2008 13th International Power Electronics and Motion Control Conference

Poznan, Poland
1-3 September 2008

Pages 515-1032

IEEE Catalog Number:	CFP0834A-PRT
ISBN 13:	978-1-4244-1741-4

Copyright © 2008 by The Institute of Electrical and Electronics Engineers, Inc.
All Rights Reserved

Copyright and Reprint Permissions: Abstracting is permitted with credit to the source. Libraries are permitted to photocopy beyond the limit of U.S. copyright law for private use of patrons those articles in this volume that carry a code at the bottom of the first page, provided the per-copy fee indicated in the code is paid through Copyright Clearance Center, 222 Rosewood Drive, Danvers, MA 01923.

For other copying, reprint or republications permission, write to IEEE Copyrights Manager, IEEE Operations Center, 445 Hoes Lane, Piscataway, New Jersey USA 08854. All rights reserved.

IEEE Catalog Number:	CFP0834A-PRT
ISBN 13:	978-1-4244-1741-4
ISSN:	2007906910

Additional Copies of This Publication Are Available from:

IEEE Service Center
445 Hoes Lane
Piscataway, NJ 08854

Phone:	(800) 678-IEEE
	(732) 981-1393
Fax:	(732) 981-9667
E-mail:	customer-service@ieee.org

2008 13th International Power Electronics and Motion Control Conference

Poznan, Poland
1-3 September 2008

IEEE Catalog Number: CFP0834A-POD
ISBN: 978-1-42441-741-4

Table of Contents

Electric Drive System for Automatic Guided Vehicles Using Contact-free Energy Transmission 1
Marcel Jufer

State-of-the-Art High Power Density and High Efficiency DC-DC Chopper Circuits for HEV and FCEV Applications... 7
Atsuo Kawamura, Martin Pavlovsky, Yukinori Tsuruta

Current-Based Condition Monitoring of Electrical Machines in Safety Critical Applications 21
Thomas G. Habetler

The Essence of Three-Phase AC/AC Converter Systems 27
J. W. Kolar, T. Friedli, F. Krismer, S. D. Round

An Analysis on Turn-off Behaviour of 1.2kV NPT-CIGBT under Clamped Inductive Load Switching........ 43
S.T. Kong, L.Ngwendson, M. Sweet, E.M. Sankara Narayanan

Turn-off behaviour of high voltage NPT- and FS-IGBT.. 48
Hans-Guenter Eckel, Karl Fleisch

Exact Circuit Power Loss Design Method for High Power Density Converters Utilizing Si-IGBT/SiC-Diode Hybrid Pairs 54
Kazuto Takao, Hiromichi Ohashi

A Forward Converter with a Monolithic Cascode Device: Design and Experimental Investigation 61
F. Chimento, S. Musumeci, A. Raciti, L. Abbatelli, S. Buonomo, R. Scollo

Switching and conducting performance of SiC-JFET and ESBT against MOSFET and IGBT 69
André Knop, W. Toke Franke, Friedrich W. Fuchs

In-Service Life Consumption Estimation in Power Modules.. 76
Mahera Musallam, C Mark Johnson, Chunyan Yin, Hua Lu, Chris Bailey

Measurement Of Temperature Sensitive Parameter Characteristics Of Semiconductor Silicon And Silicon-Carbide Power Devices 84
Mietek Nowak, Jacek Rabkowski, Roman Barlik

Unsymmetrical Gate Voltage Drive for High Power 1200V IGBT4 Modules Based on Coreless Transformer Technology Driver 88
Piotr Luniewski, Uwe Jansen

A Novel RESURFed Double Gates IGBT with Superior Performance 97
Dongming Wu, Kaihang Li, Lingling Yang

An Empiric Approach to Establishing MOSFET Failure Rate Induced by Single-Event Burnout............ 102
Jeroen van Duivenbode, Bart Smet

Comparative Study on Paralleled vs. Scaled Dc-dc Converters in High Voltage Gain Applications............ 108
Pawel Klimczak, Stig Munk-Nielsen

A Low-Loss Dc-Dc Converter For A Renewable Energy Converter 114
David S. Thompson, Otu A. Eno

A Single Active Edge-Resonant Snubber Cell-assisted ZCS Half-Bridge DC-DC Converter with Constant Frequency Asymmetrical PWM Scheme 119
Tomokazu Mishima, Mutsuo Nakaoka, Eiji Hiraki

A New Approach to High Efficiency in Isolated Boost Converters for High-Power Low-Voltage Fuel Cell Applications... 127
Morten Nymand, Michael A.E. Andersen

New Modulation Strategy with Low Switching Frequency and Minimum Baseband Distortion................... 132
N. E. Ruger, O. Schnick, W. Mathis, A. Mertens

Table of Contents

A Bit-Stream Based PWM Technique for Variable Frequency Sinewave Generation 139
N. D. Patel, U. K. Madawala

Control Strategies of the Quasi-Resonant DC-Link Inverter 144
Slawomir Mandrek, Piotr J. Chrzan

Consideration for Input Current-Ripple of Pulselink DC-AC Converter for Fuel Cells 148
Kentaro Fukushima, Tamotsu Ninomiya, Masahito Shoyama, Isami Norigoe, Yosuke Harada, Kenta Tsukakoshi

New Practical Approach to Input Current Shaping in AC-DC Power Converters 154
Kuno Janson, Viktor Bolgov, Lauri Kütt, Ants Kallaste, Heigo Mõlder

LLCC-PWM Inverter for Driving High-Power Piezoelectric Actuators 159
Rongyuan Li, Norbert Fröhleke, Joachim Böcker

Modelling and Analysis of a Matrix-Reactance Frequency Converter Based on Buck-Boost Topology by DQ0 Transformation 165
Pawel Szczeniak, Zbigniew Fedyczak, Marius Klytta

A Modular AC/DC Rectifier Based on Cascaded H-bridge Rectifier 173
H. Iman-Eini, Sh. Farhangi, JL. Schanen

Low Loss Soft Switching Boost Converter 181
So-Ri Park, Sang-Hoon Park, Chung-Yuen Won, Yong-Chae Jung

Methods for Experimental Assessment of Component Losses to Validate the Converter Loss Model 187
Yi Wang, Sjoerd de Haan, Jan Abraham Ferreira

Modified multistage semiconductor-Fitch generator topology with magnetic compression 195
Stanislaw Kalisiak, Marcin Holub

Modeling and Measuring Results of a Shunt Current Source Active Power Filter with Series Capacitor 201
P. Parkatti, M. Salo, H. Tuusa

A Multi-Drive System Based on a Two-stage Matrix Converter 207
Dinesh Kumar, Patrick W Wheeler, Jon C Clare, Lee Empringham

Characteristics of the Single Active Bridge Converter with Voltage Doubler 213
Andreas Averberg, Axel Mertens

Analysis of Capacitor Dividers for Multilevel Inverter 221
Oleg Sivkov, Jiri Pavelka

Space Vector Modulation for a Capacitor Clamped Multi-level Matrix Converter 229
Xu Lie, Jon C. Clare, Patrick W. Wheeler, Lee Empringham

New Family of Matrix-Reactance Frequency Converters Based on Unipolar PWM AC Matrix-Reactance Choppers 236
Zbigniew Fedyczak, Pawel Szczesniak, Igor Korotyeyev

Consideration of Conduction Losses for the Series Resonant Converter by Means of a Simple Extension to the SPA Approach 244
Alexander Bucher, Thomas Duerbaum, Daniel Kuebrich, Markus Schmid

Validation and Comparison of different PWM Converter Small Signal Models 250
Alexander Bucher, Markus Schmid, Lukas Bendkowski, Thomas Duerbaum

Dynamic Behaviour of a Series - Connected Multilevel Converter with Interleaved Switching 256
C. Fahrni, A. Rufer

Simple Analysis of a Flying Capacitor Converter Voltage Balance Dynamics for DC Modulation 260
A. Ruderman, B. Reznikov, M. Margaliot

Simulation of Simplified Seven Level Multilevel Converter Circuit 268
Gerardo Ceglia, Víctor Guzmán, Carlos Sánchez, Fernando Ibáñez, Julio Walter, María Giménez

Table of Contents

SEPP High-Frequency Inverter Incorporating an Auxiliary Switch and Its Performance Evaluation.......................275
H.Ogiwara,Y.Fujita, R.Urabe, M.Itoi, T.Sugai, M. kuwata, M.Nakaoka

Multiphase coupled converter models dedicated to transient response and output voltage regulation studies 281
Nadia Bouhalli, Marc Cousineau, Emmanuel Sarraute, Thierry Meynard

A 13.56 MHz Current-output-type Inverter Utilizing An Immittance Conversion Element.......................288
Yosei Sakamoto, Keiji Wada, Toshihisa Shimizu

Voltage Fed Zero-Voltage Zero-Current Switching PWM DC-DC Converter 295
Jaroslav Dudrik, Vladimír Ru1scin

PWM Spectrum Evaluation and Over-Modulation Phenomena in a Three-Phase Inverters - Analytical Approach 301
Miro Milanovic

Experimental Study of a Matrix Converter Excited Doubly-Fed Induction Machine in Generation and Motoring.......................307
Ivan Shapoval, Jon Clare, Eduard Chekhet

Effect of Type and Interconnection of DG Units in the Fault Current Level of Distribution Networks 313
H.R. Baghaee, M. Mirsalim, M. J. Sanjari, G.B. Gharehpetian

An Isolated Full-Bridge DC/DC Converter..with Bidirectional Communication Capability 320
Lon-Kou Chang, Ru-Shiuan Yang

Efficiency and Power Losses in PM BLDC Motor with Variable Bridge/half-bridge Structure Electronic Commutator 326
K. Krykowski, A. Bodora

Analysis of a device for converting a unipolar input voltage into two symmetric bidirectional output voltages with a magnetically coupled coil 331
Felix. A. Himmelstoss, Wilhelm Kraeftner

Invariant Modulation Strategy for Two-stage Direct Power Converter.......................337
Radiy Bekbudov

Experimental Study of A Multicell ac/ac Converter Balancing Circuit.......................345
Robert Stala, Andrzej Mondzik

A Comparison and Optimum Design of Reluctance-Controlled Classical Load-Resonant Converters 350
Stefan V. Mollov, Michael P. Theodoridis

Capacitor Clamped Multilevel Matrix Converter Controlled with Venturini Method.......................357
Janina Rzasa

Reliability Consideration for a High Power Zero-Voltage-Switching Flyback Power Supply 365
Arash Rahnamaee, Jafar Milimonfared, Kaveh Malekian, Mohammad Abroushan

The Traction Drive Topology Using the Matrix Converter with Middle-Frequency Transformer 372
Martin Pittermann, Pavel Drábek, Marek Cédl

Analysis of Multipulse Rectifiers with Modulation in DC Circuit in Vector Space Approach.......................377
Andrzej KAPLON and Jaroslaw ROLEK

High Efficiency Soft Switching Boost Converter for Photovoltaic System 383
Gil-Ro Cha, Sang-Hoon Park, Chung-Yuen Won, Yong-Chae Jung, Sang-Hoon Song

A Power Converter For Fault Tolerant Machine Development In Aerospace Applications 388
Liliana de Lillo, Patrick Wheeler, Lee Empringham, Chris Gerada, XiaoyanHuang

Optimal Bus Capacitance Design for System Stability in On-Board Distributed Power Architecture 393
Seiya Abe, Masahiko Hirokawa, Masahito Shoyama, Tamotsu Ninomiya

Table of Contents

Steady State Analysis of Hysteretic Control Buck Converters 400
L.K. Wong, T.K. Man

A Novel Control Method for IGBT Current Source Rectifier 405
Longcheng Tan, Yaohua Li, Ping Wang, Congwei Liu, Zixin Li, Yonggang Chen, Wei Xu

A procedure to optimize the inductor design in boost PFC applications 409
Florent Liffran

Electric Vehicle Drive Inverters Simulation Considering Parasitic Parameters 417
Wen Huiqing, Liu Jun, Zhang Xuhui, Wen Xuhui

DC-DC Converters with FPGA Control for Photovoltaic System 422
Jan Leuchter, Pavel Bauer, Vladimir Rerucha, Petr Bojda

Control of a Converter with Superconductive Energy Storage Inductor 428
Rozanov Yurie Konstantinovich, Lepanov Michail Gennadevich, Kiselev Michail Gennadevich

FPGA-based Controllers for Switching Converters 432
Karel Jezernik

Gamesa DAC converter: the way for REE grid code certification 437
Itziar Martinez, Daniel Navarro

Flatness-Based Voltage-Oriented Control of Three-Phase PWM Rectifiers 444
J. Dannehl, F.W. Fuchs

Control of a single phase H-Bridge multilevel inverter for grid-connected PV applications 451
Elena Villanueva, Pablo Correa, Jose Rodriguez

Switching and Voltage Controls for a Flyback Switch-Mode Rectifier 456
Yuan-Chih Chang, Chang-Ming Liaw

Method Of Designing ZVS Boost Converter 463
Miroslaw Luft, Elzbieta Szychta, Leszek Szychta

A New DC-DC Converter with Multi Output: Topology and Control Strategies 468
Arash A Boora, Firuz Zare, Gerard Ledwich, Arindam Ghosh

Maximum Frequency for Hysteretic Control COT Buck Converters 475
L.K. Wong, T.K. Man

Current Control Method Based on Hysteresis Control Suitable for Single Phase Active Filter with LC Output Filter 479
Yukinori Kobayashi, Hirohito Funato

Optimal Slope Compensation for step load in peak current controlled dc-dc Buck Converter 485
Susovon Samanta, Pradipta Patra, Siddhartha Mukhopadhyay, Amit Patra"

Performances of a PLL Based Digital Filter for double-conversion UPS 490
Armando Bellini, Stefano Bifaretti

10A 12V 1 chip digitally-controlled DC/DC converter IC with high resolution and high frequency DPWM 498
Kazutoshi Nakamura, Toshiyuki Naka*, Yuki Kamata*, Toyoki Taguchi, Takaaki Shimizu, Yoshiko Ikeda, Akio Nakagawa, Dragan Maksimovic*

Modelling and Modulation of Voltage Source Converter 504
Grzegorz Radomski

Sliding Mode Control of DC/DC Multiphase Power Converters 512
Vadim Utkin

A New Digital Control Method for High Performance 400 Hz Ground Power Unit 515
Zixin Li, Ping Wang, Haibin Zhu, Yaohua Li, Longcheng Tan, Yonggang Chen, Fanqiang Gao

Table of Contents

Single-phase 50-kW 16.7-Hz Four-Quadrant Line-Side Converter for Railway Traction Application 521
C. Heising, R. Bartelt, V. Staudt, A. Steimel

Technique to Improve IGBT Converter Efficiency and Transient Response 528
Robert W. Turner, Simon Walton, Richard Duke

The control of voltage converter rectifiers 536
Krzysztof Szubert

Load Voltage Regulation and Line Loss Minimization of Loop Distribution Systems Using UPFC 542
Mahmoud A. Sayed, Takaharu Takeshita

Control of Traction Single-Phase Current-Source Active Rectifier under Distorted Power Supply Voltage 550
Jan Michalík, Jan Molnár, Zdenck Peroutka

Simulation Model Of Neural Network Based Synchronous Generator Excitation Control 556
Damir Sumina, Neven Bulic, Gorislav Erceg

Predictive Current Control of a 7-level AC-DC back-to-back Converter for Universal and Flexible Power Management System 561
Stefano Bifaretti, Pericle Zanchetta, Florin Iov, Jon C. Clare

Predictive Stator Current Control For Three-Level Voltage-Source Inverters With Output LC-Filters 569
Tomasz Laczynski, Axel Mertens

Research on Dimming Control Method of Electronic Ballast for the Automotive HID Headlight 576
P. Dong, K.W.E.Cheng, S.L.Ho

Control Method for a Three-Port Interface Converter Using an Indirect Matrix Converter with an Active Snubber Circuit 581
Koji Kato, Jun-ichi Itoh

Precise Digital Control Method with Multi-rate deadbeat control for Single Phase Utility Interactive Inverter with FPGA based Hardware Controller 589
Kenta Hayashi, Tomoki Yokoyama

A Digital Current Controller for Zero-Current Transition Bidirectional Converter 595
Nobuyuki Kasa, Takahiko Iida

Control Method for a Single Phase Arbitrary Waveform-output Inverter 600
Satoshi Taniguchi, Keiji Wada, Toshihisa Shimizu

Elimination of Harmonics in Multilevel Inverters with Non-Equal DC Sources Using PSO 606
A. K. Al-Othman, Tamer H. Abdelhamid

Improved PFC Circuit Having Ladder Type Filter with Only Passive Devices 614
Kenji Ando, Keiju Matsui, Nobuhito Takeuchi, Masaru Hasegawa

Fuel Cell Current Ripple Minimization using a bi-Buck Power Interface 621
Nicu Bizon, Marian Raducu, Mihai Oproescu

Power Control Strategy of Parallel Inverter Interfaced DG Units 629
H.R. Baghaee, M.Mirsalim, M. J. Sanjari, G.B. Gharehpetian

Implementation of Nonlinear power flow controllers to control a VSC 637
Nelson L. Díaz, Fabián H. Barbosa, Cesar L. Trujillo

Harmonic Distortion Reduction Technique for Uninterruptible Power Supplies with DC Voltage Boost Technique 643
Juei Lung Shyu

Energy-based Modulation Error Control for High-Power Drives with Output LC-Filters and Synchronous Optimal Pulse Width Modulation 649
Tomasz Laczynski, Timur Werner, Axel Mertens

vii

Table of Contents

Voltage Harmonic Control of Z-source Inverter for UPS Applications 657
Arkadiusz Kulka, Tore Undeland

A Method of Optimal Control for Switched-Mode Power Converters 663
Anatoly Bekishev, Albert Iskhakov, Leonid Klyachko, Vladimir Pospelov, Sergey Skovpen

Experiment results with modified Hybrid PWM method for three phase induction motor 669
Daniel Lewandowski, Grzegorz Lisowski

Optimized Design of a Delay line based Analog to Digital Converter for Digital Power Management Applications 674
Mukti Barai, Sabyasachi Sengupta, Jayanta Biswas

Overmodulation Region of Multi-Phase Inverters 682
S. Halasz

Optimal Control of Induction Motor Using High Performance Frequency Converter 690
Jerkovic Vedrana, Spoljaric Zeljko, Valter Zdravko

Power Electronic Converter for the Reluctance Pump Drive 695
B. J. Szymanski, K. Kompa, N. Michalke, H. Kuß, U. Schuffenhauer

A Predictive Control Scheme for Current Source Rectifiers 699
Pablo Correa, Jose Rodriguez

Analysis and Design of New Switching Table for Direct Power Control of Three-Phase PWM Rectifier 703
Abdelouahab Bouafia, Jean-Paul Gaubert, Fateh Krim

Improvement of the performance for DC-DC Converter 710
X..She, Yun She

A Drive System With High-Speed Single-Phase Supplied Three-Phase Induction Motor 714
T. Binkowski, M. Grad, M. Latka, W. Malska, D. Sobczynski

A Pulse Width Modulation Technique for a Multilevel Converter in High Voltage High Frequency Applications 718
Jafar Adabi, Hamid Soltani, Firuz Zare

Bidirectional Positive Buck-Boost Converter 723
Arash A Boora, Firuz Zare, Gerard Ledwich, Arindam Ghosh

Control system of power electronics current modulator utilized in diode rectifier with sinusoidal source current 728
Michal Gwózdz, Michal Krystkowiak

Design and control of a half-bridge converter to drive piezoelectric actuators 731
Oriol Gomis-Bellmunt, Josep Rafecas-Sabate, Daniel Montesinos-Miracle, Josep-Maria Fernandez-Mola, Joan Bergas-Jane

Online Diagnosis of PEM Fuel Cell 734
Abdellah Narjiss, Daniel Depernet, Denis Candusso, Frederic Gustin, Daniel Hissel

Application of Kalman filters to the control of independent power electronic voltage sources 740
Ryszard Porada, Lukasz Nyczkowski

Verification of the load sharing characteristics in Autonomous Decentralized UPS system using FPGA based Hardware Controller 744
Nobuaki Doi, Tsuyoshi Saito, Tomoki Yokoyama

Fault Current Reduction in Distribution Systems with Distributed Generation Units by a New Dual Functional Series Compensator 750
H.R. Baghaee, M. Mirsalim, M. J. Sanjari, G.B. Gharehpetian

Dynamic Simulation of PM Motor Drive System based on Reluctance Network Analysis 758
Kenji Nakamura, Osamu Ichinokura

Table of Contents

Performance Improvement of Direct Torque Controlled Interior Permanent Magnet Synchronous Motor Drive by Considering Magnetic Saturation...763
Behrooz Majidi, Jafar Milimonfared, Kaveh Malekian

Condition Monitoring for Mechanical Faults in Fully Integrated Servo Drive Systems.................769
Jesus Arellano-Padilla, Mark Sumner, Chris Gerada

Feed-forward Compensation of Load and Parameter Variations of Electric Drive.........................776
Alon Kuperman, Yoram Horen, Saad Tapuchi, Uri Suissa

Thermal Effect of Short-Circuit Current in Low Power Induction Motors....................................782
Leo.s Beran

Generalized Model for a Class of Switched Reluctance Motors..787
Constantin Pavlitov, Yassen Gorbounov, Radoslav Rusinov, Alexandar Alexandrov, Kliment Hadjov, Dimitar Dontchev

Neural Network based Fault Detection of PMSM Stator Winding Short under Load Fluctuation.........793
J. Quiroga, D.A. Cartes, C.S. Edrington, Li Liu

Review of Electrical Machine in Downhole Applications and the Advantages.................................799
Anyuan Chen, Ravindra. B. Ummaneni, Robert Nilssen, Arne Nysveen

Broken Rotor Bar Impact on the Closed Loop and Sensorless Control of Induction Machine.........804
Piotr Kotodziejek, Elzbieta Bogalecka

Coupled Magnetic Circuit Method and Permeance Network Method Modeling of Stator Faults in Induction Machines...810
Amin Mahyob, Mohamed Y. Ould Elmoctar, Pascal Reghem, Georges Barakat

Explosion Protected Electrical Drives - Risk Assessment and Technical Diagnostics...................818
Ivica Gavranic, Drago Ban, Damirarko Zarko

The effect of subharmonics on induction machine heating..826
Piotr Gnacinski, Marcin Peplinski, Mariusz Szweda

Influence of Saturation Effects in a Transverse Flux Machine..830
M. Siatkowski, B. Orlik

A Model of Semiconductor Converter-Fed Asynchronous Machines Taking into Account Energy Losses and Thermal Processes..837
M. Pronin, O. Shonin, Y. Koskin, A. Vorontsov, P. Kalatchikov

Use of an AC Self-excited Switched Reluctance Generator as a Battery Charger.........................845
Abelardo Martínez, Estanislao Oyarbide, Javier Vicuña, Francisco Perez, Eduardo Laloya, Bonifacio Martín-del-Brío, Tomás Pollán, Beatriz Sánchez, Juan Lladó

Direct Thrust Controlled Linear Induction Motor Including End Effect..850
Berrin Susluoglu, Vedat M. Karsli

Analysis of Short-Circuit Forces at the Top of the Low Voltage U-Type and I-Type Winding in a Power Transformer..855
Leonardo Strac, Franjo Kelemen, Damir Zarko, Josipa Mokrovica

Anisotropy Comparison of Reluctance and PM synchronous Machines for Low Speed Position Sensorless Applications...859
H.W. de Kock, M.J. Kamper, R.M. Kennel

Analysis of VSI-DTC Fed 6-phase Synchronous Machines..867
Ibrahim Abuishmais, Waqas M. Arshad, Sami Kanerva

Optimal Rotor Flux Shape for Multi-phase Permanent Magnet Synchronous Motors...................874
Roberto Zanasi, Federica Grossi

Table of Contents

Modelling of Electrical Machines Using the Modelica Bond-Graph Library 880
Mieczyslaw Ronkowski

Induction Motor Parameters Identification using Genetic Algorithms for Varying Flux Levels 887
Konstantinos Kampisios, Pericle Zanchetta, Chris Gerada, Andrew Trentin, Omar Jasim

Study of the sudden symmetrical short-circuit using the mathematical models of the synchronous machine and the numerical methods 893
Petropol Serb Gabriela, Petropol Serb Ion, Campeanu Aurel, Sonia Degeratu, Anca Petrisor

Analytical Method of Calculation of the Current and Torque of a Reluctance Stepper Motor Using Fourier Complex Series 899
Pavel Zaskalicky, Maria Zaskalicka

Bearing Damage Analysis by Calculation of Capacitive Coupling between Inner and Outer Races of a Ball Bearing 903
Jafar Adabi, Firuz Zare, Gerard Ledwich, Arindam Ghosh, Robert D.Lorenz

The Model of the Squirrel Cage AC Motor including Rotor Slot Harmonics 908
Eleonora Darie, Costin Cepisca, Emanuel Darie

Identification of mathematical model induction motor's parameters with using evolutionary algorithm and multiple criteria of quality 912
Hudy Wiktor, Jaracz Kazimierz

Simulation Study on Control of Ultrahigh Speed Drives in Waste Energy Recovery Systems 916
Péter Stumpf, Miklós G. Simon, Rafael K. Járdán, István Nagy

Adaptive Back EMF Parameter Adjustment of Simplified Vector Control for Position Sensorless Permanent Magnet Synchronous Motors 924
Kiyoshi Sakamoto, Yoshitaka Iwaji, Daigo Kaneko, Toshihiro Takeuchi, Tsunehiro Endo, Atsuo Kawamura

Identification and Control of Precision XY Stages with Active Vibration Suppression System 932
Mayumi Nitta, Seiji Hashimoto

Sensitivity of the Currents Input-Output Decoupling Vector Control of the DFIM versus Current Sensors Fault 938
Meriem Abdellatif, Maria Pietrzak-David, Ilhem Slama-Belkhodja

Extended Back EMF model for PM synchronous machines with different inductances in d- and q-axis 945
Andreas Eilenberger, Manfred Schroedl

Gait generation of a two-legged robot by using adaptive network based fuzzy logic control 949
Umit Onen, Mete Kalyoncu, Mustafa Tinkir, Fatihm. Botsali

Walking robot HEXOR® II - a versatile platform for engineering education 956
M. Sajkowski, T. Stenzel, B. Grzesik

Motion Control of Steel Sheet Shears with Rocking Knife Mechanism 961
Jan Fetyko, Frantisek Durovsky, Viliam Fedak

Intelligent Adaptive Control and Monitoring Of Band Sawing 967
Ilhan Asiltürk, Ali Ünüvar

Hierarchical adaptive network based fuzzy logic controller design for a single flexible link robot manipulator 974
Mete Kalyoncu, Mustafa Tinkir

Digital Controlled High Speed Synchronous Motor 982
Zdenk Cerovský, Jaroslav Novák, Martin Novák, Marek Cambál

Analysis of combustion engine - electric Linear generator set operation 988
Jirí Pavelka

x

Table of Contents

Closed Loop Control of AC Drive with LC Filter 994
Jaroslaw Guzinski

Sensorless IPMSM based drive for reciprocating compressor 1002
Anton Dianov, Kim Young-Kwan, Lee Sang-Joon, Lee Sang-Taek, Yoon Tae-Ho

Controlling system of electrodynamic drive 1009
Josef Cernohorský

Expert System for Electric Drive Design 1017
Juhan Laugis, Valery Vodovozo

Improvement of Moving Characteristics of Cableless Micro-actuator and Consideration of Reversible Motion 1020
Hiroyuki Yaguchi, Kazumi Ishikawa, Toshihiro Zamma, Koichi Funayama

Sensorless Control of AC Machines using High-Frequency Excitation 1024
Heiko Zatocil

Adaptive PF Speed Control of SRAM Drives 1033
Laszlo Szamel

A Very Simple Fuzzy Control System for Inverter Fed Synhronous Motor 1040
Pawel Fabijanski, Ryszard Lagoda

Distributed control system of DC servomotors for six legged walking robot 1044
D. Belter, K. Walas, A. Kasinski

Optimization of Starting Process of the Frequency Controlled Induction Motor 1050
I.Ya. Braslavsky, A.V. Kostylev, D.P. Stepanyuk

3-Axes Satellite Attitude Control Based on Biased Angular Momentum 1054
Azam Ghaedi, Mohammad Ali Nekoui

Modelling and simulation of a signal injection self-sensored drive 1058
Alen Poljugan, Mark Sumner, Chris Gerada, Qiang Gao

Robust PI Cascade Control for a Multi-Mass System Optimized by Evolutionary Algorithms 1064
M. Joost, K. Zielinski, B. Orlik, R. Laur

Permanent Magnet Synchronous Servo-Drive with State Position Controller 1071
Lech M. Grzesiak, Tomasz Tarczewski, Slawomir Mandra

Closed-Loop Control of Virtual FPGA-Coded Permanent Magnet Synchronous Motor Drives using a Rapidly Prototyped Controller 1077
Christian Dufour, Vincent Lapointe, Jean Bélanger, Simon Abourida

Speed Sensorless Nonlinear Control Of Induction Motor In The Field Weakening Region 1084
Miroslaw Wlas, Haithem Abu-Rub, Joachim Holtz

Comparison of Dynamic Performances of Speed Control System Containing Time - Minimal Speed Controller with Control System Containing PI Speed Controller 1090
Andrzej Andrzejewski, Marian Roch Dubowski

Optimisation of Real-Time Complex Path Generation in Constrained Intelligent Motion Applications Based on IPM Motor Drives 1097
Silverio Bolognani, Roberto Petrella, Fabio Stefanutti, Piero Stocco

PMSM Sliding Mode Observer for Speed and Position Estimation Using Modified Back EMF 1105
Ilioudis Vasilios C., Margaris Nikolaos I.

Optimal Control of Electrical Drives with Induction Motors for Variable Torques 1111
Corneliu Botan, Marcel Ratoi, Vasile Horga

xi

Table of Contents

An Optimal Control for Saturated Interior Permanent Magnet Linear Synchronous Motors Incorporating Field Weakening 1117
Mohammad Abroshan, Jafar Milimonfared, Kaveh Malekian, Arash Rahnamaee

Improved Direct Torque Control for Induction Machine Drives using Fuzzy Logic and Particle Swarm Optimization 1123
Mohammad Mehdi Rezaei, Mojtaba Mirsalim, Kaveh Malekian

Design and Implementation of High Performance Full-Digital Spindle Drives 1128
Liu Yang, Zhao Jin

Semi hierarchical adaptive network based fuzzy logic controller design for a multi-straight-line path tracing flexible robot manipulator with rotating-prismatic joint 1132
Mete Kalyoncu, Mustafa Tinkir

Control System with the Set Point Observation 1140
Algirdas Baskys, Vitoldas Gobis, Valerijus Zlosnikas

Electropneumatic Servo System with Adaptive Force Controller 1144
Arunas Grigaitis, Vilius Antanas Gele~evicius

New fault tolerant DTC control for induction machine drives 1149
A.Ben Abdelghani Bennani, M. Ghodbane Cherif, I. Slama Belkhodja

Stability Analysis of the Natural Field Orientation Controlled Induction Machine Drive 1155
G. Mirzaeva, A. Rojas

Control of SR motor EV by instantaneous torque control using flux based commutation and phase torque distribution technique 1163
Ayumu Nishimiya, Hiroki Goto, Hai-Jiao Guo, Osamu Ichinokura

Simulation of IPM Motor by Nonlinear Magnetic Circuit Model for Comparing Direct Torque Control with Current Vector Control 1168
Hiroki Goto, Kensuke Kimura, Hai-Jiao Guo, Osamu Ichinokura

A Simplified Model for Induction Machines with Faults to Aid the Development of Fault Tolerant Drives 1173
O. Jasim, C. Gerada, M. Sumner, J. Arellano-Padilla

About the Experimental Results of an Electric Driving System Based on Asynchronous Motor and PWM Converter 1181
Petre-Marian Nicolae, Dan-Gabriel Stanescu, Ioana-Gabriela Sîrbu

Real-World Force Feedback Control for Mobile-Hapto 1187
Wataru Yamanouchi, Yuki Yokokura, Seiichiro Katsura, Kiyoshi Ohishi

The new numerical integration routine applied in sensorless drives 1193
Arkadiusz Gardecki, Krystyna Macek-Kaminska

Application of Fuzzy Logic Techniques To Robust Speed Control of PMSM 1198
Tomasz Pajchrowski, Krzysztof Zawirski

Optimal control of current commutation of high speed SRM drive 1204
Jan Deskur, Tomasz Pajchrowski, Krzysztof Zawirski

Comparison Between Direct Torque Control and Vector Control of a Permanent Magnet Synchronous Motor Drive 1209
Rafa Souad, Houcine Zeroug

Detection and self-tuning compensation of periodic disturbances by the control of DC motor 1215
Michael Ruderman, Frank Hoffmann, Johannes Krettek, Torsten Bertram

A Linear Switched Reluctance Motor Based Position Tracking System 1221
S. W. Zhao, N. C. Cheung, Y. Lu, W. C. Gan, Z. G. Sun

Table of Contents

Mobile Robot Navigation with Obstacle Avoidance Capability 1225
Anca Sorana Popa, Mircea Popa, Ioan Silea

Requirements for Power Electronics in Solid Oxide Fuel Cell System 1233
T. Riipinen, V. Väisänen, M. Kuisma, L. Seppä, P. Mustonen, P. Silventoinen

Power Supply for a IGBT-Driver with High Insulation Voltage based on a Printed Planar Transformers 1239
Günter Schmitt, Wolf Kusserow, Ralph Kennel

Variable Motor Operating Point by Integration of Power Electronic Device into Rotor 1243
Adrian Tulbure, Hans-Peter Beck, Mircea Risteiu

Magnetic Material Comparisons for High-Current Gapped and Gapless Foil Wound Inductors in High Frequency DC-DC Converters 1249
Marek S. Rylko, Brendan J. Lyons, Kevin J. Hartnett, John G. Hayes, Michael G. Egan

Feasibility Study of Half- and Full-Bridge Isolated DC/DC Converters in High-Voltage High-Power Applications 1257
Dmitri Vinnikov, Tanel Jalakas, Mikhail Egorov

Evaluation of Different Loss Calculation Methods for High-voltage IGBT-s Under Small Load Conditions 1263
T. Jalakas, D. Vinnikov, J. Laugis

Control of Power Supply Unit for Military Vehicles Based on Four-Leg Three-Phase VSI with Proportional-Resonant Controllers 1268
Tomál Glasberger, Zdenek Peroutka

Optimal Design of a Half Wave Cockroft-Walton Voltage Multiplier with Different Capacitances per Stage 1274
Ioannis C. Kobougias, Emmanuel C. Tatakis

Calculation of Leakage Inductance of Core-Type Transformers for Power Electronic Circuits 1280
Reinhard Doebbelin, Marcel Benecke, Andreas Lindemann

Enhanced Current Pulsation Smoothing Parallel Active Filter for Single Stage Grid-connected AC-PV Modules 1287
A.C. Kyritsis, N.P. Papanikolaou, E.C. Tatakis

Outline of the Design of a Cascaded H-bridge Medium Voltage STATCOM 1293
R.E. Betz, B.J. Cook, T.J. Summers, R. Fisher, A. Bastiani, S. Shao, P. Stepien, K. Willis

Investigation of High Frequency Effects on Layered Coils 1301
Georgios S. Dimitrakakis, Emmanuel C. Tatakis

Soft Switching PWM Inverter for Induction Heating Applied to Heating of Ferromagnetic Metal 1309
Sachio Kubota, Muneo Sato, Fumio Ito, Yoshihiro Shimaoka, Kunihiro Nishioka

Corona Treatment System with Resonant Inverter - Selected Proprieties 1316
Mucko Jan

Power supply unit for an electric discharge machine 1321
Wojciech Mysinski

High Power, High Voltage, High Frequency Transformer / Rectifier for HV Industrial Applications 1326
T. Filchev, D. Cook, P. Wheeler, A. Van den Bossche, J. Clare, V. Valchev

Small Power Laboratory Model and High Power Prototype of the Four-Level VSI 1332
Ryszard Michal Strzelecki, Pawel Szczepankowski, Andrzej Kasprowicz, Genady Stepanovic Zinoviev, Krzysztof Zymmer, Zbigniew Zakrzewski

AC Voltage Regulator Using PWM Technique and magnetic flux distribution 1337
A.M. Dabroom

Minimum Reactive Power Filter Design for High Power Converters 1345
Alex-Sander Amavel Luiz, Braz Jesus Cardoso Filho

xiii

Table of Contents

Injection of a carrier with higher than the PWM frequency for sensorless position detection in PM synchronous motors.. 1353
Roberto Leidhold, Peter Mutschler

Parallel Fixed Point FPGA Implementation of Sensorless Induction Motor Torque Control....................... 1359
Jacek D. Lis, Czeslaw T. Kowalski

Design of an FPGA-Based Real-Time Simulator for Electrical System.. 1365
I. Bahri, M-W. Naouar, E. Monmasson, I. Slama-Belkhodja, L.Charaabi

A New, Ultra-low-cost Power Quality and Energy Measurement Technology 1371
Alex McEachern, Andreas Eberhard

Rotor Time Constant Adaptation Using Radial Basis Function Network 1375
Pavel Brandltetter, Ondfej Skuta

Application of Speed and Load Torque Observers in High Speed Train 1382
Jaroslaw Guzinski, Marc Diguet, Zbigniew Krzeminski, Arka diusz Lewicki, Haithem Abu-Rub

Position Estimator including Saturation and Iron Losses for Encoder Fault Detection of Doubly-Fed Induction Machine.. 1390
Kai Rothenhagen, Friedrich W. Fuchs

Wide Range Low Noise Current Sensor 1398
F. Richter, C. Sourkounis

Transducerless Speed Control with Initial Position Detection for Low Cost PMSM Drives 1402
Roman Filka, Peter Balazovic, Branislav Dobrucky

Study About the Possibility of Electrodes Motion Control in the EAF Based on Adaptive Impedance Control... 1409
Manuela Panoiu, Caius Panoiu, Sorin Deaconu

Asynchronous machine stator resistance estimation using integrated PWM modulator and sampler unit as FPGA application 1416
Dag Samuelsen, Waldemar Sulkowski

Development of Monitoring System for Series HEV Bus with Touch Panel 1421
Tae-Won Chun, Quang-Vinh Tran, Uk-Don Choi, Heung-Gun Kim

A Development System for Testing Integrated Circuits Used for Power and Energy Measurements..................... 1426
Vladimir Cuk, Aleksandar Nikolic, Aleksandar Zigic

State and parameter estimation in a hydraulic system - moving horizon approach... 1432
Jerzy Baranowski, Andrzej Tutaj

Technologies of Current Sensors Suitable for Hot High Density Power Electronics... 1440
Filip Grecki, Grzegorz Iwanski, Wlodzimierz Koczara, Jozef Lastowiecki

Nonlinear dynamical feedback for motion control of magnetic levitation system 1446
Jerzy Baranowski, Pawel Piatek

Speed and position estimation of SRM 1454
Konrad Urbanski, Krzysztof Zawirski

Potential of Digital Gate Units in High Power Appliations... 1458
Harald Kuhn, Thies Koneke, Axel Mertens

Disturbance Currents of Inverters.. 1465
Petr Vrana, Jiri Javurek

Improvement of the Energy Recovery of Traction Electrical Drives using Supercapacitors 1469
Diego Iannuzzi

Table of Contents

A Multi-Core PC-based Simulator for the Hardware-In-the-Loop Testing of Modern Train and Ship Traction Systems..1475
Christian Dufour, Guillaume Dumur, Jean-Nicolas Paquin, Jean Bélanger

Energy Saving Control of Tram Motors Taking Light Signalling and City Disturbances into Account....................1481
Stanislaw Rawicki

Characterization and Improved Control of a Brushless DC Drive with In-Wheel Motor..........................1491
Manuele Bertoluzzo, Giuseppe Buja, Alessandro Pavoni

Supply of Electric Vehicles via Magnetically Coupled Air Coils..........................1497
Slawomir Judek, Krzysztof Karwowski

Sliding-Mode Approach to Control Design for Induction Motor Drive fed by a Three-Level Voltage-Source Inverter...1505
Sergey Ryvkin, Richard Schmidt-Obermoeller, Andreas Steimel

Analysis and configuration of supercapacitor based energy storage system on-board light rail vehicles.................1512
R. Barrero, X. Tackoen, J. Van Mierlo

Design of High Power Electronic Building Block based on Parallel of IGBTs for Electric Vehicle...........................1518
Wen Huiqing, Liu Jun, Zhang Xuhui, Wen Xuhui

Stability Analysis on the DC Power Distribution System of More Electric Aircraft........................1523
H. Zhang, C. Saudemont, B. Robyns, N. Huttin, R. Meuret

Design Considerations for Control of Traction Drive with Permanent Magnet Synchronous Machine....................1529
Zden..k Peroutka, Karel Zeman

Control of Primary Voltage Source Active Rectifiers for Traction Converter with Medium-Frequency Transformer..1535
Vojtech Blahník, Zdenek Peroutka, Jan Molnár, Jan Michalík

Energy management strategy for Coupling Supercapacitors and Batteries with DC-DC converters for hybrid vehicle applications..1542
M.B. Camara, F. Gustin, H. Gualous, A. Berthon

Dual-Source Fed Multiphase Traction System with Standard and Non-Standard Control Regimes Based on Synchronized PWM...1548
Valentin Oleschuk, Marian P. Kazmierkowski

Analysis of a H-NPC topology for an AC Traction Front-End Converter..........................1555
I. Etxeberria-Otadui, A. Lopez-de-Heredia, J. San-Sebastian, H. Gaztañaga, U. Viscarret, M. Caballero

Hybrid - type system of power supply for a trolleybus with an asynchronous motor.........................1562
Zygmunt Gizinski, Marcin Gasiewski, Ireneusz Mascibrodzki, Michal Zych, Krzysztof Zymmer, Marcin Zulawnik

Control of rotor flux in AC tram drive during sudden braking operation..........................1568
Andrzej Debowski, Piotr Chudzik

A New Novel Power Electronic Circuit to Reduce Stray Current and Rail Potential in DC Railway...........................1575
Reza Fotouhi, Siamak Farshad

Slip Control Upgrades for Light-Rail Electric Traction Drives..........................1581
Madis Lehtla, Hardi Hõimoja

Practical Aspects on the Improved DC Driving System Used in Electric Urban Traction.........................1585
Petre Marian Nicolae, Ioana-Gabriela Sîrbu, Ileana-Diana Nicolae, Lucian Mandache

The study of using the traction drive topology with the middle-frequency transformer..........................1593
Martin Pittermann, Pavel Drábek, Marek Cédl, Jiří Fořt

Control of a Linear Switched Reluctance Motor as a Propulsion System for Autonomous Railway Vehicles..........1598
L. Kolomeitsev, D. Kraynov, S. Pakhomin, F. Rednov, E. Kallenbach, V. Kireev, T. Schneider, J. Böcker

Table of Contents

Motion Copying System Based on Real-World Haptics in Variable Speed...1604
Yuki Yokokura, Seiichiro Katsura, Kiyoshi Ohishi

Adaptive Fuzzy Control of magnetically suspended Rotary Table 1610
Thomas Schallschmidt, Denis Draganov, Frank Palis

Wideband Force Sensing for Haptic Energy Transmission Utilizing FPGA......................................1614
Seiichiro Katsura, Masaki Kondo, Kiyoshi Ohishi

On the development of BLDC motor control run-up algorithms for aerospace application 1620
Vladimir Hubik, Martin Sveda, Vladislav Singule

Rotor Levitation by Active Magnetic Bearing Using Digital State Controller...............................1625
Chip Rinaldi Sabirin, Andreas Binder

Dynamical Torque-Speed-Curve Adaption To Damp Load Peaks Occuring In Drive Trains Of Shredding Plants...1633
Constantinos Sourkounis

Traction vehicle distributed control computer system architecture with auto reconfiguration features and extended DMA support 1638
Jiri Zdenek

Analysis and Position Control of a Linear Switched Reluctance Actuator Based on Sliding Mode Control 1646
António Espírito Santo, Maria R. A. Calado, Carlos M. P. Cabrita

Development and Control for a Reaction Wheel System Driven by Permanent Magnet Synchronous Motor 1652
Ming-Chang Chou, Chang-Ming Liaw, Sywe-Bin Chien, Fa-Hwa Shieh, Jih-Run Tsai, Hao-Chi Chang

Nonlinear control design for magnetic bearings via automatic differentiation................................1660
Stefan Palis, Mario Stamann, Thomas Schallschmidt

Design of Energy Harvesting Generator Base on Rapid Prototyping Parts.....................................1665
Zdenek Hadas, Jan Zouhar, Vladislav Singule, Cestmir Ondrusek

Control of Bouc-Wen hysteretic systems: Application to a piezoelectric actuator 1670
Oriol Gomis-Bellmunt, Faycal Ikhouane, Daniel Montesinos-Miracle

Electric drive for carding machine draft device...1676
Martin Diblík

Two-level and Multilevel Converters for Wind Energy Systems: A Comparative Study 1682
R. Melício, V. M. F. Mendes, J. P. S. Catalão

A Stand-alone Photovoltaic Supercapacitor Battery Hybrid Energy Storage System 1688
M.E. Glavin, Paul K.W. Chan, S. Armstrong, W.G Hurley

Integrated contactless power transmission systems with high positioning flexibility.......................1696
Daniel Kürschner, Christian Rathge

A Transformerless Interface Converter for a Distributed Generation System..................................1704
Tzung-Lin Lee, Zong-Jie Chen

A Comprehensive Analysis and Comparison Between Multilevel Space-Vector Modulation and Multilevel Carrier-Based PWM 1710
Constantinos Sourkounis, Ahmad Al-Diab

Identification of Electrical Parameters in a Power Network Using Genetic Algorithms and Transient Measurements 1716
Wei. Dong, Pericle Zanchetta, David W.P. Thomas

On Acoustic Noise Reduction Procedure for Inverter-Fed Induction Machines 1722
Weiss Helmut, Zaucher Peter, Xiao Jian

Table of Contents

Cascaded Doubly Fed Induction Generator for Mini and Micro Power Plants Connected to Grid 1729
Marek Adamowicz, Ryszard Strzelecki

Contactless power transmission with new secondary converter topology 1734
Matthias Dockhorn, Daniel Kürschner, Rudolf Mecke

Modeling Approach of a Generator with Non-linear Load in Embedded Electrical Network 1740
Nicolas Amelon, Mourad Ait-Ahmed, Mohamed-Fouad Benkhoris

Optimal Use of the 14 V Alternator in 42 V Automotive Supply Systems 1748
Vasile Comnac, Mihai Cernat, Adrian Mailat

New Dual Channel Quasi Resonant DC-DC Converter Topologies for Distributed Energy Utilization 1755
J. Hamar, I. Nagy, P. Stumpf, H. Ohsaki, E. Masada

Output Filtering of the Customer-end Inverter in a Low-Voltage DC Distribution Network 1763
Pasi Peltoniemi, Pasi Nuutinen, Pasi Salonen, Markku Niemelä, Juha Pyrhönen

Power Flow Control through a Multi-Level H-Bridge based Power Converter for Universal and Flexible Power Management in Future Electrical Grids 1771
Stefano Bifaretti, Pericle Zanchetta, Yue Fan, Florin Iov, Jon Clare

Energy Storage Systems The Flywheel Energy Storage 1779
Tomasz Siostrzonek, Stanislaw Piróg, Marcin Baszynski

Analysis of Wide Area Integration of Dispersed Wind Farms Using Multiple VSC-HVDC Links 1784
S. González-Hernández, E. Moreno-Goytia, O. Anaya-Lara

Generator Selection for Offshore Oscillating Water Column Wave Energy Converters 1790
D.L. O' Sullivan, A.W. Lewis

A Novel Approach To Photovoltaic Powered Water Pumping Design 1798
Michael James Case, Ernest Edward Denny

Direct Controls in Voltage-Source Converters - Generalizations and Deep Study 1803
Karoly Veszpremi, Istvan Schmidt

Multipolar double fed induction wind generator with a single phase secondary winding 1811
Leonids Ribickis, Guntis Dilevs, Nikolajs Levins, Vladislavs Pugachevs

The measurement on the solar cells in Liberec city 1815
Jiri Kubin

Rotor Turn-to-Turn Faults of doubly-fed Induction Generators in Wind Energy Plants - Modelling, Simulation and Detection 1819
Vincenz Dinkhauser, Friedrich W. Fuchs

Static and Dynamic Response of a Photovoltaic Characteristics Simulator 1827
Anastasios Ch. Nanakos, Emmanuel C. Tatakis

Modeling and Optimal Sizing of Hybrid Renewable Energy System 1834
Rachid Belfkira, Cristian Nichita, Pascal Reghem, Georges Barakat

Photovoltaic System MPPTracker Investigation and Implementation using DSP engine and Buck- Boost DC-DC converter 1840
Dimosthenis Peftitsis, Georgios Adamidis, Panagiotis Bakas, Anastasios Balouktsis

Multi Objective Distributed Generation Planning Using NSGA-II 1847
Muhammad Ahmadi, Ashkan Yousefi, Alireza Soroudi, Mehdi Ehsan

Testing of the Grid-connected Photovoltaic Systems Using FPGA-based Real-Time Model 1852
Robert Stala

xvii

Table of Contents

Output Maximization Using Direct Torque Control for Sensorless Variable Wind Generation System Employing IPMSG...1859
Yukinori Inoue, Shigeo Morimoto, Masayuki Sanada

Improving Connection and Disconnection of a Small Scale Distributed Generator Using Solid-State Controller...1866
M.M.R. Ahmed

Research control of electric systems in wind generator systems...1872
Stefan Winternheimer, Artem Kolesnikov, Evgeny Glushkin, Alexander Bukatov

Stand-alone Photovoltaic Generation System with Combined Storage using lead Battery and EDLC.........1877
Hiroaki Nakayama, Eiji Hiraki, Toshihiko Tanaka, Noriaki Koda, Nobuo Takahashi, Shuji Noda

Active Filter Action of Inverter Exciting Induction Generator for Wind Power Generation........................1884
Noriyuki Kimura, Tomoyuki Hamada, Katsunori Taniguchi, Toshimitsu Morizane

The Operation of Power Electronic Converters in Photovoltaic Drive Systems...1890
Marek Niechaj

Experimental results of a hybrid wind/hydro power system connected to isolated loads..............................1896
Mehdi Nasser, Stefan Breban, Vincent Courtecuisse, Arnaud Vergnol, Benoît Robyns, Mircea M. Radulescu

Grid Connection of Multi-Megawatt Clean Wave Energy Power Plant under Weak Grid Condition...........1904
Kai Rothenhagen, Marek Jasinski, Marian P. Kazmierkowski

Improved sizing method of storage units for hybrid wind-diesel powered system...1911
A.M. Tankari, B. Dakyo, C. Nichita

A Research Platform for a Smart-Blade Wind Generation System...1918
J. Davey, Udaya K. Madawala, R. Sharma

Soft Switching Multi-Phase Boost Converter for Photovoltaic System...1924
Joo-Hyuk Lee, Jae-Hyung Kim, Chung-Yuen Won, Su-Jin Jang, Yong-Chae Jung

Soft Switching Boost Converter for Photovoltaic Power Generation System..1929
Doo-Yong Jung, Young-Hyok Ji, Jae-Hyung Kim, Chung-Yuen Won, Yong-Chae Jung

Optimisation Of Wind Power Pmsm To Grid Conversion System...1934
Ince Kayhan, Weiss Helmut

Analysis of Wind Farm and Multilevel Converter Interactions in Medium Voltage Networks Under Steady-State and Transient Conditions...1941
J. Sosa-Ruiz, E. Moreno-Goytia, O. Anaya-Lara

A Simple, Low Cost Design Using Current Feedback to Improve the Efficiency of a MPPT-PV System for Isolated Locations..1947
Herman Fernández, Abelardo Martínez, Víctor Guzmán, María Isabel Gímenez

A Single-Phase Active Power Filter Based in a Two Stages Grid-Connected PV System...............................1951
Kleber C.A. De Souza, Denizar C. Martins

Wide Bandwidth Power Flow Control Algorithm of the Grid Connected VSI under Unbalanced Grid Voltages...1957
Zoran Ivanovic, Marko Vekic, Stevan Grabic, Evgenije Adzic, Vladimir Katic

The use of Switched Reluctance Generator in wind energy applications...1963
Eleonora Darie, Costin Cepisca, Emanuel Darie

Active Line Shaping of a Single Phase Rectifier using the Switching Function Technique............................1967
Christos Marouchos

Control of Reactive Power in Double-Fed Machine Based Wind Park...1975
Elzbieta Bogalecka, Michal Kosmecki

Table of Contents

A Novel Hybrid Modulation Method for Cascaded H-bridge Active Power Filter 1981
Yonggang Chen, Ping Wang, Yaohua Li, Zixin Li, Longcheng Tan

Apparent Power Ratio of the Shunt Active Power Filter 1987
A. Kouzou, B.S Khaldi, S. Saadi, M.O. Mahmoudi, M.S. Boucherit

Shunt Active Power Filter with Improved Dynamic Performance 1995
Krzysztof Piotr Sozanski

The Research on the Active Power Filter Based on the Cascaded H-bridge Converter 2000
Yonggang Chen, Junling Chen, Ping Wang, Yaohua Li, Longcheng Tan, Zixin Li, Wei Xu

E-laboratory in the Field of Electrical Drives 2005
H.Hõimoja, A.Rosin, T.Möller, M. Müür

Laboratory Setup for Studying Ultracapacitors in Industrial Applications 2011
I. Roasto, D. Vinnikov, T. Lehtla

Synchronous machine direct axis parameters estimation module from an iterative strategy 2015
Emile Mouni, Slim Tnani, Gérard Champenois

Determination of the Characteristic Life Time of Paper-insulated MV-Cables based on a Partial Discharge and tan(..) Diagnosis 2022
I. Mladenovic, Ch. Weindl

Elimination of Increased Excitation of Common- Mode Oscillations in Electrical Drive Systems with Active Front End and Long Motor Cables 2028
Thomas Weidinger

Internal Short Circuit in a Tooth Wound PMSM with Stranded Conductors 2037
Damien Birolleau, Christian Chillet, Laurent Albert

Implementation of a Virtual Laboratory for Low Power Electrical Drives 2043
Gh. BALUTA, V. HORGA, C. LAZAR

DQ-Transformation Approach for Modelling and Stability Analysis of AC-DC Power System with Controlled PWM Rectifier and Constant Power Loads 2049
K-N Areerak, S.V. Bozhko, G.M. Asher, D.W.P. Thomas

Genetic Identification of Parameters the Sandwich Piezoelectric Ceramic Transducers for Ultrasonic Systems 2055
Pawel Fabijanski, Ryszard Lagoda

The Impact of Higher-Order Harmonics on Tripping of Residual Current Devices 2059
Stanislaw Czapp

Estimation of the Untapped Regenerative Braking Energy in Urban Electric Transportation Network 2066
Leonards Latkovskis, Linards Grigans

Performance Evaluation of Electric Power Steering with IPM Motor and Drive System 2071
Hamidreza Akhondi, Jafar Milimonfared, Kaveh Malekian

Optimal Control: Load Frequency Control of a Large Power System 2076
Sílvio José Pinto Simões Mariano*, Luís António Fialho Marcelino Ferreira

LCL-Load Modular Converter For Induction Heating 2082
Maciej A. Dzieniakowski, Jan Fabianowski, Robert Ibach

On-line PID Controller Tuning Using Genetic Algorithm and DSP PC Board 2087
Pawel Fabijanski, Ryszard Lagoda

Regulation Properties of Pumping Station Control System In The Highest Efficiency Range 2091
Szychta Leszek

Table of Contents

Inner Gas Pressure Measurement Based Life-span Estimation of Electrolytic Capacitors..........................2096
A. Riz, D. Fodor, O. Klug, Z. Karaffy

Robust Control Methodologies for Optical Micro Electro Mechanical System - New approaches and Comparison 2102
Alireza Izadbakhsh, S.M.R. Rafiei

Modeling a Buck-Based Switching Amplifier for Sinusoid Wide Band Tracking by Using a Nonlinear Time Varying Map 2108
A. El Aroudi, E. Alarcón, E. Rodriguez, R. Leyva

Single Inductor Multiple Outputs Interleaved Converters Operating in CCM..........................2115
Luis Benadero, Vanessa Moreno-Font, Abdelali El Aroudi, Roberto Giral

Control of a two-cell dc/dc converter in presence of saturating duty cycle 2120
Moez Feki, Abdelali El Aroudi, Bruno Gerard Michel Robert, Nabil Derbel

Bifurcations and Chaotic Dynamics in a Linear Switched Reluctance Motor 2126
M.R. De Castro, B.G.M. Robert, C. Goeldel

Modular Architecture for Decentralized Hybrid Power Systems 2134
E. Ortjohann, M. Lingemann, O.Omari, A. Schmelter, N. Hmasic, A. Mohd, W. Sinsukthavorn, D. Morton

Design of a power management system for an active PV station including various storage technologies 2142
Di Lu, Tao Zhou, Hicham Fakham, Bruno Francois

Energy Management and Power Flow of Decoupled Generation System for Power Conditioning of Renewable Energy Sources 2150
Wlodzimierz Koczara, Zdzislaw Chlodnicki, Nazar Al-Khayat, Neil L.Brown

Inversion Based Control of a Diesel Fed Low Temperature Fuel Cell System 2156
Daniela Chrenko, Marie-Cecile Pera, Daniel Hissel

Power Management in an Autonomous Adjustable Speed Large Power Diesel Gensets 2164
Grzegorz Iwanski, Wlodzimierz Koczara

Cost evaluation of Generator-set with Energy Storage for 4Q-load 2170
Freek J.F.Baalbergen, Pavol Bauer

Integrating renewable energy sources and storage into isolated diesel generator supplied electric power systems 2178
Chad Abbey, Jonathan Robinson, Géza Joós

Performance comparison of different wind generator based hybrid systems 2184
Vincent Courtecuisse, Benoit Robyns, Marc Petit, Bruno Francois, Jacques Deuse

First Approach for a Fault Tolerant Power Converter Interface for Multi-Stack PEM Fuel Cell Generator in Transportation Systems 2192
Alexandre De Bernardinis, Gérard Coquery

Development of Electrical System for Hybrid Vehicles Using the Free-swinging Piston Engine and Oscillating Rotating Generator 2200
Sigitas Kudarauskas

Power flow control in different time scales for a wind/hydrogen/super-capacitors based active hybrid power system 2205
ZHOU Tao, LU Di, FAKHAM Hicham, FRANCOIS Bruno

Neuro-Fuzzy Adaptive Control of the IM Drive with Elastic Coupling 2211
Teresa Orlowska-Kowalska, Krzysztof Szabat, Mateusz Dybkowski

Control of Flexible Drive with PMSM employing Forced Dynamics..........................2219
Vittek Ján, Bris Peter, Makys Pavol, Stulrajter Marek, Vavrus Vladimír

Table of Contents

The problems of high dynamic drive control under circumstances of elastic transmission 2227
Jan Deskur, Roman Muszynski

Protective Predictive Control of Electrical Drives with Elastic Transmission 2235
Mario Vasak, Nedjeljko Peric

Low-Cost High-Performance Predictive Control of Drive Systems with Elastic Coupling 2241
Marcin Cychowski, Kieran Delaney, Krzysztof Szabat

Development of an Expert System for Identification, Commissioning and Monitoring of Drives 2248
Mario Pacas, Sebastian Villwock

Control of Axial Flux Permanent Magnet Motor by the PIPCRM Method at Standstill and at Low Speed 2254
Janusz Wisniewski, Wlodzimierz Koczara

Zero Speed Position Estimation of a Matrix Converter Fed AC PM Machine using PWM Excitation 2261
Q. Gao, G. M. Asher, M. Sumner

Sensorless Direct Torque and Flux Control of an IPM Synchronous Motor at Low Speed and Standstill 2269
Gilbert Foo, S. Sayeef, M.F. Rahman

Sensorless Control of PM Synchronous Motors Using a Predictive Current Controller with Integrated INFORM and EMF Evaluation 2275
Manfred Schrödl, Christian Simetzberger

Torque Sensorless Control of Induction Motor 2283
Karel Jezernik, Miran Rodic

Application of the induction motor torque - observer to the control of turbo - machines 2289
Andrzej Debowski, Daniel Lewandowski

Observer of induction motor speed based on exact disturbance model 2294
Zbigniew Krzeminski

Experimental Performance Evaluation for Low Speed and Regenerating Operation of Sensor-less Vector Control System of Induction Motor Using Observer Gain Tuning 2300
Kazuhiro Ohyama, Greg Asher, Mark Sumner

Application of the Stator Current-based MRAS Speed Estimator in the Sensorless Induction Motor Drive 2306
Mateusz Dybkowski, Teresa Orlowska-Kowalska

State and Parameter Estimation in Induction Motors using Sliding Modes 2312
Sachit Rao, Martin Buss, Vadim Utkin

Torque Transient Alleviation in Fixed Speed Wind Generators by Indirect Torque Control with STATCOM 2318
Marta Molinas, Jon Are Suul and Tore Undeland

Flicker Study on Variable Speed Wind Turbines with Permanent Magnet Synchronous Generator 2325
Weihao Hu, Zhe Chen, Yue Wang, Zhaoan Wang

Power Output Characteristics Analysis of Wind Energy Converter Control Methods 2331
Bingchang Ni, Constantinos Sourkounis

A Cooperative Control Method for Output Power Smoothing and Hydrogen Production by Using Variable Speed Wind Generator 2337
Rion Takahashi, Hirotaka Kinoshita, Toshiaki Murata, Junji Tamura Masatoshi Sugimasa, Akiyoshi Komura, Motoo Futami, Masaya Ichinose, Kazumasa Ide

A new interconnecting method for wind turbine/generators in a wind farm and basic characteristics of the integrated system 2343
Shoji Nishikata, Fujio Tatsuta

Educational aspects of mechatronic control course design for collaborative remote laboratory 2349
Andreja Rojko, Darko Hercog, Karel Jezernik

Table of Contents

PEMCWebLab - Distance and Virtual Laboratories in Electrical Engineering: Development and Trends.............**2354**
Pavol Bauer, Viliam Fedák, Otto Rompelman

Integrated multimedia educational program of a DC servo system for distant learning.............**2360**
Gabor Sziebig, Istvan Nagy, Rafael Kalman Jardan, Peter Korondi

Electromechanical Actuators WEB-lab.............**2368**
Dusan Maga, Jan Sitar, Juraj Dudak, Rene Hartansky, Peter Siroky, Jan Halgos, Pavol Bauer

Power Quality and Active Filters as Web-Controlled Experiment in the frame of PEMC WebLab.............**2371**
Volker Staudt, Andreas Steimel, Pavol Bauer, Vítezslav Hájek

Distant learning of Pulse Width Modulation Techniques for Voltage Source Converters.............**2378**
Bartlomiej Kamiski, Dariusz Sobczuk

Modern design optimisation exploiting field simulation.............**2383**
Jan K. Sykulski

Transmission-Line Modelling of Wave Propagation Effects in Machine Windings.............**2385**
Herbert De Gersem, Olaf Henze, Thomas Weiland, Andreas Binder

An efficient field-circuit coupling method by a dynamic lumped parameter reduction of the FE model.............**2393**
F. Henrotte, E. Lange, K. Hameyer

Coupled field-circuit-mechanical model of an electromagnetic actuator operating in error actuated control system.............**2400**
Lech Nowak

Simulation and Investigation of Magnetorheological Fluid Brake.............**2406**
Wieslaw Lyskawinski, Wojciech Szelag, Cezary Jedryczka

Field and Field-Circuit Description of Electrical Machines.............**2412**
Andrzej Demenko, Kay Hameyer

Interaction between Thermal Impedance and Parasitics in Power Sections.............**2420**
Stefan Forster, Andreas Lindemann

Discussion of Internal and External High Frequency Common Mode Noise Current on a Chopper Circuit.............**2428**
Tetsuya Mitani, Keiji Wada, Toshihisa Shimizu, Hiromichi Ohashi

A Novel Digital Control Method for DC-DC Converter.............**2434**
Fujio Kurokawa, Masashi Okamatsu, Yuichi Sumida, Yasuhiro Mimura, Masahiro Sasaki

A Novel Single/Three-phase Matrix Converter For High Power Integration.............**2439**
Makoto Saito

An Effective Design Method for High Power Density Converters.............**2445**
Yusuke Hayashi, Kazuto Takao, Toshihisa Shimizu, Hiromichi Ohashi

Power Devices in Polish National Silicon Carbide Program.............**2452**
Mariusz Sochacki, Andrzej Kubiak, Zbigniew Lisik, Jan Szmidt

SiC Power Semiconductor Devices for new Applications in Power Electronics.............**2457**
Dominique Planson, Dominique Tournier, Pascal Bevilacqua, Nicolas Dheilly, Herve Morel, Christophe Raynaud, Mihai Lazar, Dominique Bergogne, Bruno Allard, Jean-Pierre Chante

Silicon carbide Schottky diodes and MOSFETs: solutions to performance problems.............**2464**
Owen J. Guy, Michal Lodzinski, Ambroise Castaing, P. M. Igic, Amador Perez-Tomas, Michael R. Jennings, Philip A. Mawby

Characterization of the Static and Dynamic Behavior of a SiC BJT.............**2472**
M.M.R. Ahmed, N-A.Parker-Allotey, P.A. Mawby, Muhammed Nawaz, Carina Zaring

An active network control method using distributed energy resources in microgrids.............**2478**
Takayuki Tanabe, Yoshinobu Ueda, Toshihisa Funabashi, Shigeo Numata, Kimio Morino, Eisuke Shimoda

Table of Contents

Energy Management in Solar Photovoltaic Plants based on ESS 2481
M. Lafoz, L. García-Tabarés, M. Blanco

A Method of Three-Phase Balancing in Microgrid by Photovoltaic Generation Systems 2487
Masahide Hojo, Yuta Iwase, Toshihisa Funabashi, Yoshinobu Ueda

Development of HILS(Hardware In-Loop Simulation) System for MMS(Microgrid Management System) by using RTDS 2492
Jin-Hong Jeon, Jong-Yul Kim, Seul-Ki Kim, ong-Bo Ahn, JuneHo Park

Power Quality Analysis of Jeju Island Power System with Wind Farm and HVDC System 2498
Jae-Hong Kim, Eel-Hwan Kim, Se-Ho Kim, Jaeho Choi, Gil-Soo Jang, Seung-Ho Song

A New Control Method for Power Turbine Generators Using an Accurate Ship Plant System Model 2504
Nobumasa Matsui, Fujio Kurokawa, Keiichi Shiraishi

Voltage profile support in distribution networks - influence of the network R/X ratio 2510
B. Bla~ic, I. Papic

Modeling, Simulation and Analysis of Conducted Common-Mode EMI in Matrix Converters for Wind Turbine Generators 2516
S. Zhang, K.J. Tseng

Design of Frequency Shift Acceleration Contol for Anti-islanding of an Inverter-based DG 2524
Seul-Ki Kim, Jin-Hong Jeon, Heung-Kwan Choi, Jonng-Bo Ahn

Integrated Power Converter for Photovoltaic and Fuel Cell Systems in Home 2530
Yasuyuki Nishida, Shinichiro Sumiyoshi, Hideki Omori

A Comparison of Position Control Structures for Ironless Linear Synchronous Motor 2538
Martin Hrasko, Pavol Makys, Marek Franko, Jozef Kuchta

A Comparison of Sliding Mode Approaches to a Nanometre Position Control Application 2543
Paul Andreas Stadler, Stephen James Dodds

Sliding Mode Control of PMSM Drives Subject to Torsion Oscillations in the Mechanical Load 2551
Stephen J. Dodds, Jan Vittek

Sliding Mode Vector Control of PMSM Drives with Minimum Energy Position Following 2559
Stephen J. Dodds

xxiii

A New Digital Control Method for High Performance 400 Hz Ground Power Unit

Zixin Li[1,2], *Student Member, IEEE*, Ping Wang[1], Haibin Zhu[1,2], Yaohua Li[1],
Longcheng Tan[1,2], Yonggang Chen[1,2], Fanqiang Gao[1,2]
1. Institute of Electrical Engineering of Chinese Academy of Sciences
2. Graduate University of Chinese Academy of Sciences
P.O. Box 2703, 100190, Beijing, P. R. China
E-mail: zixinli@ieee.org

Abstract-In digital control, the one-sample delay for computation which is often ignored in 50/60Hz cases has great impact on the performance of a 400Hz ground power unit (GPU). Taking the delay into account, a detailed proof that it is very hard and even impossible for 400Hz GPU to maintain a low THD content in the output voltage using the conventional inductor, capacitor or load current feedback based single-loop or double-loop control is given. A new primary-plus-resonant -controller method suitable for fixed-point DSP without modifying the existing single-loop or double-loop control topology is presented. With the new approach, the computation delay is not compensated, which means that complex computation will be avoided. Meanwhile, the relationships between the discrete realization method in other literatures and the Zero Order Hold method for resonant controllers have also been discussed. Two types of tests on a 90-kVA 400Hz GPU prototype based-on a 16-bit fixed-point DSP show satisfactory performance of the new method feeding linear / nonlinear load.

I. INTRODUCTION

For inverter-based power supplies, such as 50Hz or 60Hz uninterruptible power supply (UPS) and 400Hz ground power unit (GPU) for airplanes, a main problem is to maintain a low THD content in presence of nonlinear loads. In recent years, many advanced control strategies have been proposed so as to improve the performance of power supply, such as, sliding mode control, repetitive control and dead-beat control etc. Most of them are for 50/60Hz UPS applications and few can be directly used in high power GPU as the fundamental frequency is much more higher than the UPS. [1] gives a new single-loop control method for 400Hz GPUs to avoid the employment of the double-loop control. However, the new method is not capable of controlling the harmonics and a discrete Fourier transformation (DFT) method has to be used just as proposed in [2] of which the harmonic content will be refreshed every fundamental period. [3] presents a resonant controller-based double-loop control scheme in stationary coordinates. As the resonant controller is located in the voltage loop, it has the drawback of incapability of harmonic control that will be explained in this paper. Meanwhile, many additional disturbance terms from discretization have to be compensated so as to improve the performance. As the disturbance terms are determined by the parameters of the LC filter, thus, the control effect of the algorithm is based on the accuracy of the parameters. To improve the method, [4] used a Luenberger observer to remove the influence of the

computation delay. However, parameter accuracy is still the main problem and some feed-forward terms are also used to reduce the phase lag. In [5], a three-layer voltage controller for 50Hz AC power supplies has been proposed in which the resonant controller is implemented in a DFT style. The one-sample delay in digital control will do deteriorate the performance of a 400Hz GPU which can be ignored in most 50/60Hz cases. The delay should be compensated to alleviate the impact [3], [4]. [6] and [7] offer a direct approach to lower the harmonic for a 60Hz power supply. However, the new approach requires a much higher switching frequency which is not acceptable with the current power devices.

II. MODEL OF THE GROUND POWER UNIT

Fig.1 shows the typical topology of a three-phase GPU which can represent two types. Topology I is a three-leg inverter with a Delta-Star connected transformer and Topology II is a three single-phase H-bridge inverter with a Star-Star connected transformer. The filter inductor L can be integrated in the leakage of the transformer in many cases. Generally, the unipolar-SPWM is preferred for Topology I and SVPWM is often the choice for Topology II [7].

Both of the two topologies can be modeled in the stationary α-β coordinate or in the synchronous d-q coordinate. When using d-p transformation the model of the GPU on d-axis and q-axis has coupled terms while the problem does not exit in the stationary coordinate [1], [3], and [4]. However, the coupled terms can be ignored in order to simplify the control [1]. Suppose that the transformer ratio is 1 and the continuous time model of the three-phase GPU in α-β coordinate or the simplified model in d-q coordinate can both be treated as two single-phase model as depicted in Fig.2. In the following part, Fig.2 will be taken as the model to analyze the design of the GPU controller. In Fig.2, L, C and *io* are the filter inductor, filter capacitor and load current. *io* will be taken as a disturbance of the GPU. The inductor equivalent series resistance (ESR) is *r* and capacitor ESR is ignored.

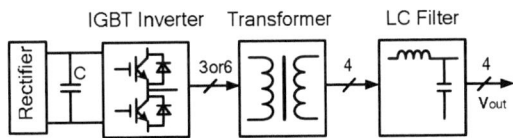

Fig. 1. Typical Topology of A Three-phase GPU

978-1-4244-1741-4/08/$25.00 ©2008 IEEE

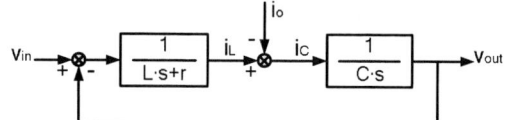

Fig. 2. Analogue Model for Analysis of GPU

III. CONTROL TOPOLOGY OPTIONS

There are various controllers to choose for the GPU. Generally, the single-loop and double-loop (or cascade) controllers are these most often used. The parameters for the GPU to be controlled is listed in TABLE I.

TABLE I
GPU PARAMETERS

Filter Inductor	L=150μH
Filter Capacitor	C=48μF
Filter Resistance:	r=0.2Ω
Sampling Frequency	fs=1/T=12kHz
Switching Frequency	fw=6kHz

A. Single-Loop Control

A single-loop voltage control seems to have a simple structure, but the increase of the voltage loop gain has a tendency to unstablize the system and this control topology may have problems with the varying load [1], [9]. A new control method is given in [1] to improve the performance of the single-loop control. The discrete model with one-sample delay of the GPU by Zero Order Hold method is

$$F(z) = \frac{1}{z} \cdot \frac{N_1 \cdot z + N_0}{(z - z_p) \cdot (z - z_{p^*})} \quad (1)$$

in which
$z_p, p^* = 0.5265 \pm j0.7859$, $N_1 = 0.42896$, and $N_0 = 0.41286$.
The controller can be expressed as [3]

$$D(z) = K_v \cdot \frac{(z - z_p) \cdot (z - z_{p^*})}{z \cdot (z - 1)} \quad (2)$$

z_p and z_p* are the two complex poles of the discrete model. The root locus of the GPU and the controller $F(z) \cdot D(z)$ with no load is shown in Fig.3. To guarantee the stability of the system with a relatively higher damping ratio, the controller gain K_v can not be very high. For instance, the damping ratio is about 0.7 when $K_v = 0.3$. The transfer function of the closed-loop system is

$$G(z) = \frac{F(z) \cdot D(z)}{1 + F(z) \cdot D(z)} = \frac{K_v \cdot N_1 \cdot z + K_v \cdot N_0}{z^3 - z^2 + K_v \cdot N_1 \cdot z + K_v \cdot N_0} \quad (3)$$

The frequency response is shown in Fig. 4 and the bandwidth of the system is about 1050Hz. There is a tradeoff between K_v and the damping ratio, so the bandwidth will not be broad enough to suppress harmonic such as 3rd, 5th and 7th introduced by nonlinear load. A DFT method is taken to control the harmonic in [1] to improve the performance of the

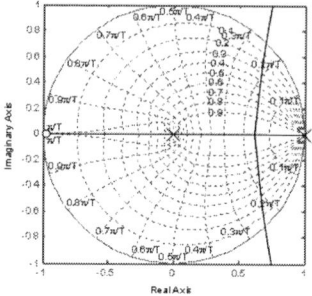

Fig. 3. Root Locus of the Single-Loop Control

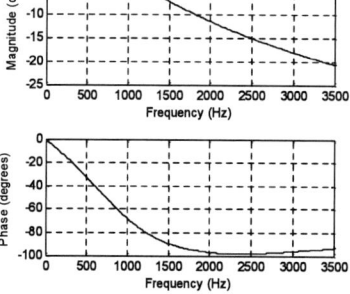

Fig.4 Frequency Response of the Single-Loop Controlled GPU

controller. However, as the harmonic detecting result refreshes every fundamental period, the response time for harmonics is long.

B. Inductor Current Based Double-Loop Control

The inclusion of an inner inductor current-loop controller in a regulation of a DC/AC power supply has many advantages: such as, alleviating the influence of the inductor parameter variation, inherent current limiting capability, decoupled design of the voltage-loop and current-loop etc [9]. However, for 400Hz GPU the one-sample delay has great impact on the inner current-loop which will limit the bandwidth of the inner loop and makes the GPU unable to suppress harmonics. A double-loop control with an inner inductor current-loop is shown in Fig.5. The following analysis is focused on the inner current-loop and the voltage-loop will not be discussed. From Fig.5, the inductor current can be expressed as

$$i_L = i_L^* \frac{K_p}{sL + (r + K_p)} - v_{out} \frac{1}{sL + (r + K_p)} \quad (4)$$

Treating v_{out} as a disturbance of the inductor current and taking the one-sample delay into account, the inner current-loop is revised into Fig.6. The discrete model of the inductor current loop with the ESR by ZOH method can be written as

$$\frac{(z-1)}{z} \cdot \frac{1}{r} \cdot \frac{z(1 - e^{-T \cdot r/L})}{(z-1)(z - e^{-T \cdot r/L})} - \frac{1}{r} \cdot \frac{(1 - e^{-T \cdot r/L})}{(z - e^{-T \cdot r/L})} \quad (5)$$

Given the parameters in TABLE I, one can get that $T \cdot r/L \ll 1$ and (5) can be approximated by

516

$$\frac{1}{r} \cdot \frac{(1-e^{-T \cdot r/L})}{(z-e^{-T \cdot r/L})} \approx \frac{1}{r} \cdot \frac{(T \cdot r/L)}{(z-1)} = \frac{T}{L} \cdot \frac{1}{(z-1)} \qquad (6)$$

The transfer function of the discrete current-loop is

$$H(z) = \frac{i_L}{i_L^*} = \frac{T}{L} \cdot \frac{K_p}{z(z-1) + K_p \cdot T/L} \qquad (7)$$

For guaranteeing the stability of the current-loop, the poles of the transfer function (7) must be all in the unit circle and then K_p must satisfy the inequality

$$K_p < L/T \qquad (8)$$

Suppose that $K_p = a \cdot L/T$ where $0 < a < 1$, the frequency response is depicted in Fig.7 in which $a = 0.3$ with the bandwidth about 1250Hz. K_p should be larger to make broader the bandwidth of the current-loop. However, for the sampling frequency is 12 kHz, at the frequency of 2000Hz

$$H(e^{j \cdot 2\pi fT}) = H(e^{j \cdot \pi/3}) = \frac{a}{a-1} < 0 \qquad (9)$$

This means that the phase lag is π at 2000Hz which is just the 5th harmonic of the fundamental frequency and the 5th harmonic will become a positive feedback. Therefore, the gain K_p has to be kept low to damp the 5th harmonic which will make the bandwidth much less than 2000Hz. Another way to broaden the bandwidth is to employ a higher switching frequency, but this is often not desirable for the limitation of the power devices especially in high power situations.

The performance of the whole system is constrained by the inner current-loop. In presence of nonlinear load, for example, the rectifier load, 5th and 7th harmonic current will come into the inner current-loop. However, as can be seen from Fig.2, the current-loop is unable to provide those harmonic currents to cancel the load current for the limitation of current-loop bandwidth and the output voltage will be deteriorated.

C. Capacitor Current Based Double-Loop Control

If a capacitor current is taken as the feedback of the inner current-loop, it seems that it has the ability of suppressing the impact of the load current. However, when the load current i_o is treated as a disturbance and ignored, the same problem as the inductor current feedback topology exits still. Namely, the capacitor current feedback has the same problem and the bandwidth of the inner current loop can not be made high enough to suppress the lower harmonics in a GPU.

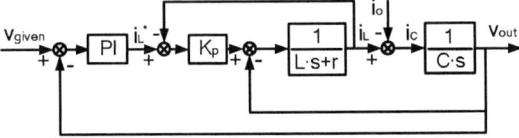

Fig. 5 Double-Loop Control with Inductor Current Feedback

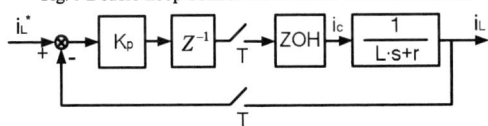

Fig.6 Inner Current Loop Model Considering One-Sample Delay

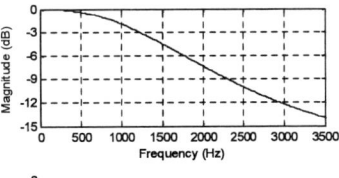

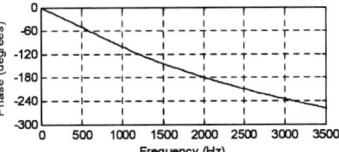

Fig.7 Frequency Response of the Current Loop

From the above analysis, it is clear that the conventional double-loop control or the single-loop control in [1] can not give a satisfactory performance and some techniques for harmonic control have to be adopted.

IV. HARMONIC CONTROL

A. The Resonant Controller

Based on the analysis in III, one can see that the conventional double-loop controller or the new single-loop controller is unable to control the harmonics. To control the voltage harmonics, resonant controller may be a good choice for its infinite gain at the resonant frequency and zero error tracking characteristic when used to control AC signals. A resonant controller can be [11]

$$R'(s) = \frac{R_{out}(s)}{R_{in}(s)} = \frac{1}{s^2 + \omega_n^2} \qquad (10)$$

or [10]

$$R''(s) = \frac{R_{out}(s)}{R_{in}(s)} = \frac{2 \cdot s}{s^2 + \omega_n^2} \qquad (11)$$

As can be seen, (10) will introduce a phase shift of 180° into the system which will reduce the phase margin greatly. However, the phase shift is only 90° in (11). Hence, (11) is preferred in practical uses. In the following analysis, the coefficient 2 will be ignored for convenience and (11) can be revised as

$$R(s) = \frac{R_{out}(s)}{R_{in}(s)} = \frac{s \cdot \cos(\theta_n) - \omega_n \cdot \sin(\theta_n)}{s^2 + \omega_n^2} \qquad (12)$$

where θ_n is the system delay compensation. The resonant controller can be descretized as (13) using the Zero Order Hold method.

$$R_{zoh}(z) = \frac{(k_{1n} - k_{2n}) \cdot z - k_{1n} - k_{2n}}{z^2 - 2\cos(\omega_n \cdot T)z + 1} \cdot \frac{1}{\omega_n} \qquad (13)$$

where

$k_{1n} = \cos(\theta_n) \cdot \sin(\omega_n \cdot T)$, $k_{2n} = \sin(\theta_n) \cdot [1 - \cos(\omega_n \cdot T)]$. As is seen from (13), the coefficients are so small that it is hard for a 16-bit fixed-point DSP to realize. The truncating errors will be great especially when the resonating frequency ω_n becomes high. [3], [4] and [11] adopt another method to make the realization of the discrete resonant controller feasible. This method is just called the virtual LC method for convenience. Suppose that there is a virtual LC filter with $L = C = 1/\omega n$ as depicted in Fig.8. If Rin is supposed to be the input voltage, the inductor current and capacitor voltage can be expressed as

$$IL(s) = Rin(s) \cdot \frac{s \cdot \omega n}{s^2 + \omega n^2} \quad VC(s) = Rin(s) \cdot \frac{\omega n^2}{s^2 + \omega n^2} \quad (14)$$

The output of the resonant controller (12) can just be expressed as

$$Rout(s) = [IL(s) \cdot \cos(\theta_n) - Vc(s) \cdot \sin(\theta_n)] / \omega n \quad (15)$$

For a specific resonant controller, ω_n is a constant and it can be synthesized in the coefficient multiplied by the input or the output of the resonant controller. Therefore, (15) can be re-written as

$$Rout(s) = [IL(s) \cdot \cos(\theta_n) - Vc(s) \cdot \sin(\theta_n)] \quad (16)$$

The continuous state space equations of the LC filter can be expressed as

$$\begin{bmatrix} \dfrac{dIL}{dt} \\[2mm] \dfrac{dVc}{dt} \end{bmatrix} = \begin{bmatrix} 0 & -\omega n \\ \omega n & 0 \end{bmatrix} \begin{bmatrix} IL \\ Vc \end{bmatrix} + \begin{bmatrix} \omega n \\ 0 \end{bmatrix} \cdot Rin \quad (17)$$

The corresponding discrete model can be calculated as

$$\begin{bmatrix} IL(k+1) \\ VC(k+1) \end{bmatrix} = \begin{bmatrix} f11n & f12n \\ f21n & f22n \end{bmatrix} \begin{bmatrix} IL(k) \\ VC(k) \end{bmatrix} + \begin{bmatrix} g1n \\ g2n \end{bmatrix} \cdot Rin(k)$$

$$\triangleq F \cdot \begin{bmatrix} IL(k) \\ VC(k) \end{bmatrix} + G \cdot Rin(k) \quad (18)$$

where

$f11n = f22n = \cos(\omega n \cdot T)$, $f21n = -f12n = \sin(\omega n \cdot T)$, $g1n = \sin(\omega n \cdot T)$ and $g2n = 1 - \cos(\omega n \cdot T)$. In digital control, the output of the discrete resonant controller can be written as

$$Rout(k) = [\cos(\theta n) \quad -\sin(\theta n)] \begin{bmatrix} IL(k) \\ VC(k) \end{bmatrix}$$

$$\triangleq C \cdot \begin{bmatrix} IL(k) \\ VC(k) \end{bmatrix} \quad (19)$$

As is seen, the resonant controller is implemented as a second order IIR filter. The abstract values of all these coefficients are less than one and have the minimum coefficient round-off errors even though the harmonic orders are different [3], [4], [11]. This method is just called the fictitious LC method.

However, the fictitious LC method is not the simplest one and it has something to do with the Zero Order Hold method. In fact, from the discrete state space model of the virtual LC filter, the z-transfer matrix or z-transfer function (for this is a SISO system) can be calculated as

$$R'(z) = C(z \cdot I - F)^{-1} G = \frac{(k1n - k2n) \cdot z - k1n - k2n}{z^2 - 2\cos(\omega n \cdot T)z + 1} \quad (20)$$

Obviously, the two discrete transfer functions in (13) and (20) are the same except for a constant coefficient. The virtual LC method just makes the small coefficients of the numerator from Zero Order Hold method multiplied by ωn. In this paper, (20) is taken directly as the discrete model of the resonant controller because it needs less computation compared with

the virtual LC method1 and the stability of the system can be analyzed using it.

B. The Proposed Method

[3] adopts the resonant controller in the outer voltage-loop to control voltage harmonics, but it does no good to the inner current-loop of which the bandwidth is greatly reduced by the computation delay and parameter-dependent terms have to be added to the current-loop. [4] adds resonant controllers to the current-loop to improve the performance and a Luenberger observer is used to remove the influence of the one-sample However, many disturbance terms have to be compensated which are still parameter dependent. A new method for harmonic control is depicted in Fig.9 in which the resonant controller is taken as the harmonic controller. The primary single-loop or double-loop controller works either in the synchronous d-q or stationary α-β coordinate; the harmonic controllers can be located in the α-β, d-q or the stationary a-b-c frame.

The new scheme does not affect the single-loop or double-loop controller and the structure is simple. The primary controller—the single-loop or double-loop controller is used to control the fundamental voltage and the harmonic controller is used to control the harmonic voltage respectively.

As the primary controller is set to control the fundamental voltage, the bandwidth of it does not need to be high. What's more, resonant controller is capable of tracking sinusoidal signals with zero errors although its bandwidth is very narrow. Therefore, a new control method for GPU is proposed further in which resonant controllers are adopted both for fundamental control and harmonic control in stationary a-b-c frame or in α-β frame. The block diagram for fundamental and harmonic controller is shown in Fig.10. The computation delay is considered in the discrete model in which $Rh(z)$ is the resonant controller.

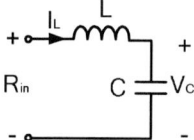

Fig.8 The Virtual LC Filter for Resonant Controller

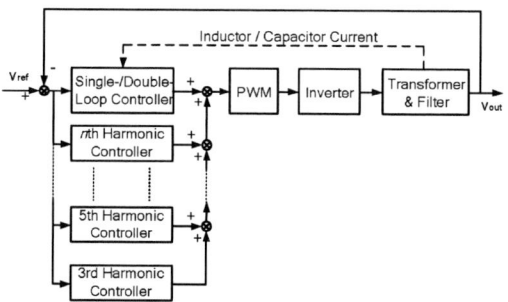

Fig.9 The Proposed New Method

The fundamental and harmonic output voltage of the discrete model can be expressed as

$$Vo\_n = Vref\_n \cdot \frac{Rh(z) \cdot H1(z)}{1 + Rh(z) \cdot H1(z)} - io \cdot \frac{H2(z)}{1 + Rh(z) \cdot H1(z)} \quad (21)$$

where $n = 1, 2, 3, \ldots$ and $H1(z)$ and $H2(z)$ are created in the discretization. At the resonant frequency, the resonant controller $Rh(z)$ has infinite gain while $H1(z)$ and $H2(z)$ have finite gain, so (21) becomes

$$Vo\_n = Vref\_n \quad (22)$$

For fundamental, the output voltage will equals the reference signal exactly. For harmonics, as the reference signal is zero, the output voltage harmonics will be eliminated completely. Given every controller works at different frequencies, the parameters of them can be designed respectively.

B. *Design of the Harmonic Controller*

The phase-lag of different harmonic mainly caused by the one-sample delay and the inverter with the LC filter are not the same and they must be compensated respectively. The harmonic controller can be written as

$$R(z) = Kn \cdot \frac{(k1n - k2n) \cdot z - k1n - k2n}{z^2 - 2\cos(\omega n \cdot T)z + 1} \quad (23)$$

where Kn is the coefficient for each harmonic controller. Not liking the discrete resonant controller by the virtual LC method, Kn can be calculated to ensure the stability of the system under different load conditions. Obviously, (24) is always true no matter what value is chosen for Kn. The phase-lag θ_n for each harmonic is mainly caused by the LC filter and the computation delay. For instance, the phase-lag of 5th harmonic caused by LC filter and the one-sample delay are -140° and -60° respectively. Then the total phase-lag to be compensated is about 200°. In terms of the newly proposed control scheme, the above analysis is also suitable for the design of the fundamental resonant controller.

V. Experimental Results

The proposed methods have been tested on a 90kVA GPU prototype of Topology I. The unipolar-SPWM is used with the sampling frequency of 12 kHz and the switching frequency of 6 kHz. The DC bus is connected to a three-phase diode rectifier with the input voltage 380V/50Hz. Other parameters are listed in TABLE I. Two types of experiments have been done to verify the proposed schemes. The first type is the single-loop control proposed in [1] in synchronous frame and the second type is the newly proposed method in this paper. The harmonic controllers are in stationary a-b-c frame in both cases and the computation is implemented on a 16-bit fixed-point TMS320LF2407 DSP.

Fig.11-Fig.14 are the results of the method in [1] feeding a 22kW three-phase resistance load. The THD is 2.4% when using only the single-loop control, which may be satisfactory. However, the THD decreased to 1.0% when the 3rd and 5th harmonic controllers are adopted. The dead time effect can deteriorate the performance of an inverter, but it is not compensated in all the experiments. It is clear that the

nonlinearity of the inverter can be reduced greatly by the new method. The test results on a three-phase rectifier load are shown in Fig.15-Fig.18. As is seen in Fig.15 and Fig.16, the output voltage has been deteriorated greatly by the harmonic current although the nonlinear load is not heavy and the THD is up to 4.5%. On the contrast, in Fig.17 and Fig.18, the THD is only 1.4% with the help of the 3rd, 5th and 7th harmonic controller. The output performance has been improved substantially by the proposed scheme under nonlinear load.

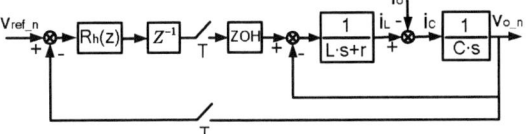

Fig. 10 Model for Fundamental and Harmonic Controller

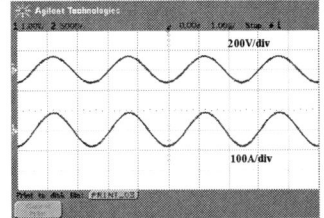

Fig. 11 Output Voltage and Current without Harmonic Controller
(Resistance Load)

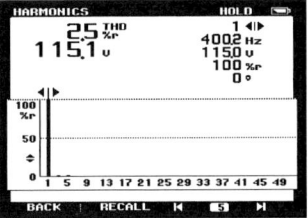

Fig. 12 Output Voltage THD without Harmonic Controller
(Resistance Load)

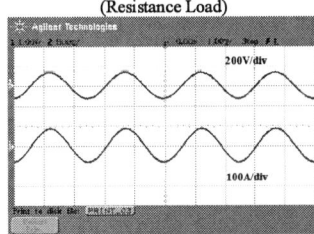

Fig. 13 Output Voltage and Current with Harmonic Controller
(Resistance Load)

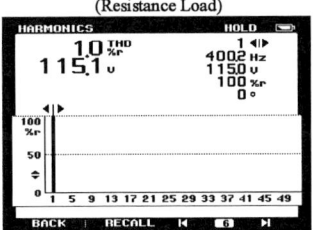

Fig. 14 Output Voltage THD with Harmonic Controller
(Resistance Load)

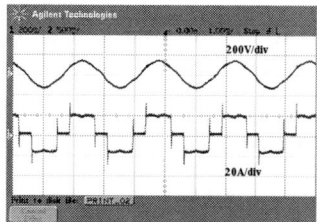

Fig. 15 Output Voltage and Current without Harmonic Controller (Rectifier Load)

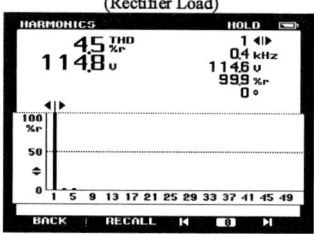

Fig. 16 THD of Output Voltage without Harmonic Controller (Rectifier Load)

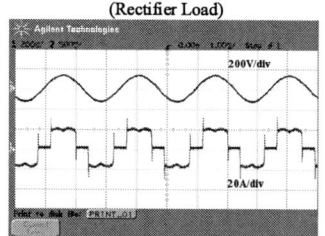

Fig. 17 Output Voltage and Current with Harmonic Controller (Rectifier Load)

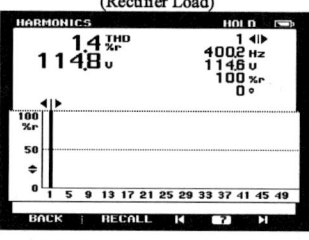

Fig. 18 Output Voltage THD with Harmonic Controller (Rectifier Load)

Fig.19-Fig.20 show the results of the newly proposed method using resonant controllers both for fundamental and harmonic control. As the steady state performances of this method are almost the same as the first experiments, only the waveforms at load change are given. Fig.19 is the three-phase output voltages and current of phase A from no load to a 13kW resistance load change of this phase while Fig.20 is the result of a 39kW three-phase resistance load change. Both cases show good performance of the proposed method.

VI. CONCLUSION

In this paper, a detailed proof is given to show that it is hard for a 400Hz GPU to reach high performance with the conventional control topology considering the one-sample delay. A new method without affecting the conventional control has been proposed to enhance the performance of

GPU. The relationships between the discrete realization method in other literatures and the Zero Order Hold method for resonant controllers have been analyzed in detail. Experiments of the new scheme with the single-loop control on a 90kVA prototype show its simplicity and validity. Another advantage of the new approach is that it can give satisfactory performance even on 16-bit fixed-point DSP.

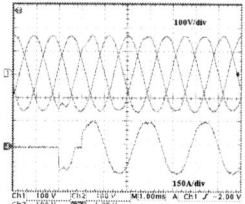

Fig. 19 13kW Phase A Load Change

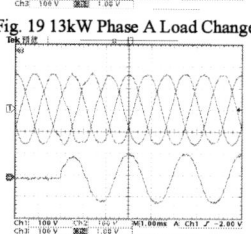

Fig. 20 39kW Three-Phase Load Change

REFERENCES

[1] Uffe Borup Jensen, Frede Blaabjerg, John K. Pedersen, "A New Control Method for 400-Hz Ground Power Units for Airplanes," *IEEE Trans. Ind. Applicat.*, Vol. 36, Jan.-Feb., 2000, pp. 180-187.

[2] Annette von Jouanne, Prasad N. Enjeti, Donald J. Lucas, "DSP Control of High-Power UPS Sytems Feeding Nonlinear Loads," *IEEE Trans. Ind. Electron.* vol. 43, Feb, 1996, pp. 121-125.

[3] Liviu Mihalache, "DSP Control of 400 Hz Inverters for Aircraft Applications," *Industry Applications Conference, 2002. 37th IAS Annual Meeting*, vol.3, pp. 1564 – 1571.

[4] Liviu Mihalache, "Improved Load Disturbance Rejection Method for 400 Hz GPU Inverters," *Applied Power Electronics Conference and Exposition, 2004. APEC '04. Nineteenth Annual IEEE* vol. 1, pp.95 – 101.

[5] Paolo Mattavelli, "Synchronous-Frame Harmonic Control for High-Performance AC Power Supplies," *IEEE Trans. Ind. Applica.*, vol. 37, May-June, 2001, pp. 864 – 872.

[6] Uffe Borup Jensen, Prasad N. Enjeti, Frede Blaabjerg, "A New Space-Vector-Based Control Method for UPS Systems Powering Nonlinear and Unbalanced Loads," *IEEE Trans. Ind. Applicat.*, vol. 37, Nov.-Dec., 2001, pp. 1864-1870..

[7] Uffe Borup Jensen, "Design and Control of a Ground Power Unit," Doctoral dissertation, Aalborg University, 2000.

[8] Micheal J. Ryan, William E. Brumsickle, Robert D. Lorenz, "Control Topology Options for Single-Phase UPS Inverters," *IEEE Trans. Ind. Applicat.*, vol. 33, Mar.-Apr., 1997, pp. 493-500.

[9] Ying-Yu Tzou, "DSP-Based Fully Digital Control of a PWM DC-AC Converter," in *Proc. IEEE PESC '95*, 1995, pp. 138-144.

[10] Daniel Nahum Zmood, Donald Grahame Holmes, "Stationary Frame Current Regulation of PWM Inverters With Zero Steady-State Error," *IEEE Transactions on Power Electronics*, vol. 18, No. 3, May, 2003.

[11] Sato, Y., Ishizuka, T., Nezu, K., Kataoka, T.," A new control strategy for voltage-type PWM rectifiers to realize zero steady-state control error in input current." *IEEE Trans. Ind. Applicat.*, vol. 34, May-June, 1998, pp. 480-486.

[12] Paolo Mattavelli, "A Closed-Loop Selective Harmonic Compensation for Active Filters," *IEEE Trans. Ind. Applicat.*, vol. 37, Jan.-Feb.,2001, pp. 81-89.

Single-phase 50-kW 16.7-Hz Four-Quadrant Line-Side Converter for Railway Traction Application

C. Heising, R. Bartelt, V. Staudt, A. Steimel

Ruhr-University of Bochum
D-44780 Bochum
Germany
Tel.: +49 234 32 23890
Fax.: +49 234 32 14597
Email: heising@eele.rub.de, bartelt@eele.rub.de, staudt@eele.rub.de, steimel@eele.rub.de

Abstract—Railway grid interaction is a major concern when developing the power chain of a railway traction vehicle and its control. Especially line-current harmonics and grid stability have to be considered.

Developing an optimised control scheme for the four-quadrant line-side converter of a railway traction vehicle operating at the 15-kV, 16.7-Hz grid requires advanced simulation tools, combined with suitable laboratory experiments for verification. This paper describes the simulation concept, the control structure, the realisation of a four-quadrant line-side converter test bench and measurement results for stationary and dynamic behaviour, verifying excellent operation.

Keywords—Four-Quadrant Converter, Railway Power Train, Converter Control

I. INTRODUCTION

Modern rail traction vehicles operating at the Central European single-phase 16.7-Hz AC overhead line employ a four-quadrant converter at the line side to feed a DC link, to which the machine-side inverter is connected, feeding the traction motors. This paper focuses on the control of the four-quadrant converter. The overall project, into which this research activity is embedded, aims at reducing grid harmonics and increasing grid stability, both major issues for the operation of rail vehicles [1].

Developing an optimised control scheme for the four-quadrant line-side converter [2] operating at the 15-kV, 16.7-Hz grid requires advanced simulation tools, which shall be backed by suitable laboratory experiments. This paper describes the simulation concept, the control concept, the realisation of a four-quadrant line-side converter test bench and measurement results for stationary and dynamic behaviour. The power of the laboratory set-up is rated at 50 kW.

The DC-link capacitor is supplemented by a resonant circuit tuned to twice the line frequency (33.4 Hz), which is to take up the power pulsation of the single-phase grid; it requires special attention, especially regarding dynamic operation. The performance of the control results from a combination of observer-based feed-forward and feed-back control schemes.

A locomotive − representing electric traction vehicles in general − on a track can be modeled in an abstract way by analysing the interaction of the electrical components of the locomotive and the surrounding substations [3]. The substations feed the locomotive via the overhead contact line (catenary) and the pantograph (Fig. 1). Because the locomotive is moving, the impedance between locomotive and substations varies. Therefore the modelling parameters of the grid representation have to be variable and the control of the four-quadrant converter has to be robust concerning grid-parameter variation.

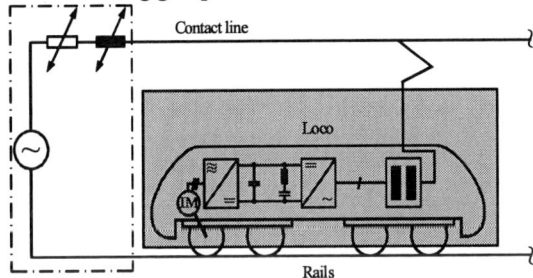

Fig. 1. Representation of substation and locomotive (first step)

The substations and the variable position of the locomotive are represented by a voltage source, on request containing harmonics, realised by a suitably controlled IGBT inverter, operated at 16.7 Hz fundamental frequency and emulating the impedance and the harmonics already present in the overhead contact-line voltage.

II. STRUCTURE OF SIMULATION AND TEST BENCH

To design a suitable test bench it is necessary to generate a 16.7-Hz single-phase voltage, which typically is not available in lab, by an inverter. This grid inverter then feeds the four-quadrant input converter of the locomotive via a transformer, followed by the three-phase motor inverter, leading to a complex structure of the simulation and the test bench (Fig. 2).

The three-phase 50-Hz lab grid feeds the DC-link capacitor of the grid inverter by a line-friendly 12-pulse diode rectifier with autotransformers [4]. A H-bridge IGBT PWM-inverter generates the 16.7-Hz grid voltage. The control of this inverter includes strategies to emulate

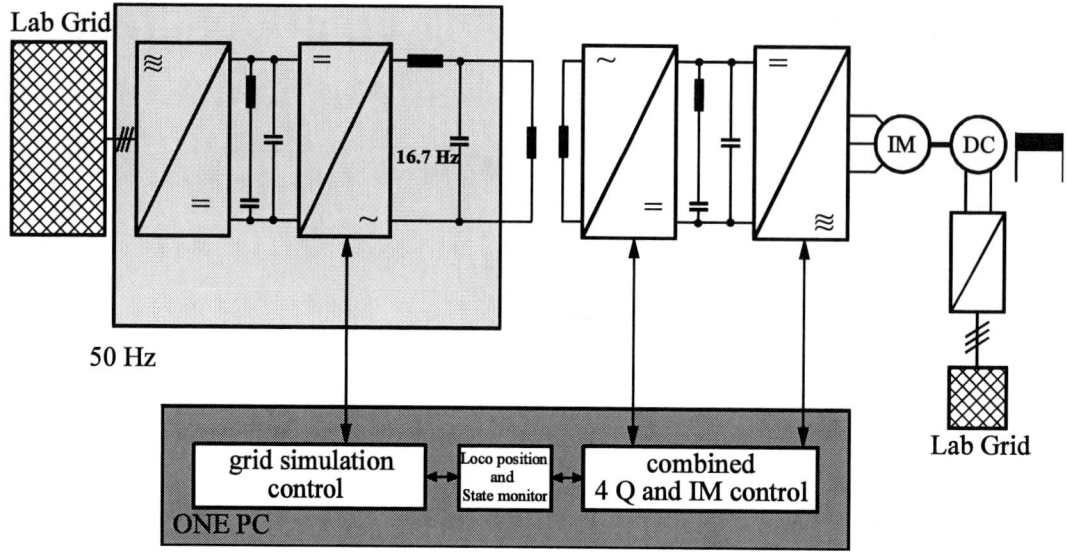

Fig. 2. Structure used for simulation and test bench

the variable grid characteristics. Because the generated 16.7-Hz voltage is single-phase, a power pulsation at double line frequency is inevitable. This is taken up by a resonant circuit tuned to 33.4 Hz, producing a smooth DC-link voltage. An IGBT brake chopper (not shown in Fig. 2) takes up dynamic brake power of the drive. To eliminate PWM voltage harmonics of the grid representation H-bridge voltage a second-order low-pass filter (LC filter) is inserted, with a cut-off frequency of 410 Hz. This value is high enough to allow impressing low-frequency line-voltage harmonics up to the 25th order [9].

The transformer decouples the potential of the grid-representation inverter and the locomotive four-quadrant converter, which is in the focus of this paper. It feeds □ in a mirrored manner □ a DC-link capacitor and a resonant circuit. The overall real-time control is realised using one PC-based hardware [5].

III. STRUCTURE OF THE SIMULATION

The simulation has to deal with continuous elements (for example voltage sources, inductivities and capacitors) as well as with discrete parts, like switching semiconductor valves. The state-space notation is used to model the continous part. The set of n first-order differential equations is solved by a Bulirsch-Stoer implementation of the Richardson Extrapolation [6], supported by an adaptive step-size algorithm. For the discrete part (e.g a converter phase leg, Fig. 3) Petri Nets are used (Fig. 4). The 'places' p_i stand for the 5 possible states of the phase leg (each two polarities of output voltage and current and all off), the t_i for possible transitions [7]. Each phase leg is described by its own Petri Net because otherwise $5^7 = 78125$ different states would be necessary to describe the test bench. Furthermore the Petri Net modulation based on the net-array description is easier to compute with modern personal computers.

This simulation framework – programmed in 'C' language – allows very fast simulation of the overall system for relatively long times. Typically about 10 seconds real-time per dominating dynamic are needed,

because the grid frequency and the resonance frequencies are low, leading to relatively long times to stationarity. They usually require long calculation time, because each switching of each of the 7 phase legs (in this example) has to be accounted for. The control algorithm can be used identically in simulation and real-time control.

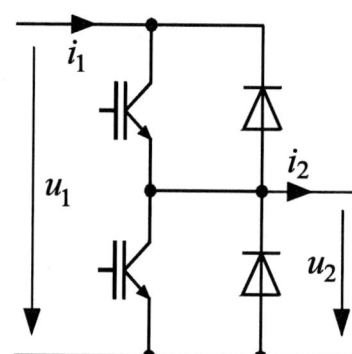

Fig. 3. Circuit diagram of converter phase leg

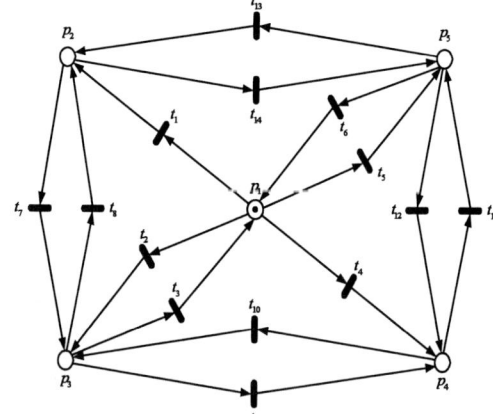

Fig. 4. Petri Net representation of a converter phase leg

522

IV. FOUR-QUADRANT LINE-SIDE CONVERTER AND ITS CONTROL IN DETAIL

The test bench of the four-quadrant converter is rated at 50 kVA; the working point of the converter has been initially chosen at fundamental oscillation characterized by 250 V (peak) input voltage, 400 A (peak) input current and 500 V DC-link voltage due to limitations set by the available transformer. The final rated values will be 300 V and 333 A. The device switching frequency of 1 kHz has been chosen to adapt to the actual value of medium-power traction drives and leads to a resulting pulse frequency of the H-bridge of 2 kHz. To model high-power locomotive converters, values down to 250 Hz per valve can be set.

A. Grid-voltage analysis

The control is a time-domain scheme with a three-part structure. First the measured grid voltage is decomposed by a discrete Fourier analysis (fig. 5). Based on the fundamental a phase-locked loop detects phase angle and frequency of the generally disturbed grid voltage. This generated data is necessary to guarantee a high power factor in steady-state operation. For this a PI-controller based on a Luenberger observer [8] is used. It observes all state quantities of leakage inductance, DC-link capacitor and inductance and capacitor of the resonant circuit and computes a suitable voltage for the four-quadrant converter. The switching instants and the switching sequence follow from regularly-sampled pulse width modulation.

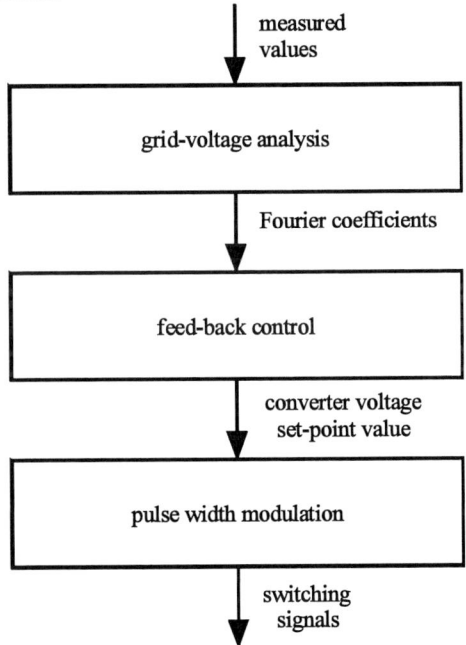

Fig. 5. Three-part structure of time-domain control

The input signals of the PI-controller for the grid voltage analysis are the fundamental coefficients F_{01} and F_{11} of the Fourier series, reduced to the phase angle ξ_1 of the fundamental:

$$F_{0k} = \sum_{n=0}^{N-1} f_n \cos\left(\frac{2\pi}{N}kn\right)$$

$$F_{1k} = \sum_{n=0}^{N-1} f_n \sin\left(\frac{2\pi}{N}kn\right)$$

$$\xi_1 = -\arctan\left(\frac{F_{11}}{F_{01}}\right)$$

The only output of the PI controller is the sampling time of the data acquisition. By changing the actual sampling time the controller is able to identify the frequency and at the same time to fix the phase angle to zero (fig. 6).

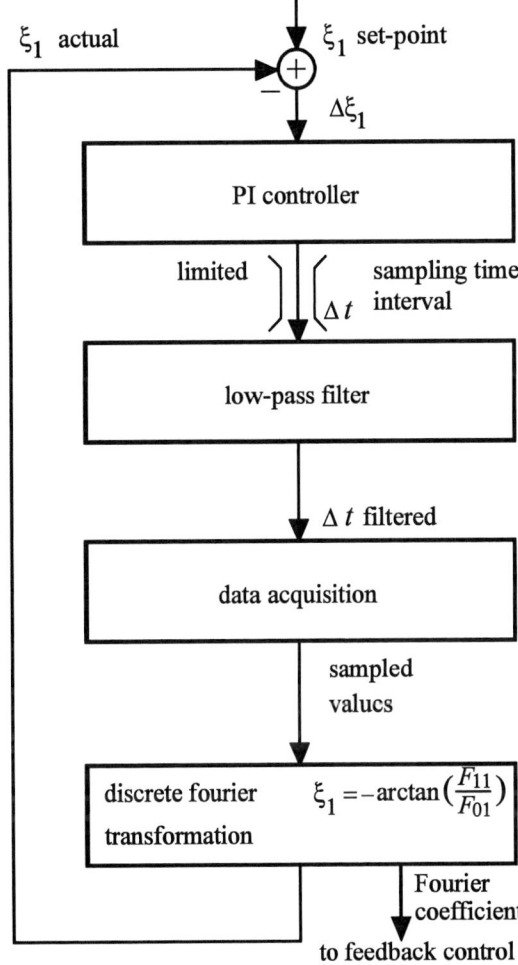

Fig. 6. Scheme of the grid-voltage analysis

A fast change in the sampling time has a strong influence on other control elements with a destabilising effect. To prevent this negative impact, the rate of change of sampling time is limited and additionally filtered by a low pass. In the majority of cases the turn-on procedure ("synchronisation") takes more than one second which is not satisfying. To shorten the procedure the multivariable control is supported by a dedicated feed-forward control. After the decay of the transient response the fundamental oscillation and the harmonics (with their phase angles) of the disturbed grid voltage are available.

In the following three figures the dynamic behaviour of the described interaction of discrete Fourier analysis and

523

phase-locked loop can be seen. To increase the demand on the controlled grid-voltage analysis an unusual frequency of 18 Hz was chosen. Initially the algorithm expects, of course, 16.7 Hz. It is important to guarantee proper functionality even if the variation of the frequency of the grid voltage is high, because the frequency variation of the railway power supply in Central Europe according to EN 50163, Table 1, is larger than that of public power supply grids.

Figure 7 illustrates the turn-on procedure of the grid voltage analysis. In the first 1.5 periods after the start the signal ξ is calculated by the discrete Fourier analysis as given in Figure 5. Based on the results of the calculation of $\xi(T)$ and $\xi(3/2\ T)$ the phase angle and the frequency of the fundamental oscillation are approximated with $T=1/16.7$ Hz. With these values the grid-voltage analysis is reset and started with new initial conditions. With these improved initial values the control eliminates the error in 3 periods. This results in a cumulated transient response time of 4.5 periods. In normal operation with the expected frequency of 16.7 Hz the response time is only 2.5 periods.

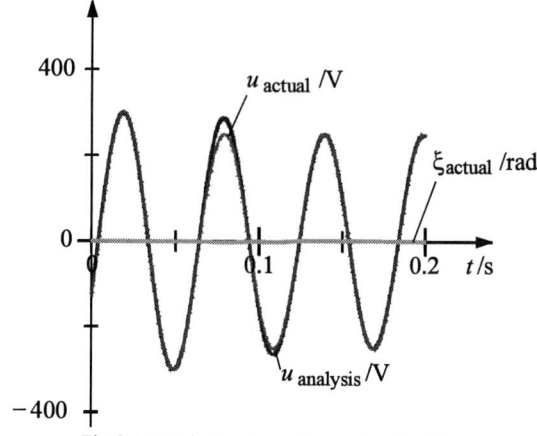

Fig. 8. Dynamic reaction on changes in grid voltage

Another important demand is operation under the influence of pantograph arcing. Figure 9 demonstrates the performance with a 100 ms line-voltage interruption. The grid-voltage analysis recovers steady-state operation mode within 1.5 periods, after the return of the voltage.

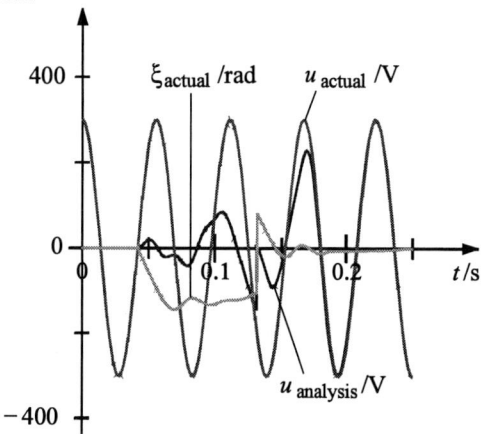

Fig. 7. Turn-on procedure of the grid-voltage analysis

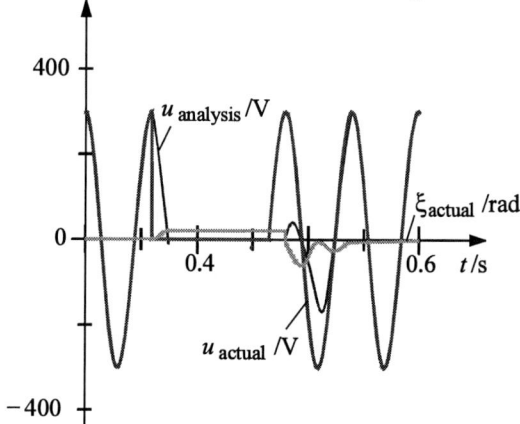

Fig. 9. Dynamic reaction on pantograph arcing

Besides a quick turn-on procedure the dynamic reaction upon changes of the grid voltage is a core requirement. Because of the high ratio of rated locomotive power to substation power ('weak grid') the amplitude of the grid-voltage can change significantly. Figure 8 shows the reaction of the grid-voltage analysis upon a –15% change in the supply voltage. It has been proven, that the change can be followed by grid-voltage analysis within one single period.

B. Feed-back and feed-forward control

The feed-back control is realized by a well-known PI-type controller for the DC-link voltage, Figure 10. The grid-voltage provided by the grid-voltage analysis serves as reference signal because it can be easily calculated for the next samples. The PI-controller is supported by a powerful feed-forward control, based on a very detailed model of the plant. This allows to increase the loop gain of the system, so that the transient response time is minimized. Besides the computed grid voltage, both feed-back and feed-forward control receive the set-point value of the DC-link voltage. In addition the feed-back control uses the output of the Luenberger observer. Though all necessary actual values like DC-link voltage and grid current are measured in a sufficient sampling time, the Luenberger observer serves as a filter and – with accurate knowledge of the system structure and parameters – allows to predict the system behaviour for the following few sampling steps. Feed-back and feed-forward control sum up to the desired output voltage of the four-quadrant converter.

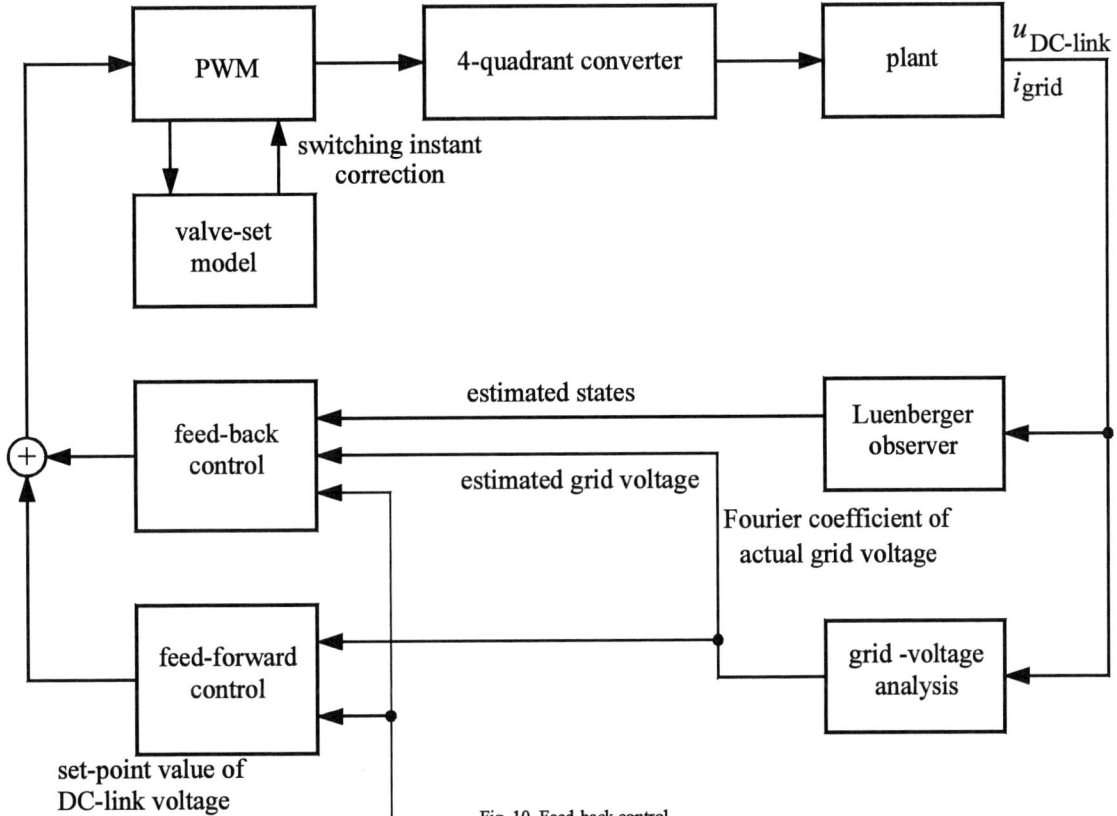

Fig. 10. Feed-back control

This voltage has to be converted into switching signals for the H-bridge IGBT converter. For this a standard double-side, regularly sampled pulse width modulation is used. Valve-voltage drops and switching delays/interlocking times cannot be neglected in case of high-power IGBT modules. Based on a detailed valve-set model – considering the actual state of the four valves and the load current of the H-bridge – switching times are adapted to correct these errors [10].

The following four figures show simulation results computed with the simulator described in the previous section. Figure 11 demonstrates the steady-state operation of the four-quadrant converter. The set-point value of the DC-link voltage is 500 V. As known from power theory the four-quadrant converter allows an operation with any power factor, preferred values are $\lambda = \pm 1$. The time functions of grid voltage and grid current are strictly proportional.

As it can be seen in Fig. 12 the controller is even able to hold this ratio between grid voltage and grid current during a DC load current step of 30 A at $t = 0$ s. In this case the four-quadrant converter behaves like a resistor with a variable parameter changing only in grid-voltage zero crossings. The resonant-tank inductance and capacitor are exited transiently at the parallel resonance frequency of DC-link capacitor and resonant tank.

$$f_{\text{resonant}} = \frac{1}{2\pi} \sqrt{\frac{C_{\text{DC}} + C_{\text{res}}}{L_{\text{res}} C_{\text{DC}} C_{\text{res}}}}$$

This can be seen in the DC-link voltage. The resonant-tank voltage (u_{resonant}) and current (i_{resonant}) are shown in Fig. 13 for the DC-load current using identical scales.

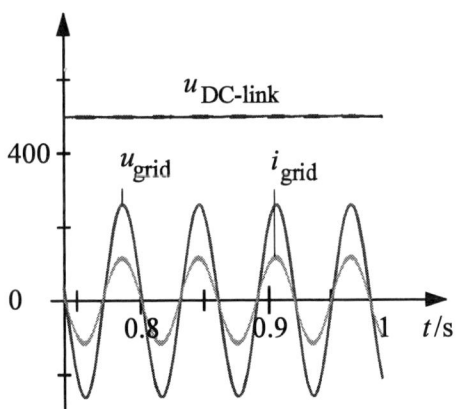

Fig. 11. Simulation results of steady-state operation

To verify the full functionality of the whole control concept the grid-voltage amplitude was modulated with a frequency of 0,167 Hz and a magnitude of 20 V. The steady-state behaviour of the system can be seen in Fig. 14. The instantaneous power is constant because the DC-link voltage of 500 V and the DC-load current of 30 A are kept constant. Thus the amplitude oscillation the grid-current is in counter-phase to the amplitude modulation of the grid-voltage.

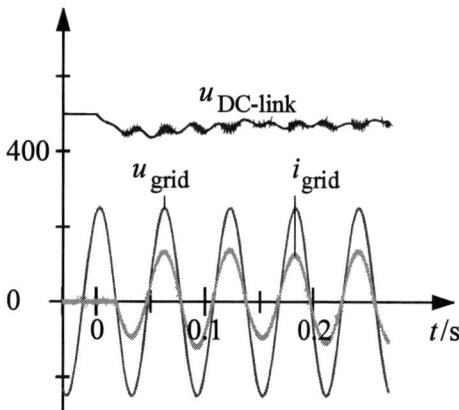

Fig. 12. Simulation results after DC load current step of 30 A

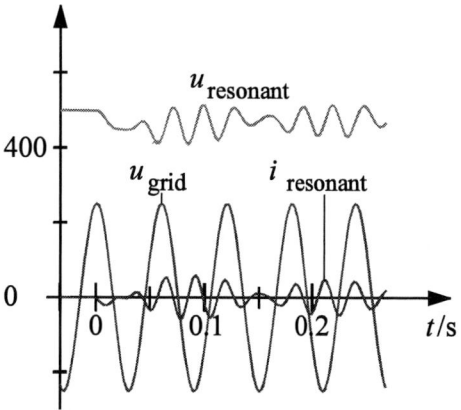

Fig. 13. Simulation results of resonant tank voltage and current after DC-load current step of 30 A

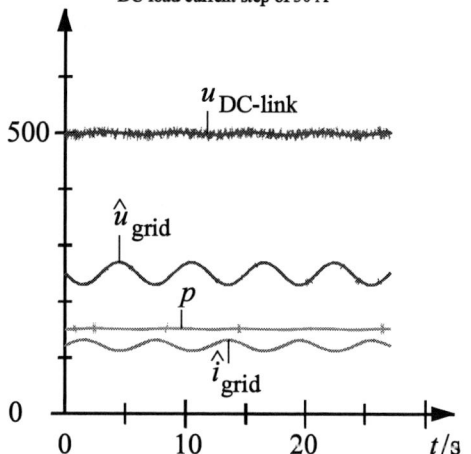

Fig. 14. Simulation results of reaction on 20-V grid-voltage amplitude modulation

C. Measurements

The following three figures document stationary and dynamic behaviour of the new control scheme as measured on the test bench. The stationary behaviour is depicted in figure 15.

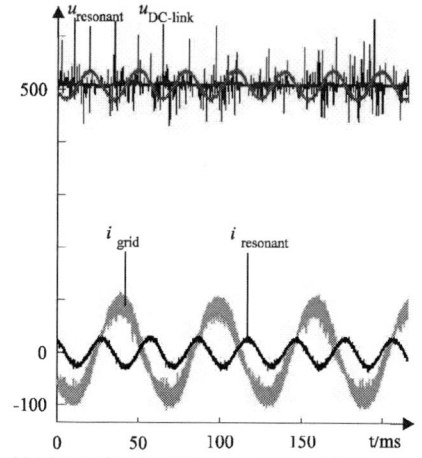

Fig. 15. Measured stationary behaviour at about 90 A peak current (P= 15 kW)

In steady-state (here with an AC current, green, of about 90 A peak) the DC-link voltage (blue, nearly straight line at about 500 V) contains a slight 33.4-Hz ripple (due to the resistive component of the resonant inductor impedance). The oscillation of power is taken up by the resonant circuit. Voltage (red, oscillating around about 500 V) and current (black, peak value about 22 A) of the resonant circuit are shown. The calculated steady-state amplitude of the resonant-tank current is [3]

$$\hat{i}_{res} = \frac{1}{2} \frac{\hat{u}_{grid}}{u_{DC-link}} \hat{i}_{grid} = 22.5A$$

Figure 16 gives the reaction to a change of the DC-link voltage set-point value from about 410 V (idle) to about 500 V (nominal value). It can be seen that the DC-link voltage (blue) rises quickly. The AC portion of the voltage at the capacitor of the resonant circuit (red, larger oscillation) shows little overshoot. All quantities quickly become stationary matching set-point conditions. The grid current (green) flows for about one period only, charging the capacitor and slightly exciting the resonant circuit, which is then quickly damped by the control.

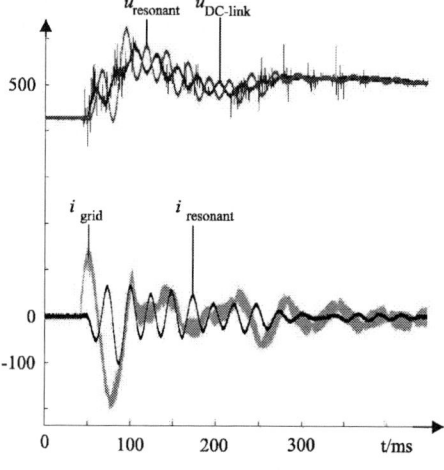

Fig. 16. Measured set-point value change (410 V to 510 V)

Figure 17 illustrates the reaction to a stepwise load change, realised by switching on a resistor on the DC side. The AC-side current rises from idle to about 90 A peak. It can be seen that again only about one to two periods of 16.7 Hz are needed to completely reach new steady-state conditions. The DC-link voltage is temporarily reduced by about 20 %, which is good taking into account the limited possibilities of a single phase four-quadrant converter operating at 16.7 Hz. Also, the resonant circuit quantities show nearly no overshoot, neither the voltage at the capacitor nor the current of the inductance. This allows to select elements with limit values not too far from stationary operating values, reducing cost and weight of these components.

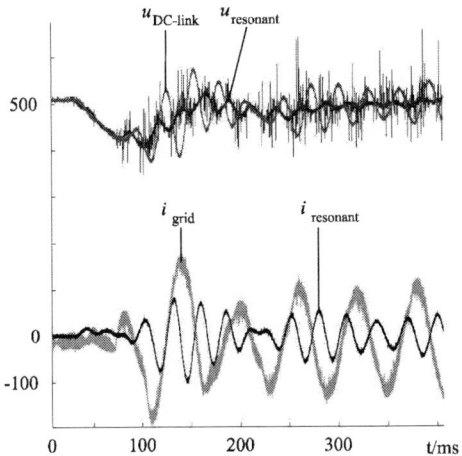

Fig. 17. Measured reaction to load step with 15 kW

V. CONCLUSION AND PERSPECTIVES

In this paper the structure of a simulation and a test-bench realisation of a 50-kW four-quadrant converter for traction application is given. The control of the four-quadrant converter, model-based and including feed-forward and feed-back strategies, is explained. Excellent control characteristics, especially with regard to the challenge resulting from the dominant oscillation of the 33.4-Hz resonant circuit under dynamic conditions, are proven by experimental results.

On this firm footing, more sophisticated control algorithms can be developed. Integrating four-quadrant line-side converter, resonant circuit, DC-link capacitor and the three-phase inverter feeding the induction machine into one overall control scheme is expected to further enhance the dynamic performance, improve railway grid stability and allow for effective mitigation of grid-current harmonics.

TABLE I.
RATED DATA OF THE TEST BENCH

$\hat{u}_{grid}$	300 V
$\hat{i}_{grid}$	333 A
P	50 kW
L_T	6.3 mH
$u_{DC-link}$	500 V
$C_{resonant}$	5 mF
$L_{resonant}$	4.53 mH
$R_{L-resonant}$	53 mΩ
$C_{DC-link}$	9.2 mF

REFERENCES

[1] M. Meyer and J. Schöning, "Netzstabilität in großen Bahnnetzen,". *Eisenbahn-Revue*, 7–8/1999, pp. 312–317

[2] M. Depenbrock, "Einphasen-Stromrichter mit sinusförmigem Netzstrom und gut geglätteten Gleichgrößen," *ETZ-A* 94 (1973), Nr. 8, pp. 466–471

[3] A. Steimel, "Electric Traction – Motive Power and Energy Supply," *Oldenbourg-Verlag*, München 2008

[4] C. Niermann "New Rectifier Circuits With Low Mains Pollution And Additional Low Cost Inverter For Energy Recovery," *3rd European Power Electronics Conference (EPE)*, Aachen 1989, vol. III, pp.1131 – 1136

[5] M. Kail, V. Staudt and A. Steimel, "PC-based control hardware for 200-kW double three-level converter," *10th International Power Electronics and Motion Control Conference*, Cavtat/Croatia 2002

[6] W. H. Press, S. A. Teukolsky, W. T. Vetterling, and B. P. Flannery, "Numerical Recipes in C," *Cambridge University Press* 1988

[7] V. Staudt, C. Heising and A. Steimel "Advanced Simulation Concept for Power Train of Loco and its Verification," *ICPE 07 Conference*, Daegu 2007

[8] J. Lunze "Regelungstechnik 2 (Mehrgrößensysteme, Digitale Regelung)," *Springer*, Berlin 2006

[9] C. Heising, M. Gorski, V. Staudt and A. Steimel, „Single-phase 50-kW, 16.7-Hz railway-grid representation featuring variable grid parameters," PESC 2008, Rhodos

[10] C. Foerth, „Traktionsantrieb ohne Messwertgeber mit minimiertem Messaufwand", Fortschritt-Ber. VDI Rh. 8 Nr. 936, Düsseldorf 2002

Technique to Improve IGBT Converter Efficiency and Transient Response

Robert W. Turner*, Simon Walton*, Richard Duke[†]

*Vectek Electronics Ltd, Napier, New Zealand, e-mail: *robert.turner@vectek.com*
[†]University of Canterbury, Electrical and Computer Engineering, New Zealand, e-mail: *richard.duke@canterbury.ac.nz*

Abstract—For medium to large power converters (1kW to 100MW) the switching frequency is limited to ensure acceptable losses at the cost of converter performance (bandwidth). A technique is presented that dynamically changes the switching frequency to maintain a low average switching frequency but for transient events shorter than the thermal time constant of the device, the switching frequency and bandwidth are increased. Relationships between the switching frequency and bandwidth are developed, and a method for determining when and how to change bandwidths is introduced. The technique is shown to give improvements in THD compared to a fixed frequency system of the same average switching frequency (same losses) in a harmonic filter application.

Index Terms—AC/AC Converter, Converter Control, Efficiency, IGBT, Parallel Operation, Power Quality, PWM

I. INTRODUCTION

For switching converters the switching frequency is directly related to the converter efficiency due to switching losses in the switching devices. Inherently for medium to high power converters (1kW to 100MW) this becomes a concern as the relative switching losses increase for increasing power devices. Therefore, conventionally in hard-switched constant switching frequency medium to high power converters, such as power line conditioners, the switching frequency is typically around 10kHz when using IGBTs, or lower for thyristor systems [1] - [3].

For medium to high power converters the switching modulation used is typically fixed frequency Pulse Width Modulation (PWM) or hysteretic [4], [5]. Hysteretic systems with no delay element theoretically have an infinite bandwidth. Their switching frequency is controlled by choosing appropriate hysteresis bands. As a result, hysteretic systems typically have no defined fully determinable switching frequency. PWM on the other hand provides a set switching frequency [6] - [9]. Natural PWM bandwidths do not have to be limited by sample time but can cause the switching frequency to increase beyond the the carrier frequency, becoming indeterminate. Regular PWM has a fixed switching frequency at all times with a fixed constant sampling delay. Regular PWM is typically preferred as it implies a fixed switching frequency which is useful when designing filters and paralleling modules.

A fixed switching frequency imposes a maximum controller bandwidth. This paper focuses on techniques using regular PWM as it allow an accurate means of defining the switching frequency and hence losses. Regular PWM modulators are also the most commonly realized modulator in a digital controllers.

When put in a closed loop the delay introduced by PWM imposes a theoretical maximum bandwidth to ensure to stability. For high power systems with low switching frequencies the maximum bandwidth is constrained, in turn constraining the performance of the system.

The limiting factor on the switching frequency is the switching loss in the device. The switching frequency is typically chosen to ensure the switching losses are kept within the device's thermal limits. However, devices have a thermal time constant (typically in the order of seconds or tens of cycles for a 50Hz system) for which thermal losses must only be within the device's limit on *average*.

For time periods shorter than the thermal time constant the switching losses may vary, provided the losses over the device time constant are acceptable. This introduces the idea that, for compact-in-time events shorter than the thermal time constant, the device can tolerate greater losses. This allows increased switching frequency as long as acceptable average thermal losses are maintained, and hence increased controller bandwidth. As compact-in-time events are not compact in frequency, an increased bandwidth would allow the controller to react to an event faster and with a smaller output error (such as overshoot). Such compact-in-time events typically include transients such as load changes or supply disturbances but also include repetitive events such as discontinuous conduction conduction periods of diode rectifiers [10].

The Dynamic Frequency Scaling (DFS) technique presents a method of dynamically changing the switching frequency and bandwidth to maintain a low average switching frequency, but increase the bandwidth and switching frequency when compact-in-time events are detected to deliver improved performance.

A. Dynamic Frequency Scaling operation

Figure 1 shows the block diagram layout of a DFS current controller. The DFS system is an additional control block that dynamically changes the loop bandwidth and switching frequency as a function of the output error. The DFS Frequency Chooser block is responsible for determining when to change switching frequency and the appropriate frequency and bandwidth to change to.

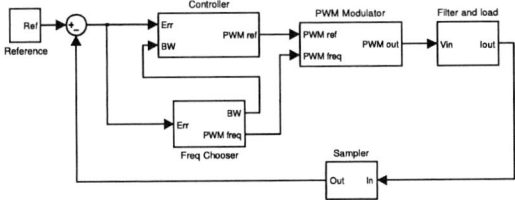

Fig. 1. DFS controller block diagram.

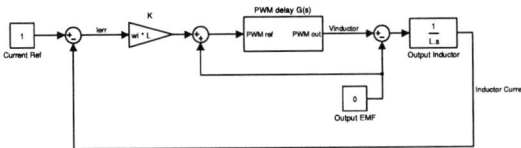

Fig. 2. Closed loop current (single order) controller with PWM.

II. GLOBAL STABILITY

A DFS controller is inherently a variable structure type system. To ensure global stability of the system, stability must be ensured for every possible switching frequency and bandwidth. As a result, general stability criteria relating the bandwidth to the switching frequency were developed. The bandwidth and switching frequency can then be dynamically adjusted while maintaining stability.

Regular PWM introduces a delay which is a result of the internal Zero Order Hold (ZOH). The delay introduced by the ZOH manifests as an effective phase change which can affect stability. For a continuous system a "pure delay" is modelled as e^{-sT_s} and can be shown to have a phase shift of $-\omega T_s$ and a magnitude of one. The Laplace Transform of a ZOH module is $\frac{1-e^{-sT_s}}{sT_s}$ with a phase delay of $\frac{-\omega T_s}{2}$ and a magnitude of $\text{sinc}\left(\frac{\omega T_s}{2\pi}\right)$, where T_s is the ZOH sample time. As the ZOH is a discrete time element the system is best examined in the discrete time domain - this is confirmed by the accurate representation and matching results presented in Zero Order Hold Stability, section II-C.

Two methods will be used to examine systems with loop delays:

1) Determine an acceptable phase or gain margin of the open loop system that will ensure stable closed loop stability.
2) The application of the Lambert W function to solve for the dominant poles in the closed loop equation of system with a delay modelled as e^{-sT_s}.

The two methods will first be examined for the pure transport delay and then for the discrete ZOH.

For this paper the bandwidth to switching frequency relationship is developed for a Current Source Inverter (CSI) system as it provides a simple first order control loop. The same procedure can also be applied to higher order systems at the cost of increased complexity.

A. Phase margin stability

A bandwidth (BW) to PWM sample time relationship for a given phase margin is developed by looking at the open loop response. Figure 2 shows the typical configuration for a CSI which will be used as the reference system. The open loop system is given by (1). The substitution $K = \omega_i L$ is made to define the system in terms of a closed loop bandwidth ω_i, and taking the simple example of driving into a short circuit ($V_{emf} = 0$) gives the open loop transfer function (2). To determine the BW to delay relationship for a given phase margin (PM), the BW is solved for the condition where the open loop gain is one (3).

$$I_{out} = \frac{KG(s)I_{err} + (G(s) - 1)V_{emf}}{Ls} \tag{1}$$

$$V_{emf} = 0 \Rightarrow H_{OL}(s) = \frac{I_{out}}{I_{err}} = \frac{\omega_i G(s)}{s} \tag{2}$$

For a pure delay system that only introduces a phase shift (unity magnitude), for a given phase margin the relationship between bandwidth (ω_i) and sample time (T_s) is given by (7).

$$H(s) = 1\angle(-180° + PM) = \frac{\omega_i G(s)}{s} \tag{3}$$

$$1\angle(-180° + PM) = \frac{\omega_i}{\omega}\angle\frac{-\omega T_s}{90°} \tag{4}$$

$$\Rightarrow \omega_i = \omega, \text{ and} \tag{5}$$

$$(PM/180° - 1)\pi = -\omega T_s - \frac{\pi}{2} \tag{6}$$

Equating (5) and (6) and solving for ω_i as a function of T_s:

$$\omega_i = \frac{1}{T_s}\frac{\pi(90° - PM)}{180°} \tag{7}$$

$$\begin{matrix} PM = 0° \\ \text{(Marginally stable)} \end{matrix} \Rightarrow \omega_i = \frac{\pi}{2T_s} \tag{8}$$

$$PM = 45° \Rightarrow \omega_i = \frac{\pi}{4T_s} \tag{9}$$

Equations (8) and (9) show the bandwidth to sample time relationship for phase margins of 0° and 45°, respectively.

B. Lambert W function stability

The Lambert W function provides a method for determining the pole positions and hence response for a closed loop system with a delay [11] - [12]. The closed loop transfer function for (2) is shown in (11).

$$H_{CL}(s) = \frac{H_{OL}}{1 + H_{OL}} \tag{10}$$

$$H_{CL}(s) = \frac{\omega_i e^{-sT_s}}{s + \omega_i e^{-sT_s}} \tag{11}$$

The response of a transfer function and also its stability is governed by the closed loop poles. To determine the poles the denominator of the closed loop transfer function is equated to zero as shown in (13).

$$0 = Poles(H_{CL}) \tag{12}$$

$$\Rightarrow 0 = s + \omega_i e^{-sT_s} \tag{13}$$

Using the Lambert W function, (13) can be solved for (14), where W is the Lambert W function. Although the Lambert W function has an infinite number of complex solutions, the primary result is the dominant pole that governs the response of the system. Equations (15) and (16) give the under-damped natural frequency and damping ratio, respectively.

$$s = \frac{W(-\omega_i T_s)}{T_s} \qquad (14)$$

$$\omega_n^2 = \Re(s)^2 + \Im(s)^2 = |s|^2 \qquad (15)$$

$$\zeta = \frac{\Re(s)}{\omega_n} = -\cos(\angle s) \qquad (16)$$

The bandwidth to delay relationship for zero phase margin as shown in (8) substituted into (14) has a zero damping ratio with a natural frequency given by (19).

$$PM = 0°, \omega_i = \frac{\pi}{2T_s} \Rightarrow s = \frac{W(-\frac{\pi}{2})}{T_s} \qquad (17)$$

$$s = \frac{0 + 1.5708i}{T_s} \qquad (18)$$

$$\Rightarrow \omega_n = \frac{1.5708}{T_s} \qquad (19)$$

$$\zeta = 0 \qquad (20)$$

As the phase margin is increased the damping ratio increases and the natural frequency decreases. The system is critically damped for $\omega_i = \frac{1}{eT_s}$, with a phase margin of 111°. Increasing the damping ratio increases the delay. A phase margin of 45° provides an acceptable tradeoff between system damping and bandwidth to delay ratio. A 45° phase margin is typically used for closed loop systems and is used for the remainder of this paper.

C. Zero Order Hold stability

Having introduced the procedure for determining bandwidth to sample time relationships for pure delays, a similar approach can be applied to the ZOH nature of a PWM modulator. The ZOH sample time (T_s) for PWM is the inverse of the switching frequency (f_{sw}). As the ZOH is a discrete time device, discrete time analysis must be applied to the system. For a single order feed back system, the open loop discrete transfer function of the reference system shown in Figure 2 with open loop transfer function (2) is given by (21). The closed loop transfer function is given by (22).

$$H_{OL}(z) = \frac{\omega_i T_s}{z - 1} \qquad (21)$$

$$H_{CL}(z) = \frac{\omega_i T_s}{z + (\omega_i T_s - 1)} \qquad (22)$$

The equivalent damping ratio, natural frequency and phase margin are obtained by solving $z = e^{sT_s}$ for time domain poles. Equations (26) and (28) give the bandwidth to switching frequency relationship for phase margins of 0° and 45°, respectively.

$$|H_{OL}(e^{sT_s})| = 1 = \left| \frac{\omega_i T_s}{e^{sT_s} - 1} \right| \qquad (23)$$

$$\omega(Mag = 0dB) = \frac{\cos^{-1}\left(1 - \frac{(\omega_i T_s)^2}{2}\right)}{T_s} \qquad (24)$$

$$PM = \tan^{-1}\left(\frac{\sin(\omega T_s)}{\cos(\omega T_s) - 1}\right) \qquad (25)$$

$$PM = 0° \Rightarrow \omega_i = \frac{2}{T_s} \qquad (26)$$

$$\Rightarrow f_{sw} = \frac{\omega_i}{2} \qquad (27)$$

$$PM = 45° \Rightarrow \omega_i = \frac{\sqrt{2}}{T_s} \qquad (28)$$

$$\Rightarrow f_{sw} = \frac{\omega_i}{\sqrt{2}} \qquad (29)$$

The system is critically damped for a ZOH sampled system when $\omega_i = \frac{1}{T_s}$ (phase margin of 60°). This occurs as a dead-beat response where the output error is reduced to zero in a single sample period.

The dead-beat response is usually the preferred response for a discrete system although is often not realizable as it requires an instantaneous sample, calculation and update time, often not possible in DSP controllers. A phase margin of 45° is used as it also provides a closer ratio between sampling time and controller bandwidth.

III. FREQUENCY CHOOSER

Having developed a bandwidth to switching frequency relationship, the criteria to determine when to change from one switching frequency to another is required. The Frequency Chooser unit, as its name suggests, decides when to change switching frequencies and which frequency to change to. To conserve the closed loop phase margin, when the switching frequency is changed so too is the controller bandwidth.

The operational logic of the frequency chooser is to change to a higher switching frequency and bandwidth for transient events (compact-in-time), while maintaining a low average switching frequency. The effect of a transient appears as a suddenly increasing output error. Typically the error propagates through the controller which, at the rate of the controller bandwidth, compensates for this error. A typical trajectory for a transient event such as a sudden load change or input reference change for the first order reference system in Figure 2 will give an output error as shown in Figure 3.

The rate of change (derivative) of the output error provides a method of indicating when a large transient has occurred. Output error and error derivative trajectories for a pure delay and a ZOH for the reference system current controller in Figure 2 are shown in Figures 4(a) and 4(b), respectively. The coordinate that the trajectory begins to move in and encircle the zero point as indicated by the dashed lines is $(1 - T_s, -\omega_i)$, scaled by the size of the step magnitude. Figure 4(b) illustrates the dead-beat response of the critically damped trajectory.

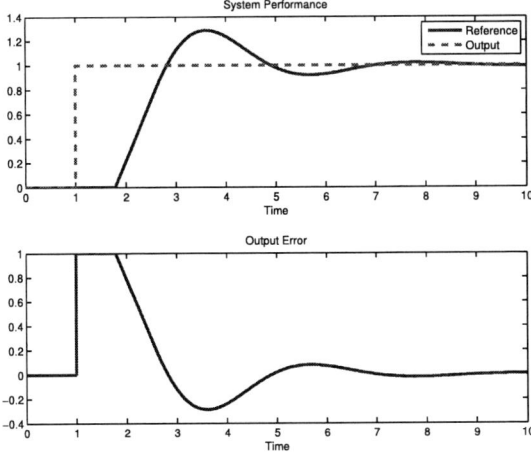

Fig. 3. Single order system step response reference. $\omega_i = 1$, PM = $45°$.

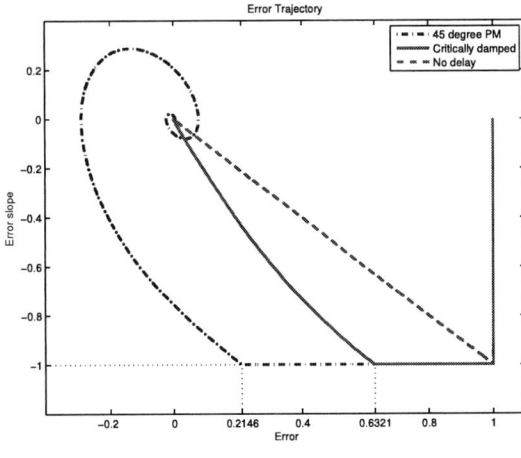

(a) Pure delay trajectories.

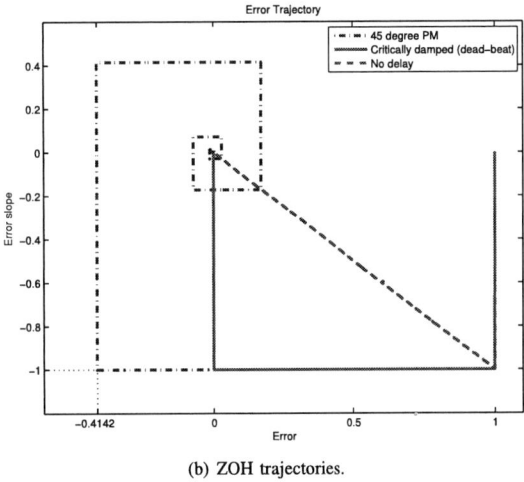

(b) ZOH trajectories.

Fig. 4. Output error and error derivative trajectories for $\omega_i = 1$.

Given the regular trajectory for an output error, transition bands can be applied that will define when to change to a higher bandwidth with a lower switching delay. As mentioned, the error trajectories scale accordingly with step size. The conditions which determine when to switch to a higher frequency are:

- The error is large and the rate of change of the error is small, and
- The error is small but the rate of change of the error is large.

Given the above criteria, a circular transition band over the error (Err) and error slope $(\dot{Err})$ trajectory is introduced. For a transition band radius of Err^* the circular transition band is given in (30). The transition band for the first order system reference system is shown in Figure 5.

$$Err^{*2} = Err^2 + \left(\frac{\dot{Err}}{\omega_i}\right)^2 \qquad (30)$$

The radius and number of transitions bands are chosen for a given application. Evaluation of the intended application must be done to choose the number of transition bands and each band's error tolerance radius. The error tolerance radius directly affects the average switching frequency and therefore must be chosen to ensure the average switching frequency is acceptable. This paper continues to investigate the use of two switching frequencies with a single transition band.

IV. APPLICATION

As DFS is a technique for increasing converter performance while maintaining a low average switching frequency, the application of interest is for high power converters. This comes as a result of high power converters being limited by their switching frequency for efficiency but with the requirement of high bandwidths. In all DFS and non-DFS applications explored, the bandwidth to

switching frequency ratio has been fixed for a phase margin of 45°. All current, voltage, time, bandwidth and component values are in per unit (pu). The ac reference signal angular frequency is normalised to one $\left(\frac{1}{2\pi}Hz\right)$. Voltages and currents are normalised to one pu at their peaks (so a one pu pure sinusoid has a RMS of 0.707).

DFS is first applied to the reference CSI in Figure 2. The CSI has been simulated with an ac reference driving into a short circuit to demonstrate and evaluate the operation of the DFS technique. The DFS configuration is compared against a fixed constant frequency configuration with the *same average bandwidth*. Both pure delays (transport delay) and ZOH modules have been examined presenting similar results.

The second application is to a CSI configured as an Active Harmonic Filter. The filter is evaluated filtering harmonics for a single phase rectifier load such as a computer power supply.

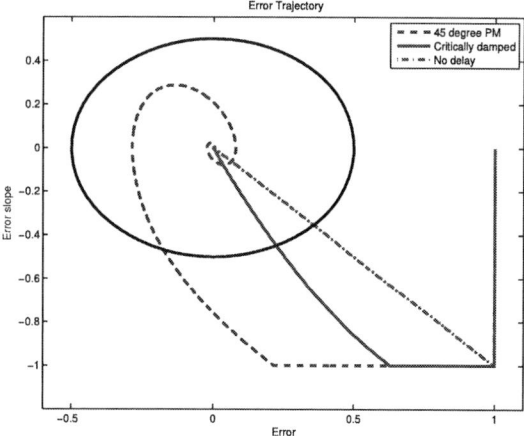

Fig. 5. Transition band.

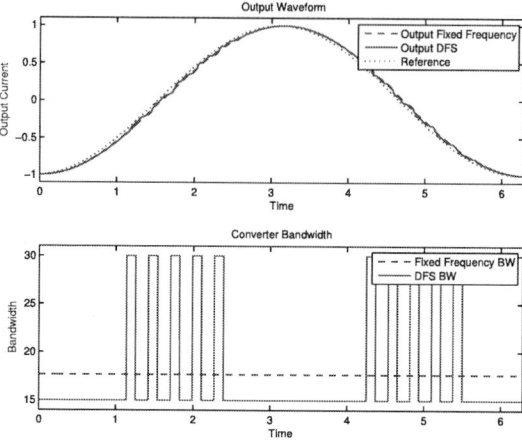

Fig. 6. CSI DFS operation for CSI with AC reference.

A. Performance Criteria

Each of the presented DFS applications is to be compared in terms of 'performance' against a fixed frequency system with the same average switching frequency. To do this a consistent and valid performance measure must be used that compares how close the output is to the reference input. For the CSI inverter DFS evaluation the method used is the RMS of the output error normalised to the reference RMS, shown in (31). This provides a measure of the relative output error for a given input signal, referred to as a converter's tracking accuracy.

$$\text{Tracking Error} = \frac{Err_{RMS}}{Ref_{RMS}} \quad (31)$$

For the Harmonic Filter application the performance will be evaluated by how well the filter is able to remove the load current harmonics. The performance of the filter is measured as the THD of the filtered load current. The IEC THD measure is used as the harmonic RMS over the fundamental RMS [13].

CSI inverters are measured in terms of performance as their ability to accurately produce the reference current and also their output impedance. An ideal current source inverter will perfectly produce the reference current, and has an infinite output impedance which allows it to reject any voltage deviation. In practice the bandwidth of the converter limits both the ability to accurately produce the reference current and the output impedance. As the bandwidth has an effect on both of these performance measures, only the reference tracking performance will be analyzed to provide a measure of the DFS relative performance to that of a fixed frequency system.

B. Current Source Inverter performance

Figure 6 shows the operation of the DFS CSI and a fixed frequency CSI. The results shown are for a reference system with two bandwidth levels of 15 and 30 (relative to the fundamental) with a transition radius of 0.06, or 6% (for a CSI the radius units are Amps). The figure shows

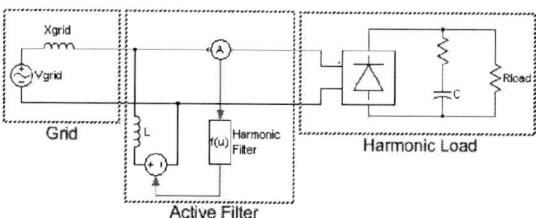

Fig. 7. Harmonic filter block layout.

the DFS and fixed frequency outputs superimposed on the reference signal. For an ac reference CSI the output error is greatest at the zero crossings, at which point the DFS system transitions to the higher bandwidth. The peaks of the waveform correspond to low output error points where the DFS system reverts to the lower bandwidth. The fixed frequency bandwidth was set as the average DFS bandwidth of 18.

From Figure 6 the fixed frequency system has a cleaner output waveform than the DFS system but has a greater tracking error about the zero crossing points of the waveform. As a result, the DFS system has a lower tracking error than the fixed frequency system (5% vs 6%) but the output ripple for DFS gives a greater THD than the fixed system.

C. Harmonic Filter Application

The main drive behind the DFS technique is the ability to get higher performance versus a fixed frequency system with the same average switching frequency. This typically becomes useful in medium to high power systems. A typical application where high bandwidth is required is active harmonic filters where the converter bandwidth directly corresponds to the harmonics it can filter [1]. Figure 7 shows the typical system configuration for an active harmonic filter with a shunt connected CSI [14]. The DFS technique has been applied to an active harmonic filter and simulated using Simulink and Plecs.

532

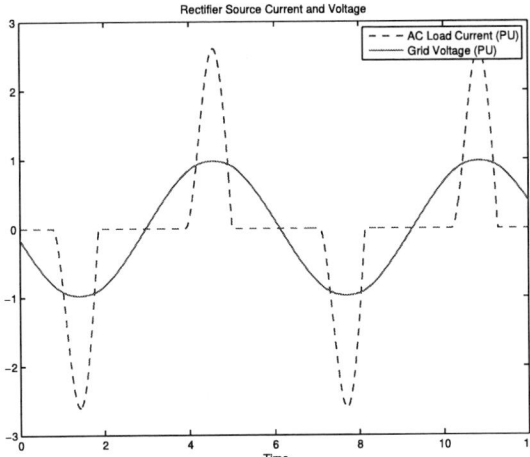

Fig. 8. Rectifier source current and voltage.

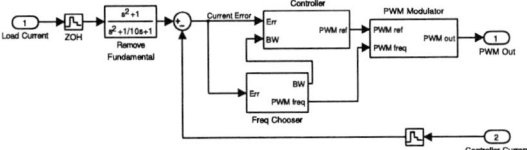

Fig. 9. Harmonic filter controller block diagram.

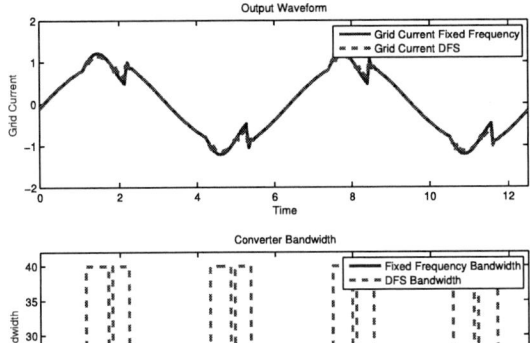

Fig. 10. Active Harmonic Filter DFS Operation at 100% load.

As suggested in Figure 7, a rectifier load is used as a typical harmonic load. An office PC was used as the reference load. A model of the harmonic load was created in Plecs with characteristics in Table I. The load is normalised for full load where the fundamental current is one pu for one pu supply voltage. Figure 8 shows the model rectifier harmonic source current for a one pu AC source voltage.

TABLE I
RECTIFIER LOAD CHARACTERISTICS

Specification	Value (Per Unit)
Supply Voltage Peak	1
Supply Voltage RMS	0.707
Current Fundamental Peak	1
Current Fundamental RMS	0.707
Current Total Peak	2.6
Current Total RMS	1.01
Current THD	103%
Crest factor	2.6

Figure 9 shows the block diagram layout for the Active Filter controller in Figure 7. The active harmonic filter was simulated both with and without DFS, and with a ZOH module and triangle-carrier PWM module. When using the PWM module the voltage and current measurements were sampled synchronously to remove the effect of the inductor current ripple. When comparing the DFS system with the fixed frequency system, the fixed frequency bandwidth and switching frequency are set as the average of the DFS system. For both systems the maximum converter output voltage is ±1.5.

The configuration has been simulated with several different bandwidth pairs. For a typical system a lower bandwidth of 20 was used with an upper bandwidth of 40 [15] (herein called a 20/40 bandwidth pair). This allows the system to consistently compensate for bandwidths up to the 20th harmonic roll-off point while being able to correct up to the 40th harmonic roll-off point in transient conditions. For a system with a fundamental of 50Hz, a

bandwidth of 20 corresponds to 1kHz ($6283rads^{-1}$). For a 45° phase margin, equation (29) specifies for a 1kHz bandwidth a switching frequency of 4.44kHz. Bandwidth pairs of 10 and 20 for the lower and upper bandwidths respectively were also tested.

The transition band is chosen as an error/error slope radius of 5% resulting in an average bandwidth of 26.2 for the 20/40 bandwidth pair configuration. The simulation results shown are using a ZOH module unless specified due to the very close correlation between the ZOH results and the PWM results. The only primary difference between the ZOH and PWM results are higher spectral components as a result of the PWM switching.

Figure 10 shows the filtered grid current for a DFS system and the equivalent fixed frequency system (the unfiltered load current is shown in Figure 8). The high bandwidth correction performed by the DFS system is immediately apparent in the figure where the overshoots that occur when the rectifier is conducting are less prominent than the fixed frequency system. The THD for the fixed frequency system is reduced from 103% (Table I) to 18%, while the DFS system reduces the THD to 13%. This illustrates a nearly one third reduction in THD for a DFS system versus a fixed frequency system for the same average switching frequency.

Figures 11, 12 and 13 each show the harmonic components of the grid current before filtering, fixed frequency filtered and DFS filtered. For each of the configurations shown the DFS system performs better than the fixed frequency system for the same average switching frequency.

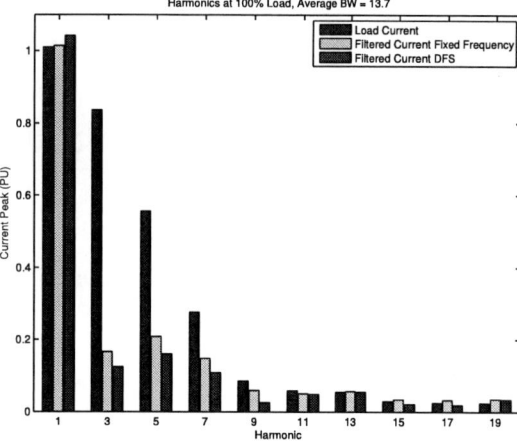

Fig. 11. Harmonics at 100% load. Average bandwidth = 13.7

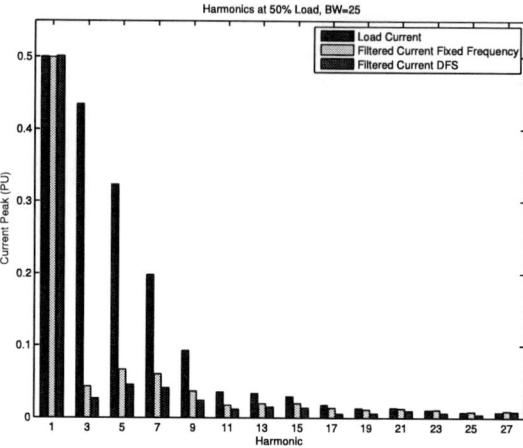

Fig. 13. Harmonics at 50% load. Average bandwidth = 25

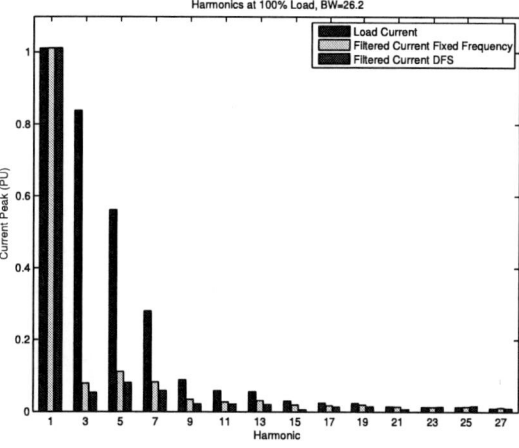

Fig. 12. Harmonics at 100% load. Average bandwidth = 26.2

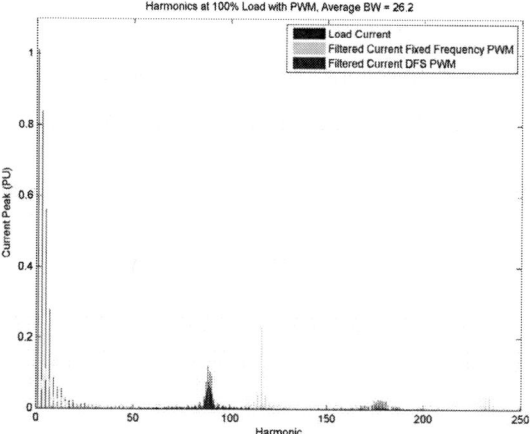

Fig. 14. Harmonics at 100% load using PWM module. Average bandwidth = 26.2

Figure 14 shows the harmonic spectrum of the grid currents using a PWM module rather than a ZOH. For both the fixed frequency and DFS system no output filter has been used. From the figure it is apparent that the peaks of the DFS carrier frequency peaks (harmonics 88.7 and 177.7) are smaller than the fixed frequency carrier peaks (harmonic 116.4).

V. ISSUES AND FUTURE WORK

The results presented show that DFS to be an effective solution (through simulation at least) but there are still several issues that are still being addressed at time of writing.

A. Changing Switching Frequency

When changing switching frequency for a PWM system the inductor current ripple will change proportionally to the switching frequency. Figure 15 shows an example of an arbitrary system with and with out current ripple correction when changing switching frequency. The system with ripple correction substitutes the set point PWM

reference (duty cycle) with an adjusted value for a short duration (in this case one period of the new frequency) to maintain the same *average* current. In most cases the shorter switching period of the two may be used to perform the correction but not in all cases. The transition duty cycle value and also the required period length to use has been proven but is beyond the scope of this paper. Without change correction the deviation in the inductor current can cause large deviations in the voltage ripple.

B. Choosing Appropriate Bandwidth Levels and Transition Radii

Ongoing investigation is being done on how to choose appropriate bandwidth levels and transition radii. By defining more transition radii many bandwidth levels may be used. Choosing the different levels becomes an optimisation problem. Optimizing the number and choice of levels involves a detailed knowledge of the relationship between performance and required bandwidths for a given

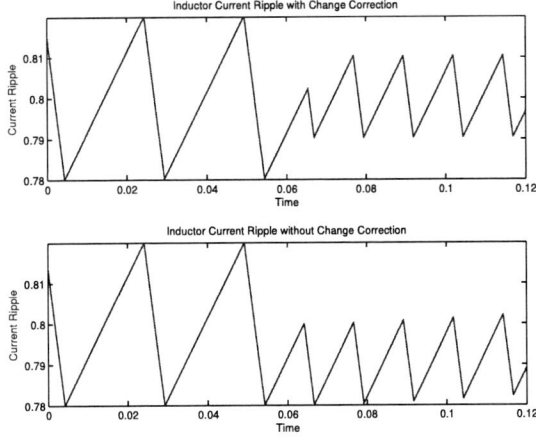

Fig. 15. Point of switching frequency change with and with out ripple correction.

application (including load type), and the relationship between switching frequency and efficiency. For example, there are limits on using high switching frequencies even for short durations as the typical high power switching devices (IGBTs, IGCTs, Thyristors etc.) have a slew rate limit. Different types of loads for a given transition radius will also result in different average switching frequencies, although still limited to the minimum and maximum.

VI. CONCLUSION

In this paper a switching technique targeted towards medium to high power converters is presented to achieve an improved performance for a given average switching frequency than a fixed frequency system. The DFS system provides a suitable method to temporarily increase the system bandwidth in the presence of a larger than acceptable output error to improve the system response.

The paper developed the relationship between bandwidth and switching frequency for a current source inverter for a chosen phase margin. The bandwidth to switching frequency relationship ensures that for any bandwidth (or conversely, switching frequency) the specified phased margin is conserved, ensuring global stability. A method for choosing when to transition between bandwidth levels is introduced based on the output error. The DFS technique is evaluated using a CSI against fixed frequency systems with the same average switching frequency.

A simple CSI inverter driving a short circuit is simulated and a more practical application of an active harmonic filter is demonstrated. The CSI inverter driving into a short circuit illustrates the DFS ability to reduce the tracking error for a given average switching frequency. The harmonic filter example demonstrates THD improvements of 30% compared to a fixed frequency system with the same average switching frequency at the cost of the increased complexity DFS involves.

REFERENCES

[1] H. Akagi, "Trends in Active Power Line Conditioners," *IEEE Transactions on Power Electronics.* vol. 9, num 2, May 1994, pp. 263 - 268.

[2] C. Y. Jeong, J. 6. Cho, Y. Kang, G. H. Rim, and E. H. Song, "A 100kVA Power Conditioner for Three-Phase Four-Wire Emergency Generators", *IEEE Power Electronics Specialists Conference*, vol. 2, 17 - 22 May 1998, pp. 1906 - 1911.

[3] D. J. Tschirhart, P. K. Jain, "Maximizing Resonant Converter Efficiency through Optimal Switch Selection", *IEEE Industrial Electronics Society IECON*, November 2005, pp. 622 - 627.

[4] J. J. Spangler, A. K. Behera, "A Comparison Between Hysteretic and Fixed Frequency Boost Converters Used For Power Factor Correction", *Applied Power Electronics Conference and Exposition*, March 1993, pp. 281 - 286.

[5] W. Gu, I. Batarseh, "Interleaved Synchronous Buck Regulator with Hysteretic Voltage Control", *Power Electronics Specialists Conference*, vol. 3, 17 - 21 June 2001, pp. 1512 - 1516.

[6] S. R. Bowes, "Advanced Regular-Sampled PWM Control Techniques for Drives and Static Power Converters", *IEEE Transactions on Industrial Electronics*, vol. 42, no. 4, August 1995, pp. 367 - 373.

[7] S. R. Bowes, D. Holliday, "Comparison of pulse-width-modulation control strategies for three-phase inverter systems", *IEE Proceedings - Electric Power Applications*, vol. 153, issue 4, July 2006, pp. 575 - 584.

[8] S. R. Bowes, D. Holliday, "Optimal Regular-Sampled PWM Inverter Control Techniques", *IEEE Transactions on Industrial Electronics*, vol. 54, no. 3, June 2007, pp. 1547 - 1559.

[9] C. Hua, "High Switching Frequency DSP Controlled PWM Inverter", *Second IEEE Conference on Control Applications*, vol. 1, 13 - 16 September 1993, pp. 273 - 283.

[10] SP Systems Pte Ltd, "Developments in Power Quality", *PowerCon 2000 presentation*, Perth Australia, 7 December 2000.

[11] Y. C. Cheng, C. Hwang, "Use of the Lambert W function for time-domain analysis of feedback fractional delay systems", *IEE Proceedings - Control Theory and Applications*, vol. 153, issue 2, 13 March 2006, pp. 167 - 174.

[12] F. Maghami Asl, A. Galip Ulsoy, "Analytical Solution of a System of Homogeneous Delay Differential Equations via the Lambert Function", *Proceedings of the 2000 American Control Conference*, vol. 4, 28 - 30 June 2000, pp. 2496 - 2500.

[13] *IEV 551-20-13 Total Harmonic Distortion*, IEC 60051-551 International Electrotechnical Vocabulary Chapter 551: Power Electronics, Nov 1, 1998.

[14] H. Akagi, E. H. Watanabe, M. Aredes, "Basic Principles of Harmonic Compensation," in *Instantaneous Power Theory and Applications to Power Conditioning.* IEEE Press, 2007, pp. 11

[15] A. M. Al-Zamil, D. A. Torrey, "A Passive Series, Active Shunt Filter for High Power Applications", *IEEE Transactions on Power Electronics*, vol. 16, no. 16, January 2001, pp. 101 - 109.

The control of voltage converter rectifiers

Krzysztof Szubert

Poznan University of Technology, Institute of Electric Power Engineering
Ul. Piotrowo 3A, Poznan, Poland
E-Mail: krzysztof.szubert@put.poznan.pl

Summary: It is talked, the rule of voltage converter control, which is used to introducing of added voltage, in this article. It is given procedure of switching converter rectifiers to hold received voltage comes from controller how most approximate to set voltage.

Keywords: voltage source converter (VSC), converter control, harmonics

1. Introduction.

As a voltage converters are accepted AC/DC converters, which are built from torn-off elements (thyristors GTO, transistors IGBT) with capacitor in DC circuit. They are wide applied in fast compensators of passive power (STATCOM), as well as in parallel (current) active high harmonic filters. They characterize possibility direct, smooth control of current on attached inductivity of filter by step change of voltage.

The control of introduced voltage by voltage converters is well work out, at foundation of sinusoidal shape of current and voltage [4]. When courses are unsinusoidal, it is very difficult make voltage drop estimation on filter coil inductivity [2], so smooth control of introduced voltage for voltage converters is difficult. Sometimes one decides to skip filter [3], or one decides to apply parallel capacitor in filter circuit (similar as in current converter) [1,5]. At the last solution, introduced voltage is formed on filter capacitor (there are applied transformers with small leakage reactance). Mutual correlation among capacity of DC circuit capacitor, and inductivity and capacity of filter elements causes fast overloading of filter capacitor when filter series inductance is small, so in this solution, real regulating does not occur. When this inductivity value is big, the changes of current have not step character and voltage readjusting on capacitor occurs.

Current converters (with inductor in DC circuit) have not higher exchanged defects. For that reason, these converters are chosen to introducing of series adding voltages. However these converters inflict similar problem with control of current at parallel circuit. So when it is demanded complex series - parallel regulation (to control both as current as voltage) by application two converters, which are joined by DC circuits, there are often used voltage converter, and very rarely current converter.

Moreover, voltage converters in relate to current converters mark smaller power loss. From here they are used even in series active filters despite problems with readjusting of voltage.

Author recognized, that it is important problem and it is unsatisfactory dissolved so far. Therefore this problem is worth of investigation.

2. The hysteresis classic PWM method and the constant frequency switch with prediction classic PWM method

The rule of converter work is possible to explain by using fig. 1 (application as series active filter). The value of set voltage U_Z of filter capacitor C_F is calculated on basis of required voltage of energy receipts as well as measured voltage of power system and transformation ratio (in different applications e.g. SSSC, voltage U_Z can be marked on basis of transmitted power etc).

978-1-4244-1741-4/08/$25.00 ©2008 IEEE

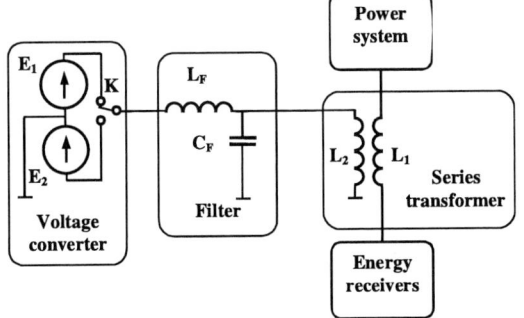

Fig.1 Simplify one-phase diagram of main circuit of series voltage converter to theoretical considerations

When capacitor input current I_{LF} is larger from capacitor output current I_{L2}, the capacitor voltage U_{CF} enlarges, in opposite this voltage diminishes (1).

$$\frac{d\,u_{CF}}{d\,t} = C_F\left(i_{LF} - i_{L2}\right) \qquad (1)$$

The value of current I_{L2} is depended on receipts group load. Current I_{LF} is controlled by switching of converter rectifiers (2).

$$\frac{d\,i_{LF}}{d\,t} = \frac{E_{1(2)} - u_{CF}}{L_F} \qquad (2)$$

By turn-on rectifier at hi voltage side of DC circuit capacitor of converter (it is similar to switch key K onto source E_1 in Fig. 1), the increase of current I_{LF} is caused. In other hand by turn-on rectifier at lo voltage side of DC circuit capacitor of converter (it is similar to switch key K onto source E_2 in Fig. 1) the decrease of current I_{LF} is caused.
Selection of impedance value L_F and capacitance value C_F is very essential, because in fact it is not only filter, but also arrangement to formation of voltage shape. Optimization of selection of these parameters is very difficult, because controller voltage U_{CF} should keep up set value U_Z, and voltage U_{CF} should contain value of high harmonic as very small as it is possible (about frequencies approximate to frequency of switches), and value load current derivative di_{L2}/dt is dependent on energy receipts.
Rule of hysteresis classic PWM method is showed in figure 2.

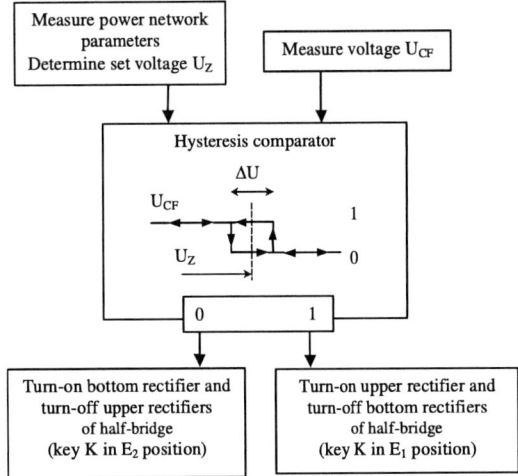

Fig.2 Block diagram of controller rectifiers work rule for hysteresis classic PWM method

In most of hysteresis arrangements, width of hysteresis loop marks range of changeability of regulated factor around set value (in approximation). Here (Fig. 2) it is possible to notice, that factor ΔU in hysteresis module accelerates switch, and in opposite case very strong oscillations would occur. In spite this oscillations still are considerable.

Factor ΔU does not limit deviations of voltage but introduces rectifiers temporary switch prediction. From here "classic" does not mark, that, this method is similar to different hysteresis methods, but that it as first created standard of voltage control by voltage converter.

The hysteresis classic PWM method is applied in analog controllers (in easier realization point of view) and constant frequency switch with prediction classic PWM method is applied in digital controllers (in easier delimitation of processor power point of view – computational cycle is comparable with constant switching cycle).

The rule of constant frequency switch with prediction classic PWM method is showed in Fig. 3. In this method is also tried of earlier switching of rectifiers – exactly one cycle before. Unfortunately oscillations also occur.

In equations that are given in block diagram (Fig.3, Fig.4, Fig.5) upper indexes concern switching cycles (n) - current, (n-1) - previous, (n+1) - foreseen next. The prediction exactitude is possible to improve by enlarging number of measurements between switching cycles, but costs of measure cards as well as processor would be larger.

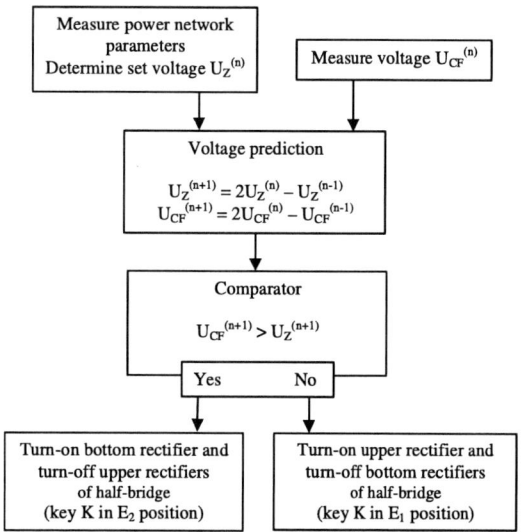

Fig.3 Block diagram of controller rectifiers work rule for constant frequency switch with prediction classic PWM method

The constant frequency switch with prediction classic PWM method depends on calculation of approximate value of capacitor voltage $U_{CF}^{(n+1)}$ and set voltage $U_Z^{(n+1)}$ in next switching cycle. The decision about switch one cycle earlier is accomplished on this basis. In effect current oscillations are smaller. It causes, that time between moment of switch (moment of decision about change of sign +/- du/dt of voltage U_{CF}), and moment of change of sign +/- du/dt of voltage U_{CF} (moment of achievement by filter current I_{LF} the value of load current I_{L2}) is shorted. So it gets smaller oscillations of voltage U_{CF} around the value of set voltage U_Z. Similarly to hysteresis classic PWM method oscillations are considerable.

3. Methods with additional current feedback loop

As it was noticed in previous chapter very important for value of oscillation of voltage U_{CF} around set voltage value U_Z has inertia – time between decision about change of sign +/- du/dt and real change of this sign (when current I_{LF} reaches value of current I_{L2}). Introducing additional feedback loop to control of current ILF can reduce this time.

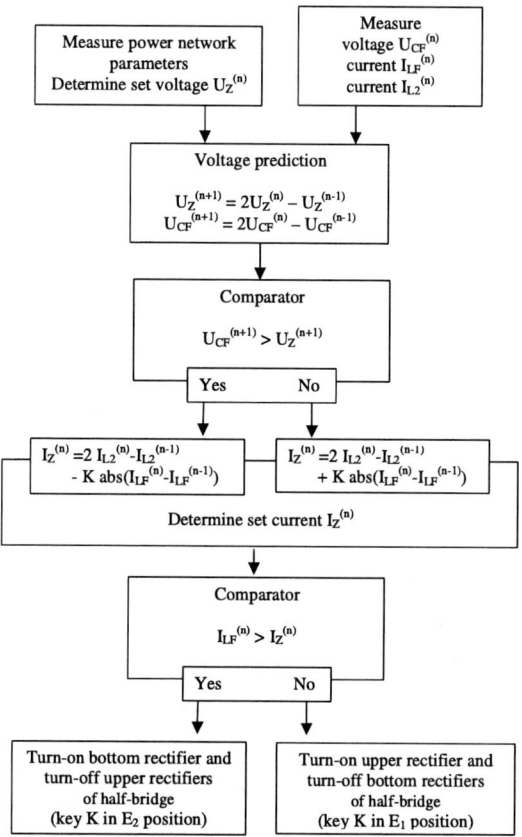

Fig.4 Block diagram of controller rectifiers work rule for constant frequency switching and current feedback with constant value PWM method

Value of current I_{LF} is possible to regulate by twofold way. In first, it is current feedback with constant value method. In this method, the set value of current is calculated as $I_Z = I_{L2} \pm \Delta I$, when about sign $\pm$ before ΔI it is decided by compare

538

predicted voltage of capacitor and predicted set voltage. In second it is current feedback with prediction method. In this method, set value of current is calculated in this manner, that if in next switching cycle value $I_{L2}^{(n+1)}$ achieves value $I_Z^{(n)}$, also voltages $U_{CF}^{(n+1)}$ and $U_Z^{(n+1)}$ would be identical.

The works rule of current feedback with constant value method is showed in Fig. 4. The value of coefficient K, in equation that described set current $I_Z^{(n)}$ (Fig.4), is from 0.5 to 1 and it decides about damping of changes of voltage U_{CF}. It is also possible to notice, that it is introduced prediction of load current $I_{L2}^{(n+1)}=2\,I_{L2}^{(n)}-I_{L2}^{(n-1)}$ as well as that ΔI is not really constant as in name of method, but is depended on last change of current I_{LF} ($I_{LF}^{(n)}-I_{LF}^{(n-1)}$). In this meter the voltage U_{CF} influence on changeability of current I_{LF} is taken in attention.

In figures 5 and 7 the voltages and currents courses for constant frequency switch PWM method were introduced, from here perpendicular lines marked possible moments of rectifiers switches. Set voltage U_Z as well as load current I_{L2} are accepted same value in these pictures (Fig5, Fig.6). Also value of initial filter voltage U_{CF} and filter current I_{LF} are identical in these pictures.

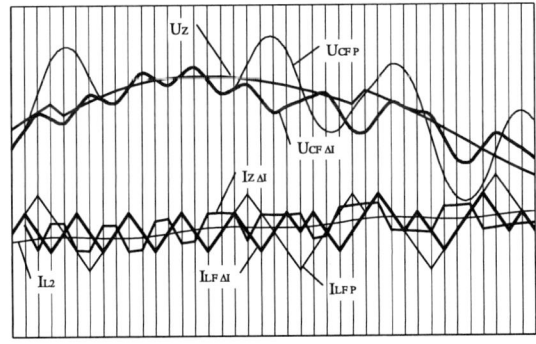

Fig.5 The voltage and current curves for constant frequency switching PWM method as with classic prediction as with current feedback about constant value

The constant frequency switching with prediction classic PWM method and the constant frequency switching and current feedback with constant value PWM method is compared, in Fig. 5. First from these methods followed decision about switching of rectifiers on basis of appointed with prediction voltages. Second method, on this basis changes set current $I_{Z\,\Delta I}$ (in relation to value of load current I_{L2} the reversal of sign +/- at ΔI follows). The comparison of filter current $I_{LF\,\Delta I}$ with set current $I_{Z\,\Delta I}$ decides about switching of rectifiers. By limitation of current hesitations to value less then $2\Delta I$ (compare $I_{LF\,\Delta I}$ and $I_{LF\,P}$), limitation of voltage hesitations followed also (compare $U_{CF\,\Delta I}$ and $U_{CF\,P}$). This meaner of calculation set current value limits speed of voltage changing, what brings additional inertia at sudden changes of set voltage value U_Z.

To suppression of introduced voltage oscillation around set value, and simultaneously do not increase inertia, it would be allowed large deviation of current I_{LF} from value I_{L2} for large difference between U_{CF} and U_Z, however for small this difference, the deviation of current I_{LF} from value I_{L2} should be limited.

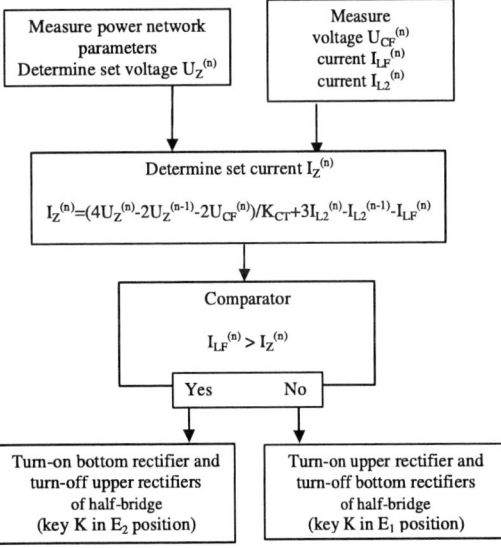

Fig.6 Block diagram of controller rectifiers work rule for constant frequency switching and current feedback with prediction PWM method

539

On this basis, it was worked out the constant frequency switching and current feedback with prediction PWM method. The works rule of current feedback with prediction method is showed in Fig. 6.

In this method, on basis of set value of voltage $U_Z^{(n)}$ as well as present value of capacitor voltage $U_{CF}^{(n)}$ and present value of load current $I_{L2}^{(n)}$ and present value of current witch flows from converter to filter $I_{LF}^{(n)}$, it is marked such set value of current $I_Z^{(n)}$, that if in next switching cycle value of current $I_{LF}^{(n+1)}$ would reach this value of current $I_Z^{(n)}$, the value $U_{CF}^{(n+1)}$ would be even value $U_Z^{(n+1)}$. In next step values $I_Z^{(n)}$ and $I_{LF}^{(n)}$ are compared, which of them is bigger it determines about switching of rectifiers.

Coefficient K_{CT} in equation which defining current value $I_Z^{(n)}$ (Fig. 6) depends from filter capacitor capacity C_F, as well as from time T between possible moments of switches. Value of set current I_Z was calculated on basis of assumption, that change value of capacitor voltage from present value $U_{CF}^{(n)}$ to appointed from prediction of set voltage value $(U_Z^{(n+1)}=2U_Z^{(n)}-U_Z^{(n-1)})$ is caused by charging of capacitor by average current from present value $(I_{L2}^{(n)}-I_{LF}^{(n)})$ to appointed from prediction $(I_{L2}^{(n+1)}-I_Z^{(n)}$ where $I_{L2}^{(n+1)}=2I_{L2}^{(n)}-I_{L2}^{(n-1)})$.

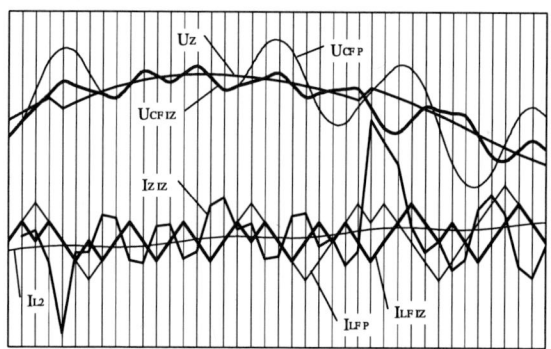

Fig.7 The voltage and current curves for constant frequency switching PWM method as with classic prediction as with prediction in current feedback

In Fig. 7 the constant frequency switching with prediction classic PWM method and the constant frequency switching and current feedback with prediction PWM method is compared. Prediction of current method is exacter. If it would be possible to match parameters of filter so, that changes of value I_Z would be slower then changes of value I_{LF}, and these last changes would have not large values at switching cycle, and simultaneously changes U_{CF} would be little faster then U_Z, then it would be possible to affirm, that controller works best. Even yet, when controller is remote from optimum work, delimitation of set current brings better results, because in voltage converter is possible to get exacter control current then voltage.

This method similarly how "constant frequency switching and current feedback with constant value PWM method" dumps oscillations of voltage great, simultaneously similarly how "constant frequency switching with prediction classic PWM method" does not limit possible speed of changes of add voltage. Moreover selection of filter parameters is not so critical, because these parameters are taken into consideration in selection of regulator parameters

4. Conclusions

It is difficult, voltage formation by voltage converter, because switching of rectifier is not equivalent of changing of sign of voltage derivate. At this moment, the sign of converter current derivate is changed. Just, when converter current value I_{LF} achieves load current value I_{L2}, the changing of sign of voltage derivate occurs. The over-regulations of voltage are as a result of inertia, which is introduced to steering circuit. For reduction of voltage over-regulations one should change way of steering from voltage to current, so one should calculate set voltage value onto set current value. It is joined with introduction of additional current feedback, so it increases cost (additional measurement of current in two places).

The limitation of converter current deviation in relation to load current (as in constant frequency switching and current feedback with constant value PWM method) suppresses add voltage oscillations strongly. However it provokes inertia between set voltage and received voltage at fast changes of set voltage.

The constant frequency switching and current feedback with prediction PWM method has not this defect; simultaneously this method keeps above advantages [6].

Reference

[1] Flota M., Alvarez R., Nunez C.: „Sliding Mode Observer-Based Control for a Series Active Filter" Proc. 2nd International Conference on Electrical and Electronics Engineering, Mexico City, Mexico, 2005 pp. 254-257

[2] Lu B., Ooi B.: „Nonlinear Control of Voltage-Source Converter Systems" IEEE Transactions On Power Electronics, Vol. 22, No. 4, 2007, pp 1186-1195

[3] Rolux A., Mouton H., Akagi H.: „Digital Control of an Integrated Series Active Filter and Diode Rectifier With Voltage Regulation" IEEE Transactions On Industry Applications, Vol. 39, No. 6, 2003 pp 1814-1820

[4] Ruszczyk A., Sikorski A.: „Przekształtnik sieciowy AC/DC przy sterowaniu napięciowym i prądowym – analiza porównawcza" („The power converter AC/DC at voltage and current control - comparative analysis") Przegląd Elektrotechniczny, 7-8/2006, pp. 52-56

[5] Szubert K.: „Control of voltage converter which includes add voltage" Proc. X Symposium Fundamental Problems of Power Electronic and Electromechanics, Wisla, Poland, 2003, pp. 39-42

[6] Szubert K.: „Sposób sterowania przekształtnikiem napięciowym" (The voltage converter method of steering). Patent notification no PL 383727

Krzysztof Szubert finished Poznan University of Technology (Poland) with specialization automatic protection in 1990. In 1998 he made doctoral dissertation and from this time he has worked as a tutor in this university. He patented method of control STATCOM to compensate reactive component, high harmonics component and sub-harmonic component of current. He is also author of few patent notifications. He is interested in application power electronic to power system. He worked in team, which in grants of Polish Committee of Scientific Investigations, worked out and executed prototype of anti-failure converter controller and prototype of unified power electronic system to control of distribution power network parameters.

Load Voltage Regulation and Line Loss Minimization of Loop Distribution Systems Using UPFC

Mahmoud A. Sayed*
mahmoud@motion.elcom.nitech.ac.jp

Takaharu Takeshita*
take@motion.elcom.nitech.ac.jp

*Dept. of Electrical and Computer Engineering,
Nagoya Institute of Technology
Gokiso, Showa, Nagoya 466-8555, JAPAN.

Abstract—This paper presents a new method for achieving line loss minimization and voltage regulation in the loop distribution systems, simultaneously. First, the line loss minimum conditions in the loop distribution systems are presented. Then, load voltage regulation by using the shunt compensation is presented. In order to achieve these two objectives, simultaneously, the unified power flow controller (UPFC) is used. The UPFC shunt converter is used as a shunt compensator to regulate the load voltage, and the UPFC series converter is used to control the power flow to achieve line loss minimization. Also, the proposed control schemes of the UPFC series and shunt converters are investigated. The effectiveness of the proposed control schemes of the UPFC has been verified experimentally using laboratory prototype in a 200V, 6 kVA system.

Index Terms—loop distribution system, voltage regulation, line loss minimization, series and shunt compensation, FACTS, unified power flow controller (UPFC).

I. INTRODUCTION

Distribution networks may be classified as either radial or loop. It is well understood that radial distribution systems are more desirable than loop distribution systems, and distribution engineers have preferred them because they use simple, inexpensive protection schemes. Radial distribution systems are used in Japan because when a fault occurs in the distribution system, the part of the fault can be isolated fast from the distribution system to avoid the influence of the fault. However, in [1], a new method of distribution system configuration was proposed. This method based on reconfiguration of the radial distribution system to a loop distribution system using the existing infrastructure of the distribution system. In this method the reliable operation of the loop distribution system was achieved by installing a high speed protective relay to prevent the expansion of the fault in the closed loop system. From the background, many papers dealing with the loop distribution systems and their merits have been published [1-5].

Much of the recent research on distribution systems has been focused on voltage regulation and minimization of the power loss. Many researches used distributed generation, series capacitors, and shunt capacitors, connected in strategic location, to regulate the load voltage and

minimize line loss by compensating the reactive power required by the loads [6,7]. Other researches minimize line loss and regulate the voltage in distribution systems by reconfiguring the existing systems using the sectionalizing switches [8]. Also, many papers dealing with loss reduction and voltage regulation using FACTS devices, that have been introduced first by Hingorani in 1988 [9], have been published. Most of these papers used STATCOM, shunt active filter and series-shunt power converter to regulate and balance the voltage at the customer side and reduce the losses by reactive power injection [10-16]. One of the most promising FACTS devices is UPFC (Unified Power Flow Controller), which has been introduced by Gyugyi in 1991 [17].

In recent years, UPFC has been proposed to increase power flow as well as an aid for system stability through the proper design of its controller. It is becoming one of the most important FACTS devices since it can provide various types of compensation, i.e. voltage regulation, phase shifting regulation, impedance compensation and reactive compensation. In [3-5], the authors have proposed line loss minimum conditions of the loop distribution systems and experimentally achieved these conditions using the UPFC. Minimization of the total line loss in loop distribution system can be realized by eliminating the loop current from the loop system, and that can be achieved if one condition of these two conditions is realized:

- The ratio of R/L of each line constituting the loop system is equal.
- The summation of the reactance voltage drops in the loop system is zero.

In this paper, achieving voltage regulation and total line loss minimization, simultaneously, in the loop distribution systems is investigated by using the UPFC series and shunt converters. The shunt converter is used to regulate the load voltage, whereas the series converter is used to minimize the total line loss of the loop distribution system. Also, the proposed control schemes of the UPFC series and shunt converters are presented. Voltage regulation and total line loss minimization in the loop system are investigated experimentally by using laboratory prototype in a 200V, 6kVA system.

978-1-4244-1741-4/08/$25.00 ©2008 IEEE

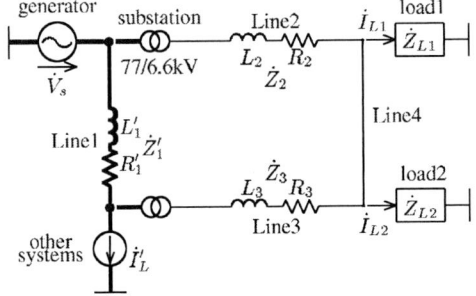

Fig. 1. Model of loop distribution system

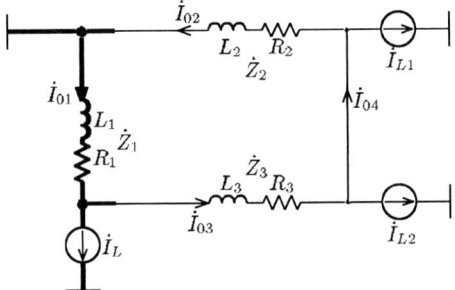

Fig. 2. Approximate model of loop distribution system

II. LINE LOSS MINIMUM CONDITIONS [3,5]

Fig. 1 shows a simple model of loop distribution system. In this model, impedances of line 1, 2, 3, and 4 are $\dot{Z}'_1=R'_1+j\omega L'_1$, $\dot{Z}_2=R_2+j\omega L_2$, $\dot{Z}_3=R_3+j\omega L_3$, and $\dot{Z}_4=0$, respectively. The load impedances are $\dot{Z}_{L1}$ and $\dot{Z}_{L2}$. The other systems, that connected to this loop system, are represented by a current source $\dot{I}'_L$. Fig. 2 shows the approximate model of the system shown in Fig. 1. The load currents $\dot{I}_{L1}$ and $\dot{I}_{L2}$, and the other systems current $\dot{I}_L$ are assumed to be constant.

Total line loss minimum conditions can be obtained in the loop distribution system from the total line loss equation that can be formulated using the line currents that flow in the loop system lines. First, the line currents and the total power loss equations are formulated. Then, the total line loss minimum conditions can be obtained from the total power loss equation. In [3,5], the authors have published the mathematical derivation process of the total line loss minimum conditions in the loop distribution system in detail. The change in the line currents is defined as the loop current $\dot{I}_{loop}$, which circulate in the loop distribution system in the same direction (counter clockwise), and can be formulated as follows:

$$\dot{I}_{loop} = -\frac{\sum_{i=1}^{3} j\omega L_i \dot{I}_{0i}}{R_{loop}} \qquad (1)$$

Total line loss minimum conditions of the loop systems can be realized by eliminating the loop current $\dot{I}_{loop}$ from the loop system. Two conditions can be obtained from equating the loop current with zero.

The first condition is:

$$\frac{R_1}{L_1} = \frac{R_2}{L_2} = \frac{R_3}{L_3} \qquad (2)$$

In other words, if the lines of the loop distribution system are constructed by the same line type, the total line loss minimum is realized without using any controller.

The second condition is:

$$\sum_{i=1}^{3} j\omega L_i \dot{I}_{0i} = 0 \qquad (3)$$

In other words, if the summation of the reactance voltage drop in the loop system is zero, the total line loss minimum is realized. Equations (2) and (3) have the same object because both of them eliminate the loop current $\dot{I}_{loop}$ from the loop system and hence minimize the total line loss. The loop current can be eliminated by using the UPFC series converter[3].

III. LOAD VOLTAGE REGULATION

The problem of the voltage and reactive power control in distribution systems is demanding great effort for the system engineers in the definition and in realization of sophisticated control schemes. Voltage regulation in the distribution system can be achieved by series or shunt compensation techniques. Both of them depend on reactive power compensation.

In series compensation technique, a device that that is connected in series with a distribution line is called a series compensator. It is referred to as a compensator since it compensates for the reactive power in the ac system. Series capacitor, Thyristor Controlled Series Compensator (TCSC), and Static Synchronous Series Compensator (SSSC) can be used as series compensator to regulate the load voltage. In [4], the authors have proposed the load voltage regulation and line loss minimization, simultaneously, in the loop distribution system by using the UPFC series converter only. Also, the proposed technique has been investigated experimentally.

In shunt compensation technique, a device that that is connected in parallel with a transmission line is called a shunt compensator. Shunt compensators have been installed on distribution feeders to supply the type of reactive power or current to counteract the out-of-phase component of current required by the inductive loads. Shunt capacitor, Static VAR Compensator, and STAT-COM can be used as a shunt compensator to regulate the load voltage.

Fig. 3(a) shows the distribution system, represented by single phase Thevenin equivalent circuit, from the load terminal, with the shunt compensator. The shunt compensator is represented by a current source that injects a current I_c. V_s and V_r are source voltage and load voltage, respectively. The load current I_l, which equals the source current I_s before installation of the shunt compensator, is assumed to be constant. Fig. 3(b) shows

543

the phasor diagram of the whole system before installation of the shunt compensator. The load voltage is assumed to be the reference voltage. The voltage drop (ΔV) at the load terminal and the load current I_l can be calculated as follows [18]:

$$\left.\begin{array}{l} \Delta \dot{V} = \dot{V}_s - \dot{V}_r = (R_s + jX_s)\,\dot{I}_l \\ \dot{V}_r = |\dot{V}_r| \angle 0.0° \\ \dot{I}_l = \dfrac{P_l - jQ_l}{\dot{V}_r} \end{array}\right\} \quad (4)$$

So that,

$$\begin{aligned} \Delta \dot{V} &= (R_s + jX_s) \times \frac{P_l - jQ_l}{\dot{V}_r} \\ &= \frac{(R_s P_l + X_s Q_l)}{\dot{V}_r} + j\frac{(X_s P_l - R_s Q_l)}{\dot{V}_r} \quad (5) \\ &= \Delta V_R + j\Delta V_X \quad (6) \end{aligned}$$

As shown in (6), the voltage drop can be divided into two components, ΔV_R in phase with the load voltage, and ΔV_X in quadrature with the load voltage.

After installing the shunt compensator, the source current can be calculated as follows:

$$\dot{I}_s = \dot{I}_l + \dot{I}_c \quad (7)$$

From (4) and (5)

$$|\dot{V}_s|^2 = \left|\dot{V}_r + \frac{(R_s P_l + X_s Q_s)}{\dot{V}_r}\right|^2 + \left|\frac{(X_s P_l - R_s Q_s)}{\dot{V}_r}\right|^2 \quad (8)$$

where:

$$Q_s = Q_l + Q_c \quad (9)$$

Fig. 3(c) shows the phasor diagram of the whole system after installing the shunt compensator. By controlling the shunt compensator current I_c, it is possible to make the load voltage to be equal in magnitude to the source voltage , $|V_r|=|V_s|$. The required reactive power of the compensator can be calculated by solving (8) for Q_s with $|V_r|=|V_s|$; then the compensator reactive power, Q_c, can be calculated.

The main object of this paper is to minimize the total line loss and to regulate the load voltages of the distribution systems, simultaneously. The line loss minimum conditions can be achieved if the loop current eliminated from the loop system. Under this condition, the load voltage can be controlled by using reactive power compensation in order to maintain it at constant value. Load voltage regulation problems in the distribution systems are commonly solved by using D-STATCOM that connected in parallel with the distribution lines to inject a controlled leading or lagging reactive power. The STATCOM is one of the FACTS devices, which can compensate reactive power in an efficient fast way. The basic electronic block of a STATCOM is a voltage sourced converter that converts a dc voltage at its input terminals into a three-phase set of ac voltages at fundamental frequency with controllable magnitude and phase angle, and it is used as a reactive power compensator by absorbing or supplying reactive

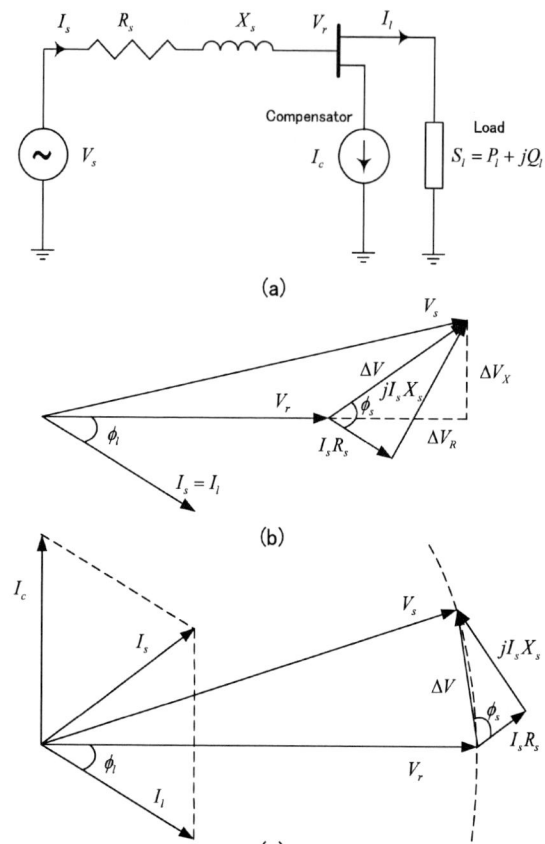

Fig. 3. (a) Equivalent circuit of the whole system, (b)Phasor diagram before shunt compensation, (c)Phasor diagram after shunt compensation.

power. Under light load conditions, the controller is used to minimize or completely diminish line over voltage; on the other hand, it can be also used to maintain certain voltage levels under heavy loading conditions.

In this paper, since the UPFC shunt converter has the same circuit configuration of the D-STATCOM, the function of the D-STATCOM to regulate the load voltage is achieved by using the UPFC shunt converter. The UPFC shunt converter is used to regulate the dc link voltage and load voltage, simultaneously. The load voltage is regulated to be equal in magnitude to the source voltage.

IV. GENERAL CONFIGURATION OF UPFC

Fig. 4 shows the general configuration of the UPFC. It consists of combined series and shunt converters connected back to back to each other through a common dc-link capacitor. The series converter, which acts as a controllable voltage source v_c, is used to inject a controlled voltage in series with the line and thereby to force the power flow to a desired value. In general, the series converter exchanges active and reactive power with the line while performing this duty. The reactive power is electronically provided by the series inverter itself, and

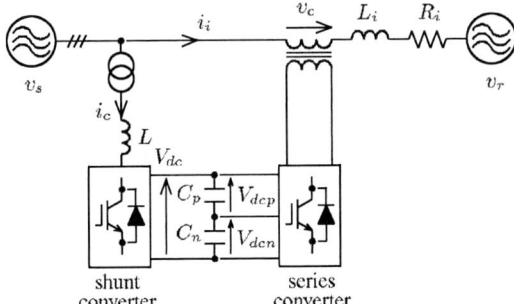

Fig. 4. Configuration of UPFC

the active power is transmitted to the dc terminals. The main function of the shunt converter, which acts as a controllable current source i_c, is to regulate the dc link voltage by adjusting the amount of active power drawn from the transmission line to meet the real power needed by the series converter. In addition, the shunt converter has the capability of controlling the reactive power.

In this paper, the main function of the UPFC series converter is to control the power flow in order to eliminate the loop current and hence minimize the total line loss of loop distribution system. The main function of the shunt converter is to regulate the dc link voltage and to regulate the load voltage to be equal in magnitude to the source voltage using reactive power injection, simultaneously.

V. CONTROL SCHEMES

The UPFC series and shunt converter control schemes are proposed in order to achieve total line loss minimization and load voltage regulation, simultaneously, in the loop distribution systems. In this paper, the system line parameters are assumed to be constant. Actually in the practical distribution systems, the system line parameters are changed according to the change of the system configuration by the sectionalizing switches, that used to isolate the faults from the system. This change in the distribution system lines can be known and the new values of the system line parameters can be considered in the UPFC control equations.

A. Series converter control scheme

Two power flow control schemes are proposed to obtain the reference voltage $\dot{V}_c$ of the UPFC series converter in order to realize the total line loss minimum conditions of the loop distribution systems. These schemes are Line Inductance Compensation and Line Voltage Compensation. These control schemes can be applied to the loop distribution system according to the system line parameters.

1) Line Inductance Compensation scheme: Line Inductance Compensation scheme is used to compensate the line parameters of the loop distribution system to realize the relation shown in (2). This control method can be used

if the condition shown in (2) is realized for all loop lines except one. The relation of the line impedance parameters of the loop system shown in Fig. 2 is:

$$\frac{R_1}{L_1} \neq \frac{R_2}{L_2} = \frac{R_3}{L_3} \qquad (10)$$

In this case, the UPFC can be controlled to compensate line 1 inductance by inserting a series inductance, L_c in order to realize same ratios of R/L in each line constituting the loop system. The value of L_c can be calculated as follows:

$$\frac{R_1}{L_1 + L_c} = \frac{R_2}{L_2} = \frac{R_3}{L_3} \qquad (11)$$

$$L_c = \frac{R_1}{R_2}L_2 - L_1 \qquad (12)$$

Considering the value of the inserted series inductance, L_c, the reference voltage of the UPFC series converter, $\dot{V}_c$, can be calculated, in the steady state, as follows:

$$\dot{V}_c = -j\omega L_c \dot{I}_1 \qquad (13)$$

In order to acheive fast and accurate response of the UPFC, the reference voltage of the UPFC series converter will be formulated as differential equation in the transient state as follows:

$$v_c = -L_c \frac{di_1}{dt} \qquad (14)$$

2) Line Voltage Compensation Scheme: Line Voltage Compensation scheme is used to compensate the inductance voltage drop in each line of the loop system in order to realize the condition shown in (3). This control scheme can be used if the relation of the line impedance parameters of the loop system shown in Fig. 2 is:

$$\frac{R_1}{L_1} \neq \frac{R_2}{L_2} \neq \frac{R_3}{L_3} \qquad (15)$$

In this case, the UPFC can be controlled to cancel the summation of the reactance voltage drop by inserting a series voltage, $\dot{V}_c$, equal in magnitude to the summation of the reactance voltage drop, but in the opposite direction. The value of the reference voltage of the UPFC series converter, $\dot{V}_c$, can be calculated using the condition shown in (3) as follows:

$$-\sum_{i=1}^{3} j\omega L_i \dot{I}_i + \dot{V}_c = 0 \qquad (16)$$

$$\dot{V}_c = \sum_{i=1}^{3} j\omega L_i \dot{I}_i \qquad (17)$$

In order to acheive fast and accurate response of the UPFC, the reference voltage of the UPFC series converter will be formulated as differential equation in the transient state as follows:

$$v_c = \sum_{i=1}^{3} L_i \frac{di_i}{dt} \qquad (18)$$

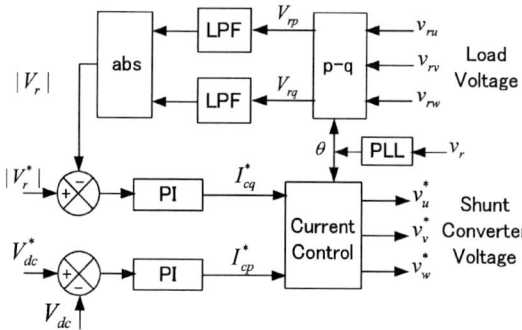

Fig. 5. Control circuit of UPFC shunt converter.

B. Shunt converter control Scheme

Regulating the reactive power injected by the shunt converter is used to achieve a constant regulated voltage at its bus. Fig. 5 shows the block diagram of the proposed direct output voltage control scheme in which the AC and DC voltage regulation are realized by PI controllers. The commanded real power current I_d^* is determined by a conventional PI controller which regulate the converter DC link voltage. The commanded reactive power current I_q^* is determined by a conventional PI controller which regulates the magnitude of the bus voltage that the shunt converter connected to. The currents I_d^* and I_q^* can be used to calculate the reference output voltage of the shunt converter.

VI. EXPERIMENTAL SYSTEM CONFIGURATION

In order to demonstrate the effectiveness of using UPFC to realize total line loss minimization and load voltage regulation of the loop distribution systems, simultaneously, a simple laboratory model of the distribution system, consisting of two lines, is used. Fig. 6 shows the 6kVA, 200V laboratory model of the distribution system and the UPFC. The distribution system consists of two sets three phase lines, line 1 and 2. The load supplied by line 1 and line 2 is a pure resistance. R_1, L_1 and R_2, L_2 are the parameters of line 1 and line 2, respectively. The parameters of the whole system are listed in Table I. The line parameters shown in Table I are chosen in order to obtain large difference between the resistance to inductance ratio of line 1 and 2, which causes a large loop current to flow in the loop system. In the practical distribution systems, the resistance to inductance ratio of each line is slightly different.

The UPFC series and shunt converters have been built as a three phase PWM converter with IGBT SKM100GB124D as the power device. The DSP TMS320VC33 is selected as the controller for both converters. The shunt converter connected in parallel with the distribution line via a three phase transformer with turns ratio of 2:1 to regulated the dc-link voltage as V_{dc}=250 V and to regulate the load voltage to be equal in magnitude to the source voltage. The series converter consists of

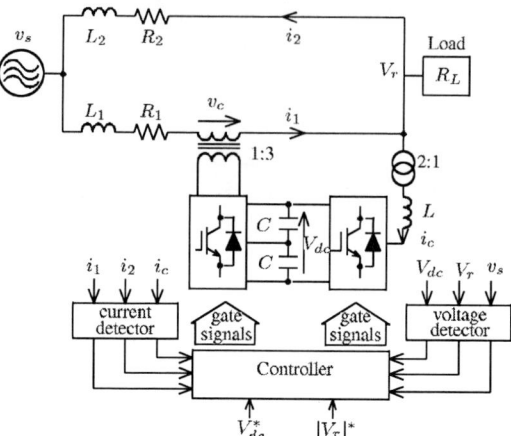

Fig. 6. Experimental system configuration

TABLE I
SYSTEM PARAMETERS(6 KVA BASE)

Source voltage v_s	200 V, 60 Hz
Load R_L	10 Ω (1.5 p.u.)
Input L of shunt conv.	3.0 mH (0.16 p.u.)
Line 1 L_1	6.0 mH (0.34 p.u.)
R_1	0.15 Ω (0.02 p.u.)
Line 2 L_2	3.0 mH (0.16 p.u.)
R_2	0.85 Ω (0.13 p.u.)
Capacitor C	3000 μF
DC link voltage V_{dc}	250 V
Transformation ratio	1:3 (system : series)
Transformation ratio	2:1 (system : shunt)
Switching time T_s	102 μs
Control parameters of shunt conv.	
AC voltage regulator k_p	0.30 A/V
k_I	0.03 A/Vs
DC voltage regulator k_p	0.30 A/V
k_I	0.30 mA/Vs

three single-phase H-bridge voltage source converters. The ac terminals of each H-bridge converter are connected in series to the distribution line through a single phase transformer with a turns ratio of 3:1 to minimize the total line loss of the loop system. The switching and sampling frequency for the series and shunt converters are 4.9 and 9.8 kHz, respectively.

VII. EXPERIMENTAL RESULT

A. Before Installing UPFC

Experimental measurements are carried out to the laboratory distribution system, shown in Fig. 6, that connected as loop distribution system before installation of the UPFC. Table II shows a comparison between the theoretical and the experimental values of each line current i_1, i_2, loop current i_{loop}, line load voltage v_r, UPFC shunt converter current i_c, UPFC series voltage v_c, each line loss P_{l1}, P_{l2}, total line loss P_l. The comparison shows that the theoretical and experimental results agree well with each other. Also, from 0 to $T1$, Fig. 7 and Fig. 8 show the experimental waveforms of line 1 current, line 2 current, loop current i_{loop}, UPFC shunt converter

TABLE II
LOOP SYSTEM BEFORE INSTALLATION OF UPFC

		Theoretical	Experimental
I_1	[A]	4.38	4.12
I_2		7.0	7.2
I_{loop}		5.42	5.43
V_r	[V]	192.2	190.5
I_c	[A]	-	0.2
V_c	[V]	-	-
P_{l1}	[W]	8.63	7.64
P_{l2}		124.95	132.19
P_l		133.58	139.83

TABLE III
EXPERIMENTAL RESULTS AFTER INSTALLATION OF UPFC SHUNT
AND SERIES CONVERTERS

		Shunt Conv. only	Series and Shunt converter	
			Line Inductance Compensation	Line Voltage Compensation
I_1	[A]	5.6	11.95	12.1
I_2		9.8	2.4	2.28
I_{loop}		7.28	0.0	0.0
V_r	[V]	199.8	200.1	200.2
I_c	[A]	18.1	13.1	13.8
V_c	[V]	0.0	27.44	27.06
P_{l1}	[W]	14.1	64.26	65.88
P_{l2}		244.9	14.69	13.26
P_l		259.0	78.95	79.14

current, actual and reference line to line load voltage V_r, load voltage v_r, voltage reference of the UPFC series converter v_c, and dc link voltage v_{dc} in the loop system before installing the UPFC. From 0 to T1, only the UPFC shunt converter is used to regulate the dc link voltage, and its current is very small that can be neglected.

B. After Installing UPFC shunt converter

In this period, The shunt converter is used to regulate the dc link voltage and to regulate the load voltage to be equal in magnitude to the source voltage by using reactive power control. Experimental measurements are carried out to the laboratory distribution system, shown in Fig. 6 after installation of the shunt converter only. Table III shows the experimental results of each line current i_1, i_2, loop current i_{loop}, line load voltage v_r, UPFC shunt converter current i_c, UPFC series voltage v_c, each line loss P_{l1}, P_{l2}, total line loss P_l. From $T1$ to $T2$, Fig. 7 and Fig. 8 show the experimental waveforms of the currents and voltages in the loop system with the effect of the UPFC shunt converter only. In this period, the shunt converter current is increased to compensate the load voltage. From these results shown from $T1$ to $T2$ it is cleared that, the UPFC shunt converter regulates the DC-link voltage to be 250 V and regulates the load voltage to be equal in magnitude to the source voltage, but the loop current increases to be 7.28 A which causes the total line loss to increase to be 259 W.

The UPFC shunt converter works here as a D-STATCOM to regulate the load voltage. Although the UPFC shunt converter regulates the load voltage, it causes the loop current to increase by 34.1% and the total line loss to increase by 85.2% .

C. After Installing UPFC shunt and series converters

In this period, the UPFC series converter is inserted, simultaneously with the shunt converter, to eliminate the loop current and hence to achieve total line loss in the loop system. The Line Inductance Compensation scheme and the Line Voltage Compensation scheme are used to control the UPFC series converter.

1) Controlling series converter with Line Inductance Compensation scheme: In the experimental system shown in Fig. 6, in order to apply the Line Inductance Compensation scheme to control the UPFC series converter, the

values of the line 1 current and L_c are needed to calculate the reference voltage of the UPFC series converter, $\dot{V}_c$, shown in (14). The value of the line 1 current can be detected using a current sensor. The value of L_c can be calculated from (12) as follows:

$$
\begin{aligned}
L_c &= \frac{R_1}{R_2}L_2 - L_1 \\
&= \frac{0.15}{0.85} \times 3.0 - 6.0 = -5.47\ [\text{mH}] \quad (19)
\end{aligned}
$$

From (13) it is cleared that the phase shift angle between the UPFC series converter voltage and its current is 90°. According to the value of the L_C, shown in (19), the UPFC series converter voltage lead the current of line 1 by 90°. In this case the UPFC series converter output power is theoretically a reactive power.

Fig. 7 shows the experimental waveforms of of the currents and voltages in the loop system before and after installation of the UPFC. From $T2$ to $T3$, the UPFC series converter is used to eliminate the loop current and achieve the line loss minimum condition by using the Line Inductance Compensation scheme, while the shunt converter is used to regulate the dc link voltage and the load voltage. In this period, the loop current is eliminated from the loop system and the total line loss is reduced to be 78.95 W by using the UPFC series converter, and the load voltage is still equal in magnitude to the source voltage by using the UPFC shunt converter. From the experimental results, shown from $T2$ to $T3$ in Fig. 7, it is cleared that, the series converter reduces the total line loss of the loop system by 43.54% from its original value before installation of the UPFC shunt converter, 139.83 W.

2) Controlling series converter with Line Voltage Compensation scheme: In the experimental system shown in Fig. 6, in order to apply the Line Voltage Compensation scheme to control the UPFC series converter, the values of the line 1 and line 2 currents are needed to calculate the reference voltage of the UPFC series converter, $\dot{V}_c$. The currents of line 1 and 2 can be detected using the current sensors. The reference voltage of the UPFC series

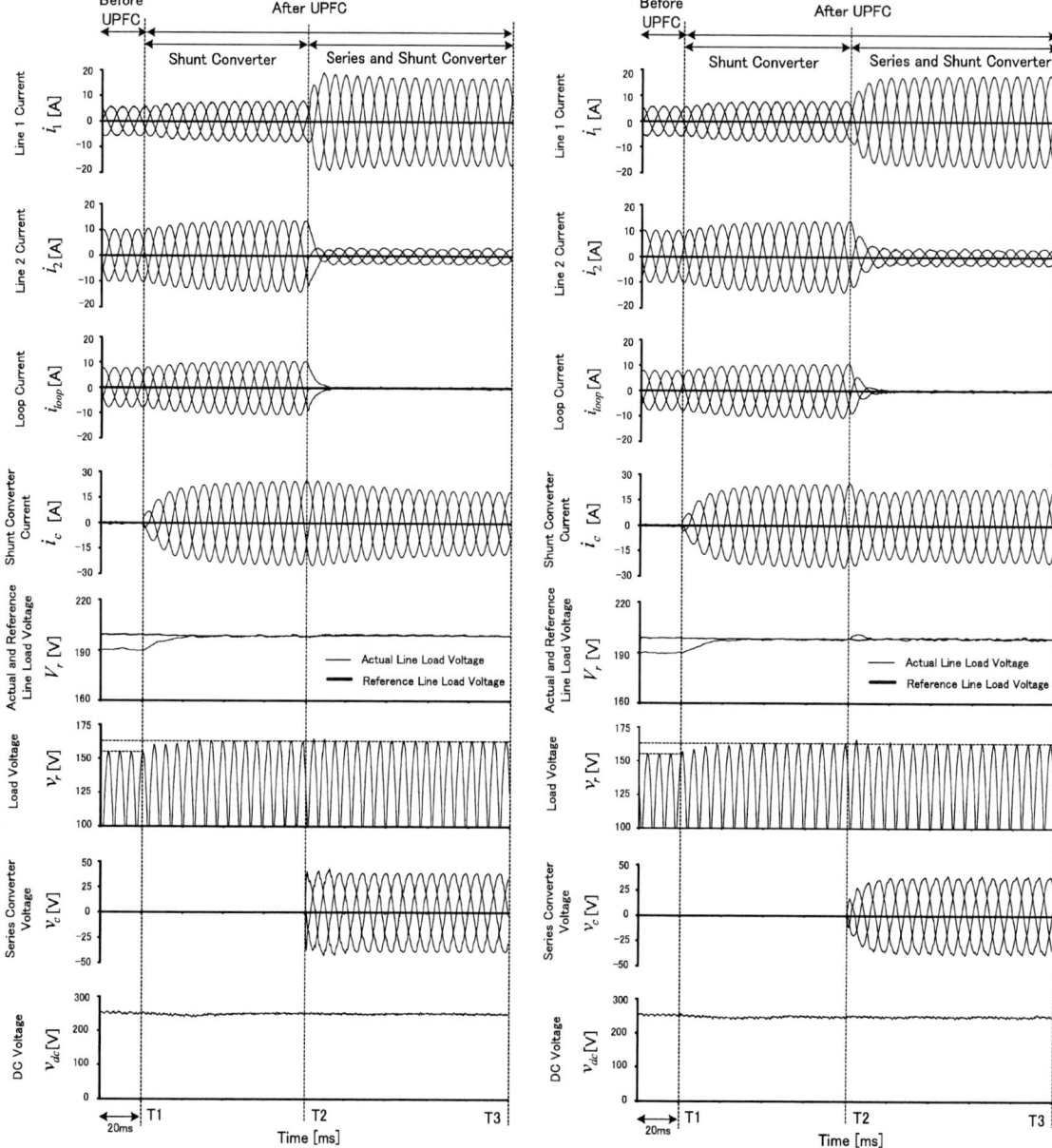

Fig. 7. Experimental waveforms of loop distribution system before and after installation of the UPFC controlled with Line Inductance Compensation scheme.

Fig. 8. Experimental waveforms of loop distribution system before and after installation of the UPFC controlled with Line Voltage Compensation scheme.

converter, $\dot{V}_e$, can be calculated as follows:

$$v_c = L_1 \frac{di_1}{dt} + L_2 \frac{di_2}{dt} \quad (20)$$

Fig. 8 shows the experimental waveforms of the currents and voltages in the loop system before and after installation of the UPFC. From $T2$ to $T3$, the UPFC series converter is used to eliminate the loop current and to achieve the line loss minimum condition by using the Line Voltage Compensation scheme, while the shunt converter

is used to regulate the dc link voltage and to regulate the load voltage to be equal in magnitude to the source voltage. In this period, the loop current is eliminated from the loop system and the total line loss is reduced to be 79.14 W by using the UPFC series converter, and the load voltage is still equal in magnitude to the source voltage by using the UPFC shunt converter. From the experimental results, shown from $T2$ to $T3$ in Fig. 8, it is cleared that, the series converter reduces the total line loss of the loop system by 43.4% from its original value before installation of the UPFC shunt converter, 139.83 W.

Table III shows a comparison between the experimental results that obtained from controlling the UPFC series converter by Line Inductance Compensation scheme and Line Voltage Compensation scheme. This comparison is in each line current i_1, i_2, loop current i_{loop}, line load voltage v_r, UPFC shunt converter current i_c, UPFC series voltage v_c, each line loss P_{l1}, P_{l2}, total line loss P_l. The comparison shows that the experimental results of both schemes agree well with each other. Also, the Line Inductance Compensation scheme and Line Voltage Compensation scheme have a great capability of controlling the UPFC series converter to achieve total line loss minimization in the loop distribution system.

From the experimental results it is cleared that, the UPFC shunt converter has a great capability to regulate the load voltage and the UPFC series converter has a great capability to minimize the total line loss in the loop distribution system, simultaneously.

VIII. CONCLUSION

This paper has presented the line loss minimum conditions and the power flow control scheme of the UPFC to realize load voltage regulation and total line loss minimization in the loop distribution systems simultaneously, along with a comprehensive analysis. The load voltage has been controlled to be equal in magnitude to the source voltage by using UPFC shunt converter, which works as D-STATCOM. Althought it regulates the load voltage, it causes the loop current and the total line loss to increase. The UPFC series converter has been inserted in the loop distribution system to eliminate the loop current and hence to achieve total line loss minimum condition in the loop distribution system. Two control schemes have been proposed to control the UPFC series converter, Line Inductance Compensation scheme and Line Voltage Compensation scheme. The effectiveness of the proposed control schemes of UPFC series and shunt converter have been verified by experiments using laboratory prototype rated 6 kVA, 200V. Experimental results prove that, the UPFC has a great capability to regulate the load voltage and minimize the total line loss of the loop distribution systems, simultaneously.

REFERENCES

[1] K. Tsukahara, C. Ishibashi, A. Kontani, Y. Yamada, "Present Situation of Distribution System in Japan and a Proposal for a New Method of Distribution System Configuration", IERE 2006 North America Workshop, July 2006.

[2] Naotaka Okada, Hiromu Kobayashi, Kiyoshi Takigawa, Masahide Ichikawa, Kosuke Kurokawa, "Loop Power Flow Control and Voltage Characteristics of Distribution System for Distributed Generation Including PV System", 3rd World Conference on Photovoltaic Energy Conversion, Osako-Japan, pp. 2284-2287, May 2003.

[3] Mahmoud A. Sayed, Nobuyuki Inayoshi, Takaharu Takeshita, Fukashi Ueda, "Line Loss Minimum Control of Loop Distribution Systems Using UPFC", Fourth Power Conversion Conference(PCC), Nagoya-Japan, pp. 2284-2287, April 2007.

[4] Mahmoud A. Sayed, Nobuyuki Inayoshi, Takaharu Takeshita, "Series Compensation for Voltage Regulation and Line Loss Minimization of Loop Distribution Systems Using UPFC", Technical Meeting on Semiconductor Power Converter and Industry Electric and Electronic Application, IEE Japan ,SPC-07-112/IEA-07-35 (2007).

[5] Mahmoud A. Sayed, Nobuyuki Inayoshi, Takaharu Takeshita, Fukashi Ueda, "Line Loss Minimization of Loop Distribution Systems Using UPFC", Trans. IEE Japan, Vol. 128-D, No. 4,2008.

[6] Ljubomir A. Kojovic, "Coordination of Distributed Generation and Step Voltage Regulator Operations for Improved Distribution System Voltage Regulation", IEEE Power Engineering Society General Meeting, June 2006.

[7] S. A. Miske, "Considerations for the Application of Series Capacitors to Radial power Distribution Circuits", IEEE Trans. Power Delivery, Vol. 16, No. 2, pp. 306-318, April 2001.

[8] Joon-Ho Cho, Jae-Chul Kim, "Network Reconfiguration at the Power Distribution System With Dispersed Generations for Loss Reduction", IEEE Power Engineering Society Winter Meeting, Vol. 4, pp. 2363-2367, Jan. 2000.

[9] N.G.Hingorani, "High Power Electronics and Flexible AC Transmission System", IEEE power Engineering Review, Vol. 8, No.7, pp. 3-4, July 1988.

[10] Takaharu Takeshita, Yuji Hayashi, Nobuyuki Matsui, "Load Voltage Compensation Using Series-Shunt Power Converter", Trans. IEE Japan, Vol. 120-B, No. 5, pp. 725-732, May 2000.

[11] Seyed Hosseini, Mohammed Reza Banaei, "Performance of Active Power Line Conditioner for Loss Reduction in the Power Distribution System", TENCON 2004, Vol. 4, pp. 97-100, November 2004.

[12] Amit K. Jain, Aman Behal, Ximing T. Zhang, Darren M. Dawson, and Ned Mohan, "Nonlinear Controllers for Fast Voltage Regulation Using STATCOMs", IEEE Trans. Control System Technology, Vol. 12, No. 6, pp. 827-842, November 2004.

[13] E. Twining, and D.G. Holmes, "Voltage Compensation in Weak Distribution Networks Using Multiple Shunt Connected Voltage Source Inverters", IEEE Bologna Power Tech Conference, Vol. 4, June 2003.

[14] Hideaki Fujita, Hirofumi Akagi, "Voltage-Regulation Performance of a Shunt Active Filter Intended for Installation on a Power Distribution System", IEEE Trans. Power Electronics, Vol. 22, No. 3, pp. 10461053-842, May 2007.

[15] Kannan Karthik, J.E. Quaicoe, "Voltage Compensation and harmonic Suppression Using Series Active and Shunt Passive Filters", Canadian Conference on Electrical and Computer Engineering, Vol. 1, pp. 582-586 , 2000.

[16] Seok-Woo Han, Seung-Yo Lee, Gyu-Ha Choe, "A 3-Phase Series Active Power Filter with Compensate Voltage Drop and Voltage Unbalance", IEEE International Symposium on Industrial Electronics (ISIE 2001), Vol. 2, pp. 1032-1037, 2001.

[17] L.Gyugyi, "Unified Power Flow control concept for Flexible AC Transmission Systems", IEE PROCEEDINGS-C, Vol. 139, No. 4, July 1992.

[18] T.J.E.Miller, "Reactive Power Control in Electric Systems", Wiley, New York, 1982.

Control of Traction Single-Phase Current-Source Active Rectifier under Distorted Power Supply Voltage

Jan Michalík, Jan Molnár and Zdeněk Peroutka

University of West Bohemia in Pilsen / Dept. of Electromechanics and Power electronics, Plzeň, Czech Republic

e-mail: *jmichali@kev.zcu.cz, jmolnar@kev.zcu.cz, peroutka@ieee.org*

Abstract— This research has been motivated by industrial demand for single-phase current-source active rectifier (CSAR) dedicated for reconstruction of older types of dc machine locomotives. This contribution presents new control strategy of CSAR using controller of phase shift angle (φ) between trolley wire voltage and current supplemented by compensation of trolley wire current waveform under supply voltage distortion. This control is able to keep the trolley wire current in demanded phase shift against supply voltage and preserve sinusoidal waveform of consumed current even under distorted trolley wire voltage. Simulation results and theoretical conclusions are verified on designed laboratory prototype of 7kVA.

Keywords— Active filter, Converter control, DSP, Harmonics, Locomotive, Modulation strategy, Pulse Width Modulation (PWM), Traction application

I. INTRODUCTION

This research has been motivated by industrial demand for design of single phase current-source active rectifier (CSAR) dedicated for reconstruction of older types of dc machine locomotives. Converter must be able to operate on both trolley wire voltages: 25kV/50 Hz and 15kV/16⅔ Hz. The goal of this contribution is the improvement of CSAR control structure in order to change and keep the phase shift (φ) between trolley wire current and voltage on demand (due to ability of cos φ correction) and also to compensate the shape of trolley wire current waveform under distorted trolley-wire voltage. The control strategy with phase shift controller proposed in [1] is shown in Fig.2. This contribution is a follow-up to this research and introduces the improvement of this control in order to achieve better converter behavior under distorted power supply voltage. Proper function of presented control was verified on the laboratory prototype of this converter.

Development was divided into three stages: control structure design, simulation, implementation and its practical function verification. The goal of this paper is also to show and describe practical problems and experiences appeared during control structure actuating.

Topic of CSAR is not very common and the published papers mostly deal with three phase version. Publications are rather theoretical, researches go into the problems connected with control using hysteresis control or PWM modulation in particular (e.g. [3]-[5]). Unfortunately, there are only a few papers dealing with single-phase CSAR for traction applications. We did not find any implementation of CSAR in passenger operating traction vehicle.

II. PROPOSED CONTROL STRATEGY OF TRACTION CURRENT-SOURCE ACTIVE RECTIFIER

The power circuit of investigated traction converter is shown in the Fig.1. The major objective of designed control was to keep the control independent on circuit parameters and measured additional values and also keep trolley wire current under distorted voltage as near to harmonic shape as possible in order to reach the improved Power Factor. Therefore, control based on phase shift (φ) controller able to work even under distorted power grid voltage was developed and is shown on Fig.2. Its principle is to keep the ϑ angle value directly depending on real phase shift between trolley wire current (I) and trolley wire voltage (U). Significant improvement of this control consists particularly of the possibility of setting phase shift (φ) within the required range (e.g. ± 50° → comparing to the conventional control making possible only zero phase shift φ=0°) and also compensation of inharmonic shape of trolley wire current under distorted supply voltage. Due to this control feature the rectifier can also be operated as a reactive power compensator working with capacitive (see Fig.3) or inductive (see Fig.5) cos φ.

Compensative function of this control strategy is explained in the following part. Conventional PWM is based on comparison of two waveforms: carrier and usually sinusoidal modulating waveform. This principal (similar to analog) supposes the same shape of supply voltage (sinusoidal) as the shape of the modulating curve. In other words, sinusoidal supply voltage causes sinusoidal trolley wire current. If the voltage is distorted, the shape of current will be distorted as well depending on this voltage distortion. Therefore, demand on Power Factor λ → 1 cannot be kept.

The goal of proposed control strategy has been to keep trolley wire current under distorted voltage as near to harmonic shape as possible. We have considered several possibilities of keeping harmonic shape of I current. The presented control was the best choice in terms of robustness, stability and waveform shape. It is based on common PWM modulator supplemented by φ angle controller and I current correction. The modulating waveform correction is based on so-called "differential waveform" (see Fig.6). This waveform represents a difference between the I current waveform and demanded (sinusoidal) waveform outgoing from modulator.

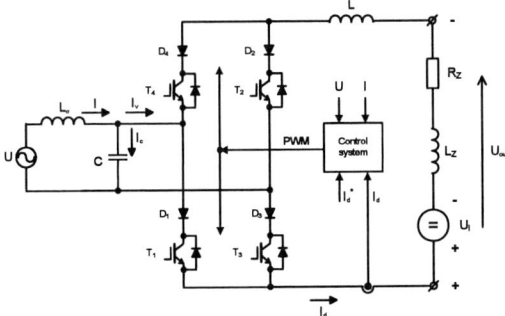

Fig.1 Power circuit configuration of single-phase current-source active rectifier

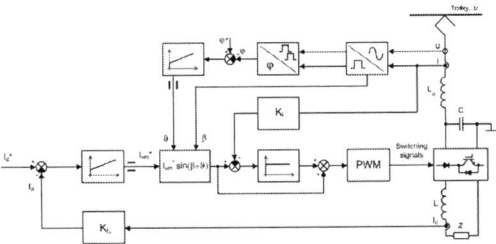

Fig.2 CSAR control with phase shift controller supplemented by trolley wire current shape correction

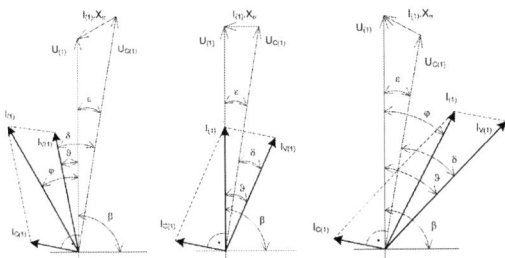

Fig.3 Vector diagram for phase shift φ in capacitive cos φ	Fig.4 Vector diagram for phase shift φ=0°	Fig.5 Vector diagram for phase shift φ in inductive cos φ

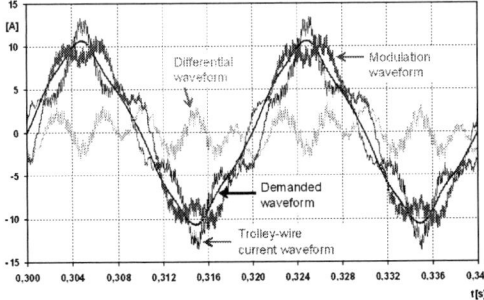

Fig.6 Principle of differential and modulation waveforms acquisition

As it can be seen, the demanded waveform is not purely sinusoidal (Fig.6). This behavior is caused by the I_d current controller which was in this particular case set to a fast response because the control of the load current has higher priority then keeping the trolley-wire current I in

purely sinusoidal shape. Due to this faster response the controller tries to follow the shape of the load current in order to decrease its ripple. Lower gain of this controller would cause more sinusoidal demanded current waveform but at the expense of load current controller response to step changes (which has higher priority).

The differential waveform obtained as a difference between demanded grid current and the measured current is leaded into proportional controller in order to adapt its amplitude according to expected distortion. This technique has been employed due to the relationship between distortion and gain. While having smaller distortion, greater gain is preferable due to the more effective compensation. When the distortion is bigger, quality compensation could easily caused limitation of modulating curve. Therefore, the gain needs to be adapted (decreased). In fact, we are usually able to find constant gain value of this proportional controller, in other words – to employ non-adaptive controller. After this adaptation the differential waveform is added to the demanded modulating waveform (sinusoidal). Final modulating waveform is distorted inversely to trolley wire current I and symmetrically according to sinusoidal waveform. This modulating waveform is leaded to PWM modulator and is called "corrected modulating waveform".

As above mentioned, vector diagram on Fig.3 shows rectifiers behavior in capacitive and Fig.5 in inductive cos φ, where behavior of current vectors $I_{V(1)}$ and $I_{(1)}$ under the change of the phase shift (φ) can be seen. Despite constant value of demanded load current I_d, during phase shift (φ) increase also the grid current $I_{(1)}$ grows due to I_d controller, which wants to keep constant value of this current by means of $I_{V(1)}$ current increase (amplitude of $I_{C(1)}$ does not change and its position depends only on $X_\sigma.I_{(1)}$). Consequently, ϑ angle reaches the area (particularly in inductive area), where average load voltage U_{out} decreases significantly which causes load current I_d ripple increase (depending also on load inductance L value). These suppositions were also verified by experimental evidence.

III. SIMULATION RESULTS

Simulation results shown in Fig.7 – Fig.14 confirm proper function of phase shift controller based control strategy under both steady-state and transient conditions. Its main advantages consist in a possibility of changing phase shift (φ) depending on user or master controller demand. Another positive feature is phase shift (φ) independent on trolley-wire voltage (U) distortion – controller is always able to keep the real phase shift φ in phase with demanded regardless of the trolley-wire voltage distortion. The behavior under distorted trolley-wire voltage can be observed in Fig.10 and can be easily compared with behavior of the control without presented correction shown in Fig.9. It is apparent that the waveform of the uncorrected trolley-wire current (I) is much more distorted. This distortion also corresponds with amount of harmonics and in case of Fig.9 is particularly significant. Comparison of both figures confirms the significant improvement of the control behavior under distorted trolley-wire voltage conditions.

The simulations were realized in programming language Pascal, graphical charts in MS Excel.

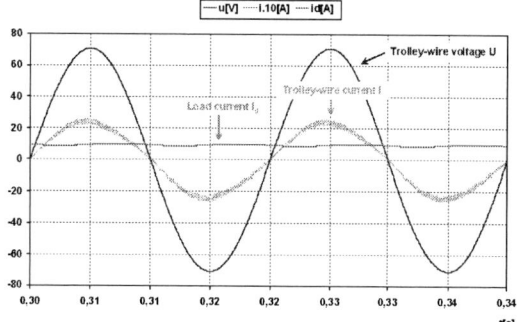

Fig.7 Steady state, load current I_d=9A, φ=0°, U=50V$_{rms}$/50Hz

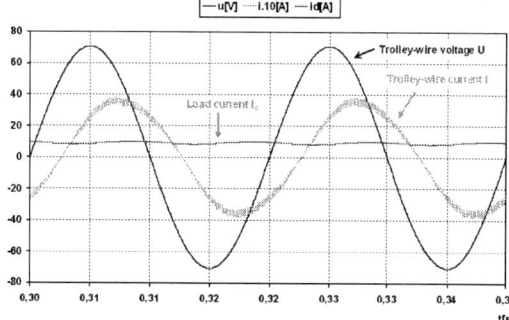

Fig.8 Steady state, load current I_d=9A, phase shift φ=-50°, U=50V$_{rms}$/50Hz

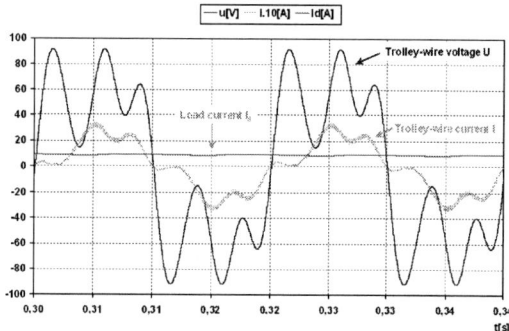

Fig.9 Steady state, I_d=9A, φ=0°, U=50V$_{rms}$/50Hz, distorted trolley wire voltage, without trolley-wire current waveform correction

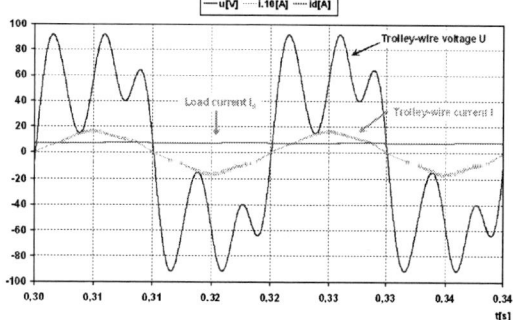

Fig.10 Steady state, I_d=9A, φ=0°, U=50V$_{rms}$/50Hz, distorted trolley wire voltage, with trolley-wire current waveform correction

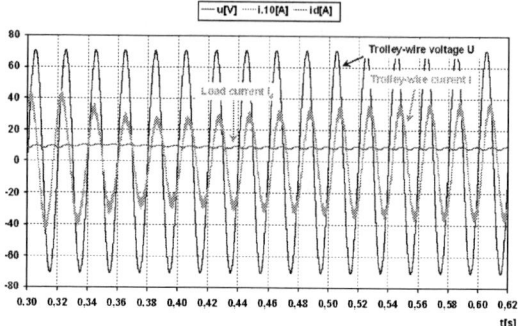

Fig.11 Transient, φ step change 50 → -50°, I_d=9A, U=50V$_{rms}$/50Hz

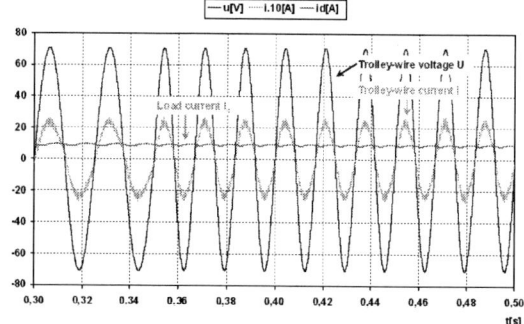

Fig.12 Transient, frequency step change 40 → 60Hz, φ=0°, load current I_d=9A, U=50V$_{rms}$/50Hz

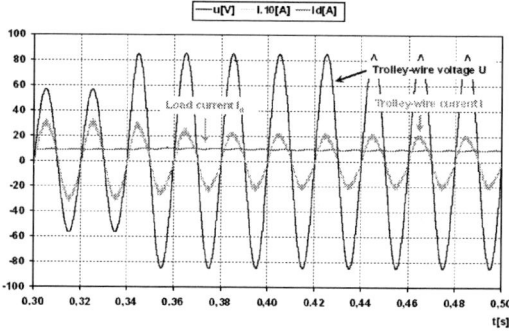

Fig.13 Transient, trolley-wire voltage step change 40 → 60V, φ=0°, load current I_d=9A

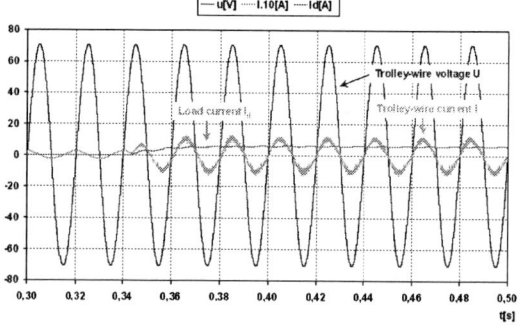

Fig.14 Transient, load current I_d step change 0 → 6A, φ=0°, U=50V$_{rms}$/50Hz

IV. EXPERIMENTAL EVIDENCE: LOW POWER LABORATORY PROTOTYPE OF TRACTION CSAR OF RATED POWER OF 7kVA

Fig.15 - Fig.26 show experimental results of properly working proposed control strategy based on phase shift controller with improved immunity against distorted trolley-wire voltage. There are analyzed both transient and steady states under harmonic and distorted trolley-wire voltage conditions. The comparison behavior between control with and without grid current correction by mean of waveforms and harmonic analysis is also provided in order to verify proper function. Every transient state presents (except for trolley wire voltage) reactions of both trolley wire current I and load current I_d in particular.

Fig.15 presents load current step change. During this transient both current and phase shift controller (which moves the phase shift to a zero value in this particular case) try to reach their demanded values. Changing phase shift also changes the trolley-wire current value. Therefore, the load current controller must also respond to the phase shift controller behavior and keep the demanded load current value even under different phase shifts trolley-wire current values.

Fig.16 - Fig.18 present transients under the distorted trolley-wire voltage. Fig.16 introduces trolley-wire frequency step change. Response of load current in this case is very good and is almost not affected by this step change as well as trolley wire current waveform (with respect to a trolley-wire voltage) and phase shift (demanded φ=0°). Step change of phase shift φ between 50° and -50° is documented in Fig.17. Slower phase shift controller response was set on purpose due to stability. Faster response caused oscillations and instability of phase shift around its demanded value. Very influential in this case is the synchronization principle. The more accurate it is, the more stable the phase shift value control is. Fig.18 shows trolley-wire voltage step change between 40 - 60V. As it can be seen, despite the significant voltage step change load current becomes stabilized quickly after approximately four periods. Important result is the trolley wire current staying in phase with trolley wire voltage during transition. Another important behavior is shown in Fig.19 where step change of trolley-wire voltage waveform is presented, in this case from sinusoidal to distorted shape. The distortion may be various but significant are value of the load current, shape of the trolley-wire current and also the phase shift. It is apparent that behavior of all these values is stable. Trolley-wire current waveform responds immediately to this change and demanded (zero) phase shift is also kept.

Figures Fig.20 - Fig.24 presents steady states of designed control strategy for both sinusoidal and distorted trolley-wire conditions. We would like to point out shape of trolley wire current (i) and load current I_d in particular. As it can be seen, with increased phase shift (in both positive and negative polarity) also amplitude of trolley wire current grows. As the phase shift increases, mean value of load voltage u_{out} drops and energy accumulated in load inductance L is not sufficient enough to keep invariable load current I_d ripple and its mean value. Hence, the load current ripple grows and the trolley wire current amplitude as well (significantly) due to effort of current controller to keep demanded load current value. In spite of the phase shift the shape of the trolley-wire

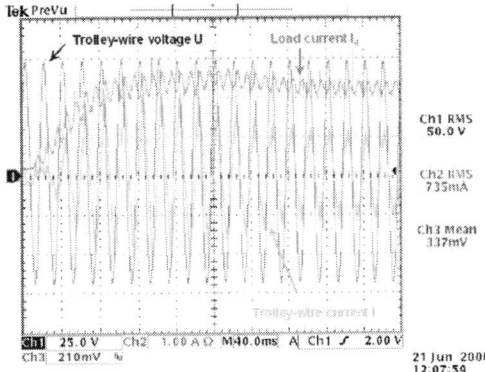

Fig.15 Load current step change, 0 → 6A, U=50V$_{rms}$/50Hz
(Ch1-u, Ch2-I, Ch3-I$_d$: 14,3A/V)

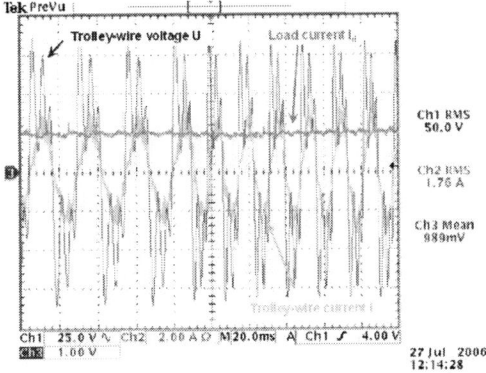

Fig.16 Trolley-wire frequency step change: 40 → 60 Hz, I$_d$=9A,
U=50V$_{rms}$/50Hz (Ch1-u, Ch2-I, Ch3-I$_d$: 9A/V)

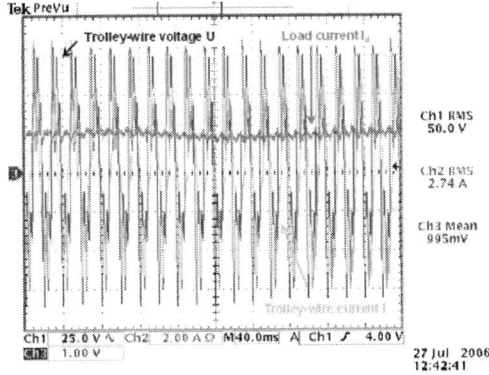

Fig.17 Phase shift (φ) step change: 50 → -50°, I$_d$=9A,
U=50V$_{rms}$/50Hz (Ch1-u, Ch2-I, Ch3-I$_d$: 9A/V)

current is very close to sinusoidal. Nevertheless, the range of the phase shift between trolley-wire voltage and current is limited due to the phase shift between PWM modulation waveform and trolley wire voltage, because the average value of load voltage decreases with growing phase shift. Considering phase shift value over ±50°, distortion of trolley wire current become significant

553

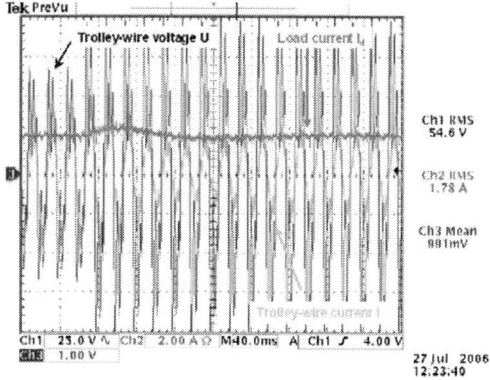

Fig.18 Trolley-wire voltage step change: 40 → 60 V, I$_d$=9A
(Ch1-u, Ch2-I, Ch3-I$_d$: 9A/V)

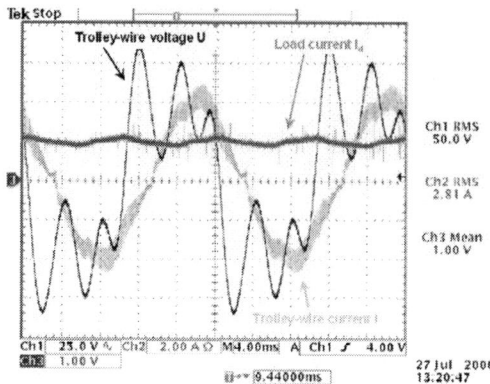

Fig.21 Steady state, φ=-50°, I$_d$=9A, U=50V$_{rms}$/50Hz , distorted
trolley-wire voltage (Ch1-u, Ch2-i, Ch3-I$_d$: 9A/V)

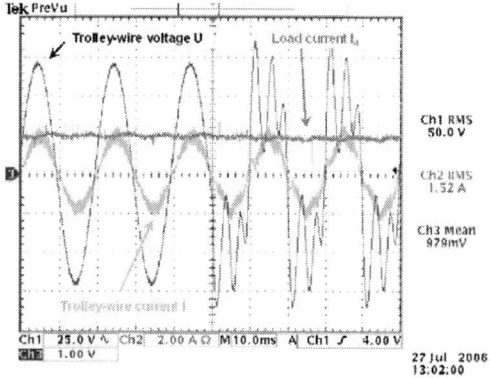

Fig.19 Trolley-wire voltage waveform step change, I$_d$=9A,
U=50V$_{rms}$/50Hz (Ch1-u, Ch2-I, Ch3-I$_d$: 9A/V)

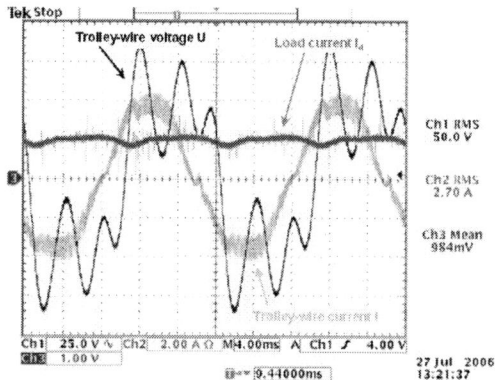

Fig.22 Steady state, φ=50°, I$_d$=9A, U=50V$_{rms}$/50Hz , distorted
trolley-wire voltage (Ch1-u, Ch2-i, Ch3-I$_d$: 9A/V)

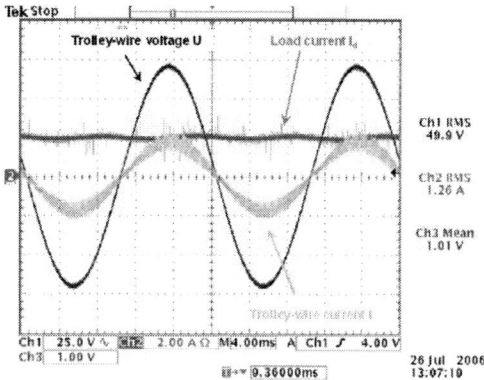

Fig.20 Steady state, phase shift φ=0°, I$_d$=9A, U=50V$_{rms}$/50Hz
(Ch1 u, Ch2 i, Ch3 I$_d$: 9A/V)

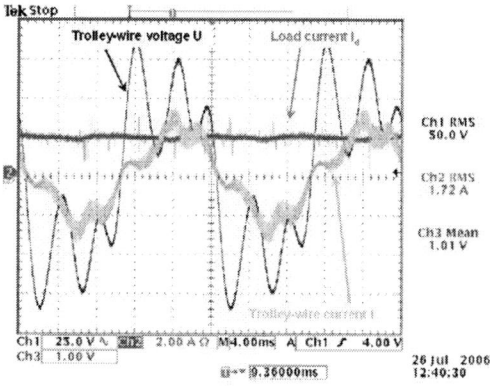

Fig.23 Steady state, φ=0°, I$_d$=9A, U=50V$_{rms}$/50Hz , distorted
trolley wire voltage, without correction
(Ch1-u, Ch2-i, Ch3-I$_d$: 9A/V)

and leads up to system breakdown. This behavior becomes significant close to phase shift of ±90°. Fig.23 - Fig.26 present the comparison between the control strategy with and without correction of trolley-wire current waveform. Comparing the trolley-wire current shape in Fig.23 and Fig.24, it is apparent that even under significant voltage distortion the control with correction behaves significantly

better and the waveform approximates sinusoidal curve with higher fidelity. This assumption is also supported by Fig.25 and Fig.26 where harmonic analysis of both considered trolley-wire current waveforms is provided. As it can be seen, the amplitude of harmonics in case of corrected waveform is half then in case of uncorrected

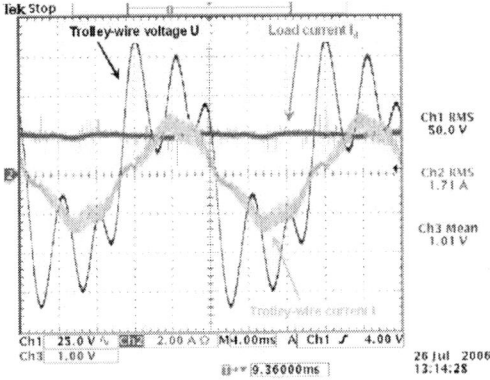

Fig.24 Steady state, φ=0°, I$_d$=9A, U=50V$_{rms}$/50Hz , distorted trolley-wire voltage, with correction (Ch1-u, Ch2-i, Ch3-I$_d$: 9A/V)

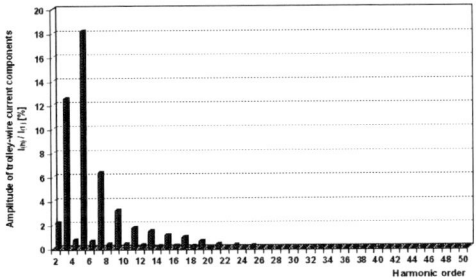

Fig.25 Harmonic analysis of trolley-wire current shown in Fig.23, control structure without correction

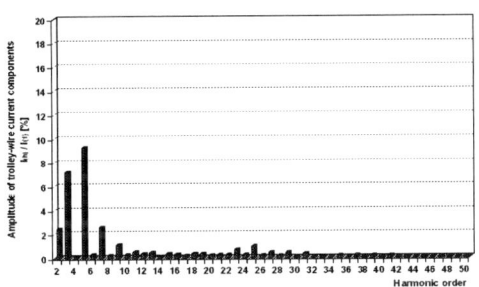

Fig.26 Harmonic analysis of trolley-wire current shown in Fig.24, control structure with correction

waveform. This analysis verifies the significant improvement of presented control behavior in case of inharmonic and distorted supply voltage. A photo of designed laboratory prototype of traction current-source active rectifier is shown in the Fig.27.

V. CONCLUSIONS / ORIGINAL RESULTS

The designed 7kVA laboratory prototype of traction CSAR employs phase shift based control strategy with synchronous PWM supplemented with trolley wire current shape correction. The synchronous PWM has been selected regarding possible interaction with railway control signaling, because disturbance produced by constant switching frequency (especially in case of

Fig.27 Designed laboratory prototype of traction current-source active rectifier

synchronous modulation) is much easier to eliminate than in case of hysteresis control. Moreover, the PWM control makes possible to employ shifted carriers, which is eligible for high power systems with low switching frequency.

The main goal of this research was to design the control strategy able to control phase shift (φ) between trolley-wire voltage and current. The presented control was designed due to industrial demand on control strategy able to work independently on circuit parameters, without need of measuring RMS values and able to work even under the distorted trolley-wire conditions. Another task was the possibility of phase shift (φ) change due to cos φ control which allows the rectifier to work as reactive power compensator. Due to the trolley wire current shape correction the controller is able to improve significantly the current shape under distorted trolley-wire voltage conditions. This solution makes possible to avoid general negative converter behavior (in terms of taken current) controlled by PWM when the supply voltage is not sinusoidal.

The presented simulation and experimental results confirm proper function of designed converter control with zero phase shift (φ=0°) as well as with nonzero phase shift. A big advantage is an ability of keeping accurate value of demanded phase shift φ even under the distorted grid conditions and independently on all others quantities. Both simulation and experimental results also confirmed that the proposed control strategy is able to improve significantly the shape of trolley wire current when the supply voltage is not sinusoidal.

VI. REFERENCES

[1] Michalík, J., Molnár, J., Peroutka, Z.: Single-Phase Current-Source Active Rectifier for Traction Applications: New Control Strategy based on Phase Shift Controller. In EPE 2007. Aalborg, Denmark. 2007.

[2] Michalík, J., Molnár, J., Peroutka, Z.: Single Phase Current-Source Active Rectifier: design, simulation and practical problems. In ISIE 2007, Vigo, Spain, 2008.

[3] Damec, V., Chlebiš, P.: New concept of single-phase traction supply converters, EDPE 2003, TU Košice, Slovakia

[4] Pou, J., Wu, B.: High Power Converters: Topologies, Controls and Applications, Tutorial, IEEE IECON 2006, Paris, France.

[5] Wang, X.; Ooi, B.: Unity PF current-source rectifier based on dynamic trilogic PWM, IEEE Transactions on Power Electronics, Volume: 8 , Issue: 3 , 1993, pages 288–294

SIMULATION MODEL OF NEURAL NETWORK BASED SYNCHRONOUS GENERATOR EXCITATION CONTROL

Damir Sumina[1], Neven Bulic[2], Gorislav Erceg[3]
Faculty of Electrical Engineering and Computing
Unska 3, Zagreb, Croatia
Tel. / Fax: + (385 1) 6129-999/ 6129-705
[1]damir.sumina@fer.hr, [2]neven.bulic@fer.hr, [3]gorislav.erceg@fer.hr

Abstract. Usage of Neural Network (NN) based excitation control on single machine infinite bus and its simulation studies are reported in this paper. The proposed feed forward neural network integrates a voltage regulator and a power system stabilizer. It is trained on-line from input and output signals of a synchronous generator. A modified error function used for training the neural network by the back propagation algorithm uses the reference and terminal voltage as controlling voltage and active power deviation to provide stabilization. The complete algorithm is simulated in Matlab Simulink. Synchronous generator (83 kVA, 50 Hz, 400V) is connected over transmission lines to AC power system. The proposed algorithm shows advantages of this method and satisfactory results.

Keywords: excitation system, synchronous generator, neural network

1. INTRODUCTION

Different types of Neural Networks (NN) have been tested for controlling synchronous generator till now. A simple structure with only one neuron for voltage control is studied in [1] and [1]. Different types of NN for improving stability with more then one neuron are studied in [3].

In this case the NN based excitation controller is placed on the position of the classical PI voltage controller, whereas the excitation current controller is kept. Some modifications are made in the modified error function to scale every signal. Therefore its influence on changing weights can be modified. The back propagation algorithm (BP) is used to update weights online.

The complete control algorithm with synchronous generator connected via transmissions lines to a power system is simulated in Matlab Simulink.

2. PROPOSED NEURAL NETWORK

The NN used in this paper is shown in Fig. 1. The NN has three inputs, six neurons in hidden layer and one neuron in output layer. The inputs of NN are reference voltage U_{ref}, terminal voltage U_g and previous output $y_{(t-1)}$. By applying $y_{(t-1)}$ at the input dynamical neural network is created.

The transig function is used as activation function for neurons in a hidden layer and for a neuron in an output layer.

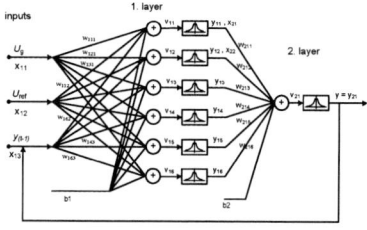

Fig. 1. Proposed neural network

The numerical representation is described in (1) and its derivation in (2).

$$\psi(v) = \frac{1}{1+e^{-g_a v}} - 1 \qquad (1)$$

$$\psi'(v) = g_a \frac{4e^{-2g_a v}}{(1+e^{-2g_a v})^2} = g_a \cdot (1 - \psi^2) \qquad (2)$$

In (1) and (2) $\psi(v)$ is the nonlinear activation function and g_a has constant value.

The NN uses a procedure to update weights on-line and there is no need for any off-line training. Furthermore, there is no need for an identifier and neither for reference model. In an online mode from the inputs and outputs of the generator the NN is trained directly and there is no need to determine the states of the system. Sampled values of the machine parameters are used by NN to calculate the error with a modified error function. This error is back propagated through the NN to update its weights after adjusting weights, the output of the NN is calculated.

3. MODIFIED ERROR FUNCTION AND BP ALGORITHM

Inputs and outputs of one neuron can be determined by:

$$y_{ki} = \psi\left(\sum_k w_{kij} \cdot x_{kj} + b_1\right), \qquad (3)$$

where y_{ki} is the output of the neuron i in layer k, ψ is the nonlinear activation function of neuron i, w is the neuron weight, x is input signal and b is bias.

The BP algorithm is an iterative gradient algorithm designed to minimize the mean square error between the actual output and NN desired output. This is a recursive algorithm starting at the output neuron and working back to the hidden layer by adjusting the weights according to:

$$w_{kij}(t+1) = w_{kij}(t) + \Delta w_{kij}(t), \qquad (4)$$

$$\Delta w_{ji}(n) = \eta \cdot \delta_j(n) \cdot y_i(n), \qquad (5)$$

$$\delta_j(n) = e_j(n) \cdot \varphi'_j(v_j(n)), \tag{6}$$

where w_{kij} is weight of neuron i in layer k of previous neuron j, η is learning rate, δ is local gradient, e is output from error function, φ'_j is derivation of activation function.

As error function for the BP algorithm commonly is used:

$$\Im = \frac{1}{2}(t_{ki} - y_{ki})^2. \tag{7}$$

The desired output of the neuron i in layer k is t_{ki}. If the neuron is in the output layer, the error function is:

$$\frac{\partial \Im}{\partial y_{ki}} = t_{ki} - y_{ki}. \tag{8}$$

If the neuron is in the hidden layer, the error function is calculated recursively:

$$\frac{\partial \Im}{\partial y_{ki}} = \sum_{p=1}^{n(k+1)} \frac{\partial \Im}{\partial y_{k+1,p}} \cdot \psi'_{k+1,p} \cdot w_{k+1,1,i}. \tag{9}$$

In the NN controlling the synchronous generator changing weights is not just based on the error between output and desired output, but also on the derivative of the error dy_{ki}/dt:

$$\frac{\partial \Im}{\partial y_{ki}} = (t_{ki} - y_{ki}) - \frac{dy_{ki}}{dt}. \tag{10}$$

Such a modified error function speeds up the BP algorithm and gives it a faster convergence. That way the NN gets appropriate for an on-line learning implementation. The error function is:

$$\frac{\partial \Im}{\partial y_{ki}} = K(U_{ref} - U_g) - k_u \frac{dU_g}{dt}, \tag{11}$$

where K and k_u are scaling factors for adjusting the influence of changing weights.

In order to perform power system stabilization, the active power deviation ΔP_{el} and the derivation of the active power in the modified error function P_{el} were considered. The complete modified error function is:

$$\frac{\partial \Im}{\partial y_{ki}} = \left[K(U_{ref} - U_g) - k_1 \frac{dU_g}{dt} \right] - \left[k_3(\Delta P_{el}) + k_2 \frac{dP_{el}}{dt} \right] \tag{12}$$

where the coefficients k_2 and k_3 are scaling factors.

The modified error function (12) is divided in two parts: the voltage control and the power system stabilizer.

4. CONTROL STRUCTURE

Classic excitation control structure include two control loops: the excitation current loop with P controller and the terminal voltage loop with PI controller. Usually, a power system stabilizer with its output is added to the summation point in front of the voltage controller. In this structure two phase currents and two line voltages are measured. The output signal of the excitation current controller is a duty cycle for the PWM input of an AC/DC IGBT converter.

The neural network based control structure is shown in Fig.2. The NN replaces the voltage controller in front of the excitation current controller.

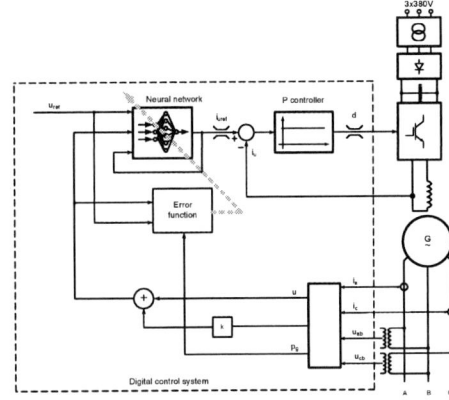

Fig. 2.   NN control structure for synchronous generator

5. DIFERENTIAL EQUATION BASED SIMULATION MODEL

Simulation model is made in Matlab Simulink. The synchronous generator parameters are shown in TABLE I and TABLE II. The generator is connected over transmission lines to the power system (Fig. 3).

Mathematical model is represented in d-q coordinate system. Based on that it is necessary to perform transformation from abc coordinate system to dq coordinate system. Assumption is that voltages are symmetrical in all phases and there is only one harmonic of magnetic flux in air gap. Equations in this model are differential in dq coordinate system represented in per unit system. Time is absolute.

$$-u_d = r \cdot i_d + \frac{1}{\omega_s} \cdot \frac{d\psi_d}{dt} + \omega \cdot \psi_q \tag{13}$$

$$-u_q = r \cdot i_q + \frac{1}{\omega_s} \cdot \frac{d\psi_q}{dt} - \omega \cdot \psi_d \tag{14}$$

$$u_u = r_u \cdot i_u + \frac{1}{\omega_s} \cdot \frac{d\psi_u}{dt} \tag{15}$$

$$0 = r_D \cdot i_D + \frac{1}{\omega_s} \cdot \frac{d\psi_D}{dt} \tag{16}$$

$$0 = r_Q \cdot i_Q + \frac{1}{\omega_s} \cdot \frac{d\psi_Q}{dt} \tag{17}$$

Equations for flux linkage and currents:

$$\psi_d = x_d \cdot i_d + x_{ud} \cdot i_u + x_{dD} \cdot i_D \tag{18}$$

$$\psi_q = x_q \cdot i_q + x_{qQ} \cdot i_Q \tag{19}$$

$$\psi_u = x_{ud} \cdot i_d + x_u \cdot i_u + x_{uD} \cdot i_D \tag{20}$$

$$\psi_D = x_{dD} \cdot i_d + x_{uD} \cdot i_u + x_D \cdot i_D \tag{21}$$

$$\psi_Q = x_{qQ} \cdot i_q + x_Q \cdot i_Q \tag{22}$$

557

Matrix representation:

$$
\begin{vmatrix} \dfrac{di_d}{dt} \\ \dfrac{di_1}{dt} \\ \dfrac{di_D}{dt} \end{vmatrix} = \omega_s \cdot \begin{vmatrix} x_d & x_{ud} & x_{dD} \\ x_{ud} & x_u & x_{uD} \\ x_{dD} & x_{uD} & x_D \end{vmatrix}^{-1} \cdot \begin{vmatrix} A_d \\ B_d \\ C_d \end{vmatrix} \tag{23}
$$

$$
\begin{vmatrix} \dfrac{di_q}{dt} \\ \dfrac{di_Q}{dt} \end{vmatrix} = \omega_s \cdot \begin{vmatrix} x_q & x_{qQ} \\ x_{qQ} & x_Q \end{vmatrix}^{-1} \cdot \begin{vmatrix} A_q \\ B_q \end{vmatrix} \tag{24}
$$

$$ A_d = -u_d - \omega \cdot \psi_q - r \cdot i_d \tag{25} $$

$$ B_d = u_u - r_u \cdot i_u \tag{26} $$

$$ C_d = -r_D \cdot i_D \tag{27} $$

$$ A_q = -u_q + \omega \cdot \psi_d - r \cdot i_q \tag{28} $$

$$ B_q = -r_Q \cdot i_Q \tag{29} $$

For the purpose of simulation it is necessary to compute inverse of matrix in (23) and (24). Results are elements of reactance matrix in d axis.

$$ Y_d(3,3) = \frac{K}{L} \tag{30} $$

$$ Y_d(3,2) = -\frac{M}{L} \tag{31} $$

$$ Y_d(3,1) = \frac{-\dfrac{x_{dD}}{x_d} \cdot K + \dfrac{x_{ud}}{x_d} \cdot M}{L} \tag{32} $$

$$ Y_d(2,2) = \frac{x_d}{I} - N \cdot Y_d(3,2) \tag{33} $$

$$ Y_d(2,1) = -\frac{x_{ud}}{I} - N \cdot Y_d(3,1) \tag{34} $$

$$ Y_d(1,1) = \frac{1}{x_d} - \frac{x_{dD}}{x_d} \cdot Y_d(3,1) - \frac{x_{uD}}{x_d} \cdot Y_d(2,1) \tag{35} $$

$$ Y_d(2,3) = Y_d(3,2) \tag{36} $$

$$ Y_d(1,3) = Y_d(3,1) \tag{37} $$

$$ Y_d(1,2) = Y_d(2,1) \tag{38} $$

$$ K = x_u - \frac{x_{ud}^2}{x_d} \tag{39} $$

$$ L = \left(x_D - \frac{x_{dD}^2}{x_d}\right) \cdot \left(x_u - \frac{x_{ud}^2}{x_d}\right) - \left(x_{uD} - \frac{x_{ud} \cdot x_{dD}}{x_d}\right)^2 \tag{40} $$

$$ M = x_{uD} - \frac{x_{ud} \cdot x_{dD}}{x_d} \tag{41} $$

$$ N = \frac{M}{K} \tag{42} $$

$$ I = x_u \cdot x_d - x_{ud}^2 \tag{43} $$

Elements of reactance matrix in q axis are:

$$ Y_q(1,1) = \frac{1}{x_q} \cdot \left(1 + \frac{x_{qQ}^2}{P}\right) \tag{44} $$

$$ Y_q(1,2) = -\frac{x_{qQ}}{P} \tag{45} $$

$$ Y_q(2,1) = Y_q(1,2) \tag{46} $$

$$ Y_q(2,2) = \frac{x_q}{P} \tag{47} $$

$$ P = x_q \cdot x_Q - x_{qQ}^2 \tag{48} $$

Simulation model has been written in relative units.
For the purpose of simulation it is necessary to know following parameters of generator: r, r_u, r_D, r_Q, x_d, x_{ud}, x_{dD}, x_u, x_{uD}, x_D, x_q, x_{qQ}, x_Q. Standard parameters that are usually known for synchronous machine are x_d, x_q, x_l, r, T_m, x_d', x_d'', x_q'', T_d'', T_q''. Relationships of those two groups of parameters are following:

$$ x_{ud} = x_d - x_l \tag{49} $$

$$ x_{uD} = x_{ud} \tag{50} $$

$$ x_u = \frac{(x_d - x_l)^2}{x_d - x_d'} \tag{51} $$

$$ x_D = x_{ud} + \frac{(x_d' - x_l) \cdot (x_d'' - x_l)}{x_d' - x_d''} \tag{52} $$

$$ x_{qQ} = x_q - x_l \tag{53} $$

$$ x_Q = \frac{(x_q - x_l)^2}{x_q - x_q''} \tag{54} $$

$$ r_u = \frac{x_u}{T_{d0} \cdot \omega_s} \tag{55} $$

$$ r_D = \frac{(x_d' - x_l)^2}{x_d' - x_d''} \cdot \frac{x_d''}{x_d'} \cdot \frac{1}{T_d'' \cdot \omega_s} \tag{56} $$

$$ r_Q = \frac{(x_q - x_l)^2}{x_q - x_q''} \cdot \frac{x_q''}{x_q} \cdot \frac{1}{T_q'' \cdot \omega_s} \tag{57} $$

Equations of motion are:

$$ \frac{d\varphi}{dt} \cdot \frac{1}{\omega_s} = \omega \tag{58} $$

$$ \frac{d\delta}{dt} \cdot \frac{1}{\omega_s} = 1 - \omega \tag{59} $$

$$ \frac{d\omega}{dt} = \frac{1}{T_m} \cdot (m_{meh} + m_{elm}) \tag{60} $$

$$m_{elm} = \psi_q \cdot i_d - \psi_d \cdot i_q \qquad (61)$$

SMIB system is simulated. System is consisted of synchronous generator connected on AC grid over transmission lines with reactance x_v and resistance r_v.

$$u_{dv} = i_{dv} \cdot r_v + \frac{x_v}{\omega_s} \cdot \frac{di_{dv}}{dt} + \omega \cdot x_v \cdot i_{qv} + u_{kmd} \qquad (62)$$

$$u_{qv} = i_{qv} \cdot r_v + \frac{x_v}{\omega_s} \cdot \frac{di_{qv}}{dt} - \omega \cdot x_v \cdot i_{dv} + u_{kmq} \qquad (63)$$

$$u_{kmd} = u_{km} \cdot \sin \delta \qquad (64)$$

$$u_{kmq} = u_{km} \cdot \cos \delta \qquad (65)$$

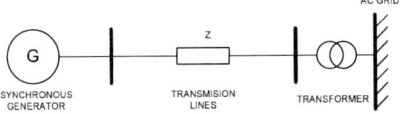

Fig. 3. Synchronous generator connection

Control structure simulated in Matlab Simulink is shown on the Fig. 4).

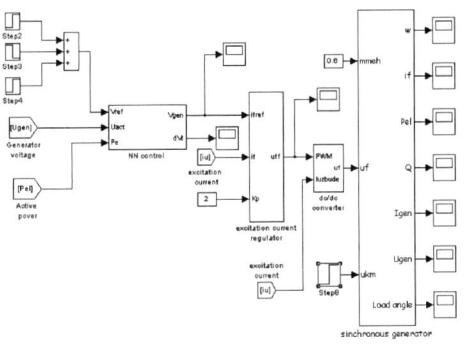

Fig. 4. Simulation model

TABLE I.
SYNCHRONOUS GENERATOR PARAMETERS

Terminal voltage	400 V
Phase current	120 A
Power	83 kVA
Freqvency	50 Hz
Speed	600 r/min
Power factor	0,8
Excitation voltage	100 V
Excitation current	11.8 A

TABLE II.
SIMULATION MODEL PARAMETERS

X_d	X_q	X_l	T_m	T_f	X_d'
0,8 (p.u.)	0,51 (p.u.)	0,04 (p.u.)	2,6 (s)	0,55 (s)	0,35 (p.u.)
X_d''	X_q''	T_d''	T_q''	r_v	x_v
0,15 (p.u.)	0,15 (p.u.)	0,054 (s)	0,054 (s)	0,05 (p.u.)	0,35 (p.u.)

6. SIMULATION RESULTS

Various tests with a step down of 0.1 p.u. of the reference voltage were made.

In the Fig. 5 voltage and active power of synchronous generator are shown for the NN control structure. The stabilization effect in the modified error function is not simulated. The generator operates with approximately 0.5 p.u. of active power.

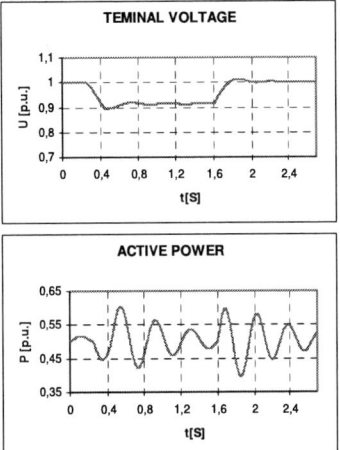

Fig. 5. Synchronous generator terminal voltage and active power with NN control structure without stabilization operating at 0.5. p.u. of active power

The results with the NN control structure without stabilization at operating conditions of 0.8 p.u. active power are presented in Fig. 6.

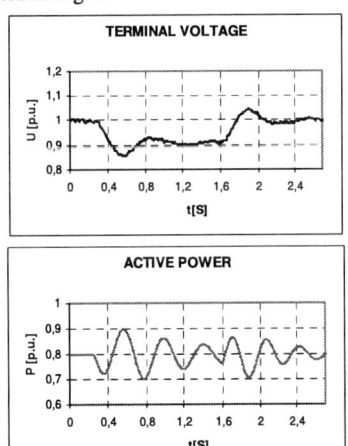

Fig. 6. Synchronous generator terminal voltage and active power with NN control structure without stabilization operating at 0.8. p.u. of active power

In Fig.8 is shown behaviours of the generator's voltage and active power for the NN control structure with stabilization. The stabilization effect in the modified error function is simulated and the generator operates at 0.5 p.u. and 0.8 p.u. of active power. The results are shown in Fig.7 and Fig.8.

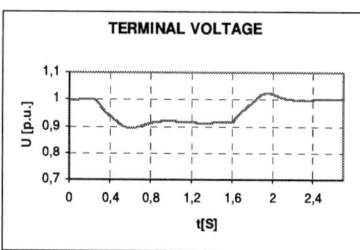

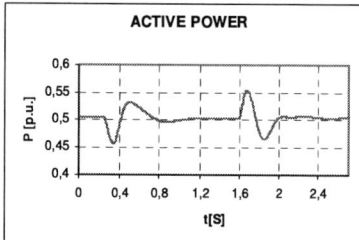

Fig. 7. Synchronous generator terminal voltage and active power with NN control structure with stabilization operating at 0.5. p.u. of active power

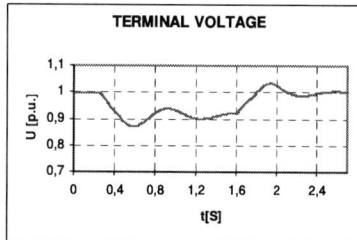

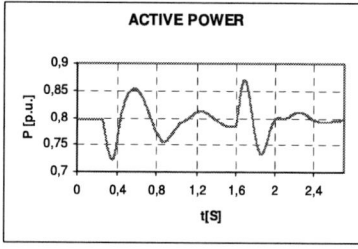

Fig. 8. Synchronous generator terminal voltage and active power with NN control structure with stabilization operating at 0.8. p.u. of active power

The NN control structure with stabilization effect in the modified error function significantly damps active power oscillations. In the NN control structure without stabilization the terminal voltage control also remains good.

7. CONCLUSIONS

The neural network based excitation control structure with synchronous generator connected via transmission lines to the power system were simulated in Matlab Simulink. A back propagation algorithm with modified error function is used to train the neural network online. The modified error function uses sampled input and output values of the synchronous generator. Therefore is no need to determine the states of the system. Test results show that the NN with stabilization effect effectively damp down active power oscillations and provide a good terminal voltage control.

REFERENCES

[1] O.P.Malik,.M.M.Salem, A.M. Zaki, O.A. Mahgoub and E. Abu El-Zahab, *"Experimental studies with simple neuro-controller based excitation controller"*, IEEE Proc.-Gener. Transm. Distrib. Vol. 149. No I. January 2002

[2] Bulic N., Erceg G., Idzotic T, *Comparison of different methods of excitation control for a synchronous generator*, EPE-PEMC, Riga, 2004

[3] M.M.Salem, O.P.Malik,. A.M. Zaki, O.A. Mahgoub and E. Abu El-Zahab *"Simple neuro-controller with modified error function for a synchronous generator"*, Electrical Power and Energy Systems 25 (2003) 759-771

[4] Sumina, D.; Idzotic, T.; Erceg, G., *"The appliance of the estimated load angle in the fuzzy power system stabilizer"*, Electrotechnical Conference, 2006. MELECON 2006

APPENDIX: NOMENCLATURE

φ	generator current voltage angle
ϑ	Load angle
$\Im$	Criteria function
ψ	activation function of neuron
ψ_d, ψ_d	flux linkage in d q axes
ψ_D, ψ_Q	damping flux linkage in d q axes
ω	generator speed (electrical)
ω_{meh}	angle speed (mechanical)
E_0	Generator internal voltage
I	Generator current amplitude
i_d, i_q	generator current in dq axes
i_u	excitation current
i_α, i_β	generator currents in α β axes
m_{elm}	electrical torque
m_{meh}	mechanical torque
P, P_{el}	generator active power
P_{max}	maximal generator power
Q	reactive power
R	armature resistance per phase
R_D, R_Q	damping windings resistance in d q axes
R_u	exciter resistance in d q axes
r_v	infinite bus resistance
S	total power of generator
T_d''	sub transient time constant in d axes
T_q''	sub transient time constant in q axes
T_m	mechanical time constant
T_u	excitation time constant
U	generator voltage
U_{ref}	generator voltage reference value
u_d, u_q	generator voltage in d q *axes*
u_{km}	ac grid voltage
u_{kmd}, u_{kmq}	ac grid voltage in d q axes
u_α, u_β	generator voltage in α, β axes
w_{ijk}	weights of neurons
x	neuron input
x_d, x_q	synchronous reactance in d q axes
x_D, x_Q	damping reactance in d q axes
x_d'	transient reactance in d axes
x_d''	sub transient reactance in d axes
x_q''	sub transient reactance in q axes
x_u	exciter reactance
x_v	infinite bus reactance
y_{ki}	neuron output after activation function
v_{ki}	neuron output before activation function

Predictive Current Control of a 7-level AC-DC back-to-back Converter for Universal and Flexible Power Management System

Stefano Bifaretti*, Pericle Zanchetta[†], Florin Iov[††] and Jon C. Clare[†]

* University of Rome Tor Vergata, Dept. of Electronic Engineering, Rome, Italy, *email: bifaretti@ing.uniroma2.it*
[†] University of Nottingham, School of Electrical and Electronic Engineering, Nottingham, UK,
email:Pericle.Zanchetta@nottingham.ac.uk
[††] Aalborg University, Institute of Energy Technology, Aalborg, Denmark, *email: fi@iet.aau.dk*

Abstract— **The paper proposes a novel power conversion system for Universal and Flexible Power Management (UNIFLEX-PM) in Future Electricity Network. Its structure is based on a back-to-back three-phase AC-DC 7-level converter; each AC side is connected to a different PCC, representing the main grid and/or various distributed generation systems. Effective and accurate power flow control is demonstrated through simulation in Matlab-Simulink environment on a model based on a two-port structure and using a Predictive Control technique. Control of different Power flow profiles has been successfully tested in numerous network conditions such as voltage unbalance, frequency excursions and harmonic distortion.**

*Keywords—***Converter Control, Distributed power, HVDC, Multilevel converters.**

I. INTRODUCTION

With the present architecture of the electricity network, most of the electricity is generated in large power stations and transmitted, using a passive transmission line, through high voltage transmission systems. Power is then delivered to consumers via medium and low-voltage distribution systems [1]. The power flow in this arrangement is only in one direction: from the central power stations to the consumers. Nowadays most of the European countries have started to liberalize the electricity market. In order to enable the electricity market, different distribution system operators (DSOs) will operate on the electricity network transparently and without discrimination under the governance of a regulator [1]. This scenario requires an increase penetration of the renewable energy resources (RES) and other distributed generation (DG) and an active role for DSOs in controlling the network stability, optimising central and distributed power inputs into the network. Moreover, in order to reach this goal, the entire architecture of the electricity network must be redesigned on the basis of models, such as Micro-grids, an Internet model and Active Networks, derived by the information and communication technologies (ICT) that will transform the existing electrical grid into a smart one [2]. According to [1], active networks technically and economically may be the best way to facilitate DG initially in a deregulated market. Its architecture employs an increased number of power input nodes, as a result of DG, bi-directional energy flow

is possible and new technologies are emerging that can enable the direct routing of electricity.

New power electronics systems offer ways of controlling the routing of electricity and also provide flexible DG interfaces to the network. Additionally, power flow control using power electronic converters is needed to ensure proper and secure functioning of the grid. In such framework, the paper analyses a new power conversion structure to be employed for Universal and Flexible Power Management (UNIFLEX-PM) proposing a Predictive Control technique suitable to manage the energy exchanges between two different electricity networks and to obtain high power quality at both grids. The paper points out the behavior of the conversion system in different grid conditions, in particular when the grid voltages are unbalanced and distorted.

II. UNIFLEX-PM STRUCTURE

The main objective of the Uniflex-PM system is to provide a flexible and modular power electronic interface able to connect different kind of sources and loads including MV electrical networks, RES and energy storage systems [3]. The foreseen structure of UNIFLEX-PM conversion system is shown in Fig. 1.

The conversion system is composed by three power converters, each one connected to a PCC. Port One and Port Two are used for MV electrical networks, while Port Three is used mainly to connect the conversion structure to an energy storage system and to a low voltage electricity network.

The conversion structure must be able to satisfy the following requirements:

- Bi-directional power flow operation in all ports with active and reactive power control capabilities;
- Compliance to most of European and International grid standards for DG connection in terms of injected harmonics, robustness to grid voltage distortions and excursions;
- Galvanic isolation among the ports;
- Modular architecture providing high reliability and easy maintenance.

A basic schematic block of the conversion structure, based on a multi-stage architecture, that can provide all the above mentioned capabilities, is illustrated in Fig. 2.

978-1-4244-1741-4/08/$25.00 ©2008 IEEE

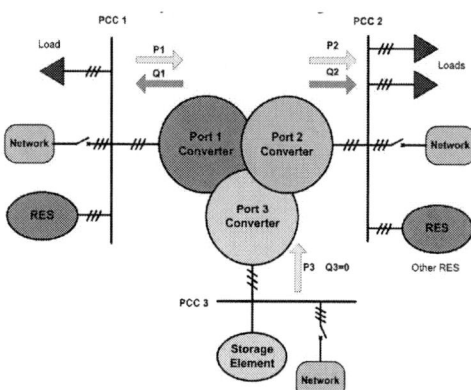

Fig. 1. General Structure of UNIFLEX model.

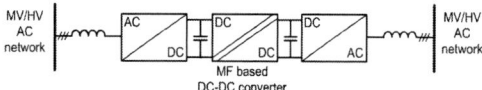

Fig. 2. Basic schematic block of the power conversion structure.

Such architecture consists of two AC/DC power conversion stages and a DC/DC conversion stage, based on Medium Frequency (MF) transformer to achieve the galvanic isolation between the AC terminals.

Different topologies of power converters can be considered for medium and high voltage grid applications [4]-[7]. However, due to the recent developments in power semiconductor devices, particularly in the IGBT technology, there is an increasing interest in the last years, in multi-level power converters especially for medium to high-power, high-voltage [8]-[10]. Multi-level converters present different advantages such as reduced harmonics content in the input and output voltage, reduced switching losses at the same harmonic performance as for a two-level converter. Among the different topologies, to assure also a modular architecture, a multi-level cascaded single phase H-bridge topology has been selected for the implementation of the Uniflex-PM system.

To simplify the control implementation, the paper accounts a model based on two-port structure, as shown in Fig. 3. Each phase of Port 1 and Port 2 of the UNIFLEX-PM model employs a conversion structure based on a three-phase 7-level AC-DC cascaded converter. A more detailed schematic of the structure is shown in Fig. 4, where only Phase A is represented. A series input filter L_f is included to separate the converter from the grid and provides a suitable attenuation to current harmonic.

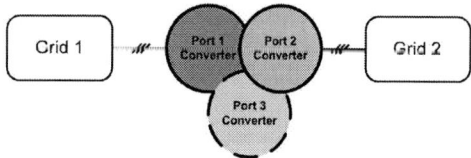

Fig. 3. Two-Port Structure of UNIFLEX model.

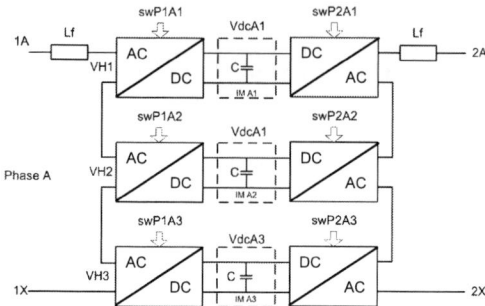

Fig. 4. Structure of one phase of the 2-Port UNFILEX-PM system.

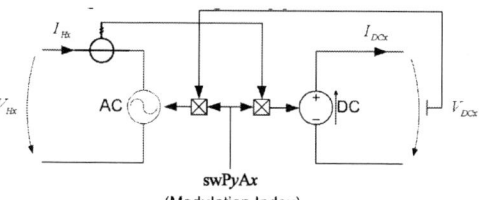

Fig. 5. Model of an H-Bridge converter.

Since the isolated DC/DC conversion stage has an independent control for each branch, an equivalent capacitor C is instead used in the model. In Fig. 4, the control input of each converter swPyAx, where y=1..2 and x=1..3 denotes, respectively, the Port number and the branch number of H-Bridge converters, has been represented.

In order to use the same model for both average and switching simulations, a model of each H-Bridge converter, able to support continuous and switching signals is used. The H-Bridge model is shown in Fig. 5 where are highlighted the electrical variables on both AC and DC side and the control input.

Referring to Phase A Port1 structure and considering a neutral-connected conversion system, the instantaneous phase voltage is obtained using the following relation:

$$v_{ha} = \sum_{x=1}^{3} VH_x = \sum_{x=1}^{3} V_{dc}A_x \cdot swP1A_x$$

III. CONTROL SYSTEM

Different control systems for grid connected converters, based on natural, stationary or synchronous reference frame, have been proposed in literature [11]-[17]. Synchronous reference frame (dq) control uses Park transformation to obtain dc values for voltages and current and makes the control system easy to implement. Their control structure becomes instead very complex if unbalanced grid voltage conditions have to be accounted [11]-[13]. Stationary reference frame ($\alpha\beta$) control uses sinusoidal control variables so, to avoid steady-state errors, PI controllers cannot be used; more complex kind of controllers, such as resonant controllers, have therefore to be employed [14]. Finally, hysteresis [15] or predictive [16], [17] controls based on a natural reference frame (abc) present very fast response and are particularly

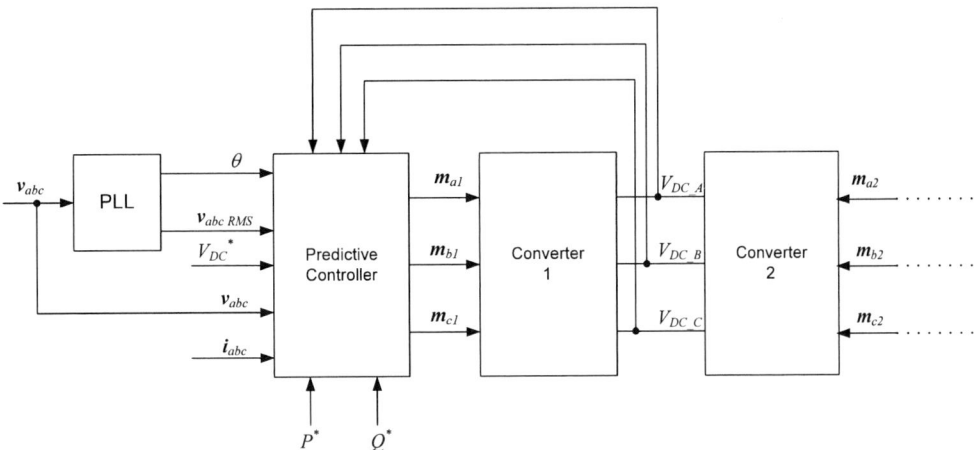

Fig. 6 - Block diagram of Predictive Control

suitable for an easy microprocessor implementation. This paper proposes the application of a predictive current control to the UNIFLEX conversion structure to control the grid currents and the DC-link voltages.

The overall block diagram of the Controller is shown in Fig. 6, where only Port 1 side is illustrated in detail. For Port 2, a control scheme similar to Port 1, but without the outer DC-link control loop, is used. A multi loop structure is selected to design the controller to regulate the power/DC-link voltage and input line currents at the same time. The outer loop is the power/DC-link voltage loop while the inner loop, is responsible for the current control. In Port 1, the regulation of active power P is obtained through the DC-link voltage balance, performed by a PI controller. The predictive controller needs the calculation of the amplitude and phase of the current references. Active power P contributes to generate only the amplitude of the current references while the reactive power Q acts on both the amplitude and phase of the current references.

For Port 2, the controller is similar to that of Port 1 except for active power regulation. In fact, both the active and reactive powers are controlled through PI controllers without any interaction of DC-link voltage. Then, the calculation of the amplitude and phase of the current references, necessary for Predictive controller, is the same.

A. Predictive Controller Description

The basic equation of the Predictive controller model is obtained starting from the following state equation for the inductor L_f:

$$i(t+t_0) = i(t_0) + \frac{1}{L_f} \int_{t_0}^{t} (v_j - v_{hj}) dt \qquad (1)$$

where j represents the phase a, b or c. Applying a time discretisation to (1) with a sampling period T_s and supposing v_{xj} constant during T_s, the following expression for voltage v_{xj} is obtained:

$$v_{xj_k} = \frac{L_f}{T_s}(i_k - i_{k+1}) + \frac{1}{T_s} \int_{t_k}^{t_k+T_s} v_j dt \qquad (2)$$

In order to zero the current error, the control system imposes at each sampling period i_{k+1} equal to the reference current. To reduce the effects of the harmonics in the grid voltages, the integral of v_j is calculated using the method proposed in [17].

The block diagram of the predictive controller employed for Phase a of Port 1 is shown in Fig. 6.

Reference current i_a^*, necessary to the predictive controller to zero the error at each sampling period, is calculated on the base of the following sinusoidal function:

$$i_a^* = I_a^* \sin \alpha = I_a^* \sin(\theta + \omega T_s - \varphi_a^*)$$

whose amplitude I_a^* and phase α depend on actual values of active power reference P^*, reactive power Q^* and the output a PI regulator used to control the of DC-link voltage.

In particular, amplitude I_a^* is obtained on the basis of the following relationship:

$$I_a^* = \frac{\sqrt{2}P_a^*}{V_{aRMS} \cdot \cos \varphi_a^*}$$

where P_a^* is the active power reference for Phase a obtained modifying the overall active power reference, divided by 3, on the basis of DC-link error on the same phase; V_{aRMS} is the RMS of grid voltage useful in case of distorted grid voltages.

φ_a^* is the phase difference between voltage and current at PCC needed to produce the desired power factor and calculated as:

$$\varphi_a^* = \text{at an} \frac{Q_a^*}{P_a^*}$$

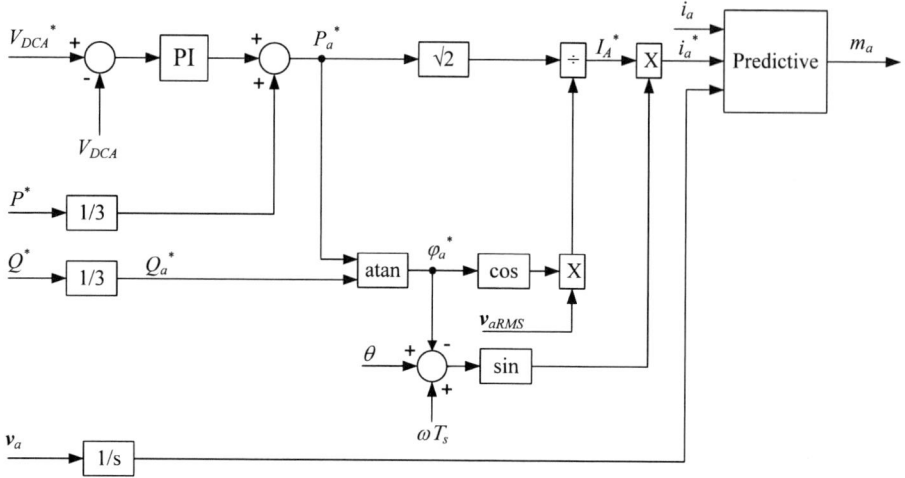

Fig. 7. Block diagram of Predictive Controller for Phase A Port 1.

To avoid that the systems overcomes the desired apparent power, φ_a^* modifies both the values of I_a^* and phase α of reference current i_a^*.

Phase angle α of reference current i_a^* is achieved by subtracting from the actual voltage grid angle θ, provided by a Phase Locked Loop (PLL), the angle φ_a^*; moreover, a term ωT_S is added to account that i_a^* value, calculated at the generic sampling instant k, is applied at next sampling period $k+1$. Since the conversion structure presents a neutral connection, the control system uses three different PLL, based on SOGI filter [18]. Such method allows an easy implementation on DSPs, very good dynamic performances, fast and accurate response based on the tuning and it is frequency independent. However, this approach is very sensitive to the discretization method of the integrators.

Finally, the block 1/s in Fig. 7 provides the voltage grid integral necessary to calculate the integral term in (2).

IV. SIMULATION RESULTS

The proposed control strategy was evaluated under different power flow profiles, grid conditions and situations such as voltage unbalances, frequency excursions, and harmonic distortion. The simulations, performed on Matlab-Simulink for 5 different study cases, accounts the following parameters for the power conversion structure: rated power 300 kVA, rated line-to-line voltage 3.3 kV, DC-link voltage on each capacitor 1100 V, DC-link capacitor 6.2 mF, input filter inductance 16 mH. A bidirectional power flow is simulated using an average model of the UNIFLEX system for the first 3 cases; in case E harmonics on the grid voltages are introduced and in case D a switching model of the converter with suitable PWM modulation is used.

A. Power Flow

The waveforms of active and reactive power (blue line), shown respectively in Fig. 8a and Fig. 8b, tracks the reference (green line) with satisfying dynamics and steady state error. Moreover, as illustrated in Fig. 9, the DC-link

voltage presents insignificant ripple where voltage on the capacitor of Phase B Port 1 is considered for example.

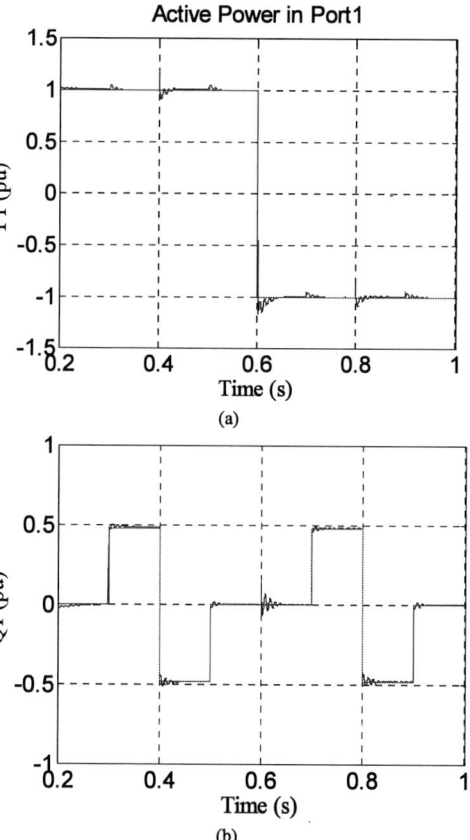

Fig. 8. Simulation results for power flow study in Port 1:
(a) Active Power, (b) Reactive Power.

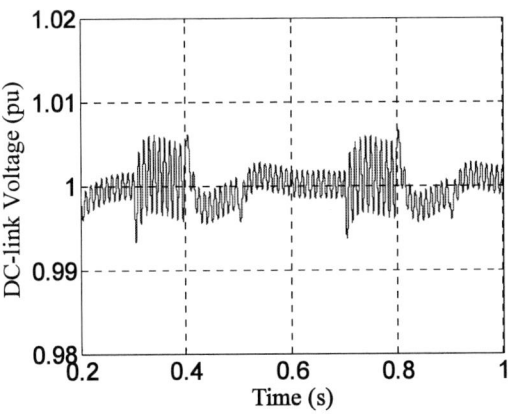

Fig. 9. DC-link Voltage on Phase B Port 1

B. Voltages Unbalance

Further tests are carried out when 3% voltages unbalance, defined as the ratio between the negative and the positive sequence of the voltage, is considered in the grid at Port 1. Fig. 10a and Fig. 10b show simulation results for active and reactive power flow control, while Fig. 11a and Fig. 11a show the grid currents and the DC-link voltage on phase B.

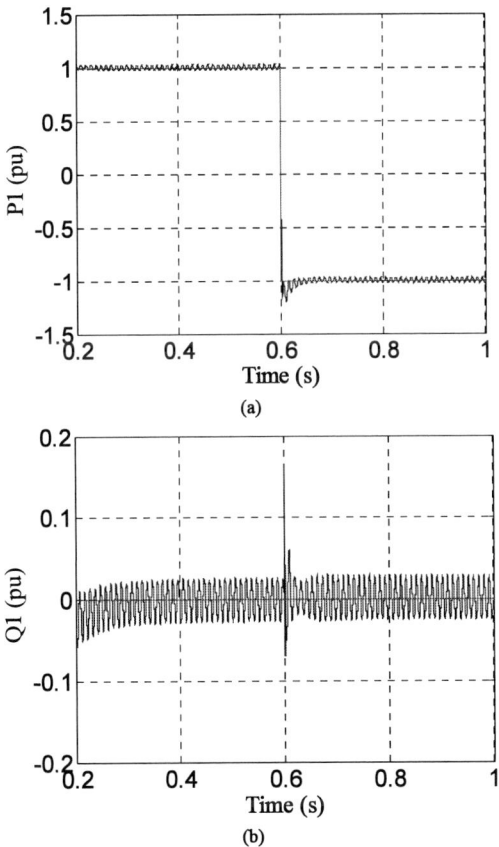

Fig. 10. Power Flow under voltage unbalances in Port 1:
(a) Active Power, (b) Reactive Power.

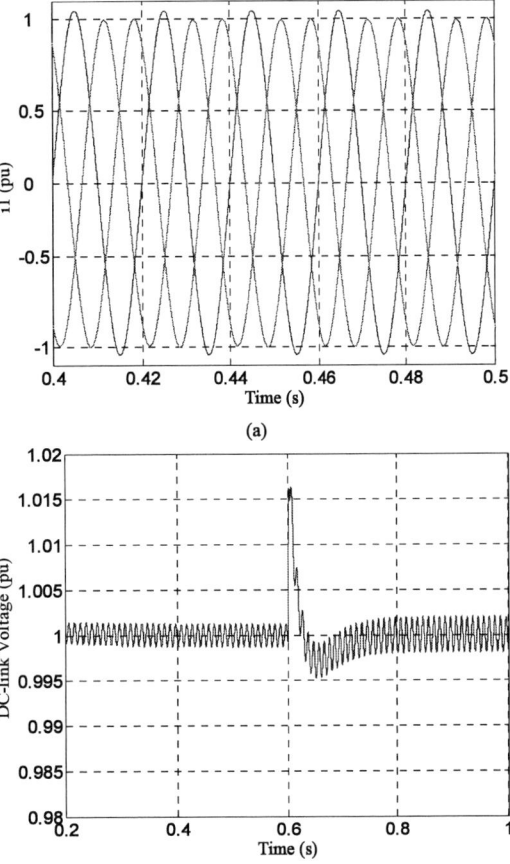

Fig. 11. Simulation results for voltage unbalances in Port 1:
(a) Input currents Port 1, (b) DC-link Voltage Phase B Port 1

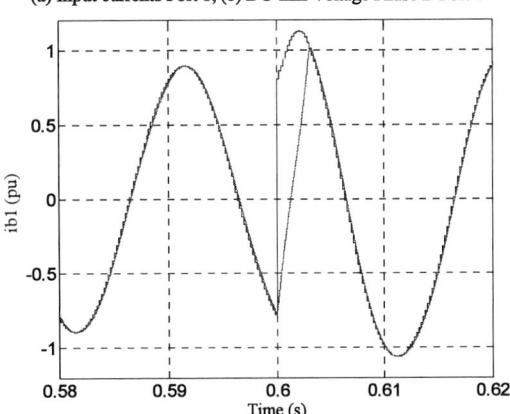

Fig. 12. Phase B Current tracking during Active Power step.

It is to note that the grid currents are essentially sinusoidal and the active power is well regulated. The effect of the unbalance causes ripples on the reactive power which is not possible to eliminate. Moreover, the DC-link voltage presents a maximum variation of about 1.5% of the nominal value.

The effectiveness of predictive current controller can be observed by Phase B current tracking, illustrated in Fig. 12, when the step variation in Active power shown in Fig.

10a is applied; in the figure the blue line is the current reference and the red line is the phase current. It is to note that the current transient duration, due to the step change occurred at 0.6s, is limited to about 1/5 period; before and after that interval the predictive controller is able to track very accurately the reference current.

C. Frequency Excursions

The behavior of the control systems has been also evaluated when a ±6% frequency variation, with the profile shown in Fig. 13a, of grid voltages is applied.

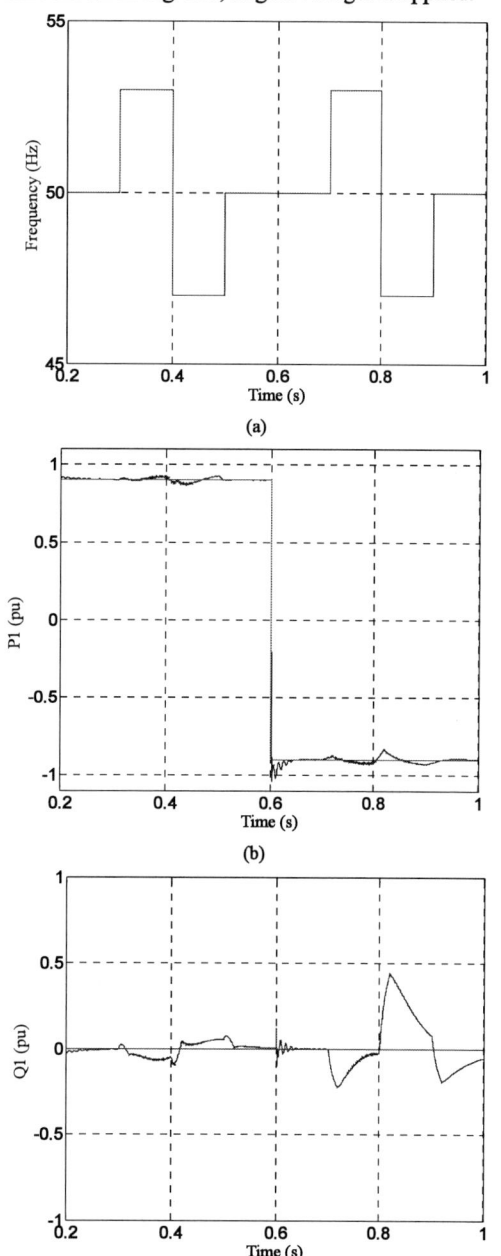

Fig. 13. Power Flow under frequency excursions in Port 1:
(a) Frequency profile, (b) Active Power, (c) Reactive Power.

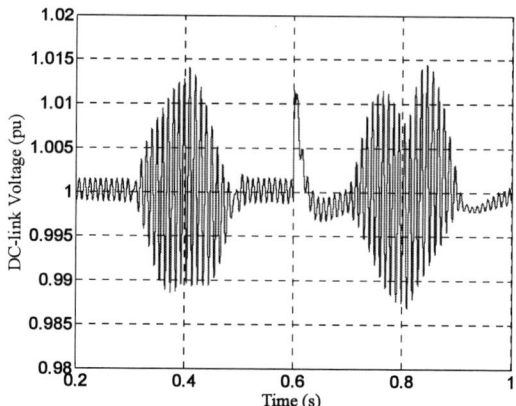

Fig. 14. Simulation results for frequency in Port 1:
DC-link Voltage

Fig. 13b and Fig. 13c show the simulation results obtained for active and reactive power, while Fig. 14 illustrates the DC-link voltage on phase B; also in this case the DC-link voltage is very well regulated.

D. Harmonic content

Harmonic study tests were performed to evaluate the current harmonics injected into the grid by the converter modulation. The analysis considers that the power modules deliver 100% active power into the PCC while the reactive power is set to zero.

For the current harmonics analysis a 1800 Hz switching frequency was imposed to the converter and undistorted grid voltages are considered. Fig. 15 and Fig. 16 show, respectively, the output voltage produced on Port1 Phase A by the converter and the corresponding phase current in the mentioned operating conditions.

The complete harmonic contents and the THD are summarized in Fig. 17 that shows individual harmonics values for a phase current up to 50th harmonic vs EN 61000-2-4 harmonic levels, represented in the figure respectively with a green bar and a red bar.

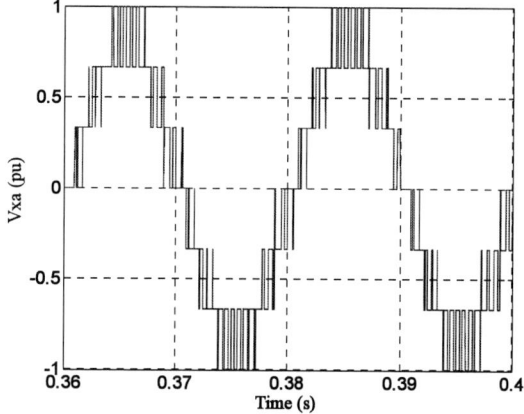

Fig. 15. Modulated voltage produced by the converter.

566

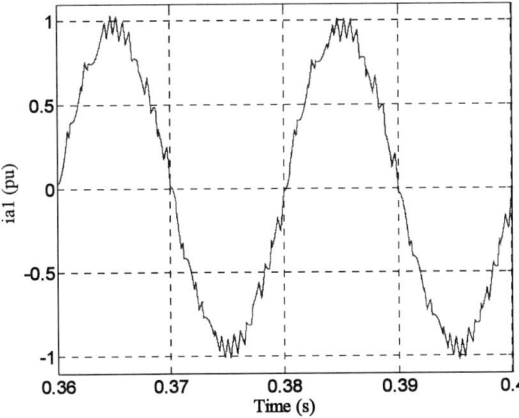

Fig. 16. Phase current produced applying modulated voltage.

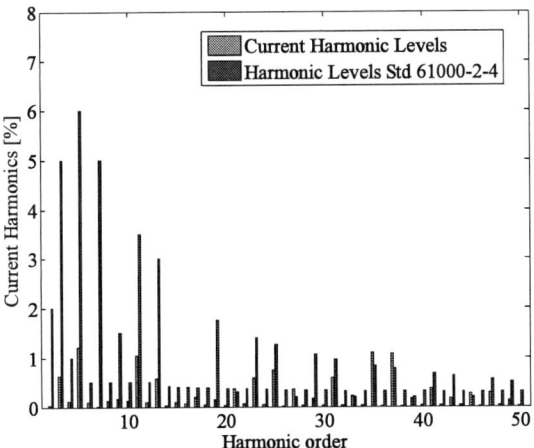

Fig. 17. Harmonic content of a phase current.

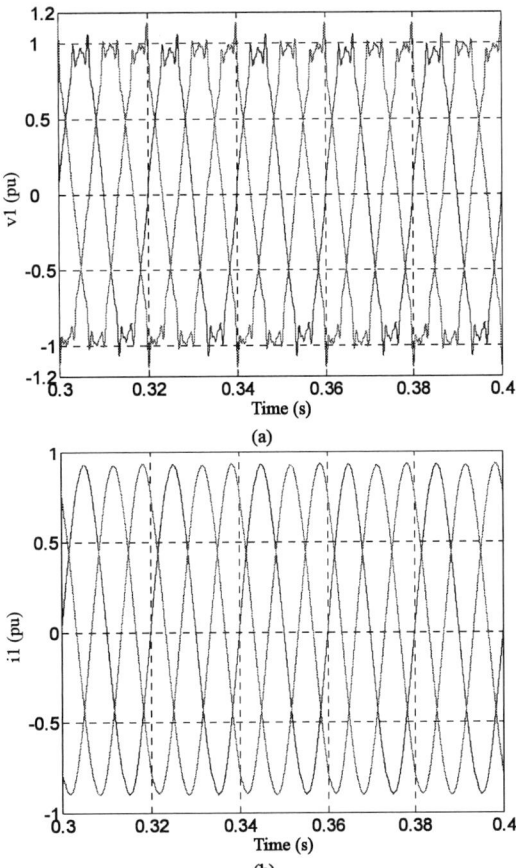

Fig. 18. Effects of distorted grid voltages in Port 1:
(a) grid voltages, (b) grid currents.

Table I. Harmonic content of grid voltages and currents.

	5th(%)	7th(%)	11th(%)	13th(%)
V1	6	5	3.5	3
I1	0.697	0.282	0.210	0.156

The current harmonic levels that exceeds the standard values are positioned at harmonic orders 35th and 37th, i.e. around the switching frequency. Important value, equal to about 0.6%, of the 3rd harmonic is present due to the neutral connection required by this control strategy; however, such value results widely lesser than 5% required by standard EN 61000-2-4. Finally, a value of 5.55% is obtained for the current THD that is lower than 8% limit imposed by the considered standard.

E. Effects of Distorted Grid

Another interesting aspect is the performance evaluation of the converter control when the grid voltages are distorted. In this case, the analysis considers that the power modules deliver 90% active power into the PCC while the reactive power is set to zero.

The grid voltages are polluted with 5th, 7th, 11th and 13th harmonics according to the amplitudes furnished by standard EN 61000-2-4.

Fig. 18a and Fig. 18b show respectively the waveforms of the grid voltages and the grid currents in Port 1, while Table I summarises the voltage and current harmonic contents. It can be noticed that the proposed predictive controller produces almost sinusoidal currents even if the supply voltages are strongly distorted.

V. CONCLUSIONS

The paper proposes the application of a Predictive controller to a back-to-back three-phase AC-DC 7-level converter suitable for Universal and Flexible Power Management (UNIFLEX-PM) in Future Electricity Network. The main goal of the system is to provide a flexible and modular power electronic interface able to connect different kinds of sources and loads including MV electrical networks, RES and energy storage systems.

The effectiveness of the Predictive controller in a natural reference frame is verified through different study cases. The analysis was performed in a Matlab/Simulink environment using a model of the converter that suitable for both average and switching simulations.

The simulation results have shown that the control system can trace the active and reactive power under different power reference profiles and grid operating conditions; moreover, the control system presents a fast and accurate current tracking. A good decoupling of the active and reactive power control is obtained in all considered cases. The supply currents are essentially sinusoidal even under unbalanced supply voltage conditions. The predictive controller can reject significantly also the harmonic distortion in the supply voltage without using any additional filter in the control loop; thus, the produced currents present a good harmonic content.

The overall system performance relies on the angular information; therefore, the PLL plays an important role in the control.

The average DC-link voltage has small variations; the voltage peaks reach the maximum value of about 1.5% of the rated one.

ACKNOWLEDGMENT

The authors acknowledge the support from European Commission through Contract no. 019794 SES6.

REFERENCES

[1] European Commission, "New ERA for electricity in Europe. Distributed Generation: Key Issues, Challenges and Proposed Solutions," EUR 20901, 2003, ISBN 92-894-6262-0.

[2] European Commission, "Towards Smart Power Networks. Lessons learned from European research FP5 projects," EUR 21970, 2005, ISBN 92-79-00554-5.

[3] F. Iov, F. Blaabjerg, R. Bassett, J. Clare, A. Rufer, S. Savio, P. Biller, P. Taylor and B. Sneyers, "Advanced Power Converter for Universal and Flexible Power Management in Future Electricity Network," in *Proc. CIRED 2007*, Vienna, Austria, May 2007.

[4] M. Marchesoni and M. Mazzucchelli, "Multilevel converters for high power ac drives: a review," in *Proc. of IEEE International Symposium on Industrial Electronics*, pp. 38-43, 1993.

[5] D. Soto and T.C. Green, "A comparison of high-power converter topologies for the implementation of FACTS controllers," *IEEE Trans. Ind. Electron.*, vol.49, no.5, pp. 1072-1080, Oct. 2002.

[6] Y. Cheng, C. Qian, M.L. Crow, S.Pekarek and S. Atcitty, "A comparison of diode-clamped and cascaded multilevel converters for STATCOM with energy storage," *IEEE Trans. Ind. Electron.*, vol.53, no.5, pp. 1512-1521, Oct. 2006.

[7] F. Blaabjerg and F. Iov, "Wind power – a power source now enabled by power electronics," in *Proc of 9th Brazilian Power Electronics Conference* COBEP 07, Blumenau, Santa Catarina, Brazil, ISBN 978-85-99195-02-4, October 2007.

[8] Bin Wu, *High-Power Converters and AC Drives*, IEEE Press, Wiley Interscience 2006, ISBN 10-0-471-73171-4.

[9] V.G. Agelidis, G.D. Demetriades and N. Flourentzou, "Recent Advances in High-Voltage Direct-Current Power Transmission Systems", in *Proc. IEEE Int. Conf. on Industrial Technology ICIT 2006, Dec. 2006*.

[10] A. Rufer, "Today's and Tomorrow's Meaning of Power Electronics within the grid interconnection," keynote paper presented at *The 12th European Conf. on Power Electronics, EPE 2007*, Aalborg, Denmark, September 2007.

[11] P. Rodriguez, J. Pou, J. Bergas, J.I. Candela, R.P. Burgos and D. Boroyevich, "Decoupled Double Synchronous Reference Frame PLL for Power Converters Control Power Electronics," *IEEE Trans. on Power Electron.*, vol. 22, no. 2, pp.584 – 592, March 2007.

[12] R. Teodorescu and F. Blaabjerg, "Flexible control of small wind turbines with grid failure detection operating in stand-alone or grid-connected mode," *IEEE Trans. Power Electron.*, vol. 19, no. 5, pp. 1323–1332, Sep. 2004.

[13] R. Teodorescu, F. Iov and F. Blaabjerg, "Flexible development and test system for 11 kW wind turbine," in *Proc. of IEEE PESC*, vol. 1, pp. 67–72, June 2003.

[14] W. Lenwari, M. Sumner, P. Zanchetta and M. Culea, "A High Performance Harmonic Current Control for Shunt Active Filters Based on Resonant Compensators," in *Proc. of IECON 2006*, pp. 2109 – 2114, Nov. 2006.

[15] A. Bellini, S. Bifaretti and S. Costantini, "A Hysteresis Modulation Technique for NPC Inverter in Digitally Controlled Induction Motor Drives," in *Proc. of 10th Int. Conf on POWER ELECTRONICS and MOTION CONTROL, EPE-PEMC 2002*, Sept. 2002.

[16] J. Rodriguez, J. Pontt, C.A. Silva, P. Correa, P. Lezana, P. Cortes and U. Ammann, "Predictive Current Control of a Voltage Source Inverter" *IEEE Trans. on Ind. Electron.*, vol. 54, no. 1, pp. 495 – 503, Feb. 2007.

[17] P. Zanchetta, D.B. Gerry, V.G. Monopoli, J.C. Clare, P.W. Wheeler, "Predictive Current Control for Multilevel Active Rectifiers with Reduced Switching Frequency," *IEEE Trans. on Ind. Electron.*, vol. 55, no.1, pp. 163-172, Jan. 2008.

[18] M. Ciobotaru, R. Teodorescu, and F. Blaabjerg, "A New Single-Phase PLL Structure Based on Second Order Generalized Integrator," in *Proc. of 37th PESC*, pp. 1-6, June 2006.

Predictive Stator Current Control For Three-Level Voltage-Source Inverters With Output LC-Filters

Tomasz Laczynski, Axel Mertens

Institute for Drive Systems and Power Electronics
LEIBNIZ UNIVERSITY OF HANNOVER
Hannover, Germany
E-mail: laczynski@ial.uni-hannover.de

Abstract — In high power medium voltage drives the switching frequency of power semiconductor devices is limited because of high switching losses. These drives often consist of a three-level NPC inverter with an LC output filter to protect the motor from harmonics. Application of the filter creates a resonant circuit that may be excited for instance by switching harmonics or control transients. However, state of the art damping control methods require switching frequencies well above the filter resonance. This paper presents improvements to a predictive stator current control method which was recently proposed by the authors. The improvement consists of an optimized definition of the inverter's switching transitions and a control method of the neutral point potential. The predictive stator current controller avoids the excitation of the filter resonance and enables very fast stator current control while maintaining low switching frequency. Simulation results of the proposed current control method with a 2.4 kV induction motor drive model show good dynamic and stationary performance.

Index Terms — LC filter, medium voltage inverter, predictive current control, three-level inverter, neutral point potential.

I. INTRODUCTION

In order to minimize the switching losses of the power semiconductors, high power medium voltage three-level Neutral Point Clamped (NPC) inverters typically operate at a switching frequency lower than 500 Hz [1-6]. They are frequently applied to the retrofit of existing fixed speed induction motors supplied by mains with medium voltage variable speed drives so as to achieve energy savings. Because existing motors usually are not designed for inverter supply, the use of *LC* output filters becomes necessary in order to avoid isolation problems and bearing currents [7-10]. The filter creates sinusoidal output voltages and thereupon conditions for the machine similar to operation from the grid. The inverter switching frequency that is acceptable for thermal reasons lies only slightly above the resonance frequency of the LC filter, which causes problems relating to dynamic stator current control.

The PWM is used in medium voltage three-level NPC inverters only for low modulation indices because of the demand for low switching frequencies [11,12]. For higher modulation indices optimised pulse patterns are utilised in order to minimise the current harmonics. Compared to other modulation methods, optimised pulse patterns using

very low switching frequencies create the lowest harmonic distortions at stationary operation. When these two modulation methods are used, a dynamic stator current control can be achieved only after solving several issues. For instance, the application of the PWM with a switching frequency of even 600 Hz creates at stationary operation considerable harmonic distortions of the stator current. In this case spectral side bands around the switching frequency are created and some of them coincide with the resonance of the filter. Furthermore, the sampling frequency of the stator current controller is limited to 1200 Hz when its reference voltage is generated by the PWM. A stable control operation at such a low sampling frequency and inherent dead time of one sampling period due to PWM can be achieved, but only with strong distorted stator currents. For that reason, only at low modulation indices does the dynamic stator current control give acceptable results, as in this case the sampling frequency can be four times higher than the switching frequency.

At stationary operation optimised pulse patterns provide very low distortion currents, but a dynamic stator current control is not possible with them without additional measures. Every change of the modulation index or pulse number creates weakly damped transients which excite the filters resonance. The current controller is not able to control these transients and creates by its intervention additional current distortions by the excitation of the filter. In [13] was principally shown, that a dynamic operation with optimised pulse patterns could be possible by the application of a supplementary controller for the harmonic voltages and currents of the filter and motor.

The control approach presented in this contribution shows a different solution to that problem. For a predictive stator current control scheme a neutral point potential control method is proposed. The potential control method makes use of the possibility of the choice between complementary vectors. The predictive control, which was recently presented by the authors in [14,15], is an alternative to the DTC method [6] and in simulations shows similar performance. The control method takes the nonlinear nature of the inverter into account by computing the future stator current trajectories for possible inverter switching states and choosing a vector with a minimal performance index. The approach avoids the excitation of the filter resonance and makes very fast stator current control possible while maintaining low average switching frequency. The control method was verified by simulations of a 2.4 kV medium voltage three-level NPC inverter with an LC filter and an induction motor.

II. STATOR CURRENT CONTROL

A. System Description

The signal flow graph of the proposed controlling system for the inverter fed induction machine with *LC* filter is shown in Fig. 1.

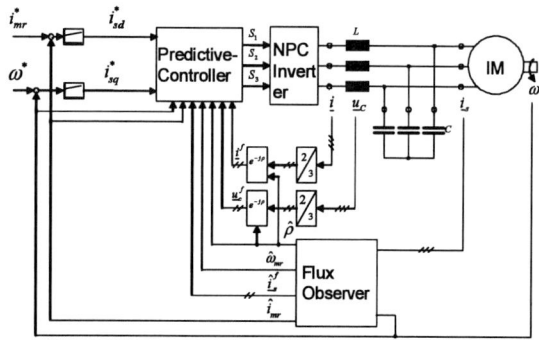

Fig. 1: Signal flow graph of the control structure

The predictive current controller receives its reference value for the complex stator current vector from the flux and speed PI controllers, and computes the switching functions for the inverter using the measured or estimated instantaneous state variables of the motor and filter. The controller makes use of the instantaneous values of the stator current, the choke current, the capacitor voltage and the rotational speed. Additionally, a flux observer provides instantaneous values for the magnetizing current, the angle and the angular velocity of the rotor flux. A three-level NPC inverter generates 19 different voltage space vectors (Fig. 2). The machine stator currents are the controlled variables. The controller predicts possible stator current trajectories which come with each of the voltage space vectors by solving a discrete-time state space model of the filter and motor. By means of a weighting function taking into consideration the deviation of the current trajectory from the stator reference current, a performance index is computed. For every sample interval, the switching state having the minimal performance index is computed, and is then generated by the inverter.

B. System Model

The dynamic behavior of an induction machine can be described in field coordinates by the following equations [16]:

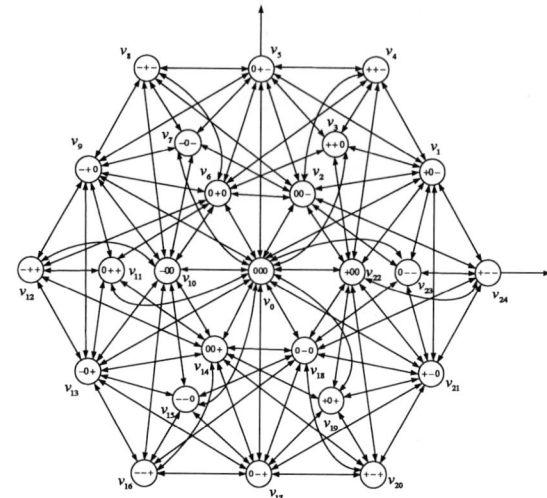

Fig. 2: Voltage space vectors of the three-level NPC inverter and possible transitions

$$u_{sd} = R_s i_{sd} + \sigma L_s \frac{di_{sd}}{dt} - \sigma L_s \omega_{mr} i_{sq} + (1-\sigma)L_s \frac{di_{mr}}{dt}, \quad (1)$$

$$u_{sq} = R_s i_{sq} + \sigma L_s \frac{di_{sq}}{dt} + \sigma L_s \omega_{mr} i_{sd} + (1-\sigma)L_s i_{mr} \omega_{mr}, \quad (2)$$

$$i_{sd} = T_r \frac{di_{mr}}{dt} + i_{mr}, \quad (3)$$

$$\omega_{mr} = \frac{i_{sq}}{T_r i_{mr}} + p\omega. \quad (4)$$

The nomenclature used is shown in Tab. I. The equation of motion is:

$$J\frac{d\omega}{dt} = \frac{3}{2}p(1-\sigma)L_s i_{mr} i_{sq} - m_l. \quad (5)$$

In the above equations an amplitude invariant space vector transformation is used:

$$x_\alpha + x_\beta = \frac{2}{3}\begin{bmatrix} 1 & e^{j\frac{2}{3}\pi} & e^{j\frac{4}{3}\pi} \end{bmatrix}\begin{bmatrix} x_1 \\ x_2 \\ x_3 \end{bmatrix}. \quad (6)$$

The differential equations of the *LC* filter inductor current space vector are:

$$\frac{di_d}{dt} = \frac{1}{L}u_d - \frac{1}{L}u_{cd} + \omega_{mr}i_q - \frac{R_l + R_c}{L}i_d + \frac{R_c}{L}i_{sd} \quad (7)$$

and

$$\frac{di_q}{dt} = \frac{1}{L}u_q - \frac{1}{L}u_{cq} - \omega_{mr}i_d - \frac{R_l + R_c}{L}i_q + \frac{R_c}{L}i_{sq}. \quad (8)$$

The capacitor voltages are:

$$\frac{du_{cd}}{dt} = \omega_{mr} u_{cq} + \frac{1}{C} i_d - \frac{1}{C} i_{sd} \qquad (9)$$

and

$$\frac{du_{cq}}{dt} = -\omega_{mr} u_{cd} + \frac{1}{C} i_q - \frac{1}{C} i_{sq} . \qquad (10)$$

Using the above equations a continuous state space system can be formed (11), which allows the computation of the stator current trajectories. The inputs of the model are the inverter output voltages u_d, u_q and the magnetizing current i_{mr}. The prediction is carried out based on the assumption that the motor speed ω, the rotor flux velocity ω_{mr} and magnetising current i_{mr} are constant during the prediction period.

$$\frac{d}{dt}\begin{bmatrix} i_{sd} \\ i_{sq} \\ i_d \\ i_q \\ u_{cd} \\ u_{cq} \end{bmatrix} = \begin{bmatrix} a_{11} & a_{12} & a_{13} & 0 & a_{15} & 0 \\ -a_{12} & a_{11} & 0 & a_{13} & 0 & a_{15} \\ a_{31} & 0 & a_{33} & a_{12} & a_{35} & 0 \\ 0 & a_{31} & -a_{12} & a_{33} & 0 & -a_{35} \\ a_{51} & 0 & -a_{51} & 0 & 0 & a_{12} \\ 0 & a_{51} & 0 & -a_{51} & -a_{12} & 0 \end{bmatrix}\begin{bmatrix} i_{sd} \\ i_{sq} \\ i_d \\ i_q \\ u_{cd} \\ u_{cq} \end{bmatrix}$$
$$+ \begin{bmatrix} 0 & 0 & b_{13} \\ 0 & 0 & b_{23} \\ b_{31} & 0 & 0 \\ 0 & b_{31} & 0 \\ 0 & 0 & 0 \\ 0 & 0 & 0 \end{bmatrix}\begin{bmatrix} u_d \\ u_q \\ i_{mr} \end{bmatrix} = \underline{A}\underline{x} + \underline{B}\underline{u} \qquad (11)$$

The matrix elements of the above state space model are:

$$a_{11} = -R_e/(\sigma L_s) - 1/(T_s) - (1-\sigma)/(\sigma T_r)$$
$$a_{12} = \omega_{mr}$$
$$a_{13} = -R_e/(\sigma L_s)$$
$$a_{15} = -1/(\sigma L_s)$$
$$a_{31} = -R_e/L$$
$$a_{33} = -(R_f + R_e)/L$$
$$a_{35} = -1/L$$
$$a_{51} = -1/C$$
$$b_{13} = (1-\sigma)/(\sigma T_r)$$
$$b_{23} = -(1-\sigma)/\sigma \, p\omega$$
$$b_{31} = -a_{35}$$

In order to reduce the computational demand, the current trajectories are predicted by means of a time discrete state space model:

$$\underline{x}(k+1) = \underline{\Phi}\underline{x}(k) + \underline{\Gamma}\underline{u}(k) . \qquad (12)$$

The discrete state space model is obtained through discretization of the state space model (11) using the following series:

$$\underline{\Phi}(T) = \sum_{i=0}^{\infty} \frac{\underline{A}^i T_{Step}^i}{i!} , \qquad (13)$$

$$\underline{\Gamma}(T) = \sum_{i=1}^{\infty} \frac{\underline{A}^{i-1} T_{Step}^i}{i!} \underline{B} . \qquad (14)$$

T_{Step} is the sample period of the discrete prediction model.

C. Predictive stator current controller

A three-level NPC inverter generates a total of 28 voltage space vectors, but only 19 vectors ($z = 1,...,19$) cause different voltages for the motor with filter and therefore a specific stator current trajectory. At the beginning of every sample interval $t(i) = T^*i$ ($i = 0,1,2,3,...$) the algorithm determines the set of permitted transitions according to the Fig. 2 and the actual generated vector by the inverter. A set of possible space vectors for the next sampling period results from the set of permitted transitions of the actual vector. Next, stator current trajectories for the set of possible space vectors are predicted by the solving of Eq. (12). Trajectories of the complementary vector pairs need to be computed only once, as they provide similar results. In this way the computational demand for prediction is reduced.

The measurement of the instantaneous values and the calculation of the control algorithm demand a finite time amount and create that way a dead time T_d. Therefore the measured instantaneous values at the end of the calculation period differ from the actual one. The difference is compensated using the Eq. (12-14) by prediction of the instantaneous values with prediction horizon equal to the dead time T_d.

The compensated instantaneous values of the stator current, the filter inductor current and the filter capacitor voltage provide the initial values for the trajectories prediction. Computation of possible current trajectories $i_{sd}(z,k)$ and $i_{sq}(z,k)$ is carried out for discrete time instants $t_{step}(k) = T_{step}^*k$ ($k = 0,1,2,3,...,k_{max}$). The parameter k_{max} defines the prediction horizon $T_{pred} = T_{step}^*k_{max}$ which has a significant effect on the controller performance. The optimal prediction horizon varies slightly depending on load torque and speed. However, good results were obtained in the entire operation range, when the optimal prediction horizon values valid for the rated operation point were used. Analysis of simulation results with equal parameters but varying values for the prediction horizon have shown that the optimal prediction horizon for a defined operation point is the global minimum of a top opened parabolic-like function. The operation point dependant functions are wide enough i.e. the magnitude of the gradient is not too high, so as ensure in that way robustness in respect of the optimal prediction horizon. It is also possible to control the optimal prediction horizon in order achieve the best performance and especially the lowest harmonics of motor torque in all operating points.

From the current trajectories the resulting trajectory errors

$$e(z,k+1)=\sqrt{\left(f_d\left(i_{sd}^*(k)-i_{sd}(z,k+1)\right)\right)^2+\left(i_{sq}^*(k)-i_{sq}(z,k+1)\right)^2} \quad (15)$$

are computed for the entire prediction horizon. The parameter f_d is a weighting factor and allows giving a higher control priority to the quadrature current in order to reduce torque harmonics. The average area between the error function and the abscissa for a time instant k results from:

$$e_A(z,k)=\frac{T_{step}}{2}\left(e(z,k)+e(z,k-1)\right)\cdot \quad (16)$$

The summation of the average areas yields the performance index for the vector z:

$$e_{A,tot}(z)=\sum_{k=0}^{k_{max}}e_A(z,k)\cdot \quad (17)$$

Finally, in the next sampling period the vector with the minimal performance index $e_{A,tot}(z)$ will be generated by the inverter.

III. NEUTRAL POINT POTENTIAL CONTROL

In this part a new neutral point (NP) potential control method for the predictive stator current control is presented. During transient operations increasing NP offsets can build up necessitating a fast compensation. The resulting NP potential errors lead to higher voltage stress on semiconductor devices and on capacitors. The errors can also increase or decrease the actual output voltages of the inverter. The circuit diagram in Fig. 3 shows a three-level NPC inverter comprising two dc-link capacitors C_1 and C_2.

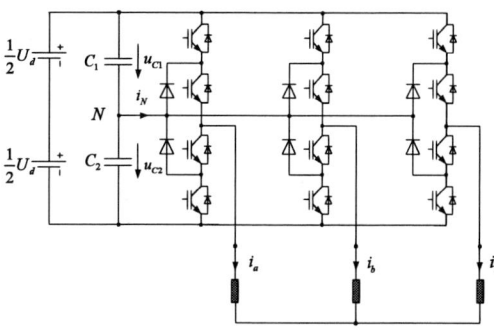

Fig. 3: Circuit of a three-level NPC inverter

The capacitors work as a voltage divider but their voltages u_{c1} and u_{c2} are not stabilized by external sources. Therefore the potential of the NP is floating in proportion to the integral of the NP current. The magnitude of the NP current and ints sign are determined by the switching state of the inverter and the load conditions.

A three-level inverter can create 27 switching states which are shown in Fig. 2. The notation (- 0 +) means that the phase a is linked to the negative dc rail, the phase b to

the NP and the phase c to the positive dc rail. The switching state vectors can be summarised to four sets.

1) Large vectors, like (- + +), link the phase terminals only with positive or negative dc rail. Therefore they do not cause an NP current and do not affect the NP potential.

2) Medium voltages, like (- 0 +), link one phase terminal to the NP. The resulting NP current affects the NP potential.

3) Small vectors make up redundant pairs. For example, an equal output voltage is generated by a positive small vector (+ + 0) and the corresponding negative small vector (0 0 -). For both vectors the resulting NP current has the same magnitude but opposite sign. In this way the NP potential increases or decreases.

4) The three zero vectors (- - -), (0 0 0) and (+ + +) do not create an NP current and do not affect the NP potential.

The above discussion shows that the small and medium switching state vectors cause imbalances of the NP potential which can be described by a NP potential error:

$$\Delta u_N = u_{c1} - u_{c2}\cdot \quad (18)$$

For that reason an NP potential has to be controlled during the operation of the stator current controller. This is done by appropriate selection of complementary small vectors. The algorithm makes use of the fact that the complementary voltage vectors generate for the machine equal voltages, but influence the neutral point in opposite way. A positive NP current generally increases the NP potential error, whereas a negative one results in a reduction of the error. The actual NP current can be described as a linear combination of the three phase currents i_a, i_b and i_c. The switching state of the inverter determines how the phase currents contribute to the NP current. The selection of the voltage vectors by the predictive stator control can be easily adjusted, which simplifies the integration of the NP control. However, a higher priority is given to the current controller. This means, that the predictive controller first selects according to the performance function (17) an optimal vector from the set of the possible transitions which belongs to the actual generated voltage. Subsequently, if the predictive controller has selected a small vector, the NP controller checks whether the appropriate complement vector is contained in the set of possible transitions of this selected vector. If this is the case, the NP potential will be regulated by a calculated selection of one of the both complementary vectors. Otherwise, when no complementary vector is available, the controller will select the optimal vector, even if the vector will increase the asymmetry of the NP potential. For that, by the selection of the small vectors the NP controller makes use of the fact, that complementary vectors create an NP current with the same magnitude but of opposite sign. Furthermore, when small vectors are used, the magnitude of the NP current depends always only on a single phase current. For

572

example, the vectors pairs V_{10}, V_{11} and V_{22}, V_{23} create an NP current which depends on the phase current i_a. On the other hand, the NP current depends on the phase current i_b when vector pairs V_6, V_7 and V_{18}, V_{19} are used. The remaining four small vectors generate an NP current, which depends on the phase current i_c. Furthermore, the outside vectors of the inner hexagon (V_3, V_7, V_{11}, V_{15}, V_{19} and V_{23}) create NP currents of the same sign as the respective phase currents, whereas the complementary inside vectors create NP currents with the opposite sign to the phase currents. The vectors V_{11} and V_{23} for example generate an NP current m$i_N = i_a$ and vectors V_{10} and V_{22} cause $i_N = -i_a$.

The proposed NP potential control method takes the above discussed features into consideration and has therefore the following logical algorithm. If the condition

$$\Delta u_N * i_a > 0 \quad \text{or} \quad \Delta u_N * i_b > 0 \quad \text{or} \quad \Delta u_N * i_c > 0 \qquad (19)$$

is valid, then the NP potential controller gives preference to the inside vectors of the inside hexagon, whereas for

$$\Delta u_N * i_a < 0 \quad \text{or} \quad \Delta u_N * i_b < 0 \quad \text{or} \quad \Delta u_N * i_c < 0 \qquad (20)$$

priority is given to the outside vectors of the inside hexagon. The simulation results of the NP potential control are shown in the next section.

IV. SIMULATION RESULTS

Simulation results for a 2.4 kV medium voltage drive system with an *LC* filter were obtained using MATLAB/Simulink software and the PLECS toolbox. The a.c. network, the 12-pulse diodes rectifier, the dc-link, the three-level NPC inverter and the *LC* filter were modelled using the PLECS toolbox. The toolbox models the power semiconductors as ideal switches in order to speed up the simulation. The induction motor and the time time-discrete controller were modelled in MATLAB/Simulink. The parameters of the simulated model are given in the appendix in Table II.

A sampling frequency of $f = 1/T = 6$ kHz was used and the prediction horizon T_{pred} chosen to 900 μs. The dead time was $T_d = 1/(2T)$. Fig. 4 and Fig. 5 shows the step response of the current controller to a step command in speed at $t = 2$ s, when the motor was operating at 20 % of rated speed and 10 % rated torque with the rated magnetising current. The quadrature current reaches its maximal value in 5 ms from the step command. This demonstrates the dynamic performance of the current controller. The stator current is depicted in Fig. 5 and shows no excitation of the filter after the step command was applied. The dc-link capacitor voltages and the NP potential error are presented in Fig. 6 respectively in Fig. 7.

The drive has been simulated at rated frequency, rated flux and rated torque and the results are presented in Fig. 8 to Fig. 11.

The average switching frequency by operation at nominal values is about $f_s = 500$ Hz and lies near the resonance frequency of 460 Hz. A further reduction of the average switching frequency is possible by reduction of the controller's sampling frequency. However, with the reduction of the sampling frequency comes a degradation of the stationary performance, especially an increasing of the THD values.

V. CONCLUSIONS

The simulation results presented show that the predictive stator current controller combined with the proposed NP control make possible a dynamic operation of drive systems with an *LC* output filter while using a low average switching frequency. The filter resonance is not excited, though a fast stator control is achieved. At nominal values of the drive the average switching frequency is about $f_s = 500$ Hz and lies near the resonance frequency. The proper operation principle of the controller was verified by simulation of a 2.4 kV medium voltage drive.

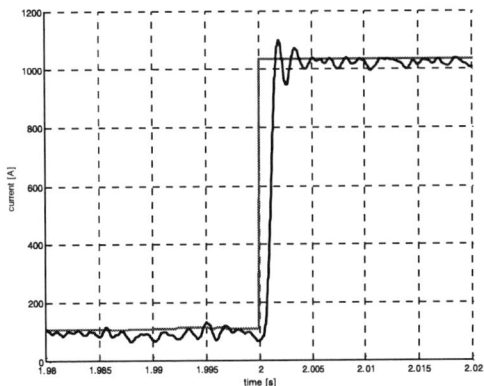

Fig. 4: Quadrature current i_{sq} (reference speed step, simulation)

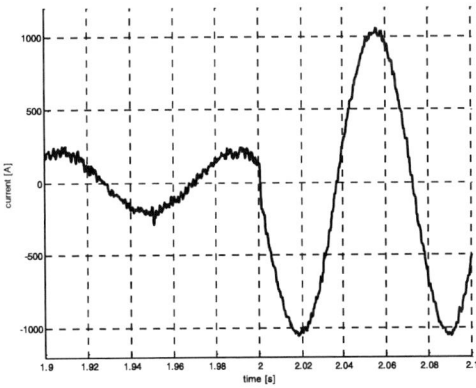

Fig. 5: Stator current (reference speed step, simulation)

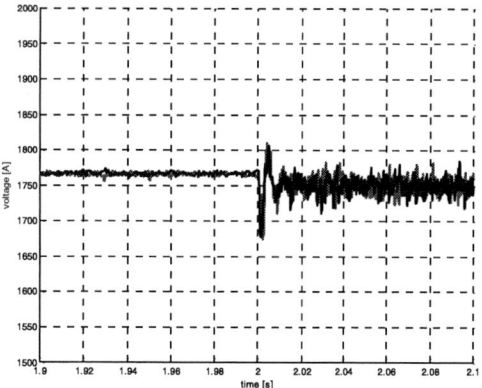

Fig. 6: Dc-link capacitor voltages u_{c1} (red) and u_{c2} (black) (reference speed step, simulation)

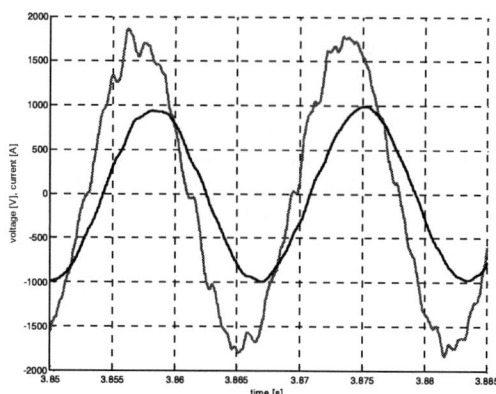

Fig. 9: Motor phase voltage (red) and stator current (black) (simulation)

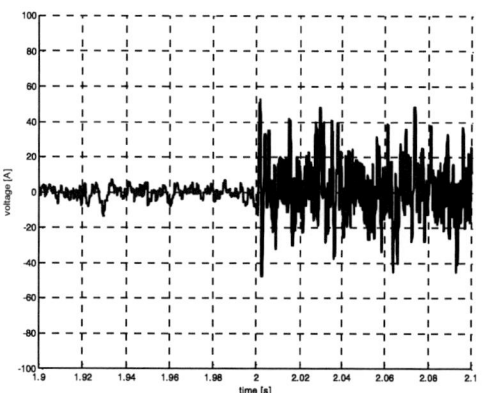

Fig. 7: Neutral Point potential error Δu_N (reference speed step, simulation)

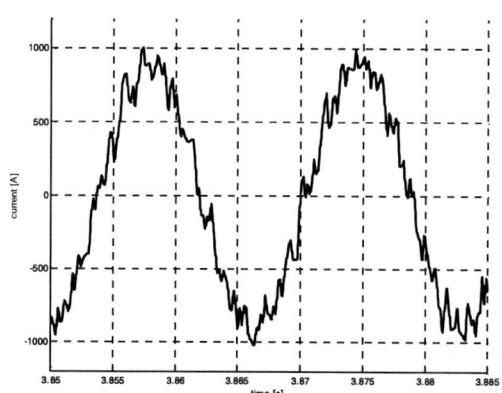

Fig. 10: Filter choke current (simulation)

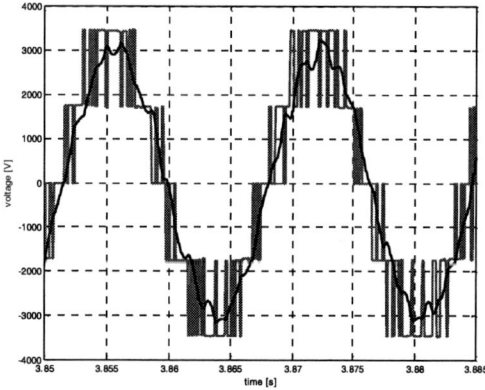

Fig. 8: Inverter (red) and motor (black) line-to-line voltages (simulation)

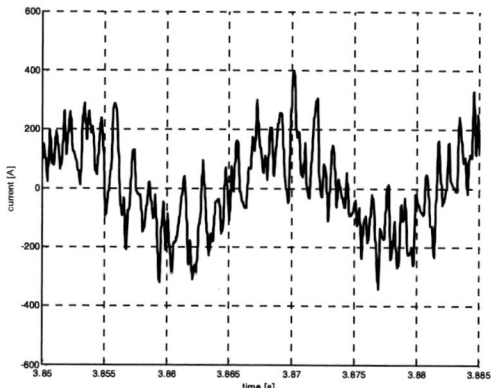

Fig. 11: Filter capacitor current (simulation)

APPENDIX

The used nomenclature and the parameters of the simulated drive systems are listed in Table I and Table II.

TABLE I
NOMENCLATURE

R_s, R_r	stator and rotor resistance
L_s, L_r	stator and rotor self inductance
L_h	magnetizing inductance
T_s	stator time constant, $T_s = L_s/R_s$
T_r	rotor time constant, $T_r = L_r/R_r$
σ	leakage coefficient, $\sigma = 1 - L_h^2/(L_s L_r)$
ω_{mr}	angular velocity of the rotor field
ω	rotational speed
p	number of pole pairs
J	total moment of inertia
m_l	load torque
R_l, R_c	filter inductor and filter capacitance resistance
L, C	filter inductance and capacitance

TABLE II
PARAMETERS OF THE 2.4 KV DRIVE

U	2.4 kV
I	600 A
n	1190 min^{-1}
f	60 Hz

REFERENCES

[1] A. Nabae, I. Takahashi, H, Akagi, "A New Neutral Point Clamped PWM Inverter", IEEE Trans. Ind. Appl., vol. IA-17, pp. 518-522, 1981.

[2] A. Mertens, M. Bruckmann, R. Sommer, "Medium Voltage Inverter Using High-Voltage IGBTs", EPE 1999, Lausanne, Switzerland.

[3] R. Sommer, A. Mertens, M. Griggs, H.-J. Conraths, M. Bruckmann, T. Greif, "New Medium Voltage Drive Systems Using Three-Level Neutral Point Clamped Inverter With High Voltage IGBT", IEEE IAS 1999, Phoenix, AZ.

[4] P.N. Enjeti, R. Jakkli, " Optimal Power Control Strategies for Neutral Point Clamped (NPC) Inverter Topology", IEEE Trans. Ind. Appl., vol. 28, no. 3, pp. 558-566, 1992.

[5] J. Holtz, "Pulsewidth Modulation for Electronic Power Converters", IEEE Trans. Ind. Appl., vol. 31, no. 5, pp. 1110-1120, 1995.

[6] A. Sapin, P.K. Steimer, J.-J. Simond, "Modeling, Simulation and Test of a Three-Level Voltage-Source Inverter With Output LC Filter and Direct Torque Control", IEEE Trans. Ind. Appl., vol. 43, no. 2, pp. 469-475, 2007.

[7] H.-J. Conraths, F. Giessler, H.-D. Heining, "Shaft Voltages and Bearing Currents – New Phenomena in Inverter Driven Induction Machines", EPE 1999, Lausanne, Switzerland.

[8] B.P. Schmitt, R. Sommer, "Retrofit of Fixed Speed Induction Motors With Medium Voltage Drive Converters Using NPC Three-Level Inverter High-Voltage IGBT Based Topology", IEEE ISIE 2001, Pusan, South Korea.

[9] J.K. Steinke, "Use of an LC Filter to Achieve a Motor-friendly Performance of the PWM Voltage Source Inverter", IEEE Trans. Energy Conv., vol. 14, no. 3, pp. 649-654, 1999.

[10] S. Pöhler, A. Mertens, R. Sommer, "Optimisation of Output Filters for Inverter Fed Drives", IEEE IECON 2006, Paris, France.

[11] J. Holtz, B. Beyer, "Fast Current Trajectory Control Based on Synchronous Optimal Pulsewidth Modulation", IEEE Trans. Ind. Appl., vol. 31, no. 5, pp. 1110-1120, 1995.

[12] J. Holtz, N. Oikonomou, "Neutral Point Balancing Algorithm at Low Modulation Index for Three-Level Inverter Medium-Voltage Drives", IEEE Trans. Ind. Appl., vol. 43, no. 3, pp. 761-768, 2007.

[13] T. Laczynski, T. Werner, A. Mertens, "Active Damping of LC-Filters for High Power Drives using Synchronous Pulsewidth Modulation", IEEE PESC 2008, Rhodes, Greece.

[14] T. Laczynski, A. Mertens, "New Current Controller for Inverter Fed Medium Voltage Drives with LC Filter", IEEE ICPE 2007, Daegu, South Korea.

[15] T. Laczynski, A. Mertens, "Predictive Current Controller for Inverter Fed Medium Voltage Drives with LC Filter", IEEE PEDS 2007, Bangkok, Thailand.

[16] W. Leonhard, "Control of Electrical Drives", Springer-Verlag Berlin, 1996.

Research on Dimming Control Method of Electronic Ballast for the Automotive HID Headlight

P. Dong *, K.W.E.Cheng [†] and S.L.Ho[††]

The Hong Kong Polytechnic University, Department of Electric Engineering, Hung Hum, Hong Kong, China, e-mail: *
05900654r@polyu.edu.hk [†] *eeecheng@polyu.edu.hk* [††] *eeslho@polyu.edu.hk*

Abstract—It is the purpose of this paper to explore the problem of dimming the HID lamp used in the automotive headlight system. To investigate the relationship between the lamp power and control parameters, a power-dependent HID lamp model is established in this paper. The two methods, dynamic duty-ratio control and dynamic bus voltage control, are analyze and applied in the system. The dynamic bus voltage control which can be realized according to the structure of controller is stressed in this paper. The simulation results and experimental results give the difference between the different dimming control methods.

Keywords—Car Xenon headlamp, dimming control, dynamic duty-ratio, dynamic bus voltage

I. INTRODUCTION

Dimmable control for light is a necessary method for energy saving. The dimming control for incandescent lamp, fluorescent has been developed for a long time. [1-3]. Many different techniques have been used in dimming control and they can be divided into three different modes. For the lamp that needs to use a power converter to operate, the control parameter can be classified variable switching frequency control, dynamic duty-ratio control and dynamic bus voltage control. The variable switching frequency control is popularly used in practical implementation because of its simplicity [4-8]. When the switching frequency is located further away from the resonance of the tank, the resonance energy stored in the tank is less and hence less energy is coupled to the lamps, and the lamps dim. There is also an inherent unstable region exists due to the interaction between the negative impedance of the lamps and the output characteristics of the electronic ballast with frequency control. Controlling the duty cycle of the switches can control the load power. Actually, with the asymmetrical duty ratio control, a small DC-biased lamp current is resettled [9-11]. But in half-bridge inverter, the maximal duty-ratio is 0.5. Using dynamic duty-ratio control will change its operation from ZVS to ZCS, if the duty-ratio is small.

Today, HID lamps are being fitted in both high beam and low beam. The low-beam position may only need lower intensity. A dimmer should be proposed for this purpose. In addition, the other motivating factor for this work is to include some flexibility in the control of the

HID lamp to suit the various driving environments in the modern cities and country roads. Many methods have been used in dimming control. There are mainly three different modes: variable switching frequency control, dynamic duty-ratio control and dynamic bus voltage control. The dimming control happened in steady state normally and the equivalent circuit of the system is shown in Fig. 1. The converter is a DC-DC converter and a flyback converter is used. It works in a fixed high frequency mode (usually equal to 100 kHz), and the full-bridge inverter works in a fixed low frequency mode. The inductor L_p is the primary inductance of the transformer, i_0 and v_C is the output current and voltage of the flyback converter, respectively. The resonant circuit is not implemented and it is feasible to use the dynamic duty-ratio control and dynamic bus voltage control in the dimming control of automotive system.

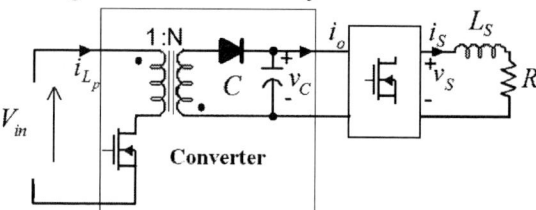

Fig. 1 Circuit Schematic of the electronic ballast in the steady state

It is the purpose of this paper to explore the problem of dimming the HID lamp used in the automotive headlight system. To investigate the relationship between the lamp power and control parameters, a power-dependent HID lamp model is established. The two methods, dynamic duty-ratio control and dynamic bus voltage control, are analyzed and applied in the system.

II. MODELING OF HID LAMPS UNDER LAMP POWER CONTROL

The physical characteristic of this type HID lamp in dimming condition is the same with [12]. An mean model is now used for the analysis in this section. The expression is an extension of the lamp mean power concept as in Ref [10].

$$V_{lamp} = cV_{lamp}^2 R_{lamp} + d$$

$$R(P_{lamp}) = \frac{V_{lamp}^2}{P_{lamp}} \tag{1}$$

where V_{lamp} is the lamp voltage, P_{lamp} is the lamp power and R_{lamp} is the equivalent lamp resistance. The values of c and d could be obtained by linear regression using the measurement. Fig.2 shows the simulation result and the experiment result using this model.

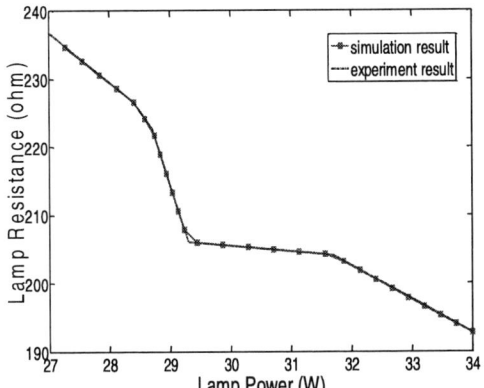

Fig. 2 Agreement of the simulation and experimental results of lamp Resistance under dimming condition

III. DYNAMIC DUTY-RATIO CONTROL FIR DIMMING

Power tracking control is now implemented using the duty ratio method. The power is regulated through the control method for the duty ratio.

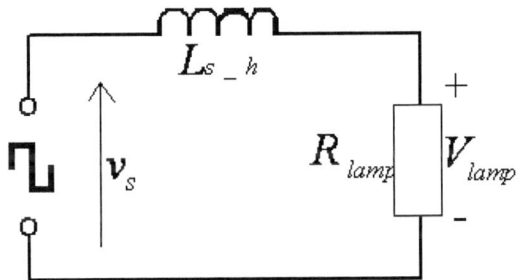

Fig. 3 Equivalent circuit of the inverter bridge for the lamp circuit

The source voltage is provided by an inverter that is a H-bridge or half-bridge configuration. The DC link voltage of the inverter is V_c and hence the switching waveforms output form the inverter is $\pm V_C$ that is to excite the lamp. The switching frequency of the inverter is ω_1

The duty ratio of the transistor is fixed. The other parameters used are: the inductor L_{S_h} is the secondary inductor of high voltage transformer used in the electronic ballast. R_{lamp} is the equivalent resistance of the lamp.

The duty ratio that gives the lamp power P_{lamp} is determined by

$$d = \cfrac{1}{1 + \cfrac{NV_{battery}}{V_{lamp}} \sqrt{\cfrac{\pi}{2} \left| \cfrac{R_{lamp}}{Z_s + R_{lamp}} \right|}} \tag{2}$$

Fig. 4 shows the calculated P_{lamp}-d curves with different Z_S values, and with $V_{battery}$=12V, turns ratio of main transformer of flyback converter is 5.5.

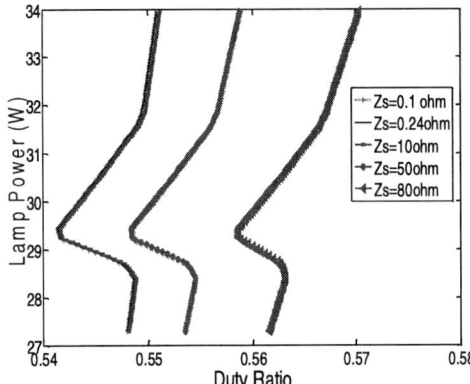

Fig. 4 Lamp Power versus duty ratio under various Z_s

Fig 4 shows the characteristics of the lamp power with duty ratio and Inductive impedance Z_s. Z_s varies with the duty ratio under given lap power. The lamp voltage and current does not vary with inductance.

The constant power is used in the dimming condition. In addition, the duty ratio varies with the dimmed power. So, the varied duty ratio dimming methods are dependent on the constant power control methods.

IV. DYNAMIC BUS VOLTAGE CONTROL FOR DIMMING

The feedback values from the sensor of the lamp current and voltage was used in the dynamic bus voltage control for the adjusting the HID lamp input voltage.

Two methods used in the constant power control circuit, namely alternative Adder method (used in UCC2305 chip) and a multiplier (used in the microprocessor or DSP chip) and are displayed in Figs. 6(a) and 6(b). The voltage gain k_v and current gain k_i are used in the following analysis. Normally, the peak current comparison method is used to obtain the switching signal. The duty ratio is adjusted by the comparison of the power control signal v_{level} with the power reference signal v_{ref}. The power reference signal v_{ref} is a constant level that is used to control the output power. If v_{ref} is fixed to a fixed power, the current sensor and voltage sensor is fixed. Different power levels are obtained by changing the current sensor and voltage sensor to get the variable bus voltages.

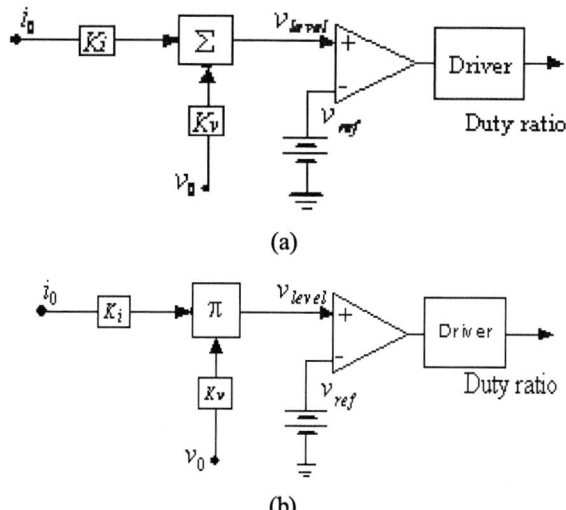

(a)

(b)

Fig.6 (a) Adder Method to Obtain Constant (b) Power Multiplier Method to Obtain Constant Power

A. Adder Method

Fig. 6(a) implied that the power control signal v_{level} is now developed into

$$v_{level} = k_i \cdot i_0 + k_v \cdot v_0 \qquad (7)$$

Peak current modulation method defines the peak current of the lamp to be:

$$i_{0p}(v_0) = \frac{1}{k_i}(v_{ref} - k_v v_0) \qquad (8)$$

The output power of the flyback is the lamp power, under low circuit loss condition is:

$$P_0 = V_{in} N i_{0p} v_0 \cdot (1 - N \frac{i_{0p}}{v_0} L_p N f_s) \qquad (9)$$

The lamp power is examined in Fig 7 and 8 with the variation of k_i and k_v. The condition is based on a 35HID lamp that operated between 60V to 110V, source voltage 12V and power reference signal $v_{ref} = 2.5$.

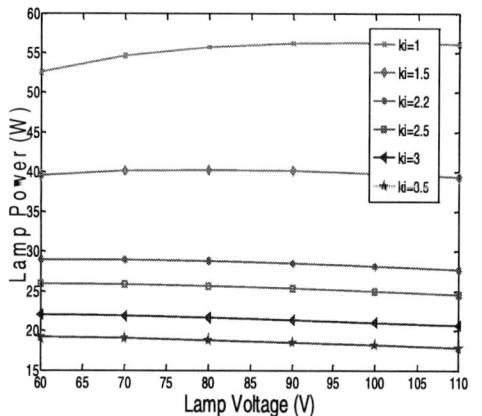

Fig. 7 Lamp power versus current gain ($k_v = 0.00485$)

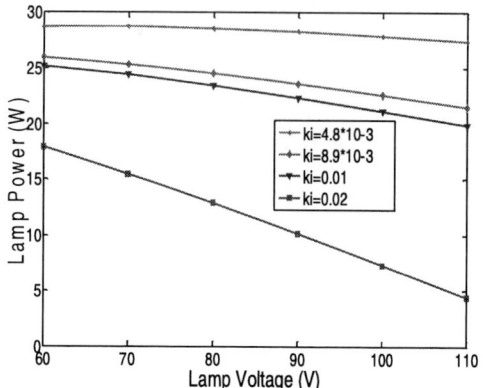

Fig. 8 Lamp power versus current gain ($k_i = 2.224$)

There are two points that can be concluded from Figs. 7 and 8, that is: (1) the lamp output power can be changed by varying the current gain or voltage gain. (2) In dimming operation, the current gain is used to change the lamp power, because it can keep the lamp power constant at large lamp voltage scope.

B. Multiplier Method

The power control signal v_{level} can expressed that based on Fig 6b as

$$v_{level} = k_i i_0 \times k_v v_0 \qquad (10)$$

This can be simplified as:

$$\frac{v_{level}}{v_0} = k_i i_0 k_v \qquad (11)$$

The peak current modulation method allows the peak current of the lamp written as:

$$i_{0p} = \frac{1}{k_v k_i} \frac{v_{ref}}{v_0} \qquad (11)$$

The power reference signal $v_{ref} = 2.5$ V, input battery voltage is 12V. The relationship between the current gain, voltage gain and power are shown in Fig. 9 and Fig.10.

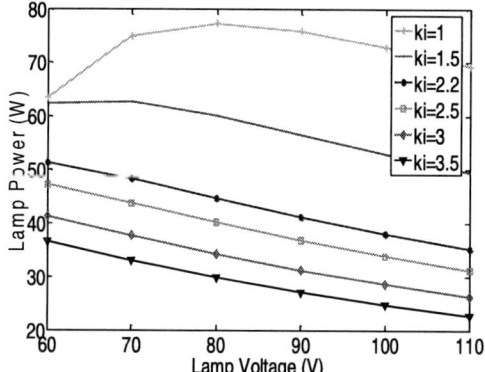

Fig. 9 Lamp power versus current gain ($k_v = 0.008924$)

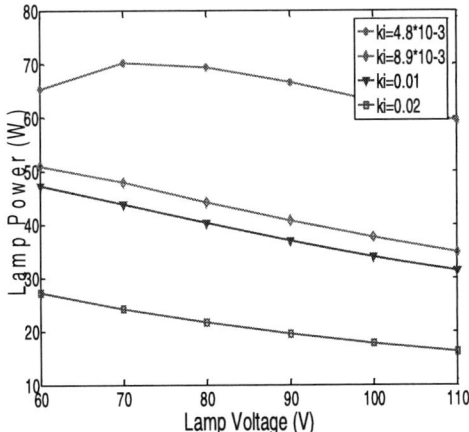

Fig. 10 Lamp Power versus current gain ($k_i = 2.224$)

Figs. 9 and 10 show lamp power with the k_v and k_i respectively. The variable current gain or voltage gain would change the lamp power and the current gain is recommended due to small variable power scope. In addition, comparing the two methods, Adder method and multiplier method, it could be concluded that: a) If the current gain is same ($k_i = 2.224$), larger voltage gain is needed in the multiplier method. b) A larger variable scope happened in the multiplier method when the lamp voltage is changed from 60V to 110V, so the dimming power scope is larger in using Adder method than the multiplier method. This conclusion is verified by the experimental result. Using Adder method to get constant power control, 34% power of 35W-HID lamp can be dimmed. 23% power of 35W-HID lamp can be dimmed by using the multiplier method.

V. EXPERIMENT RESULTS

A. Power tracking duty ratio dimming method

Fig. 11 gives the lamp current and lamp voltage waveform under dimming control whose output power is 28 W. Using the power tracking dimming control method, the output power varies with the lamp current obviously and the output light will reduce evidently and the energy is saved. The lamp voltage changes in a narrow range under the conditioning of dimming,

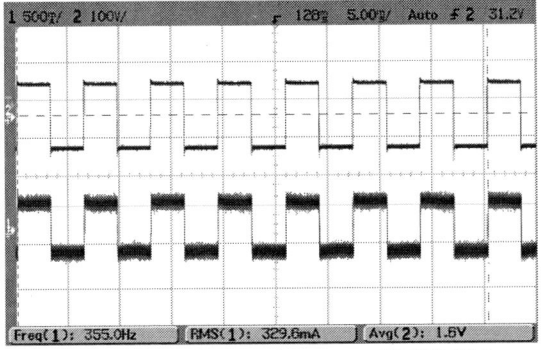

Fig.11 Lamp voltage and current at 29.2W (Voltage: 85.56V, Current: 0.329A) (ch1: lamp current, 500mA/div, 2ms/div; ch2: lamp voltage, 100V/div, 2ms/div

B. Dynamic bus dimming method

In this controller, the Adder method is used, so a larger dimming scope is obtained. Fig. 12 shows the lamp current and lamp voltage under 28W output power.

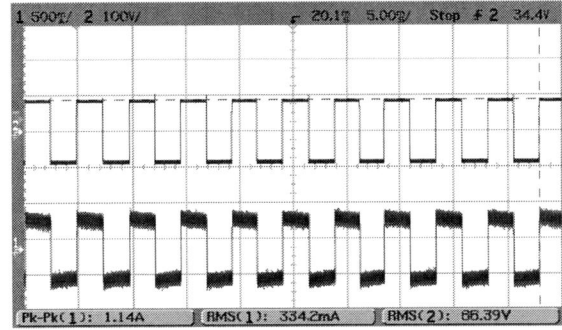

Fig.12 Lamp voltage and current at 28W (Voltage: 86.39V, Current: 0.334A) (ch1: lamp current, 500mA/div, 2ms/div; ch2: lamp voltage, 100V/div, 2ms/div)

VI. CONCLUSION

The dynamic bus voltage method is feasible in dimming control used in automotive headlight system. According to the simulation results, the change of current gain is better than the voltage gain in dimming control. Because the lamp voltage can varied in a large scope but the lamp power is nearly constant. The Adder method is better than the multiplier method and the reasons are the Adder method is easy to realize. Dynamic duty ratio dimming method and dynamic bus voltage dimming method are described. Their both mathematical analysis, simulation and experiment results are conducted. These two methods could be used in the dimming control. Contrasting two methods, dynamic bus voltage dimming method and dynamic duty ratio dimming method, a larger dimming scope could be realized in dynamic bus dimming method.

ACKNOWLEDGMENT

The authors gratefully acknowledge the support of the Hong Kong Polytechnic University. This work was also supported by ITF under the project GHS/073/04.

REFERENCES

[1] M.-S. Lin, M.-C. Lee, 13.Y. Chen and W.-S. Feng, Synchronous dimming control for a cold- cathode fluorescent lamp driver, Electronics Letters 20th June 1996 Vol. 32 No. 13. pp: 1151-1153.

[2] Chin S. Moo Hung L. Cheng Tsai F. Lin, Designing a Dimmable Electronic Ballast with Voltage Control for Fluorescent Lamp, ISIE'99 - Bled, Slovenia, pp: 786-791.

[3] C. S. Moo, Y. C. Chuang, Y. H. Huang, and H. N. Chen, Modeling of fluorescent lamps for dimmable electronic ballasts, IEEE Industry Applications Society IAS Annual Meeting 1996, pp: 2231-2236.

[4] J. Rozenboom. The electronic ballast circuit and low pressure lamps. Int. Journal Electronics, Vol. 82, No.3, 1997, pp. 269-294.

[5] J. Ribas, J. M. Alonso, E. L. Corominas, et al. Design Considerations for Optimum Ignition and Dimming of Fluorescent Lamps Using a Resonant Inverter Operating Open Loop. Conference Record of IEEE-IAS 1998, pp: 2068-2075.

[6] C. S. Moo, H. L. Chung, et al. Design Dimmable Electronic Ballast with frequency control. Conference Record of IEEE-APEC 1999, pp: 727-733.

[7] T. F. Wu, T. H. Yu, et al. Single-stage Electronic ballast with dimming feature and Unity Power Factor. IEEE Trans. On Power Electronics. Vol.13, No.3, May 1998. pp: 586-597.

[8] T. J. Ribarich and J. J. Ribarich. A new Control Method for Dimmable High frequency Electronic Ballasts. Conference Record of IEEE-IAS 1998, pp: 2038-2043.

[9] J. H. Reijnaerts. Circuit Arrangement for Reducing Striations in a low-pressure Mercury Discharge Lamp. US Patent No. 5,369,339, Nov. 29, 1994.

[10] C. R. Sullivan. Control System for Proving Power to a Gas Charge Lamp. US Patent No. 5,864,212, Jan. 26, 1999.

[11] R.L. Steigerward and L. D. Stevanovic. Elimination of Striations in Fluorescent Lamps driven by High-frequency Ballast. US Patent No. 5,701,059, Dec. 23, 1997.

[12] Y. Wei, S. Y. Hui. An analysis into the dimming control and characteristic of Discharge Lamps. IEEE Trans. On Power Electronics, Vol. 20, No.6, Nov. 2005, pp: 1432-1440.

[13] Ho, Y.K.E., Lee, S.T.S, Chung, H.S.H., and Hui, S.Y.: A comparative study on dimming control methods for electronic ballasts. IEEE Trans. Power Electronic, 2001, 16, (6), pp: 828-836.

Control Method for a Three-Port Interface Converter Using an Indirect Matrix Converter with an Active Snubber Circuit

Koji Kato*, Jun-ichi Itoh*

*Nagaoka University of Technology, Nagaoka City Niigata, Japan, e-mail: *itoh@vos.nagaokaut.ac.jp*

Abstract— This paper proposes a novel control method for an interface converter using an indirect matrix converter with three power supplies, which consist of two AC and one DC supplies. An active snubber circuit in the indirect matrix converter is used as an interface with the DC power supply. The proposed control method is based on the indirect control method with a triangular carrier wave [10]. Therefore the proposed method can be easily applied to a DC/DC converter. In addition, this paper proposes reducing the number of switching during one carrier period for the inverter stage converter. In the proposed method, the inverter stage carrier is controlled asymmetry to reduce the switching times in the inverter stage. To apply the proposed method, the switching loss of the inverter stage can be decreased to 2/3 times in comparison with a conventional method. The validity of the proposed method is confirmed by simulation and experimental results.

Keywords—Matrix converter, Converter control, Power factor correction, Energy converters for HEV.

I. INTRODUCTION

Recently, renewable energies and hybrid electric vehicle (HEV) systems are receiving significant interest, with consideration of global warming and environmental problems. There are two types of power sources for renewable energies; AC power sources such as wind power generator, and DC power sources such a photovoltaic cells, batteries, and fuel cells. Therefore, to implement these applications, interface power converters have been intensely studied. A conventional power converter system, which consists of a pulse width modulation (PWM) rectifier, a DC/DC converter and an inverter, requires a large energy buffer, such as an electrolytic capacitor. However, the electrolytic capacitor in a conventional system has disadvantages such as large size, short-life time and high costs.

On the other hand, there is an AC/AC direct converter with a DC link, which is referred to as an "indirect matrix converter" [1-5]. The indirect matrix converter is composed by the current source rectifier and voltage source inverter without energy buffer in DC link part. The utilization of an indirect matrix converter, which has no large energy buffer such as an electrolytic capacitor, can bring advantages such as size reduction, long-life time and cost reduction. We have already proposed an interface converter using an indirect matrix converter [6]. The proposed interface converter connects a DC/DC converter to DC link of the indirect matrix converter. With regard to the voltage relationship between the battery and the DC

link, the proposed circuit works as a boost up converter for the battery. However, the proposed converter has the same problem due to an indirect matrix converter. That is, the voltage transfer ratio of indirect matrix converters, which defines the ratio between the output and input voltage, is well known as being constrained to 0.866. Improvement of the voltage transfer ratio for indirect matrix converters has been widely discussed [7,8]. However, it seems that the proposed methods increase the number of components, due to the insertion of voltage transfer ratio compensator, which consists of an H-bridge inverter and a capacitor, to the DC link of the indirect matrix converter.

Meanwhile, a snubber circuit is required in order to protect the indirect matrix converter from over-voltage and over-current. In Ref. [9], a snubber circuit was used in order to compensate the imbalance in the supply voltage. However there was no discussion of the DC power supply interface using a snubber circuit.

This paper proposes a novel control method for multi-power supply interface system using an indirect matrix converter with a step-down chopper. The proposed converter is constructed based on an indirect matrix converter that does not have a large energy buffer, such as an electrolytic capacitor. The active snubber circuit of the indirect matrix converter is used as a step-down chopper for a DC power supply. In addition, the voltage transfer ratio can be improved by the DC power supply. An indirect control method with a triangular carrier wave [10] is expanded into the proposed control method. In the proposed control system, a DC/DC converter can be easily added to the indirect matrix converter.

In addition, this paper also proposes a control method reducing the number of switching for the inverter stage converter. In the proposed converter which is composed by a current source rectifier, a voltage source inverter and a step-down chopper, the switching frequency of the inverter stage increases in comparison with rectifier stage. Because the zero current switching operation of the rectifier stage can be achieved when the inverter output zero voltage vectors, which are generated by every upper or lower peak of the inverter carrier. Therefore, in the proposed method, the switching frequency is reduced by using a transformed asymmetry inverter carrier in order to control the zero voltage vector timing.

The basic operation of the proposed method is confirmed by the simulation and experimental results; the total harmonic distortion (THD) of the input, output and dc output currents are 7.4%, 4.8%, 1.9%, respectively, and the input power factor is 99% and efficiency is 93.8%. In addition, bidirectional energy flow characteristic among the power supplies in this circuit is confirmed.

II. CIRCUIT TOPOLOGY

Figure1 shows AC and DC power supply interface systems. A conventional interface system consists of a PWM rectifier, a DC/DC converter and an inverter, as shown in Fig. 1(a). This system requires a large electrolytic capacitor in DC link part in order to smooth DC link voltage. This system is very flexible in term of voltage condition among the AC input AC output side and DC power source because of using voltage type converters. However, the problem for an electrolytic capacitor in DC link part is large volume, short lifetime in high temperature and high cost.

Figure 1(b) shows the proposed AC and DC power

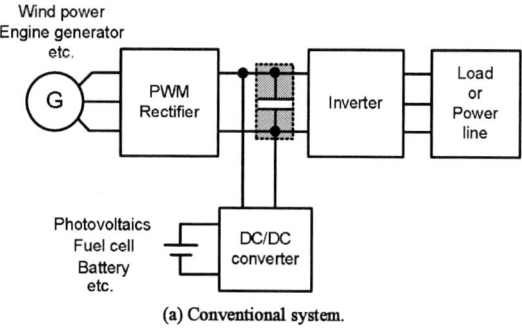

(a) Conventional system.

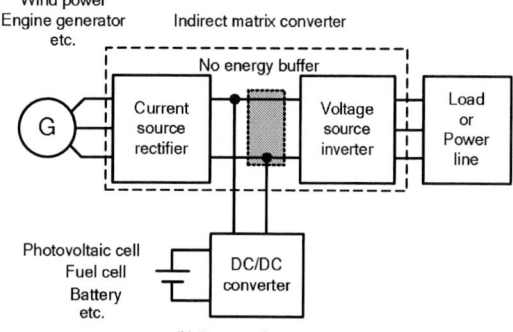

(b) Proposed system.

Fig. 1. Block diagrams of AC and DC power supply interface system.

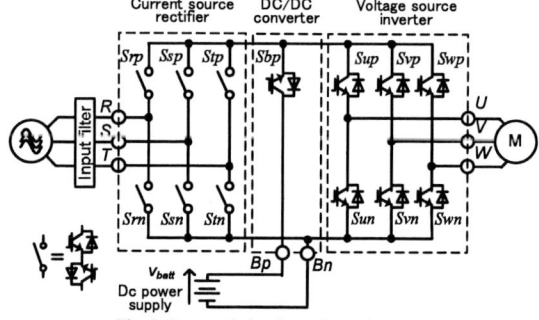

Fig. 2. Proposed circuit configuration.

supply direct interface converters for the energy management system. The proposed interface converter is constructed based on an indirect matrix converter without a large energy buffer such as an electrolytic capacitor. The DC/DC converter connects to DC link of the indirect matrix converter.

Figure 2 shows the specific main circuit of the proposed interface converter. The DC/DC converter is built as an active snubber circuit in the indirect matrix converter. An IGBT is connected anti-parallel to the snubber circuit diode. This snubber circuit with the IGBT is used as a step-down chopper of the DC power supply. In this case, the DC power supply voltage of the snubber circuit is higher than the peak of the AC input line voltage, because a rush current occurs between the AC and the DC input power supplies when the peak of the AC input line voltage is higher than the DC power supply voltage. Therefore, this converter is referred to as a "step-down type AC/DC/AC direct converter".

Figure 3 shows the equivalent circuits of the proposed converter. The rectifier stage converter is similar to a four phase current source rectifier including the DC/DC converter. Thus, the switches in the three-phase PWM rectifier and DC/DC converter must be separately turned on in order to avoid a short circuit of the AC and DC power supplies. When the DC/DC converter switch S_{bp} is turned off, the proposed converter operates as a conventional indirect matrix converter. In this case, the DC/DC converter becomes a conventional snubber circuit, as shown in Fig. 3(a). On the other hand, the proposed converter operates as a conventional inverter when the DC/DC converter switch S_{bp} is turned on, when all switches in the rectifier are turned off. In this case, the DC/DC converter is similar to a DC power supply, as shown in Fig. 3(b). That is, the proposed circuit operates as an indirect matrix converter or inverter alternately. In

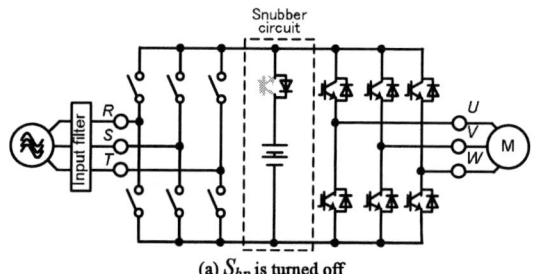

(a) S_{bp} is turned off

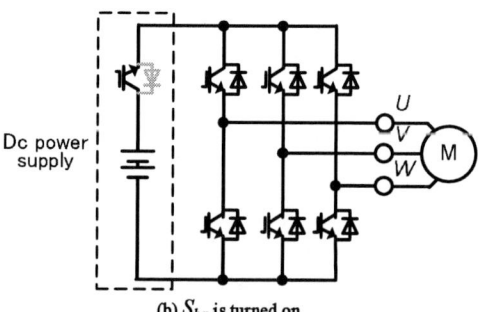

(b) S_{bp} is turned on

Fig. 3. Equivalent circuits of the proposed converter.

addition, the input power ratio between the AC and DC power supplies and the voltage transfer ratio between the input and the output voltage are controlled by the duty ratio of the DC/DC converter.

III. CONTROL STRATEGY

The proposed control strategy, which is based on an indirect control method with a triangular carrier wave, can easily realize the addition of the DC/DC converter. The conventional indirect control strategy, which uses space vector modulation, must calculate the pulse width for each switch, including the DC/DC converter. In addition, it is difficult to define the input current command vector, because it is four-phase, consisting of the three-phase AC power supply and the DC power supply. Therefore, the control in this paper can be achieved by using an independent command for the DC/DC converter based on carrier comparison modulation.

A. Control method for the rectifier stage and DC/DC converter

Figure 4 shows a block diagram of the step-down type AC/DC/AC direct converter. The rectifier stage converter in the proposed interface converter functions as a four-phase current source rectifier including the DC/DC converter. The relation between the input voltage $[v_r\ v_s\ v_t\ v_b']^t$ and the output voltage $[v_u\ v_v\ v_w]^t$ can be expressed as

$$\begin{bmatrix} v_u \\ v_v \\ v_w \end{bmatrix} = \begin{bmatrix} S_{up} & S_{un} \\ S_{vp} & S_{vn} \\ S_{wp} & S_{wn} \end{bmatrix} \begin{bmatrix} S_{rp} & S_{sp} & S_{tp} & S_{bp} \\ S_{rn} & S_{sn} & S_{tn} & 0 \end{bmatrix} \begin{bmatrix} v_r \\ v_s \\ v_t \\ v_b' \end{bmatrix} \quad (1),$$

where the input voltage v_b' is the DC power supply voltage v_{batt} based on the neutral point of the input voltage. To avoid a short circuit of the AC and DC power supplies, the switches in the rectifier stage converter and the DC/DC converter are not turned on at the same time. As a

result, the current command for each phase is decreased by the working time of the DC chopper. Therefore, the input current command i_{rec}^{**} for one phase is converted by

$$i_{rec}^{**} = i_{rec}^{*} \cdot (1 - i_b^{*}) \quad (2),$$

where i_{rec}^{*} is rectifier stage current command, and i_b^{*} is the DC/DC converter input current command.

Figure 5 shows the relationship between the DC link voltage and the DC/DC converter duty ratio for the proposed converter. The DC link voltage is decided by the average value of the output voltage between the rectifier stage and the DC/DC converter. Therefore the average value of the DC link voltage E_{dc} in the proposed circuit can be expressed as

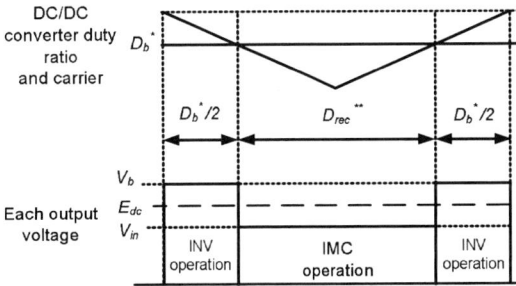

Fig. 5. The relationship between the DC link voltage and the DC/DC converter duty ratio.

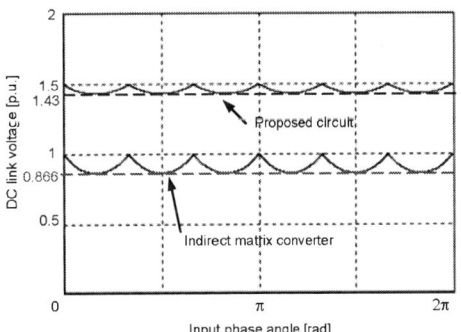

Fig. 6. DC link voltage waveform.

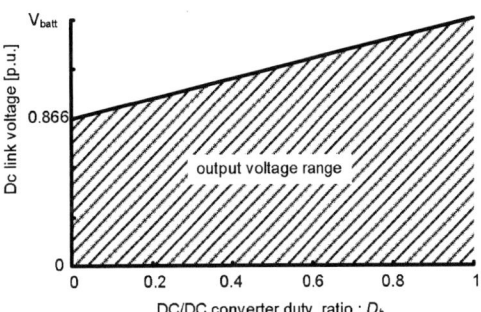

Fig. 7. Output voltage range for proposed converter.

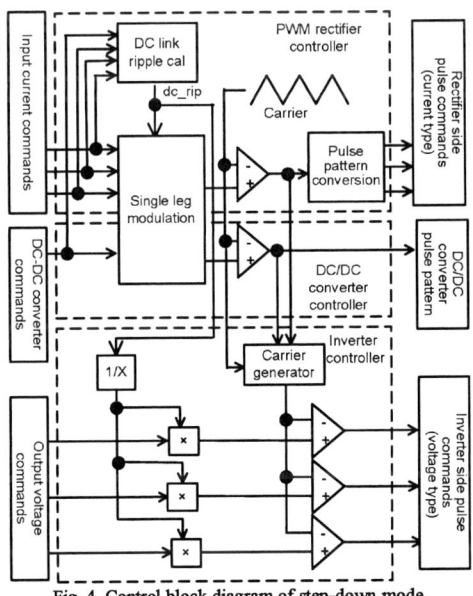

Fig. 4. Control block diagram of step-down mode.

$$E_{dc} = D_{rec}^{**} v_{in} + D_b^{*} v_b \qquad (3),$$

where D_{rec}^{**} is the rectifier stage converter duty ratio based on input current command converted by equation (2), and D_b^{*} is the DC/DC converter duty ratio, and v_{in} is input voltage for maximum value, and v_b is DC power supply voltage.

Figure 6 shows a comparison between the DC link voltage of the proposed circuit and the indirect matrix converter, where the maximum value of the indirect matrix converter is defined as 1 [p.u.]. The DC link voltage contains a ripple, which is constrained by the sixth order component of the AC power supply frequency. Therefore, the output voltage is limited to 0.866 times that of the maximum DC link voltage. On the other hand, the proposed circuit can operate alternately between the indirect matrix converter and inverter as discussed and shown early in Fig. 3. The DC link voltage waveform of the proposed circuit is shown in Fig. 6 when D_b^{*} is set to 0.5 and the DC power supply voltage v_b is assumed to be 2 times that of the indirect matrix converter. In this case, the DC link voltage of the proposed circuit obtains 1.43 times that of the indirect matrix converter. Thus, the proposed circuit can improve the voltage transfer ratio from the AC input side to the AC output side by using the DC power supply voltage.

Figure 7 shows the DC link voltage range for the proposed converter. In the proposed converter, the DC power supply voltage of the snubber circuit has to keep higher than the peak of the AC input line voltage in order to prevent a rush current. Thus, the range of the DC link voltage is defined from the output voltage of rectifier stage to battery voltage, as expressed by equation (3). Therefore, the range of the output voltage is defined from zero volts to battery voltage as show in Fig. 7.

It should be noted that the ratio between the rectifier duty D_{rec}^{*} and the DC/DC converter duty D_b^{*} is the same as the input power ratio between the AC and DC power supplies.

B. Basic control method of the inverter stage

Figure 8(a) shows the relation of carrier signals between the inverter and the rectifier stages. The rectifier stage is controlled under the condition of constant DC link current. However, the actual DC link current must be zero when the inverter stage controller selects the zero voltage vectors. In the proposed method, this is achieved by controlling the slope of the inverter carrier signal, as shown in Fig. 6. By adopting this method, the zero current period of the DC link is distributed to each input current by the same ratio, as expressed by equation (4).

$$\frac{T_{r0}}{T_r} = \frac{T_{s0}}{T_s} = \frac{T_{b0}}{T_b} = \frac{T_{r0} + T_{s0} + T_{b0}}{T_r + T_s + T_b} \qquad (4),$$

where T_r is turn on period in S_{rp}, T_s is tern on period in S_{sp}, T_b is tern on period in S_{bp}, T_{r0} is zero current period in T_r, T_{s0} is zero current period in T_s, T_{b0} is zero current period in T_b.

C. Proposed switching loss reduction method for inverter stage

In the Fig. 8(a), two-phase modulation is introduced in order to reduce the switching loss in the inverter stage besides the zero current period distribution. However, the

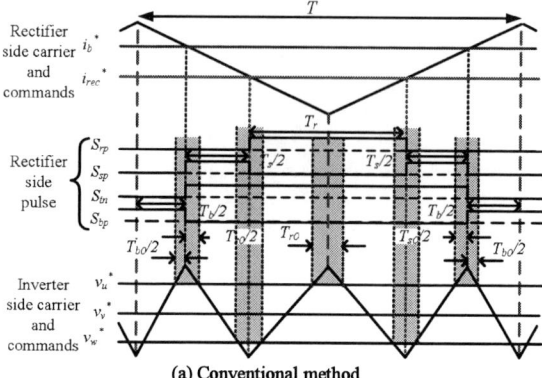

(a) Conventional method

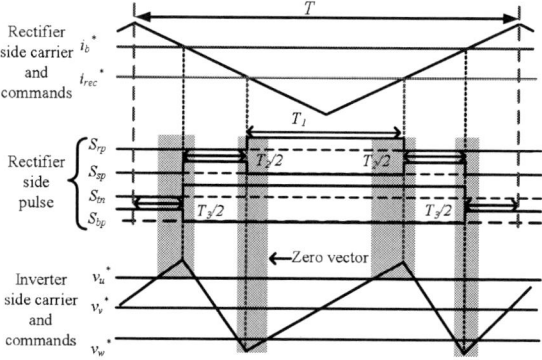

(b) Proposed method

Fig.8. Relation between inverter carrier and rectifier pulse.

number of switching of the inverter stage increases to 3 times in compared with rectifier stage. For example, the U-phase switching devices in the inverter stage are changed in one rectifier carrier period 6 times though that of the rectifier stage is 2 times. The switch timings of the rectifier and DC/DC converter stage are only agree with the bottom of the inverter carrier in order to achieve the zero current switching. As a result, the switching loss of the inverter stage increases.

Figure 8(b) shows proposed switching loss method using the carrier signals. In the proposed method, the inverter stage carrier is controlled asymmetry to reduce the number of switching of the inverter stage, as shown in Fig. 8(b). The switch timings of the rectifier and DC/DC converter stage are agree with both bottom and top of the inverter carrier. In this case, the U-phase switching devices in the inverter stage are changed 4 times. It is should be noted that the zero current period of the DC link is also distributed to each input current by the same ratio in the proposed method, as expressed by equation (4).

As a result, the switching loss of the inverter stage can be decreased to 2/3 times in the comparison with a conventional method because the number of the switching in one carrier period is 4 times though that of the conventional method 6 times. Note that the proposed method requires the zero voltage vectors at both top and bottom of the inverter carrier. This means that the two phase modulation can not be applied to the inverter voltage command. That is, the proposed method is effective in low voltage aria such as low speed in motor drive applications because the three phase modulation is used in low speed aria.

D. Commutation method

In indirect matrix converter, the rectifier stage converter can achieve zero current switching when the inverter generates the zero voltage vectors [11]. In other words, the DC link current must be zero when the inverter controller selects the zero voltage vectors. The proposed circuit alternately operates between an indirect matrix converter and an inverter. The switching of S_{bp} in the DC/DC converter is also achieved when the inverter outputs the zero vectors. Therefore, switching losses do not occur in the rectifier stage converter and the DC/DC converter. It is should be noted that a conventional dead time is applied for the commutation of the inverter stage. The conventional dead time for the inverter stage causes output voltage error and input current error. To overcome it, the commutation error compensation method [6] is applied for proposed circuit. As a result, the influence of the commutation for the input and output current can be compensated.

IV. SIMULATION AND EXPERIMENTAL RESULS

Table 1 provides the simulation and experimental circuit parameters and conditions. The operation of the proposed circuit is demonstrated by the simulation and experimental results.

A. Basic operation simulation results for the proposed converter

Figure 9 shows the simulation results of the proposed converter. In the simulation, an ideal current source load and a main circuit are used to check the proposed control strategy. Good sinusoidal waveforms are obtained for the input current and the output voltage. The THD of the input and the output current are less than 1%, respectively. In addition, a good DC waveform, without low frequency ripple, is obtained for the DC power supply current.

In the waveform of Fig. 9(a), power grid generation, battery discharge, and motoring operations are obtained. Similarly, power grid regeneration, battery charge, and generating operations are obtained for waveform of Fig. 9(b). As a result, bidirectional energy flow among the power supplies in this circuit is confirmed.

These simulation results confirm the basic operation of the proposed circuit.

B. Basic operation experimental results for the proposed converter

This chapter shows the basic operation of the proposed converter. The proposed switching loss reduction method is not applied in experimental results of this chapter as rated voltage range. However the validity of the proposed switching number reduction method is confirmed by next chapter.

Figure 10 shows the input current, the output current and the DC current waveforms. Good sinusoidal and dc waveform are obtained for input and output current. In the waveform of Fig. 9(a), power grid generation, battery discharge, and motoring operations are obtained.

Figure 11 shows the DC link voltage waveform for the conventional indirect matrix converter and proposed converter. In consideration of the low frequency

Table 1. Experimental parameters.

Input voltage	200 [V]	LC filter	2 [mH]
Input frequency	50 [Hz]		6.6 [□F]
Carrier frequency	10 [kHz]	Cut-off frequency	1.3 [kHz]
Output frequency	30 [Hz]	load	R-L
DC power supply	300 [V]	Commutation time	2.5 [□s]
Power ratio (AC:DC)		2:1	

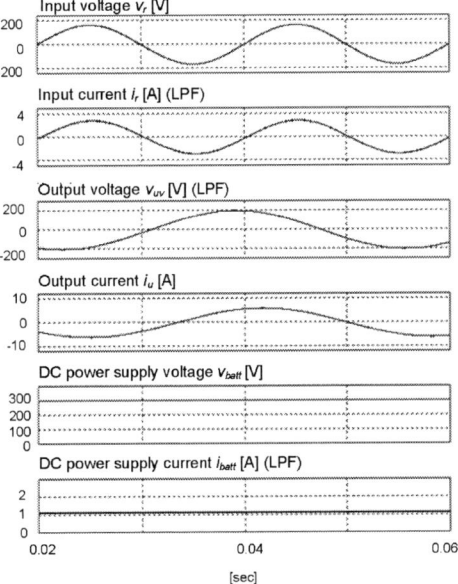

Waveform when power grid generation, battery discharge, and motoring operations

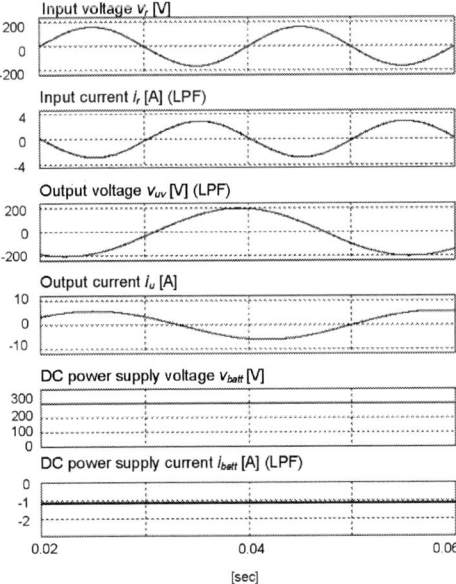

(b) Waveform when power grid regeneration, battery charge, and generating operations

Fig. 9. Simulation results.

components, a low pass filter was applied to the lower waveform in Fig. 10. As a result, the DC link voltage of the conventional indirect matrix converter is 245 [V]. On the other hands, the results shown in Fig.11 (b) confirm that the DC power supply voltage and the input voltage are alternated to the DC link voltage. As a result, the minimum average value of the DC link voltage is 260 [V], as shown by the upper waveform of Fig. 11(b). The DC link voltage of the conventional indirect matrix converter is 245 [V]; therefore, the proposed circuit can improve the voltage transfer ratio.

Figure 12 shows the efficiency and the input power factor of the proposed circuit. An input power factor (P.F.) over 99% and high efficiency of 93.8% were obtained. In comparison, the efficiency of a conventional multi-power supply interface converter with a large electrolytic capacitor is approximately 90%. In addition, the efficiency can be improved by the applying of reverse blocking IGBTs to the rectifier stage.

Figure 13 shows the THD of the input and output current confirming that the input, output and DC input current THD are 7.5%, 3.7% and 1.9%, respectively.

These experimental results confirm the validity and basic operation of the proposed circuit.

C. The proposed switching loss reduction method

This chapter shows the validity of the proposed switching loss reduction method. The validity of proposed method is confirmed by the indirect matrix converter. This means that the DC/DC converter duty ratio is just zero. The reason is that the effect of switching loss ruction is depends on the DC/DC converter duty ratio. Therefore, to clear the effect of the proposed method, the DC/DC converter stage was stopped.

Figure 14 shows the input current and the output current waveforms of the proposed method. Good sinusoidal waveform are obtained for input and output current. In addition, it is confirmed in Figure 15 that the input and output current THD are 1.8% and 1.3% respectively. In the input and output waveform, almost a similar waveform is obtained in comparison with the conventional method. Thus, the good experimental waveforms are confirmed, in case of the reducing of switching number for inverter stage.

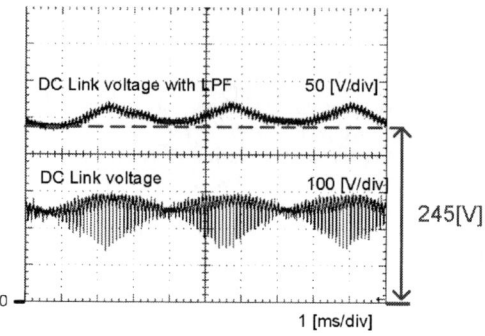

(a) Conventional indirect matrix converter.

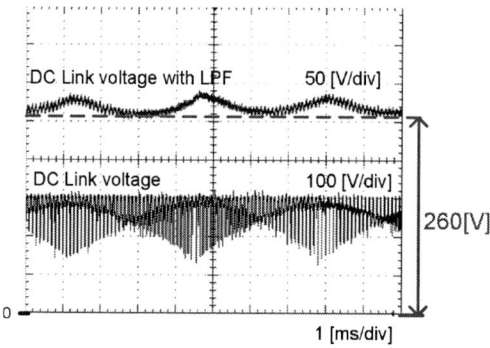

(b) Proposed converter.
Fig. 11. DC link voltage waveform.

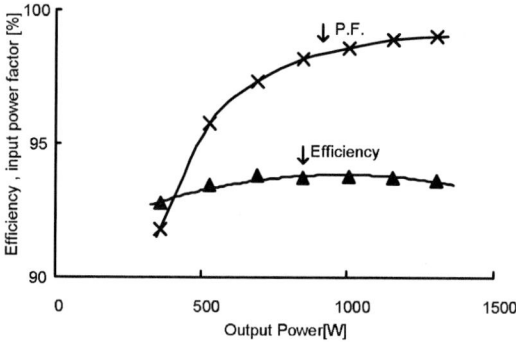

Fig. 12. Efficiency and input power factor. (IMC=indirect matrix converter; INV=inverter.)

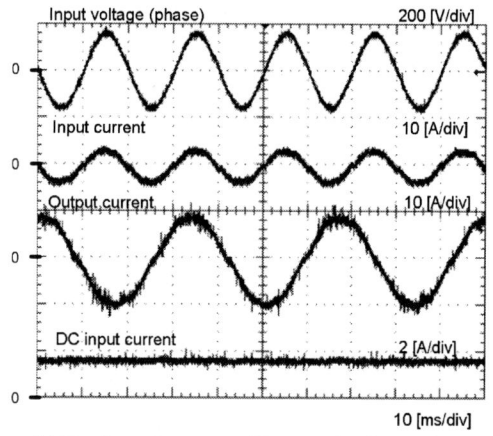

(a) Waveform when power grid regeneration, battery charge, and generating operations.
Fig. 10. Experimental waveform.

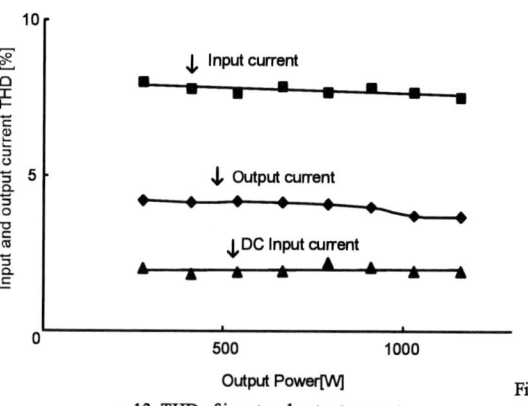

g. 13. THD of input and output current.

586

Figure 16 shows the efficiency and the input power factor of the proposed circuit. In the conventional method, the input power factor over 99% and high efficiency of 93.8% were obtained. On the other hand, the efficiency is improved by about 0.5 point by applying the proposed method. In addition, the input power factor over 99% is obtained.

Figure 17 shows the loss analysis for the proposed circuit using circuit simulator (PSIM, Powersim Technologis Inc) and DLL file (Dynamic Link Library)[12]. The three phase modulation, two phase modulation, and proposed switching loss reduction method are applied for the inverter stage control. The two phase modulation decreases the switching loss of inverter stage by about 2/3 in compared with a three phase modulation. On the other hands, the switching loss of the inverter stage is improved by about 1/2 by applying of the proposed switching loss reduction method. In the view of loss on rectifier stage, there is no difference in each method, since the switching losses do not occur in the rectifier stage during zero current switching. In case of the heavy load, the conduction loss is dominant to the switching loss. In contrast, the switching loss is dominant to conduction loss when the light load is used. Therefore, proposed method is valid in the low speed and light load area, because the two-phase modulation can not apply in the low speed area to avoid the output current bias to one switching device and to keep the output voltage accuracy.

These experimental results confirm the validity of the proposed circuit and proposed switching loss reduction method.

V. CONCLUSIONS

This paper proposes a novel control method for a multi-power supply interface converter using an indirect matrix converter. An active snubber circuit in the indirect matrix converter is used as the interface of the DC power supply. The proposed control method can be easily applied to multi-port systems, due to the use of carrier comparison modulation. Therefore the proposed method can be applied easily to a DC/DC converter. In addition, this paper proposes a control method of switching loss reduction for the inverter stage converter. Applying the proposed method, the switching loss of the inverter stage can be decreased to 2/3 times in the comparison with conventional method. The validity of the proposed method was confirmed by the simulation and experimental results.

The experimental results of the proposed system using 1kW class prototype circuit are obtained as follows,

i) The bidirectional energy flow among the power supplies in this circuit is confirmed.

ii) The input, output and dc output current THD are 7.5%, 3.7%, 1.9%, the input power factor is 99%, and the efficiency is 93.8%.

iii) The efficiency is improved by about 0.5 point by application of the proposed method.

iv) The proposed switching number reduction method is valid in the low speed and light load area.

This study was supported by Industrial Technology Grant Program in 2005 from New Energy and Industrial Technology Development Organization (NEDO) of Japan.

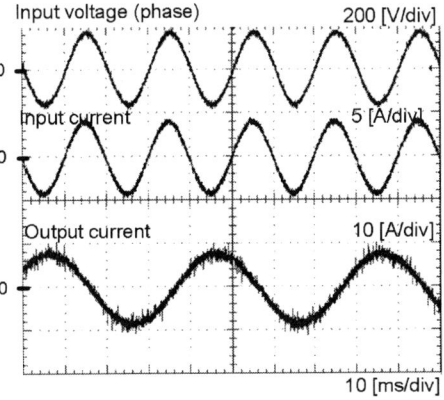

(a) Conventional method.

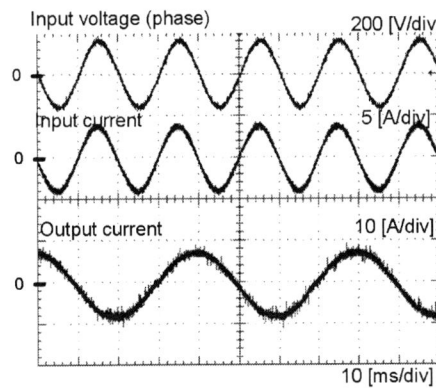

(b) Proposed method.

Fig. 14. Experimental waveform.

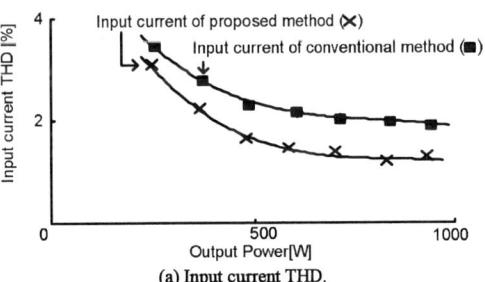

(a) Input current THD.

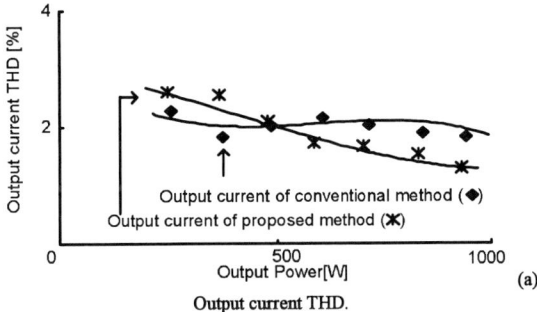

Output current THD.

Fig.15. THD of input and output current.

REFERENCES

[1] J.W.Kolar, M.Baumann, F.Schafmeister, H.Ertl: "Novel Three Phase AC-DC-AC Sparse Matrix Converter", IEEE APEC 2002

[2] L.Wei, Y.Matsusita, T.A.Lipo: "Investigation of Dual-bridge Matrix Converter Operating under Unbalanced Source Voltage" IEEE PESC2003, 1293(2003)

[3] C.Klumpner, F. Blaabjerg: "A new generalized two-stage direct power conversion topology to independently supply multiple AC loads from multiple power grids with adjustable power loading" Proc. of IEEE PESC2004, pp.2028-2084 (2004)

[4] Y. Minari, K. Shinohara, R. Ueda: "PWM-rectifier/voltage-source inverter without DC link components for induction motor drive", IEE Proc. on Electric Power Applications, Vol. 140, No. 6, pp. 363 .368, Nov. 1993.

[5] L. Wei, T.A. Lipo: "A novel matrix converter topology with simple commutation", Proc. of IAS'01, Vol. 3, pp. 1749-1754, 2001.

[6] K. Kato, J. Jtoh: "A Novel Control Method for Direct Interface Converters used for DC and AC Power Supplies", Proc. of EPE'07, 448, 2007.

[7] S. Mariethoz, T. Wijekoon, P. Wheeler: "Analysis, control and comparison of hybrid two-stage matrix converters for increased voltage transfer ratio and unity power factor", PCC-NAGOYA 2007

[8] C. Klumpner, T. Wijekoon, and P. Wheeler: "New methods for the active compensation of unbalanced supply voltages for two-stage direct power converters." Proc. IEEJ IPEC'05.

[9] C. Klumpner, T. Wijekoon, P. Wheeler: "Active Compensation of Unbalanced Supply Voltage for Two-Stage Direct Power Converters Using the Clamp Capacitor" PESC'05, pp.2376 - 2382, 2005

[10] J.Itoh, I.Sato, A.Odaka, H.Ohguchi, K.Kodachi: "A Novel Approach to Practical Matrix Converter Motor drive System with RB-IGBT" Power Electronics Specialists Conference 2004

[11] K.Iimori, K.shinohara, M.Muroya, H.kitanaka: "Characteristics of New Current Controlled PWM Rectifier-Voltage Source Inverter without DC Link Components for Induction Motor Drive" IEEJ Vol.119-D No.2,1999(in Japanese)

[12] J. Itoh, T. Iida, A. Odaka:" Realization of High Efficiency AC link Converter System based on AC/AC Direct Conversion Techniques with RB-IGBT" Industrial Electronics Conference, Paris, ,PF-012149,2006

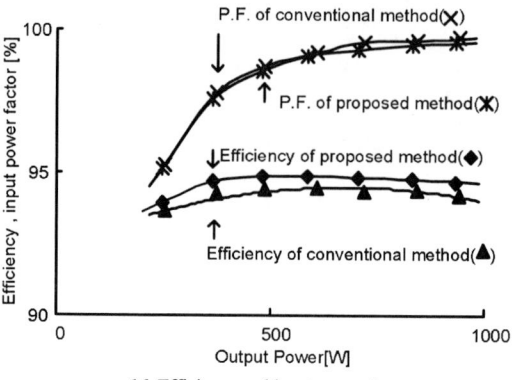

Fig. 16. Efficiency and input power factor.

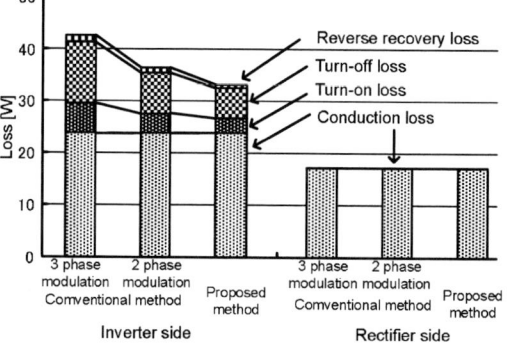

Loss analysis of heavy load. (a)

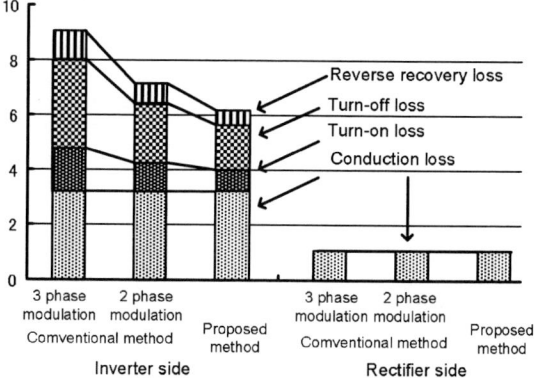

(b) Loss analysis of light load.
Fig. 17. Loss analysis.

Precise Digital Control Method with Multi-rate deadbeat control

for Single Phase Utility Interactive Inverter with FPGA based Hardware Controller

Kenta Hayashi*, Tomoki Yokoyama*
*Tokyo Denki University, Ishizaka Hatoyama, Hiki-gun, Saitama, Japan 350-0394

Abstract—A new current control method for the utility interactive inverter based on multi-rate deadbeat control with FPGA based hardware controller is proposed. Multi-rate deadbeat control improves the tracking accuracy to the reference compared with the conventional single rate deadbeat control. As a result, superior characteristics were derived with very small LC filter component. Also to utilize the capability of FPGA based hardware controller, all the control circuit for the proposed current control method using voltage deadbeat control and PLL control with quasi-dq transformation are implemented in FPGA based hardware controller. The design concept of the controller is also described, and the advantages and the disadvantage are discussed through simulations and experiments.

I. INTRODUCTION

Nowadays, FPGA technology makes it possible to realize a new control system for power electronics with its superior high speed calculation capability. Authors have already proposed FPGA based hardware controller for power electronics control and applied to PWM inverter for CVCF operation and utility interactive inverter system [3]-[7].

Generally, the control system of the utility interactive inverter is constructed based on the conventional PI control method. Improvement of the FPGA technology realizes very high-speed calculation capability. In the case of DSP based software controller, because of the limitation of A/D conversion time and the calculation time of the DSP, it was hard to use the sampling point data for the control output of the same sampling period. So the state observer is introduced to derive the next sampling data using pre-sampled data, then the calculation of the control law is performed using the observed data. Proposed method presented before in [4] are based on such techniques. In that case, the control characteristics was much depends on the accuracy of the state observer. If the control system was implemented using FPGA based hardware controller, parallel processing control circuit can be realized, and all the calculation circuit, A/D interface circuit, reference generation circuit, PLL control circuit and pulse pattern generation circuit can be implemented in one FPGA chip. As the result, total calculation time for the control law can be dramatically reduced compared with the DSP based software controller.

In [6] and [7], authors were proposed a current control methods for the utility interactive inverter based on single rate deadbeat control using FPGA based hardware controller. The remand item is the precise discrete time system model and the robustness for the LC filter parameter variation.

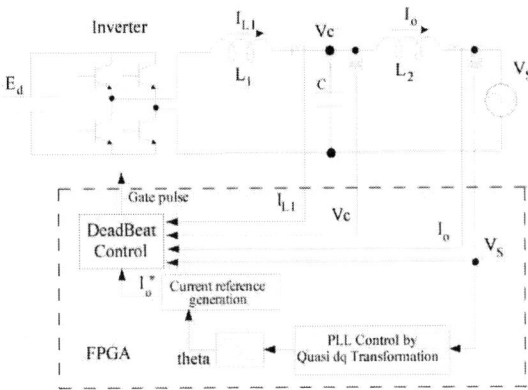

Fig. 1. Configuration of proposed utility interactive inverter system

In this paper, a new discrete time system model based on multi-rate deadbeat control for the utility interactive inverter is proposed. Adopting multi-rate deadbeat control, sampling frequency becomes one third of single rate deadbeat control. Even in that situation, tracking accuracy to the reference is much improved compared with single rate deadbeat control. Throughout simulations and experiments, the advantages and the disadvantages are discussed.

II. CONTROL METHOD

Fig.1 shows the diagram of the proposed system configuration. PWM inverter bridge with the DC voltage E_d outputs V_i, then connected to the output filter L_1 and C, then connected to the utility voltage V_S via inductor L_2. The inductor current I_{L1}, the capacitor voltage V_c, the output current I_o and the utility voltage V_S are measured as the state variables respectively. The output current I_o is depend on the difference voltage between the inverter output voltage V_c and the utility voltage V_S. As the utility voltage is stable, so if the output voltage V_c is determined, the output current can be controlled, and the current regulation depends on the parameter of the inductor L_2.

A. Current Control for the I_o

Fig.2 shows the equivalent circuit of the single phase utility interactive system when the inverter is replaced as the sinusoidal voltage source. In this stage, there are three state variables, which are I_o, V_s, $\dot{V}_s$. It is assumed that the utility voltage V_s is a sinusoidal waveform. The state

978-1-4244-1741-4/08/$25.00 ©2008 IEEE

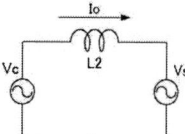

Fig. 2. Equivalent circuit for Single phase utility interactive system

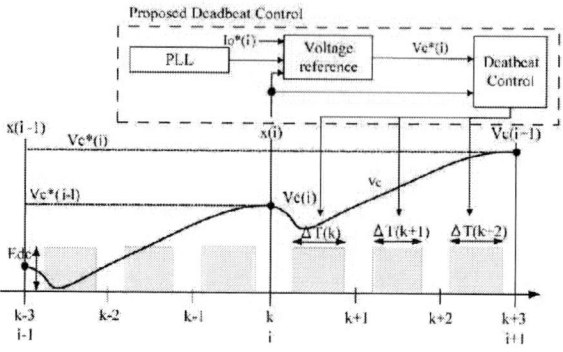

Fig. 3. Timing chart of multi-rate deadbeat control

equation becomes as follows,

$$\dot{x}_i = A_i x_i + B_i u_i \qquad (1)$$

$$y_i = C_i x_i \qquad (2)$$

$$, x_i = \begin{bmatrix} I_o \\ V_s \\ \dot{V}_s \end{bmatrix}, A_i = \begin{bmatrix} 0 & -\frac{1}{L_2} & 0 \\ 0 & 0 & 1 \\ 0 & -\omega^2 & 0 \end{bmatrix}$$

$$, B_i = \begin{bmatrix} \frac{1}{L_2} \\ 0 \\ 0 \end{bmatrix}, C_i = \begin{bmatrix} 1 & 0 & 0 \\ 0 & 1 & 0 \\ 0 & 0 & 0 \end{bmatrix}, u_i = \begin{bmatrix} V_c \end{bmatrix}.$$

The state equation for discrete time system becomes as follows [1],

$$x_i(i+1) = F_i x_i(i) + G_i u_i(i) \qquad (3)$$

$$y_i(i+1) = C_i x_i(i) \qquad (4)$$

$$, x_i(i) = \begin{bmatrix} I_o(i) \\ V_s(i) \\ \dot{V}_s(i) \end{bmatrix}, F_i = \begin{bmatrix} f_{i11} & f_{i12} & f_{i13} \\ f_{i21} & f_{i22} & f_{i23} \\ f_{i31} & f_{i32} & f_{i33} \end{bmatrix}$$

$$, G_i = \begin{bmatrix} g_{i11} \\ g_{i21} \\ g_{i31} \end{bmatrix}, C_i = \begin{bmatrix} 1 & 0 & 0 \\ 0 & 1 & 0 \\ 0 & 0 & 0 \end{bmatrix}, u_i(i) = V_c(i).$$

To derive the reference $x_v(i+1)$ in the next section, the output voltage V_c in (3) and (4) can be rewritten as follows.

$$V_c(i) = (I_o(i+1) - f_{i11} * I_o(i) - \\ f_{i12} * V_s(i) - f_{i13} * \dot{V}_s(i))/g_{i11} \qquad (5)$$

Here, the derivative of the output voltage $\dot{V}_s(k)$ is assumed as follows.

$$\dot{V}_s(i) = \frac{V_s(i) - V_s(i-1)}{T_s} \qquad (6)$$

Here T_s is the sampling interval. (5) indicates that if the inverter output voltage V_c is determined, the output current I_o matches to the reference $I_o(i+1)$ at the next sampling instant.

The equation for the capacitor current I_c in (5) can be rewritten as follows,

$$I_c(i) = C\dot{V}_c(i). \qquad (7)$$

The inductor current I_{L1} can be derived by the capacitor current I_c and the reference of output current I_o^*,

$$I_{L1}(i) = I_o^* + I_c(i). \qquad (8)$$

B. Deadbeat Control for the V_c

The inverter output voltage V_c is depend on the condition of the PWM inverter bridge voltage V_i and the utility voltage V_s. V_i is the PWM pulse waveforms and the V_s is a continuous sinusoidal waveform in the normal utility line. In the proposed method, the V_i and the V_s are divided into different terms in (9), the state equation becomes as follows,

$$\dot{x}_v = A_v x_v + B_{v1} u_v + B_{v2} V_s \qquad (9)$$

$$y_v = C_v x_v \qquad (10)$$

$$, x_v = \begin{bmatrix} I_{L1} \\ V_c \\ I_o \end{bmatrix}, A_v = \begin{bmatrix} 0 & -\frac{1}{L_1} & 0 \\ \frac{1}{C} & 0 & -\frac{1}{C} \\ 0 & \frac{1}{L_2} & 0 \end{bmatrix}, B_{v1} = \begin{bmatrix} \frac{1}{L_1} \\ 0 \\ 0 \end{bmatrix}$$

$$, B_{v2} = \begin{bmatrix} 0 \\ 0 \\ -\frac{1}{L_2} \end{bmatrix}, C_v = \begin{bmatrix} 1 & 0 & 0 \\ 0 & 1 & 0 \\ 0 & 0 & 1 \end{bmatrix}, u_v(t) = \begin{bmatrix} V_i(t) \end{bmatrix}.$$

The state equation for discrete time system becomes (11), (12), (13) and (14).

$$x_v(k+3) = F_v x(k+2) + G_v \Delta T(k+2) + H_v V_S(k+2) \quad (11)$$

$$x_v(k+2) = F_v x(k+1) + G_v \Delta T(k+1) + H_v V_S(k+1) \quad (12)$$

$$x_v(k+1) = F_v x(k+2) + G_v \Delta T(k) + H_v V_S(k) \quad (13)$$

$$y_v(k+3) = C_v x_v(k) \quad (14)$$

Here, to derive the discrete time equation for the V_i, it is assumed that the PWM pulse is outputted in the center of the carrier interval, G_v can be derived as (15) , where T_c is the sampling interval and T_p is the time interval of the pulse width $\Delta T(k)$.

$$G_v = \int_{(T_c-T_p)/2.0}^{(T_c+T_p)/2.0} e^{A_v(T_c-\tau)} B_{v1} d\tau \qquad (15)$$

On the other hand, the V_s is the infinite utility line, so it is assumed that the V_s is constant while the one carrier interval, H_v can be derived as (16).

$$H_v = \int_0^{T_c} e^{A_v \tau} B_{v2} d\tau \qquad (16)$$

The equation for $x_v(k+3)$ in (11), (12) and (13)

can be rewritten as follows,

$$x_v(k+3) = F_v^3 x(k) + \begin{bmatrix} F_v^2 G_v & F_v G_v & G_v \end{bmatrix} \begin{bmatrix} \Delta T(k) \\ \Delta T(k+1) \\ \Delta T(k+2) \end{bmatrix}$$
$$+ \begin{bmatrix} F_v^2 H_v & F_v H_v & H_v \end{bmatrix} \begin{bmatrix} V_S(k) \\ V_S(k+1) \\ V_S(k+2) \end{bmatrix}. \quad (17)$$

(17) can be rewritten based on the sampling period i as (18).

$$x_x(i+1) = F_x x_x(i) + G_x u_x(i) + H_x V_S(i) \quad (18)$$

$$, x_x(i) = \begin{bmatrix} I_{L1}(k) \\ V_c(k) \\ I_o(k) \end{bmatrix}, F_x = F_z^3, G_x = \begin{bmatrix} F_z^2 G_z & F_z G_z & G_z \end{bmatrix}$$

$$, H_x = \begin{bmatrix} F_z^2 H_z & F_z H_z & H_z \end{bmatrix}, u_x(i) = \begin{bmatrix} \Delta T(k) \\ \Delta T(k+1) \\ \Delta T(k+2) \end{bmatrix}$$

$$, V_S(i) = \begin{bmatrix} V_S(k) \\ V_S(k+1) \\ V_S(k+2) \end{bmatrix}.$$

Finally, the derived pulse width to control the output current I_o matches to the reference becomes as (19).

$$\begin{bmatrix} \Delta T(k) \\ \Delta T(k+1) \\ \Delta T(k+2) \end{bmatrix} = G_x^{-1} \begin{bmatrix} I_{L1}(k+3) \\ V_c(k+3) \\ I_o(k+3) \end{bmatrix} - G_x^{-1} F_x \begin{bmatrix} I_{L1}(k) \\ V_c(k) \\ I_o(k) \end{bmatrix}$$
$$- G_x^{-1} H_x \begin{bmatrix} V_S(k) \\ V_S(k+1) \\ V_S(k+2) \end{bmatrix} \quad (19)$$

$V_S(k)$, $V_S(k+1)$ and $V_S(k+2)$ in (19) can be replaced to the reference value, the pulse width $\Delta T(k)$, $\Delta T(k+1)$ and $\Delta T(k+2)$ can be derived. Finally, I_o matches to the reference $I_o(k+1)$ at next sampling instant.

III. SIMULATION

Simulations were carried out for 1kW class utility interactive inverter system. Detailed conditions are summarized in Table I and Table II.

To clarify the control accuracy, simulations of single rate deadbeat control for the utility interactive inverter presented in [8] were also applied for comparison.

TABLE I

SYSTEM PARAMETERS

E_{dc}	$200.0(V)$	System rating	$1(kW)$
V_S	$100.0(Vrms)$	Utility frequency	$50(Hz)$

TABLE II

FILTER PARAMETERS

Carrier frequency	$20.0(kHz)$	$30.0(kHz)$	$60.0(kHz)$
L_1	$0.79(mH)$ $2.48(\%Z)$	$0.54(mH)$ $1.69(\%Z)$	$0.29(mH)$ $0.88(\%Z)$
C	$4.03(\mu H)$ $1.27(\%Z)$	$2.74(\mu H)$ $0.86(\%Z)$	$1.46(\mu H)$ $0.46(\%Z)$
L_2	$2.03(mH)$ $6.39(\%Z)$	$1.39(mH)$ $4.35(\%Z)$	$0.74(mH)$ $2.31(\%Z)$

A. Tracking accuracy for the I_o^*

Simulation was carried out to verify the tracking accuracy for the I_o^*. Fig.4 shows simulation result for the single rate deadbeat control.

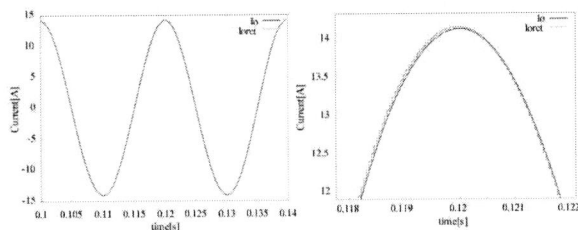

Fig. 4. Single rate deadbeat control (conventional method)

Fig.5 shows simulation result for the multi-rate deadbeat control.

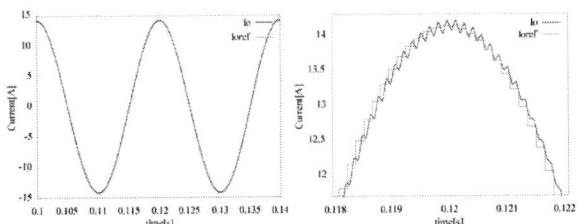

Fig. 5. Multi-rate deadbeat control (proposed method)

To clarify the tracking accuracy to the reference, (20) is defined to evaluate the tracking error. From Table III, it is obvious that the tracking accuracy is much improved in the proposed method.

TABLE III

TRACKING ERROR OF OUTPUT CURRENT I_o TO THE REFERENCE

	tracking error (%)
Single rate DB	0.208
Mulit-rate DB	0.187

$$tracking\ error = \frac{|I_{o\_fund}^* - I_{o\_fund}|}{I_{o\_fund}^*} * 100 \quad (20)$$

Here, I_{o_fund} is the fundamental amplitude of the output current I_o which is derived by FFT calculation.

B. Steady-State Response with Small LC Filter Component

To confirm the affection for the LC filter parameter selection, simulations were carried out in the case that the filter parameter of L_1, L_2 and C are varied from 30% to 100% of the nominal value in Table I.

Fig.6 shows the tracking error for the LC filter selection. It is obvious that the tracking error characteristics is much improved with small LC filter component in multi-rate deadbeat control. Fig.7 and Fig.8 show the waveform when the nominal parameter of L_1, L_2 and C are varied to 50% of nominal value (point (A) in Fig.6).

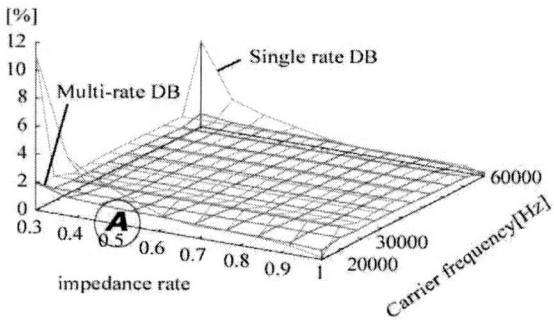

Fig. 6. Tracking error for small LC filter component

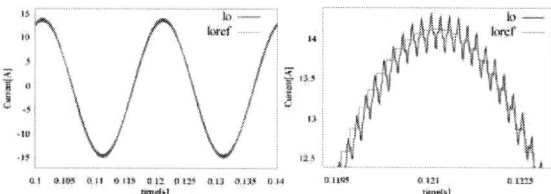

Fig. 7. Tracking error with small LC filter component (50%) (Multi-rate deadbeat control) (point (A) in Fig.6)

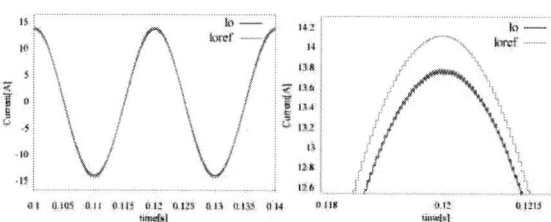

Fig. 8. Tracking error with small LC filter component (50%) (Single rate deadbeat control) (point (A) in Fig.6)

C. Step Response

Based on the result of Fig.6, the system parameter for the experimental set up is decided as shown in Table IV. To clarify the step responce, current reference I_o^* is changed from 10[Arms] to 5[Arms] in 100[ms]. Fig.9 shows the waveforms of the step response for single rate deadbeat control and Fig.10 shows the waveforms of the step response for multi-rate deadbeat control.

TABLE IV

SYSTEM PARAMETERS

E_d	100.0(V)	L1	1.0(mH) 3.14(%Z)
V_S	50.0(V)	C	1.0(μF) 0.314(%Z)
System rating	1(kW)	L2	3.0(mH) 9.42(%Z)
Utility frequency	50(Hz)	Carrier frequency	20.0(kHz)

Table V shows the settling time of I_o when the current reference is changed. The settling time is much improved

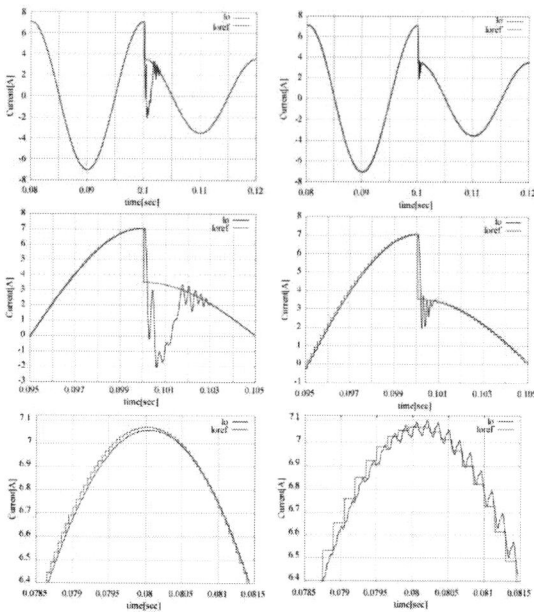

Fig. 9. Step response for single rate deadbeat control

Fig. 10. Step response for multi-rate deadbeat control

in the case of multi-rate deadbeat control.

TABLE V

SETTLING TIME OF OUTPUT CURRENT I_o

	Settling time (ms)
Single rate DB	2.83
Mulit-rate DB	0.75

D. Steady-State Response with distorted utility voltage

Fig.11 and Fig.12 shows the waveforms in the case that the distorted utility voltage V_s is applied for single rate deadbeat control and multi-rate deadbeat control. In Fig.11 and Fig.12, based on the fundamental frequency, 2% of 3rd order harmonics, 2% of 5th order harmonics, 2% of 7th order harmonics and 1% of 11th order harmonics are included to the V_s. Even in such condition, the output current I_o follows the current reference precisely.

IV. FPGA IMPLEMENTATION

To implement the proposed method, Stratix EP2S60 (ALTERA Corp.) is applied. EP2S60 has built-in 36-blocks DSP core, which can perform four multiplication, and addition or subtraction can be performed for the multiplication result in 1 clock, and can operate up to 450MHz clock frequency. FPGA board is driven by the 50MHz oscillator. Fig.13 shows the FPGA and AD converter board.

Fig.14 shows the system configuration and the module block diagram of the FPGA chip, which is constructed with the quasi dq transformation and amplitude/phase detection block, power detection/control block, I_o reference

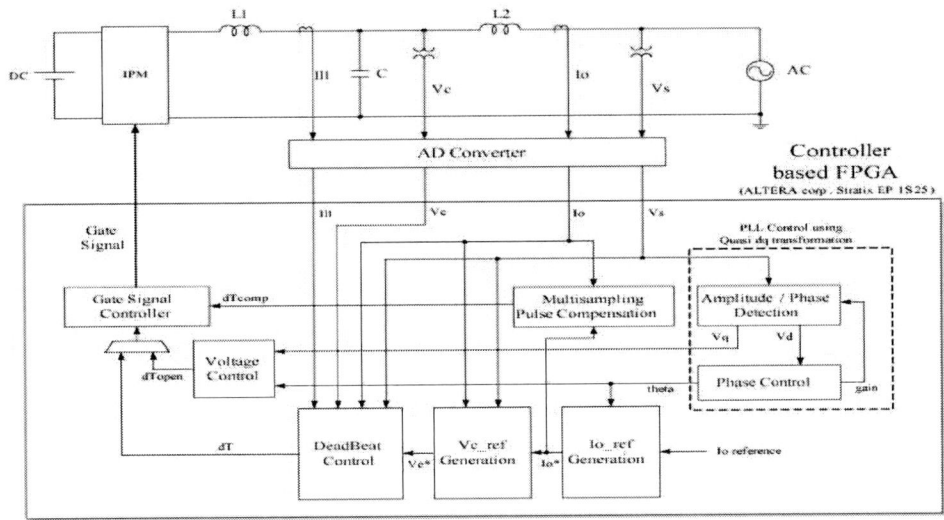

Fig. 14. System configuration and diagram of FPGA

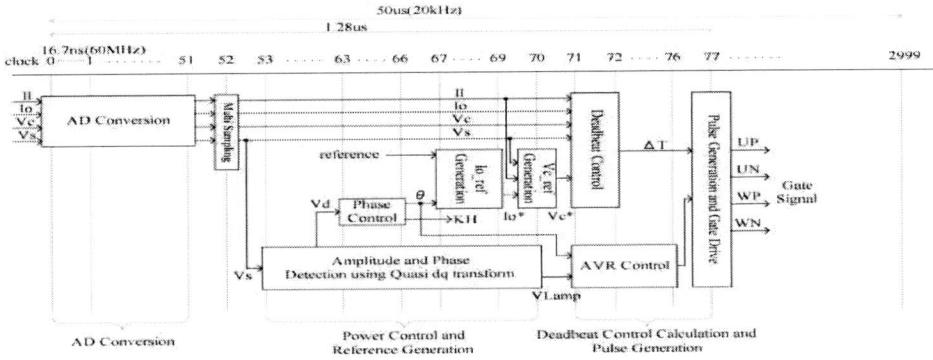

Fig. 15. Timing chart of the FPGA logic

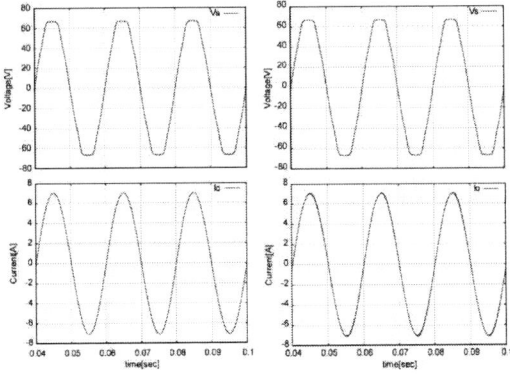

Fig. 11. Distorted utility voltage V_s and output current I_o(single rate deadbeat control)

Fig. 12. Distorted utility voltage V_s and output current I_o(multi-rate deadbeat control)

Fig. 13. FPGA based hardware controller

generation block, V_C reference generation block, deadbeat control block and gate signal control block.

Fig.13 shows the FPGA based hardware controller with 4ch AD converters and 8ch AD/DA unit board. Fig.15 shows the timing chart of the FPGA based hardware controller. AD conversion is finished in $835ns$ (50 $clocks$), and all the calculation is finished in $1.28\mu s$ (77 $clocks$), which is shorter than the dead time of the IGBT, so the control output can be reflected immediately to the PWM gate signal.

V. EXPERIMENT

Experiments were carried out for multi-rate deadbeat control. Fig16 shows the experimental result of the multi-rate deadbeat in the steady state conditions.

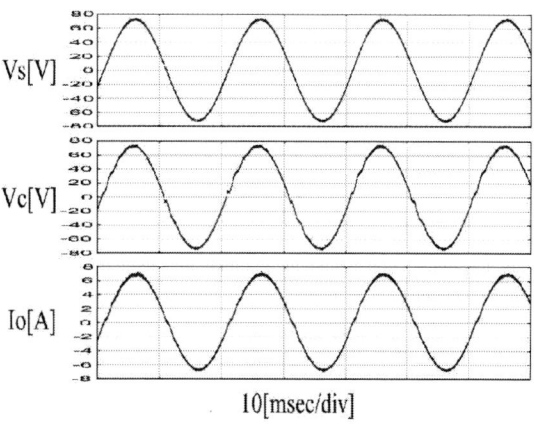

Fig. 16. Experimental result(multi-rate deadbeat control)

In Fig16, the output current I_o follows the reference of the output current I_o^*. The output current I_o is stable in the steady state condition for multi-rate deadbeat control.

Fig.17 shows the experimental result in the case that the distorted utility voltage V_s is applied for multi-rate deadbeat control. It is obvious that the output current I_o follows the reference of the output current I_o^*.

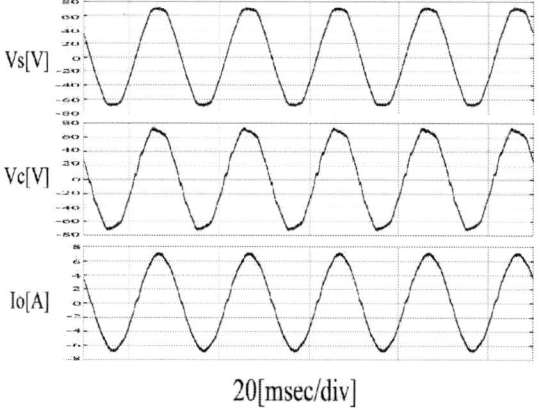

Fig. 17. Experimental result with distorted

VI. CONCLUSION

A new approach of digital control method for the utility interactive inverter to improve the regulation accuracy of the utility current is proposed. Applying multi-rate deadbeat control method, precise control accuracy can be achieved with FPGA based hardware controller.

REFERENCES

[1] K.P.Gokhale, A.Kawamura and R.G.Hoft "Deadbeat Microprocessor Controlof PWM Inverter for Sisusoidal Output Waveform Synthesis", ibid.IA-23,901, (1987) (originally published at PESC'85,1985)

[2] S.Hamasaki, A.Kawamura: "Experimental Verification of Line Current Detection type Active Filter Combined with Disturbance-observer", PCC-Osaka, Vol.1 O7-3, 2002

[3] T.Komiyama, T.Kojima, T.Yokoyama: "Implementation of High-Speed Frequency Detection for Single Phase Utility Interactive Inverter System", EPE-PEMC 2004 DS3.5 Topic 3-13

[4] K.Aoki, N.Uemura, E.Shimada, T.Yokoyama: "A Study of Current Control Using Deadbeat Control for Utility Interactive Inverter", EPE-PEMC 2004 LS2.2 Topic 2+3-7

[5] E.Shimada, K.Aoki, T.Komiyama, T.Yokoyama: "Implementation of Deadbeat Control for Single Phase Utility Interactive Inverter Using FPGA", IPEC 2005 S13-4

[6] E.Shimada, K.Aoki, T.Komiyama, T.Yokoyama: "Implementation of Single Phase Utility Interactive Inverter using FPGA based Hardware Controller", EPE 2005 T14-2

[7] E.Shimada, T.Yokoyama: "Current Control Method Using Voltage Deadbeat Control with Multi Sampling Pulse Compensation for Single Phase Utility Interactive Inverter with FPGA based Hardware Controller", INTELEC 2005

[8] S.Takamatsu, E.Shimada, T.Yokoyama: "Digital Control method for Single Phase Utility Interactive Inverter using Deadbeat Control with FPGA based Hardware Controller", EPE-PEMC 2006

[9] K.Hayashi, S.Takamatsu, T.Yokoyama: "Precise Digital Control Method with Sinusoid based Model for Single Phase Utility Interactive Inverter with FPGA based Hardware Controller", EPE 2007

[10] A.Kawamura, H.Fujimoto, T.Yokoyama: "Survey on the real time digital feedback control ofPWM inverter and the extension to multi-ratesampling and FPGA based inverter control", IECON 2007

A Digital Current Controller for Zero-Current Transition Bidirectional Converter

Nobuyuki Kasa* and Takahiko Iida*

* Okayama University of Science, Okayama, Japan, e-mail: *kasa@ee.ous.ac.jp*

Abstract—This paper presents a digital current controller for zero-current transition (ZCT) bidirectional converters. In order to reduce switching losses and noises, a soft-switching technique is adopted. It is suitable for the pulse-width modulation (PWM), because the resonant transition is very short and it is independent of the switching frequency. A field programmable gate array (FPGA) and a 2.5MHz analog to digital (A/D) converter can control the inductor current within 2us. From the experimental results, the current is controlled precisely even if the duty factor is more than 50%. Also, the zero-current transition operation of the main switch is confirmed and it is expected that the efficiency of the converter can be improved. The proposed converter is suitable for an energy transformer between batteries and motors, or generators.

Keywords—Soft switching, ZCS converters, Converter control.

I. INTRODUCTION

In a wind power generation system as shown in Fig.1, a bidirectional converter charges batteries or supplies the energy to the load. Therefore, efficiency and response of the bidirectional converter are required to be high level.

This paper presents a zero-current transition (ZCT) bidirectional converter adopted by a digital current controller. In a conventional converter, insulated gate bipolar transistors (IGBT) are used as the semiconductor main switches. However, the power losses, caused by the IGBT tail currents and their resister-capacitor-diode (RCD) snubber circuits, make the system efficiency low. In order to improve the efficiency, soft switching techniques are adapted[1-4].

In addition, a digital controller for the inductor current of the converter is proposed. An FPGA and a 2.5MHz A/D converter can control the inductor current within 2 us. To avoid subharmonic instability in a continuous current mode (CCM), a down counter is used as the sawtooth ramp circuit, which is used in a conventional analog current controller.

A 48.8 kHz ZCT bidirectional converter is assembled in the laboratory.

II. ZCT BIDIRECTIONAL CONVERTER

The efficiency of the bidirectional converter is greatly influenced by the losses caused by the hard switching of

This work was supported by the Ministry of Education, Culture, Sports, Science and Technology of Japan through a Financial Assistance Program of the Social Cooperation Study (2006-2010)

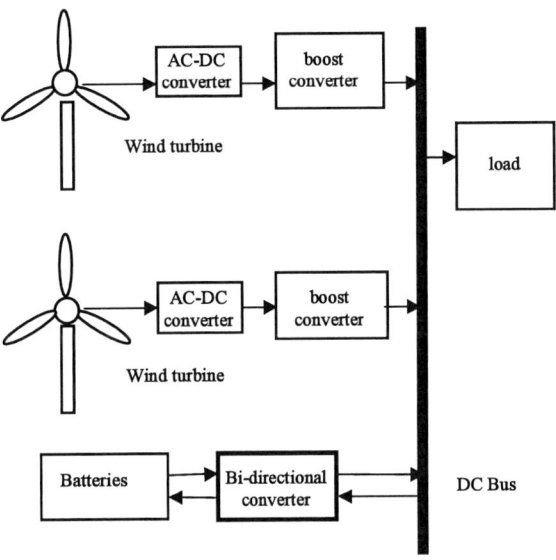

Fig. 1 Application of proposed bidirectional converter

the main switch Sm and the RCD snubber across it. Therefore, to solve these problems, we propose a ZCT circuit for the converter.

Fig. 2 shows the main circuit configuration for a prototyped ZCT bidirectional converter. The ZCT circuit consists of a resonant capacitor C_r and inductor L_r, two auxiliary switches Sa1 and Sa2, six diodes and two auxiliary capacitors. The ZCT circuit is connected to the main switch in parallel, and then it is useful for any topologies. The bi-directional converter consists of the boost converter in which the IGBT Sm1 and diode Dm2 are used and the buck converter in which the IGBT Sm2 and diode Dm1 are used.

In this system, the efficiency is greatly influenced due to losses by the hard switching of the main switches Sm1 and Sm2 and the RCD snubber circuits across each switch. Therefore, to solve these problems we adapt the ZCT circuit for the proposed bi-directional converter. The same resonant components of inductor L_r and capacitor C_r can be used for the ZCT operation of both boost and buck converters. Therefore the ZCT circuit configuration becomes quite simple as shown in Fig. 2.

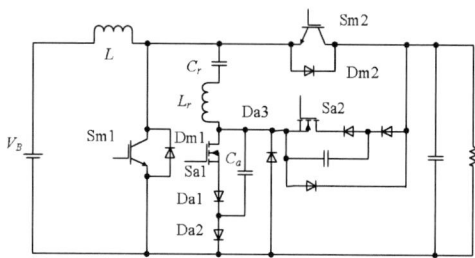

Fig. 2 Proposed ZCT bidirectional converter.

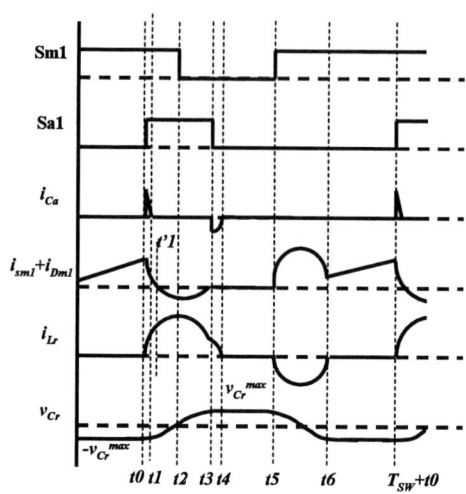

Fig. 3 Typical waveforms in one switching duration or T_{SW} to illustrate the principle of operation of ZCT converter.

Fig. 3 shows the key voltage and current waveforms of the converter, when it is operated in the boost mode. T_{SW} is the one switching duration. There are seven intervals of operation for the converter and Fig. 4 shows equivalent circuits for each interval when the converter is operated in the boost mode. The inductor current of the converter is controlled in CCM. The ZCT circuit reduces the power losses caused by the IGBT tail current when Sm1 is turned-off. If the inductor current is controlled in the continuous current mode, it reduces the switching losses only when Sm1 is turned-off.

To simplify the analysis, following assumptions are used: The circuit is operating in steady state. The switches, diodes and all the components are ideal.

Interval 1 ($t_0<t<t_1$): When the auxiliary switch Sa1 is turned on at t_0, the discharge current from the capacitor C_a begins to flow through Sa1 and Da1. In the mean time, the capacitor C_r with $-v_{Cr}^{max}$ at t_0 begins to discharge due to the resonance between C_r and L_r. The C_a current i_{Ca} decreases to zero at t_1.

Interval 2 ($t_1<t<t_2$): Since the Sm1 current i_{Sm1} is the difference between the source current i_{in} and the resonance current i_{Lr}, and i_{Sm1} decreases to zero at t'_1, because i_{Lr} equals i_{in}. That is, Sm1 turns-off with zero current at t'_1 and its anti-parallel diode Dm1 starts conducting carrying current in the opposite direction.

$$i_{Lr}^{peak} = \frac{v_{Cr}^{max}}{Z_r} \tag{1}$$

where, $Z_r = \sqrt{L_r / C_r}$

The gating signal for Sm1 is removed at t_2.

$$t_2 - t_0 = \frac{2\pi\sqrt{L_r C_r}}{4} \tag{2}$$

Interval 3 ($t_2<t<t_3$): The resonant current i_{Lr} charges C_r with the polarity shown in Fig. 4(iii). Since the resonant current i_{Lr} is larger than source current i_{in}, Dm1 continues to conduct. This mode ends when Sa is turned-off at t_3.

$$t_3 - t_2 \leq \frac{2\pi\sqrt{L_r C_r}}{4} \tag{3}$$

Interval 4 ($t_3<t<t_4$): Since Sa1 is turned-off, the electromagnetic energy stored in L_r transfers to the capacitor C_a.

Interval 5 ($t_4<t<t_5$): Sm1 and Sa1 are both in off state. The source current i_{in} supplies the load and the smoothing capacitor C_o through the boost diode Dm2. The electromagnetic energy stored in L transfers to them. In this interval, C_r voltage remains at v_{Cr}^{max}.

Interval 6 ($t_5<t<t_6$): When Sm1 is turned on at t_5, both the source current i_{in} and the resonant current i_{Lr} through L_r and C_r begins to flow into Sm1. When the resonant current i_{Lr} reaches the crest value in negative direction, the Sm1 current i_{Sm1} becomes the maximum value. After a half of resonant cycle, i_{Lr} decreases to zero, i_{Sm1} becomes to the same level as i_{in} at t_6.

Interval 7 ($t_6<t<T_{SW}+t_0$): In this interval, C_r voltage v_{Cr} is not varied, because there is no resonant current i_{Lr}. Only the source current i_{in} flows through Sm1, and electromagnetic energy is stored in L. After the end of *Interval 7*, the operation returns to *Interval 1* again.

III.   DIGITAL CURRENT CONTROLLER

When the inductor current is controlled in the CCM, an external sawtooth ramp circuit is added to the current controller to stabilize the current feedback loop[5,6]. In the proposed system, a digital controller is adopted to control the inductor current in the CCM.

Fig. 5 shows the digital controller which is

596

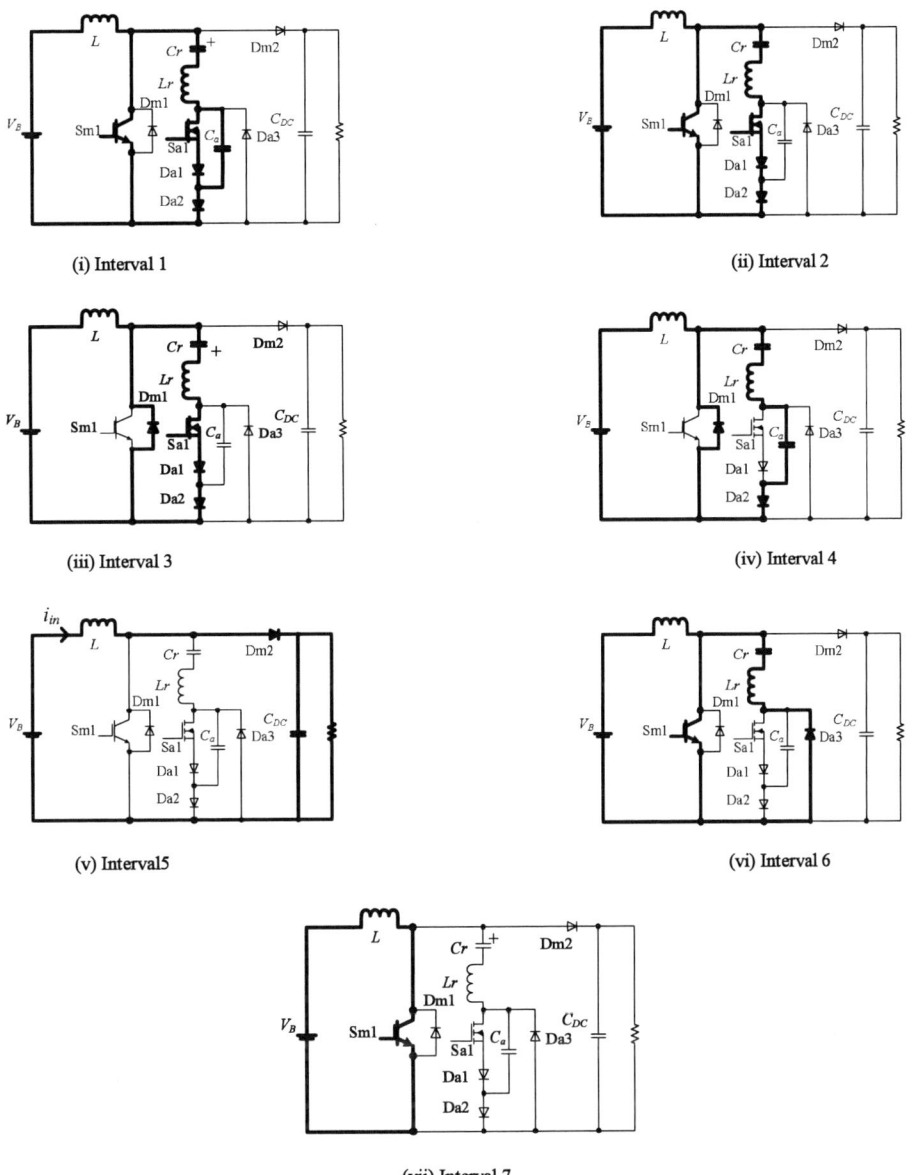

(i) Interval 1

(ii) Interval 2

(iii) Interval 3

(iv) Interval 4

(v) Interval5

(vi) Interval 6

(vii) Interval 7

Fig. 4 The equivalent circuits in each interval.

implemented by a FPGA. Fig. 5 (a) shows the block diagram of the current controller and (b) shows the related timing chart. The frequency of the main clock CLK is 25 MHz. The switching frequency of the converter is obtained by counting the CLK to 511 (=2^9-1) and then the frequency becomes approximately 49 kHz. The inductor current AD_DATA is measured by a shunt resister through a 2.5MHz A/D converter. A commanded value of the peak inductor current CMD_DATA is calculated in DSP and the commanded current is send as 12 bit data from DSP to FPGA through a parallel port. To achieve the external ramp in FPGA, a down counter is added to the controller. The down counter counts from the commanded current CMD during one switching term. A comparator compares the output from the down

counter CM to the inductor current AD. When AD is bigger than the output CM, the comparator outputs 1. In the conventional converter, the switch is turned off when the output of the comparator is changed. However, in the proposed ZCT converter, the aux switch must be turned on before the main switch is turned off as explained in section II. A Johnson counter and a RS flip flop are used for the switches to achieve the ZCT.

Fig 6 shows how a down counter affects the system's PWM pulses and inductor currents. When there is no down counter, the current cannot be controlled in the same pulse width as shown in fig.6 (a). When a down counter is built in FPGA, the current is controlled in the same pulse width as shown in fig.6 (b).

The stability of the system depends on the down

597

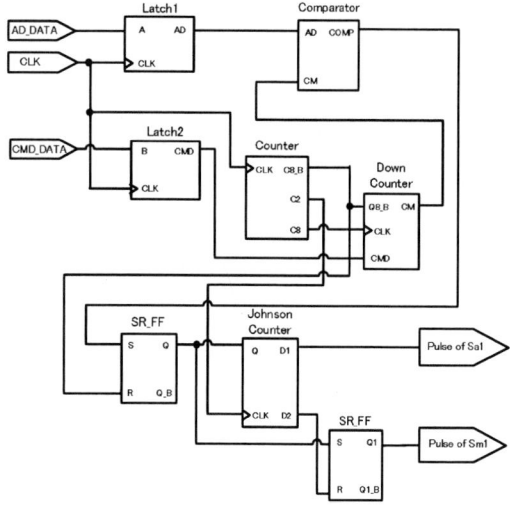

(a) block diagram

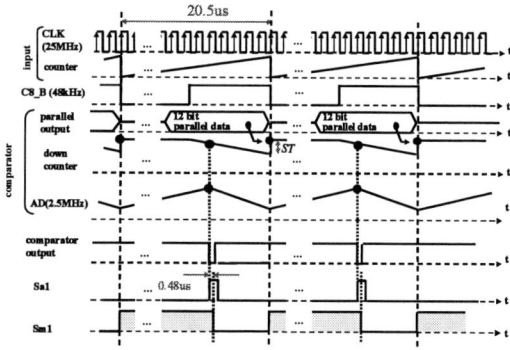

(b) timing chart

Fig. 5 Proposed digital current controller in FPGA.

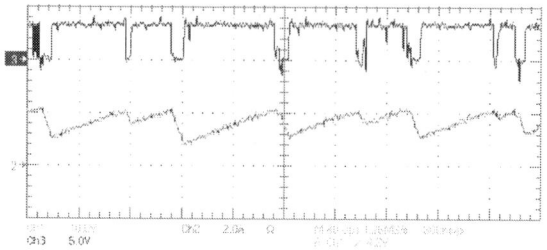

(a) without down counter

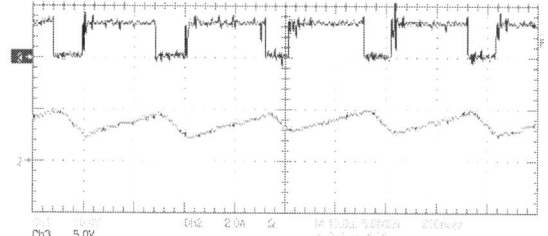

(b) with down counter

Fig. 6 Inductor current without down counter and with down counter

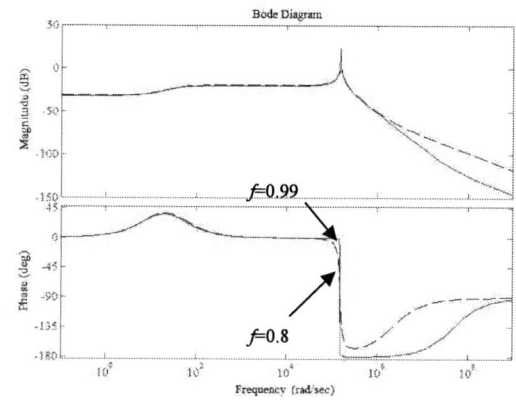

Fig. 7 bode diagram.

counter in the digital controller. The state equation in the discrete time is expressed as follows[6]:

$$x[k+1] = (I + TA_{dk})x[k] + TB_{dk}v_{in}[k] \qquad (4)$$

where, x is the vector $[iL \; vC]^T$ and T is the sampling period.

$$I + TA_{dk} = \begin{bmatrix} 1 - \dfrac{V_C}{LS + V_{in}} & -TD'/L \\ (TD'/C) + \dfrac{I_L}{C}\dfrac{1}{S + \dfrac{V_{in}}{L}} & 1 - (T/RC) \end{bmatrix} \qquad (5)$$

where, S is the value of down count per second and D' is the duty ratio of the off state (1-D).

The characteristic equation is deriving from (5) as:

$$G(z) = \left\{ z - \left(1 - \frac{V_C}{LS + V_{in}} \right) \right\} \left(z - 1 + \frac{T}{RC} \right) + \frac{TD'I_L}{LC\left(S + \dfrac{V_{in}}{L} \right)} + \frac{T^2D'^2}{LC}$$

$$(6)$$

When the sampling time is short enough, the second and third terms of the right hand in (6) can be neglected. Therefore, the stability depends as follows:

$$f(S) = \left| -\left(1 - \frac{Vc}{LS + V_{in}} \right) \right| < 1 \qquad (7)$$

When the system has not a down counter and the boost ratio is larger than 2, $f(S)$ is not smaller than 1 because S is 0. Therefore the inductor current cannot be controlled

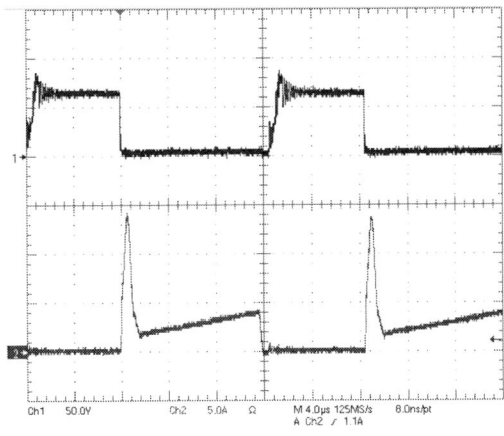

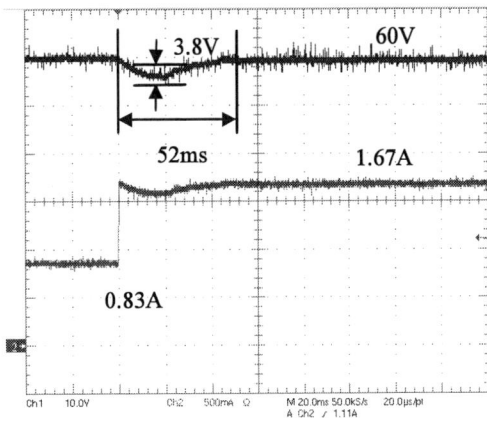

Fig. 8 Voltage and current waveforms of the main switch Sm1.

Fig. 9 Transient characteristics.

in the same pulse width as shown in Fig.6 (a). Although the boost ratio is larger than 2, a value of down count makes $f(S)$ smaller than unity when the system has the down counter. Fig. 7 shows bode diagrams which are obtained by Matlab. The phase margin in $f(S)$ is 0.8 is larger than the one in $f(S)$ is 0.99. Therefore, this result shows that the down counter in FPGA can stabilize the inductor current.

IV. EXPERIMENTAL RESULTS

Based on the foregoing design method, an experimental system was assembled in the laboratory. The bidirectional converter consists of the main switch Sm1 and Sm2 (IGBT: IRG4PC40UD) and the auxiliary switches Sa1 and Sa2 (MOSFET: IRF8010 made by IR). The system controller consists of a mother board (DSK6713 made by TI: DSP TMS670C6713) and a daughter board (AED-109 made by Signalware: FPGA VertexE200 made by Xilinx).

In the experimental system, the peak of the inductor current can be controlled from -10A to +10 A. The input voltage is 24V in the boost mode. To obtain more power, the input voltage will be set at higher value.

Fig. 8 shows the switching transient waveforms of Sm1 in the boost mode. In this figure, the voltage across Sm1 and the current of Sm1 are shown, where the output power is 50 W. The input voltage is 24 V and the output voltage command is set at 60V, and then the duty ratio is approximately 60%. When the current i_{Sm1} becomes equal to zero or a negative value, Sm1 can be turned-off with ZCS.

Fig. 9 shows the characteristics of the digital controller for the output voltage in the boost mode. The output voltage can be controlled to the commanded value, even if the output power is changed rapidly from 50W to 100W. Where, the gains K_P and K_I of the digital PI controller are set at 0.07 and 0.07, respectively. Under the condition where the output power is 100W, the efficiency of the converter is 88%.

V. CONCLUSIONS

The ZCT bidirectional converter has been proposed. To achieve ZCS of the main switch, ZCT circuits have been adapted to the bi-directional converter. From the experimental results, we confirmed ZCS turn-off of the main switch in the boost mode. It is expected that the efficiency of the converter can be increased by the soft-switching technology.

In addition, a digital controller for the inductor current of the converter has been proposed. To avoid the subharmonic instability in CCM, a down counter is used in FPGA. From the experimental results, the response of the converter can be increased by the current controller.

Therefore, the proposed converter is suitable for the bidirectional converter in wind power generation systems or other renewable generation systems.

REFERENCES

[1] G.Hua, E.X.Yang, Y.Jiang and F.C.Lee, "Novel zero-current-transition PWM converters", IEEE PESC, p.538 (1993)

[2] B. Ray and A. Romney-Diaz, "Constant Frequency Resonant Topologies for Bidirectional DC/DC Power Conversion," IEEE APEC, pp.1031-1037(1993)

[3] O. Dranga, B. Buti and I. Nagy, "Stability Analysis of a Feedback-Controlled Resonant DC-DC Converter," IEEE Trans. on Industrial Electronics, vol. 50, no. 1, pp.141-152, February 2003

[4] P. Chen and A.K.S. Bhat, "A Soft-Switched AC to DC Converter Operating in DCM: Analysis, Design, Simulation and Experimental Results," IEEE APEC (2004)

[5] R. Ridley, "A New, Continuous-Time Model For Current-Mode Control, IEEE Trans. on Power Electronics, vol. 6, no. 2, April, pp.271-280, 1991

[6] J.G. Kassakian, M.F. Schlecht, and G.C. Verghese, Principles of Power Electronic, Addison-Wesley, Massachusetts, 1991

[7] N. Kasa, Y. Harada, T. Iida and A. Bhat, "Zero-current transition converters for independent small scale power network system using lower power wind turbines," SPEEDAM 2006

[8] T. Iida, N. Hino and N. Kasa, "Soft-switching boost converter with current-minor-loop by FPGA," Annual meeting reports of Japan Society of Power Electronics, 2007

Control Method for a Single Phase Arbitrary Waveform-output Inverter

Satoshi Taniguchi*, Keiji Wada* Toshihisa Shimizu*
*Tokyo Metropolitan University / Department of Electrical and Electronics Engineering
1-1 Minami Osawa, Hachioji, Tokyo, Japan
e-mail: *shimizu@eei.metro-u.ac.jp*

Abstract—A method for the control of a single phase voltage type (PWM) inverter that can generate arbitrary waveforms in a wide frequency region is presented. The feature of this method is that single-phase rotation coordinate transformation is applied. A 90 degrees phase shifter is required, although there is no ideal phase shifter. The operational characteristics are experimentally compared using two kinds of quasi phase shifters. Generation of sinusoidal and rectangular waves was verified for the inverter using both methods. However, some problems of both phase shifters are identified for the generation of arbitrary waveforms.

Keywords—Single phase system, Converter control, Switched-mode power supply

I. INTRODUCTION

Power supply systems that generate an arbitrary voltage waveform in a wide frequency range are used in the power quality testing process of electronics products. Analogue amplifier systems have been used for this purpose; therefore, some serious problems such as low conversion efficiency and large weight and volume have been unavoidable. In order to overcome these problems, a high-frequency pulse width modulation (PWM) inverter system with adequate control methods was studied.

In recent years, a number of papers regarding PWM amplifiers that utilize high frequency switching technology have been reported [1]-[3]. In order to realize an inverter system that can generate arbitrary waveforms, we have investigated a control method suitable for a single-phase voltage source PWM inverter that utilizes rotating frame transformation in the control circuit [4]. The rotating frame transformation has been widely used for three-phase inverters, because three-phase to $\alpha\beta$-phase transformation can be easily realized [5]. However, in the case of a single phase inverter, the imaginary (β) axis value on the $\alpha\beta$-frame cannot be determined, because the output voltage has only a single-phase signal. Therefore, rotation coordinate transformation (dq-frame transformation) cannot be realized for the single-phase system. Therefore, we have proposed an approximate $\alpha\beta$-frame transformation method for the single-phase system using an imaginary axis value that is generated by a 90 degrees phase shifter. In this system, instantaneous voltage control can be realized based on the dq-frame. Since the amplitude and phase value on the dq-frame can be controlled adequately, the control characteristics of the arbitrary waveform generation are improved. The

major concern with this method is how to realize an ideal 90 degrees phase shifter, and a study has been reported regarding this [6]. This method is effective for the generation of a single frequency waveform, but some problems still remain for generation of an arbitrary waveform that includes many frequency components. This paper focuses on two kinds of 90 degrees phase shifter, and discusses the operational characteristics of the arbitrary waveform controller for the PWM inverter system.

II. CIRCUIT CONFIGURATION

Fig. 1 shows the main circuit configuration of the proposed inverter system. A single-phase voltage source PWM inverter is used for the main circuit, and an LC filter is used for elimination of the switching ripple voltage in the output.

Fig. 2 shows a control block diagram of the syetem. The inverter output voltage v_o is divided into a real-axis component v_α and an imaginary-axis component v_β by the 90 degrees phase shifter, and the rotating frame transformation is processed based on a phase angular reference θ^*, which is given individually. The output voltage is transformed into v_{od} and v_{oq}. These values are filtered by the first-order lag low pass filter (LPF), and are compared with V_d^* and V_q^*, which are the amplitude reference values of v_o. These results are then multiplied by the gain, Kp. It should be noted that V_d^* is given as an amplitude signal of the output voltage v_o on the d-axis (Fig. 3). V_q^* is given as zero in order for the phase angle of v_o and the d-axis to coincide. Finally, the values v_{id} and v_{iq} on the rotational frame are transformed to the value v_α^* on the stationary frame, and this value is used for the voltage command of the inverter. It is well known that a band-pass filter formed by placing the LPF on a dq-frame passes around the frequency of the rotating frame. Since the frequency of the rotating frame is controlled individually, the LPF operates as an adjustable frequency resonator and the control error can be minimized if the frequency component of the control signal is coincident with the rotating frame frequency. In the case of an arbitrary wave form, the rotating frame frequency is adjusted to the instantaneous frequency of the arbitrary waveform. The control error at the arbitrary waveform condition can then also be minimized. For example, it is possible to generate a sinusoidal voltage if the frame angular frequency, ω^*, is constant. The arbitrary waveform out put can also be

978-1-4244-1741-4/08/$25.00 ©2008 IEEE

generated by instantaneously changing the frame angular frequency ω^*, depending on the waveform command.

III. INTRODUCTION OF 90-DEGREE PHASE SHIFTER

An ideal 90 degrees phase shifter is required in this system. However, an ideal 90 degrees phase shifter does not really exist. The Hilbert transformer is one of the well known 90 degrees phase shifters in the communications field. However, the inherent time delay with the Hilbert transformer is relatively large, and causes instability in the feedback systems. Therefore, we have attempted to apply another quasi-90 degrees phase shifter that does not have an inherent time delay. In this investigation, two kinds of quasi 90 degrees phase shifters were selected, a differentiator and an all-pass filter, and the effectiveness of these shifters was examined.

A. Differentiator

A differentiator is considered as the first quasi phase shifter. The relation between v_α and v_β can be expressed by Eq. (1), and is shown in Fig. 4. The bode diagram of the relation between v_α and v_β is also shown in Fig. 4. ω_d is the gain-crossover angular frequency of the differentiator. The feature of the differentiator is that the phase shift is fixed to 90 degrees, regardless of the frequency. However, the amplitude of v_β increases as the frequency increases, and unity gain can only be obtained at the gain-crossover angular frequency. Hence, the high frequency components caused by the switching operation are amplified and these undesirable signals impose serious problems on the control characteristics.

$$V_{\beta-d}(s) = -\frac{s}{\omega_d}V_\alpha(s) \qquad (1)$$

B. All-pass filter

An all-pass filter is considered as the second quasi phase shifter. The relation between v_α and v_β is expressed as Eq. (2), and the bode diagram for this filter is shown in Fig. 5. The feature of the all-pass filter is that the gain is kept constant, regardless of the frequency, but the phase delay changes with the frequency and only becomes 90 degrees at the corner angular frequency, ω_{all}.

$$V_{\beta-all}(s) = -\frac{s - \omega_{all}}{s + \omega_{all}}V_\alpha(s) \qquad (2)$$

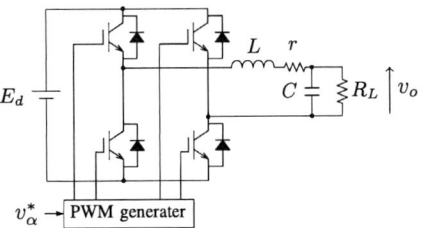

Fig. 1. Main circuit configuration.

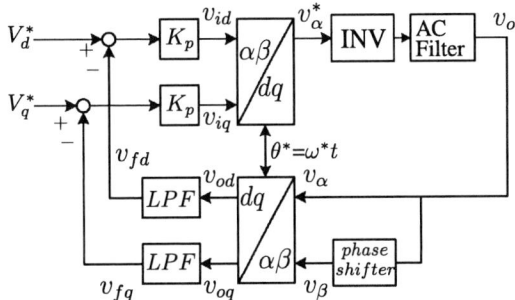

Fig. 2. Block diagram of control circuit.

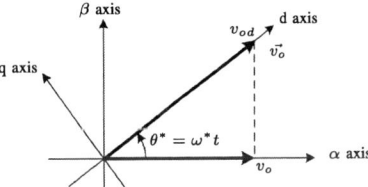

Fig. 3. Relation between output voltage vecter and the rotating frame system.

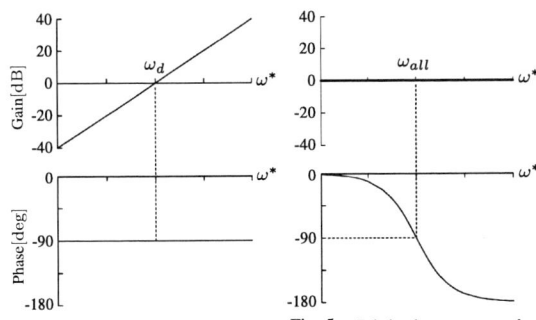

Fig. 4. Relation between v_α and v_β in a differentiator. Fig. 5. Relation between v_α and v_β in an all-pass filter.

IV. Operating Characteristic

Sinusoidal and rectangular waveforms were experimentally generated as representative of arbitrary waveforms. The operating characteristics when utilizing each phase shifter were examined. Table I shows the circuit parameters of the experimental setup.

A. Sinusoidal wave output

The sinusoidal wave, as one of the arbitrary waveforms, was considered. Figs. 6 and 7 show the experimental waveforms of the sinusoidal waveform output in the case for the differentiator and all-pass filter, respectively. In this case, the amplitude command is kept to a constant value (DC component), whereas the rotating phase angle of the rotation frame is increased at a fixed rate of $d\theta^*/dt$ $(=\omega^*) = 200 \times 2\pi$ [rad/s]. In this experiment, the gain-crossover angular frequency, ω_d, of the differentiator and the corner angular frequency, ω_{all}, of the all-pass filter were adjusted to the rotational angular frequency, ω^*, of the rotational frame. This means each phase shifter is operated as an ideal 90 degrees phase shifter at the rotational angular frequency component. When using either quasi phase shifter, the output voltage waveform generates a sinusoidal wave voltage. However, near the zero crossing of the output voltage v_o, the output waveform is distorted under the influence of the dead time, thereby, v_β is distorted. In the case of the differentiator and the all-pass filter, the distortion of v_β is different. Because the distortion v_o near the zero crossing contains high frequency component, it is not amplified using the all-pass filter, but it is amplified when using the differentiator. Therefore, the distortion of v_β has little direction when using the all-pass filter. However, there are few differences agreement if both output waveforms are compared. Therefore, it can be said that the distortion of v_β does not affect the output waveform. This is because the influence of the distortion of v_β is reduced by the LPF (which forms a resonator for a stillness coordinate system) inserted on the dq-axis.

B. Rectangule wave output

The rectangular wave, which is the another arbitrary waveform, is considered. Figs. 8 and 9 show the experimental rectangular waveform outputs in the case when the differentiator and the all-pass filter is used, respectively. The amplitude command is kept to a constant value, as in the case of the sinusoidal wave generation. However, the phase angle of the rotational frame is controlled as follows:
(a) For the period where the instantaneous output voltage is constant, the phase angle reference θ^* is kept constant.
(b) For the period where the instantaneous output voltage is changing, increasing to high voltage and decreasing to low voltage, the phase angle, $\omega^*=d\theta^*/dt$, is changed as $200 \times 2\pi$ [rad/s].
Fig. 8 shows that a rectangular wave voltage can be generated when the differentiator is used.

TABLE I
Experimental Circuit Paramerters.

Inductance of Main Circuit L	2 mH
Capacitance of Main Circuit C	2 μF
Load Resistance R_L	25 Ω
Resistance of Inductance r	1 Ω
DC Voltage Source E_d	100 V
Switching Frequency f_S	10 kHz

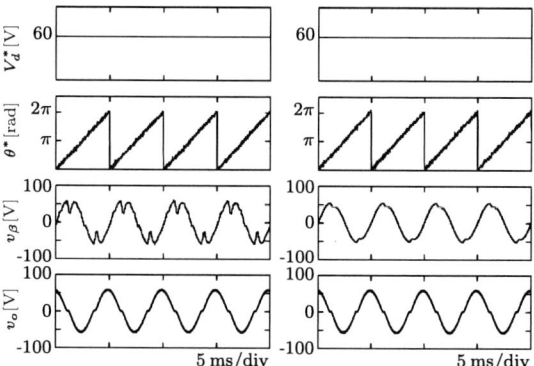

Fig. 6. Operation waveforms when differentiator is used (ω^*:200×2π [rad/s]).

Fig. 7. Operation waveforms when all-pas filter is used (ω^*:200×2π [rad/s]).

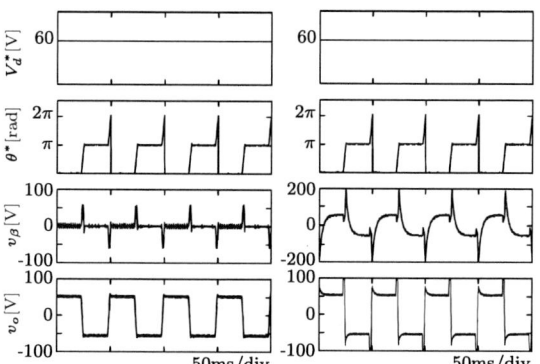

Fig. 8. Operation waveforms when differentiator is used.

Fig. 9. Operation waveforms when all-pass filter is used.

However, as shown in Fig. 9, when the all-pass filter is used, the waveform has an overshoot superimposed on the generated rectangular wave. On the duration where ω^* is $200\times2\pi$[rad/s] for the case of the differentiator and the case of the all-pass filter, a small difference can be observed the difference is in the amount of change of v_β. On the other hand, on the duration where ω^* is zero, for both the cases of using the differentiator and the all-pass filter, v_β is significantly different.

When using the differentiator, v_α and v_β always have a phase difference of 90 degrees. However, when the all-pass filter is used, and ω^* is zero, neither the amplitude nor the phase is changed, and the 90 degrees phase shift for v_β cannot be achieved and v_β becomes the same as v_α.

When ω^* is zero, $\sin\omega^*t$ is zero; therefore, v_{od} and v_{oq} are expressed as Eq. (3)

$$v_{od} = v_\alpha \cos\omega^*t + v_\beta \sin\omega^*t \rightarrow v_\alpha \qquad (3)$$
$$v_{oq} = -v_\alpha \sin\omega^*t + v_\beta \sin\omega^*t \rightarrow v_\beta$$

Therefore, v_β only affects the q-axis and does not affect d-axis. Moreover, v_α^*, which is on the stationary frame and is the voltage command of the inverter, is expressed as Eq. (4).

$$v_\alpha^* = K_P(V_d^* - v_{fd})\cos\omega^*t - K_P(V_q^* - v_{fq})\sin\omega^*t \qquad (4)$$

$$\rightarrow K_P(V_d^* - v_{fd})$$

Therefore, the value of the q-axis does not affect the output waveform, when ω^* is zero. Additionally, a closed loop is not formed on the q-axis, so the value of the q-axis is not controlled. This means that, when ω^* is zero, v_β does not influence the output waveform. Therefore, when ω^* is zero, for both cases of using the differentiator or the all-pass filter, v_o is not different. However, if ω^* changes to $200\times2\pi$ [rad/s], a closed loop is formed also on the q-axis; therefore, the value of the q-axis at that time is an initial deviation, and the output waveform is affected.

In the case when the all-pass filter is used, the closed loop on the q-axis is not formed during ω^* is zero, and v_β comes to the same value of v_α. After that, when the closed loop on the q-axis control is formed, the value of v_β affects as a large initial deviation. Hence, the generated rectangular waveform has an overshoot.

On the other hand, in the case when the differentiator is used, when ω^* comes to zero, the value of v_β becomes zero even if the closed loop is not formed on the q-axis control. Thus, even if a closed loop is formed, there is almost no initial deviation, and a rectangular waveform without overshoot can be generated.

C. Sinusoidal wave output($\omega^* \neq \omega_d$ and ω_{all})

In an arbitrary waveform, the output frequency variably changes. Therefore, it is necessary to consider that ω_{all} and ω_d are not adjusted to ω^*. Figs. 10 and 11 show the experimental sinusoidal waveform outputs for the

differentiator and all-pass filter, respectively. At this time, ω^* is $200\times2\pi$ [rad/s], and ω_d and ω_{all} are $50\times2\pi$ [rad/s]. For either phase shifter, the required sinusoidal wave cannot be generated. In particular, the amplitude error is quite large when the differentiator is used. This indicates that when either phase shifter is used, it is necessary to adjust ω_d and ω_{all} to ω^*.

V. THE CHARACTERISTIC OF SYSTEM, WHEN ω_d AND ω_{all} ARE ADJUSTED TO ω^*

In Section IV, the requirement for the adjustment of ω^* at ω_d and ω_{all} was checked. Figs. 10 and 11 show the experimental waveforms of the sinusoidal waveform output, when $\omega^*(=200\times2\pi$ [rad/s]) is not adjusted to ω_d and ω_{all}, which are $50\times2\pi$ [rad/s], whereas Figs. 12 and 13 show the experimental sinusoidal waveform outputs when ω^* is adjusted to ω_d and ω_{all}, which are $50\times2\pi$ [rad/s].

The output waveforms when the differentiator is used in the controller are sinusoidal waveforms, as shown in Fig. 12. However, the output waveforms when the all-pass filter is used in the controller do not generate sinusoidal waveforms, but include a large amount of DC component, as shown Fig. 13.

When the all-pass filter is used, it is not sufficient to simply adjust ω^* and ω_{all}. In order to verify the characteristics of the control system, the open-loop transfer function of the whole system is considered.

A. Frequency characteristic

In the case when the differentiator is used, the transfer function of the control circuit is given by,

$$\frac{V_\alpha^*(s)}{V_\alpha(s)} = \frac{K_P(1+2s\tau)}{(1+s\tau)^2 + (\omega^*\tau)^2}. \qquad (5)$$

where τ is the time constant of the LPF. On the other hand, the transfer function of the control circuit when the all-pass filter is used is given by,

$$\frac{V_\alpha^*(s)}{V_\alpha(s)} = \frac{K_P(1+s\tau+j\omega^*\tau)}{(1+s\tau)^2 + (\omega^*\tau)^2}. \qquad (6)$$

The transfer function of the main circuit (LC filter and load resistance)is expressed as follows.

$$\frac{V_o(s)}{V_\alpha^*(s)} = \frac{R_L}{CR_LLs^2 + (L+CR_Lr)s + R_L + r} \qquad (7)$$

Figs. 14 and 15 show bode diagrams of the open-loop transfer function for each case. Bode diagrams are a solid line and a dotted line when making rotating frame angular frequency ω^* into $50\times2\pi$[rad/s] and $200\times2\pi$[rad/s], respectively. The phase shift decreases to zero at low frequency region when the differentiator is used, indicating that the stability of the control system is warranted. On the other hand, the phase shift is increased to 180 degrees at low frequency region when the all-pass filter is used. This indicates that some stability problems will occur in the system when the all-pass filter is applied; therefore, stability evaluation should be considered in the low frequency region.

B. Direct-current characteristic

A/D converters are used in the inverter system in order to detect the output voltage and current signals; therefore, it is unavoidable that some DC component is contained in the detected signal. Moreover, a few DC components may be contained in the inverter output voltage. Therefore, the DC component V_{DC}, is regularly contained in the control signal of the inverter, and this functions as a disturbance. This section discusses the influence of the DC component upon the control stability on both the differentiator and the all-pass filter. Fig. 16 shows a block diagram of the inverter system with focus on the DC component, V_{DC}. H_{DC} is the transfer function of the control circuit focused on the DC component. This is the positive feed back system for the DC component, so that the DC gain of this closed loop is expressed as,

$$v_\alpha^* = \frac{H_{DC}}{1 - H_{DC}} V_{DC} . \tag{8}$$

Equation (8) indicates that the DC component contained in v_α^* increases when the value of H_{DC} is close to 1. Therefore, the value of H_{DC} should be much smaller or much higher than 1.

1) Differentiator: In the case when the differentiator is used as the 90 degrees phase shifter, the DC component of the differentiator output becomes zero. Therefore, H_{DC} is expressed by Eq. (9),

$$H_{DC} = -K_p G \cos \Delta \tag{9}$$

where G and Δ are the gain and the phase characteristic of the LPF, respectively. G and Δ are given by

$$G = \frac{\omega_c/\omega^*}{\sqrt{1 + (\omega_c/\omega^*)^2}} \tag{10}$$

$$\Delta = \tan^{-1} \left(\frac{1}{\omega_c/\omega^*} \right). \tag{11}$$

ω_c is the cutoff angular frequency of LPF.

Fig. 17 shows the amplitude of H_{DC} in the case when the angular frequency ratio, ω_c/ω^*, is changed, and the cutoff angular frequency of the LPF on the control circuit is kept constant at $10 \times 2\pi$[rad/s]. It can be seen that H_{DC} always takes a value smaller than 1. The DC component in v_α^* can then be suppressed at any output frequency condition.

2) All-pass-filter: When the all-pass filter is used, the DC component remains in the output of the all-pass filter as it is. Therefore H_{DC} is expressed by Eq. (12):

$$H_{DC} = \sqrt{2} K_p G \sin(\Delta - \pi/4). \tag{12}$$

Fig. 18 shows the amplitude of H_{DC} in the case when the frequency ratio, ω_c/ω^*, is changed. It is clear that the amplitude characteristic shown in Fig 18 is rather different from that shown in Fig. 17. The amplitude of H_{DC} takes a value close to 1 in the range from $\omega_c/\omega^* =$

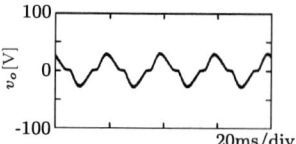

Fig. 10. Output-waveform when the differentiator is used (ω^*:$200 \times 2\pi$ [rad/s]).

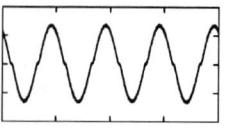

Fig. 11. Output-waveform when the all-pass filter is used (ω^*:$200 \times 2\pi$ [rad/s]).

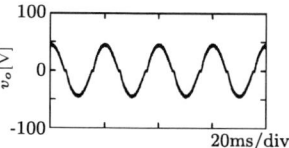

Fig. 12. Output-waveform when the differentiator is used (ω^*:$50 \times 2\pi$ [rad/s]).

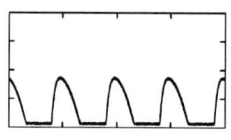

Fig. 13. Output-waveform when the all-pass filter is used (ω^*:$50 \times 2\pi$ [rad/s]).

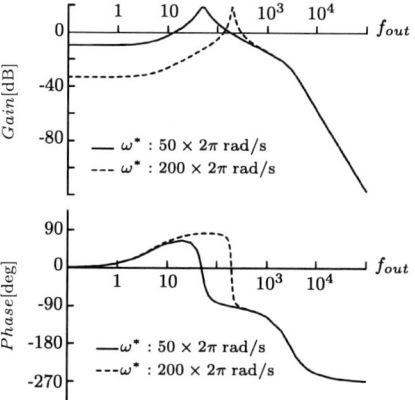

Fig. 14. Bode diagram when the differentiator is applied.

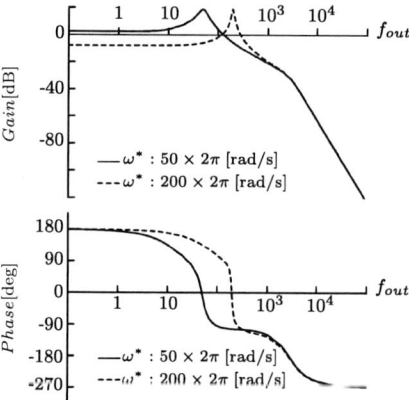

Fig. 15. Bode diagram when the all-pass filter is applied.

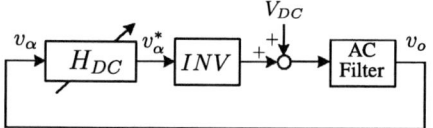

Fig. 16. Block diagram for the DC component.

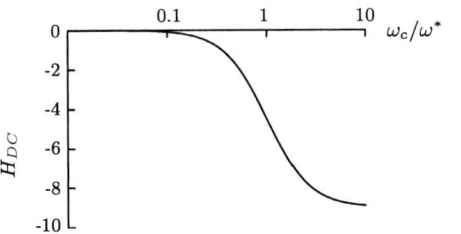

Fig. 17. Characteristics of H_{DC} for application of the differentiator (applying differentiator).

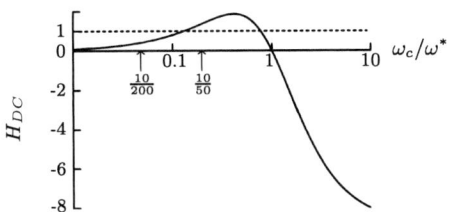

Fig. 18. Characteristics of H_{DC} for application of the differentiator (applying all-pass filter).

0.13 to 0.76. In the case when the rotating frame angular frequency, ω^*, is $50\times2\pi$[rad/s], the frequency ratio, ω_c/ω^* is 0.2. The amplitude of H_{DC} is then close to 1, and the resultant DC component contained in v_α^* is significantly increased, as shown in Fig. 13. On the contrary, the amplitude of H_{DC} is close to zero at an rotating frame angular frequency of $200\times2\pi$[rad/s] ($\omega_c/\omega^*= 0.05$), and no DC component is included, as shown in Fig. 7. From the above discussion, care should be taken in that a large DC voltage component is included in the output voltage at a specific rotating frame angular frequency when the all-pass filter is used as the 90 degrees phase shifter. In the case of the differentiator, the system can generate a sinusoidal waveform, regardless of the rotating frame angular frequency. On the other hand, in the case of all-pass filter, the stability of the DC component depends on the rotating frame angular frequency and the cutoff angular frequency of the LPF.

VI. CONCLUSION

A control method was proposed for a single-phase voltage source PWM inverter, which utilizes a rotating frame transformation. In order to realize practical rotation frame transformation in the single-phase system, the transfer characteristics of a differentiator and an all-pass filter, used as the 90 degrees phase shifter, were examined. The performance of the waveform generation was studied for both rectangular and sinusoidal waveforms. The influence of the DC component included in the control signal was also discussed.

When the differentiator is used, it is necessary to adjust the gain-crossover angular frequency and the rotating frame angular frequency. When the all-pass filter is used, it is not only required to adjust the corner angular frequency and the rotating frame angular frequency, but also to redesign the cutoff frequency of the LPF, with output frequency. However, since the cutoff angular frequency of the LPF affects the transient response of the amplitude information, changing the cutoff angular frequency is not realistic. Therefore, when using the all-pass filter, the frequency band that can be generated is limited. In conclusion, the differentiator excels the all-pass filter as a 90 degrees phase shifter in this system.

ACKNOWLEDGMENT

This research project was supported by KAKENHI (Grant-in-Aid for Scientific Research)(No.19360132)

REFERENCES

[1] K. Iwaya, and I. Takahashi : "Switching Type Power Amplifier Using Multilevel Inverter" , Trans on IEEJ, Vol.123-D, No.11, pp. 1339-1344, 2003 (in Japanese)

[2] N. Zomood, and G. Holmes : "Stationary Fraqme Current Regulation of PWM Inverters With Zero Steady-State Error," *IEEE Trans. on Ind. Appl.*, Vol. 18, No. 3, pp. 814-822, May, 2003

[3] H. Abe, and H. Fujimoto : "Multirate Perfect Tracking Control of Single-phase Inverter with Inter Sampling for Arbitrary Waveform" , in Proc. The Fourth Power Conversion Conference, Nagoya, pp. 810-815, 2007.

[4] S. Hashino, K. Wada, and T. Shimizu : "A Generation Control of Arbitrary AC Waveforms for the Single-phase Voltage Source PWM Inverter Utilizing an Adaptive Frequency Loss-less Resonator" ,IEEJ, Vol. 127-D, No. 2, pp. 103-111, 2007 (in Japanese)

[5] S. Fukuda, and T. Yoda : "A Novel Current-Tracking Method for Active Filters Based on a Sinusoidal Internal Model" *IEEE Trans. on Ind. Appl.*, Vol. 37, No. 3, MAY/JUNE 2001

[6] M. Ciobotaru, R. Teodorescu, and F. Blaabjerg : "A New Single-Phase PLL Structure Based on Second Order Generalize Integrator" , 37th IEEE Power Electronics Specialists Conference 2006 TuD2-2

Elimination of Harmonics in Multilevel Inverters with Non-Equal DC Sources Using PSO

A. K. Al-Othman [*], and Tamer H. Abdelhamid [†]

[*] College of Technological Studies, Electrical Engineering Department, Kuwait, e-mail: ak.alothman@*paaet.edu.kw*
[†] College of Technological Studies, Electrical Engineering Department, Kuwait, e-mail: th.hassan@*paaet.edu.kw*

Abstract—**Multilevel inverters supplied from equal and constant dc sources almost don't exist in practical applications. The variation of the dc sources affects the values of the switching angles required for each specific harmonic profile, as well as increases the difficulty of the harmonic elimination's equations. This paper presents an extremely fast optimal solution of harmonic elimination of multilevel inverters with non-equal dc sources using a novel Particle Swarm Optimization (PSO) algorithm. The overall system is suitable for large variable speed drives, UPS systems, and on-line utility applications such as static var compensation. A set of mathematical equations describing the general output waveform of the multilevel inverter with non-equal dc sources is formulated. PSO is then employed to compute the optimal solution set of switching angles, if it exists for each required harmonic profile. Theoretical studies for different case studies regarding the number of levels and harmonic profile are carried out to show the effectiveness and robustness of the proposed technique, and validated through both simulations and laboratory experimentation.**

Keywords—**Multilevel converters, harmonics, pulse width modulation (PWM), optimal control.**

I. INTRODUCTION

Multilevel inverter is considered as one of the most significant recent advances in power electronics. They have drawn tremendous interest in the field of high-voltage high-power applications such as laminators, mills, compressors, large induction motor drives, UPS systems, and static var compensation [1]. Its concept is based on producing small output voltage steps, resulting in better power quality. Despite the need for more power transistors, they operate at low voltage levels and also at low switching frequency so that the switching losses are also reduced. Some of the fundamental multilevel topologies include the diode-clamped, flying capacitor, and cascaded H-bridge structures [2]. Multilevel inverters are mostly supplied from dc sources obtained from fuel cells, ultra capacitors, ect. It is worth noting that in most of the reported work, it was assumed that the dc sources were all equal, which will probably not be the case in applications even if the sources are nominally equal.

The key issue in designing an effective multilevel inverter is to ensure that the total harmonic distortion (THD) of the output voltage waveform is within acceptable limits. Selective harmonic elimination pulse width modulation has been intensively studied in order to achieve low THD [3]. The output voltage waveform analysis using Fourier theory produces a set of non-linear transcendental equations. The solution of these equations, if exists, gives the switching angles required for certain fundamental component and selected harmonic profile. Iterative procedures such as *Newton-Raphson* method has been used to solve these sets of equations [4]. This method is derivative-dependent and may end in local optima, and a judicious choice of the initial values alone guarantees conversion. Another approach based on converting the transcendental equation into polynomial equations is presented in [5], where resultant theory is applied to determine the switching angles to eliminate specific harmonics. This approach, however, appears to be unattractive because as the number of inverter levels increases, so does the degree of the polynomials of the mathematical model. This is likely to lead to numerical difficulty and substantial computational burden as well.

Genetic algorithms (GA) have been used to obtain optimal solutions for inverter circuits supplied from constant dc sources [6]. Despite their effectiveness in selective harmonic elimination, they are complicated and their parameters such as crossover and mutation probability, population size and number of generations are usually selected as common values given in literature or by means of a trial and error process to achieve the best solution set.

Heuristic algorithms such as Particle Swarm Optimization (PSO) [7] have the ability to combat the above drawbacks. As an optimization technique, PSO is much less dependent on the start values of the variables in the optimization problem when compared with the widely used *Newton-Raphson*. In addition, PSO doesn't rely on the guidance of the gradient information, such as the Jacobian matrix, hence it is more capable of determining the global optimum solution. PSO deal with all problems that usually considered very hard for researchers, such as integer variables, non-convex functions, non-differentiable functions, domains not connected, badly behaved functions, multiple local optima, and multiple objectives [8]. For these reasons, PSO has been adopted in this study.

This paper presents an optimal minimization technique assisted with PSO algorithm in order to highly reduce the computational burden associated with the solution of the nonlinear transcendental equations of the harmonic elimination problem of a cascaded H-bridge inverter with non-equal dc sources. The presented method has the advantage of its extremely short time required to reach at the optimal solution, if existed, as this is essential for online updates such that the algorithm can cope with any sudden variations of the voltage levels of the dc sources. The presented method can be extended to other multilevel inverter topologies, where an accurate and fast solution is

978-1-4244-1741-4/08/$25.00 ©2008 IEEE

guaranteed even for large number of levels and switching angles. Problem formulation and analysis are presented, simulations of the overall system for different case studies are carried out, and experimental verifications are also conducted and compared to those from both simulations and conventional methods, where the superiority of the presented algorithm is reported.

II. CASCADED H-BRIDGE MULTILEVEL INVERETR

The cascaded H-bridge multilevel inverter consists of a series of single-phase H-bridge inverter units, as shown in Fig. 1. It is modular in nature and can be extended to any required number of levels. It is supplied from several separate dc sources (SDCSs), which may be obtained from batteries, solar cells, or ultra-capacitors [9]. Each SDCS is connected to a single-phase H-bridge inverter and can generate three different voltage outputs, $+V_{dc}$, 0, and $-V_{dc}$. This is accomplished by connecting the dc source to the ac output side by using different combinations of the four switches Q_1, Q_2, Q_3, and Q_4. The ac outputs of the modular H-bridge inverters are connected in series such that the synthesized voltage waveform is the sum of all of the individual inverter outputs. All semiconductor devices of the H-bridges are only switching at the fundamental frequency, and consequently this is referred to as the fundamental switching scheme. Also, each H-bridge unit generates a quasi-square waveform by phase-shifting its positive and negative phase legs' switching timings. The number of output voltage levels in a cascaded multilevel inverter is then $2s+1$, where s is the number of dc sources. Three-phase version of this circuit is also available by adding another two phases and connecting their neutral point together. Fig. 2 shows the generalized output voltage of cascaded H-bridge multilevel inverter with non-equal dc sources. The total output voltage is given by $v_o= v_1+ v_2+ v_i+......+ v_s$. With enough levels and an appropriate

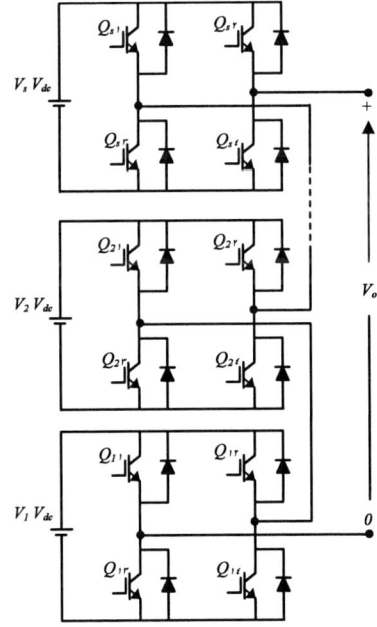

Fig. 1. Single-phase structure of a multilevel cascaded H-bridge inverter

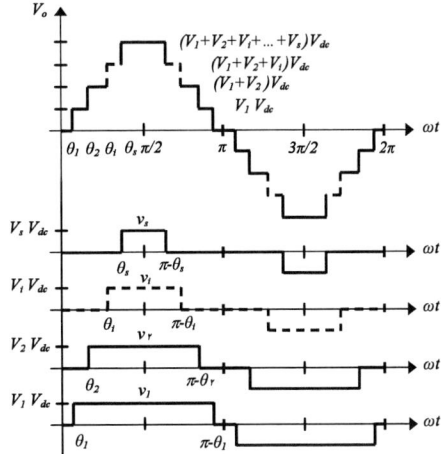

Fig. 2. Generalized output waveform of a multilevel inverters

switching angles θ_1, θ_2, θ_i, ..., and θ_s, the multilevel inverter results in an output voltage that is almost sinusoidal with a low THD (<5%) with each of the active devices subjected to a single dc source, and only switching at the fundamental frequency. This reduces both the voltage stress and the switching losses of the semiconductor devices, resulting in a better utilization and high overall efficiency.

III. PROBLEM FORMULATION

Assuming that the non-equal dc sources are known, and taking into consideration the characteristics of the inverter waveform shown in Fig. 2, from its odd nature, half- and quarter-wave symmetry, Fourier series expansion of the stepped output voltage waveform of the multilevel inverter with non-equal dc sources can be expressed as [10]:

$$V_o(\omega t) = \sum_{n=1,3,5,...}^{\infty} \frac{4V_{dc}}{n\pi}\{V_1 \cos(n\theta_1) + V_2 \cos(n\theta_2) + \cdots$$

$$\cdots +V_s \cos(n\theta_s)\}\sin(n\omega t) \quad (1)$$

Where the product V_iV_{dc} is the value of the ith dc source. Equation (1) has s variables (θ_1, θ_2, θ_3,......, θ_s), where $0 \leq \theta_1 < \theta_2 < \cdots < \theta_s \leq \pi/2$, and a set of solutions is obtainable by equating $s-1$ harmonics to zero and assigning a specific value to the fundamental component, as given below:

$$\left. \begin{array}{l} V_1 \cos(\theta_1) + V_2 \cos(\theta_2) + \cdots +V_s \cos(\theta_s) = m \\ V_1 \cos(3\theta_1) + V_2 \cos(3\theta_2) + \cdots +V_s \cos(3\theta_s) = 0 \\ V_1 \cos(5\theta_1) + V_2 \cos(5\theta_2) + \cdots +V_s \cos(5\theta_s) = 0 \\ \vdots \\ V_1 \cos(n\theta_1) + V_2 \cos(n\theta_2) + \cdots +V_s \cos(n\theta_s) = 0 \end{array} \right\} \quad (2)$$

Where $m=V_1/(4V_{dc}/\pi)$, and the modulation index m_a is given by $m_a=m/s$.

An objective function is then needed for the optimization procedure, which is selected as a measure of effectiveness of eliminating selected order of harmonics while maintaining the fundamental component at pre-specified value. Therefore, this objective function to be minimized is defined as:

$$F\left(\theta_1,\theta_2,\cdots,\theta_3\right)=\left[\sum_{n=1}^{s}V_1\cos(\theta_n)-m\right]^2+$$

$$+\left[\sum_{n=1}^{s}V_2\cos(3\theta_n)\right]^2+\cdots+\left[\sum_{n=1}^{s}V_2\cos\left((2s-1)\theta_s\right)\right]^2 \quad (3)$$

The optimal switching angles are obtained by minimizing (3) subject to the constraint $0\leq\theta_1<\theta_2<\cdots<\theta_s\leq\pi/2$, and consequently the required harmonic profile is achieved. The main challenge is the non linearity of the transcendental set of equations (2), as most iterative techniques suffer from convergence problems and other techniques such as elimination using resultant and GA are complicated.

IV. SOLUTION USING PSO

PSO has recently received much attention as robust stochastic search algorithms for optimization problem. PSO combines social psychology principles in socio-cognition human agents and evolutionary operations. Based and inspired by social behavior of bird flocking or fish schooling, PSO conduct the searching process using a population of particles. A particle represents a potential solution to the problem under investigation. Each particle in a given population adjusts its position by flying in a multi-dimensional search space until an unchanging position of the fittest particle is encountered.

Generally, PSO has the advantage of being very simple in concept, easy to implement, and computationally efficient algorithm. Unlike other heuristic algorithms, PSO possesses flexible and well-balanced operators to enhance and adapt the global and fine tune local search. PSO has been applied to various power system problems in which it proved to be extremely efficient relative to other evolutionary computation technique [11-13].

Like GA, PSO is a population based optimization tool. The system is initialized with a population of random solutions and searches for optima by updating generations. However, unlike GA, PSO has no evolution operators such as crossover and mutation. In PSO, the potential solutions, called particles, are flown through the problem space by following the current optimum particles. Each particle keeps tracking its coordinates in the problem space which is associated with the best solution (fitness) it has achieved so far. The fitness of each particle is also stored or memorized. This best fitness is called *personal best*. Another best value that is tracked by the particle swarm optimizer is the best value obtained so far by any particle in the neighbors of the particle. When a particle takes all the population as its topological neighbors, the best value is a best of personal bests and is called *global best*.

Generally, the advantages of PSO may be summarized as in the following:

1. PSO is fairly easy to apply, and only few parameters are required to be adjusted before application.
2. PSO has no evolution intermediate operators. (i.e. no crossover and mutation like the GA)
3. In PSO, the Global best particle (*Gbest*) leads the population by giving out information to the others potential solutions. Unlike GA, the whole population moves like one group [14].
4. PSO frequently converges to the solutions in fewer objective function evaluations than those required by GAs [14]. Therefore, PSO appears to be more efficient than GA.
5. PSO uses payoff (performance index) information along with memory to help and assist the search in the problem space.
6. In order to avoid premature convergence, PSO utilizes a distinctive feature of controlling a balance between global and local exploration of the search space. Such capability doesn't exist in GA.

PSO can be modeled mathematically by velocity and position equations as flows:

$$v_{i,j}(k+1)=\phi(k)v_{i,j}(k)+\alpha_1\lfloor\gamma_{1,j}(p_{i,j}-x_{i,j}(k)\rfloor$$
$$+\alpha_2\lfloor\gamma_{2,j}(G_{i,j}-x_{i,j}(k)\rfloor \quad (4)$$

$$x_{i,j}(k+1)=x_{i,j}(k)+v_{i,j}(k+1) \quad (5)$$

where:

i : Particle index.
j : Index of parameter of concern to be optimized.
x : The position of the ith particle and j^{th} parameter.
k : The discrete time index.
v : The velocity of the ith particle and j^{th} parameter.
P : The best position found by the ith particle and j^{th} parameter. (*personal best*)
G : The best position found by swarm. (*global best*)
γ : A random uniform number between [0,1] applied to the ith particle.
φ : Inertia function.
α : Acceleration constants.

The first term represents the inertia or habit, where each particle continues moving in the direction it had previously moved. As training progresses, the influence of the past velocity becomes smaller. In this study, a decreasing linear inertia function $\phi(k)$ is used. For the early stages of the search, a relatively large inertia is used to enhance the global exploration. On the other hand, when reaching the last stages, the inertia is reduced for better local exploitation. The second term represents the memory where each particle is attracted to the best point in its trajectory P. The third term represents a cooperation or information exchange, in which each particle is attracted to the best point found by all particles G. Fig. 3 illustrates the concept of particle movement influenced by three terms.

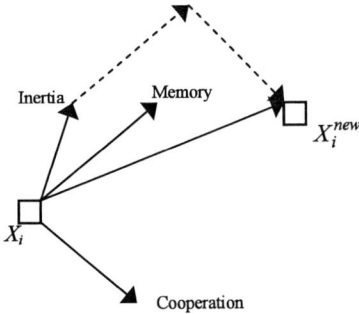

Fig. 3. Ilustration of a particle movement influenced by three terms

PSO starts with a random initialization of N particles (positions). As an analogy with GA, N would best represents the population size [15, 16]. Each particle is a row vector of a dimension l of randomly generated real values, where l is the number of parameters to be optimized. This is similar to chromosome length of a real-code GA. In this study, the stopping criterion is either reaching the maximum number of iterations (1000 epoch), or reaching the minimum global error gradient which has to satisfy $|G(k) - G(k+1)| < 10^{-9}$ to exit the training loop. It has been suggested that the acceleration constants α_1 and α_2 are both set to a fixed value of 2 [26]. The inertia function is a decreasing liner function with initial inertia weight of 0.9 and final weight of 0.2. The population size was chosen to be 40 particles.

V. SIMULATION RESULTS

The generalized transcendental equations of multilevel inverter (2) are solved using the described PSO algorithm. The proposed technique has been applied to different study cases in order to confirm its ruggedness. The simulation results are obtained accordingly using Matlab [17]. It is assumed that the level of the non-equal dc sources can be measured, and V_{dc} has a nominal value of 1 p.u. and so does V_1, while the following sources will acquire different given values less than 1 p.u. For each inverter topology with a specific number of levels, a large number of solution sets can be obtained according to the values of m and the dc sources V_1, V_2, V_3,...ect. Therefore, different study cases will be presented for 5-, 7- and 9-level inverters to ensure the feasibility of the presented algorithm.

A. Case 1: 5-level inverter; V_1=1 p.u., V_2=0.9 p.u.

The proposed technique is applied to minimize the defined cost function for the above stated case. The convergence characteristic of the proposed PSO algorithm is depicted in Fig. 4. It is obvious that a near optimal solution is achieved by the PSO algorithm in about 11 iterations. The CPU execution time required for convergence is 1.27 sec.

The PSO algorithm is used to find the switching angles for the abovementioned case. However, the solution exists for a limited range of m, where $0.84 \leq m \leq 1.59$. Despite this is a natural phenomenon of multilevel inverters even

with equal dc sources, the obtained range of m is wider than that obtained from conventional techniques. Fig. 5 illustrates the variation of the switching angles θ_1 and θ_2 versus m. As an example, an operating point when m=1.5 was chosen which sets the fundamental output voltage, V_f, to be 1.9 p.u. (s =2, m=1.5, V_{dc}=1 p.u., $V_f = (4mV_{dc}/\pi)$ =1.9 p.u.). For this point, the optimum values of the switching angles are: θ_1=9.815° and θ_2=55.122°. Fig. 6 shows the inverter output voltage and the corresponding harmonic spectrum at the abovementioned operating point. It is clear that the targeted 3rd harmonic is eliminated and the fundamental component is equal to 1.9 p.u. as desired.

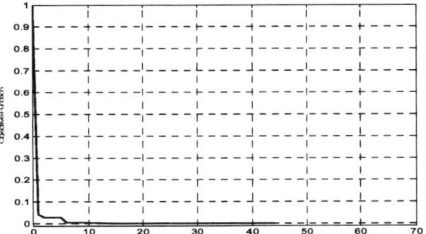

Fig. 4. Convergence characteristic of case 1

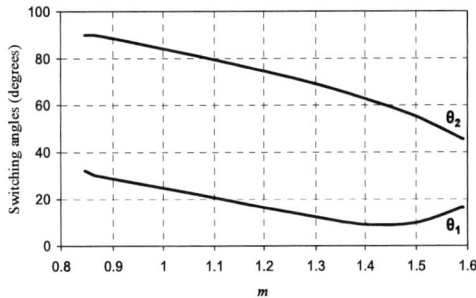

Fig. 5. Solutions for 2 angles versus m for Case 1

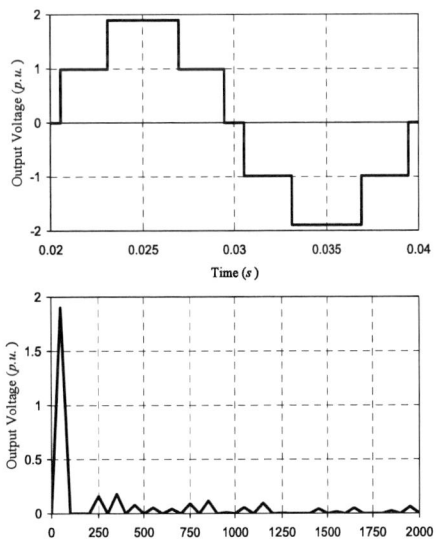

Fig. 6. Output voltage and Corresponding FFT of case 1 at m=1.5

B. Case 2: 7-level inverter; V_1=1 p.u., V_2=0.9 p.u., V_3=0.8 p.u.

The proposed technique was applied to minimize the defined cost function for the above stated case. The convergence characteristic of the proposed method is depicted in Fig. 7. It is obvious from the convergence that a near optimal solution was achieved by PSO in about 18 iterations and approximately other 30 iterations was required to refine the solution. The total CPU execution time of PSO is 1.55 sec.

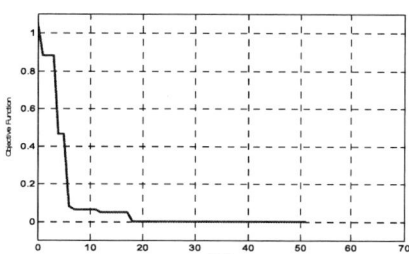

Fig. 7. Convergence characteristic of case 2

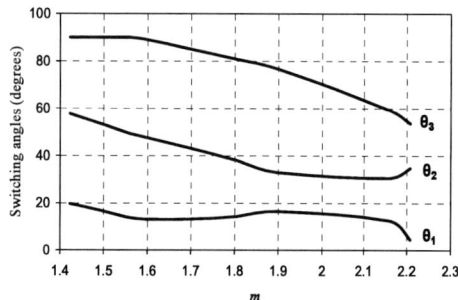

Fig. 8. Solutions for 3 angles versus *m* for Case 2

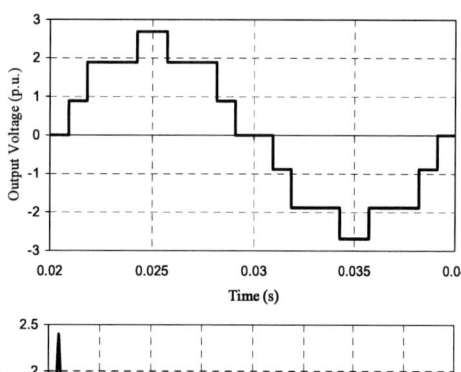

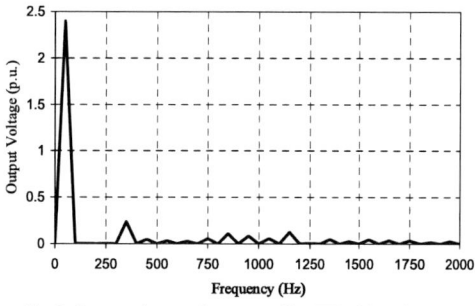

Fig. 9. Output voltage and corresponding FFT of Case 2 at *m*=1.9

Fig. 8 illustrates the variation of the switching angles θ_1, θ_2, and θ_3 versus *m*, where it can be seen that solution exists in the range $1.82 \leq m \leq 2.22$. One particular operating point was chosen to demonstrate the effectiveness of the proposed method, *m*=1.9, which sets the fundamental component V_f to 2.4 p.u. (*s*=3, *m*=1.9, V_{dc}=1 p.u., $V_f = (4mV_{dc}/\pi)$ =2.419 p.u.). For this point, the optimum values of the switching angles are: θ_1 =16.5014°, θ_2=32.849°, and θ_3=76.6228°. The inverter output voltage and its corresponding harmonic spectrum at the abovementioned operating point are depicted in Fig. 9, where it is clear that the 3rd and 5th harmonics are totally eliminated and the desired fundamental of 2.4 p.u. is achieved.

C. Case 3: 9-level inverter; V_1=1 p.u., V_2=0.9 p.u., V_3=0.8 p.u., V_4=0.7 p.u

In this case, four dc sources are considered to verify the feasibility and ruggedness of the proposed technique. The available four degrees of freedom offer the elimination of three low order harmonics and maintaining the fundamental at specific value. The proposed PSO technique was applied to minimize the cost function for the above stated case. The convergence characteristic is shown in Fig. 10, where the optimal solution is reached after 62 iterations in 3.14 sec.

Fig. 11 illustrates the variation of switching angles θ_1, θ_2, θ_3, and θ_4 over the defined range of *m* , $2.02 \leq m \leq 2.66$. An operating point of *m*=2.4 was chosen and applied to the inverter to indicate the effectiveness of the proposed technique to eliminate the 3rd, 5th, and 7th harmonics while maintaining the fundamental component V_f at 3.05 p.u. (*s*=4, *m*=2.4, V_{dc}=1 p.u., $V_f = (4mV_{dc}/\pi)$ =3.05 p.u.). For this point, the optimum values of the switching angles are: θ_1=7.457°, θ_2=30.4133°, and θ_3 =47.4292°, and θ_4=82.3501°. Fig. 12 shows the inverter output voltage and its corresponding harmonic spectrum at the abovementioned operating point, where the elimination of targeted harmonics is clearly evident and the fundamental is maintained at 3.05 p.u.

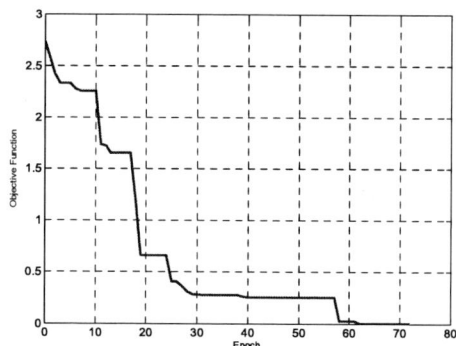

Fig. 10. Convergence characteristic of case 3

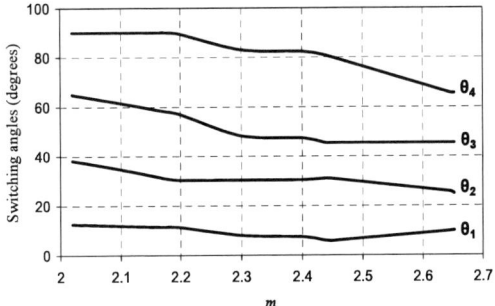

Fig. 11. Solutions for 4 angles versus m for Case 3

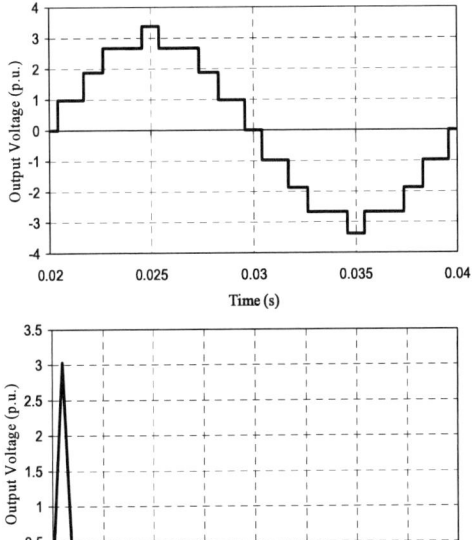

Fig. 12. Output voltage and corresponding FFT of Case 3 at m=2.4

D. Performance Index

In order to indicate the usefulness and effectiveness of the presented method, a quality factor is chosen as a performance index. The THD is very useful parameter to evaluate the performance of the inverter, and therefore it is considered in this work. The THD is defined as the total amount of harmonics relative to the fundamental, and can be calculated using (6) up to the 31st harmonic, where the inverter low pass filter typically eliminates other higher order harmonics.

$$\%THD = \frac{\sqrt{\sum_{n=3,5,\cdots}^{31} V_n^2}}{V_1} \times 100 \qquad (6)$$

Fig. 13 depicts the variation of the %THD versus m for the study cases under consideration. The obtained values of the %THD are better or at least the same as those obtained from other techniques [5]. Moreover, the

computational time is also chosen as a secondary performance index to indicate the superiority of the proposed technique over conventional iterative techniques. The conventional Newton-Raphson iterative technique was used and applied to the same study cases mentioned earlier. The degree of accuracy was kept at 0.00001 while it was 10^{-9} in the proposed PSO algorithm, and a limit of 1000 iterations was also set for halting as no optimal solution situation is highly expected as the inequality of the dc sources increases the complexity of the transcendental equations, and convergence problems are highly arise. The comparison between the conventional Newton-Raphson iterative method and the proposed PSO technique for the given study cases, regarding the computational time and corresponding %THD are tabulated in Table 1. It is clearly evident that the proposed PSO technique is both extremely faster and optimal than the conventional Newton-Raphson method, which reflects the superiority of the introduced technique as far as computational burdens and quality of the output waveform are concerned.

VI. EXPERIMENTAL VERIFICATION

A prototype single-phase cascaded H-bridge inverter was built using IRF630 (200V, 9A) MOSFETs as the switching devices, and MUR820G (200V, 8A) as fast recovery diodes. A battery bank of 4 SDCSs of 60V dc (nominal) was used to individually supply each inverter level. This prototype was configured to work as 5-, 7-or 9-level cascaded H-bridge inverter according to the number of activated levels. The switching angles which were obtained from the PSO algorithm were converted into time intervals and stored in an EPROM in the form of look-up tables for all possible harmonic profiles and SDCSs values over their defined range of modulation

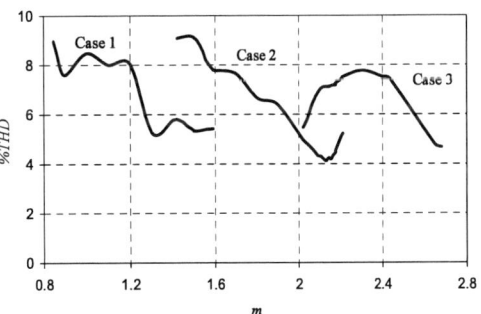

Fig. 13. The %THD versus m for different study cases

TABLE 1.
Comparison between iterative Newton-Raphson technique and the proposed PSO technique

Case study	Iterative Newton-Raphson technique		Proposed PSO technique	
	Computational time	%THD	Computational time	%THD
Case 1	18.02s	8.01	1.27 s	5.44
Case 2	20.45s	9.52	1.55s	6.40
Case 3	24.26s	10.26	3.14s	7.20

indices. A real-time controller based on the MCB-1A Hampden microprocessor kit was used to implement the harmonic elimination PWM method. The required output voltage of any application such as SVG, UPS, or variable speed drive is mapped to a modulation index, then the appropriate set of switching angles was obtained by cycling through the tables. The switching angles were converted into PWM switching patterns using down-counters and some logic operations, and then were are interfaced to the inverter power switches through optocoupler isolators and drivers. A digital oscilloscope was used to display and capture the output waveforms, and with its FFT built-in feature, the spectrum of the output voltage is obtained for different study cases. In order to verify the presented simulation results, the hardware implementation was developed for the same study cases of 5-, 7-, and 9-level inverters with the same values of SDCSs as indicated in cases 1, 2, and 3. The experimental inverter output voltage and its corresponding FFT of case 1 regarding 5-level inverter are shown in Fig. 14. The selected value of V_{dc}= 60V sets V_1=60V and V_2=54V corresponding to 1 p.u. and 0.9 p.u. respectively. Clearly, the spectrum shown in Fig. 14-b confirms that V_f is about 110V ($\approx$1.9 p.u.) and the targeted 3rd harmonic is eliminated. The small deviation between the measured value of V_f and that from simulations is due to the voltage drop on the semiconductor devices, as ideal switches were used in simulations. The experimental results of case 2 regarding 7-level inverter are shown in Fig. 15, where V_1=60V, V_2=54V, and V_3=48V. Three switching angles per quarter cycle can be seen, and a fundamental voltage V_f of about 142V ($\approx$ 2.419 p.u.) is achieved while the

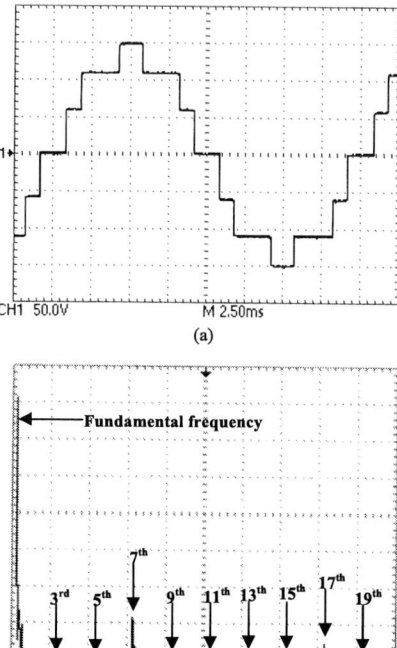

(a)

(b)

Fig. 15. (a) Output voltage, (b) Corresponding FFT of Case 2
(V_{dc}= 60V, m= 1.9)

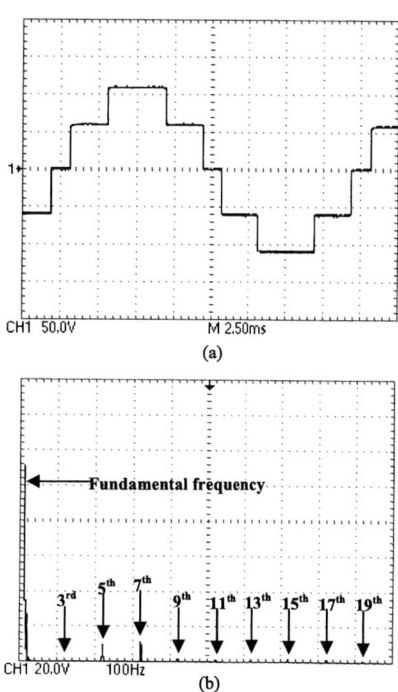

(a)

(b)

Fig. 14. (a) Output voltage, (b) Corresponding FFT of Case 1
(V_{dc}= 60V, m= 1.5)

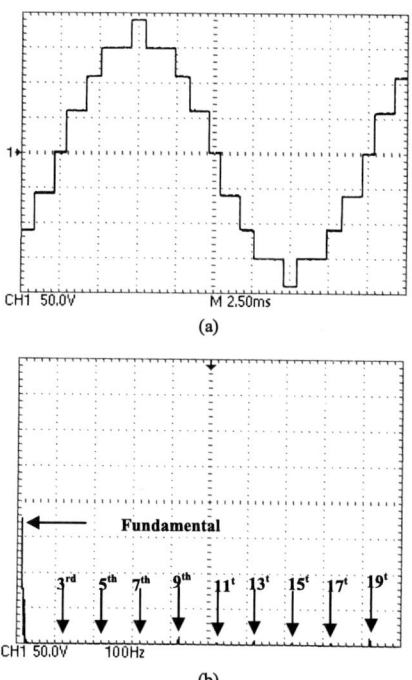

(a)

(b)

Fig. 16. (a) Output voltage, (b) Corresponding FFT of Case 3
(V_{dc}= 60V, m= 2.4)

targeted 3rd and 5th harmonics are eliminated. The experimental verification of case 3 regarding 9-level inverter with four switching angles per quarter cycle is shown in Fig. 16. The fundamental voltage V_f is about 178V (≈ 3.05 p.u.) and the targeted 3rd, 5th, and 7th harmonics are totally are eliminated. It is clearly evident that all experimental results are in close agreement with those from simulations, which reflect the solidness and effectiveness of the presented PSO algorithm. The %THD of the experiments was found a little higher than those of the simulation because the control resolution of the microprocessor is limited $8\mu s$, and the switches are not ideal as in simulations.

VII. CONCLUSIONS

Harmonic elimination of multilevel inverters with non-equal dc voltage sources using a PSO algorithm has been presented. The algorithm is easy to implement as it requires few parameters and no evolution intermediate operators. It overcomes the complicated computations associated with conventional iterative techniques, and the large number of parameters required for GA. It also reduces both the computational burden and running time, and ensures the accuracy and quality of the calculated angles. This method was found superior to conventional techniques that my fail to converge if higher levels with non-equal dc sources are sought. In order to prove the feasibility and effectiveness of the proposed algorithm, it is applied to different study cases regarding the number of inverter levels and targeted harmonics to be eliminated. For each case defined by the values of the dc sources and the required harmonic profile, optimal solution can be found over a definite range of modulation index. The PSO technique proved its effectiveness in finding optimal solutions in extremely short time for different values of the dc sources, where THD was taken as a performance index to examine the effectiveness of the solution. Some of these sets of solutions couldn't be obtained before using traditional techniques. Experimental implementations proved the effectiveness of the proposed method and were in a good agreement with the simulation results. A comparison between the proposed technique and the conventional Newton-Raphson method in terms of computational times and resulted %THD is also reported, where it reveals that the algorithm can be effectively used for selective harmonic elimination of multilevel inverters and results in a dramatic decrease in both the computational times and the output voltage %THD. This method is expected to have widespread applications especially in on-line static var compensation systems, as practical multilevel inverters have non-equal dc sources, and on-line updates of the power system harmonics is highly required.

REFERENCES

[1] J. Lai and F. Peng, "Multilevel converters—a new breed of power converters," *IEEE Trans. Ind. Appl.*, vol. 32, pp. 509-517, May/Jun. 1996.

[2] J. Rodriguez, J.-S. Lai, and F. Z. Peng, "Multilevel inverter: a survey of topologies, controls, and applications," *IEEE Trans. Power Electron.*, vol. 49, pp. 724–738, July 2000.

[3] P. N. Enjeti, P. D. Ziogas, and J. F. Lindsay, "Programmed PWM techniques to eliminate harmonics: a critical evaluation," *IEEE Trans Ind Appl* vol. 26, pp. 302-316, Mar./Apr. 1990.

[4] H. S. Patel and R. G. Hoft, "Generalized technique of harmonic elimination and voltage control in thyristor inverters—Part I: harmonic elimination," *IEEE Trans. Ind. Applicat.*, vol. IA-9, pp. 310–317, May/June 1973.

[5] J. N. Chiasson, L. M. Tolbert, K. J. McKenzie, and Du Z., "A complete solution to the harmonic elimination problem," *IEEE Trans. Power Electron.*, vol. 19, pp. 491-499, Mar. 2004.

[6] K. El-Naggar, T. H. Abdelhamid, "Selective harmonic elimination of a new family of multilevel inverters using genetic algorithms," *Energy Conversion and Management*, vol. 49, pp. 89-95, Jan. 2008.

[7] J. Kennedy, "The particle swarm: social adaptation of knowledge," *IEEE Int. Conference on Evolutionary Computation* ICEC'97, Indianapolis, USA, 1997, pp. 303-308.

[8] V. Miranda, D. Srinivasan, and L. M. Proenca, "Evolutionary computation in power systems," International Journal of Electrical Power & Energy Systems, vol. 20, pp. 89-98, Jan. 1997.

[9] D. G. Holmes and B. P. McGrath, "Opportunities for harmonic cancellation with carrier-based pwm for two-level and multilevel cascaded inverters," *IEEE Trans. Ind. Appl.*, vol. 37, pp. 574–582, Mar.Apr. 2001.

[10] G. J. Su, "Multilevel DC-link inverter," *IEEE Trans. Ind. Appl.*, vol. 41, pp. 848–854, May/Jun. 2005.

[11] S. Kannan, S. M. Slochanal, P. Subbaraj, and N. P. Padhy, "Application of particle swarm optimization technique and its variants to generation expansion planning problem," *Electric Power Systems Research*, vol. 70, pp. 203-210, 2004.

[12] M. A. Abido, "Optimal power flow using particle swarm optimization," *International Journal of Electrical Power & Energy Systems*, vol. 24, pp. 563-571, 2002.

[13] X. Yu, X. Xiong, and Y. A. Wu, "PSO-based approach to optimal capacitor placement with harmonic distortion consideration," Electric Power Systems Research, vol. 71, pp. 27-33, Jan. 2002.

[14] R. Eberhart, and Y. Shi, "Comparison between genetic algorithms and particle swarm optimization," *IEEE International Conference on Evolutionary Computation*, 1998, pp. 611-616.

[15] D. E. Goldberg, *Genetic algorithms in search, optimization, and machine learning*. Reading, Mass. Harlow: Addison-Wesley; 1989.

[16] Z. Michalewicz, *Genetic algorithms + data structures = evolution programs*. 3rd rev. and extended ed. Berlin ; New York: Springer-Verlag; 1996.

[17] MATLAB 6.0 software package, 2006. http://www.Mathworks.com.

Improved PFC Circuit Having Ladder Type Filter with Only Passive Devices

Kenji Ando,

NITTO KOGYO CORPORATION
Nagakute, Aichi, JAPAN

Keiju Matsui, Nobuhito Takeuchi
Masaru Hasegawa
Chubu University
1200, Matsumoto, Kasugai, JAPAN
keiju@isc.chubu.ac.jp

Abstract - It is well known that non-linear circuits such as various rectifiers generate harmonics in the power system. To improve those harmonic problems, various PFC circuits have been proposed so far. Among these, Takahashi have proposed a PFC scheme using a LC filter without switching devices. This method makes a PFC effect by widening conduction period using the operations of capacitors and inductor. On the basis of this scheme, we propose another PFC circuit. This method attempts to improve the input current, towards satisfactory waveform by using a novel ladder type filter.

Keywords - PFC, ladder type filter, passive device, harmonics, rectifier.

I.   INTRODUCTION

It is well known that non-linear circuits such as rectifier circuits generate various harmonics in the power system. Hence, with the spread of applications incorporating such rectifiers, the occurrence of harmonics in the power system is likely to increase. Because of this phenomenon, various power factor correction circuits have been proposed and studied for both single-phase and three-phase rectifiers [1-3]. These methods use switching devices to attempt to modify the input current into a sinusoidal waveform. However, these methods can have an adverse effect on the cost performance, or introduce EMI noise problems. Some high-performance electronic devices must avoid such switching noise to obtain high-quality noise characteristics. From this point of view, it is appropriate to study harmonic reduction methods that function without the use of switching devices. Many researchers have previously proposed several novel rectifier circuits.

Amongst these devices, some circuits have achieved power factor correction without the use of switching devices [4-9]. Since switching devices, such as power transistors are not used, switching losses and switching noise, such as EMI, are not generated. Since a control circuit for the switching device is also unnecessary, the system cost is reduced. Amongst the many studies, a novel PFC circuit without switching devices has been proposed by Takahashi, et al [9]. Against this background of PFC schemes without switching devices, we have proposed a novel PFC circuit, which is different from Takahashi's. The original aspects of the proposed circuit have been identified and analyzed.

II.   OPERATIONAL   PRINCIPLE

A. PFC Circuit by Takahashi

Fig.1 shows the PFC scheme proposed by Takahashi. The operational principle is explained as follows; an auxiliary LC resonant circuit is connected to an intermediate dc terminal. The input supply current flows into LC circuit when $v_a < v_o$. When $v_a > v_o$, the current flows toward the output from the input supply and the LC circuit. In this way, the conduction period is divided into two sections, permitting the conduction angle to be widened, producing an improved input current waveform. This is an excellent circuit which offers a unique and simple method. On a basis of this circuit topology, we will seek a still more simple solution by a schematic analysis. Compared to the 'conventional' circuit above, the circuit construction is analogous, but the operational concept is different. Because of the simple construction, the circuit could be expected to develop in the direction of a three-phase circuit configuration.

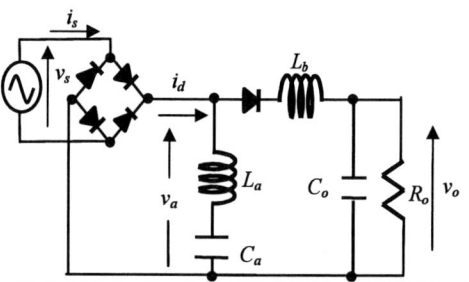

Fig.1. PFC circuit with auxiliary LC resonant by Takahashi

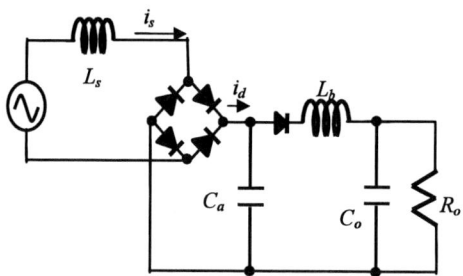

Fig.2. Proposed PFC with ladder type circuit

978-1-4244-1741-4/08/$25.00 ©2008 IEEE

B. Ladder Type PFC scheme

Fig.2 shows the proposed fundamental PFC circuit, termed a 'ladder type PFC circuit'. In the dc circuit, C_a, L_b, C_o are connected as the π type filter. The circuit obtained, however, includes an additional inductor, L_s, which is sometimes eliminated from the line, so that only the line inductance remains, to function effectively as a filter. From the above reasoning, the proposed circuit is termed a 'Ladder Type Filter'. C_o is a large capacitance for smoothing the output voltage and C_a is a somewhat reduced capacitance. Because of reduced value of C_a, this voltage can be rapidly discharged to zero as the input voltage is decreasing. Because of this, the input current can be varied according to the variation of the input voltage waveform. L_b is relatively large inductor, which plays a role in maintaining an independence between the input capacitance, C_a, and output capacitance, C_o, together with a series diode preventing the reverse current.

The difference from the conventional circuit is that the auxiliary inductor, L_a, in the conventional circuit is removed, and its role is transferred to the line inductance, L_s, or an additional inductor in series. The number of circuit components can be reduced by one, in a case of no additional line inductor being required. This series line inductance has an advantage, in as much as, since the inductor current is increased compared to the auxiliary current i_a, even an inductor of reduced size can work sufficiently. Usually, this additional series inductor can be eliminated and be subsumed into the line inductance, as explained later. In such way, the series inductor can be somewhat reduced to as small as the line inductance.

C. Explanation of operational principle
in a schematic modeling way

In order to make clear the thought process behind the proposed circuit topology, consider the operation in terms of the circuit model with a schematic wave in Fig.3. The output smoothing inductor is assumed to be very large, to achieve almost zero ripple current and render the circuit analysis very simple. In order to realize an ideal input current, it is necessary to understand the process of conversion on the capacitor C_a. For this purpose, the input current is taken as an ideal current source, and the conversion process to an ideal sinusoidal wave is then examined. The manner of realizing an adequate conversion will now be discussed. This operational waveform is shown schematically in Fig.3 (b).

Fig.4 shows operational circuits, including the schematic waves. The purpose of these figures is to explain the thought behind the proposed circuit.

(1) Period I($i_d < i_b$)

The inductor current, i_b, which is sufficiently smoothed by sufficiently larger inductor, flows through the input power supply and the auxiliary capacitor in parallel path. If v_a attempts to decrease towards a negative value, ($v_a < 0$), for instance, when the input power supply or the load are suddenly changed, the current flows through the double diodes in the diode bridge like a free wheel phenomenon. Thus, v_a does not become negative, but is clamped at $v_a=0$. In case of commonly used capacitor input type, the circuit capacitor is sufficiently charged to a fairly large voltage, so the previously mentioned current prevention operation occurs. In the proposed method at the steady state, on the other hand, the capacitor is forced to be progressively discharged by the inductor current, i_b. Because of this, the current prevention phenomenon observed in the usual capacitor input type does not occurred in the proposed case. The current waveform, screened by various patterns shown in the figure, is distinguished by its own pattern. The same analytical technique is used in the following:

(2) period II($i_d > i_b$)

In contrast to a constant dc current, i_b, an ideal input current rises sinusoidally, and the incremental component, i_d-i_b, flows into the auxiliary capacitance, C_a, as a charging current in parallel.

(3) period III($i_d < i_b$)

When i_d again reduces below i_b, the accumulated charge on C_a is forced to be discharged by i_b and is falls towards zero. The discharge current is i_b-i_d. At the end of the period, v_a reaches zero, and operation returns to the first period, I, and the whole operation is repeated. The idea of the proposed circuit is such that the charge accumulated on C_a is forced to be discharged, and C_a is then prepared to accept incoming charge due to input current. If the above can be accomplished, an ideal converter could be realized. Actually, however, a small size L_b is desirable, and the input current waveform is not from a current source, but a voltage source with additional inductor or a line inductance of less than 1 mH. Because of this, the actual waveform is distorted from the ideal shape.

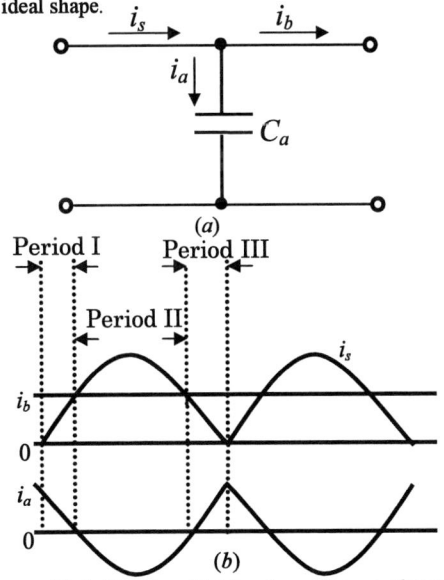

Fig.3. Circuit model (*a*), and operating waveforms (*b*)
due to current source

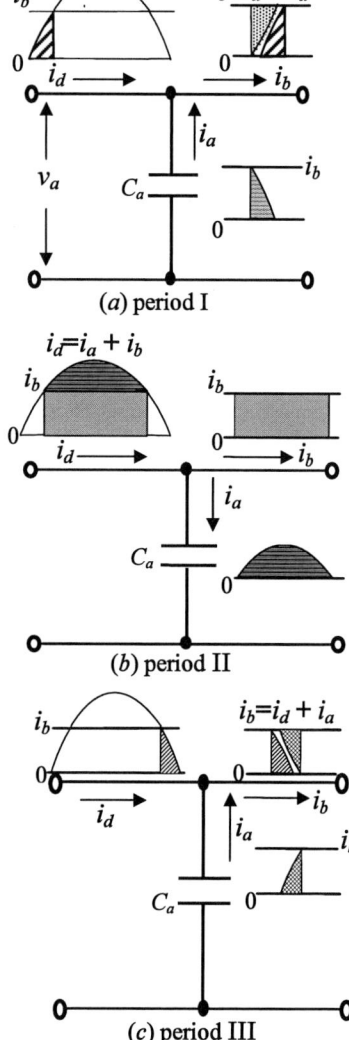

Fig.4. Circuits in schematic operation

(a) period I

(b) period II

(c) period III

III. SIMULATION RESULTS

A. Operation waveforms by ideal current source

The purpose of this paper is to resolve a satisfactory input current waveform. Because of this, circuit simulation will be executed using an ideal model circuit. With such technique, the reason for waveform distortion will be examined and identified. The circuit simulation is executed by using an ideal sinusoidal current source described above, and by using a sufficiently large L_d to achieve a smooth dc current without ripple and renders the system operation simple. Fig.5 shows various waveforms obtained from such a circuit simulation. Harmonic components of the rectified full wave of sinusoidal input current wave, i_s, are eliminated at this stage by the

auxiliary capacitor, and the wave is towards output without ripple, as in Fig.5 *(a)*.

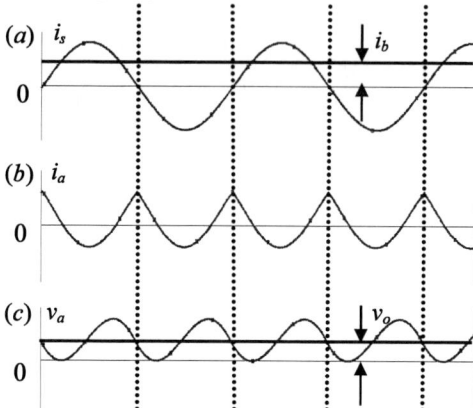

Fig.5. Waveforms by ideal current source

Consequently, the filtered harmonic component flows through the auxiliary capacitor as a charge and discharge current (Fig.5 *(b)*). The Integrated currents provide capacitor voltage in Fig.5 *(c)*. The mean value of the capacitor voltage is then transmitted towards the output as v_o. The current, i_a, is symmetry over one quarter cycle, as a point symmetry, relative to an origin on one-quarter period axis. As a result, the peak value becomes $v_a=2v_o$.

These waveforms are obtained using idealized circuit model and simulation. By observing both waveforms, v_a and i_a, the phases are shifted according to the fundamental ac circuit theory. Namely, in both waves, i.e, the capacitor current i_a, in Fig.5 *(b)* that charges and discharges the harmonic components of rectified ideal input sinusoidal wave through C_a, and the capacitor voltage, v_a, obtained from integration of i_a in Fig.5 (c), the relationship of both is seen to be always distorted. Intrinsically, we can not obtain both to be sinusoidal waves at the same time. We can see that the auxiliary capacitor, which combines both waveforms, plays a very important role. By an adequate selection of circuit constants, such as LC component, the purpose is to approach an effective sinusoidal wave as possible.

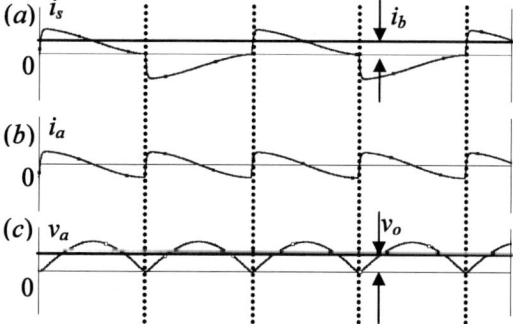

Fig.6. Waveforms by ideal voltage source

616

B. Operation waveforms by ideal voltage source

In order to realize an effective waveform, circuit simulation is executed with the circuit being modeled from a voltage source in the manner of the previously mentioned technique. Other techniques employed are the same as previously mentioned. i.e., L_b is sufficiently large, v_a has momentary zero point at every half cycle, etc. Each of the voltage and current waveforms are shown in the same arrangement as in the former Fig.5. It is found that the obtained waveforms are quite different from current source waveforms. The reason for this may be explained as follows; at t=0 in Fig.6, when the positive voltage, v_a starts to increase, the discharge current, i_a, is instantaneously commutated to the power supply. At the same time, the output inductor current, i_b, is also commutated to input power supply. Whilst increasing the input supply voltage, C_a is gradually being charged. The following relationship is established between each of the variables.

$$i_s = C_a \frac{dv_a}{dt} + i_b \ldots\ldots\ldots\ldots\ldots(1)$$

$$i_a = -C_a \frac{dv_a}{dt} \ldots\ldots\ldots\ldots\ldots\ldots(2)$$

where i_b is assumed to be constant, due to the sufficiently large inductor, and is added to the input supply current. v_s is given by $V_{sm}\sin\omega t$. Assuming that $v_s=v_a$ for simplicity of analysis, then,

$$i_s = \omega C_a V_{sm} \cos \omega t + i_b \ldots\ldots\ldots\ldots(3)$$

$$i_a = -\omega C_a V_{sm} \cos \omega t \ldots\ldots\ldots\ldots(4)$$

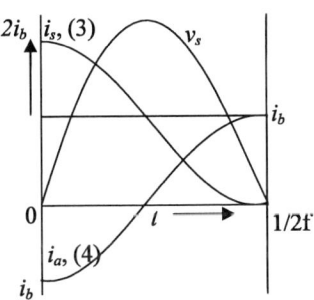

Fig.7.   Relationship between each waveform by voltage source

Fig.7 shows results, calculated from (3) and (4). As in the previously mentioned simulation results in Fig.6, the reasons of waveform generation can be explained by inspection of Fig.7. Consequently, in the transient state when v_a is increasing from zero, rising waveforms of i_a or i_s abruptly changing in nature. Because of this, it is expected that a series inductor, inserted in the line of the input supply, would be effective to suppress the input current rising, even if only a small inductance.

IV.   DISCUSSION OF THE SIMULATION

Fig.8 shows various waveforms of the proposed circuit. Employed circuit constants are shown in Table 1. To resolve the ideal operation and waveforms, various considerations and possibilities have been examined, where the auxiliary capacitor plays an important role in waveform improvement. In practice, however, the smoothing inductor, L_a, and the like can not be much increased. Hence, the input current waveform becomes distorted to some extent.

To reduce the harmonics of the input current waveform, it is very important to select an optimum value of C_a as mentioned above. Fig.9 (a) shows a relationship between THD and auxiliary capacitance, C_a. For low values of C_a, the flow of input current is prevented due to small C_a, deteriorating THD. The line inductance, L_s (=0.4mH, 1mH, 2mH), are given as a parameter, and calculated from the results of simulation. If L_s is increased from L_s=0.4mH to 2mH, THD is not influenced very much. We can say that, even a reduced value, such as the purely line inductance, is sufficient for an improvement in the waveform. Fig.9 (b) shows harmonic components for the calculation of THD in Fig.9 (a). It can be seen that the THD characteristic is significantly affected by the third order harmonic component. The results from the number of charges and discharges in one half period. Resultant waveforms are analogous to the those of the third order harmonic. These harmonic components are entirely cleared with respect to IEC standard. Fig.10 shows the relationship between the THD and smoothing inductor, L_d. In a region of reduced L_d, the quantity of the accumulated charge on C_a can not effectively be withdrawn towards the output, so that current inflow is prevented by already accumulated charge. In a region of larger L_d, the influence on THD is not much large with varying L_d, so that the characteristics can not be improved. Fig.8 can be produced by techniques in which an optimum waveform having minimum THD is obtained. Compared to the voltage waveform in Fig.8 (a), the current waveform in Fig.8 (b) includes short zero voltage periods, but these are synchronized to the voltage waveform. The reasons for the appearance of zero periods is that the v_a voltage does not have zero voltage period, as shown in Fig.8 (e), that results in a current prevention effect, as with the capacitor input type. In order to obtain the optimum input current waveform having minimum THD, the circuit constants are examined. This v_a wave does not have a zero region. In these waveforms, however, the minimum THD can be obtained, Fig.11 shows the conventional waveforms from Takahashi. By comparison between the conventional and the proposed circuit, it can be said of the proposed circuit that the number of circuit components can be reduced by one; nevertheless, we can obtain an input current waveform representing an equal level of quality. From simulation results in Fig.8 and Fig.11, THD is 14.19% in the Takahashi circuit and 13.71% in the proposed circuit. In regard to the optimum THD, the proposed circuit is a little more advantageous than the conventional one. Table 2 shows these THDs of both waveforms. From inspection of both waveforms, we can see that analogous results and, hence, same quality level THD can be obtained.

Table.1 Circuit constants

L_s: Line inductance	0.4 mH
R_s: Line resistance	0.6 Ω
L_b: Filter inductance	5 mH
C_a: Auxiliary capacitance	
Proposed	400 μF
Conventional	150 μF
C_o: Filter capacitance	6000 μF
R_o: Load resistance	8 Ω
L_a: Auxiliary inductance	2.2 mH
v_s: Supply voltage	100 V
f_s: Supply frequency	60 Hz

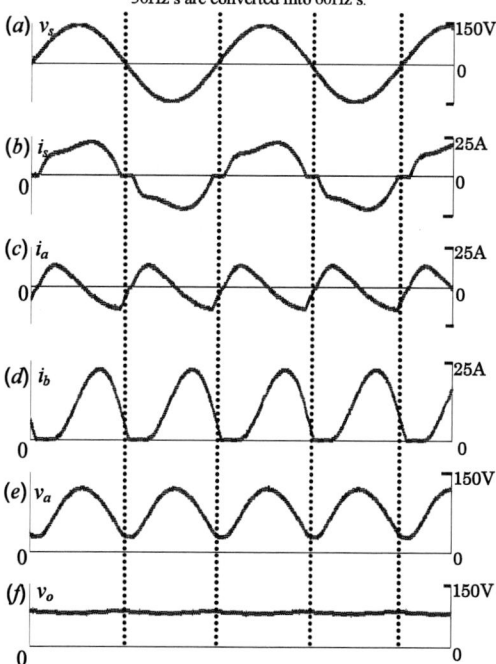

50Hz's are converted into 60Hz's.

Fig.8. Waveforms in proposed circuit (Simulation)

Table.2 Comparison of THDs.

THD	
Conventional	14.19 %
Proposed	13.71 %

$$\text{THD} = \sqrt{\frac{I_e^{\,2} - I_1^{\,2}}{I_1^{\,2}}}$$

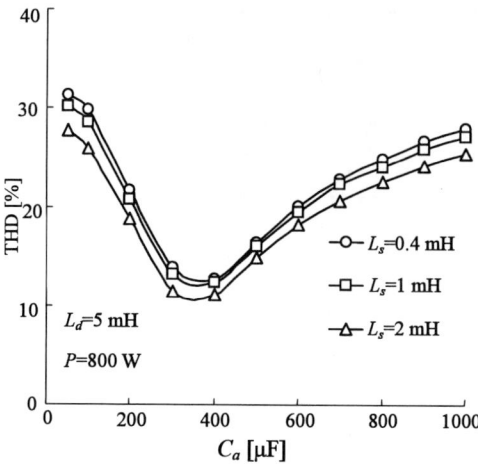

Fig.9.(*a*) Relationship between THD and auxiliary capacitance

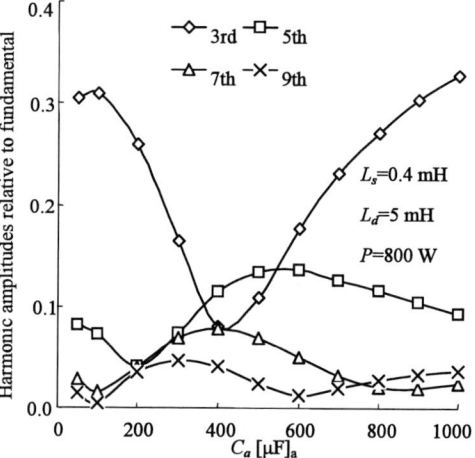

Fig.9.(*b*) Relationship between harmonics and auxiliary capacitance

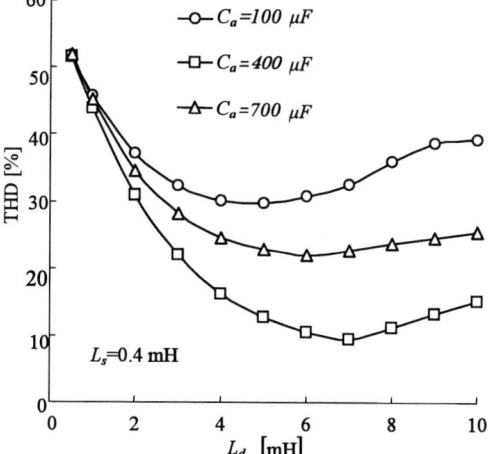

Fig.10. Relationship between THD and output inductance

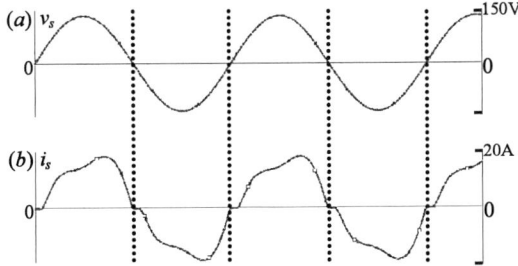

Fig.11. Conventional waveforms

Fig.12 shows the THD characteristic with varying L_a, for the conventional circuit. The auxiliary capacitance, C_a, is given as a circuit parameter. In the case of C_a=150㎌, which represents the Takahashi solution [9], an increase in L_a brings about an improvement in THD. Such large size of L_a, in addition to a large L_d, is inadvisable, so that reasonably reduced value of about 2.0mH is utilized and recommended. In the case of an auxiliary capacitance of C_a=400㎌, when L_a is increased, THD has, also, a tendency to increase. The reason may be explained, similarly to that as shown in Fig.9 (b), the third order harmonic is increased and the waveforms are apt to be distorted.

With C_a=300㎌, on the other hand, when L_a is decreased, THD is also decreased. However, at around 2mH, the effect of decreasing L_a has no effect and the same quality level THD is obtained as the proposed value. The value in the case of C_a=400㎌ shows for the proposed one, whose THD can be compared favorably. Fig.13 shows relationship between THD and the line inductance L_s. It might have been thought that increased L_s would bring about an improvement in THD, however such effect could not be observed and almost nearly constant characteristics were obtained. In the proposed method, current flowing in to C_a has a significant influence, but reduced L_s does not influence the characteristics. In case of larger L_s values than those represented in the figure, the voltage drop due to the main current is significant, and such a large inductor can not be used in practice.

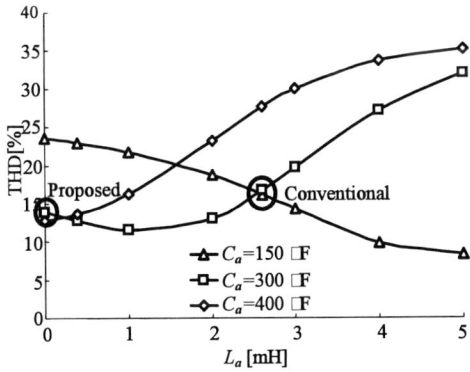

Fig.12. Relationship between THD and L_a in conventional

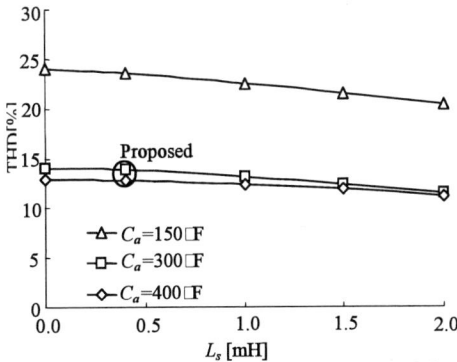

Fig.13. Relationship between THD and L_s in proposed

V. EXPERIMENT RESULTS

Fig.14 shows various waveforms in the experiment. Both waveforms are agreeing well each other. It can be seen that the analytical results by simulation are confirmed by these experiments.

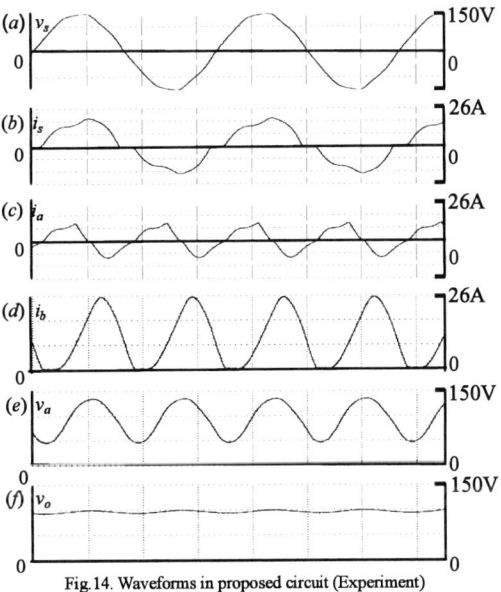

Fig.14. Waveforms in proposed circuit (Experiment)

VI. CONCLUSION

The proposed method could have been evaluated, merely considering the feature of reducing the number of components. However, the thought process for this idea is a little different, and yet the characteristics are a still little improved, using a reduced number of components. The problem with a capacitor input type rectifier is the prevention of input current flow by virtue of, the previously accumulated charge, so that the input current is distorted. In the relationship between the input current waveform into an auxiliary capacitor and the capacitor voltage, the phases are shifted and hence distort each other. Consequently, it is very difficult to maintain both waveforms satisfactory. In the distortion relationship, a compromise is to obtain an effective waveform having a minimum THD.

The idea of the proposed method hinges on continuously removing the auxiliary capacitor charge. In this way, the proposed circuit topology can be envisaged. The secondary stage smoothing inductor plays a role of continuously removing the stored charge. Consequently, the influence of the stored charge becomes fairly ineffective on the input current. It can be seen that the proposed characteristics provide almost the same quality level as in the conventional circuit, even with less circuit components. Reduced number of components could easily give rise to a novel three phase circuit.

ACKNOWLEDGEMENT

This research was partly supported by grants for scientific research from Monbu Kagakusho (Ministry of Education, Culture, Sports, Science and Technology).

REFERENCES

[1] JISC: "Electromagnetic compatibility (EMC) - Part 3-2: Limits" JISC61000-3-20 (2005-3)

[2] Isao Takahashi, Wataru Ikeshita, "Improvement of Input Current Waveforms of a Single-Phase Rectifier Circuit" *IEEJ Trans.* Vol.105-B, No.2,p.82 (1985-2)

[3] PFC Circuit Investigation Committee on IEEJ, "Current Circumstances and Trends of PFC Converter-circuit and control strategies", Technical Report of Institute of Electrical Engineers in Japan, no.785 (2000-6)

[4] Keiju Matsui, Kazuo Tsuboi, Atsushi Kobayashi, Tomonori Fukuda, Saburo Muto, "Discussion on Single-Phase Rectifier to Reduce Lower Order Harmonics", National Convention Record of IEE Japan No, 579,(1989)

[5] Kenichiro Fujiwara, Hiroshi Nomura, "A new operating principle of voltage-doubler diode rectifiers to meet the harmonic guide lines" " *IEEJ Trans.* Vol.119-D, No.1, pp.103 - 108 (1999-1) (in Japanese)

[6] K.Fujiwara and H.Nomura,"A Power Factor Correction for Single-Phase Diode Rectifiers without employing PWM Strategy", IPEC-Yokohama '95, pp.1501 -1506 (1995).

[7] Isamu Yamamoto, Keiju Matsui, "A Power Factor Correction with Two-Input Current Mode using Voltage Doubler Rectifier" *IEEJ Trans. IA,*Vol.121-D, No.2, pp.225 – 230 (2001-2)

[8] Isamu Yamamoto, Keiju Matsui, "A Power Factor Correction with Two-Input Current Mode using Voltage Doubler Rectifier" *IEEJ Trans. IA,*Vol.121-D, No.2, pp.225 - 230 (2001-2)

[9] Isao Takahashi, Kazuhiro Hori, "Improvement of Input Current Waveforms of a Single Phase Diode Rectifier by Passive Devices" *IEEJ Trans.* Vol.117-D, No.1, pp.13 - 18 (1997-1)

[10] Nobuhito Takeuchi, Keiju Matsui, Hiroo Kojima, Isamu Yamamoto, Masaru Hasegawa, "A Discussion on PFC Circuit Using Ladder Type Filter" , Industry Applications Society of the Institute of Electrical Engineers of Japan (2007-8)

[11] Kazutaka Hori, Isao Takahashi, "PFC Rectifier Having Charge Pumps", Hokuriku Section Joint Conference Record of Institute of Electrical and Related Engineers (1993-10)

Fuel Cell Current Ripple Minimization using a bi-Buck Power Interface

Nicu BIZON*, Marian RADUCU*, Mihai OPROESCU*

*University of Pitesti/Electronics, Communication and Computer Science Department, Pitesti, Romania, e-mail:
nicubizon@yahoo.com; moproescu@yahoo.com; marian_raducu1963@yahoo.com;

Abstract— **This paper proposes a method to minimize the inverter ripple by active compensation of the low frequency harmonics using a buck current source with a bang-bang control. Remain power spectrum is spreading by chaotification of the buck voltage source command using a nonlinear transfer function for the control loop. Simulation results and designing methodology are presented.**

Keywords—**Fuel cell, inverter current ripple, power interface, chaotification.**

I. INTRODUCTION

Power electronics is an interdisciplinary green technology with the main goal to convert electrical energy at high conversion efficiency from different (renewal) energy sources to loads. One problem appears when fuel cells (FC) as energy sources supply equipments which has been designed for alternating current (ac); because FC generate direct current (dc) an inverter system (dc-ac converter) must be used as power interface. Inverter input current ripple has been reported as main factor for performance degradation of the FC proton exchange membrane (PEMFC) and PEMFC life cycle reduction, if current ripple are not limited by an adequate controlled [1,2,3]. The PEMFC low frequency (LF) current ripple affects in much measure the PEMFC life cycle, so some restrictions are specified [4,5,6]. The proposed techniques to meet these requirements are few, and propose innovative control for LF current ripple decreasing [7,8] or HF current ripple spreading [9-13] in the high frequencies (HF) band.

Passive filters (with inductors and/or capacitors) for high PEMFC output current are bulky, expensive and inherently unreliable) are usually specified to reduce current ripple in the intermediate dc-link of an energy generation system (EGS). Recently there has been renewed interest in the EGS complex behavior [14-17], especially in hybrid-electrical vehicle (HEV), where an efficient energy management must be done [18-24]. In order to optimize the power interface (multi-port converter) between energy source, PEMFC, energy storage devices (ESD), batteries stack, devices for energy compensation (DEC), ultracapacitors stack, and dynamic load (usually ac electrical motors in HEV), the maximum power point (MPP) control principle has been extended to PEMFC control [25]. The MPP control for PEMFC power interface with EGS must integrate the abouve mention techniques concerning the PEMFC current ripple

spectrum. In this paper is proposed a PEMFC power interface of bi-buck type, with a new nonlinear control that integrate the following objectives: MPP control; PEMFC LF current ripple decreasing; PEMFC HF current ripple spreading; maintaining of the performance parameters for buck output voltage.

An equivalent electrical model of a fuel cell is used in simulation. This can simulate the PEMFC i-u characteristic with a PEMFC appropriate dynamic [26], and the effects of inverter input current ripple on the PEMFC performance [27]. The proposed nonlinear control [13] is here used for spreading the PEMFC HF current ripple. For decreasing the PEMFC LF current ripple is proposed a buck current source with an adaptive bang-bang control. The MPP control [25] is not detailed here; only the control performances improvement regarding the PEMFC current ripple is shown. The PEMFC reference current is estimate for a nominal load.

The remainder of the paper is organized as follows. Section 2 presents the bi-buck Matlab model and gives details about of the used models. Section 3 briefly discusses the control loops and presents some design relations. Section 4 presents the bang-bang control of the buck current source and some selected simulation results. Section 5 presents the simulation results for bi-buck interface. Last section concludes the paper.

II. PROBLEM STATEMENT

The proposed PEMFC power interface is a bi-buck topology shown in figure 1. The dc Controlled Voltage Source, CVSref, is controlled by the reference voltage, V_{ref}, of the Buck Voltage Source, so $V_{ref}=V_{out(AV)}$. The ac Controlled Voltage Source (CVSource) is controlled by the Buck Current Source output current, I_L, which try to following the LF PEMF current ripple ($I_{FC}-I_{ref}$) in the bang-bang control loop. If the PEMFC reference current (I_{ref}) is estimate by MPP control techniques for a nominal load current, I_{out}, then I_{out2} current is a estimation of the I_{out} current ripple, and $I_{out1(AV)} \cong I_{out(AV)}$.

In this paper different kind of load current ripple shapes are used (figure 2 shows a superposition of the first three 100Hz inverter harmonics over nominal load current, and figure 3 show a pulse load model). The PEMFC as Vin1 energy source in simulation has the parameters mention in figure 4. The second energy sources, Vin2, which assure the power flow for compensate the LF inverter ripple, can be a low energy source. For example, can be a ultracapacitors stack, charged from a energy source (PEMFC) or with the

978-1-4244-1741-4/08/$25.00 ©2008 IEEE

recovered energy in the HEV braking process. It is shown that proposed bang-bang current control is independent by the Vin2 source level (for demonstration, a simple controlled voltage source is used).

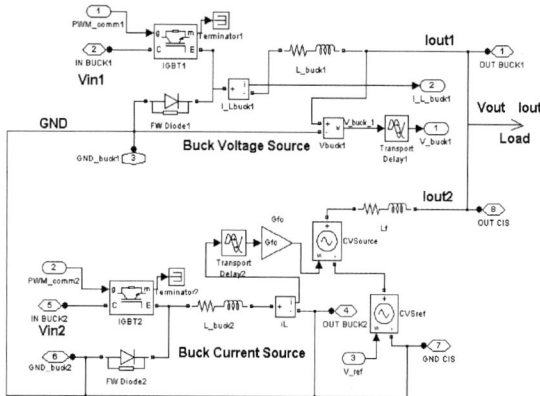

Fig. 1. Bi-buck topology – Matlab diagram

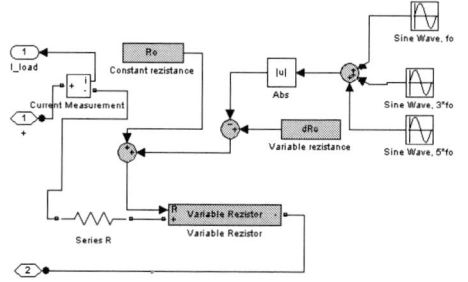

Fig. 2. Variable load

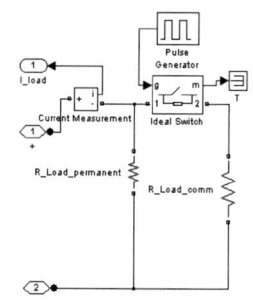

Fig. 3. Pulse load

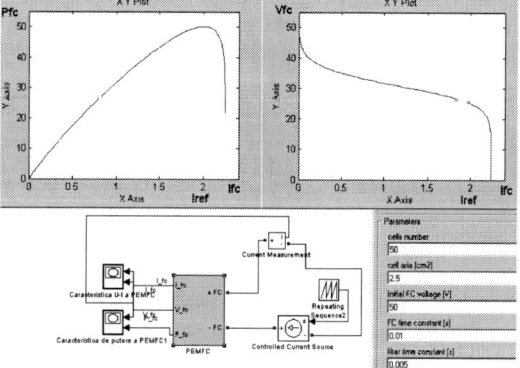

Fig. 4. PEMFC model characteristics and parameters

III. CONTROL LOOPS AND SOME DESIGNING ASPECTS

The complete Matlab diagram of the proposed PEMFC power interface, including control loops, is presented in figure 5.

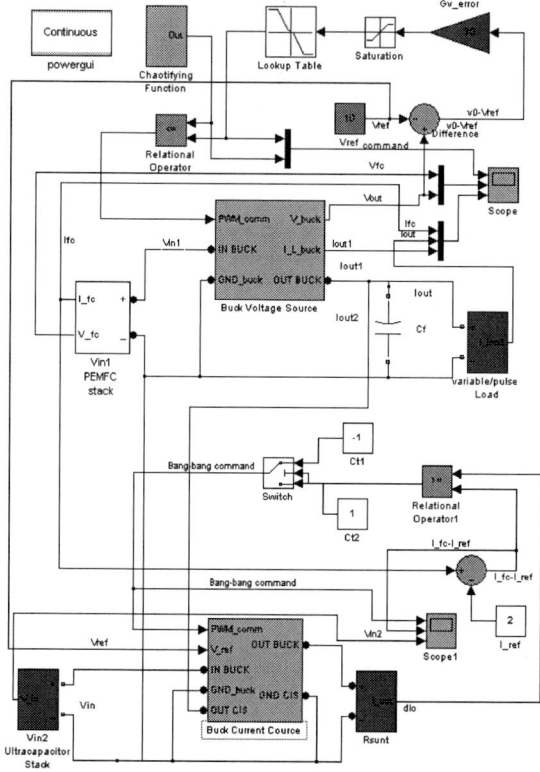

Fig. 5. PEMFC power interface – Matlab diagram of the control loops

The control loop for buck voltage source has been first analyzed in [28] and designing aspects are specified in [29], so only few comments will be included here. The nonlinear characteristic law of the control loop is modeled by a look-up table, and is optimized for a small output voltage ripple:

$$
v_{\substack{\text{Look-up} \\ \text{table out}}}\left(\Delta v_{og}\right) = \begin{cases} s_0 \cdot \Delta v_{og} + \dfrac{1}{2} \cdot \left(s_1 - s_0\right) \cdot \\ \cdot \left(\left|\Delta v_{og} + \Delta v_{og(min)}\right| - \left|\Delta v_{og} - \Delta v_{go(min)}\right|\right), \\ \quad for \quad \left|\Delta v_{og}\right| < \Delta v_{og(max)} \\ - \operatorname{sign}(\Delta v_{og}) \cdot V_{max}, \quad \left|\Delta v_{og}\right| \geq \Delta v_{og(max)} \end{cases} \quad (1)
$$

where:

$$
s_o = \frac{-1-1}{0.15-(-0.15)} = -\frac{20}{3}
$$

is the slope in the inner zone $\left|\Delta v_{og}\right| < \Delta v_{og(min)}$,

$$
s_1 = \frac{1-10}{0.6-0.15} = -20
$$

is the slope in the intermediate zone $\Delta v_{og(min)} < \left|\Delta v_{og}\right| < \Delta v_{og(max)}$, $\pm V_{max}$ are the voltage limits in the outer zone $\left|\Delta v_{og}\right| \geq \Delta v_{og(max)}$,

$$\left|\Delta v_{og}\right| = G_{V\_error} \cdot \left|v_o - V_{ref}\right| \qquad (2)$$

is the gained output voltage ripple (G_{V_error}=30), and $\pm\Delta v_{o(min)}$, $\pm\Delta v_{o(max)}$ denote the breakpoints ($\Delta v_{o(min)}$=150mV, $\Delta v_{o(max)}$=600mV).

This paper presents also a new way to chaotifying the voltage buck converter behavior using a simple signal as a chaotifying function; for ex., we chosen the saw-tooth signal as a signal for buck behavior chaotification. The switching command is obtained by comparison of the two signals: signal used for buck PWM control chaotification, v_{sw}, and look-up table output signal, $v_{look\text{-}up\ table\ out}$.

The bang-bang control loop for buck current source is simple, and was chosen because assure a short time response, limited only by buck time constants. For a dynamic compensation of the LF inverter current ripple harmonics (up to 5-th harmonics – 500Hz) the buck response time must be shortly than 2 ms. If nominal output power is 100W and V_{ref}=10V, results I_{out}=10A and R_{load}=1Ω. As a consequence, the appropriate filter inductance value, L_f, must be in range 1mH to 5mH. A filter capacitance value in range 10µF to 47µF gives a filter resonance frequency over to 500 Hz. The buck inductance value used in simulation will be in range 0.1mH to 1 mH, in order to assure a continuous conduction mode for buck operation, with acceptable HF current ripple for PEMFC. The gain for ac controlled voltage source, G_{fc}, is equal with $\dfrac{I_{out(AV)}}{I_{FC(AV)}} \cdot \dfrac{L_1}{L_f}$ ratio. The rectangular pulse load is used for testing the response time in the next section.

IV. Time Response of the Buck Current Source

Matlab diagram of the buck current source (with bang-bang command) and some simulation results are shown in figure 6. From simulation results we conclude that a value of L_2 buck inductance up to 1 mH can assure a well following of the low frequency current ripple shape by buck current source (figure 6.d).

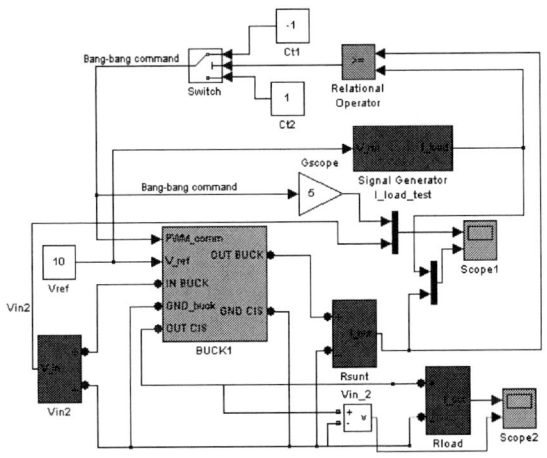

a) Matlab diagram of buck current source with bang-bang command

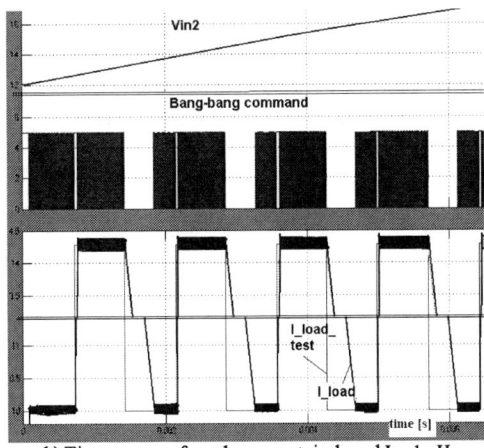

b) Time response for pulse current ripple and L_2=1mH

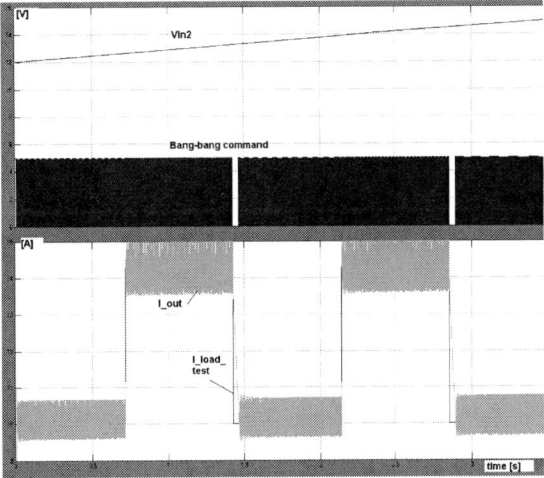

c) Time response for pulse current ripple and L_2=0.1mH

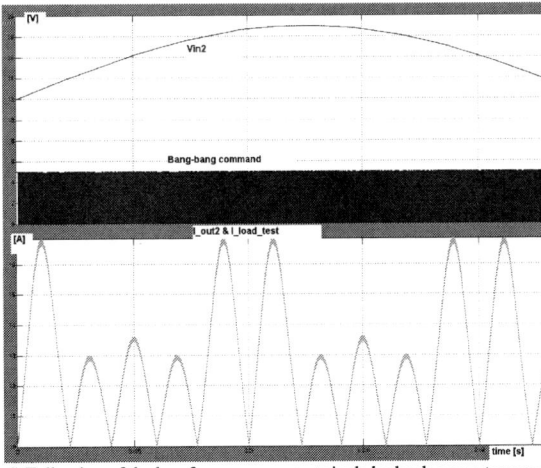

d) Following of the low frequency current ripple by buck current source when L2=1mH

Fig. 6. Buck current source. Matlab diagram and simulation results

A L_2 buck inductance up to 0.1 mH gives big HF current ripple for the buck current source. So, for the next simulations we choose L_1=L_2=0.1mH and L_f=0.5mH, that gives G_{fc}=1

V. SIMULATION RESULTS

The output voltage reference is, for example, V_{ref}=10V, and different shapes of the current ripple are chosen for the load current. This simulated inverter current ripple includes the LF harmonics (up to 500Hz) with different level of amplitudes (see figures 7 and 8) and is simulated using a variable load (see figure 2).

The simulated inverter current, I_{out}, the obtained PEMFC output current, I_{FC}, and the bi-buck interface currents, I_{out1} and I_{out2}, are presented in the figure 7.a. The simulated inverter current harmonics and the obtained PEMFC output current harmonics are presented in the figure 7.b, and figure 7.c, respectively.

The reduction of the harmonics ratio, Rk=(simulated inverter current harmonic level)/(PEMFC output current harmonic level), is shown in table 1 for the first set of amplitudes used in simulation of the inverter current.

TABLE I.
REDUCTION OF THE HARMONICS RATIO, R_K - FIRST SET OF AMPLITUDES

Harmonic Freq. [Hz]	0	100	200	300	400	500
R_k relation	$R_0 = \dfrac{11.29}{2.052}$	$\dfrac{8.4}{2.5} \times \times R_0$	$\dfrac{3.2}{0.73} \times \times R_0$	$\dfrac{5.7}{1.3} \times \times R_0$	$\dfrac{7.2}{1.53} \times \times R_0$	$\dfrac{12.8}{2.6} \times \times R_0$
Rk value	5.5019	18.49	24.12	24.12	25.89	27.09

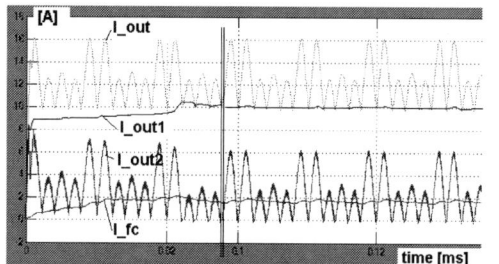

a) Dynamic compensation of the low frequency inverter current ripple by buck current source

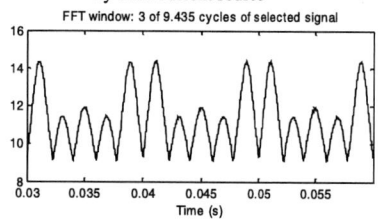

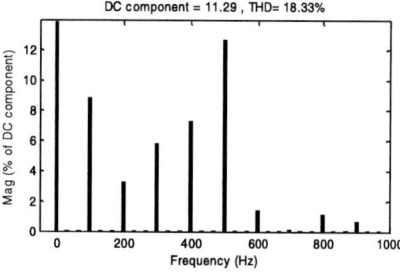

b) Harmonics for the simulated inverter current

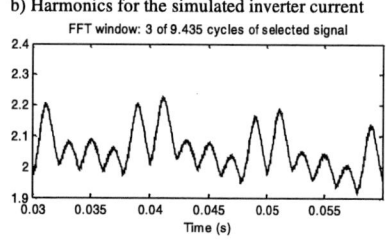

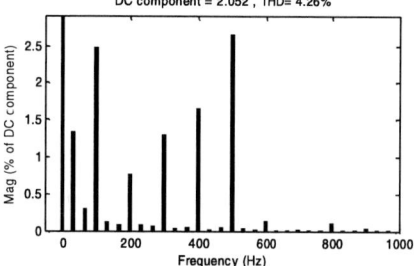

c) Harmonics for the obtained PEMFC output current

Fig. 7. Simulation results for a simulated inverter current – superposition of five harmonics: first set of amplitudes

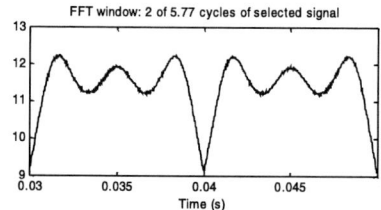

a) Harmonics for the simulated inverter current

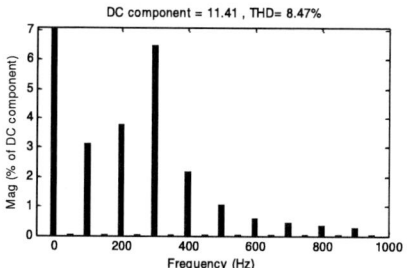

b) Harmonics for the obtained PEMFC output current

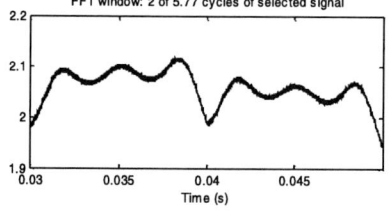

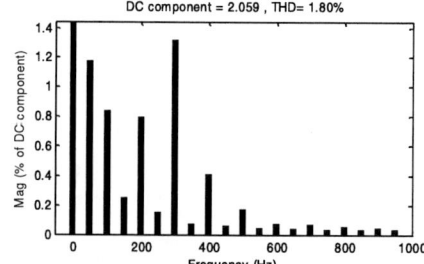

Fig. 8. Simulation results for a simulated inverter current – superposition of five harmonics: second set of amplitudes

The reduction of the harmonics ratio, R_k, is shown in table 2 for the second set of amplitudes used in simulation of the inverter current by superposition.

TABLE II.
R_k REDUCTION - SECOND SET OF AMPLITUDES

Harmonic Freq. [Hz]	0	100	200	300	400	500
R_k relation	$R_0 =$ $\dfrac{11.41}{2.059}$	$\dfrac{3.1}{0.82} \times$ $\times R_0$	$\dfrac{3.7}{0.78} \times$ $\times R_0$	$\dfrac{6.4}{1.3} \times$ $\times R_0$	$\dfrac{2.1}{0.4} \times$ $\times R_0$	$\dfrac{1.2}{0.18} \times$ $\times R_0$
Rk value	5.4515	20.95	26.29	27.28	29.09	36.94

The output voltage ripple includes the inverter ripple and buck ripple (usually given by the associate on-off control of the transistor), and this is spread in a large frequencies band by proposed chaotifying technique (figure 9). For used filter capacitor, $C_f = 47\mu F$, the output voltage ripple is quite well as level.

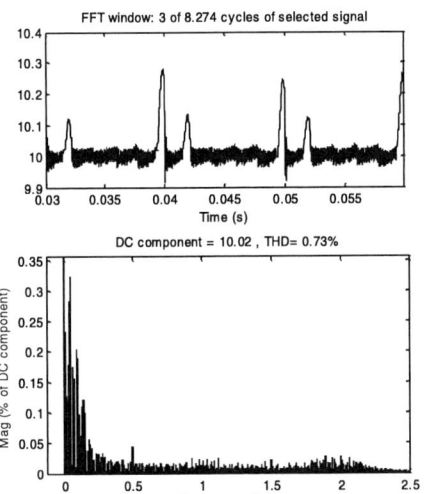

Fig. 9. Output voltage of bi-buck power interface (top) and associate frequencies spectrum (bottom)

Figure 10 present a circuit implementation using the output currents addition. In this case we must have $V_{in2} > V_{ref}$.

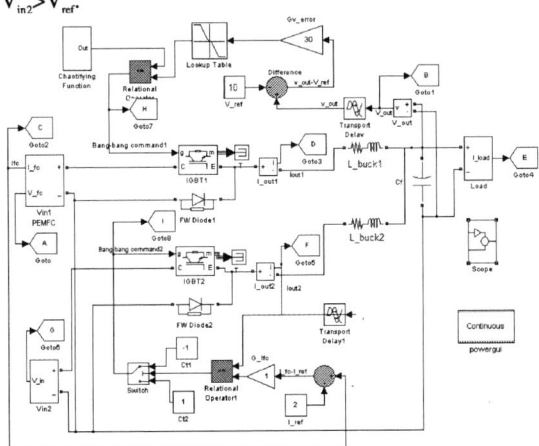

Fig. 10. Bi-buck power interface using current superposition

For a half cell number and a double cell aria and the same load parameters ($V_{ref}=10V$ and variable load) the PEMFC current is double, so we fix Iref=4A. The gain of the PEMFC current ripple, G_{Ifc}, plays a significant role in the active compensation of the low frequency inverter current harmonics with a buck current source using a bang-bang current control. Figure 11 presents the PEMFC current with buck current source (figure 10) and using $G_{Ifc}=20$, and the reduction of the harmonics ratio is shown in figure 12, which presents the PEMFC current without buck current source.

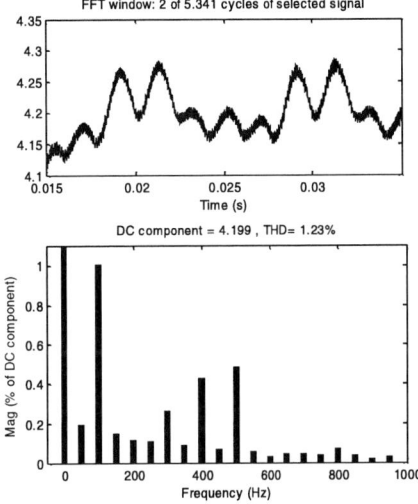

Fig. 11. PEMFC current with buck current source

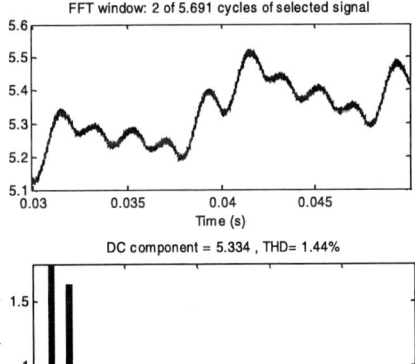

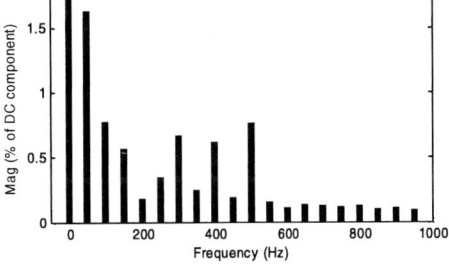

Fig. 12. PEMFC current without buck current source

Table 3 presents the PEMFC current harmonics with (*)/ without (**) buck current source as % of dc component. Last row present the reduction ratio of these harmonics calculate with relation: Ratio = (PEMFC current harmonic level**) / (PEMFC current harmonic level*). These results show more clearly the active compensation effect of the inverter current harmonics with a buck current source. For LF inverter current harmonics the reduction ratio is in range 1.5÷2.

TABLE III.
HARMONICS LEVEL OF PEMFC CURRENT FOR CIRCUIT IMPLEMENTATION ON THE FIGURE 10 WITH (*) AND WITHOUT (**) BUCK CURRENT SOURCE

Harmonic Freq. [Hz]	100	200	300	400	500
Harmonic Level* [%]	1	0.11	0.28	0.45	0.51
Harmonic Level ** [%]	1.65	0.2	0.54	0.52	0.6
Ratio =**/*	1.65	1.82	1.93	1.16	1.18

The active compensation start-up of the inverter current harmonics, I_{out}, with buck current source, I_{out2}, is shown in figure 13.

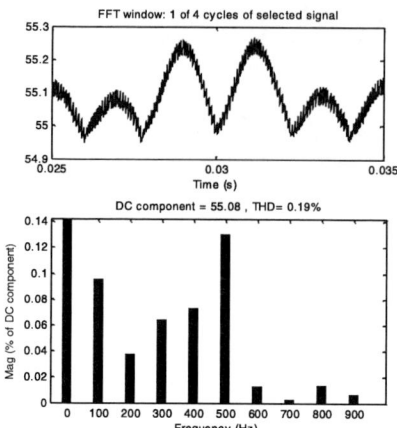

Fig. 14. PEMFC current (cells number=60; cell aria=100cm^2) with buck current source

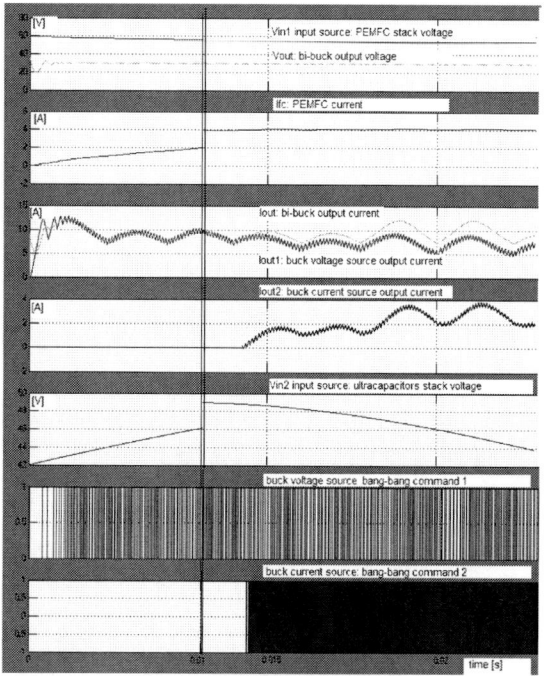

Fig. 13. Bi-buck interface start-up

The proposed active compensation technique was tested for high PEMFC power (cells number=60; cell aria=100cm^2), and the results are shown in figures 14-15. The load current is shown in figure 16. The ripple factor, RF, is defined by relation:

$$RF\ (\%) = 100 \times (I_{fc(max)} - I_{fc(min)})/ I_{fc(max)} \quad (3)$$

We can calculate RF for load current and for PEMFC current:

$RF_{load\_current}$=100 x (88-58)/88=34%;
$RF_{PEMFC\_current}$=100 x (55.25-54.95)/55.25=0.5%.

It is difficult to compare the obtained results, even if the load power is the same, because all latest active/passive compensation proposed methods use a single input voltage source, PEMFC stack.

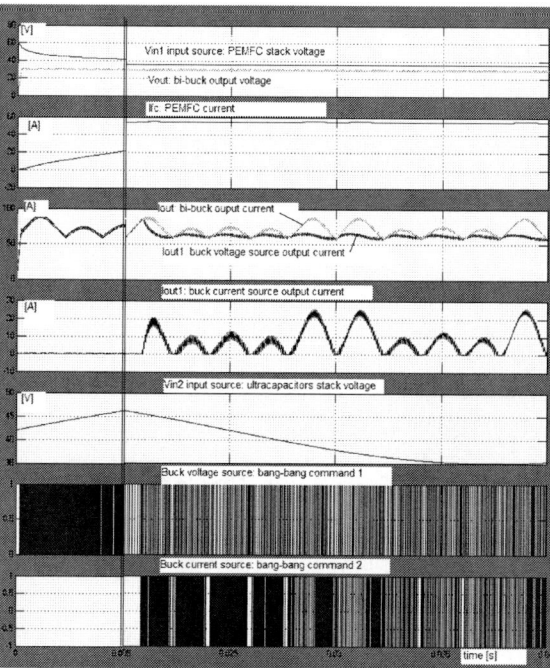

Fig. 15. Bi-buck interface start-up with high PEMFC power (cells number=60; cell aria=100cm^2), and G_{Ifc}=100

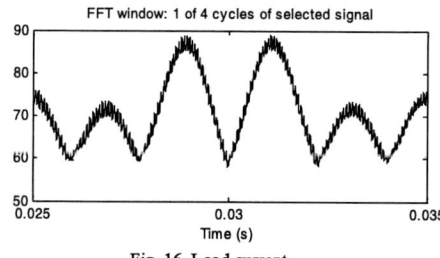

Fig. 16. Load current

For example, an advanced active control technique is proposed in [30] that incorporate a current control loop in the dc-dc converter for inverter current ripple reduction. In [31], a high performance fuel cell power conditioning system is presented in order to show superior efficiency and high-frequency ripple current elimination using a multiphase dc-dc converter and low-frequency inverter current ripple current elimination using a multiple loop control technique. The active harmonic filter proposed in [32] provides an alternate path for the 120 Hz ripple, thereby, preventing the ripple from circulating through the source. The difference of low frequency current ripples between simple resistor load condition and actual residential load conditions is compared in [33]. A design methodology for interleaved boost converters is studied in [34] in order to design high efficiency and high performance converters for fuel cell applications (including input current ripple, output voltage ripple, and losses analysis). All these control method eliminates the inverter current ripple going back to fuel cell with a cost effective control design, and report a ripple factor up to 4% for a load power in the same range.

Other bi-buck interface that uses the Matlab model presented in figure 1 is shown in figure 17. That design is a little bite more complex, but can use a second input voltage source lower than V_{ref}.

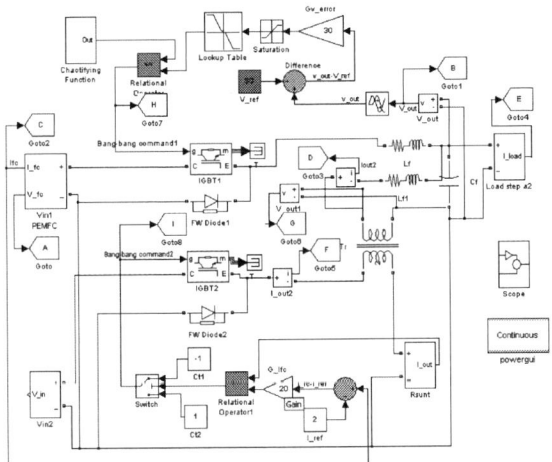

Fig. 17. Bi-buck power interface using model on Fig. 1

VI. CONCLUSION

This paper proposes a method to minimize the inverter ripple effect on the PEMFC source by an active compensation of the low frequency current harmonics with a buck current source supplied by an auxiliary source. The proposed control takes the advantages of the bang-bang control (high dynamic, simplicity of design and reliability). Remaining power spectrum is spread by chaotification of the buck voltage source behavior using a nonlinear transfer function in the feedback voltage control loop. Simulation results using Matlab model are promising, so two practical solutions for bi-buck topology implementation are proposed.

ACKNOWLEDGMENT

The CNCSIS Grant #570, CEEX #226 and CEEX #310 of the National Research Council (MEC) has supported part of the research for this paper.

REFERENCES

[1]. R.S. Gemmen, "Analysis for the effect of the ripple current on fuel cell operating condition", *ASME* 369 (4), pp. 279–289, 2001.

[2]. C. Woojin, J. Gyubum, N.E. Prasad, and W.H. Jo, "An Experimental Evaluation of the Effects of Ripple Current Generated by the Power Conditioning Stage on a Proton Exchange Membrane Fuel Cell Stack", *JMEPEG* 13, pp.257-264, 2004.

[3]. R.M. James, J. Faryar, B. Jacob, L.M. Josh, and G.S. Samuelsen, "Analysis of stationary fuel cell dynamic ramping capabilities and ultracapacitor energy storage using high resolution demand data", *Journal of Power Sources* 156, pp. 472–479, 2006.

[4]. B. Suddhasatwa, *Recent Trends in Fuel Cell Science and Technology*, Springer, ISBN 0-387-35537-5, 2007.

[5]. NETL Group, "NETL published fuel cell specifications", in *Future Energy Challenge 2005 Competition* (www.netl.doe.gov; www.energychallenge.org), 2005.

[6]. S. Wajiha, A.K. Rahul, and M. Arefeen, "Analysis and minimization of input ripple current in PWM inverters for designing reliable fuel cell power systems", *Journal of Power Sources* 156, pp. 448–454, 2006.

[7]. N. Bizon, and M. Oproescu, "Hysteretic Fuzzy Control of the Power Interface Converter", *Fuzzy Systems and AI*, Publishing House of the Romanian Academy, unpublished.

[8]. G. Fontes, C. Turpin, S. Astier, and T Meynard, "Interactions between fuel cells and power converters : Influence of current harmonics on a fuel cell stack", *IEEE Transaction on Power Electronics*, vol. 22, no.°2, pp. 670-678, 2007.

[9]. K.K. Tse, R.W.M. Ng, H.S.H. Chung, et al., "An evaluation of the spectral characteristics of switching converters with chaotic carrier-frequency modulation", *IEEE Trans. on Industrial Electronics*, vol. 50, no. 1, pp. 171-182, 2003.

[10]. C. Morel, M. Bourcerie, and F. C. Blondeau, "Improvement of power supply electromagnetic compatibility by extension of chaos anticontrol", *Journal of Circuits, Systems, and Computers*, vol. 14, no. 4, pp. 757–770, 2005.

[11]. F. Mihali, and D. Kos, "Reduced Conductive EMI in Switched-Mode DC–DC Power Converters Without EMI Filters: PWM Versus Randomized PWM", *IEEE Transactions on Power Electronics*, vol. 21, no. 6, pp. 1783 – 1794, 2006.

[12]. Z-Z. Li, S-S, Qiu, and Y-F. Chen, "Experimental Study on the Suppressing Emi Level of DC-DC Converter with Chaotic Map", *Proceedings of the Chinese society for electrical engineering*, vol. 26, no.5, pp. 76-81, 2006.

[13]. N. Bizon, "EMI Spectral Optimization using a Nonlinear Controller for Buck Output Voltage Stabilization", unpublished.

[14]. B. Choi, D. Kim, D. Lee, S. Choi, and J. Sun, "Analyze on input filter interaction in switching power converter", *IEEE Trans. on Power Electronics*, vol. 22, no. 2, pp. 452 – 460, 2007.

[15]. C.K.Tse, *Complex behavior of switching power converters*, CRC Press, 2003.

[16]. N. Bizon, and M. Oproescu, "Energy Generation System Behaviour using a Clocked Fuzzy Peak Current Control", *in 12th European Conference on Power Electronics and Applications*, IEEE Catalog Number 07EX1656C, 2007.

[17]. S. Banerjee, and G. C. Verghese, *Nonlinear phenomena in power electronics: attractors, bifurcations, chaos, and nonlinear control*, N. Y.: IEEE Press, 2001.

[18]. N. Bizon, "Intelligent Integrated Control Of The Power Flows Into An Energy Generation System", *Mediterranean Journal of Measurement and Control*, Vol. 3, No. 3, 2007, pp. 59-71, 2007.

[19]. W.A. Keith, W.T. Keith, M.J. Marcel, T.A. Haskew, W.S. Shepard, and B.A. Todd, "Experimental investigation of fuel cell dynamic response and control", *Journal of Power Sources 163*, pp. 971–985, 2007

[20]. I. Valero, S. Bacha, and E. Rulliere, "Comparison of energy management controls for fuel cell applications", *Journal of Power Sources 156*, pp. 50–56, 2006.

[21]. K.-S. Jeong, W.-Y. Lee, and C.-S. Kim, "Energy management strategies of a fuel cell/battery hybrid system using fuzzy logics", *Journal of Power Sources 145*, pp. 319–326, 2005.

[22]. P. Corbo, F.E. Corcione, F. Migliardini, and O. Veneri, "Energy management in fuel cell power trains", *Energy Conversion and Management 47*, pp. 3255–3271, 2006

[23]. N. Bizon, E. Sofron, and M. Oproescu, "Intelligent Control of the Power Flows on an Energy Generation System", *The Fifth Multi-Conference on Systemics, Cybernetics and Informatics – EIC2007*, vol. 5, pp. 314-319, ed. by International Institute of Informatics and Systemics (IIIS), USA, 2007.

[24]. N. Bizon, E. Lefter, and M. Oproescu, "Modeling and Control of the Energy Sources Power Interface for Automotive Hybrid Electrical System", *in 21st JUMV International Automotive Conference Science and Motor Vehicles 2007*, proceedings on CD ed. by the Society of Automotive Engineers of Serbia under FISITA patronage, 2007.

[25]. Z.-D. Zhong, H.-B. Huo, X.-J. Zhu, G.-Y. Cao, and Y. Ren, "Adaptive maximum power point tracking control of fuel cell power plants", *Journal of Power Sources*, in press, doi:10.1016/j.jpowsour.2007.10.080.

[26]. N. Bizon, A. Zafiu, and M. Oproescu, "PEMFC modeling using genetics algorithm for parameters tuning", *in Int. conference ISEE 2005*, pp. 163-169, 2005.

[27]. W. Choi, J.W. Howze, and P. Enjeti, "Development of an equivalent circuit model of a fuel cell to evaluate the effects of inverter ripple current", *Journal of Power Sources 158*, pp. 1324–1332, 2006.

[28]. N. Bizon, "Buck Supplies Output Voltage Ripple Reduction Using Fuzzy Control", in the *Annals of "Dunarea De Jos" University of Galati - Electrotechnics, Electronics, Automatic Control, Informatics*, Fascicle III, ISSN 1221-454X, pp. 102-106, 2007 (http://www.ann/ugal.eeai/index.html).

[29]. N. Bizon, "Chaotification of the buck converter that use a modified Chua's diode in the feedback loop", *Advances in Intelligent Systems and Technologies – 5th European Conference on Intelligent Systems and Technologies*, in press.

[30]. C. Liu, and J. S. Lai, "Low Frequency Current Ripple Reduction Technique with Active Control in a Fuel Cell Power System with Inverter Load", *Power Electronics Specialists Conference – PESC'05*, Vol. IEEE 36th, pp. 2905 – 2911, 2005.

[31]. J. S. Lai, "A high-performance V6 converter for fuel cell power conditioning system", *Vehicle Power and Propulsion - IEEE Conference*, pp. 7, 2005.

[32]. S. Wajiha, and N. Hrishikesh, "Active Filtering of Input Ripple Current to Obtain Efficient and Reliable Power from Fuel Cell Sources", *Telecommunications Energy Conference - INTELEC '06*, pp. 1-6, 2006.

[33]. J.-S. Kim, H.-S. Kang, B.-K. Lee, and W.-Y. Lee, "Analysis of low frequency current ripples of fuel cell systems based on a residential loads modeling", *Electrical Machines and Systems – ICEMS07*, pp. 282-287, 2007.

[34]. G.-Y. Choe, H.-S. Kang, B.-K. Lee, and W.-Y. Lee, "Design consideration of interleaved converters for fuel cell applications", *Electrical Machines and Systems – ICEMS07*, pp. 238-243, 2007.

Power Control Strategy of Parallel Inverter Interfaced DG Units

H.R. Baghaee, M.Mirsalim, M. J. Sanjari, G.B. Gharehpetian

Center of Excellence on Power system, Amirkabir University of Technology, Tehran, Iran

E-mails: hrbaghaee@aut.ac.ir, mirsalim@aut.ac.ir, m_j_sanjari@aut.ac.ir, grptian@aut.ac.ir,

Abstract—In this paper, a power control strategy for parallel inverters of DGs, which are connected to harmonic polluted grids, is proposed. In order to optimize the controller performance, the dual-time sampling scheme is implemented. Also, the parameters of the controller are optimized by Harmony Search Algorithm. The simulation results show that by using this scheme the system control delay is minimized, and the proposed controller presents a high performance even under presence of system harmonics or fault.

Keywords— Parallel Inverters, Distributed Generation, Harmonics, Space Vector Pulse Width Modulation.

I. INTRODUCTION

The Installation of a variety of small-size DG is changing the traditional structure of distribution systems. The integration of DG units within the existing infrastructure requires a full understanding of their impact on power flow and power quality at both customer and utility sides [1-3]. Depending on the distribution system operating characteristics and the DG characteristics, the impacts might be positive or negative. DGs have different characteristics, and therefore, their impact on voltage control, stability, and system protection will also be different [4-7]. DGs should meet various operating requirements of the utilities or the power system operators [8-10]. Considering the inherent benefits of power electronic (PE) interfaces [11], they have widely used for DG sources. Also, different aspects of DG interconnection like power quality and protection issues have been discussed, too [12-14].

As the system load grows, the Uninterruptible Power Supply (UPS) needs to be replaced with a higher capacity one. Also, if the UPS fails, the entire system is affected. To increase the reliability as well as power capability of the supply system, a single UPS unit can be replaced with a multiple smaller UPS in parallel. This system has many advantages like expandability, modularity, maintainability, redundancy, and increased reliability [15].

To realize the above mentioned goals, different approaches have been discussed in [16-39]. The load sharing of parallel three phase inverters has been discussed in [20-24]. The current sharing control strategy for parallel multi-inverter systems to achieve a weighted output current distribution has been introduced in [25, 26]. The wireless load sharing and control of parallel three phase inverters has also been presented [27-30]. Some other aspects like maximum current control strategy, failure isolation and hot-swap features and bidirectional power flow in grid connected inverters have been investigated in [31-35]. In [30,36-38], the interconnection of DGs to the system using this scheme has been proposed.

With a pulse width modulation (PWM) based switching strategy, the converter connecting DGs to the grid will contains low-order harmonics. To decrease the current harmonic content of the DG interconnection, a new approach based on space vector pulse width modulation (SVPWM) switching strategy is used in this paper. In the proposed approach, a power control strategy for parallel inverter interfaced DG units, shown in Fig. 1, is presented, too. In order to optimize the parameters of the controller, Harmony Search Algorithm (HSA) is used.

II. SVM SWITCHING STRATEGY

All modulation techniques try to obtain variable output voltage with maximum fundamental component and minimum harmonics. Many PWM techniques have been developed to let the inverters to posses desired output characteristics, achieve a wide linear modulation range and higher efficiency, and have less switching and commutation losses and lower Total Harmonic Distortion (THD) [39-43].

The Space Vector Modulation (SVM) technique is more popular than conventional techniques because it has lower base band harmonics than regular PWM or sinusoidal PWM (SPWM). SVM increases the output capability of SPWM without distorting line to line output voltage waveform. Moreover, SVM has 15% higher fundamental output voltage than conventional modulation methods and leads to better DC link utilization [43].

In SVM technique, the three phase quantities can be transformed to their equivalent two phase quantity either in synchronously rotating or stationary reference frame [44, 45]. The reference vector magnitude can be found using this two phase component and used to modulate the inverter output. A three phase sinusoidal voltage set is assumed to be as follows:

$$V_a = V_m \sin(\omega t)$$
$$V_b = V_m \sin\left(\omega t - \frac{2\pi}{3}\right)$$
$$V_c = V_m \sin\left(\omega t + \frac{2\pi}{3}\right)$$

(1)

978-1-4244-1741-4/08/$25.00 ©2008 IEEE   629

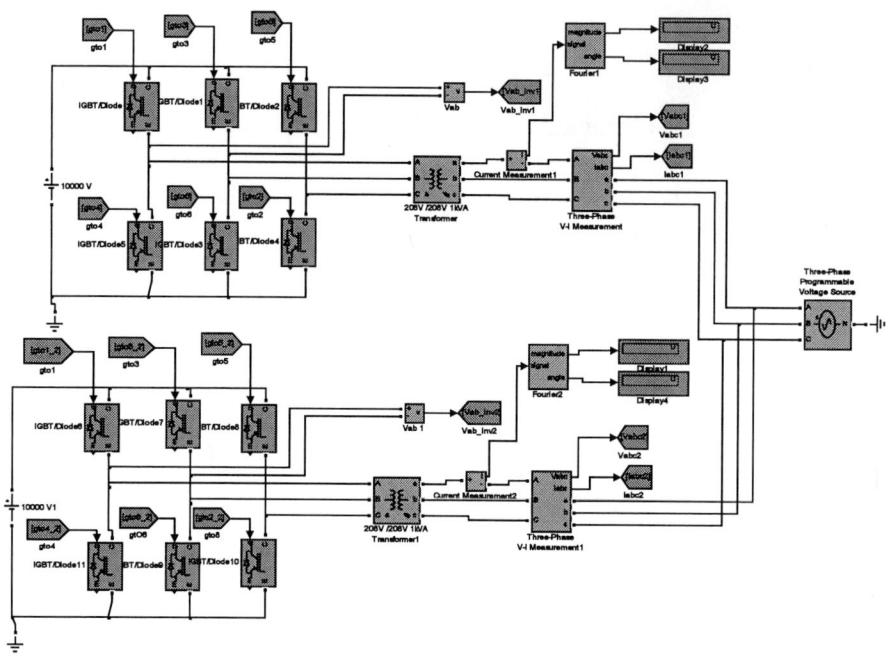

Fig.1 Configuration of system under study

When this three phase voltage set is applied to the AC machine, it produces a rotating flux in the air gap of the AC machine. This rotating flux component can be represented as a single rotating voltage vector. The magnitude and angle of the rotating vector can be found by means of Clark's transformation [44-45]. The representation of the rotating vector in the complex plane has been shown in Fig. 2.

Space vector representation of the 3 phase quantity can be expressed as follows.

$$\vec{V}^* = V_\alpha + jV_\beta = \frac{2}{3}\left(V_a + aV_b + a^2V_c\right) \tag{2}$$

where,

$$a = 1\angle120° \tag{3}$$

$$|V| = \sqrt{V_\alpha^2 + V_\beta^2} \tag{4}$$

$$\alpha = \tan^{-1}\left(\frac{V_\beta}{V_\alpha}\right) \tag{5}$$

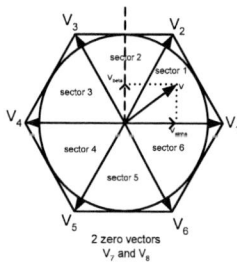

Fig. 2 Representation of rotating vector in complex plane

Hence, we have:

$$
\begin{aligned}
V_\alpha + jV_\beta &= \frac{2}{3}\left(V_a + aV_b + a^2V_c\right) \\
&= \frac{2}{3}\left(V_a + \cos\left(\frac{2\pi}{3}\right)V_b + \cos\left(\frac{2\pi}{3}\right)V_c\right) + j\frac{2}{3}\left(\sin\left(\frac{2\pi}{3}\right)V_b - \sin\left(\frac{2\pi}{3}\right)V_c\right)
\end{aligned} \tag{6}
$$

Separating the real and imaginary parts of (6), results in the following equations:

$$V_\alpha = \frac{2}{3}\left(V_a + \cos\left(\frac{2\pi}{3}\right)V_b + \cos\left(\frac{2\pi}{3}\right)V_c\right) \tag{7}$$

$$V_\beta = \frac{2}{3}\left(0V_a + \sin\left(\frac{2\pi}{3}\right)V_b - \sin\left(\frac{2\pi}{3}\right)V_c\right) \tag{8}$$

$$
\begin{aligned}
\begin{bmatrix} V_\alpha \\ V_\beta \end{bmatrix} &= \frac{2}{3}\begin{bmatrix} 1 & \cos\left(\frac{2\pi}{3}\right) & \cos\left(\frac{2\pi}{3}\right) \\ 0 & \sin\left(\frac{2\pi}{3}\right) & -\sin\left(\frac{2\pi}{3}\right) \end{bmatrix}\begin{bmatrix} V_a \\ V_b \\ V_c \end{bmatrix} \\
&= \frac{2}{3}\begin{bmatrix} 1 & -\frac{1}{2} & -\frac{1}{2} \\ 0 & \frac{\sqrt{3}}{2} & -\frac{\sqrt{3}}{2} \end{bmatrix}\begin{bmatrix} V_a \\ V_b \\ V_c \end{bmatrix}
\end{aligned} \tag{9}
$$

In the SVM switching strategy, a sinusoidal voltage is treated as a constant amplitude vector rotating at constant frequency. The reference voltage, V_{ref} is usually approximated by a combination of the eight switching patterns (V_0 to V_7). Then, three-phase voltage vectors are transformed into a vector in the stationary d-q coordinate frame which represents the spatial vector sum of the three-phase voltages [14].

To realize the SVM switching strategy, direct and quadrature axis voltage components, V_d and V_q, and the reference voltage and angle α must be obtained using (10).

$$\begin{bmatrix} V_d \\ V_q \end{bmatrix} = \frac{2}{3}\begin{bmatrix} 1 & -\frac{1}{2} & -\frac{1}{2} \\ 0 & \frac{\sqrt{3}}{2} & -\frac{\sqrt{3}}{2} \end{bmatrix}\begin{bmatrix} V_{an} \\ V_{bn} \\ V_{cn} \end{bmatrix}$$

$$\left|\overline{V_{ref}}\right| = \sqrt{V_d{}^2 + V_d{}^2}$$

$$\alpha = \tan^{-1}\left(\frac{V_q}{V_d}\right) = \omega_s t = 2\pi f_s t$$

(10)

T_1, T_2 and T_0 represent the time widths for vectors V_1, V_2

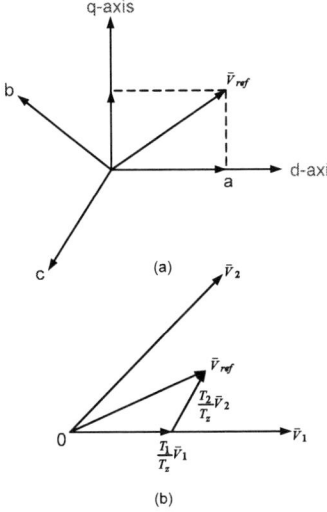

Fig. 3. a) Voltage Space Vector and its components in (d,q) axis
b) Reference vector as a combination of adjacent vectors at sector 1

TABLE. 1 SWITCHING SEQUENCE

Sector	Upper Switches (S₁, S₃, S₅)	lower Switches (S₄, S₆, S₂)
1	$S_1 = T_1 + T_2 + T_0/2$ $S_3 = T_2 + T_0/2$ $S_5 = T_0/2$	$S_4 = T_0/2$ $S_6 = T_2 + T_0/2$ $S_2 = T_1 + T_2 + T_0/2$
2	$S_1 = T_1 + T_0/2$ $S_3 = T_1 + T_2 + T_0/2$ $S_5 = T_0/2$	$S_4 = T_2 + T_0/2$ $S_6 = T_0/2$ $S_2 = T_1 + T_2 + T_0/2$
3	$S_1 = T_0/2$ $S_3 = T_1 + T_2 + T_0/2$ $S_5 = T_2 + T_0/2$	$S_4 = T_1 + T_2 + T_0/2$ $S_6 = T_0/2$ $S_2 = T_1 + T_0/2$
4	$S_1 = T_0/2$ $S_3 = T_1 + T_0/2$ $S_5 = T_1 + T_2 + T_0/2$	$S_4 = T_1 + T_2 + T_0/2$ $S_6 = T_2 + T_0/2$ $S_2 = T_0/2$
5	$S_1 = T_2 + T_0/2$ $S_3 = T_0/2$ $S_5 = T_1 + T_2 + T_0/2$	$S_4 = T_1 + T_0/2$ $S_6 = T_1 + T_2 + T_0/2$ $S_2 = T_0/2$
6	$S_1 = T_1 + T_2 + T_0/2$ $S_3 = T_0/2$ $S_5 = T_1 + T_0/2$	$S_4 = T_0/2$ $S_6 = T_1 + T_2 + T_0/2$ $S_2 = T_2 + T_0/2$

and V_0. T_0 is the period in a sampling period for null vectors should be filled. As each switching period (half of sampling period) T_z must start and end with zero vectors, i.e., there will be two zero vectors per T_z or four null vectors per T_s, the duration of each null vector is $T_s/4$ [43].Therefore, the time duration T_1, T_2 and T_0 can be calculated as following:

$$\int_0^{T_z}\overline{V_{ref}}dt = \int_0^{T_1}\overline{V_1}dt + \int_{T_1}^{T_1+T_2}\overline{V_2}dt + \int_{T_1+T_2}^{T_2}\overline{V_2}dt$$

$$\therefore T_z\overline{V_{ref}} = \left(T_1\overline{V_1} + T_2\overline{V_2}\right)$$

$$T_z\left|\overline{V_{ref}}\right|\begin{bmatrix} \cos(\alpha) \\ \sin(\alpha) \end{bmatrix} = \frac{2}{3}.T_1.V_{dc}\begin{bmatrix} 1 \\ 0 \end{bmatrix} + \frac{2}{3}.T_2.V_{dc}\begin{bmatrix} \cos\left(\frac{\pi}{3}\right) \\ \sin\left(\frac{\pi}{3}\right) \end{bmatrix}$$

$where\ 0 < \alpha < 60$

(11)

$$\therefore T_1 = T_z a\frac{\sin\left(\frac{\pi}{3} - \alpha\right)}{\sin\left(\frac{\pi}{3}\right)}$$

$$\therefore T_2 = T_z a\frac{\sin(\alpha)}{\sin\left(\frac{\pi}{3}\right)}$$

$$T_0 = T_z - (T_1 + T_2), where, T_z = \frac{1}{f_s} and\ a = \frac{\left|\overline{V_{ref}}\right|}{\frac{2}{3}.V_{dc}}$$

(12)

Then, switching time of each switch (S₁ to S₆) must be determined. The voltage space vector and its components in dq plane is show in Fig. 3a. The switching sequence for the lower and upper switches has been shown in Table. 1.

The above mentioned symmetrical pulse pattern for two consecutive T_z intervals are shown and $T_s=2$ and $T_z=1/f_s$ is the sampling time, where f_s is switching frequency. Note that the null time has been conveniently distributed between the V_0 and V_7 vectors to describe the symmetrical pulse width. Studies have been shown that a symmetrical pulse pattern gives minimal output harmonics [39].

III. CONTROLLER DESIGN

The line currents and the grid voltage at the Point of Common Coupling (PCC) are feedback variables. These variables are transformed to the dqo frame by the Park transform as [44] follows:

$$\begin{bmatrix} f_q \\ f_d \\ f_0 \end{bmatrix} = \frac{2}{3}\begin{bmatrix} \cos(\theta_g) & \cos\left(\theta_g - \frac{2\pi}{3}\right) & \cos\left(\theta_g + \frac{2\pi}{3}\right) \\ \sin(\theta_g) & \sin\left(\theta_g - \frac{2\pi}{3}\right) & \sin\left(\theta_g + \frac{2\pi}{3}\right) \\ \frac{1}{2} & \frac{1}{2} & \frac{1}{2} \end{bmatrix}\begin{bmatrix} f_a \\ f_b \\ f_c \end{bmatrix}$$

(13)

Where,

$$\theta_g = \int_0^t \omega_g(t)dt + \alpha(0)$$

(14)

Using this transform, the line currents and the PCC voltage are transformed to the $dq0$ frame. Then, the reference currents of q and d axis can be calculated by the following equations:

$$I_{q,ref} = \frac{P_{ref}}{\left(\frac{2}{3}\right) \times V_{PCC,q}} \tag{15}$$

$$I_{d,ref} = \frac{Q_{ref}}{\left(\frac{2}{3}\right) \times V_{PCC,q}} \tag{16}$$

The difference between reference and actual currents of q and d axis are the input of the proportional integral (PI) controller. The outputs of PI controllers are added to the actual q and d axis voltages. The result is the reference q and d axis voltages.

$$V_{d,ref} = V_d + K_{P,d}\left(I_{d,ref} - I_d\right) + K_{I,d} \int_0^{T_{end}} \left(I_{d,ref} - I_d\right) dt \tag{17}$$

$$V_{q,ref} = V_q + K_{P,q}\left(I_{q,ref} - I_q\right) + K_{I,q} \int_0^{T_{end}} \left(I_{q,ref} - I_q\right) dt \tag{18}$$

where, T_{end} is the simulation time. These reference voltages are the input of SVM system.

In the normal operating condition of parallel inverters, the converter should compensate the harmonics of the grid. The controller output signals are used as the input to the SVM pulse generating module.

IV. HARMONY SEARCH ALGORITHM

The parameters of the controller has been also optimized using HAS described in the next section. The objective function is integral of time square of total error (ITSE) namely:

$$ITSE = \int_0^{T_{end}} t.\left(\left(I_{d,ref} - I_d\right)^2 + \left(I_{q,ref} - I_q\right)^2\right) dt \tag{19}$$

The steps in the procedure of HSA are as follows [46-47]:
Step 1: Initialize the problem and algorithm parameters.
Step 2: Initialize the harmony memory.
Step 3: Improvise a new harmony.
Step 4: Update the harmony memory.
Step 5: Check the stopping criterion.
These steps will be described in the next five subsections.

A. Initialize the problem and algorithm parameters

The optimization problem is specified as follows:
$$\min\{f(x) \mid x \in X\} \text{ subject to } g(x) \geq 0 \text{ and } h(x) = 0$$
where, $f(x)$ is the objective function and $g(x)$ is the inequality constraint function; $h(x)$ is the equality constraint function. x is the set of each decision variable, x_i, and X is the set of the possible range of values for each decision variable, that is $X_{i,min} \leq X_i \leq X_{i,max}$. where $X_{i,min}$ and $X_{i,max}$ are the lower and upper bounds for each decision variable. The HSA parameters are also specified in this step. These are the harmony memory size

(HMS), or the number of solution vectors in the harmony memory; harmony memory considering rate (HMCR); pitch adjusting rate (PAR); number of decision variables (N) and the number of improvisations (NI), or stopping criterion. The harmony memory (HM) is a memory location where all the solution vectors (sets of decision variables) are stored. This HM is similar to the genetic pool in the GA [48]. Here, HMCR and PAR are parameters that are used to improve the solution vector and both defined in Step 3.

B. Initialize the harmony memory
The HM matrix is filled with as many randomly generated solution vectors as the HMS.

$$HM = \begin{bmatrix} x_1^1 & x_2^1 & \cdots & x_{N-1}^1 & x_N^1 \\ x_1^2 & x_2^2 & \cdots & x_{N-1}^2 & x_N^2 \\ \vdots & \vdots & \vdots & \vdots & \vdots \\ x_1^{HMS-1} & x_2^{HMS-1} & \cdots & x_{N-1}^{HMS-1} & x_N^{HMS-1} \\ x_1^{HMS} & x_2^{HMS} & \cdots & x_{N-1}^{HMS} & x_N^{HMS} \end{bmatrix} \tag{20}$$

C. Improvise a new harmony
A new harmony vector, $x_1' = (x_1', x_2', \ldots, x_N')$, is generated based on three rules: (1) memory consideration, (2) pitch adjustment and (3) random selection. Generating a new harmony is called 'improvisation' [48]. In the memory consideration, the value of the first decision variable x_1' for the new vector is chosen from any value in the specified HM range $(x_1^1 - x_1^{HMS})$. Values of the other decision variables $(x_2', x_3', \ldots, x_N')$ are chosen in the same manner. The HMCR, which varies between 0 and 1, is the rate of choosing one value from the historical values stored in the HM, while (1 _ HMCR) is the rate of randomly selecting one value from the possible range of values.

$$x_i' \leftarrow \begin{cases} x_i' \in \{x_i^1, x_i^2, \ldots, x_i^{HMS}\} & \text{with probability HMCR} \\ x_i' \in X_i & \text{with probability } (1-HMCR) \end{cases} \tag{21}$$

For example, a HMCR of 0.85 indicates that the HS algorithm will choose the decision variable value from historically stored values in the HM with the 85% probability or from the entire possible range with the 100–85% probability. Every component obtained by the memory consideration is examined to determine whether it should be pitch-adjusted. This operation uses the PAR parameter, which is the rate of pitch adjustment as follows:

$$x_i' \leftarrow \begin{cases} Yes & \text{with probability PAR} \\ No & \text{with probability } (1-PAR) \end{cases} \tag{22}$$

The value of (1 _ PAR) sets the rate of doing nothing. If the pitch adjustment decision for x_i' is Yes, x_i' is replaced as follows:
$$x_i' \leftarrow x_i' \pm rand() * b_w \tag{23}$$
Where, b_w is an arbitrary distance bandwidth, rand () is a random number between 0 and 1.

In Step 3, HM consideration, pitch adjustment or random selection is applied to each variable of the new harmony vector in turn.

D. Update harmony memory

If the new harmony vector, $x'_i = (x'_1, x'_2, ..., x'_N)$ is better than the worst harmony in the HM', judged in terms of the objective function value, the new harmony is included in the HM and the existing worst harmony is excluded from the HM.

E. Check stopping criterion

If the stopping criterion (maximum number of improvisations) is satisfied, computation is terminated. Otherwise, Steps 3 and 4 should be repeated.

V. SIMULATION RESULTS

The proposed parallel inverter interfaced DG units has been simulated for four different cases:

Case1: Normal condition

In this case, both DG units of parallel inverter interfaced are microturbine generator (MTG) discussed in [49]. The PCC voltage, harmonic spectrum and active and reactive power of inverters have been shown in Fig. 4. in this case THD of PCC voltage is equal to 4.06%.

Case2: Normal condition, DG units are not the same

In this case, both DG units of parallel inverter interfaced are micro turbine generator (MTG) discussed in [49] and the fuel cell discussed in [50]. The PCC voltage, harmonic spectrum and active and reactive power of the inverters have been shown in Fig. 5. In this case, THD of PCC voltage is equal to 6.63%.

Case3: Fault condition of case1

In this case, a three phase fault has been simulated in the case 1. The PCC voltage and active and reactive power of the inverters have been shown in Fig. 6.

Case4: Fault condition of case 2

In this case, the three phase fault has been applied to the case 2. The PCC voltage and active and reactive power of the inverters has been shown in Fig. 7. Simulation results indicate that the proposed controller presents a good performance for normal and fault conditions.

VI. CONCLUSION

In this paper, a new control strategy for parallel inverter interfaced distributed generation units have been proposed. The space vector pulse width modulation has also used to generate switching signals of parallel

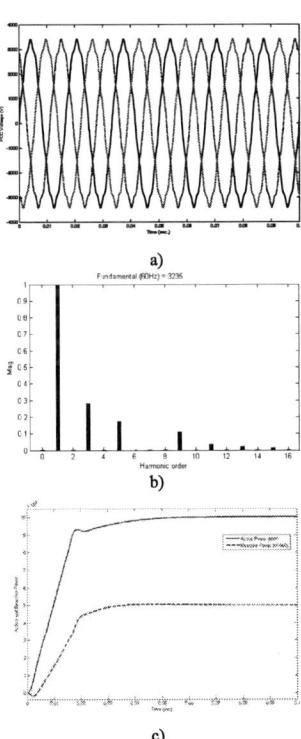

a)

b)

c)

Fig. 4. a) PCC voltage, b) Harmonic spectrum and c) active and reactive power for case1

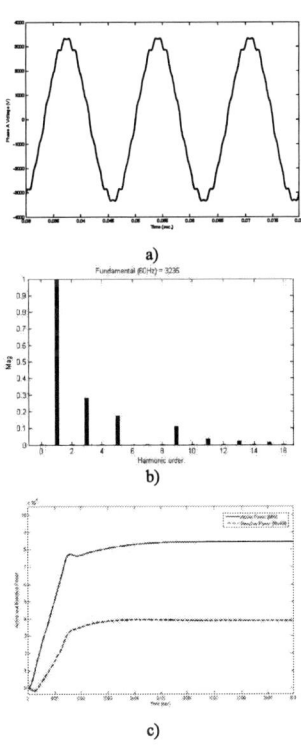

a)

b)

c)

Fig. 5. a) PCC voltage, b) Harmonic spectrum and c) active and reactive power for case2

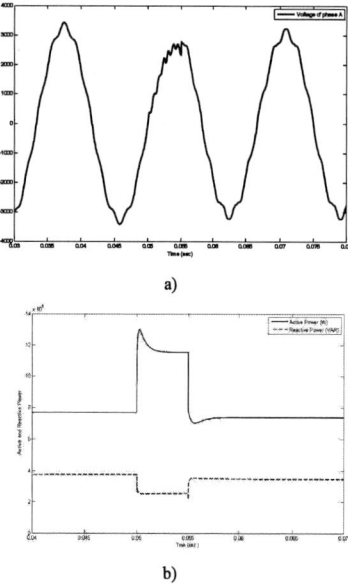

a)

b)

Fig. 6. a) PCC voltage, b) Harmonic
spectrum and c) active and reactive power
for case3

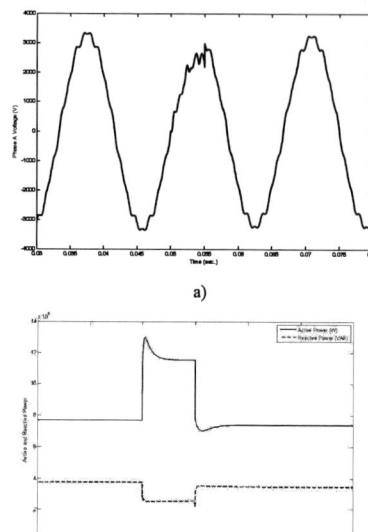

a)

b)

Fig. 7. a) PCC voltage, b) Harmonic
spectrum and c) active and reactive power
for case4

inverters. Also, the control system has been tested in normal and fault conditions. Simulation results indicate that the system performance is good from viewpoint of harmonics and power flow control.

REFERENCES

[1] T. Ackermann, G. Andersson and L. Soder, "Distributed generation: a definition", Elsevier Electric Power System Research, 2001,Vol. 57, pp.894-895.

[2] R. C. Dugan and T. E. McDermott, "Distributed generation," IEEE Ind. Appl. Mag., vol. 18, no. 2, pp. 19–25, Apr./May 2002.

[3] F. V. Edwards, G. J. W. Dudgeon, J. R. McDonald and W. E. Leithead, "Dynamics of distribution network with Distributed Generation", IEEE Power Engineering Society Summer Meeting, 2000, Vol. 2, pp. 1032-1037

[4] P. Barker and R.W. DeMello, "Determining the impact of DG on power systems, radial distribution," in Proc. IEEE Power Eng. Soc. Summer Meeting, 2000, pp. 1645–1656.

[5] M. T. Doyle, "Reviewing the impact of distributed generation on distribution system protection," in Proc. IEEE Power Eng. Soc. Summer Meeting, 2002, pp. 103–105.

[6] R. A. Walling, R. Saint, R. C. Dugan, J. Burke, L. A. Kojovic, "Summary of Distributed Resources Impact on Power Delivery Systems", Power Delivery, IEEE Transactions on : Accepted for future publication Volume PP, Issue 99, 2007 Page(s):1 – 10

[7] R. C. Dugan and T. E. McDermott, "Operating Conflicts for Distributed Generation Interconnected with Utility Distribution Systems", IEEE Industry Applications Magazine, 2002, Vol. 8, No. 2, pp. 19–25.

[8] T. Ackermann and V. Knyazkin, "Interaction between distributed generation and the distribution network: Operation aspects," in Proc. IEEE T&D Conf., 2002, pp. 1357–1362.

[9] IEEE Standard for Interconnecting Distributed Resources with Electric Power Systems, IEEE Std 1547-2003.

[10] S.A. Papathanassiou, "A Technical Evaluation Framework for the Connection of DG to the Distribution Network", Elsevier Electric Power System Research, Vol 77, January 1 2007, pp. 24–34.

[11] B. Kroposki, C. Pink, R. DeBlasio, H. Thomas, M. Simoes and P.K. Sen, "Benefits of power electronic interfaces for distributed energy systems", IEEE Power Engineering Society General Meeting, June 2006, pp. 18-22.

[12] J. Liang, T. C. Green, G. Weiss, and Q.-C. Zhong, "Evaluation of repetitive control for power quality improvement of distributed generation," in Proc. IEEE-PESC'02 Conf., 2002, pp. 1803–1808.

[13] J. C. Gomez and M. M. Morcos, "Coordinating over current protection and voltage sags in distributed generation systems," IEEE Power Eng Rev., vol. 22, no. 2, pp. 16–19, Feb. 2002.

[14] H.R. Baghaee, M. Mirsalim, .M.Ale-Emran, M. Abedi and G.B. Gharehpetian, "Power Factor Improvement of DC/DC Converter of Micro-Turbines ", International Conference on Renewable Energies and Power Quality, ICREPQ'08 , March 12-14, 2008, Santander, Spain available online at: www.icrepq.com/icrepq-08/324-baghaee.pdf

[15] A. Tuladhar, T. Hua Jin Unger, K. Mauch, "Control of parallel inverters in distributed AC power systems with consideration of line impedance effect", Industry Applications, IEEE Transactions on Volume 36, Issue 1, Jan/Feb 2000 Page(s):131 - 138

[16] Wu T F , Huang Y H , Chen Y K, et al.A 3C strategy for multi-module inverters in parallel operation to achieve an equal current distribution ,IEEE Transaction IE , 2000 , 47(2)

[17] Yu-Kai Chen; Yu-En Wu; Tsai-Fu Wu; Chung-Ping Ku, "ACSS for paralleled multi-inverter systems with DSP-based robust controls", Aerospace and Electronic Systems, IEEE Transactions on Volume 39, Issue 3, July 2003 Page(s):1002 -1015

[18] Liangliang Chen; Lan Xiao; Chunying Gong; Yangguang Yan, "Circulating current's characteristics analysis and the control strategy of parallel system based on double close-loop controlled VSI", Power Electronics Specialists Conference, 2004. PESC 04.2004 IEEE 35th Annual 1-1

[19] Woo-Cheol Lee; Taeck-Ki Lee; Sang-Hoon Lee; Kyung-Hwan Kim; Dong-Seok Hyun; In-Young Suh, "A master and slave control strategy for parallel operation of three-phase UPS systems with different ratings", Applied Power Electronics Conference and Exposition, 2004. APEC '04. Nineteenth Annual IEEE Volume 1, 2004 Page(s):456 - 462 Vol. 1

634

[20] U. Borup, F. Blaabjerg, and P. N. Enjeti, "Sharing of nonlinear load in parallel-connected three-phase converters," IEEE Trans. Ind. Applicat., vol. 37, pp. 1817–1823, Nov./Dec. 2001.

[21] Yan Xing; Lipei Huang; Sun, S.; Yangguang Yan, "Novel control for redundant parallel UPSs with instantaneous current sharing", Power Conversion Conference, 2002. PCC Osaka 2002. Proceedings of the , Volume: 3 , 2-5 April 2002 Pages:959 – 963 vol.3

[22] Xiao Sun; Yim-Shu Lee; Dehong Xu, "Modeling, analysis, and implementation of parallel multi-inverter systems with instantaneous average-current-sharing scheme", Power Electronics, IEEE Transactions on Volume 18, Issue 3, May 2003 Page(s):844- 856

[23] K. Jiarong, X. Shaojun, "Research on the Power sharing of the Parallel Inverters without Control Interconnection Basing on Droop Characteristic Parameters unbalance, Power Electronics and Motion Control Conference, 2006. IPEMC apos;06. CES/IEEE 5th International Volume 3, Issue , 14-16 Aug. 2006 Page(s):1 - 5

[24] Z. Qinglin, C. Zhongying, W. Weiyang, "Improved Control for Parallel Inverter with Current-Sharing Control Scheme power sharing" Power Electronics and Motion Control Conference, 2006. IPEMC apos; 06. CES/IEEE 5th International Volume 3, Issue , 14-16 Aug. 2006 Page(s):1 - 5

[25] Y.K. Chen, T.F. Wu, Y.E. Wu, and C.P. Ku, "CWDC strategy for paralleled multi-inverter systems achieving a weighted output current distribution", Applied Power Electronics Conference and Exposition, 2002. APEC 2002. Seventeenth Annual IEEE Volume 2, 10-14 March 2002 Page(s):1018 - 1023 vol.

[26] Y.K. Chen, T.F. Wu, Y.E. Wu, and C.P. Ku, "CWDC current-sharing control strategy for paralleled multi-inverter systems achieving a weighted output current distribution," in *Proc. IEEE Appl. Power Electron. Conf.*, Mar. 2002, pp. 1018–1023.

[27]Y.B. Byun , T.Y. Joe , E.S. Kim , J.I. Seo , D.H. Kim."Parallel operation of Three-Phase UPS Inverter by Wireless Load Sharing Control, IEEE Trans.on Indus.Appl.,2000,29(1):526-532

[28] J.M. Guerrero, L.Garcia de Vicuna, J.Miret, M.Castilla.A Wireless Load Sharing Controller to Improve Dynamic Performance of Parallel-Connected UPS Inverters, Power Electronics Specialist Conference, 2003. PESC '03. IEEE 34th Annual

[29] J. Hongxin Ding, M. Su, J. Du, Y. Chang, Liuchen, "Communicationless Parallel Inverters Based on Inductor Current Feedback Control", Applied Power Electronics Conference, APEC 2007 - Twenty Second Annual IEEE Publication Date: Feb. 25 2007-March 1 2007, On page(s): 1385-1389

[30] J.M. Guerrero, L.G. de Vicuna, J. Matas, M. Castilla, J. Miret, "A Wireless Controller to Enhance Dynamic Performance of Parallel Inverters in Distributed Generation Systems", Power Electronics, IEEE Transactions on Volume 19, Issue 5, Sept. 2004 Page(s): 1205 - 1213

[31] X. Rihui, L. Yunfeng, Z. Jixiang, "Modeling and Analysis of Stability for Parallel Inverters Operated with Instantaneous Maximum Current Control Strategy", Computational Engineering in Systems Applications, IMACS Multiconference on Volume , Issue , Oct. 2006 Page(s):1701 - 1706

[32] T.-F. Wu et al., "Design and implementation of a paralleled inverter system with failure isolation and hot-swap feature," in Proc. IEEE Appl. Power Electron. Conf., Feb. 2005, pp. 531–536.

[33] W. Tsai-Fu, H. Hui-Ming, W. Yu-En, C. Yu-Kai, "Parallel-Inverter System With Failure Isolation and Hot-Swap Features", Industry Applications, IEEE Transactions on Volume 43, Issue 5, Sept.-oct. 2007 Page(s):1329 - 1340

[34] H. Matthias and S. Helmut, "Control of a three phase inverter feeding an unbalanced load and operating in parallel with other power sources," in Proc. EPE-PEMC'02 Conf., 2002, pp. 1–10.

[35] Seshadri Sivakuma;, Tom Parsons and Shyamalt Sivakuma, "Modeling analysis and control of Bidirectional Power Flow in Grid Connected Inverter Systems", IEEE conference of PCC, Osaka, 2002. pp. 1015-1019.

[36] C.-C. Hua, K.-A. Liao, and J.-R. Lin, "Parallel operation of inverters for distributed photovoltaic power supply system," in Proc. IEEE PESC'02 Conf., 2002, pp. 1979–1983.

[37] S. R. Wall, "Performance of inverter interfaced distributed generation," in Proc. IEEE/PES-Transmission and Distribution Conf. Expo., 2001, pp. 945–950.

[38] K. De Brabandere, B. Bolsens, J. Van den Keybus, A. Woyte, J. Driesen, R. Belmans,"A Voltage and Frequency Droop Control Method

for Parallel Inverters", Power Electronics, IEEE Transactions on Volume 22, Issue 4, July 2007 Page(s):1107 - 1115

[39] H. W. v. d. Brocker, H. C. Skudenly, and G. Stanke, "Analysis and realization of a pulse width modulator based on the voltage space vectors, "in Conf. Rec. IEEE-IAS Annu. Meeting, Denver, CO, pp. 244–251, 1986.

[40] H. W. Vander Brocek, H. C. Skudelny, and G. Stank, "Analysis and Realization of a Pulse Width Modulator based on Voltage Space Vectors," Proc. IEEE-IAS'86, Vol. 1, pp. 12-17, 1986.

[41] Y. H. Lee, B. S. Suh, and D. S. Hyun, "A novel PWM scheme for a three-level voltage source inverter with GTO thyristors," IEEE Trans. On Industry Applications, Vol. 32, No. 2, pp. 260-268, April 1996.

[42] H. Nikkhajoei, M. Saeedifard, R. Iravani, "A three-level converter based micro-turbine distributed generation system", Power Engineering Society General Meeting, 2006. IEEE Volume, Issue, 18-22 June 2006 Page(s): 6 pp. 1-6.

[43] D. Rathnakumar, J. LakshmanaPerumal, T. Srinivasan, "A new software implementation of space vector PWM", SoutheastCon, 2005. Proceedings. IEEE Publication Date: 8-10 April 2005 On page(s): 131-136

[44] P. C. Krause,"Analysis of electric machinery", McGraw-Hill Publications, 1986.

[45] C. M. Ong, "Dynamic Simulation of Electric Machinery usin MATLAB/Simulink", Prentice-Hall, 1998.

[46] ZW.Geem, C. Tseng and Y. Park, 'Harmony search for generalized orienteering problem: best touring in China", Springer Lecture Notes for Computer Science, 2005, Vol. 341, pp.741–50.

[47] ZW. Geem, JH. Kim, GV. Loganathan, "Harmony search optimization: application to pipe network design", International Journal of Modeling and Simulation, 2002, Vol., 22, pp.125–33.

[48] KS. Lee, ZW. Geem, "A new structural optimization method based on the harmony search algorithm", Computer Structure, 2004, Vol. 82, pp. 781–98.

[49] D.N. Gaonkar, R.N. Patel, "Modeling and simulation of microturbine based distributed generation system", Power India Conference, 2006 IEEE Volume, Issue, 10-12 April 2006 pp. 1-5

[50] C. Wang, H. Nehrir, G. Hongwei, "Control of PEM fuel cell distributed generation systems", Power Engineering Society General Meeting, 2006. IEEE Publication Date: 18-22 June 2006 On page(s): 1 pp. 586-595

BIOGRAPHIES

Hamid Reza Baghaee *(IEEE Student Member' 2008)* received the BSc degree in Electrical Engineering from Kashan University in 2006. Currently he is graduate student of Power Engineering in Amirkabir University of Technology. His research interests are power system dynamic and control, HVDC & FACTS devices, Distributed Generation (DG) and application of Artificial Intelligence in power systems.

Mojtaba Mirsalim *(IEEE Senior Member' 2004)* was born in Tehran, Iran, on February 14, 1956. He received his B.S. degree in EECS/NE, M.S. degree in Nuclear Engineering from the University of California, Berkeley in 1978 and 1980 respectively, and the PhD in Electrical Engineering from Oregon State University, Corvallis in 1986. Since 1987 he has been at Amirkabir University of Technology, has served 5 years as the Vice Chairman and more than 7 years as the General Director in Charge of Academic Assessments, and currently is a Full Professor in the department of Electrical Engineering where he teaches courses and conducts research in energy conversion and CAD, among others.

His special fields of interest include the design, analysis, prototyping, and optimization of electric machines, renewable energy, FEM, and hybrid vehicles. Mirsalim is the author of more than 100 international journal and conference papers and three books on electric machinery and FEM. He is the founder and at present, the director of the Electrical Machines & Transformers Research Laboratory at http://ele.aut.ac.ir/EMTRL/Homepage.htm

Mohammad Javad Sanjari received the BSc degree in Electrical Engineering from Amirkabir University of Technology in 2006. Currently he is graduate student of Power Engineering in Amirkabir University of Technology. His research interests are power system dynamic and control, power system security assessment, HVDC & FACTS devices, Distributed Generation (DG) and application of Artificial Intelligence in power systems.

G.B. Gharehpetian *(IEEE Member)* was born in Tehran, in 1962. He received his BS and MS degrees in electrical engineering in 1987 and 1989 from Tabriz University, Tabriz, Iran and Amirkabir University of Technology (AUT), Tehran, Iran, respectively, graduating with First Class Honors. In 1989 he joined the Electrical Engineering Department of AUT as a lecturer. He received the Ph.D. degree in electrical engineering from Tehran University, Tehran, Iran, in 1996. As a Ph.D. student he has received scholarship from DAAD (German Academic Exchange Service) from 1993 to 1996 and he was with High Voltage Institute of RWTH Aachen, Aachen, Germany. He held the position of Assistant Professor in AUT from 1997 to 2003, and has been Associate Professor since 2004. Dr. Gharehpetian is a Senior Member of Iranian Association of Electrical and Electronics Engineers (IAEEE), member of IEEE and member of central board of IAEEE. Since 2004 he is the Editor-in-Chief of the Journal of IAEEE. The power engineering group of AUT has been selected as a Center of Excellence on Power Systems in Iran since 2001. He is a member of this center and since 2004 the Research Deputy of this center. Since November 2005 he is the director of the industrial relation office of AUT. He is the author of more than 200 journal and conference papers. His teaching and research interest include power system and transformers transients, FACTS devices and HVDC transmission.

Implementation of Nonlinear power flow controllers to control a VSC

Nelson L. Díaz*, Fabián H. Barbosa*, Cesar L. Trujillo*
* Laboratorio de Investigación en Fuentes Alternativas de Energía
Universidad Distrital Francisco José de Caldas
Bogotá DC-Colombia

Abstract-- The voltage source converters VSC's are power converters highly used in systems of distributed generation, HVDC's and motor drivers, thanks to the feasibility and facility to control the active and reactive power independently, in a fast and effective way. This paper presents the analysis, design and results obtained in digital simulation and in physical implementation, of nonlinear controllers applied to a VSC prototype, to control independently the active and reactive power flow.

There are two control strategies. Firstly a fuzzy control based on linguistic rules was proposed, however it present a high computational cost in time, hence was proposed an Artificial Neural Network as a dynamic emulator of the fuzzy control. The computational time in simulation was compared between the two intelligent controllers. All the simulations were done in (MATLAB®/SIMULINK®). Finally the system composed by the VSC and the controllers was implemented.

Index Terms-- Direct power control, Fuzzy control, Neuronal control, Voltage Source Converter (VSC).

I. INTRODUCTION

The Voltage Source Converter (VSC) is the base of the emergent technology VSC-HVDC, this converter topology has more advantages in control strategies and permits an easy and flexible interconnections with AC networks in comparison with conventional HVDCs based on line-commutated Current Source Converter (CSC) [1][2]. A VSC interconnected with an AC network, is a nonlinear coupled double-input double-output control object, where the power flow between the DC side and the AC side can be controlled by means of an intervention of the phase and amplitude of the AC voltage, generated by the converter [5][9]. Then a mathematical analysis to obtain a model and subsequent design of independents controllers for the active and reactive power flow might result in a complex procedure, instead of independents fuzzy controllers can be easily designed, because of they are based on a qualitative knowledge about the dynamic response of the converter, as was shown in precedent works [8].

A VSC-HVDC link is composed at least by two VSC stations, each one in both sides of the DC link with the same configuration [5]. Usually is necessary for a station adopts the constant DC voltage control, while the other can adopt the control of the active power flow between the dc link and the Ac network [5], the reactive power can be controlled independently in each station. In this

This work was supported in part by the laboratory of investigation on alternative energy sources of the Universidad Distrital.

work will be analyzed the station responsible of the power flow control, then one important assumption in this work for design propose is that the DC voltage is regulated.

This paper presents the analysis of a VSC operation and its power flow control, followed by design and results in simulation and in physical implementation, of nonlinear controllers based on Knowledge-based fuzzy control and artificial neural networks applied to a Voltage Source Convert prototype, to control independently the active and reactive power.

II. ANALYSIS OF THE VSC

A VSC is composed of a 6-pulse bridge whose kernel is based on Insulated Gate Bipolar Transistor (IGBTs), which present a better performance in high voltage and long distances than others self-commutating power semiconductors such as (MOSFETs, GTOs,) [1][3]. However the VSC compared to the CSC can not operate in high voltage levels. The VSC uses the Pulse Wide Modulation (PWM) technique to become a DC voltage into an AC current and vice versa, the VSC can operate as a inverter or as a rectifier [1][4], The PWM permits a fast and easy control of the AC voltage generated from the converter as well as flexible control of the active and reactive power flow between the converter and an active load, this can be achieved by a correct intervention over the phase and amplitude of the AC voltage generated by the VSC.

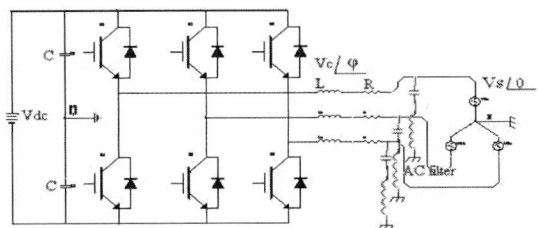

Fig. 1 . Basic Model of a VSC.

In addition of the 6-pulse bridge the VSCs are consisted of diodes connected in anti-parallel, a converter-reactor which is the linkage between the VSC and the AC network or the passive load, DC capacitors that have the function of energy storage, and shunt AC filters to reduce the harmonic components in the AC side (Fig.1) [1][6]. A sinusoidal AC voltage can be virtually generated at any angle and amplitude by using Sinusoidal

978-1-4244-1741-4/08/$25.00 ©2008 IEEE   637

Pulse Wide Modulation (SPWM) techniques at forced commutation up to 2000 Hz [1].

III. POWER FLOW CONTROL ANALYSIS

Some authors have mentioned that the active and reactive power flow can be independently controlled by mean of intervention of the phase and amplitude of the Ac voltage generated from the converter respectively, with respect to the AC network voltage [1][3][5][7]. This assumption is based on the fact that the VSC connected to an active AC network acts as a synchronous machine, therefore, it can control active and reactive power almost instantaneously [1], then the active power P and reactive power Q exchanged between the converter and the AC network are expressed as follow in (1) and (2) seen from the VSC.

$$P = 3 * \frac{V_c V_s}{X} \sin \varphi \tag{1}$$

$$Q = 3 * \frac{V_s(-V_s + V_c \cos \varphi)}{X} \tag{2}$$

Where Vc is the fundamental component of the converter voltage (Van(t), Vbn(t), Vcn(t)), Vs is the fundamental component of the AC network in each phase, X is the reactance of the reactor that connects the two voltage sources, in addition R which is the resistance of the line, it will be ignored in order to simplify the analysis [4]; this assumption can be done taking into account that typically the loss of the VSC is usually les than 5% of the rated capacity, and it determines the value of the resistor R [5]. Analysis previously done in a precedents work shows that in this range the expressions (1) and (2) are a good approximation [8].

Formulas (1) and (2) show that P depends mainly of the angle φ and Q depends mainly of the magnitude of Vc which is controlled by means of the modulation index m of the SPWM which is the relationship between the peak AC output voltage an the DC voltage in each pole [5][6]. However the VSC is a nonlinear double input double output coupled control system, this relationship between the two control variables P and Q and its respective manipulate variables φ and Vc also can be seen in (1) and (2) [9]. So the system cannot be controlled easily using only linear controllers as the conventional PI controllers. Another important fact to be in account is that the powers converters present important variations in the values of their principal parameters [12], as will be showed later in this document. A nonlinear fuzzy controller is a more general type of controller than a PID, and can work easily and directly whit these kind of nonlinear plants and with system that does not have constant parameters, what might limit the linear control strategies performance, because of the linear controllers are generally based on Linear Invariant Time (LTI) models while the fuzzy controllers does not use the parameters of the plant for their design [8][10] [12].

IV. DESIGN AND IMPLEMENTATION OF A VSC.

A VSC prototype was designed an implemented with the aid of probe the fuzzy controllers in a real application, based on their simulation results their implementation would be desirable, the prototype designed present the follow characteristics: is a VSC fed by a DC voltage V_{dc} = 390V, designed for $S_{(trifasic)}$=400VA, m=0.9, Vs (Phase) =120V@60Hz (which is the voltage of the AC network); where V_{dc} >Vs(line), that is necessary for normal operation of the VSC [5]. And an inductance of the reactor in each line X=0.2(p.u)*Z_{base} where, Z_{base}=(V_{base})²/S_{base}=(0.9)*(195/√2)²/(400/3) [5][11]. Then the value of the inductance of the reactor is 63mH this parameter is very important in the analysis and implementation of the converter because the reactor permits the interconnection between to Ac systems and the interchange of power. However it is important to be in account, that in a physical implementation the value of this parameter does not remains constant because the value of the inductance, depends of the value of the current that going through the coil. This can be seen in Fig. 2. Which presents the inductance (L μH) of the coils that were used in this application in variations of the current that going through the coil. The coils were built by using toroidal cores.

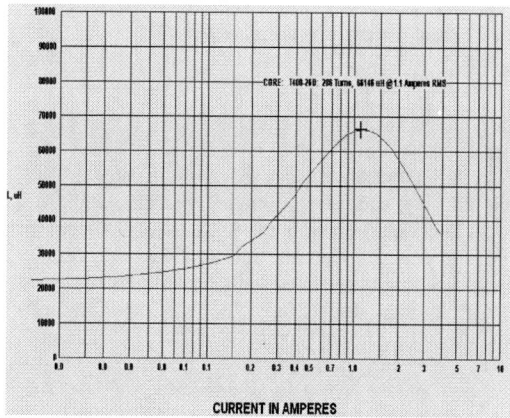

Taken from design of inductors for power filter, MICROMETALS, INC.
Fig. 2. Inductance Vs current.

An analysis of the dynamic response of P as a function of φ and Q in function of m was done [8]. This analysis was focused on understanding the dynamic behavior of the system, in this analysis was seen that the system present a rapid response to a step excitation then a sample time of 1.4ms was selected, in this time each controller must apply its control action, since the controller will be applied in a digital resource.

In spite of the design of the converter was done for the previous specifications, the range of operation chosen for P and Q includes -100 and 100 W for P and -30 and 30 VAr for Q in each phase; these values permit a comfortable range to work. At this point, and as is showed in Fig 2. is evident that the parameters of the plan

638

like the reactor inductance do not remain constants, this is the reason why the fuzzy control was selected for a physical implementation over conventional controllers based on LTI models because of the fuzzy controllers are not based on the values of the parameters and only need a qualitative knowledge about the dynamic behavior of the system [10][14]. Then a nonlinear fuzzy control might have a desirable response in control of the active and reactive power independently.

V. NONLINEAR FUZZY CONTROLLER.

In a precedent work "[8]" were designed independents fuzzy controllers to control the active and reactive power flow between a VSC and an active load [7] which is the urban AC network (120V@60Hz). Whit the fuzzy controllers were possible an independent control of the active and reactive power flow, regarding that the VSC is a nonlinear coupled system [8].

The fuzzy controllers designed were Mandani these kinds of controllers are usually used in feedback control mode, because they are computationally simples [10][15], presents low sensibility to noise in the input, what is important in power systems and can represent easily the knowledge about what an action control must o how the controller would be act in an specific case, this knowledge is represent by means of rules in the form if-then [10] and synthesized in form of an input-output mapping between the antecedent and the consequent variables [10].

The controllers are a kind of Fuzzy PD controller in counterpart of linear Proportional Derivative controllers [10], the rule base of each controller has two inputs as antecedents the error (e) and the error change (Δe), in order to obtain a derivative effect, and the consequents are the phase shift (shift_φ) and the amplitude shift (shift_m) for FCP (Fuzzy Control of P) and FCQ (Fuzzy Control of Q) respectively [8]. The inputs and outputs of the fuzzy controllers are summarized in table I.

The input-output mapping for FCP and FCQ are showed in the control surface of Fig. 3. and Fig. 4. respectively, they were designed in MATLAB® using the fuzzy inference system toolbox.

TABLE I
INPUTS AND OUTPUTS OF THE FUZZY CONTROLLERS

	Input 1	input 2	Output
FCP	Error (e)	Δe	shift_φ
FCQ	Error (e)	Δe	shift_m

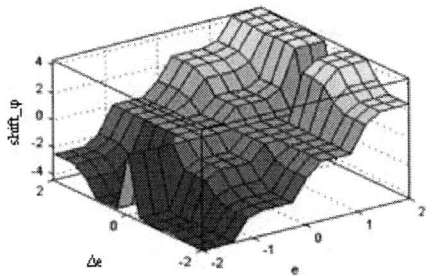

Fig. 3. Control surface of FCP.

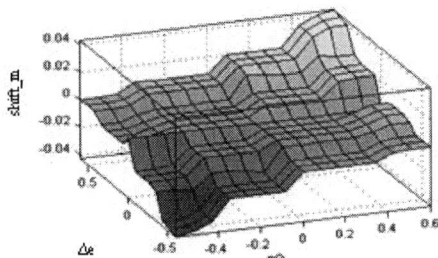

Fig. 4. Control surface of FCQ.

In addition to the Mandani inference system the fuzzy controllers are also composed of dynamic pre and post-filters the pre-filter used is an approximation of the derivative effect in the domain of discrete time, this approximation is often used an well accepted for computational implementation [10], the propose of the derivative effect is obtain a fast response to perturbations and an excessive overshot [16]. To improve the precision in steady state was used an integrator as dynamic pos-filter in the output of the fuzzy controller the diagram of the fuzzy controller is shown in Fig. 5.

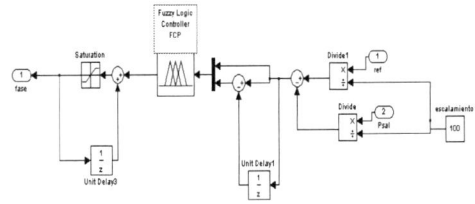

Fig. 5. Structure of the fuzzy controller.

Digital simulation were made using the dynamic model of the VSC that is presented in [8], this model is composed by the equations (3) (4) and (5) and was built in Simulink® Linking these equations.

$$
\begin{bmatrix} \dfrac{di_a}{dt} \\[2mm] \dfrac{di_b}{dt} \\[2mm] \dfrac{di_c}{dt} \end{bmatrix} = -\frac{1}{L}\begin{bmatrix} R & 0 & 0 \\ 0 & R & 0 \\ 0 & 0 & R \end{bmatrix} * \begin{bmatrix} i_a \\ i_b \\ i_c \end{bmatrix} - \frac{1}{L}\begin{bmatrix} \frac{2}{3} & -\frac{1}{3} & -\frac{1}{3} \\ -\frac{1}{3} & \frac{2}{3} & -\frac{1}{3} \\ -\frac{1}{3} & -\frac{1}{3} & \frac{2}{3} \end{bmatrix} * \begin{bmatrix} v_{xan} \\ v_{xbn} \\ v_{xcn} \end{bmatrix}
$$

$$
+ \frac{E \cdot m}{L}\begin{bmatrix} \sin(\theta) & 0 & 0 \\ 0 & \sin(\theta - \frac{2\pi}{3}) & 0 \\ 0 & 0 & \sin(\theta + \frac{2\pi}{3}) \end{bmatrix} \tag{3}
$$

$$
P = V_{sn} i \cos \beta \tag{4}
$$

$$
Q = V_{sn} i \sin \beta \tag{5}
$$

Where: θ= ωt+φ, i_a, i_b and i_c: are the line currents in each phase, V_{sn} the phase voltage of the AC network, and β the phase between the voltage and current.

The response of the system composed by the model of the VSC and the fuzzy controllers in close loop can be seen in Fig. 6. for different reference values in the operation range of the VSC. The simulations made in Simulink® shows that the fuzzy control achieves to control the active and reactive power flow easily; only is required a knowledge about the behavior of the system, and the need to resort to complicated mathematical models that are required with linear controllers is not necessary [10]. In addition the response of the fuzzy controller was compared with linear PI controllers that were designed using the mathematic model of the VSC, the response of the independent linear controllers is shown in Fig.7., where is evident that the objective of independent control of the active and reactive power flow was well achieved however the design of the linear controllers will not be treated in deep in this paper. In the Table II and III is summarized a comparison about the step response over the invariant model for the control of P and Q respectively, where is evident that the linear controllers present a better response that the fuzzy controllers.

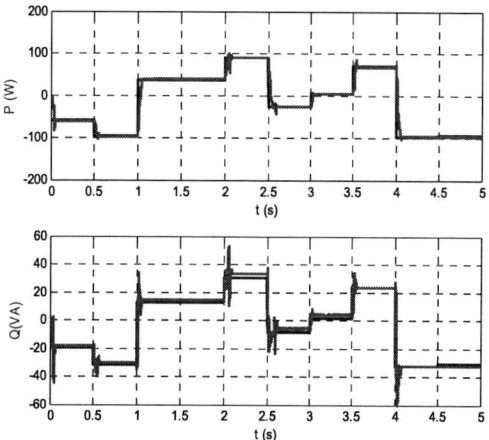

Fig. 6. Response fuzzy controller.

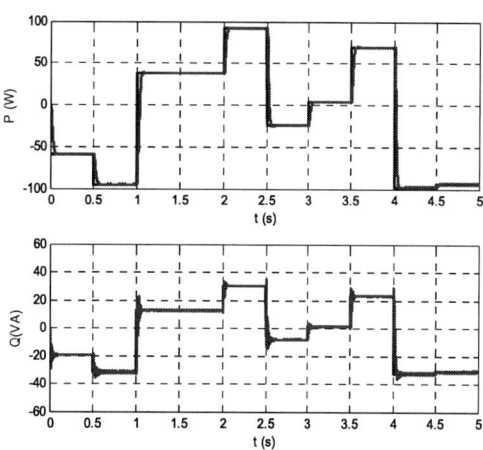

Fig. 7. Response linear controllers.

TABLE II
CHARACTERISTICS STEP RESPONSE FOR P WITH A REFERENCE OF 100

	Linear control of P	FCP
Setting time (s) 5%	0.0462	0.1232
Overshot (%)	2.3	18.3

TABLE III
CHARACTERISTICS STEP RESPONSE FOR Q WITH A REFERENCE OF 30

	Linear control of Q	FCQ
Setting time (s) 5%	0.056	0.133
Overshot (%)	15	65

At this point would be desirable the implementation of linear controllers because of they are computationally simpler, however it is important to take in account that the real system is not an invariant model because of their mains parameters does not remains constant (i.e. the reactor inductance as was showed in Fig.2.), then the response of the linear controller, that was based on an invariant model might not have a good behavior, when change the parameters of design, this fact was probed by reducing the value of the reactor inductance by 33% with both control strategies. The response of the system using the linear controllers is shown in Fig.8, while the response using the fuzzy controllers is depicted in Fig.9. By comparing both responses is evident that the fuzzy controllers have a more robust behavior that the linear controllers, because of that the fuzzy controllers do not used the parameters of the model for their design, only are based on qualitative knowledge about the dynamic behavior of the system as was mentioned before. This is the reason why the fuzzy controllers were selected for an implementation with a real model.

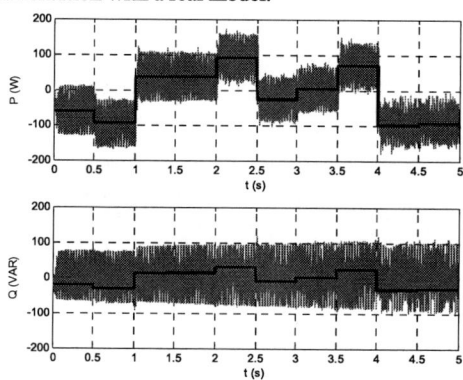

Fig. 8. Response linear controller when L is changed.

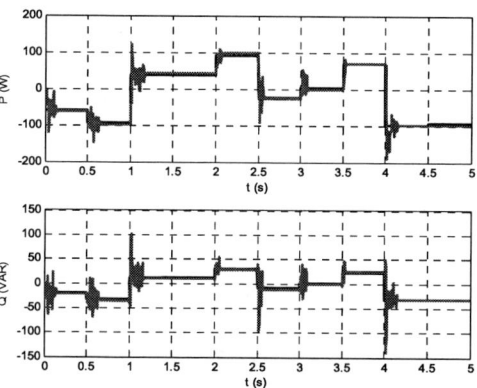

Fig. 9. Response fuzzy controller when L is changed.

The fuzzy controllers were implemented in MATLAB by using the Real-Time Workshop and were integrated to the system through a data acquisition target PCI 6024-E of National Instrument. However the sample time was too small to execute the two controllers (1.4mS), take in account that MATLAB generate a sequential code in C then the computational time of the two controllers seems to be higher than the sample time. The result was that the computers cannot execute the fuzzy controllers in real time by the computer; the fuzzy control this proves was made using a computer with the follow characteristics: (Inter. Pentium(R) 4, CPU 3.00 GHz, 512 MB of RAM) with the processor at the 100% disposed for this application.

However the fuzzy controllers have a robust behavior when the parameters of the plant do not remain constants and works correctly in nonlinear plants [10]. Then their use is desirable in this application, taking in account that the converter will be exposed to load variations, in this sense if an Artificial Neural Network (ANN) that emulate the dynamic behavior of the fuzzy controllers is used, it would diminish the computational time of the controller, in the implementation, since the artificial neural network is characterized for being a system with a high degree of parallelism [12] then with a control that operates in the same way that the fuzzy one is possible to reduce the computational time whit similar results in the implementation.

VI. EMULATOR BASED ON ANN.

An ANN can easily emulate any linear or nonlinear function or relationship between input and output vectors, since it can be used as a universal approximation of a function and can work with different data sets of inputs and outputs with the purpose of emulating any system [12], it is possible to approximate the input-output mapping developed by the fuzzy inference system of the fuzzy controllers which are represented in Fig.3. and Fig.4. for FCP and FCQ respectively . It is important to recall that the obtained fuzzy systems are Mandani controllers [10], then they are not memory system, reason why the suitable identifier to use is of parallel form [12], this is advisable since in this way we did not require of the presence of the fuzzy controller in the implementation of the emulator like controller. The ANN architecture used is a Feed forward Network, it is composed by 6 neurons in the hidden layer with tansig transfer function that lets the network work with positive and negative input vectors and a neuron in the output layer with purelin transfer function to obtain an linear output layer that lets the network produce values outside the range -1 to +1 [13]. The emulators based on neural network were implemented on SIMULINK® and is shown in Fig.10. this emulator only replace the fuzzy inference system reason why the ANN must be accompanied for the dynamic pre and post filter that accompanying the fuzzy controllers as is showed in Fig.5.

The BP Methodology [13] was used for the training of the networks, which was done using the Neural Network TOOLBOX of MATLAB®, to train the network were used Input vectors composed by the error, Δe and 4 delays of each variable and the corresponding target

vectors that was shift_φ for FCP and shift_m for FCQ. and the results was a mean quadratic error between the output of the fuzzy inference system and the output of the ANN of $6.7*10^{-3}$ for FCP and $5.9*10^{-7}$ for FCQ, the results show that the ANN has a similar behavior than the original fuzzy control.

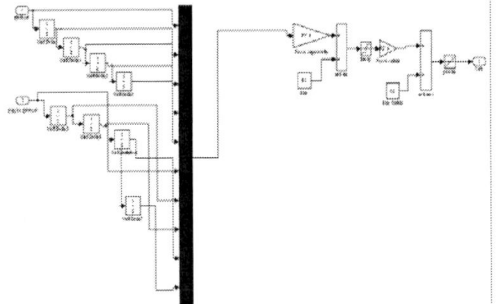

Fig. 10. ANN structure.

This application of the based on ANN has the robust of the fuzzy controller with a smaller computational time in than the fuzzy controller, fact which is showed in a benchmarking [12][17][18], that compare the computational time in simulation between the two controllers in a model of the converter built in Simulink® using the Simpowersystems toolbox [8]. The results of the benchmark comparing the simulation time in minutes for a simulated time of 1 second is summarized in Table IV, where is evident that the computational time with the ANNs is almost a third of the computational time with the fuzzy controllers.

TABLE IV
BENCHMARK COMPUTATIONAL TIME IN SIMULATION

Fuzzy controllers	Emulators based on ANNs.
43.95 min	18.1833 min

VII. RESULTS.

The ANN presents a smaller computing time in simulation than the fuzzy controller, which was showed before by means a benchmarking [12] between the two controllers in simulations. Finally the complete system was implemented through a data acquisition target PCI 6024-E of National Instrument where the objective of the control of following the reference was achieved.

Some responses of the system in real time are shown in Fig.11. Fig.12. and Fig.13 using as controller the emulator based on ANN where the horizontal axes represent the number of periods sampled (the sampling time is 1ms). Into the figures we can be seen that the response of the system follow its reference. However is also evident that the output signals present high degree of oscillation that is common in power systems due to a variety of interactions among components and the heavy power transfer across interconnections that often create damping problems [19].

In Fig.11 is possible to see the VSC transfering100 W in each phase to the AC network at the same time is forced the AC network to present a capacitive behavior, in Fig.12 is shown the otherwise where the AC network transfers power to the VSC and the AC network is forced to present an inductive behavior. In Fig.13 is showed the VSC transferring power to the AC network with a power factor close to the unit.

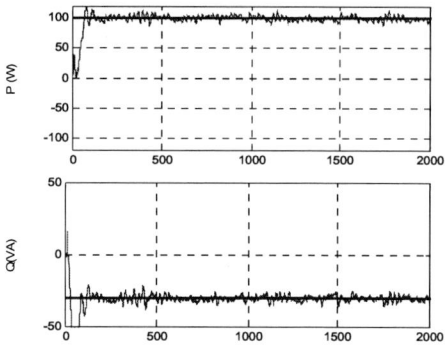

Fig. 11. Active and reactive power controlled to Q Reference in -30 and P reference in 100.

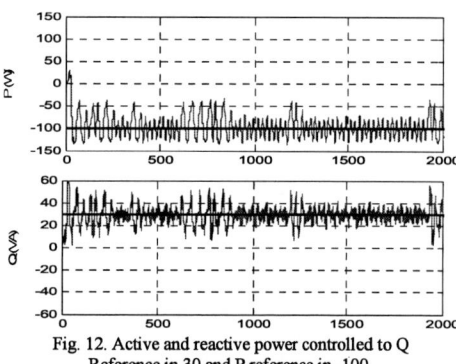

Fig. 12. Active and reactive power controlled to Q Reference in 30 and P reference in -100.

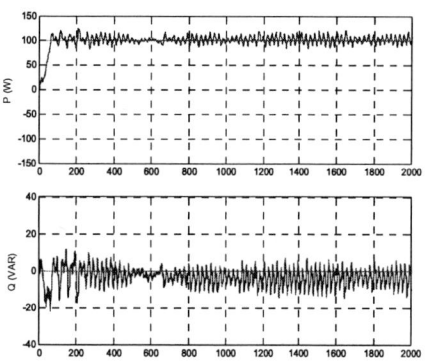

Fig. 13. Active and reactive power controlled to Q Reference in 0 and P reference in 100.

VIII. Conclusions

In spite of the parameters of the plant do not remain constant, the fuzzy controllers presents a robust behavior due to they are not based on these parameters, only is required a single knowledge about the behavior of the system, without the need to resort to complicated mathematical models that are required in linear controllers, however the fuzzy control has a greater computational cost in time, then its implementation could be limited To the availability of more powerful computational tools. With the ANN we obtain a good response emulating the fuzzy control with a smaller computational time than the fuzzy control reason why the ANN can be implemented. In spite of the oscillation present in the real implementation, the objective of the control has been achieved.

References

[1]. Vijay, K. Sood "HVDC and FACTS Controllers" Kluwer power electronics and power editions series, 2004.

[2]. Michael, P., Johnson, B., and B " the ABC of HVDC transmission technologies", Power and energy magazine, vol 5, Masrch/Aplil 2007.

[3]. Gengyin Li., et al, "power Flow Calculation of Power System Incorporating VSC-HVDC" Paper published in POWERCOM 2004, 21-24 November, Singapore.

[4]. Acha, E., Agelidis, V.,Anaya, O.,Millar, T. "Power electronic control in electrical systems" Newnes Power engineering series, 2002.

[5]. Guibin Z., Zheng Xu., "Steady-State Model for VSC Based HVDC and Its Controller Design", Department of Electrical Engineering, Zhejiang University Hangzhou, P.R.China, IEEE. 2001.

[6]. Guangkai, Li., et al "Research on Dynamic Characteristics of VSC-HVDC System" Power Engineering Society General Meeting, IEEE, 2006.

[7]. Padiyar, K. R., Prabhu, N., "Modeling, Control desing and Analysis of VSC based HVDC transmission" Paper published in POWERCOM 2004, 21-24 November, Singapore.

[8]. Diaz, N., Barbosa, F., Trujillo, C., "Analysis and Design of a Nonlinear Fuzzy Controller Applied to a VSC to Control the Active and Reactive Power Flow" IEEE Proceedings, CERMA 2007, September de 2007 Cuernavaca Mexico.

[9]. Song, R., et al. "VSCs based HVDC and its control strategy" Electronic Power Research Institute, Beijin, China, IEEE/PES transmission and Distribution Conference 2005.

[10] BABUSKA, R. "Fuzzy and Neuronal Control" Delft University of technology, October 2001.

[11] Peña, R., Trujillo, C. "Análisis y simulación de conversores VSC con modulación PWM para uso en sistemas HVDC" Paper published in Revista de Ingeniería. Vol.: 11 No. 2 Pag.: 30-39, Bogotá, Colombia, 2006.

[12]. Diaz, N., Soriano, J. "Study of Two Control Strategies Based in Fuzzy Logic and Artificial Neural Network Compared with an Optimal Control Strategy Applied to a Buck Converter" IEEE proceedings publications, NAFIPS 2007.

[13]. Soriano, J., Guacaneme, J.,Guarnizo, J., "General inverse neural control Nnarmax structure for Buck Converter using state variable ", ACCA,2007.

[14]. Muhammad, R. "Power Electronics Handbook" Canada, Academic press 2001.

[15]. Li-Xin Wang, "A course in fuzzy systems and control", Ed Pearson, 1996.

[16]. Ogata, K., "Ingeniería de control moderna " Prentice Hall, tercera edición, 1998.

[17]. http://www.des.udc.es/uploads/media/prac2.pdf, febrero de 2007.

[18]. http://es.wikipedia.org/wiki/Benchmark Category: Benchmarks, February of 2007.

[19]. Paserba, J., "Robust control (improving system dynamic performance)", Book reviews, IEEE power & energy magazine, September/October 2007.

Harmonic Distortion Reduction Technique for Uninterruptible Power Supplies with DC Voltage Boost Technique

Juei Lung Shyu

Department of Electrical Engineering, Kao Yuan University

No.1821,Chung Shan Rd., Lu Chu Hsiang, Kaohsiung County 821,Taiwan R.O.C.

Abstract—This paper presents a new control strategy for UPS inverter, which is based on the addition of the DC voltage boost compensation for keeping the fundamental output voltage at the preset value and suppressing the generation of low-order harmonics caused by nonlinear load. This solution involves the combination of a total harmonic distortion (THD) reduction controller and a power factor correction (PFC) controller, such that the harmonics contaminated in the output voltage can be eliminated despite the highly distorted load current while maintaining an acceptable dynamic performance. Thus, the proposed strategy is viewed as a refinement term added to the outer voltage control loop. Besides, the DC voltage of the inverter under linear loads is adaptively reduced so that low switching losses is achieved, when compared with conventional constant DC voltage approaches. The new strategy is quite simple and requires only the measurement of the output voltage to compute the THD. Simulation results are presented to exhibit the improved performance of the proposed approach especially under rectifier loads.

Keyword—UPS, total harmonic distortion, power factor correction.

I. INTRODUCTION

Uninterruptible power supplies are widely used as standby power for critical loads in case of utility power failure, such as computers, medical/life support systems, and communication systems. Among different UPS topologies, on-line UPS system has been commonly adopted since it can provide continuous power to the load with seamless transition from normal mode to backup mode during utility power failures and vice versa. Generally, the UPS is designed to produce high quality AC voltage with low distortion, particularly under nonlinear loads and sudden change in load. But the UPS system has high THD for nonlinear loads, for example, the rectifier load drawing a rectangular-shaped current

causes voltage dips and notches in the AC output. In practice, even though it is filtered, the actual output voltage may still be distorted due to nonlinear load injected harmonic current. In order to achieve fast dynamic response and eliminate output voltage distortion under nonlinear loads, many improved control systems have been proposed such as sliding mode control, multi-loop, optimal state feedback, repetitive-based control, deadbeat control, and many others [1]-[7]. Although such types of feedback control approaches have fast transient response, the distortion is still not compensated completely for load disturbances. The disadvantage is that the maximum PWM voltage area of the pulse is limited by the computation and sampling time. This limitation results in relative large harmonic components occurring for nonlinear loads. In addition, all such control strategies are slightly complex to implement. For conventional PWM methods with constant DC input voltage, this can cause the inverter output voltage to contain large amounts of low order harmonics and significantly affect the inverter performance. Hence, a large PWM voltage area of each pulse is required to keep the distortion of output voltage low. This issue is completely ignored in previous works.

In this paper, a simple THD reduction technique with addition of DC voltage boost compensation is proposed. In this technique, a DC voltage boost signal which is proportional to the THD value of the inverter output voltage is injected into the conventional PFC outer voltage loop. Then, the switching instance can be determined that the PWM voltage areas of the inverter are proportional to the instantaneous THD value. The THD value is obtained at every sampling cycle from the feedback of inverter output voltage. The desired DC voltage boost signal is determined through THD reduction control loop. As a result, the proposed control scheme leads to improve the THD of the inverter output voltage while keeping an acceptable dynamic response.

978-1-4244-1741-4/08/$25.00 ©2008 IEEE

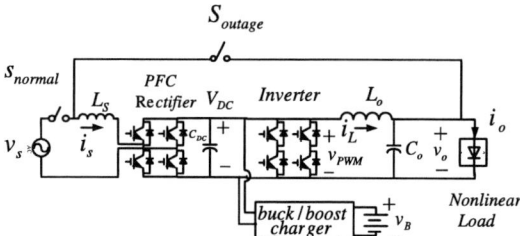

Fig. 1(a). Circuit configuration of the single-phase on-line UPS system.

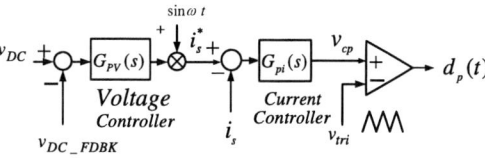

Fig. 1(b). Conventional PFC multi-loop control block diagram.

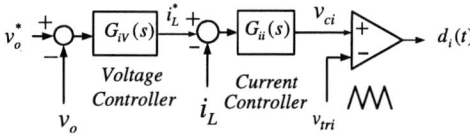

Fig. 1(c). Basic multi-loop control scheme of inverter.

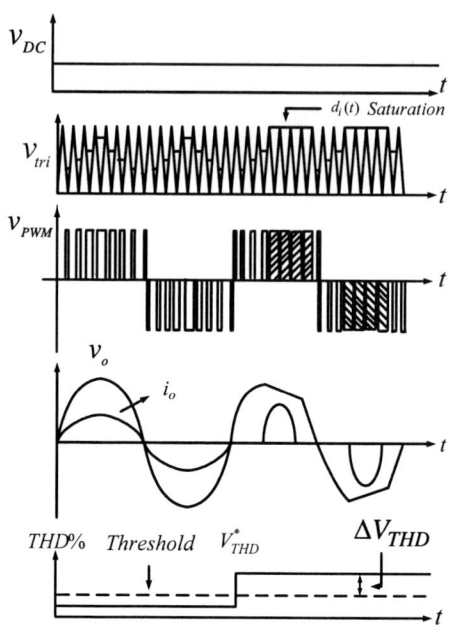

Fig. 1(d) Key waveforms of a conventional inverter without DC voltage boost compensation.

To further achieve a good transient response performance, an inner current loop is introduced to regulate the inductor current and generate necessary PWM signals. With such a multi-loop arrangement, the PWM inverter possesses the features of easy implementation, good transient response, and insensitivity to the load variation. Moreover, by the addition of DC voltage boost compensation, the DC voltage supplied to inverter at various loads is automatically compensated and lower switching loss is achieved for resistive load. Compared to previous works, a new approach is used to develop the DC voltage boost control, which results in a new feed-forward signal that has an additional term proportional to the THD of the inverter output voltage. The additional DC voltage boost signal is shown to be capable of reducing total harmonic distortion significantly, especially nonlinear load. The performance of the proposed control approach will be validated by simulation results.

II. SYSTEM DESCRIPTION

A. Description of the Circuit Topology

Fig. 1(a) shows the typical schematic diagram of the on-line UPS system. The multi-loop control schemes are adopted in the designed UPS system. The UPS can be operated in two modes: one is the normal mode, and the other is outage mode. In the normal mode, the PFC rectifier converts the AC supply voltage into DC voltage

for the inverter and battery charging. Moreover it generates input current with low THD as well as nearly unity power factor. The buck/boost battery charger is operated to provide backup energy in the outage mode. Meanwhile, to produce a stable power with high reliability during outage mode, the inverter should be modulated such that the output voltage is regulated well with low THD. Besides, bypass switches (S_{normal} , S_{outage}) change the operating mode in accordance with the line condition. As to the conventional PFC control system shown in Fig. 1(b), there are two control loops for PFC rectifier, one outer voltage loop and one inner current loop. The outer voltage loop uses the sensed DC voltage v_{DC} as a feedback signal, which is compared with reference signal v_{DC}^*. The output of the voltage controller $G_{pv}(s)$ is then multiplied by a unity supply voltage v_s to produce a control voltage i_s^* as a reference signal for the inner current loop. Current error is then used to produce a control voltage v_{cp} through the current controller $G_{pi}(s)$. The PWM generator generates required PWM signals to regulate the DC output voltage with the line current shaping.

For an UPS inverter system with sinusoidal PWM technique, the output voltage v_o is given by

$$v_o = (d_i \cos \omega_o t) V_{DC} \qquad (1)$$

Where d_i is the duty ratio of a pulse-width modulation (PWM) switching sequence at a relatively high frequency, which varies between 0 and 1. ω_o is the output frequency. To guarantee a good tracking performance

644

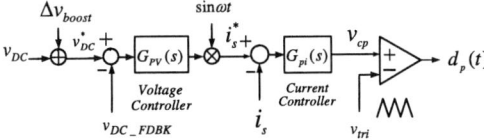

Fig. 2 (a) Block diagram of the proposed DC voltage boost compensation with THD reduction.

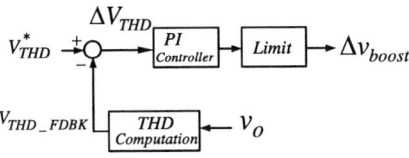

Fig. 2 (b) THD reduction control scheme.

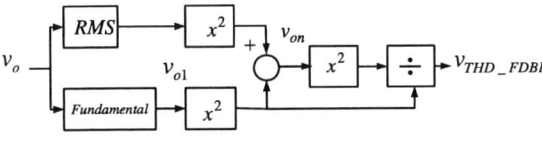

Fig. 2 (c) Block diagram for THD computation.

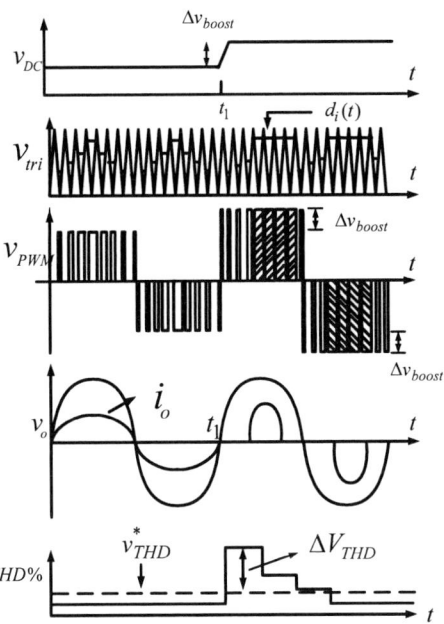

Fig. 2 (d) Key waveforms of inverter with DC voltage boost compensation.

of inverter output voltage towards its pure sinusoidal reference command, the DC voltage should be increased properly if d_i reaches to upper limit. A general multi-loop control block diagram, as shown in Fig. 1(c), consists of two loops. The outer voltage loop is mainly responsible for the output voltage regulation, which is achieved by comparing the voltage feedback signal v_o and reference command v_o^*. The error signal is then generated as a reference command for the inner current loop. However, if a UPS inverter is digitally implemented with a microprocessor system, the maximum duty ratio is inevitably limited by A/D sampling and the required computation time to determine the pulse width of switching. As shown in Fig. 1(d), if the inverter with constant DC-link voltage connected to a nonlinear load, saturation of the duty ratio $d_i(t)$ results in large distortion of the output voltage. Hence, it is necessary to increase the DC voltage to respond to the higher THD instantaneously and reduce the harmonic distortion through varying the PWM voltage area of each pulse.

B. Control Strategy

In order to provide precise voltage control of inverter, any harmonic distortion must be properly compensated. However, conventional PFC control with constant DC voltage can not achieve good performance under nonlinear loads disturbance. By introducing the DC

voltage boost technique in the conventional voltage control loop of PFC rectifier, the performance of the single-phase PWM inverter can be optimized so that the output voltage is controlled with low THD. The proposed DC voltage boost scheme with improved steady-state performance is shown in Fig. 2(a). An extra THD reduction control loop in Fig. 2(b) is then added at the input of the conventional PFC voltage-loop to yield satisfactory inverter performance. In this technique, instead of a fixed DC voltage, a boost DC voltage Δv_{boost} which is proportional to the THD value v_{THD_FDBK} is injected at the input of the PFC rectifier. The output voltage v_o of the UPS inverter is sampled and calculated to a detected THD value v_{THD_FDBK} that is required for THD reduction control loop. The error signal between the THD threshold value v_{THD}^* and detected THD value v_{THD_FDBK} is then passed to a PI controller and the desirable boost signal Δv_{boost} is thus generated. Due to the negative feedback in the THD reduction loop, the boost signal Δv_{boost} changes in such a manner that the inverter output voltage is modulated so that the minimum THD is obtained. Finally, DC voltage boost signal is combined with voltage command v_{DC} of conventional PFC control to form the new command signal v_{DC}^*. From the description above, one can derive the following key equations for resistive load and nonlinear load, respectively:

645

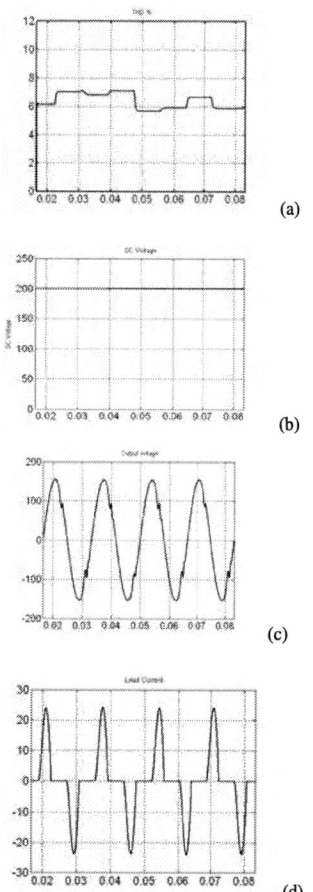

Fig. 3 Simulated results of rectifier load operation without compensation. (a) THD of output voltage (b) DC voltage v_{DC} waveform (c) output voltage v_o waveform (d) load current i_o waveform. (time: 01s/div)

Resistive Load:

$$v_{DC}^* = v_{DC} \text{ and } \Delta v_{boost} = 0, \text{ if } v_{THD\_FDBK} \le v_{THD}^* \quad (2)$$

Nonlinear Load:

$$v_{DC}^* = v_{DC} + \Delta v_{boost}, \quad \text{if } v_{THD\_FDBK} \ge v_{THD}^* \quad (3)$$

Total harmonic distortion (THD) is an important figure of merit used to quantify the level of harmonics in voltage or current waveforms. THD can be defined as the ratio of the root-mean-square value of total harmonic voltage and fundamental voltage, as indicated in formula (4):

$$THD = \sqrt{\left. v_{o,rms}^2 - v_{o1,rms}^2 \middle/ v_{o1,rms}^2 \right.} = \sqrt{\left. \sum_{n=2}^{\infty} v_{on}^2 \middle/ v_{o1}^2 \right.} \quad (4)$$

where n is the harmonic order. $v_{o,rms}$ is RMS value of output voltage. $v_{o1,rms}$ -RMS value of fundamental component of output voltage and v_{on} is the harmonic order n component. The THD is obtained by the calculations with samples of the output voltage v_o at sample frequency equal to integer multiple of the nominal frequency of v_o. THD computation block is

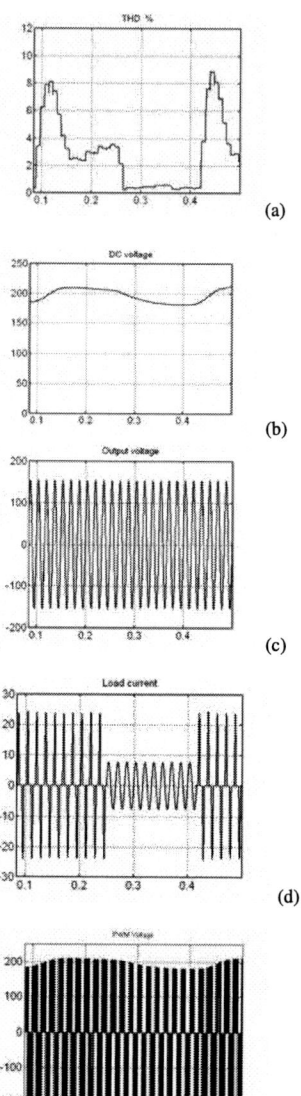

Fig. 4 Simulated results with compensation when load changes from rectifier load to resistive load. (a) THD of output voltage (b) DC voltage v_{DC} waveform (c) output voltage v_o waveform (d) load current i_o waveform. (e) Inverter output voltage v_{PWM} waveform. (time: 01s/div)

described in Fig.2(c).

Fig. 2(d) shows key waveforms of the UPS inverter during a load step transient. As can be seen from Fig.2 (d), at $t = t_1$, as sensed distorted output voltage, THD value experiences a positive step change, exceeding the threshold value v_{THD}^*. The Δv_{boost} is then summed up with v_{DC} and generate a new voltage command v_{DC}^* at the input of the voltage loop. Since the desired voltage boost command Δv_{boost} is determined by comparing the reference command v_{THD}^* with measured THD value v_{THD_FDBK}, thus, as expected, any waveform distortion in the output voltage leads to immediate Δv_{boost} increase within one switching cycle. As shown in Fig. 2(d), the boost voltage Δv_{boost} is proportional to

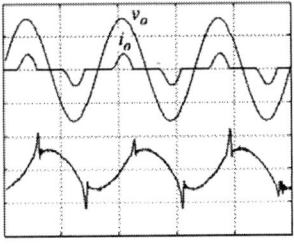
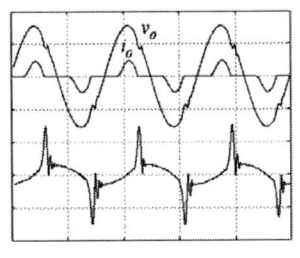

(a)　　　　　　　　　　　　　　　　　　　(b)

(1) output voltage v_o (100V/div)　　　　　　(1) output voltage v_o (100V/div)

(2) load current i_o (50A/div)　　　　　　　(2) load current i_o (50A/div)

(3)Tracking error(10V/div)　　　　　　　　(3)Tracking error(20V/div)

Fig. 5 Nonlinear load operation. (a) $V_{THD}^* = 0.01$ (b) $V_{THD}^* = 0.05$. (time: 0.01ms/div)

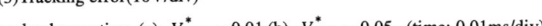

TABLE I THD PERFORMANCE FOR NONLINEAR LOAD					
v_{THD}^*	1%	2%	3%	4%	5%
v_{DC}	220V	216V	212V	208V	203V

the error voltage ΔV_{THD} . After the DC voltage is increased at $t = t_1$, because of an immediate increase of the PWM voltage area of pulse, the THD of the output voltage caused by the nonlinear load is can be significantly reduced. Clearly, a high performance UPS inverter could be obtained, if a smaller v_{THD}^* is chosen. It should be noted that the DC voltage supplied to inverter does not increase immediately after the THD sudden change because of the assumed low bandwidth of the voltage loop.

III. SIMULATION RESULTS

To verify the performance of the proposed control technique, some simulated results were carried out with Matlab/Simulink programming language. The proposed UPS inverter was tested without and with DC voltage boost technique. Nominal values of circuit parameters are listed as:

$v_s = 110 V_{rms}$, $V_{DC} = 200 V$ $v_o = 110 V_{rms}$ $C_{DC} = 940 \mu F$
$L_s = 3 mH$, $L_o = 1 mH$, $C_o = 30 \mu F$, and switching frequency $f_s = 18 KHz$ .The performance of the UPS inverter with and without are shown in Fig. 3 and Fig. 4, respectively. As can be seen from Fig. 3, the output voltage THD is higher up to 7% while keeping DC voltage constant. Fig. 4 shows the key waveforms with compensation for load step change. For the transient response test, the DC voltage was stepped from 200V to 220V while nonlinear load applied, and then stepped back to 180V at resistive load. As can be seen from Fig. 4, the maximum THD during transient response is approximately 8%, whereas the corresponding THD under steady state is below 3% with compensation. As shown in Fig. 4(d), DC voltage experiences a positive step change. However, because of slow voltage control loop, v_{DC} starts going up slowly some time after load change. The simulated results indicate that the proposed control scheme offers low THD with output voltage well

regulated even under a highly distorted condition.

By the addition of the THD reduction control loop, the DC voltage required for resistive load is much lower than that for nonlinear loads in terms of the THD. Since the switching loss of the UPS inverter is proportional to the DC voltage, it is easily seen that higher efficiency for resistive load is obtained with the proposed control scheme. On the other hand, the proposed approach does not increase more voltage stress of the power switches for inverter implementation.

Table I shows the simulated DC voltage at same nonlinear load with different threshold values v_{THD}^* . The harmonic distortion for nonlinear load could be reduced if an extra DC voltage boost compensation is added to the control strategy. As expected, the DC voltage increases as v_{THD}^* is increased.

Fig. 5 compares the output voltage waveforms of the inverter that is controlled by the proposed method for same nonlinear load when v_{THD}^* was set at 0.01 and 0.05. This clearly shows that lower v_{THD}^* value can achieve excellent steady-state performance with acceptable tracking error of output voltage. The dynamic response of the system for a step change in the reference voltage v_{THD}^* between 0.01 and 0.05, under same rectifier load, are shown in Fig. 6. It shows that the dynamic response is seen to be reasonable with THD feedback v_{THD_FDBK} reaching to the reference command v_{THD}^* fast.

IV. CONCLUSIONS

A simple and effective control strategy for the THD reduction of inverter output voltage is proposed. To improve the performance of the UPS inverter, a DC voltage boost technique is added to significantly suppress the output voltage distortion caused by nonlinear load and thus nearly sinusoidal output voltage waveform is obtained. Transient response to load step changes is presented to indicate that good output voltage regulation

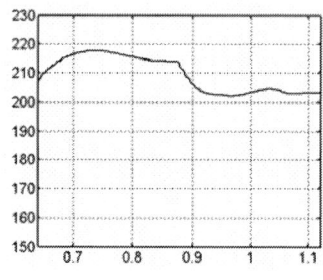
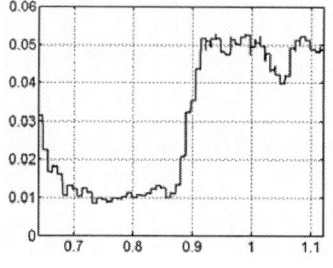

(a) DC voltage v_{DC} (b)THD value v_{THD_FDBK}

Fig. 6 THD step response when v_{THD}^{*} changes from 0.01 to 0.05. (time: 01s/div)

with low THD are obtained simultaneously. Besides, the steady state performance is rather insensitive to DC voltage fluctuation and load variations. The present work is still on-going. The author is encouraged to ask for additional information on experimental results.

REFERENCES

[1] A. Kawamura and T. Haneyoshi, "Deatbeat controlled PWM inverter with parameter estimation using only voltage sensor," *IEEE Trans. Power Electron.*, vol. 3, no. 2, pp. 118-125, Apr. 1988.

[2] T. Kawabata, T. Miyashita, and Y. Yamamoto, "Deat beat control of three phase PWM inverter," *IEEE Trans. Power Electron.*, vol. 5, no. 1, pp. 21-28, Jan. 1990.

[3] S. –L. Jung, H.-S. Huang, M.-Y. Chang, and Y.-Y. Tzou, "DSP-based multi-loop control strategy for single-phase inverters used in AC power sources," in proc. *IEEE Power Electronics specialists Conf.*, 1997, pp. 706-712.

[4] N. M. Abel-Rahim and J. E. Quaicoe, "Analysis and design of a multiple feedback loop control strategy for single-phase voltage-source UPS inverters," *IEEE Trans. Power Electron.*, vol. 11, no. 4, pp. 532-541,Jul. 1996.

[5] K. Zhang, Y. Kang, J. Xiong, and J. Chen, "Direct repetitive control of SPWM inverter for UPS purpose " *IEEE Trans. Power Electron.*, vol.18, no. 3, pp. 784-792, May. 2003.

[6] Heng Deng, Ramesh Oruganti, and Dipti Srinivasan, " Analysis snd Design of Iterative Learning Control Strategies for UPS Inverters," *IEEE Trans. Ind. Electron.*, vol. 54, no. 3, pp. 1739-1751, Jun.2007..

[7] Gustavo Willmann, Daniel Ferreira Coutinho, Luis Fernando, and Fausto Bastos, "Multi-Loop H-Infinity Control Design for Uninterruptible Power Supplies," *IEEE Trans. Ind. Electron.*, vol. 54, no. 3, pp. 1591-1602, Jun.2007.

Energy-based Modulation Error Control for High-Power Drives with Output LC-Filters and Synchronous Optimal Pulse Width Modulation

Tomasz Laczynski, Timur Werner and Axel Mertens

Institute for Drive Systems and Power Electronics
LEIBNIZ UNIVERSITY OF HANNOVER
Hannover, Germany
E-mail: laczynski@ial.uni-hannover.de

Abstract— In order to reduce the switching frequencies of power semiconductors, synchronous optimal pulse width modulation provides an appropriate solution. The switching angles result from complex offline calculations, assuming steady-state operation. Medium voltage drives when applied to the retrofit of existing fixed speed induction motors often include an output LC-filter which introduces a resonant circuit. Therefore, only in steady-state operations the combination of a high-power drive with a LC-filter and optimized pulse width modulation delivers satisfactory behavior. Changes in operating conditions cause weakly damped electrical oscillations, which can be described as dynamic modulation errors. In this paper, a novel energy-based control method is proposed, in order to damp the LC-filter excitation without increasing the switching frequency. The introduced control technique has been verified by simulations of a 2.4 kV induction motor drive.

Keywords—Active damping, control methods for electrical drives, induction motor, passive filter, pulse width modulation (PWM), variable speed drive, voltage source inverters (VSI).

I. INTRODUCTION

Due to their reliable and economic design induction motors are used in many present medium voltage drives. Originally supplied by the mains these machines are preferably retrofitted by medium voltage variable speed drives to improve the overall efficiency of the system. In order to minimize the switching losses and increase the maximum output power of the inverter at the same time, the switching frequency is kept at very low level below 500 Hz [1-3, 5- 6].

The optimal pulse width modulation represents the best possible solution for rectangular shaped voltage waveforms with minimal harmonic distortions at low switching frequencies [6,9]. Due to the algorithm complexity, they are calculated offline and only for stationary operations.

Usually the already installed motors are not suitable to run with inverters because of weaker motor winding isolation and bearing currents [4,5]. Consequently, LC-filters are often applied between inverter and motor, providing almost sinusoidal output voltages with conditions similar to the grid [10,11]. In order to reduce the switch-

ing losses and thus increase the maximum output power, the switching frequency is kept very low [1,8,15]. This demand increases the possibility of excitation of filter resonance. For this purpose, offline-optimized pulse patterns provide appropriate results, but only for steady-state operations. This fact does not allow fast dynamic operations like fast load or pulse pattern changes without the creation of weakly damped oscillating transient currents [14]. Therefore, an additional suitable control method is required, which can quickly and precisely recognize these transients and actively damp them without increasing the switching frequency.

This paper presents a new control method which analyses the energy of transients that appear in the filter and motor. To calculate its actuating variable, the controller uses an energy function which describes the behavior of the transients. The actuating variable is then generated by direct time shifting of the pulse pattern edges and therefore considers the non-linear behavior of the inverter. These control actions actively damp filter oscillations and allow fast dynamic operations with switching frequencies down to 280 Hz.

II. MODELING

The design and simulation of the control method was done in MATLAB/Simulink® using appropriate dynamic models of the NPC three-level inverter, the control logic, the output LC-filter and the induction motor. Fig. 1 shows the principle structure of the considered drive system.

A. Three-Level NPC Inverter

In order to convert the reference signals of the controller and the control logic into corresponding output voltages, a three-level NPC inverter has been applied. It can realize three distinct voltages $U_d/2$, $0V$ and $U_d/2$ relatively to the neutral point. U_d is the dc link voltage. The combination of these voltages allows 27 switching states leading to 19 different switching vectors. In order to gain an easier insight in relevant dynamics, the inverter has been modeled ideally assuming constant dc link voltage and delay free behavior of the switches.

978-1-4244-1741-4/08/$25.00 ©2008 IEEE

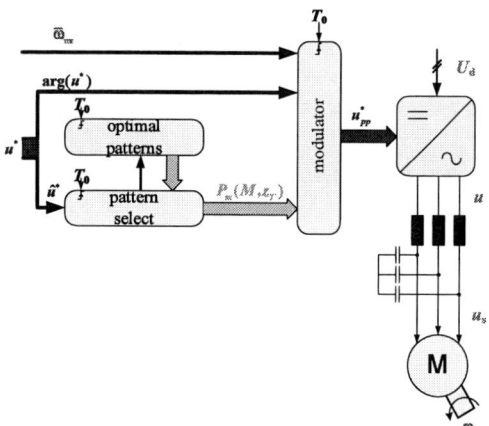

Fig. 1: Scheme of the drive system

The modulation control unit transforms the reference values of the outer fundamental stator current controllers into switching signals for each phase of the inverter. The main component of the modulation control unit is the modulator, which is designed for the use with the optimized pulse patterns. The information on all available pulse patterns is stored in tables within the modulator's control logic. The control unit chooses the actually applied pulse pattern with respect to the reference variables of the controller (Fig. 1).

B. Modulator and Optimized Pulse Patterns

The applied pulse patterns were offline computed using the method of selective harmonic elimination [3]. The selective harmonic elimination can be used to avoid harmonics which would cause resonance excitations of the filter. It is also possible to optimize the pulse patterns in respect of the THD of the motor current or related characteristics. Since the optimization of the steady-state quality of the pulse patterns is not the main topic of this work, no further analyses have been done in this direction yet.

The offline optimized pulse patterns are synchronous with the fundamental frequency [6]. Therefore, the switching frequency is an integer multiple

$$z_C = \frac{f_s}{f_{,1}} = \text{pulse number} \qquad (1)$$

of the fundamental frequency. Due to the synchronism, the switching function, the output voltage and the stator currents contain only integer harmonics. Another important feature of the pulse patterns is their quarter-wave symmetry. Only notch angles for the first quarter period need to be stored in the control logic unit. The pulse patterns are stored in the control logic unit as characteristic curves of the modulation index. The modulation index is defined as the ratio of the fundamental voltage amplitude and $U_d/2$. The actual switching frequency needs to be controlled and limited, in order to minimize the switching losses, because an increase of the fundamental frequency results in an increase to the switching frequency. The limitation of the switching frequency is provided by changing from a pulse pattern with a higher pulse number to one with a lower pulse number using a hysteresis. Since every change of the pulse patterns leads to distortions in the voltages and currents of the motor and the LC-filter, the change frequency must be kept at low level.

The amplitudes of the harmonics can be computed from [6,9]:

$$b_v = \frac{U_d}{2} \frac{4}{v\pi} \sum_{n=1}^{q} (-1)^{n+1} \cos(v\alpha_n) \qquad (2)$$

with

$$n = 1,5,7,11\ldots$$

and

q: number of switching edges of a pulse pattern within the first quarter of the period.

The variable q can be calculated by

$$q = \frac{z_C - (3 - W)}{4 - W} \quad \text{with} \quad W \in \{2,3\}. \qquad (3)$$

with W as the number of the inverter's levels.

The spectrum of the pulse patterns comprises only sinusoidal harmonics without any phase shift. The number of eliminated harmonics is proportional to the pulse number. Therefore, the quality of the optimization also depends on the pulse number of the pulse pattern.

C. Induction Motor with LC-Filter

The simulated induction motor contains a squirrel-cage rotor and is star-connected with open neutral point.

1) Harmonic Model: The harmonic behavior of the stator current space vector can be described with

$$\vec{u}_{s,v}^s = R_\sigma \vec{i}_{s,v}^s + L_\sigma \frac{d\vec{i}_{s,v}^s}{dt} \quad \text{with} \quad v \gg 1, \qquad (4)$$

where

$$R_\sigma = \left(R_s + \left(\frac{L_h}{L_r'} \right)^2 R_r' \right) \qquad (5)$$

and

$$L_\sigma = \left(L_s + \left(\frac{L_h}{L_r'} \right)^2 L_r' \right) \qquad (6)$$

with

$\vec{u}_{s,v}^s$: v-order harmonic of the filter output voltage.

Eq. (4) can be derived from the differential equations of a fundamental wave motor model which were presented in [12].

The harmonic model for an induction motor with a LC-filter is shown in Fig. 2.

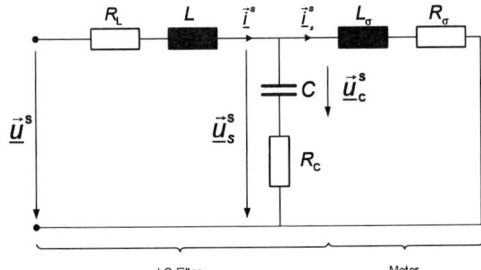

Fig. 2: Structure of the transient model of the induction motor with LC-filter

For the two orthogonal space vector components, the following state space model is valid, because the dynamics between the components are uncoupled

$$\frac{d}{dt}\underbrace{\begin{pmatrix} i_{,h} \\ i_{s,h} \\ u_{c,h} \end{pmatrix}}_{\dot{\vec{x}}_{,h}} = \underbrace{\begin{bmatrix} \dfrac{R_L + R_C}{L} & \dfrac{R_C}{L} & \dfrac{1}{L} \\ \dfrac{R_C}{L_\sigma} & \dfrac{R_\sigma + R_C}{L_\sigma} & \dfrac{1}{L_\sigma} \\ \dfrac{1}{C} & \dfrac{1}{C} & 0 \end{bmatrix}}_{A} \underbrace{\begin{pmatrix} i_{,h} \\ i_{s,h} \\ u_{c,h} \end{pmatrix}}_{\vec{x}_{,h}} + \underbrace{\begin{bmatrix} \dfrac{1}{L} \\ 0 \\ 0 \end{bmatrix}}_{B} u_{,h}\,. \quad (7)$$

With the above state space model, the transfer function between the stator current and the filter input voltage can be computed. The resonance frequency is by 457 Hz.

III. CONTROL METHOD

Fig. 4 presents the basic structure of the complete control system. The control system comprises above all the fundamental stator current controllers and the modulation error controller. Both controllers use a sampling frequency of 2 kHz.

A. Fundamental Control

1) Estimation of the Fundamental Components: High amplitudes and frequencies of the distortions in the choke current, stator current and capacitor voltage make fundamental wave instant value measurements by using conventional filtering methods difficult. The measured fundamentals contain considerable phase delays and distortions. Therefore, to deal with this problem, the proposed control method uses a predictive model based estimation of the distortions. This estimation allows a delay-free determination of distortions and thus improves the controller's performance. The determination of the fundamentals is done by subtraction of the estimated distortions from the measured instantaneous values. The future shape of the pulse pattern in the next sampling period is already known to the controllers in the actual sampling period before its generation by the inverter. For this reason, the distortions, which would first be effective in the next sampling period,

can be numerically predicted by solving the harmonic model (Fig. 3).

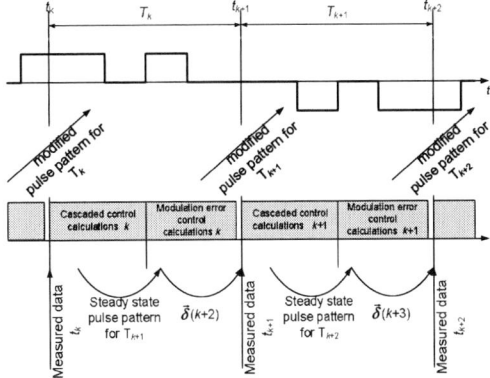

Fig. 3: Sequence of the digital modulation error control

In order to perform the prediction, the harmonic model is solved for the next sampling period with the distortion voltages

$$u_{,h}(k+1) = u^*(k+1) - \hat{u}_{,1}^*(k+1)\sin(\varphi_{u,1}^*(k+1)) \quad (8)$$

as the input. The fundamental voltage reference values $\hat{u}_{,1}^*(k+1)$ and $\varphi_{u,1}^*(k+1)$ come from the current controllers (Fig. 4). The voltage $u^*(k+1)$ describes the pulse pattern, already including the edge shifts for the next sampling period. As a result, the reference voltages are used instead of the measured inverter output voltages. The estimation method assumes ideal switching characteristics of the inverter. The determination of the distortions is done by numerically solving the harmonic state space model in (7). For this purpose, the continuous state space model is discretized. The discrete system matrices result from [13]:

$$\mathbf{A_d} = \mathbf{\Phi}(T_d) = e^{\mathbf{A}T_d}\,, \quad (9)$$

$$\mathbf{B_d} = \int_0^{T_d} \mathbf{\Phi}(T_d - \tau)\,\mathbf{B}\,d\tau = \int_0^{T_d} e^{\mathbf{A}(T_d - \tau)}\,\mathbf{B}\,d\tau \quad (10)$$

with

$$T_d = 0.83\,\mu s\,.$$

2) Cascaded Control: The PI controllers have a cascaded structure. The inner PI controllers regulate the fundamental of the stator current space vector. The magnetizing current and the mechanical speed are regulated by the outer PI controllers. The dimensioning of the controllers' parameters was done with classical frequency domain methods by considering a dead time of one sampling period T_0. The controllers are modeled in the discrete time domain.

651

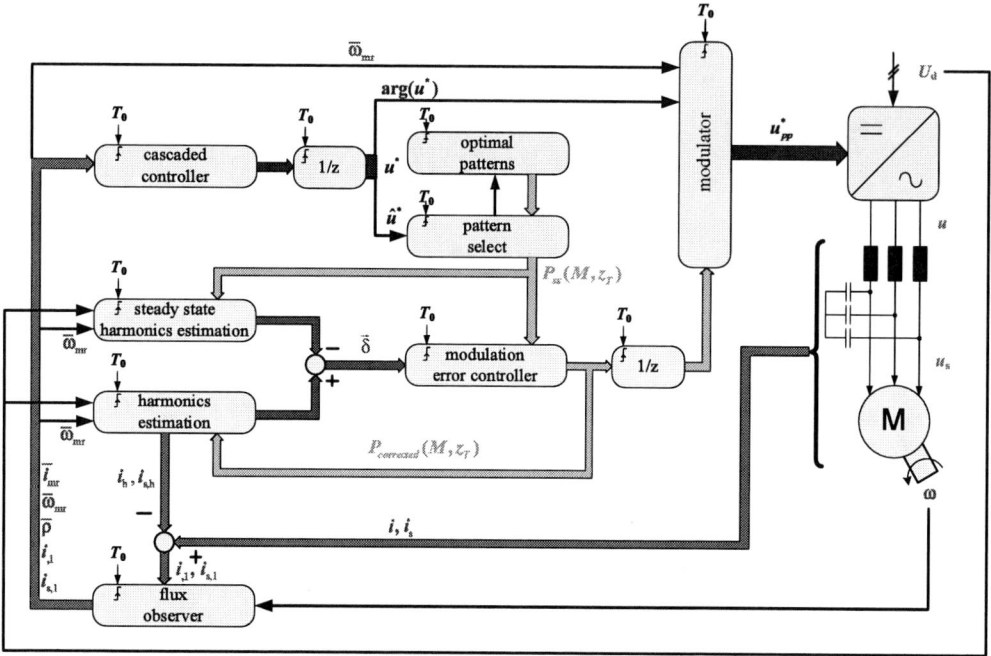

Fig. 4: Structure of the complete control scheme

B. Modulation Error Control

1) Definition and Calculation of Modulation Errors:
The offline computed pulse patterns are optimized with respect to the accurate mapping of the fundamental and the reduction or elimination of undesired harmonics. The computation of each pulse pattern is performed for a stationary operation and for a defined value of the modulation index. Each pulse pattern shows a specific distribution of its voltage harmonics with respect to the amplitudes and frequencies. During a dynamic operation of the drive, fast changes of the modulation index or the fundamental frequency give rise to inevitable pulse pattern changes. This results in discontinuous changes of the harmonic voltage components of the pulse patterns and therefore causes oscillating and weakly damped transients.

The modulation error is defined as the deviation of the instant harmonic states from their stationary reference values [14,16,17]. For a drive system with an output LC-filter and an induction motor, the modulation error can be defined as:

$$\vec{\delta} = \begin{pmatrix} \delta \\ \delta_s \\ \delta_c \end{pmatrix} = \begin{pmatrix} i_{,h} - i_{h\_ss}^* \\ i_{s,h} - i_{s,h\_ss}^* \\ u_{c,h} - u_{c,h\_ss}^* \end{pmatrix} . \tag{11}$$

The reference values are the stationary harmonic voltages and currents of the filter and the motor. The dynamics of the modulation error can thus be described using the above definition of the modulation error and the harmonic state space model (7) as:

$$\dot{\vec{\delta}} = \mathbf{A}\vec{\delta} + \mathbf{B}\Delta u_h . \tag{12}$$

The harmonic input voltage Δu_h consists of the difference between the stationary harmonic voltage of a defined pulse pattern and the actually generated harmonic voltage

$$\Delta u_h = u_h - u_{h\_ss}. \tag{13}$$

The modulation error controller receives the actual values of the harmonic states after their prediction for the determination of the fundamentals with

$$\vec{x}_{,h}(k+1) = \mathbf{A}_d \vec{x}_{,h}(k) + \mathbf{B}_d u_{,h}(k) . \tag{15}$$

In order to reduce the calculation requirements, the computation of the stationary harmonic states is performed analytically. For this purpose, the transfer functions $\underline{Z}_i(j\omega_v)$, $\underline{Z}_{i_s}(j\omega_v)$ and $\underline{Z}_{u_c}(j\omega_v)$ are computed with the state space system (7), describing the steady-state transfer characteristics between the harmonic input voltage and the harmonic states. The phasor of the stationary harmonic choke current for a defined frequency ω_v results for instance analytically from:

$$\underline{I}(j\omega_v) = \underline{Z}_i(j\omega_v)\underline{U}(j\omega_v) \tag{15}$$

and

$$\underline{I}(j\omega_v) = \hat{z}_i(j\omega_v)\,\hat{u}_v\,e^{j\left(\arg \underline{Z}_i(j\omega_v)+\varphi_{u,v}\right)}. \tag{16}$$

In the time domain, the corresponding harmonic choke current is given by:

$$i_{,v\_ss} = \hat{i}_{,v\_ss} \sin(\omega_{,v} t + \varphi_{i,v}). \qquad (17)$$

The distortion current can thus be approximated by a sum of its harmonics:

$$i_{,h\_ss}(t) = \sum_{v \geq 5}^{\infty} i_{,v\_ss}(t). \qquad (18)$$

The spectrum of the pulse pattern is composed of sinusoidal harmonics, which have no phase shift. As a result, the amplitudes of the harmonics can be computed analytically from Eq. (2). This allows calculation of the steady-state distortion current in the first phase from:

$$i_{1,h\_ss}(t) \approx \sum_{v \geq 5}^{v_{max}} \underbrace{\left(\frac{U_d}{2} \frac{4}{v\pi} \sum_{n=1}^{q} (-1)^{n+1} \cos(v\alpha_n) \right)}_{b_v} \cdots \\ \cdots \hat{z}_i(j\omega_{,v}) \sin\left(\omega_{1,v} t + \arg \underline{Z}_i(j\omega_{1,v}) \right). \qquad (19)$$

A phase shift of $2\pi/3$ or $4\pi/3$ must be considered for the calculation of stationary distortion currents in the other phases. The same procedure is also valid for the computation of the other two state distortions.

2) Energy-based Modulation Error Control: The modulation error occurs as high-frequency and weakly damped states oscillations of the system's energy storage devices (L_σ, L und C). An active damping of these oscillations can be achieved by a decrease of the modulation error energy, which corresponds with a Lyapunov control approach. Using the modulation error vector (11) the following energy space vector can be defined:

$$E_{\alpha,\beta} = \frac{1}{2}\left[L\delta_{\alpha,\beta}^2 + L_\sigma\delta_{\alpha,\beta,s}^2 + C\delta_{\alpha,\beta,c}^2 \right] = \vec{\delta}_{\alpha,\beta}^T \mathbf{P} \vec{\delta}_{\alpha,\beta} \qquad (20)$$

with

$$\mathbf{P} = \left\{ \begin{matrix} \frac{1}{2}L & 0 & 0 \\ 0 & \frac{1}{2}L_\sigma & 0 \\ 0 & 0 & \frac{1}{2}C \end{matrix} \right\}. \qquad (21)$$

$E_{\alpha,\beta}$ is a positive definite quadratic function with a global minimum at zero. Thus, the modulation error can be decreased using the command signal for the inverter that keeps the derivative of the energy function negative

$$\dot{E}_{\alpha,\beta} = \vec{\delta}_{\alpha,\beta}^T \operatorname{grad} E_{\alpha,\beta}(\vec{\delta}_{\alpha,\beta}) < 0. \qquad (22)$$

The approximation of the energy derivative by the difference quotient gives:

$$\dot{E}_{\alpha,\beta}(kT_0) \approx \frac{E_{\alpha,\beta}\big((k+1)T_0\big) - E_{\alpha,\beta}(kT_0)}{T_0} \\ \approx \frac{1}{2T_0}\begin{pmatrix} \vec{\delta}_{\alpha,\beta}^T\big((k+1)T_0\big)\mathbf{P}\vec{\delta}_{\alpha,\beta}\big((k+1)T_0\big)\cdots \\ \cdots - \vec{\delta}_{\alpha,\beta}^T(kT_0)\mathbf{P}\vec{\delta}_{\alpha,\beta}(kT_0) \end{pmatrix}. \qquad (23)$$

Applying the time-discrete state space system for the modulation error space vector

$$\vec{\delta}_{\alpha,\beta}(k+1) = \mathbf{A}_0\vec{\delta}_{\alpha,\beta}(k) + \mathbf{B}_0\Delta u_{\alpha,\beta}(k). \qquad (24)$$

to the Eq. (23) provides a scalar quadratic function for the energy derivative

$$\dot{E}_{\alpha,\beta}(k) \approx a\left(k, \vec{\delta}_{\alpha,\beta}(k)\right) + b\left(k, \vec{\delta}_{\alpha,\beta}(k)\right)\Delta u_{\alpha,\beta}(k) + \cdots \\ \cdots c\Delta u_{\alpha,\beta}^2(k) \qquad (25)$$

with the coefficients

$$a\left(k, \vec{\delta}_{\alpha,\beta}(k)\right) = \frac{1}{2T_0}\begin{pmatrix} \left(\mathbf{A}_0\vec{\delta}_{\alpha,\beta}(k)\right)^T \mathbf{P}\mathbf{A}_0\vec{\delta}_{\alpha,\beta}(k) - \cdots \\ \cdots \vec{\delta}_{\alpha,\beta}^T(k)\mathbf{P}\vec{\delta}_{\alpha,\beta}(k) \end{pmatrix} \\ = \frac{1}{2T_0}\vec{\delta}_{\alpha,\beta}^T(k)\left(\mathbf{A}_0^T\mathbf{P}\mathbf{A}_0 - \mathbf{P}\right)\vec{\delta}_{\alpha,\beta}(k), \qquad (26)$$

$$b\left(k, \vec{\delta}_{\alpha,\beta}(k)\right) = \frac{1}{2T_0}\left(\mathbf{B}_0^T\mathbf{P}\mathbf{A}_0\vec{\delta}_{\alpha,\beta}(k) + \left(\mathbf{A}_0\vec{\delta}_{\alpha,\beta}(k)\right)^T \mathbf{P}\mathbf{B}_0\right) \\ = \frac{1}{T_0}\left(\mathbf{B}_0^T\mathbf{P}\mathbf{A}_0\vec{\delta}_{\alpha,\beta}(k)\right), \qquad (27)$$

$$c = \frac{1}{2T_0}\left(\mathbf{B}_0^T\mathbf{P}\mathbf{B}_0\right). \qquad (28)$$

The coefficient a is independent from the input $\Delta u_{\alpha,\beta}(k)$ and negative because the modulation error state space system Eq. (24) is stable. The coefficient c is positive, since the matrix $\mathbf{P}$ is positive definite. Therefore, assuming constant coefficients a and b in Eq. (25) during a sampling period, the energy derivative represents a top opened parable which has a defined global minimum (Fig. 5).

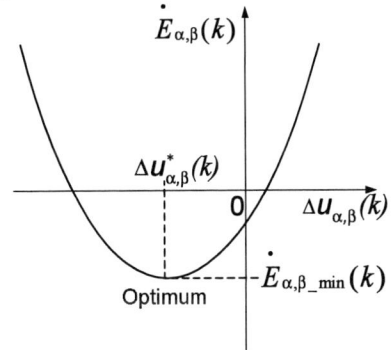

Fig. 5: Discretized derivative of the modulation energy as a function of $\Delta u^*(k)$ within one sampling period

The calculation of the control variable $\Delta u_{\alpha,\beta}(k)$, which minimises the derivative of the modulation error energy is performed by finding the minimum of Eq. (25) with

$$\frac{d\left(\dot{E}_{\alpha,\beta}(k)\right)}{d\Delta u_{\alpha,\beta}(k)} = b\left(k,\vec{\delta}_{\alpha,\beta}(k)\right) + 2c\Delta u_{\alpha,\beta}(k) = 0. \qquad (29)$$

The solution of the above Eq. gives the time-discrete control law

$$\Delta u_{\alpha,\beta}(k) = -\frac{b\left(k,\vec{\delta}_{\alpha,\beta}(k)\right)}{2c} = -\frac{\mathbf{B}_0^T \mathbf{P} \mathbf{A}_0}{\mathbf{B}_0^T \mathbf{P} \mathbf{B}_0}\vec{\delta}_{\alpha,\beta}(k) = \qquad (30)$$
$$= \mathbf{K}\vec{\delta}_{\alpha,\beta}(k).$$

The control matrix $\mathbf{K} = [k_1\ k_2\ k_3]$ can be pre-calculated offline, which reduces the demands on the online calculation. The voltage value $\Delta u_{\alpha,\beta}(k)$ is the modulation error controller's command signal, which has to be realized in the next sampling period by shifting of the pulse pattern edges.

3) Generation of the Reference Voltage: The modulation error controller works in a time-discrete manner and generates the reference space vector voltage within the next sampling period by shifting of the pulse pattern edges. The shifting of the edges causes changes of the voltage time integral, which can be approximated by voltage mean value changes. Hence, the reference voltage space vectors of the modulation error controller can be described as the voltage mean value changes. The effective change of the voltage mean value within a sampling period T_0 goes along with a time shift ΔT of the edges:

$$\Delta u = \left(\frac{U_d}{2}\frac{1}{T_0}\right)(-s)\Delta T \qquad (31)$$

with

$$s = \begin{cases} 1, & \text{for switches from lower to upper voltage state} \\ -1, & \text{for switches from upper to lower voltage state.} \end{cases}$$

The length of the sampling period and the location of the edge, which can occur at any point, restrict the maximal possible range of edge shift. This results in unsymmetrical limits for the possible changes of the voltage mean values. The limits are also time-variant and they change with every period. During some sampling periods no switching edges occur and therefore it is not always possible to generate the required reference voltage space vector. For this reason, for most of the sampling periods the controller can only approximately generate its reference variable. This can be explained by existence of time-variant limits and a random number of the switching edges within the sampling periods. The generation of the reference voltage is therefore performed in iterative steps, which are defined by the number of the switching edges.

The shifting of the edges starts with the first phase and is done within each phase sequentially. The future influence of the actual possible edge shift on the energy of the modulation error is always verified before its execution by employing (20) and (24). In this way, in the next sampling period only those shifts will be carried out, for which the prediction shows a reduction of the modulation's error energy. Effects of the former shifts on the modulation error are taken into account during computation of the future shifts. This means, that a new calculation of the reference voltage is executed every time the controller decided to shift an edge in the next period.

IV. SIMULATION RESULTS

The control algorithm has been verified by simulations of a 2.4 kV medium voltage drive system with a LC-filter fed by an NPC three-level inverter. The simulations were executed with MATLAB/Simulink®.

A. Steady-state Operation with Ideal Control Variable

The performance of the control method can be verified by the analysis of the modulation error energy, after applying a pulse pattern change during the operation with a constant modulation level. Therefore, the cascaded stator fundamental control was not involved at this stage. As a benchmark, the absolute value of the modulation error energy space vector was used

$$|E| = \sqrt{E_\alpha^2 + E_\beta^2}\ . \qquad (32)$$

The operation with an ideal control variable allows direct control of the system without any shifting of the switching edges (Fig. 6). Thus, the basic effects of the control algorithms can be analysed, as the actuating variable is not disturbed. In Fig. 6 a change of the pulse pattern with the pulse number $z_T = 7$ to $z_T = 4$ at $t = 0.6005$ s was simulated employing an ideal control variable. Such a high pulse number change is unusual in real operations, but is appropriate to check the performance of the energy controller. The modulation index was kept at a constant value of M = 0.8. Primarily, the controller minimizes the modulation error energy and thereby the error of the stator current. The energy-based controller delivers almost a steady reduction of the error energy without causing any excitations and with reasonable speed (Fig. 6).

B. Steady-state Operation with Shifts of the Pulse Pattern Edges

In comparison to Fig. 6, Fig. 7 shows the changes of the pulse pattern (performed by the controller) and the modulation error energy which results from pulse pattern shifts. The sampling intervals have been colored depending on the amount of shifted switching edges to visualize their effects. The results confirm the accuracy and dynamic performance of the control method. The application of the edge shifts reduces the speed of the controller at about factor 2. Nevertheless, it allows a fast reduction of the torque excitations caused by the modulation error (Fig. 8 and 9)

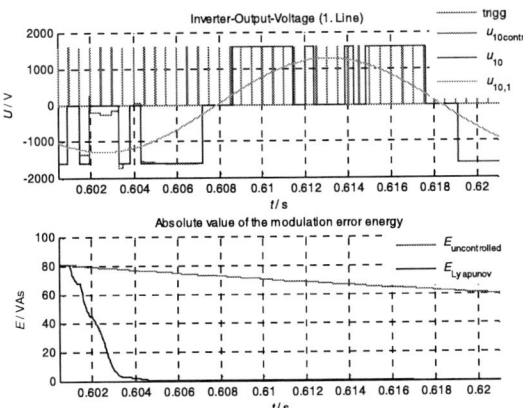

Fig. 6: Resulting absolute value of the modulation error during the ideal control

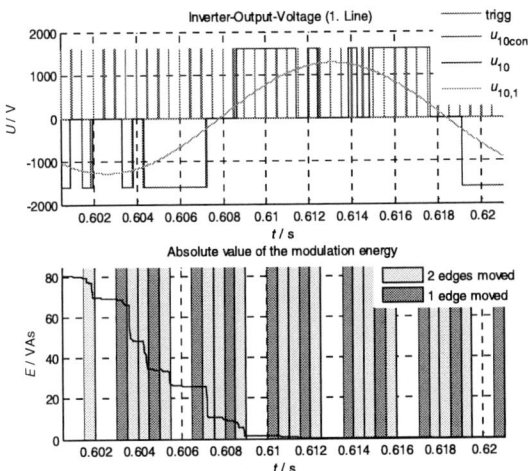

Fig. 7: Edge shifts and the resulting absolute value of the modulation error

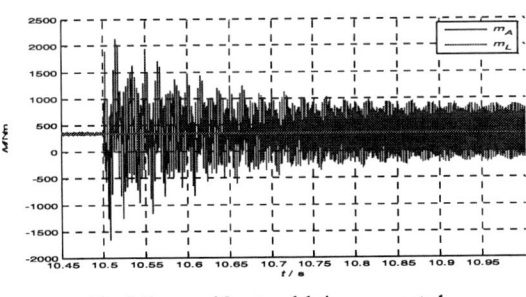

Fig. 8: Torque **without** modulation error control

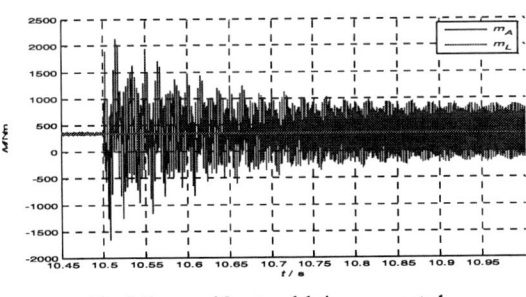

Fig. 9: Torque **with** modulation error control

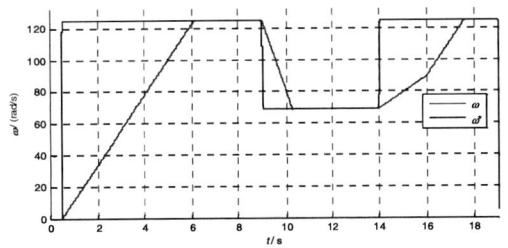

Fig. 10: Mechanical speed

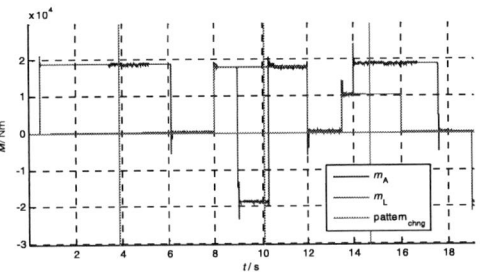

Fig. 11: Torque

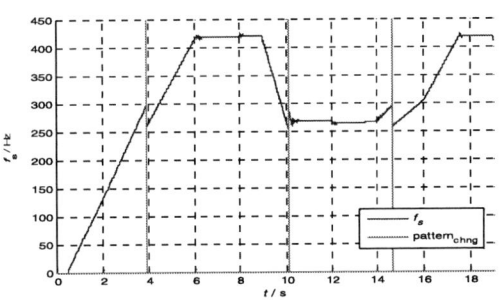

Fig. 12: Switching frequency

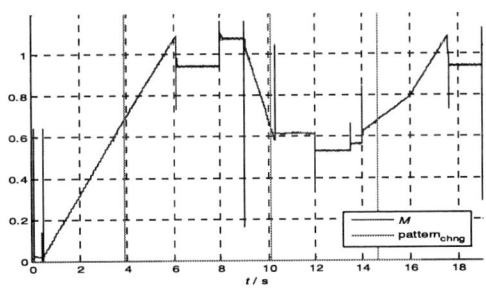

Fig. 13: Modulation index

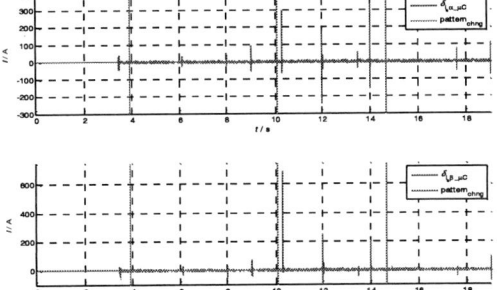

Fig. 14: Modulation error of the stator current space vector

C. Dynamic Operation with Modulation Error Control

The modulation error control works well together with the cascaded PI-control and allows also dynamic operations with IGBT switching frequencies down to 280 Hz (sampling frequency $f_0 = 2$ kHz), as shown in Fig. 10, 11 and 12. The simulations were done within a broad range of the modulation index including the upper limit of the space vector modulation (Fig. 13). The modulation error achieves its highest values during discontinuous or fast changes of the modulation index, because it implies changes of the pulse pattern. Nevertheless, it is possible to reduce the modulation error rapidly by employing the proposed energy-based control (Fig. 14).

V. CONCLUSION

This contribution introduces a control method for medium voltage drives with LC-filters and three-level NPC inverters which use optimal pulse patterns. Simulations with MATLAB/Simulink® of the presented control method show that a dynamic control of the drive is possible at very low switching frequencies. The proposed control method consists of two main control units. The space vector of the stator current fundamental is controlled using a PI-controller working with a sampling frequency of 2 kHz. A delay-free computation of the stator current fundamental is performed by a predictive, model-based estimation of the stator current distortion. The underlayed modulation error controller is based on discrete evaluation of the modulation error energy of all energy storage devices and regulates the transients in the output LC-filter and of the stator current. The modulation error is defined as the deviation of the transient distortions from the steady-state distortions that occur during ideal stationary supply with the pulse patterns. The modulation error is computed by means of a predictive, model-based estimation of actual distortions and an analytical computation of the steady-state distortions. The realization of the corrective space vector reference voltage $\Delta u_{\alpha,\beta}$ is done iteratively by shifting the switching edges within the sampling periods. Before the execution of each edge shift, the controller checks by prediction, if the resulting energy of the modulation error will decrease. This corrective direct intervention in the shape of the pulse patterns makes possible a fast settling time of the modulation error without increasing the switching frequency and without degradation of the patterns' optimal quality.

REFERENCES

[1] A. Nabae, I. Takahashi, H, Akagi, "A New Neutral Point Clamped PWM Inverter", IEEE Trans. Ind. Appl., vol. IA-17, pp. 518-522, 1981.

[2] A. Mertens, M. Bruckmann, R. Sommer, "Medium Voltage Inverter Using High Voltage IGBTs", IEEE EPE 1999, Lausanne, CH.

[3] R. Sommer, A. Mertens, M. Griggs, H.-J. Conraths, M. Bruckmann, T. Greif, "New Medium Voltage Drive Systems Using Three-Level Neutral Point Clamped Inverter With High Voltage IGBT", IEEE IAS 1999, Phoenix, AZ, pp. 1513-1519.

[4] H.-J. Conrath, F. Gießler, H.-D. Heining, "Shaft Voltages and Bearing Currents – New Phenomena in Inverter Driven Induction Machines", IEEE EPE 1999, Lausanne, CH.

[5] B.P. Schmitt, R. Sommer, "Retrofit of Fixed Speed Induction Motors With Medium Voltage Drive Converters Using NPC Three-Level Inverter High-Voltage IGBT Based Topology", IEEE ISIE 2001, Pusan, Korea.

[6] P.N. Enjeti, R. Jakkli, " Optimal Power Control Strategies for Neutral Point Clamped (NPC) Inverter Topology", IEEE Trans. Ind. Appl., vol. 28, no. 3, pp. 558-566, 1992.

[7] J. Holtz, "Pulsewidth Modulation for Electronic Power Converters", IEEE Trans. Ind. Appl., vol. 31, no. 5, pp. 1110-1120, 1995.

[8] A. Sapin, P.K. Steimer, J.-J. Simond, " Modeling, Simulation and Test of a Three-Level Voltage-Source Inverter With Output LC Filter and Direct Torque Control", IEEE Trans. Ind. Appl., vol. 43, no. 2, pp. 469-475, 2007.

[9] H. L. Liu, G. H. Cho, S. S. Park, "Optimal PWM Design for High Power Three-Level Inverter Through Comparative Studies", IEEE Trans. Power Electronics, vol. 10, no. 1, pp. 38-47, 1995.

[10] J. K. Steinke, "Use of an LC Filter to Achieve a Motor-friendly Performance of the PWM Voltage Source Inverter", IEEE Trans. Energy Conv., vol. 14, no. 3, pp. 649-654, 1999.

[11] S. Pöhler, A. Mertens, R. Sommer, "Optimisation of Output Filters for Inverter Fed Drives", IEEE IECON 2006, Paris, France.

[12] W. Leonhard, "Control of Electrical Drives", Springer-Verlag, Berlin, 2001.

[13] G. F. Franklin, J. D. Powell, M. Workmann, "Digital Control of Dynamic Systems", Addison-Wesley, Menlo Park, CA, 1997.

[14] J. Holtz, B. Beyer, "Fast Current Trajectory Control Based on Synchronous Optimal Pulsewidth Modulation", IEEE Trans. Ind. Appl., vol. 31, no. 5, pp. 1110-1120, 1995.

[15] J. Holtz, N. Oikonomou, "Synchronous Optimal Pulsewidth Modulation and Stator Flux Trajectory Control for Medium-Voltage Drives", IEEE Trans. Ind. Appl., vol. 43, no. 2, pp. 600-608, 2007.

[16] T. Laczynski, T. Werner, A. Mertens, "Active Damping of LC-Filters for High Power Drives using Synchronous Pulsewidth Modulation", IEEE PESC 2008, Rhodos, Greece.

[17] T. Laczynski, T. Werner, A. Mertens, "Modulation Error Control for Medium Voltage Drives with LC-Filters and Synchronous Optimal Pulse Width Modulation", IEEE IAS 2008, Edmonton, Alberta, Canada.

APPENDIX

TABLE 1

PARAMETERS OF THE DRIVE

U	2,4 kV
I	600 A
n	1190 min^{-1}
f	60 Hz

Voltage Harmonic Control of Z-source Inverter for UPS Applications

Arkadiusz Kulka, Tore Undeland

Norwegian University of Science and Technology
O.S. Bragstadsplass 2E, Trondheim, Norway
+47 73594241
e-mail: {arkadiusz.kulka, tore.undeland}@elkraft.ntnu.no
URL: http://www.elkraft.ntnu.no/eno

Abstract—This paper presents a control method for obtaining sinusoidal output voltage regardless of the nonlinear and unbalanced loads. Control of the DC boost stage and capacitor voltage is presented. The resonant regulators are used for selective harmonic cancelation of the output AC voltage. The Z-source inverter is able to provide higher AC voltage related to the DC link voltage than in conventional VSI, possessing embedded property of boost converter. This work presents the optimal control of boost factor and capacitor voltage, reducing the voltage transistor stress under desired AC voltage level. Experiment implementation on TMS320F2812 DSP show possibility of accommodating blanking time DSP circuits for controlling shoot-through duty ratio without any additional external logic. Modified space vector modulation gives only two transistor switching per cycle, thus minimizing the switching loses as much as possible.

Keywords—uninterruptible power supply, Z source inverter, distributed power generation, voltage control, renewable energy.

I. INTRODUCTION

Z-source voltage-type inverter (ZSI) has been proven experimentally and in the literature as an attractive single-stage solution for buck-boost, three phase dc-ac power conversion [1]. The general layout is shown in Fig. 1.

The ZSI provides special features which can't be observed in the traditional VSI inverter:

- The ZSI is a boost converter for dc-ac power conversion and higher peak to peak ac output voltage can be obtained than available input voltage.

- A short circuit across any phase legs is allowed, so the dead time is not necessary. The cross conductive short circuit is called shoot through state and is similar to those in Current Source Inverter.

- Shorting of any phase legs provide a boost up capability, thus must be carefully controlled (similar to step-up converter).

The proposed AC voltage control scheme is suited for ZSI with LC output filter. It use resonant regulators (1) and it is suited for UPS or standalone power generation where sine wave output voltage is to be maintained. The proposed controller is able to compensate voltage distortion from unbalanced and nonlinear loads, thus controlling negative and positive voltage sequence and its harmonics. For control purpose only two phase to phase output voltage measurements are required and one measurement of the C_2 capacitor voltage (pseudo dc link).

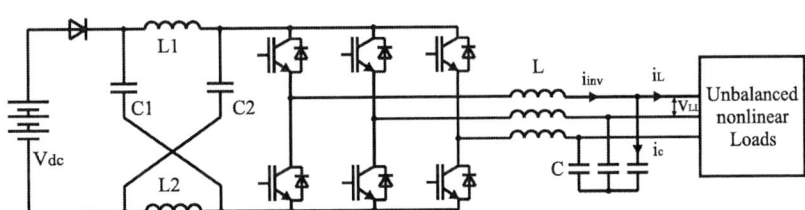

Fig. 1. Three phase Z source inverter.

II. AC VOLTAGE CONTROL TOPOLOGY

The proposed control topology is depicted in Fig. 2 and is based on [2]. Filter capacitor current control is used for selective harmonic voltage rejection. As an alternative to sensing the capacitor current a sensor less scheme is used based on derivative of output voltage. The derivative introduces higher amplitude gain for higher harmonics. Normally the filter capacitor current contain large amount of switching noises and the derivative in the digital system will additionally introduce delay and even more noise. It is found that resonant controller handle well this type of signal. The delay associated with modulator and discrete derivative can be compensated by adjusting the leading angle γ of resonant controller (1) for given harmonic. The so called leading angle can change phase relation between input and output which can be adjusted due the fact that (1) is composed of two orthogonal components.

$$Hac(s) = K_i \cdot \frac{s \cdot \cos(\gamma) - \omega \cdot \sin(\gamma)}{s^2 + \omega^2} \qquad (1)$$

The system delay is the same for all harmonics, but for each separate harmonic there is different compensation (leading) angle. Note, that with increased order of harmonic the given leading angle is increasing. E.g. for fundamental harmonic of 50Hz, one sample delay (e.g.100us) is just 1.8 degree, but for 11th harmonics it is almost 20 degree. In order to achieve high quality voltage output the 5th, 7th, 11th and 13th harmonic must be included. Without correcting the leading angle for the last two harmonics the quality would be not satisfactory.

The reference for capacitor current $i_{Cref,\alpha}$, $i_{Cref,\beta}$ (2) is easily obtained from voltage reference $v_{ref,\alpha}$, $v_{ref,\beta}$ (3) by interchanging axes and thus obtaining 90° advanced reference angle. It should be noted that in current reference definition capacitor value is included, but later on is cancelled out as can be seen in Fig. 6, avoiding the C parameter uncertainty.

$$\begin{bmatrix} i_{C,ref,\alpha} \\ i_{C,ref,\beta} \end{bmatrix} = \begin{bmatrix} -v_{ref,\beta} \\ v_{ref,\alpha} \end{bmatrix} [\omega \cdot C] \qquad (2)$$

$$\begin{bmatrix} v_{ref,\alpha} \\ v_{ref,\beta} \end{bmatrix} = \begin{bmatrix} \cos(\omega t) \\ \sin(\omega t) \end{bmatrix} \cdot v_{d,ref} \qquad (3)$$

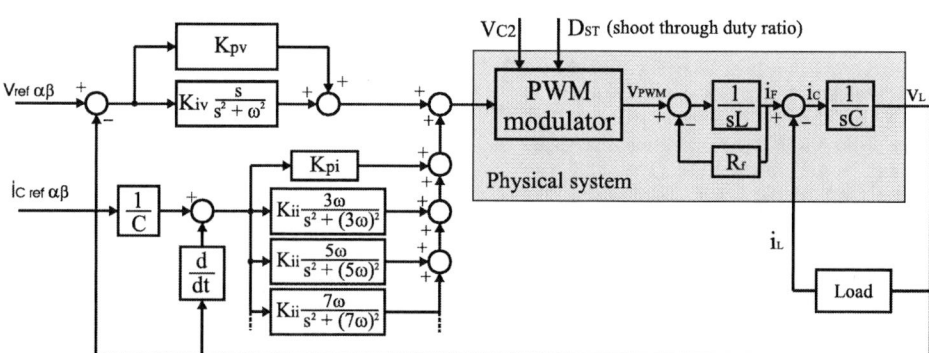

Fig. 2. Voltage harmonic controller layout.

III. MODULATOR WITH SHOOT THROUGH STATES

ZSI uses modified modulation strategy that insert shoot through states into standard space vector modulation, SVM [4]. These shoot-through states boost the dc link capacitor voltages and can be placed instead the null states without altering the normalized volt-sec average voltage. The duration of each active state in a switching cycle is kept the same as in traditional SPWM. Therefore, the output waveform will still be kept sinusoidal. The generation of switching signals is shown in Fig. 3.

The first shoot through interval $T_{ST}/3$ is inserted between two active states (common point of a and b line in Fig. 3). The active states are left/right shifted accordingly by $T_{ST}/6$ with their time intervals kept constant, and the remaining two (most left and most right) shoot-through states ($T_{ST}/3$) are lastly inserted

within the null intervals, at the beginning and end of active states. The modulation is symmetrical (left and right side of Fig. 3 is the mirror image). This way of sequencing inverter states also ensures a single device switching at all transitions and also allows simultaneously use of shoot through states

The reference signals of the inverter legs for upper and lower transistors are shown on Fig. 4. Very important implementation detail is that during the saturation (usually transients), the highest priority is given to the shoot through states (Fig. 5), so the active states are clamped first. This allow boost up the voltage first and then the modulation index can come back to not saturated level.

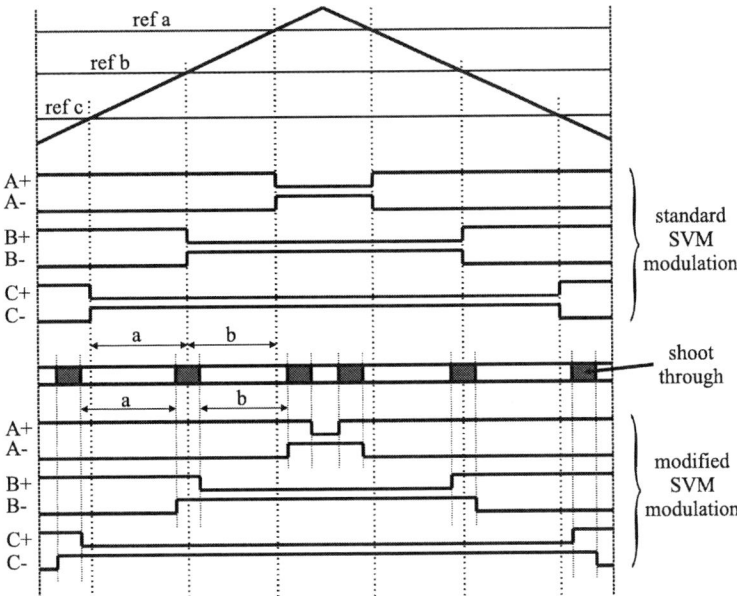

Fig. 3. Generation of switching signals, with shoot through states (red).

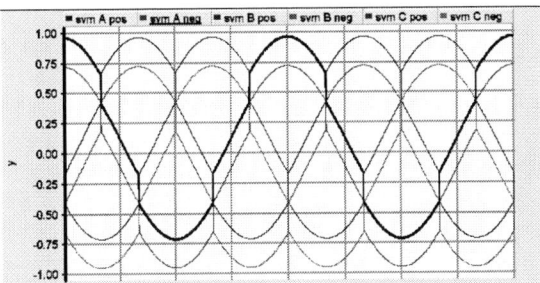

Fig. 4. Reference signal for modulator during steady state operation. The bands limited by parallel lines are the shoot through states.

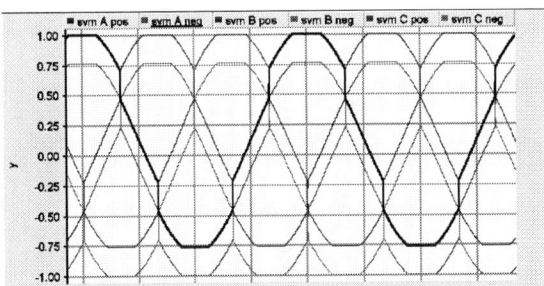

Fig. 5. Reference signal for modulator during transients. The modulated signal become saturated but not the shoot through signal.

For the simulation and the experiment the switching frequency was set to 10 kHz. The shoot through zero state (ST) was populated among the three phase legs, achieving equivalent switching frequency of 60 kHz from the view of Z-source network.

In experiment the selection of inductance was based on ripple current and maximum transistor current.

Since the transistor peak current is high (60 A), the converter rated current was defined as 10 A, the DSP cannot output long ST states (and we don't want use long time for ST during one PWM period), relatively small inductor in size and value can be designed. With this specific hardware it was convenient to have high current ripple (utilizing fully transistors and short ST time), but this is not usual case.

Assumed maximum capacitor voltage of 100 V, maximum ripple current of 40 A, the maximum shoot through time is limited by the DSP to 3 µs.

$$L = \frac{T_{ST}V_{CAP}}{\Delta I_{MAX}} = 30\mu H$$

IV. VIRTUAL DC LINK VOLTAGE CONTROLLER

The average of pseudo dc-link voltage across the inverter bridge is identical to the capacitor voltage because the average of an inductor voltage is zero. The capacitor voltage (C_1 and C^2) is dependent on the shoot through time, and it can be stepped up by increasing shoot-through time. Reducing the transistor voltage stress under a desired load is important, it should ensure that there is no high boost ratio and simultaneously the modulation index is not fully used (and the dc voltage as well). As has been analyzed in [1] the voltage gain (boost) is defined as $B=(1-2T_{ST}/T)^{-1}$, B is always ≥ 1, where T_{ST} is the half of shoot through time during the switching period T.

The AC voltage output relation is, $v_{peak,phase}=M\,B\,V_{DC}$ where M is the modulation index. Therefore, to minimize the voltage stress for any given voltage gain, we have to maximize the range of modulation index M by using as much of available DC link voltage and leaving enough time for shoot through state. Defining the shoot through duty ratio by $D_{ST}=T_{ST}/T$ resulting that during normal operation. $M+D_{ST}\leq 1$ In case of using boost property, minimum voltage stress appears for $M+D_{ST}=1$.

A discrete-time PI voltage controller based on trapezoidal method of approximation is used to regulate the average voltage of DC-link, V_{C2} (Fig. 6). It is important to include the wind up protection, thus limiting the maximum shoot through duty ratio.

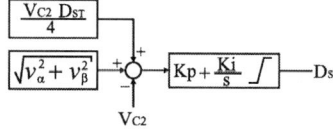

Fig. 6. Optimal DC-link voltage controller

As a result the aim of this controller is to keep the maximum peak modulation index as close to one as possible. The v_α and v_β are the output signals of the voltage controller, and the peak value of that vector is calculated. The other feed-forward term is calculation of equivalent voltage which would be "taken out" by shoot through duty ratio and not available for active vectors. It is very important to add this term, as the D_{ST} is increasing (consuming equivalent time of zero vectors) there is less available time for the active vectors.

V. SIMULATION RESULTS

In Fig. 7 the simulation shows the load voltages and currents under a rectifier load. The DC source voltage is 400 V. The output reference AC voltage is set to 600 V peak.

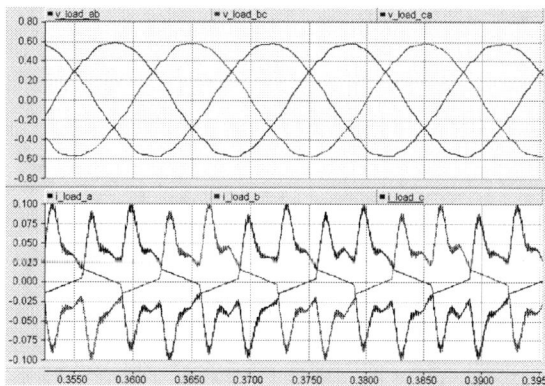

Fig. 7. Inverter Output voltage and load currents, no boost, voltage harmonic enabled.

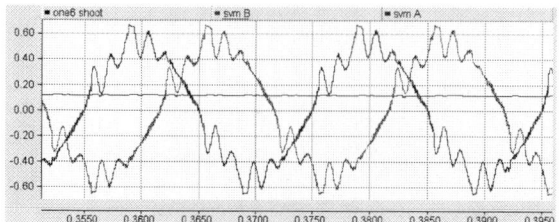

Fig. 8. Duty ration of modulation signal, when voltage harmonic algorithm is enabled.

Figure 8 shows the controller output modulation in phase A and B which is given to SVM. The visible ripples are cancelling the influence of rectifier current on output filter. Because the goal is to achieve sinusoidal voltage, they compensate voltage droop over a filter inductance. The blue line shows the 1/3 of the shoot through time, so this explain why modulation index at peaks is not reaching one, summing up gives, $0.65 + 3 * 0.11 = 0.98$.

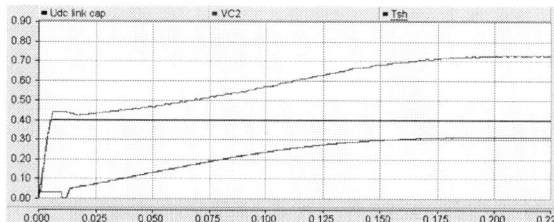

Fig. 9. Response of the optimal dc-link controller.

Figure 9 shows the start up response of dc-ling boost controller to a given set-point value. Dc link voltage at C_2 (green), DC supply voltage (blue), and shoot through time (red).

The problem can arise when very nonlinear current is drawn (crest factor >2). This leads to high spikes in reference voltage calculated by set of resonant controllers. The spikes "consume" the dc-link voltage, thus leading to even higher boost factor. The problem in real implementation was solved by inserting a peak detector and large time constant low-pass filter. The insertion took place in Fig. 6 between the instantaneous reference v_α, v_β and module calculation and the summation block.

VI. EXPERIMENTAL RESULTS

Since the standard VSI converter cannot be used due to significant changes in dc-link layout and gate signal interlocking, the 3 kW prototype of Z-source inverter was design and constructed. The used transistors rated current is 60 A, and voltage class of 1200 V. The gate driver is based on monolithic, opto isolated integrated circuit HCPL 316J which require galvanic small power supply for each transistor. Two level over voltage dc-link protection is also designed, since it easy to boost voltage to dangerous level for dc-link capacitors and transistors. There is no hardware interlocking protection for upper and lower transistors, so the gate signals from DSP card are directly connected to the gate drivers enabling shoot-through. The capacitor over voltage protection act in two steps. First the transistor which damp the power to the resistor is activated, secondly if the voltage is still increasing and pass certain level the signals to the drivers are cut down. The protection circuit is visible on the right side of the board at Fig. 10.

Fig. 10. Photograph of the setup: control board, inverter, Z network.

The controller is DSP, TMS320F2812 which has all necessary circuits, like A/D converter, PWM generator and embedded hardwire to control dead time. The PWM outputs also can be adjusted to be active high or low. Exploiting those features enables direct DSP control of duty ratio and shoot through factor. The dead time unit act as a shoot through time generator and is adaptively changed during operation. A found limitation in DSP is that the shoot through can be only 3,2 µs for a given PWM resolution, and there are always six shoot through per switching period, giving a total of 19,2 µs. For 100 µs switching period and PWM resolution of 6,66 ns the step up duty ratio is limited to 19 %. By lowering the resolution the boost factor can be further increased.

The death time generator from the DSP act as a shoot through generator, however the death time logic actually consumes the average volt-sec of the modulator duty. In order to prevent it a software correction function is introduced which shifts the active states by adding duration which is proportional to ST time. With varying ST time (the band on Fig. 4) the active states are shifted accordingly. The function is similar to one used for correcting death time effect in conventional VSI where current control is used. The death time correction is especially important e.g. when virtual flux is estimated [9].

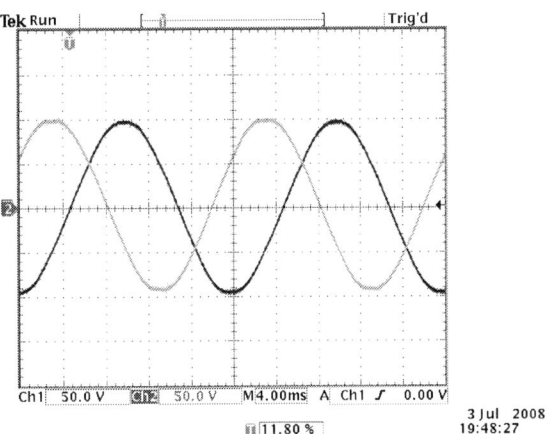

Fig. 11. Two phase sinusoidal output voltage with harmonic cancellation enabled, no load current.

The influence of enabling selective voltage harmonic cancellation algorithm is illustrated in Fig. 12 and Fig. 13.

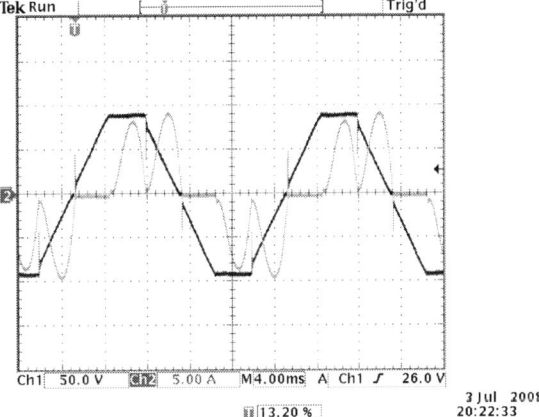

Fig. 12. Voltage and current on rectifier type of load connected to the output as a nonlinear load. The voltage harmonic cancellation is disabled.

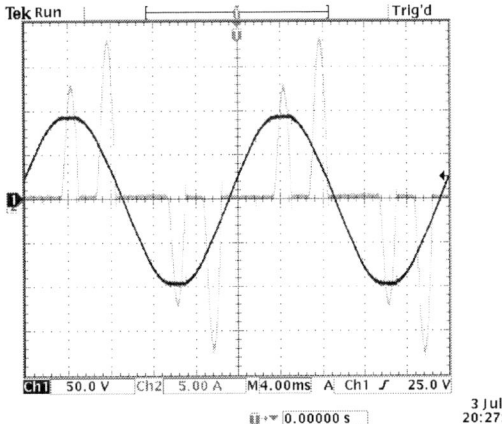

Fig. 13. Voltage and current on rectifier type of load connected to the output as a nonlinear load. The voltage harmonic cancellation is enabled.

In both cases the supply voltage is 80 V, and the reference output peak voltage is 100V. The boost factor is 25%. The three phase rectifier output is connected parallel to 4700 μF capacitor and to 25 Ω load resistor.

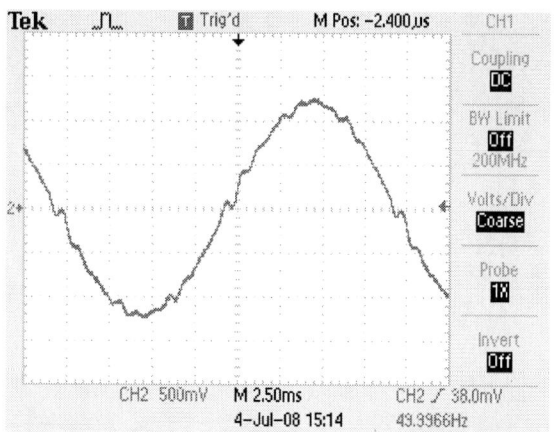

Fig. 14. Inverter output voltage where no load is applied, the boost ration is 300 %, the voltage harmonic cancellation is disabled. The reference output voltage is set to 240 V peak, the source is 80 V.

With very high boost ratio (the dc-link voltage three times higher than available supply voltage) the second order effect is easily visible. Due to harmonics coming from the shoot through state, even when the reference is set to sinusoidal the output voltage is distorted. The maximum available ST time is used in Fig. 14 (3.2 μs).

Experiments measurements shows that operation of Z-source inverter with high boos ratios (higher than 2) are inefficient. The voltage harmonic cancellation algorithm can be enabled, improving significantly the voltage shape but with increased nonlinear load current the shape will be distorted quickly. In other words, with higher output voltage (high boost factor) less current can be drawn. The current capacity can be improved by increasing the ST time. Increasing the ST time is not feasible since the maximum transistor current can be reach and redesign of inductor would be required.

VII. CONCLUSION

This paper has presented a Z-source inverter for implementing UPS or stand alone power generation system. It can boost the input voltage by a practical factor 1.5 to 2 not scarifying the efficiency, reducing cost and minimized component count. The voltage and current transistor class for Z-source inverter must be higher compared to VSI of the same rated power. For the high boost ratios (>2) the efficiency compared to standard VSI with boost stage is lower. The boost property can be vital where the input voltage is not changing in wide range and can decrease with the load. Example is the draining battery of UPS, or variable speed PM generator where for conventional VSI would be lack of DC link voltage. It can be used in variable speed diesel based systems where the speed changes would be in the range of 2. At minimum load, the diesel could run with lower speed, saving the fuel, and the required AC voltage would be still achieved. It can be used for renewably energy sources where is need for small boost up capability but it should not exchange transformers where is need for high boost up (like PV generation). In practice to comply with standards for EMI radiation it can be problem since large amount of components is under high dV/dt.

VIII. REFERENCES

[1] Fang Z. Peng "Z-Source Inverter" *IEEE Transaction on Industry Application*, Vol. 39, No. 2, March/April 2003.

[2] Arkadiusz Kulka, Tore Undeland, Vazquez S., Franquelo S. "*Stationary Frame Voltage Harmonic Controller for Standalone Power Generation*" Proceedings of EPE 2007 Aalborg/Denmark.

[3] Fang Zheng Peng, Miaosen S.,Zhaoming Q, "Maximum Boost Control of the Z-Source Inverter" *IEEE Transaction on Power Electronics*, Vol. 20, No. 4, July 2005.

[4] Poh Chiang Loh, D. Mahinda Vilathgamuwa, Yue Sen Lai, Geok Tin Chua and Yunwei Li "Pulse-Width Modulation of Z-Source Inverters" *IEEE Transaction on Power Electronics*, Vol. 20, No. 6, November 2005.

[5] Miaosen Shen, Alan Joseph, Jin Wang, Fang Z. Peng1, Donald J. Adams "*Comparison of Traditional Inverters and Z-Source Inverter for Fuel Cell Vehicles*" IEEE 0-7803-8538-1, 2004.

[6] Jacek Rabkowski "*The bidirectional Z-source inverter for energy storage application*" Proceedings of EPE 2007, Aalborg/Denmark.

[7] Jin-Woo Jung, Ali Keyhani "Control of a Fuel Cell Based Z-Source Converter" IEEE *Transaction on Energy Conversion*, Vol. 22, No. 2, June 2007.

[8] Poh Chiang Loh, Feng Gao, Pee-Chin Tan, Frede Blaabjerg "*Three-Level AC-DC-AC Z-Source Converter Using Reduced Passive Component Count*" IEEE 1-4244-0655-2, 2007.

[9] Arkadiusz Kulka, Tore Undeland "*Double Frame Virtual Flux, Voltage Sensor-less Algorithm for Three Phase VSC in Unbalanced Condition – Experimental study*" unpublished.

A Method of Optimal Control for Switched-Mode Power Converters

Anatoly Bekishev[1], Albert Iskhakov[1], Leonid Klyachko[2], Vladimir Pospelov[3] and Sergey Skovpen[4]

[1] Open Society "Concern "Morinformsistema – Agat", Moscow, Russia, e-mail: *asiskhakov@mail.ru*
[2] Central Scientific Research Institution, Moscow, Russia, e-mail: *asiskhakov@mail.ru*
[3] Military-Industrial Commission of the Russian Federation, Moscow, Russia, e-mail: *asiskhakov@mail.ru*
[4] State Marine Technical University "Sevmashvtuz", Severodvinsk, Russia, e-mail: *skovpen@atnet.ru*

Abstract—An optimal control method allowing to achieve the high dynamic performances of switched-mode power converters described by the second-order difference equation is offered. To this purpose, control vector is formed as a linear combination of differences of system variables, and matrix transformation of the linear equation to the nilpotent form is implemented. This control approach gives an ability to achieve a convergence of transient in a fixed point within two steps and to remain this property under regulation and perturbation of system parameters. Matrix transformation of the equation to the nilpotent form is realized by the system solution of the linear algebraic equations regarding the multipliers at the differences of system state variables. A proposed control method can be used as a basis for development of systems with as much as possible high dynamic behavior. To confirm the efficiency of this method, the system for load current control of a PWM buck converter is resulted as an example of the switched-mode power supply.

Keywords—converter control, optimal control, pulsed power converter, pulse width modulation (PWM), switched-mode power supply.

I. INTRODUCTION

Switched-mode power converters currently are one of the most widespread technical devices in industrial applications. Dynamic processes in such converters have complex nonlinear nature. To simplify their mathematical description different approximate methods are used. In particular, averaging methods are widely applied which not allow to provide maximum possible dynamic characteristics of converters.

In practice, the most often way to achieve desired dynamic parameters of control system is using regulators in the form of proportional-integral-differential circuits and their discrete analogues. Coefficients of these regulators are calculated and adjusted for specific operating conditions only, more often for nominal operation mode. If any change in the operating conditions of the system occurs, dynamic parameters deteriorate and become worse than values, which can be achieved by means of adjustment for other operation mode. This is a fundamental shortcoming of regulators with constant parameters.

The proposed optimal control method is applied for the second-order discrete system. The special attention this individual problem is given, because a lot of electrotechnical devices (in particular, controlled semiconductor power converters) containing active switches can be described by the second-order difference equation. Current regulation systems that are used to current control in superconducting solenoids, in the windings of high power electromagnetic suspensions, in electric-arc processes are examples of these devices.

In the paper, an optimal control method for system that are described by linear difference equation of the second order, allowing to transfer the system from arbitrary condition into the specific point in the space of system parameters within two steps is considered. Theoretical results are confirmed by a concrete application example of the method to solve the problem of a current control of PWM buck converter with resistive-inductive load (RL-load).

II. MATHEMATICAL THEORY

A. Formulation of a problem

A problem of control synthesis with the maximum speed of response for linear discrete system of the second order is considered. An offered control method allows to lead the system from any state condition into specified point within two steps.

A special attention to this particular problem is given in order to obviously demonstrate a solution and wide practical application of this approach. A current regulation in windings of electromagnetic devices (electric machines and others) can be referring to this problem. A current regulation system is considered as an example of switched-mode power converter.

Let the system is described by the linear difference equation

$$\mathbf{X}_{n+1} = \mathbf{A}\mathbf{X}_n + \mathbf{B}, \qquad (1)$$

where $\mathbf{X}_n = (x_{1,n}, x_{2,n})^T$ is a vector of solution, $\mathbf{B} = (b_1, b_2)^T$ is a vector of constant coefficients, $n = 0, 1, 2, ..., T$ is a transposing index, $\mathbf{A} = \begin{bmatrix} a_{11} & a_{12} \\ a_{21} & a_{22} \end{bmatrix}$.

If matrix $\mathbf{A}$ is stable, the solution $\mathbf{X}_n$ of the equation (1) tends to the steady-state value, which called the fixed point

$$\mathbf{X} = \mathbf{X}_{n+1} = \mathbf{X}_n = (\mathbf{I} - \mathbf{A})^{-1}\mathbf{B}, \qquad (2)$$

where $\mathbf{I}$ is an identity matrix of the second order.

According to the solution (1)

$$\mathbf{X}_n = \mathbf{A}^n \mathbf{X}_0 + \sum_{i=0}^{n-1} \mathbf{A}^i \mathbf{B}, \qquad (3)$$

where $\mathbf{X}_0$ is an initial vector. In order to lead the system (1) into specified point (2) within two steps with any $\mathbf{X}_0$, two conditions must be fulfilled:

$$1)\ \mathbf{A}^2 = 0,$$

$$2)\ \sum_{i=0}^{n-1} \mathbf{A}^i \mathbf{B} = \mathbf{X}. \tag{4}$$

The first condition of (4) is satisfied if matrix $\mathbf{A}$ is nilpotent [1]. Let's show, that the second condition is a consequence of the first one, i.e. it is satisfied for nilpotent matrix $\mathbf{A}$. In fact, having set any initial value of a solution vector $\mathbf{X}_0 = \mathbf{X}_n$ $(n = 0)$, as a result of two calculations by the formula (3), we obtain

$$\mathbf{X}_1 = \mathbf{A}\mathbf{X}_0 + \mathbf{B},$$

$$\mathbf{X}_2 = \mathbf{A}\mathbf{X}_1 + \mathbf{B} = \mathbf{A}^2\mathbf{X}_0 + \mathbf{A}\mathbf{B} + \mathbf{B}. \tag{5}$$

The first term of $\mathbf{X}_2$ gives a zero, therefore

$$\mathbf{X}_2 = \mathbf{A}\mathbf{B} + \mathbf{B} = (\mathbf{A} + \mathbf{I})\mathbf{B}. \tag{6}$$

If the right part of (6) to multiply by product of a matrix $(\mathbf{I} - \mathbf{A})$ and inverse matrix $(\mathbf{I} - \mathbf{A})^{-1}$

$$\mathbf{X}_2 = (\mathbf{I} - \mathbf{A})^{-1}(\mathbf{I} - \mathbf{A})(\mathbf{A} + \mathbf{I})\mathbf{B} \tag{7}$$

and to consider, that product of matrixes in the right part with nilpotent matrix $\mathbf{A}$ gives an identity matrix

$$(\mathbf{I} - \mathbf{A})(\mathbf{A} + \mathbf{I}) = (\mathbf{I}^2 - \mathbf{A}^2) = \mathbf{I},$$

the equality (7) means achievement a fixed point (2)

$$\mathbf{X}_2 = (\mathbf{I} - \mathbf{A})^{-1}\mathbf{B} = \mathbf{X}. \tag{8}$$

Thus, in order to a system of the second order will reached a specified point (2) from any condition $\mathbf{X}_0$ within two steps it is necessary, that the matrix of difference equation was a nilpotent. However, the matrix of the initial equation does not satisfy this property; therefore, there is a problem of transformation the equation (1) with a normal matrix $\mathbf{A}$ to an equation with a nilpotent matrix $\mathbf{N}$, i.e. the following condition must be satisfied

$$\mathbf{N}^2 = 0. \tag{9}$$

This problem is not trivial in the sense, that the specified transformation cannot be fulfilled in the usual manner, i.e. by finding a matrix with a required form using elementary transformations [2], as a matrix $\mathbf{N}$ is degenerate, and an original matrix does not belong them. It is, therefore, necessary to use other way that not falls into a technique of matrix transformations.

It should be noted that even in special mathematical literature, including quoted here [1, 2] and other references, it is resulted a few information about nilpotent matrixes. Anyway, the mentioned problem of transformation, as far as it is known, has not been formulated before.

In this paper, the transformation, which allows to lead a matrix of the second order to the nilpotent form by means of the special matrix, whose elements are determined through the original matrix elements, is considered. It gives the possibility of convergence of the transient into the fixed point after two steps and one enables to remain this property under regulating and perturbations of the system parameters.

B. Transformation of the second-order matrix to nilpotent form

The second-order matrix

$$\mathbf{N} = \begin{bmatrix} n_{11} & n_{12} \\ n_{21} & n_{22} \end{bmatrix} \tag{10}$$

is nilpotent as it can be seen by equating the coefficients of the characteristic equation

$$\lambda^2 - (n_{11} + n_{22})\lambda + n_{11}n_{22} - n_{12}n_{21} = 0$$

to zero, i.e.

$$n_{11} + n_{22} = 0, \quad n_{11}n_{22} - n_{12}n_{21} = 0. \tag{11}$$

Expressions (11) are called the conditions of nilpotency.

Transformation of matrix $\mathbf{A}$ to a nilpotent matrix $\mathbf{N}$ can be implemented differently. Here the transformation satisfying the two conditions is considered. These conditions are invariance of the fixed point (2) and additivity. The first condition has a semantic meaning, namely, transformation of equation should not change the fixed point (2). The additivity condition is formal. It defines the special matrix $\mathbf{H}_A$, called as nilpotenting, from additive operation of original $\mathbf{A}$ and nilpotent $\mathbf{N}_A$ matrixes

$$\mathbf{N}_A = \mathbf{A} + \mathbf{H}_A \quad \text{or} \quad \mathbf{H}_A = \mathbf{N}_A - \mathbf{A}. \tag{12}$$

Thus, it is necessary to determine the elements of matrix $\mathbf{H}_A$ as functions of elements of the second-order matrix $\mathbf{A}$ under condition $\mathbf{N}_A^2 = 0$. There are two nilpotency conditions (11). Considering them as the equations regarding the four elements $\mathbf{H}_A$, it is possible to choose two elements by any way, in particular, to accept zero. We designate two required items as h_1 and h_2. It is possible to arrange these elements by various ways, and we will consider two variants.

Let the matrix $\mathbf{H}_A$ has the diagonal form

$$\mathbf{H}_A = \begin{bmatrix} h_1 & 0 \\ 0 & h_2 \end{bmatrix}. \tag{13}$$

Substituting (13) in (12) and using conditions (11), we obtain a system of the nonlinear algebraic equations regarding the coefficients h_1 and h_2

$$\begin{cases} h_1 + h_2 = -p; \\ a_{22}h_1 + a_{11}h_2 + h_1h_2 = -q \end{cases} \tag{14}$$

with the solution

$$h_1 = -a_{11} \pm d, \quad h_2 = -a_{22} - d, \tag{15}$$

where $p = a_{11} + a_{22}$, $q = a_{11}a_{22} - a_{12}a_{21}$, $d = \sqrt{-a_{12}a_{21}}$, and, accordingly, the two nilpotent matrixes

$$\mathbf{N}_{A1} = \begin{bmatrix} d & a_{12} \\ a_{21} & -d \end{bmatrix}, \quad \mathbf{N}_{A2} = \begin{bmatrix} -d & a_{12} \\ a_{21} & d \end{bmatrix}. \tag{16}$$

If nonzero elements are located in row of matrix, for example, in the first row

$$\mathbf{H}_A = \begin{bmatrix} h_1 & h_2 \\ 0 & 0 \end{bmatrix}, \tag{17}$$

we have a system of the linear equations

$$\begin{cases} h_1 = -p; \\ a_{22}h_1 + a_{21}h_2 = -q \end{cases} \qquad (18)$$

with the solution

$$h_1 = -a_{11} - a_{22}, \quad h_2 = -a_{12} - a^2{}_{22}/a_{21}, \qquad (19)$$

defining the unique nilpotent matrix

$$\mathbf{N}_A = \begin{bmatrix} -a_{22} & -a^2{}_{22}/a_{21} \\ a_{21} & a_{22} \end{bmatrix}. \qquad (20)$$

So, depending on the chosen structure of nilpotenting matrix $\mathbf{H}_A$ (13) or (17), the calculation of its elements can be reduced to the solution of system both linear and nonlinear equations. For system of the second order this circumstance does not practically influences the complexity of calculation. However, if a system dimension increases, this question becomes basic.

C. Transformation of the linear difference equation to form with a nilpotent matrix

Transformation of the equation (1) to the form with a nilpotent matrix has the following feature. If procedure (12) is used directly, when the product $\mathbf{HX}_n$ is added to the right part of (1) to ensure an invariance of a fixed point it is necessary to subtract product $\mathbf{HX}$, that leads (1) to the form

$$\mathbf{X}_{n+1} = \mathbf{AX}_n + \mathbf{B} + \mathbf{H}(\mathbf{X}_n - \mathbf{X}),$$

where $\mathbf{H}$ is a 2 by 2 matrix whose elements are the coefficients.

However, the vector $\mathbf{X}$ appears here. In computational problems this vector is unknown, in control problems its elements, generally, are known, but a memory is necessary for their storage. For possibility to apply procedure (12) without using $\mathbf{X}$, the equation (1) is added the product $\mathbf{H}\Delta\mathbf{X}_n$, which not changes a fixed point, according to the expression

$$\mathbf{X}_{n+1} = \mathbf{AX}_n + \mathbf{B} + \mathbf{H}\Delta\mathbf{X}_n, \qquad (21)$$

where $\Delta\mathbf{X}_n = \mathbf{X}_{n+1} - \mathbf{X}_n$.

The same way is also used for transformation a system of the linear algebraic equations to the iterative form. In [3] the same operation has been called a linear invariant transformation of a fixed point.

As was noted in [4], conformably to a control system, the expression (21) has the clear physical sense, which consists that the additional term is a feedback on differences of co-ordinates of the space vector $\mathbf{X}_n$.

Further transformation of the equation (21) is yielded as follows. For matrix $\mathbf{H}$ we choose, for example, the form (17), then in the first row (21)

$$x_{1,\,n+1} = a_{11}x_{1,\,n} + a_{12}x_{2,\,n} + b_1 + h_1\Delta x_{1,\,n} + h_2\Delta x_{2,\,n} \qquad (22)$$

we substitute the second one and after collecting terms we can write down it in the form

$$x_{1,n+1} = a^*_{11}x_{1,n} + a^*_{12}x_{2,n} + b^*_1, \qquad (23)$$

where $a^*_{11} = \left(a_{11} - h_1 + a_{21}h_2\right)/c$, $a^*_{12} = \left[a_{12} + (a_{22} - 1)h_2\right]/c$, $b^*_1 = \left(b_1 + b_2h_2\right)/c$, $c = 1 - h_1$.

Having replaced the first equation of system (1) by (23), we obtain the transformed equation

$$\mathbf{X}_{n+1} = \mathbf{A}^*\mathbf{X}_n + \mathbf{B}^*, \qquad (24)$$

where $\mathbf{A}^* = \begin{bmatrix} a^*_{11} & a^*_{12} \\ a_{21} & a_{22} \end{bmatrix}$, $\mathbf{B}^* = \begin{bmatrix} b^*_1 \\ b_2 \end{bmatrix}$, with the same fixed point (2)

$$\mathbf{X} = (\mathbf{I} - \mathbf{A}^*)^{-1}\mathbf{B}^* = (\mathbf{I} - \mathbf{A})^{-1}\mathbf{B}. \qquad (25)$$

A convergence of the solution (24) to a fixed point (25) depends on the coefficients h_1 and h_2. Following the derivation of the expressions (19), we obtain the system similar to (18)

$$\begin{bmatrix} 1+a_{22} & -a_{21} \\ a_{22} & -a_{21} \end{bmatrix} \begin{bmatrix} h_1 \\ h_2 \end{bmatrix} = \begin{bmatrix} p \\ q \end{bmatrix}. \qquad (26)$$

Solution of (26) gives the elements of nilpotenting matrix

$$h_1 = p - q, \quad h_2 = [a_{22}p - q(1 + a_{22})]/a_{21}. \qquad (27)$$

The matrix $\mathbf{A}^*$ becomes a nilpotent by substituting the coefficients (27) into elements of this matrix in (24), since $\left(\mathbf{A}^*\right)^2 = 0$. According to (12) we designate

$$\mathbf{A}^* = \mathbf{N}_A. \qquad (28)$$

Let us name the transformed equation (24) with a nilpotent matrix as a nilpotent. Solution of this equation is

$$\begin{aligned} \mathbf{X}_1 &= \mathbf{N}_A\mathbf{X}_0 + \mathbf{B}^*, \\ \mathbf{X}_2 &= \mathbf{N}_A\mathbf{X}_1 + \mathbf{B}^* = \mathbf{N}_A^2\mathbf{X}_0 + \mathbf{N}_A\mathbf{B}^* + \mathbf{B}^*. \end{aligned} \qquad (29)$$

According to (8) at the second iteration it gives a fixed point (2)

$$\mathbf{X}_2 = (\mathbf{N}_A + \mathbf{I})\mathbf{B}^* = \mathbf{X}. \qquad (30)$$

Thus, transformation of the second order linear difference equation to the form with a nilpotent matrix enables one to achieve a convergence of the solution into a fixed point within two steps. It should be noted that this property remains irrespective of a matrix stability at a zero vector $\mathbf{B}$, when the solution converges to zero. Let us consider more in detail the last case, because this result, as was noted in [4], is not matched to a treatment of discrete system controllability.

D. Controllability of a system described by the nilpotent equation

In control theory [5], the term "controllability" is used to formal description of the system, which is called the controllable (controllable by Kalman), if it can be transferred from specified state to another one in finite time [6]. The devices ensuring this system property are called the aperiodic regulators [7].

In [4] it is noted that the term "controllability" is not suitable when it is used in the specific technical area to describe a system, which is able to transfer from one state to another one within finite number of steps. The results obtained here confirm the unsuitability of this term in the formal aspect also.

Indeed, the equation (1) describes a system with linear combination of control [6]. At $\mathbf{B} = 0$ the equation (1)

665

$$\mathbf{X}_{n+1} = \mathbf{A}\mathbf{X}_n \qquad (31)$$

describes an uncontrollable system in the substantial sense, because it has not control, and in the formal sense, because controllability criterion is not satisfied.

However, if the matrix $\mathbf{A}$ is a nilpotent, how it is shown here, the solution converges to zero within two steps with any initial vector $\mathbf{X}_0$, i.e. system (31) has a property of controllability while it has not a control. Thus, there is a conflict when the same system can be belong to controllable and uncontrollable simultaneously. Certainly, this conflict situation has the formal nature, since in order to lead a system matrix to the nilpotent form at $\mathbf{B} = 0$ in accordance with (21), system should have an input and must satisfy the controllability criterion.

This conflict can be resolved by adding up the definition of controllability of the discrete system with the property of "automatic" controllability of the system (with a nilpotent matrix) without control action.

Let us note the fundamental difference between discrete and continuous systems in a context of the realization of finite-duration processes [5]. For continuous systems these processes are achieved by altering a sign of the right part of the differential equation, thus alteration of an initial vector leads to change the switching moments and duration of process. In discrete system the right part does not vary or is equal a zero for system (31), and alteration of an initial vector does not influence duration of process.

E. Nonlinear discrete system

Systems with switches are basically being described by the nonlinear difference equations. Mathematical nonlinear discrete problems are solved using a special method for hit into convergence region. Then a solution is found inside of this region by the well-known method with finite number of iterations (for example, conjugate gradient method [1]). This strategy gives the maximum effect for control of nonlinear discrete system, when a convergence process to a fixed point occurs practically within minimum number of steps. To this purpose, the feedback signal on differences of space vector co-ordinates is formed, and multipliers are chosen by reduction of the linearized system equation to the nilpotent form. Let us examine this way in more detail.

Let the system is described by the nonlinear difference equation in the following form

$$\mathbf{F}(\mathbf{X}_{n+1}, \mathbf{X}_n) = 0 \qquad (32)$$

with a fixed point $\mathbf{X} = \mathbf{X}_{n+1} = \mathbf{X}_n$. By analogy to the linear equation (21) the equation (32) is supplemented with product $\mathbf{H}\Delta\mathbf{X}_n$, which does not change a fixed point

$$\mathbf{F}(\mathbf{X}_{n+1}, \mathbf{X}_n) + \mathbf{H}\Delta\mathbf{X}_n = 0, \qquad (33)$$

where $\mathbf{F}$ is a nonlinear vector function with elements $F_1(\mathbf{X}_{n+1}, \mathbf{X}_n)$ and $F_2(\mathbf{X}_{n+1}, \mathbf{X}_n)$.

The further actions are carried out in the following order. At first, the equations (33) are linearized in the neighborhood of the fixed point $\mathbf{X}$. Then the obtained equation is transformed to the form with a nilpotent matrix. Lastly, the elements of nilpotenting matrix $\mathbf{H}$ are calculated using the expression (27) through the elements of matrixes of linearized system. In [3] the elements of nilpotenting matrix are called the correction coefficients.

A linearized equation (33) is expressed as

$$\mathbf{A}_2\delta\mathbf{X}_{n+1} + \mathbf{A}_1\delta\mathbf{X}_n + \mathbf{H}\Delta\mathbf{X}_n = 0, \qquad (34)$$

where $\delta\mathbf{X}_n = \mathbf{X}_n - \mathbf{X}$, $|\delta| \ll 1$, $\mathbf{A}_2 = \partial\mathbf{F}/\partial\mathbf{X}_{n+1}$, $\mathbf{A}_1 = \partial\mathbf{F}/\partial\mathbf{X}_n$, which is led to the form (21)

$$\delta\mathbf{X}_{n+1} = \mathbf{A}\delta\mathbf{X}_n + \mathbf{G}\Delta\mathbf{X}_n, \qquad (35)$$

where $\mathbf{A} = -\mathbf{A}_2^{-1}$, $\mathbf{G} = -\mathbf{A}_2^{-1}\mathbf{H}$.

The elements of matrix $\mathbf{H}$ ensures a nilpotency of matrix $\mathbf{A}$ in (35) and ones are defined by analogy to derivation of the expressions (27).

In term of the control theory, a mathematical operation of transformation of original equation can be interpreted both as the system correction [9] and as the control.

Here the theoretical part of the method is completed. Further it is shown how the obtained results can be used for achievement of practically extremely possible dynamic properties of the switched-mode power converter as an example of concrete technical system with switches.

III. CURRENT REGULATION SYSTEM

A buck converter simplified topology with RL-load is shown in Fig. 1a. An average load current I_d is a controllable variable of the system. The current value is changed by the control system (it is not shown on the circuit) of the transistor switch S. We assume that the circuit elements are ideal, and converter operates under continuous-current condition. The timing diagrams are depicted in Fig. 1b to make it easier to explain the operating principle of the system.

The control system containing the PWM modulator defines the switching period T. At the instant t_n switch S is compulsory closed. According to the control law the control system forms the pulse duration τ_n and it opens the switch at the instant t_n'.

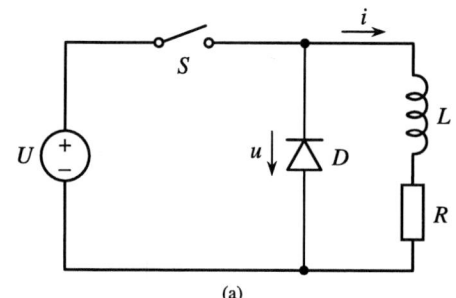

(a)

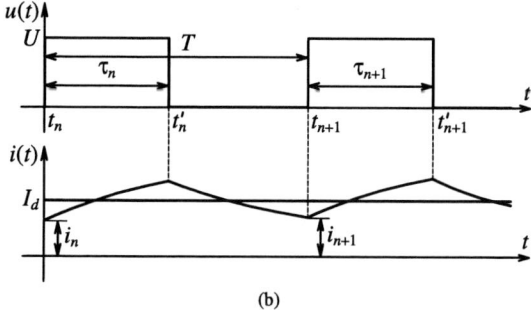

(b)

Fig. 1. Current regulation system: (a) buck converter topology, (b) PWM pattern and load current waveform.

At the n-th time interval the differential equation regarding the load current i is

$$L\frac{di}{dt} + Ri = \begin{cases} U, & t \in [t_n, t_n']; \\ 0, & t \in [t_n', t_{n+1}]. \end{cases} \quad (36)$$

It has the solution

$$i = \begin{cases} I + C_{1n}\exp(st), & t \in [t_n, t_n']; \\ C_{2n}\exp(st), & t \in [t_n', t_{n+1}], \end{cases} \quad (37)$$

where U is an input dc voltage, $I = U/R$, $t_n = nT$, $T = 1/f$ is a discrete interval (switching period); f is a switching frequency; $t_n' = t_n + \tau_n$, τ_n is a pulse duration, C_{1n} and C_{2n} are constants, $s = -R/L$.

Having designated $i_n = i(t_n)$ and using conditions of a continuity of a current during the moments of the switching, allowing to express C_{1n} and C_{2n} through i_n, we obtain the difference equation of a load current

$$i_{n+1} = gi_n - I(g - e_n), \quad (38)$$

where $g = \exp(sT)$, $e_n = \exp[s(T - \tau_n)]$.

In practice, a number of different control systems with current integration can be used to obtain proportionality between a reference signal x and an average current value. In this case, the variant with current averaging during a discrete interval is considered according to the expression

$$k\tau_{n+1} = x - \frac{1}{T_I}\int_{t_n}^{t_{n+1}} i(t)dt + k\tau_n, \quad (39)$$

where k defines a slope of the ramp generator; T_I is an integration constant. Substituting a current i from (37) into (39), we find the control difference equation

$$\tau_{n+1} = \frac{x}{k} - \frac{I\tau_n}{kT_I} - \frac{i_n(g-1) + I(e_n - g)}{skT_I} + \tau_n. \quad (40)$$

The equations (38) and (40) characterize nonlinear discrete model of the second-order system, where duration τ_n and current i_n are system variables. Let us represent this system in the form of (32)

$$\begin{cases} \tau_{n+1} = \dfrac{x}{k} - \dfrac{I\tau_n}{kT_I} - \dfrac{i_n(g-1) + I(e_n - g)}{skT_I} + \tau_n; \\ i_{n+1} = gi_n - I(g - e_n). \end{cases} \quad (41)$$

The main advantage of the proposed method is possibility to physical realization of the procedure (33), which describes the operation of summation and, hence, assumes to use the summer with an input for connection the signal $\mathbf{H}\Delta\mathbf{X}_n$. In control systems, as a rule, the summer with several inputs for control signals and feedbacks is used to control of switches. Summer output usually is connected to the comparator, which alters the state at the instant when changing a sign of the input sum occurs. Comparator operation leads to alteration of the switch state. To this purpose the signal is formed as a linear combination of the differences of system variables

$$z = h_1(\tau_{n+1} - \tau_n) + h_2(i_{n+1} - i_n) = h_1\Delta\tau_n + h_2\Delta i_n \quad (42)$$

and is added to the control signal x in the (39). As a result, the first equation of (41) changes as follows

$$\tau_{n+1} = \frac{x}{k} - \frac{I\tau_n}{kT_I} - \frac{i_n(g-1) + I(e_n - g)}{skT_I} + \tau_n + \\ + \frac{h_1}{k}\Delta\tau_n + \frac{h_2}{k}\Delta i_n. \quad (43)$$

By linearizing the equations (43) and (38) in the neighborhood of a fixed point with co-ordinates $\tau_{n+1} = \tau_n = \tau$ and $i_{n+1} = i_n = i$ we obtain a linear system in the form of (34) with identity matrix $\mathbf{A}_2$

$$\begin{cases} \delta\tau_{n+1} = a_{11}\delta\tau_n + a_{12}\delta i_n + \dfrac{h_1}{k}\Delta\tau_n + \dfrac{h_2}{k}\Delta i_n; \\ \delta i_{n+1} = a_{21}\delta\tau_n + a_{22}\delta i_n, \end{cases} \quad (44)$$

where $a_{11} = \partial F_1/\partial\tau_n = \dfrac{I(e-1)}{kT_I} + 1$, $a_{12} = \partial F_1/\partial i_n = \dfrac{1-g}{skT_I}$, $a_{21} = \partial F_2/\partial\tau_n = -esI$, $a_{22} = \partial F_2/\partial i_n = g$, $e = \exp[s(T - \tau)]$.

Solution of the system (44) converges always after two steps independently of the initial conditions.

The first equation of (44) differs from (22) by presence an identical multiplier $1/k$ at coefficients h_1 and h_2, therefore it is not necessary to calculate them. Thus, using the expressions (27) we can write down at once

$$\begin{cases} h_1 = k(a_{11} + a_{22} - a_{11}a_{22} + a_{12}a_{21}); \\ h_2 = k\left[a_{22}^2(1 - a_{11}) + a_{12}a_{21}(1 + a_{22})\right]/a_{21}. \end{cases} \quad (45)$$

The obtained coefficients are functions of all system parameters. Some parameters (for example, pulse duration τ) can vary during the system operation. System variable τ relates with reference signal x by an expression for an average load current, following from (39)

$$I_d = \frac{1}{T}\int_0^T i(t)dt = I\frac{\tau}{T} = x\frac{T_I}{T}. \quad (46)$$

By taking into account (46) relations (45) can be presented in the form

$$h_1(I_d) = 0, \quad h_2(I_d) = 0. \quad (47)$$

Relations (47) can be realized by software or hardware way. The simplest way of reproducing these relations are to write their in a memory unit. Implementation of these dependences allows to achieve a convergence of transient in a neighborhood of a fixed point within two steps, thereby, to provide system speed of response close to the theoretical maximum even at the high deviations.

IV. COMPUTATIONAL EXPERIMENT

Theoretical statements are confirmed using MATLAB environment to make computational experiment. The special program has been developed to solve a system of nonlinear difference equations. The converter parameters used for modeling are given in the table I.

TABLE I.
CONVERTER PARAMETERS

U (V)	L (H)	R (Ω)	T (s)	T_I (s)	k	I_d (A)
12	10^{-3}	1	10^{-4}	10^{-3}	10^5	6

The results of common solution (38) and (43) are presented in the form of curves which are shown in the Fig. 2 and Fig. 3 as the functions $i(n)$, where i is a load current at the instant t_n (see Fig. 1b), i.e. $i(n) = i_n$. These curves correspond to a transient resulting from step of the reference signal x.

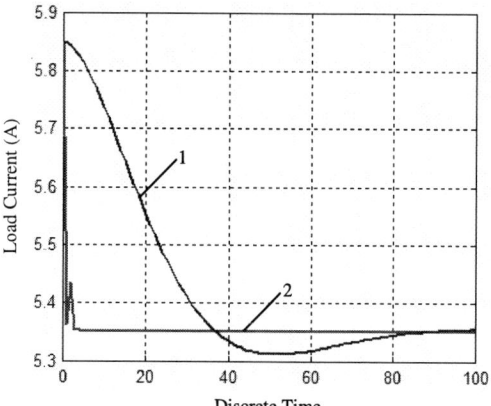

Fig. 2. Load current at step of reference signal from $x = 0,6$ to $x = 0,55$.

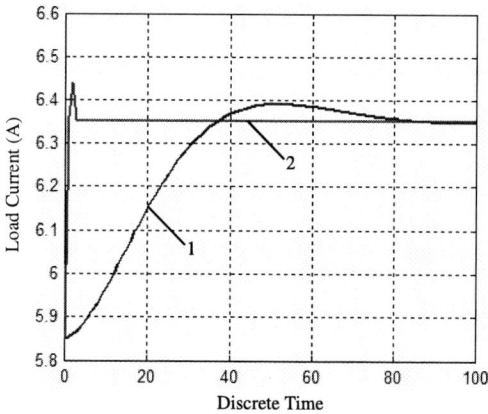

Fig. 3. Load current at step of reference signal from $x = 0,6$ to $x = 0,65$.

The curve of solution of original system (41) is designated by digit 1. The curve of common solution of equations (38) and (43) labeled by digit 2 corresponds to the system with optimal control.

In accordance with (46) desired load current value $I_d = 6$ (A) is provided by setting the reference signal $x = 0,6$. In the plots the current value is lesser than average value because $i(n)$ is used instead of $I_d(n)$, which value can be recalculated for n-th interval by the formula

$$I_d(n) = I_{dn} = I\frac{\tau_n}{T} + \frac{i_n(g-1) + I(e_n - g)}{sT}.$$

The plotted curves show that the proposed control method provides fast time response under small perturbations from steady-state condition. In particular, for system with optimal control the transient completes after 3 discrete time intervals, while for original system the transient period is about 80 discrete time intervals.

Such perturbations are small since the duration of control signal τ_n remains inside the interval $[0; T]$ during entire transient.

This control method can be also used to control of systems, which operate under large perturbations. For that aim, the range of variation of the control signal τ_n should be limited in a forced manner.

V. CONCLUSIONS

A control method for a system described by the second-order linear difference equation with formation a feedback signal in the form of a linear combination of differences of variables, allows to achieve convergence of transient to a fixed point within two steps, for what an equation matrix should be a nilpotent. A transformation of an equation matrix to a nilpotent form can be implemented by various ways. One of them is reduced to a problem of the solution of system of the linear algebraic equations regarding the multipliers of differences.

Application of this principle for control of system with switches, described by the nonlinear difference equation, also gives the same effect that is the process of convergence to a fixed point in its neighborhood occurs within minimum number of steps practically. In this case, multipliers of differences are chosen by transformation of the matrix of a linearized system equation to the nilpotent form.

A control principle for system with switches, in which a matrix is transformed to the nilpotent form, is a basis for design regulators forming course of transients with the maximum speed of response. This property remains under regulation and perturbations of system parameters.

The described system has an unique combination of high static accuracy and practically maximum speed of response in any point of a regulation range. In view of this fact, it is possible to recommend control systems for application in advanced devices for which fast response and accuracy of current control is necessary, for example, for electro-arc installations, electromagnetic suspension systems and others.

The results of computational experiment validated the theoretical statements.

REFERENCES

[1] V. V. Voevodin and Yu. A. Kuznetsov, *Matrixes and calculations.* Nauka, Moscow, 1984.

[2] R. Horn and C. Johnson, *Matrix Analysis*, MIR, Moscow, 1989.

[3] A. S. Iskhakov, V. Ya. Pospelov and S. M. Skovpen', "A method of invariant transformation of difference equations applied to the problem of stabilizing the voltage of controlled rectifiers," *Electrical Technology Russia*, no. 2, pp. 35-49, February 2004.

[4] A. S. Iskhakov, "A new control method for a system with switches," *Electrichestvo*, no. 12, pp. 50-58, December 2005.

[5] A. Strashak, *Controllability.* Sov. Entsiklopediya, Moscow, 1972.

[6] A. A. Krasovsky, *The Automatic Control Theory Handbook.* Nauka, Moscow, 1987.

[7] R. Izerman, *Digital Control Systems.* MIR, Moscow, 1984.

[8] E. Jury, *Pulsed Automatic Regulation Systems.* Fizmatgiz, Moscow, 1963.

[9] A. S. Iskhakov, E. S. Novikov, V. Ya. Pospelov, S. M. Skovpen' and L. M. Klyachko, "A mathematical description of control method on the basis of difference correction," *Proc. 12th European Conference on Power Electronics and Applications EPE 2007*, Aalborg, Denmark, CD-ROM Paper 0051, 2007.

Experiment results with modified Hybrid PWM method for three phase induction motor

Daniel Lewandowski, Grzegorz Lisowski

Technical University of Lodz/Institute of Automation Control, Łódź, Poland

e-mail: *daniel.lewandowski@p.lodz.pl*, *grzeslis@p.lodz.pl*

Abstract—This paper describes the experiment results with noise reduction in induction motor drive fed by the voltage inverter. The tests were conducted using the Hybrid PWM method proposed in [1]. The HPWM reduces computational overhead in space vector computation and introduces k_0 factor. The proper shaping of k_0 allows either reducing switching losses in power devices or minimizing current ripple in the load. However in some cases modulation of k_0 is ineffective. The authors proposed modification in PWM realization to improve silencing of the induction motor.

Index Terms—Induction motors, Pulse width modulated inverters, Pulse shaping methods, Random noise, Acoustic noise

I. INTRODUCTION

The voltage inverters supplying induction motor are very common in an industry. High reliability, economic usage and almost service-free running are the positive characteristic of this drive system. The space-vector method (SVPWM) operates in a complex plane divided in the six sectors separated by the switching state vectors $v_{x,x\in\{0,..,7\}}$ - figure 1. These vectors are defined by a combination of conducting / non-conducting state in the power modules of the inverter – figure 2. The voltage vector in a complex plane is used to select two adjacent switching – state vectors. On figure 1 the voltage space vector v is located in the sector 1 between vectors v_4 and v_6. Based on this information coefficients α, β, γ are calculated to fulfill the equations (1). The factors α, β are the duration of modulation the proper switching - state vectors during PWM interval. The factor γ is the total duration of the zero vectors $v_0 + v_7$. The final result of the method is the vector of pulse widths p.

$$v = \begin{bmatrix} v_\alpha \\ v_\beta \end{bmatrix} = \alpha v_4 + \beta v_6 + \frac{1}{2}\gamma v_0 + \frac{1}{2}\gamma v_7$$

$$p = \begin{bmatrix} \alpha + \beta + \frac{1}{2}\gamma \\ \beta + \frac{1}{2}\gamma \\ \frac{1}{2}\gamma \end{bmatrix} \quad (1)$$

$$\alpha + \beta + \gamma = 1$$

$$\alpha, \beta, \gamma \geq 0$$

The Hybrid PWM method (HPWM) [1] allows fast calculation of p vector. In comparison to the standard SVPWM method it also introduces k_0 factor which controls the ratio of duration of the vector v_7 and v_0. The sum of these durations remains constant. The pulse

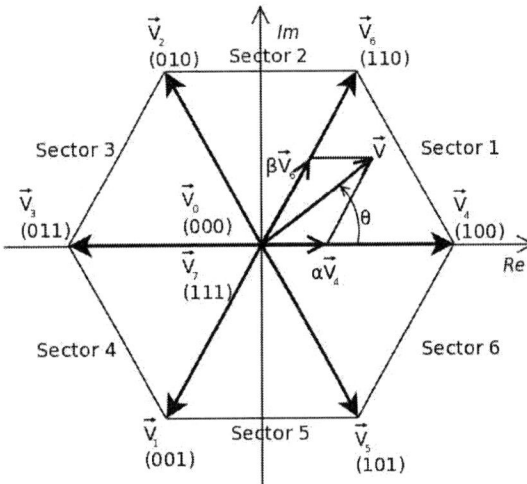

Fig. 1: The definition of switching-state vectors in the complex plane

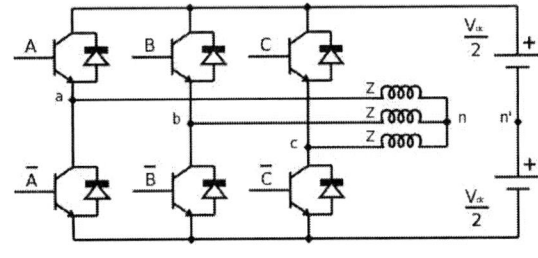

Fig. 2: The three phase inverter with induction load controlled by three signals A, B, C.

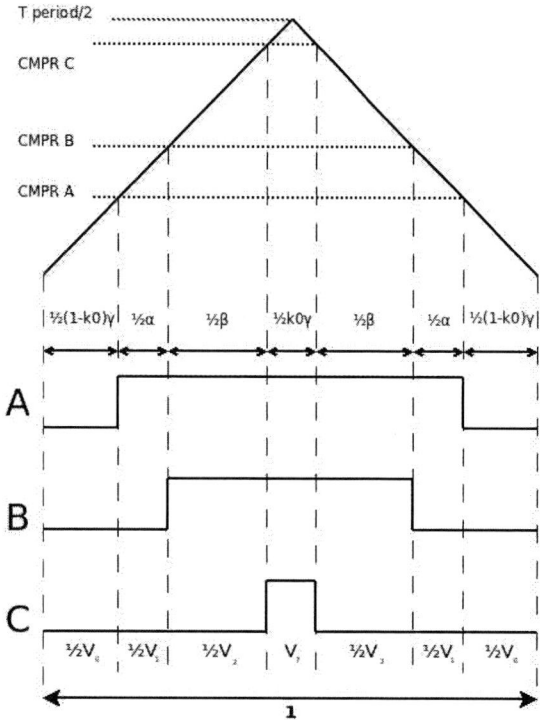

Fig. 3: The relation between p and α, β, γ

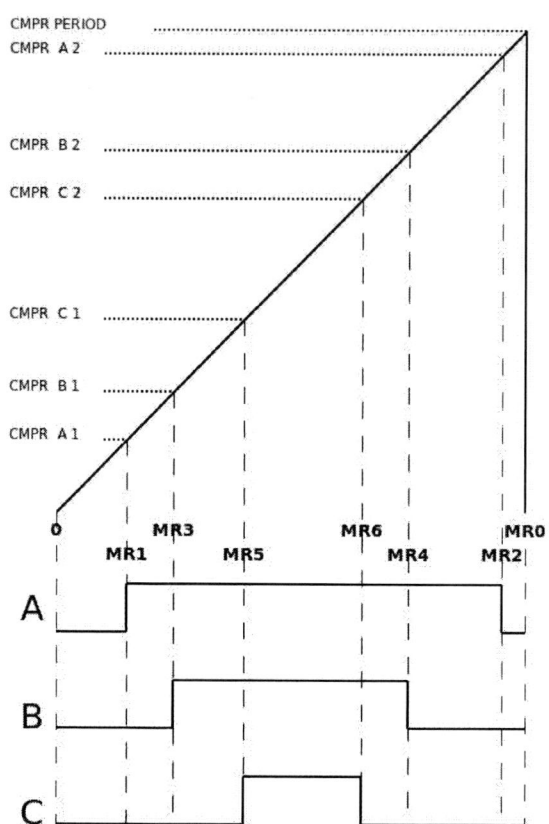

Fig. 4: Dual Edge PWM in LPC2148 microprocessor

widths vector p for the space vector v in first sector can be calculated according to the equations (2). In this paper, classical SVPWM and HPWM was compared to new algorithm Dual Edge PWM (DEPWM) with random vector r.

$$V = \begin{bmatrix} V_a \\ V_b \\ V_c \end{bmatrix} = \begin{bmatrix} v_\alpha \\ -\frac{1}{2}v_\alpha + \frac{\sqrt{3}}{2}v_\beta \\ -\frac{1}{2}v_\alpha - \frac{\sqrt{3}}{2}v_\beta \end{bmatrix}$$

$$v_{zs} = 2k_0 - 1 - k_0 \max_{i \in [a,b,c]} V_i - (1 - k_0) \min_{i \in [a,b,c]} V_i$$

$$p = \begin{bmatrix} p_A \\ p_B \\ p_C \end{bmatrix} = \begin{bmatrix} V_a + v_{zs} \\ V_b + v_{zs} \\ V_c + v_{zs} \end{bmatrix}$$

$$k_0 \in [0, 1]$$

(2)

II. HPWM

The majority of microprocessors gives the programmer choice between symmetric or asymmetric PWM with register controlling filling of pulse. In case of the constant PWM frequency the only parameter is the pulse width vector p. The relation between p and α, β, γ is shown on figure 3.

HPWM method proposed in [1] gives k_0 parameter which controls the duration ratio of zero vectors. This parameter can be changed every PWM cycle and it can even have the random value from the unity range $[0, 1]$ [2].

However HPWM has some limitation in over-modulation condition of inverter.

III. DUAL EDGE PWM (DEPWM)

Some microprocessors like LPC2148 has interesting hardware solution of PWM generator. One PWM channel has two match registers [5]. Three channels give six match registers (CMPR_A1, CMPR_A2, CMPR_B1, CMPR_B2, CMPR_C1, CMPR_C2) and seventh one to control PWM duration (CMPR_PERIOD). The match register event can be coupled with actions like set, clear or toggle PWM channel output. This solution gives more capabilities in realising PWM scheme, because it allows flexible placing of pulses in PWM cycle. The exemplary DEPWM cycle is shown on figure 4.

In general Dual Edge PWM can be described by the pulse widths vector, defined by equation (1) or (2), and the pulse positions vector (3). In case when v is in the first sector there is fulfilled inequality $p_A \geq p_B \geq p_C$ and the match registers can be written according to the equation (4). For the rest of the sectors the similar equations can be written.

670

$$r = \begin{bmatrix} r_A \\ r_B \\ r_C \end{bmatrix} \qquad (3)$$

$$r_A, r_B, r_C \in [0, 1]$$

$$MR1 = (1 - p_A) r_A MR0$$
$$MR3 = MR1 + (p_A - p_B) r_B MR0$$
$$MR5 = MR3 + (p_B - p_C) r_C MR0$$
$$MR2 = MR1 + p_A MR0 \qquad (4)$$
$$MR4 = MR3 + p_B MR0$$
$$MR6 = MR5 + p_C MR0$$

The p vector is remarkably determined by the space vector v regardless of used method (1, 2). The advantage of the Dual Edge PWM is the flexibility of setting the components of the pulse positions vector r. The proper voltage vector v is always generated independently from the value of the r. This freedom of r implies that r can have random values from unity range.

It is obvious that DEPWM can emulate SVPWM ($k_0 = 0.5$ and $r = [0.5, 0.5, 0.5]^T$) and HPWM ($k_0 = rand()$ and $r = [0.5, 0.5, 0.5]^T$). This property of DEPWM allows comparison of all these method in silencing the induction motor.

IV. Experiments

Experiments were conducted with 0.55kW induction motor fed by 5kW voltage inverter. The inverter worked at 2.5kHz frequency with $1\mu s$ dead time. The load was the steel wheel mounted on the motor shaft. The authors considered four conditions of the induction motor work:

- accelerating with torque $\tau = 1$
- 0.8 nominal speed with $\tau = 0$
- braking with $\tau = -1$
- 0.8 nominal speed with $\tau = 0$ and 50% link voltage (over-modulation region)

DEPWM capabilities allows for comparison of three PWM generation schemes:

- SVPWM

$$k_0 = 0.5 \quad r = \begin{bmatrix} 0.5 \\ 0.5 \\ 0.5 \end{bmatrix} \qquad (5)$$

- HPWM with random k_0

$$k_0 = rand() \quad r = \begin{bmatrix} 0.5 \\ 0.5 \\ 0.5 \end{bmatrix} \qquad (6)$$

- DEPWM with random k_0 and r

$$k_0 = rand() \quad r = \begin{bmatrix} rand() \\ rand() \\ rand() \end{bmatrix} \qquad (7)$$

The standard SVPWM is commonly used in the industry inverters. However it uses fixed switching frequency which results in predetermined multiples harmonic components. As it is shown on figures 5a, 6a, 7a, 8a there is a lot of the harmonic components which are present at all considered conditions. The components below 20kHz are responsible for audible noise. The total energy of the noise can not be minimised but it can be spread to wider spectrum.

The HPWM scheme (6) introduced random components which leads in variation of base switching frequency. This strategy spread energy of harmonic components into a new ones. The total energy carried by all components is not changed but the peaks of them are smaller. This effect is seen on figures 5b, 6b, 7b as flattening of phase A current FFT. In audible form the HPWM improvements can be heard as noise without clear frequency.

However HPWM effectivenes is limited when the inverter is working in over-modulation region. In case of narrow DC link voltage there is often a situation when PWM pulses does not have zero vectors. This results in no effect of k_0 variation. On figure 8b can be seen a very little change of current spectrum in comparison with figure 8a.

The authors propose the scheme (7) to overcome this problem. During over-modulation DEPWM changes the position of middle pulse which results in flattening of the current spectrum shown on figure 8c. In comparison to figure 8b there is only small first harmonic component. In rest of considered conditions the results of DEPWM are slightly better in comparision to HPWM - figures 5c, 6c, 7c.

V. Conclusion

Space vector PWM algorithm was modified to enable introduce random vector r (DEPWM). This algorithm is useful to easily achieve different PWM schemes described in literature. It made PWM universal and flexible, but DEPWM is suitable only for microprocessors with two compare register per phase. The experiments has proved that randomizing of r vector and/or k_0 leads to flattening of frequency spectrum of current in comparison with standard SVPWM method (5). In contrast to HPWM proposed modification changes the spectrum of the current even in over-modulation region.

References

[1] Vladimir Blaskos, *Analysis of Hybrid PWM Based on Modified Space-Vector and Triangle-Comparison Methods*, IEE Transactions on industry application, vol. 33, no. 3, May/June 1997

[2] Vladimir Blaskos, Michael M. Bech, Frede Blaabjerg and John K. Pedersen *A New Hybrid Random Pulse Width Modulator for Industrial Drives* Applied Power Electronics Conference and Exposition, 2000. APEC 2000. Fifteenth Annual IEEE vol. 2,

[3] Henrik Kragh, Frede Blaabjerg, John K. Pedersen, *Reduce of the acoustic noise effect from PWM-VSI inverter controlled ac-drives by music and random modulation*, PCC – Yokohama 1993

[4] S.-H. Na, Y.-G. Jung, Y.-C. Lim, S.-H. Yang, *Reduction of audible switching noise in induction motor drives using random position space vector PWM* IEE Proceedings Electrical Power Applications, vol. 149, no. 3, May 2002

[5] Philips Electronics, *LPC214x User Manual*, August 2005

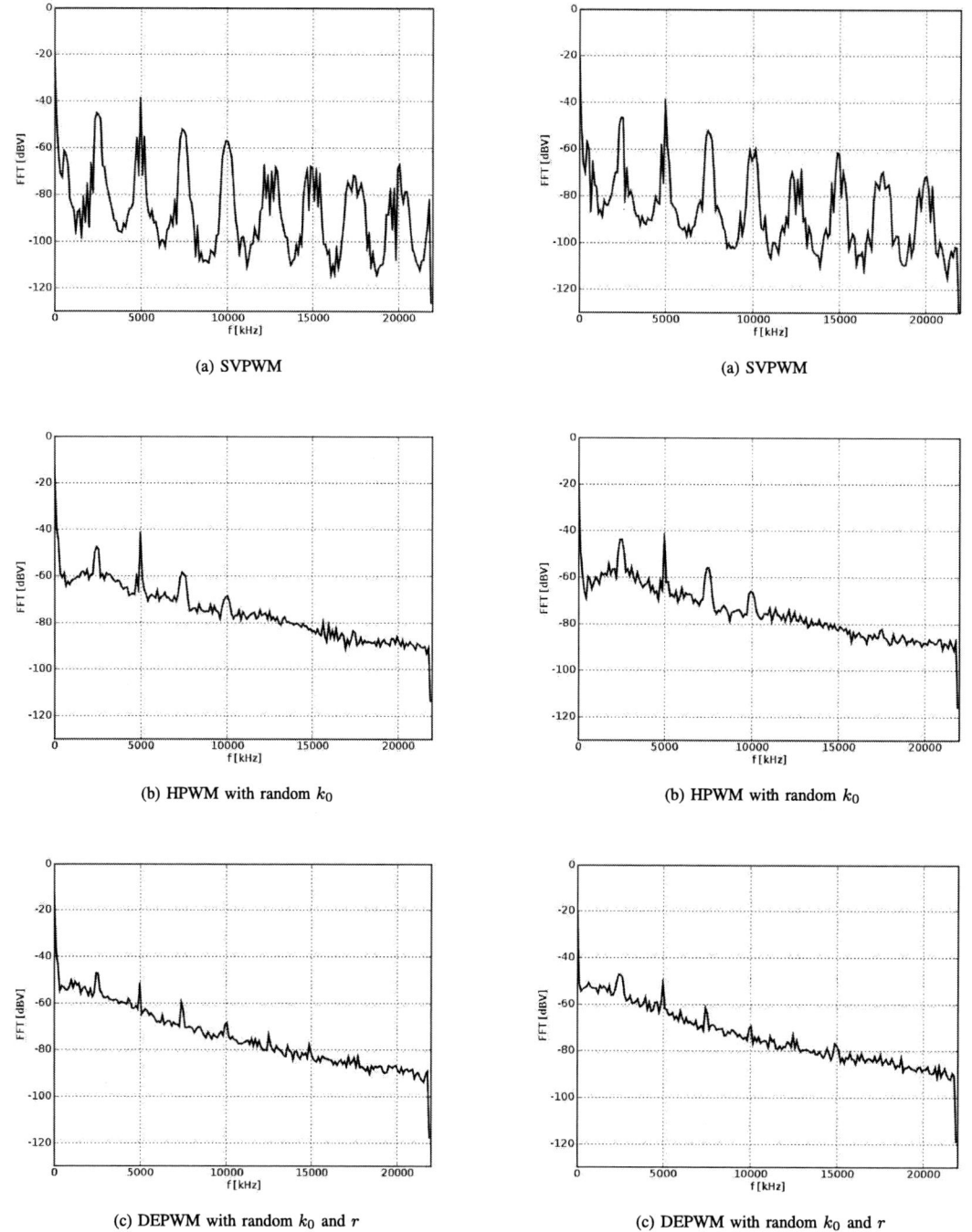

(a) SVPWM

(a) SVPWM

(b) HPWM with random k_0

(b) HPWM with random k_0

(c) DEPWM with random k_0 and r

(c) DEPWM with random k_0 and r

Fig. 5: Phase A current FFT during accelerating with $\tau = 1.0$

Fig. 6: Phase A current FFT at 0.8 nominal speed with $\tau = 0$

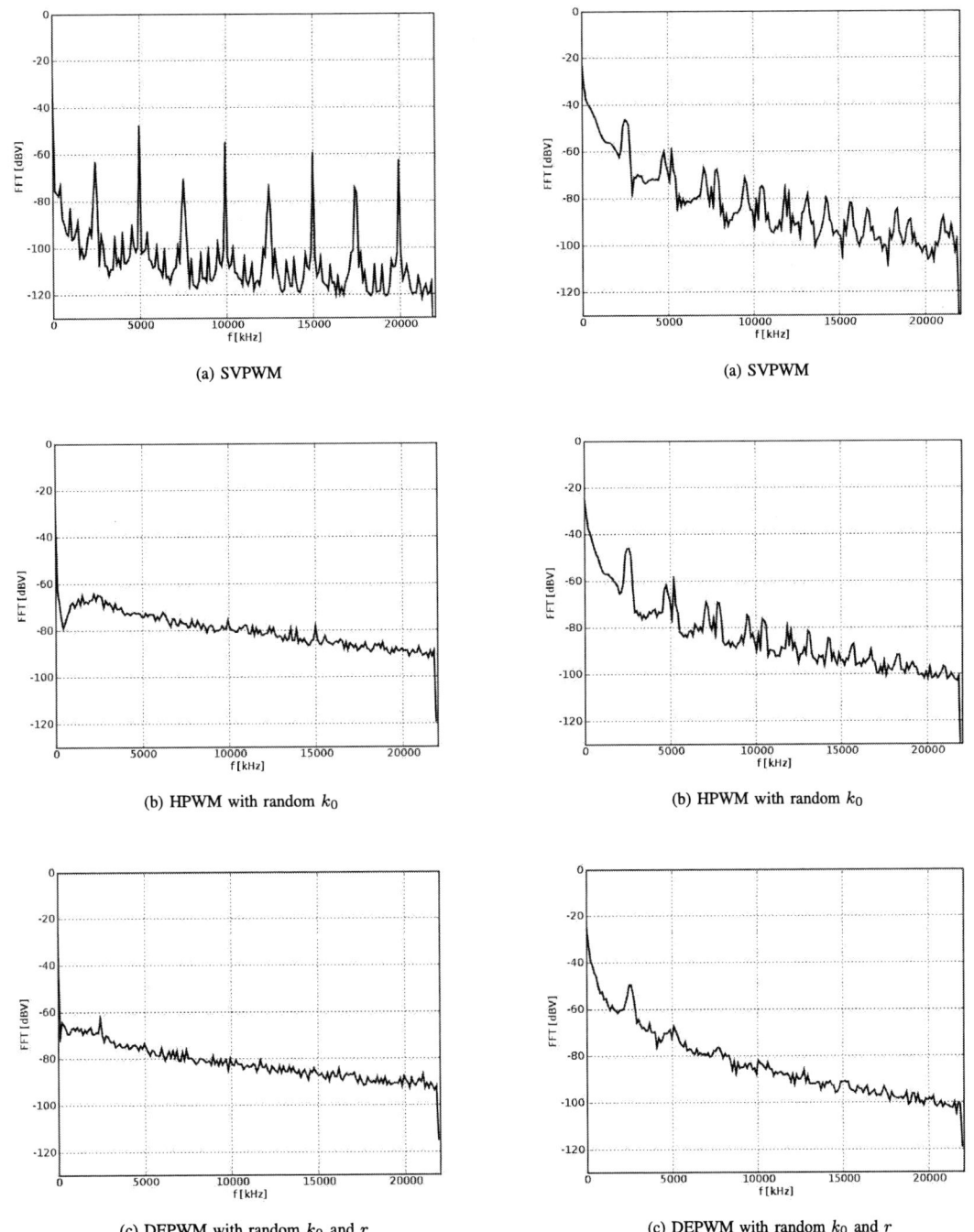

(a) SVPWM

(a) SVPWM

(b) HPWM with random k_0

(b) HPWM with random k_0

(c) DEPWM with random k_0 and r

(c) DEPWM with random k_0 and r

Fig. 7: Phase A current FFT during braking with $\tau = -1.0$

Fig. 8: Phase A current FFT at 0.8 nominal speed with $\tau = 0$ and 50% of link voltage

Optimized Design of a Delay line based Analog to Digital Converter for Digital Power Management Applications

Mukti Barai, Sabyasachi Sengupta
Department of Electrical Engineering
Indian Institute of Technology Kharagpur
{mukti, ssg} @ee.iitkgp.ernet.in

Jayanta Biswas
CEM Solutions
Bangalore
jayanta@cem-solutions.net

Abstract—The proliferation of mobile electronic equipment is driving the need for aggressive real-time power management techniques, beyond the incremental efficiency improvements in DC-DC switching converters. In *dynamic voltage scaling (DVS)* power managament technique, performance, *i.e.* the operating clock frequency is adjusted with the time-varying workload and the supply voltage is scaled down dynamically with the clock frequency to meet the specific performance requirements. A new class of Analog-to-Digital Converter (ADC) architecture is a challenging requirement in DVS power management implementation.

This paper presents an optimized hardware design of a delay line based ADC to meet the requirements of DVS power management, under direct performance control over a wide range of clock frequency. A novel formulation of digital error value based on target clock frequency and the corresponding regulated output voltage is presented with optimum hardware. Support for process, voltage, temperature (PVT) variations is incorporated in the design framework.

I. INTRODUCTION

Integration of different fast processing applications in a single portable device, such as laptops with specialized computing, requires the portable system to run with higher clock frequency. Consequently, the portable device needs high supply current and power consumption of the portable device increases. Power consumption in digital circuits is proportional to the square of the supply voltage V_{dd} [1]. *Dynamic Voltage Scaling* (DVS) is a popular power management technique used in digital systems. DVS scales down the processor supply voltage until the processor just meets the specific performance requirements. Popularity of DVS for digital processing ICs in portable applications, presents a set of technical challenges to the digital power controller design, such as efficiency, static and dynamic regulation accuracy and fast dynamic response. New architectures of digital power controllers are needed for low power, high frequency portable applications to meet these challenges.

Digital power controller in a closed loop with DC-DC converter is shown in Fig. 1. In this digital control implementation, an Analog to Digital Converter (ADC) samples regulated output voltage, V_{out} in every switching period. Digital error value, $e[n]$, is generated by comparing sampled values for V_{out} with the voltage which corresponds to the

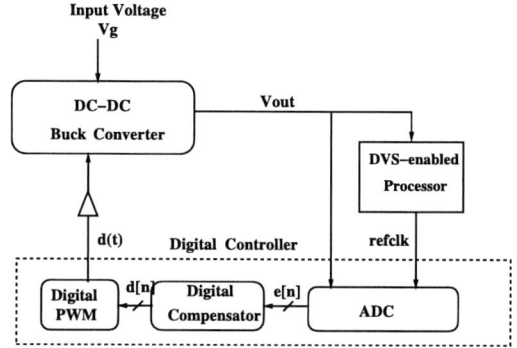

Fig. 1. Digital controller in closed loop with DC-DC Converter

frequency of *refclk* signal. A fast digital compensator block computes the digital control command, $d[n]$ from $e[n]$ and a Digital Pulse-Width Modulator (DPWM) converts $d[n]$ into an analog pulse width modulated signal, $d(t)$. Each module of the digital controller architecture needs to be optimized in terms of chip area and power consumption, so that integrated digital controller becomes faster and efficient in power management digital applications.

In a voltage mode digitally controlled switched-mode converter, the resolution of the ADC determines the precision of the converter regulated output voltage, V_{out}. In practical situations, the controlled output voltage V_{out} remains normally close to the reference. This calls for an ADC architecture which has high resolution in a small window, specially around the reference. This architecture is known as *Window ADC*. Concentration of high resolution only around the reference permits the *Window ADC* to have a small footprint. Consequently, switching power consumption is reduced, which is desirable for high frequency applications. *Window ADC* is also fast and robust against switching noise.

Window flash ADC is the fastest ADC as the conversion takes place in one cycle [2]. However, Window flash ADC is inefficient in terms of power consumption and area requirement for higher resolution. The basic window architecture of delay-line ADC as reported in [3], [4] are not suitable for DVS enabled digital systems architecture where input ref-

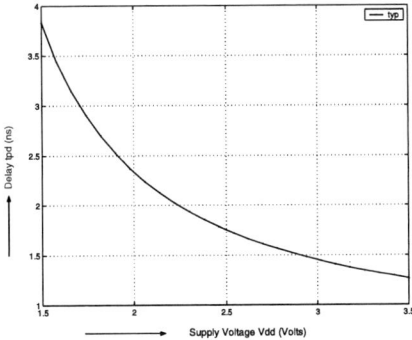

Fig. 2. Delay vs. V_{dd} Characteristics for one delay cell in $0.5\mu m$ CMOS

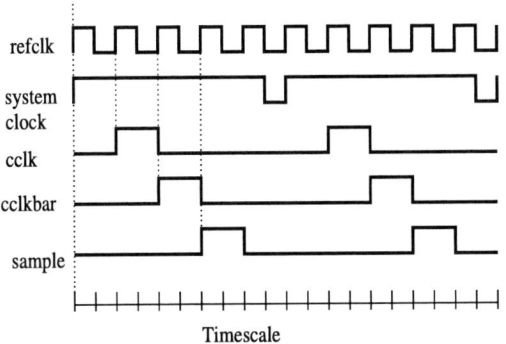

Fig. 3. Timing diagram of derived signals from *refclk*

erence corresponds to the frequency of control signal. The ring-oscillator based ADC structure [5] comprises of an analog and a digital block, consumes more power and silicon area and is not completely synthesizable. The work reported in [6] implements PVT support for finite state, delay-line based window ADC, accepting input command as reference clock and compromises with the maximum clock frequency of the controller.

In this paper, a novel delay-line based *Window ADC* is proposed where input reference is taken from a signal with varying frequency. Optimization is carried out to reduce the ADC footprint and to increase the maximum operating frequency of the proposed ADC. DVS enabled processor outputs a signal whose frequency is the reference input to the delay-line based ADC. The proposed ADC compares the converter regulated output voltage with the voltage which corresponds to the frequency of the reference signal. An efficient, polarity-sensitive digital error computation is executed which minimizes delay-line ADC latency. Simplification of conversion and comparison permits significant hardware reduction. Based on the piecewise-linear delay characteristic of the delay line with respect to delay line supply voltage V_{dd}, multiple band concept is introduced in this work to improve the output voltage regulation in steady state.

The principle of delay line ADC operation is described in section II. Section III explains design implementation and a novel digital error formulation. Section IV illustrates post

layout simulation and experimental results. Section V provides performance comparison with existing literature and section VI concludes the presentation.

II. PRINCIPLE OF FREQUENCY INPUT DELAY LINE ADC OPERATION

The conversion characteristics of a delay line ADC is strongly influenced by the voltage-dependent propagation delay of a standard CMOS logic gate. The propagation delay time, t_{pd} of the logic gate is a function of its supply voltage V_{dd} and can be described by

$$ t_{pd} = \frac{K.V_{dd}}{(V_{dd} - V_{th})^\alpha} \qquad (1) $$

where V_{th} is the CMOS device threshold voltage, and K is a constant that depends on the device/process parameters and the capacitive loading of the gate. The value of α also depends on process parameters. The t_{pd} of a logic gate increases, if the gate supply voltage is reduced. Increasing supply voltage V_{dd} results in a shorter delay. The delay versus voltage characteristic of a delay cell used for the proposed delay line ADC is shown in Fig. 2. In delay line based ADCs, the digital conversion utilizes this variation in propagation delay across delay cells with respect to the supply voltage V_{dd}. Initially, a test signal pulse propagates along the delay line. After a fixed time, delay line taps are sampled to obtain digital output in thermometer code (sequence of 1's followed by 0's). The sampled value obtained, corresponds to the supply voltage, V_{dd} of the delay line. The sampled values of the delay line ADC depend on the delay versus voltage characteristics of the delay cell and the length of the delay line. Here, the power converter output voltage, V_{out} is the supply voltage, V_{dd} of the delay line.

The power converter output voltage, V_{out} is so regulated that the propagation delay that a signal takes to reach the middle tap of a delay line, is made equal to single period of the *refclk*. As shown in Fig. 1, digital controller accepts a clock frequency *refclk* as a reference input, obtained from the DVS processor. The supply voltage, V_{out} of the delay line is precisely regulated to provide the voltage corresponding to *refclk* frequency. The delay to reach the middle tap of delay line, is made equal to one clock period of *refclk* and the V_{out} is regulated to maintain the same delay for a given *refclk* frequency. The delay versus supply voltage V_{dd} characteristics of one delay cell for the proposed design in $0.5\mu m$ CMOS process is shown in Fig. 2.

Timing diagram of the signals derived from the reference clock *refclk* signal is shown in Fig. 3. The DVS processor is projected to generate different *refclk* depending on the activity of the portable system. jay

The proposed delay line ADC is self-calibrated for output regulated voltage variations since reference clock frequency variation is considered in delay line design. Temperature variation is monitored by the processor and reference frequency input command is adjusted for temperature variations. In effect, the delay of the proposed delay line ADC scales linearly with load and frequency in presence of PVT variation.

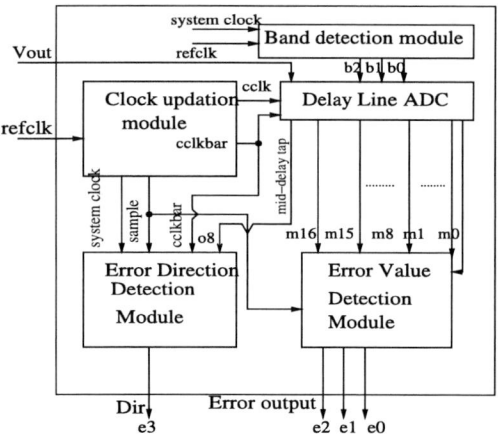

Fig. 4. Block Diagram of ADC

A. Piecewise-linear Delay Characteristics

Non-linear delay characteristic of a delay line is adequately approximated by piecewise-linear segments on a bounded domain of regulation window. The width of these segments is sufficiently small. Let us consider that the delay T_d of a delay line is approximated in this way, i.e.

$$T_d = m.t_d + k \qquad (2)$$

where, t_d is delay of each delay cell and m is the number of delay cells in each segment of the delay line, k represents constant delay for zero regulation window. For each possible values of m, the piecewise-linear approximation matches T_d within regulation window and is continuous. Given this piecewise-linear approximation of T_d, the corresponding piecewise difference of T_d simply depends on m, the number of delay cells. The value of the design parameter m depends implicitly on T_d in this formulation. Evaluation of T_d would then require just one multiplication with number of delay cells and one addition for initial delay. Hence, approximate evaluation of difference in T_d value does not require any computation. It is natural to consider the use of the piecewise linear expressions to estimate delay of a delay line.

Therefore, the piecewise linear expressions for delay of the delay line within a small range of voltage change in presence of PVT variation is appropriate. Piecewise linear characteristics of the delay line offers the opportunity to employ multiple overlapped band concept to increase the resolution of a delay line window ADC. Multiple overlapped band improves the output voltage regulation in steady state. Each band corresponds to a small range of voltage. ADC interface remains unchanged as multiple band concept is implemented within the ADC itself. This concept also overcomes the worst case non-linear delay effect of the delay line.

III. DESIGN IMPLEMENTATION

The block diagram of the proposed ADC is shown in Fig. 4. It contains five sub-modules namely the *clock updation*, *band detection*, *delay line ADC*, *error direction detection* and *error amount detection* modules. The sub-modules are described below.

A. Clock Updation Module

The *clock updation* module accepts the frequency input command, *refclk*, of 50% duty ratio from the DVS processor. The *refclk* frequency is measured utilizing a standard high frequency clock of 50 MHz or higher and a *system clock* signal is derived by simple adjustable division which is equal to the constant switching frequency of the converter, here at 1 MHz. The *system clock* signal synchronizes the complete operation of the digital controller. The measurement of *refclk* frequency is performed over a few *refclk* cycles when *system clock*, whose frequency is significantly lower, remains low and *refclk* remains unchanged. The frequency measurement accuracy is realized by performing it over a multi-pulse period of *refclk*. If the frequency of *system clock* nears *refclk* frequency, single cycle measurement is used for calculating period of the *refclk*. The *clock updation* module also derives the timing signals *cclk*, *cclkbar* and *sample* for proposed delay line ADC operation. The *cclk*, *cclkbar* and *sample* signals contain one high pulse during one period of *system clock*. *cclk* is the derived signal from *refclk* and positive edge of *cclk*, *cclkbar*, *sample* matches with the positive edge of *refclk* as shown in timing diagram Fig.. 3. The difference between the positive edge of *cclk* and *cclkbar* is equal to one clock period of *refclk* signal. Similarly, difference between the positive edge of *cclkbar* and *sample* is equal to one clock period of *refclk* signal. The width of both *cclk* and *cclkbar* derived signals equals to the one clock period of the *refclk* signal. The *cclk* signal is used as an ADC *start* pulse and traverses through the delay line. Delay tap outputs are sampled at the positive edge of the *cclkbar*. The ADC conversion is performed only once in every switching period for the calculation of digital error.

B. Band Detection Module

The *band detection* module identifies the reference band corresponding to the target input *refclk*. Multiple overlapped-band concept is introduced based on the piecewise-linear characteristics of the delay line. The *band detection module* is activated when *system clock* is low. The band detector output signals (b2 b1 b0) are fed to the *delay line ADC* module along with the *cclk* signal. The length of the delay line between the delay taps are adjusted to provide linear resolution based on (b2 b1 b0) across all the bands, whereas the number of delay tap outputs remains constant. In this proposal, seventeen delay taps are used for implementing eight overlapped bands for a target frequency range of 6 - 16 MHz.

Multiple overlapped-band concept helps to achieve precise quantization step for the proposed window ADC design over a wide regulated voltage range of 1.6 - 3.2 V. Eight overlapped bands are implemented to differentiate a minimum voltage variation of 12.5 mV over a wide range (1.6 -

676

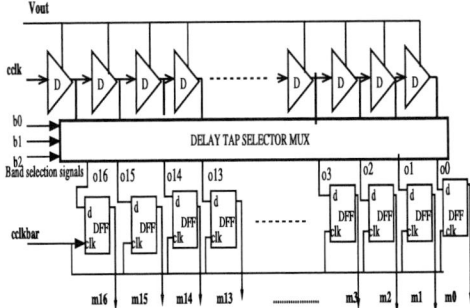

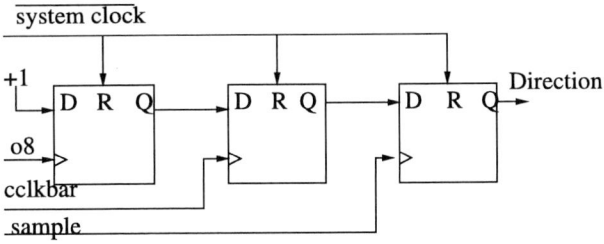

Fig. 7. Logic circuit of the error direction detection

Fig. 5. Schematic of delay line ADC

3.2 V) of regulated output voltage. Seventeen 8:1 multiplexers receive delay-tap outputs corresponding to the seventeen taps of the delay line as shown in Fig. 6. Multiplexer outputs (o16....o0) are sampled with *cclkbar* signal and the sampled outputs (m16.....m0) are fed to the *error value detection module*.

C. Delay line ADC Module

The delay corresponding to the regulated output voltage variation in the window of (1.6 - 3.2 V) is the critical path delay. This delay is affected by delay line loading, which reflects as an additional capacitive loading of the delay line. An *adjustable delay line module* continuously monitors the critical path delay with respect to load, target input clock frequency and PVT variations. Interface of the ADC, however, remains unaffected as the bands are embedded within the ADC itself. The proposed multiple-band delay line ADC provides higher resolution to the regulation of the output voltage. The schematic of the proposed delay line ADC is shown in Fig. 5.

The mid-length delay of the delay line is made equal to the one clock period of the target frequency, *refclk*. The converter output voltage is such regulated that the propagation delay that a signal takes to reach the middle tap, *o8* of the delay line, is equal to the period of *refclk* signal. The length of the delay taps is adjusted for each band based on *band detection* block output.

D. Error Direction Detection Module

The error calculation module consists of *error direction detection module* and *error value detection module*. The logic circuit of the *error direction detection* module and *error value detection* module are shown in Fig. 7 and 8 respectively. In each period of the *system clock*, the *error value detection* module computes 4-bit digital error with 3-bits value *e2, e1, e0* and 1-bit direction *e3*.

The error direction detection logic is an important part in generating the error sample values. Here, zero error condition corresponds to the situation when the rising edge of *cclk* signal arrives at the mid-tap of the delay line output of multiplexer number 8 - (*o8*), at the sampling instant, *i.e.* at the

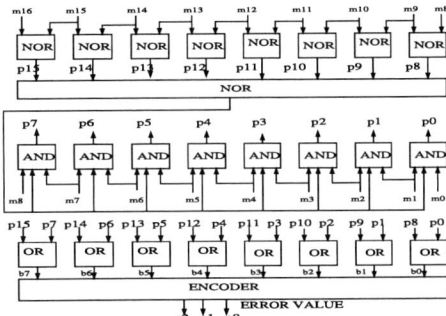

Fig. 8. Logic circuit for error value detection

rising edge of *cclkbar* signal. This implies that a regulated output voltage, V_{out} is achieved for the corresponding target frequency input *refclk*. The supply voltage, V_{dd} determines the delay of the delay line and therefore any change in V_{out} with respect to the target clock frequency, *refclk*, is indicated as an error. The rising edge detection at the middle tap, *o8* output solves the most challenging part of the error direction detection logic. There is a possibility when no rising edge is detected from the delay tap outputs and all the sampled values of tap outputs remain low. The following two cases, when no rising edge is detected.

1) delay line supply voltage V_{dd} is within regulated voltage range, but *refclk* frequency is very low.
2) delay line supply voltage V_{dd} is low compared to the threshold voltage V_{th} of 0.6 - 0.7 V.

To separate out the above two cases, the *error direction detection* module takes *system clock, sample, cclkbar* and *o8* as input and generates the *error direction* signal. The proposed solution detects and latches the rising edge of the middle tap, *o8*, output, when the *system clock* signal remains high for every switching period. This logic identifies whether the rising edge of *cclk* signal passed *o8* tap in the current switching period *i.e* after system clock becomes high. We can separate out the above two cases with this novel logic. This is very important in generating the error sample values. This module requires less than 20 logic gates and optimizes the hardware burden.

677

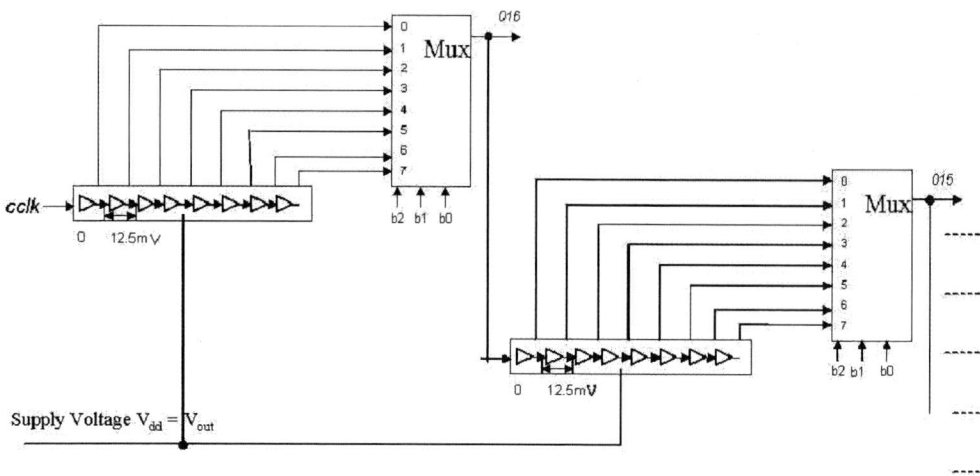

Fig. 6. Multiple Band Delay Tap Selector

TABLE I

INPUT AND OUTPUT VALUES OF THE ERROR VALUE DETECTION
MODULE

$x7$	$x6$	$x5$	$x4$	$x3$	$x2$	$x1$	$x0$	$e2$	$e1$	$e0$
0	0	0	0	0	0	0	0	0	0	0
0	0	0	0	0	0	0	1	0	0	1
0	0	0	0	0	0	1	1	0	1	0
0	0	0	0	0	1	1	1	0	1	1
0	0	0	0	1	1	1	1	1	0	0
0	0	0	1	1	1	1	1	1	0	0
0	0	1	1	1	1	1	1	1	0	0
0	1	1	1	1	1	1	1	1	1	1
1	1	1	1	1	1	1	1	1	1	1

E. Error Value Detection Module

When the frequency of *refclk* becomes high, with V_{out} unchanged, the rising edge of the *cclk* gets detected by the rising edge of *cclkbar* before the mid-length point, *o8*, of the delay line because the period between the rising edges of the *cclk* and *cclkbar* gets smaller with higher clock frequency and the pulse has a smaller time to propagate. This indicates a positive error and any two consecutive sampled outputs from *m16* to *m8* are set to high. Sampled outputs values from *m8* to *m0* need not be attended for the positive error case. If, on the contrary, when the output voltage is higher than the corresponding value of the input target frequency, all of the sampled outputs from *m16* to *m8* remain low, and depending on the error value, the positive edge is detected in between *m8* to *m0*. The negative error value is measured from sampled outputs *m8* to *m0*. If no positive edge is obtained within a system clock period, it corresponds to a maximum value of +/-7. Consequently all higher error values are truncated to a value of +/-7. Digital error values of +/-5 and

+/-6 are truncated to +/-4 value. Zero error condition occurs only when the rising edge of *cclk* is detected by *cclkbar* at the middle *o8* delay tap. Digital error values between 0 to 4 are retained as such. The *error direction detection* module discussed above, decides the polarity of the error. The generated intermediate error values are a stream of 0's and followed by 1's. Intermediate error values are represented by eight bits and the number of 1's in the value represents the error value. Intermediate error values are fed to the three bit binary encoder. Possible input combinations to the encoder used in *error value detection* module of Fig. 8 and the corresponding output is shown in Table I.

F. Process, voltage and temperature (PVT) variation support

In an implementation of the design, process and temperature variations may cause deviations in the output voltage and they may exceed the specified static and dynamic voltage regulation requirement of the power converter. The proposed delay line ADC is self-calibrated for such voltage variations since PVT influenced target frequency variation is considered in delay-line design. The DVS-enabled processor monitors temperature variation and adapts the target frequency. PVT variation monitoring is set as a software task of the DVS-enabled processor where such parameters variations are periodically monitored and target frequency command, *refclk* correspondingly updated. This monitoring is performed once in every second which is much slower compared to *refclk* and hence the processor load overhead does not increase substantially. All temperature related processes having comparatively larger time constants, this solution does not affect overall system accuracy. An ADC is thus realized which scales linearly with load and frequency in presence of PVT variation. The advantage of this approach is that the ADC hardware is not loaded with unnecessary counters, permitting high frequency operation. The pro-

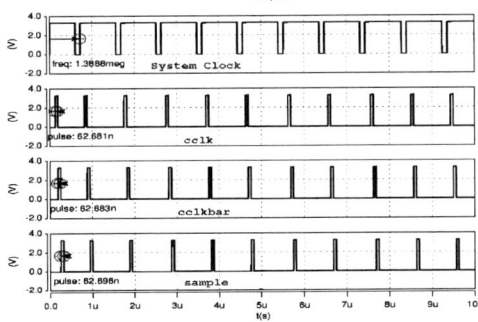

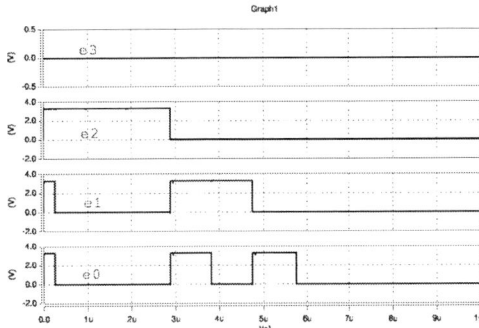

Fig. 9. Derived signals obtained from post layout simulation

posed ADC hardware can operate at a maximum clock frequency of 100 MHz. The finite state based digital controller reported in [6], puts the burden of PVT support in the error detection hardware and compromises maximum frequency of the ADC.

IV. PERFORMANCE RESULTS OF THE IMPLEMENTATION

The proposed optimized design architecture of delay line ADC has been implemented standard 0.5 μm CMOS technology. Synopsis [9] synthesis (Design Vision), timing verification (Prime Time), chip layout (Astro) and post layout chip simulation (Nanosim) tools have been used to implement and verify the functionality of the complete digital controller chip design in 0.5 μm CMOS using technology library of National Semiconductor. Active chip area of the proposed delay line ADC is 0.08 mm^2. The optimized design of delay line ADC has also been realized on an experimental prototype and verified in an integrated digital controller architecture with an 1 MHz synchronous buck converter. The experimental prototype is shown in Fig. 17. The delay line has been fabricated with discrete components. Altera DE2 cyclone-II FPGA platform has been used for multiple band logic, digital error calculation, LUT based PID compensator and an 8-bit edge-triggered hybrid DPWM implementation. FPGA board has been connected to delay line ADC and synchronous buck converter by its I/O interface. The closed loop prototype is tested for 1 Watt, 1 MHz synchronous buck converter with input voltage range of 4-6 Volts and a wide regulated output voltage range of 1.6-3.2 Volts over a target frequency range of 6 - 16 MHz. The output capacitor and inductor value used are 22 μF and 4.7 μH respectively. The synchronous buck converter operates at constant switching frequency to improve its stability over a wide range of target frequency.

A. Simulation results

The proposed ADC is implemented in 0.5 μm CMOS technology using Cadence library of National Semiconductor and Synopsis tools. Proposed implementation takes 0.08 mm^2 chip area.

Fig. 9 depicts the *system clock, cclk, cclkbar* and *sample* timing signals for the proposed delay line ADC operation

Fig. 10. Digital error Output of proposed ADC from post layout simulation

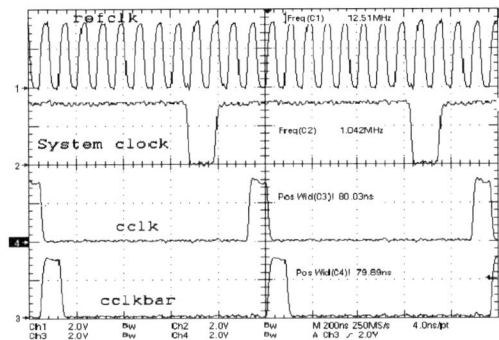

Fig. 11. Signals *System Clock, cclk, cclkbar* derived experimentally from 12.5 MHz *refclk*

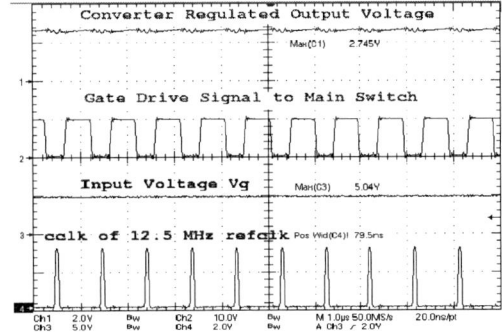

Fig. 12. Experimental results in closed loop steady state operation: Ch1:Converter Regulated Output Voltage; Ch2: Gate drive signal for main switch in synchronous buck converter; Ch3: Input Voltage Vg; Ch4: *cclk* of 12.5 MHz *refclk*

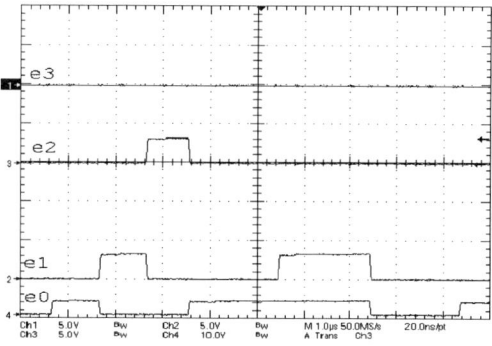

Fig. 13. Digital Error output from the proposed delay line ADC

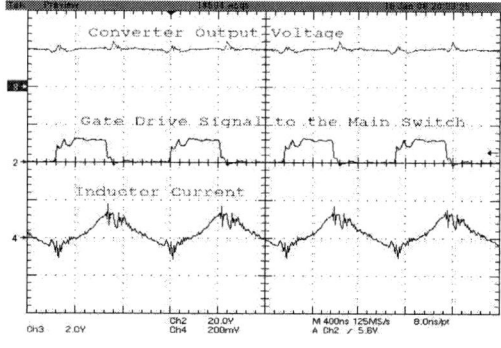

Fig. 14. Experimental results in closed loop steady state operation: Ch3:Converter Regulated Output Voltage; Ch2: Gate drive signal for main switch in synchronous buck converter; Ch4: Inductor current waveform

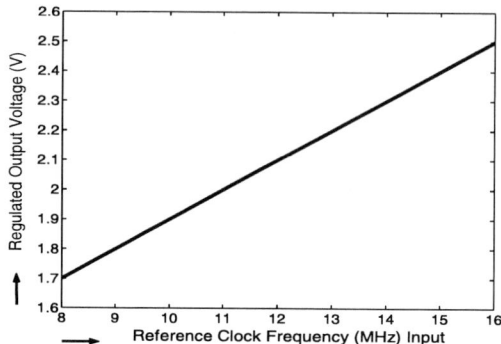

Fig. 15. Delay-line ADC characteristic curve of regulated supply voltage and reference clock frequency input

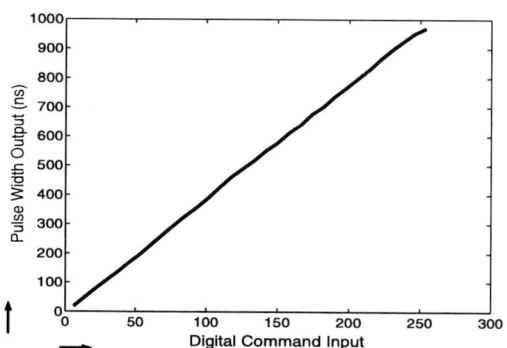

Fig. 16. Measured Pulse-width (ns) versus Digital command input obtained from hybrid DPWM

for a 16 MHz target frequency. The positive pulse width of *cclk*, *cclkbar* and *sample* equals approximately to one clock period 16 MHz frequency, 62.68 ns. The digital error output obtained from the proposed optimized delay line ADC is shown in Fig. 10, which demonstrates the fast computation of digital error for a output voltage change from 2.2 to 2.4 V and error reaches zero at 2.4 V regulated voltage. The optimized delay line ADC occupies only 0.08 mm^2 silicon area and consumes an average current of 6 μA at 2.4 V for the target frequency of 16 MHz. Operation details and performance verification of the proposed optimized delay line based ADC design has been demonstrated experimentally with an integrated digital controller in the next section.

B. Experimental results

Fig. 11 demonstrates the derived timing signals obtained from *refclk* of 12.5 MHz. *cclk*, *cclkbar* and *sample* signals are shifted monoshot version of the input reference command *refclk*.

Fig. 12 demonstrates the steady state closed loop operation for 12.5 MHz *refclk* signal and the corresponding regulated output voltage of 2.352 Volts respectively. Gate drive signal for the main switch of the synchronous buck converter and input voltage are also shown in this figure. The controller provides regulated output voltage without limit cycle

oscillations over a voltage range of 1.6 - 3.2 V. Fig. 13 shows the digital error output obtained from the proposed delay line ADC.

Fig. 14 shows converter output voltage and the corresponding inductor current waveform with main switch control signal. Prototype delay line ADC provides zero error output at 2.1 V of converter regulated output voltage for a reference frequency of 10 MHz.

Fig. 15 demonstrates the linear characteristics between reference clock frequency and the regulated output voltage of the proposed delay line ADC measured from the experimental results. Linearity of an edge triggered hybrid PWM is shown in Fig. 16 with the input digital command value. Prototype experimental setup is shown in Fig. 17.

V. PERFORMANCE COMPARISON

Previous work [7] uses mixed-mode circuitry of analog and digital components to implement DVS and provides non-linear regulated output voltage for target clock frequency. The sampled output of a 3 -tap delay line based ADC is used with a mixed-mode glue logic to generate the controller timing signals for the main switch, auxiliary switch and the output switch of a Watkins-Johnson switching converter accordingly. However, this proposal does not achieve smooth output regulation.

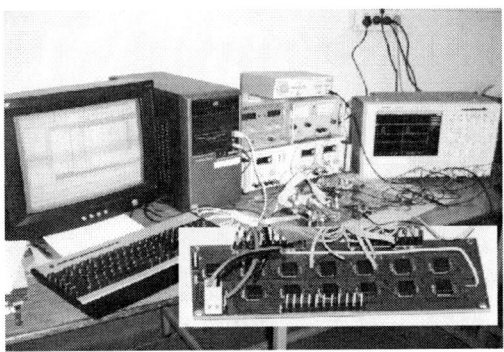

Fig. 17. Experimental Setup with Delay line PCB

TABLE II
DESIGN PERFORMANCE SUMMARY

Parameter	Value
refclk frequency	6-16 MHz
Regulated output voltage	1.6-3.2 Volt
ADC chip area	0.08 mm^2
ADC conversion time	2 periods of *refclk*
ADC Quantization step	less than 10 mV
ADC power consumption	$6\mu A$/MHz
Output Inductor and Capacitor	4.7μH and 22 μF
Converter Switching frequency	1 MHz

A voltage controlled oscillator (VCO) is used in [8] to generate a frequency signal for the sensed converter output voltage. The generated frequency signal and the target input frequency signal is passed through a phase detector block which generates the corresponding error difference value. Phase detector logic is implemented with a very high frequency clock signal and two digital counters. The use of very high frequency clock signal makes this approach inefficient for power management applications. This work uses compensator and DPWM in the digital controller framework.

A delay line based approach is used in [4] for DVS implementation. Three delay line tap outputs are passed to the digital glue logic to to differentiate between steady state and transient state. When the sampled values of the tap outputs for all the three taps are same, this implies transient state and the edge of the reference clock signal is not captured in the sampled values. The edge of the reference clock is captured in the delay line sampled values in steady state. State based delay line controller initiates controller action based on the current state. The exact error amount value is not determined in this scheme and hence the output regulation is not smooth. In steady state, a pulse train is fed as the control signal for the switching converter and the width of the pulse train is proportional to the regulated voltage. In transient state, maximum duty ratio signal (always high) is provided if the output voltage is less than the corresponding voltage of the target frequency. When the output voltage is more than the corresponding voltage of target frequency, minimum duty ratio signal (always low) is provided to the

switching converter. This approach does not provide smooth regulation of converter output voltage.

The work reported in [6] uses delay line and finite state machine (FSM) approach to generate PWM control signal for the switching converter for a given input target frequency. This work implements PVT variation support for the target frequency command and provides error compensation, frequency compensation using a false low level detection logic in the critical path of the state based controller. Therefore, the maximum clock frequency of the digital controller is compromised as the delay of the critical path increases. This finite state based approach provides good performance for a medium range (upto 600 KHz) of switching frequency.

Design performance summary in closed loop operation are listed in Table II. In this proposal, exact error value is generated from the input target frequency and classical digital controller framework is used. This proposal achieves smooth regulation, reduces chip area and provides regulated output voltage over a wide range of input frequency command values. Proposed ADC can work with 100 MHz clock and the complete digital controller can be synthesized with HDL tools.

VI. CONCLUSION

This work presents an optimized hardware design of a delay line based ADC with frequency reference input and optimizes delay line ADC latency. A novel concept of multiple band is introduced to improve steady state regulation. The complete digital PWM controller is verified in a closed loop with dc-dc buck converter. Any variation in V_{out} or in *refclk* is handled with good accuracy and fast response. The objective of linear characteristic curve between the input *refclk* frequency and the corresponding regulated voltage is achieved.

REFERENCES

[1] Thomas D. Burd, Trevor A.Pering, Anthony J. Stratakos, and Robert W. Brodersen, "A Dynamic Voltage Scaled Microprocessor System" *IEEE Journal of Solid-State Circuits*, vol. 35, No. 11, pp. 1571-1580, November. 2000.

[2] A.V.Peterchev, J. Xiao, and S.R. Sanders. "Architecture and IC implementation of a digital VRM controller" *IEEE Transactions on Power Electronics*, vol. 18, pp. 301-308, Jan. 2003.

[3] Patella, B.J. Prodic, A. Zirger, A. Maksimovic, D. "High-frequency digital PWM controller IC for dc-dc converters," *IEEE Transactions on Power Electronics*, vol. 18, Issue 1, Part 2, Jan.2003 pp. 438-446, Special Issue on Digital Control.

[4] C. Zhang, D. Ma, A. Srivastava. "Integrated adaptive dc/dc conversion with adaptive pulse-train technique for low-ripple fast-response regulation " *In Proceedings of IEEE ISLPED Conference, 2004*, pp. 257-262.

[5] Jinwen Xiao, Angel V. Peterchev, Jianhui Zhang, and Seth R. Sanders. "A 4- μA Quiescent-Current Dual-Mode Digitally Controlled Buck Converter IC for Cellular Phone Applications" *IEEE Journal of Solid-State Circuits*, vol. 39, No. 12, pp. 2342-2348, December. 2004.

[6] D. Kang, Y. Kim, and J.T. Doyle, "A High-Efficiency fully digital synchronous Buck converter power delivery system based on a Finite-State machine " *IEEE Transactions on VLSI Systems*, vol. 14, no. 3, pp. 229-240, March 2006.

[7] D. Dhar, and D. Maksimovic, "Switching Regulator with Dynamically Adjustable Supply Voltage For Low Power VLSI", *Proceedings of IEEE Industrial Electronics Society*, pp. 1874-1879, 2001.

[8] G. Wei, and M. Horowitz, "A fully digital, Energy-Efficient adaptive Power-Supply Regulator", *IEEE Journal of Solid-State Circuits*, vol. 34, no. 4, pp. 520-528, April, 1999.

[9] "Synopsys EDA tools", Available: http://www.synopsys.com

Overmodulation Region of Multi-Phase Inverters

S. Halász

Electric Power Engineering Department, Budapest, Hungary, shalasz@eik.bme.hu

Abstract—**Overmodulation region control of multi-phase inverters is investigated from point of view of ac motor harmonic losses. It is shown that in full overmodulation region neither PWM inverter control nor the optimization of switching angles of inverter can ensure the decrease of motor harmonic losses to acceptable values. In the first half of overmodulation region these losses permanently increase leading to considerable values. Therefore the motor voltage region above 85-87% of the inverter maximal possible value is virtually restricted at least for steady state operation. The theoretical results are checked by simulation.**

Keywords: multi-phase drive, converter control, voltage source inverters, AC machine.

I. INTRODUCTION

The structure of multi-phase inverter-fed ac drive is presented in Fig. 1. In this case the phase number of motor and inverter is bigger than 3. The multi-phase inverter-motor system has better fault tolerance, lower motor torque ripples, smaller power rating of converter semiconductors and lower phase current for a given voltage rating [1-2].

These advantages may be very useful in some areas of industry. Therefore the wider application of multi-phase system is expected.

II. STATE OF THE ART

In case of three phase system the overmodulation region is $\sqrt{3}\pi/6 \le U_1/U_{1\max} \le 1$ where U_1 the motor fundamental voltage is, $U_{1\max} = 2U_{dc}/\pi$ is the maximum possible value of this fundamental voltage and U_{dc} is the dc voltage of the inverter. Usually the motor harmonic losses are characterized by square of rms. value of the stator harmonic flux:

$$\Delta\Psi^2 = \sum_{\nu>1}^{\infty} (\frac{U_\nu}{U_1\nu})^2, \qquad (1)$$

where U_ν is the voltage harmonic of order ν.

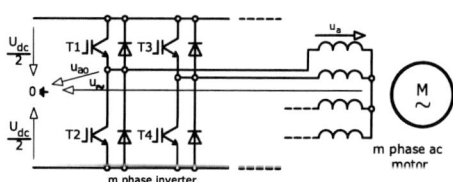

Fig. 1. Structure of multi-phase inverter-fed ac drive

In case of six step 3-phase inverter in p.u. :

$$\Delta\Psi_3^2 = \frac{1}{5^4} + \frac{1}{7^4} + \frac{1}{11^2}... = \frac{5\pi^4}{486} - 1 = 0.00215.$$

It is well known that in the operational point with fundamental frequency $f_1 = 50Hz$ the motor harmonic losses usually are approximately 10% in the stator and 50-60% in the rotor (related to the rated motor coil losses in the stator and in the rotor, respectively) when the rotor skin effect is taken into account. Usually, it is better to characterize the motor harmonic losses by a loss-factor which is the relation of $\Delta\Psi^2$ to 0.00215:

$$k_\Psi = \Delta\Psi^2 / 0.00215. \qquad (2)$$

When $k_\Psi = 1$ approximately 15% motor power degradation is necessary while for case $k_\Psi = 0.1$ the motor degradation is usually unnecessary.

For 3-phase system the Park-vector investigation gives very sensible advantages but for multi-phase system probably the vector method is less obvious. Therefore the investigation of the phase values (voltages, current, flux etc.) can produce a good result. In latter case the basic point will be the investigation of $u_{a0}(t)$ phase end voltage function for different modulation techniques.

For a multi-phase system $(m > 3)$ the maximum possible value of the motor fundamental voltage is the same as in case of 3-phase system, but the span of

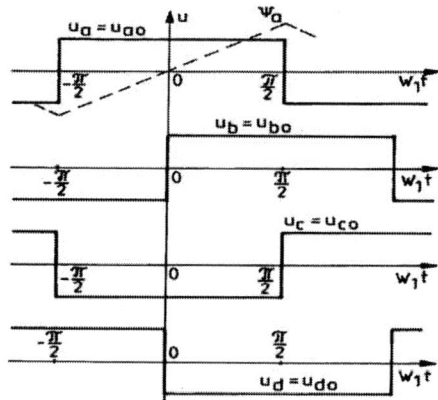

Fig. 2. Phase end voltage and phase voltages, m=4, $\gamma = 1$.

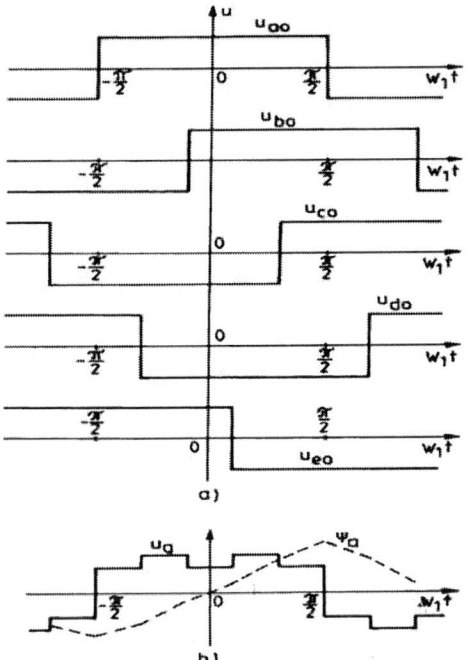

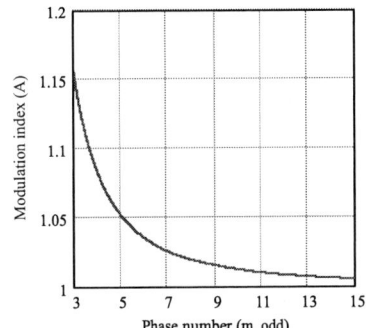

Fig. 4. Modulation index at boundary
of linear and overmodulation regions

Fig. 3. Phase end voltages (*a*) and phase voltage (*b*), *m*=5, $\gamma = 1$.

overmodulation region is considerably bigger. It is clear from Fig. 2 and Fig. 3 where the motor phase voltages and as well the motor voltages to the middle point of dc (phase end voltages) are presented for phase number $m = 4$ and $m = 5$. Figs. 2-3 are valid for $\gamma = 1$ (full inverter control) where the switching number of one transistor during one fundamental period is γ.

In case of 4 phases the motor star point voltage in any time is zero:

$$u_{\sim} = \frac{u_{a0} + u_{b0} + u_{c0} + u_{d0}}{4} = 0, \qquad (3)$$

thus the phase voltage and the motor phase end voltage to the middle of dc circuit are the same. This statement is valid for any even number of phases except of $m = 2$. Since when $m > 2$ (Fig. 2) there are two phase voltages with opposite phase it is impossible to increase the modulation index $A = U_1 / (U_{dc} / 2)$ above 1 without overmodulation. Therefore for any even number of phases the overmodulation region begins from $U_1 = U_{dc} / 2 = \pi U_{1\max} / 4 = 0.785 U_{1\max}$.

In case of odd number of phases the star point voltage $u_{\sim} = \pm U_{dc} / (2m)$ and it changes the sign every π / m (Fig. 3b). Therefore the phase voltage $u_a = u_{a0} - u_{\sim}$ has the stepped shape (Fig. 3b). The overmodulation region of multi-phase inverters with odd m begins at $A = 1 / \cos(\pi / (2m))$ (Fig. 4). From Fig. 4 it is seen that overmodulation region of multi-phase inverters virtually begins at $A = 1$ and only for $m = 5$ it is possible to increase the normal region up to still sensible value A=1.051.

III. HARMONIC LOSSES OF MULTIPHASE AC DRIVES

There are two methods for the voltage control of overmodulation region. In the first one the PWM overmodulation technique is used (Fig. 5) and switching angles are determined by intersection points of the reference wave and triangular carrier one. This method

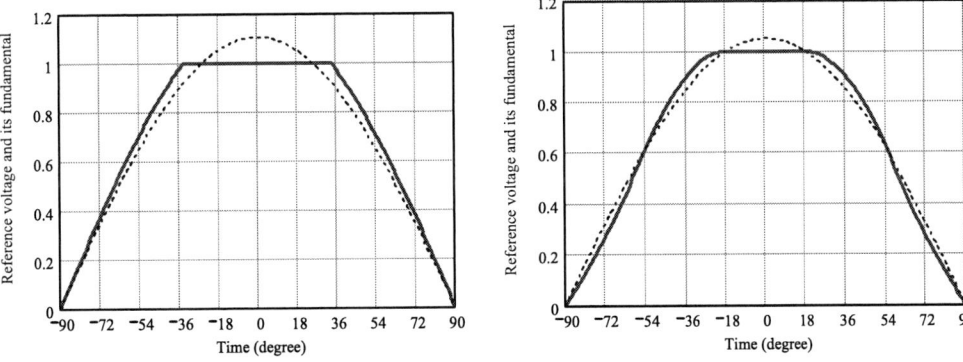

a) Continuous PWM, ($A = 1.3$, $U_{aref1} = 1.14 U_{dc} / 2$)   b) Discontinuous PWM, *m*=5 (A=1.051, $U_{1ref} = 1.051 U_{dc} / 2$

Fig. 5. Reference wave and its fundamental component

683

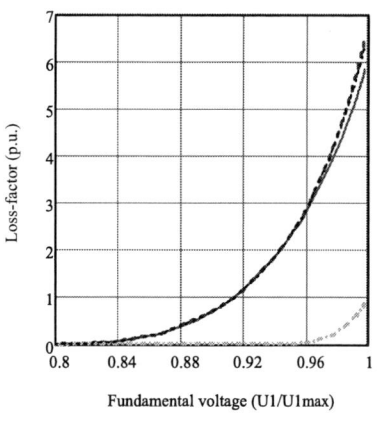 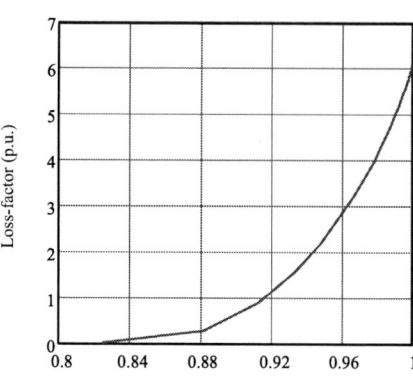

a) Continuous PWM, ·—·· $m=3$, --- $m=4$, —$m=5$, ··$m=7$ b) Discontinuous PWM, $m=5$

Fig. 6. Loss-factor of overmodulation region

can be applied for both the continuous and discontinuous PWMs. It is well known that the linear control region can be increased above $A=1$ when the space vector method or a PWM method with injection of m order harmonic is applied instead a sinusoidal PWM. The same result is valid for any discontinuous PWM. In any case the maximum of the reference wave will be at $t=\pi/(2m)$.

In the second method the simple inverter with additional switching is applied. In latter case the switching instants are selected by Newton-Rapson method providing the minimum of harmonic losses for a given fundamental motor voltage.

III.1. PWM overmodulation technique

Assuming a very high modulation frequency it may be supposed that the motor harmonic losses are determined only by reference wave low order harmonics while the current harmonics of modulation frequency can be omitted. The voltage harmonics of reference wave $u_{aref}(t)$ (Fig. 5.) are calculated by Fourier equations. In case of high modulation frequency the harmonic content of a both reference wave and motor voltage is the same except the zero voltage harmonics (e.g. in case of $m=5$ the harmonics of order 5., 15., 25.,... are cancelled by 5 wire system). Therefore the loss-factor is determined by (1) and (2).

In case of the continuous PWM (Fig. 5a) the simplest method of overmodulation voltage control is the increase of the modulation index above 1. In this case there is no switching when the reference wave is bigger than 1.

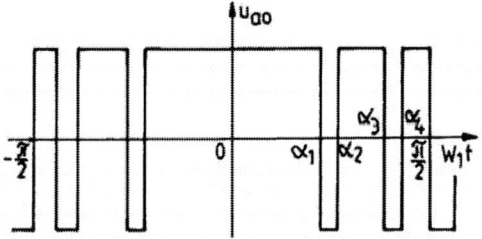

Fig. 7. Phase end voltage with additional switching, $\gamma=9$.

Therefore this part of reference wave is taken equal to 1, as shown in Fig. 5a.

In case of discontinuous PWM the shape of a reference wave depends on number of phases. Therefore the number of phases $m=5$ is investigated when the each phase is fixed to positive and to negative dc bar on π/m time as shown in Fig. 5b for the boundary of linear (normal) and overmodulation regions. In this case the modulation index is $A=1/\cos(\pi/(2m))=1/\cos(\pi/10)=1.051$. The further increase of inverter voltage can be worked out by multiplying the reference wave shown in Fig. 5b by a rising factor $k>1$.

The loss-factor for different phase number is shown in Fig. 6a for continuous PWM and in Fig. 6b for $m=5$ and discontinuous PWM. It is seen that when $U_1>0.91U_{1\max}$ the motor harmonic losses become bigger than those for full inverter control of 3-phase inverters. This means that the harmonic loss level of this part of overmodulation region cannot be acceptable for steady state operation.

It should be noted that in case of 3-phase system the overmodulation region is usually restricted by $U_1=(0.95\div0.97)U_{1\max}$. The loss-factor for this fundamental voltage is $k_\Psi\approx0.1$ therefore the motor harmonic losses and low order voltage harmonics are at acceptable values. For the same loss-factor the multiphase system must be restricted by $U_1=(0.85\div0.87)U_{1\max}$ but this leads to very sensitive restriction of motor voltage.

III.2. Simple inverter with additional switching

In case of full voltage control the number of transistor switching on the fundamental period is $\gamma=1$ (Figs. 2-3). The loss-factor can be computed from $\Psi_a(t)$ time function or from the summation $\sum_{\nu>1}^{\infty}1/\nu^4$, where ν is order of existing voltage harmonics. The result is given in Table 1. In this case the motor voltage must be controlled by dc voltage control.

684

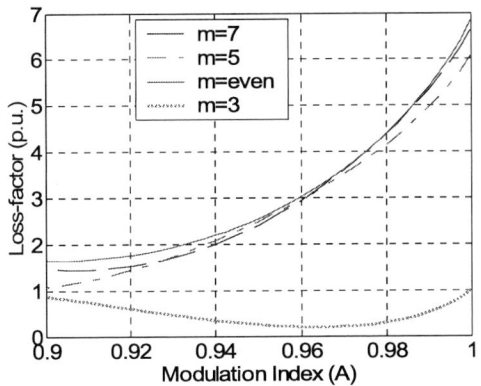

Fig. 8. Loss-factor for different phase number, $\gamma = 5$

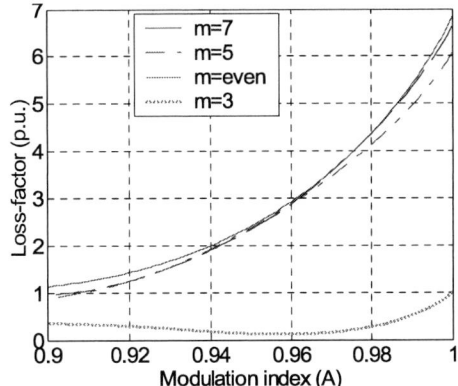

Fig. 9. Loss-factor for different phase number, $\gamma = 7$

Table 1. Loss-factor for $\gamma = 1$

Phase number m	3	4 (even)	5	7
$\Delta\Psi^2$ (p.u)	$\dfrac{5\pi^4}{486}-1$ $=0.00215$	$\dfrac{\pi^4}{96}-1$ $=0.0147$	$\dfrac{27\pi^4}{5^44}-1$ $=0.0131$	$\dfrac{25\pi^4}{7^4}-1$ $=0.0143$
k_Ψ (p.u)	1	6.823	6.169	6.627

It is seen that for $\gamma = 1$ the loss-factor of multi-phase inverters is more than 6 times bigger than for $m = 3$. Mainly it is explained by existence of third order current since the third order voltage harmonic has the amplitude $U_3 = U_1/3$.

This unacceptable value of harmonic losses can be lowered by inverter voltage control with additional switching. For different motor fundamental voltage the additional switching angles α_i (Fig. 7) are determined for the smallest value of loss-factor. The result for $\gamma = 5$ and $\gamma = 7$ is shown in Figs. 8-9. In case of fundamental voltage decreasing by 2-3% e.g., the loss-factor of 3-phase

system can be diminished to $k_{3\Psi} = 0.17 - 0.22$. Nevertheless, for $\gamma = 5$ and $\gamma = 7$ and $m > 3$ the loss-factor of this voltage region cannot be pressed to an acceptable level.

Naturally, the number of switching can be increased and this leads to further moderation of the harmonic losses. However, the loss-factor value is restricted similarly by $u_{ref}(t)$ function in Fig. 5 and therefore considerable decreasing of the loss-factor is impossible (as it is shown by simulation results later on).

III.3. Decrease of the low order harmonics

The motor low order voltage harmonics are prescribed by a low order harmonics of reference wave. For the continuous PWM they are harmonics of the reference wave shown in Fig. 5a. The 3rd, 5th and 7th order harmonics of reference wave are presented in Fig. 10a. It is seen that the most unpleasant 3rd harmonic continuously grows with increase of the fundamental voltage. The same situation can be observed for the discontinuous PWM. In Fig. 10b the 3rd and 7th order harmonics of the reference wave (Fig. 5b) are given for number of phase $m = 5$ (the 5th order harmonic is cancelled). In value of 3rd order harmonic between continuous and discontinuous PWMs

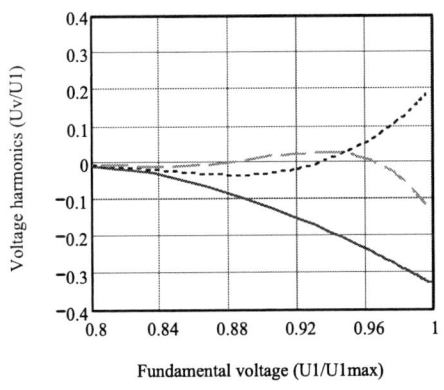

a) Continuous PWM

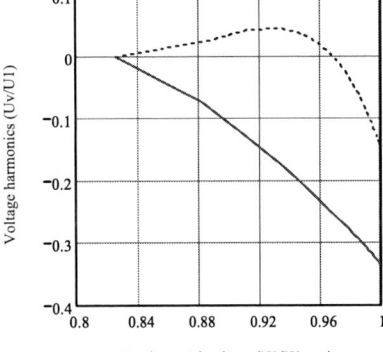

b) Discontinuous PWM

Fig. 10. Low order voltage harmonics of reference wave of overmodulation region
——— 3rd , ——— 5th , --- 7th order

there is no sensible difference. This result is also valid for additional switching control since the additional switching can be located only in those part of reference wave where

$$u_{ref} \leq 1.$$

It is seen that the 3rd order harmonic really dominates in overmodulation region. Therefore another way to decrease harmonic losses is minimization of 3$^{rd.}$ order voltage harmonic for a given motor fundamental voltage. However, in overmodulation region the optimization of third order voltage harmonic shows that the considerable decrease of the amplitude of this harmonic is impossible and therefore this method can decrease the motor harmonic losses only in a very small rate.

The considerable decreasing of the harmonic losses can be reached only by cancellation of 3$^{rd.}$ order harmonic. But this is possible only for e.g. when 6-phase inverter is controlled as two three-phase ones and the star points of two 3-phase system are isolated.

IV. SIMULATION RESULTS

The motor current is determined by simulation of the system. For no-load condition, $U_1 = 0.9 U_{1\max}$ and two methods of overmodulation control the motor current vs. time functions are presented in Fig. 11. Taking into

consideration the rotor skin effect the motor transient inductance was lowered to $L' = 0.12 = const$. The rated no-load current was $I_{no-load} = 0.5$.

It is known [1] that assuming a sinusoidal flux distribution in the air gap only the current harmonics of order $km \pm 1$ (k=0, 2, 4, 6..., m odd or semi-even [1]) are able to produce the counter rotor currents and torque. The other current harmonics give zero value of the resultant MMF and therefore they are restricted only by stator leakage inductance and stator resistance. The stator leakage inductance is lower than the transient one therefore many current harmonics will be bigger than in case of 3-phase system. On the other side the rms. value of the rotor harmonic current will be considerably lower. For a rough approximation the rms. value of stator harmonic current is computed with $L' = 0.12 = const$ neglecting resistance influence on harmonic current. These approximations gives only qualitative information of performances but considerably simplifies the computations:

$$\Delta i = \Delta \Psi / L' \qquad (4)$$

Fig. 11a shows the current time function in case of optimized switching angles with $\gamma = 13$ while in Fig. 11b

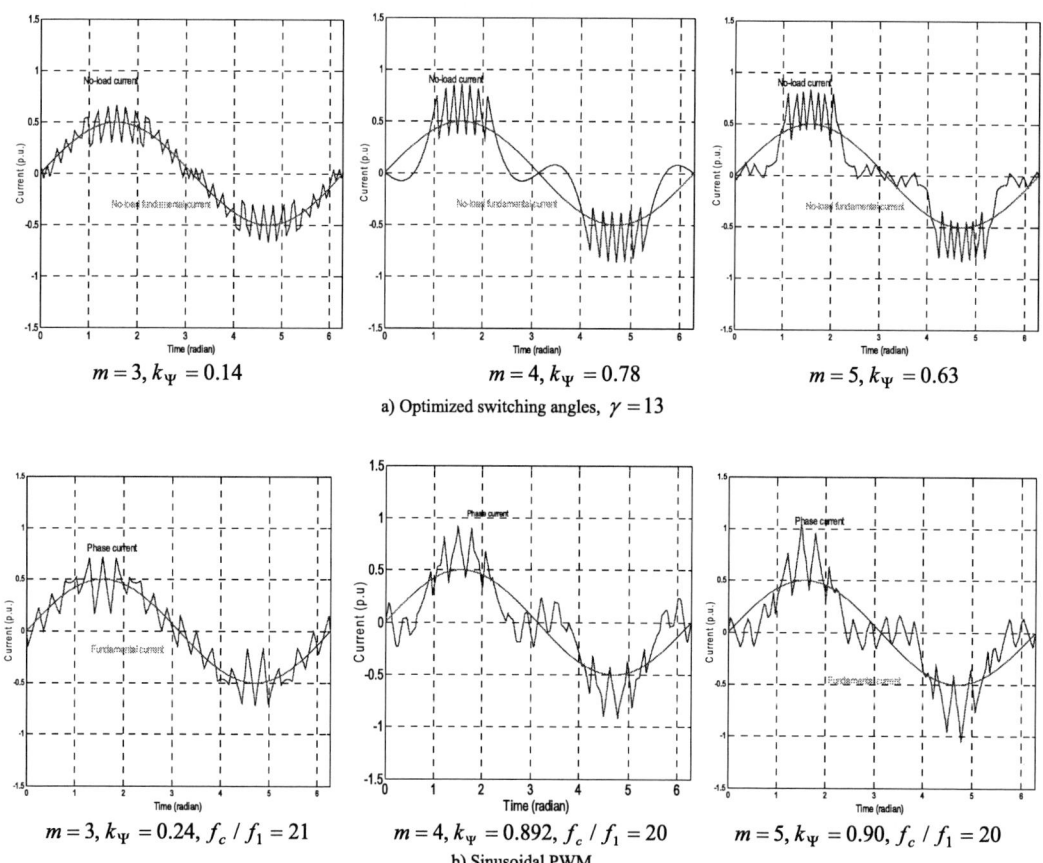

$m = 3, k_\Psi = 0.14$ $m = 4, k_\Psi = 0.78$ $m = 5, k_\Psi = 0.63$

a) Optimized switching angles, $\gamma = 13$

$m = 3, k_\Psi = 0.24, f_c / f_1 = 21$ $m = 4, k_\Psi = 0.892, f_c / f_1 = 20$ $m = 5, k_\Psi = 0.90, f_c / f_1 = 20$

b) Sinusoidal PWM

Fig. 11. Motor fundamental current and phase current, no-load, $U_1 = 0.9 U_{1\max}$

Table II. Overmodulation parameters of different phase number.

Phase number m	PWM	k_Ψ	A	U_1/U_{max}	k_Ψ	A	U_1/U_{max}	k_Ψ	A	U_1/U_{max}
3	Sin	0.1	2.27	0.967	0.25	2..85	0.979	0.5	4.01	0.990
	SPV		1.52	0.967		1.90	0.979		2.68	0.990
	Discont.		k=1.45	0.966		k=1.78	0.979		k=2.45	0.990
5	Sin		1.141	0.850		2.13	0.871		1.30	0.890
	SPV		1.19	0.867		1.28	0.881		1.40	0.897
	Discont.		k=1.13	0.863		k=1.19	0.879		k=1.29	0.895
even	Sin		1.12	0.848		1.21	0.869		1.29	0.889

the current time function is given for sinusoidal natural PWM overmodulation technique. In latter case the carrier frequency was selected in order to have the number of switching close to the first case. Therefore the carrier frequency was $f_c = 21 f_1$ in case of three-phase system and $f_c = 20 f_1$ in case of four- and five-phase system.

From Fig. 11 it is seen that the loss-factor in case of optimized switching angles is lower than that in case of PWM. It is explained by different voltage spectra, i.e. in case of optimized switching angles the spectrum is spread on all harmonics but the big amplitudes are lower than for carrier-based methods.

In case of 3-phase system it should be noted that space vector PWM or PWM with 3-rd harmonic addition have normal voltage control region $0 < U_1 < 0.907 U_{1max}$ therefore in such case in the investigated operational point with $U_1 = 0.9 U_{1max}$ the loss-factor can be close to zero. This statement is not true for 5-phase system where $U_1 = 0.9 U_{1max}$ lies in overmodulation region and therefore the loss-factor can be only lowered to $k_\Psi = 0.68$ (in the face of $k_\Psi = 0.90$ for sinusoidal PWM).

In Table II the overmodulation parameters of reference

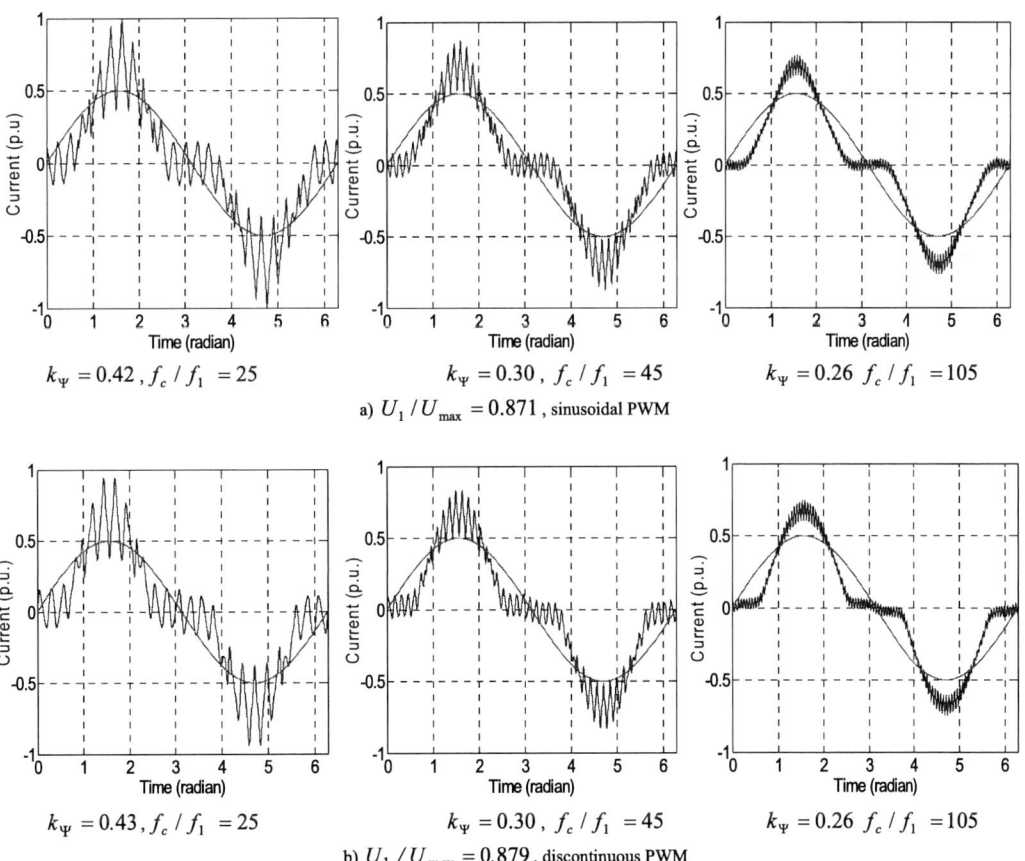

$k_\Psi = 0.42 , f_c / f_1 = 25$ $k_\Psi = 0.30 , f_c / f_1 = 45$ $k_\Psi = 0.26 \ f_c / f_1 = 105$

a) $U_1 / U_{max} = 0.871$, sinusoidal PWM

$k_\Psi = 0.43 , f_c / f_1 = 25$ $k_\Psi = 0.30 , f_c / f_1 = 45$ $k_\Psi = 0.26 \ f_c / f_1 = 105$

b) $U_1 / U_{max} = 0.879$, discontinuous PWM

Fig. 12. Motor fundamental and phase current, no-load, $m = 5$

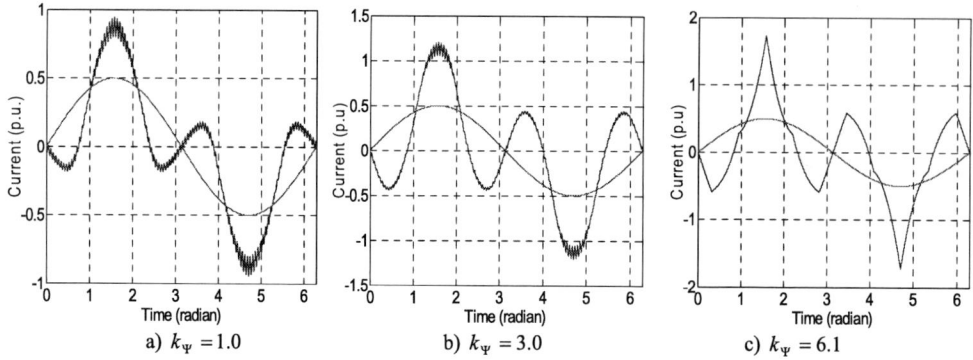

Fig. 13. Motor fundamental and phase current, no-load, $f_c / f_1 = 105$, $m = 5$ (sinusoidal PWM)

waves shown in Fig. 5 are gathered for different value of loss-factor and for sinusoidal, space vector (SPV) and discontinuous PWMs. The loss-factor $k_\psi = 0.1$ determines the overmodulation region which does not require the deration of the steady-state motor power. It is seen that in this case for 3-phase system 97% of motor voltage can be used while in case of multi-phase system this value lowered to 85-87%. The loss-factor $k_\psi = 0.25$ and $k_\psi = 0.5$ on the one hand require derating of the motor, on otherwise give very small increase of motor voltage in comparison with case of $k_\psi = 0.1$. Thus this overmodulation region can be used only for short time operation. At the same time the current peak can be considerably bigger than in linear voltage region. This phenomenon is illustrated by Fig.12 where the motor current vs. time function is given for $m = 5$ (sinusoidal and discontinuous PWMs) and different carrier frequencies. It is seen that the loss-factor with increase of the carrier frequency decreases to the loss-factor of reference wave $k_\psi = 0.25$ (Table II) and that 3rd current harmonic produces the predominating value of motor harmonic losses. Finally, in Fig. 13 the motor current is given for unacceptable value of the loss-factor $k_\psi = 1.0, 3.0$ and 6.1 (full voltage, simple inverter).

V. TORQUE PULSATION

One important advantage of multi-phase drives is the considerably lower value of torque pulsation. This is quite clear from comprasing the first torque component-time functions in Fig. 14. These are drawn for full motor voltage and for case when $m = 3$ and $m = 5$. In spite of high 3rd order current harmonic (Fig. 13c) in case of $m = 5$ the amplitude of the first torque harmonic of order 10th is approximately times lower than the 6th order torque harmonic amplitude in case of $m = 3$. Due to higher frequency the speed oscillation in case of $m = 5$ will be by ≈ 8.0 times lower than for $m = 3$.

At the same time the square of rms. value of the stator and rotor currents will be:

$$\Delta i^2 = \frac{\Delta \Psi_s^2}{(L')^2} = \frac{0.0131}{0.12^2} = 0.91,$$

$$(5)$$

$$\Delta i_r^2 = \frac{1}{0.12^2}(\frac{1}{9^4} + \frac{1}{11^4} + \frac{1}{19^4} + \frac{1}{21^4} + ...) = 0.0165,$$

respectively. These currents in case of $m = 3$ are equal:

$$\Delta i^2 \approx \Delta i_r^2 = 0.15.$$

CONCLUSION

Different PWM strategies of the overmodulation region for multi-phase inverters are investigated from point of view of ac motor harmonic losses. It is shown that in overmodulation region the 3rd order voltage harmonic has considerable value and therefore the multi-phase motor harmonic losses are significantly bigger than in case of 3-phase system. The carrier-based PWM inverter control (natural sinusoidal PWM, space vector PWM and discontinuous PWM) or the optimization of switching angles of inverter give approximately the same value of harmonic losses for a given value of motor voltage. Therefore the multi-phase motor voltage region above 85-87% of the inverter maximal possible value is virtually restricted at least for steady state operation. In case of a carrier-based PWM a better solution is the limitation of the modulation index to $A \approx 1.2$ while in case of a discontinuous PWM the rising factor

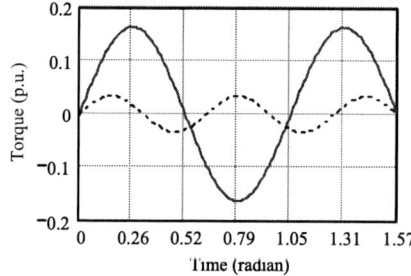

Fig. 14. Torque pulsation,
— 6th order harmonic, $m = 3$, – – – 10th order harmonic, $m = 5$.

must be limited to $k \approx 1.15 - 1.16$. In these cases the decrease of the network voltage cannot produce motor operation in the unwanted part of overmodulation region. In comparison with 3-phase machines especially the stator harmonic losses increase in a big rate while the rotor harmonic losses of multi-phase machine are considerably lower. The theoretical results are checked by simulation.

ACKNOWLEDGMENT

The research was supported by the Hungarian National Scientific Fund (OTKA #T 042866) for which author expresses his sincere gratitude.

REFERENCES

[1] E. A. Klingshirn, "High phase order induction motors, Part I, Experimental results", *IEEE Transaction on Power Apparatus and Systems*, No. 1, January 1983, pp.47-53.

[2] E. A. Klingshirn, "High phase order induction motors, Part II, Description and theoretical considerations", *IEEE Transaction on Power Apparatus and Systems*, No. 1, January 1983, pp.47-53

[3] A. Igbal, E. Levi, "Space vector modulation schemes for a five-phase voltage source inverter", *European Power Electronics and Appl. Conference*, EPE, Dresden, Germany, 2005, CD ROM, paper 0006.

[4] V. Oleschuk, G. Griva, F. Pofumo, A. Tenconi, "Synchronized PWM control of symmetrical six-phase drives", *The 7th International Conference on Power Electronics*, October 22-26, 2007, Daegu, Korea, CD ROM.

[5] S. Halász, I. Varjasi, A. Zacharov, "Overmodulation region of inverter- fed ac drives", *3rd IAS Power Conversion Conference*, 2-5 April 2002, Osaka, Japan, vol. III, pp.: 1346-1351.

[6] X.F. ZhangZhang, F. Yu. , H. S. Li, Q. G. Song, "A novel discontinuous space vector PWM control for multiphase inverter",, *International Symposium on Power Electronics, Electrical Drives and Motion (SPEEDAM 2006)* , May 23-26, 2006, Taormina, Italy, CD ROM.

[7] S. Williamson, S. Smith, "Pulsating torque and losses in multiphase induction motor", *IEEE Transaction on Industry Application,* No. 4, July-August 2003, pp. 986-993.

Optimal Control of Induction Motor Using High Performance Frequency Converter

Jerkovic Vedrana*, Spoljaric Zeljko* and Valter Zdravko*

* Faculty of Electrical Engineering / Department of Electro-mechanical Engineering, Osijek, Croatia, e-mail:
zeljko.spoljaric@etfos.hr

Abstract—This paper discusses optimal configuring and commissioning of laboratory system that consists of FC-302 high performance frequency converter and induction motor. In the beginning frequency converter control of induction motor is explained. Furthermore, the measurements of slip compensation parameter change influence on motor speed and motor load dependence are performed, as well as measurements of power factor correction attained by frequency converter. It is possible to compensate slip completely, so that motor speed is constant, regardless of motor load, which provides optimal control. Frequency converter partially compensates power factor of converter – motor system with capacitors of passive filter.

Keywords—converter control, induction motor, power factor correction, optimal control.

I. INTRODUCTION

Latest development of high performance frequency converters opens more possibilities for optimal control of induction and synchronous motors. In this paper optimal control of induction motor is considered. Induction motors have wide spread application in industry because of their variable speed and power range. The control and estimation of AC drives in general are considerably more complex than those of DC drives, and this complexity increases substantially if high performances are demanded [1]. Speed of induction motor can be economically controlled if source voltage has variable frequency [2]. The operating speed of induction motor can be determined by relation:

$$n = \frac{60f}{p}(1-s) \qquad (1)$$

where n is shaft speed in revolutions per minute, f is the supplied frequency in Hertz, p is the number of pole pairs and s is the operating slip. Speed of induction motor depends upon applied frequency, number of pole pairs and load torque. The most common control principle for induction motors is the constant volts per hertz (*V/Hz*) principle [3]. If the ratio (*V/Hz*) remains constant with the change of frequency than stator flux (Φ) also remains constant and the torque (T) is independent of the supply frequency. Relation that connects these values in relative terms is:

$$\overline{T} = \overline{\Phi}^2 = \left(\frac{\overline{U}}{\overline{f}}\right)^2 \qquad (2)$$

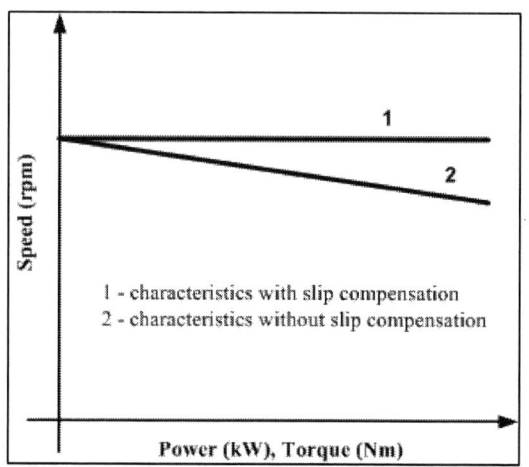

Fig. 1. Dependence of speed, load power and torque with and without slip compensation.

More advanced control principle from standard volts per hertz ratio control is voltage vector control (VVCplus) which uses frequency converter considered in this paper.

Voltage vector control improves the dynamics and stability, both when the speed reference is changed and in relation to the load torque. Vector control simplifies motor control using d-q or Park-Transformations which convert current and voltage from a-b-c to d^e-q^e or synchronously rotating reference frame [4].The main problem that occurs in induction motor drives is drop of speed because the speed depends on motor load. This can be solved by slip compensation which will be shown further. With slip compensation speed of induction motor remains constant regardless to load power or load moment change, as it is shown in Fig. 1. Influence of frequency converter on power factor is also considered in paper. It will be proved that using high performance frequency converter improves power factor and efficiency of used power source.

II. VECTOR CONTROL

Vector or field oriented control is based on control of magnitude and phase alignment of vector variables. Requests of very dynamic drives can only be accomplished by using vector control because the scalar control is too slow. Vector control of induction motor is based on torque control of separately excited DC motor [5]. DC motor can be controlled by controlling field or

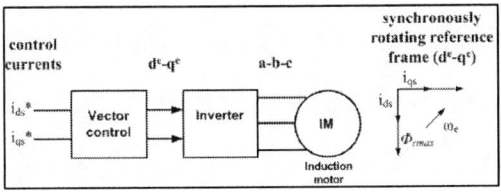

Fig. 2. Vector-controlled induction motor [1].

armature circuit. These two circuits of DC motor are completely separate. This means that, when torque is controlled by controlling the armature current (I_a), the field flux (Φ_f) is not affected and we get the fast transient response and high torque/ampere ratio with the rated field flux [1]. When field current (I_f) is controlled, it affects the field flux only, but not the armature flux (Φ_a) because of decoupling. DC-machine like performance can also be implemented to an induction motor if the machine control is considered in a synchronously rotating reference frame (d^e-q^e), where the sinusoidal variables appear as DC quantities in steady state [5]. Fig. 2. shows the simplified scheme of vector controlled induction motor which consists of inverter and vector control in the front. Vector control is shown with two control current inputs, $i_{ds}{}^*$ and $i_{qs}{}^*$. These currents are the direct axis component and quadrature axis component of the stator current, respectively, in synchronously rotating reference frame [1]. Field current is analogous to i_{ds} and armature current is analogous to i_{qs}. Torque can be expressed as given in [1]:

$$T = K_t\,\Phi_{rmax}\,i_{qs} = K_t{}'\,i_{ds}\,i_{qs}\,, \qquad (2)$$

where Φ_{rmax} is the peak value of flux, K_t and $K_t{}'$ are constant values. Current i_{ds} is oriented in direction of flux and i_{qs} is established perpendicular to it, as shown in space-vector diagram on the right of Fig. 2. This means that when $i_{qs}{}^*$ is controlled, it affects the actual i_{qs} current only, but does not affect the flux [5]. Also, when $i_{ds}{}^*$ is controlled, it controls the flux only and does not affect the i_{qs} component of current. As mentioned before in paper, voltage vector control principle is used for motor control.

A. Voltage Vector Control (VVCplus)

Voltage vector control will be explained from usage perspective. VVCplus control is used in open loop without feedback sensor. Control structure in VVCplus is shown on Fig. 3. This type of control is adaptive to

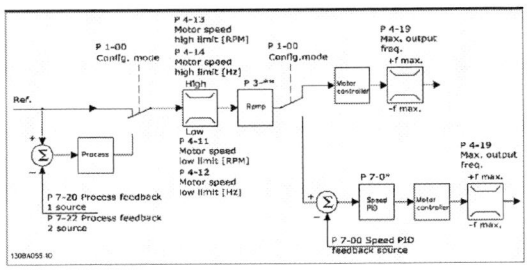

Fig. 3. Control structure in VVCplus.

motor load and adaptation to speed and torque changes is less then three milliseconds [5]. Motor torque can remain constant regardless to speed changes. VVCplus control principle is suitable for most applications. The main benefit of this control principle is robust motor model. This control principle can be used with or without slip compensation which will be described in the next chapter.

III. SLIP COMPENSATION

Relative difference between synchronous speed and rotor speed of induction motor is given in [2] with slip parameter s:

$$s = \frac{n_s - n}{n_s}\,, \qquad (3)$$

where n_s is synchronous speed and n is rotor speed. Hence, slip is equal to ratio of relative speed between rotating stator field and rotor winding, and synchronous speed. Induction motor has two limits of speed: motor standstill (short circuit) and synchronism (ideal idle motion). At motor standstill rotor speed is zero revolutions per minute and slip factor equals one. On the other hand, at ideal idle motion, rotor speed equals synchronous speed, and slip is zero. Synchronism is not possible as stationary state, because there would be no current induced in rotor and, therefore, there would be no torque to support this state. In real idle motion there is a weak current induced in the rotor to overcome the friction resistance and iron loss, so slip is small, but grater than zero. Open loop control of induction motor can be improved by slip regulation.

A. Slip Compensation in Scalar Control

To regulate slip, the slip command ω_{sl} is needed. Slip command is yielded from speed loop error, processed in PI controller and brought through limiter. The estimated slip is than added to the feedback speed signal and frequency command is generated. For the Volts per Herz control principle, voltage command is derived from frequency command and processed in Volts per Herz function generator, which incorporates low-frequency

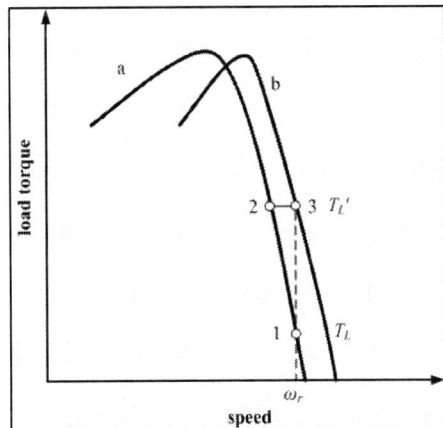

Fig. 4. Effects of load torque variation and speed compensation.

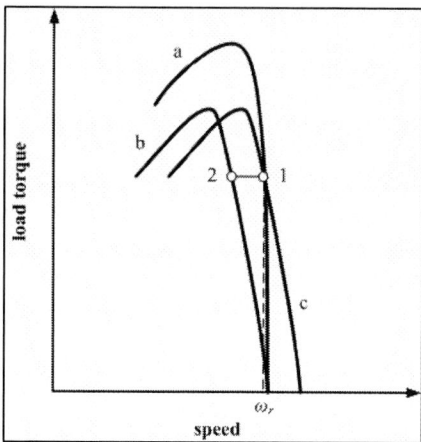

Fig. 5. Effects of supply voltage variation and speed compensation.

voltage drop compensation. At low frequencies voltage drop is present due to the resistance of stator windings. Since slip is proportional to developed torque at constant flux, slip compensation can be considered as an open loop torque control within a speed control loop. The effect of load torque variation is explained on the fig. 4. With the torque increase, speed will tend to drop. However, the speed control loop will increase the frequency until the original speed is restored. If the line voltage changes, the flux does also change. The line voltage drop is considered on the fig. 5. Decrease of line voltage leads to flux reduction and therefore to the drop of speed. Resulting speed drop will act on the speed loop to raise frequency and restore the original speed.

B. Slip compensation in vector control

There are several methods used to estimate speed of induction motor, such as: direct synthesis from state equations, model referencing adaptive system (MRAS), speed adaptive flux observer (Lienberger observer), extended Kalman filter and slip calculation. If slip calculation is chosen as a method for speed estimation, speed is calculated from slip frequency ω_{sl}. As given in [1], slip frequency can be calculated in stator flux-oriented direct vector control:

$$\omega_{sl} = \frac{(1 + \sigma s T_r) L_s i_{qs}}{T_r (\psi_{ds} - \sigma L_s i_{ds})}, \qquad (4)$$

where $T_r = L_r / R_r$ is rotor electrical time constant given as ratio of rotor inductance L_r and rotor resistance R_r, and i_{ds}, i_{qs}, ψ_{ds} are the signals corresponding to stator flux orientation. σ represents following expression:

$$\sigma = 1 - \frac{L_m^2}{L_s L_r}, \qquad (5)$$

where L_m is magnetizing inductance, L_s stator inductance, and L_r rotor inductance. For sensorless vector control, synchronous angular frequency ω_e is control variable. Hence, actual speed of induction motor can be derived from rotor angular frequency given with the following relation in [1]:

$$\omega_r = \omega_e - \omega_{sl} . \qquad (6)$$

An accurate calculation of slip frequency for high-efficiency machines, especially near synchronous speed, is difficult because the signal magnitude is small and highly dependent on machine parameters. There is also problem of direct integration of machine terminal voltages at low speed, which are used to synthesize ω_e and ω_{sl} signals.

IV. EXPERIMENTAL MODEL

Experimental model (Fig. 6) consists of induction motor (4 kW) and frequency converter FC-302, which controls motor performance. Induction motor is configured in delta connection, with DC generator as load. Direct current that excites magnetic field in DC generator is variable.

A. Frequency converter FC-302

FC-302 is a high performance frequency converter for demanding applications. It can handle various kinds of motor control principles. Beside normal squirrel cage induction motors, FC-302 can also handle permanent magnet synchronous motors. In Fig. 7. scheme of FC-302 power transmission part is shown. Converter consists of diode rectifier, LC filter and three-phase bridge inverter. Short circuit behaviour on converter depends on the 3 current transducers in motor phases. LC filter indicates that FC-302 is voltage-fed converter. FC-302 is capable of controlling either the speed or the torque on the motor shaft. There are two types of speed control: speed open loop (does not require any feedback) and speed closed loop (needs feedback to an input). There are four control principles that FC-302 can operate with: U/f (a special motor mode), VVCplus (Voltage Vector Control principle), Flux sensorless (i. e. Flux Vector Control without encoder feedback) and Flux with encoder feedback. FC-302 supply voltage range is 380-500 volts, supply frequency 50 Hertz, output voltage can reach

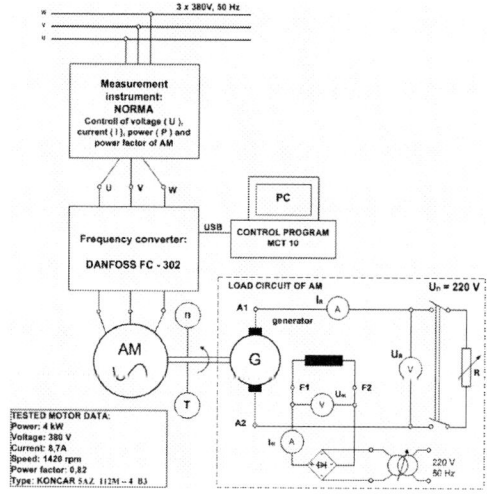

Fig. 6. Block scheme of experimental model and tested motor data.

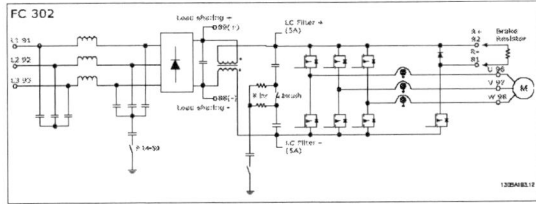

Fig. 7. Power transmission part of FC-302.

amount 0 – 100 % of supply voltage and its power size is 5.50 kilowatts. Control of induction motor is without mechanical sensor. Controlled AC drives without a mechanical sensor have in common that only terminal quantities, i.e. stator voltages and currents are measured from which the information on flux and speed of the motor must be derived, based on the nominal knowledge of the important motor parameters [6].

B. Measurements

Frequency converter can compensate motor slip by giving frequency supplement that follows the measured motor load. Slip compensation can be preformed by changing values of parameter 1-62 of frequency converter. Value for slip compensation is entered in percentages. Slip compensation is calculated automatically, on the basis of rated motor speed (speed given on the nameplate data). User can enter values of this parameter in range -500 to 500% in order to compensate tolerances in the value of motor rated speed. To find an optimal value of parameter 1-62 for the specified motor, series of measurements are made. Speed reference is kept constant and motor load varied from motor nominal load to idle motion. Value of parameter 1-62 is altered manually. Motor load is changed by changing current that excites magnetic field in DC generator. Measured values are motor speed and motor power.

Since FC-302 is a voltage-fed inverter, it is expected that it alters power factor of laboratory system. To find out how frequency converter influences on power factor of laboratory system, another series of measurements are made. Again, speed reference is kept constant and motor load varied from 6.3 kilowatts to idle motion. Motor load is changed by changing current that excites magnetic field in DC generator. Measured values are power factor, motor power and frequency. Measurements were taken first on motor without frequency converter, than on frequency converter – motor system.

V. RESULTS

A. Slip Compensation

Results of taken measurements are given on the Fig. 8. As shown in Fig. 8. value of converter parameter 1-62 must be 70% to keep motor speed constant and independent of motor load. Accordingly, it is possible to compensate slip completely and achieve induction motor performance equal to performance of synchronous motor.

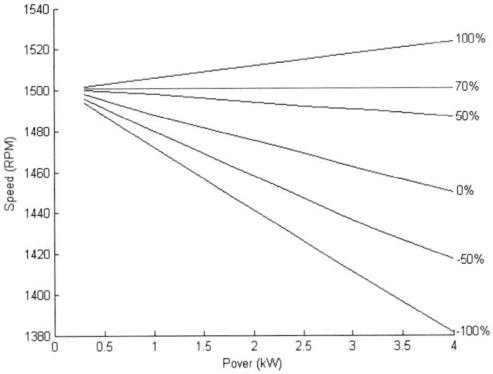

Fig. 8. Slip compensation.

As mentioned before, speed compensation principle is based on giving frequency supplement that follows the measured motor load. To find out how the frequency increases with motor load at optimal value of parameter 1-62, frequency was also measured. The result of this measurement is shown on fig. 9. Fig. 9. indicates that, when slip is totally compensated, frequency supplement is proportional to motor power. At motor nominal power, that is 4 kilowatts, frequency supplement reaches 1.7 Hertz.

B. Power Factor Corection

Results of taken measurements are given on the Fig. 10. As shown in Fig. 10. power factor of frequency converter – motor system is closer to value 1 in all power range than power factor of motor alone. Power factor of frequency converter – motor system is therefore better, because the use efficiency of mains power is greater. Frequency converter partially compensates power factor of converter – motor system. This is achieved with capacitors of passive filter placed in DC circuit. It is important to mention that power factor is equal to phase shift ($cos\varphi$) only if voltage and current wave form is sinus. If voltage and current waves are not sinus, relation between power factor (λ) and $cos\varphi$ is determined by relation [7]:

$$\lambda = \cos\varphi_1 (I_1 / I) \qquad (7)$$

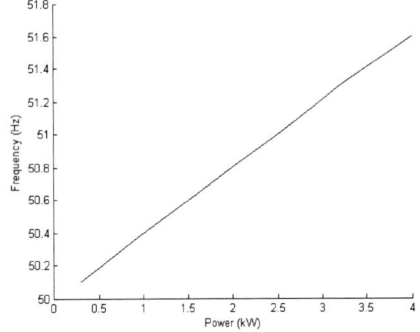

Fig. 9. Frequency supplement due to the slip compensation.

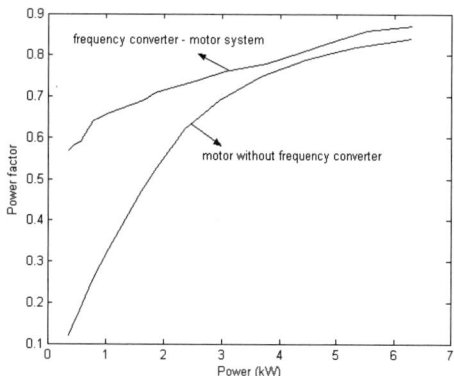

Fig. 10. Power factor correction.

where I_1 is effective value of basic current harmonic and I is total effective current.

VI. CONCLUSION

High performance frequency converters enable very effective motor control. In this paper two benefits of using such frequency converter for control of motor are described.

First benefit is possibility of slip compensation. It is proved that frequency converter can compensate motor slip completely, so that induction motor controlled with frequency converter behaves as synchronous motor. This is convenient because synchronous motors have narrow operation array comparing to induction motors, which is limited with nominal frequency at which they still can provide motor nominal torque. Slip compensation intensity is set manually, because accurate calculation of slip for high-efficiency machines is difficult since the signal magnitude is small and highly dependent on machine parameters. However, manual adjustment also means that motor can even be overcompensated or undercompensated purposely, what can be used in some special

applications. Slip compensation is gained with frequency change, that is, frequency increases with motor load. The relation between frequency supplement of slip compensation and motor power for used frequency converter is proved to be linear.

Second benefit is improvement of power factor of the system. It is proved that frequency converter compensates inductive reactive power of motor with large capacitors of DC circuit. The use of energy from power supply is therefore greater. Fig. 10. shows that the maximal improvement of power factor is gained at motor idle motion. Power factor of motor alone is than 0.12, and power factor of frequency converter – motor system is 0.57. This is especially important for drives which are not constantly loaded, as chain mortisers and so on.

All these benefits together with use of vector control enable optimal control of induction motor. Other benefits and possible imperfections of high performance frequency converters should be subject of further research.

REFERENCES

[1] B.K. Bose, Modern Power Electronics and AC Drivers. Prentice Hall, Upper Saddle River, New Jersey, USA, 2002, pp. 333-435.

[2] B. Skalicki, J. Grilec, Electrical Machines and Drivers. FSB, University of Zagreb, Zagreb, Croatia, 2005, pp. 337-407.

[3] H.A. Toliyat, S. Campbell, DSP – Based Electromechanical Motion Control. CRC Press LCC, Boca Raton, Florida, USA, 2004

[4] Z. Spoljaric, J. Jelecanin and Z. Valter, "Supervisory Control of EC Synchronous Motor Using LabView program", DAAAM International, Vienna, Austria, pp. 699-700, October 2007 [Proceedings of 18-th International DAAAM Symposium, Zadar, Croatia]

[5] V. Jerkovic, Z. Spoljaric, K. Miklosevic and Z. Valter, "Comparison of Different Motor Control Principles Using Frequency Converter", International Conference Science in Practice, SIP 2008, Osijek, Croatia, May 2008, in press

[6] W. Leonhard, Control of Electrical Drivers. Springer, Berlin, Heidelberg, NewYork, 2001, pp. 241-301.

[7] V. Siladi, "Frequency Converters and Power Factor", Elektro, vol. 4/98, pp. 30-33, June/July 1998.

Power Electronic Converter for the Reluctance Pump Drive

B. J. Szymanski *, K. Kompa **, N. Michalke ***, H. Kuß ***, U. Schuffenhauer ***

* Warsaw University of Technology / Department of Electrical Engineering, Warsaw, Poland
** Warsaw University of Technology / Department of Electronics and Information Technology, Warsaw, Poland
*** Centre of Applied Research and Development at the University of Applied Sciences, Dresden, Germany

Abstract — **In a standard application of hydrostatic drive unit, drive and pump are interconnected as a separate components. Nowadays, the trend towards miniaturization and integration of drives can be observed. The example of such integration is the drive system where the components of the hydraulic pump and electric drive are functionally integrated. The drive consists of specially designed switched reluctance machine integrated with the external gear pump and a proper power converter along with its unique control scheme. The paper focuses on the power electronics converter along with its control and describes the concept of the "reluctance pump". Control scheme of the drive is realized on the TMS320F2812 DSP from Texas Instruments.**

Keywords—**switched reluctance drive, DSP**

I. INTRODUCTION

The conventional design of hydraulic power supplies consists of electric motor, hydraulic pump and a coupling unit. The concept of the "reluctance pump", where an external gear pump is integrated with switched reluctance motor, can be one of the examples of functional integration in the electric-hydrostatic drive systems.

Today's rapid development of power electronics converters application areas along with their growing complexity forces the demand for high performance semiconductor devices as well as efficient control systems.

Nowadays, there are plenty of microprocessors able to fulfill high requirements stated by today's power electronics. There are chips, with AD/DA converters, PWM generators and interface peripherals, which allow to build converters controller on a single chip [1]. The TMS320F2812 DSP, fixed point microprocessor from Texas Instruments was chosen for the application where electric-hydrostatic drive control is considered.

The control algorithms of reluctance motors are widely described in [2] so they won't be the subject matter of further discussion. During the laboratory investigations, control algorithms were often modified what forced the need of proper power electronics software architecture, which allows in relative simple manner to adjust the system parameters or change the control algorithm structure.

II. THE CONCEPT OF THE "RRELUCTANCE PUMP"

The main aspect in the optimization and improvement of electric-hydrostatic drive systems is the integration of modules [3]. So far the efforts are being made to develop compact aggregates where the main aim is the integration of the assembly groups of the electric motor and hydraulic pump.

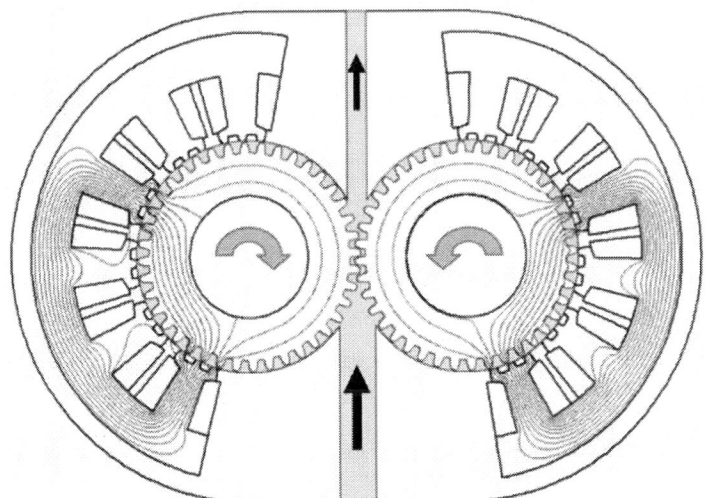

Fig. 1. Magnetic flux lines of "switched reluctance pump"

978-1-4244-1741-4/08/$25.00 ©2008 IEEE

FLUX2D 7.40/3 TR_PV1C_SINUS 12/12/07 12:03 Analysis Display Regions

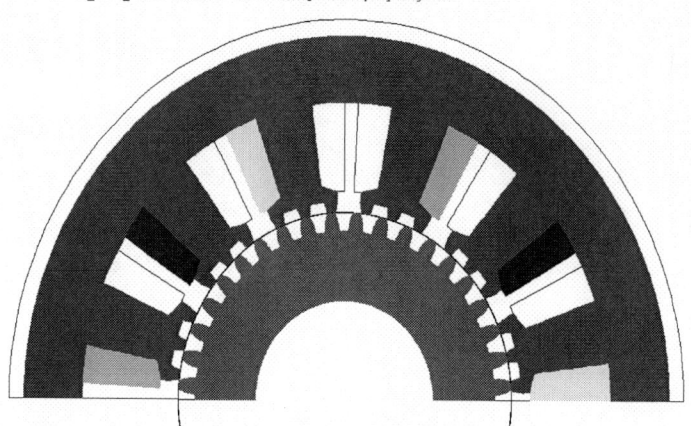

Fig.2. Cross-Section of the switched reluctance motor: stator, rotor and windings (only positive halves of the winding are highlighted)

High level of integration can only be achieved if motor and pump share essential groups. The concept of a "reluctance pump" may serve as an example of functional integration. It is realized by means of synthesis of an external gear pump with a switched reluctance motor (fig.1).

The principle of the "reluctance pump" combines the function of the teeth for oil displacement with the pole formation for the magnetic circle. The stator, which extends over 360 deg in reluctance drives, is cut in order to form two stators. The stators are distributed over 180 deg among the two gear wheels (fig. 1).

In the first phase of the experimental investigations, in order to verify the main concepts of the designed drive system, the test were made on the specially designed four pole, switched reluctance motor which cross-section is presented in the fig.2. The construction of the motor is similar to the construction of the "reluctance pump". In the final stage the "reluctance pump" depicted in fig.1 was built.

III. POWER ELECTRONIC CONVERTER FOR "RELUCTANCE PUMP"

Switched reluctance motor is a specialized form of stepper motor and has the lowest construction cost of any industrial electric motor due to its lack of magnets and simple structure. Because of the construction of the switched reluctance motor, the motor can't be fed directly from standard 3 phase mains supply.

Nowadays there are practically no power converters available on the market adjusted to be utilized in the reluctance motors drives. Because of that reason, in order to verify the control possibilities of prototyped drive system, the power electronic converter (fig.3) along with control algorithms and proper software structure was developed.

The 5kW bridge converter suitable to work with different types of 3-phase reluctance machines was built [4]. The topology of converter is depicted in the fig. 3. Converter consists of 12 IGBTs (3 independent full H-bridges) along with driver's protection circuits, hall effect current sensors for all machine phases, DC voltage measurement and encoder interface circuit.

The modular power electronics converter is presented in the fig. 4.The control system of the drive is based on TMS320F2812 DSP from Texas Instruments. Digital to analog converter gives possibility of real time observation of control algorithm's variables.

The communication with the PC is realized by means of RS-232. Properly designed Java application makes

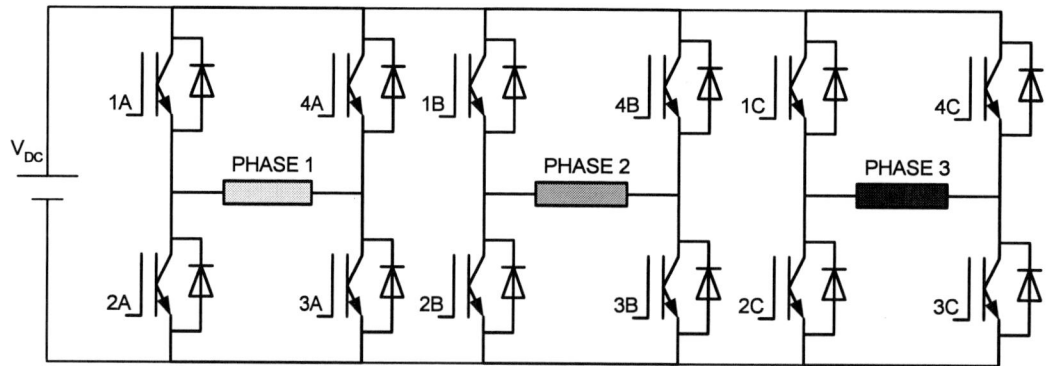

Fig.3. Topology of power electronic converter

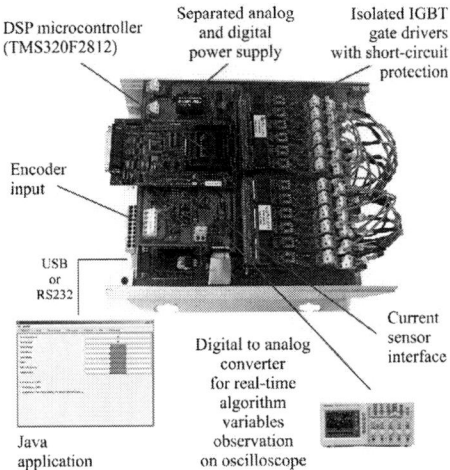

Fig.4. Power electronics converter used during
laboratory investigations

possible the data acquisition from the converter's measurement system. Furthermore, it helps to monitor the drive behavior and perform investigations with reluctance motors.

The software structure of the power electronics converter offers wide range of control algorithm procedures and protection functions and allows flexible system parameters adjustment.

IV. SIMULATION RESULTS

During the designing stage simulations based on the finite elements method (FEM) were made and the optimal stator current shape was determined.

In order to minimize the torque ripple, provide proper torque characteristic and assure proper magnetic flux spatial distribution in coupled rotors of the "reluctance pump" - bidirectional, trapezoidal stator winding current form (fig 5a) is necessary in the stator winding.

On the contrary, from the fig. 5b it can be observed that when the stator current waveform is different form the current waveform depicted in the fig. 5a, the torque's ripples increase.- the torque ripples in the fig 5b change between 2 Nm and 6 Nm, whereas in the fig. 5a the generated torque is in the range between 6 Nm and 7 Nm.

During the simulation investigations it was determined that that rated speed of 1000 rpm turned out to be most useful from of the mechanical and magnetic construction point of view.

V. EXPERIMENTAL RESULTS

Fig. 6 presents experimental results of implemented current control scheme operation. Three phase bidirectional, trapezoidal stator windings current waveforms are presented. The modulation frequency was set to 10 kHz. The amplitude of the current is equal to 15 A and the frequency of the current waveform is equal to 30 Hz. According to the simulation results, such kind of stator current waveform minimizes the torque ripple produced by the machine.

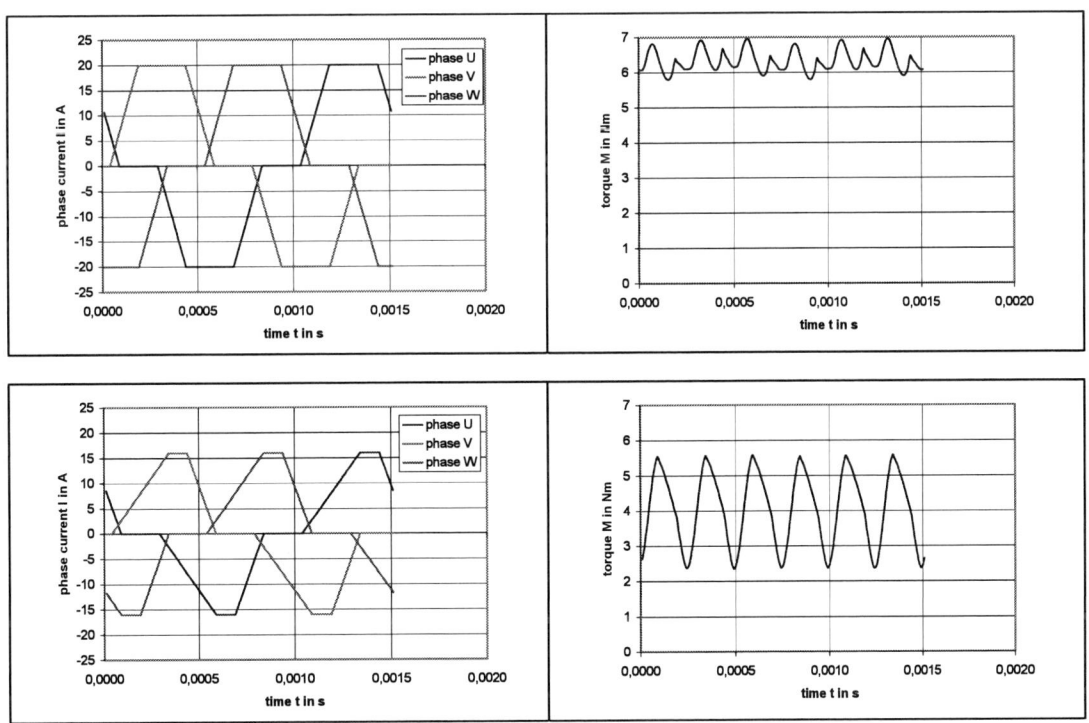

Fig.5. Simulation results of the "reluctance pump" torque for different shapes of stator windings currents.

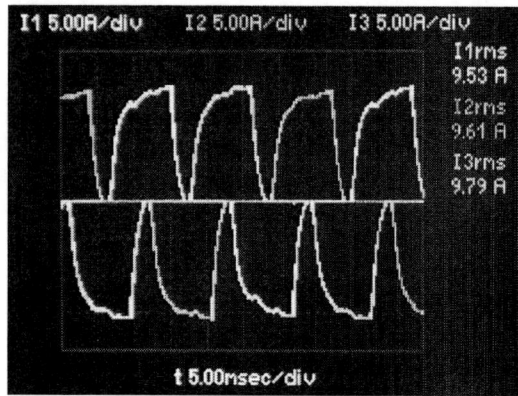

Fig.6. Experimental results – stator current waveforms
$I_{MAX} = 15$ A; $f = 30$ Hz

VI. CONCLUSION

This paper described the "reluctance pump" drive along with its components. Because of the non-conventional construction of the switched reluctance motor used as the "reluctance pump" special power electronic converter along with its control scheme was developed.

In order to verify the main concepts of the DSP control scheme, before the usage of the power electronics converter on the "reluctance pump", the investigations were made on the machine model of the "reluctance pump" with non-coupled rotors (fig.2).

Simulation and experimental results show that in order to assure proper magnetic flux spatial distribution along with optimal "reluctance pump" torque characteristic - bipolar, trapezoidal current shape should be applied to the stator windings.

It should be mentioned that so far the investigations were conducted without the usage of the position sensor. In the further researches the position of the rotor will be used what in turn should improve drive's overall performance. However, because of the construction of the pump, non-encoder operation is needed, what forces the usage of sensorless control methods.

Nevertheless, the investigations showed that functionally integrated electric-hydrostatic drives can be used only in the low-pressure applications. However, the further investigations can be carried out in the area of high-pressure application where integration idea can be realized by placing the hydraulic pump in to idle core volume of the electric motor [3]

ACKNOWLEDGMENT

The project "Development of Integrated Electro-Hydrostatic Drive Systems" is supported by the German Research Foundation. The authors would like to thank the German Research Foundation for the financial support of the research work.

REFERENCES

[1] B. Kaminski, K. Wejrzanowski, W. Koczara, "An application of PSIM simulation software for rapid prototyping of DSP based power electronics control systems", Power Electronics Specialists Conference, PESC 2004, 20-25 June 2004, Aachen, Germany

[2] Miller T.J.E. "Electronic Control of Switched Reluctance Machines", Reed Educational and Professional Publishing Ltd, 2001.

[3] W. Wustmann, S. Helduser ,U. Schuffenhauer, H. Kuß, N. Michalke, "Fully Integrated Electric-Hydrostatic Drive based on a gear pump and a switched reluctance motor", The Tenth Scandinavian International Conference on Fluid Power, SICFP'07,May 21-23,2007, Tampere, Finland

[4] B. Szymanski, H. Kuß, T. Wichert, K. Kompa, "Switched Reluctance Motor in textile machine drive", 39th Power Electronics Specialists Conference, PESC 2008, June 15 – 19, 2008, Rodos, Grece

A Predictive Control Scheme for Current Source Rectifiers

Pablo Correa[*], Jose Rodriguez[†]

[*]University Federico Santa Maria, Valparaiso, Chile, e-mail: *pablo.correa@usm.cl*

[†]University Federico Santa Maria, Valparaiso, Chile, e-mail: *jose.rodriguez@usm.cl*

Abstract: **Current source rectifiers constitute a commonly used topology in high power applications, featuring several advantages such as regeneration capability and high quality currents. This paper presents a simple predictive control scheme for a current source rectifier. Main achievements of the here presented control scheme are the simplicity in controlling the dc link current and the generation of currents with unity power displacement factor. Experimental results are presented to validate the proposed control principle.**

Keywords— **Converter Control, Active Damping**

I. INTRODUCTION

Current Source Rectifiers (CSR) are nowadays a well suited alternative to diode rectifiers and back-to-back converters. They offer on the one hand the capability of directly feeding back energy to the mains, and on the other, they generate sinusoidal input currents with unity displacement power factor. Thanks to these features CSRs are the preferred choice as current source for DC-loads or drives [1-3]. These converters are characterized by a weakly damped filter which requires suitable control methods.

Nowadays, commercial off-the-self microprocessors make possible the implementation of predictive control schemes in several converter topologies. One advantage of predictive control methods is the use of a very intuitive control law, in contrast to the controller design with the classical control theory, which can be very complex. In addition, multivariable cases, dead time compensation and the treatment of constraints can be easily dealt with [4,5]. Recently, predictive control has been introduced for matrix converters [6-8], active front ends [9], and two-level and multilevel inverters [5,10,12], featuring power factor control and high quality sinusoidal input currents.

Several control schemes have been proposed for the CSR, most of them use transformations to a rotational coordinates frame and PI controllers [13]. Additionally, all of these methods make use of Pulse Width Modulation (PWM) techniques, such as the Trapezoidal PWM [14], off-line calculated patterns with Selective Harmonic Elimination (SHE) [15], and Space Vector Modulation [16].

In this paper, a predictive control scheme for a CSR is presented. The proposed approach uses the prediction of the phase input current and of the output current with an horizon of one sampling time to calculate the best suited

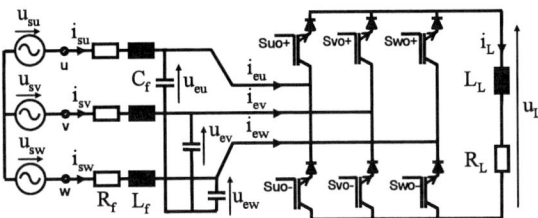

Fig. 1. Current source rectifier.

switching state for the rectifier that accomplish the dual purpose of controlling the power factor and the output current, without using any modulation scheme. The ideas presented are validated through simulation and experimental results obtained with a 2 kW converter prototype that uses IGBTs.

II. CONVERTER TOPOLOGY

The CSR consists of an array of six unidirectional power semiconductors switches which feed a single phase load, as it is shown in Fig. 1. The converter operates connecting at any time one switch of the high side and one switch of low side of the rectifier to the load. This constraint limits the operation of the rectifier to nine feasible switching states. In addition, an input filter is necessary to avoid over-voltages during the switching transitions and in case that a phase is not connected to the load.

In the following, it will be assumed that the three phase quantities of the rectifier are symmetrical and hence, can be represented by the well known two-dimensional space phasor. E.g. the phase current components i_{su}, i_{sv}, and i_{sw} will be described by the complex space vector:

$$\underline{i}_S = \underline{i}_{S\alpha} + j \cdot \underline{i}_{S\beta}, \tag{1}$$

defined as:

$$\left. \begin{array}{l} i_{S\alpha} = \dfrac{1}{3}\left(2i_{sU} - i_{sV} - i_{sW}\right), \\[2mm] i_{S\beta} = \dfrac{1}{\sqrt{3}}\left(i_{sV} - i_{sW}\right). \end{array} \right\} \tag{2}$$

This space phasor is referred to a stationary reference frame also called the α–β reference frame.

III. PRINCIPLE OF PREDICTIVE CONTROL

The predictive control scheme that is here proposed considers the discrete change of switching states at equidistant points with a constant sampling period. This state is selected out of the nine available possibilities in order to accomplish two requirements: first the line side of the rectifier must deliver mainly active power, and, secondly, in the load side the current must follow the reference. The first condition is fulfilled by minimizing the instantaneous reactive power:

$$q = \left(u_{s\alpha}^{k+1} \cdot i_{s\beta}^{k+1} - u_{s\beta}^{k+1} \cdot i_{s\alpha}^{k+1} \right), \qquad (3)$$

where k is the sample number. On the other hand, the second condition requires a minimum error between the load current and its reference:

$$\Delta i_L = i_L^* - i_L^{k+1}, \qquad (4)$$

where i_L is the load current and i_L^* corresponds to the DC-link current reference value. Both requirements are merged in a unique so called quality function:

$$g = A \cdot |q| + B \cdot |\Delta i_L| . \qquad (5)$$

The control scheme works as follows: At each sampling time, all possible switching states are used to calculate the predicted values of the load and input currents which are used to calculate the quality function g. Then, the switching state which delivers the minimum value of *g* will be used at the next sampling period. The weighting factors A and B decide whether the priority is given to those states which minimize the error in the load current or those which improve the power factor. A simplified block diagram of this control scheme is shown in Fig. 2.

IV. CALCULATION OF THE PREDICTED VALUES

A mathematical model of the converter provides the basis for the prediction of the input and output currents which are needed for evaluating the quality function. The line side of the rectifier is represented by a second order system described by the following equations:

$$\left. \begin{aligned} L_f \frac{di_s}{dt} &= u_s - u_e - R_f \cdot i_s \\ C_f \frac{du_e}{dt} &= i_s - i_e \end{aligned} \right\} \qquad (6)$$

where L_f represents the inductance and R_f the resistance of both the filter and the mains. The prediction of the input current is computed from a first order difference equation, as described in [6],

$$i_s^{k+1} = c_1 \cdot u_s^k + c_2 \cdot u_e^k + c_3 \cdot i_s^k + c_4 \cdot i_e^k . \qquad (7)$$

The real coefficients c_1, c_2, c_3, c_4 are defined so that the same values for the currents of the equivalent continuous

time system are obtained for all sampling instants. This is carried out by representing the equations (5) and (6) through a state space system with state variables i_s and u_e and inputs variables u_s and i_e:

$$\begin{bmatrix} \dot{u}_e \\ \dot{i}_s \end{bmatrix} = \begin{bmatrix} 0 & 1/C_f \\ -1/L_f & R_f \end{bmatrix} \begin{bmatrix} u_e \\ i_s \end{bmatrix} + \begin{bmatrix} 0 & -1/C_f \\ 1/L_f & 0 \end{bmatrix} \begin{bmatrix} u_s \\ i_e \end{bmatrix} \quad (8)$$

The discrete form of this system is given by:

$$\begin{bmatrix} u_e^{k+1} \\ i_s^{k+1} \end{bmatrix} = \Phi \begin{bmatrix} u_e^k \\ i_s^k \end{bmatrix} + \Gamma \begin{bmatrix} u_s^k \\ i_e^k \end{bmatrix}, \qquad (9)$$

where:

$$\Phi = e^{A \cdot Ts}, \qquad \Gamma = A^{-1}(\Phi - I)B . \qquad (10)$$

The model of the load is obtained in a similar way. Let us assume an inductive load as it is shown in Fig. 1. This system is described as follows:

$$L_L \frac{di_L}{dt} = u_L - R_L \cdot i_L \qquad (11)$$

Finally, expressing (11) as a difference equation, the predicted value of the load current is obtained:

$$i_L^{k+1} = d_1 u_L^k + d_2 i_L^k . \qquad (12)$$

IV. ACTIVE DAMPING OF THE INPUT FILTER

The described method selects the best suited switching combination that minimizes the quality function. By controlling the converter in this way, the switching frequency in the input is not fixed and rectifier could easily excite the parallel resonance mode of the input filter, especially when a low switching frequency is used. Because most current rectifiers are applied in high power, it is important to modify the method in such a way that operation at low switching frequency can be possible.

A standard approach to mitigate the resonance of the input filter is the use of a damping resistor. It is well know that a resistor in parallel with the filter capacitor could successfully accomplish with this function, at expenses of high losses which could not be acceptable in high power drives. A way to obtain the same effect is by means of the so called active damping, which has been described extensively in the literature [17]. In simple words, the active damping emulates the damping resistor by drawing a current with the rectifier which is proportional to the capacitor voltage:

$$i_{dU} = \frac{u_{eU}}{R_d}, i_{dV} = \frac{u_{eV}}{R_d}, i_{dW} = \frac{u_{eW}}{R_d}, \qquad (13)$$

or equivalently, in the α-β frame:

$$i_{d\alpha} = \frac{u_{e\alpha}}{R_d}, i_{d\beta} = \frac{u_{e\beta}}{R_d}. \qquad (14)$$

Previous works determine the output current using a reference frame rotating synchronous with the input voltage space vector. In this work, the output current is obtained out of simple relations between the instantaneous power of the damping resistor and the instantaneous power that should be drawn by the load:

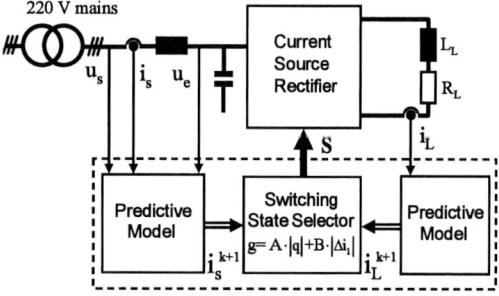

Fig. 2. Block diagram of the control scheme.

$$P_d = 1.5 \cdot \left(i_{d\alpha} u_{e\alpha} + i_{d\beta} u_{e\beta} \right) = i_{dL}^2 \cdot R_L, \quad (15)$$

In this way, the output current component that generates a equivalent instantaneous power is equal to:

$$i_d = \sqrt{\frac{1.5}{R_L \cdot R_d} \left(u_{e\alpha}^2 + u_{e\beta}^2 \right)}. \quad (16)$$

Because the predictive nature of the control scheme, the predicted value of (16) is actually used in the control scheme:

$$i_d^{k+1} = \sqrt{\frac{1.5}{R_L \cdot R_d} \left(\left(u_{e\alpha}^{k+1} \right)^2 + \left(u_{e\beta}^{k+1} \right)^2 \right)}. \quad (17)$$

In order to avoid active power flowing through the load, a more practical implementation of the active damping should damp all harmonics except the fundamental component. This selectivity in the frequency is done by applying a high pass filter for the current determined in (17), eliminating in this way the DC component responsible for the active power. A simple discrete first order high pass filter is used for this goal. Thus, the filtered value of the damping current is given by:

$$i_{df}^{k+1} = k_1 i_{df}^k + k_2 i_d^{k+1} + k_3 i_d^k. \quad (18)$$

Thus, the resulting current i_{df} is added to the original reference i_L^*, therefore eq. (4) is given now by:

$$\Delta i_L = i_{df}^{k+1} + i_L^{*\,k+1} - i_L^{k+1}. \quad (19)$$

V. RESULTS

In order to test the performance of the proposed method, simulations were carried out with Matlab Simulink. The rectifier filter and the load are selected with values which are equivalent to high power converters. The filter has parameters $C_f = 0.28$ p.u, $L_f = 0.083$ p.u, $R_f = 0.01$ p.u. and the load of the rectifier is $L_L = 0.62$ p.u. and $R_L = 1.3$ p.u. The predictive controller includes active damping with an equivalent resistor of $R_d = 1$ p.u. Fig. 3-a and 3-b shows sinusoidal input currents and unity power factor. Fig. 3-c shows a load step from $I_L^* = 1.27$ p.u. to 0.76 p.u. It can be seen that current follows the reference with accuracy.

The proposed control scheme was also experimentally validated in a 2kW CSR prototype. This set up consists of a transformer, a LC-filter, an IGBT-based rectifier with control board including a Xilinx Spartan-3 FPGA. The control scheme was implemented on a Xilinx Microblaze softprocesor of 32 bits. Results shown in Fig. 4 were obtained with the same parameters of simulations, except that the load resistor is RL=2.6 p.u. and only passive damping was used. In a similar way than simulations, the current feature unity power displacement. It should be noted that the distortion observed in the input currents is mainly produced by the 5th harmonic component present in the input voltage and it is not just a consequence of the control method proposed in this work. Fig. 4 shows that the output current follows the reference very closely.

Because the time discrete characteristic of the proposed scheme, the analytical determination of the switching frequency f_{sw} of each semiconductor is not direct. The maximum switching frequency is given $f_{sw} = f_s/2$, fortunately, the average switching frequency of each valve is much lower. Simulations and experimental measurements show a mean switching frequency lower than 2 kHz for the sampling rate previously mentioned. This switching frequency is quite suitable for modern converters with IGBTs and a rated output of the order of hundred kVAs.

VI. CONCLUSIONS

A predictive control strategy has been presented in this paper which permits a simple but effective control of a CSR. This control scheme uses a time discrete model of the converter and a quality function. The main idea relies in predicting the best suited switching state which has to be applied in the next sampling period. Both dynamic and stationary behaviour have been analyzed with simulations and experimentally. The measured output current follows the reference current very closely even under transient conditions, which corroborates the good dynamics provided with the proposal. Furthermore, the condition of

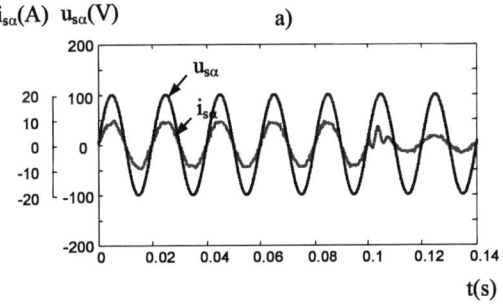

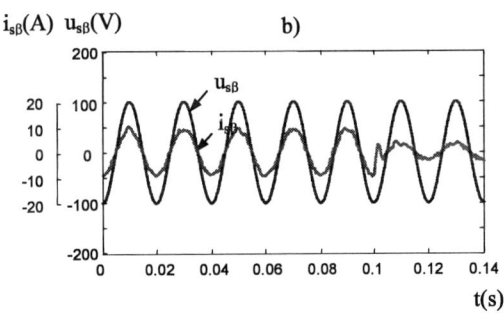

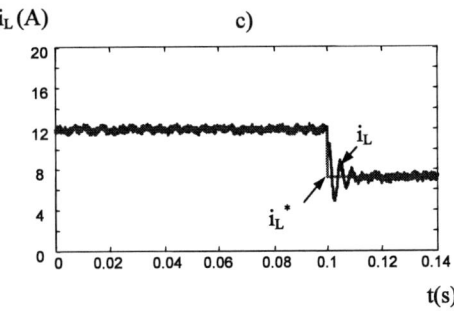

Fig. 3. Voltage and current waveforms for the rectifier: a) component α of the input voltage and current, b) component β of the input voltage and current, c) output current.

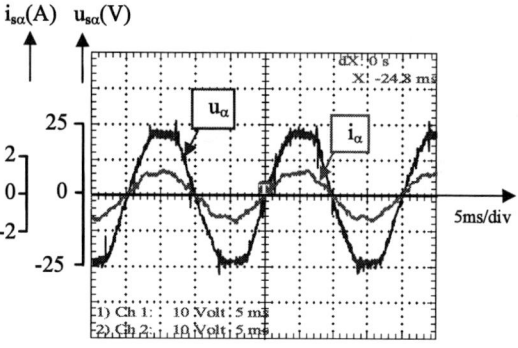

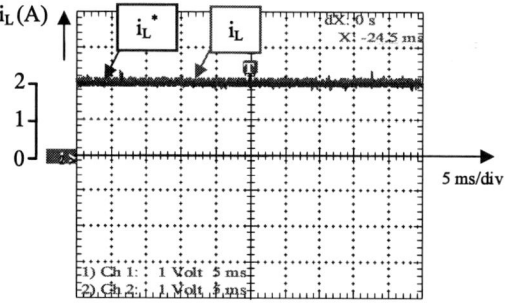

Fig. 4 Voltage and current waveforms for the rectifier (experimental): a) α component of the input voltage and current, b) output current.

unity power displacement has also been achieved. The disadvantage of this method is the parameters of the load and the input filter must be known. However, this is also a requirement for the design of other control schemes based on PI controllers and modulators.

VII. APPENDIX

The parameters of the rectifier are the following:

Nominal power: 2 kW	
Base voltage: 70 V rms	
Mains Filter :	Load :
L_f = 2 mH (0.083 p.u.)	R_L = 10 ohm (1.33 p.u.)
C_f = 120 uF (0.28 p.u.)	L_L = 15 mH (0.62 p.u.)
Rf = 1 ohm (0.013 p.u.)	

VIII. REFERENCES

[1] S. Rees "New Cascaded Control System for Current- Source Rectifiers," IEEE Transaction on Industrial Electronics, Vol. 52, N°3, June 2005, pp: 774-784.

[2] Y. Wei Li, B. Wu, N.R. Zargari, J.C. Wiseman, D. Xu; "Damping of PWM Current-Source Rectifier Using a Hybrid Combination Approach," IEEE Transactions on Power Electronics, Vol.22, issue 4, July 2007, pp: 1383-1393.

[3] T. Noguchi, D. Takeuchi, S. Nakatomi, A. Sato; "Novel Direct-Power-Control Strategy of Current-Source PWM Rectifier," in the proceedings of the 2005 international conference on Power Electronics and Drives Systems, PEDS 2005. Vol. 2, pp: 860- 865.

[4] J. Holtz, U. Boelkens, "Direct frequency converter with sinusoidal line currents for speed-variable AC motors;" IEEE Transactions

on Industrial Electronics, Vol. 36, Issue 4, Nov. 1989, pp. 475 – 479.

[5] P. Cortes, J. Rodriguez, R. Vargas, U. Ammann, "Cost Function-Based Predictive Control for Power Converters;" on the proceedings of the 32nd Annual IEEE Industrial Electronics conference, IECON 2006, Nov. 2006, pp:2268 – 2273.

[6] S. Müller, U. Ammann, S. Ress, „New Time-Discrete Modulation Scheme for Matrix Converters," *IEEE Trans. Ind. Electron*, vol. 52, no. 6, pp. 1607–1615, Dec. 2005.

[7] J. Rodriguez, J. Pontt, R. Vargas, P. Lezana, U. Ammann, P. Wheeler, F. Garcia, "Predictive direct torque control of an induction motor fed by a matrix converter;" in the proceedings of the 2007 European Conference on Power Electronics and Applications, 2-5 Sept. 2007, pp.1 – 10.

[8] M.E. Rivera, R.E, Vargas, J. Espinoza, J. Rodriguez; " Behavior of the Predictive DTC Based Matrix Converter Under Unbalanced AC Supply," in the proceedings of the 42nd IAS Industry Applications Conference, 2007. 23-27 Sept. 2007, pp: 202 – 207.

[9] H. Lamsahel, P. Mutschler, M. Reddy;. "Controlling voltage source converters with minimal energy storage," in the proceedings of the 2005 European Power Electronics and Applications Conference, 11-14 Sept. 2005. pp:9-.

[10] J. Rodriguez, J. Pontt, C.A. Silva, P. Correa, P. Lezana, P. Cortes, U. Ammann, "Predictive Current Control of a Voltage Source Inverter;" IEEE Transactions on Industrial Electronics, Vol. 54, Issue 1, Feb. 2007 Page(s):495 – 503.

[11] P. Correa, M. Pacas, J. Rodriguez, "Predictive Torque Control for Inverter-Fed Induction Machines," IEEE Transactions on Industrial Electronics, Vol. 54, Issue 2, April 2007, pp.1073 – 1079.

[12] R. Vargas, P. Cortes, U. Ammann, J. Rodriguez, J.Pontt; "Predictive Control of a Three-Phase Neutral-Point-Clamped Inverter," IEEE Transactions on Industrial Electronics, Vol. 54, Issue 5, Oct. 2007 :2697 – 2705.

[13] Bin Wu, *High Power Converters and AC-drives*, Wiley-IEEE Press, 2006, ISBN 0471731714

[14] G. Joos, J. Espinoza; "PWM control techniques in current source rectifiers" in the proceedings of the International Conference on Industrial Electronics, Control and Instrumentation, IECON 93′, Nov. 1993, pp. 1210-1214, vol. 2.

[15] B. Wu, S. B. Dewan, G. R, Slemon; "PWM-CSI inverter for induction motor drives," IEEE Transactions on Industry Applications, Vol. 28, Issue 1, Part 1, Jan.-Feb. 1992 pp. 64 – 71.

[16] D. Xu, N.R. Zargari, B. Wu, J. Wiseman, B. Yuwen, S. Rizzo; "A Medium Voltage AC Drive with Parallel Current Source Inverters For High Power Applications;" in the proceedings of 36th IEEE Power Electronics Specialists Conference,. PESC '05. 2005, pp: 2277-2283.

[17] J. C. Wiseman, B. Wu "Active damping control of a high-power PWM current-source rectifier for line-current THD reduction," IEEE Transactions on Industrial Electronics, Vol. 52, Issue 3, Jun. 2005 pp. 758 – 764.

Analysis and Design of New Switching Table for Direct Power Control of Three-Phase PWM Rectifier

Abdelouahab Bouafia [*], Jean-Paul Gaubert [†] and Fateh Krim [*]

[*] Laboratoire d'Electronique de Puissance et Commande Industrielle (LEPCI), Université de Sétif, *Algérie*
bouafia_aou@yahoo.fr   krim_f_ieee_org@yahoo.fr
[†] Laboratoire d'Automatique et d'Informatique Industrielle (LAII-ESIP), Université de Poitiers, *France*
Jean.Paul.Gaubert@univ-poitiers.fr

Abstract— in this paper, a new switching table is presented for direct power control (DPC) of three-phase PWM rectifiers. The proposed switching table is synthesized by analyzing the instantaneous active and reactive power correction. For this reason, and based on the sign and magnitude of the change in instantaneous active and reactive power, the best converter switching state allowing smooth control of active and reactive power simultaneously during each sector is selected. The dc-bus voltage is maintained at the required level by controlling active power to be constant and equal to its reference value. The unity power factor (UPF) operation of the converter is achieved by maintaining the reactive power zero during all sectors. Simulation and experimental results have proven that the proposed DPC is much better than the classical one and verify the validity and performance of the presented DPC.

Keywords— Direct power control, PWM, DC power supply, converter control, power factor correction.

I. INTRODUCTION

Various control strategies have been proposed in recent works on three-phase PWM rectifiers. They can be classified for its use of current loop controllers or active/reactive power controllers. The well-known method of indirect active and reactive power control is based on current vector orientation with respect to the line voltage vector. It is called voltage-oriented control (VOC) [1]-[5]. VOC guarantees high dynamics and static performance via internal current control loops. However, the final configuration and performance of the VOC system largely depends on the quality of the applied current control strategy.

Over the last few years an interesting emerging control technique has been direct power control (DPC) developed analogously with the well-known direct torque control (DTC) used for adjustable speed drives [5]-[12]. In DPCs schemes there are no internal current loops and the converter switching states are appropriately selected by a switching table based on the instantaneous errors, between the commanded and estimated values of active and reactive power, and voltage vector position [6] or virtual-flux vector position [8].

In this paper, a new switching table is designed for direct power control (DPC) of three-phase PWM rectifier, which makes it possible to achieve the unity power factor

(UPF) operation by directly controlling its instantaneous active and reactive power, without any power-source voltage sensors. A method, based on the study of the instantaneous active and reactive power correction, is adopted for synthesizing the new switching table, different to the one found in the control of ac machines and used in [6] and in [8], is presented. Finally, the proposed DPC is simulated and experimentally implemented. It is shown via simulation and experimental results that the proposed DPC has high performance and much better than the classical one. Good regulation of the dc-bus voltage and unity power factor operation are successfully achieved in transient and steady states.

II. PRINCIPLES OF DPC

A. System Configuration:

DPC is based on the instantaneous active and reactive power control loops. In DPC scheme, the converter switching states are selected by a switching table based on the instantaneous errors between the commanded and estimated values of active and reactive power. Fig.1 shows the configuration of the direct instantaneous active and reactive power control for three-phase PWM rectifier in which the symbols are as follows:

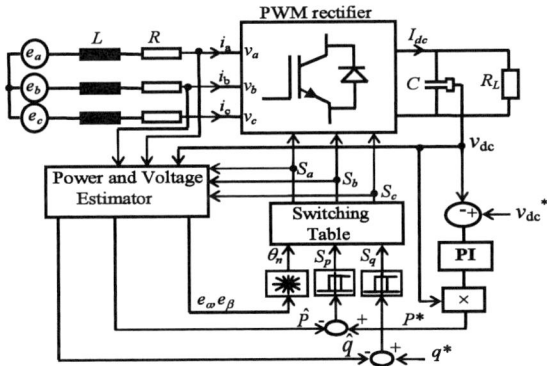

Fig.1. Configuration of DPC for three-phase PWM rectifier.

e_a, e_b, e_c three-phase power-source voltages;
v_a, v_b, v_c ac terminal voltages of the PWM rectifier;
i_a, i_b, i_c three-phase line currents;

S_a, S_b, S_c switching states of the converter;
L, R parameters of interconnecting reactors;
C, R_L dc-link capacitor and load resistance.

In DPC scheme, the dc-bus voltage is regulated by adjusting the active power, and the UPF operation is achieved by controlling the reactive power to be zero. As shown in Fig.1, the active power command, $P*$, is provided from the outer proportional-integral (PI) dc-bus voltage controller. The reactive power command, $q*$, is directly given from the outside of the controller, usually equal to zero for UPF. Errors between the commands and the estimated feedback power are input to the hysteresis comparators and digitized to the signals S_P and S_q. Also, the phase of the power-source voltage vector is converted to the digitized signal θ_n. For this purpose, the stationary coordinates are divided into 12 sectors, as shown in Fig.2.

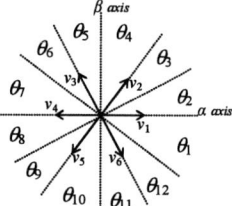

Fig.2. Sectors in stationary coordinates and rectifier voltage vectors.

The digitized error signals S_P and S_q and digitized voltage phase θ_n are input to the switching table in which every switching state, S_a, S_b, and S_c, of the converter is stored. By using this switching table, the optimum switching state of the converter can be selected uniquely in every specific moment according to the combination of the input signals.

B. New Switching Table Synthesizing:

In the stationary reference frame α-β and for a balanced three-phase system, the line currents equation can be represented as follows:

$$\frac{di_\alpha}{dt} = \frac{1}{L}\left(e_\alpha - v_\alpha - R.i_\alpha\right)$$
$$\frac{di_\beta}{dt} = \frac{1}{L}\left(e_\beta - v_\beta - R.i_\beta\right) \qquad (1)$$

From (1), line current vector $[i_\alpha\ i_\beta]^T$ can be controlled by selecting the proper rectifier voltage vector. The change in line current depends on the actual supply voltage vector, $e_{\alpha\beta}$, on the selected rectifier voltage vector, $v_{\alpha\beta}$, and in less measure on the actual line current. The parameter R can be practically neglected and a discrete first order approximation of (1) can be adopted. So the change in line current vector for the next control period is given by:

$$\Delta i_\alpha = i_\alpha(k+1) - i_\alpha(k) = \frac{T_s}{L}\left(e_\alpha(k) - v_\alpha(k)\right) \quad (2)$$
$$\Delta i_\beta = i_\beta(k+1) - i_\beta(k) = \frac{T_s}{L}\left(e_\beta(k) - v_\beta(k)\right)$$

The instantaneous active and reactive power can be represented as follows:

$$\begin{bmatrix} P \\ q \end{bmatrix} = \begin{bmatrix} e_\alpha & e_\beta \\ e_\beta & -e_\alpha \end{bmatrix}\begin{bmatrix} i_\alpha \\ i_\beta \end{bmatrix} \qquad (3)$$

As first approximation, and if the switching frequency is high enough, the change in power-source voltage can be neglected. The change in the active and reactive power can be estimated for the next control cycle as follows:

$$\begin{cases} \Delta P = e_\alpha(k).\Delta i_\alpha + e_\beta(k).\Delta i_\beta \\ \Delta q = e_\beta(k).\Delta i_\alpha - e_\alpha(k).\Delta i_\beta \end{cases} \qquad (4)$$

By replacing (2) in (4) we obtain:

$$\Delta P = \frac{T_s}{L}\left(e_\alpha^2(k) + e_\beta^2(k)\right)$$
$$- \frac{T_s}{L}\left(e_\alpha(k).v_\alpha(k) + e_\beta(k).v_\beta(k)\right) \qquad (5)$$
$$\Delta q = \frac{T_s}{L}\left(e_\alpha(k).v_\beta(k) - e_\beta(k).v_\alpha(k)\right)$$

For controlling the instantaneous active and reactive power there are six basic none zero rectifier voltage vectors and two zero rectifier voltage vectors available as shown in Fig.2. For the seven basic rectifier voltage vectors we obtain seven possible values of change in active and reactive power. As a result, there are different ways of selecting the corresponding switching state that controls the evolution in active and reactive power. For $i=(0,1,2,.....,6)$ change in the active and reactive power are given by the following expressions:

$$\Delta P_i = \frac{T_s}{L}\left(e_\alpha^2(k) + e_\beta^2(k)\right)$$
$$- \frac{T_s}{L}\left(e_\alpha(k).v_{\alpha i} + e_\beta(k).v_{\beta i}\right) \qquad i=0,....,6. \quad (6)$$
$$\Delta q_i = \frac{T_s}{L}\left(e_\alpha(k).v_{\beta i} - e_\beta(k).v_{\alpha i}\right)$$

In the stationary reference frame α-β, the power-source voltage vector is given by the following transformation:

$$e_{\alpha\beta} = \begin{bmatrix} e_\alpha \\ e_\beta \end{bmatrix} = \sqrt{\frac{2}{3}}\begin{bmatrix} 1 & -1/2 & -1/2 \\ 0 & \sqrt{3}/2 & -\sqrt{3}/2 \end{bmatrix}\begin{bmatrix} e_a \\ e_b \\ e_c \end{bmatrix} \qquad (7)$$

This vector can be represented in (α, β) plane as:

$$e_\alpha = E.cos(\theta),\ e_\beta = E.sin(\theta)\ \text{and}\ \|e_{\alpha\beta}\| = E \qquad (8)$$

Where: E is the *RMS* value of line to line power-source voltage and θ is the angular position of the power-source voltage vector in α-β coordinates defined as:
$$\theta = arctg(e_\beta / e_\alpha),\ -\pi/6 \leq \theta \leq 11\pi/6$$

For the rectifier voltage vector, v, the typical space vector representation, for each converter switching state and its corresponding, v_α and v_β components values are shown in the following Table I.

TABLE I
RECTIFIER VOLTAGE SPACE VECTORS

v_i	v_a	v_b	v_c	$v_{\alpha i}$	$v_{\beta i}$
0	0	0	0	0	0
1	$2/3 v_{dc}$	$-1/3 v_{dc}$	$-1/3 v_{dc}$	$\sqrt{2/3}\ v_{dc}$	0
2	$1/3 v_{dc}$	$1/3 v_{dc}$	$-2/3 v_{dc}$	$1/\sqrt{6}\ v_{dc}$	$1/\sqrt{2}\ v_{dc}$
3	$-1/3 v_{dc}$	$2/3 v_{dc}$	$-1/3 v_{dc}$	$-1/\sqrt{6}\ v_{dc}$	$1/\sqrt{2}\ v_{dc}$
4	$-2/3 v_{dc}$	$1/3 v_{dc}$	$1/3 v_{dc}$	$-\sqrt{2/3}\ v_{dc}$	0
5	$-1/3 v_{dc}$	$-1/3 v_{dc}$	$2/3 v_{dc}$	$-1/\sqrt{6}\ v_{dc}$	$-1/\sqrt{2}\ v_{dc}$
6	$1/3 v_{dc}$	$-2/3 v_{dc}$	$1/3 v_{dc}$	$1/\sqrt{6}\ v_{dc}$	$-1/\sqrt{2}\ v_{dc}$

The rectifier voltage vector in (α, β) plane is given by:

$$v_{\alpha\beta} = \begin{bmatrix} v_\alpha & v_\beta \end{bmatrix}^T, \quad \|v_{\alpha\beta}\| = \sqrt{2/3}.v_{dc} \qquad (9)$$

The normalised value of the change in active power and reactive power can be deduced from (6) as follows:

$$\overline{\Delta P_i} = \frac{\Delta P_i}{\frac{T_s}{L}\|e_{\alpha\beta}\|\|v_{\alpha\beta}\|} = \frac{\|e_{\alpha\beta}\|}{\|v_{\alpha\beta}\|} - \left(\cos(\theta).\overline{v}_{\alpha i} + \sin(\theta).\overline{v}_{\beta i}\right) \qquad (10)$$

$$\overline{\Delta q_i} = \frac{\Delta q_i}{\frac{T_s}{L}\|e_{\alpha\beta}\|\|v_{\alpha\beta}\|} = \cos(\theta).\overline{v}_{\beta i} - \sin(\theta).\overline{v}_{\alpha i}$$

It can be seen from (10) that the change in reactive power during all sectors has a sinusoidal waveform for all rectifier voltage vectors v_i. The change in active power has a shifted sinusoidal waveform as shown in Fig.3 and Fig.4 respectively.

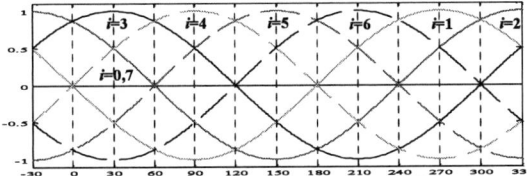

Fig.3 Change in instantaneous reactive power $\overline{\Delta q_i}$

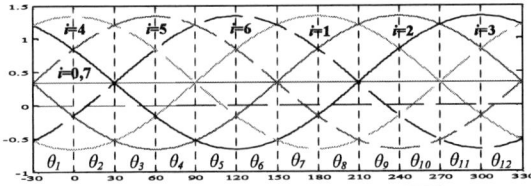

Fig.4 Change in instantaneous active power $\overline{\Delta P_i}$

The basic idea of the proposed DPC was to choose the best rectifier voltage vector among the seven possible vectors in order to ensure smooth control of instantaneous active and reactive power during each sector. To maintain the dc-bus voltage close to the reference value, and to keep the unity power factor, active power is controlled to be constant and equal to its reference value, the reactive power should be maintained zero for all sectors. For this reason, the new switching table synthesis is based on the sign and magnitude of the change in active and reactive power for each sector, as shown in Fig.3 and Fig.4. For example, for sector one the sign of the change in active and reactive power are shown in table II.

TABLE II
SIGN OF CHANGE IN ACTIVE AND REACTIVE POWER FOR SECTOR 1

$\overline{\Delta P_1}$		$\overline{\Delta q_1}$		
>0	<0	>0	=0	<0
v_4, v_3, v_5, v_0	v_1, v_6	v_2, v_3, v_1	v_0	v_5, v_6, v_4

For each combination of hysteresis output signals, S_P and S_q, rectifier voltage vectors are selected for sector one as shown in the following table:

TABLE III
SELECTED RECTIFIER VOLTAGE VECTORS FOR SECTOR 1

Sector 1		$\overline{\Delta q_1}$	
		$>0 \leftrightarrow S_q=1$	$<0 \leftrightarrow S_q=0$
$\overline{\Delta P_1}$	$>0 \leftrightarrow S_p=1$	v_3	v_4, v_5
	$<0 \leftrightarrow S_p=0$	v_1	v_6

For all sectors, the proposed new switching table is represented in the following Table IV.

TABLE IV
THE NEW SWITCHING TABLE FOR DPC

S_p	S_q	θ_1	θ_2	θ_3	θ_4	θ_5	θ_6	θ_7	θ_8	θ_9	θ_{10}	θ_{11}	θ_{12}
1	0	v_5	v_6	v_6	v_1	v_1	v_2	v_2	v_3	v_3	v_4	v_4	v_5
	1	v_3	v_4	v_4	v_5	v_5	v_6	v_6	v_1	v_1	v_2	v_2	v_3
0	0	v_6	v_1	v_1	v_2	v_2	v_3	v_3	v_4	v_4	v_5	v_5	v_6
	1	v_1	v_2	v_2	v_3	v_3	v_4	v_4	v_5	v_5	v_6	v_6	v_1

$v_1(100), v_2(110), v_3(010), v_4(011), v_5(001), v_6(101), v_0(000), v_7(111)$.

The switching table of classical DPC presented in [6] is the following:

TABLE V
THE SWITCHING TABLE OF CLASSICAL DPC

S_p	S_q	θ_1	θ_2	θ_3	θ_4	θ_5	θ_6	θ_7	θ_8	θ_9	θ_{10}	θ_{11}	θ_{12}
1	0	v_6	v_7	v_1	v_0	v_2	v_7	v_3	v_0	v_4	v_7	v_5	v_0
	1	v_7	v_7	v_0	v_0	v_7	v_7	v_0	v_0	v_7	v_7	v_0	v_0
0	0	v_6	v_1	v_1	v_2	v_2	v_3	v_3	v_4	v_4	v_5	v_5	v_6
	1	v_1	v_2	v_2	v_3	v_3	v_4	v_4	v_5	v_5	v_6	v_6	v_1

To achieve voltage sensorless operation of DPC for three-phase PWM rectifier, active and reactive powers are estimated using the switching state of the converter, the three-phase line currents, the dc-bus voltage, and the inductance of the reactors [6]. It can be derived as:

$$\hat{p} = L\left(\frac{di_a}{dt}i_a + \frac{di_b}{dt}i_b + \frac{di_c}{dt}i_c\right) + v_{dc}(S_a i_a + S_b i_b + S_c i_c) \qquad (11)$$

$$\hat{q} = \sqrt{3}L\left(\frac{di_a}{dt}i_c - \frac{di_c}{dt}i_a\right) \qquad (12)$$

$$-\frac{1}{\sqrt{3}}v_{dc}\left[S_a(i_b - i_c) + S_b(i_c - i_a) + S_c(i_a - i_b)\right]$$

The power-source voltage vector, $e_{\alpha\beta}$, can be estimated using the following equation:

$$\begin{bmatrix} \hat{e}_\alpha \\ \hat{e}_\beta \end{bmatrix} = \frac{1}{i_\alpha{}^2 + i_\beta{}^2} \begin{bmatrix} i_\alpha & -i_\beta \\ i_\beta & i_\alpha \end{bmatrix} \begin{bmatrix} \hat{p} \\ \hat{q} \end{bmatrix} \qquad (13)$$

III. SIMULATION RESULTS

The main electrical parameters of the power circuit and control data, used in simulation and implementation tests, are given in Table VI. Several tests, in simulation studies, were conducted to verify feasibility and performance of the proposed DPC. Fig.5 and Fig.6 show the simulated waveforms, under UPF operation in the steady state for purely sinusoidal power-source voltages, for the new DPC and classical one respectively.

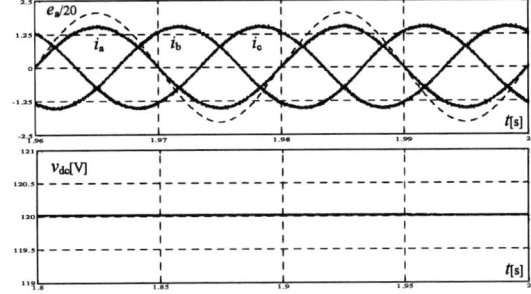

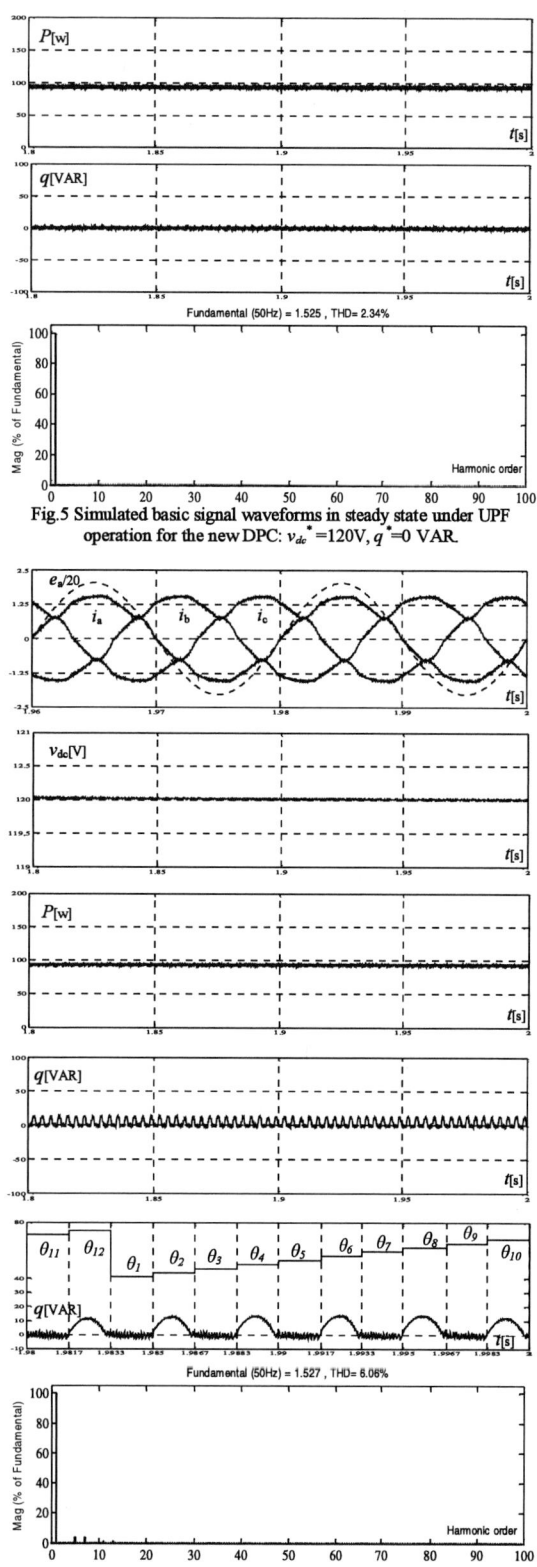

Fig.5 Simulated basic signal waveforms in steady state under UPF operation for the new DPC: v_{dc}^{*}=120V, q^{*}=0 VAR.

Fig.6 Simulated basic signal waveforms and enlargement on one main period for sectors and reactive power in steady state under UPF operation for the classical DPC: v_{dc}^{*}=120V, q^{*}=0 VAR.

It can be seen from these figures, that the new switching table ensures smooth control of active and reactive power during all sectors. Each instantaneous power tracks its reference value with a good approximation and stability. As a result, input currents are very close to sine waveform (THD=2.34%) and in phase with line voltages. The reactive power for the classical DPC is poorly controlled during even sectors (2,4,6,8,10,12), because in the first and second line of classical switching Table V, the same zero rectifier voltage vector is selected for both values of S_q which do not ensures control of reactive power as shown in Fig.3. As a result, input currents for classical DPC are distorted (THD=6.06%). Also, active power is poorly controlled during odd sectors (1,3,5,7,9,11), for classical DPC, because the selected vectors in the first line of switching Table V decrease active power as shown in the following figures, Fig.7 and Fig.8, in case of load power increasing.

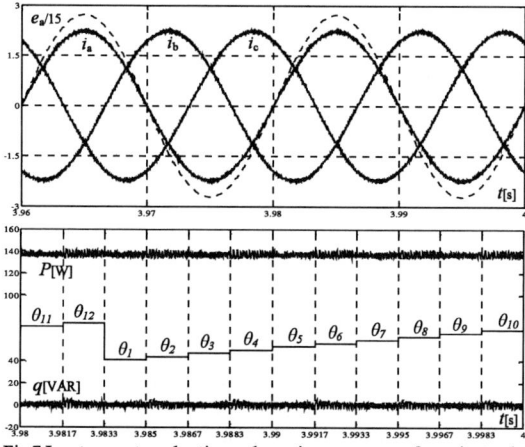

Fig.7 Input currents and active and reactive power waveforms in steady state under UPF operation of the new DPC for load power increasing (50%): v_{dc}^{*}=120V, q^{*}=0 VAR.

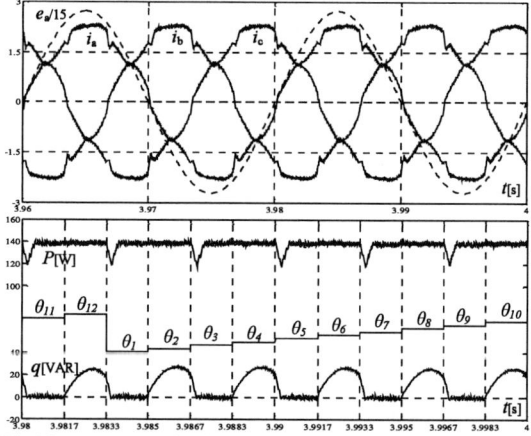

Fig.8 Input currents and active and reactive power waveforms in steady state under UPF operation of the classical DPC for load power increasing (50%): v_{dc}^{*}=120V, q^{*}=0 VAR.

Fig.9 shows a result of a step response against the disturbance load power under the unity power factor operation for the proposed DPC. The load power was increased (100%) in this test. It can be observed that the unity power factor operation is successfully maintained, even in this transient state. Notice that, after a short transient, the dc-bus voltage is maintained close to its reference value. From Fig.9, it can be found that the active power control and the reactive power control are independent of each other.

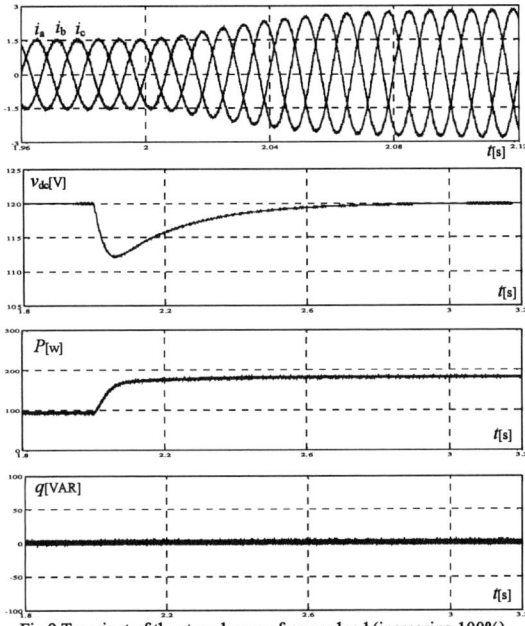

Fig.9 Transient of the step change of power load (increasing 100%).

The proposed control method can control the power factor indirectly by changing the reactive power command q*. Fig.10 and Fig.11 show examples of the operating characteristics under lagging and leading power factor operations. In these tests, the reactive power command has been changed to +80 VAR or -80 VAR, and it is confirmed that the proposed DPC can adjust the current phase indirectly with respect to the line voltage through the reactive power command.

Fig. 10 Waveforms under lagging power factor operation q*=+80 VAR.

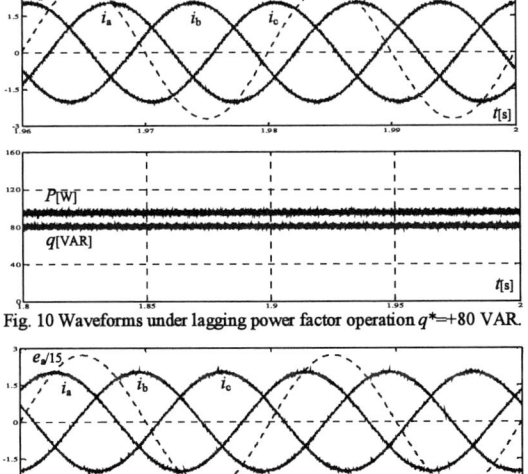

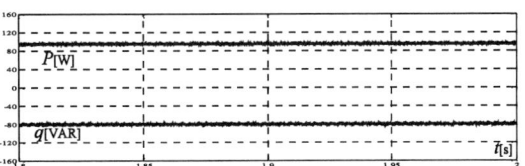

Fig. 11 Waveforms under leading power factor operation q*= -80 VAR.

TABLE VI
ELECTRICAL PARAMETERS OF POWER CIRCUIT

Sampling frequency	20 kHZ
Resistance of interconnecting reactors R	0.56 [Ω]
Inductance of interconnecting reactors L	19.5 [mH]
dc-bus capacitor C	2350 µF
Load resistance R_L	175 [Ω]
Line to line ac voltage E	50V rms
Source voltage frequency f	50 Hz
dc-bus voltage v_{dc}	120 V

IV. EXPERIMENTAL RESULTS

An experimental prototype, Fig.12, has been developed, in LAII-ESIP-laboratory France, to examine operating characteristic of the proposed DPC scheme. A three-phase IGBT based inverter is used as PWM rectifier. The development of control algorithm is performed with Matlab/Simulink™ and the real-time implementation with dSPACE (RTI1104) digital signal processor inserted in a PC-Pentium.

Fig.12 Experimental test bench and its dSPACE system control.

The experimental waveforms and measurements, under UPF operation in the steady state, for the new DPC and classical DPC, are presented in Fig.13 and Fig.14 respectively. It can be seen from these figures, that the new DPC is much better than the classical one. Input currents obtained with the new switching table are very close to sine wave (THD=3.5%) and in phase with power-source voltages (φ=1°, displacement factor DPF=1). Power factor is very close to one (PF=0.979).

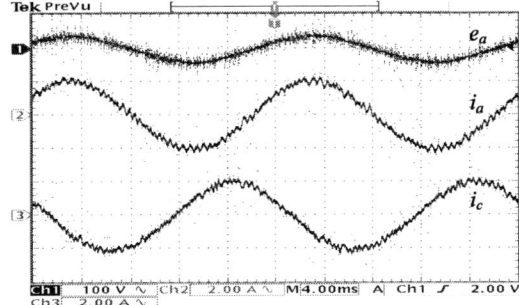

707

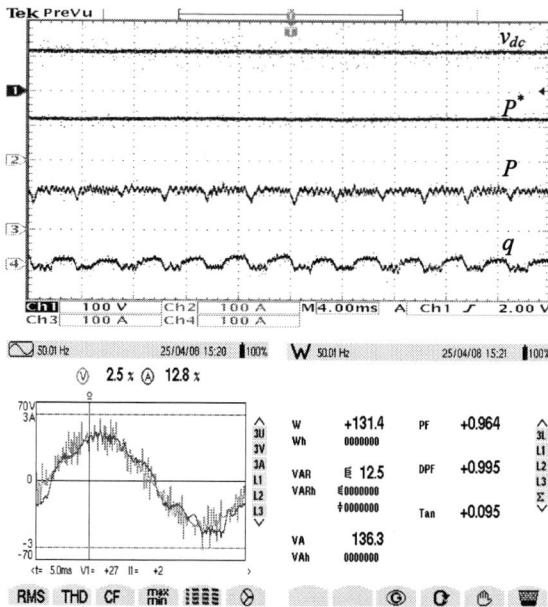

Waveforms with THD of e_a and i_a and power quality analysis

Fig.14 Experimental waveforms and measurements (using power quality analyser) under unity power factor operation in steady state for the classical DPC: $v_{dc}^* =120$V, $q^*=0$ VAR.

Other experimental tests were conducted to verify feasibility and performance of the proposed DPC in transient state, in case of step change of v_{dc}^* or step change of load power, are presented in Fig.15 and Fig.16 respectively. After a short transient, the dc-bus voltage and active power are maintained close to their new reference values with good approximation and stability. The reactive power is maintained zero even in this transient response. The input currents have nearly sinusoidal waveforms.

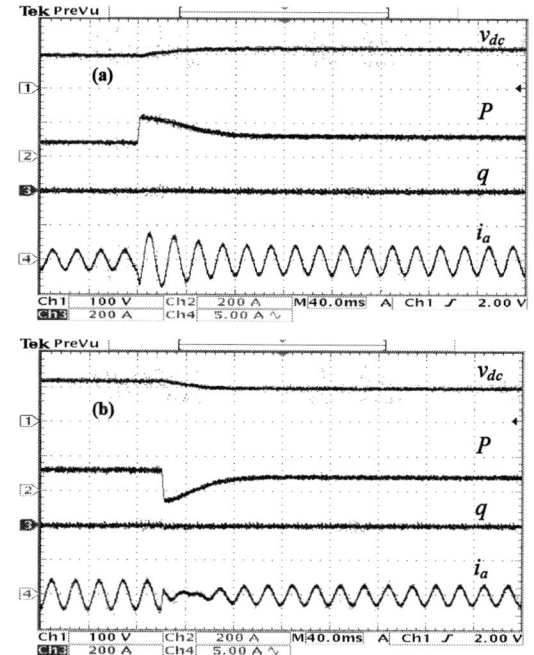

Fig.15 Transient of the step change of v_{dc}^* for $q^*=0$ VAR: (a) increasing from 100 to 120V, (b) decreasing from 120 to 100V.

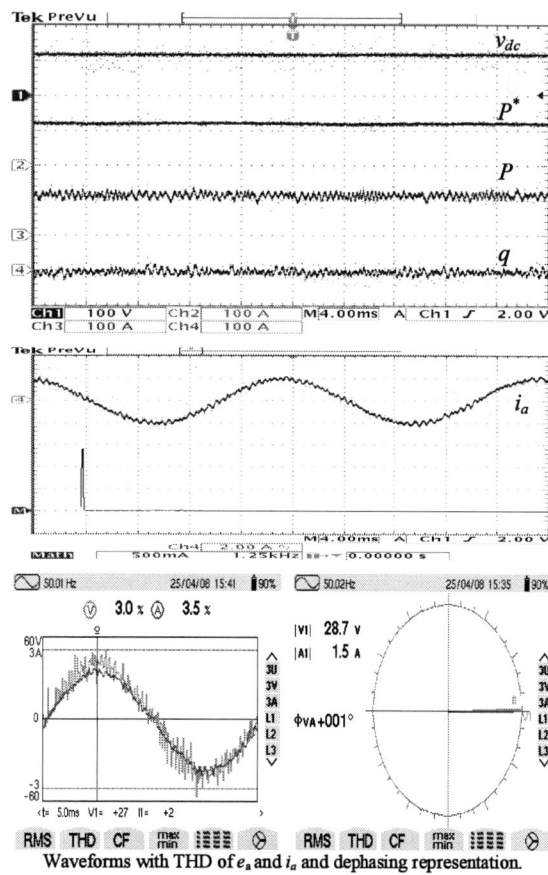

Waveforms with THD of e_a and i_a and dephasing representation.

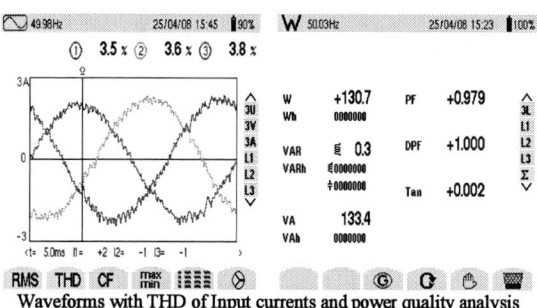

Waveforms with THD of Input currents and power quality analysis

Fig.13 Experimental waveforms, input currents spectrum and measurements (using power quality analyser) under unity power factor operation in steady state for the new DPC: $v_{dc}^* =120$V, $q^*=0$ VAR.

Experimental results and measurements of classical DPC are presented in the following figures.

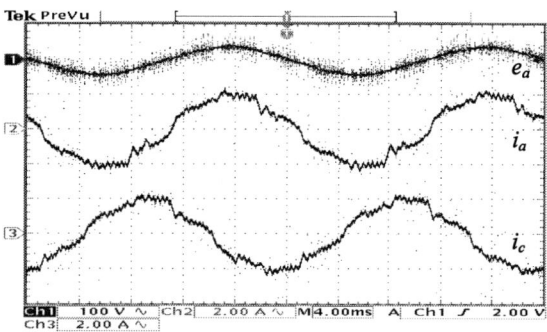

708

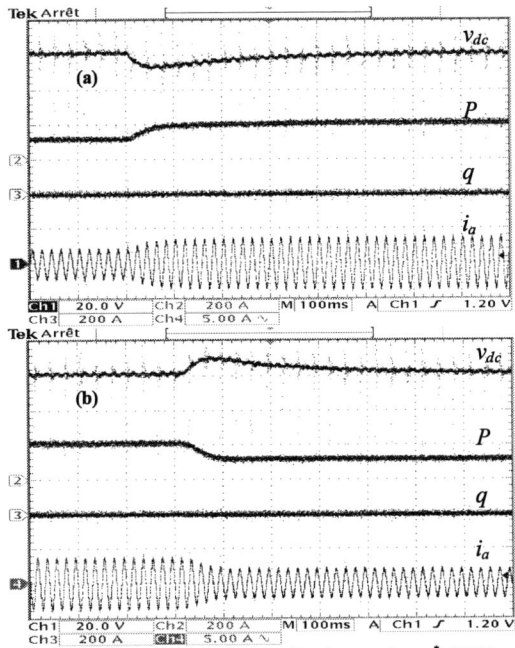

Fig.16 Transient of the step change of load power for v_{dc}^*=120V and q^*=0: (a) load power increasing 100%, (b) load power decreasing 50%.

Fig.17 shows examples of the operating characteristics under lagging (q^*=+80VAR), and leading (q^*=-80VAR) power factor respectively.

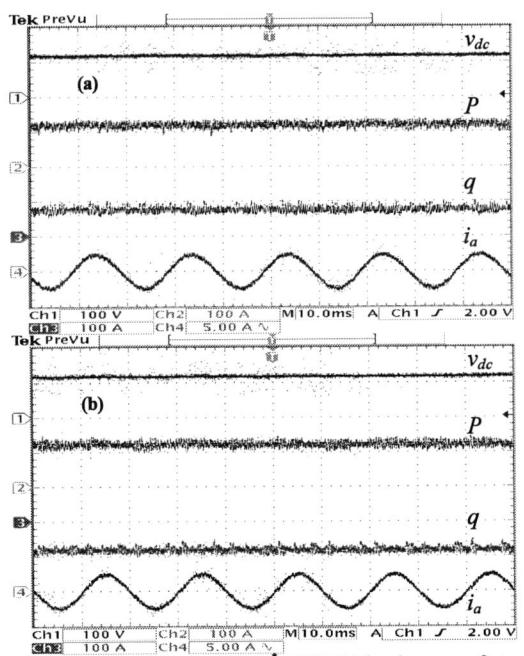

Fig.17 Experimental results for v_{dc}=120V: (a) lagging power factor operation q^*=+80VAR, (b) leading power factor operation q^*=-80VAR.

V. CONCLUSION

This paper has described a direct power control of three-phase PWM rectifier (DPC) based on new switching table. The main goal of the control system is to maintain the dc-bus voltage at the required level and satisfy the unity power factor (UPF) operation of the three-phase PWM rectifier. The converter switching states of the new switching table are chosen by analysing the instantaneous active and reactive power correction. The best rectifier vector allowing smooth control of active and reactive power during each sector is selected. The proposed DPC was simulated and implemented in real-time using conventional PI controller for dc-bus voltage regulation. The presented results indicate that the novel DPC is much better than the classical one in steady and transient states. Input currents waveforms are more sinusoidal, near unit power factor and decoupled active and reactive power control are successfully achieved using the proposed switching table.

REFERENCES

[1] S. Hansen, M. Malinowski, F. Blaabjerg, and M. P. Kazmierkowski, "Control strategies for PWM rectifiers without line voltage sensors," in Proc. IEEE APEC, vol. 2, pp. 832-839, 2000.

[2] M. P. Kazmierkowski and L. Malesani, "Current Control techniques for three-phase voltage-source PWM converter: A survey," IEEE Trans. Ind. Electron., vol. 45, pp. 691-703, Oct 1998.

[3] B. H. Kwon, J. H. Youm, and J.W. Lin, "A line voltage-sensorless synchronous rectifier," IEEE Trans. Power Electron., vol. 14, pp. 966-972, Sept 1999.

[4] M.Liserre, A.Dell'Aquila, F.Baabjeerg, "Genetic algorithm-based design of the active damping for an LCL-filter three-phase active rectifier," IEEE Trans. Power Electron., vol.19, No. 1,pp. 76-86, January 2004.

[5] M. Malinowski, M. P. Kazmierkowski, A.Trzynadlowski, "Review and comparative study of control techniques for three-phase PWM rectifiers," Electrimacs, static power converters II, Auguest 2002.

[6] T. Noguchi, H. Tomiki, S. Kondo, and I. Takahashi, "Direct power control of PWM converter without power-source voltage sensors," IEEE Trans. Ind. Applicat., vol. 34, pp. 473-479, May/June 1998.

[7] G. Escobar,A. M. Stankovic, J. M. Carrasco, and E. Galvan, and R.Ortega, "Analysis and design of Direct power control (DPC) for a three phase synchronous rectifier via output regulation subspaces," IEEE Trans. Power Electron., vol. 18, pp. 823-830, May 2003.

[8] M. Malinowski, M. P. Kazmierkowski, S. Hansen, f. Blaabjerg, and G. D. Maeques, "Virtual flux based direct power control of three phase PWM rectifiers," IEEE Trans. Ind. Applicat., vol. 37, pp. 1019-1027, july/august 2001.

[9] M. Malinowski, M. Jasinski, and M. P. Kazmierkowski, "Simple Direct power control of three phase PWM rectifier using space vector modulation (DPC-SVM)," IEEE Trans. Ind. Electron., vol. 51, pp. 447-454, April 2004.

[10] J.Restrepo, J.Viola, J.M.Aller, A.Bueno, "A simple switch selection state for SVM direct power control," IEEE ISIE2006, pp.1112-1116, July 2006, Montreal, Quebec, Canada.

[11] M.Cichowlas, M.Malinowski, P.Kazmierkowski, D.L.Sobczuk, J.Pou, "Active filtering function of three-phase PWM boost rectifier under different line voltage conditions," IEEE Trans. Ind. Electron., vol. 52, No. 2, pp. 410-419, April 2005.

[12] S. A. Larrinaga, M. A. R. Vidal, E. Oyarbide, J. R. T. Apraiz, "Predictive control strategy of DC/AC converters based on direct power control," IEEE Trans. Ind. Electron, vol. 54, No. 3, pp. 1261-1271, June 2007.

Improvement of the performance for DC-DC Converter

X. She[1], student member, IEEE, Yun She[2]

1. Huazhong University of Science and Technology, Wuhan, China, shexu8511211@gmail.com
2. Advanced Control Technology Lab, Central South University, Changsha, China
sheyuncsu@yahoo.com.cn

Abstract—This paper proposes a novel fuzzy PI controller that used in phase-shifted full bridge(PSFB) converter. Besides, a load current feed forward control is adopted to eliminate its influence to output voltage. Compared with conventional PI controller, experimental results testify that this novel controller performs much better in both steady and dynamic response.

Keywords—DC power supply, ZVS converters, Fuzzy control

I. INTRODUCTION

The main control strategy for phase-shifted full bridge DC-DC converter is PI control [1]–[2]. Although the PI control strategy has been widely used in many power electronic devices, it has many disadvantages especially in dynamic responding characters. Fuzzy control has been applied in many industrial areas since it was put forward for its nice robust character. While in the area of power electronic, fuzzy control is still a new control strategy. This paper proposes a novel fuzzy PI controller in phase-shifted full bridge DC-DC converter, besides, a load current feed forward control was also adopted to eliminate the influence of load current to the output voltage. Experimental results show that this control strategy has enhanced the performances of converter to a large extent.

II. SYSTEM ANALYSIS AND CONTROL STRATEGY [3]–[5]

A. System Analysis

Fig.1 is the main circuit of a phase-shifted full bridge converter, where Q_1 and Q_3 are switches of leading bridge, Q_2 and Q_4 are switches of lagging bridge. L_r is resonant inductor. L_f, C_f compose output filter. According to the small signal analysis, the output voltage is influenced by three parameters: input voltage $v_g(s)$, load current $i_{load}(s)$ and duty cycle d, as shown in Fig.2.

Because of the non-linear character of power electronic system, conventional PI control strategy can not perform well in dynamic response in a wide area. Fuzzy control is famous for its nice robust character and has been used in many industry control process. A novel fuzzy PI control strategy will be introduced in phase-shifted full bridge DC-DC converter.

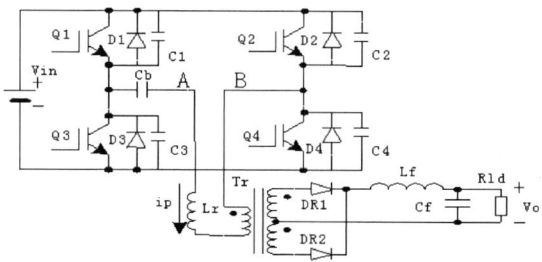

Fig.1. main circuit of phase-shifted full bridge DC-DC converter

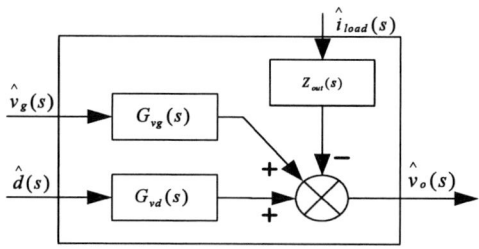

Fig.2. parameters that influence output voltage

The conventional increment mode PI control can be described as shown in (1) and (2):

$$u(k) = u(k-1) + \Delta u(k) \qquad (1)$$

$$\Delta u(k) = k_p[e(k) - e(k-1)] + k_i e(k) \qquad (2)$$

The fuzzy PI controller is composed by PI regulator and fuzzy controller. The inputs of fuzzy controller are system error and change of the error de/dt. The output of fuzzy controller are increment of K_p and K_i. It can be described as shown in (3) and (4):

$$u(k) = u(k-1) + \Delta u(k) \qquad (3)$$

$$\Delta u(k) = (k_p + \Delta k_p)[e(k) - e(k-1)] + (k_i + \Delta k_i)e(k) \qquad (4)$$

Fig.3 is the control diagram of the whole system, where $G_u(s)$ is a PI regulator, $G_{filter}(s)$ is a first order filter, K_{md} stands for phase shift PWM generator. $G_i(s)$ is current feed forward regulator, $Z_{out}(s)$ equals to the output resistance. According to the feed forward control theory, the influence of load current can be eliminated when (5) is satisfied.

$$G_i(s) * K_{sen\_i} = -\frac{Z_{out}(s)}{G_u(s) * K_{md}} \qquad (5)$$

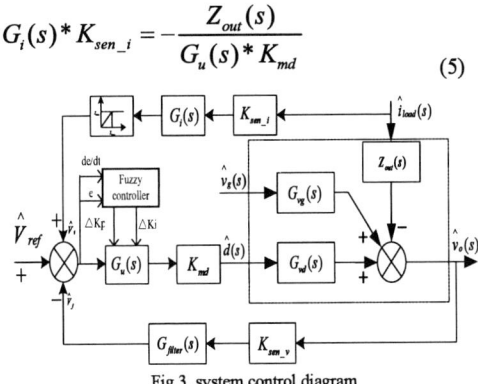

Fig.3. system control diagram

B.Control Strategy

Linguistic domain of input values(E, E_c) and output value (Δk_p, Δk_i) of fuzzy controller are described as below:

$$E \in \{-4,-3,-2,-1,0,1,2,3,4\}$$
$$E_c \in \{-4,-3,-2,-1,0,1,2,3,4\}$$
$$\Delta k_p \in \{-4,-3,-2,-1,0,1,2,3,4\}$$
$$\Delta k_i \in \{-4,-3,-2,-1,0,1,2,3,4\}$$

The linguistic variables of above four parameters can be divided into NB, NS, Z, PS, PB five levels.

Choosing Gaussian fuzzifiers to fuzzy input values, which are shown in (6).

$$\mu_A(x) = e^{-(\frac{x_1-\overset{*}{x_1}}{a_1})^2} \nabla ... \nabla e^{-(\frac{x_1-\overset{*}{x_n}}{a_n})^2} \qquad (6)$$

Where, a_i is a positive, and choose the smallest t-norm ∇, as shown in Fig.4.

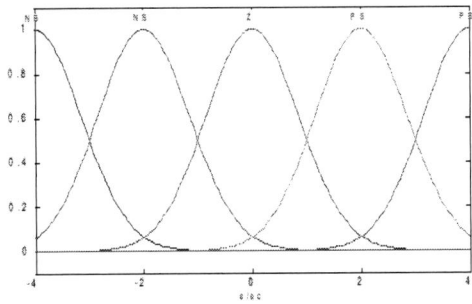

Fig.4. membership function

The rules for choosing k_p and k_i are:

1. When the absolute value of E is too large, we choose a large k_p to ensure dynamic response and choose k_i=0 to avoid large overshoot.

2. When the absolute value of E is small, we choose a larger k_i and k_p to ensure a nice steady state response.

3. When E is a middle value, we choose a smaller k_p and a suitable k_i to ensure small overshoot.

So the rules for adjusting parameters of fuzzy controller can be described as shown in Table I and II. According to the method for choosing linguistic domain and rules of regulation of each value, we can get output surface diagram of fuzzy controller, as shown in Fig.5 and Fig.6.

Table I. FUZZY RULES OF Δk_p

E \ Ec	NB	NS	Z	PS	PB
NB	PB	PB	PS	PS	Z
NS	PB	PS	PS	Z	Z
Z	PS	Z	Z	Z	NS
PS	Z	Z	NS	NS	NB
PB	Z	NS	NS	NB	NB

Table II. FUZZY RULES OF Δk_i

E \ Ec	NB	NS	Z	PS	PB
NB	NB	NB	NS	NS	Z
NS	NB	NS	NS	Z	Z
Z	NS	Z	Z	Z	PS
PS	Z	Z	PS	PS	PB
PB	Z	PS	PS	PB	PB

The rule for fuzzy composition is Max-min composition, and we adopt Mamdani inference. Choosing center of gravity defuzzifier to defuzzy the output values, as shown in (7).

$$y^* = \frac{\int_v y u_B(y)dy}{\int_v u_B(y)dy} \qquad (7)$$

Where $u_B(y)$ is membership function of output values.

According to Table I and II, we can get Δk_p and Δk_i at any time I, assuming that $k_p^{'}$ and $k_i^{'}$ are parameters of conventional PI regulator. The output value of fuzzy controller can be calculated by (8) and (9):

$$k_p = k_p^{'} + \sum_{i=1}^{k} \Delta k_p(i) G_1 \qquad (8)$$

$$k_i = k_i^{'} + \sum_{i=1}^{k} \Delta k_i(i) G_2 \qquad (9)$$

Where G_1 and G_2 can be decided as following rules:

If the real domain of $\Delta k_p(i)$ is {-a,a}, and the real domain of $\Delta k_i(i)$ is {-b,b}. Then G_1=a/4; G_2=b/4.

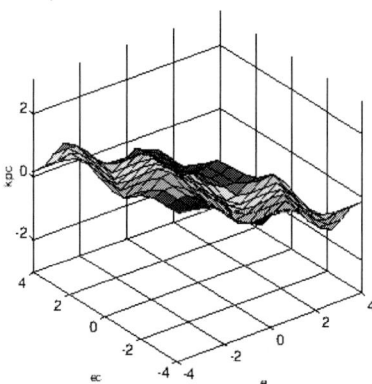

Fig.5 output surface- Δk_p as the output

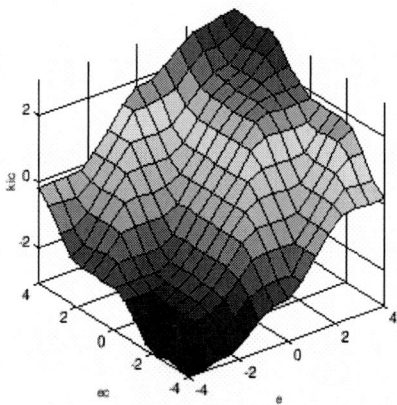

Fig.6 output surface- Δk_i as the output

III. EXPERIMENTAL RESULTS AND ANALYSIS

To testify the theory above, a 28.5v/350A prototype of phase-shifted full bridge DC-DC converter was designed by using both conventional PI controller and fuzzy PI controller. The whole system is controlled by TMS320F240 DSP at the switch frequency of 20KHz. The main parameters of system are: input voltage is 513v, output voltage is 28.5v, output current is 350A. Parallel capacitor in leading bridge is 10nF, parallel capacitor in lagging bridge is 22nF, capacitor in output filter is 0.27F, inductor in output filter is 50uH, resonant inductor is 50uH.

Fig.7 is the ripple voltage of the output DC voltage by adopting conventional PI controller; Fig.8 is the ripple

voltage of the output DC voltage by adopting fuzzy PI controller. The maximum ripple voltage has been decreased from 200mv to 90mv. It is obviously that ripple voltage has been decreased greatly by adopting fuzzy PI controller.

In order to compare the dynamic character of the two methods, adding and cutting the load abruptly: load current changes from 350A to 30A, then back to 350A. Fig.9 is the dynamic responding wave by adopting conventional PI controller. Fig.10 is the dynamic responding wave by adopting fuzzy PI controller. Comparing two waves in both overshoot value and responding time, as seen in Table III and Table IV. It is obviously that fuzzy PI controller performs much better compared with conventional PI controller.

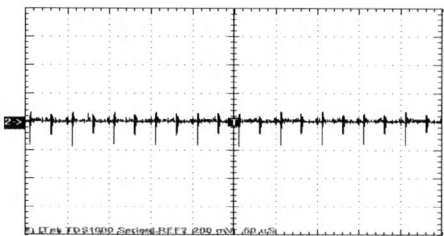

Fig.7 Ripple voltage in output DC voltage by adopting traditional
PI controller

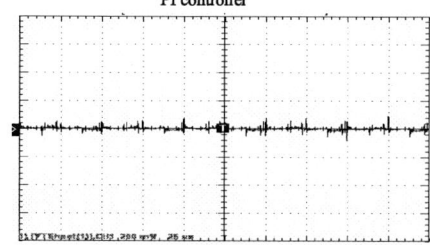

Fig.8 Ripple voltage in output DC voltage by adopting fuzzy
PI controller

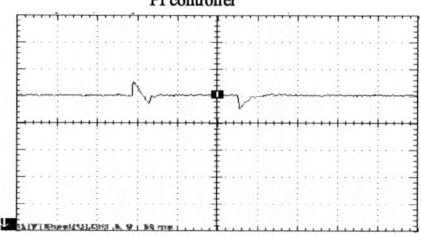

Fig.9 dynamic responding wave(350A-30A-350A) by adopting
traditional PI controller

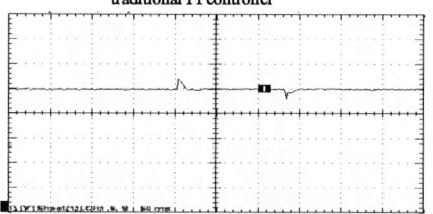

Fig.10 dynamic responding wave(350A-30A-350A) by adopting
fuzzy PI controller

Table III dynamic character of conventional PI controller

	350A-30A	30A-350A
Max overshoot voltage	2.7v	2.2v
Responding time	40ms	30ms

Table IV dynamic character of fuzzy PI controller

	350A-30A	30A-350A
Max overshoot voltage	2.0v	2.1v
Responding time	10ms	12ms

IV. CONCLUSION

This paper proposes a novel fuzzy control combined with load current feed forward control which is applied in phase-shifted full bridge converter. This control strategy combines the nice robust character of fuzzy control and accuracy of PI control to accomplish automatic adjustment of parameters of controller. Experimental results show that compared with conventional PI control, this control strategy performs much better in both steady state and dynamic response.

REFERENCES

[1] Cho J G, Sabate J A, Lee F C, "Novel full bridge zero-voltage transition PWM DC/DC converter for high power application." APEC1994[C], 1994:143-149

[2] Tabisz W A, Lee F C, "Zero-voltage switching multi-resonant technique: a novel technique to improve quasi-resonant converters." IEEE PESC Conference Record 1987[C], 1987:386-390

[3] Brown,M. , and C.Harris, "Neuro fuzzy Adaptive Modeling and Control", Prentice Hall, Ehglewood Cliffs, NJ,1994

[4] Chen,J.Q., and L.J.Chen, "Study on stability of fuzzy closed-loop control systems," Fuzzy Sets and Systems, 57, no.2,PP. 159-168

[5] Johansen. T.A, "Fuzzy model based control: stability, robustness, and performances issues", IEEE Trans. on Fuzzy Systems,2,PP.221-234,1994

A DRIVE SYSTEM WITH HIGH-SPEED SINGLE-PHASE SUPPLIED THREE-PHASE INDUCTION MOTOR

T. Binkowski, M. Grad, M. Latka, W. Malska, D. Sobczynski

Rzeszow University of Technology, Department of Power Electronics and Power Engineering,
Rzeszow, Poland
e-mail: *tbinkow@prz.edu.pl, mgrad@prz.edu.pl, mlatka@prz.edu.pl, wmalska@prz.edu.pl, dsobczyn@prz.edu.pl*

Abstract— Use of high-speed three-phase induction motors supplied through electronic power converters from single-phase network in electric drive systems used in household appliances and power tools is proposed. Such solution would result in reduction of motor weight and thus the mass of the whole drive system, possibility to regulate (set) the rotation rate within wide range, and, most importantly, increase of efficiency and improvement of reliability of drive systems of that type.

This paper presents results of analysis and simulation tests concerning considered technical solutions depending on the type of electronic power converter (single-phase power supply + inverter) feeding a three-phase induction motor, based on which preliminary parameters of the laboratory circuit were determined.

Keywords— high frequency power converter, high speed driver, variable speed driver, induction motor.

I. INTRODUCTION

High-speed drive systems with rotation rate above 314 rad/s (3000 rpm), are in most cases realised with use of classic AC commutator motors. Such solutions can be found in e.g. electric power tools (sawing machines, hand drills etc.) and in household equipment (vacuum cleaners, juice extractors etc.). Commutator motors manufacturing technology is relatively complicated, and AC commutator motors itself have numerous disadvantages, the most important of which are: presence of commutator making by itself the motor heavier and its manufacturing more expensive; limited service life of brushes – especially in case of motors operating in AC power supply network; generation of electromotive force of large value and significant electromagnetic interference; necessity for continuous maintenance; and, finally, limited maximum rotation rate value resulting from commutator's definite mechanical strength [2][3].

By contrast with AC commutator motors, a three-phase induction motor has simplest and more compact design, is more reliable, and its windings as well as the magnetic circuit are utilised better. Torque produced by induction machine is a function of its geometrical dimensions, magnetic-, and current-carrying capacity. That is represented by the following relations [8][9]:

$$T = (c_e, \phi, i) \quad \text{and} \quad P = T\omega \qquad (1)$$

where: T – torque, c_e – a machine constant (containing information about its dimensions), ϕ – magnetic carrying capacity, i – current-carrying capacity, P – output power at motor's shaft, ω – angular velocity.

It follows from the above relationships that increase of rotation rate at unchanged output power makes it possible to reduce a motor's overall dimensions.

II. A CONCEPT OF DRIVE SYSTEM FOR HOUSEHOLD APPLIANCES WITH A HIGH-SPEED THREE-PHASE INDUCTION MOTOR

Realization of a high-speed drive using a three-phase induction motor, especially that of low and medium power and with rotational speed of up to 40,000 rpm designed for use in household appliances, became possible thanks to rapid progress in electronic power control technology. Increase of both capability and diversity of electronic power circuits results from increase of semiconductor elements voltage and current operational ranges parallel to improvement of their dynamical parameters, as well as from application of up-to-date microprocessor and programmable circuits for control purposes.

Choice of topology for a drive system to be used in household appliances, including the electronic power converter, is based on the following quality- and economy-related criteria: single-phase supply and sinusoidal waveform of power network current; high efficiency and reliability; small overall dimensions; minimum number of semiconductor elements; and low price.

A block diagram of the proposed drive system solution with a high-speed three-phase induction motor supplied from single-phase network through an inverter is shown in Figure 1.

In the presented circuit, one can single out five principal blocks: blocks A, B and C constitute the drive system's power circuit, while blocks D and E are circuits regulating/controlling the power supply and the inverter, respectively. Various technical solutions are possible here, both in scope of the power circuit and the control system.

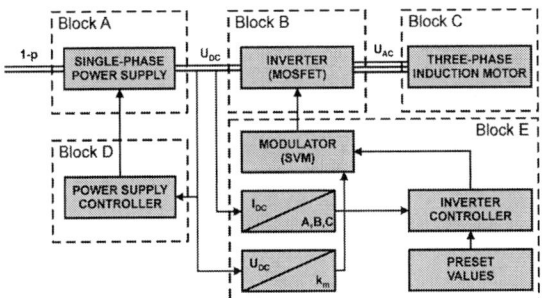

Fig. 1. Block diagram of a high-speed drive system model.

III. TOPOLOGIES OF AC/DC/AC CONVERTERS

The AC/DC/AC converter (blocks A and B) can be realised in different ways. The most classic solution consists in combination of two circuits: a single-phase AC/DC rectifier and three-phase DC/AC inverter connected by means of an intermediate circuit.

Single-phase AC/DC rectifiers are used as input circuits in many practical solutions. The decisive factors here are: high efficiency, possibility to obtain high power densities, and low cost of the system's components. However, typical circuits of AC/DC converters such as e.g. 4D diode bridge, do not allow for regulation of output voltage value and are characterised with distorted supply network current. That results in low value of power factor (PF) and presence of harmonics in the network current [5].

Disadvantages of classical rectifier-type circuits can be limited by using supply circuits containing active elements that make possible to change output voltage and synthesise the converter's input current. Among other solutions, the so-called UFP (Unity Power Factor) converter circuits are considered. These circuits, known as boost- and buck-type systems, make possible to control the rectified voltage. Advantage of buck-type circuit consists in possibility to obtain controlled voltage lower than the input voltage. Boost-type converters make it possible to obtain controlled output voltage higher than the input voltage. Input current can be easily synthesised by means of the converter's active elements with use of appropriate control method. Therefore, boost-type converters provide controlled DC voltage, with possibility to achieve Unity Power Factor conditions and low content of harmonics in input current, THD_I.

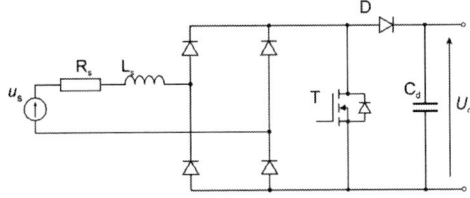

Fig. 2. A diode bridge rectifier with PWM converter synthesising sinusoidal input current.

The most frequently used topology of boost-type converters is a circuit consisting of a single-phase diode bridge rectifier (not controlled) and a tranzistor interrupter (chopper) in the voltage boosting circuit – Fig. 2. A disadvantage of circuits of that type with reference to

non controlled circuit is its lower efficiency. Properly chosen control systems makes possible to synthesise a sinusoidal input current waveform [7].

By substituting two diodes in the bridge rectifier with transistors (Fig. 3 ÷ 4) one can reduce the number of semiconductor elements connected in series which affects the system's efficiency. A disadvantage of such circuits consists in unidirectional energy flow.

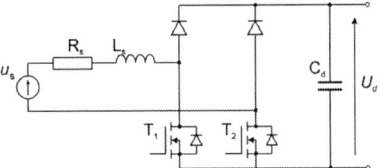

Fig. 3. Symmetric active rectifier (semi-boost) with sinusoidal line current waveform and unidirectional energy flow.

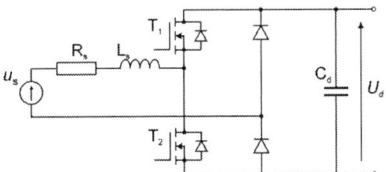

Fig. 4. Asymmetric active rectifier (semi-boost) with sinusoidal line current waveform and unidirectional energy flow.

Bidirectional energy flow is ensured by half-bridge and bridge rectifiers shown in Figures 5 ÷ 6.

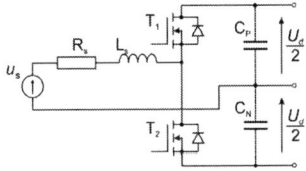

Fig. 5. Half-bridge active rectifier with sinusoidal line current waveform and bidirectional energy flow.

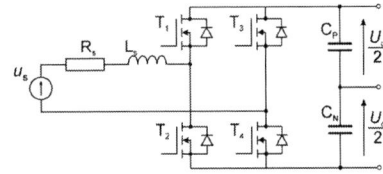

Fig. 6. Active bridge rectifier with sinusoidal line current waveform and bidirectional energy flow.

An interesting solution is represented by half-bridge converter because of lowest possible power losses (only one element is active during operation) – Fig. 5.

The bridge converter (Fig. 6) with controllable voltage is characterised with less input current distortions at higher reliability and efficiency compared to the diode rectifier circuit. Moreover, such circuit allows for bidirectional energy flow, the feature being especially desirable in drive systems with recovery braking.

An additional rationale behind use of a half-bridge circuit consist in possibility to realise the AC/DC/AC converter based only on the three-phase module of bridge inverter: using one of the inverter branches as an input rectifier, and the remaining two – as a V-connected inverter (Fig. 7).

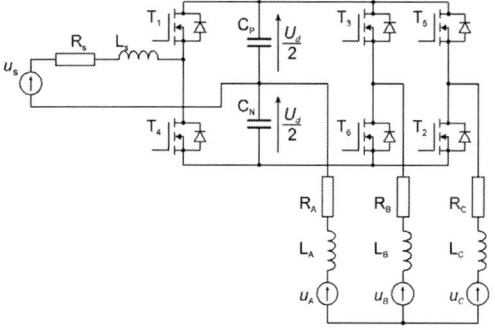

Fig. 7. Block diagram of a high-speed drive system model.

In the presented circuit, one can single out five principal blocks: blocks A, B and C constitute the drive system's power circuit, while blocks D and E are circuits regulating/controlling the power supply and the inverter, respectively. Various technical solutions are possible here, both in scope of the power circuit and the control system.

IV. Control Of High Speed Drive

The control process of high speed electric drive used in household appliances or power tools must satisfy two fundamental requirements. The first of them and the uncompromised one is the drive current control. The second requirement of drive control is the speed or supply voltage control. The critical values of a load current exciting causes damage of semiconductor switches, so in the structure of a current controller the integrator with an output limitation must be used. In the household appliances and power tools the speed control accuracy is not very important. For that reason, the master speed or voltage controller as a proportional controller is allowed. A limitation of the sensor quantity in household appliance or electric power tool drives is the result of the economical criteria and causes the speed/voltage replacement using the internal feedback. The block diagram of the considered case is shown in Figure 8.

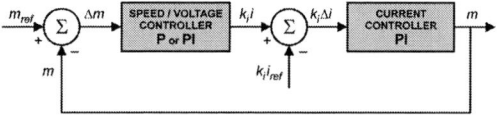

Fig. 8. Block diagram of control system

The output value m of the controller is scaled to the range from 0 to 1 and is directly equal to the set value of the voltage inverter amplification coefficient. Based on this value, the controller calculates the wave frequency and realizes the scalar control method of induction motor keeping the value of u/f ratio constant.

In relation to the considered topology of power converter, the values of output voltages and output wave frequency are realized using proper modulators [1]. In case of the induction motor supply using the three phase voltage inverter connected with the boost converter, the control unit consist of the three classical modulators with three sinusoidal references or space vector modulators. Then, the reference signals are in a three phase sinusoidal form, or direct as a space vector components. Besides, the boost converter requires an additional control unit which controls the sinusoidal current of supply source with unity power factor. In case of V-connected converter, the controllers of both the inverter and the power supply are integrated. Then the control unit uses only three modulators. Two of them realize the set values shifted by 60 electrical degrees, the third realizes sinusoidal supply current with unity power factor.

V. Simulations

For the discussed AC/DC/AC converter topologies (Fig. 2 ÷ 7) with appropriately chosen control circuits, simulation test were performed. Examples of results are presented on Figures 9 ÷ 10.

a)

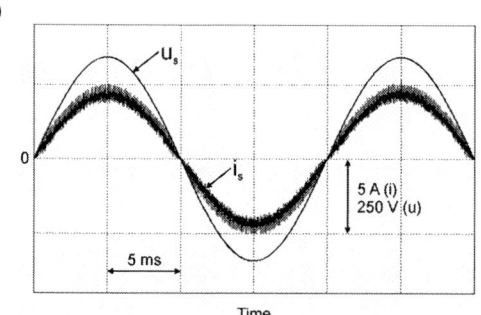

b)

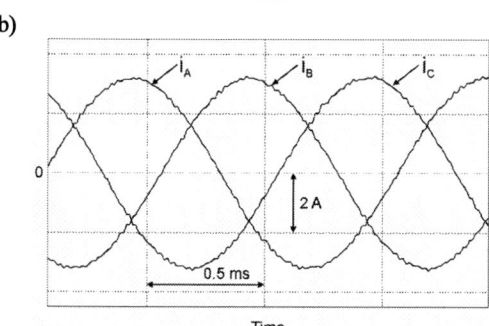

c)

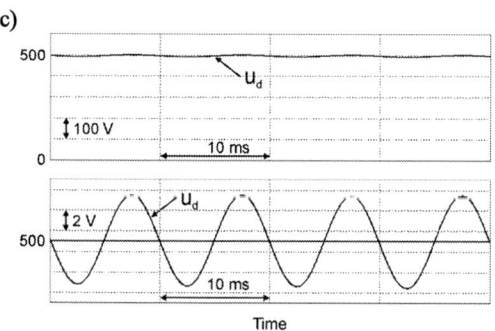

Fig. 9. Results of simulation for 1-phase UPF converter and 3-phase inverter: a) input current and voltage waveforms, b) output currents supplying high-speed induction motor (667 Hz), c) DC voltage on capacitor in the intermediate circuit.

a)

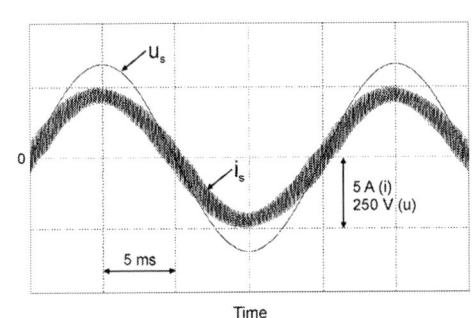

b)

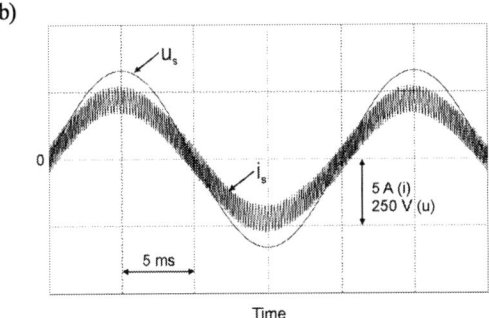

c)

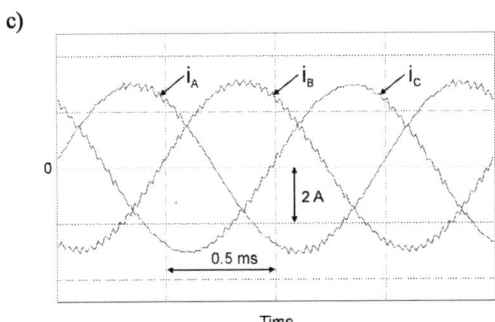

d)

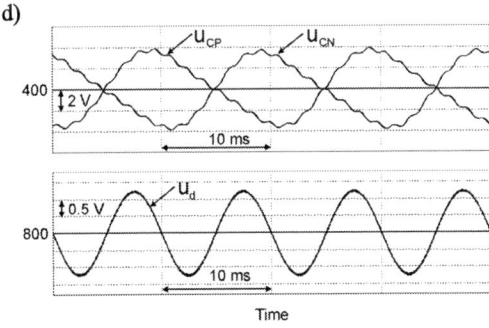

Fig. 10. Results of simulation for V-connected converter: a) input current and voltage waveforms with PI control, b) input current and voltage waveforms with hysteresis control, c) waveforms of output currents supplying high-speed induction motor (667 Hz), d) DC voltage on capacitor in the intermediate circuit.

VI. SUMMARY

On the grounds of the analysis carried out above and simulation tests, one can conclude that both proposed solutions, i.e.:

- single-phase diode bridge rectifier with PWM converter synthesising sinusoidal line current and three-phase voltage inverter;
- V-connected converter,

seem to be appropriate for use in electric drive systems for household appliances.

Carrying out tests on actual circuits will make possible to chose an appropriate converter topology that will be applied in a prototype household appliance drive system that could be used in e.g. vacuum cleaners.

ACKNOWLEDGMENT

The research work supported by the Ministry of Science and Higher Education (Poland) under Grant R01 039 02 (2007 – 2010).

REFERENCES

[1] T. Binkowski, "Modulators in 3-phase voltage inverter control", (in polish: „Modulatory w układach sterowania trójgałęziowego falownika napięcia"), *Przeglad Elektrotechniczny*, nr 2 /2007.p.6-8.

[2] K. Buczek, D. Sobczynski, "Analysis and examination of high-speed induction motor driver system". *Power Electronics and Electrical Drive*, Polish Academy of Sciences, electrical Engineering Committee, Wroclaw 2007, p 487-498.

[3] K. Buczek, D. Sobczynski, "Analysis of Induction Motor and 330 Hz Invereter Drive System", *13th International Conference on Electrical Drives and Power Ectronics*, EDPE'99, High Tatras, Slovakia, 5-7 October 1999, p. 64-68.

[4] K. Jalili, D. Krug, S. Bernet, M. Malinowski, B.J. Cardoso Filho, "Design and Characteristics of a Rotor Flux Controlled High Speed Induction Motor Drive Applying Two-Level and Three-Level NPC Voltage Source Converters", *Power Electronics Specialists Conference*, 2005. PESC '05. IEEE 36th 2005 p:1820 - 1826

[5] A. Pandey, B. Singh, P. Kothari, "Comparative evaluation of single-phase unity power factor ac-dc boost converter topologies", *IE (I) Journal-EL*, vol 85, September 2004, p.102-109.

[6] S. Piróg, "Power Electronics: systems with network and hard commutations", (in polish: „Energoelektronika: układy o komutacji sieciowej i o komutacji twardej"), *AGH Uczelniane Wydawnictwa Naukowo-Dydaktyczne*, 2006.

[7] J. Rodr´iguez, J. Dixon, J. Espinoza, P. Lezana, "PWM regenerative rectifiers: state of the art", *IEEE Transactions on Industrial Electronics*, Vol. 52, No. 1, February 2005 p.5-22.

[8] D. Sobczynski, "The analysis of properties of induction motor supplied from 300 Hz output voltage inverter", *4th International Symposium on Microelectronic Technologies and Microsystems*, Zwikau 2000, October 26-27 p. 58-65.

[9] D. Sobczyński, "High speed drive with induction motor fediing by inverter", (in polish: „Wysokoobrotowy układ napędowy z silnikiem indukcyjnym zasilanym z falownika"), *Przeglad elektrotechniczny*, nr 7-8/2006, p.24-31.

A Pulse Width Modulation Technique for a Multilevel Converter in High Voltage High Frequency Applications

Jafar Adabi, Hamid Soltani, Firuz Zare

Queensland University of Technology, School of Electrical Engineering, GPO Box 2434, Brisbane, QLD, 4001, Australia, adabi.jafar@student.qut.edu.au

Abstract— in this paper, a new hysteresis voltage control technique based on optimized harmonic elimination method is presented for a single phase multilevel inverter with flying capacitor topology in high voltage high frequency applications such as high voltage plasma systems. A main objective of the proposed technique is to generate high voltage and control capacitors voltages in all operational conditions. For this purpose, principals of optimized harmonic elimination technique with and without considering switching time are described by mathematical analysis and simulation results. Then, a hysteresis voltage control is analyzed in detail to control capacitors voltages when load impedance changes significantly due to nonlinearity of load such as a plasma reactor. A 100 times decrease in load impedance is simulated and flying capacitors voltages are controlled without any voltage collapse.

Keywords— multilevel inverter, harmonic elimination, switching time, hysteresis voltage control

I. INTRODUCTION

Multilevel converters have been broadly used in high-power applications. They have more attraction due to reduction of harmonic contents and voltage stress in power electronic applications. A general structure of the multilevel converter is to synthesize a sinusoidal voltage using several levels of voltages and a pulse width modulation technique [1]. There are three most popular configurations in multilevel converters: diode-clamp inverter, flying-capacitor topology and cascade inverter [1-4]. Generalized structures and applications of the multilevel inverters have been discussed in [5-6]. Fig.1 shows the general representation of two famous topologies of multilevel inverters.

Several pulse width modulation techniques for multilevel inverter topologies had been mentioned in literature such as: sinusoidal pulse width modulation (SPWM) [7-9], hysteresis current control technique [10], optimized harmonic stepped-waveform technique. [11-12]

In this paper, principals of optimized harmonic elimination technique with and without considering switching time are described by mathematical and simulation analysis and a new pulse position technique are presented for harmonic elimination. Then, hysteresis voltage control for a high voltage power converter in high frequency applications will be analyzed.

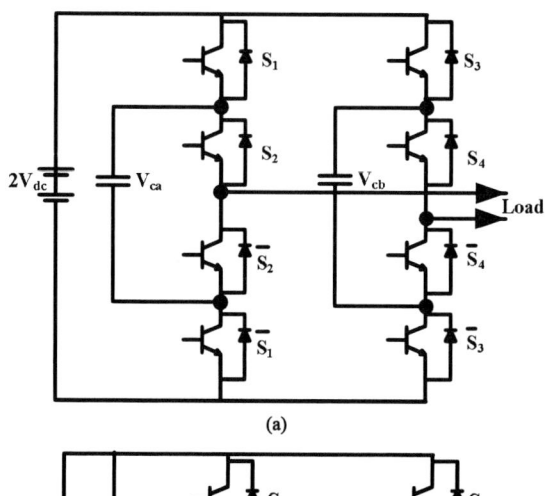

(a)

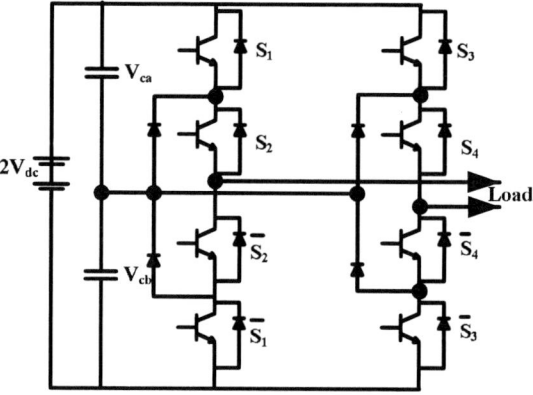

(b)

Fig.1. Single-phase Five Level Inverter Topologies: (a) Flying Capacitor (b) Diode-Clamped.

II. OPTIMISED HARMONIC ELIMINATION TECHNIQUE

Fig.2 shows a (2N+1) level inverter output voltage waveform in which the switching rise and fall times are not considered and switches are assumed ideal. Fourier expansion of the waveform is:

$$V(\omega t) = \frac{4V_{dc}}{n\pi} \sum_{n=1}^{\infty} (\cos n\alpha_1 + \cos n\alpha_2 + ... + \cos n\alpha_N) \times \sin n\omega t \quad (1)$$

Where N is the number of voltage levels and n is the harmonic order. Optimized switching angles ($\alpha_1, \alpha_2... \alpha_N$) for a single phase to achieve the fundamental voltage and eliminate the (N-1) odd harmonics ($3^{rd}, 5^{th}, ...$) can be calculated through solving the following equation.

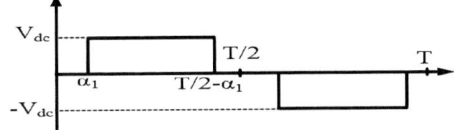

$$\begin{cases} \dfrac{4V_{dc}}{\pi}(\cos\alpha_1 + \cos\alpha_2 + \dots + \cos\alpha_N) = V_1 \\ (\cos 3\alpha_1 + \cos 3\alpha_2 + \dots + \cos 3\alpha_N) = 0 \\ \vdots \\ (\cos(2N-1)\alpha_1 \dots + \cos(2N-1)\alpha_N) = 0 \end{cases} \quad (2)$$

In the previous method, on-off switching times have been ignored in a pulse width modulated voltage. Fig.3 shows a (2N+1) level inverter output voltage waveform with considering the switching rise and fall times. The Fourier expansion of (2N+1)-level waveform is:

$$V(t) = \sum_{n=1,3,5,\dots}^{\infty} \frac{2V_{dc}T}{\pi^2 n^2 t_r} \left[\begin{array}{c} \left(\sin\dfrac{2\pi n}{T}(t_1 + t_r) \\ + \dots \sin\dfrac{2\pi n}{T}(t_N + t_r) \right) - \\ \left(\sin(\dfrac{2\pi n}{T}t_1) + \dots \sin(\dfrac{2\pi n}{T}t_N) \right) \end{array} \right] \sin n\omega t \quad (3)$$

The variations of different harmonics, b_i (i=3,5,7), versus switching time for various frequencies are shown in Fig.4. The DC power supply voltage is considered 200 volts.

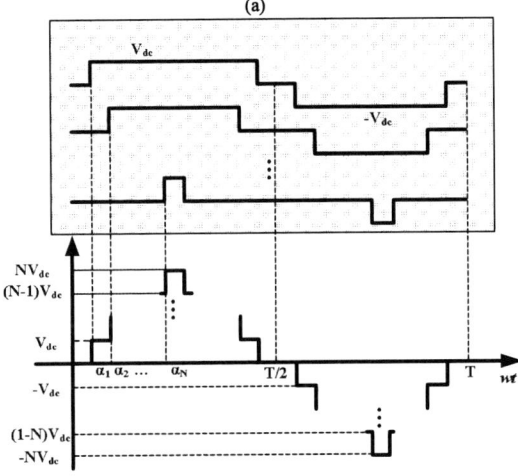

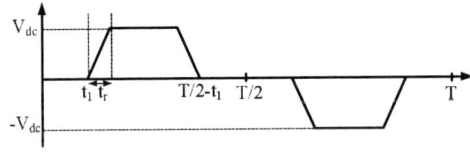

Fig.2. (a) a step waveform (b) (2N+1)-level voltage waveform

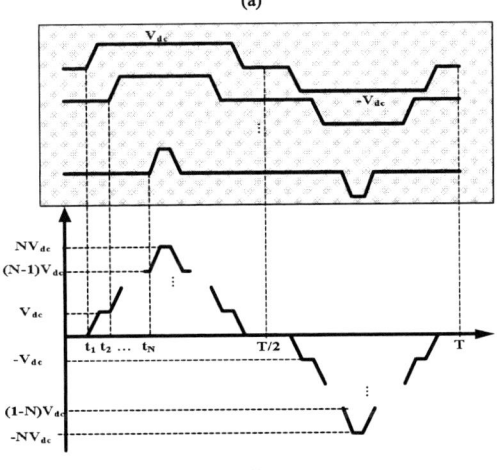

Fig.3. (a) a simple step waveform (b) a (2N+1) level inverter output waveform with consideration of t_r

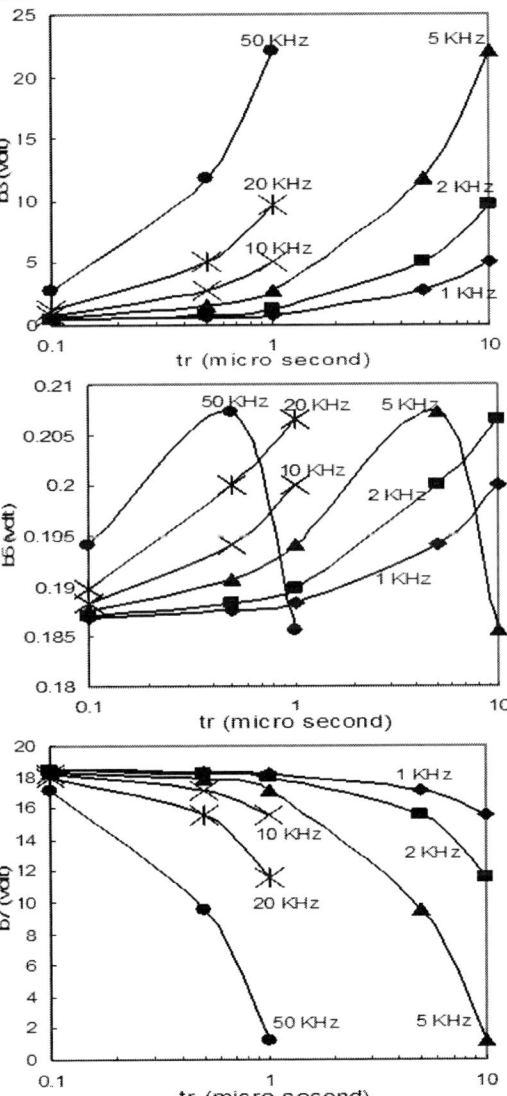

Fig.4. Harmonics Variations with respect to t_r in 5 Level Inverter at 1,2,5,10,20,50 kHz

III. PROPOSED PULSE POSITION TECHNIQUE FOR HARMONIC ELIMINATION

The simulation results show that the harmonic elimination method with considering the switching time increases the Total Harmonic Distortion (THD). In fact, for high frequency applications in which the switching time cannot be neglected, the off-line analysis based on traditional method (Eq.1) is not a proper method as it may not eliminate the expected harmonics. A simple method to solve the problem is to shift the edges of the pulse to the left and right sides by $t_r/2$ as shown in Fig.5.c. In this case, we can use the off-line switching times based on sharp pulses and turn on the switches $t_r/2$ earlier and turn off them $t_r/2$ later with respected to the traditional method (Fig.5.a). In this section, we have performed the analysis for a five-level inverter considering new switching times for different t_r and output frequencies. Fig.6 shows that the 3rd and 5th harmonics are eliminated when new switching times are employed. Harmonic contents at different frequencies are shown in Fig.7 with respect to t_r.

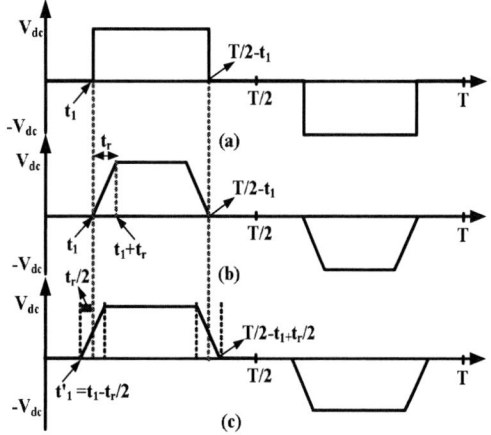

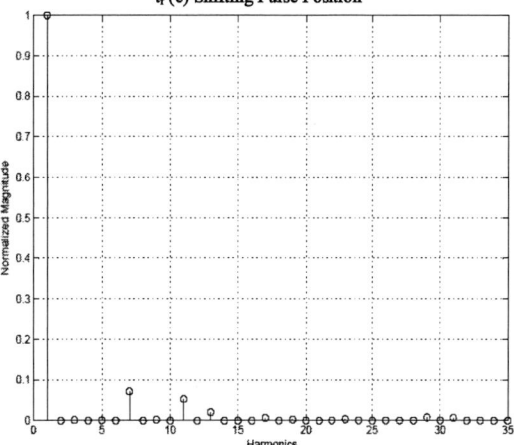

Fig.5: Voltage Waveform (a) Sharp Pulse (b) with Considering t_r (c) Shifting Pulse Position

Fig.6. Harmonic Spectrum of a 5-Level Inverter at 50 kHz and t_r=1 μs

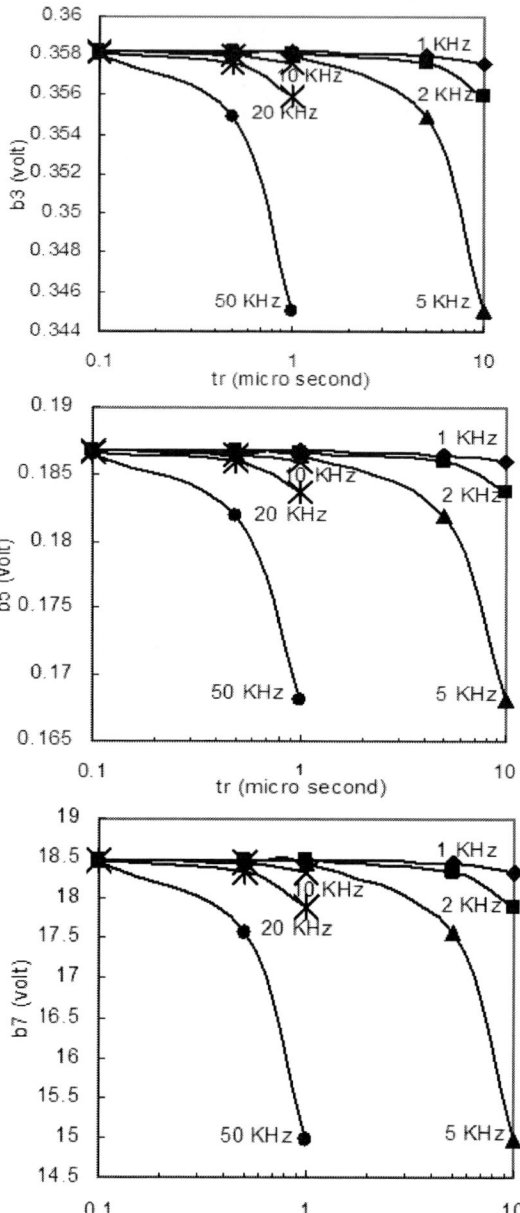

Fig.7. the 3rd, 5th, 7th and 9th Harmonics Variations with respect to t_r in a 5-level inverter

IV. HYSTERESIS BAND VOLTAGE CONTROL OF A SINGLE PHASE 5-LEVEL INVERTER

Switching states of a five level flying capacitors inverter (Fig.1.a) and charging and discharging states and output voltage levels are shown in Table. I. The most important issue in controlling the output voltage of proposed inverter is to control capacitors voltages. At first, switching angles must be calculated based on optimized harmonic elimination technique. According to Fig.8 capacitors voltages are compared with reference values and when the error reaches the upper (lower) limit, the proposed capacitor must be discharged (charged). Then a proper switching states must be selected with respect to charging or discharging states of capacitors in legs a & b, voltage level and current direction. The main objective of proposed technique is to control the output voltage and the capacitors voltage in all operational conditions such as nonlinear load and sudden short circuit. It can prevent voltage collapse when the load impedance is dramatically decreasing.

A case study for a high voltage single phase 5-level inverter with a resistive and inductive load is investigated. Fig.9 shows the load voltage and current, capacitor voltage in each phase leg based on the hysteresis voltage control. Table II shows the simulation parameters.

TABLE I. SWITCHING STATES AND CHARGING AND DISCHARGING STATES OF FLYING CAPACITIRS IN A5-LEVEL INVERTER

S_1	S_2	S_3	S_4	V_{out}	V_{ca} For i>0	V_{cb} For i>0	V_{ca} For i<0	V_{cb} For i<0
1	1	0	0	$2V_{dc}$	No Change	No Change	No Change	No Change
1	1	0	1	V_{dc}	No Change	Increase	No Change	Decrease
1	1	1	0	V_{dc}	No Change	Decrease	No Change	Increase
1	0	0	0	V_{dc}	Increase	No change	Decrease	No change
0	1	0	0	V_{dc}	Decrease	No change	Increase	No change
0	1	0	1	0	Decrease	Increase	Increase	Decrease
0	1	1	0	0	Decrease	Decrease	Increase	Increase
1	0	0	1	0	Increase	Increase	Decrease	Decrease
1	0	1	0	0	Increase	Decrease	Decrease	Increase
0	0	0	0	0	No Change	No Change	No Change	No Change
1	1	1	1	0	No Change	No Change	No Change	No Change
0	0	0	1	$-V_{dc}$	No Change	Increase	No Change	Decrease
0	0	1	0	$-V_{dc}$	No Change	Decrease	No Change	Increase
0	1	1	1	$-V_{dc}$	Decrease	No Change	Increase	No Change
1	0	1	10	$-V_{ds}$	Increase	No Change	Decrease	No Change
0	0	1	1	$-2V_{dc}$	No Change	No Change	No Change	No Change

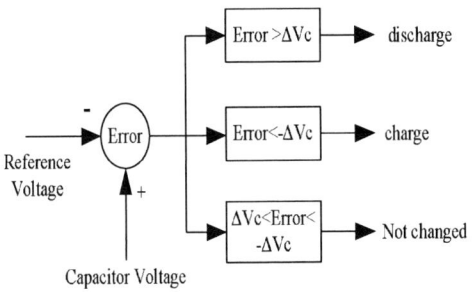

Fig.8, (a) charging and discharging states of flying capacitors in hysteresis voltage control

TABLE II. SIMULATION PARAMETERS IN HIGH VOLTAGE APPLICATIONS

Load Resistance (R)	10 ohms
Load inductance (L)	50 mH
Flying Capacitors(C_a, C_b)	1 mF
Input DC voltage	4000 volts
Capacitor reference voltage (V_{ref})	2000 volts
Hysteresis voltage bands (ΔV)	50 volts

The main objective of the proposed control method is to control the capacitors voltages when load impedance is changed significantly (in a high voltage plasma reactor due to voltage collapse in a reactor). To show the capability of proposed hysteresis technique to prevent voltage collapse across load or capacitors in all load conditions, the load impedance is decreased dramatically during specific time. Fig.10.a shows the load voltage and currents, capacitors voltages when the load impedance (1 mH &100Ω) is changed to 1mH between 0.002 & 0.0022 seconds. Simulation results show that in a new condition the proposed control technique can control the flying capacitors voltage in the hysteresis bands which can protect the power switches and devices. Fig.10.b shows the load voltage and currents, capacitors voltages when the load impedance (10 mH &100Ω) is decreased 100 times (0.1mH &1Ω) between 0.002 & 0.0022 seconds. Thus, during the short circuit, output voltage and capacitor voltage are controlled without any voltage collapse. In case of sudden short circuit of load, by charging and discharging states, the capacitor voltage is settled between the bands. An algorithm is defined in order to change the charging or discharging states of capacitors when it reaches the band and don't change the switching states when the capacitor voltage is between the bands. When the load is come back to its initial value, capacitor voltages are kept around 2000 volts.

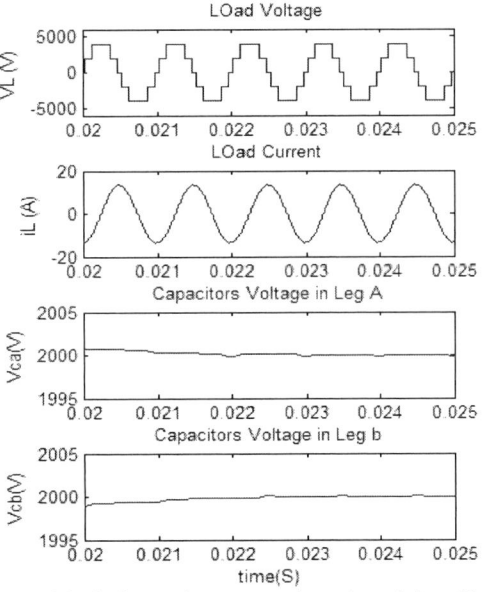

Fig.9. (a) load voltage and current, capacitors voltages in leg a &b (b) capacitors voltage

This analysis and simulation results show that the hysteresis voltage band control can easily keep the capacitors voltages between the bands in case of nonlinear loads and sudden short circuits. Defining the bands above the reference current can be used as a factor in the control strategy to apply the hysteresis voltage control for protection and capacitor voltage control. Thus, the hysteresis voltage control is not active in a normal operation when there is no inrush current and when the load current exceeds the limits, the hysteresis voltage control will be activated to protect the system and power switches.

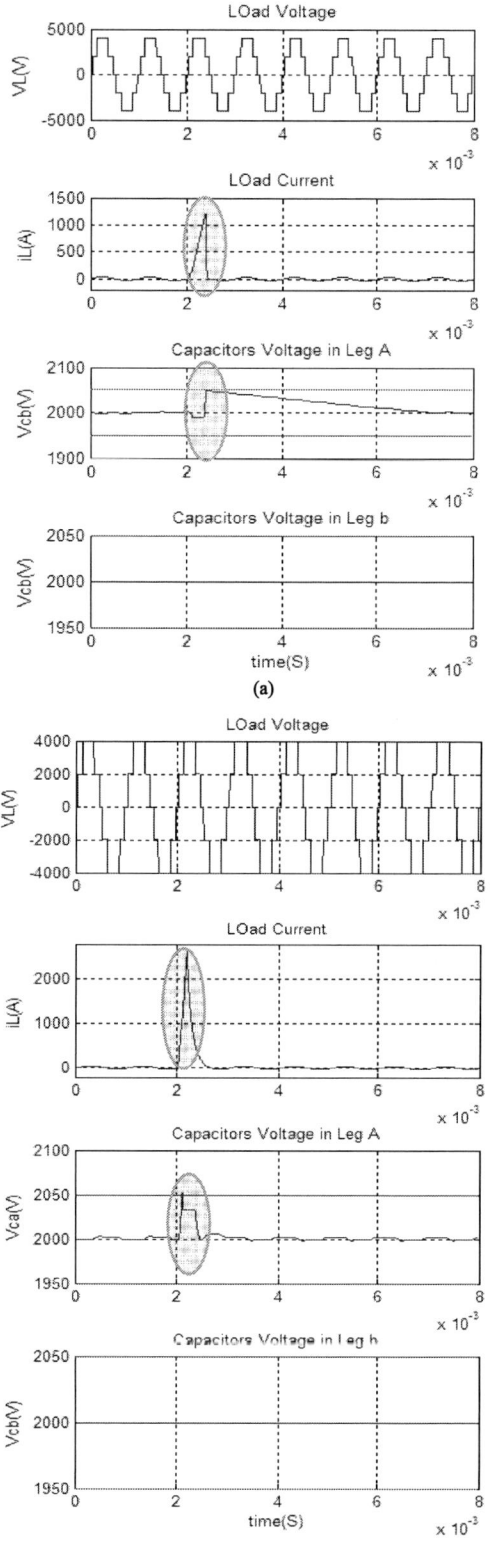

(a)

(b)

Fig.10. load voltage and current, capacitor voltage at each phase leg

V. CONCLUSIONES

The main objective of the proposed control method is to control the capacitors voltages when load impedance is changed significantly (in a high voltage plasma reactor due to voltage collapse in a reactor). Based on the proposed analysis, hysteresis voltage band control can easily keep the capacitors voltages between the bands in case of nonlinear loads and sudden short circuits. Defining the bands above the reference current can be used as a factor in the control strategy to apply the hysteresis voltage control for protection and capacitor voltage control. Also, optimized switching technique with and without consideration of rise and fall time is presented based on mathematical and simulation results.

REFERENCES

[1] A.Nabae, I.Takahashi, and H.Akagi, "A new neutral-point-clamped PWM inverter," IEEE Trans.Ind.Applicat., pp518-523, Sep./Oct.1981.

[2] T.A.Meynard and H.Foch, "Multilevel conversion: High voltage choppers and voltage source inverter" in proc.23rd Annual. IEEE Power Electron Conf vol.1, 1992, pp.397-403

[3] F.Z.Peng, J.Sheng-Lai, McKever-JW, VanCoevering-J, "a multilevel voltage source inverter with separate DC sources for static VAR generation", IEEE-Transactions-on-Industry-Applications. vol.32, no.5, Sept-Oct.1996; p.1130-8

[4] F. Zare, G. Ledwich, "A New Predictive Current Control Technique for multilevel Converters," presented at TENCON 2006. IEEE Region 10 Conference, 2006

[5] Y.Chen and B.T.Ooi, "Regulating and equalizing DC capacitor voltages in multilevel converters," IEEE Trans.Ind.Applicat., vol.12, pp.901-907, Mar./Apr.1997.

[6] J.S.Lai and F.Z.Peng, "Multilevel converters-A new breed of power converters," IEEE Trans.Ind.Applicat.,vol.32,pp.509-517,May/June 1996.

[7] H .S.Patel, R.G.Hoft, "Generalized Techniques of harmonic elimination and voltage control in thyristor inverter: part I-Harmonic elimination," IEEE Transactions on industry application, vol.IA.9,no.3,May/Jun.,1973,pp.310-317.

[8] P.M.Bhagwat and V.R.Stefanovic, "Generalized structure of a multilevel PWM inverter," IEEE Trans.Ind.Applicat., vol.19, no 6,pp.1057-1069,Nov./Dec.1983

[9] G.Carrara, S.Gardella, M.Marchesoni, R.Salutari, G.Sciutto, "A new multilevel PWM method: a theoretical analysis," in Proc, IEEE Power Electronic Specialist conf. (PESC), June, 1990, pp.363-371.

[10] F.Zare, G.Ledwich, "A hysteresis current control for single-phase multilevel voltage source inverters: PLD implementation," IEEE Transactions on power electronics, vol 17, issue 5, Sept.2002, pp. 731-738

[11] Chiasson, J.N.Tolbert, L.M, McKenzie, K.J.Zhong, Du, "A new approach to solving the harmonic elimination equations in multilevel converters", IEEE Transaction on Power Electronics, Vol.pp:478-490,March 2004

[12] Chiasson, J.N.Tolbert, L.M, McKenzie, K.J.Zhong, Du, "A new approach to solving the harmonic elimination equations for a multilevel converter;" in the Record of the Industry Applications Conference, 2003. 38th, Volume: 1, Pages: 640 647 v-ol.1,12-16 Oct.2003

Bidirectional Positive Buck-Boost Converter

Arash A Boora, *Student Member, IEEE*, Firuz Zare, *Senior member, IEEE*, Gerard Ledwich, *Senior member, IEEE*,
Arindam Ghosh, *Fellow, IEEE*
School of Engineering Systems
Queensland University of Technology
arash.boora@student.qut.edu.au

Abstract - **In a positive buck-boost (PBB) converter, inductor current and capacitor voltage can be decoupled which may improve system stability. In fact for a specific level of capacitor voltage, the inductor current can be adjusted at different levels and can be utilized to increase the robustness of the converter against input voltage and load disturbances. But when demand is a fast response with respect to step change in reference voltage, this topology needs to be modified. In this paper, a family of topologies based on a positive buck boost converter are presented which have a fast response and bidirectional power flow capability. This feature leads to some applications in hybrid vehicle systems and telecommunications.**
Simulations have been carried out to validate fast response of the proposed converters.

Keywords— **Bidirectional, DC-DC converters, Fast response**

I. INTRODUCTION

Non inverting Buck Boost converter is a known DC-DC power electronic converter which has characteristics of both buck and boost converter and can be applied in both step up and step down applications (Fig.1).

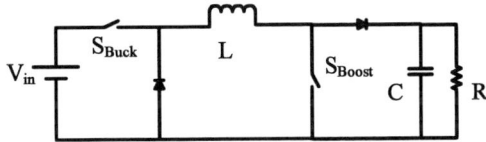

Figure 1: Positive Buck-Boost Converter

In addition it has advantage of a capacity for extra current storage in the inductor. Comparing the inductor current in Buck, Boost, inverting Buck Boost and positive Buck Boost, we have:

$$I_L = \frac{I_o}{D'} = \frac{V_C}{RD'} = \frac{V_{in}}{RD'^2} \qquad \text{Boost} \qquad (1)$$

$$I_L = \frac{I_o}{D'} = \frac{V_C}{RD'} = \frac{DV_{in}}{RD'^2} \qquad \text{Inverting Buck-Boost} \qquad (2)$$

$$I_L = I_o = \frac{V_C}{R} = \frac{DV_{in}}{R} \qquad \text{Buck} \qquad (3)$$

Equation (4) is same relationship for PBB.

$$I_L = \frac{1}{D'_{Boost}} I_o = \frac{1}{D'_{Boost}} \frac{V_C}{R} = \frac{D_{Buck}}{D'^2_{Boost}} \frac{V_{in}}{R} \qquad (4)$$

The output voltage is:

$$V_C = \frac{D_{Buck}}{D'_{Boost}} V_{in} \qquad (5)$$

So we have two controlling parameters (D_{Buck} and D_{Boost}) and two controlled quantities (i_L and v_C) which can be decoupled according to (4) and (5).

In this way the inductor of PBB can be used as an energy storage as well as energy deliverer while the amount of stored energy is independent from the level of delivered energy by D'_{Boost} in (4).

This capacity is utilized for improving the dynamic response of PBB to disturbances due to an increase in load or decrease in input voltage [8]. There are some DC-DC applications like hybrid vehicles [1-3] and hybrid power sources [4] which require bidirectional power flow. When there is a requirement for step up and step down voltage conversions or a higher degree of dynamic response to either disturbance or control reference signal a bidirectional converter topology (Fig. 1) may be promising. This converter is based on positive buck-boost converter. The other application is such as broader bandwidth where the switching frequency is restricted. For example envelop tracking in RF power amplifiers [5-7]. In this application, signal which should be amplified is divided into high and low frequency spectrums. The high frequency spectrum is amplified by a linear amplifier and the low frequency spectrum is converted by a switching circuit which reduces the size of the linear amplifier. The optimization between switching converter's band width and switching loss should be conducted to have the appropriate switching frequency. In the case of proposed converter owing to extra current capacity of positive buck-boost converter it can increase the dynamic response band width without increasing the switching frequency.

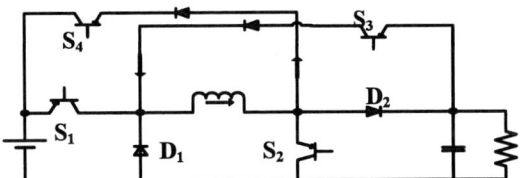

Figure 2: Bidirectional Positive Buck Boost (step up)

Using this degree of freedom of this converter another parameter is applied to decrease response time which may lead to a broader band width with lower switching frequency in communication systems [5-7]. In these low power applications there is no need to transfer small quantity capacitor energy back to input source thus the switch S_4 will be eliminated shown in Fig. 3.

978-1-4244-1741-4/08/$25.00 ©2008 IEEE

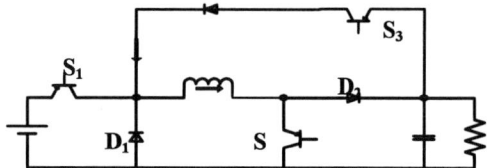

Figure 3: Topology of unidirectional Positive Buck-Boost converter with increased bandwidth.

In this paper we consider that for Inductor current control we need to recuperate the capacitor's energy in case of falling step change in reference voltage or load.

In this paper a current hysteresis, voltage hysteresis control strategy has been developed for proposed converters. The main focus is on the control strategy to increase the bandwidth of the system without increasing the switching frequency on one side and its ability to conduct bidirectional power flow on the other side.

If the bidirectional power flow be required in step down case D2 and S4 should be substituted. (Fig 4)

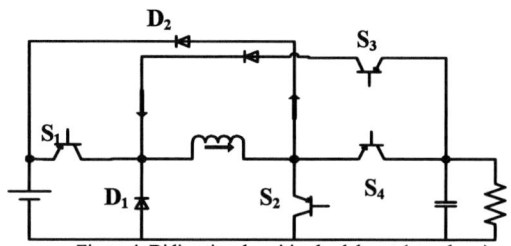

Figure 4: Bidirectional positive buck boost (step down)

Both topologies shown in Fig. 4 and Fig. 2 can perform step up and step down operations. The difference is that Fig. 2 topology can reverse the direction of energy only in step up operation and Fig. 5 topology can reverse the direction of energy only in step down operation.

The conventional way to develop a bidirectional converter is to anti parallel a diode with each switch and a switch with each diode in the unidirectional topology. Applying this method to positive buck boost converter is shown in Fig. 5.

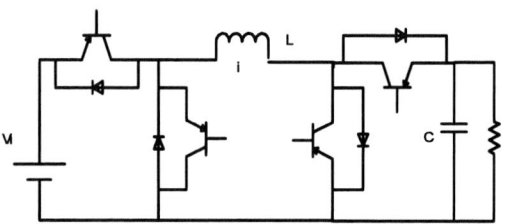

Figure5: Positive Buck-Boost Converter

Comparing this topology and the new topologies presented in this paper we can see that the number of switches and diodes are same or less.

The main difference between the new topology and the conventional bidirectional topology is that the capacitor energy can be transferred to the inductor faster in the new topology because it can direct the inductor current through the capacitor in negative way.

So the change in output voltage of the topologies presented in this paper can be performed faster.

This is more remarkable when we consider that PBB can store a level of extra current in its inductor. In the other words adding S_3 can improve the dynamic response of PBB and the dynamic response can be even more improved by storing more current in the inductor. Details of controlling PBB with a level of extra current are presented in [8] which include the effect of extra current storage on the switching frequency and switching loss of this converter.

The advantage of conventional topology is that the bidirectional energy delivery in this topology (Fig. 5) can be performed in both step down and step up cases.

In this paper we present the topologies based on PBB which are developed to be wideband and unidirectional (Fig. 3) or wideband and bidirectional (Fig. 2 and 4). All of these topologies are controlled to utilize the current storage advantage of PBB [8].

The following sections of this paper cover switching states, steady state and dynamic equations, control strategy and simulation results for proposed topologies.

II. SWITCHING STATES

To explain a general control strategy of the proposed DC-DC converter (Fig. 2) all switching states are presented in Fig.6. There are 9 switching states, 3 of which freewheels the inductor current and are called "zero states". Looking at the voltage imposed on the inductor in the other 6 states, the states can be divided into three groups with two states in each. The groups are named A, B, and C. and their states are shown in Fig.6. Because of this symmetrical availability of switching states there are options to transfer the inductor energy to the load (A, C) or to the voltage source (B`, C`), as well as options to transfer capacitor's energy to the inductor (A`, C`) and voltage source's energy to the inductor (B, C). In zero states the inductor and capacitor are disconnected and capacitor energy supplies the load while inductor keeps its current constant.

To reduce the switching loss the controller tries to decrease the switching frequency of each switch by switching from one state to the next one by changing the on or off state of only one switch. The states which can be transferred to each other by only one switching are called adjacent states. Fig. 7 shows the adjacent states.

To familiarize this converter we can see that switching between A and C states is a Buck converter and switching between B and C is a Boost converter. To discharge capacitor quickly we can apply states which direct the inductor current to the capacitor in the reverse way (A`, C`). In case of step up S_4 switch can be used to direct the energy of capacitor and inductor to the voltage source in switching states of (B` and C`).

There are three Zero states which are called so because inductor voltage is 0 when the state is 0, 0(-), or 0(+). And capacitor is discharging by load current. Because of these states we can store some extra current in the inductor while we are controlling output voltage independently.

724

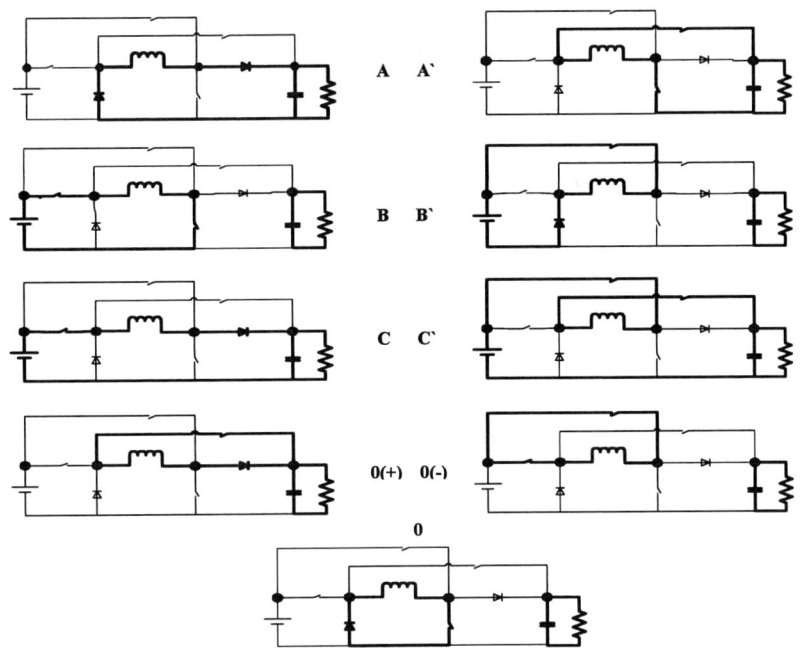

Figure 6: Switching states of Fig. 1 topology

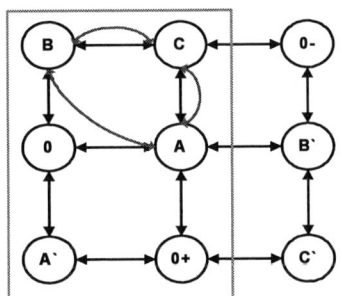

Figure 7 adjacency of switching states in bidirectional
positive buck boost converter
The states in box are the states of Fig. 3 topology

Table I: Switching states and charging condition of Capacitor and Inductor in each switching state.

$(S_1S_2S_3S_4)$	C	L	
A (0000)	Charge	Discharge	
B (1100)	Discharge	Charge	
C (1000)	Charge	Charge $V_o<V_i$	Discharge $V_o>V_i$
0 (0100)	Discharge	No Change	
A' (0110)	Discharge	Charge	
B' (1001)	Discharge	Discharge	
C' (0011)	Discharge	Charge $V_o<V_i$	Discharge $V_o>V_i$
0+ (0010)	Discharge	No Change	
0- (1001)	Discharge	No Change	

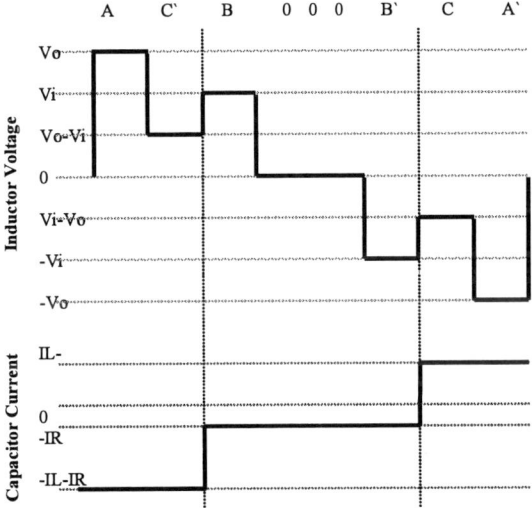

Figure 8: the effect of each switching state on Inductor and Capacitor

Fig. 8 shows the capacitor voltage and inductor current on each switching state. There are seven levels for inductor voltage and three levels for capacitor current. In capacitor current the level of $-I_L-I_R$ is identical in this topology and allows reducing the capacitor voltage faster than conventional bidirectional positive Buck Boost converter.

In drawing Fig. 8 input voltage has been considered to be less than output voltage and more than V_o-V_i.

Of course Fig 8 has been drawn simply. The detailed version should include fluctuations of output voltage and inductor current.

III. STEADY STATE AND DYNAMIC EQUATIONS

Writing dynamic equations of the inductor current we have:

$$i_L = inductor\ current$$
$$v_C = capacitor\ voltage \tag{6}$$

$$(T_A - T'_A)(-v_C) + (T_B - T'_B)V_{in} + (T_C - T'_C)(V_{in} - v_C) = L\frac{di_L}{dt}$$

Defining:

$$T_A - T'_A = \Delta T_A$$
$$T_B - T'_B = \Delta T_B \tag{7}$$
$$T_C - T'_C = \Delta T_C$$

We have:

$$\Delta T_A(-v_C) + \Delta T_B V_{in} + \Delta T_C(V_{in} - v_C) = L\frac{di_L}{dt} \tag{8}$$

$$-(\Delta T_A + \Delta T_C)v_C + (\Delta T_B + \Delta T_C)V_{in} = L\frac{di_L}{dt}$$

And writing dynamic equations for the capacitors voltage we have:

$$((T_A - T'_A) + (T_C - T'_C))i_L - T\left(-\frac{v}{R}\right) = C\frac{dv_C}{dt} \tag{9}$$

Where, T is switching cycle. With same definition:

$$(\Delta T_A + \Delta T_C)i_L - T\left(-\frac{v}{R}\right) = C\frac{dv_C}{dt} \tag{10}$$

Writing these equations in a state space form with some simplifications leads to:

$$\begin{bmatrix} L & 0 \\ 0 & C \end{bmatrix}\begin{bmatrix} \dot{i_L} \\ \dot{v_C} \end{bmatrix} = \tag{11}$$
$$\left(\frac{\Delta T_A + \Delta T_C}{T}\begin{bmatrix} 0 & -1 \\ 1 & 0 \end{bmatrix} + \begin{bmatrix} 0 & 0 \\ 0 & -\frac{1}{R} \end{bmatrix}\right)\begin{bmatrix} i_L \\ v_C \end{bmatrix} + \frac{\Delta T_B + \Delta T_C}{T}\begin{bmatrix} 1 \\ 0 \end{bmatrix}V_{in}$$

Calculating transfer functions for inductor current and capacitor voltage results in:

$$i_L = \frac{(RCs + 1)\left(\frac{\Delta T_B + \Delta T_C}{T}\right)}{RLCs^2 + Ls + \left(\frac{\Delta T_A + \Delta T_C}{T}\right)^2 R}V_{in} \tag{12}$$

$$v_C = \frac{R\left(\frac{\Delta T_B + \Delta T_C}{T}\right)\left(\frac{\Delta T_A + \Delta T_C}{T}\right)}{RLCs^2 + Ls + \left(\frac{\Delta T_A + \Delta T_C}{T}\right)^2 R}V_{in} \tag{13}$$

Steady state equation can be driven by putting $s = 0$ in transfer functions:

$$i_L = \frac{T(\Delta T_B + \Delta T_C)V_{in}}{(\Delta T_A + \Delta T_C)^2 R} = \frac{T}{(\Delta T_A + \Delta T_C)}I_{Load} \tag{14}$$

$$v_C = \frac{(\Delta T_B + \Delta T_C)}{(\Delta T_A + \Delta T_C)}V_{in} \tag{15}$$

It is interesting to mention that equations are similar to conventional positive buck boost converter equations, which are [8]:

$$i_L = \frac{T(T_B + T_C)V_{in}}{(T_A + T_C)^2 R} = \frac{T}{(T_A + T_C)}I_{Load} \tag{16}$$

$$v_C = \frac{(T_B + T_C)}{(T_A + T_C)}V_{in} \tag{17}$$

But in case of bidirectional positive buck boost converter ΔT_X can have negative values unlike T_X in case of unidirectional positive buck boost converter for states of *A*, *B*, *C*, and *0*.

IV. CONTROL STRATEGY AND SIMULATION RESULTS

Hysteresis control strategy has been developed for the family of topologies presented in this paper

Hysteresis control strategy is applied to step down case for topologies shown in Fig. 1, 3, 4. This control strategy is simple but has not exact control on switching frequency and dynamic transient of the system.

This control strategy applies two bands for capacitor voltage and two bands for inductor current. Crossing these bands makes commands to change switching configurations. Fig. 9 explains hysteresis control strategy for bidirectional positive buck boost converter in step down case. The topology is shown in Fig. 4.

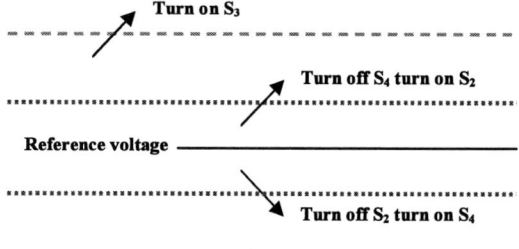

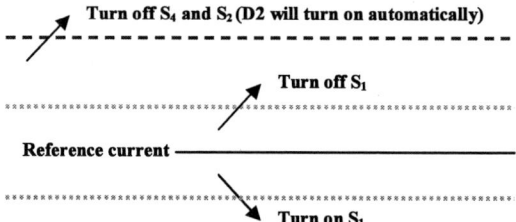

Figure 9: looking at Fig 4 the hysteresis control strategy for step down case is explained here.

Fig. 10 shows the simulation results for step change in reference voltage in conventional, wide band and bidirectional positive buck boost converters which are controlled by hysteresis strategy.

In all cases the input voltage is constantly 100V and output resistance is constantly 10 ohms.

The controller keeps the inductor current twice as minimum required current in all cases

Comparing the plots shown in Fig. 10 can be seen that the reference voltage has been changed to 80V from 20V and at 0.01sec and back to 20 from 80 at 0.03sec.

The voltage and current rise in all of the cases is same it could be quicken by increasing stored current in the inductor.

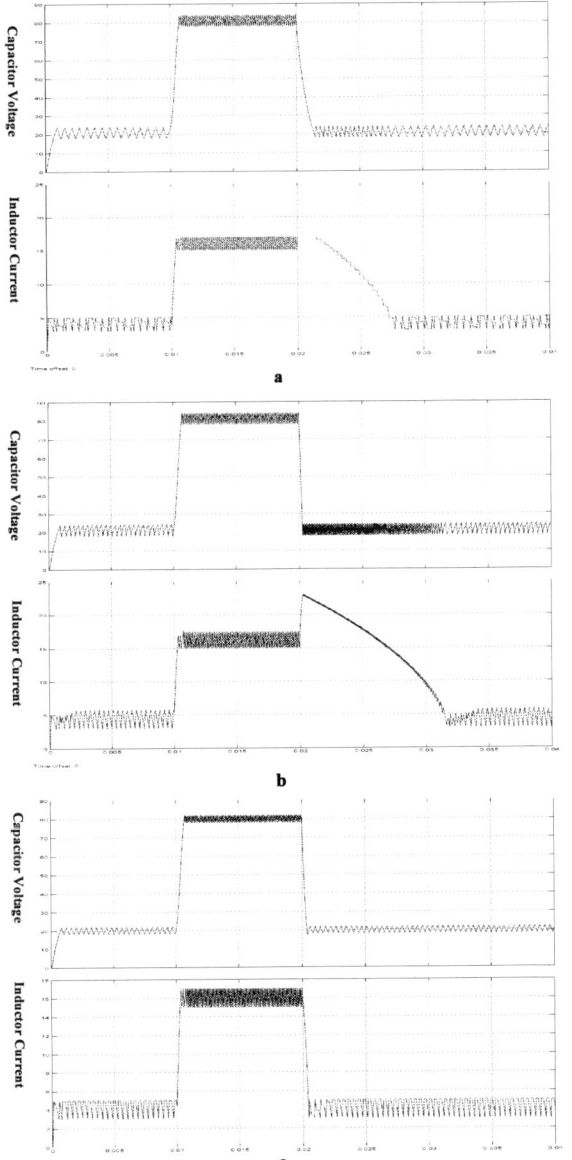

Figure 10: a) Conventional PBB (topology: Fig. 1)
b) Wide band PBB (topology: Fig. 3)
c) Bidirectional PBB (topology: Fig. 4)

Voltage drop is almost same in (a) and (b) where switch S_3 has been used. But current has risen in (b) because the topology is not bidirectional and the extra energy of the capacitor is transferred to the inductor.

In (c) the extra energy of the capacitor has been directed to the source by Diode D_2 which is included in bidirectional topology shown in Fig. 4. Looking closer at (c) the voltage drop is slightly slower than (b) the reason is transfer of energy to the source and then lower inductor current during voltage drop.

V. CONCLUSION

A family of DC-DC converter topologies based on positive Buck Boost converter are introduced. The main approach is to apply PBB's capacity for current storage to increase band width of theses converters without increasing switching frequency.

Also a bidirectional fast response topology has been derived from PBB.

These topologies have been controlled by hysteresis control strategies. Simulation results have been presented.

VI. ACKNOWLEDGMENT

The authors thank the Australian Research Council (ARC) for the financial support for this project through the ARC Linkage Grant LP0774899.

References

[1] "Superconducting Magnetic Energy Storage (SMES) for Energy Cache Control in Modular Distributed Hydrogen-Electric Energy Systems" Louie, H.; Strunz, K.;Applied Superconductivity, IEEE Transactions on Volume 17, Issue 2, Part 2, June 2007 page(s):2361 - 2364

[2] "System Integration and Power Flow Management for a Series Hybrid Electric Vehicle using Super-capacitors and Batteries" Yoo, Hyunjae; Sul, Seung-Ki; Park, Yongho; Jeong, Jongchan; Applied Power Electronics Conference, APEC 2007 - Twenty Second Annual IEEE , Feb. 25 2007-March 1 2007 Page(s):1032 – 1037

[3] "Modelling and design of super capacitors as peak power unit for hybrid electric vehicles", Timmermans, J.-M.; Zadora, P.; Cheng, Y.; Van Mierlo, J.; Lataire, Ph.; Vehicle Power and Propulsion, 2005 IEEE Conference 7-9 Sept. 2005 Page(s):8 pp.

[4] "Dynamic Performance of a Static Synchronous Compensator with Superconducting Magnetic Energy Storage" Molina, M.G.; Mercado, P.E.; Watanabe, E.H.; Power Electronics Specialists Conference, 2005. PESC '05. IEEE 36th 2005 Page(s):224 – 230

[5] "A high-efficiency linear RF power amplifier with a power-tracking dynamically adaptive buck-boost supply" Sahu, B.; Rincon-Mora, G.A.; Microwave Theory and Techniques, IEEE Transactions on Volume 52, Issue 1, Part 1, Jan. 2004 page(s):112 – 120

[6] "A DSP structure authorizing reduced-bandwidth DC/DC Converters for Dynamic Supply of RF Power Amplifiers in Wideband Applications" Cesari, Albert; Cid-Pastor, Angel; Alonso, Corinne; Dilhac, Jean-Marie; IEEE Industrial Electronics, IECON 2006 - 32nd Annual Conference on Nov. 2006 Page(s):3361 – 3366

[7] "Efficiency optimization in linear-assisted switching power converters for envelope tracking in RF power amplifiers" Yousefzadeh, V.; Alarcon, E.; Maksimovic, D.; Circuits and Systems, 2005. ISCAS 2005. IEEE International Symposium on 23-26 May 2005 Page(s):1302 - 1305 Vol. 2

[8] "A General Approach to Control a Positive Buck-Boost Converter to Achieve Robustness against Input Voltage Fluctuations and Load Changes" Arash A Boora, Student member, IEEE, Firuz Zare, Senior member, IEEE, Gerard Ledwich, Senior member, IEEE, Arindam Ghosh, Fellow, IEEE . PESC 2008 (unpublished)

Control system of power electronics current modulator utilized in diode rectifier with sinusoidal source current

Michał Gwóźdź*, Michał Krystkowiak [†]

* Poznan University of Technology, Institute of Electrical Engineering and Electronics, Poznań , Poland, e-mail:
michal.gwozdz@eranet.pl

[†] Poznan University of Technology, Institute of Electrical Engineering and Electronics, Poznań , Poland, e-mail:
Michal.Krystkowiak@put.poznan.pl

Abstract—In the article a structure of control circuit of power electronics current modulator, connected to DC circuits of diode rectifiers, is presented. This solution is very attractive in case of large power rectifiers due to obtaining close to sinusoidal power grid current and relatively low modulator power. The criterions of parameters selection of control circuit are described. Selected results of simulation researches are presented also.

Keywords— device, harmonics, sensor, simulation.

I. INTRODUCTION

The rectifiers belong to most widely utilized group of power electronics converters. Unfortunately the classical solutions of diode and thyristor rectifiers have disadvantageous harmonic spectrum of source (power grid) current [5]. One of the ways, which corrects the waveform of current, depends on utilization a current modulation in DC output circuit of these rectifiers.

These converters are built with two 6-pulse diode rectifiers, which are supplied by two 3-phase transformer about connection star-star and star-delta. With their help phase shift equal 30 degrees of voltage sources for each of diode converter is obtained. Additionally, in DC output circuit a special current modulator can be placed. The block scheme of such modified converter group is shown on Fig. 1.

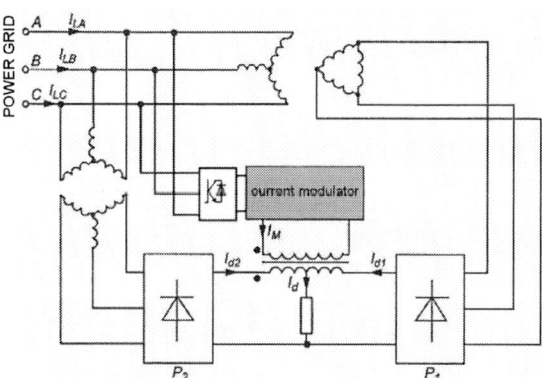

Fig. 1. The diode rectifier with current modulator in DC output circuit.

Presented solution of diode rectifier has got close to sinusoidal source current. Utilized there current modulator is kind of current source, which is connected to main circuit of rectifiers via a wideband pulse transformer with two taps on secondary site. With aid of this transformer the current of modulator is adding or subtracting to output currents of each 6-pulse diode rectifiers. In this way we can change a shape of input currents of these two rectifiers and resultant source current of converter, being sum of these. In aim of getting sinusoidal source (power grid) current, the modulator has to generate triangular waveform with fundamental frequency 300 Hz (what is equal 6 times of power grid voltage frequency). The control circuit of current modulator directly decides about the effectiveness of improvement of energy transformation quality and source current at the input of rectifiers group.

The waveforms of source current for classical 12-pulse rectifier and rectifier with current modulator are shown on Fig. 2.

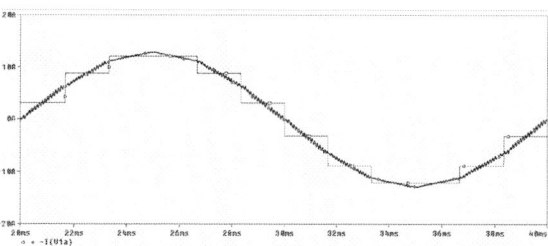

Fig. 2. The waveforms of source current of 12-pulse rectifier without current modulator (black line) and rectifier with such modulator (blue).

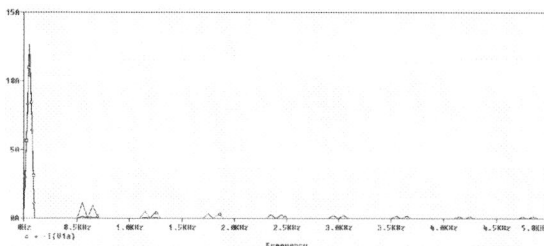

Fig. 3. The spectrum of source current of 12-pulse rectifier without current modulator (black line) and rectifier with such modulator (blue line).

The spectrum analysis of these signals is presented on Fig. 3.

As result of current modulation in DC output circuit, the current of power grid is close to sinusoidal, with THD value about 1.2%. In case of classical 12-pulse diode rectifiers, THD value is equal about 15%, what is above 10 times higher.

II. SIMULATION MODEL OF CONTROL CIRCUIT AND CURRRENT MODULATOR

A. Structure of control circuit and current modulator

Fig. 4 presents a simulation model of current modulator. The current modulator is built with IGBT full bridge inverter with passive serial RL filter at the output. This converter is connected to DC output circuits of each diode rectifiers via wideband pulse transformer with two taps on secondary site.

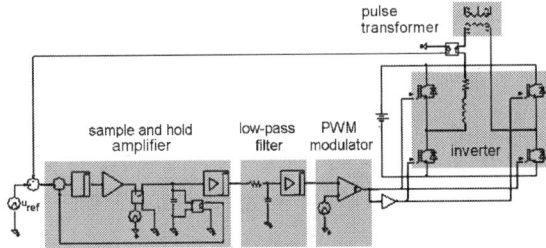

Fig. 4. The simulation model of control circuit and current modulator.

The control circuit of current modulator consists of following elements: sample and hold amplifier, low-pass filter with gain block, PWM modulator and reference signal generator. The amplitude of reference signal depends on average value of DC load current.

B. Signal model of current modulator with control circuit

The small signal (linear) model of current modulator with control circuit is shown on Fig. below.

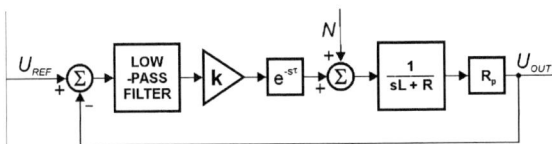

Fig. 5. The small signal (linear) model of current modulator with control circuit.

Following the low-pass filter, gain block (k) is placed. This block represents resultant gain of control system with current modulator, which is equal to product of low-pass filter gain and gain of inverter. Delay block represents time delay carried in by both, PWM modulator and inverter. Parameters R and L of low-pass block are described by following equations:

$$R = R_f + R_{Z1}^{"} + R_{Z2}^{"} \qquad (1)$$

$$L = L_f + L_{Z1}^{"} + L_{Z2}^{"} \qquad (2)$$

where:

$L_{Z1}^{"}, L_{Z2}^{"}$ – short-circuited inductances of star-stat and star-delta transformers on secondary windings terms,

$R_{Z1}^{"}, R_{Z2}^{"}$ – short-circuited resistances of star-star and star-delta transformers on secondary windings terms,

R_f, L_f – resistance and inductance of output filter in current modulator.

The short-circuited parameters of pulse transformer were disregard, because the values of these parameters are much smaller than the values of shot-circuited parameters of both main transformers (with star-star and star-delta connections). The resistance R_p represents the sensor, which is used to measure the output signal. The signal N represents voltage, which is induced on secondary windings of pulse transformer.

The transfer function of such linear model of current modulator is as follows:

$$U_{out}(s) = \frac{\dfrac{F(s)kR_p e^{-s\tau}}{R+sL}}{1+\dfrac{F(s)kR_p e^{-s\tau}}{R+sL}} U_{ref}(s) +$$

$$+ \frac{\dfrac{R_p}{R+sL}}{1+\dfrac{F(s)kR_p e^{-s\tau}}{R+sL}} N(s) \qquad (3)$$

where: $F(s)$ – transfer function of low-pass filter.

If amplitude of u_{ref} is much larger than amplitude of n, equation (3) can be reduced to following form:

$$U_{out}(s) = \frac{\dfrac{F(s)kR_p e^{-s\tau}}{R+sL}}{1+\dfrac{F(s)kR_p e^{-s\tau}}{R+sL}} U_{ref}(s) \qquad (4)$$

This situation has been analyzed mainly in most of researches.

C. Criterions of low–pass filter parameters selection

During the simulation researches different criterions of low-pass filter parameters selection were analysed.

The first of analysed criterions was the limitation of slew rate value of PWM modulator input signal. Fulfillment of this condition is necessary to assure the correct switching frequency, which must be equal to frequency of carrying signal.

This condition is described by following equation:

$$\frac{d}{dt}\left[\mathbf{L}^{-1}\left(\frac{k_F F(s)(R+sL)}{R+sL+kF(s)R_p e^{-s\tau}}U_{ref}(s)\right)\right] < \frac{d}{dt}\left[s_n(t)\right] \quad (5)$$

where: k_F – gain of low-pass filter, $s_n(t)$ – carrying signal.

The second criterion is stability of closed loop circuit. This depends on assurance of the largest possible gain value of open-circuit transfer function (4) in useful bandwidth and below 1 in bandwidth, where the shift phase is close (and above) 180 el. degrees.

The output signal u_{out} of current modulator model for rectangular reference signal and circuit stable is shown on Fig. below.

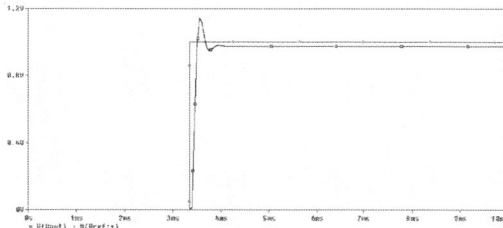

Fig. 6. The output signal (blue line) and rectangular reference signal (black line) when system is stable.

Fig. 7 and 8 below present current waveform at the output of current modulator for triangular reference signal and 2 different values of low-pass filter gain.

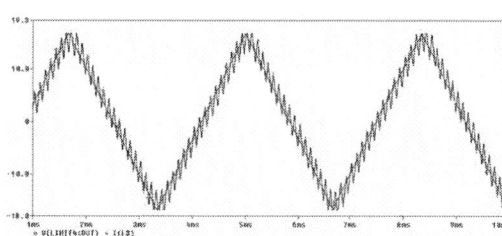

Fig. 7. The output signal of current modulator when gain of control circuit is correct (dark blue line).

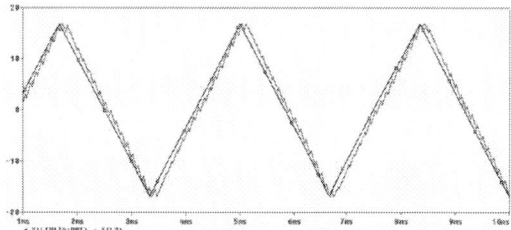

Fig. 8. The output signal of current modulator when gain of control circuit is incorrect – too small value of these.

The last criterion, which has been analyzed during the researches, is limitation of spectrum aliasing effects of output signal u_{out} [8]. Limitation of band of u_{out} is obtained thanks low-pass filter implementation mainly.

As can be seen the source current is close to sinusoidal, so higher harmonics amplitudes, characteristic for classical 12-pulse diode rectifiers, are much reduced.

III. SUMMARY

In this article one of possible ways leading to sinusoidal source current of power electronics rectifiers is presented. For achieving this, the current modulator in DC output circuit of two 6-pulses rectifiers has been utilized.

As result of these, the current of power grid is close to sinusoidal. This solution is very attractive in case of large and very large power rectifiers (for example for electrical traction) due to the power of the current modulator is fraction (only about 2,5%) of total DC power load of both rectifiers.

REFERENCES

[1] L. Frackowiak, M. Krystkowiak, "Rectifier with modulation in DC circuit", *Electrotechnical review*, pp. 61-64, July 2006.

[2] R. Strzelecki, H. Supronowicz, "The power factor of AC circuits and correction method", Warsaw, OWP, 2000, pp. 120-135.

[3] H. Tunia, B. Winiarski, "Power electronics", Warsaw: WNT, 1994, pp. 109-113.

[4] L. Frackowiak, "Power electronics", vol. II. Poznan: WPP, 1998, pp. 188-191.

[5] R. Barlik, M. Nowak, "Thyristor technician", Warsaw: WNT, 1998, pp. 174–178.

[6] L. Frackowiak, M. Krystkowiak, "Rectifier with modulation in DC circuit" in. *Non – sinusoidal currents*, Lagow, Poland, 2006.

[7] N. Mohan, T.M. Undeland, W.P. Robins: "Power Electronics, Converters, Applications and Design", New York, J.Wiley, 1989.

[8] M. Gwóźdź: "Impact of Aliasing Effect on Work of Wide Band Power Electronics Current Source", XIX Symposium Electromagnetic Phenomena in Nonlinear Circuits, Maribor, Słowenia, 28-30.06.2006, ss. 147-148.

Design and control of a half-bridge converter to drive piezoelectric actuators

Oriol Gomis-Bellmunt*, Josep Rafecas-Sabaté*, Daniel Montesinos-Miracle*,
Josep-Maria Fernández-Mola* and Joan Bergas-Jané*
*Centre d'Innovació Tecnològica en Convertidors Estàtics i Accionaments (CITCEA-UPC),
Departament d'Enginyeria Elèctrica, Universitat Politècnica de Catalunya.
ETS d'Enginyeria Industrial de Barcelona, Av. Diagonal, 647, Pl. 2. 08028 Barcelona, Spain
Tel: +34 934016727, Fax: +34 934017433 e-mail: *oriol.gomis@upc.edu*

Abstract—**The present work deals with the development and control of a converter to drive piezoelectric actuators. Due to the piezoelectric load specification, a half-bridge converter is proposed. Such a converter is analyzed considering charging and discharging modes. The converter is controlled by means of a sliding mode control strategy, employing two different sliding surfaces for the charging and discharging operation mode. The proposed scheme is validated by means of simulations.**

Index Terms—**Piezoelectric actuators, Sliding mode control.**

I. INTRODUCTION

The increasing number of applications where the piezoelectric actuators are used is motivating the development of specific converters to drive such actuators. Different switching converter topologies have been employed to drive such elements, including fly-back [1], buck-boost [2], [3], push-pull [4] and converters based on piezoelectric transformers [5]. The most common strategy to control the position and force of a piezoelectric actuator is the direct regulation of the voltage applied [4]. The non-linear relationship between displacement and voltage can be improved working with the electric charge instead of the voltage. This strategy has been used successfully by many authors [6]–[13]. The main drawback of the charge control is the necessity of sensing and integrating the current supplied to the load, with the corresponding problems involved.

Piezoelectric loads differ from the conventional loads assumed in the conventional converters. They behave as voltage sources and there is only current when the voltage is changing or when they are producing mechanical work. If the voltage is unchanged and the actuator is unloaded there is no current. Hence the converter can operate in either continuous or discontinuous conduction mode depending on the converter and load state. Furthermore, the voltage in the output capacitance (or load) cannot be considered constant and therefore the currents cannot be assumed linear. Such a complexity increase is to be introduced in the converter design in order to optimize the actuator drive performance.

This work introduces a half-bridge converter and a sliding mode control strategy. The necessity for such a control strategy arises from the fact that a PI controller

cannot deal properly with small capacitance ranges. It is important to remark that a very simple model could be used considering only a capacitance as long as the working frequency was far away from the resonance (aprox. 40 kHz). For higher frequencies more branches have to be included to describe appropriately the behavior of the actuator. Each new branch corresponds to a mechanical resonance frequency. The employed model includes a branch corresponding to the mechanical resonance, however the results considering or not such a resonance are undistinguishable as long as the frequency applied to the piezoelectric actuators is kept far away from the resonance.

II. CONVERTER

The converter analyzed is shown in Fig. 1. It consists in a half-bridge converter. The converter can operate with positive and negative i_L current but always with positive u_C voltage. It is important to note that no output capacitor is used; only the equivalent capacitance of the piezoelectric load is considered. It improves the efficiency of the converter, since no power is needed to charge and discharge the output capacitor, but it complicates the control, not allowing normal PID control.

The converter show two clear operating modes: charging and discharging, as illustrated in Figs. 2 and 3. Charging operation is controlled by S1. Inductance current is positive and when S1 is disconnected the current flows through D2. Discharging operation is controlled by S2, having negative inductance current which flows through D1 once S2 is disconnected.

III. CONTROL

The non-linearity of the converter along with the problems encountered when working with classical PI controllers for low equivalent capacitance values have motivated the election of a non-linear control strategy. Among them the sliding mode control has been chosen due to the appropriate physical interpretation of the sliding surface as an energy surface, its ease of implementation and the suitability to drive power converters as shown in [2]. Two energy based sliding surfaces have been introduced to allow the controller disconnect the switches when the

978-1-4244-1741-4/08/$25.00 ©2008 IEEE

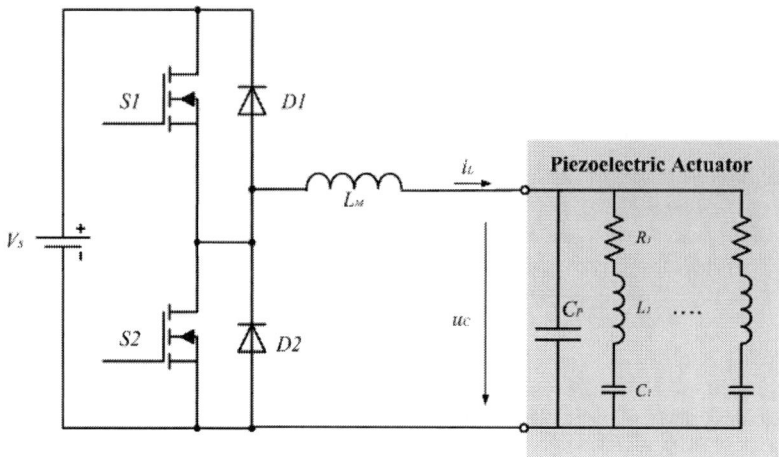

Fig. 1. Converter under study

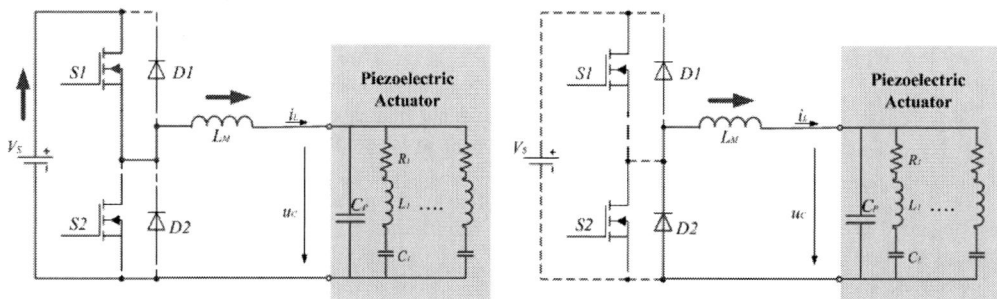

Fig. 2. Charging operation

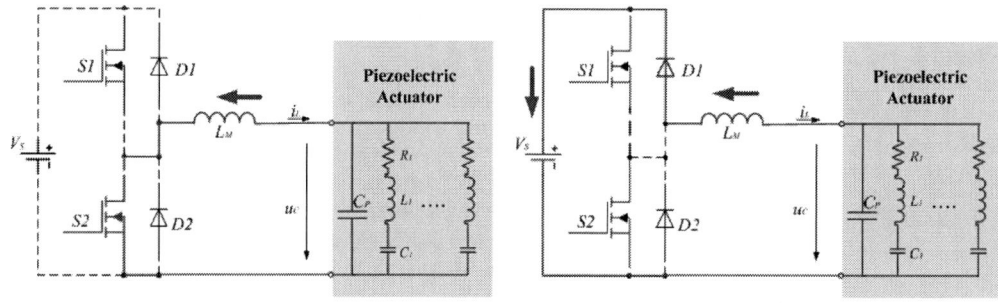

Fig. 3. Discharging operation

needed energy is in the inductance, before the voltage raise in the equivalent capacitance occurs.

Taking $x_1 = i_L$ and $x_2 = u_C$, the control law is based on the sliding surfaces:

$$S_1(e,t) = E_1^* - E_1 =$$
$$\frac{1}{2}C_P x_2^{*2} - \frac{1}{2}C_P x_2^2 - \frac{1}{2}L_M x_1^2 \quad (1)$$

$$S_2(e,t) = E_2^* - E_2 = \frac{1}{2}C_P x_2^{*2} - \frac{1}{2}C_P x_2^2 \quad (2)$$

The control law yields:

$$U_1(t) = \begin{cases} 0 & S_1(e,t) < -\varepsilon \\ 1 & S_1(e,t) > \varepsilon \end{cases}$$

$$U_2(t) = \begin{cases} 1 & S_2(e,t) < -\varepsilon \\ 0 & S_2(e,t) > \varepsilon \end{cases} \quad (3)$$

IV. SIMULATION

The proposed scheme has been validated by means of simulation. The results considering a sampling frequency of 1000 kHz, with L_M=1 mH and C_P=0.5 μF are shown in Fig. 4, where it can be noted that the piezoelectric device voltage tracks remarkably well the reference voltage.

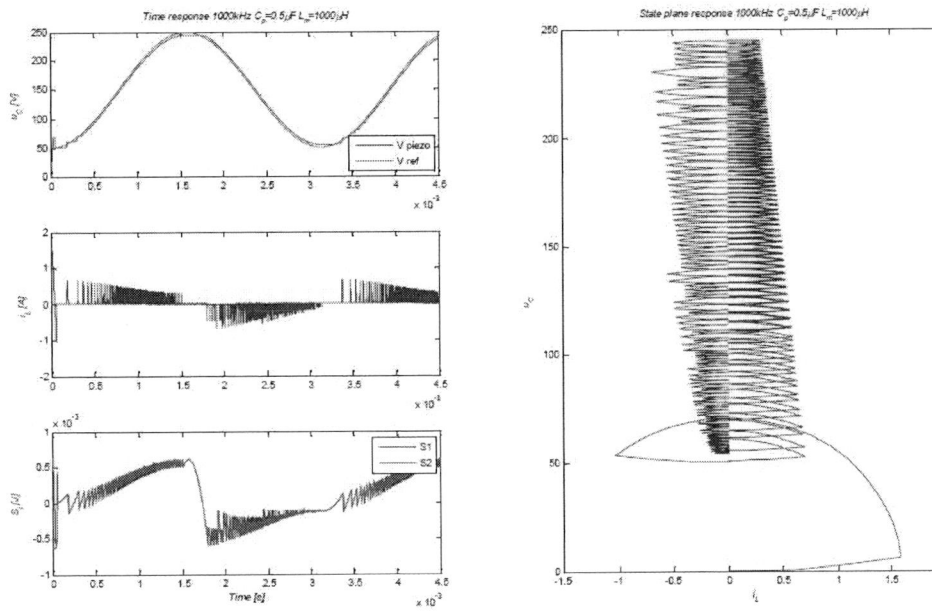

Fig. 4. Simulation results with C_P=0.5 μF, L_M=1 mH, sampling frequency = 1000 kHz

V. CONCLUSION

The present paper has presented a half-bridge converter to drive piezoelectric actuators. The converter has been analyzed considering its charging and discharging modes. The converter is controlled by a sliding mode control strategy, employing two different sliding surfaces for the charging and discharging operation mode. The proposed scheme has been validated by means of simulations.

ACKNOWLEDGMENT

Supported by CICYT through grant DPI2005-08668-C03-03.

REFERENCES

[1] S.-Y. Tseng, S.-Y. Tseng, Y.-M. Chen, Y.-K. Huang, H.-T. Hsieh, and T.-F. Wu, "Quasiresonant flyback converter for transdermal drug delivery applications," in *Proc. Nineteenth Annual IEEE Applied Power Electronics Conference and Exposition APEC '04*, Y.-M. Chen, Ed., vol. 3, 2004, pp. 1653–1659 Vol.3.

[2] O. Gomis-Bellmunt, D. Montesinos-Miracle, S. Galceran-Arellano, and J. Rull-Duran, "Sliding mode control based buck-boost bidirectional converter to drive piezoelectric loads," *Electrical Engineering (Archiv fur Elektrotechnik)*, vol. 90, no. 2, pp. 115–125, Dec. 2007. [Online]. Available: http://dx.doi.org/10.1007/s00202-007-0067-1

[3] B. Krishnamachari, B. Krishnamachari, and D. Czarkowski, "Bidirectional buck-boost converter with variable output voltage," in *Proc. IEEE International Symposium on Circuits and Systems ISCAS '98*, D. Czarkowski, Ed., vol. 6, 1998, pp. 446–449 vol.6.

[4] C. Stiebel and H. Janocha, "New concept of a hybrid amplifier for driving piezoelectric actuators," *http://www.lpa.uni-saarland.de/pdf/stiebel_ja2000.pdf*, 2000.

[5] S. Manuspiya, P. Laoratanakul, and K. Uchino, "Integration of a piezoelectric transformer and an ultrasonic motor," *Ultrasonics*, vol. 41, pp. 83–87, 2003.

[6] S. Chandrasekaran and D. K. Lindner, "Power flow through controlled piezoelectric actuators," *Journal of Intelligent Material Systems and Structures*, vol. 11, no. 6, pp. 469–481, Jun. 2000. [Online]. Available: http://jim.sagepub.com/cgi/content/abstract/11/6/469

[7] S. Chandrasekaran, D. K. Lindner, and D. J. Leo, "Effect of feedback control on the power consumption of induced-strain actuators," *Journal of Intelligent Material Systems and Structures*, vol. 13, no. 2-3, pp. 85–95, Feb. 2002. [Online]. Available: http://jim.sagepub.com/cgi/content/abstract/13/2-3/85

[8] S. Chandrasekaran, D. K. Lindner, and R. C. Smith, "Optimized design of switching amplifiers for piezoelectric actuators," *Journal of Intelligent Material Systems and Structures*, vol. 11, no. 11, pp. 887–901, Nov. 2000. [Online]. Available: http://jim.sagepub.com/cgi/content/abstract/11/11/887

[9] A. Fleming, A. Fleming, and S. Moheimani, "Precision current and charge amplifiers for driving highly capacitive piezoelectric loads," *Electronics Letters*, vol. 39, no. 3, pp. 282–284, 2003.

[10] A. J. Fleming and S. O. R. Moheimani, "Improved current and charge amplifiers for driving piezoelectric loads, and issues in signal processing design for synthesis of shunt damping circuits," *Journal of Intelligent Material Systems and Structures*, vol. 15, no. 2, pp. 77–92, Feb. 2004. [Online]. Available: http://jim.sagepub.com/cgi/content/abstract/15/2/77

[11] A. Gandelli, A. Gandelli, and R. Ottoboni, "Charge amplifiers for piezoelectric sensors," in *IEEE Instrumentation and Measurement Technology Conference IMTC/93. Conference Record*, R. Ottoboni, Ed., 1993, pp. 465–468.

[12] J. A. Main, D. V. Newton, L. Massengill, and E. Garcia, "Efficient power amplifiers for piezoelectric applications," *Smart Mater. Struct.*, vol. 5, no. 766-775, pp. 282–284, Aug 1996.

[13] K. Yi, K. Yi, and R. Veillette, "A charge controller for linear operation of a piezoelectric stack actuator," *IEEE Trans. Contr. Syst. Technol.*, vol. 13, no. 4, pp. 517–526, 2005.

Online Diagnosis of PEM Fuel Cell

Abdellah NARJISS*, Daniel DEPERNET*, Denis CANDUSSO+, Frederic GUSTIN*, Daniel HISSEL*

*FC-LAB, FEMTO-ST/ ENISYS Department, Belfort, France, e-mail: *anarjiss@edu.univ-fcomte.fr*
daniel.depernet@utbm.fr, frederic.gustin@univ-fcomte.fr, daniel.hissel@univ-fcomte.fr
+FC-LAB, INRETS, Belfort, France, e-mail: *denis.candusso@inrets.fr*

Abstract—**This study consists on online detection of fuel cell dysfunction in embedded applications without additional hardware component. The power converter usually coupled with the fuel cell is used to perform the diagnosis strategy. The main interest of the method presented in this paper is to simultaneously optimize the performances, the cost and the size of the Proton Exchange Membrane Fuel Cell (PEMFC). In this work, we focus on a system including a fuel cell stack coupled with an isolated DC/DC power converter and controlled by a Digital Signal Processor (DSP). These elements associated with measurement sensors and adequate programmings allow the control of the power conversion and the determination of fuel cell health at the same time. The fuel cell harmonic impedance is measured using spectroscopy method for a large frequency range. Specific cases of fault detections related to membrane humidification and reactive gas feeding are treated in this paper.**

Keywords—**Fuel cell, online diagnosis, DC/DC converter, control, DSP controller, impedance spectroscopy**

I. INTRODUCTION:

For Polymer Electrolyte Fuel Cell (PEFC), the performance is linked to operating conditions and among them, water management has a significant part [1]. It is well-known now that the electrolyte ionic conductivity depends on membrane water content [2] which is function of different parameters as gas pressure, fuel cell temperature and fuel cell load. The control of these parameters allows better humidification of membrane which signifies higher fuel cell performance [3]. But to ensure enhanced PEFC operation, the challenge is to estimate the membrane state and the impact of water management on the global fuel cell performances. In this work, we demonstrate how to determine state of fuel cell in real time thanks to impedance spectroscopy procedure and by using only fuel cell power converter associated to innovative control. On the one hand this diagnosis method allows detecting miscellaneous phenomena characterized by spectral impedance analysis as for example anther (??) phenomena. On the other hand it allows limiting the overall cost and size of the power transfer unit and monitoring devices.

In this work, a reduced scale fuel cell system is used as well as a power converter to study behavior of PEFC dedicated to usual transport applications. The set-up consists on a stack of 20 cells and 1kW power, an isolated high frequency DC/DC converter and a DSP controller [4]. We study how fuel cell performances can be optimized by exploiting results acquired from impedance spectroscopy. One issue, selected as an example of significant performance factor for the fuel cell, is

membrane water content. This system of diagnosis can also include supervision of voltage for each individual cell in stack.

II. DC/DC CONVERTER

A. Topology of converter

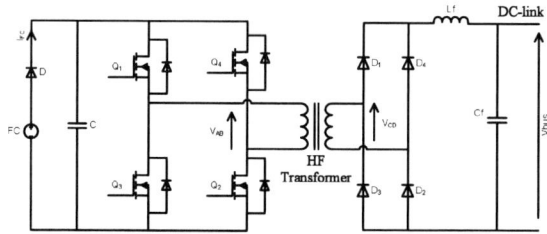

Fig. 1: Topology of DC/DC converter

The DC-DC Full-bridge converter (Fig. 1) is a two stage converter with an intermediate high frequency transformer. The first stage consists of a single phase inverter based on four switches associated to an adequate control law. The intermediate high frequency transformer ensures voltage gain and transmits power from primary to secondary. The second stage is based on a single phase rectifier associated with an LC filter to supply the DC link. General system is controlled by DSP controller and control interface.

A capacitor must be added in parallel with the fuel cell stack to reduce the current harmonics.

This structure presents following main advantages:

* no high-current inductance is required in the primary stage of the converter,

* the switch constraints are reduced (i.e. voltage and current constraints) regarding the previous structures,

* the two stages are isolated.

B. Design of converter

Presentation of fuel cell and DC-link parameters:

	Voltage	Current
Fuel cell stack (FC)	20V	0A
	13V	50A
DC-link	110V	0 to 9A
	$\Delta V = \pm 10\%$ of 110V	

These parameters allow designing semi-conductors for the converter as described in the following table.

Component	characteristics
MOSFET IXFN180N07	$V_{DSS} = 70V$ $I_D = 180A$ $R_{DS} = 4m\Omega$
Diode BYV34-400M	$V = 400V$ $I = 20A$ $T_{rr} = 35ns$

The originality of this converter consists in the technology of its transformer which operates at high-frequency and includes PAYTON planar technology. This one allows a better integration and compact structure of the power system than classical converters with high transformation ratio.

A working frequency of 50 kHz appears as a good compromise between power losses and transformer volume.

Output filter is calculated for reducing current and voltage ripples on DC-link. The inductance of filter can be determined from Eq. 1:

$$L = \frac{V_s}{\Delta I_s} \cdot \alpha \cdot \frac{1}{2.f} \qquad \text{Eq. 1}$$

Where:

$$\alpha = 0,5 \qquad \text{and} \qquad \Delta I_s = I_s * 10\%$$

Eq. 2 is used to determine output capacitor:

$$C = \frac{I_s}{\Delta V_s} \cdot \alpha \cdot \frac{1}{2.f} \qquad \text{Eq. 2}$$

Where:

$$\Delta V_s = V_s * 1\%$$

These two equations lead to the following results:

| Lf = 0,6mH |
| Cf = 10uF |

$$f_0(Hz) = \frac{1}{2.\pi.\sqrt{L_f.C_f}} \qquad ; \qquad f_0 = 2056Hz$$

In the same manner as for the output filter, we can calculate the input filter (capacitor) using Eq. 2:

| C = 2,5mF |

C. Control of converter

The semi-conductor control law is designed to modulate the conduction time of each diagonal switch ("Q_1, Q2" and "Q_3, Q4") of the inverter stage, as shown in

Fig. 2. The control of the parameter α allows controlling the average value of the DC link output.

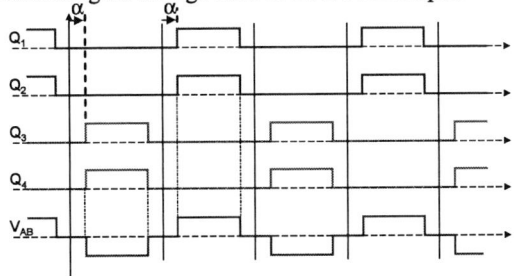

Fig. 2: Control strategy

Proportional-Integral (PI) controller has been added to the system. This PI controller allows regulation of the DC-link level at reference voltage (in our case: 110V).

III. PEFC ON LINE DIAGNOSIS USING IMPEDANCE SPECTROSCOPY

A. General method

To evaluate on-line state-of-health of fuel cell stack, impedance spectroscopy principle is used and implemented by injecting a low sinusoidal current signal, at different frequencies, by controlling our converter. This method allows having the real-time impedance spectroscopy of the fuel cell stack, without additional measurement and spectrometry apparatus (just using our converter and DSP controller).

The relation between all parts of our system is given in Fig. 3. Function of acquisition board is to transmit filtered and adapted information acquired from load and PEFC stack to the DSP controller. We can point out two sets of measurements assumed by the acquisition board. The first one, which comes from the load, is used to regulate load voltage and power transit while the second consists to acquire electrical parameters of fuel cell. This makes possible for the DSP controller to fix the load voltage, to analyze the acquired signals, to control the DC/DC converter, to realize current injection and impedance spectroscopy at the same time.

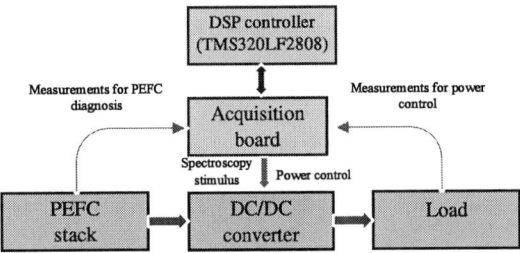

Fig. 3: General functional scheme

In our case, the difficulty is to extract the signals resulting from injection and used for impedance spectroscopy. Indeed, on the one hand the injected signals must be reduced to not affect performances and DC link stability and on the other hand, they must be extracted from a high ripple due to switching semi-conductors and from a high continuous component due to the average values of fuel cell current and voltage.

B. Calibrating of measures

Measured signals consist on continuous and alternative components. To obtain alternative one, the continuous one is removed by an electronic board including programmable gain and differentiator control by the controller (Fig. 4).

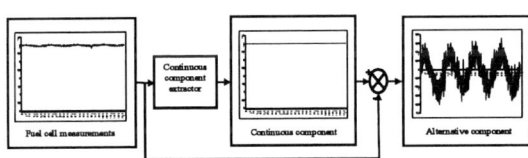

Fig. 4: Extraction of ripple signal

735

To understand how the offset system operates, the example of fuel cell voltage is selected and it is shown how its alternative component can be extracted.

The first step is to define the K-factor to obtain better resolution of alternative signal.

A PWM digital output is filtered to create an analogue signal whose average value is controlled. From this value, the offset value can be compensated as seen in Fig. 5.

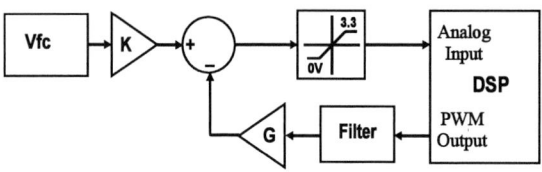

Fig. 5: Offset compensation

C. Current ripple injection for spectroscopy

To realize impedance spectroscopy, DSP controller has to inject electrical stimulus in PEFC. This is done by means of a sinusoidal modulation of duty cycle around its nominal value in order to create a very low sinusoidal signal for different frequencies. From this control, it results a sinusoidal voltage superposed on the load voltage, and a sinusoidal current in fuel cell superposed on the load current.

The injection principle and impedance spectroscopy are applied only when the dynamical conditions of the load are sufficiently low to disable the voltage controller with respect of voltage stability.

The current ripple injection is based on the control of the ripple peak in the PEFC in order to ensure at the same time non-deteriorating current and significant signal measurement for spectroscopy as described in Fig. 6.

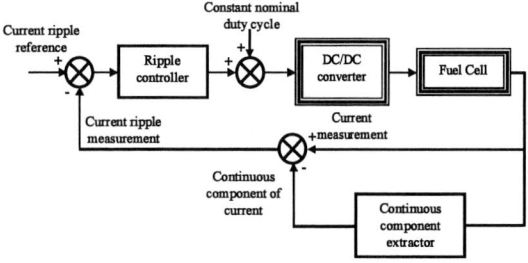

Fig. 6: current ripple injection

D. Measure of impedance spectroscopy

During the current ripple injection, ripple current and ripple voltage in the PEFC are extracted from their absolute values by using programmable gain and offset circuits. Then, they are digitally converted in order to compute the PEFC impedance for each injection frequency. The spectral impedance measurement consists on several steps.

As a first step, given the use of the DC/DC converter, current and voltage ripples are naturally mixed with high frequency noise due to Pulse Width Modulation (PWM). This noise is totally rejected by choosing a sample

frequency adjusted to a division of the PWM frequency. Then the analogical digital conversion results in the extraction of desired components of current and voltage ripples.

To see the effect of this sample frequency on signals, results obtained through the DSP controller are plotted in Fig. 7, and these curves are compared with those measured by the oscilloscope and shown in Fig. 8.

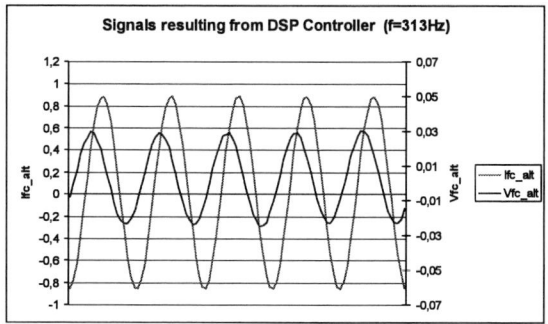

Fig. 7: Ripple measurement with PWM rejection

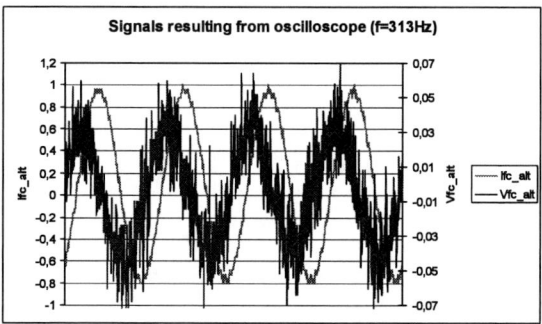

Fig. 8: Ripple measurement without PWM rejection

We deduce from these results that the DSP controller sampling method efficiently rejects PWM noise.

As a second step, complex expression of the current and voltage ripples are computed by using the Discrete Fourier Transform (DFT) algorithm described by Eq. 3 which gives real and imaginary parts of the measured signals.

This method is used to have a good accuracy even if the measured signal presents a residual perturbation.

As a third step, spectral impedance is computed from current and voltage expressions, and compared to the correct spectral impedance for PEFC diagnosis.

$$X(k) = \sum_{n=0}^{N-1} x(n)\, e^{-j 2\pi kn / N}, \quad 0 \leq k \leq N-1 \qquad \text{Eq. 3}$$

$$real[X(k)] = \sum_{n=0}^{N-1} x(n) \cdot \cos(2\pi kn / N) \qquad \text{Eq. 4}$$

$$imag[X(k)] = \sum_{n=0}^{N-1} -x(n) \cdot \sin(2\pi kn / N) \qquad \text{Eq. 5}$$

The entire process of impedance measurement is resumed in the functional diagram of Fig. 9.

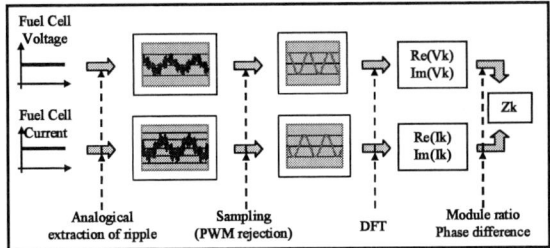

Fig. 9: Functional diagram of impedance measurement

Using this method and injecting current for various frequencies, impedance spectra of fuel cell can be recorded.

IV. EXPERIMENTALS RESULTS

A. Description of test bench

The following photo shows different parts of our test bench where we can observe DC/DC converter, DSP controller, measurement board, fuel cell stack, and dynamic load.

This reduced scale test bench is conceived in compliance with characteristics of applications dedicated to electrical vehicles. It has been tested and validated with conclusive operating results.

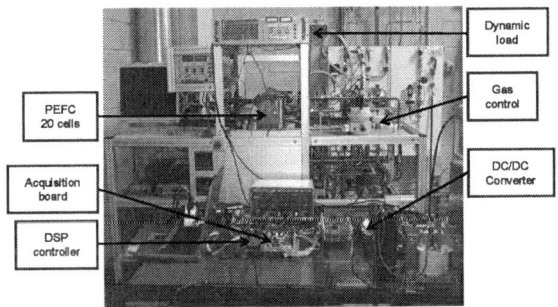

Fig. 10: test bench photo

To acquire different measurement results from the interface board and from the DSP controller, a serial communication interface between the DSP controller and a computer has been used.

B. Power transit with converter

The dynamic load used in the test bench emulates a typical reference of vehicle electrical power as shown in Fig. 11 (a). The proposed DC/DC converter has to control its output voltage i.e. feeding the DC link of the vehicle, at the reference value. Fig. 11 (a) illustrates the good behavior of the controller.

Waveforms of fuel cell current and voltage resulting from this power request are shown in Fig. 11 (c). Evolutions of fuel cell electrical parameters due to load variations do not disturb bus voltage regulation as shown in Fig. 11 (a). The controller compensates voltage variations due to current variations in passive components

and semi-conductors and also fuel cell internal voltage variations.

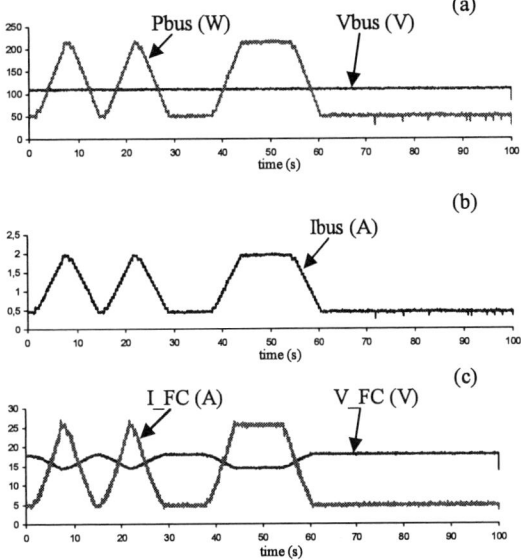

Fig. 11: (a) Electrical power (Pbus) and output voltage (Vbus) - (b) Load current - (c) Fuel cell stack voltage (V_FC) and current (I_FC)

C. Experimental spectroscopy

The following figure illustrates two impedance spectra measured for two different fuel cell currents to demonstrate the ability of converter and control to realize all the steps described earlier. These results validate the online spectroscopy method. Variation of the fuel cell impedance spectrum versus fuel cell current is here directly highlighted by online spectroscopy.

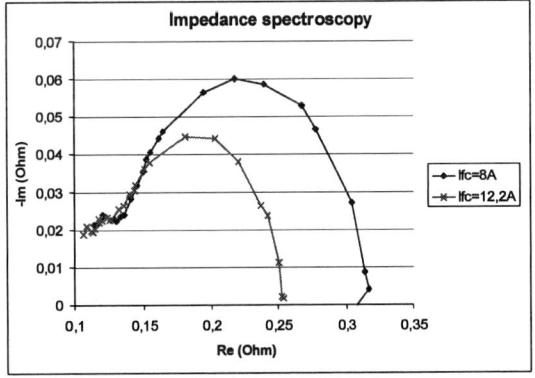

Fig. 12: Impedance spectroscopy of fuel cell stack

Now, several diagnoses based on impedance spectroscopy can be proceeded in order to analyze fuel cell state of health, evaluate performances of gas feeding or influence of environmental parameters. An example is provided in the following section.

On the other hand, it is necessary to validate the ability of the converter to preserve DC link stability during impedance spectroscopy. Fig. 13 illustrates variations of output voltage for spectroscopy at three different frequencies (2, 30 and 360 Hz). As planned, the greatest influence of spectroscopy on DC voltage is observed for lower frequencies due to output LC filter. However, injection method which controls amplitude of injected ripples allows the output voltage variations to not exceed 6% of rated value for DC link voltage.

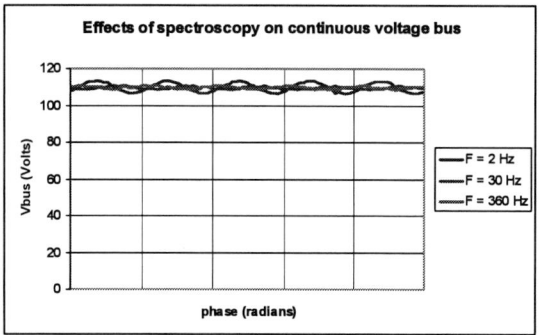

Fig. 13: Effects of impedance spectroscopy on DC link stability

For global stability of DC link output, voltage control is performed when power consumption fluctuates. Online impedance spectroscopy is performed when DC link stability is detected and then voltage control is disabled. Spectroscopy is aborted and voltage control enabled if DC voltage variations out of specifications are detected during impedance measurement.

D. Examples of diagnosis

The conductivity of the fuel cell membrane depends on its humidification rate. To have an idea about the effect of the membrane state on the system performance, Fig. 14 is presented. It shows fuel cell voltage as a function of load and air hygrometry rate (close to 100%, 75% and 50%).

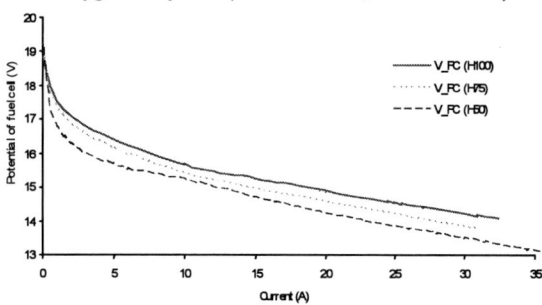

Fig. 14: Humidification effect on the fuel cell voltage

The membrane is humidified in this test by air hydration upstream of the fuel cell stack inlet, and the effect of humidification on the fuel cell voltage can be observed in this figure. Indeed, when the hygrometry rate decreases, fuel cell power decreases. The best operation is obtained when hygrometry rate is close to 100%.

Hygrometry variations cause impedance variations. Impedance spectroscopy and then impedance spectral analysis allow detecting the beginning of the phenomenon. Correlating this study with the monitoring of cell voltages allows the detection, the localization and if possible the correction of the problem. The method is applicable to all other phenomena involving a variation of the impedance.

This first example illustrated by Fig. 15 reflects the problem of humidification mentioned above. The impact of air dew point temperature (Tr) on the fuel cell impedance can be seen. The increase of dew point temperature allows better humidification which means a better behavior of the fuel cell and thus lower impedance. This phenomenon may be highlighted by impedance spectral analysis as shown in Fig. 15.

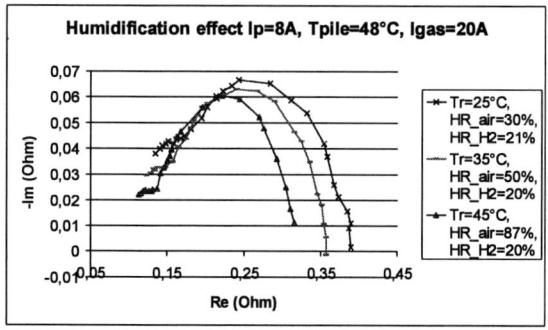

Fig. 15: Humidification effect on the impedance spectroscopy

The second example demonstrates that the state of fuel cell regresses when amounts of gas flows (computed for currents equal to 20A, 15A and 10A) decrease for a constant load as shown in Fig. 16. In this case, impedance increases, what is detected by our spectroscopy strategy.

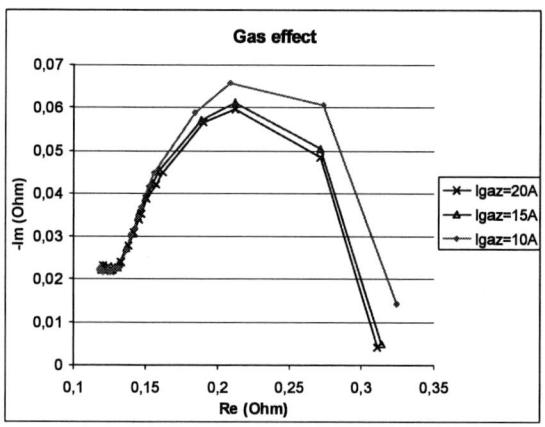

Fig. 16: Effect of gas flow on the impedance spectroscopy

V. CONCLUSION

Diagnosis of energy supply system and especially PEFC diagnosis are necessary to ensure good performances, reliability and autonomy of electrical vehicles. However, classical diagnosis methods require complex hardware and software. Their main disadvantages are their high cost, their large size and the difficulty of implementation. The diagnosis proposed in this paper allows detecting some

PEFC dysfunctions by online impedance spectroscopy. No additional cost and place is required for PEFC online diagnosis using the proposed power DC/DC converter topology and control.

REFERENCES

[1] "Performance of a PEM Fuel Cell Water Management System Using Static Output Feedback". by Amey Y. Karnik, Julia H. Buckland and Jing Sun. Proceedings of the 2007 American Control Conference. New York City, USA, July 11-13, 2007. IEEE.

[2] "MIMO Nonlinear Identification pf a PEM Fuel Cell Humidifier Using Wavelet Networks". by Xian-Rui Deng and Ying-Fei Xiong. Proceedings of the 6th world Congress on Intelligent Control and automation. June 21-23, 2006.Dalian. China. IEEE

[3] "Heat and Mass Transfer Effects in PEM Fuel Cell". By N. E. Vanderborgh, J. R. Huff and J. Hedstrom. Energy Conversion Engineering Conference, 1989. IECEC-89. Proceedings of the 24th Intersociety. 6-11 Aug. 1989. IEEE.

[4] "Design and Control of a Fuel Cell DC/DC Converter for Embedded Applications". By Abdellah Narjiss, Daniel Depernet, Frédéric Gustin, Daniel Hissel, Alain Berthon. EPE2007. 12th European Conference on Power Electronics and application 2-5 September 2007, Aalborg, Denmark.

[5] "Design of a High Efficiency Fuel Cell DC/DC Converter Dedicated to Transportation Applications". By Abdellah Narjiss, Daniel Depernet, Frédéric Gustin, Daniel Hissel, Alain Berthon, FC-06-1051, Journal of Fuel Cell Science and Technology

Application of Kalman filters to the control of independent power electronic voltage sources

Ryszard Porada[*], Łukasz Nyczkowski[†]

Poznan University of Technology, Institute of Electrical Engineering and Electronics, Poznań, Poland,
[*]e-mail: ryszard.porada@put.poznan.pl
[†]e-mail: lukasz.nyczkowski@put.poznan.pl

Abstract—Power electronic independent voltage sources are used i.a. in electroacoustics and different special spheres; as systems realizing eg. the optimal control of electric drives, and also as executive blocks in active compensation systems. In the paper we discusse possibility of realization of a broadband power electronic voltage source with application of Kalman filter and basic quantities describing the working system. Selected simulation results of the independent voltage source for different reference signals are also included.

Keywords: power converters, controlled voltage source, adaptive control, Kalman filter

I. Introduction

The basic task of power electronic systems is transformation of electrical energy received from technically accessible sources of energy with speciefied voltage and frequency into voltage and the frequency required by receivers of the electrical energy, and also flow control of this energy. Converters should not disturb work of energy sources, should only consume energy necessary for the receiver to perform expected work through suitable signals of voltage or current, and also shaping optimal output voltage signals in accordance with tasks realized by the receiver.

Power electronics independent voltage and current sources find application in electroacoustics, different special spheres (i.a. power generators of references waveforms), as systems realizing eg. the optimal control of converter drives and also as executive blocks in active compensation systems [1,2,6,7]. Such systems have to meet higher requirements connected with mapping of reference specified signals in defined frequency band, both in static and dynamic states. It ia also essential to assure required quality of voltage of controlled sources, feeding receivers of the electrical energy.

The paper presents possibility of realization of power electronic broadband voltage source as a system controlled with application of Kalman filters. We discussed influence of elements and parameters of the system with assumed constant carrier frequency of PWM modulation on accuracy and stability of reference signal on the energy output. Some selected results of simulation researches for different reference signals are also presented.

II. General Structure Of Broadband Voltage Source

Broadband power electronic voltage source should generate voltage of desired parameters on energetic output. Such task can fulfill inverter which enables linear transfer of a reference signal [3,7,9,11] (in specified frequency band with assumed level of map error and good static and dynamic proprieties). General structure of such system is shown in figure 1.

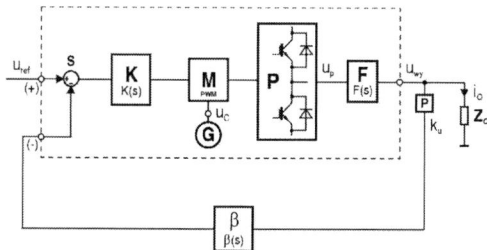

Fig. 1. General block diagram of power electronic independant controlled energy source

Structure of such power electronic voltage source is a closed system, controlled in PWM modulation with constant carrier frequency. Feedback loops fulfill required task (depending on destination of the system) in shaping output signals of desired proprieties. In general structure of the system can be specified by the following elements: adder (S), modulator PWM (M), power electronic inverter (P) with control system and protections as well as measure transducer (P-k_u) of signals required for realization of suitable feedback loops. The most essential elements of the system in view of control quality are: filter $K(s)$ in the main circuit, passive output filter $F(s)$ which structure depends on destination and filter $\beta(s)$ in loop feedback which shapes the frequency characteristic of the system to desired form.

The system contains strongly nonlinear elements, like modulator (M) and inverter (P). It has unfavorable influences on stability of the closed system. These nonlinearities have to be compensated in an optimal way by use of two filters: input K (in the main circuit) and output passive filter F.

Task of the passive filter F is effective suppression of

components with higher frequencies, particularly products of PWM modulation in the waveforms of output voltage. Structure and frequency characteristic of this filter depends on application of the system – as a voltage or current source. Application of filters F with possibly simple structure has a practical reason – it decreases its dimensions and weight.

To provide maximum fidelity in the mapping of reference signals, in the possible wide frequency band, the independent power electronic voltage source should have closed structure with transmittance copying frequency charcteristic of the low-pass ideal filter. To aproksymate this characteristic we can use Butterworth or Czebyszev filters [10,11]. However because of delay caused by the switching elements of inverter we proposed aproksymation of transmittance of the ideal filter in the form

$$G(s) = \frac{\exp(-sT_c)}{(1+s\tau) + \beta(s)\exp(-sT_c)} \qquad (1)$$

where: T_c – the period of the PWM modulation, τ – the time-constant. Period T_c has essential influence on the transfer band, while time-constant τ – on the shape of the frequency characteristics.

Taking into account the structure of the source (figure 1) and desired transmittance of the system (1), transmittance of the filter $K(s)$ of the main circuit is in general form

$$K(s) = \frac{1}{1+s\tau}\frac{1}{F(s)} \qquad (2)$$

where $F(s)$ – transmittance of the exit filter including the influence of parameters of the receiver.

A task of the filter $K(s)$ is to compensate influence of delays caused by inverter (P), output passive filter F and a receiver (in general case nonlinear and nonstationary) on transmittance of the closed system. Parameters of this filter should change in a way which provides the desired shape of attenuation and phase characteristic and could ensure possible wide transfer band at possibly large value of gain and suitable margin of stability of the closed system. Fulfilling the above requirements enables to shape respons of the system and to achieve repeatability of the output waveform in relation to reference signal, thus reduction of undesired spectrum components in range of lower frequences.

Simulate investigation of system were carried out for transistor inverter working in closed structure, for evaluation of the efficiency of harmonical reductions in output voltage, with constant carrier frequency of PWM modulation equal 15 kHz, for various reference signal. Researches were performed in OrCAD program.

Figure 2 shows waveforms of reference and output voltages, generated by the voltage source for references signals: sinusoidal ($f_o = 50\,\text{Hz}$), rectangular ($f_o = 100\,\text{Hz}$), triangular ($f_o = 200\,\text{Hz}$) and amplitudes of signals of nominal system parameters (fig. 2a) as well

as 0,01 values of these amplitudes (fig. 2b). For each presented case becomes visible a very good mapping of the reference signal.

a)

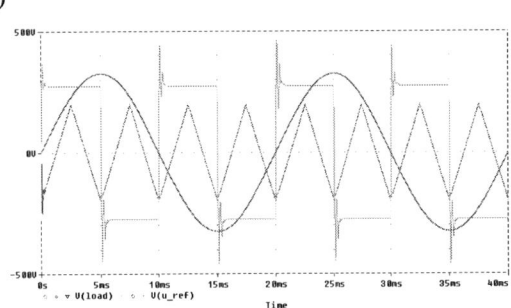

b)

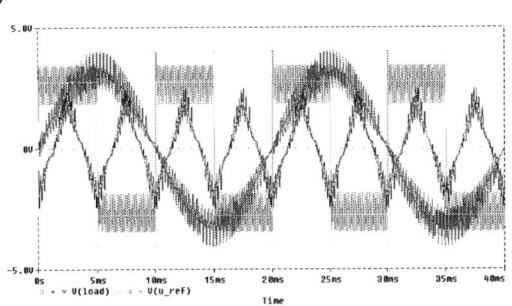

Fig. 2. Waveforms of output voltage of voltage source
for reference signals with different parameters

The presented waveforms show that there is possibility of practical realization of broadband power electronic voltage source of very good qualitative parameters of the output voltage, for different shapes of reference signals, in a wide range of regulation of the voltage (amplitudes – 1:100, frequencies – from 0 to 200 Hz).

The presented above methode of work of linear power electronic source of energy [6,7], in the form of the linear current source, was applied in the laboratory model of generator of the spatial magnetic field to supply of coil of field applicator [1,2].

III. APPLICATION OF THE KALMAN FILTERS

The obtainment of the required quality of parameters of inverter output signals at changing parameters of component elements of the system (e.g. the passive filter F or the receiver) demands the adaptive correction of parameters of the control filter K. This assignment, inclusive realization of the lineal transfer of signals by the power electronic voltage source can be realized by the use of adaptive digital filters, i.e. the Kalman filter.

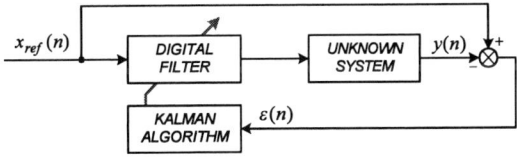

Fig. 3. Adaptive control structure based on Kalman Filter

The Kalman filter is based on classic adaptive digital filter structure [8,9,10,12] – figure 3. The unknown system in this case is a voltage source.

An adaptive Kalman filter works as a one step predictor of dynamic system state [8,12]. It can be described by equations

$$x_k = Ax_{k-1} + Bu_{k-1} + w_{k-1} \qquad (3)$$

$$z_k = Hx_k + v_k \qquad (4)$$

where: x, u, z – are vectors of: state, input and output (of measured signals); A, B, H – matrixes of: state, control and output. Random variable vectors w and v represent the process and measurement noise. They are assumed to be independent (of each other), white and with normal probability distribution [8,12].

There are two groups of equations that describe the Kalman filter. The first is used to predict the system state at k step, based on state estimate at $k-1$ step

$$\hat{x}_k^- = A\hat{x}_{k-1} + Bu_{k-1} \qquad (5)$$

$$P_k^- = AP_{k-1}A^\mathrm{T} + Q \qquad (6)$$

where: $\hat{x}_k^-$ is *a priori* state estimate at k step.

The second group of equations modifies the *a posteriori* state estimate and filter structure in way to minimise the estimation

$$K_k = P_k^- H^\mathrm{T} \left(H P_k^- H^\mathrm{T} + R \right)^{-1} \qquad (7)$$

$$\hat{x}_k = \hat{x}_k^- + K_k \left(z_k - H\hat{x}_k^- \right) \qquad (8)$$

$$P_k = \left(I - K_k H \right) P_k^- \qquad (9)$$

The Q matrix (6) and R matrix (7) are covariance matrixes, respectively to the process noise and measurement noise variables. In practice, they might change with each time or measurement step, however here we assume they are constant [8]. The K is called the Kalman gain matrix and decides of measurement or computation priority.

Both cycles of Kalman algorithm are linked up with *a priori* estimate error covariance of state estimate

$$e_k^- = x_k - \hat{x}_k^- \qquad (10)$$

$$P_k^- = E\left[e_k^- (e_k^-)^\mathrm{T} \right] \qquad (11)$$

and *a posteriori*

$$e_k = x_k - \hat{x}_k \qquad (12)$$

$$P_k = E\left[e_k (e_k)^\mathrm{T} \right] \qquad (13)$$

The task for Kalman filter as an adaptive control algorithm is to generate a proper reference signal for PWM inverter for the purpose of mapping references signal on output system.

IV. SIMULATION RESULTS

Preliminary simulate investigation of system were carried out for 1-phase transistor inverter controlled with application of Kalman filter, with constant carrier

frequency of PWM, for various reference signal. Researches were performed in Matlab®/Simulink®.

All parts included in a voltage source shows figure 4. The signal $u_{ref}(t)$ is generated by Kalman filter; the voltage $u_o(t)$ is the output voltage of voltage source. Researches concentrate on effectiveness of mapping of references signals, related to adaptive controll of voltage source based on Kalman filters.

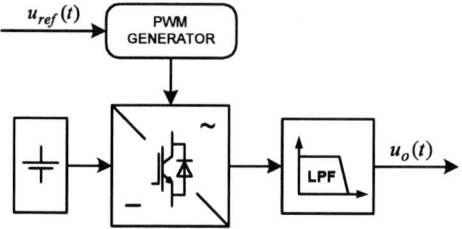

Fig. 4. Block structure of voltage source

The simulations were passed for DC voltage equal 400V, the PWM modulation frequency $f_{PWM} = 15\,\mathrm{kHz}$, signal sampling frequency $f_s = 60\,\mathrm{kHz}$ and the algorithm running frequency (comparing the error signal to the reference) – $f_A = 60\,\mathrm{kHz}$, the same for all presented cases. Researches were carried out to estimate the three most significant parameters having influence on system work: the additive voltage PWM (f_{PWM}), data sampling frequency (f_s) and order of Kalman filter (N), mostly for evaluate dynamic and delay of output voltage.

a)

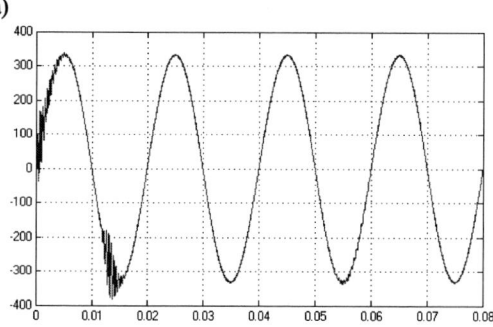

b)

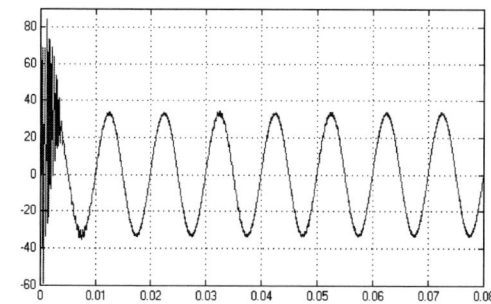

Fig. 5. Waveforms of output voltage of voltage source for sinusoidal reference signals: a) nominal RMS value, $f_o = 50\,\mathrm{Hz}$,

b) 0,1 nominal RMS value, $f_o = 100\,\mathrm{Hz}$

a)

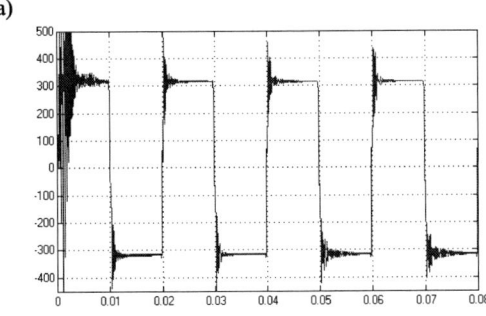

b)

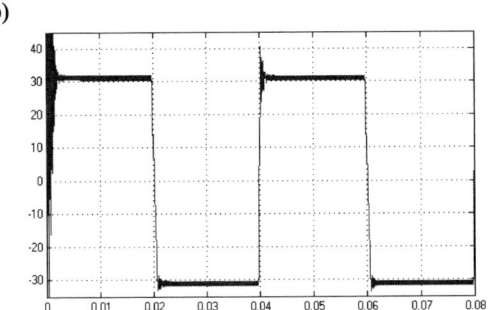

Fig. 6. Waveforms of output voltage of voltage source for rectangular reference signals: a) nominal RMS value, $f_o = 50\,\text{Hz}$,

b) 0,1 nominal RMS value, $f_o = 12,5\,\text{Hz}$

a)

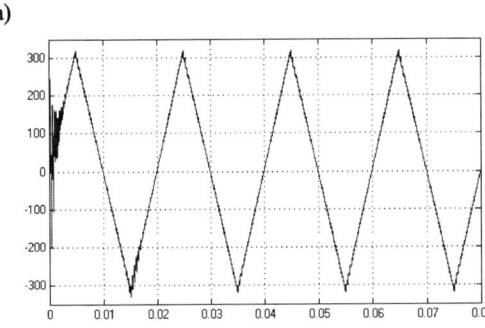

b)

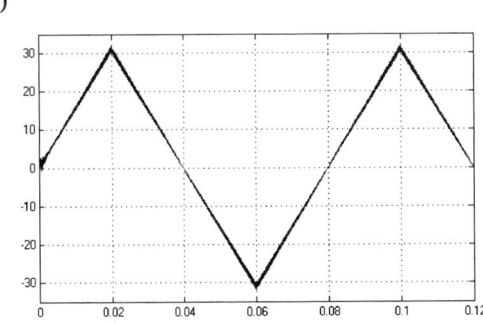

Fig. 7. Waveforms of output voltage of voltage source for triangular reference signals: a) nominal RMS value, $f_o = 50\,\text{Hz}$,

b) 0,1 nominal RMS value, $f_o = 12,5\,\text{Hz}$

Figures 5, 6 and 7 show waveforms of 1-phase sinusoidal voltages: reference and output generated by the

voltage source for nominal RMS value (figure 5a,6a,7a) and 0,1 values of their amplitude (figure 5b,6b,7b) and different frequency. We can notice good mapping of reference signals – the deformation of the voltage signal does not exceed 2 % in the frequency band approx. 2,5 kHz, while the statical exactitude is approx. 1,5 %.

Conclusion of the simulations is, that using the Kalman filter to control voltage source gives satisfying results. Executed simulations show that there is a possibility to achieve better dynamic characteristics by using Extended Kalman Filter [4,8,10]. Better modeling of system may lower required number and rate of computations that are necesary to estimate optimal system state and improvement of its dynamic.

V. CONCLUSION

At present we can observe considerable interest in modern, effective methods and systems which decrease level of deformations of voltages in receivers of electric energy. This can be achieved by application converters with sinusoidal output voltage, which provide work conditions comparable to power supply from standard power network of alternating current. These are converters controlled usually with application of different methods of PWM modulation. It gives opportunity of maximal utilization of transfer band, limited by carrier frequency of PWM modulation, with preservation of indispensable margin of stability of the closed system. The presented in the paper broadband voltage source controlled by Kalman filter in high degree realize correctly task of shaping output voltages with amplitude and frequency changing in wide range.

REFERENCES

[1] M. Gwóźdź and R. Porada, "The generator of the spatial magnetic field", *Int. Conf. on "Power Electronics and Inteligent Control for Energy Conservation"*, PELINCEC'2005, Warszawa, Poland, October 2005, (paper no. 113).

[2] M. Gwóźdź and R. Porada, "Utilization of Low Distortion Power Electronics Current Sources in Generator of Spatial Magnetic Field", *Proc. of 12th International Power Electronics & Motion Control*, EPE-PEMC'06, Portorož, Slovenia, September 2006, pp. 852-856.

[3] N. Mohan, *Power elektronics: Converters, Aplications, and Design*. John Wiley&Sons, New York 1989.

[4] A. Niederliński, J. Mościński and Z. Ogonowski, *Regulacja adaptacyjna*, PWN, Warszawa, 1995.

[5] M. Nowak and R. Barlik, *Poradnik inżyniera energoelektronika*, WNT, Warszawa 1998.

[6] R. Porada and M. Gwóźdź, „Optymalizacja czasowo-częstotliwościowych parametrów niezależnych źródeł energii w strukturach zamkniętych", *Proc. of XXVII IC-SPETO'04*, 2004, t. 2, ss. 293-296.

[7] R. Porada, "Simulation researches of power electronics broadband voltage source", in „*Computer Application in Electrical Engineering*", ALWERS, Poznań 2006, pp. 258-266.

[8] M. I. Ribeiro, *Kalman and Extended Kalman Filters: Concept, Derivation and Properties*, Wiley&Sons 2004.

[9] L. Rutkowski, *Filtry adaptacyjne i adaptacyjne przetwarzanie sygnałów*, WNT, Warszawa 1994.

[10] B. Stewart, *Adaptive Signal Processing*, University of Strathclyde, Glasgow 1999.

[11] J. Szabatin, *Podstawy teorii sygnałów*, WKŁ, Warszawa 2000.

[12] G. Welch and G. Bishop, *An Introduction to the Kalman Filter*, UNC-Chapel Hill, TR 95-041.

Verification of the load sharing characteristics in Autonomous Decentralized UPS system using FPGA based Hardware Controller

Nobuaki Doi*, Tsuyoshi Saito*, Tomoki Yokoyama*

*TOKYO DENKI UNIVERSITY / Ishizaka, Hatoyama, Hiki-gun, Saitama, Japan

Abstract— Autonomous decentralized control system based on FPGA based hardware controller is proposed. Progress of FPGA technology makes it possible to include the software macro CPU core into the FPGA chip, a high flexibility can be realized for the construction of the control processor in power electronics application.

In the proposed method, all the control system is implemented in one FPGA chip. Complicated calculations are assigned to hardware calculation logic, and the parallel processing circuit makes it possible to realize minimizing the calculation time. Also Nios II CPU core is implemented in the same FPGA chip, and the software development can be applied for non-time critical calculations. Two parallel connected UPS system was constructed and load sharing characteristics were verified through simlations and experiments.

I. INTRODUCTION

To improve the reliability of the UPS system, parallel connected redundant system is realized. The autonomous decenrtalized UPS system has advantages to easy plug-in/out features. Generally DSP based software controller is used for the UPS controller, DSP treats the digital control schemes and ASIC or PLD treat the PWM logic control. Nowadays DSP has many features to treat power electronics applications such as PWM generation logic and other gate control logic. On the other hand, FPGA has the hardware configurable features using Hardware Description Language (HDL). The required circuit only can be built into the chip, the optimized control circuit can be realized. But if all logics with complicated control schemes are made into hardware, the total circuit will become huge, and the gate size of FPGA is limited in some cases. So it is important to balance the process sharing with software and hardware. In this paper, an autonomous decentralized control UPS system is constructed with Hardware/Software (HW/SW) codesign procedure using FPGA based hardware controller.

In the proposed system, the calculation logic for a quasi dq transformation, PLL control and gate pulse generation block were implemented in one FPGA chip. The calculation for a droop control is implemented in software for Nios II CPU. Therefore each operation are processed by hardware and software, the optimal control system can be realized. Two parallel connected UPS system was constructed with FPGA based hardware controller and load sharing characteristics were verified through simlations and experiments.

II. SYSTEM MODELING AND CONTROL METHOD

Fig. 1 shows a system configuration of parallel-connected UPS system, and Fig. 2 shows a basic parallel-connected UPS system where the voltage v_i is the output voltage for the ith UPS unit as $v_i = E_i \cos(\omega t + \phi_i)$. Each UPS is controlled independently with the same control law, and it is assumed that each UPS can observe only its output voltage and current.

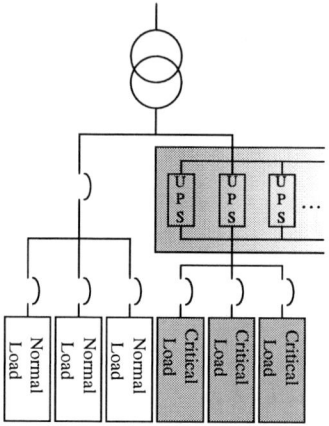

Fig. 1. Autonomous Decentralized Control UPS system

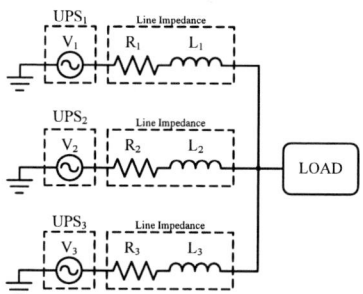

Fig. 2. Model of parallel-connected single-phase UPS system

A. Droop Control

The phase and the voltage amplitude are controlled based on a droop characteristics in this system. Control factor for the output voltage are the phase ϕ_i and the voltage amplitude E_i. The reference of the phase $\phi_i{}^*$ and the reference of the voltage amplitude $E_i{}^*$ are determined according to the active power of ith UPS P_i and the reactive power of ith UPS Q_i. Fig.3 shows how to decide the phase $\phi_i{}^*$ and the voltage amplitude $E_i{}^*$ based on the droop characteristics. The active element and the reactive element of the output voltage, and the active element and the reactive element of the output current can be detected in every sampling period by quasi dq transformation [1].

$$\phi_i^* = \phi_0 - m_i (P_{0i} - P_i) \qquad (1)$$
$$E_i^* = E_0 - n_i (Q_{0i} - Q_i) \qquad (2)$$

Equation (1) and (2) indicate how to decide the phase and the voltage amplitude from the active power and the reactive power. Differentiate the rated active power of the

978-1-4244-1741-4/08/$25.00 ©2008 IEEE

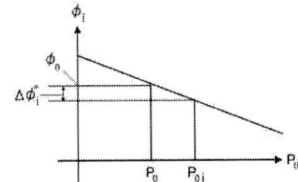

Fig. 3. Droop characteristics

E_i^*	:	reference of voltage amplitude
E_0	:	nominal voltage amplitude
ϕ_i^*	:	reference of phase
ϕ_0	:	nominal phase
P_{0i}	:	rated active power for ith UPS
Q_{0i}	:	rated reactive power for ith UPS
n_i, m_i	:	droop characteristics gain
Vd_{0i}	:	d phase voltage for ith UPS
Id_{0i}	:	d phase current for ith UPS
Iq_{0i}	:	q phase current for ith UPS

ith UPS P_{0i} and the active power of the ith UPS P_i, then multiplying by the droop characteristics gain m_i, subtract from the nominal phase ϕ_0, desired phase reference ϕ_i^* is derived. Also differentiate the rated reactive power of the ith UPS Q_{0i} and the reactive power of the ith UPS Q_i, then multiplying by the droop characteristics gain n_i, subtract from the nominal voltage amplitude E_0, desired voltage amplitude reference E_i^* is derived. P_{0i} and Q_{0i} are the power reference of each UPS. Instant output power is derived by the phase and amplitude of the output voltage and current obtained from the quasi dq transformation. By changing the power reference P_{0i} and Q_{0i} of each UPS, shared ratio of the load current becomes proportional to the ratio of the power reference.

B. Quasi dq Transformation

Quasi dq transformation was used to detect an instantaneous phase and an amplitude of a single phase voltage. Target waveform V_s is assumed as the normal line voltage, and is sampled in every sampling period T_s, so the present sampled data $Vs(k)$ can be described as equation (3). Also pre-sampled data $Vs(k-1)$ and $Vs(k-2)$ can be expressed as (4) and (5). Here, V_L is the maximum peak value of Vs, and $V_s(k-1)$ is defined as V_{re} [8].

$$V_s(k) = V_L \cos(\omega t) \tag{3}$$

$$V_s(k-1) = V_L \cos\{\omega(t-T_s)\} = V_{re} \tag{4}$$

$$V_s(k-2) = V_L \cos\{\omega(t-2T_s)\} \tag{5}$$

Differentiate (3) and (5), adapting the formula of trigonometric function, (6) can be derived.

$$
\begin{aligned}
\frac{V_s(k-2) - V_s(k)}{2T_s\omega} &= \frac{V_L}{2T_s\omega}[\cos\{\omega(t-2T_s)\} - \cos\omega t] \\
&= \frac{V_L}{2T_s\omega}[-2\sin\omega(-T_s)\sin\{\omega(t-T_s)\}] \\
&\cong V_L \sin\{\omega(t-T_s)\} = V_{im} \tag{6}
\end{aligned}
$$

Here, it is obvious that the phase of V_{im} delays 90° than the phase of V_{re}, so using only three sampled data of the single phase waveform, V_{re} and V_{im} form the rectangular coordinates system, and dq transformation can be adopted to the single phase system. Therefore, the phase and the voltage amplitude can be detected from three sampling data using

quasi dq transformation. In order to reduce the influence of the noise, three kind of sampling frequencies are adopted for the implementation, which are 20[kHz], 10[kHz] and 6.7[kHz] respectively. Three parallel calculation circuit blocks are prepared, each block is driven by three different sampling frequencies, averaging the calculated values of every circuit to derive dq elements of the target waveform.

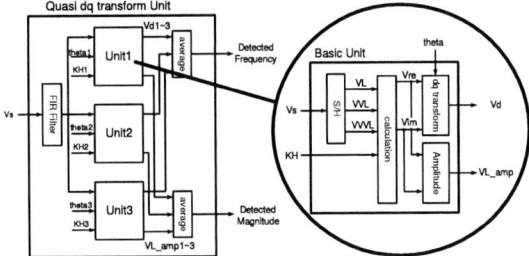

Fig. 4. Quasi dq Transformation

C. Control Block Diagram for Single Phase UPS system

Fig.5 shows the proposed control block diagram of the single phase UPS system. Single phase UPS is controlled by the droop control and the PLL control based on the quasi dq transformation [1]. It is assumed that each UPS can observe only its output voltage and current. The phase and the voltage amplitude can be detected by the quasi dq transformation. The operation frequency is detected by the PLL control using the phase ϕ_{V_i} of the ith parallel connected UPS output voltage.

Also, the active power of the ith UPS can be obtained by multiplying the voltage amplitude and the current amplitude, and the reactive power of the ith UPS can be obtained by multiplying the voltage amplitude and the phase of the output current, which is nominalized to calculate the reactive power. The phase reference ϕ_i^* and the voltage amplitude reference E_i^* can be decided by applying the phase and the voltage amplitude control based on the droop characteristics using the active power P_{0i} and the reactive power Q_{0i}. Therefore the voltage of each UPS can be controlled and each UPS supplies a balanced power to the load.

III. SIMULATION

Simulations were carried out for the load fluctuate condition. The parameters for the simulations are listed in Table.I.

TABLE I
SIMULATION CONDITIONS

rated active power of ith UPS P_{0i}	1000 [W]
rated reactive power of ith UPS Q_{0i}	0 [Var]
nominal frequency f	50.0 [Hz]
sampling frequency	20.0 [kHz]
reference of voltage amplitude E_{0i}	$100\sqrt{2}$ [V]
reference of phase difference ϕ_{0i}	0 [rad]
maximum impedance R	0.1+j0.037 [Ω]
droop characteristic gain m_{0i}	-0.0016785
droop characteristic gain n_{0i}	-0.00021362

TABLE II
LOAD PARAMETERS

load R	10 [Ω]
change load R	5 [Ω]
variation range for voltage amplitude	±3 [%]
variation range for phase angle	±1 [%]

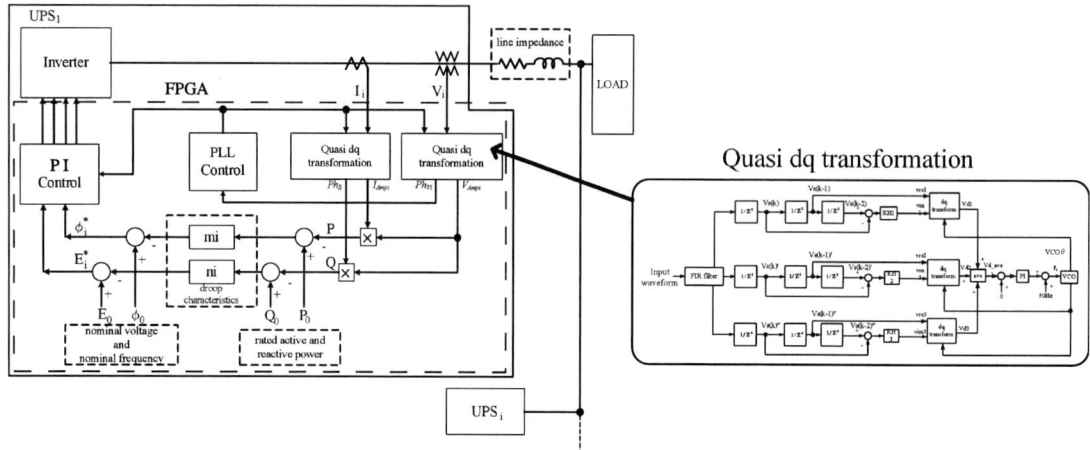

Fig. 5. Control block diagram for one single phase UPS system

1) Load Variation: The load is changed at 0.5 seconds with two parallel-connected UPSs, each UPS is settled in the same output power references. Fig.6 shows the active power, the reactive power, the output current of each UPS, and load current respectively. From this figure, it is confirmed that the active power and the reactive power of the load is supplied equally balanced from the two UPSs while the load is changed.

2) Rapid load change in one Cycle: The stability of the system is verified when rapid load change occurs five times in one cycle. Fig7 shows the load current and the output current of each UPS. Even in such conditions, each UPS supplies the balanced current to the load.

IV. EXPERIMENT

A. System Design

The parameters for the experimental setup are listed in Table.III. The proposed single phase UPS system was implemented using FPGA with a 32bit Nios II processor[11] as shown in Fig.5. StratixII EP2S60(ALTERA Corp.) is applied in this system. The Nios II CPU has a pipelined RISC architecture and reconfigurable features. In the experiments of the proposed single phase UPS system, a quasi dq transformation, PLL control and gate pulse generation block were implemented in FPGA logic. The operations such as droop control were implemented in Nios II CPU software.

B. Experimental System

Fig.8 shows the detailed module block diagram of the proposed single phase UPS system in the FPGA with a 32bit Nios II processor. The FPGA blocks are constructed with quasi dq transformation block, PI control block and VCO block. In the Nios II CPU blocks, power calculation block, droop control block and reference of output voltage calculation block are implemented. Two quasi dq transformation blocks are constructed in parallel, which calculates in the different sampling frequency, and the phase difference and the amplitude are calculated for every control stage. In the FPGA, the phase difference ϕ_{V_i}, ϕ_{I_i} and the voltage amplitude V_{amp_i}, I_{amp_i} can be calculated by the quasi dq transformation using the ith UPS output voltage and the ith UPS output current. Also the

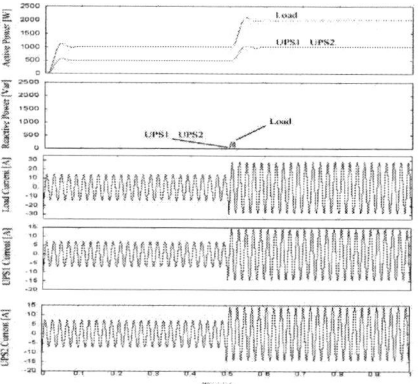

Fig. 6. Step response for the load changing in two parallel connected UPSs

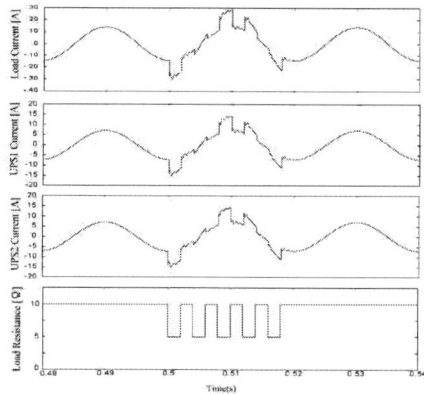

Fig. 7. Rapid load change in one cycle in two parallel connected UPSs

operation frequency f is detected by the PLL control using the phase difference of the ith UPS output voltage ϕ_{V_i}. These values ($V_{amp_i}, I_{amp_i}, \phi_{I_i}, f$) are sent to Nios II CPU. In Nios II CPU, the reference of the output voltage is decided by the droop control.

Fig.9 shows the timing chart of the FPGA based hardware controller. AD conversion and hardware calculation is finished

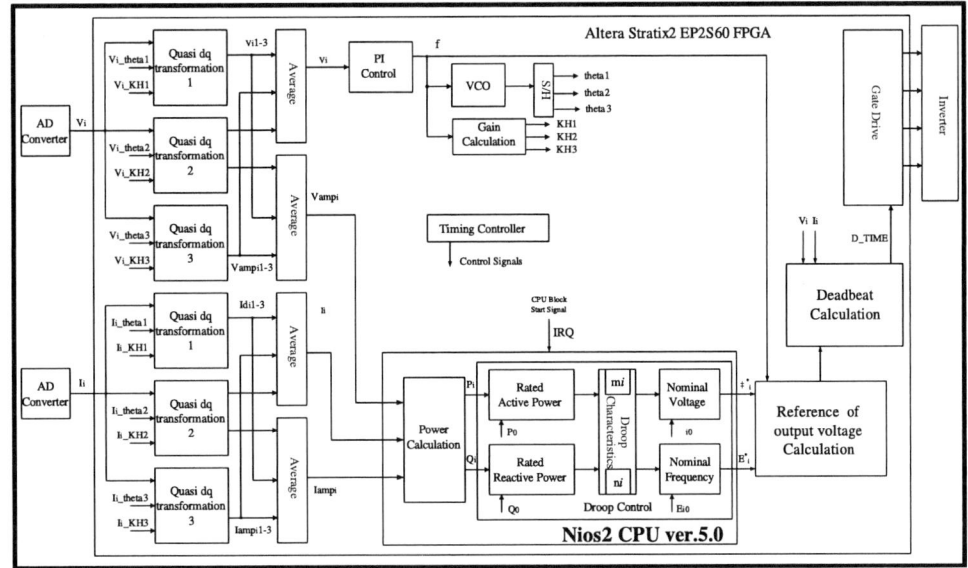

Fig. 8. Module block diagram of FPGA controller

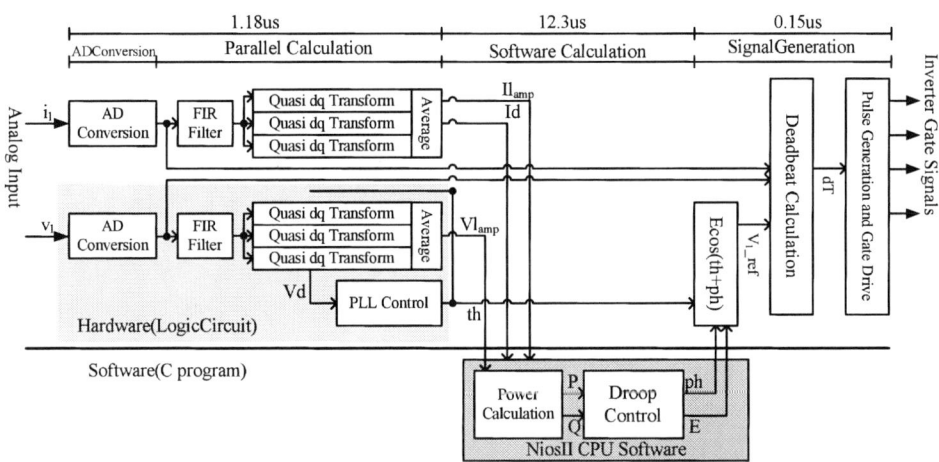

Fig. 9. Timing Chart of the Hardware and Software

TABLE III

EXPERIMENTAL CONDITIONS

Controller Board	Altera NiosII Development Kit
FPGA Device	Altera Stratix2 2S60
CPU Core	Altera NiosII CPU
rated active power of ith UPS P_{0i}	62.5 $[W]$
rated reactive power of ith UPS Q_{0i}	0 $[Var]$
nominal frequency f	50.0 $[Hz]$
sampling frequency	20.0 $[kHz]$
reference of voltage amplitude E_{0i}	$25\sqrt{2}$ $[V]$
reference of phase difference ϕ_{0i}	0 $[rad]$

Fig. 10. FPGA based hardware controller

in 1.18us(73clocks), the software calculation is finished in 12.3us, and all the calculation is finished in 13.6us.

Stratix II (EP2S60) is used for the FPGA controller, which has 60,440LE (logic elements), and 20% of the LE is used for the proposed control logic including Nios II CPU core.

Fig.10 shows the experimental setup of FPGA based hardware controller.

C. Experimental Results

In the experimental system, two parallel connected UPS condition was verified. In the condition that UPS1 supplied the power to the load, UPS2 was connected and disconnected to the load line.

In the case of share ratio 1:1, P_{01} and P_{02} are settled equally to 62.5[W]. From Fig.13 and Fig.14, two UPSs shared the load current equally.

TABLE IV shows the load share characteristics when the share ratio P_{01} and P_{02} were changed as shown in the table. Fig.15 shows power balance characteristics. It is confirmed that the load share balance can be controlled by the ratio of the rated power gain P_{01} and P_{02}.

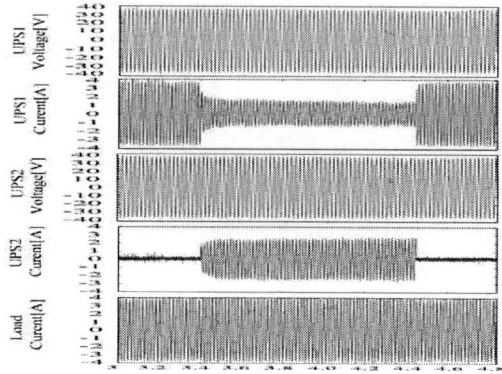

Fig. 13. Voltage and current waveform for UPS1 and UPS2 (1:2)

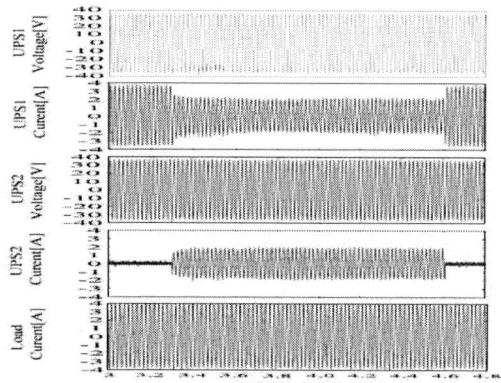

Fig. 11. Voltage and current waveform for UPS1 and UPS2 (1:1)

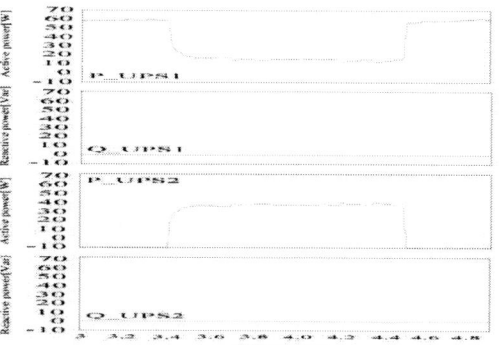

Fig. 14. Active Power and Reactive Power for UPS1 and UPS2 (1:2)

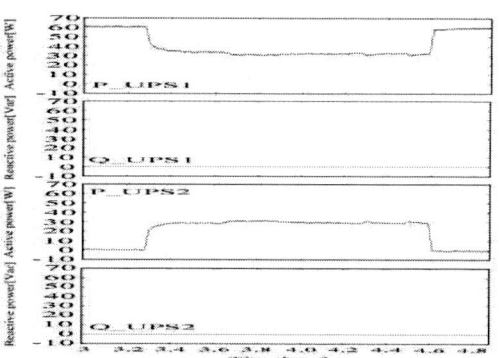

Fig. 12. Active Power and Reactive Power for UPS1 and UPS2 (1:1)

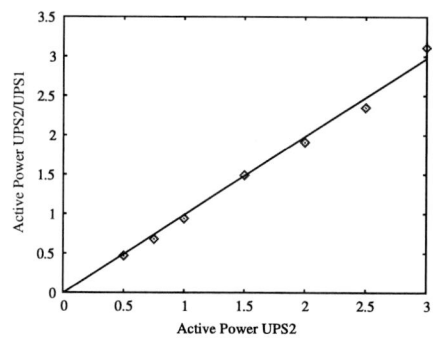

Fig. 15. Power balance characteristics

TABLE IV
LOAD SHARE CHARACTERISTICS

Share ratio		P0i[W]	Q0i[Var]	P[W]	Q[Var]	Power balance
1:0.5	UPS1	62.5	0.0	44.64	0.0	1.00:0.47
	UPS2	31.25	0.0	20.78	0.0	
1:0.75	UPS1	62.5	0.0	37.65	0.0	1.00:0.68
	UPS2	46.875	0.0	25.64	0.0	
1:1	UPS1	62.5	0.0	31.92	0.0	1.00:0.94
	UPS2	62.5	0.0	30.23	0.0	
1:1.5	UPS1	62.5	0.0	25.06	0.0	1.00:1.49
	UPS2	93.75	0.0	37.27	0.0	
1:2	UPS1	62.5	0.0	21.98	0.0	1.00:1.91
	UPS2	125.0	0.0	42.02	0.0	
1:2.5	UPS1	62.5	0.0	18.65	0.0	1.00:2.35
	UPS2	156.25	0.0	43.74	0.0	
1:3	UPS1	62.5	0.0	15.07	0.0	1.00:3.11
	UPS2	187.5	0.0	46.96	0.0	

V. CONCLUSION

In this paper, an autonomous decentralized control system for single phase UPS inverter with FPGA based hardware controller using software CPU core is proposed. By applying the proposed method in parallel-connected single phase UPS system, each UPS shares the power of the load equally when the power of the load changed or the number of the connected UPS is changed, an autonomous decentralized control for single phase UPS system can be realized. Constructing the control system using FPGA, and HW/SW codesign procedure is applied to realize the optimal control system. Operation of autonomous decentralized UPS was verified through simulations and experimental results.

REFERENCES

[1] T.Komiyama, T.Kojima, T.Yokoyama : "Implementation of High-speed Frequency Detection for Single Phase Utility Interactive Inverter System", EPE-PEMC' 04

[2] T.Yokoyama, S.Simogata, M.Horiuchi, T.Ide : "Instantaneous Deadbeat Control for Single Phase and Three Phase PWM Inverter Using FPGA Based Hardware Controller", EPE-PEMC' 04

[3] H.Takeda, T.Yokoyama : "A Study of Autonomous Decentralized Control for Single Phase UPS System based on quasi dq transformation", PESC' 04

[4] M.Horiuchi, T.Fujii, T.Yokoyama : "Comparison of Multirate Deadbeat Control for Single Phase PWM Inverter", IPEC' 05

[5] E.Shimada, K.Aoki, T.Komiyama, T.Yokoyama : "Implementation of Deadbeat Control for Single Phase Utility Interactive Inverter Using FPGA", IPEC' 05

[6] T.Fujii, T.Yokoyama : "FPGA based Multi-Sampling Pulse Compensation Method with Deadbeat Control for Single Phase PWM Inverter", EPE-PEMC' 05

[7] E.Shimada, K.Aoki, T.Komiyama, T.Yokoyama : "Implementation of Deadbeat Control for Single Phase Utility Interactive Inverter Using FPGA based Hardware Controller", EPE-PEMC' 05

[8] A.Takeuchi, S.Kondo : "Quick Voltage Amplitude Control on Parallel Operation of Single Phase Inverters", 2002 National Convention Record I.E.E of Japan p122 (in Japanese)

[9] W.Fujii, T.Yokoyama : "Construction of FPGA based hardware controller for autonomous decentralized control for UPS application", EPE-PEMC 2006

[10] Myway labs -PSIM- http://www.myway-labs.co.jp/psim/

[11] ALTERA http://www.altera.com/

[12] T.Saito,N.Doi,T.Yokoyama : "Verification of Autonomous Decentralized Control UPS system using FPGA based Hardware Controller", ICPE 2007

Fault Current Reduction in Distribution Systems with Distributed Generation Units by a New Dual Functional Series Compensator

H.R. Baghaee [*], M. Mirsalim [*][+] (IEEE Senior Member), M. J. Sanjari [*], and G.B. Gharehpetian [*]

[*] Center of Excellence in Power Engineering, Amirkabir University of Technology, Tehran, Iran
[+] Also, with the Department of Engineering, St. Mary's University, San Antonio, TX, USA
, e-mails: hrbaghaee@aut.ac.ir, mmirsalim@aut.ac.ir, m_j_sanjari@aut.ac.ir, and grptian@aut.ac.ir

Abstract— In this paper, a new control strategy is proposed to limit the fault current in power systems which include distributed generation units. In the normal mode of operation, the series compensator acts as a line compensator. As soon as the fault is sensed, the new control mode is activated to limit the fault current, and properly interrupt the breaker. Harmony search algorithm has also been used to optimize the parameters of series compensator controller. Simulations performed in MATLAB/Simulink environment indicate that the proposed control strategy performs well to limit the fault currents of distribution systems and restore the voltage at the point of common coupling to its preset value.

Keywords— fault current limiter, series compensator, distributed generation, harmony search algorithm.

I. INTRODUCTION

The distributed generation (DG) interconnection with power systems can significantly impact power flow and power quality for both customers and the utilities. These impacts may have strengths and weaknesses depending on both the power system operating characteristics and the DG characteristics. The benefits provided by DGs are not only in improved power quality and system reliability, but also in loss reduction. If certain minimum standards for control, installation and protection are not maintained, power system operations may adversely be affected by the introduction of DGs. Thus, DGs should meet some operating requirements of the utilities or the power system operators [1-3].

The widespread use of DG units connected to networks creates new problems for electric utilities due to the planning, control, and management of the system, and might also increase fault currents, cause malfunction in relay coordination, and decrease power quality (PQ) [4]. Another significant problem is that the short circuit capacity (SCC) of the buses increases and may exceed the breaker interrupting capability. Thus, the existing protection system could not cope with the increase in fault level. Consequently, the interrupting capacity of the circuit breaker exceeds the rated one, and the instantaneous voltage sag occurs at the normal line. Therefore, it is important to limit fault currents which are affected by DG units. Different approaches, in particular passive devices (i.e., fault current limiters (FCLs)) have been proposed to reduce fault currents and mitigate voltage dips [5, 6].

As the concern for power quality grows, much research attention has been directed towards the design of custom power devices. These devices such as series compensators (SC) are designed to mitigate voltage perturbation caused by faults. The mitigation function is achieved by SC through the injection of a voltage V_{sc} generated from a voltage source converter (VSC) as shown in Fig 1.

In distribution systems faults usually happen in the downstream of the SC; anywhere on the feeder connecting the SC and the sensitive load. And because the SC is in series with the faulty feeder, a large current will pass through the supply transformer and the VSC. Depending on the location and the type of fault, and the design of the protection system, the fault clearing action may last more than 10 cycles, which can damage the VSC. In addition, the loads connected to the point of common coupling (PCC) may become affected by the voltage sag caused by the fault, i.e. the equipment connected to the feeder might fail, malfunction, or shut down) [7].

In [8], a bypass scheme has been implemented to protect SCs against large currents exceeding the converter ratings. When a downstream fault occurs, a protection scheme can be performed to close the secondary terminals of the injection transformer by implementing a solid-state shorting-switch thyristor. Fault current limiting methods can be categorized into two types; non-superconducting [9-15] and superconducting [16-23].

With the progress of power semiconductor devices and power conversion technologies towards handling larger currents, new applications of power-electronic-based controllers have started to be investigated [9-12]. The basic operation and effectiveness of the advanced FCL with harmonic compensation has been presented in [13]. Also, the operating characteristics of FCLs have been discussed and their practical designs verified by experimental tests in [14-15].

In [16-23] the superconducting fault current limiters (SFCLs) have been used to solve the issues caused by the interconnection of DG units to grids.

978-1-4244-1741-4/08/$25.00 ©2008 IEEE

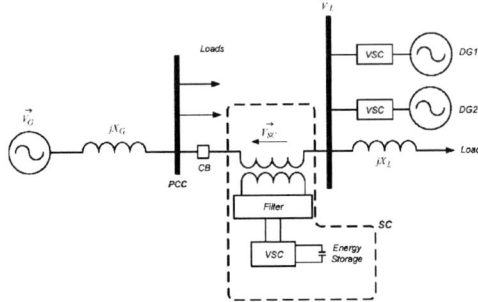

Fig. 1 The distribution system under study

In this paper, the authors are presenting a new control strategy to limit the fault currents of distribution systems having DG units. In the normal mode of operation, the SC is controlled for line compensation. As the fault is detected, the new control mode is activated to limit the fault currents flowing through the breaker. This strategy leads to lower fault levels and as a consequence, to lower rating lower-cost circuit breakers. Harmony search (HS) algorithm based on minimization of a proper objective function has also been used to optimize the parameters of the SC controller. The proposed control scheme can be implemented in medium voltage (MV) networks either in the form of VSC-based custom power devices, or in the interconnection of DGs to a network operating as a flexible distributed generation (FDG).

Next, the proposed control strategy is applied to a 2-bus distribution system and the results are presented. Simulation results indicate that the proposed scheme can effectively limit fault currents of the grid.

The operation modes of a series compensator are briefly discussed as in the following.

A. Voltage Restoration Mode

Fig. 1 shows the system under study which consists of a synchronous generator, a compensated transmission line and two DG sources connected to bus 2 through multi-level VSCs. Implementation of these converters can improve power quality by reducing harmonic content of the output voltage which results in lower costs for the filters. In this study, it is assumed that the series reactance of injection transformer is negligible and that load has a constant power factor. Moreover, the network is assumed to be lossless and the converter is working in the linear mode.

With the above assumptions and without any loss of generality, the load voltage equation can be written as:

$$\vec{V}_L = \vec{V}_G - jX_G \vec{I}_L \qquad (1)$$

where $\vec{I}_L$ is the line current phasor under normal condition and X_G is the equivalent reactance of the generator. Under an upstream fault, the voltage sag can be modeled as a decrease in generator voltage to $\vec{V}_{sag}$. As shown in Fig.2, the PCC voltage under the upstream fault condition and without series compensation can be obtained as:

$$\vec{V}_{PCC,f} = \vec{V}'_L = \vec{V}_{sag} - jX_G \vec{I}'_L \qquad (2)$$

where $\vec{I}'_L$, $\vec{V}_{PCC,f}$ and $\vec{V}'_L$ are the line current and the PCC and load voltages under upstream fault condition, respectively. Now, the SC must inject a voltage $\vec{V}_{SC}$ (represented by a voltage source in Fig. 2) to restore the load voltage to its pre-fault value $\vec{V}_L$, and protect the load against voltage sag. Thus:

$$\vec{V}_L = \vec{V}_{sag} - jX_G \vec{I}_L + \vec{V}_{SC} \qquad (3)$$

$\vec{V}_{SC}$ should be obtained such that the dynamic restoration of $\vec{V}_L$ be realized [8-9]. Setting $\vec{V}_L$ in (3) equal to $\vec{V}_{PCC}$, $\vec{V}_{sag} - jX_G \vec{I}_L$ is the voltage on source side of SC and $\vec{V}_{SC}$ is:

$$\vec{V}_{SC} = \vec{V}_{PCC} - \vec{V}'_{PCC,f} \qquad (4)$$

where $\vec{V}'_{PCC,f}$, is the measured source side voltage of SC.

B. Fault Current Limiting Mode

The principle of SC operation as a fault current limiter is illustrated in the single-line diagram of Fig. 3a, where X is the effective reactance between the PCC and the fault location f. Under the pre-fault condition, the relationships between the line current and voltages are:

$$\vec{V}_G = \vec{V}_{PCC} + jX_G \vec{I}_L \qquad (5)$$

$$\vec{V}_{PCC} = \vec{V}_f + jX \vec{I}_L \qquad (6)$$

where, $\vec{V}_{PCC}$ and $\vec{V}_f$ are the pre-fault voltages at PCC and node f, respectively. Fig. 3b demonstrates the phasor diagram representation of Fig.3a. When the downstream fault occurs at f, $\vec{V}_f = 0$, and if the equivalent voltage source of generator $\vec{V}_G$, and the reactance X_G, are assumed unchanged, then:

$$\vec{V}_G = \vec{V}_{PCC,f} + jX_G \vec{I}_f \qquad (7)$$

$$\vec{V}_{PCC,f} = jX \vec{I}_f \qquad (8)$$

where $\vec{V}_{PCC,f}$ and $\vec{I}_f$ are the PCC voltage and fault current during the fault without the presence of SC, respectively. The network and the corresponding phasor diagram under a downstream fault is shown in Fig. 4. Rearranging (4) and (6), the pre-fault and the fault currents are as in below:

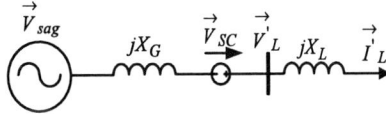

Fig. 2 Network model under an upstream fault condition

$$\vec{I}_L = \frac{\vec{V}_G - \vec{V}_{PCC}}{jX_G} \tag{9}$$

$$\vec{I}_f = \frac{\vec{V}_G - \vec{V}_{PCC,f}}{jX_G} \tag{10}$$

If $\left|\vec{V}_{PCC}\right|\cos\theta > \left|\vec{V}_{PCC,f}\right|$, where θ is phase angle between $\vec{V}_{PCC}$ and $\vec{V}_G$, then the fault current $\left|\vec{I}_f\right|$ will be greater than the pre-fault $\left|\vec{I}_L\right|$, and this represents a downstream fault.

The proposed scheme is as in the following. As soon as the fault is detected, the SC injects a series voltage $\vec{V}_{SC}$ thorough the injection transformer as shown in Fig.5. $\vec{V}_{SC}$ has a value to restore $\vec{V}_{PCC}$ to its pre-fault value so that $\vec{V}_{PCC} = \vec{V}_{PCC,f}$, and the fault current is limited to its pre-fault level namely, $\vec{I}_f = \vec{I}_L$. Thus:

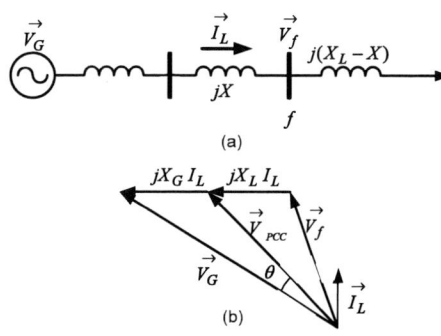

(a)

(b)

Fig. 3 The network and its phasor diagram for pre-fault condition

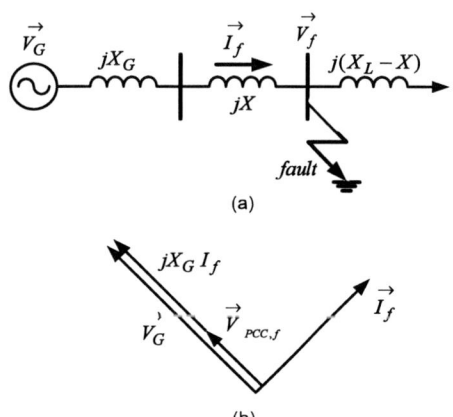

(a)

(b)

Fig. 4 The network and its phasor diagram under a downstream

$$\vec{V}_G = \vec{V}_{PCC} + jX_G \vec{I}_L \tag{11}$$

$$\vec{V}_{PCC} = \vec{V}_{SC} + jX \vec{I}_L \tag{12}$$

From phasor diagrams of figures 5b and 3b, we have

$$\vec{V}_{SC} = \vec{V}_f \tag{13}$$

If the injected voltage is equal to the value of the pre-fault voltage phasor at node f, then the fault current will be limited to the pre-fault load current. As the fault location is random and unknown, hence $\vec{V}_f$ is variable. Rewriting (12), the injected voltage can be expressed as:

$$\vec{V}_{SC} = \vec{V}_{PCC} - jX \vec{I}_L \tag{14}$$

Thus, if the pre-fault PCC voltage and the load current are known, the injected voltage can be obtained. $\vec{V}_{PCC}$ can be measured using the phase lock loop (PLL) technique. The voltage component $jX\vec{I}_L$ is the load voltage $\vec{V}_L$ and can also be measured. Therefore, the injected voltage can be generated continuously. The dynamic process of voltage injection has also been depicted in Fig. 6.

Based on the above explanation, the fault current can be limited to the pre-fault load current. The maximum injected voltage occurs for $X=0$ (i.e. the three phase fault at a load bus). Thus, the primary winding voltage and the volt-ampere rating of the injection transformer are $\left|\vec{V}_{PCC}\right|$ and $\left|\vec{V}_f \vec{I}^*_L\right|$ (approximately the downstream load capacity), respectively.

II. THE PROPOSED METHOD

A. Controller design

Fig. 7a shows the SC-connected distribution system. The PCC voltage is restored by the SC through the injection of voltage $\vec{V}_{SC}$. An inductance–capacitance (LC) filter is used to block all but the desired output voltage fundamental frequency component. The SC harmonic filter shown in the figure, has an inductance L_{filter}, an internal resistance r_{filter}, and a capacitance C_{filter} [7]. The load is represented by an impedance with a constant power factor. The resistance and leakage inductance of the injection transformer with a turns-ratio of 1:n are assumed to be negligible. The block diagram of the control system can be obtained based on the following equations:

$$V_{conv} = V_c + r_{filter} I_{filter} + L_{filter}\frac{dI_{filter}}{dt} \tag{15}$$

$$I_{filter} = I_c + nI_L \tag{16}$$

$$I_c = C_{filter}\frac{dV_c}{dt} \tag{17}$$

$$V_{SC} = nV_c \tag{18}$$

where, V_{conv} and V_c are the output voltage of the converter and the capacitor voltage, respectively. Also, I_{filter}, I_c and I_L are the filter current, capacitor current, and load current, respectively. Based on the above equations, the open-loop block diagram of the control system is shown in Fig. 7b where, $V^*_{sc}(s)$ is the reference injection voltage i.e. the actual SC output voltage. It is obvious from this figure that:

$$V_{SC}(s) = G_1 V^*_{SC}(s) + G_2 I_L(s) \qquad (19)$$

$$G_1 = \frac{n k_i}{L_{filter} C_{filter} s^2 + r_{filter} C_{filter} s + 1} \qquad (20)$$

$$G_2 = \frac{-n^2 (L_{filter} s + r_f)}{L_{filter} C_{filter} s^2 + r_{filter} C_{filter} s + 1} \qquad (21)$$

where, $G_1(s)$ and $G_2(s)$ are the transfer functions from $V^*_{sc}(s)$ to $V_{sc}(s)$ and I_L to $V_{sc}(s)$, respectively.

For these second order transfer functions, the open loop natural damping frequency $\omega_{n,OL}$ and the damping constant $\xi_{n,OL}$, are:

$$\omega_{n,OL} = \sqrt{\frac{1}{L_{filter} C_{filter}}} \qquad (22)$$

$$\xi_{n,OL} = \frac{r_{filter}}{2 L_{filter} \omega_{n,OL}} = \frac{r_{filter}}{2 X_{filter} K_f} \qquad (23)$$

where X_{filter} is inductive reactance of the filter at the base frequency of the network, and K_f is the ratio of the natural damping frequency to the network base frequency. K_f should be chosen much greater than unity to eliminate the harmonics produced by the power converter. The goal is to have a high quality filter (quality factor more than 10) and hence, the ratio of r_{filter}/X_{filter} should be less than 0.1 ($\xi_{n,OL} \ll 1$ and $K_f \gg 1$). Now, as the SC operates in the open loop control mode, it has a low damping level.

To enhance the damping level of the SC controller, the control system shown in Fig. 7b is proposed. Because $V_{SC}(s)$ is proportional to $I_c(s)$ to control $V_{SC}(s)$ in the event of load current changes, $I_c(s)$ must be regulated. Furthermore, because $V_{SC}(s)$ should track $V^*_{SC}(s)$, the outer voltage loop is included.

The controllers have been chosen as simple proportional regulators k_c and k_v. This is because, when the output voltage varies sinusoidally, the integral part of the PI controller causes additional phased shift and thus it is not considered.

As shown in Fig. 7b, the close loop transfer function between $V^*_{SC}(s)$ and $V_{SC}(s)$ can be written as:

$$V_{SC}(s) = G_{1c} V^*_{SC}(s) + G_{2c} I_L(s)$$

where, G_{1c} and G_{2c} are the closed loop transfer functions from $V^*_{sc}(s)$ to $V_{sc}(s)$ and I_L to $V_{sc}(s)$, respectively. $\qquad (24)$

$$G_{1c} = \frac{n k_v k_c k_i}{L_{filter} C_{filter} s^2 + C_{filter} (r_{filter} + k_c k_i) s + n k_v k_c k_i +} \quad (25)$$

$$G_{2c} = \frac{-n^2 (L_{filter} s + r_f)}{L_{filter} C_{filter} s^2 + C_{filter} (r_{filter} + k_c k_i) s + n k_v k_i k_c +} \quad (26)$$

The closed loop natural damping frequency ($\omega_{n,CL}$) and the damping constant ($\xi_{n,CL}$) can also be obtained as below:

$$\omega_{n,CL} = \sqrt{n k_v k_c k_i + n} \sqrt{\frac{1}{L_{filter} C_{filter}}} \qquad (27)$$

$$\xi_{n,CL} = \frac{r_{filter} + k_i k_c}{2 L_{filter} \omega_{n,CL}} = \frac{r_{filter} + k_i k_c}{2 X_{filter} K_f \sqrt{n + n k_v k_c k_i}} \qquad (28)$$

k_c and k_v are determined by solving the following optimization problem with harmony search algorithm.

$$\min objective\ finction = \int_0^{T_{end}} (I_f - I_L)^2 \exp(-\alpha t) dt$$

$$subject\ to: \qquad (29)$$

$$\omega_{n,CL} = \omega_{n,CL,desired}$$

$$\xi_{n,CL} = \xi_{n,CL,desired}$$

where T_{end} is the simulation time and α is an arbitrary constant. $\omega_{n,CL,desired}$ and $\xi_{n,CL,desired}$ can be determined from the required closed-loop system-response settling time and overshoot.

B. Harmony search algorithm

Harmony search algorithm (HAS) has recently been developed in an analogy with music improvisation process where music players improvise the pitches of their instruments to obtain better harmony [24, 25]. The steps in the procedure of harmony search (HS) are described in the next five subsections.

Step 1: Initialize the problem and algorithm parameters

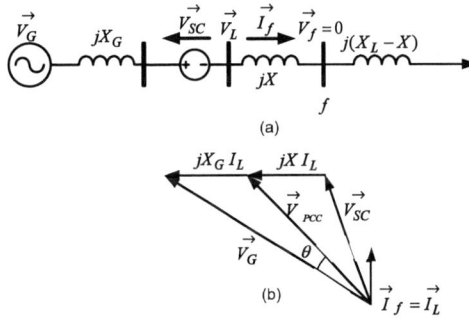

(a)

(b)

Fig. 5 Downstream fault condition with the FCL

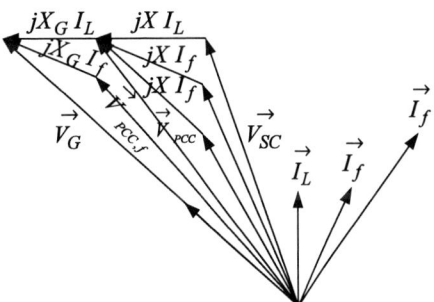

Fig. 6 Dynamic process of voltage injection

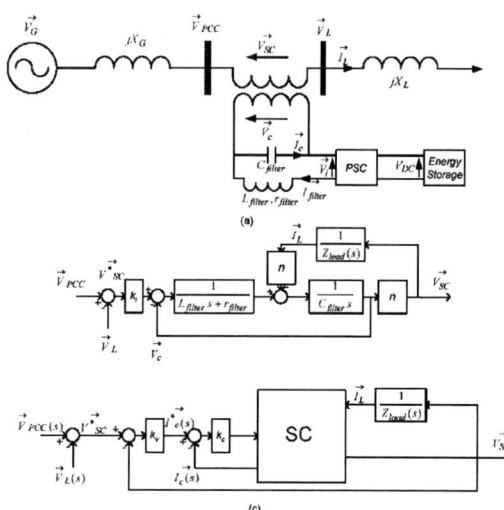

Fig. 7 a) Typical SC-connected distribution system b) The control system for the FCL

The optimization problem is specified as: $\min\{f(x)\mid x\in X\}$ subject to $g(x)\geq 0$ and $h(x)=0$ where, $f(x)$, $g(x)$, and $h(x)$ are the objective function, the inequality constraint function, and the equality constraint function, respectively. x is the set of each decision variable, x_i, and X is the set of the possible range of values for each decision variable, that is $X_{i,\min}\leq X_i\leq X_{i,\max}$, where $X_{i,\min}$ and $X_{i,\max}$ are the lower and upper bounds for each decision variable. The parameters of the HSA are also specified in this step. The parameters are the harmony memory size (HMS) or the number of solution vectors in the harmony memory, harmony memory considering rate (HMCR), pitch adjusting rate (PAR), the number of decision variables (N), and the number of improvisations (NI) or the stopping criterion. The harmony memory is a memory location where all the solution vectors (sets of decision variables) are stored, and is similar to the genetic pool in the genetic algorithm [26]. Here, HMCR and PAR are the parameters that are used to improve the solution vector, and are defined in step 3.

Step 2: Initialize the harmony memory

In step 2, the HM matrix is filled with as many randomly generated solution vectors as the HMS

$$HM=\begin{bmatrix} x_1^1 & x_2^1 & \cdots & x_{N-1}^1 & x_N^1 \\ x_1^2 & x_2^2 & \cdots & x_{N-1}^2 & x_N^2 \\ \vdots & \vdots & \vdots & \vdots & \vdots \\ x_1^{HMS-1} & x_2^{HMS-1} & \cdots & x_{N-1}^{HMS-1} & x_N^{HMS-1} \\ x_1^{HMS} & x_2^{HMS} & \cdots & x_{N-1}^{HMS} & x_N^{HMS} \end{bmatrix} \quad (30)$$

Step 3: Improvise a new harmony

A new harmony vector, $x_1'=(x_1',x_2',...,x_N')$, is generated based on three rules: (1) memory consideration, (2) pitch adjustment and (3) random selection. Generating a new harmony is called 'improvisation' [26]. In the memory consideration, the value of the first decision variable x_1' for the new vector is chosen from any value in the specified HM range $(x_1^1-x_1^{HMS})$. Values of the other decision variables $(x_2',x_3',...,x_N')$ are chosen in the same manner. The HMCR, which varies between 0 and 1, is the rate of choosing one value from the historical values stored in the HM, while (1 _ HMCR) is the rate of randomly selecting one value from the possible range of values.

$$x_i' \leftarrow \begin{cases} x_i' \in \{x_i^1,x_i^2,...,x_i^{HMS}\} & with\ probability\ HMCR \\ x_i' \in X_i & with\ probability\ (1-HMCR) \end{cases} \quad (31)$$

For example, a HMCR of 0.85 indicates that the HS algorithm will choose the decision variable value from historically stored values in the HM with the 85% probability or from the entire possible range with the 100–85% probability. Every component obtained by the memory consideration is examined to determine whether it should be pitch-adjusted. This operation uses the PAR parameter, which is the rate of pitch adjustment as follows:

$$x_i' \leftarrow \begin{cases} Yes & with\ probability\ PAR \\ No & with\ probability\ (1-PAR) \end{cases} \quad (32)$$

The value of (1 _ PAR) sets the rate of doing nothing. If the pitch adjustment decision for x_i' is Yes, x_i' is replaced as follows:

$$x_i' \, x_i' \leftarrow x_i' \pm rand()*b_w \quad (33)$$

where, b_w is an arbitrary distance bandwidth, and rand () is a random number between 0 and 1.

In step 3, HM consideration, pitch adjustment or random selection is applied to each variable of the new harmony vector in turn.

Step 4: Update harmony memory

If the new harmony vector $x_1'=(x_1',x_2',...,x_N')$, is better than the worst harmony in the harmony memory (HM), judged in terms of the objective function value, the new harmony is included in the HM and the existing worst harmony is excluded from the HM.

Step 5: Check stopping criterion

If the stopping criterion (maximum number of improvisations) is satisfied, computation is terminated. Otherwise, steps 3 and 4 are repeated. In this paper, HS is implemented to optimize the parameters of the controller based on the minimization criterion given in (29).

III. SIMULATION RESULTS

The proposed control strategy has been applied to the system under study with parameters given in Table I. By choosing $\omega_{n,CL,desired}$ and $\xi_{n,CL,desired}$ equal to 3 kHz and 0.75 respectively, the minimization of the objective function defined by (27) yields $k_v=0.696$ and $k_c=71.54$. The parameters of the HSA are tabulated in Table II.

The open loop and closed loop root locus and Nichols diagrams have been depicted in Figs. 8 and 9, respectively.

The system introduced in section II is subjected to a single line to ground (SLG) fault at node f at t=0.05 sec. The PCC voltage and the line current without and with the proposed FCL strategy are demonstrated in Fig. 10 and Fig. 11, respectively. In the second scenario, a three-phase fault at node f at t=0.05 sec is imposed on the system. The PCC voltage and the line current without and with the proposed FCL strategy are shown in Fig. 12 and Fig. 13, respectively. The simulation results show that the proposed control strategy performs well in limiting the fault current, and restoring the PCC voltage to its preset value.

TABLE I.
THE SYSTEM PARAMETERS

Parameters	Values
Distributed supply voltage	11 kV
Equivalent source reactance	30 Ω
Injection transformer turns-ratio	1:1
DC-link voltage	15.5 kV
Filter capacitor	48 µF
Filter inductance	0.2 mH
Filter resistance	2.2 Ω

TABLE II.
THE PARAMETERS OF THE HARMONY SEARCH ALGORITHM

HMS	10
HMCR	0.9
PAR	0.3
b_w	0.01
Number of iterations	1000

IV. CONLUSION

In this paper, a new control strategy was presented for a series compensation scheme to limit fault current contribution of distributed generation units in distribution systems. Harmony search algorithm was also used to optimize parameters of the series compensator controller based on minimization of a proper objective function.

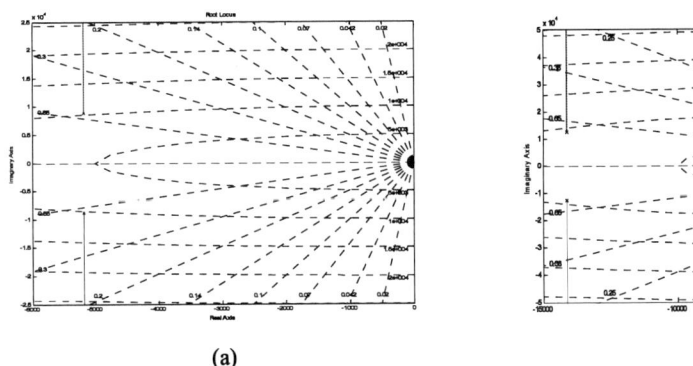

(a)   (b)

Fig. 8 a) Open-loop and b) Closed-loop root locus diagrams of the control system

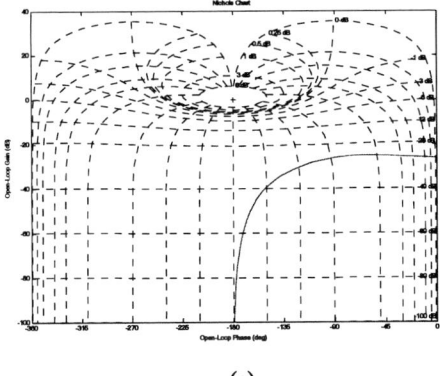 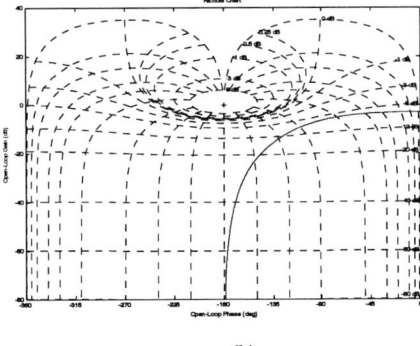

(a)   (b)

Fig. 9 a) Open loop and b) Closed loop Nichols diagrams of the control system

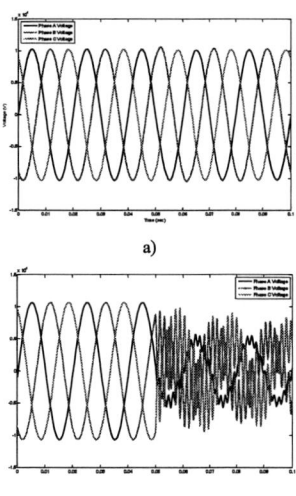

a)

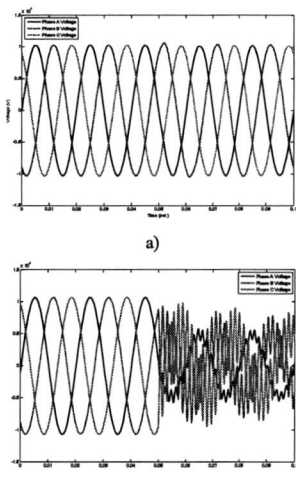

b)

Fig. 10 a) PCC voltage and b) line current without FCL strategy

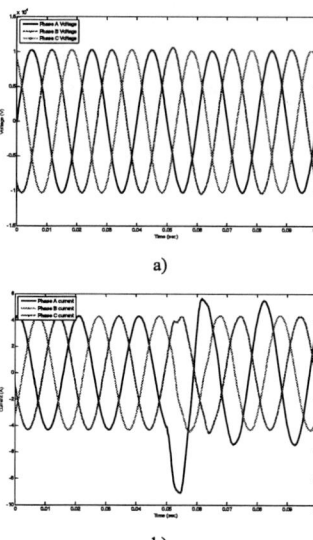

a)

b)

Fig. 11 a) PCC voltage and   b) line current with FCL strategy

a)

b)

Fig. 12 a) PCC voltage and   b) line current without FCL

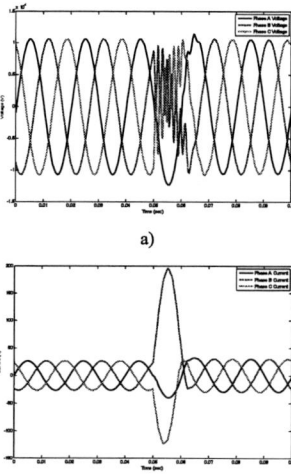

a)

b)

Fig. 13 a) PCC voltage and   b) line current with FCL strategy

The proposed control scheme was implemented on a medium voltage test system. The simulation results indicate that by proper control of the converter, the downstream fault currents of the grid can effectively be reduced.

REFERENCES

[1] T. Ackermann, G. Andersson and L. Soder, "distributed generation: a definition", *Elsevier Electric Power System Research*, vol. 57, pp. 894-895, 2001.

[2] F. V. Edwards, G. J. W. Dudgeon, J. R. McDonald and W. E. Leithead, "dynamics of distribution network with distributed generation", *IEEE Power Engineering Society Summer Meeting*, Vol. 2, pp. 1032-1037, 2000.

[3] R. A. Walling, R. Saint, R. C. Dugan, J. Burke, L. A. Kojovic, "summary of distributed resources impact on power delivery systems", *IEEE. Trans. Power Del*, Accepted for future publication, 2007

[4] R. C. Dugan and T. E. McDermott, "operating conflicts for distributed generation interconnected with utility distribution systems", *IEEE Ind. Appl. Magazine*, vol. 8, No. 2, pp. 19–25, 2002.

[5] A. Cataliotti, G. Cocchiara, M. G. Ippolito, G. Morana, "applications of the fault decoupling device to improve the operation of LV distribution networks", *IEEE Trans. Power Delivery*, 2007.

[6] A. Campoccia, G. Cocchiara, M. G. Ippolito, and G. Morana, "fault decoupling device: a new device to reduce the impact of distributed generation on electrical distribution systems" *Proc. of IEEE Power Tech conference*, pp. 123–128, 2003.

[7] S. S. Choi, B. H. Li, and D. M. Vilathagmuwa, "design and analysis of the inverter-side filter used in the dynamic voltage restorer", *IEEE Trans. on Power Del*, vol. 17, No. 3, pp. 857–864, 2002,.

[8] S. S. Choi, B. H. Li, and D. M. Vilathgamuwa, "dynamic voltage restoration with minimum energy injection", *IEEE Trans. Power System*, vol. 15, no. 1, pp. 51–57, Feb. 2000.

[9] N. H. Woodley, L. Morgan, and A. Sundaram, "experience with an inverter-based dynamic voltage restorer," *IEEE Trans. Power Del*, vol. 14,no. 3, pp. 1181–1186, Jul. 1999.

[10] F. Tosato and S. Quaia, "reducing voltage sags through fault current limitation," *IEEE Trans. Power Del*, vol. 16, no. 1, pp. 12–17, Jan. 2001.

[11] M. Steurer, K. Fröhlich,W. Holaus, and K. Kaltenegger, "a novel hybrid current-limiting circuit breaker for medium voltage: Principle and test results," *IEEE Trans. Power Del*, vol. 18, no. 2, pp. 460–467, Apr. 2003.

[12] S. S. Choi, T. X. Wang, and D. M. Vilathgamuwa: A series compensator with fault current limiting function, *IEEE Trans. on Power Del*, vol. 20, pp. 2248-2256, 2005.

[13] M. Hojo, N. Kuroe and T. ohnishi, "fault current limiter with harmonics compensation based on voltage source inverter", *Proc. of Intl. Conf. on Electrical Eng*, 2005.

[14] M. Hojo, N. Kuroe and T. Ohnishi, "fault current limiter by series connected voltage source inverter", *IEEJ Trans. on Ind. App*, special issue on IPEC-Niigata, vol.126, No.4, pp.438-443, 2006.

[15] M. Hojo, Y. Fujimura, T. Ohnishi and T. Funabashi, "design of fault current limiter by series-connected voltage source inverter", *proc. of Intl. Conf. on advanced power system automation and protection* 2007.

[16] T. Nomura, M. Yamaguchi, S. Fukui, K. Yokoyama, T. Satoh, and K. Usui, "single DC reactor type fault current limiter for 6.6 kV Power Systems", *IEEE Trans. Applied Superconductivity*, vol. 11, no. 1, pp. 2090–2093, March 2001.

[17] K. Usui, T. Nomura, T. Satoh, M. Yamaguchi, S. Fukui, K. Yokoyama, and T. Nagasawa, "a single dc reactor type fault current limiting interrupter for three-phase power system," *IEEE Trans. Applied Superconductivity*, vol. 11, no. 1, pp. 2126–2129, March 2001.

[18] K. Duangkamol, Y. Mitani and K. Tsuji, "power system stabilizing control and current limiting by a SMES with a series phase compensator", *IEEE Trans. on Applied Superconductivity*, Vol.11, No.1, pp.1753-1756, 2001.

[19] L. Ye, L. Z. Lin, and K. Juengst, "application studies of superconducting fault current limiters in electric power system," *IEEE Trans. Appl. Superco.*, vol. 12, no. 1, pp. 900–904, Mar. 2002.

[20] H. Shao, M. Yamaguchi, S. Fukui, J. Ogawa, T. Sato, and H. Ishikawa, "performance comparison of a DC hybrid type FCLI with other types", *IEEE Trans. Applied Superconductivity*, vol. 15, no. 2, pp. 2126–2129, June 2003.

[21] H. Kang, M. C. Ahn, Y. K. Kim, D. K. Bae, D. K. Park, Y. S. Yoon, T. K. Ko, J. H. Kim, and J. Joo, "design, fabrication and testing of type superconducting DC reactor for 1.2 kV/80 A inductive fault current limiter," *IEEE Trans. Applied Superconductivity*, vol. 13, no. 2, pp. 2008–2011, June 2003.

[22] M. C. Ahn, H. Kang, D. K. Bae, D. K. Park, Y. S. Yoon, S. J. Lee, and T. K. Ko, "The short-circuit characteristics of a DC reactor type superconducting fault current limiter with fault detection and signal control of the power converter," *IEEE Trans. Applied superconductivity*, vol. 15, no. 2, pp. 2102–2105, June 2005.

[23] Y. Shirai, A. Mochida, T. Morimoto, M. Shiotsu, T. Oide, M. Chiba, and T. Nitta, "repetitive operation of three-phase superconducting fault current limiter in a model power system", *IEEE Trans. Applied Superconductivity*, vol. 15, no. 2, pp. 2110–2113, June 2005.

[24] ZW.Geem, C. Tseng and Y. Park, "harmony search for generalized orienteering problem: best touring in China", *Springer Lecture Notes for Computer Science*, Vol. 341, pp.741–50, 2005.

[25] ZW. Geem, JH. Kim, GV. Loganathan, "harmony search optimization: application to pipe network design", *International Journal of Modeling and Simulation*, Vol., 22, pp.125–33, 2002.

[26] KS. Lee, ZW. Geem, "a new structural optimization method based on the harmony search algorithm", *Computer Structure*, Vol. 82, pp. 781–98, 2004.

BIOGRAPHIES

Hamid Reza Baghaee *(IEEE Student Member' 2008)* received the BSc degree in Electrical Engineering from Kashan University in 2006. Currently he is graduate student of Power Engineering in Amirkabir University of Technology. His research interests are power system dynamic and control, HVDC & FACTS devices, Distributed Generation (DG) and application of Artificial Intelligence in power systems.

Mojtaba Mirsalim *(IEEE Senior Member' 2004)* was born in Tehran, Iran, on February 14, 1956. He received his B.S. degree in EECS/NE, M.S. degree in Nuclear Engineering from the University of California, Berkeley in 1978 and 1980 respectively, and the PhD in Electrical Engineering from Oregon State University, Corvallis in 1986. Since 1987 he has been at Amirkabir University of Technology, has served 5 years as the Vice Chairman and more than 7 years as the General Director in Charge of Academic Assessments, and currently is a Full Professor in the department of Electrical Engineering where he teaches courses and conducts research in energy conversion and CAD, among others.

His special fields of interest include the design, analysis, prototyping, and optimization of electric machines, renewable energy, FEM, and hybrid vehicles. Mirsalim is the author of more than 100 international journal and conference papers and three books on electric machinery and FEM. He is the founder and at present, the director of the Electrical Machines & Transformers Research Laboratory at http://ele.aut.ac.ir/EMTRL/Homepage.htm

Mohammad Javad Sanjari received the BSc degree in Electrical Engineering from Amirkabir University of Technology in 2006. Currently he is graduate student of Power Engineering in Amirkabir University of Technology. His research interests are power system dynamic and control, power system security assessment, HVDC & FACTS devices, Distributed Generation (DG) and application of Artificial Intelligence in power systems.

G.B. Gharehpetian *(IEEE Member)* was born in Tehran, in 1962. He received his BS and MS degrees in electrical engineering in 1987 and 1989 from Tabriz University, Tabriz, Iran and Amirkabir University of Technology (AUT), Tehran, Iran, respectively, graduating with First Class Honors. In 1989 he joined the Electrical Engineering Department of AUT as a lecturer. He received the Ph.D. degree in electrical engineering from Tehran University, Tehran, Iran, in 1996. As a Ph.D. student he has received scholarship from DAAD (German Academic Exchange Service) from 1993 to 1996 and he was with High Voltage Institute of RWTH Aachen, Aachen, Germany. He held the position of Assistant Professor in AUT from 1997 to 2003, and has been Associate Professor since 2004. Dr. Gharehpetian is a Senior Member of Iranian Association of Electrical and Electronics Engineers (IAEEE), member of IEEE and member of central board of IAEEE. Since 2004 he is the Editor-in-Chief of the Journal of IAEEE. The power engineering group of AUT has been selected as a Center of Excellence on Power Systems in Iran since 2001. He is a member of this center and since 2004 the Research Deputy of this center. Since November 2005 he is the director of the industrial relation office of AUT. He is the author of more than 200 journal and conference papers. His teaching and research interest include power system and transformers transients, FACTS devices and HVDC transmission.

Dynamic Simulation of PM Motor Drive System based on Reluctance Network Analysis

Kenji NAKAMURA* and Osamu ICHINOKURA*

* Tohoku University/Graduate School of Engineering, Sendai, Japan, e-mail: *nakaken@ecei.tohoku.ac.jp*

Abstract— **This paper presents a method for dynamic simulation of a permanent magnet (PM) motor including its drive circuit based on reluctance network analysis (RNA). The RNA, which is based on the magnetic circuit method, has some advantages for simulating electric motors as follow: simple analytical model, high calculation accuracy, and easy to combine with external electric circuits, motion and thermal dynamics. First, the basis of the magnetic circuit method is described briefly. Next, a case study of dynamic simulation of a PM motor drive system is presented. The validity of the proposed method is proved by comparison with experimental results.**

Keywords—**Permanent magnet (PM) motor, Reluctance Network Analysis (RNA), Dynamic analysis**

I. INTRODUCTION

In order to reduce the development period and cost, computer aided engineering (CAE) is being applied to design and analysis of electrical machinery. The finite element analysis (FEA) is one of the powerful CAE tools, and a number of general-purpose FEA programs have been placed on the market nowadays. The most conventional optimum design for electrical machinery is shape optimization. Recent electrical machinery, however, particularly an electric motor is controlled by a power electronics circuit to achieve high power and efficiency. Therefore, the optimization of not only the shape but also the whole system including the power electronics circuit is necessary. Although the FEA based coupled analysis being examined, it takes much time to obtain the calculation results in general.

The magnetic circuit method is also one of the CAE tools and more practical than FEA, because an analytical model is very simple, calculation accuracy is reasonably high, and it is easy to combine with an external electric circuit, motion and thermal dynamics [1]-[3]. Since the electric motor has generally complicated structure or flux distribution however, it is difficult to express it in a simple magnetic circuit.

To overcome the above problem, a more detailed model namely, reluctance network analysis (RNA) model is proposed in this paper. First, the basis of the magnetic circuit method is described, and then a case study of dynamic simulation of a PM motor drive system based on RNA is presented.

II. THE BASIS OF THE MAGNETIC CIRCUIT METHOD

Fig. 1(a) illustrates a ring core with a winding of N turns. The cross section, magnetic path length, and permeability of the core are S, l, and μ, respectively. In this circuit, when the current i flows in the winding, a relationship between magnetomotive force (MMF) Ni and flux ϕ can be expressed in the following equation [4]:

$$Ni = R_m\phi, \text{ where } R_m = \frac{l}{\mu S}. \tag{1}$$

In Eq. (1), if let Ni and ϕ correspond to voltage v and current i of the electric circuit, it is understood that the same relationship as the Ohm's low exists between the MMF and flux. Accordingly, the ring core of Fig. 1(a) can be expressed in the magnetic circuit shown in Fig. 1(b). In the figure, R_m which corresponds to a resistance of the electric circuit is called *reluctance*. Based on this magnetic circuit, the flux ϕ can be calculated when the current i is given.

If magnetic nonlinearity is taken into account, a *B-H* curve of core material is approximately expressed by the following nonlinear function:

$$H = \alpha_1 B + \alpha_n B^n, \tag{2}$$

where the coefficients are α_1 and α_n. The order n is determined by the strength of nonlinearity. When the *B-H* curve of core material used in the ring core of Fig. 1 is represented in Fig. 2, the order n is 31. Using the cross section S and magnetic path length l, Eq. (2) can be rewritten as follows:

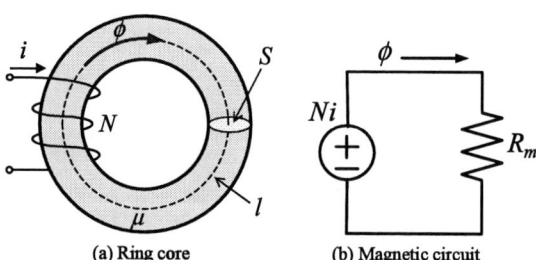

(a) Ring core   (b) Magnetic circuit

Fig. 1 Ring core with a winding of N turns and its magnetic circuit.

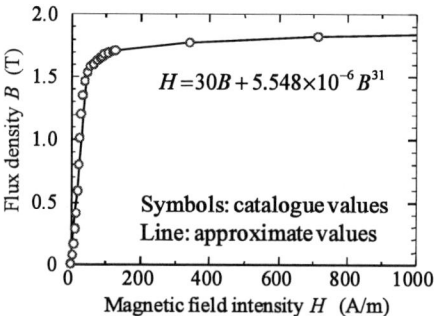

$$H = 30B + 5.548\times10^{-6}B^{31}$$

Symbols: catalogue values
Line: approximate values

Fig. 2 B-H curve of core material and its approximate curve.

$$\frac{Ni}{l} = \alpha_1 \frac{\phi}{S} + \alpha_n \left(\frac{\phi}{S}\right)^n. \qquad (3)$$

Consequently,

$$Ni = \left(\frac{\alpha_1 l}{S} + \frac{\alpha_n l}{S^n} \phi^{n-1}\right)\phi. \qquad (4)$$

In Eq. (4), the nonlinear formula enclosed in the parenthesis indicates the reluctance R_m.

Next, let us consider the circuit shown in Fig. 3. The ac power source supplies alternating current to the winding of the ring core. In the circuit, the following equation applies:

$$v = ri + N\frac{d\phi}{dt}, \qquad (5)$$

where the circuit resistance including the winding is r. Substituting Eq. (4) into Eq. (5), the differential equation for the flux is given by,

$$N\frac{d\phi}{dt} + \frac{r}{N}\left(\frac{\alpha_1 l}{S} + \frac{\alpha_n l}{S^n} \phi^{n-1}\right)\phi = v \qquad (6)$$

Thus, if the voltage v is given, the flux ϕ can be obtained by solving the above equation with Runge- Kutta method, and then the current i is calculated when the obtained flux is substituted into Eq. (4). In the case of the simple magnetic circuit mentioned above, it can be easily solved. However, it is difficult to solve it with Runge- Kutta method if the magnetic circuit is more complicated or has the strong magnetic nonlinearity. In that case, a general purpose circuit simulator like "SPICE" becomes a useful tool.

Fig. 4 illustrates the SPICE model of nonlinear reluctance. As shown in the figure, the linear term in Eq. (4) can be expressed as the linear resistance and the nonlinear one is represented by the nonlinear controlled voltage source. Furthermore, using SPICE, the nonlinear inductor circuit shown in Fig. 3 can be expressed as the electric- and magnetic-coupled model shown in Fig. 5. As shown in the figure, the electric circuit is combined with the nonlinear magnetic circuit using the two controlled voltage sources E_1 and H_1. In this coupled model, when the voltage v is given, the winding current i is calculated based on the electric circuit. The winding current i gives the MMF Ni of the controlled source H_1, and then the flux ϕ in the magnetic circuit is calculated. Accordingly, the counter electromotive force e' of the controlled source E_1

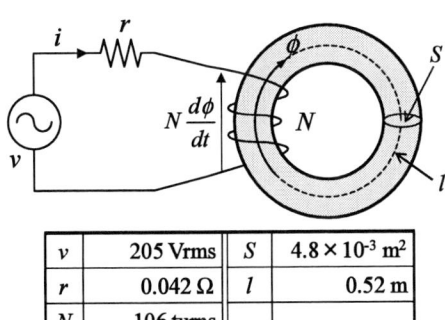

Fig. 3 Circuit configuration of a nonlinear inductor.

v	205 Vrms	S	4.8×10^{-3} m^2
r	0.042 Ω	l	0.52 m
N	106 turns		

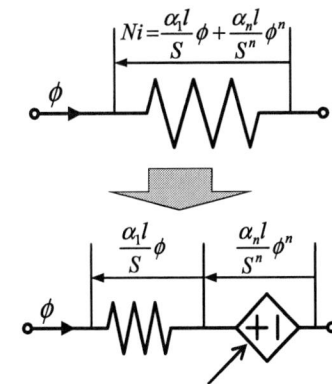

Fig. 4 SPICE model of nonlinear reluctance.

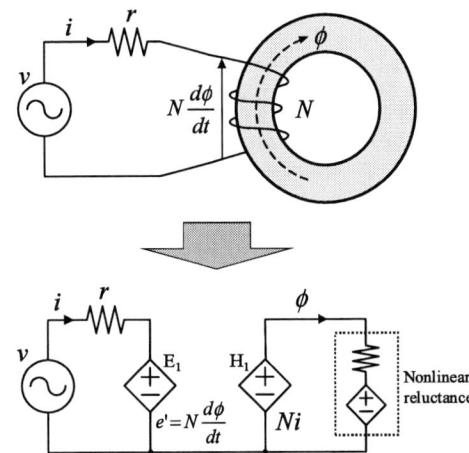

Fig. 5 Electric and magnetic coupled model of the nonlinear inductor.

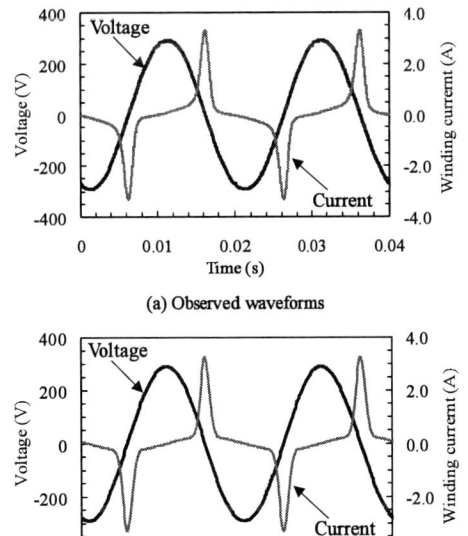

(a) Observed waveforms

(b) Calculated waveforms.

Fig. 6 Exciting voltage and winding current of the nonlinear inductor.

is obtained from the differential of flux linkage. All the above calculation can be carried out simultaneously on SPICE.

Fig. 6 shows the observed and calculated waveforms of the exciting voltage and winding current. From the figure, it is seen that the calculation result agrees well with the observed one. The peak of the winding current caused by magnetic saturation is reproduced by the SPICE model. It is understood that the consideration of magnetic nonlinearity and the coupled analysis are successful when SPICE is used as the solver for the magnetic circuit. In the following paragraph, a case study of dynamic simulation of a PM motor drive system based on RNA is presented.

III. DYNAMIC SIMULATION OF PM MOTOR DRIVE SYSTEM BASED ON RNA

Fig. 7 shows the specifications of the three-phase four-pole PM motor used in the consideration. The core material is non-oriented silicon steel with a thickness of 0.5 mm. The material of the permanent magnet is ferrite with the residual magnetic flux density B_r and coercive force H_c are 0.38 T and 294 kA/m, respectively. Fig. 8 is the photograph of the stator and rotor as viewed from top. It is seen that the test PM motor has an overhang structure that the axial length of the rotor is 17.8 mm longer than that of the stator stack.

First of all, the PM motor is divided into multiple elements as shown in Fig. 9(a). The numbers of stator and rotor divisions in a circular direction are 48 and 96, respectively. Each divided element is expressed in a two-dimensional unit magnetic circuit as shown in Fig. 9(b). The reluctances R_r and R_θ in the unit magnetic circuit are found by the dimensions of the element and permeability of core material [5].

Fig. 10(a) illustrates the arrangement of the stator windings. It is seen that the five stator poles are wound together with one U-phase coil. Accordingly, the MMFs by the winding currents are arranged in these stator poles including slot spaces according to the winding distribution as shown in Fig. 10(b).

Fig. 11 shows the expanded view of the RNA model obtained in the way described above. In the figure, MMFs by the winding currents are Ni_u, Ni_v, and Ni_w, respectively. The internal reluctance of the magnet is R_p, which is found

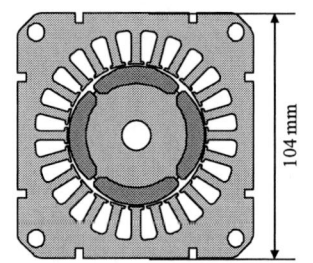

Number of turns/phase:	480 turns
Winding resistance/phase:	8.46 Ω
Stator and rotor stacks:	35.0 mm
Axial length of the magnet:	52.8 mm
Permanent magnet (ferrite)	
Coercive force:	294 kA/m
Residual magnetic flux density:	0.38 T

Fig. 7 Specifications of the three-phase four-pole PM motor.

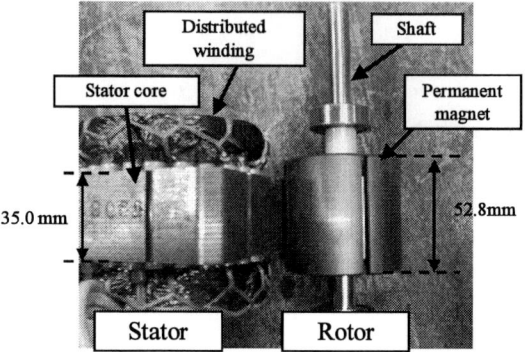

Fig. 8 Photograph of the stator and rotor as viewed from top.

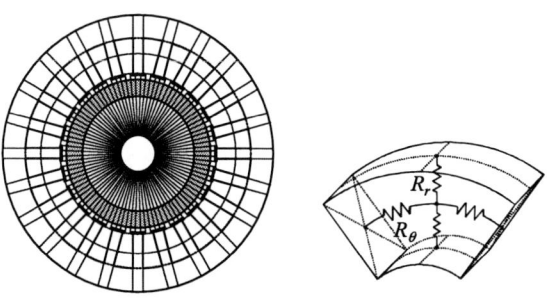

(a) Division of the motor   (b) Unit magnetic circuit.

Fig. 9 Divisions on the PM motor and the unit magnetic circuit.

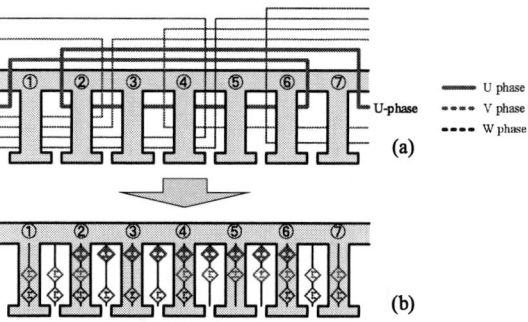

Fig. 10 Schematic diagram of the stator windings and the arrangement of MMFs by winding currents.

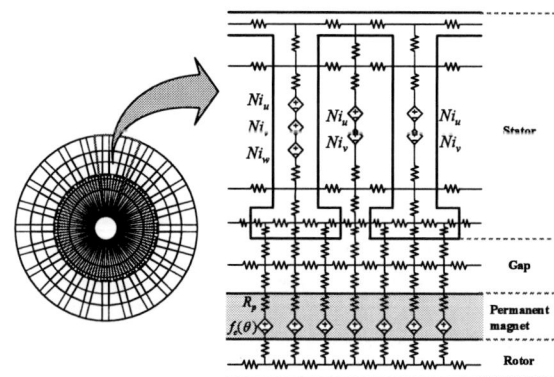

Fig. 11 Expanded view of the RNA model of the PM motor.

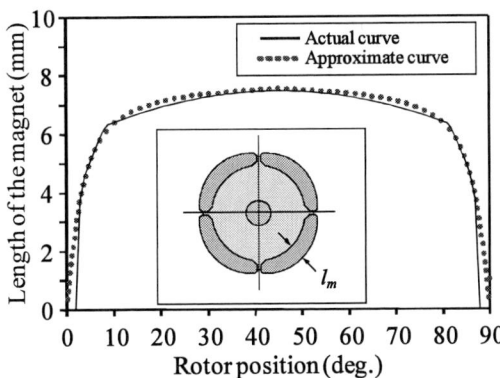

Fig. 12 Length of the magnet corresponding to the rotor position.

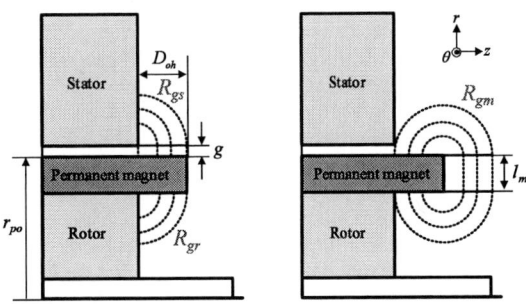

Fig. 13 Magnetic paths around the permanent magnet (left: magnet to both cores, right: magnet to the air).

by the dimensions and recoil permeability of the permanent magnet. The MMFs by the magnet are defined by

$$f_c = H_c l_m , \qquad (7)$$

where the length of the magnet is l_m, which changes with the rotor position θ as shown in Fig. 12. Thus, the MMF f_c is given by the following function in order to adopt the rotary motion in the RNA model:

$$f_c(\theta) = H_c l_m(\theta)$$
$$= H_c \times a \tan^{-1}(b \sin 2\theta) , \qquad (8)$$

where the coefficients a and b are 4.99 and 9.0, respectively.

Three-dimensional flux distribution around the permanent magnet must be considered in the RNA model because the test PM motor has the overhang structure. In this paper, let us assume three magnetic paths expressed approximately by a circular arc and straight line as shown in Fig. 13 [5], [6]. Consequently, the reluctances R_{gs}, R_{gr}, and R_{gm} are obtained from:

$$R_{gs} = \frac{n_\theta g}{2\pi\mu_0 r_{po} D_{oh}} + \frac{n_\theta}{8\mu_0(r_{po} + g)} \qquad (9)$$

$$R_{gr} = \frac{n_\theta}{8\mu_0(r_{po} - l_m)} \qquad (10)$$

$$R_{gm} = \frac{n_\theta l_m}{2\pi\mu_0 D_{oh}\left(r_{po} - l_m/2\right)} + \frac{n_\theta(2r_{po} - l_m)}{4\mu_0 r_{po}(r_{po} - l_m)} \qquad (11)$$

where the number of rotor divisions in a circular direction is n_θ.

Fig. 14 shows the electromagnetic and motion coupled model of the PM motor. The proposed coupled model consists of three different blocks as follows: *Motor drive circuit*, *RNA model of PM motor*, and *Motion calculation circuits*. In the coupled model, when the rotor position θ is given, the controller generates gate signals of the motor drive circuit, and the phase currents i_u, i_v, and i_w are calculated. The phase currents give MMFs Ni_u, Ni_v, Ni_w in the RNA model, and then flux distribution in the PM motor is calculated. The counter electromotive forces e_u', e_v', and e_w' are obtained from the flux linkage. The motor

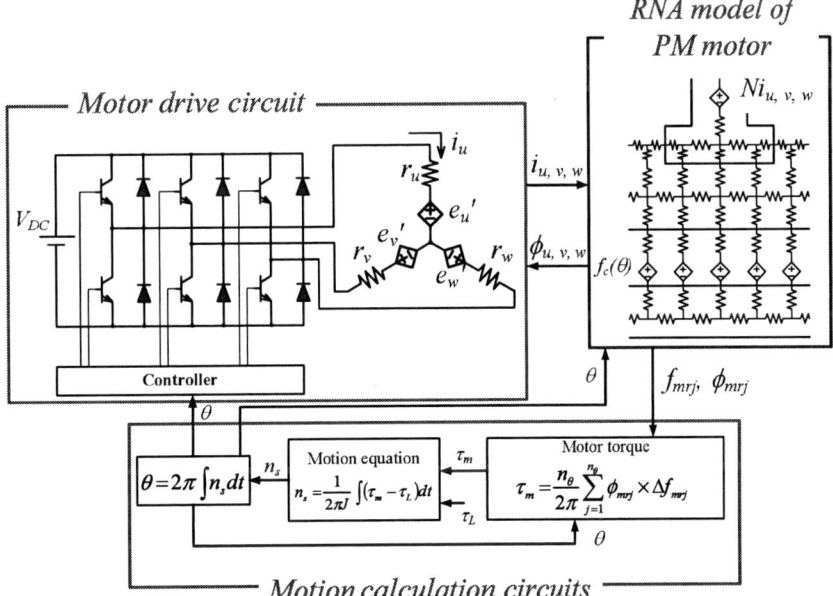

Fig. 14 Electromagnetic and motion coupled model of the PM motor.

761

torque τ_m is calculated based on magnetic energy [7]. When the load torque τ_L is given, the rotational speed n_s is obtained from the motion equation. Finally, an integral of the rotational speed gives the rotor position. All the above calculation can be performed simultaneously on SPICE.

Fig. 15 shows the calculated starting characteristics of the PM motor. A load torque increases linearly from 0 to 2.0 N·m between 150 and 250 ms. It is seen from the figure that the phase currents, motor torque, and the rotational speed increase sharply after starting the motor, and then these are changed gradually according to the load torque, and come closer to a steady state. Using the proposed coupled model, the dynamic characteristics including the starting and sudden change in load can be calculated.

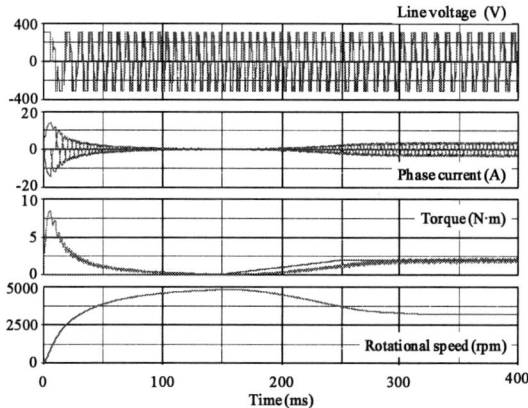

Fig. 15 Staring characteristics of the PM motor.

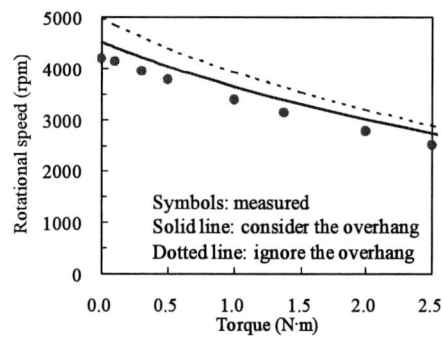

(a) Rotational speed characteristics

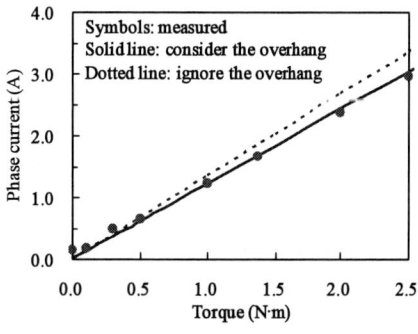

(b) Phase current characteristics.

Fig. 16 Measured and calculated characteristics of the PM motor.

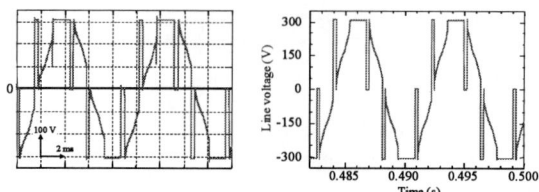

(a) Waveforms of exciting voltage.

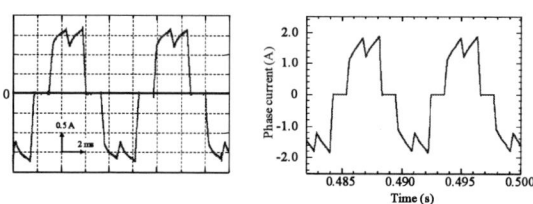

(b) Waveforms of phase current.

Fig. 17 Observed (left) and calculated (right) waveforms at a load torque of 1.0 N·m.

Fig. 16(a) shows the rotational speed versus torque characteristics. In the figure, the symbols show the measured values, and the solid and broken lines indicate the calculated ones with and without considering the overhang. Fig. 16(b) shows the phase current versus torque characteristics. From the figures, it is clear that the calculated values with considering the overhang agree well with the measured ones. Fig. 17 shows the observed and calculated waveforms of the exciting voltage and phase current at a load torque of 1.0 N·m.

IV. Conclusion

This paper presented a method for dynamic simulation of a PM motor drive system based on RNA. It was clearly indicated that the dynamic behavior of the PM motor including the starting and sudden change in load can be calculated by RNA and the calculation accuracy is also high.

The proposed method will be one of the powerful CAE tools for optimization of the electric motors including its drive circuit.

References

[1] J. M. Kokernak, D. A. Torrey, Magnetic circuit model for the mutually coupled switched-reluctance machine, IEEE Trans. Magn., 36, (2000), 500-507.

[2] M. Cheng, K. T. Chau, C. C. Chan, E. Zhou, X. Huang, Nonlinear varying-network magnetic circuit analysis for doubly salient permanent-magnet motors, IEEE Trans. Magn., 36, (2000), 339-348.

[3] K. Nakamura, K. Kimura, O. Ichinokura, Electromagnetic and Motion Coupled Analysis for Switched Reluctance Motor Using a Nonlinear Magnetic Circuit Model, 11th Power Electronics and Motion Control Conference (EPE-PEMC 2004), Riga, Latvia, (2004), A52267.

[4] V. Karapetoff, The Magnetic Circuit, McGraw-Hill, New York, (1911).

[5] K. Nakamura, M. Ishihara, O. Ichinokura, Reluctance Network Analysis Model of a Permanent Magnet Generator Considering an Overhang Structure and Iron loss, Proc. 17th Int. Conf. on Electrical Machines, Chania, Greece, (2006), PSA1-16.

[6] H. C. Roters, Electromagnetic Devices, John Wiley & Sons, Inc., New York, (1941).

[7] T. Sashida, T. Kenjo, An Introduction to Ultrasonic Motors, Oxford University Press, (1993), 191-228.

Performance Improvement of Direct Torque Controlled Interior Permanent Magnet Synchronous Motor Drive by Considering Magnetic Saturation

Behrooz Majidi[*†]   Jafar Milimonfared[*]   Kaveh Malekian[*]

*Amirkabir University of Technology (Tehran Polytechnic),
Center of Excellence in Power Systems, No. 424, Hafez Ave., Tehran 15914, Iran.

† Corresponding author: bmx@aut.ac.ir

Abstract—The influence of magnetic saturation on the maximum torque per ampere strategy in constant torque region and field weakening strategy in constant power region for a direct torque controlled interior permanent magnet synchronous motor drive are discussed in this paper. In other words, by considering magnetic saturation, all optimal strategies and motor-inverter limitations are derived in the $T - |\psi_s|$ plane to apply in the direct torque control method. These strategies that take magnetic saturation into account and determine the optimal torque and flux commands are derived and implemented in direct torque control method. Simulation results on a prototype interior permanent magnet motor are included to validate the usefulness of the work.

Index Terms—Direct torque control, field weakening strategy, interior permanent magnet synchronous motor, magnetic saturation, maximum torque per ampere strategy.

I. Introduction

RECENTLY, the Interior Permanent Magnet Synchronous Motor (IPMSM) is widely used in high performance applications due to some its advantageous such as high torque to current ratio, high power to current ratio, high efficiency, low noise, and robustness [1]-[3]. In high performance variable speed drive systems, the motor speed should closely follow specified reference trajectory regardless any load disturbances, parameter variations and any model uncertainties [4].

In addition, in some applications, wide-speed operation of electric drives is desired [5]-[8]. In other words, operation in both constant torque region and constant power region is needed. The optimal behavior of drive can be achieved by considering Maximum Torque per Ampere (MTPA) strategy below the base speed and Field Weakening (FW) strategy above the base speed. It should be noted that the MTPA and FW strategies must be modified in order to consider magnetic saturation. This paper focuses on optimal torque control of direct torque controlled interior permanent magnet synchronous machines capable of operating in both the constant torque and the constant power region.

In the Interior Permanent Magnet (IPM) motor, since the effective air-gap length on the d-axis is large and the relative magnetic permeability of the Permanent Magnet (PM) is close to unity, the variation of the corresponding magnetizing inductance, L_d, due to magnetic saturation, is minimal. On the contrary, the effective air-gap length on the q-axis is small, and, therefore, the saturation is significant [9], [10].

II. Motor Dynamic by Considering Magnetic Saturation

The mathematic model of IPMSM in the rotor reference frame can be obtained from synchronous machine model. Due to constant field produced by permanent magnets, the field variation is zero. In this paper it is assumed that:

- Magnetic saturation is considerable;
- Core losses are negligible;
- The induced EMF is sinusoidal; and
- There is no dampener winding on rotor.

Using these assumptions, the voltages and torque equations could be written as,

$$\begin{pmatrix} V_q \\ V_d \end{pmatrix} = \begin{pmatrix} R & \omega_{re}L_d \\ -\omega_{re}L_q & R \end{pmatrix}\begin{pmatrix} i_q \\ i_d \end{pmatrix} + \begin{pmatrix} i_q & 0 \\ 0 & i_q \end{pmatrix}\begin{pmatrix} pL_q \\ pL_d \end{pmatrix}$$
$$+ \begin{pmatrix} L_q & 0 \\ 0 & L_d \end{pmatrix}\begin{pmatrix} pi_q \\ pi_d \end{pmatrix} + \begin{pmatrix} \omega_{re}\psi_f \\ 0 \end{pmatrix} \tag{1}$$

$$T_e = \frac{3P\psi_s}{4L_dL_q}[2\psi_f L_q \sin\delta - \psi_s(L_q - L_d)\sin 2\delta] \tag{2}$$

$$T_e = T_L + B\omega_{rm} + Jp\omega_{rm} \tag{3}$$

$$\omega_{rm} = \frac{1}{P}\omega_{re} \tag{4}$$

where, V_q, V_d are the q- and d-axis voltages; L_q, L_d are the q- and d-axis inductances; i_q, i_d are the q- and d-axis currents; R is the stator resistance per phase; ψ_f is the constant flux linkage due to rotor permanent magnet; ψ_s is the stator flux

978-1-4244-1741-4/08/$25.00 ©2008 IEEE

linkage; δ is the load angle; ω_{rm}, ω_{re} are the mechanical and electrical rotor angular speed, respectively; P is the number of pair poles; p is the differential operator; T_e is the electro-magnetic torque; T_L is the load torque; B is the viscous coefficient; and J is the inertia constant.

III. MAGNETIC SATURATION IN IPMSM

In IPM synchronous motor, the magnetic saturation on the high-inductance axis (q-axis) is significant. The L_q inductance varies depending on the q-axis component of the magnetizing current i_{mq}, which is identical to i_q current, and can be represented by [10]-[12],

$$L_q = L_q(i_q) \qquad (5)$$

where $L_q(i_q)$ is a nonlinear function of the i_q current that models the saturation effects. An acceptable approximation is that, two inductances are defined for the q-axis: an unsaturated inductance which applies for i_q current lower than i_{qs}, and a saturated one (linearly varied with the i_q current) for higher i_q current [13], [14].

$$L_q = \begin{cases} L_{qo}, & i_q < i_{qs} \\ L_{qo} - \beta(i_q - i_{qs}), & i_q > i_{qs} \end{cases} \qquad (6)$$

where β is a coefficient expressing the saturation effect. Unlike the L_q inductance, the variation of L_d according to i_d is negligible and, therefore, L_d can be considered constant. Furthermore, mutual coupling effects between q- and d-axis could be assumed negligible [10].

The variation of L_q with respect to i_q, considered in a three phase IPM motor as illustrated in Fig. 1. The parameters of the IPM motor are given in Table I.

IV. DIRECT TORQUE CONTROL PRINCIPLE BY CONSIDERING MAGNETIC SATURATION

Fig. 2 shows the block diagram of a DTC-based ac motor drive. As shown in this figure, a switching table is used for inverter control such that the torque and flux errors are kept within the specified bands.

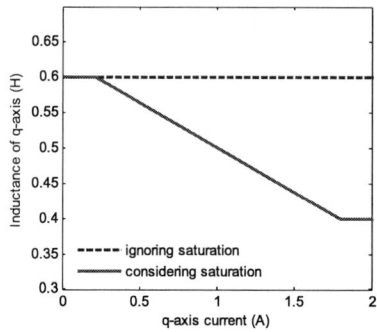

Fig. 1. The variation of q–axis inductance (L_q) with respect to the q–axis current (i_q).

TABLE I
MOTOR PARAMETERS

Rated voltage	240 V
Rated frequency	60 Hz
Number of pair poles	2
d-axis inductance	0.375 H
q-axis inductance	$\begin{cases} 0.601, H & i_q < (0.21A) \\ 0.601 - 0.1258(i_q - 0.21), H & i_q > (0.21A) \end{cases}$
Stator resistance	19.4 Ω
Magnetic flux constant	0.447 v/rad/s
Rotor inertia	0.8×10^{-3} kgm^2

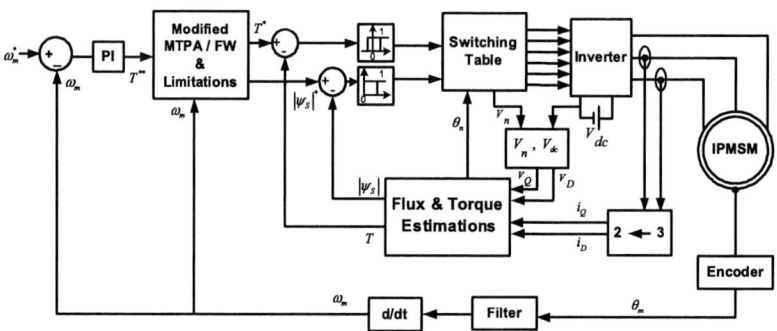

Fig. 2. Block diagram of the proposed direct torque controlled IPMSM drive.

Compared to the conventional current vector control method, the DTC scheme has the following features.

- There are no current control loops, hence, the current is not regulated directly;
- Coordinate transformation is not required;
- There is no separate voltage pulsewidth modulator; and
- Stator flux vector and torque estimation is required.

In the current vector control method, since q- and d-axis current components are directly controlled, all control strategies and motor-inverter limitations are considered in the $i_q - i_d$ plane. Unlike current vector control, torque and stator flux are directly controlled in the direct torque control method. As a result, all control strategies and limitations in the $i_q - i_d$ plane must be transmitted into the $T - |\psi_s|$ plane. Because each point in the $i_q - i_d$ plane can be transmitted into its corresponding point in the $T - |\psi_s|$ plane using (7) and (8), all control strategies and limitations can be mapped into the $T - |\psi_s|$ plane point by point.

$$T = \frac{3}{2} P \left[\psi_f + \left(L_d - L_q(i_q) \right) i_d \right] i_q \qquad (7)$$

$$|\psi_s| = \sqrt{\left(L_d i_d + \psi_f \right)^2 + \left(L_q(i_q) i_q \right)^2} \qquad (8)$$

It should be noted that saturation of the q-axis inductance should be considered in (7) and (8) to achieve correct mapping. In the following sections, the current and voltage constraints as well as the optimal strategies in the constant torque and constant power regions are explained.

A. Constraints

Considering the voltage and current constraints, the armature voltages and currents could be written as,

$$I_a = \sqrt{i_d^2 + i_q^2} \leq I_{am} \qquad (9)$$

$$V_a = \sqrt{v_d^2 + v_q^2} \leq V_{am} \qquad (10)$$

where I_{am} is continuous armature current rating in continuous operation or maximum available current of the inverter. The maximum voltage V_{am} is the maximum available output voltage of the inverter. The critical condition of (9) (i.e., $I_a = I_{am}$) is given by the current limit circle in the $i_q - i_d$ plane, which is independent with respect to the magnetic saturation, as shown in Fig. 3(a). For each (i_q, i_d) pair satisfying (9), torque and $|\psi_s|$ can be found and plotted in the $T - |\psi_s|$ plane. Figs. 3(b)-(c) indicate the current limit trajectory by ignoring and considering saturation while mapping, respectively.

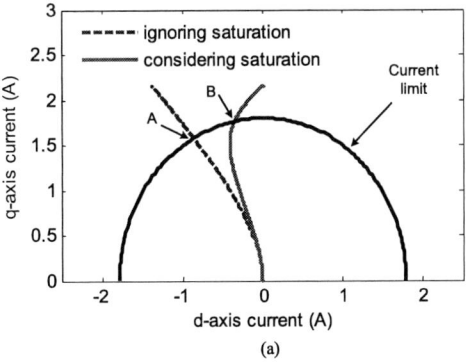

(a)

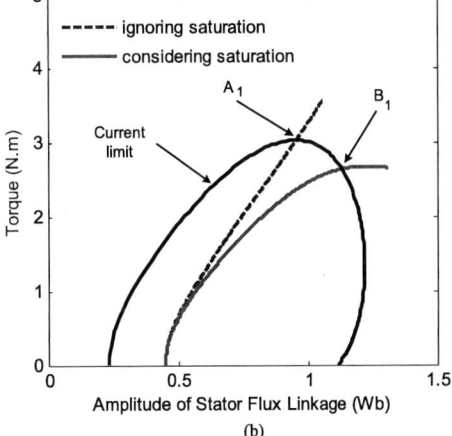

(b)

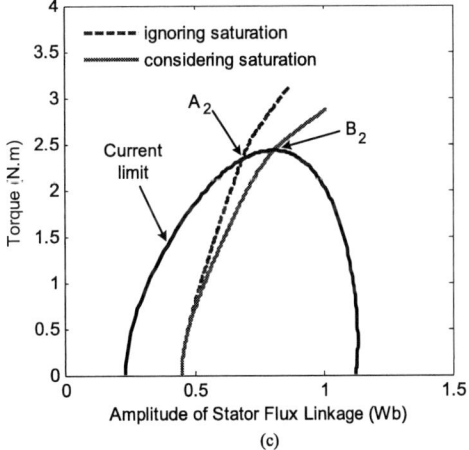

(c)

Fig. 3. The conventional MTPA and modified MTPA strategies as well as the current limit: (a) in the $i_q - i_d$ plane; (b) mapped into the $T - |\psi_s|$ plane by ignoring saturation while mapping; and (c) mapped into the $T - |\psi_s|$ plane by considering saturation while mapping.

B. Control in Constant Torque Region

The most popular strategy in the constant torque region is maximum torque per ampere. The MTPA strategy must be modified in order to consider magnetic saturation. The relationships between i_q and i_d which satisfy MTPA by ignoring and considering saturation are given by (10) and (11), respectively [8], [15].

$$i_d = \frac{\psi_f}{2(L_q - L_d)} - \sqrt{\frac{\psi_f^2}{4(L_q - L_d)^2} + i_q^2} \tag{10}$$

$$i_d = \frac{\psi_f - \sqrt{\psi_f^2 + 4i_q^2(L_q - L_d)\left[(L_q - L_d) - (L_q')i_q\right]}}{2\left[(L_q - L_d) - (L_q')i_q\right]} \tag{11}$$

Considering (6), L_q' could be written as

$$L_q' = \begin{cases} 0 & i_q < i_{qs} \\ -\beta & i_q > i_{qs} \end{cases} \tag{12}$$

Fig. 3(a) illustrates the MTPA trajectories by considering saturation and ignoring saturation in the $i_q - i_d$ plane. Figs. 3(b)-(c) show both conventional and modified MTPA trajectories mapped into the $T - |\psi_s|$ plane by ignoring and considering saturation while mapping, respectively.

The command vector producing maximum torque, T_{max}, is the cross point of the MTPA trajectory and the current-limit, which corresponds to points A and B for traditional MTPA and modified MTPA strategies, respectively, in all Figs. 3(a)-(c). The maximum torque corresponding to the conventional MTPA strategy mapped by ignoring saturation (corresponding to point A_1) can not be produced because a real IPMSM magnetically saturates, however, it is greater than the maximum torque corresponding to the modified MTPA mapped by considering saturation (corresponding to point B_2). In other words, by considering effect of saturation on both derivation of MTPA relationship and map of this relationship, the maximum possible torque can be achieved.

C. Control in Field Weakening Region

The torque capability in field weakening region is determined by both of the voltage and the current limits. In the steady state, the voltage constraint could be expressed as

$$V_o = \sqrt{v_{do}^2 + v_{qo}^2} \leq V_{om} \tag{13}$$

where,

$$v_{do} = -\omega_{re} L_q(i_q) i_q, \; v_{qo} = \omega_{re} L_d i_d + \omega_{re} \psi_f \tag{14}$$

$$V_{om} = V_{am} - RI_{am} \tag{15}$$

The "o" subscripts are defined to simplify the control algorithm. Fig. 4 illustrates voltage limits in both considering saturation and ignoring saturation in $i_q - i_d$ plane. Figs. 4(b)-(c) show both conventional and modified voltage limit

trajectories mapped into the $T - |\psi_s|$ plane by ignoring and considering saturation while mapping, respectively.

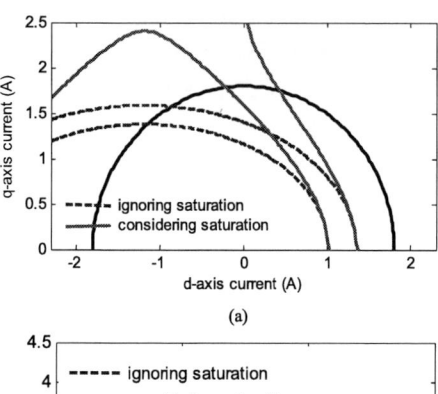

(a)

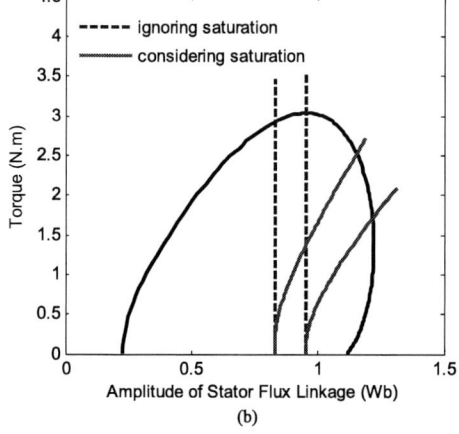

(b)

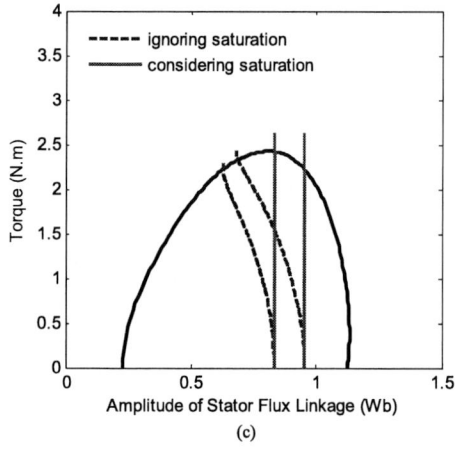

(c)

Fig. 4. The conventional and modified voltage limits as well as the current limit: (a) in the $i_q - i_d$ plane; (b) mapped into the $T - |\psi_s|$ plane by ignoring saturation while mapping; and (c) mapped into the $T - |\psi_s|$ plane by considering saturation while mapping.

As shown in Fig. 4(c), the modified voltage limits mapped by considering saturation are vertical lines in the $T-|\psi_s|$ plane. Also, a given relationship in the $T-|\psi_s|$ plane can be derived for the voltage limit, which verifies the modified voltage limit trajectories in Fig. 4(c).

Combining (13) and (14) yields,

$$\left(-L_q(i_q)i_q\right)^2 + \left(L_d i_d + \psi_f\right)^2 = \left(\frac{V_{om}}{\omega_{re}}\right)^2 \quad (16)$$

So

$$|\psi_s| = \frac{V_{om}}{\omega_{re}} \quad (17)$$

This equation is independent with respect to the magnetic saturation. As a result, the modified voltage limit are, always, vertical lines in the $T-|\psi_s|$ plane, regardless of motor saturates or how variation of L_q with respect to i_q is.

When the rotor speed is below the base speed, the voltage limit line is on the right side of the intersection of the MTPA and current limit trajectories (point B$_2$) and, therefore, the voltage limit is always satisfied with MTPA trajectory control. When the rotor speed is increased above the base speed, voltage limit line moves to left and the stator flux linkage should be reduced according to (17) for FW operation. In other words, for operation above the base speed, the amplitude of the stator flux linkage is approximately inversely proportional to the rotor speed.

V. SIMULATION RESULTS

The complete proposed IPMSM drive, as shown in Fig. 2, has been simulated using Matlab/Simuink for the prototype IPMSM of Table I.

The direct torque controlled IPMSM drive incorporating traditional MTPA and FW strategies has been also simulated in order to compare the performance to those obtained from the proposed drive system. In order to make a fair comparison, the same limitations are considered for both of them. Also, several tests have been performed to evaluate the performance of the proposed IPMSM drive system.

The simulated responses are shown in Figs. 5 (a)-(b) for both of the conventional and modified direct toque controlled drive systems, to see the starting performance as well as the response with a step change in the command speed. The drive system is started with the speed reference set at 400 rpm. It is seen from Fig. 5(a) that the proposed drive can follow the command speed within 0.1 Sec., whereas the conventional direct toque controlled drive follow the command speed within 0.12 Sec. because, as shown in Fig. 5(b), the maximum available torque is less in the conventional system. At $t = 0.3$ Sec., command speed changed into 800 rpm. In this case, the modified system has better performance as well.

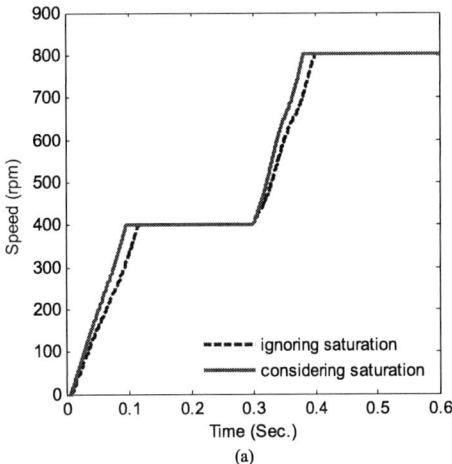

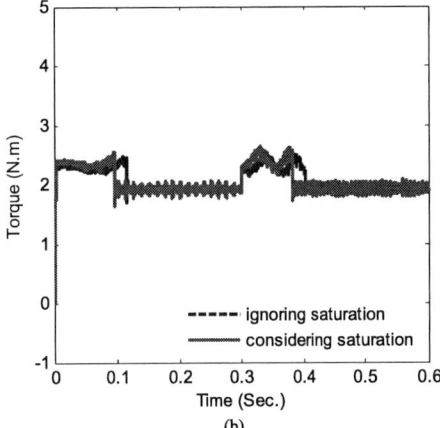

Fig. 5. Simulated responses of the both conventional and modified drive systems: (a) speed responses and (b) torque responses.

In another simulation test, the proposed drive performance including speed, torque are observed under different operating conditions such as sudden change in load and step change in command speed over wide speed range. The proposed drive system is started at no load condition with the rotor speed of 800 rpm. At $t = 0.1$ Sec., a step change in load torque (1.1 N.m) is applied to the motor shaft, and at $t = 0.2$ Sec., the command speed changes into 2000 rpm (above the base speed). The proposed drive responses are shown in Fig. 6(a)-(b). It is shown that both drive systems are also capable of following the command speed above the base speed, but dynamic of responses for the modified drive system is better than ones obtained for the conventional system.

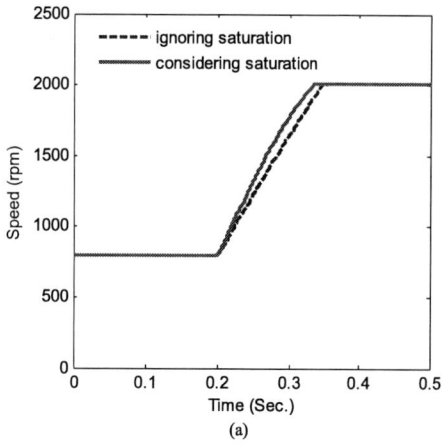

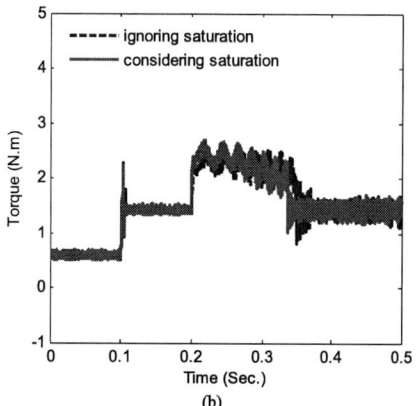

Fig. 6. Simulated responses of the both conventional and modified drive systems with respect to sudden change in load and step change in speed command above the base speed: (a) speed responses and (b) torque responses.

VI.   CONCLUSION

Considering magnetic saturation, the modified MTPA and FW strategies as well as drive limitations have been derived in the $T-|\psi_s|$ plane to apply in the direct torque control method. The proposed direct torque controlled IPMSM drive has been simulated for an IPM motor. The validity of the proposed IPMSM drive has been established in simulation at different operating conditions by considering saturation. In order to prove the superiority of the proposed controller, a performance comparison with the conventional direct torque controlled IPMSM drive has also been provided. The simulation results show ability of the proposed technique at different operation condition such as sudden load change and step change of speed (over wide speed range).

REFERENCES

[1]   G.R. Slemon, *Electric Machines and drives*, Addison-Wesley Publication Company, 1992, pp. 503-511.

[2]   J. K. Gieras and M.Wing, *Permanent Magnet Motor Technology: Design and Applications*, New York: Marcel-Dekker, 1997.

[3]   P. Vas, *Sensorless Vector and Direct Torque Control*, Oxford, 1998, pp. 87-90.

[4]   K. Malekian and J. Monfared, "A Genetic Based Fuzzy Logic Controller for IPMSM Drive over Wide Speed Range," *Electric Machines & Drives Conference, IEMDC '07, IEEE International*, vol. 1, pp. 847–853, 3-5 May 2007.

[5]   T. M. Jahns, "Flux- weakening Regime Operation of an Interior Permanent-magnet Synchronous motor Drive," *IEEE Trans. Ind. Appl.*, vol. IA-23, no.4, pp.681~689, 1986.

[6]   B. K. Bose, "A High-Performance Inverter-Fed Drive System of an Interior Permanent Magnet Synchronous Machine", *IEEE Trans. Ind. Appl.*, vol. IA-24, no.6, pp.987~997, 1988.

[7]   S.R. MacMinn and T. m. Jahns, "Control Techniques for Improved High-Speed Performance of Interior PM Synchronous Motor Drive," *IEEE Trans. Ind. Appl.*, vol. IA-27, no.4, pp.997~1004, 1991.

[8]   S.Morimoto, M Sanada and Y. Taketa, "Wide-Speed Operation of Interior Permanent Magnet Synchronous Motor with High-Performance Current Regulator," *IEEE Trans. Appl.*, vol. IA-30. no.4, pp.920-926, 1994.

[9]   S. A. Nasar, I. Boldea, and L. E. Unnewehr, *Permanent Magnet, Reluctance, and Self-Synchronous Motors. Florida*: CRC Press, Inc., 1993.

[10] B. J. Chalmers, S. A. Hamed, and G. D. Baines, "Parameters and performance of a high-field permanent-magnet synchronous motor for variable-frequency operation," *IEE Proc.*, pt. B., vol. 132, no. 3, pp.117–124, May 1985.

[11] N. Bianchi and S. Bolognani, "Parameters and volt–ampere rating of a synchronous motor drive for flux-weakening applications," *IEEE Trans. Power Electronics*, vol. 12, no. 5, pp. 895–903, Sept. 1997.

[12] S. Morimoto, Y. Takeda, T. Hirasa, and K. Taniguchi, "Expansion of operating limits for permanent magnet motor by current vector control considering inverter capacity," *IEEE Trans. Ind. Appl.*, vol. 26, no. 5, pp. 866–871, Sept./Oct. 1990.

[13] B. J. Chalmers, "Influence of saturation in brushless permanent-magnet motor drives," *IEE Proc.*, pt. B, vol. 139, no. 1, pp. 51–52, Jan. 1992.

[14] B. J. Chalmers, R. Akmese, and L. Musaba, "Validation of procedure for prediction of field-weakening performance of brushless synchronous machines," *in Proc. ICEM 1998*, vol. 1, Istanbul, Turkey, pp. 320–323.

[15] C. Mademlis and V. G. Agelidis, "On Considering Magnetic Saturation with Maximum Torque to Current Control in Interior Permanent Magnet Synchronous Motor Drives," *IEEE Trans. on Energy Conversion*, vol. 16, no. 3, Sep. 2001.

Condition Monitoring for Mechanical Faults in Fully Integrated Servo Drive Systems

Jesus Arellano-Padilla*, Mark Sumner and Chris Gerada

University of Nottingham/School of Electrical & Electronic Engineering, Nottingham, UK,

e-mail: * *eexja@nottingham.ac.uk, Mark.Sumner@nottingham.ac.uk, Chris.Gerada@nottingham.ac.uk*

Abstract—A modified motor current signature analysis (MCSA) approach to monitor the performance of a servo drive system using only one set of current transducers at the input side of the rectifier is proposed. The paper introduces a mathematical analysis to determine how the harmonic frequencies related to a fault propagate from the motor to the input inverter, and includes the effect of the DC Link components. Experimental results are presented for a Permanent Magnet servo drive system with faulty bearings to validate the proposed scheme.

Keywords—Prognosis, Diagnostics, Integrated adjustable speed drive.

I. INTRODUCTION

Typical motor drive systems consist mainly of two components as integral units: an electrical machine and a drive which is a voltage-source-input (VSI)-fed PWM inverter. In general, faults in motor drives can be classified into three different groups: Growing faults with only small effects on the operation, partial non-catastrophic faults with emergency operation possible, and catastrophic faults with total motor drive breakdown. These faults are either mechanical or electrical in origin. Electrical faults are usually related to the drive and the machine, while mechanical faults are related exclusively to the machine.

Condition monitoring avoids severe economical losses resulting from unexpected failures and improves the system reliability by providing warnings when incipient faults are detected. This opens the possibility of scheduling future preventive maintenance and repair work before a catastrophic failure occurs. According to an IEEE, sponsored survey [1], three quarter of all motor failure related problems can be prevented with early diagnostics and proper corrective actions. Motor-signature-current-analysis or MSCA is traditionally used to detect problems in the machine using measured motor currents; however it is not suitable for monitoring the power converters. This paper proposes a modified MSCA scheme to detect mechanical/electrical problems in a drives system using measured supply currents (rectifier currents).

II. FAULTS IN MOTOR DRIVE SYSTEMS

Although modern electrical machines are reliable and simple in construction, faults can occur in any of their three main components: stator, rotor and bearings. These faults produce air-gap eccentricity problems [2]. In the static air-gap eccentricity the position of the minimal radial air-gap length is fixed in space, i.e., is only space

dependant. Examples of this are: ovality of the core, and incorrect positioning of the stator and rotor. For the case of dynamic air-gap eccentricity, the centre of the rotor is not at the centre of the rotation and the minimum air-gap rotates with the rotor. It is said that dynamic eccentricity is time and space dependent. Dynamic eccentricity can be caused by a bent rotor shaft, wear of bearings, misalignment of bearings and mechanical resonances at critical speeds. Problems related to air-gap eccentricity produce unbalanced air-gap voltages and line currents which produces an increment in torque pulsations, reducing generated torque and efficiency [2].

III. CONDITION MONITORING FOR SERVO DRIVE SYSTEMS

Traditionally fault detection and prediction has been aimed mainly at the machine where several approaches are been proposed to detect faults such as broken bars, stator windings inter-turn problems, phase to phase or phase to ground short circuits, bearings failures, air gap eccentricity [3-6]. Special attention has been devoted to non-invasive methods, which are capable of detecting faults using measured data without interfering with the machine or its structural parts [2].

MSCA is considered one of the most promising fault detection methods, as it permits the detection of several common machines faults related to air-gap eccentricity. The MSCA approach is based in detecting the stator line currents and by using a frequency spectrum analyzer, detect the presence of air-gap asymmetries [5]. The conventional MSCA scheme is presented in Fig.1a. It is observed that the approach requires a set of current transducers mounted on the motor side, providing this way the ability to monitor the machine. The basic principle of MSCA is that when a machine presents air-gap eccentricity, side band components appear around the slot harmonics in the stator line current frequency spectra [2].

In general the frequency components in the stator currents of an induction machine which are due to the air-gap eccentricity can be obtained as

$$f_e = f_1 [(kZ_2 \pm n_d)(1-s)/P \pm v] \qquad (1)$$

where f_1 is the fundamental stator frequency, k is any integer, Z_2 is the number of rotor slots, and n_d is the eccentricity order number, which for static eccentricity is $n_d = 0$ and for dynamic eccentricity is $n_d = 1$. Furthermore, in (1) s is the slip, P is the number of pole-pairs, and v is the harmonic of the stator m.m.f. time harmonics, $v = \pm 1$, $\pm 3, \pm 5, \pm 7$, etc. One of the most common reasons for failure in electrical machines (related to dynamic eccentricity) is related to faults in bearings. An incipient fault in a rolling-element bearing can be for example a

978-1-4244-1741-4/08/$25.00 ©2008 IEEE

crack on the inner or outer race, or in the rolling element itself. This crack produces small impulses at every instant when one of the rolling elements passes over it [7]. The Characteristic fault frequencies which can be detected in the motor currents [8] are modulated by the electrical supply and are predicted by:

$$F_{BNG} = \left| F_E \pm m \cdot F_V \right| \qquad (2)$$

where F_{BNG} is the resulting fault frequency component in the stator current, F_E is the electrical supply frequency, F_V is one four characteristic faults frequencies and m is an integer. The major disadvantage of current-based bearing monitoring is that typically bearing fault signatures are very small components in the stator current [9].

Literature related to the condition monitoring for drives (power converters) is practically inexistent. Modern inverters have fault protection system built in which offer very basic protection, however drive monitoring systems that can detect incipient faults and reasonably prevent a total drive breakdown with system condition monitoring are still required. A novel condition monitoring scheme for a motor drive system including power converters was proposed in [10]. This approach requires several voltage transducers to monitor the output voltages and DC link, in addition to the current transducers to determine motor phase currents. The main disadvantage of the scheme proposed in [10] compared to traditional MSCA and the scheme proposed in this paper, is that it requires a large number of transducers which complicates the monitoring scheme.

IV. Considerations For the Proposed Approach

Traditional MSCA requires a set of current transducers mounted on the motor side, providing the ability to monitor the machine (see Fig.1a). Unfortunately, this approach is not suitable for monitoring problems in the power converters since current controllers in the system can reduce the effect of faults. Therefore problems such as supply unbalance, open circuit faults in diodes and/or switches may pass undetected for certain time until a major fault happens as consequence. This paper proposes a condition monitoring scheme for a motor drive system based in MSCA which can monitor both, the electrical machine and the power converter.

The proposed approach is based in the propagation of harmonic content through the drive, which is now explained.

It is well known that power converters generate significant amount of harmonics and inter harmonic currents on the input and the output side, and they transfer from one side to the other. Inter harmonics are defined as steady-state currents which are not integer multiples of the fundamental frequency. It is therefore assumed that in a VSI-fed adjustable speed drive, the frequency content at the supply utility side (input of the rectifier) will consists of a contribution of characteristic harmonics and inter harmonics of both converters (rectifier and inverter) and the machine. As shown in Fig.1b, a traditional motor drive system contains a three-phase diode bridge rectifier, a dc-link filter consisting of a large shunt capacitor and a series inductance, and a three-phase VSI-fed PWM inverter. The link filter is normally selected to reduce the ripple in the dc-voltage and to improve current quality in the supply [11, 12]. The inverter produces three-phase PWM wave forms supplying the electrical machine for speed control. A simple 6-pulse diode rectifier is normally used. Due to the normal operation of the inverter, a high switching frequency is generated at the input of the inverter and the current becomes discontinuous with high harmonic content. Since the switching frequency is much higher than the input and output frequencies, the effects of the harmonics on the DC link voltage and supply are small. The dc-link filter should be designed to block or minimise the propagation of the switching frequency harmonics through the converters. This is discussed further with more detail.

Since the proposed approach requires continuous monitoring of the rectifier input currents to detect abnormalities which will appear in the current spectra together with the characteristic harmonics (generated by the static converters), it is necessary to define first the magnitude and order of those expected harmonics appearing due to the normal operation of the converters (i.e. with no fault present). It should be pointed out that the calculation of harmonic content at the input of a servo drive system is very complex; however literature states that a good approximation can be obtained by using the equivalent circuit depicted in Fig.2a, where the inverter and machine can be represented as an impedance and voltage supply [12].

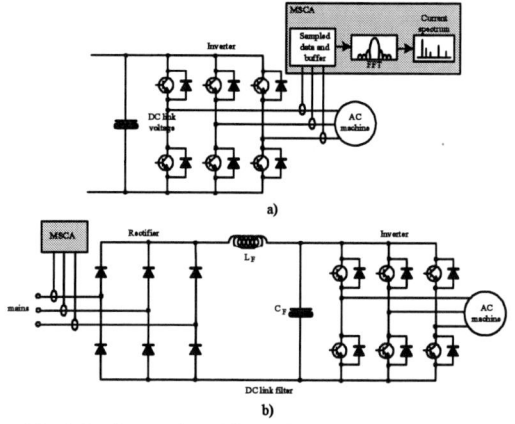

Fig. 1. Implementation of the MSCA approach. a) traditional implementation, b) Basic circuit of the VSI-fed motor drive system and proposed condition monitoring scheme.

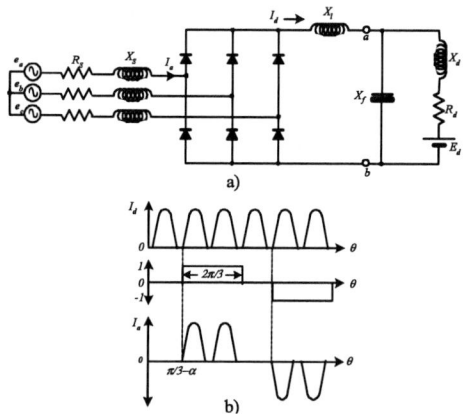

Fig. 2 Equivalent rectifier circuit. a) Diode bridge circuit with a L-C filter on the dc side, b) Current waveforms and rectifier switching function for discontinuous operation.

Symbols X_d, X_f, X_l, and X_s represent reactances with respect to the source frequency. The dc side current I_d in Fig.1a may be continuous or discontinuous, depending upon the values of circuit parameters.

A. Propagation of harmonics through the rectifier

As shown in Fig.2b, the ac side current waveforms are not sinusoidal as in the case of input current I_a. References [13, 14] propose a frequency domain analysis for the equivalent circuit by using rectifier switching functions. In [13] the case of the rectifier operating under continuous mode is considered, while in [14] both continuous and discontinuous modes are accounted. Note that most of the adjustable speed drive systems available operate under the discontinuous mode, therefore this is the operating mode considered in this paper. From inspection of Fig.2, the ac output side current I_a can be represented as:

$$I_a = S_a i_d = S_a (I_d + i_{dr}) \qquad (3)$$

where i_d is the discontinuous dc side current in Fig.2b, i_{dr} is the ac component injected into the rectifier circuit [14], while the rectifier switching function S_a which can be represented by the Fourier series is given by:

$$S_a = \sum_{n=1}^{\infty} \left(A_{sn} \cos n\theta + B_{sn} \sin n\theta \right) \qquad (4)$$

where

$$A_{sn} = \frac{2\sqrt{3}(-1)^l}{\pi} \cdot \frac{\sin n(u+\phi) + \sin n\phi}{n}$$

$$\qquad (5)$$

$$B_{sn} = \frac{2\sqrt{3}(-1)^l}{\pi} \cdot \frac{\cos n(u+\phi) + \cos n\phi}{n}$$

From (5) we have: $n = 6l \pm 1$ and $l = 0,1,2,\cdots, n > 0$, while u corresponds to the overlap angle and ϕ is the firing angle given by (6) where E_m is the peak value between terminals a-b in Fig.2a.

$$\phi = \sin^{-1} \left(\frac{I_d R_s}{\sqrt{3} E_m} \right) \qquad (6)$$

Note that from Fig.2b, the current i_d begins to flow at $(\pi/3 - \alpha)$, where α is equal to (7) and V_d is the ripple-free dc voltage between terminals a-b [14].

$$\alpha = \cos^{-1} \left(\frac{V_d}{\sqrt{3} E_m} \right) \qquad (7)$$

According to [14], the current i_a at the input of the rectifier is derived from (3) and (4) as:

$$i_a = \sum_{n=1}^{\infty} \left\{ (A_{in} + \Delta A_{in}) \cos n\theta + (B_{in} + \Delta B_{in}) \sin n\theta \right\} \qquad (8)$$

Where we have:

$$A_{in} = \frac{2\sqrt{3} I_d (-1)^{l+1}}{\pi} \cdot \frac{\sin n\gamma}{n}, \quad B_{in} = \frac{2\sqrt{3} I_d (-1)^l}{\pi} \cdot \frac{\cos n\gamma}{n}$$

$$\Delta A_{in} = \sum_{m=6}^{\infty} \frac{\sqrt{6} I_{dm}(-1)^{k+l+1}}{\pi} \left[\frac{\sin\{(m+n)\gamma - \lambda_m\}}{m+n} + \frac{\sin\{(m-n)\gamma - \lambda_m\}}{m-n} \right]$$

$$\Delta B_{in} = \sum_{m=6}^{\infty} \frac{\sqrt{6} I_{dm}(-1)^{k+l}}{\pi} \left[\frac{\cos\{(m+n)\gamma - \lambda_m\}}{m+n} - \frac{\cos\{(m-n)\gamma - \lambda_m\}}{m-n} \right]$$

with:

$$\gamma = \alpha - \pi/6$$

$$m = 6k \qquad (k = 1,2,3,\cdots)$$

$$n = 6l \pm 1 \qquad (l = 0,1,2,\cdots, n > 0) \qquad (9)$$

From (8), I_{dm} and λ_m are harmonic components of the dc side current which are defined in [14]. From equations above, it is possible to obtain some expressions representing the magnitude and order (9) of harmonics for the case of the system in Fig.2a, which is a simplified version of the servo drive unit of Fig.1b -with no faults. The rms value of the nth harmonic component of i_a in discontinuous mode is obtained from (8) and (9) and is given by:

$$I_n = \sqrt{\frac{(A_{in} + \Delta A_{in})^2 + (B_{in} + \Delta B_{in})^2}{2}} \qquad (10)$$

For the case of mechanical faults (bearings) considered in this paper, the magnitude of the characteristic harmonics given by (10) are not important, instead the harmonic's order stated by (9) are relevant since this helps to determine the location of possible bearing faults in the machine since *fault frequencies appear as side-band harmonics around the characteristic harmonics of the rectifier*. Note that for the case of monitoring electrical faults in the converters, the magnitude of the harmonics given by (10) should be taken in account and monitored.

B. Propagation of harmonics through the inverter

The inverter (Fig.3) can be considered to be ideal switches operating at very high frequency. A simplified equivalent model can be obtained [11].

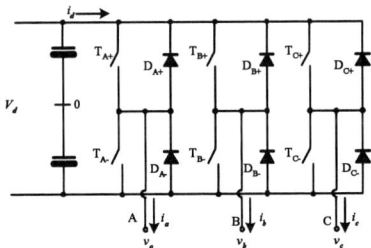

Fig. 3 Three phase inverter

Since the converter itself has no energy storage elements, the instantaneous power input must equal the instantaneous power output. Also it is assumed the output voltages are pure sine waveforms at fundamental frequency and for the case of phase a, we have:

$$v_a(t) = \sqrt{2}V_0 \sin(\omega t) \tag{11}$$

where V_o is the rms value of the phase voltage at the output of the inverter. The output current I_a is defined by:

$$i_a(t) = \sqrt{2}I_0 \sin(\omega t - \phi) \tag{12}$$

Assuming high switching frequency, i_d only consists of low frequency and dc components, and the instantaneous ac power output can be expressed in terms of fundamental frequency output voltages and currents [11]. By equating the instantaneous power input to the instantaneous power output, we get:

$$V_d i_d^* = v_a(t)i_a(t) + v_b(t)i_b(t) + v_c(t)i_c(t) \tag{13}$$

where i_d^* would only consists of the low frequency and dc components. Assuming $\sqrt{2}V_0$ and $\sqrt{2}I_0$ are the amplitudes of the phase voltages and currents respectively, yields:

$$i_d^* = \frac{2V_0 I_0}{V_d}[\cos\omega t \cos(\omega t - \phi) + \cos(\omega t - \tfrac{2}{3}\pi)\cos(\omega t - \tfrac{2}{3}\pi - \phi)$$
$$+ \cos(\omega t + \tfrac{2}{3}\pi)\cos(\omega t + \tfrac{2}{3}\pi - \phi)] \tag{14}$$

$$= \frac{3V_0 I_0}{V_d}\cos\phi = Id \tag{15}$$

It should be noted that in reality, i_d consists of high frequency switching components in addition to the dc-component i_d^* however for this analysis, the approximation given by (15) is reasonable [11]. It is well known that the magnitude (V_0) of a VSI-fed inverter is adjusted according with the modulation index M as:

$$V_0 = \frac{M}{2\sqrt{2}}V_d \tag{16}$$

By combining (15) and (16) the current inverter gain is obtained, which is valid at fundamental frequency:

$$I_d = \frac{3M}{2\sqrt{2}}I_0 \cos\phi \tag{17}$$

Note that the inverter gain depends of the modulation index M and the angle between voltage and current, ϕ. Expression (17) can be used to determine the inverter propagation gain for fault signatures generated in the machine and appearing at the input of the inverter. Since air-gap eccentricities produce torque disturbances in the machine with higher frequencies than that of the fundamental, and with lower frequencies than that of the switching frequency; we can assume ϕ is irrelevant. Therefore, a new expression for the harmonic propagation gain through the inverter can be obtained:

$$|I_{dh}| = \frac{3M}{2\sqrt{2}}I_{0h} \tag{18}$$

According to (18), any balanced disturbance appearing on the motor currents I_{0h}, (e.g. those generated by faulty bearings) will appear as an individual frequency at the input side of the inverter I_{dh} where the magnitude is related to the inverter modulation index.

C. Propagation of harmonics through the DC-link filter

It has been noted that for most of the published work related to the design of VSI-fed drives, the design of the dc-link filter has been given little attention and the selection of the filter is normally determined only by the necessity of having a constant DC-link voltage to reduce ripple and to improve current quality [11]. Further, literature states that to improve power factor, a serial inductance is required. Unfortunately, no relevant information is provided respect to the design criteria [12]. In the case of the proposed monitoring approach, the effects of the switching frequency at the input of the rectifier are not desirable, therefore the dc-link filter should be designed to block or minimize propagation of the switching frequency through the system. This can be done by a proper selection of the components of the LC filter, which should have a resonance response close to the switching frequency for maximum attenuation. A good approximation for the resonance frequency of the filter is obtained by:

$$2\pi F_R = \frac{1}{\sqrt{L_F C_F}} \tag{19}$$

where F_R corresponds to resonance frequency while L_F and C_F are the values of the inductor and capacitor. As it was stated, a correct selection of the resonance frequency F_R is important to reduce switching frequency propagation effects and to encourage propagation of other frequencies of interest as signatures generated for faulty components as the machine and power converters.

V. IMPLEMENTATION OF THE PROPOSED APPROACH

For purposes of this work, only mechanical disturbances are considered. As the machine is vector controlled, when a torque disturbance T_{dist} with frequency F_{dist} appears in the machine, a balanced current disturbance appears in each phase with the same frequency of F_{dist}. This frequency (signature) is transferred to the input of the inverter and appears as an individual component with the same frequency as the disturbance and with a magnitude given by (18). Note that the magnitude of this signature depends of the machine conditions given by M.

This signature must propagate through the DC Link to the rectifier output with minimal attenuation. At the rectifier stage, the fault signature is transferred to the rectifier input due to natural propagation. Note that the disturbance signatures will appear at the input side of the rectifier in conjunction with the rectifier characteristic harmonics given by (9) as side-band frequencies i.e., $\pm F_{dist}$. The magnitude of these frequencies is considerably smaller than those of the characteristic harmonics in the inverter, therefore the sampling frequency and system resolution should be carefully considered.

VI. EXPERIMENTAL RESULTS

Experimental results are presented for the case of a servo drive system consisting of a 4kw vector controlled PM machine and a commercial three phase inverter. The switching and sampling frequency for the currents controllers is 10 kHz, while the speed controller is sampled at 2kHz.

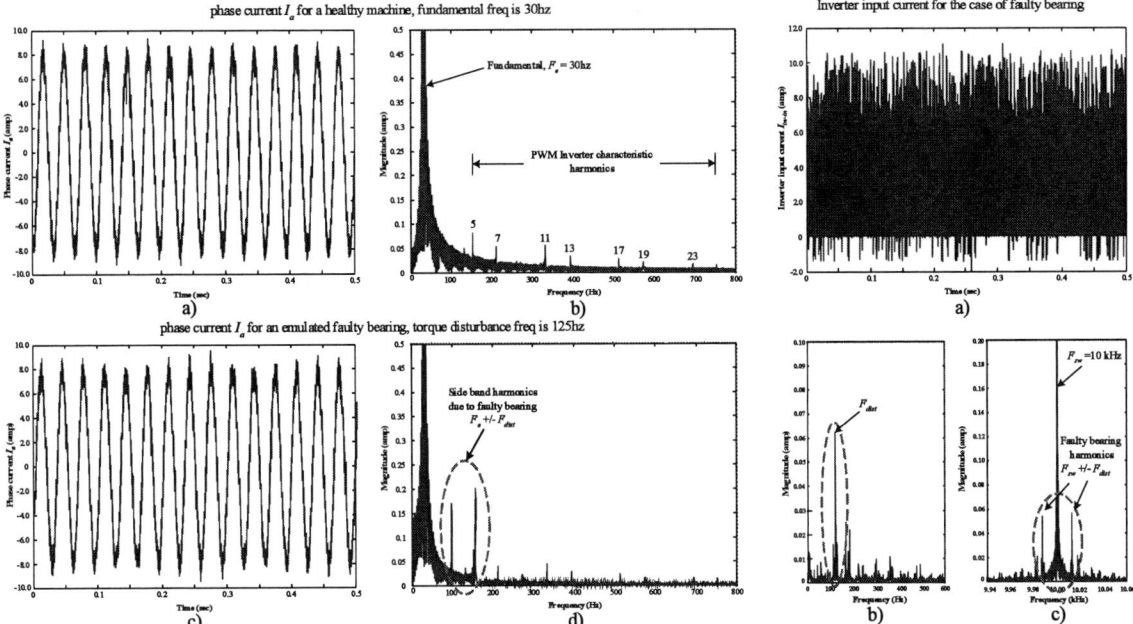

Fig.4. Application of traditional MSCA and detection of a bearing problem. Top graphs: healthy machine. a) machine phase current I_a, b) current spectrum of I_a. Bottom graphs: machine with faulty bearing, c) phase current, d) current spectrum.

Fig.5. Propagation of fault signature (bearings) trough the inverter. a) Current at the input of the inverter. b) and c) Fault signatures in the current spectrum.

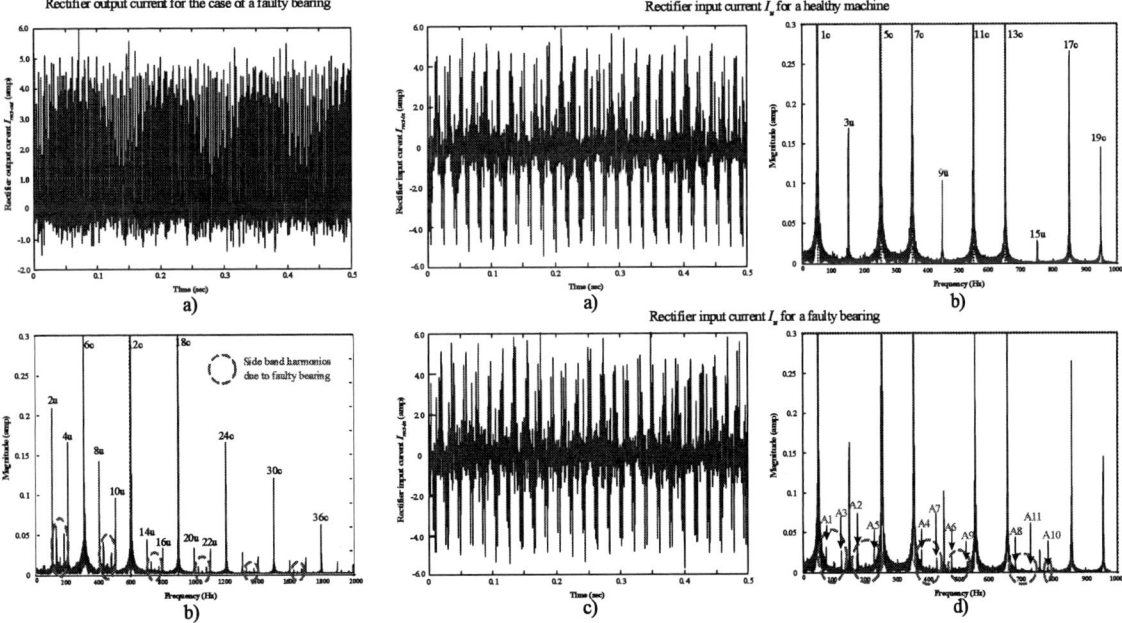

Fig.6. Propagation of fault signature (bearings) trough the DC-link filter. a) Current at the output side of the rectifier. b) Current spectrum showing characteristic "c", unbalances "u", and fault harmonics.

Fig.7. Application of proposed MSCA for the detection of a bearing problem. Top graphs: healthy machine. a) rectifier input current I_u, b) current spectrum of I_u. Bottom graphs: machine with faulty bearing, c) phase current, d) current spectrum where signature harmonics appear as side band of characteristic harmonics.

773

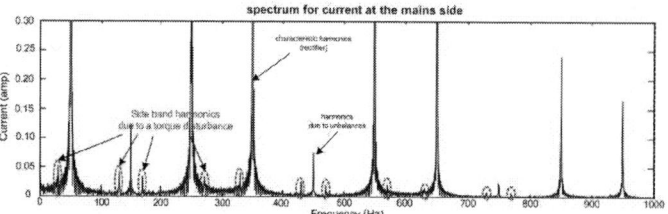

Fig.8. Spectrum analysis for the current at the rectifier side (mains). Note that the special harmonics from the disturbance (80Hz) have propagated through the DC link to the rectifier input and appear as side band harmonics around the characteristic ones

The DC-link filter has a resonant frequency of 10.53kHz. The condition monitoring approach is independent of the drive control and has been implemented with a 10-bit resolution data acquisition system with a sampling frequency of 200 kHz. Sampled data are stored and analysed using Matlab.

A DC machine is used to simulate faulty bearings by using a dynamic emulation [15], where a small torque disturbance representing a bearing fault (2% of the PM machine rated torque) is applied. Note that the magnitude of this disturbance is very small and can be compared with that of a typical disturbance found in a machine with a real bearing problem. The applied torque disturbance F_{dist} was set to 125Hz while the fundamental frequency for the machine was set to 30Hz. Note that the fundamental frequency of the mains is 50Hz. For this combination of frequencies the signature frequencies predicted by (9) and (18) at the rectifier input are: characteristic harmonics: $1c$=50Hz, $5c$=250Hz, $7c$=350Hz, and so on. The fault signature harmonics (which appear as side bands around characteristic ones) are: 75Hz and 175Hz around $1c$, 125Hz and 375Hz around $5c$, 225Hz and 475Hz around $7c$, etc.

Fig.4 shows a comparison for the measured motor currents for the machine running at a speed of 30Hz and 80% load for the case of a *healthy machine* (Figures 4a and 4b), and the case of a *faulty machine* (Figures 4c and 4d). Fig.4.b shows the current spectrum for a healthy machine, note that fundamental frequency (30Hz) dominates as expected however, additional small magnitude harmonics are observed (5, 7, 11 ...). These harmonics are related to the converter and always appear in VSI-fed PWM inverters. No other harmonics are observed, therefore by observing Fig.4.b we can conclude the machine is healthy. For the case of the faulty machine, the current spectrum in Fig.4d shows the introduction of side band harmonics around the fundamental frequency which are clearly differentiated from the characteristic harmonics. Note that traditional MSCA is based on the results observed in these graphs.

Fig.5 presents the current at the input side of the inverter for the case of the faulty machine. The current at the input of the inverter is discontinuous and contains high frequency switching components as observed in the current spectrum of Fig.5a. Note however that the disturbance frequency generated for the faulty bearing (125Hz) is easily recognized, appearing as a individual inter harmonic in the low part of the spectrum (Fig.5b), and as sideband harmonics in the higher part of the spectrum, around the switching frequency as observed in Fig.5c. Since the impedance of the DC-link filter is high

for frequencies close to the switching frequencies, these latter harmonics will not be detected at the input side of the rectifier.

Fig.6 shows the wave form at the output of the rectifier for the case of the faulty machine. Characteristic harmonics at the output are dominant as expected ($6c$, $12c$, $18c$) as shown in Fig.6b, where the subscript symbol "c" is included to indicate "characteristic" harmonics. Is observed that other dominant harmonics exist besides those at "c", and these are indentified with "u" meaning uncharacteristic harmonics. The "u" harmonics are a result of small unbalances present in the three phase supply and that is why they are identified as "unbalance harmonics". Note that for a perfectly balanced system these harmonics are not present however, their presence has little effect on the performance of the drive or the proposed MSCA scheme since they do not affect the location of the signature frequencies highlighted in Fig.6b. In this case they appear as side bands of the characteristic frequencies "c" (and are still present due to the transparency of the filter).

Fig.7 represents the application of the proposed MSCA monitoring approach for the two cases considered, the healthy machine (top figure), and the faulty machine (bottom figures). Again, "c" and "u" harmonics are indicated and appear for both conditions. The signature frequencies again, appear as sideband harmonics around harmonics "c". $A1$ and $A2$ are side band harmonics around the fundamental $1c$, while $A3$ and $A4$ are located around component $5c$. Note that these are entirely consistent with the theoretical values calculated from eqns. (9) and (18). It should be pointed out that despite to their magnitude, fault signature signals are easy to be identified at the input rectifier currents which validate the approach proposed.

For visualization purposes, Fig.8 shows with more detail the current spectrum at the input of the rectifier for a different disturbance frequency, in this case 80Hz for the machine operating under same conditions as above. Note again that the fault signature frequencies are easily identified.

VII. CONCLUSIONS

A theoretical analysis of the propagation of currents through a power converter has been presented together with experimental results to demonstrate the feasibility of using only the supply currents (rectifier input) for condition monitoring of electric motor drive systems. In this paper, mechanical faults have been covered. Contrary to conventional MSCA, the approach presented can be used to monitor the status of the electrical machine and also the status of the converter and may be achieved independent of the drive controller. The design of the dc-link filter is important to allow correct propagation of the disturbance frequencies to the input side of the drive; therefore resonance response close to the switching frequency is required.

The presented approach may be relevant for applications where access to internal signals may not be desirable or possible such an in the case of fully integrated motor drives. In the aerospace field the application of these servo drives is of particular interest and a mechanism to monitor for faults which is independent of the drive controller is also advantageous.

ACKNOWLEDGEMENT

The authors would like to acknowledge the support provided for this work by GE Aviation and EPSRC as part of the SMARTPACT University Technology Strategic Partnership.

REFERENCES

[1] IEEE Petro-Chemical Paper PCIC-94-01.

[2] Peter Vas. "Parameter estimation, condition monitoring, and diagnosis of electrical machines". *Monographs in electrical and electronic engineering*. Oxford science publications. 1993.

[3] Betta, G.; Liguori, C.; Paolillo, A.; Pietrosanto, A. (Maio 2001), "A DSP-based FFT-analyzer for the fault diagnosis of rotating machine based on vibration analysis". *Instrumentation and Measurement Technology Conference*. Proceedings of the 18th IEEE, Vol.1, pp.572.

[4] M. Y. Chow. "Methodologies of using neural network and fuzzy logic technologies for motor incipient fault detection". Singapore, World Scientific Publishing. 1997.

[5] Thomson, W.T. "On-line MCSA to diagnose shorted turns in low voltage stator windings of a 3-phase induction motors prior to failure", *Electrical machines and drives conference*, IEMDC 2001. IEEE International, pp.891-898.

[6] Thorsen, O; Dalva, M. "Condition monitoring methods, failure identification and analysis for high voltages motors in petrochemical industry". *Eighth International Conference on Electrical Machines and Drives*. No. 444, 1-3, pp.109-113., 1997.

[7] R. A. Callot, "Vibration, Monitoring and Diagnosis". New York: Wiley, 1979.

[8] Jason R. Stack, Thomas G. Hebetler and Ronald G. Harley, "Fault classification and fault signature production for rolling element bearings in electrical machines". *IEEE Trans. Ind. Applicat.*, Vol 40, No.3, May/June 2004.

[9] Wei Zhou, Thomas G. Hebetler and Ronald G. Harley, "Bearing condition monitoring methods for electric machines: a general review".

[10] Abdul Kadir, Talib Alukaidey, Omar Al-Ayasrah, R. Salman. "Embedded control with predictive diagnostic algorithm of an induction machine drive system", *Proceedings of the 2006 American Control Conference*. USA.

[11] Mohan, N., Undeland, T.M., and Robbins, W.P. "Power Electronics: Converters, Applications and Design". John Wiley & Sons, 1995.

[12] Arthur W. Kelley, and William F. Yadusky, "Rectifier design for minimum line current harmonics and maximum power factor". *IEEE Trans. on Power Electronics*. Vol.7, No.2, April 1992.

[13] Masaaki Sakui, Hiroshi Fujita and Mitsuo Shioya "A method for calculating harmonic currents of a three-phase bridge uncontrolled rectifier with dc filter". *IEEE Trans. on Industrial Electronics*. Vol.36, No.3, August 1989.

[14] Masaaki Sakui and Hiroshi Fujita, "An analytical method for calculating harmonic currents of a three-phase diode-bridge rectifier with dc filter", *IEEE transactions on Power Electronics*, Vol.9, No.6, November 1994.

[15] J. Arellano, G. Asher and M. Sumner. "Control of a dynamometer for dynamic emulation of mechanical loads with stiff and flexible shafts" *IEEE Trans. on Industrial electronics*, vol.53, No.4, 2006.

Feed-forward Compensation of Load and Parameter Variations of Electric Drive

Alon Kuperman*, Yoram Horen , Saad Tapuchi and Uri Suissa

Sami Shamoon College of Engineering, Beer-Sheva, Israel, *Corresponding Author e-mail: *alonk@sce.ac.il*

Abstract—**The paper presents a method to compensate effects caused by slow varying loads and plant parameters drift as well as by detuned controller using a simple yet robust algorithm in voltage controlled electric drives. In case of known variations an analytical expression of pre-computed feed-forward compensation voltage is derived, while in presence of unknown disturbances the control algorithm uses a simple Luenberger observer fed by data from a low cost encoder to estimate the feed-forward voltage command reflecting the deviation of the model from its nominal value. It is shown that the feed-forward command reflects parameter changes as well as load torque and slow varying actuator and measurement noises. Simulation and experimental results are given to describe the control algorithm performance and limitations.**

Keywords—**Control of Drive, DSP, Robustness**

I. INTRODUCTION

Control algorithm software executed by a digital signal processor (DSP) is the "soul" of a controlled electric drive [1]. The main goal of the control algorithm is to assure that the electric drive performs as planned when the system is exposed to mixture of disturbances such as measurement noises, loads and parameter drifts. Modern control theory focuses on a variety of subjects, related to motion control. Repetitive control theory deals with tracking and rejecting periodic non sinusoidal inputs [2], while active disturbance rejection theory focuses on rejection load signals of any type [3]. Input loop shaping theory tries to improve performance of torque-limited drives [4]. Robust [5] and quantitative feedback [6] control theories concentrate their efforts on reducing the influence of parameter variations and noises on drive performance, while stochastic optimal control theory assures optimal operation in presence of input, output- and state-dependent noises [7, 8]. In order to experimentally verify any control algorithm, an actual disturbance should be applied. A real mechanical load, for example, is quite difficult to implement even it is as simple as a constant torque one. The real mechanical loading becomes even more cumbersome when the load torque should change in time, e.g., a sinusoidal change. The use of torque-controlled load dynamometers, which is common in engine test beds or in the testing of electrical machines [9]-[12], is often impossible in an educational laboratory, due to the high price and

complexity. In addition, there are servo drives applications where the inertia has a large variation, e.g., in robotics the inertia range could be as wide as 1:10 [13]; moreover, phase resistance may change as a result of high temperature. Additionally, the plant model used to derive the controller is never an exact model of the physical process, resulting in unmodeled dynamics. These changes and deviations are difficult to be mechanically implemented in order to verify the control algorithm. In order to overcome these difficulties, a method for obtaining effects similar to those caused by the physical disturbance inputs and motor parameters variation without any mechanical parts supplementary to the electric motor of the studied electric drive and without actually changing any electrical or mechanical parameter of the motor, but using DSP software only, was recently developed [14-16]. Using a digital signal processing software, a virtual feed-forward voltage signal is created, forcing electric motor to behave as in the real variation case, while the motor parameters maintain their nominal values and no actual torque load is applied. Reversing the proposed method, a motor drive where parameter variations occurred or a disturbance such as actuator noise or load torque was applied can be viewed as a nominal motor drive, where a feed-forward voltage signal, reflecting all the disturbances, was added to the voltage command input to the motor. Hence, by estimating this so-called disturbance voltage and subtracting it from the command input to the motor, disturbance effects could be cancelled.

II. APPROACH PRESENTATION

A general closed-loop Brushless DC (BLDC) motor speed control system, which consists of the motor itself and a controller, is shown in Fig.1 [17, 18], where ω_{ref} is the reference speed, ω is the actual motor speed, ω_m is the measured motor speed, U_e is the speed error, U_c is the controller output voltage, U_p is the motor input voltage, $G_c(s)$ is the controller transfer function, $G_p(s)$ is the single-phase equivalent motor transfer function, T_L is the load torque, U_n is the actuator noise, ω_n is the measurement noise and s is the Laplace operator. The speed control loop could be replaced by position or torque loop without loss of generality as long as a single loop configuration is preserved. Moreover, in a multi loop case, which is out of the present paper scope, each loop can be treated as an independent single loop using the proposed approach. According to the BLDC motor electrical equivalent circuit shown in Fig. 2 (where L is phase inductance, R is phase resistance, J is rotor inertia, B is viscous friction constant, U_p is applied phase voltage and T_L is applied load torque), the following equations are derived,

$$U_P - e = \frac{d(Li)}{dt} + Ri$$
$$T_E - T_L = \frac{d(J\omega)}{dt} + B_m\omega \quad , \qquad (1)$$
$$e = K_b \cdot \omega = K\omega$$
$$T_E = K_t \cdot i = Ki$$

where K_t and K_b are equal torque and back electromotive force (emf) constants, respectively, referred as K through the paper, e is back emf voltage and T_E is electrical torque.

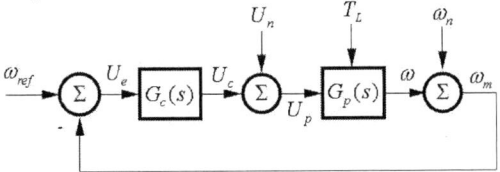

Fig. 1. Speed controlled BLDC drive block diagram

The phase inductance and the rotor inertia, usually assumed constant, may vary in general case. Hence, (1) is rearranged into

$$U_P - K\omega = i\frac{dL}{dt} + L\frac{di}{dt} + Ri = L\frac{di}{dt} + (R + \frac{dL}{dt})i \qquad .(2)$$
$$Ki - T_L = \omega\frac{dJ}{dt} + J\frac{d\omega}{dt} + B_m\omega = J\frac{d\omega}{dt} + (B_m + \frac{dJ}{dt})\omega$$

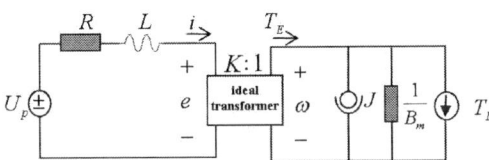

Fig. 2. BLDC motor electrical equivalent circuit

According to (2), block diagram shown in Fig. 3 is used throughout the paper to represent the BLDC motor.

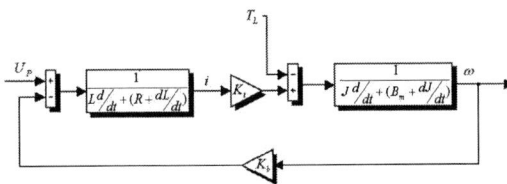

Fig. 3. Block diagram model of a BLDC motor

The loop controller $G_c(s)$ (mostly PID type with Feed-Forward branch) tuning is based on the nominal plant model, including nominal electrical and mechanical

parameters. However, besides parameter drift, unpredicted disturbances such as load torque changes, actuator and measurement noises are always present. Hence the nominal controller is never truly tuned, because the plant is rarely nominal. The goal of the paper, presented in the introduction, can be reformulated using Fig. 4: create a feed-forward voltage command U_{dist}, reflecting all possible deviations of the plant from its nominal form such that adding it to the controller output makes the controller "see" a nominal plant despite the disturbances and hence remain the desired performance. Only low frequency disturbances are treated in the paper, hence actuator and measurement noises are not included.

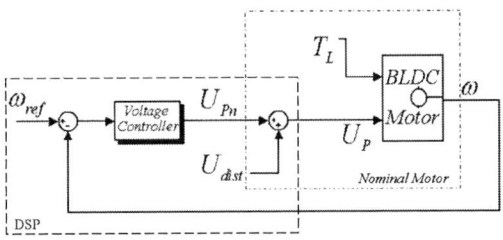

Fig.4. Voltage controlled BLDC drive with disturbance and parameter variations

III. ANALYTICAL DERIVATION OF THE FEED-FORWARD VOLTAGE COMMAND

In case of quantifiable or measurable disturbances, the compensating feed-forward voltage command can be exactly calculated as follows. Eq. (2) can be rearranged into state space form as

$$\begin{pmatrix} \dot{i} \\ \dot{\omega} \end{pmatrix} = \begin{pmatrix} -\dfrac{R+\dot{L}}{L} & -\dfrac{K}{L} \\ \dfrac{K}{J} & -\dfrac{B_m+\dot{J}}{J} \end{pmatrix} \begin{pmatrix} i \\ \omega \end{pmatrix} + \begin{pmatrix} \dfrac{1}{L} & 0 \\ 0 & -\dfrac{1}{J} \end{pmatrix} \begin{pmatrix} U_P \\ T_L \end{pmatrix} =$$

$$= \begin{pmatrix} -\dfrac{R+\dot{L}}{L} & -\dfrac{K}{L} \\ \dfrac{K}{J} & -\dfrac{B_m+\dot{J}}{J} \end{pmatrix} \begin{pmatrix} i \\ \omega \end{pmatrix} + \begin{pmatrix} \dfrac{1}{L} \\ 0 \end{pmatrix} \begin{pmatrix} U_P \\ 0 \end{pmatrix} + \begin{pmatrix} 0 \\ -\dfrac{1}{J} \end{pmatrix} \begin{pmatrix} 0 \\ T_L \end{pmatrix}$$

(3)

or

(4)
$$\dot{X} = AX + BU = AX + [\mathbf{B_1} \quad \mathbf{B_2}]U = AX + [\mathbf{B_1} \quad 0]\begin{bmatrix} U_1 \\ 0 \end{bmatrix} + [0 \quad \mathbf{B_2}]\begin{bmatrix} 0 \\ U_2 \end{bmatrix}$$

in a compact parametric form.

Consider the two following systems:

the first, representing the nominal motor (consisting of the nominal input voltage, parameters and load torque; subscripted n), given by

$$(5)\ \dot{X}_n = \mathbf{A_n} X_n + [\mathbf{B}_{1n}\ \ 0]\begin{bmatrix} U_{1n} \\ 0 \end{bmatrix} + [0\ \ \mathbf{B}_{2n}]\begin{bmatrix} 0 \\ U_{2n} \end{bmatrix}$$

where

$J = J_n,\ \ L = L_n,\ \ R = R_n,\ \ B_m = B_{mn},\ \ K = K_n,\ \ i = i_n,\ \ \omega = \omega_n$
$T_L = T_{Ln}$ and $U_P = U_{Pn}$ (nonzero nominal load torque can be created by impeller attached to the motor, for example);

and the second, representing the disturbed motor (consisting of nominal input voltage with parameters and load torque different from nominal, subscripted d), given by

$$\dot{X}_d = \mathbf{A_d} X_d + [\mathbf{B}_{1d}\ \ 0]\begin{bmatrix} U_{1d} \\ 0 \end{bmatrix} + [0\ \ \mathbf{B}_{2d}]\begin{bmatrix} 0 \\ U_{2d} \end{bmatrix} \quad (6)$$

where

$J_d = J_n + \Delta J,\ \ L_d = L_n + \Delta L,\ \ R_d = R_n + \Delta R,$
$B_{md} = B_{mn} + \Delta B_m,\ i_d = i_n + \Delta i, \omega_d = \omega_n + \Delta\omega,\ \ K = K_n + \Delta K,$
$T_{Ld} = T_{Ln} + \Delta T_L,\ \ \ U_{Pd} = U_{Pn} + U_{dist}$

Subtracting (5) from (6), the following system is derived,

$$(7)$$

$$\dot{X}_d - \dot{X}_n = \begin{pmatrix} \dot{\Delta i} \\ \dot{\Delta\omega} \end{pmatrix} = (\mathbf{A_n} + \Delta\mathbf{A})\begin{pmatrix} i_n + \Delta i \\ \omega_n + \Delta\omega \end{pmatrix} + (\mathbf{B}_{1n} + \Delta\mathbf{B}_1)\begin{pmatrix} U_{Pn} + U_{dist} \\ 0 \end{pmatrix} +$$

$$+ (\mathbf{B}_{2n} + \Delta\mathbf{B}_2)\begin{pmatrix} 0 \\ T_{Ln} + \Delta T_L \end{pmatrix} - \mathbf{A_n}\begin{pmatrix} i_n \\ \omega_n \end{pmatrix} - \mathbf{B}_{1n}\begin{pmatrix} U_{Pn} \\ 0 \end{pmatrix} - \mathbf{B}_{2n}\begin{pmatrix} 0 \\ T_{Ln} \end{pmatrix}$$

where

$\Delta\mathbf{A} = \mathbf{A_d} - \mathbf{A_n},\ \ \Delta\mathbf{B}_1 = \mathbf{B}_{1d} - \mathbf{B}_{1n},\ \ \Delta\mathbf{B}_2 = \mathbf{B}_{2d} - \mathbf{B}_{2n},\ \ \Delta i = i_d - i_n,$
$\Delta\omega = \omega_d - \omega_n$

Substituting $\Delta\omega = \dot{\Delta\omega} = 0$ into (7) in order to make the speeds of (6) and (7) to behave alike and performing some algebraic manipulations, an expression of feed-forward compensation voltage for the case of known disturbances is derived:

$$(8)$$

$$U_{dist} = \frac{1}{B_{1n_{11}}}\left[\dot{\Delta i} - \left(A_{n_{11}} + \Delta A_{11} \right)\Delta i - \Delta A_{11} i_v - \Delta A_{12}\omega_v - \Delta\mathbf{B}_1 U_{Pn} \right]$$

IV. ESTIMATING THE FEED-FORWARD VOLTAGE COMMAND

In case of unknown or non-measurable disturbances, the compensating feed-forward voltage command can be estimated as follows. Eq (2) can be rearranged into single equation form as

$$(9)$$

$$\frac{LJ}{K}\ddot{\omega} + \frac{LB_m' + R'J}{K}\dot{\omega} + \frac{R'B_m' + K^2}{K}\omega = U_P - \frac{1}{K}(R'T_L + L\dot{T}_L)$$

where $B_m' = B_m + \dfrac{dJ}{dt}$ and $R' = R + \dfrac{dL}{dt}$. Consider the following parameter deviations from nominal values:

$$J_d = J_n + \Delta J,\ \ L_d = L_n + \Delta L,\ \ R_d' = R_n' + \Delta R', \\ B_{md}' = B_{mn}' + \Delta B',\ \ \omega_d = \omega_n + \Delta\omega,\ \ T_{Ld} = T_{Ln} + \Delta T_L. \quad (10)$$

By substituting (10) into (9) and rearranging, it is possible to represent the disturbed motor as a nominal motor with an additional input term U_{dist} reflecting all the changes:

$$(11)$$

$$\frac{L_n J_n}{K_n}\ddot{\omega}_n + \frac{L_n B_{mn}' + R_n' J_n}{K_n}\dot{\omega}_n + \frac{R_n' B_{mn}' + K_n^2}{K_n}\omega_n =$$

$$U_{Pn} - \frac{1}{K_n}(R_n' T_{Ln} + L_n \dot{T}_{Ln}) -$$

$$- U_{dist}(J_n, \Delta J, L_n, \Delta L, R_n', \Delta R', B_{mn}', \Delta B', \omega_n,$$

$$\Delta\omega, \dot{\omega}_n, \Delta\dot{\omega}, \ddot{\omega}_n, \Delta\ddot{\omega}, T_{Ln}, \Delta T_L, \dot{T}_{Ln}, \Delta\dot{T}_L, U_{Pn})$$

Hence, estimating U_{dist} and adding it to the controller output would cancel all the disturbance effects. Eq. (11) can be rearranged into state space form as

$$(12)$$

$$\begin{pmatrix} \dot{\omega}_n \\ \ddot{\omega}_n \\ \dot{U}_{dist} \end{pmatrix} = \begin{pmatrix} 0 & 1 & 0 \\ -\dfrac{L_n B_{mn}' + R_n' J_n}{L_n J_n} & -\dfrac{R_n' B_{mn}' + K_n^2}{L_n J_n} & -\dfrac{K_n}{L_n J_n} \\ 0 & 0 & 0 \end{pmatrix}\begin{pmatrix} \omega_n \\ \dot{\omega}_n \\ U_{dist} \end{pmatrix}$$

$$+ \begin{pmatrix} 0 \\ \dfrac{K_n}{L_n J_n} \\ 0 \end{pmatrix} U_{Pn}'$$

where $U_{Pn}' = U_{Pn} - \dfrac{1}{K_n}(R_n' T_{Ln} + L_n \dot{T}_{Ln}).$

Representation of (12) is valid only in case of low frequency disturbances. The estimation of U_{dist} is performed using Luenberger observer [19] which proved to be suitable for disturbance torque estimation [20]. In addition, the observer (13) estimates instantaneous speed for speed control systems and position for position control system using data from low cost encoder, which is the only sensor used. In practice, this kind of observer is capable of estimating quantities varying slower than the observer time constant (limited by sample rate, as explained further on). Hence, good performance is expected estimating constant and slow varying load torques and parameter drifts.

The observer is a two-input two-output model, driven by estimated speed signal obtained from a low cost encoder (usually a stair-like non-continuous wave) and motor voltage command. Its outputs are estimated instantaneous speed and the compensating feed-forward voltage command U_{dist}, as shown in Fig. 5.

The proposed observer is of the form

$$
\begin{pmatrix} \dot{\hat{\omega}} \\ \dot{\hat{\omega}} \\ \hat{U}_{dist} \end{pmatrix} = \begin{pmatrix} 0 & 1 & 0 \\ -\dfrac{L_n B_{mn}{'} + R_n{'} J_n}{L_n J_n} & -\dfrac{R_n{'} B_{mn}{'} + K_n{}^2}{L_n J_n} & -\dfrac{K_n}{L_n J_n} \\ 0 & 0 & 0 \end{pmatrix} \begin{pmatrix} \hat{\omega} \\ \dot{\hat{\omega}} \\ \hat{U}_{dist} \end{pmatrix}
$$
(13)
$$
+ \begin{pmatrix} 0 \\ \dfrac{K_n}{L_n J_n} \\ 0 \end{pmatrix} U_{Pn}{'} + \begin{pmatrix} L_1 \\ L_2 \\ L_3 \end{pmatrix} (\omega_n - \hat{\omega})
$$

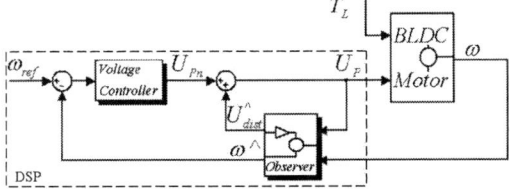

Fig. 5. Observer-based disturbance compensation system

Pole placement strategy is used in order to determine the gains as follows. The characteristic polynomial is calculated,

(14)
$$
\left| sI - (A - LC) \right| = s^3 + (L_1 + \frac{R_n{'} B_{mn}{'} + K_n{}^2}{L_n J_n}) s^2
$$
$$
+ (L_1 \cdot \frac{R_n{'} B_{mn}{'} + K_n{}^2}{L_n J_n} + \frac{L_n B_{mn}{'} + R_n{'} J_n}{L_n J_n} + L_2) s - L_3 \cdot \frac{K_n}{L_n J_n}
$$

and then compared to the desired characteristic polynomial based on the chosen poles p_1, p_2 and p_3:

(15)
$$
\left| sI - (A - LC) \right| = (s - p_1)(s - p_2)(s - p_3) =
$$
$$
s_3 - (p_1 + p_2 + p_3) s^2 + (p_1 p_2 + p_1 p_3 + p_2 p_3) s - p_1 p_2 p_3
$$

The resulting observer gains are

(16)
$$
L_1 = -(p_1 + p_2 + p_3) - \frac{R_n{'} B_{mn}{'} + K_n{}^2}{L_n J_n}
$$
$$
L_2 = p_1 p_2 + p_2 p_3 + p_1 p_3 - (L_1 \cdot \frac{R_n{'} B_{mn}{'} + K_n{}^2}{L_n J_n} + \frac{L_n B_{mn}{'} + R_n{'} J_n}{L_n J_n})
$$
$$
L_3 = p_1 p_2 p_3 \frac{L_n J_n}{K_n}
$$

In order to force the observer states to quickly converge to the motor states, poles p_1, p_2 and p_3 must be chosen as large as possible. However, the largest pole is limited by the sample rate,

$$
\max(p_i) < \frac{F_s}{2}; \quad i = 1,2,3 , \tag{17}
$$

where F_s is the sampling frequency of the observer executing algorithm.

V. RESULTS AND DISCUSSION

Consider a brushless motor with nominal parameters given in Table I and zero nominal torque. The motor is a part of MCK240 motion control kit. The MCK240 is a complete stand-alone system that allows to experiment with and use the TMS320F240 ('F240) DSP controller for digital motion control (DMC) applications. The 'F240 DSP controller is designed to implement advanced DMC applications, by integrating high performance of a DSP core and the on-chip peripherals of a microcontroller into a single chip solution. The MCK240 has an 'F240 DSP controller on board, as well as a 3-phase inverter and as mentioned before a brushless motor to allow complete verification of the DSP code for basic DMC evaluation applications. The MCK240 allows to evaluate all the specific DMC features of the 'F240 DSP controller and to directly control a brushless motor in DC or AC modes. External power modules may be easily interfaced with the DSP board through a universal motion control bus (Fig. 6).

TABLE I

BRUSHLESS MOTOR PARAMETERS

parameter	units	Value
R	Ω	7.5
L	μH	480
K_b	V/Krpm	2.1
K_t	Nm/A	$20e^{-3}$
J	Kgm^2	$4.6e^{-7}$
B_m	N/m/s	~0

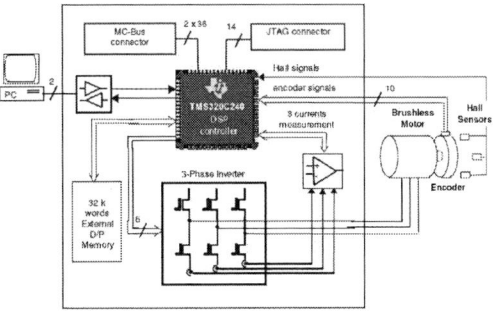

Fig. 6. MCK240 block diagram [21]

In order to theoretically demonstrate the observer-based control approach, a simulation was performed for two different sets of observer gains without taking into consideration any sampling, sensor and finite word length limitations. Then the same simulation was performed, while the feed-forward compensation voltage was calculated according to (8) without using observer, assuming the disturbance was a priori known. The drive was commanded to follow a constant speed reference of 25 lines/sample (750 RPM), while a step load torque of 5mNm was applied at $t = 0.05s$ and removed at $t = 0.15s$, as shown in Fig. 7a. Figs. 7b, 7c describe the drive speed response and the compensating feed-forward voltage for the two following observer gain sets, respectively: $p_i = 500$, $p_i = 5000$; $i = 1,2,3$. Fig. 7d illustrates the drive speed response and the compensating feed-forward voltage in case of ahead known variations. It is clear that the higher the observer gains are, the faster the observer converges to the motor states and compensates the disturbance torque. The case of the a priori known disturbances may be considered as a theoretically ideal compensation. However, practical bounds of (17) apply, limiting the achievable performance.

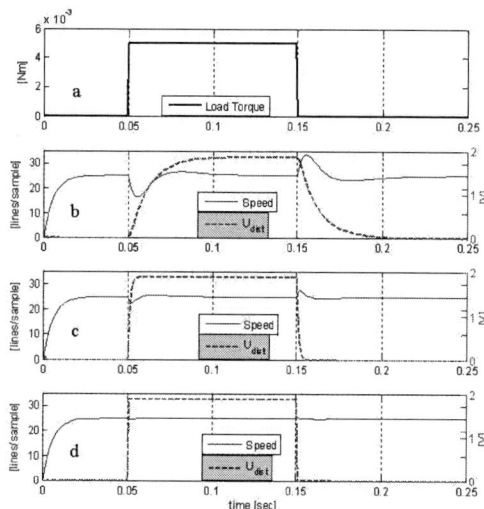

Fig. 7. Feed-forward voltage influence on performance: (a) applied load torque; (b) observer estimated compensating voltage and motor speed for $p_i = 500$, $i = 1,2,3$ (c) observer estimated compensating voltage and motor speed for $p_i = 5000$, $i = 1,2,3$ (d) compensating voltage, pre-computed according to (8) and motor speed.

In an effort to experimentally verify the proposed method, a load having the following characteristics was applied to the motor: moment of inertia J (the same as the motor moment of inertia), friction described by the torque-speed characteristics of

$$T_{f_L}(\omega) = 3 \cdot 10^{-5} \cdot \omega + sign(\omega) \cdot (4 \cdot 10^{-3} + 2.5 \cdot 10^{-3} \cdot e^{-|\omega|})$$ (18)

and shown in Fig. 8, composed of stiction, viscous and coulomb frictions [22]; torque of 5mNm step, starting at $t = 100ms$. The drive was commanded to follow a speed reference of 25 lines/sample (750 RPM) for $0 < t < 350$ ms and 50 lines/sample (1500 RPM) for $350 < t < 500$ ms.

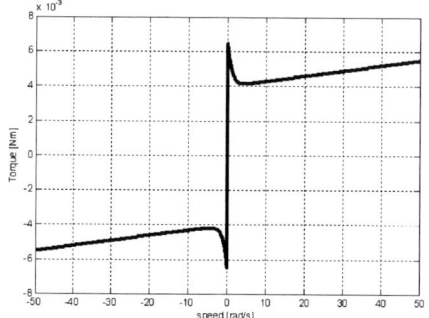

Fig. 8. Friction torque characteristics of the load.

First, a digital simulation (sample time of 1ms, observer gains $p_i = 500$, $i = 1,2,3$) was performed without taking into account fixed-point effects of the DSP in order to predict the system behavior. Fig. 9 presents the simulated speed (lines/sample) and feed-forward compensating voltage (Q15 format). It is clear that before a load torque is applied ($t < 100ms$), inertia and friction mismatches with the nominal values are reflected by the observer and a corresponding compensation voltage is created. Following the load torque application ($t = 100ms$) the feed-forward voltage command is updated in a step-like manner. The update rate is set by the observer gains. When the reference speed command rises to 75 lines/sample at t = 350ms, the friction torque also rises according to (26) and, hence, the feed-forward voltage command is further renewed and settles at a value slightly higher than the one before the command change occurred. Another useful feature of the method is the fact that the feed-forward voltage command U_{dist}, reflects all possible deviations of the plant from its nominal form such that adding it to the controller output makes the controller "see" a nominal plant despite the disturbances and hence remain the desired performance. The PI controller outputs are shown in Fig. 10 for three following set-ups: unloaded nominal drive with PI controller only; loaded drive with PI controller only (with no observer) and loaded drive with an observer in addition to the PI controller. The first set-up might be considered as a "nominal" one, because the PI controller "sees" a nominal plant only. Comparing the outcomes of the other two set-ups, it is obvious that in case of observer-based system the output of the PI controller converges to the "nominal" one in steady state, while the other one reflects all the plant changes.

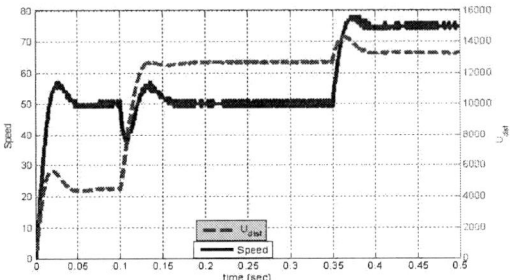

Fig. 9. Simulation results – speed and feed-forward compensating voltage command

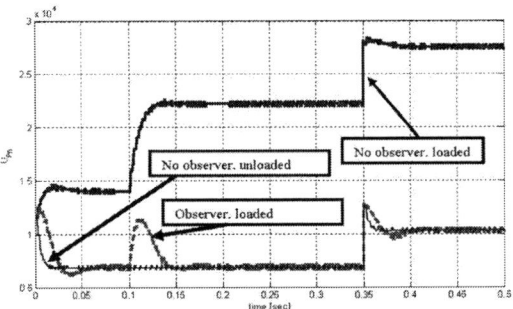

Fig. 10. Simulation results – PI controller outputs for a nominal unloaded, nominal loaded and observer-based loaded cases.

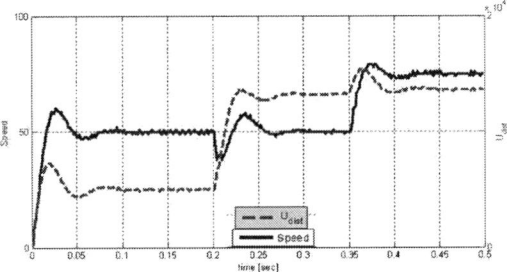

Fig. 11. Experimental results – speed and feed-forward compensating voltage command

The experimental results having the same setup as the simulations except the load torque application time ($t = 200ms$), presenting motor speed and U_{dist} are shown in Fig. 11. The experimental results slightly differ from the simulations because of the fixed-point effects reflected by higher overshoots and small steady-state errors. However, the conclusions from the simulations are fully valid from the experimental point of view.

VI. CONCLUSION

Motor drive, where parameter variations occurred or a disturbance such as actuator noise or load torque was applied can be viewed as a nominal motor drive having a feed-forward voltage signal, reflecting all the disturbances, added to the nominal voltage command input to the motor. Hence, by estimating this so-called disturbance voltage and subtracting it from the command input to the motor, disturbance effects can be cancelled.

Creating a feed-forward voltage command, which reflects all possible deviations of the plant from its nominal form and adding it to the controller output makes the controller "see" a nominal plant despite the disturbances and hence maintain the desired performance. The disturbance estimation is possible using Luenberger observer which proved to be suitable for disturbance torque estimation. In addition, the observer estimates instantaneous speed for speed control systems and position for position control system using data from low cost encoder, which is the only sensor used. Simulation and experimental results show excellent agreement with the proposed method theory.

REFERENCES

[1] Bose, B.K, *Microcomputer control of power electronics and drives*, New York, IEEE Press, 1987

[2] Weiss, G. and Haefele, M., "Repetitive control of MIMO systems using H^infinity design." *Automatica*, Vol. 35, pp. 1185-1199, 1999

[3] Wu, D., Xiankui W., Tong Z. and Weilong L., "Application of active disturbance rejection to tracking control of a fast tool servo system ," *In Proc. IEEE-CCA*, pp. 547-552, 2005

[4] Singhose, W., Derezinski, S., and Singer, N., "Input shapers for improving the throughput of torque-limited systems," *In Proc. IEEE-CCA*, pp. 1517-1522, 1994

[5] Zhou K., Doyle J. and Glover K., *Robust and Optimal Control*, Prentice Hall, 1996

[6] Yaniv O., *Quantitative feedback design of linear and nonlinear control systems*, Springer, 1999

[7] Bertsekas D.P. and Shreve S.E., *Stochastic Optimal Control: The Discrete-Time Case, Athena Scientific*, 1996

[8] Todorov E., "Stochastic optimal control and estimation methods adapted to the noise characteristics of the sensorimotor system," *Neural Computation*, vol. 17(5), pp. 1084-1108, 2005

[9] Wasko, C.R., "A universal AC dynamometer for testing motor drive systems", in Conf. Rec. *IEEE-IAS Annu. Meeting*, 1987, pp.409-412

[10] Collins, E.R. and Huang, Y., "A programmable dynamometer for testing rotating machinery using a three-phase induction machine", *IEEE Trans. Energy Conversion*, vol. 9, pp. 521-527, Sept. 1994

[11] Newton, R.W.,Betz, R.E. and Penfold, H.B., "Emulating dynamic load characteristics using a dynamic dynamometer", in *Proc. Int. Conf. Power Electronics and Drive Systems*, 1995, vol. 1, pp. 465-470

[12] Akpolat, Z.H., Asher, G.M. and Clare, J.C., "Experimental dynamometer emulation of nonlinear mechanical loads", *IEEE Trans. on Ind. Appl.*, v. 35, no. 6, 1999, pp. 1367-1373[

[13] Leonhard, W., *Control of electric drives*, Springer-Verlag, Berlin, 1997

[14] Rabinovici, R. and Kuperman, A., "Virtual loading of electric drive", in *Proc. ICEM* , Belgium, 2002

[15] Kuperman A. and Rabinovici R., "Virtual torque and inertia loading of controlled electric drive," *IEEE Trans. Education*, vol. 48(1), pp. 47-52, 2005

[16] Kuperman A., "HIL-Based Virtual Disturbance and Parameter Variations of Controlled Electric Drive," Technical Report,   2007

[17] Dorf, R.C. and Bishop, R.H., *Modern control systems*, Addison-Wesley, 1998

[18] Chen, J. and Rodriguez, F., "SPICE modeling of a resolver-to-digital converter for closed loop simulations of brushless DC motors", in *Proc. IECEC*, Boston, USA, 1991

[19] Luenberger D. G., "Observers for multivariable systems," *IEEE Trans. Automatic Control*, vol. AC-11(2), pp. 190-196, 1966

[20] Rabinovici R., Andreescu G.D., "Generalized approach of the disturbance torque control," In *Proc. ELECTROMOTION'*, pp. 255-260, Greece, 1999

[21] *MCK240 v1.0 User Manual*, Technosoft DSP Motion Solutions, 2001

[22] Armstrong-Helouvry B. and Dupont P., "Friction Modeling for Control," *in Proc. 1993 American Control Conference*, San Francisco, CA, pp. 1905-1909, 1993.

Thermal Effect of Short-Circuit Current in Low Power Induction Motors

Leoš Beran*

*Technical university of Liberec, Liberec, Czech Republic, e-mail: *leos.beran@tul.cz*

Abstract—Diagnostic of reliability of induction motors (IM) is very important for industry. The article deals with the analyse of short-circuit currents in stator winding of low power induction motors which can strongly affected reliability of IMs. The occurrence of short-circuit arises when some parts of the insulation system of a stator winding breakdown. The stator winding loses a part of active coil and its inductivity decreases. The short-circuit current warms up the surrounding stator winding with its heat effect at point of the short-circuited stator winding.

Theoretical part describes two models to simulate different short-circuits that can occur in the stator winding. The first model is a single-phase linear model and the second is a numerical model in SymPowerSystems Toolbox of Matlab software.

Chapter III analyses thermal activity on separate parts of the stator winding. Thermal effects were verified on a set up model that consists of a specially designed experimental motor (EM) with the temperature monitoring in slots of the stator.

I. INTRODUCTION

Contemporary trends in technical diagnostic are focused upon a high power induction motors (IM) with regard its price. High power IM can be diagnosed via expensive on-line or off-line expert systems. The price of low power IMs powered by frequency converters (FC) is too close price to be able to develop some expert system for its diagnostic.

A. The main conception

Our research was established on this conception. The price of diagnostic system doesn't need to be as low as the price of low power IM. The price of a IM is relative to price a machine or a technology which operates. If the low power IM break down, it can cause a lot of economic losses. Therefore it is necessary to watch these cheap but very important IM in a production technology. (textile industries, automotive industries, glass industries, power plant technology etc.)

New diagnostic method should be independent on type of feed. The most known methods used for diagnostic high power IM depend on their feed. There are usually problems with feeding by FC. With subsequent research is possible to develop method independent on feeding the IMs. This article presents first attempt which invoke a lot of next research steps in the future.

B. The biggest problem of low power IM

The most critical part of IM is the stator winding. There are the most reasons to brake down all IMs. Therefore our effort is focused upon these parts of the IMs. Heat produced by a little short-circuit currents has the greatest degradation effect of insulation materials used in the stator winding. Monitoring a field of temperature in the IM could by susceptible to monitor short-circuits in the stator winding. Such method would be independent on feeding.

II. MODELS OF SHORT-CIRCUIT IN STATOR WINDING

Models was chosen with emphasis on simplicity to carry out results very fast by different setting of constrains. Opted model for turn to turn short-circuit is in the Fig.1 and for turn to the ground short-circuit numerical model in the Fig.5. Both of these models were performed by MATLAB software. Even if the IM is powered by FC, it can be supposed harmonic signal of voltage and current for the case of turn to turn short-cirucuit.

Therefore the first model was performed by the help analytical computing with using the Kirhoff's laws. In case of turn to the ground short-circuit must be the problem solved numerically. There is no way how to describe voltage and current with analytical equations. The Fig.6 shows real record in time of current to the earth.

At assemblage of above mentioned models had to be taken some simplification for reducing the computing time. It was supersaturation of stator core and dislocation of stator winding in stator core.

A. Model of short-circuit turn to turn in one phase

The circuit model is intended as one phase linear model. There was assembled the linear circuit with the inductance and the resistance of serial arrangement. When the short circuit occurs this simply RL serial combination is divided to three basic parts.

The first represents non short-circuited part, the second represents shorted part and the last part includes resistance of short-circuit between turns in the stator winding. The first and second is RL serial combination also. The third part represents physical short-circuit – it consists of the resistance.

All parameters of the model were measured in our EM (see III-A). The measurement of resistance of point short-circuit is not easy . It is usualy non-linerar with regard to pressure, current, temperature etc.

The solution of the model which represents circuit on Fig. 1. was divided to two steps of superposition. The Fig. 2. shows the first step of superposition. There is short-circuited electric voltage U_{r2} and U_{inz}. The electric

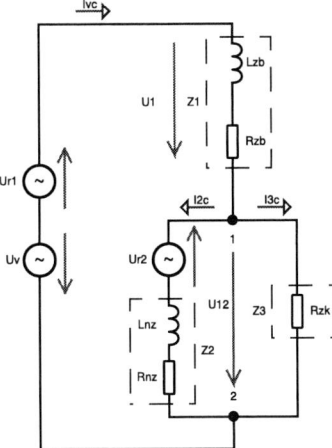

Fig. 1.   Model of short-circuit turn to turn

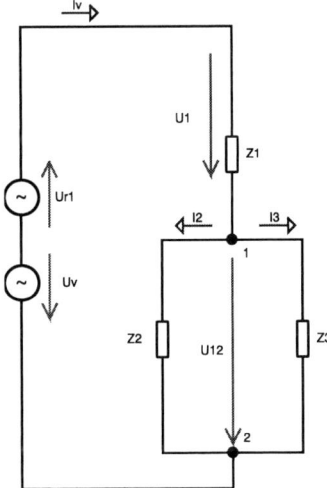

Fig. 2.   Model of short-circuit turn to turn superposition step 1st

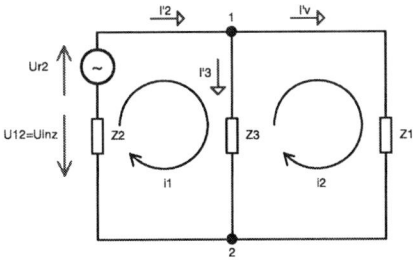

Fig. 3.   Model of short-circuit turn to turn superposition step 2nd

The most important information about power losses is in the part 2. It is located in stator winding in one place. There is not enough heat sink to reduce increasing temperature which rise steeply up to danger values for insulation materials.

Equations (2), (3) and (4) describe currents in the first step of superposition.

$$\hat{I}_v = \frac{\hat{U}_v - \hat{U}_{r1}\,(s)}{\hat{Z}} = \frac{(\hat{U}_v - \hat{U}_{r1}\,(s))(\hat{Z}_2 + \hat{Z}_3)}{\hat{Z}_1(\hat{Z}_2 + \hat{Z}_3) + \hat{Z}_2\hat{Z}_3} \quad (2)$$

$$\hat{I}_2 = \frac{\hat{U}_{12}\,(s)}{\hat{Z}_2} \quad (3)$$

$$\hat{I}_3 = \frac{\hat{U}_{12}\,(s)}{\hat{Z}_3} \quad (4)$$

Equations (5), (6) and (7) describe currents in the second step of superposition.

$$\hat{i}_1 = \frac{(\hat{U}_{inz} - \hat{U}_{r2}\,(s))(\hat{Z}_1 + \hat{Z}_3)}{\hat{Z}_2\hat{Z}_3 + \hat{Z}_1(\hat{Z}_2 + \hat{Z}_3)} \quad (5)$$

$$\hat{i}_2 = \frac{(\hat{U}_{inz} - \hat{U}_{r2}\,(s))\hat{Z}_3}{\hat{Z}_2\hat{Z}_3 + \hat{Z}_1(\hat{Z}_2 + \hat{Z}_3)} \quad (6)$$

$$\hat{I}'_v = \hat{i}_2 \qquad \hat{I}'_2 = \hat{i}_1 \qquad \hat{I}'_3 = \hat{i}_1 - \hat{i}_2 \quad (7)$$

When it is possible to evaluate all currents and voltage in the model of short-circuit, than the power losses can be compute according equations (8).

$$P_1 = R_{zb}|\hat{I}_{vc}|^2 \quad P_2 = R_{nz}|\hat{I}_{2c}|^2 \quad P_3 = R_{zk}|\hat{I}_{3c}|^2 \quad (8)$$

The main result of all above mentioned equations is present on the Fig. 4. There are power losses in stator winding of short-circuited part two. More results can be computed with the help of all derived equations. Result from area in our focus was opted in this case.

The Fig. 4. presents the three dimensional graph where s is the slip of rotor, N_z/n is ratio short-circuited to non-short-circuited count of turns of the stator winding and P_z/P_N is ratio power losses to nominal power of EM.

The Fig. 4. shows how the power losses steeply rise with load of IM (increasing s). The power losses rise

voltage U_{r2} is induced voltage from rotor. The U_{inz} is induced voltage in stator winding from power supply but only the part belongs to short-circuited turns of stator coil.

The Fig. 3. presents the second step of superposition. There is short-circuited U_v and U_{r1}. The U_v is the power supply voltage of current phase and the U_{r2} is induced voltage from rotor the part belongs to short-circuited turns of stator coil again. For both steps are solved all partial currents which are necessary for computing overall currents in stator winding.

Now all currents and voltages can be evaluate in each branch of our model. With utilization the values of resistances and acquired currents can be determine power losses according (1) in our model.

$$P = \frac{1}{T}\int_0^T R \cdot i^2(t)\,\mathrm{d}t \quad (1)$$

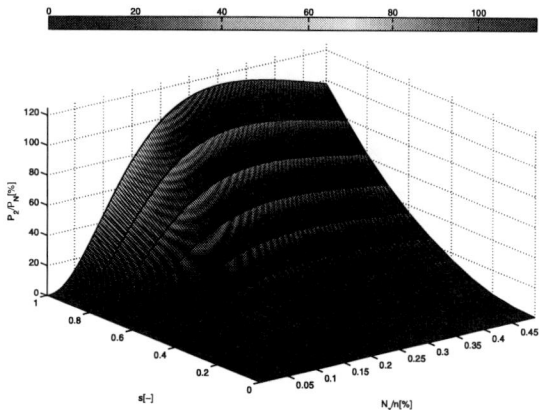

Fig. 4. Power losses in stator winding of short-circuited part

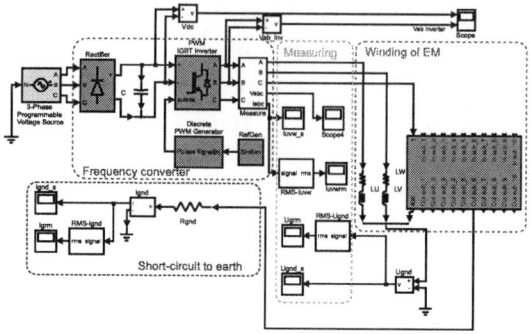

Fig. 5. Model of short-circuit to the earth of induction motor

with raising ratio N_z/n also but only up to 0.25. Then the power losses go down because of rising impedance in short-circuited part of the stator winding.

Overall electric power losses in motor can be computed by the help equation (9) and (10). For example the power losses can be computed for $s = 0.2$, it corresponds to motor torque $M = 4.8Nm$. This constrains represent nominal load for our EM. The results of equation (9) and (10) highlight that power losses are 13 % to the nominal power of the IM.

$$\frac{P_1}{P_N} = 6,2\% \quad \frac{P_2}{P_N} = 4,55\% \quad \frac{P_3}{P_N} = 2,25\% \quad (9)$$

$$\frac{P_z}{P_N} = 13\% \quad (10)$$

It is not anything surprising at the first sight because it could be common value of power losses for a low power IMs without bugs.

This case is very danger for insulation materials in stator winding with regard to location of producing heat by the power losses. Here must be noticed that power losses are for instance only in one or two slots of stator core. It depend on the type of short-circuit and on the type of stator winding.

In the first time the short-circuit can be small, for instance 5 turns of all winding in one phase. It is not danger at the genesis of this bug. Motor can work without problems for long time but later the short-current produce localized heat which warms the surrounding turns and isolation materials up. Than can be small short-circuit extended at more turns of the stator winding. After extension of short-circuit could be a motor damaged in short running time.

This problem is stronger because producers of IMs save materials nowadays. The stator core is smaller and smaller. It rises the problem with heat capacity of stator core. When the stator core is smaller than surface for motor cooling is smaller too. These above mentioned

aspects get higher demands to insulation materials that must fulfill higher requirement.

B. Model of short-circuit to the earth

The second model of short-circuit to the earth was carry out by toolbox SimPowerSystems. This way was chosen because by the help linear circuit is this problem insurmountable. It is the best way of solution when the motor is powered by frequency converter. This model is a little bit time-consuming.

Some unusable results was acquired with this model. The results are not comparable with the real drive because of the unknown algorithm and structure in frequency converter. Fig. 5. shows the real measured currents which flow to the earth. For selected frequencies the current has different time behaviour and amplitude. There was no possibility to find some rules even in long time window of measurement.

It is secret know-how of SIEMENS how to drive our motors. If is not known algorithms that FCs use, it is not possible to assembly correct model. This model must be improved in co-operation with company SIEMENS in the future. According above mentioned reasons the evaluation of short-current to the earth is not easy.

III. FIELD OF TEMPERATURE IN STATOR WINDING

For verification the results of our models was necessary to design special workplace with experimental motor and monitoring unit of temperature. The other important instruments which caried out all our workplace are described in chapter IV.

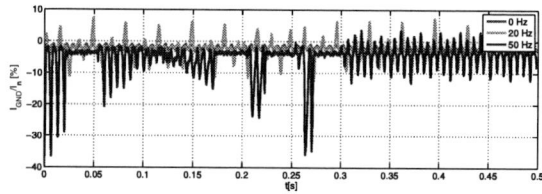

Fig. 6. Current to the earth when the short-circuit to the earth occur for 0 Hz, 20 Hz and 50 Hz

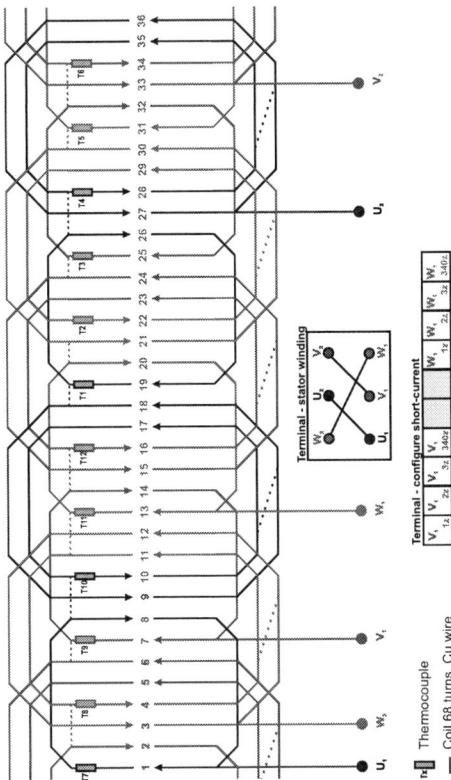

Fig. 7. Winding plan of experimental motor

A. Experimantal motor EM

The experimental induction motor (EM) had to be carry out for monitoring field of temperature and for simulation artificial short-currents in stator winding. The IM had to allow us to create an artificial short circuits in stator winding. There is no easy way how to monitor field of temperature at line produced motors. There is no way to put directly a thermo sensor in stator winding (in a gap of stator coil) and create short circuits too. Therefore was designed special and original stator winding including both possibilites.

It consists of three-pahase winding U, V and W. There are 4 wire branches in the winding V and W which are connected to terminal outside the EM. This terminal allow to configure different short-circuits in stator winding. The coil of phase U is without wire branches. Twelve thermocouple are placed in each third gap in stator core. The complete design of stator winding is shown in the Fig. 7.

The thermocouple was opted because of their small dimensions. It takes small volume in stator winding. With this type sensor of temperature can be monitor many places where is not enough space. For this design was used 12 thermocouple. In the future is designed motor with 24 thermocouple. It will be experimental motor second generation. With it will be possibility to monitor more places in the EM even in the surface.

B. Measuring unit TEMP_12

Monitoring field of temperature carries out special designed unit. The unit TEMP_12 was designed and assembled at Technical university of Liberec. It can measure temperature in twelve diverse points where are thermocouple type J placed.

The unit TEMP_12 monitors all points with multiplex. There are used amplifiers ADG594 for amplifying thermoelectric voltage form thermocouple. All is driven with ADUc824 CPU by Analog Devices. The TEMP_12 sends acquired data via RS232 to PC where they are analysed.

C. Results of measuring field of temperature

The unbalance in field of temperature was verified at experimental workplace with special designed motor for this purpose. The Fig. 8. illustrates measured field of temperature in chosen slots of the stator. There we can see the distribution of heat or more precisely temperature in stator core.

The numbers of T_x present twelve thermocouple in the Fig. 8. They have radial arrangement in stator core. For instance when short-circuit of stator winding occur in phase V, than the place number 6 is the hottest place in motor. It is caused by the distribution of stator winding in slots.

The temperature is distributed from the hottest place towards the other places in the stator core. The field of temperature is outstandingly susceptible to changes of configuration in the stator winding. When the motor is ready, it can be even identify in field of temperature the mechanical part of stator as feet (thermocouple 2,3 and 10, 11).

The results are independent on feeding, speed, load, ambient temperature etc. There are important the difference between temperature of diverse places in our results not absolute values.

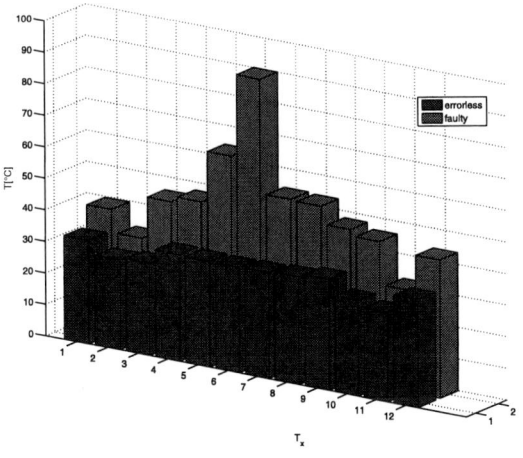

Fig. 8. The field of temperature EM for short-circuited (blue) and for non-short-circuited (red) winding.

These results are very important and interesting but in this case is the problem with places of acquired data. In practise is not possible measure so many places directly in the stator winding. The field of temperature could be measured very easy on the surface of a motor, because heat produced by short-current penetrates via stator core to the surface.

Now is necessary to continue with new design of EM and new workplace which carry out comparison of results in stator winding and on the surface of measured motor. Than it will be possible to verify mutual relation between temperature on the surface of motor and in the stator winding.

IV. EXPERIMENTAL WORKPLACE

In this paragraph is shortly described our workplace for experiments which were carried out at EM.

The experimental workplace consists of two main parts. The first of them is special motor (EM) for simulation different types of short-circuits. EM is powered by frequency converter SIEMENS Micromaster 440. EM has special designed the stator winding with twelve thermocouples described in chapter III-A. These thermocouples perform by the help of measuring unit TEMP_12 monitoring field of temperature in stator winding.

The second part consists of a synchronous servomotor and the FC Simovert Masterdrive MC. The software USS ControlPanel wraps up it to compact easy controlled load for EM. We can with this loading apply many different type of load that can in industry occur. Fig. 9 shows the the arrangement and real realization of experimental workplace.

V. CONCLUSION

There is described incidence of short-circuit current in stator winding of induction motor in this article. Mainly it was focused upon production of power losses (Joules heat) by short-current.

Our experiments show us that some configuration of short-circuit is fatal for running motors in short time of running time. The time to damage depend on immediate loading too. On the other side some configuration of short-cicruit can be detected as different distribution of heat in stator winding of a motor.

If is possible to monitor penetration of heat to the surface of induction motor, the short-circuit will be identify more easily without changes in construction of current IMs. Then can be possible to evaluate unbalance in field of temperature on surface and decide whether the stator winding is in order or damaged. For this state must be done a lot of experiments in laboratory. It is not only task of penetration of heat. It must be consider that unbalance in the field of temperature can affect cover and cooler of used IM. Next factor is type of stator winding (2 poles, 4..8 poles).

The area for experiments is very wide. Therefore is long time research for using intended method in industries. In the end of our research the number of thermal sensors will be reduced with regard experiences that were collected by experiments.

At the end it must be mentioned that configuration of short circuit which don't endanger our motor can occur. For example up to 3 % of stator winding can motor operate without the marks of bugs and problem for long time. The detection in the field of temperature is not almost possible for low unbalance of temperatures.

ACKNOWLEDGMENT

This works is solved with financial contribution of the project 1M0553 "Research center of textile II" and grant No. 102/08/1118 Grant Agency of Czech Republic.

REFERENCES

[1] ARTBAUER, J.- ŠEDOVIČ, J.- ADAMEC A. *Izolanty a izolácie*. 1. vyd. ALFA 1969.
[2] J.F.Chavez, J.J. Martinez Vega. *The High Electric Field Behavior in PET*. 2004 IEEE International Conference on solid Dielectric. Vol.1., page 47-50.. ISBN: 0-7803-8348-6, 2004.
[3] CIGÁNEK, Ladislav. *Stavba elektrických strojů*. Vydání I. Praha: Státní nakladatelství technické literatury, 1958. 716 s. Kapitola 51., Vinutí trojfázového statoru, s. 217-228.
[4] PEROUTKA, Zdeněk. *Dizertační práce: Přechodná přepětí ve střídavých regulovaných pohonech a jejich vliv na izolaci motoru*. Plzeň: Západočeská univerzita v Plzni. Fakutla elektrotechnická. Katedra Elektroniky. 2004. 109 s., 1 s. příloh. Vedoucí dizertační práce Doc. Ing. Václav Kůs, CSc.
[5] UTSUMI, T. – Yamaguchi, I. Detection and Location of Inter–Turn Short Circuit in Linear Induction Motor. In $3^{t}h$ *IEEE Symposium on Diagnostics for Electric Machines, Power Electronics and Drives*. Atlanta, GA, USA, 2003. p.63–68. ISBN 0-7803-7838-5/03.
[6] LEBEY, T. – CASTELAN, P., – KANDEV, N. at al. Testing o low-voltage motor turn insulation intended for pulse-width modulated applications. In *IEEE Transactions on dielectrics and electrical insulation*. 2000. p. 783–789. Vol. 7 No.6. ISBN. 1070-9878.
[7] PATOČKA, Miroslav. *Vybrané stati z výkonové elektroniky – Svazek II: Pulsní měniče bez vf. impulsního transformátoru*. Vydání první. Brno: Ústav el. pohonů a výkonové elektroniky FEI VUT , 1997. Kapitola 6.3. Řízení střídačů – sinusová PWM. s. 138-145. ISBN 80-214-0883-9.
[8] BERAN, Leoš. *Dizertační práce: ANALÝZA ZKRATOVÉHO PROUDU A JEHO TEPELNÉHO ČINKU V MALÝCH ASYN-CHRONNÍCH MOTORECH*. Liberec: Technická univerzita v Liberci. Fakulta mechatroniky a mezioborových inženýrských studií. Ústav Mechatroniky a technické informatiky. 2007. 122 s., 1 s. příloh. Vedoucí dizertační práce Prof. Ing. Aleš Richter, CSc.

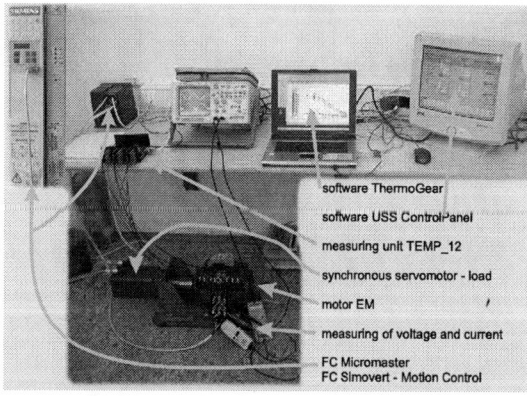

Fig. 9. Experimental workplace for monitoring field of temperature in stator winding of IM

Generalized Model for a Class of Switched Reluctance Motors

Constantin Pavlitov[*], Yassen Gorbounov[*], Radoslav Rusinov[*], Alexandar Alexandrov[**], Kliment Hadjov[**], Dimitar Dontchev[**]

* Department of Electrical Drives Automation, Technical University of Sofia, Bulgaria
** Department of Applied Mechanics, University of Chemical Technology and Metallurgy, Bulgaria
knp@tu-sofia.bg, gorbounov@gmail.com, r_rusinov@yahoo.com, alexa@uctm.edu, klm@uctm.edu, dontchev@uctm.edu

Abstract -- The paper deals with the problems of identification of a class of Switched Reluctance Motors (SRM). A generalized mathematical description of different SRMs with central geometrical symmetry of active poles is suggested. Convenient simulation cellular model of the motor is presented here that is of extreme importance in order to build up a proper SRM drive. The identification of Switched Reluctance Motors is rather sophisticated task due to the fact that the stator inductance depends on both position of the rotor and the stator current. The problem is further complicated due to the influence of the air gap and the highly nonlinear characteristic of the magnetic circuit. The mathematical model of the SRM is of a high order and is described by partial differential equations. In order to avoid this complex mathematical description, artificial neural network substitutions are utilized. They simplify the process of identification and make it easy - using Matlab Simulink. A suggestion for future work is also made in the paper.

Index Terms – Artificial neural networks, Rotating machine nonlinear analysis, Reluctance motor drives.

I. INTRODUCTION

The purpose of the paper is to point out an approach that is based on a superposition of a simple hypothetical motor model described by an artificial neural network [4], which gives an opportunity for easy understanding of the final SRM model. The only constraint of the approach is the motor to be with central geometrical symmetry of the active stator poles. As an example the following motors can be identified using this method: SRM6-4, SRM4-6, SRM8-6, SRM6-8, SRM12-8 and etc. [2]. The suggested approach helps to avoid the sophisticated mathematical description, provoked by the fact that the stator inductance depends on both, position of the rotor and the stator current value [2, 3]. The last dependence leads to partial differential equation description, which sometimes is rather complex task, even for contemporary simulation facilities. Furthermore, neural network description helps to reflect the nonlinearity character of the magnetic circuits, as well as the influence of the air gap.

II. THE APPROACH ESSENCE

A. Hypothetical nonlinear neural network motor model

The starting point of this approach is the hypothetical motor model illustrated in Fig. 1. In fact it is a SRM2-2 motor that has 2 stator and 2 rotor poles. This model is not of practical use but it is a base point for further development of different SRM models.

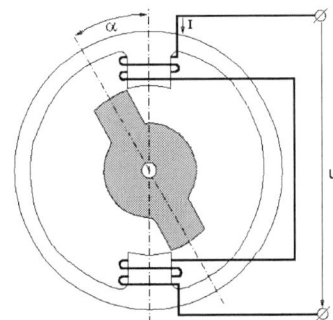

Fig. 1. Hypothetical SRM2-2.

The mathematical description of this hypothetical single-phase motor is given below (system of equations (1)). It is a result of a previous research of the authors [1].

$$L(I,\alpha) = L(I).N(\alpha) \tag{1-a}$$

$$N(\alpha) = NetN(\alpha) \tag{1-b}$$

$$\frac{dN(\alpha)}{d\alpha} = NetN'(\alpha) \tag{1-c}$$

$$L(I) = NetL(I) \tag{1-d}$$

$$\frac{dL(I)}{dI} = NetL'(I) \tag{1-e}$$

$$\frac{d\alpha}{dt} = \omega \tag{1-f}$$

$$\frac{d\omega}{dt} = \varepsilon \tag{1-g}$$

$$\frac{dI}{dt} = \frac{I.[\dfrac{U}{I} - R - L(I).\dfrac{dN(\alpha)}{d\alpha}.\omega]}{L(I,\alpha) + I.N(\alpha).\dfrac{dL(I)}{dI}} \tag{1-h}$$

$$\varepsilon = \frac{1}{J}(\frac{1}{2}.I^2.L(I).\frac{dN(\alpha)}{d\alpha} - c.\omega - T_L) \tag{1-i}$$

978-1-4244-1741-4/08/$25.00 ©2008 IEEE

where: $L(I,\alpha)$ – is the current inductance of the phase, depending on both phase current and rotor position; $L(I)$ – is the inductance in the aligned position of the rotor which depends only from the phase current; $N(\alpha)$ – is the normalized inductance of the phase in zero current conditions or when the motor is not loaded; U is the stator voltage; I is the stator current; R is the active resistance of the stator winding; c is the friction coefficient (bearing and ventilating); J is the moment of inertia; ω is the angular speed and ε is the angular acceleration. Due to the description of $N(\alpha)$, $dN(\alpha)/d\alpha$, $L(I)$, $dL(I)/dI$ with backpropagation artificial neural networks which are demonstrated on the graphics in Fig. 2 and Fig. 3, the system (1) consists only of algebraic and ordinary differential equations, otherwise it would has been described by partial derivatives. The modeling of the system is not hard to be done by MATLAB Simulink.

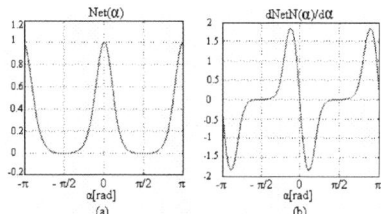

Fig. 2. Normalized stator inductance and its derivative in respect to the angle α.

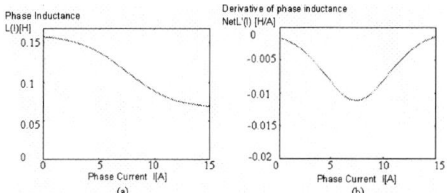

Fig. 3. Aligned phase inductance in function of current (a) and phase inductance derivative (b).

The Neural Networks descriptions in the hypothetical motor model are shown in Fig. 4.

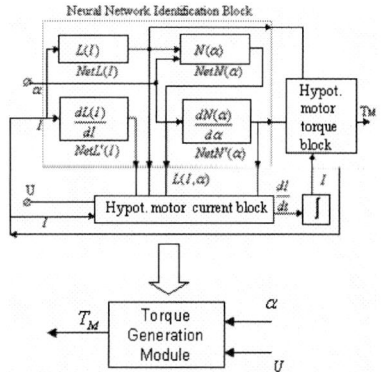

Fig. 4. Neural network descriptions in the hypothetical motor model: torque generation module and its neural network ingredients.

They take part as ingredients of the torque generation module. The structure of the acceleration block is depicted in Fig. 5.

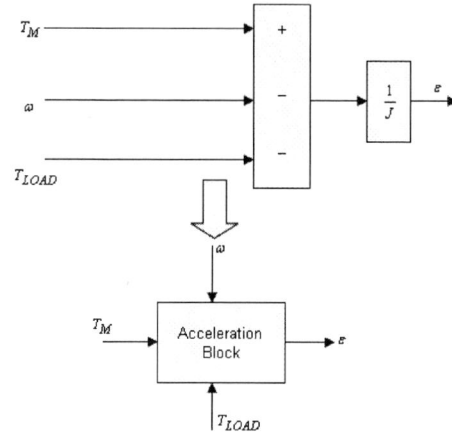

Fig. 5. Acceleration Block Reflected by Equation (1-i)

It is seen from it that tribo (friction and bearings) effects are also taking into account. The complete block diagram of the hypothetical motor which is described by the whole system – from (1-a) to (1-i), can be seen in Fig. 6.

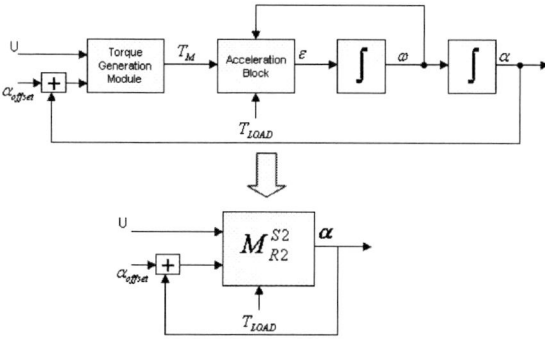

Fig. 6. Block diagram of the hypothetical motor SRM2-2 (upper one). Down in the picture the simplified SRM2-2 representation is shown. The position feedback and offset angle give opportunity for a variation of the rotor positioning.

B. Superposition of the hypothetical model

If there is a need to describe the hypothetical motor shown in Fig. 7, then the suggested convenient description will be like (2):

$$M_{R4}^{S2}(\alpha,U) = M_{R2}^{S2}(\alpha,U) + M_{R2}^{S2}\left(\alpha + \frac{\pi}{2},U\right) \qquad (2)$$

The last model is obtained as a superposition of the model that is pictured in Fig. 1, taken two times and with phase difference of $\pi/2$. As it is denoted by the superscript and the subscript (M_{R4}^{S2}) the motor is a single phased SRM2-4 motor with 2 stator poles and 4 rotor poles. The simplified model can be seen in Fig. 8. The position feedback gives opportunity for 4 stable aligned rotor locations. Evidently this description is very close to a single phased SRM6-4 motor type.

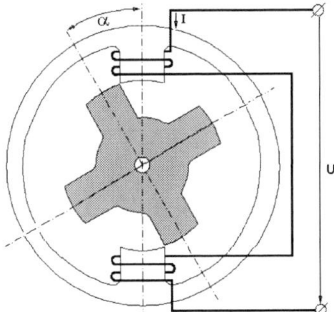

Fig. 7. Hypothetical SRM2-4

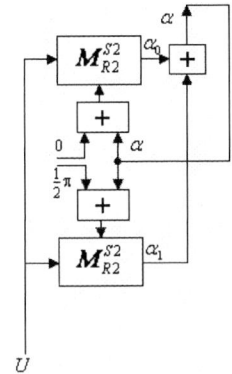

Fig. 8. The SRM2-4 is presented as a superposition of two hypothetical SRM2-2 with phase angle deviation of π/2

C. Mathematical model of SRM6-4

What happens if the motor consists of three phases like the one shown in Fig. 9?

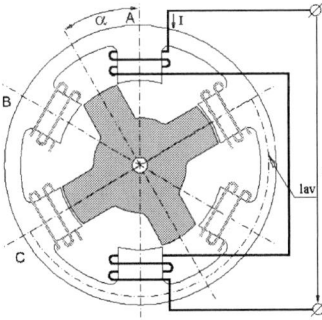

Fig. 9. Three phase SRM6-4

Obviously the equation of the kind (2) will be tripled. For every phase there will be a π/3 radians phase difference, as in (3):

$$PhA \rightarrow M_{R4}^{S2}(\alpha, U_A) = M_{R2}^{S2}(\alpha, U_A) + M_{R2}^{S2}(\alpha + \frac{\pi}{2}, U_A)$$

$$PhB \rightarrow M_{R4}^{S2}(\alpha + \frac{\pi}{3}, U_B) = M_{R2}^{S2}(\alpha + \frac{\pi}{3}, U_B) + M_{R2}^{S2}(\alpha + \frac{\pi}{2} + \frac{\pi}{3}, U_B)$$

$$PhC \rightarrow M_{R4}^{S2}(\alpha + \frac{2}{3}\pi, U_C) = M_{R2}^{S2}(\alpha + \frac{2}{3}\pi, U_C) + M_{R2}^{S2}(\alpha + \frac{\pi}{2} + \frac{2}{3}\pi, U_C)$$

(3)

After some transformations of (3) having in mind the periodical character of the neural network functions [0-π], the following model of SRM6-4 can be obtained, as in (4):

$$PhA \rightarrow M_{R4}^{S2}(\alpha, U_A) = M_{R2}^{S2}(\alpha + \frac{0}{6}\pi, U_A) + M_{R2}^{S2}(\alpha + \frac{3}{6}\pi, U_A)$$

$$PhB \rightarrow M_{R4}^{S2}(\alpha + \frac{\pi}{3}, U_B) = M_{R2}^{S2}(\alpha + \frac{2}{6}\pi, U_B) + M_{R2}^{S2}(\alpha + \frac{5}{6}\pi, U_B)$$

$$PhC \rightarrow M_{R4}^{S2}(\alpha + \frac{2}{3}\pi, U_C) = M_{R2}^{S2}(\alpha + \frac{4}{6}\pi, U_C) + M_{R2}^{S2}(\alpha + \frac{1}{6}\pi, U_C)$$

(4)

The mathematical description (4) points out how to obtain the mathematical model of the SRM6-4 in Fig. 9, using only the description (1), i.e. M_{R2}^{S2}. Superposing it 6 times with different initial rotor angles the system is ready to be used. There is also more compact way to describe the same system (5):

$$PhA, B, C \triangleright \alpha = \sum_{K=A,B,C} M_{R4}^{S2}(\alpha, U_K) = \sum_{i=0}^{5} M_{R2}^{S2}(\alpha + i\frac{\pi}{6}, U_K)$$

$$\begin{matrix} case \\ (i=0) \oplus (i=3) \rightarrow K=A \\ (i=1) \oplus (i=4) \rightarrow K=C \\ (i=2) \oplus (i=5) \rightarrow K=B \end{matrix}$$

(5)

A graphical representation of (5) is pictured in Fig. 10.

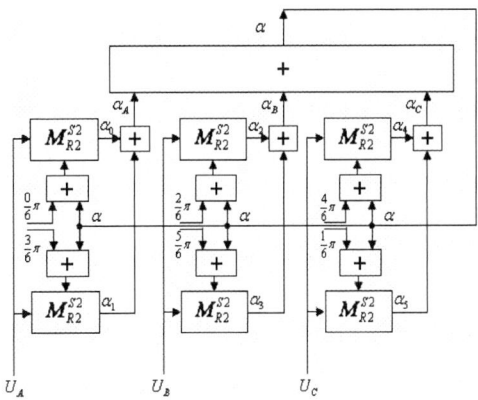

Fig. 10. The model of SRM6-4 based on a hypothetical model superposition

There are 12 stable aligned positions of the rotor in this model representation. In other words it is equal to 0.52[rad] step size. These steps can be seen by the stepper working mode of the motor, demonstrated in Fig.11.

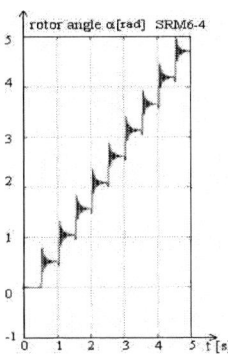

Fig. 11. Stepper mode for SRM6-4. Step of 0.52[rad].

Phase currents and inductances are shown in Fig. 12.

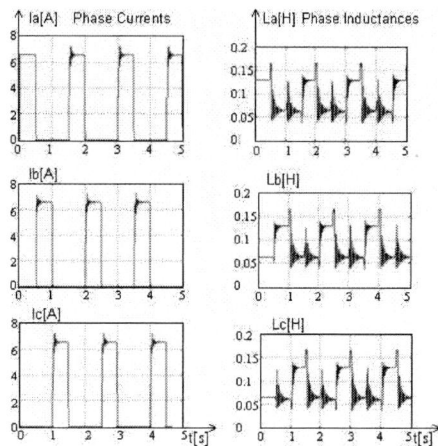

Fig. 12. Phase currents and inductances of SRM6-4

The inductance peak marks the switching off instance of the phase. This peak will not exist if motor is not loaded and phase current is near to zero. The continuous mode where stator coils activation is synchronized to the position of the rotor is depicted in Fig. 13.

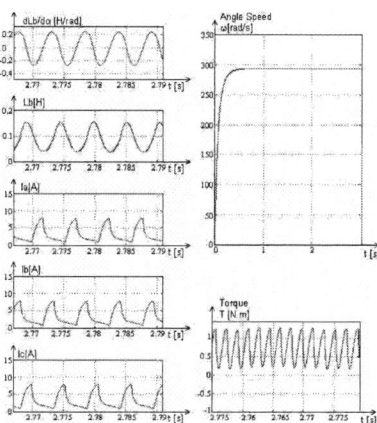

Fig. 13. Continuous mode

The commutations are made in certain advance angle, before rotor reaches the align position. That advance angle can be explored by the suggested model. Some results are given in Fig. 14.

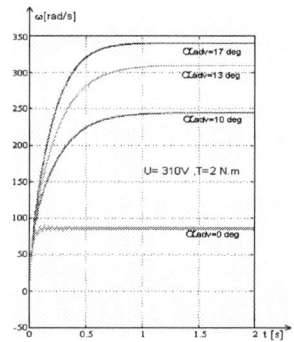

Fig. 14. Advance angle influence on the motion speed

By the help of this model the mechanical characteristics of the SRM6-4 can be also explored. They are pointed in Fig. 15.

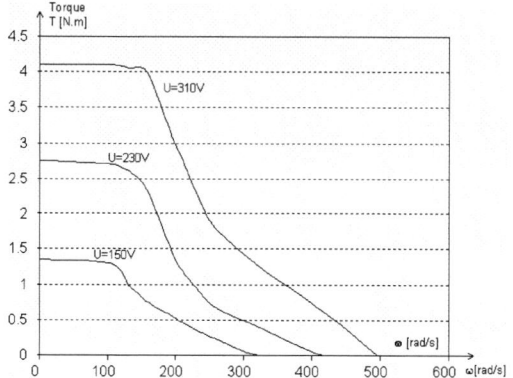

Fig.15.Mechanical characteristics of the SRM6-4 obtained by the mathematical model

D. Mathematical model for SRM8-6

Analogically to (5) the following relation for SRM8-6 motor can be written (6):

$$PhA, B, C, D \triangleright \alpha = \sum_{K=A,B,C,D} M_{R6}^{S2}(\alpha, U_K) = \sum_{i=0}^{11} M_{R2}^{S2}(\alpha + i\frac{\pi}{12}, U_K) \tag{6}$$

$$\begin{aligned} case \\ (i=0)\Theta(i=4)\Theta(i=8) \to K=A \\ (i=1)\Theta(i=5)\Theta(i=9) \to K=D \\ (i=2)\Theta(i=6)\Theta(i=10) \to K=B \\ (i=3)\Theta(i=7)\Theta(i=11) \to K=C \end{aligned}$$

The cell offset in that case is $\pi/12$. The 4 neural networks described in the first case should be trained again but the equations for the hypothetical model remain untouched. The cellular (superposed) model of the motor will be like in Fig. 16:

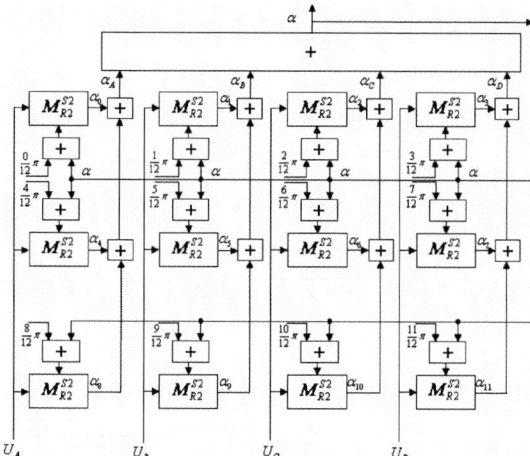

Fig. 16. SRM8-6 cellular model based on a hypothetical motor superposition

It is seen from Fig. 16 that every phase consists of three cells which offset (vertically) is $\pi/3$ from each other. This offset is determined by the rotor form which has symmetrical structure with six poles. The phase offset is $\pi/4$. The cell offset (horizontally) is equal to $\pi/3-\pi/4=\pi/12$. Stepper mode of work is pictured in Fig. 17. It is seen from it that the value of the step is equal to 0.26[rad] which is equal to $\pi/12$. The two times smaller step of SRM8-6 compared to SRM6-4, means two times bigger frequency of control on equal speeds.

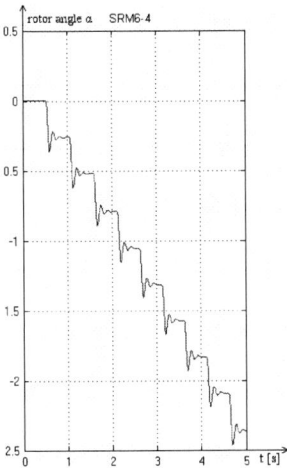

Fig. 17. Stepper Mode of SRM8-6

The cellular method can be applied the same way to SRM12-8, SRM6-8 etc. In order to do so, two things have to be kept in mind:

• The 4 neural networks must be properly trained for each motor. They have to be obtained experimentally or in theoretical way [1]. The hypothetical motor equations are always the same.

• The cell offset should be found in advance. It is equal to rotor offset minus phase offset. The cellular offsets (demonstrated in Fig. 10 and Fig. 16) are incremented (by one step) from left to right starting with the upper left cell. After finishing the row, the increment continues one row down also from left to right.

E. Mathematical model for SRMn-m

Generally any switched reluctance motor model with central symmetry of the active poles can be described with the generalized equation as follows (7):

$$Phases A_1 A_2 ... A_n \rightarrow \alpha = \sum_{K=A_1,A_2,...A_n} M_{Rm}^{S2}(\alpha, U_K) =$$

$$= \sum_{i=0}^{\frac{m.n}{2(n-m)}-1} M_{R2}^{S2}(\alpha + 2.\pi.i\frac{n-m}{m.n}, U_K) \quad (7)$$

In this expression the index K is determined as follows (8):

$$(i=0) \oplus (i=\frac{m}{2}+1) \cdots (i=\frac{n.m}{2(n-m)}-1-\frac{m}{2}) \rightarrow K = A_1$$

$$(i=1) \oplus (i=\frac{m}{2}+2) \cdots (i=\frac{n.m}{2(n-m)}-1-\frac{m}{2}+1) \rightarrow K = A_2$$

$$\vdots$$

$$(i=\frac{n}{2}-1) \oplus (i=\frac{m}{2}+\frac{n}{2}) \cdots (i=\frac{n.m}{2(n-m)}-1-\frac{m}{2}+\frac{n}{2}-1) \rightarrow K = A_n$$

$$(8)$$

Where $n.m/(n-m)$ is the number of the steps in the stepper mode (for revolution) and $2.\pi.n.m/(n-m)$ is the size of these steps.

III. PROSPECTIVE

The generalized mathematical identification model can be further applied to the sensorless control of SRM motors which eliminates the need of the expensive and environmentally sensitive optical encoder. In this case the model of the flux linkage is necessary to be obtained. It is not hard to be done by the aid of the identification model and especially this part of it that concerns the neural network inductance description. An example for the SRM6-4 flux linkage in regard to the phase current and advanced angle, obtained using the model from chapter II.C is given in Fig. 18.

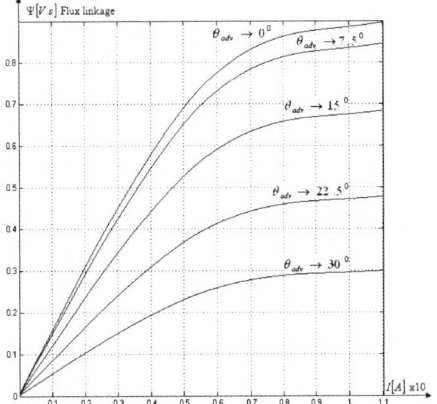

Fig. 18. Flux linkage in regard to phase current and advanced angle, obtained by the SRM6-4 model.

If measuring the current through each of the phases, and knowing the advance angle, the exact instant of the switching-off time can be easily derived. The general block diagram of the sensorless control for a single phase is given in Fig. 19.

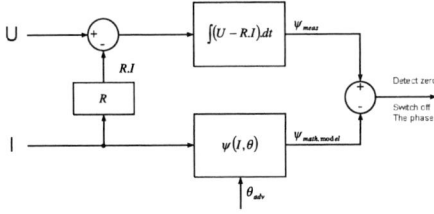

Fig. 19. Flux linkage in sensorless control.

At the bottom of the picture the etalon model of the flux linkage is given. This model has been obtained by the aid of the neuronal SRM mathematical model. At the top of the Fig. 19, an observer of the flux linkage has been built. Subtracting the model flux linkage from the observed one, the switching-off angle for the phase can be determined. This comparison can be made using a zero zone detector.

This sensorless control algorithm is intended to be practically developed in the future.

IV. CONCLUSIONS

A generalized mathematical description of various SRMs with central geometrical symmetry of active poles was suggested. Convenient hypothetical cellular simulation model of the motor has been presented in the paper that has showed to be of extreme importance in order to build up a proper SRM drive.

The usage of neural networks description of the phase inductance and its derivatives, simplifies the mathematical model. The neural networks help to describe the model with ordinary derivatives instead of expressing it with partial ones, which makes the solution much easier. The four neural networks have been trained using the *Levenberg-Marquardt algorithm*. Performance goal has been met in 50% of the cases for the *NetL(I)*, 39% for the *NetL(I)/dI*, 83% for the *NetN(α)* and 91% for the *NetN(α)/dα*. The values of the mean square errors for these networks have been found to be $1,3742.10^{-5}$, $2,4476.10^{-6}$, 0.0073 and 0.0344 respectively.

The model has been initially created for *the SRM2-2 hypothetical motor (called cellular block)* and afterwards it has been transformed by means of superposition to SRM6-4, SRM8-6 and etc. Besides, this model can be spread easily over various VRM motor types by obtaining different neural network descriptions over phase inductance and its derivatives. In this model the nonlinearity of the magnetic circuits and the air gap influence has also been taken into account [1].

Finally, a general overview of continuing further work for implementing sensorless motor control of SRMs that eliminates the need of the expensive and environmentally sensitive optical encoder has been proposed.

All of the above-mentioned items give opportunities for rather precise and adequate mathematical description that can help further for the optimal solution of the motor control task.

REFERENCES

[1.] Pavlitov C.; Gorbounov Y.; Georgiev Tz.: *Application of Artificial Neural Networks for Identification of Variable Reluctance Motors*, EDPE 24-26 September 2007, Podbanske, The High Tatras, Slovakia.

[2.] Miller T. J. E.: *Electronic Control of Switched Reluctance Machines*, ISBN 0 7506 50737, Reed Educational and Professional Publishing Ltd., 2001.

[3.] Krishnan R.: *Switched Reluctance Motor Drives: Modelling, Simulation, Analysis, Design, and Applications*, Virginia Tech, CRC Press, 2001.

[4.] Demuth H., Beale M., *Neural Network Toolbox For Use with MATLAB, Users Guide v.4*, www.mathworks.com

[5.] Nagy I.; Sütő Z.: *Nonlinear Phenomena in the Current Control of Induction Machines*, in nonlinear phenomena in power electronics Attractors, Chaos, Bifurcation and Nonlinear Control edited by S. Banerjee, G. Verghese, IEEE Press USA, 2001

Neural Network based Fault Detection of PMSM Stator Winding Short under Load Fluctuation

J. Quiroga[*], D.A. Cartes[*], C.S. Edrington[*] and Li Liu[*]

[*] Center for Advanced Power Systems-Florida State University /Tallahassee, USA, e-mails: jeq06@fsu.edu, dave@eng.fsu.edu, edrinch@eng.fsu.edu, lliu@caps.fsu.edu

Abstract— A negative sequence analysis coupled with a neural network based approach is applied to fault detection of a single phase winding short in a PMSM. A multilayer network provides a near term current prediction as input to the fault detection system. The fault detection is performed using the negative sequence analysis of the residuals (difference between the actual and predicted values of currents). The negative sequence component of the residuals provides the detection of the fault and a measurement of the level of severity of the winding short. The method is validated using a 15 hp PMSM experimental setup.

Keywords— Fault Detection, PMSM, Stator Winding Short, Neural Network.

I. INTRODUCTION

Early detection of a stator short-circuit fault in permanent magnet synchronous motors (PMSM) is critical in preventing damage to the machine. Fast and accurate diagnosis of these types of faults allows actions to be taken to protect the power system and the machine. Recently, there has been considerable interest in fault detection and diagnosis techniques for condition based maintenance (CBM) [1-2]. These approaches rely on information provided by condition monitoring systems that assess the system condition continuously. The proposed technique in this work forms a basis for implementing CBM in a physical system.

In literature, work on fault detection and diagnosis (FDD) of electric machines is primarily focused on induction motors [3-5]. The absence of publications in detection of stator winding faults in PMSMs prevents a systematic comparison of the proposed method. However, recent success in using the negative sequence as a fault indicator for induction motors, as presented in [3, 6] may be extended as an approach to the PMSM.

In [7], it is proposed to use a dynamic recurrent network in the form of an IIR filter as a multi-step predictor for complex systems. Present and delayed observations of the measured system inputs and outputs, are utilized as inputs to the network. The proposed architecture includes local and global feedback. In [8] a FDD system for mechanical and electrical faults is developed for induction motors based on neuropredictors and wavelets; however, the load fluctuation is not addressed.

This work was supported in part by the Office of Naval Research (ONR) under grant N000140210623.

Additionally, experiments with a PMSM testbed show a significant increase in the negative sequence components under load fluctuation, which can trigger false alarms in condition monitoring systems currently operating under healthy conditions. Although the effect of other sources of asymmetries such as inherent motor imbalance and instrumentation is not considered in this paper, the experimental NN model developed does permit a real assessment of the machine's condition; because during the process of modeling (training) the effect of other sources of asymmetries different from the faults are embedded in the model.

II. METHODOLOGY

A. Negative Sequence Component

The method of symmetrical components is a mathematical technique to describe unbalanced power systems. For a three phase system, the sequence components of the current are described by (1)

$$I^0 = \frac{1}{3}\left(I_a + I_b + I_c\right)$$

$$I^+ = \frac{1}{3}\left(I_a + \alpha I_b + \alpha^2 I_c\right) \qquad (1)$$

$$I^- = \frac{1}{3}\left(I_a + \alpha^2 I_b + \alpha I_c\right)$$

where I^0, I^+, and I^- are the zero, positive and negative sequence currents, respectively, and α is a phase rotation operator equivalent to $e^{j2\pi/3}$ or 120°. A balanced set of three phase currents only contains the positive sequence current component, the negative sequence current is an indication of the amount of unbalance (asymmetry) in the system and the zero sequence current is a measure of the amount of current not returning through the phase conductors. The positive sequence components determine the torque produce by the motor and the direction of rotation of the machine [9].

Asymmetries in the 3φ quantities of the PMSM may arise due to a variety of reasons: 1) stator winding short circuits, 2) inherent machine and instrumentation asymmetries, 3) load fluctuations and 4) unbalanced supply voltages. Such asymmetries are reflected in the negative sequence component. In this study, the negative sequence of the residual signal generated by the comparison between the NN simulated fundamental currents and the actual fundamental currents is employed as a fault indicator, which is obtained for the PMSM using (2) [7]:

$$r_{a2} = \frac{1}{3}\left(r_a + \alpha^2 r_b + \alpha r_c\right) \tag{2}$$

where r_a, r_b and r_c are the magnitudes of the residual signals for the three fundamental phase-currents and r_{a2} is the negative sequence of the residuals.

B. Neural Network Model Development

Because of the complexity of the dynamic behavior associated with the PMSM under load fluctuation and the difficulty associated in establishing an exact mathematical formulation to develop an explicit model of the PMSM with conventional methods, a nonlinear empirical model using a NN is developed. In this paper, it is proposed to utilize a multi-layer dynamic recurrent NN with local feedback of the hidden nodes and global feedback, as shown in Fig. 1. Local feedback implies use of delayed hidden node outputs as hidden node inputs, whereas global feedback is produced by the connection of delayed networks outputs as network inputs. This architecture provides a network in the form of a nonlinear infinite impulse response (IIR) filter.

The operation of a recurrent NN predictor that employs global feedback can be represented as (3):

$$\hat{y} = \Phi \begin{pmatrix} u(k), u(k-1), u(k-2), ..., u(k-p) \\ \hat{y}(k-1), ..., \hat{y}(k-q), W \end{pmatrix} \tag{3}$$

where $\Phi(\bullet)$ represents the nonlinear mapping of the NN, u is the inputs, $\hat{y}$ is the simulated values and W is the parameters associated to the NN.

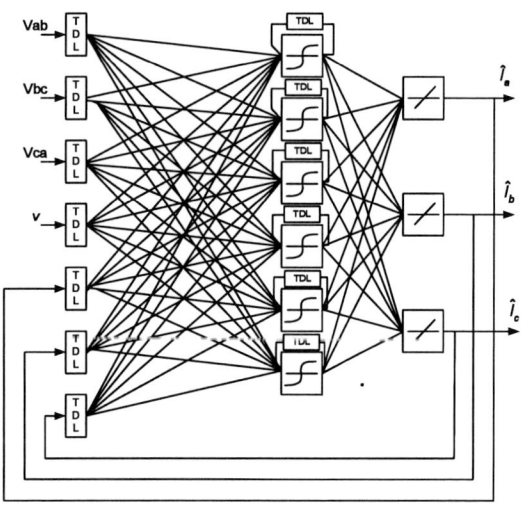

Fig. 1 General structure of multilayer NN

This NN architecture provides the capability to predict the output several steps into the future without the availability of actual outputs. Empirical models with predictive capabilities are desirable in fault monitoring and diagnosis applications. The implemented NN consists of an input layer, a hidden layer, and an output layer. Each of the processing elements of a MLP network is governed by (4). (1)

$$x_{[l,i]} = \sigma_{[l,i]}\left(\sum_{j=1}^{N_{[l-1]}} w_{[l-1,i][l,i]} x_{[l-1,j]} + b_{[l,i]}\right) \tag{4}$$

for $i = 1,...,N_{[l]}$ (the node index), and $l = 1,....,l$ (the layer index), where $x_{[l,i]}$ is the ith node output of the lth layer, $b[l,i]$ is the bias, and $\sigma_{[l,i]}(\bullet)$ is the activation function of the ith node in the lth layer. The relationship between inputs and outputs in a multilayer NN can be expressed using a general nonlinear input-output model:

$$\hat{y}(k,W) = f\left(u(k);W\right) \tag{5}$$

where W is the weight matrix determined by the learning algorithm, $f(\cdot)$ represents the nonlinear mapping of the vector input using any activation function. In this study, the *tansig* function is used in the hidden layer and *purelin* is used in the output layer. The input vector is defined as:

$$U(t) = \begin{bmatrix} V_f^{NS}(t), V_f^{NS}(t-1), V_f^{NS}(t-2), \\ v^{NS}(t-1), v^{NS}(t-2), \hat{I}_f^{NS}(t-1), \\ \hat{I}_f^{NS}(t-2) \end{bmatrix} \tag{6}$$

where NS represents a non-stationary signal, V_f^{NS} are the actual normalized values of the 3 phases line voltages:

$$V_f^{NS} = \left[V_{ab_f}^{NS}, V_{bc_f}^{NS} V_{ca_f}^{NS}\right] \tag{7}$$

The vector $\hat{I}_f^{NS}$ is the three normalized predicted phase currents:

$$\hat{I}_f^{NS} = \left[\hat{I}_{a_f}^{NS}, \hat{I}_{b_f}^{NS}, \hat{I}_{c_f}^{NS}\right] \tag{8}$$

The variable v^{NS} is the normalized rotational velocity of the rotor. The hidden layer is composed of 6 neurons with delayed local feedback employed in each neuron. The hidden layer node number is chosen considering the balance of accuracy and network size. The output layer, with global feedback, has 3 nodes, which correspond to the three phase-current predictions as shown in (9).

$$\hat{y}(t+1|t) = \left[\hat{I}_a(t+1|t), \hat{I}_b(t+1|t), \hat{I}_c(t+1|t)\right] \tag{9}$$

C. Model Training and Validation

Generally, training recurrent dynamic networks is computationally intensive and in this work has been difficult due to the time dependencies present in their architectures. Recurrent networks exhibit complex error surfaces characterized by very narrow valleys which bottoms are often cusps. Additionally, initial conditions assigned in the training stage and variations in the input sequence can produce spurious valleys in the error surface [10].

The goal of the NN training is to produce a network, which yield small errors on the training set, but which will also respond properly to novel inputs (regulation). Therefore, in order to provide appropriate training of the model consideration of issues such as: regulation, initial values of the parameters, as well as the need to train the NN several times, must be addressed in order to achieve optimal results.

The neural network model proposed is trained using Bayesian regulation conveniently implemented within the framework of the Levenberg-Marquardt algorithm.

Regulation is used to avoid an over fitted network and to produce a network that generalizes well [11]. This approach constrains the size of the network weights, adding a penalty term proportional to the sum of the squares of the weights (*msw*) and biases to the performance function (mean sum of squares of the neural network). As can be seen, the objective function becomes a maximum penalty likelihood estimation procedure as shown in (10).

$$msereg = \beta mse + \alpha msw \qquad (10)$$

where α and β are objective function parameters. This approach provides to the neural network a smooth response. The values of α and β determine the response of the NN. When $\alpha \ll \beta$, over fitting of the NN occurs. If $\alpha \gg \beta$ the NN does not adequately fit the training data. In [12] one approach is proposed to determine the optimal regulation parameter based on a *Bayesian framework*. In this framework, the weights and biases of the network are assumed to be random variables with specified distributions. The regularization parameters are related to the unknown variances associated with these distributions. These parameters can be estimated using statistical techniques.

The Levenberg-Marquardt algorithm is a variation of Newton's method and was designed for minimizing functions that are sums of squares of other nonlinear functions [12]. The algorithm speeds up the training by employing an approximation of the Hessian matrix (11). The gradient is computed via (12),

$$H = J^T J \qquad (11)$$

$$g = J^T e \qquad (12)$$

where J is the Jacobian matrix that contains first derivatives of the network errors with respect to the weights and biases, and e is a vector of network errors.

The Jacobian matrix is much less complex than computing the Hessian matrix. The Levenberg-Marquardt algorithm uses this approximation to the Hessian matrix in the following Newton-like update:

$$x_{k+1} = x_k - \left[J^T J + \mu I \right]^{-1} J^T e \qquad (13)$$

where x_k is a vector of current weights and biases. The Levenberg-Marquardt algorithm is an accommodative approach between Gauss-Newton's method (faster and more accurate near an error minimum) and gradient descent method (guaranteed convergence) based on the adaptive value of μ. If scalar μ is zero then the update process described by (13) resembles Newton's method using the approximate Hessian matrix. If μ is large, then this process becomes gradient descent method with a small step size.

A complete description of the Levenberg-Marquardt Backpropagation (LMBP) algorithm can be found in [12]. A detailed discussion about the implementation of Bayesian Regulation in combination with Levenberg-Marquardt training is presented in [11].

As mentioned previously, the initial weight and bias values assigned during the training stage affect the performance of the neural network. During the NN model training each layer's weights and biases are setting according to the method proposed for Nguyen and Widrow in [14]. This method for setting the initial weights of hidden layers of a multilayer neural network provides a considerable reduction in training time. Using the Nguyen and Widrow initialization algorithm, the values of weights and bias are assigned at the beginning of the training, then the network is trained and each hidden neuron still has the freedom to adjust its own values during the training process.

The proposed NN is trained offline using values collected at a 625 Hz sampling frequency followed by scaling in the range of [0:1] of the magnitude of the fundamental components of the voltage and the rotational velocity as inputs. The targets are the normalized values of the magnitudes of the fundamental components of the three phase currents. The training data set consists of 5487 samples; the number of parameters to be calculated (weights and biases) during the training stage is 579 for the neural network proposed. The training data set is comprised of measurements taken by loading the motor between no-load and 30% of the rated torque applied by ramping for 2 seconds, followed by 2 seconds of constant 30 % of rated torque and then a ramp down for 2 seconds to no-load, as shown in Fig. 2. The criterion established for checking if the network has been trained for a sufficient number of iterations to ensure convergence is obtained when the values of the Sum Squared Error (*SSE*) is relatively constant over several iterations. In the case of the performed training, the neural network achieved convergence when the value of *SSE* was approximately 1.

The use of regulation avoids the need to use a validation data set.

In testing the performance of the developed network, the normalized mean square error and the absolute mean error are used. The testing data set consists of measurements (no-load, 10%, 20%, 30%, 40%, and 45% of variation with respect to the rated torque) entirely different than the ones used in the training data set in order to evaluate the generalization performance of the network. The performance evaluation results for the testing set are summarized in Table I in terms of mse and mean error and demonstrate the generalization performance of the network up to 45% of the rated torque.

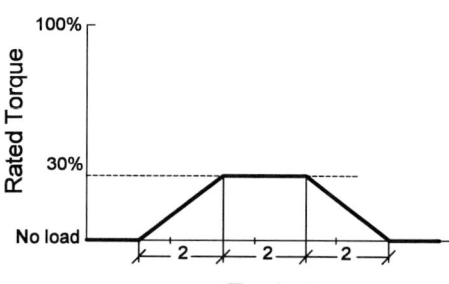

Fig. 2 Load applied to obtain the training set

Table I. GENERALIZATION PERFORMANCE OF THE NETWORK FROM NO TORQUE UP TO 45% OF THE RATED TORQUE

Rated torque (%)	NMSE (%)			Mean |Error (%)|		
	$I_A(t)$	$I_B(t)$	$I_C(t)$	$I_A(t)$	$I_B(t)$	$I_C(t)$
0	1.44	1.43	1.47	0.12	0.21	0.01
10	1.13	1.16	1.17	0.10	0.20	0.01
20	0.68	0.71	0.73	0.05	0.11	0.02
30	0.35	0.41	0.35	0.04	0.08	0.04
40	1.16	1.2	1.17	0.14	0.14	0.03
45	2.26	2.20	2.25	0.15	0.16	0.03

D. Residual Generation

The residuals are the basis of the fault detection strategy and are produced by comparing the three phase current predictions and the actual values of the three phase currents. Theoretically, the residuals are small values during normal operation even under load fluctuation. However, the residuals generated under a fault condition, deviate from the nominal value. The residual for phase A (phases B and C are similar) is expressed in (12):

$$r_a(t) = \left[I_a(t) - \hat{I}_a(t) \right] \tag{12}$$

where I_a is the actual value of current in phase A at time t and $\hat{I}_a$ is the predicted value of current in phase A at time t.

As mentioned in section A, the fault indicator proposed is based on the observation of the magnitude of the negative sequence component of the residuals. The rise in magnitude in this indicator, above some baseline

established previously by observation of the motor under normal conditions, is then the result of the presence of a fault condition. The near term current predictor proposed is able to track the variations in amplitude in each one of the fundamental phase currents in the motor under a changing load condition, one of the sources of the asymmetry in the negative sequence current. Therefore, information carried by the negative sequence current of residuals is immune to the asymmetry produced by the load fluctuation.

E. Experimental Approach

In the proposed fault detection system, the data acquisition system allows the sampling of $V^{NS}(t)$, $I_f^{NS}(t)$ and $v(t)$. The signals are sampled at 625 Hz and the voltage and current signals are filtered using bandpass filters to obtain the fundamental components of each phase i.e. $V_f^{NS}(t)$ and $I_f^{NS}(t)$. A NN based motor model is used to produce the residuals. The block diagram of the overall system is illustrated in Fig. 3.

The proposed fault detection system is experimentally validated using a system which consists of a 28.8 kVA variable frequency drive connected to an 11.25 kW, 640 V, 60 Hz, Y-connected 8-pole PMSM. A dc motor is mechanically coupled to the PMSM to serve as a load (see Fig. 4).

During the experiments, the load is changed by varying the armature resistance of the dc motor, in order to emulate a load fluctuation condition, e.g. increasing or decreasing the load from 0% up to 45%.

In this experimental setup, the PMSM is specifically designed to imitate turn-to-turn fault conditions in the stator windings. A schematic illustration of this is shown in Fig. 5, note that monitoring of the fault current generated in the short circuit loop is provided. Currently, only two taps are placed in phase A resulting in the percentage of shorted windings as 6.25% and 12.5%, respectively. To emulate a less severe fault condition, an adjustable resistor is used as shown in Fig. 5. The severity level of the stator winding fault can be adjusted by varying the value of this resistor, which is reflected in the variation of the loop current as shown in Table II.

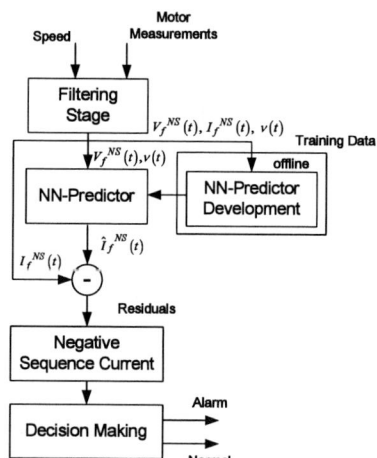

Fig. 3 Overview of the fault detection system

The system is developed using MATLAB®/Simulink with dSPACE® as an interface to the data acquisition hardware and PMSM drive system. The fully tested algorithm is applied to the electrical system and performance can be studied in dSPACE®, which is used to display and record the line voltages, line currents, the fault detection output and the torque signal.

Fig. 4 PMSM experimental test bed

Table II. VALUES OF LOOP CURRENT COMPARED WITH THE PERCENTAGE OF SHORTED WINDINGS IN PHASE A OF THE PMSM

Implementation	Loop current (A)	Percentage of shorted windings (%)
Taps	115	12.5
	75	6.25
Variable resistor	65	$6.25+R_1$
	41	$6.25+R_2$
	25	$6.25+R_3$

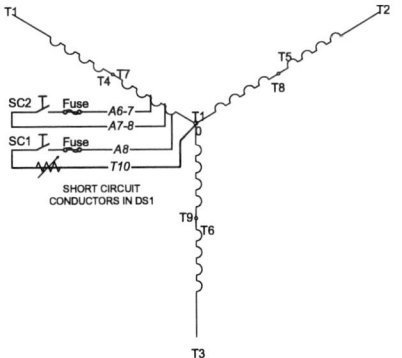

Fig. 5 Stator winding short circuit setup with external variable resistor

III. EXPERIMENTAL RESULTS

This section presents the experimental results obtained using the neural network approach to fault detection of stator winding short circuit in a PMSM. A series of tests are designed to demonstrate detectability and capacity to evaluate the grade of severity of the emulated stator winding short circuit fault using the proposed system, both with and without load fluctuation.

A series of case studies were performed in order to cover a wide variety of operating conditions at different fault and load levels. Two typical cases are presented here for each approach.

Case 1. Stator winding short with 25 A loop current and 30% load variation

This experiment is performed to demonstrate detectability under an incipient fault (less than 6.25% of the winding short in phase A) and under fluctuating load conditions. The fault is applied two times and the negative sequence component of the residuals is obtained as shown in Fig. 6. It is noted in Fig. 6 that the system is effective in distinguishing between the normal and fault condition under load fluctuation. Additionally, the magnitude of the negative sequence components of the residuals under fault conditions presents slight changes in magnitude mainly caused by deviations in the prediction of the currents in the neural network. Nevertheless the variations are not significant and the system performs robustly and satisfactorily in detecting and evaluating the level of severity.

Case 2. Stator winding short with 41.5 A loop current and 45% load variation

A second experiment is performed under load fluctuations of no-load up to 45% of the rated torque at a loop current of 41.5 A in the fault. In this case the fault is applied three times. The values of the negative sequence component of the residuals obtained are illustrated in Fig. 7.

An increase in the magnitude of the negative sequence current of residuals can be observed as a result of a more severe fault condition.

Additionally, as in the previous cases, the proposed indicator shows a strong correlation between its magnitude and the level of severity. This is illustrated in Table III, and Fig. 8 from whence it can be observed that the fault indicator, negative sequence current of residuals, decrease proportionally as the fault severity decreases. The percent increase in negative sequence current of residuals is calculated using (13)

$$\% = \frac{x - x_0}{x_0} \times 100\% \tag{13}$$

where x is the actual value and x_0 is the normal condition.

IV. CONCLUSIONS

The fault detection proposed shows its efficacy in performing fault detection when the PMSM is running under different conditions of load and under different grades of severity of turn-to-turn stator winding short circuits. It is noted, that experimentally the load fluctuation condition does not produce any significant increase in the values of negative sequence components of the residuals. Therefore, the most important factor in reducing the uncertainty in evaluating the stator winding failures under load fluctuation is eliminated. A model for current prediction is provided, which can expand the scope of this method to study the performance of the motor under aging process.

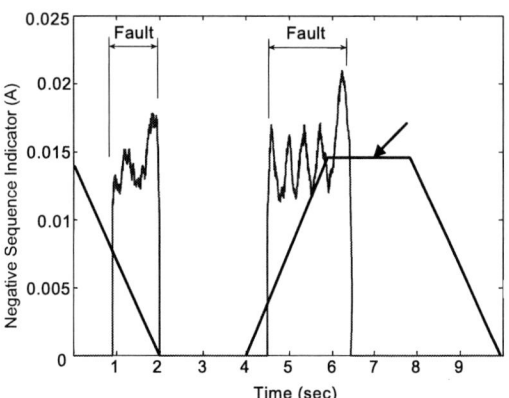

Fig. 6 Output of the NN fault detection proposed (loop current 25 A, 0-30% load)

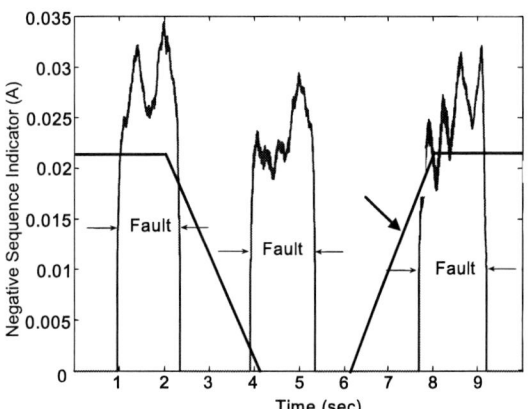

Fig. 7 Output of the NN fault detection purposed (loop current 41.5 A, 0-45% load)

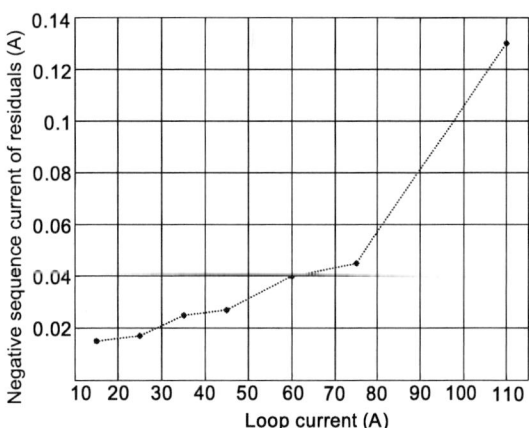

Fig. 8 Negative sequence current of residuals for different levels of fault severity

Table III. CORRELATION OF LOOP CURRENT, PERCENTAGE OF SHORTED WINDING AND PERCENTAGE OF R_{A2} INCREASE

Implementation	Loop current (A)	Shorted windings (%)	r_{a2} (A)	Increase in r_{a2} (%)
Taps	115	12.5	0.13	2500
	75	6.25	0.045	800
Variable Resistor	60	6.25+R$_1$	0.04	700
	45	6.25+R$_2$	0.027	440
	35	6.25+R$_3$	0.025	400
	25	6.25+R$_4$	0.017	240
	15	6.25+R$_5$	0.015	200
Normal condition	0	0	0.005	0

REFERENCES

[1] Li Liu, K.P. Logan, D.A. Cartes, and S.K. Srivastava, "Fault Detection, Diagnostics, and Prognostics: Software Agent Solutions," IEEE Trans. on Vehicular Technology, Volume 56, Issue. 4, Part 1, pp. 1613-1622, July 2007.

[2] M.E.H. Benbouzid, "A review of induction motors signatures analysis as a medium for fault detection", Industrial Electronics, IEEE Transactions on Volume 47, Issue 5, pp. 984 – 993, Oct.

[3] Trutt, F.C.; Sottile, J.; Kohler, J.L.;, "Online condition monitoring of induction motors" Industry Applications, IEEE Transactions on Volume 38, Issue 6, pp 1627 – 1632, Nov.-Dec. 2002.

[4] F. Filippetti, G. Franceschini, C. Tassoni, and P. Vas, "Recent developments of induction motor drives fault diagnosis using AI techniques", IEEE Trans. Industrial Electronics, vol.47, no. 5, pp. 994–1004, Oct. 2000.

[5] S. M. A. Cruz, and A. J. M Cardoso, "Diagnosis of stator inter-turn short circuits in DTC induction motor drives", IEEE Trans. Ind. Applicat., vol. 40, Issue 5, pp. 1349–1360, 2004.

[6] Albizu, I. Zamora, A.J. Mazon and A. Tapia," Techniques for Online Diagnosis of Stator Shorted Turns in Induction Motors", Electric Power Components and Systems, 34:1, 97-114.

[7] Alexander G. Parlos, Omar T. Rais and Amir F. Atiya, "Multi-Step-Ahead Prediction using Dynamic Recurrent Neural Networks", Neural Networks, 1999. IJNN'99, Vol 1.pp. 349 - 352 July 1999.

[8] Kyusung Kim, Alexander G. Parlos, "Induction Motor Fault Diagnosis Based on Neuropredictors and Wavelet Signal Processing", IEEE/ASME Transactions on Mechatronics, Vol. 7, No 2, pp 201-219, June 2002.

[9] Bollen, M. H. J.Gu, I. Signal Processing of Power Quality Disturbances, pp.861

[10] De Jesus, O.; Horn, J.M.; Hagan, M.T., "Analysis of recurrent network training and suggestions for improvements," Neural Networks, 2001. Proceedings. IJCNN '01. International Joint Conference on , vol.4, no., pp.2632-2637 vol.4, 2001.

[11] Foresee, F.D., and M.T. Hagan, "Gauss-Newton approximation to Bayesian regularization," Proceedings of the 1997 International Joint Conference on Neural Networks, 1997, pp. 1930–1935.

[12] MacKay, D.J.C., "Bayesian interpolation," Neural Computation, Vol. 4, No. 3, 1992, pp. 415–447.

[13] Hagan, M.T., H.B. Demuth, and M.H. Beale, Neural Network Design, Boston, MA: PWS Publishing, 1996.

[14] Nguyen, D., and B. Widrow, "Improving the learning speed of 2-layer neural networks by choosing initial values of the adaptive weights," Proceedings of the International Joint Conference on Neural Networks, Vol. 3, 1990, pp. 21–26.

Review of Electrical Machine in Downhole Applications and the Advantages

Anyuan Chen, Ravindra. B. Ummaneni, Robert Nilssen and Arne Nysveen

Electrical Power Engineering Department

Norwegian University of Science and Technology, Trondheim, Norway, e-mail: *anyuan.chen@elkraft.ntnu.no*

Abstract—To enable safe, economical and environmentally acceptable solutions for improving oil and gas recovery from mature fields and for exploitation of deep and ultra-deep offshore reservoirs, new downhole technologies are required. Electrification of downhole applications has proven to be promising. This paper presents electrical machine in currently available or emerging downhole applications and their advantages. Permanent magnet machines are the trend for downhole applications.

Keywords—Efficiency, permanent magnet motor, induction motor and environment.

I. INTRODUCTION

The majority of oil and gas increasing in coming decades will be from two sources: extracting more from mature fields and the exploitation of deep and ultra deep water reservoirs [1]. Today the production from the mature fields is gradually declining and the fields will be shut down prematurely if new technologies for improving recovery and more economical operation and maintenance are not put in place. For deep offshore oil and gas reservoirs, the cost of traditionally fixed or floating facilities for processing, which are today widely used, increases significantly. Many deepwater fields may therefore be uneconomical to develop. Downhole processing, by moving the processing from topside or onshore to downhole, can generally improve production, ultimate recovery and processing efficiency whilst decreases environmental impact. Currently downhole applications are powered by four types of system from topside: mechanical rod connection, pressurized gas, hydraulic power and electrical power systems. Of them electrically bottom-driven downhole applications have proven to be promising for deep wells, particularly for deepwater offshore wells, which is the exploitation trend today. This paper reviews electrical machine in currently available and emerging downhole applications, their advantages and the new trend for electrical downhole machines.

II. ELECTRICAL MACHINE IN DOWNHOLE APPLICATIONS

A. Operating Subsurface Valve (SSV)

Fig. 1 illustrates currently available hydraulic and electro-hydraulic SSVs as well as emerging electric cabled SSVs and future cable free electric SSVs.

Hydraulic SSV systems require two hydraulic lines for each valve and an instrument cable for the gauges and flow meter. Benefits to this approach are simple valve designs and well proven technology. The main disadvantage is the slow response time for long hydraulic lines. To overcome the drawbacks, an electro-hydraulic system uses electrical signals to operate solenoids that control the delivery of signal pressures to shuttle control valves, which in turn release hydraulic power to activate the tool functions. This system increases the reliability, shortens the response time and reduces the vast number of lines required in purely hydraulic systems. But both hydraulic and electro-hydraulic SSV systems need a hydraulic power unit (HPU), comprising of a low-pressure fluid storage reservoir to store control fluid, a high-pressure pump and a high-pressure storage reservoir (accumulator). The HPU needs a relatively large top space that is a critical issue for offshore applications. Moreover, the hydraulic lines leading from the surface to control values not only frequently cause failure by high fluid leakage rate, but also result in high cost to build up.

In the emerging electric cabled SSVs and cable free electric SSVs (Tubing is used to transmit signals and power) shown in Fig. 1 c and d, the valves are directly driven by high torque electrical motors, the response speeds are therefore faster and large volume on the topside occupied by the HPU is released. A valve prototype of a 50W motor driving one-inch variable orifice valve has been presented in [2].

Advantages of electric SSVs:

- No mechanical force, tension or compression required in the tubing to operate valve.
- Valve can be changed to any position at any time regardless of the pressure in the tubing or wellbore or whether the tubing is in tension or compression.
- Fast response and less transmission lines.
- More efficient systems and easier maintenance.
- More environment-friendly.

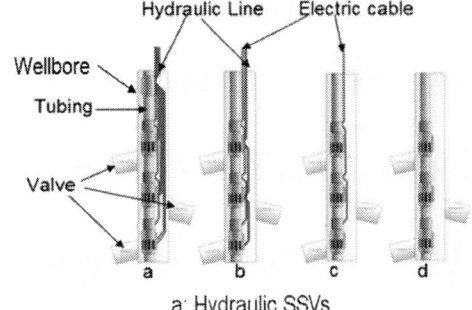

a: Hydraulic SSVs
b: Electro- hydraulic SSVs
c: Electric cabled SSVs
d: Cable free electric SSVs

Fig. 1. Different SSV technologies [2].

B. Driving Rod Pumps and Progressing Cavity Pumps

It has long been recognized that production of oil by means of sucker rod pumps or progressing cavity pumps (PCPs) driven by a motor at the surface via a long flexible rod system, is very inefficient. Not only are the pumps and the rod connecting the downhole pump very expensive, the rod also usually rubs against the tubing in a number of places and therefore often breaks since most wells are not really "straight". By respectively using downhole linear electrical motors to drive rod pumps and rotate electrical motor to drive PCPs, the abovementioned drawbacks can easily be overcome. The electrically bottom-driven pumps not only eliminate the rod break problem and significantly reduce the tubing wear, but also increase the production by eliminating the couplings, centralizers and rod strings required in surface-driven pumps [3]. Fig. 2 shows a simulation comparison between the tubing flow losses for an electrical submersible PCP and a rod-driven PCP.

C. Driving Electrical Submersible Pumps (ESPs)

ESP system, consisting of an electrical motor and a centrifugal pump, can be flexibly located in vertical, deviated even horizontal wells. Due to the robustness, flexibility and reliability, it has been widely installed for artificial lift and downhole oil / water separation (DOWS).

Fig. 3 gives the typical application ranges of the five most prevalent artificial lifts: rod pump, progressing-cavity pump, gas lift, hydraulic lift and ESP lift. ESP lift and gas lift have higher capacity and deeper capability than the others and are therefore the standard solution for offshore when artificial lift is required [4]. However,

natural gas shortage sometimes limits the use of gas lift, and tubing size and long flow lines also limit the system pressure and hence restrict its efficiency. Moreover, gas lift has reduced benefits as reservoir pressures approach abandonment levels. ESP lift is then normally required to extend field life as shown in Fig. 4 [4] [5]. The green area represents production extracted by natural flow, the orange area by gas lift and the blue area by ESP lift. Another important reason to employ ESP lift for offshore wells is that it has less footprint topside while offering the highest yield from most deep-water wells.

A relatively new technology of ESP DOWS has been developed based on ESP lift systems to reduce the cost of handling produced water as shown in Fig. 5, wherein there are two ESPs driven by an electrical motor. One ESP pumps oil-rich stream to the surface while the other one injects water-rich stream to an underground formation so that the cost of treatment and disposal of produced water is significantly reduced. Fig. 6 shows a practical production history from an ESP DOWS trial. It was clearly shown that the oil production almost kept constant whilst water production decreased from 300m³/day to 10m³/day when an ESP DOWS was installed.

Advantages of ESP [6][7] [8]:

- Easy to install and low maintenance.
- More compact and deepwater capability.
- High efficiency over 500-1000 barrel/day.
- Good in deviated wells.
- Minor surface equipment needed.
- Increased production and recovery.
- Less environmental impact and cost.

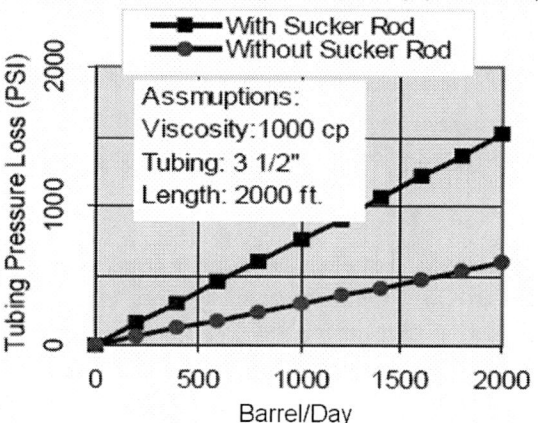

Fig. 2. Tubing flow losses [3].

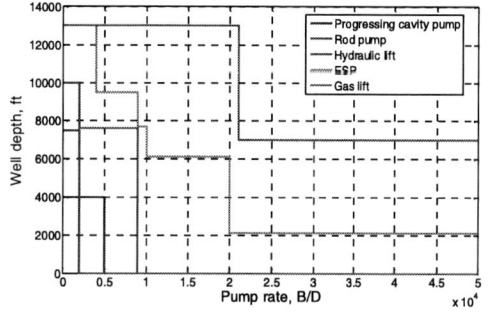

Fig. 3. Typical artificial application ranges [5].

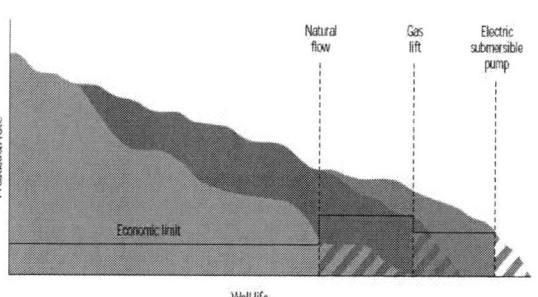

Fig. 4. Well life extended by gas & ESP lift [4].

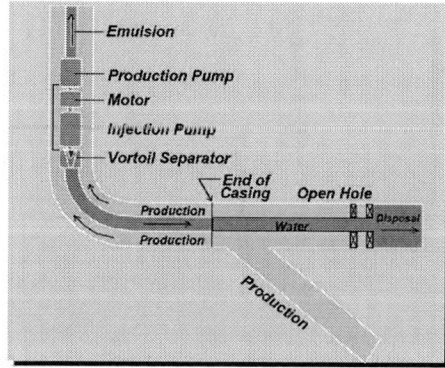

Fig. 5. An ESP DOWS system

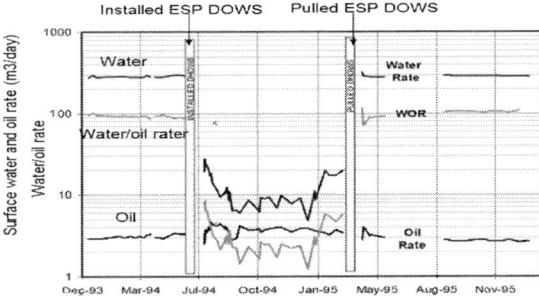

Fig. 6. Production history of an ESP DOWS [17].

D. Driving Downhole Gas Compressors

Downhole gas compression is to boost gas pressure by placing a compressor in close proximity to the source gas reservoir. This can have major economical gains in hydrocarbon recovery in addition to using conventional central gas compression. Fig. 7 shows a simulation result of potential yield improvement by downhole gas compression [9]. These applications demand the shaft speeds in the range of 5,000 to 20,000 rpm that are much faster than the rated speeds of standard induction downhole motors. To match the application speeds, standard electrical induction motors have to be coupled with a speed-increasing gearbox. This is also problematic as it is extremely difficult to make a reliable gearbox in the small borehole diameter, and it is also expensive. To eliminate the gearbox, a high-speed permanent magnet (PM) motor as shown in Fig. 10 has been designed and tested for downhole gas compressors.

Advantage of PM motor driven compressor [9]:

- More compact.
- Increased reliability.
- Increased production.

E. As Drilling Motor

In the vast majority of drilling services, downhole power is provided by positive displacement motors (PDMs). However, PDMs have many weaknesses, such as short motor run life, poor performance in high temperature operations, a limited choice of drilling media and the need to compromise the fluid program between drilling and formation requirements. This restricts the operational effectiveness of PDMs in many areas whereas an electrical drilling motor offers an efficient and reliable alternative. In [10] and [11], two PM drilling motors have been presented.

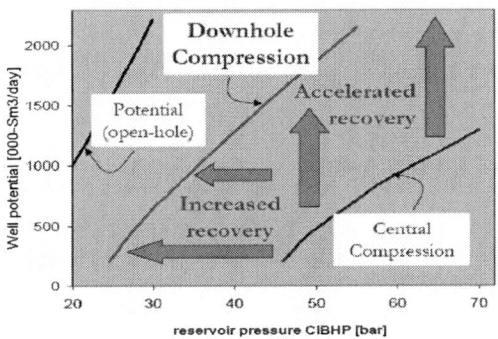

Fig. 7. Potential yield improvement by downhole gas compression [9].

Advantages of electrical drilling [10] [11] [12]:

- Increased reliability of drilling motor extends mean time between failures of drilling string and reduces drilling costs.
- Electrical motors increase the control and flexibility of the bottom-hole assembly (BHA) since drilling bit speed is maintained independent of fluid flow rate through direct operator control.
- Suitable for a wider range of drilling environments, particularly for deepwater operations. Electrical motors can operate with a wider range of drilling media than PDMs, such as energized fluids. This makes electrical motors ideal for aggressive under-balanced drilling applications and in deepwater operations.
- Electrical motors could tolerate temperatures up to 200°C with new materials.
- Possible higher rate of penetration.

F. As Downhole Generator

Today most downhole tools operate based on battery power, mainly lithium-ion that is limited to 180°C [13]. For very deep wells, the instrumented BHA is critical, but the battery life and higher downhole temperatures constrain their use. The most practical and promising solution to power intelligent downhole tools in high pressure and high temperature (HPHT) environments is a turbine generator, wherein the turbine coverts a small amount of the hydraulic energy in the mud stream to rotary energy, which is then converted to electrical power by an electrical generator.

Advantage of power supply from a HTHP downhole generator:

- More reliable in HTHP environment.
- Suitable for deep wells.

G. Intervention tools

Besides the abovementioned applications, electrical motors can also be combined with other different equipments working as downhole intervention tools. Normally these electrical motors are in a relatively lower power range with high output torque, and their characteristics are significantly dependent on specific applications.

III. STANDARD INDUCTION AND DOWNHOLE MOTORS.

Due to the harsh conditions downhole and the extremely high cost of failure replacement in offshore downhole applications, an electrical downhole machine must have robust construction and high reliability. Today standard electrical downhole motor is three-phase, squirrel-cage induction motor (IM) due to its high reliability, robustness and simple construction. Fig. 8 shows structure of a typical multi-rotor induction downhole motor. The stator is wound as a single unit and the rotor consists of a number of electrically discrete rotors with bearings between them to accommodate the slender construction. Normally downhole motors are oil-filled, and the oil, having low compressibility, makes it compatible with the high external ambient pressures existing downhole. It also provides bearing lubrication and effective heat transfer for dissipation of the losses radially outward through the motor housing. The outer diameter is typically between 100 to 300 mm and the length is normally from 5 m to 10m, even up to as long as 20 m or longer. The most common power range is from 40 to 200 kW. More power can be achieved with the inclusion of additional motors. The working temperatures are usually limited to 180°C, some specially designed motors can withstand up to 218°C. Their efficiencies are from 70% to 89% for high-speed applications, and from 60% to 73% for low-speed, high-torque applications (see Table 1). However, induction motors are inherently unsuited to low-speed, high-torque applications, which is the case for driving PCPs. Thus currently most installations rely on a gearbox to match the normal motor running speed and torque to the pump characteristics. The re-introduced gear system is not only costly and different to make, but also further decreases the efficiencies of the whole systems.

IV. NEW DOWNHOLE MACHINES.

PM machines, having almost the same robustness and reliability as conventional induction motors whilst with higher efficiencies, higher torque densities and smaller volumes, have widely employed in industrial applications to replace conventional machines, but few have been developed for HPHT downhole applications due to the low temperature stability of PM materials in past time. Today with the development of advanced technologies and applications of high temperature magnets, it is increasingly interesting for oil and gap industries to develop PM machines for downhole applications.

A study shows that PM downhole motors consume an average of 23% less energy than IMs [15]. It is obviously seen from Table 1 that the efficiencies of PM motor, 91-93% for high-speed applications and 85%-89% for high-torque applications, are much higher than IMs. Furthermore, PM motors have the capability of delivering high torque over whole operation ranges including start-up. Fig. 9 shows a permanent magnet DC (PMDC) downhole motor with brushes. Its speed is controlled by simply varying the DC current from topside, and no variable speed drive (VSD) is required. The main disadvantage of this motor is that the brushes re-introduce extra maintenance that is extremely costly for offshore applications. Therefore, this type of motor is mainly limited to onshore applications.

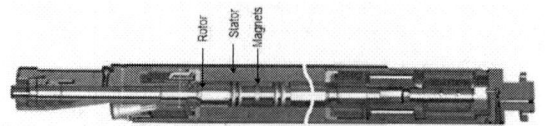
Fig. 9. A PMDC downhole motor [15].

TABLE 1
EFFICIENCY COMPARISON OF ELECTRICAL DOWNHOLE MOTORS

Company	Motor type	Efficiency	Application
Borets	IM	77-85%	Driving ESP
Borets	IM	60-73%	Driving PCP
Woodgroup	IM	70-89%	Driving ESP
Borets	PMDC	91-93%	Driving ESP
Borets	PMDC	85-89%	Driving PCP
(Research)	PMSM	95%	Driving PCP
(Research)	PMSM	92%	

With the technology development in VSD, PM synchronous machine (PMSM) and PM brushless DC machine (BLDC) become the trend for downhole applications. In [10] and [12], a high-torque BLDC capable of withstanding temperatures up to 230°C has been designed and tested for electrical drilling. The motor is 2.7 m long and has an outer diameter of 80 mm. Its operating torque is 420 Nm and the peak power is around 20 kW. Fig. 10 shows a high-speed BLDC motor for downhole gas compression [9]. In [16], a 10-pole interior-mounted PMSM shown in Fig. 11 is presented for driving PCPs and one research project focusing on linear PM synchronous machines for downhole drilling is also going on according to [11]. PM machine for downhole applications holds tremendous promise, but is still in its infancy.

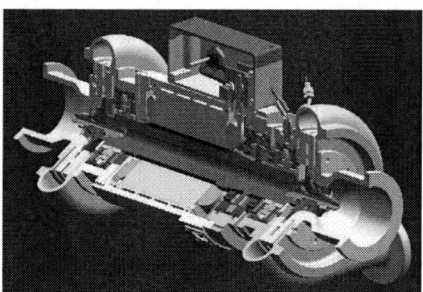
Fig. 10. A high speed BLDC downhole motor [9].

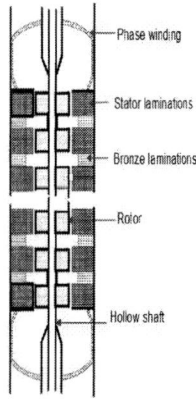
Fig. 8. Structure of a typical multi-rotor induction motor [14]

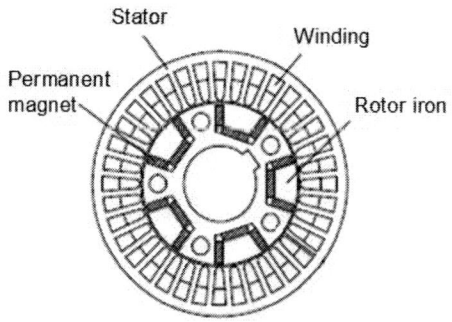
Fig. 11. A downhole PMSM [16].

V. CONCLUSION

This paper has reviewed electrical machine in downhole applications and their advantages. Electrification of downhole applications has presented lots of benefits, particularly for deep-water offshore reservoirs that are the exploitation trend for oil and gas industries now. Today's standard electrical downhole motors are not energy efficient. PM motor becomes the trend for downhole applications.

ACKNOWLEDGMENT

The authors are grateful to NFR for the financial support.

REFERENCES

[1] "Oil and Gas Industries Technology master plan", European oil and gas innovation forum, February 2004.

[2] Electricwell.com

[3] "High Volume Down-Hole Progressing Cavity Pumps with Electrical Submersible Motors", R&M energy systems, a unit of tobbins and myers inc, Paper 18, ESP WORKSHOP, 1998

[4] Roy Fleshman and Harryson Obren Lekic, "Artificial lift for High-Volume Production", Oil review, spring 1999.

[5] Ron Bates, Charlie Cosad, Lance fielder, Alex Kosmala, Steve Hudson, George Romero and Valli Shanmugam, "Taking the Pulse of Producing Wells-ESP Surveillance", 2004.

[6] John A Veil, Bruce G. Langhus and Stan Belieu, "Feasibility Evaluation of Downhole Oil/Water separator (DOWS) Technology", report for U.S department of energy, office of fossil energy and national petroleum technology office, January 1999.

[7] Jesus R. Rodriguez, Fathi Finaish, Shari Dunn-Norman, "Parametric study of Motor/Shroud Heat Transfer Performance in an Electrical Submersible Pump", Transactions of the ASME, Vol.122, September 2000.

[8] William D. Holmes, Monty D. Corbett, Daniel B. Wells, "Successful Downhole Oil Water Separator Installation", 2003 ESP Workshop, May 9, 2003.

[9] "Downhole gas compression", PTAC technical information session, 27, November 2003.

[10] Dan Turner, Philip Head and Mike Yuratich, "New DC motor for Downhole Drilling and Pumping Applications", Society of petroleum engineers paper 68489, March 2001, Houston, Texas.

[11] Ummaneni Ravindra Babu, Nilssen Robert and Brennvall, J.E, "Force Analysis in Design of High Power Linear Permanent Magnet Actuator with Gas Springs in Drilling Applications", Electric Machines & Drives Conference, 2007, IEMDC'07, IEEE International.

[12] D.R Turner, T.W.R Harris, M. Slater, M.A. Yuratich and P.F. Head, "Electric Coiled Tubing: A Smart CT Drilling System", Society of petroleum engineers paper 52791, March 1999, Amsterdam, Holland.

[13] Timothy F. Price, "Development of a High-Pressure/High-Temperature Downhole Turbine Generator", Dexter Magnetic Technologies Inc., DE-FC26-05NT42655, March 27, 2007.

[14] J.R.Smith, D.M.Grant, A-AI-Mashgari and R.D.Slater, Operation of subsea electrical submersible pumps supplied over extended length cable systems, IEE Proc.-Electr, Power Appl., Vol. 147, No. 6, November 2000.

[15] NTK-BP Technology magazine, October-November 2007, Russian, pp.27-31.

[16] Zhang Binyi, Liang Binxue, Feng Guihong and Zhuang Fuyu, "Research of Multipolar permanent Magnet Synchronous Submersible Motor for Screw pump", Proceedings of 2007 IEEE International Conference on Mechatronics and Automation, August 5-8, 2007 Harbin, China, pp.1011-1016, 2007.

[17] Kelly Piers, "Coping with Water from Oil & Gas Wells", Presentation, June 14, 2005.

Broken Rotor Bar Impact on the Closed Loop and Sensorless Control of Induction Machine

Piotr Kołodziejek[*], Elżbieta Bogalecka [†]

[*†] Gdańsk University of Technology /Department of Electrical and Control Engineering, Gdańsk, Poland,
e-mail: *pkolod@ely.pg.gda.pl, e.bogalecka@ely.pg.gda.pl*

Abstract— **This paper presents analysis of closed loop and sensorless control of induction machine with broken rotor bar for on-line diagnostics purposes. Closed loop control system synthesis of induction motor is based on symmetrical model of the machine. In case of induction machine operation with broken rotor bars, rotor speed pulsation is compensated by the control system. Simulation of Multiscalar Control with the fault machine shows that decoupling variables can be used as detection signals. Analysis of speed observer frequency response in open loop operation gives information about disturbance impact on estimation error. Further analysis indicates that machine asymmetry can affect stability of sensorless control system with speed observer.**

Keywords— **Diagnostics, Induction motor, Sensorless control, Modelling.**

I. INTRODUCTION

Development of sensorless control system with AC machine sets a new problem for fault detection and diagnostics. Many rotor fault modelling methods and detection techniques have been developed and classified. Most of them require setting specified operation state of machine e.g. MCSA and EPVA methods reported in [1] and [2] require steady-state operation. Other require transient state i.e. start up in wavelet motor current analysis or inertia braking in voltage space vector analysis after supply disconnection method [3,4]. Closed loop and sensorless control with fault machine require exact model of asymmetric machine with space vector representation. Rotor fault detection method in FOC driven induction machine has been proposed in [5] and [6]. Model-based detection methods are not popular due to nonlinear and nonstationary model of the machine that requires estimation of parameters. Sensorless control requires exact identification of the machine to limit estimation error and provide stability of the control system, therefore model-based approach in this case is reasonable. This paper refer to Multiscalar control as a generalized space-vector control where state variables are reference frame independent [7]. Simulated sensorless control system is based on multiscalar model of the machine with selected speed observer structure. Three structures of state observers reported in [8-11] has been taken into consideration.

Recent years researches in induction machine rotor fault diagnostics were focused on application of Fourier transformation and wavelet analysis of specific electric signal i.e. stator current. Acquisition of the signal is simple but spectra includes fundamental frequency of supply voltage and other harmonic components related to different phenomena besides rotor fault components. Broken rotor bar characteristic symptoms in stator current spectra are sideband frequency components which can exist also in case of load torque pulsation or natural asymmetry of high-power machine. Towards uniqueness of spectra analysis problem advanced rotor fault isolation techniques are investigated in [12] and [13]. Practical methods for induction motor rotor fault diagnostics are analysis of power spectra analysis at reduced voltage short-circuit, stator current space vector hodograph, stator phase current, stator voltage after supply disconnection and mechanical variables analysis. Increasing applications of sensorless controlled inverter fed drives with induction motors is the reason for developing new diagnostics methods. Machine asymmetry related to rotor is cause of additional components appearance in control system variables and residuals of state observers. Analysis of signals in the control system may be useful for fault detection, isolation and diagnostics purposes. This paper presents analysis of closed loop and sensorless control of induction machine with broken rotor bar for on-line diagnostics purposes.

II. ROTOR FAULT MODELLING

Broken rotor modelling methods has been reported in [6] and [14-18]. Authors have considered feasibility of using T transformation proposed in [18], where rotor current was modified to calculate torque pulsation in open-loop model and diagnose rotor fault. Analysis of instantaneous rotor currents impact on d-q-0 components shows that T transformation can be also used in closed-loop modelling algorithm of the machine with assumption of specific reference frame and state variables choice to calculate model differential equations. Taking stator current and rotor flux as state variables is suitable, but in this case rotor current vector does not exist directly in model equations to use the T transformation. Solution for this problem is taking rotor flux vector as modified variable or indirect variables calculation with rotor current modification. Simulation analysis confirmed uniqueness of results comparing to rotor current modification with assumption of rotor and stator flux as state variables and linear main magnetic path of the machine. Proposed broken rotor modelling algorithm with stator current and rotor flux as state variables with rotor current modification is shown in fig. 1. This case requires variables transformation to rotor reference frame

978-1-4244-1741-4/08/$25.00 ©2008 IEEE

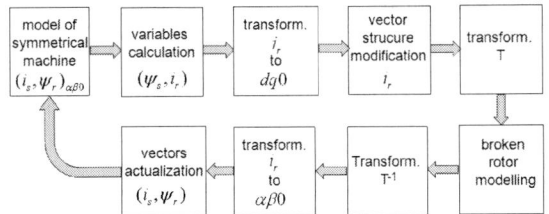

Fig. 1. Rotor fault modelling in closed-loop algorithm.

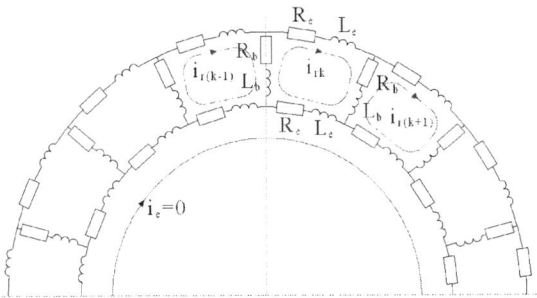

Fig. 2. Rotor circuit model for symmetrical machine.

and indirect calculation of stator flux and rotor current vectors. Assuming instantaneous rotor currents as in fig. 2 and insulated rotor bars electric circuit theory is used for simplified description of fault machine as in [18]:

- broken bar and end ring equations:

$$i_{rk}' = \frac{i_{rk} + i_{r(k+1)}}{2} \quad (1) \quad i_{r(k+1)}' = \frac{i_{rk} + i_{r(k+1)}}{2} \quad (2)$$

$$i_{rk}' = -\frac{i_{rk}}{N_b - 1} \quad (3)$$

- partially broken rotor bar with higher impedance:

$$i_{rk}' = \frac{1}{2}ai_{rk} + (1-a)i_{r(k+1)} \quad (4)$$

$$i_{r(k+1)}' = (1-a)i_{rk} + ai_{r(k+1)} \quad (5)$$

$$i_{rk}' = \frac{1}{2}ai_{rk} + (1-a)i_{r(k+1)} \quad (6)$$

$$i_{r(k+1)}' = (1-a)i_{rk} + ai_{r(k+1)} \quad (7)$$

$$a = \left(\frac{\exp^{\frac{1}{N_b}}}{\exp^{\left(\frac{R_b'}{N_b R_b}\right)}} + 1 \right) \quad (8)$$

where i_{rk}' is modified rotor current vector element related to k-th rotor circuit (Fig. 2), R_b is rotor bar resistance and N_b is number of rotor bars.

III. MULTISCALAR CONTROL SYSTEM ANALYSIS

Multiscalar based control system reported in [7] is generalized space vector control method where new state variables are reference frame independent. New state variables are calculated as follows:

$$x_{11} = \omega_r \quad (9) \qquad x_{12} = \psi_{rx} i_{sy} - \psi_{ry} i_{sx} \quad (11)$$

$$x_{21} = \psi_r^2 \quad (10) \qquad x_{22} = \psi_{rx} i_{sx} + \psi_{ry} i_{sy} \quad (12)$$

where x_{11} is rotor speed, x_{12} is proportional to electromagnetic torque, x_{21} is square rotor flux and x_{22} is scalar product of rotor flux and stator current vectors. Feedback linearization is applied to compensate nonlinear components in multiscalar model of the machine:

$$u_1 = \frac{w_\delta}{L_r}\left(x_{11}\left(x_{22} + \frac{L_m}{w_\delta}x_{21}\right) + \frac{1}{T_v}m_1 \right) \quad (13)$$

$$u_2 = \frac{w_\delta}{L_r}\left(-x_{11}x_{12} - \frac{R_r L_m}{L_r w_\delta}x_{21} - \frac{R_r L_m}{L_r}\frac{x_{11}^2 + x_{22}^2}{x_{21}} + \frac{1}{T_v}m_2 \right) \quad (14)$$

where R_r is rotor resistance, L_r is rotor inductance, Lm is magnetizing inductance, w_σ is dissipation coefficient, m_1, m_2 are new control variables for linearized model of the machine and u_1, u_2 are nonlinear control variables. Stator voltage space vector components in stationary reference frame are calculated as follows:

$$u_{s\alpha} = \frac{\psi_{rx}u_2 - \psi_{ry}u_1}{\psi_r^2} \quad (15) \quad u_{s\beta} = \frac{\psi_{rx}u_1 + \psi_{ry}u_2}{\psi_r^2} \quad (16)$$

Multiscalar control system has been simulated with presented asymmetric model of the machine as shown in fig. 3. In simulated Multiscalar Control System rotor speed, rotor flux and speed signals are fed from model of the machine. Pulsation of rotor speed and stator current has lower amplitude comparing to steady-state open-loop broken machine operation (as shown in fig. 4). This is result of compensating effect of the control system which depends on parameters of PI controllers. Frequency

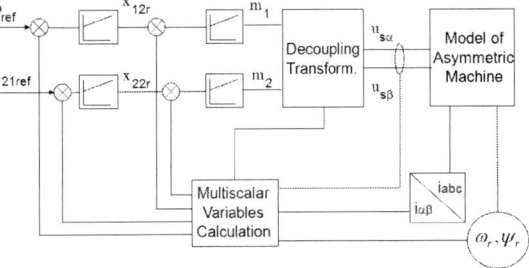

Fig. 3. Multiscalar control system with fault of the machine.

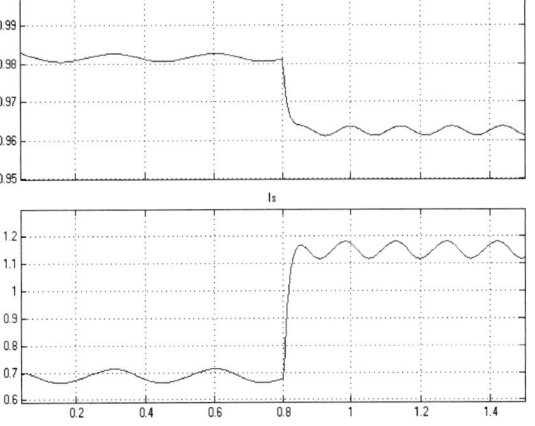

Fig. 4. Stator current module and speed at step torque load.

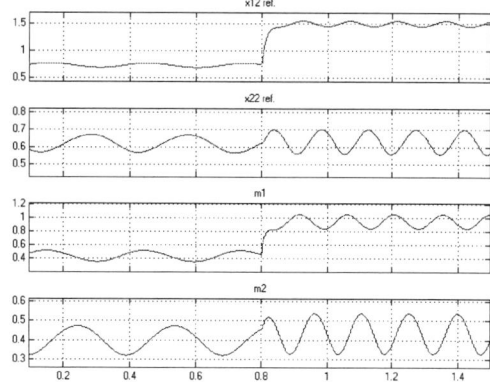

Fig. 5. Output signals of PI controllers at step load torque.

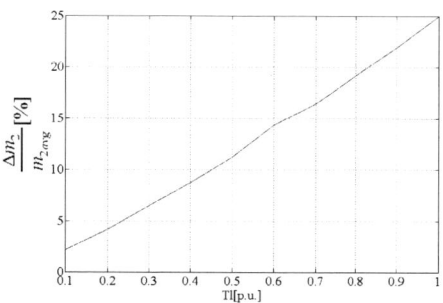

Fig. 6. Relative amplitude difference of m_2 control variable as function of torque load for broken end-ring.

of the pulsation is related to slip of the machine and supply voltage frequency. Output signals of PI controllers in steady-state operation contain harmonic components as shown in fig. 5. Amplified harmonic component is visible in decoupling variable m_2. This feature can be useful in detection of the broken rotor. Relative amplitude of the harmonic component is up to 25% for m_2 variable (as in fig. 6) comparing to approximately 5% for state variables of the machine at nominal torque load. This feature can be used in detection techniques for control system with speed measurement. Amplification of the oscillation give chances for earlier broken rotor detection, which is a common problem of known diagnostics methods. Detection techniques in this case can be based on filtering signal methods, instantaneous slip calculation or characteristic oscillations frequency estimation from asymmetric model of the machine.

IV. OPEN LOOP OBSERVERS FREQUENCY ANALYSIS

Analysis of speed observers frequency characteristics in open-loop operation is needed for determination of observer structure influence on estimated harmonic signal and discrimination of control system and observer structure influence on estimated signals. Three structures of speed observers has been taken into consideration. Equations of the observers are presented in orthogonal vector components notation in stator reference frame as follows:

a) Kubota speed observer reported in [11]:

$$\frac{d\hat{i}_{s\alpha}}{dt} = a_1\hat{i}_{s\alpha} + a_2\hat{\psi}_{r\alpha} + \hat{\omega}_r a_3\hat{\psi}_{r\beta} + \qquad (17)$$
$$+ a_4 u_{s\alpha} + k_f(i_{s\alpha} - \hat{i}_{s\alpha})$$

$$\frac{d\hat{i}_{s\beta}}{dt} = a_1\hat{i}_{s\beta} + a_2\hat{\psi}_{r\beta} + \hat{\omega}_r a_3\hat{\psi}_{r\alpha} + \qquad (18)$$
$$+ a_4 u_{s\beta} + k_f(i_{s\beta} - \hat{i}_{s\beta})$$

$$\frac{d\hat{\psi}_{r\alpha}}{dt} = a_5\hat{\psi}_{r\alpha} + a_6\hat{i}_{s\alpha} - \hat{\omega}_r\hat{\psi}_{r\beta} + \qquad (19)$$
$$+ k_{f1}(i_{s\alpha} - \hat{i}_{s\alpha}) - k_{f2}\hat{\omega}_r(i_{s\beta} - \hat{i}_{s\beta})$$

$$\frac{d\hat{\psi}_{r\beta}}{dt} = a_5\hat{\psi}_{r\alpha} + a_6\hat{i}_{s\beta} + \hat{\omega}_r\hat{\psi}_{r\alpha} + \qquad (20)$$
$$+ k_{f1}(i_{s\beta} - \hat{i}_{s\beta}) + k_{f2}\hat{\omega}_r(i_{s\alpha} - \hat{i}_{s\alpha})$$

$$\frac{d\hat{\omega}_r}{dt} = (i_{s\alpha} - \hat{i}_{s\alpha})a_3\hat{\psi}_{r\beta} - (i_{s\beta} - \hat{i}_{s\beta})a_3\hat{\psi}_{r\alpha} \qquad (21)$$

b) Krzemiński speed observer reported in [8-9]:

$$\frac{d\hat{i}_{s\alpha}}{d\tau} = a_1\hat{i}_{s\alpha} + a_2\hat{\psi}_{r\alpha} + a_3\zeta_\beta + bu_{s\alpha} + \qquad (22)$$
$$+ k_3\left(k_1(i_{s\alpha} - \hat{i}_{s\alpha}) - \hat{\omega}_r\xi_\alpha\right)$$

$$\frac{d\hat{i}_{s\beta}}{d\tau} = a_1\hat{i}_{s\beta} + a_2\hat{\psi}_{r\beta} - a_3\zeta_\alpha + bu_{s\beta} + \qquad (23)$$
$$+ k_3\left(k_1(i_{s\beta} - \hat{i}_{s\beta}) - \hat{\omega}_r\xi_\beta\right)$$

$$\frac{d\hat{\psi}_{r\alpha}}{d\tau} = a_6\hat{i}_{s\alpha} + a_5\hat{\psi}_{r\alpha} - \zeta_\beta - k_2\left(\hat{\omega}_r\hat{\psi}_{r\beta} - \xi_\beta\right) \quad (24)$$

$$\frac{d\hat{\psi}_{r\beta}}{d\tau} = a_6\hat{i}_{s\beta} + a_5\hat{\psi}_{r\beta} + \zeta_\alpha - k_2\left(\hat{\omega}_r\hat{\psi}_{r\alpha} - \xi_\alpha\right) \quad (25)$$

$$\frac{d\zeta_\alpha}{d\tau} = k_1(i_{s\beta} - \hat{i}_{s\beta}) \quad (26) \qquad \frac{d\zeta_\beta}{d\tau} = k_1(i_{s\alpha} - \hat{i}_{s\alpha}) \quad (27)$$

$$\frac{dV_f}{d\tau} = k_5(V - V_f) \quad (28) \qquad V = \zeta_\beta\hat{\psi}_{r\alpha} - \zeta_\alpha\hat{\psi}_{r\beta} \quad (29)$$

$$S = \hat{\psi}_{r\beta}\zeta_\beta - \hat{\psi}_{r\alpha}\zeta_\alpha \qquad (30)$$

$$\hat{\omega}_r = S\left(\sqrt{\frac{\zeta_\alpha^2 + \zeta_\beta^2}{\hat{\psi}_{r\alpha}^2 + \hat{\psi}_{r\beta}^2}} + k_4(V - V_f)\right) \qquad (31)$$

c) New speed observer of Krzemiński reported in [10]:

$$\frac{d\hat{i}_{s\alpha}}{d\tau} = a_1\hat{i}_{s\alpha} + a_2\hat{\psi}_{r\alpha} + a_3\zeta_\beta + bu_{s\alpha} + k_1(i_{s\alpha} - \hat{i}_{s\alpha}) \quad (32)$$

$$\frac{d\hat{i}_{s\beta}}{d\tau} = a_1\hat{i}_{s\beta} + a_2\hat{\psi}_{r\beta} - a_3\zeta_\alpha + bu_{s\beta} + k_1(i_{s\beta} - \hat{i}_{s\beta}) \quad (33)$$

806

$$\frac{d\hat{\psi}_{r\alpha}}{d\tau} = a_6\hat{i}_{s\alpha} + a_5\hat{\psi}_{r\alpha} - \zeta_\beta - k_2V\hat{\psi}_{r\alpha} + k_3S\hat{\psi}_{r\beta}\left(V - V_f\right) \quad (34)$$

$$\frac{d\hat{\psi}_{r\beta}}{d\tau} = a_6\hat{i}_{s\beta} + a_5\hat{\psi}_{r\beta} + \zeta_\alpha - k_2V\hat{\psi}_{r\beta} - k_3S\hat{\psi}_{r\alpha}\left(V - V_f\right) \quad (35)$$

$$\frac{d\zeta_\alpha}{d\tau} = -\hat{\omega}_{\psi r}\zeta_\beta - k_4\left(i_{s\beta} - \hat{i}_{s\beta}\right) \quad (36)$$

$$\frac{d\zeta_\beta}{d\tau} = \hat{\omega}_{\psi r}\zeta_\alpha + k_4\left(i_{s\alpha} - \hat{i}_{s\alpha}\right) \quad (37)$$

$$\frac{dV_f}{d\tau} = k_5\left(V - V_f\right) \quad \frac{d\hat{\omega}_{rf}}{d\tau} = k_6\left(\hat{\omega}_r - \hat{\omega}_{rf}\right) \quad (38)$$

$$\hat{\omega}_r = \frac{\zeta_\alpha\hat{\psi}_{r\alpha} + \zeta_\beta\hat{\psi}_{r\beta}}{\hat{\psi}_r^2} \quad (39)$$

$$V = \zeta_\alpha\hat{\psi}_{r\beta} - \zeta_\beta\hat{\psi}_{r\alpha} \quad (40) \quad S = \begin{cases} 1 & \text{if } \hat{\omega}_{\psi r} \geq 0 \\ -1 & \text{if } \hat{\omega}_{\psi r} < 0 \end{cases} \quad (41)$$

$$\hat{\omega}_{\psi r} = \hat{\omega}_{rf} + a_6\frac{\hat{\psi}_{r\alpha}\hat{i}_{s\beta} - \hat{\psi}_{r\beta}\hat{i}_{s\alpha}}{\hat{\psi}_r^2} \quad (42)$$

where a_1-a_6 are coefficients of machine parameters, k_1-k_5 are parameters of observer gain, $i_{s\alpha}$, $i_{s\beta}$, $u_{s\alpha}$, $u_{s\beta}$, $\psi_{r\alpha}$, $\psi_{r\beta}$, ζ_α, ζ_β, are respectively stator current, supply voltage, rotor flux and disturbance space vector components, ω_r is rotor speed, "^" and "f" means estimated and filtered state variables. Presented speed observers are used for rotor flux and speed estimation. Equations of presented observers are based on the same symmetrical model of the machine. Structural difference between Krzemiński observers is new additional disturbance model. Method of the analysis is based on additional low-amplitude sinusoidal signal, which is added to rotor speed output as shown in fig. 7. The additional signal frequency range is 1-500 Hz with 1 Hz resolution. Magnitude relative difference between rotor speed, rotor flux, stator current and their estimated values were calculated to obtain frequency characteristics and presented in fig. 8-10. Parameters of speed observers were optimized for transient state operation and stability was verified by pole configuration characteristics. For rotor speed and stator current estimation all structures of speed observers have frequency characteristics similar to low-pass filters, but with different cut-off frequency value as shown in fig. 8 and 10. For rotor speed estimation Kubota observer has the lowest cut-off frequency near 10 Hz but there is no resonance frequency characteristic as shown in fig. 8. Low frequency slip pulsation range disturbances related to rotor asymmetry at steady state operation is enough for signal

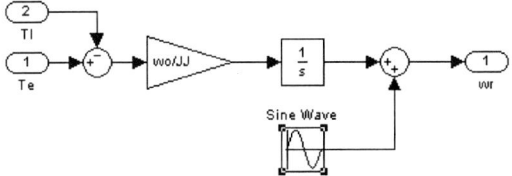

Fig. 7. External disturbance signal simulation scheme.

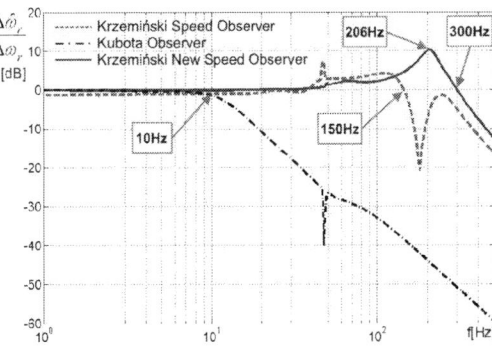

Fig. 8. Frequency characteristics for rotor speed estimation.

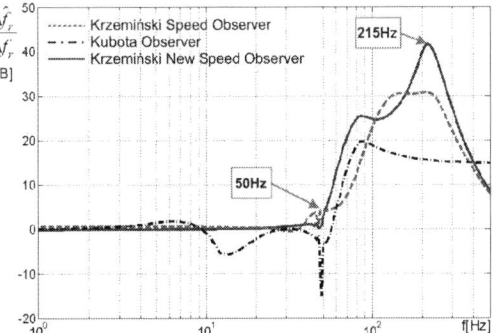

Fig. 9. Frequency characteristics for rotor flux estimation.

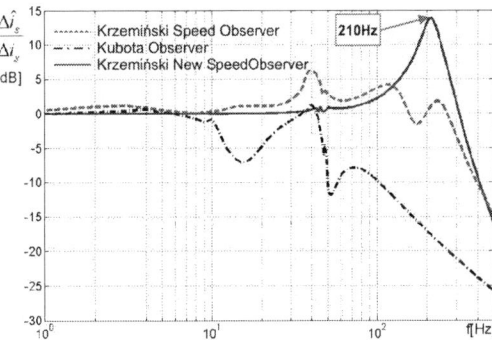

Fig. 10. Frequency characteristics for stator current estimation.

reconstruction. In this case higher frequencies related Krzemiński observers have both higher cut-off frequency at approximately 150 Hz and 300 Hz. This can be useful for transient state analysis e.g. start-up and for other symptoms detection, but in this case other frequencies not related to rotor asymmetry can be also reconstructed. For rotor flux estimation observer characteristics have good characteristics up to 50 Hz which is the voltage supply frequency. Adequacy of frequency diagrams has been verified in time domain. Consistency of time and frequency domain characteristics is shown in fig. 11 and 12, where examples of time domain diagrams has been showed for resonance frequency of 206 Hz. Frequency response analysis of state observers is complex task because observer must operate with state variables from model or plant. Results show that all investigated

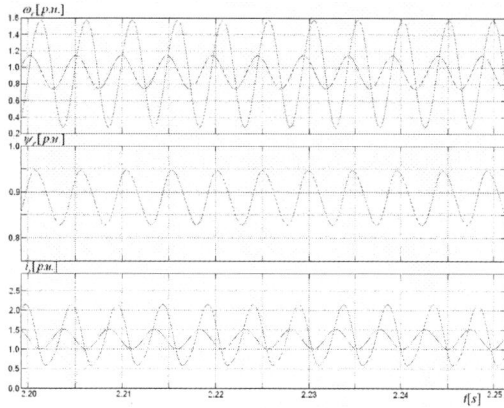

Fig. 11. Estimated and real state variables in new Krzemiński observer for open-loop operation.

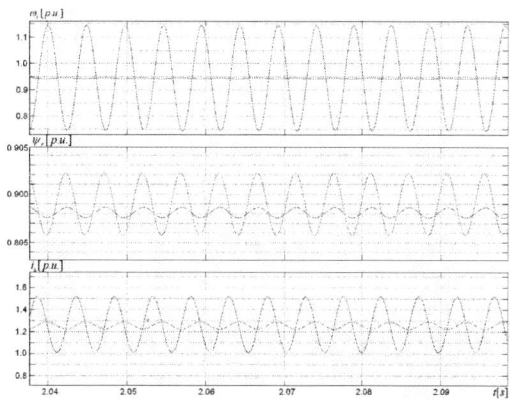

Fig. 12. Estimated and real state variables in new Kubota observer for open-loop operation.

observers can be used for diagnostic purpose for symptoms of limited frequency range because of low-pass filtering features (Fig. 1). Filtering high frequency disturbances is very important advantage comparing to popular MCSA rotor fault diagnostic method, where frequencies related to rotor asymmetry require high resolution FFT and high quality measurement devices to isolate stator current band frequencies with amplitude near to measurement disturbance at low motor slip.

V. SENSORLESS CONTROL SYSTEM ANALYSIS

For sensorless control operation rotor speed and flux are estimated state variables in dedicated state observers. Scheme of sensorless control system based on multiscalar control with presented broken rotor model is shown in fig. 13. Fig. 14 and 15 show real and estimated values of rotor speed, modulus of rotor flux and stator current in sensorless control operation with new Krzemiński and Kubota observers. The figures show transients and quasi-steady-state after dynamical rotor fault. Harmonic components of rotor speed and rotor flux are reconstructed with mean value error but control system is stable in both cases. In fig 16 and 17 case of instability as been showed for first version of Krzemiński observer. This phenomena is due to little error of speed estimation

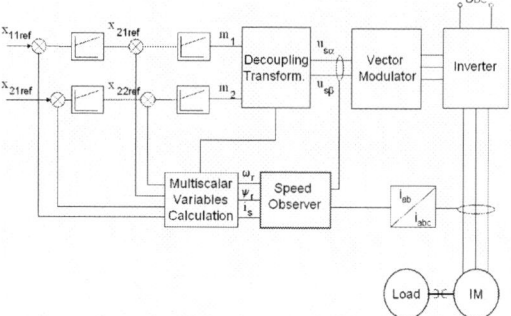

Fig. 13. Scheme of multiscalar-based sensorless control system.

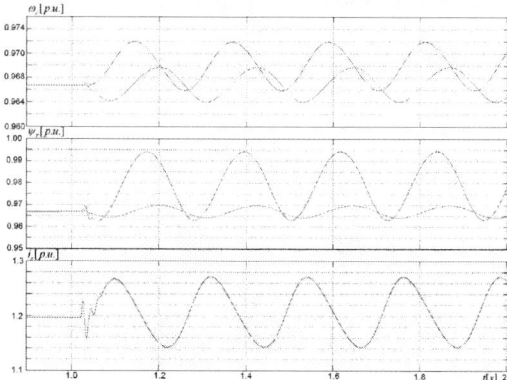

Fig. 14. Real and estimated state variables in new Krzemiński observer for sensorless closed-loop operation.

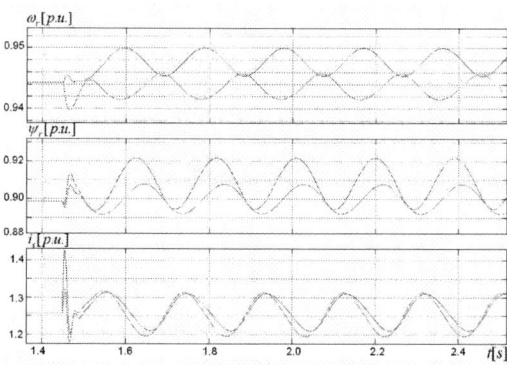

15. Real and estimated state variables in Kubota observer for sensorless closed-loop operation.

in low frequency range as in fig. 8. The shape of the characteristics is related to impact of control system time constant and slip of the machine, which determines low frequency pulsations due to broken rotor. It is clearly visible after step torque load. In this case control system is generating electromagnetic torque pulsation with limitation set in the PI controllers.

VI. CONCLUSIONS

Broken rotor impact on the sensorless control system of induction machine requires extensive simulation analysis for different operation states and configuration. For closed-loop operation with speed measurement output

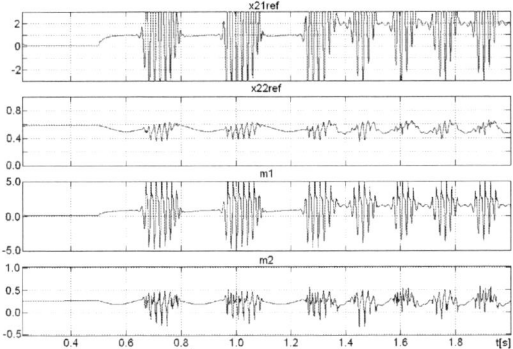

Fig. 16. Control system outputs with Krzemiński observer for sensorless closed-loop operation.

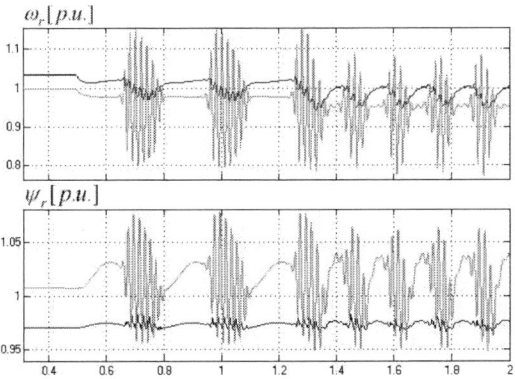

Fig. 17. Rotor speed and flux real and estimated values in Krzemiński observer for sensorless closed-loop operation.

signals of PI controllers in steady-state operation contain amplified harmonic components which can be used for diagnostic purposes. Open-loop observers analysis in frequency domain show that low frequency slip pulsation range disturbances related to rotor asymmetry at steady state operation are correctly reconstructed, but observer structure determines pulsation frequency range and quality reconstruction. Next sensorless operation with broken rotor simulation analysis confirmed that pulsations can be reconstructed using speed observer, which states basis for developing new diagnostics methods, but in this case further analysis of stability issues is required. Simulation analysis show that two structures of observers are able for stable operation with serious asymmetry of the machine while the third structure sensitivity features may be useful for early detection of the fault.

APPENDIX

Simulations has been prepared in Matlab/Simulink environment with following parameters: Rs=0.045, Rr=0.045, Ls=1.927, Lr=1.927, Lm=1.85, JJ=177.

ACKNOWLEDGMENT

This work has been supported by polish science fund as research project (2007 – 2009). Software equipment was provided by CI TASK.

REFERENCES

[1] A. Bellini, G. Franceschini, C. Tassoni, R. Passaglia i M. Saottini, "MCSA in Inverter Fed Machines:Pitfall and Fallacies," *16th International Conference on Electrical Machines*, Cracow, Poland, 2004, pp. 847-848

[2] M. Cruz, A. J. Cardoso, "Rotor cage fault diagnosis in three-phase induction motors by Extended Park's Vector Approach," *Electric Machines and Power Systems*, Vol. 28, No. 4, pp. 289-299, April 2000

[3] H. Dougles, P. Pillay, A. Ziarani,"A New Algorithm for Transient Motor Current Signature Analysis using Wavelets," *IEEE Transactions on Industry Applications*, vol. 40, no. 5, 2004, pp.1361-1367

[4] F. Cupertino, E. de Vanna, L. Salvatore, S. Stasi,"Analysis Techniques for Detection of IM Broken Rotor Bars After Supply Disconnection," *IEEE Transactions on Industry Applications*, vol. 40, no. 2, pp. 523-533, March/April 2004

[5] A. Bellini, F. Filippetti, G. Franceschini, C. Tassoni,"Closed Loop Control Impact on the Diagnosis of Induction Motor Faults," IEEE 1999

[6] H. Rodriguez-Cortes, Ch.N. Hadjicostis, A. M. Stankovic, "Model-Based Broken Rotor Bar Detection on an IFOC Driven Squirrel Cage Induction Motor," *Proceeding of the 2004 American Control Conference*, vol. 15, no. 2, pp. 3094-3099, Boston, July 2004

[7] Z. Krzeminski,"Multiscalar model based control system for induction motor fed by CSI," *The 2nd Inter. Conf. on Electr. Drives*, Poiana, Romania 1988

[8] Z. Krzeminski,"Sensorless Control System of the Induction Motor Based on New Observer" *PCIM'00*, Nurnberg, Germany

[9] Z. Krzemiński," A new speed observer for control system of induction motor. *IEEE Int. Conference on Power Electronics and Drive Systems*, PESC'99, Hong Kong, 1999

[10] Z. Krzemiński, "Struktura obserwatora prędkości silnika indukcyjnego z modelem zakłóceń," *Modelowanie i Symulacja 2006*, Kościelisko, 2006

[11] H. Kubota, K. Matsuse and T. Nakano, "DSP Based Adaptive Flux Observer of Induction Motor," *IEEE Transactions on Industrial Applications*, 1993, s. 344-348.

[12] F. Filippetti, G. Franceschini, C. Tassoni,"Neural networks aided on-line diagnostics of induction motor faults," *in Proc. IEEE Transactions on Industry Applications* Soc. Annual Meeting Conf., vol. 1, Toronto, ON, Canada, 1993, pp. 316–323.

[13] F. Filippetti, G. Franceschini, C. Tassoni,"Recent Developements of Induction Motor Drives Fault Diagnosis Using AI Techniques," *IEEE Transactions Ind. Electronics*, vol. 47, no. 5, pp. 994–1002 Oct. 2000.

[14] J. Manolas, J. A. Tegopoulos,"Analysis of Squirrel Cage Induction Motors with Broken Bars and Rings," *IEEE Transactions On. Energy Convercion*, vol. 14, no. 4, Dec. 1999

[15] A. Muiioz-Garcia, T. A. Lipo,"Complex vector model of squirrel cage induction machine including instantaneous rotor bar currents," *IEEE Trans.Ind. Appl* 1998

[16] G. Didier, H. Razik, A. Rezzoug,"An induction motor model including the first space harmonics for broken rotor bar diagnosis," *European Transactions on Electrical Power*, pp. 229-243, 2005

[17] A. Munoz-Garcia, T. A. Lipo,"Complex Vector Model of the Squirrel Cage Induction Machine including Instantaneous Rotor Bar Currents," *IEEE 1998*, pp. 57-64

[18] Cunha, Soares, Oliveira, Filho: "A new method to simulate induction machines with Rotor Assymetries" IEEE IECON 2002

Coupled Magnetic Circuit Method and Permeance Network Method Modeling of Stator Faults in Induction Machines

Amin Mahyob[*], Mohamed Y. Ould Elmoctar, Pascal Reghem and Georges Barakat[†]

University of Le Havre / GREAH, Groupe de Recherche en Electrotechnique et Automatique du Havre,
76058 Le Havre, France,

[*]e-mail: *amin.mahyob@univ-lehavre.fr*
[†]e-mail: *georges.barakat@univ-lehavre.fr*

Abstract— **The aim of this paper is to present and compare two modeling methods of the inter-turn short circuit fault in the stator winding of a cage induction machine. The first method is a Coupled Magnetic Circuit Method (CMCM) and the second method is a Permeance Network Method (PNM) which allows taking into account the saturation effect on the fault signature. The simulation results show that by proper modeling of induction machine it is possible to detect the effect of inter-turn short circuit faults on the machine line currents. The effect of saturation on the machine line current signature is discussed. The detailed equations describing the performance of the machine in the two cases of modeling under inter-turns short circuit faults are presented. Simulation results are compared for both healthy and faulty cases**

Keywords— **AC machine, asynchronous motor, diagnostics, modeling, simulation.**

I. INTRODUCTION

The stator winding inter-turn short circuit fault is one of the most common machine faults. It is usually related to failure of the turn-to-turn insulation. This fault belong that class of faults that may often have a negligible influence on the machine performance but the presence of such fault leads to generate major ones such as phase to phase faults or phase to ground faults that cause catastrophic failure of the machine[1]–[2]. Therefore, early detection of inter-turn shorts especially, turn-to-turn shorts of winding during motor operation would eliminate subsequent damage to adjacent coils and the stator core, reducing repair cost and motor outage time[3].

Recent years, the studies on fault diagnosis of electrical machines including, chemical analysis, mechanical and magnetic measurement and motor current signature analysis (MCSA) have been squared [4]-[5]-[6]. The success of the technique depends not only on its ability to distinguish between healthy and faulty states but also on its ability to discriminate between various faults. So, this requires the development of a complete and accurate modeling methods for electrical machines able to simulate electrical faults within reasonable consumed time in order to study a large number of situations and build databases contains sufficient information on the machine components related to the various faults.

In this paper, two methods of modeling to simulate the faulty induction machine are presented. The first method is based on coupled magnetic circuit theory. This method takes into account all the space harmonics in the machine thank to winding function theory and dedicated function modeling of doubly slotted air-gap permeance but it neglects the effect of the magnetomotive force drop (m.m.f.) in the iron, and by the way the effect of saturation on the fault signature [4]-[7]. In fact In the case of the inter-turn short circuit fault, the short circuit current reaches several times the rated value of the phase current in the healthy case and generates an important local saturation around the fault zone. This phenomenon impacts strongly the harmonic contents of the stator current and imposes the use of a nonlinear model for the calculation of this fault signature in the phase currents.

The second method is based on permeance network of the machine. This method allows taking into account the saturation effect as well as all the space harmonics of doubly slotted machine. Simulation results show that the PNM seems to be the more adequate modeling approach helping to simulate a saturated electrical machine under an inter turn short circuit stator fault with moderated computing time compared to the FEM.

II. CMCM MODELING OF INDUCTION MACHINES UNDER STATOR FAULTS

A. Case of Healthy Machine

1) Case delta connected stator phase: Consider initially a general *m*-phase delta connected induction machine with *q* rotor bars. The proposed model is derived from a well-known method [7] [8] consisting to replacing the *q* squirrel cage bars by an equivalent circuit containing *q+1* magnetically coupled meshes as shown in Fig. 1. The current in each mesh of the rotor cage is an independent variable. The following assumptions are to be considered:

i) The permeance of iron is infinite,

ii) The eddy current and windage losses are negligible,

iii) There are no inter-bar currents.

With these assumptions, the equations describing the general induction machine with m stator phases and q rotor bars can be written in vector-matrix form and in a compact manner as follows:

$$[V] = [R][I] + \frac{d}{dt}[\Phi] \qquad (1)$$

where:

$$[R] = \begin{bmatrix} [R_{SS}] & [0] \\ [0] & [R_{RR}] \end{bmatrix} \quad (2)$$

$$[\Phi] = \begin{bmatrix} [\Phi_S] \\ [\Phi_R] \end{bmatrix} = [L] \cdot [I] = \begin{bmatrix} [L_{SS}] & [L_{SR}] \\ [L_{RS}] & [L_{RR}] \end{bmatrix} \cdot \begin{bmatrix} [I_S] \\ [I_R] \end{bmatrix} \quad (3)$$

$$[V_S]^T = [v_{S1} \ v_{S2} \ v_{S3} \ldots\ldots v_{Sm}]^T \quad (4)$$

$$[I_S]^T = [i_{S1} \ i_{S2} \ i_{S3} \ldots\ldots i_{Sm}]^T \quad (5)$$

$$[I_R]^T = [i_{R1} \ i_{R2} \ \ldots\ldots \ i_{Rq} \ i_e]^T \quad (6)$$

where T represents the transpose of a vector.

The resistance matrix $[R_{SS}]$ in (2) is a $m*m$ diagonal matrix of stator phases resistances. The inductance matrix $[L_{SS}]$ in (3) is a $m*m$ symmetric matrix of stator phases inductances. Also, $[L_{SR}]$ is the $m*(q+1)$ mutual inductances matrix between stator phases and rotor cage meshes with $[L_{SR}] = [L_{RS}]^T$. The $(q+1)*(q+1)$ resistances $[R_{RR}]$ and inductances $[L_{RR}]$ matrices in (2) and (3) respectively, are derived from the equivalent circuit of Fig 1 and are given in equations (14) and (15), where R_b is the rotor bar resistance and R_e is the end ring portion resistance, L_{ob} is the bar leakage inductance, L_{oe} is the end ring portion inductance and L_{RRii} is the magnetizing inductance of the ith mesh and L_{RRij} is the mutual inductance between the ith mesh and the jth one.

To the set of the equations (1)-(3), one must add the mechanical equation of the shaft:

$$J \frac{d\Omega}{dt} = T_{em} - f\Omega - T_L \quad (7)$$

in which Ω is the rotor angular speed and T_{em} the electromagnetic torque expressed as :

$$T_{em} = \frac{1}{2} [I]^t \left\{ \frac{d}{d\theta_m} [L] \right\} [I] \quad (8)$$

with θ_m the rotor position.

At this stage and taking into account the torque expression, one can easily remark that the accuracy of the coupled magnetic circuit based modeling method depends greatly on the accuracy of the inductances calculation which is discussed in the next section C.

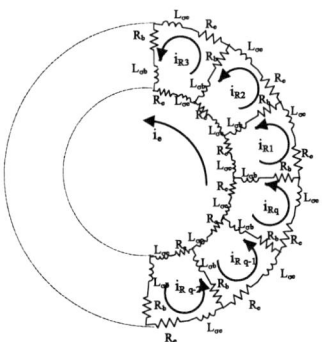

Fig.1. Equivalent circuit of the rotor cage showing meshes

2) Case star connected stator phase: In the case of the stator phases star connected without a neutral line, the Kirchoff's law applied to the phase currents yields:

$$i_{Sm} = -\sum_{n=1}^{m-1} i_{Sn} \quad (9)$$

Replacing of the mth current by its new expression in equation (1) results in the obtainment of a new equation where the stator currents and voltages vectors are as follows:

$$[V_S^Y]^T = [v_{S1} - v_{Sm} \ \ v_{S2} - v_{Sm} \ \ \ldots \ \ v_{Sm-1} - v_{Sm}]^T \quad (10)$$

$$[I_S^Y]^T = [i_{S1} \ i_{S2} \ \ldots \ i_{Sm-1}]^T \quad (11)$$

and the $[R_{SS}]$, $[L_{SS}]$ and $[L_{SR}]$ matrices of the star case are deduced from the delta case ones by the following relations:

for $k=1..m-1; j=1..q$

$$\begin{cases} L_{SR}^Y(k,j) = L_{SR}^\Delta(k,j) - L_{SR}^\Delta(m,j) \\ L_{RS}^Y(j,k) = L_{RS}^\Delta(j,k) - L_{RS}^\Delta(j,m) \end{cases} \quad (12)$$

and for $k,j=1..m-1$

$$\begin{cases} R_{SS}^Y(k,k) = R_{SS}^\Delta(k,k) + R_{SS}^\Delta(m,m) \\ R_{SS}^Y(k,j) = R_{SS}^\Delta(m,m) \quad with \quad k \neq j \\ L_{SS}^Y(k,j) = L_{SS}^\Delta(k,j) - L_{SS}^\Delta(k,m) - L_{SS}^\Delta(m,j) + L_S^\Delta(m,m) \end{cases} \quad (13)$$

where the superscript Y and Δ indicate the quantities related to the star connection and the delta connection of stator phases respectively[7].

The rotor resistances and inductances matrices as well as the rotor voltages and currents vectors remain unchanged

B. Case of Inter-Turn Short Circuit Faults

In electrical machines, the insulation between stator turns can be represented by an electrical resistance which is infinite in the healthy case. Generally, inter-turn short circuit in the stator phases is due to a local insulation failure which induces this fault to spread in degrees causing the shutdown of the machine. So, the contact point between the shorted turns and the faulty phase can be represented by a variable resistance R_{CC} in parallel with the shorted turns as indicated in fig. 2 and its value depends on the severity of the fault.

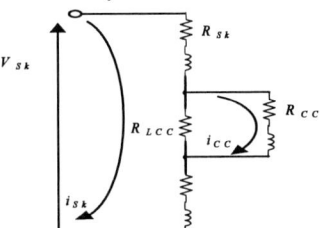

Fig. 2: Electric scheme of a phase with short circuited turns

The occurrence of inter-turn short circuit in N_{CC} coils belonging to the stator phases results in adding N_{CC} dimensions to the stator related vectors of the differential equations system (9). In other words, one must write the above differential equations system with $m+N_{CC}$ stator phases where the N_{CC} additional stator phases represent the N_{CC} coils of shorted turns. The additional elements of the matrix resistances R_{SS} corresponding to the 'new' N_{CC} phases are determined as a multiple of the resistance value of a mean length turn. The magnetizing and mutual inductances of the additional N_{CC} phases are carried out by the use of their respective winding functions. The rotor matrices R_{RR} and L_{RR} remain unchanged [7]. The star connection case of stator phases can be treated in the same manner as that for a healthy machine.

$$[R_{RR}] = \begin{bmatrix} 2(R_b+R_e) & -R_b & 0 & 0 & . & . & . & . & -R_b & -R_e \\ -R_b & 2(R_b+R_e) & -R_b & 0 & . & . & . & . & 0 & -R_e \\ 0 & & & . & & . & & & . & . \\ . & & . & & . & & . & & . & . \\ . & & & . & & . & & & . & . \\ . & & . & & . & & . & & . & . \\ . & & & . & & . & & & . & . \\ . & & . & & . & & . & & . & . \\ -R_b & 0 & & . & . & . & . & -R_b & 2(R_b+R_e) & -R_e \\ -R_e & -R_e & . & & . & . & . & . & -R_e & qR_e \end{bmatrix} \quad (14)$$

$$[L_{RR}] = \begin{bmatrix} L_{RR11}+2(L_{\sigma b}+L_{\sigma e}) & L_{RR12}-L_{\sigma b} & L_{RR13} & . & . & . & . & L_{RR1q}-L_{\sigma b} & -L_{\sigma e} \\ L_{RR21}-L_{\sigma b} & L_{RR22}+2(L_{\sigma b}+L_{\sigma e}) & L_{RR23}-L_{\sigma b} & . & . & . & . & L_{RR2q} & -L_{\sigma e} \\ . & . & & & & & . & & -L_{\sigma e} \\ . & . & & & & & . & & . \\ . & . & & & & & . & & . \\ . & . & & & & & . & & . \\ . & . & & & & & . & & . \\ L_{RRq1}-L_{\sigma b} & . & & . & . & . & . & L_{RRqq}+2(L_{\sigma b}+L_{\sigma e}) & -L_{\sigma e} \\ -L_{\sigma e} & -L_{\sigma e} & . & & . & . & . & -L_{\sigma e} & qL_{\sigma e} \end{bmatrix} \quad (15)$$

C. Calculation of Inductances Under Inter-Turn Short Circuit Faults

1) Inductances Expression Driven From Energy Consideration: the calculation of all the relevant inductances for the induction machine is based on the winding function theory developed in [13]. In some previous study [9,]-[10], the inductances expressions were derived from the calculation of the air-gap flux across by the windings. The authors of this study [9] have made a distinction between the variation of L_{SR} and L_{RS} with the rotor position in the case of air-gap eccentricity. This distinction results from the use of the distribution function of the windings in the calculation of mutual inductances. In this paper, the calculation of inductances is derived from the magnetic energy stored in the air-gap. Assuming an infinite permeability of the iron, the air-gap flux density due to a winding 'i' is given by the well-known expression:

$$B_i(\theta,\theta_s,z) = F_i(\theta,\theta_s,z)P(\theta,\theta_s,z) \quad (16)$$

where $F_i(\theta, \theta_s, z)$ is the magnetomotive force (m.m.f) of winding 'i', $P(\theta, \theta_s, z)$ is the air-gap permeance, θ is the angular position of the rotor with respect to some stator frame, θ_s is the angular position in the stator frame and z is the axial position in the case of skewed rotor bars. If

I_i is the current flowing in winding 'i', the m.m.f. can be expressed by :

$$F_i(\theta,\theta_s,z) = F_{w_i}(\theta,\theta_s,z)I_i \quad (17)$$

where $F_{wi}(\theta, \theta_s, z)$ is the winding function of winding 'i'. In effect, F_{wi} is the m.m.f. distribution along the air-gap for a unit current in winding 'i'. The magnetic energy stored by winding 'i' in the air-gap is given by

$$W_i(\theta) = \frac{1}{2}L_{ii}(\theta)I_i^2 = \frac{1}{2\mu_0}\iiint_{\tau_{ag}}B_i^2(\theta,\theta_s,z)d\tau_{ag} \quad (18)$$

where L_{ii} is the magnetizing inductance of winding i and τ_{ag} is the air-gap volume. Incorporating (16) and (17) in (18) and considering that the air-gap flux density is constant in the radial direction, the inductance L_{ii} can be computed by

$$L_{ii}(\theta) = \int_0^L\int_0^{2\pi} F_{w_i}^2(\theta,\theta_s,z)P^2(\theta,\theta_s,z) \\ R_{av}(\theta,\theta_s,z)e(\theta,\theta_s,z)d\theta_s dz \quad (19)$$

where R_{av} is the average radius of the air-gap and $e(\theta, \theta_s, z)$ is the effective air-gap function. In the same manner, the mutual inductance between winding '*i*' and winding '*j*' can be computed by

$$L_{ij}(\theta) = \int_{0}^{L}\int_{0}^{2\pi} F_{w_i}(\theta,\theta_s,z)F_{w_j}(\theta,\theta_s,z)P^2(\theta,\theta_s,z)$$

$$R_{av}(\theta,\theta_s,z)e(\theta,\theta_s,z)d\theta_s dz \quad (20)$$

From the expression (20), one can easily remark that L_{ij} is equal to L_{ji} unlike the winding embraced air-gap flux based approach.

2) Calculation of the Air-Gap Permeance Function: a sophisticated permeance model has been developed in [11] and [12]. It is based on a statistical study of the drop of flux density waveform in the air-gap in front of a slot by means of finite elements method. This study was made essentially for the more widely used slot geometry shapes in induction machines. The expressions of the air-gap permeance function for a slot pitch τ_e are given as follows:

$$P(x) = \begin{cases} P_{\max} & for \quad \tau_2 - \tau_3/2 \le |x| \le \tau_e/2 \\[2em] \dfrac{P_{\max}+P_{moy}}{2} + \dfrac{P_{\max}-P_{moy}}{2}\dfrac{\tan^{-1}\left(a(-|x|+\tau_2/2)\right)}{\tan^{-1}\left(\dfrac{a}{2}(\tau_3-\tau_2)\right)} \\[1em] \qquad\qquad for \quad \tau_3/2 \le |x| \le \tau_2 - \tau_3/2 \\[2em] \dfrac{(P_{\min}-P_{moy})e^{-fx^2}+P_{moy}-P_{\min}e^{-f\tau_3^2}}{1-e^{-f\tau_3^2}} \\[1em] \qquad\qquad for \quad -\tau_3/2 \le x \le \tau_3/2 \end{cases} \quad (21)$$

The different parameters in the air-gap permeance expressions are expressed in terms of the slot dimensions. These expressions are given in [12].

Using the permeance model of (21) , one can easily construct the permeance of a single slotted air-gap. If $P_s(\theta, \theta_s)$ denotes the air-gap permeance for a slotted stator and a smooth rotor and $P_r(\theta, \theta_r)$ the air-gap permeance for a slotted rotor and a smooth stator, the air-gap permeance of a doubly slotted air-gap is then given by:

$$P(\theta,\theta_s) = \frac{P_s(\theta,\theta_s).P_r(\theta,\theta_r)}{P_s(\theta,\theta_s)+P_r(\theta,\theta_r)} \quad (22)$$

Fig 3 illustrates the air-gap permeance of the studied machine, whose parameters are given in the appendix.

III. PNM MODELING OF INDUCTION MACHINES UNDER STATOR FAULTS

A. Case of Healthy Machine

An electrical machine can be represented as a set of flux tubes (Fig. 4) characterized by their magnetic permeances. These permeances are expressed as functions of the machine geometry and the instantaneous fluxes flowing in each one of them [13]-[14].

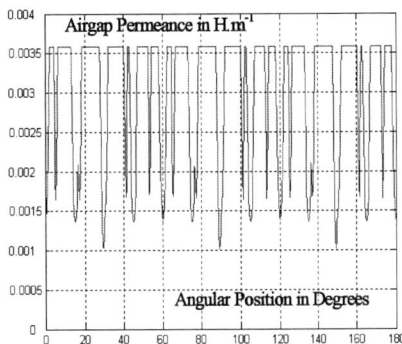

Fig. 3. Air-gap permeance of the studied machine.

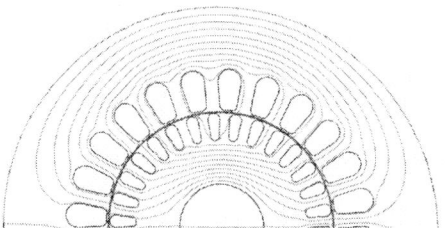

Fig. 4. Flux tubes for a half part of the studied machine

Exploiting the flux tubes of fig. 4, one can deduce the magnetic equivalent circuit of the induction machine as shown in Fig. 5 where one can distinguish the stator magnetic circuit region, the air-gap region as well as the rotor magnetic circuit region.

The stator slot currents are modeled by magnetomotive force (m.m.f.) sources in series with the tooth permeances. To establish the relation between these m.m.f and the phase currents we consider a closed contour around the slot as shown in Fig. 6 and we associate a magnetomotive force (F_{tsi}) to the slot *i*. Using the magnetic and electric laws the following equation are deduced:

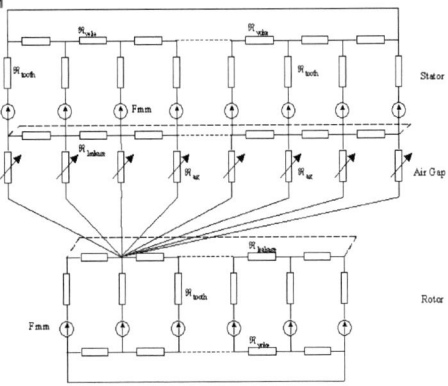

Fig. 5. Permeance network of studied machine

$$\begin{cases} \int H * dl = \sum n_k I_k = \sum_j \dfrac{\phi}{P_j} \\ \sum_j \dfrac{\phi}{P_j} - F_{sti} + F_{sti+1} = 0 \\ F_{sti} - F_{sti+1} = \sum n_k I_k \end{cases} \quad (23)$$

By generalizing the above equations for all the teeth we can obtain the following compact matrix form:

$$\left[F_{st} \right] = \left[M_{is} \right] . \left[I_s \right] \quad (24)$$

where $[F_{st}]$ is the vector of tooth m.m.f., $[I_s]$ is the vector of the phase currents $[M_{is}]$ is the matrix that relates the tooth m.m.f. to the phase currents and n_k is the number of the turns inside the slot carrying the current i_k.

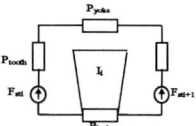

Fig. 6 Closed contour around a slot

In the same manner, the rotor bar currents are modeled by m.m.f. sources in series with the rotor tooth permeances. The relation between these m;m.f and mesh currents is calculated with same way as in the stator case by considering the number of turns inside the slot equals 1.

The air-gap region is subdivided in a set of air-gap elements each one of them is modeled by a permeance which depend on the rotor position in order to connect the stator tooth fluxes to the rotor tooth fluxes. By this manner, one can take into consideration the rotor motion. As an example, equation (26) [13] gives a possible air-gap element permeance expression relating the stator tooth 'i' to the rotor tooth 'j' versus the rotor relative position θ :

$$P_{i,j} = \begin{cases} P_{max} & 0 \le \theta \le \gamma_t^{'} \; et \; 2\pi - \gamma_t^{'} \le \theta \le 2\pi \\ P_{max} \left(1 + \cos\pi \dfrac{\theta_{ij} - \gamma_t^{'}}{\gamma_t - \gamma_t^{'}} \right) \Big/ 2 & \gamma_t^{'} \le \theta_{ij} \le \gamma_t \\ P_{max} \left(1 + \cos\pi \dfrac{\theta_{ij} - 2\pi + \gamma_t^{'}}{\gamma_t - \gamma_t^{'}} \right) \Big/ 2 & 2\pi - \gamma_t \le \theta_{ij} \le 2\pi - \gamma_t^{'} \\ 0 & else \end{cases} \quad (25)$$

The different parameters in the air-gap permeances expression are expressed in terms of the stator and rotor slot dimensions and air-gap length.

To take into account the magnetic circuit saturation, the B-H curve can be modeled by the Marrocco's formula [15] or also numerically by the use of Spline functions fitting the measured B-H curve.

The system of magnetic equations governing the machine is obtained by treating the permeance network of Fig. 5 in the same manner as an electric circuit. In

our case, node and branch equations are written considering stator phase and rotor loop currents as the entry of the system of equations. These equations are given by the following general form:

$$\begin{bmatrix} [\Re] & [N^t] \\ [N] & [P] \end{bmatrix} . \begin{bmatrix} [\Phi_t] \\ [U] \end{bmatrix} = \begin{bmatrix} [M_i] \\ [0] \end{bmatrix} . [I] \quad (26)$$

where $[\Re]$ is a diagonal reluctance matrix, $[P]$ is a permeance matrix, $[\Phi_t]$ is the vector of stator and rotor tooth fluxes, $[U]$ is the vector of stator and rotor node scalar magnetic potential, $[M_i]$ is a matrix that relates the tooth m.m.f. to the phase currents and rotor mesh currents, N is a matrix its general element equals 1 if the flux coming out of to the node , - 1 if flux entries to the node and zero else , and $[I]$ is the vector of stator and rotor currents.

The developed magnetic network is related to the electrical differential equations of the stator phases by the calculation of the linkage fluxes issued from the tooth fluxes. In order to establish the relation between the linkage fluxes and tooth fluxes, we consider a coil of n_c turns embarrasses y teeth, as shown in fig. 7, the linkage flux of this coil is given by:

$$\psi_{ck} = \sum_{i=j+1}^{y+j} n_c \Phi_{Si} \quad (27)$$

and the linkage flux of the phase is equal to the some of the linkage fluxes of the coils constituting this phase; i.e:

$$\psi_{pi} = \sum_k \psi_{ck} \quad (28)$$

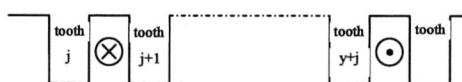

Fig. 7 Linkage flux of a coil

where y is the coil pitch, Φ_S is the stator tooth flux, ψ_{ck} is the linkage flux of the coli k and ψ_{pi} is the linkage flux of the phase i.

By repeating the above equations for each one of the stator phases we get the whole stator equations system witch is given by:

$$\left[\psi_s \right] = \left[M_{\phi s} \right] . \left[\Phi_s \right] \quad (29)$$

where $[M_\Phi]$ is the matrix that relates the stator phase linkage fluxes to stator tooth fluxes

As the studied machine is a squirrel cage induction machine and as the rotor meshes are defined by two bars consecutive, the flux linkage of a mesh is equal to the flux of the tooth embarrassed by these two bars.

As in the coupled magnetic circuit method, the electrical equations describing the delta connected general induction machine with m stator phases and q rotor bars can be written by replacing the q bars of squirrel cage by an equivalent circuit containing $q+1$ magnetically coupled circuits [4]. The compact matrix form of these equations is given by:

$$[V] = [R].[I] + [L_\sigma].\frac{d}{dt}[I] + \frac{d}{dt}[\psi] \qquad (30)$$

with

$$[R] = \begin{bmatrix} [R_{SS}] & [0] \\ [0] & [R_{RR}] \end{bmatrix}, [L_\sigma] = \begin{bmatrix} [L_{\sigma S}] & [0] \\ [0] & [L_{\sigma R}] \end{bmatrix}$$

and

$$[\psi] = \begin{bmatrix} [\psi_S] \\ [\psi_R] \end{bmatrix} = \begin{bmatrix} [M_{\phi S}] & [0] \\ 0 & [M_{\phi R}] \end{bmatrix} . \begin{bmatrix} [\Phi_S] \\ [\Phi_R] \end{bmatrix}$$

where $[L_{\sigma S}]$ is a $m*m$ diagonal matrix of stator winding end leakage inductances, $[L_{\sigma R}]$ is a $(q+1)*(q+1)$ end ring leakage inductance matrix driven as $[L_{RR}]$ by considering only $L_{\sigma e}$ and setting the other elements to zero, $[\psi]$ is the linkage flux vector and $[M_\phi]$ is the matrix that relates the stator phase and rotor mesh linkage fluxes to tooth fluxes

To the set of magnetic and electric equations one must add the mechanical equation of the shaft. The torque equation is established by deriving the magnetic co-energy with respect to the rotor angular position; thus the electromagnetic torque obtained is given by:

$$T_{em} = \frac{1}{2} \sum_{i=1}^{Ns} \sum_{j=1}^{q} \frac{dP_{i,j}}{d\theta} U_{i,j}^2 \qquad (31)$$

and the mechanical equation is given by:

$$T_{em} = T_L + J\frac{d\Omega}{dt} + f\Omega \qquad (32)$$

where Ω is the rotor angular speed and θ is the rotor position. $U_{i,j}$ is the scalar magnetic potential between the stator tooth 'i' and the rotor tooth 'j'.

In the case of star connected induction machine the stator equations are written as function of stator line currents and voltages. Two new matrices then developed, the first relates the phase currents to the new line currents considering the relation (9), and the second relates electromotive forces to the mesh electromotive forces, in other words the mesh linkage fluxes to phase linkage.

Then each one of $[M_{is}]$ and $[M_\phi]$ is multiplied by the matrix corresponding. The stator resistance and leakage inductance matrices are treated with the same way as in the case of coupled magnetic circuit. The rotor matrices remain unchanged.

B. Case of Inter-Turn Short Circuit Faults

In the presence of inter-turn short circuit fault the electrical equation are treated as in the case of coupled magnetic circuit method. The additional elements of the matrix resistances and the matrix of the end winding inductances corresponding to the 'new' N_{CC} phases are calculated as described in the case of coupled magnetic circuit while the N_{CC} additional elements of the matrices $[M_{is}]$ and $[M_{\phi s}]$ are carried out with the same manner as

the main phases by considering their respective turns. The rotor resistance and end ring leakage inductance matrices remain unchanged.

IV. SIMULATION RESULTS AND DISCUSSION

The proposed model was used to simulate a 4 kW, 230/400 V, 14.2/8.2 A, 2840 rpm, 2 poles, 24 stator slots, 30 rotor bars, Y connected, standard squirrel cage induction whose detailed parameters are given in the appendix . Each stator phase contains 4 coils in series and each coil is constituted of 31 turns. The simulations were carried out for 100% of rated load in the case: of healthy machine and in the case 3.33% phase (a) short circuited turns for the two methods of modeling, CMCM and PNM with zero contact resistance (R_{CC}=0).

Fig. 8 and 9 show the simulation results of stator current spectrum and Concordia's pattern (diagram) respectively for the healthy machine for the two methods of modeling. One can remark that in the healthy case the results are closed to each other and the form of the Concordia's pattern is circle, but the dimension of the Concordia's patterns in the case of PNM method is greater than that in the case of CMCM due to the drop of m.m.f. in the iron, consequently the induced electromotive force is reduced which leads to the increase of the current absorbed.

It is known that rotor slot harmonics reflected in the time domain of the stator currents are boosted by the machine asymmetries [4]. These harmonics are induced in the short circuited turns coil which induce them in the stator faulty phase current at the frequencies

$$f_{rh} = (kq\frac{(1-S)}{p} \pm v)f_s \qquad (33)$$

where q is the number of rotor slots and p is the pole pair number, S is the slip and v is the supply harmonics.

Furthermore, the effect of saturation can be introduced to expression (33) simply as additional harmonics representing the fluctuation of the air-gap flux at twice the number of pole pairs and twice the frequency of the fundamental wave, then expression (34) becomes [1]:

$$f_{rh} = (kq\frac{(1-S)}{p} \pm 2m_{sat} \pm v)f_s \qquad (34)$$

where m_{sat} is an integer representing the indices of saturation harmonics.

Fig. 10 shows the simulation results of the phase (a) stator current for 4 shorted turns (3.33% of phase turns).

It is seen that the contribution of the principal slot harmonic (PSH) (k=1, v=1) increases in the case of the PNM model than that in the case of CMCM model.

The third harmonic is also presented in the two models because of the speed oscillations [9]-[16] induced by the torque ripple. In fact, one of the consequences of the inter-turn short circuit fault is the appearance of the negative m.m.f due to the negative sequence current in the stator phases. The negative sequence m.m.f reacts with the fundamental component of rotor bar currents giving a torque spectral at twice the supply frequency as shown in fig. 11. This pulsating

815

torque produces a speed ripple at the same frequency. Finally, the pulsation of the air-gap flux at twice the supply frequency induces the third harmonic in the stator currents. It is known that the contribution of this third harmonic in phase current increases with increasing the severity of the fault.

Some recent studies report the detection of third harmonics components in line currents as signature for stator faults; however this will be confused with the voltage unbalance present in the line voltage, structural asymmetry and the saturation [16].

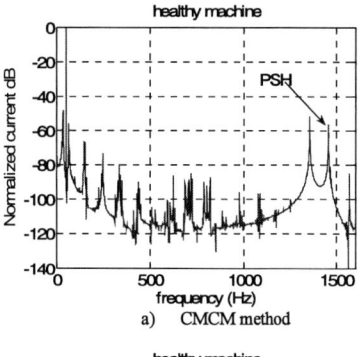

a) CMCM method

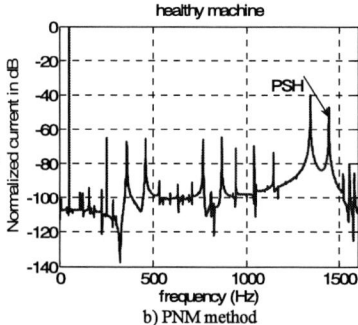

b) PNM method

Fig. 8 Faulty phase stator current spectrum for healthy machine

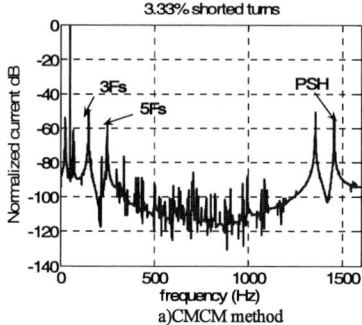

a)CMCM method

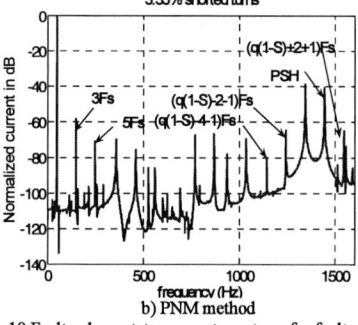

b) PNM method

Fig. 10 Faulty phase stator current spectrum for faulty machine

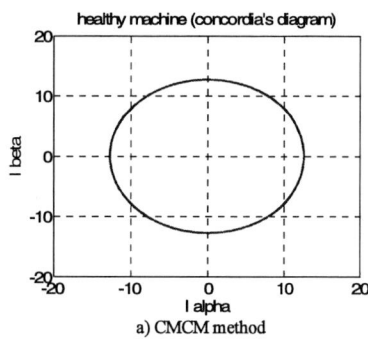

a) CMCM method

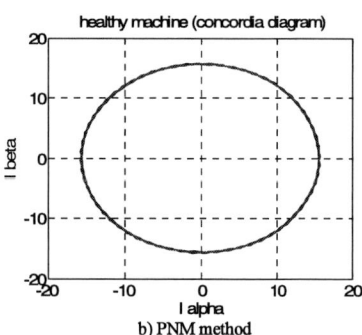

b) PNM method

Fig. 9 Concordia's patterns for healthy machine

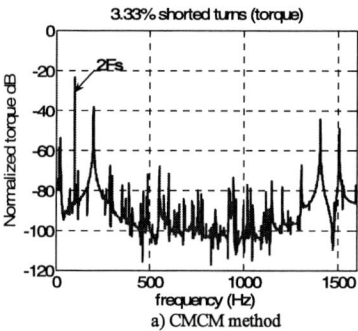

a) CMCM method

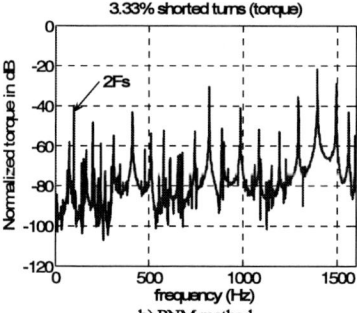

b) PNM method

Fig. 11 Torque spectrum for faulty machine

816

Regarding fig. 10 (a) and (b) one can clearly remark that the harmonic contents of phase currents spectrum differ from the linear model to nonlinear model. In fact, short circuit current saturates locally and strongly the magnetic circuit of the machine, so the operating point lies on the nonlinear part of B(H) curve used to simulate the machine causing the apparition of new in the harmonics case of PNM based model representing the saturation effect according to the relation (34)

V. CONCLUSION

In this paper, two methods of modeling for the simulation and diagnosis of induction machines have been presented and compared. The simulation results show a good correspondence between the electromagnetic phenomena and the frequency components of the stator currents in the two models for the healthy case. In the case of inter-turn short circuit faults, it was demonstrated that the saturation impact can not be neglected which leads to unsatisfactory results with the MCMC based model. In spite of the flexibility of CMCM method, the PNM is more adequate to study the saturated electrical machines and give a robust model of simulation for both model based diagnosis and AI-based diagnosis.

APPENDIX

Studied motor parameters:
4 kW, 230/400 V, 14.2/8.2 A, 2840 rpm, 2 poles, 24 stator slots, 30 rotor bars

$D_{so} = 145$ mm	stator outer diameter
$D_{si} = 75.4$ mm	stator inner diameter
$L_s = 125$ mm	stack length
$e = 0.35$ mm	effective airgap
$R_r = 37.35$ mm	rotor radius
$\beta = 1.4$	bar skew coefficient
$s_s = 2.6$ mm	stator slot opening width
$b_s = 5.78$ mm	stator slot base length
$d_s = 0.6$ mm	stator slot isthmus width
$s_r = 1.1$ mm	rotor slot opening width
$b_r = 4.0$ mm	rotor slot base length
$d_r = 0.9$ mm	rotor slot isthmus width
$N_s = 124$	# of stator phase turns in series
$R_s = 1.595\ \Omega$	stator phase resistance
$L_s = 0.004$ H	stator phase leakage inductance
$R_b = 3.04\text{E-}04\ \Omega$	rotor bar resistance
$L_b = 5.16\text{E-}07$ H	rotor bar leakage inductance
$R_e = 8.75\text{E-}07\ \Omega$	rotor end ring segment resistance
$L_e = 1.59\text{E-}09$ H	rotor end ring segment leakage inductance
$J = 0.045$ kg.m2	drive inertia
$f_v = 0.0038$ kg.m^2.s^{-1}	friction coefficient

REFERENCES

[1] Andreas Stavrou, Hawrad G. Sedding and and Jams Penman," Current monitoring for detecting inter-turns short circuit in induction motors", *IEEE Trans. Energy Conv.* Vol. 16. No. 1, March., 2001.

[2] Sang Bin Lee, karim Younsi and Gerald Kliman, "An online technique for monitoring the insulation condition of ac machine stator winding", *IEEE Trans. Energy Conv.* Vol. 2. No. 4, Dec., 2005.

[3] Hyung-Woo Lee, Tae-Hyung Kim and Changho Choi "A Novel Internal Fault Analysis of a Brushless DC Motor Using Winding Function Theory " , *IEMDC* 2005, May 15-18, 2005, Sant Antonio, Texas, USA

[4] O. A. Mohammed, Z. Liu, S. Liu and N. Y. Abed," Internal short circuit fault diagnosis for PM machine using FE-based phase variable model and wavelet analysis", *IEEE Trans. Mag..* Vol. 43. No. 4, April, 2007.

[5] Arafat Siddique, G. S Yadava and Bhim Sigh, "A Review of stator fault monitoring techniques of induction motors", *IEEE trans. Energy conv.* Vol. 20. No. 1, march, 2005.

[6] M; E. H. Benbouzid, "Bibliography on induction motors faults detection and diagnosis", *IEEE trans. Energy conv.* Vol. 14. No. 4, Dec, 1999.

[7] G. Houdouin, G. Barakat, B. Dakyo, E. Destobbeleer and C. Nichita, "A Coupled Magnetic Circuit Based Global Method for the Simulation of Squirrel Cage Induction Machines Under Rotor and Stator Faults", in *Proc. Electrimacs 2002*, Montreal, Canada, 18-21 Aug. 2002

[8] M. Poloujadoff, "The theory of three phase induction squirrel cage machine", *Electric Machines and Power Systems* , 13, pp245-264, 1967..

[9] G.M. Joksimovic and J. Penman, "The detection of inter-turn short circuits in the stator windings of operating motors, " *IEEE Trans. Ind. Electronics*, vol. 47, no. 5, pp. 1078-1084, Oct. 2000.

[10] A. Toliyat and T. A. Lipo, "Transient analysis of cage induction machines under stator, rotor bar and end ring faults," *IEEE Trans. Energy Conversion*, vol. 10, pp. 241-247, June 1995.

[11] G. Houdouin, G. Barakat, T. Derrey and E. Destobbeleer, "A simple analytical model for the calculation of harmonics due to slotting in the air-gap flux density waveform of an electrical machine," in *Proc. EPE1997*, Vol. 2, pp. 2601-2605.

[12] G. Houdouin, G. Barakat, B. Dakyo and E; Destobbeleer, "An improved method for dynamic simulation of air-gap eccentricity in induction machines," *in Proc. IEEE SDEMPED2001*, pp.133-138.

[13] V.Ostovic, *Dynamics of Saturated Electric Machines*, spring -Verlag, New York, 1989.

[14] Wan Shuting, Wang Aimeng, Li Yongang and Wang Yi, " Reluctance Network Model of Turbo-Generator and its Application in Rotor Winding Inter-Turn Short Circuit Fault," *IEMDC 2005*, San Antonio, Texas, USA, 15-18, May, 2005.

[15] A. Marrocco, "Analyse numérique de problèmes d'électrotechnique", *Ann.Sc.math*, Québec, vol. 1, pp. 271-296, 1977.

[16] Subhasis Nandi," Detection of stator faults in induction machines using residual saturation harmonics", *IEMDC2005*,,15-18,May, 2005, San Antonio, Texas, USA

Explosion Protected Electrical Drives
- Risk Assessment and Technical Diagnostics

Ivica Gavranić*, Drago Ban† and Damir Žarko†

* Ex - Agency, Zagreb, Croatia, e-mail: *i.gavranic@ex-agencija.hr*
† University of Zagreb, Faculty of Electrical Engineering and Computing, Zagreb, Croatia,
e-mail: *drago.ban@fer.hr, damir.zarko@fer.hr*

Abstract—In this paper the application of motors in plants at risk of explosive atmospheres, and the special requirements of such applications from the standpoint of explosion protection are considered. The application of motors with type of protection increased safety, marked as "EEx e", with and without associated safety systems, are analyzed for zone 2. Risk assessment of such applications, based on the requirements of ATEX Directive, is the main part of this paper. There is a specific example which illustrates the results of selection and application of motors with type of protection "EEx e" without the application of additional safety systems, which is the traditional approach. Possible application, in special cases, of the mentioned motor with the application of additional safety systems, which is the new approach, is also analyzed. The concept of the new approach is used in cases where diagnostic methods have determined that the basic explosion protection of an "EEx e" motor has been undermined. To establish the actual condition of a rotor squirrel cage, a diagnostic method based on the spectrum analysis of the stator current is used. This diagnostic test can ascertain the deteriorated condition of the rotor cage which directly affects its explosion protection. Damage to the rotor cage can be a serious source of ignition of explosive atmospheres.

Keywords—Explosive Atmosphere, Electrical Drives, Explosion Protected Motor, Safety and Risk Assessment, Technical Diagnostics.

I. INTRODUCTION

Because of widespread application, maintenance problems and intensive construction of new and increasingly larger electrical drives (ED) in oil refineries, oil pipelines, gas pipelines and other similar facilities, it is essential to perfect known methods and procedures and develop new methods of risk assessment of these drives in areas at risk of explosive atmospheres.

The simultaneous occurrence of an explosive atmosphere (e.g. mixture of flammable gas and air) and an effective source of ignition, such as electrical and mechanical sparks or hot surfaces, presents a serious risk of explosion with these motor drives.

One possible source of ignition is the motor. Malfunctions of the rotor cage and stator winding present a danger of electrical and mechanical sparks, increased heating and hot surfaces.

Lead by the idea of avoiding and/or reducing the risk of explosions (by preventing the simultaneous occurrence of an explosive atmosphere and an ignition source), today's

methodology, with respect to the approach to insuring explosion protection in these types of motor drives, is based on the application of electrical equipment specially designed and built for use in areas where the occurrence of explosive atmospheres is possible.

In Europe the prevention of explosions is based on the European Directives (ATEX Directives) of which the most significant are ATEX 95 (Directive 94/9/EC of the European Parliament and the Council) [2] and ATEX 137 (Directive 1999/92/EC of the European Parliament and the Council) [1].

The traditional approach to insuring explosion protection of motor drives in areas at risk of explosive atmospheres, specified in the mentioned ATEX Directives, is based on the application of motors that ensure certain Equipment Protection Levels (EPLs Ga, Gb or Gc), that is, that they fulfill the requirements of a certain category (1, 2 or 3).

The compliance of a motor with the needed category and Equipment Protection Levels (EPLs) is achieved by constructing the motor with a certain type of protection.

The characteristics of a hazardous area define the requirements of the application of motors with the appropriate Equipment Protection Levels (EPLs) or category.

Hazardous areas are divided into three zones; zone 0, zone 1 and zone 2 [5]. The basic characteristics of each hazardous area (with regard to the possible occurrence of an explosive atmosphere and its expected duration) are briefly illustrated in table I.

TABLE I.
CHARACTERISTICS OF ZONES 0, 1 AND 2.

Zone	Possible occurrence of explosive atmosphere	Expected duration
0	Normal operation	Continuously, for long periods or frequently
1	Likely to occur in normal operation	Occasionally
2	Not likely to occur in normal operation	Short period only

The safety levels provided by motors with certain Equipment Protection Levels (categories) can be basically demonstrated according to table II.

Usually motors are not applied in zone 0 and are not manufactured in category 1, EPLs Ga, therefore this safety level is not analyzed within the scope of this paper.

TABLE II.
LEVEL OF PROTECTION OF MOTOR

EPLs (Cat.)	Level of protection	Source of ignition
Ga (1)	Very high	Not in normal operation and not in expected rare malfunctions
Gb (2)	High	Not in normal operation and not in expected malfunctions which are taken into consideration
Gc (3)	Enhanced	Not in normal operation

The selection of motor depending on hazardous area and permitted Equipment Protection Level (category) are illustrated in table III.

TABLE III.
PERMITTED EQUIPMENT CATEGORIES / EQUIPMENT PROTECTION LEVELS (EPLS) DEPENDING ON ZONES

Zone	Permitted Equipment Protection Levels - EPLs (Permitted categories)	Permitted Type of Protection of Motor
0	Ga (1)	Theoretical: Two independent types of protection. (e.g. "EEx e" + "EEx d")
1	Ga or Gb (1 or 2)	"EEx d", "EEx e" "EEx px", "EEx py"
2	Ga, Gb or Gc (1, 2 or 3)	"EEx d", "EEx e" "EEx px", "EEx py" "EEx pz", "EEx n"

The marks of the Type of Protection of Motor in table III signify the following: "EEx d" is Flameproof enclosures, "EEx e" is Increased safety, "EEx p (x, y or z)" is Pressurized enclosures and "EEx n" is Non sparking.

The demonstrated approach is based on the selection of the permitted equipment category for a particular zone. For example, a motor with type of protection "Increased safety - EEx e" which belongs to category 2 is permitted for application in zone 1, but a motor with type of protection "Non sparking - EEx n" which belongs to category 3 is not permitted for application in this zone.

The conformity of the explosion protection on a motor, designed for application in areas at risk of explosive atmospheres, with requirements of the safety level needed, is ascertained through the procedure of testing and certification of the motor during and after production, but before use.

The current standard documents in the field of explosion protection precisely specify the basic testing that is necessary to perform in the certification procedure of explosion protected motors. These tests, depending on the type of explosion protection, can include temperature measurement, testing of mechanical protection (IP) of the motor, insulation system of the stator winding, impact resistance testing, testing with increased pressure and non-transmission test, and determining permitted stop time with starting current - t_E.

The new approach is being intensively developed for the application of electrical motors in hazardous areas [6]. According to this approach the application of motor for general purposes (without category) is also acceptable in zone 2, but with the insurance of an associated safety device (system) with the corresponding Safety Integrity Level - SIL, which completes the essential safety level of equipment under control - EUC. Thus the required category of the combined equipment (1, 2 or 3) is achieved.

An ongoing research has been conducted where the concept of the "new" approach is applied in analysis of eventual further application of a motor with type of protection increased safety, "EEx e", in conditions when such a motor, after a longer period of use, has been diagnosed with "questionable" basic explosion protection. A specific example illustrates the results of diagnostic testing to determine the actual condition of a motor after a 30-year period of use.

One of many possible associated safety systems is also demonstrated, which supplement the deteriorated basic protection of an "EEx e" motor, and with its application the required Category of the combined equipment is insured.

Of course, monitoring further "progress" of diagnosed malfunctions and their elimination, when they reach an unacceptable level is essential. The application of the "new" approach and introduction of safety systems intend to insure and supplement the explosion protection of equipment in conditions when basic explosion protection is "questionable".

It is not the purpose of the "new" approach to apply motors with unacceptable malfunctions in view of explosion protection. The goal of this concept is to diagnose the condition of a motor that indicates deteriorated basic explosion protection that can lead to the occurrence of ignition sources, and in these cases supplement the basic explosion protection up to the final elimination of the deficiencies. From the standpoint of explosion protection, undiagnosed malfunctions or deficiencies in the motor (e.g. "questionable condition of rotor squirrel cage) can be hidden sources of ignition. These are not possible to diagnose through basic tests on the motor.

The installation of additional safety systems allows, at least temporarily, the application of a motor in areas at risk of explosive atmospheres in a way that is different from traditional, described in table III, but which fulfills the requirements of ATEX Directive with regard to preventing explosions, which is the final goal.

II. ELECTRICAL DRIVE IN HAZARDOUS AREA - OIL PIPELINE

The structure of a modern electrical drive (ED) is illustrated in Fig 1. In a specific application, for explosive atmospheres, it is necessary, wherever possible, to install all ED components (Power Electronics Converter System, Transformer, Electromechanical Separator and Protection System) outside of the hazardous area, so that the only connection between a motor in the hazardous area and the other ED components, which are outside of the hazardous area, are the electrical cables.

819

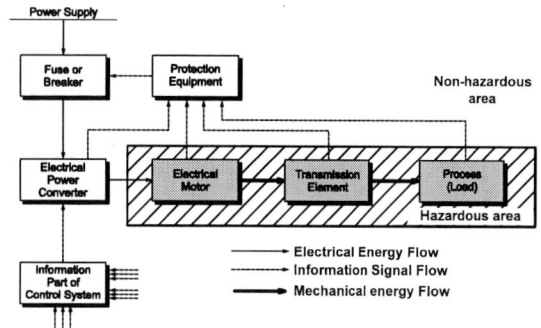

Fig. 1. Electrical drive in hazardous area.

As an example ED in a hazardous area, in this paper, one of the drives from a Croatian oil pipeline has been selected.

In this electrical drive for oil transport, squirrel cage motors are used (6 kV, 50 Hz, 1900 kW) with increased safety protection, "EEx e". The electrical drive, demonstrated in Fig 2., is installed in a closed and naturally ventilated pump station with an interior that is at risk of an explosive atmosphere and classified as zone 2. In accordance with the earlier described principle of selection and application of motors in hazardous areas (table III), the application of a motor with protection "EEx e" in this pump station, zone 2, is acceptable.

The principle mentioned, as well as all of today's methodology of explosion protection insurance, assume that the condition of the motor's explosion protection will be maintained at the same level as it was when brand new (or when the motor was certified) for the entire life-span of the motor. A motor's explosion protection during its life-span is based on this assumption, based on which the proper selection of motor and selection of protection are assessed.

Fig. 2. Electrical drive in hazardous area, zone 2.

However, during exploitation of the motor its explosion protection can degrade, which can cause the occurrence of effective sources of ignition inside the motor, mechanical and electrical sparks and increased heating that were not present when the motor was new.

Based on this it can be concluded that it is not possible, without performing detailed tests or diagnostics of the motor's current condition, to verify with certainty the acceptability of further application of the motor with regard to explosion protection.

Along with basic tests of the explosion protection elements it is also essential to perform special tests on the motor. These special tests, as equally important as the basic tests, can include diagnostic of the condition of the rotor squirrel cage, stator winding, gaps and bearings.

With these diagnostic tests it is possible to detect malfunctions in the motor which can become serious sources of ignition of explosive atmospheres. Malfunctions that can be detected by application of new diagnostic methods on the motor's condition have not been sufficiently analyzed in today's methodology of insuring explosion protection on older motor drives. The proceeding is a brief description of one of the diagnostic methods.

A. Diagnostic testing of explosion protected "EEx e" motors

To illustrate the importance of diagnostic testing on motors, some test results of motors in an oil transport system are demonstrated here.

The basic data on tested motors with type of protection "EEx e" are shown in table IV [9].

TABLE IV.
BASIC DATA ON TESTED "EEx e" MOTORS

Rated voltage, power, current	6000 V, 1900 kW, 210 A
Rated frequency, speed	50 Hz, 2980 min⁻¹
I_A/I_N, t_E	5.5, 8.5 s
Number of rotor/stator slots	52/60

I_A/I_N is the ratio between initial starting current I_A and rated I_N.
t_E is the permitted stop time with starting current.

A diagnostic method based on spectrum analysis of the stator current is used to check the electromechanical condition of the rotor squirrel cage. This diagnostic test of the rotor squirrel cage is based on the conversion of current signals from time domain to frequency domain. This conversion is achieved with the aid of a spectrum analyzer using the Fourier transformation. Based on the value of each component in the frequency spectrum of the stator current, and with knowledge of the motor's construction parameters, it is possible to arrive at a conclusion about the type and degree of rotor malfunction in a squirrel cage motor. This method is well-known and thoroughly treated in literature [22].

The results of testing two identical motors in normal drive conditions are shown in Fig. 3 and Fig. 4. [9].

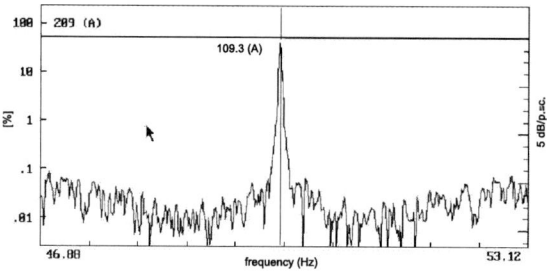

Fig. 3. Spectrum analysis of the stator current of the motor ED1, 6 kV, 1.9 MW, in a frequency range of 50 Hz.

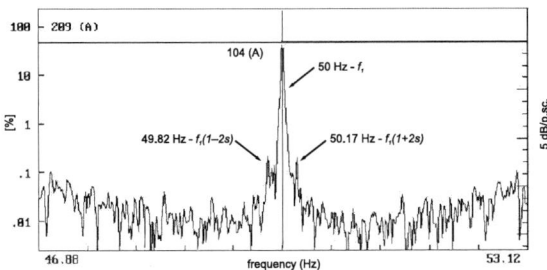

Fig. 4. Spectrum analysis of the stator current of the motor ED2, 6 kV, 1.9 MW, in a frequency range of 50 Hz.

For the diagnostic of the condition, the commercial program "Motormonitor" is used [22], based on the spectrum analysis of the stator current.

The diagnostic assessment of the condition of motor ED1 was:

"The rotor is in proper condition and a repeat test in 12 months from today's date or after 200 start-ups, whichever comes first, is recommended." In Fig. 3 it is evident that there are no additional notable components at frequencies that indicate damaged condition of the rotor squirrel cage [$f_1(1-2s)$ and $f_1(1+2s)$].

The diagnostic assessment of motor ED2 was:

"The rotor shows indications of increased contact resistance and should be re-tested within 9 months. If the drive with this motor experiences cycles of "heavy drive", testing should be performed within 6 months or after 50 start-ups, whichever comes first". In Fig. 4 there is evident increase of additional components at frequencies that indicate the beginning of malfunctions and non-symmetry of the rotor squirrel cage [$f_1(1-2s)$ and $f_1(1+2s)$].

Based on the rotor squirrel cage test performed on motor ED2, it can be concluded that it is no longer possible to determine with certainty that the condition of this motor, with regard to explosion protection, is the same as when the motor was new and certified. Test results indicate that the occurrence of sources of ignition, electrical sparks and increased heating are possible inside the motor.

With the application of these modern diagnostic methods it is possible, in an early phase, to detect initial malfunctions of the rotor that precede larger malfunctions, which may become sources of ignition of explosive atmospheres. At the same time these methods contribute to increased quality of the system of maintenance and to the planning of overhaul work, which reduces expenses.

These motors have also undergone diagnostic testing of the condition of the stator winding insulation systems. Testing was performed on the insulation between the coils by applying the shock voltage response analysis method. Measurement of the stator winding insulation resistance and a test with increased voltage were also performed. The results of those tests are not provided in this paper since they did not indicate any damage to the stator winding, and therefore are not significant in the analysis of explosion protection with regard to the concept of this paper.

Only the implementation of regulated tests and application of diagnostic assessments can insure a realistic picture of a motor's condition in areas at risk of explosive atmospheres. Based on these results a decision can be made on the necessary service interventions or other safety activities.

III. TECHNICAL DIAGNOSTICS AND RISK ASSESSMENT

The diagnostic test results demonstrated in the previous chapter have shown that the explosion protection of motor ED2 can no longer be verified with certainty. This formally means that the motor, from the standpoint of explosion protection, no longer conforms to the condition it was in when it was certified. Furthermore, it can be formally concluded that this motor can no longer fulfill the requirements of explosion protection for which it was designed and manufactured. In other words, it can no longer guarantee any Equipment Protection Levels (EPLs). Therefore this motor, with this newly-determined condition of explosion protection, can only be considered as a "motor for general purposes", that is, a motor that can be a source of ignition in normal operation.

A. "New" approach and risk assessment

According to the "new" approach concept, which for zone 2 is illustrated by table V [6], the application of motor ED2, at least temporarily, is possible under certain conditions.

TABLE V.
REQUIREMENTS OF SAFETY INTEGRITY LEVEL OF SAFETY DEVICE (SYSTEM) MONITORING POTENTIAL IGNITION SOURCES IN EQUIPMENT UNDER CONTROL (EUC)

Hazardous areas	Zone 2	
Fault tolerance of EUC (electrical motor)	0	-1
Fault tolerance of the safety device	---	0
Safety Integrity Level of the safety device *	---	SIL 1
Category of the combined equipment **	3	

	"-1" indicates that the EUC becomes a source of ignition in normal operation (e.g. motor for general purposes i.e. motor without applicable basic construction measures of explosion protection, in other words a motor that does not comply with EN 60079-0).
	"0" indicates that the EUC is assessed as safe in normal operation in zone 2 (e.g. motor type of protection "EEx n"). "0" indicates that one single fault may cause potential ignition sources in EUC and indicates that one single fault may cause the safety device to fail.
*	SIL (according to EN 61508 series).
**	Category (according to Directive 94/9/EC - ATEX 95).

A motor with type of protection "EEx e" can, when fully functional from the standpoint of explosion protection, be given a "Fault tolerance of EUC" value of 1 [6]. This association is based on the "new" approach and the fact that a motor with type of protection "EEx e" is an apparatus with EPL Gb or category 2. This is not shown in table V since it refers only to zone 2 which does not require category 2 motors, that is, motors that have "Fault tolerance of EUC" value of 1 ("1" indicates that two faults may cause the potential ignition sources in EUC). In zone 2 the application of milder requirements is sufficient, that is, motors with "Fault tolerance of EUC" value of 0.

However, it is clear that zone 2 allows the application of an "EEx e" motor (since motors that have "Fault tolerance of EUC" value of 1 comply with applications in zone 2 without additional safety systems).

The important question is which value of "Fault tolerance of EUC" can be given to a motor that has, through diagnostic testing, been assessed with a deteriorated condition of the rotor squirrel cage, as was shown for motor ED2. The answer is "Fault tolerance of EUC" value of -1.

In table V it is evident that motors which have "Fault tolerance of EUC" value of 0 (also those with value of 1) can also be used in zone 2 without the application of a safety device (system). These are, for instance, motors with type of protection "EEx e". This possibility conforms to the "new" approach as well as the traditional solution.

In table V it is also evident that it is possible to use motors that have "Fault tolerance of EUC" value of -1 (motor ED2 with deteriorated condition of rotor squirrel cage). However in this case it is necessary to ensure the application of a safety device or system which fulfills the requirements of SIL 1. The safety system foreseen is briefly illustrated below.

The explosion protection of motor ED2 should be supplemented and insured with additional explosion protection measures, e.g. effective forced ventilation which turns the surrounding area in the vicinity of the motor, zone 2, into a safe area. In order to ensure this, two ventilators should be installed of which one operating is sufficient and the other one is for back-up. Fig. 5 illustrates the concept mentioned.

The photo in Fig. 2 shows an ED in zone 2, where the concept of forced ventilation can be applied (the motor ED2 is designed with type of protection "EEx e").

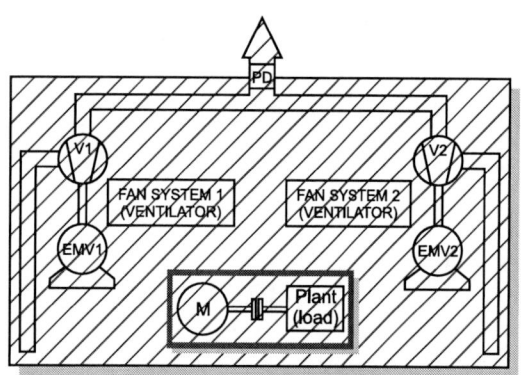

Fig. 5. Electrical drive with M motor and forced ventilation system, zone 2

In a specific example the required safety system is ensured with two separate safety systems as illustrated in Fig. 6 and Fig. 8 [7]. The first system, illustrated in Fig. 6 and Fig. 7, monitors the continuous presence of forced ventilation with a pressure device PD.

The information "there is / there is no" ventilation is sent to the logic subsystem which, in the event there is no ventilation (e.g. fault in ventilator V1-EMV1 which is in continuous operation) activates contactor V2 and turns on the back-up ventilator V2-EMV2. The same system simultaneously activates contactor V1 which shuts down the motor to ventilator V1-EMV1.

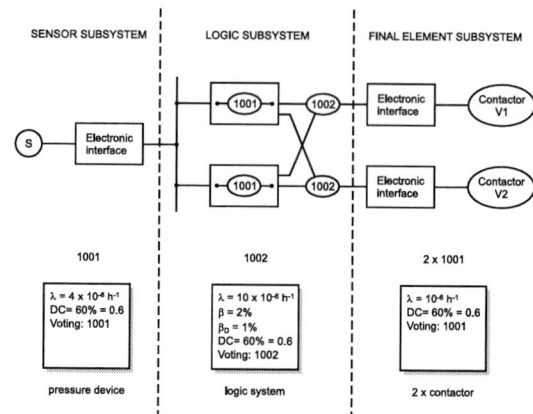

Fig. 6. Architecture of safety system 1.

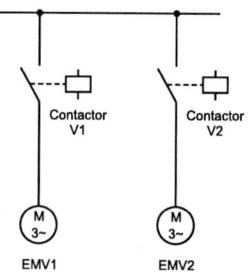

Fig. 7. Contactors V1 and V2.

With a comparative solenoid, the second safety system monitors the start-up of ventilator V2 when ventilator V1 shuts down. The information on "switched-on" or "switched-off" of the ventilator V2 is sent to the logic subsystem which, in the event that ventilator V2 does not switch on, activates contactor KM and disconnects the "EEx e" motor from its power supply while simultaneously activating contactor ALARM, which turns on the audio alarm. The description of the architecture is illustrated in Fig. 8 while the function of the contactors is shown in Fig. 9.

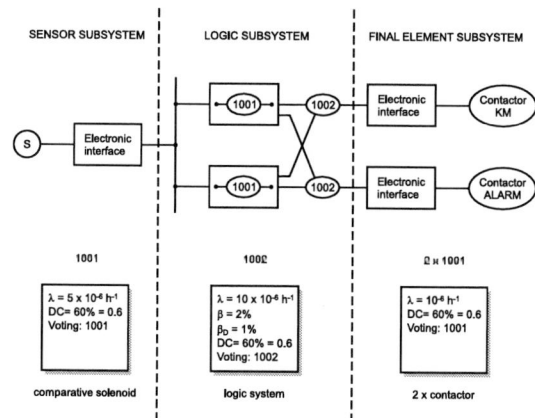

Fig. 8. Architecture of safety system 2.

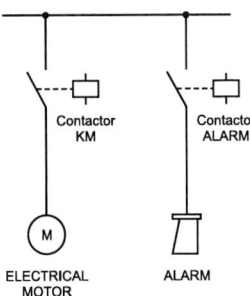

Fig. 9. Contactors KM and ALARM.

In order to establish which SIL is fulfilled by the designed safety systems, it is necessary to calculate PFD_{SYS} according to [10]

$$PFD_{SYS} = PFD_S + PFD_L + PFD_{FE} \qquad (1)$$

where PFD_{SYS} is the average probability of failure on demand of a safety function for the safety system, PFD_S is the average probability of failure on demand for the sensor subsystem, PFD_L is the average probability of failure on demand for the logic subsystem and PFD_{FE} is the average probability of failure on demand for the final element subsystem.

For architecture 1001 in Fig. 6. and Fig. 8. the following calculation is made according to [10]:

$$PFD = (\lambda_{DU} + \lambda_{DD}) \cdot t_{CE} \qquad (2)$$

For architecture 1002 according to [10] one has:

$$PFD = 2\big((1-\beta_D)\lambda_{DD} + (1-\beta)\lambda_{DU} \big)^2 t_{CE} t_{GE} + \\ \beta_D \lambda_{DD} MTTR + \beta \lambda_{DU} \left(\frac{T_1}{2} + MMTR \right) \qquad (3)$$

$$t_{GE} = \frac{\lambda_{DU}}{\lambda_D}\left(\frac{T_1}{3} + MTTR \right) + \frac{\lambda_{DD}}{\lambda_D} MTTR \qquad (4)$$

For architectures 1001 and 1002 according to [10]:

$$\lambda_{DU} = \frac{\lambda}{2}(1 - DC) \qquad (5)$$

$$\lambda_{DD} = \frac{\lambda}{2} \cdot DC \qquad (6)$$

$$t_{CE} = \frac{\lambda_{DU}}{\lambda_D}\left(\frac{T_1}{2} + MTTR \right) + \frac{\lambda_{DD}}{\lambda_D} \cdot MTTR \qquad (7)$$

where λ is the failure rate (per hour) in a subsystem, λ_{DU} is the undetected dangerous failure rate (per hour) in a subsystem, λ_{DD} is the detected dangerous failure rate (per hour) in a subsystem, t_{CE} is the channel equivalent mean down time (hour) - combined down time for all the components in the channel of the subsystem, t_{GE} is the voted group equivalent mean down time (hour) - combined down time for all channels in the voted group, β is the fraction of undetected failures that have common cause ($\beta=2\beta_D$), β_D is of those failures that are detected by diagnostic tests, $MTTR$ is the mean time to restoration (hour), DC is the diagnostic coverage and T_1 is the proof-test interval (hour).

The entry parameters used for safety system 1 and safety system 2 (λ, β, β_D i DC) obtained according to [10], [11] and $MTTR$ and T_1 are shown in table VI.

The final calculation results for safety system 1 and safety system 2 obtained by applying (1) to (7) are also shown in table VI .

TABLE VI.
ENTRY PARAMETERS USED (FOR SAFETY SYSTEM 1 AND SAFETY SYSTEM 2) AND CALCULATION RESULTS

Parameter	Safety system 1			Safety system 2		
	Sensor subsystem	Logic subsystem	Final element subsystem	Sensor subsystem	Logic subsystem	Final element subsystem
Entry parameters used (for safety system 1 and safety system 2)						
λ	4×10^{-6} h^{-1}	10×10^{-6} h^{-1}	10^{-6} h^{-1}	5×10^{-6} h^{-1}	10×10^{-6} h^{-1}	10^{-6} h^{-1}
β	---	0.02	---	---	0.02	---
β_D	---	0.01	---	---	0.01	---
DC	0.6	0.6	0.6	0.6	0.6	0.6
$MTTR$	8 h	8 h	8 h	8 h	8 h	8 h
T_1	17520 h	17520 h	17520 h	17520 h	17520 h	17520 h
Calculation results (for safety system 1 and safety system 2)						
λ_{DU}	0.8×10^{-6} h^{-1}	3×10^{-6} h^{-1}	0.2×10^{-6} h^{-1}	---	---	---
λ_{DD}	1.2×10^{-6} h^{-1}	2×10^{-6} h^{-1}	0.3×10^{-6} h^{-1}	---	---	---
t_{CE}	3512 h	3512 h	3512 h	---	---	---
t_{GE}	---	2344 h	---	---	---	---
PFD	7.1×10^{-3}	7.5×10^{-4}	$2 \times 1.8 \times 10^{-3}$	8.8×10^{-3}	7.5×10^{-4}	$2 \times 1.8 \times 10^{-3}$
PFD_{SYS}	11.5×10^{-3}			13×10^{-3}		
SIL	SIL 1			SIL 1		

Part of the interim result/parameter for *PFD* with safety system 2, for the purpose of simplification, has been obtained from [10] (for cases in which the same has been available). The connection between *PFD* and the corresponding Safety Integrity Level is shown in table VII according to [6].

TABLE VII.
CONNECTION BETWEEN *PFD* AND THE CORRESPONDING SAFETY INTEGRITY LEVEL OF THE SAFETY DEVICE SYSTEM

Safety Integrity Level SIL	Low demand mode of operation *PFD*
SIL 4	$\geq 10^{-5} < 10^{-4}$
SIL 3	$\geq 10^{-4} < 10^{-3}$
SIL 2	$\geq 10^{-3} < 10^{-2}$
SIL 1	$\geq 10^{-2} < 10^{-1}$

After the analyses and calculations have been performed, and based on the acquired results shown in table VI, it can be concluded that the application concept, in zone 2, of a motor with type of protection "EEx e" which has been diagnosed with a questionable condition of the rotor squirrel cage, that is, deteriorated condition of explosion protection ("Fault tolerance of EUC" value of -1), with the application of safety systems that fulfill SIL 1, is acceptable.

IV. APPLICATION OF RISK ASSESSMENT AND TECHNICAL DIAGNOSTICS FOR OLDER DRIVES

Electrical drives designed for operation in areas at risk of explosive atmospheres are subject to high safety requirements, and because of this the costs of investment and maintenance can be significantly high. Opposition to the safety and economic requirements can be, to a lesser or greater degree, settled by performing an assessment of technological risks. A risk assessment in hazardous areas must recognize and analyze every possible danger or event that can result in an explosion.

Today's legal standard regulations in the field of explosion protection (ATEX 95 and 137) only basically refer to risk assessments "as a general category", do not regulate general methodology, required analyses and do not provide criteria for the assessment of results.

In this paper the importance and necessity of the application (implementation) of all-inclusive and complete risk assessments as a scientific discipline, which significantly contributes to improvements in safety, but also critically considers the technical and economic justification of the application of safety measures, is emphasized.

Emphasis is also placed on the necessity and usefulness of performing diagnostic "on-line" methods for ascertaining the actual and current condition of a motor's explosion protection, particularly important after a longer period of use. The application of the traditional practice of insuring explosion protection, without implementing risk assessments and without diagnostic testing, does not guarantee an acceptable and complete level of safety.

Earlier shown results of diagnostic testing of the condition of the rotor in the ED2 motor have indicated initial malfunctions in the rotor squirrel cage.

Since the ED2 motor, with type of protection "EEx e", is installed in an area at risk of explosive atmospheres, zone 2, its further application without additional safety systems is unacceptable. Requirements to re-test within 6 months, for areas at risk of explosive atmospheres where rotor malfunctions can become sources of ignition, are not sufficient and therefore unacceptable.

One of the solutions for overcoming these deficiencies is to apply the "new" approach by installing additional safety systems, described earlier, which will supplement or replace the deteriorated explosion protection of the motor ED2.

With this approach, from the standpoint of explosion protection, the further application of this motor is acceptable, however it is clear that measures should be undertaken to eliminate the noted deficiencies from the rotor squirrel cage.

The application of the "new approach" concept presents temporary solutions and facilitates safe operation of the motor until its deficiencies can be eliminated. Deficiencies detected in the condition of the rotor squirrel cage do not necessarily mean that there are also functional deficiencies in the motor's operation. In this sense, the application of the "new" approach insures time needed for additional inspections and tests, while the motor is still operational, but not putting the electrical drive at risk with regard to explosion protection. After the service activities have been performed and deficiencies eliminated, additional safety systems will no longer be necessary. It is important to point out that the costs of designing, construction and later removal of the additional safety systems are negligible in comparison with the price of a motor of 6 kV, 1900 kW, with type of protection "EEx e". The costs mentioned are also negligible in comparison with the losses that would occur during stoppage of the electrical drive in oil transport.

The possibility of further application of a motor, based on diagnostic test results and risk analysis are shown in table VIII.

From the results demonstrated, the possibilities (combinations) of applying motors in zone 2 are clearly evident.

TABLE VIII.
ANALYSIS RESULTS OF APPLICATION OF "EEx e" MOTORS WITH AND WITHOUT ADDITIONAL SAFETY SYSTEMS AND ASSESSMENTS OF SAFETY/RISKS IN AREAS AT RISK OF EXPLOSIVE ATMOSPHERES, ZONE 2.

Assessed Condition of basic Explosion Protection of "EEx e" Motor (diagnostic tests)	Satisfactory (functional)	Unsatisfactory (questionable/deteriorated)	Unsatisfactory (questionable/deteriorated)
Fault tolerance of Electrical Motor	1	-1	-1
Safety Integrity Level of the safety system (SIL)	---	---	SIL 1
Category of the combined equipment	3 (2)	--- (No category)	3
Satisfactory: Y (yes)/N (no)	Y	N	Y

V. CONCLUSIONS

The effective regulations in the field of explosion protection (ATEX 95 and 137) treat risk assessment only as a general category, do not regulate general methodology and required analyses, and do not provide criteria for the assessment of results.

Because of widespread application, maintenance problems and intensive construction of new and increasingly larger electrical drives in oil refineries, pipelines, gas pipelines and other similar facilities, it is essential to perfect the known methods and procedures and develop new methods of risk assessment.

The application of motors with type of protection increased safety, "EEx e", in areas at risk of explosive atmospheres, zone 2, in the oil transport system has been analyzed. By implementing diagnostic tests the actual condition of the rotor cage and stator winding of a high voltage motor has been determined, which is of the utmost importance for the condition of a motor's explosion protection.

Based on the results acquired the risk assessment has been performed. Also, an assessment of further possible application of the motor in the drive under consideration has been performed. The possibility of introducing additional safety systems has been analyzed, which supplement the deteriorated protection on the motor.

REFERENCES

[1] Directive 1999/92/EC of the European Parliament and the Council.

[2] Directive 94/9/EC of the European Parliament and the Council.

[3] N. Marinović, "Electrotechnology in Mining," *Elsevier, Amsterdam*, 1990.

[4] R. L. Rogers, (Co-ordinator) "EU Project No: SMT4-CT97-2169 (Methodology for Risk Assessment of Unit Operations and Equipment for Use in Potentially Explosive Atmospheres," Co-ordinator Dr. R. L. Rogers, *Inburex GmbH, Hamm*, 2000.

[5] IEC 60079, Electrical Apparatus for Explosive Gas Atmospheres.

[6] prEN 50495 (August 2006), Safety devices required for the safe functioning of equipment with respect to explosion risks.

[7] I. Gavranić, "Diagnostics of explosion protected electrical drives," M.S. thesis, *FER Zgreb, Zagreb*, 2002. (Croatian language only).

[8] P. J. Tavner, "Condition Monitoring of Electrical Machines," *Research studies press LTD, Letchwort*, 1987.

[9] I. Gavranić, "Testing and maintenance of explosion protected EDs in older plant," *VI International Conference Ex 2005*, Dubrovnik, 2005. (Croatian language only).

[10] IEC 61508, Functional safety of electrical /electronic/ programmable electronic safety-related systems.

[11] Application of IEC 61508 and IEC 61511 in the Norwegian Petroleum Industry, OLF, 2004.

[12] I. Gavranić, D. Ban and D. Žarko, "Electrical Drives for Explosive Atmospheres - Motor Selection and Risk Assessment," *EDPE*, High Tatras, 2007.

[13] Practical Assistance for the Preparation of an Explosion Protection Document, ISSA, Mannheim, 2006.

[14] R. Wilcox, M. Burrows, S. Ghosh, and B. M. Ayyub, "Risk-based Technology Methodology for the Safety Assessment Compressed Natural Gas Fuel Systems," *ICMES/SNAME*, New York, 2000.

[15] I. Boldea and S. A. Nasar, "The Induction Machine Handbook," *CRC Press, New York*, 2002.

[16] I. Boldea and S. A. Nasar, "Electric Drives," *Taylor & Francis*, 2006.

[17] C. Petitfrere and C. Proust, "Analysis of ignition risk on mechanical equpment in ATEX," *PCIC Europe*, Paris, 2007.

[18] I. Gavranić, "Methodology of Risk Assessment in Explosive Atmosphere/Hazardous Areas," *Ex-Journal 2007*, ISSN 1845-0172, Zagreb, 2007.

[19] J. Tixier, G. Dusserre, O. Salvi, and D. Gaston, "Review of 62 risk analysis methodologies of industrial plans," *Journal of Loss Prevention in the Process Industries*, Elsevier, 2002.

[20] M. Lanphier, P. K. Sen, and J. P. Nelson, "An updateon surge protection of medium voltage motors: A comparison of the standards and applications," *PCIC Europe*, Paris, 2007.

[21] E. J. Thornton, and J. K. Armintor, "The fundamentals of AC electric induction motor desing and application," *Proceedings of the twentieth international pump users symposium*, 2003.

[22] Motormonitor - operating manual, ENTEK, Ohio, 1992.

The effect of subharmonics on induction machine heating

Piotr Gnacinski, Marcin. Peplinski, Mariusz Szweda

Gdynia Maritime University, Department of Ship Electrical Power Engineering, Gdynia, Poland,
e-mails: piotrg@am.gdynia.pl, marcinpe@atol.am.gdynia.pl, szweda@atol.wsm.gdynia.pl

Abstract: One of the most rarely discussed power quality disturbances are subharmonics – voltage components of frequency less than the fundamental one. They may appear in both land and in ship power systems. This work deals with the influence of voltage subharmonics on temperature-rise distribution in an induction cage machine. The results of experimental investigations are presented for a totally-enclosed, four poles 3-kW induction machine with built-in thermocouples.

Keywords: **induction motor, power quality, thermal stress**

I. INTRODUCTION

Induction cage machines are commonly used in industrial applications. It is estimated that they consume about 2/3 of the produced electric energy in the word. The common usage of induction cage machines results from their main advantages: low price and comparatively high reliability and durability. Induction cage machines have also some drawbacks. One of them is susceptibility to the lowered voltage quality. Power quality disturbances appearing in the supply network causes additional power losses in induction machines, and consequently, they may lead to a significant increase in windings temperature [4-7], [10]. The higher windings temperature leads to faster aging of insulation system. For the most common insulating materials 8-11 K increase in working temperature causes double reduction of its life time. Consequently, power quality disturbances appearing in the supply network may significantly reduce the machine reliability and operational life [1-3], [5], [7], [10].

One of the most rarely discussed power quality disturbances are subharmonics – voltage components of frequency less than the fundamental one. They may appear in both land and in ship power systems. Subharmonics are caused by arc furnaces, cycloconverters, automated spot welders, rectifiers feeding fluctuating or cyclic load, motors working with cyclic load and wind generators [3].

The effect of voltage subharmonics on induction cage machine was presented in works [1-4], [8,11]. The effect of voltage subharmonics on currents and electromagnetic torque of induction machine was analyzed by Lazim *et al* [11]. The thermal loss of life due to subharmonics was investigated by de Abreu *et al* on the basis of computer simulation [1-3]. Additionally, in the works of Abreu *et al*

a proposal of limiting subharmonics to 0.1% of the fundamental voltage was made [3] and the effect of subharmonics on magnetic flux was investigated with analytical method [2]. Fuchs *et al* [4] analyzed the subharmonic torque, additional power losses and temperature-rise due to harmonics, interharmonics and subharmonics (of frequency ≥50% of the fundamental voltage frequency). The work [4] also suggests limiting the subharmonics injection to 0.1%. In the authors' work [8] the effect of voltage subharmonics on flux in an induction machine was investigated using an experimental method.

This work deals with the influence of voltage subharmonics on temperature-rise distribution in an induction cage machine. The results of experimental investigations are presented for a 3-kW induction cage machine and subharmonic frequency equal to 10% of the fundamental voltage frequency.

II. THE EXPERIMENTAL STAND

The experimental stand consists of two synchronous generators, a transformer, a system for power quality analysis, a totally-enclosed four poles induction cage machine mSZJe34a type (3 kW) with built-in thermocouples, loaded with a DC generator.

The thermocouples are embedded in different parts of the investigated induction machine: in end-windings, slot windings, in the stator core, in a stator tooth, in the rotor and on the casing. The thermocouples location is shown in Fig. 1 and described in Tab. I. The machine parameters are presented in [6].

For the purpose of subharmonics generation, a two-frequency method presented in [9] was applied. The method requires two synchronous generators. One of the generators produces the voltage of fundamental frequency, and the other of subharmonic frequency. Owing to coupling via transformer, both the frequencies impose on each other.

The voltage subharmonic content was measured with a PC-based power quality analyzer. It consists of the PC plug-in data acquisition board (DAQ) PCI-703-16A Eagle Technology and the signal conditioning unit. The DAQ is fitted with 16 Channel Simultaneous Sample and Hold analog inputs and 14-bit A/D converter. The presented below signals were sampled with sampling frequency 10504 Hz. The cut-off frequency of

978-1-4244-1741-4/08/$25.00 ©2008 IEEE

antialinsing filter was equal to 3.5 kHz. All frequency components in registered signals were obtained by using Discrete Fourier Transform (DFT) with the frequency resolution equal to 1/100 of the fundamental frequency. The rectangular window with width equal exactly to 100 periods of measured voltage was used for the analysis.

The diagram of the laboratory stand is shown in Fig. 2.

TABLE I
THE THERMOCOUPLE LOCATION IN THE INVESTIGATED INDUCTION
CAGE MSCHINE MSZJE34A (A DESCRIPTION TO FIG. 1)

number of thermocouples	location
1-2	end-windings (driving end)
3-4	end-windings (non-driving end)
5	slot windings
6	stator tooth
7	stator core
8	bearing
9-10	casing

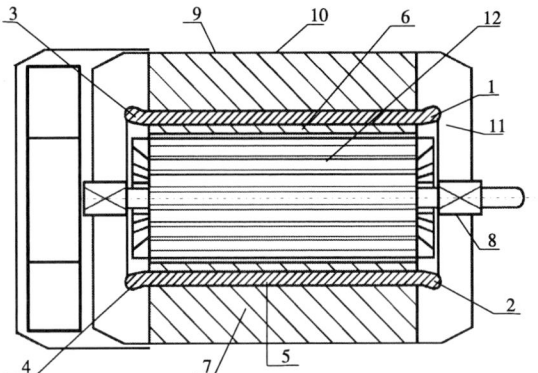

Fig. 1. The thermocouples location in the investigated induction machine mSZJe34a

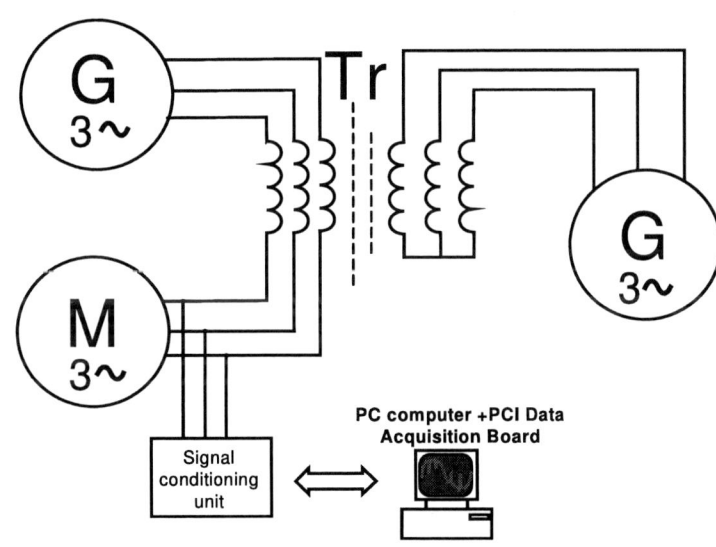

Fig. 2. The diagram of the laboratory stand

III. THE RESULTS OF INVESTIGATIONS

For the purpose of investigation of subharmonics on induction machine heating two thermal tests were carried out – with subharmonic injection and without it (a reference test). In the former one the machine was supplied with voltage containing a positive-sequence subharmonic of frequency equal to 5 Hz (10% of rated frequency) and amplitude equal to 2.8% of the fundamental voltage amplitude. For both the thermal tests the fundamental voltage harmonic, voltage frequency and

load torque were kept equal to their rated values. It is also worth mentioning that both the tests were carried out at the same values of ambient temperature and atmospheric pressure (both the agents slightly influence the temperature-rise distribution).

Fig. 3 and 4 show the recorded voltage and current waveforms during the test with subharmonic injection. Additionally, the subharmonic and harmonic content of this current waveform is shown in Fig. 5. As can be seen, merely 2.8% voltage subharmonic causes about 40% current subharmonic. This results from the fact that low-frequency current subharmonics are attenuated practically by the windings resistance only, whereas higher harmonic currents are limited mostly by windings reactance (see [2], [4]).

The 40% subharmonic current flowing through the windings leads to significant additional power losses. Additionally, supplying with voltage containing subharmonics produces additional torques [4] that increases the load torque value and losses inside a motor. Consequently, even apparently inconsiderable subharmonics injection may expose a machine to a risk of serious overhheating.

A comparison of the measured temperature-rise distributions in the investigated machine for the reference test and the supply voltage with subharmonic injection are presented in Fig. 6 and 7, respectively. As it can be seen in Fig. 6 and Fig. 7, the subharmonic injection caused the most significant increase in temperature-rise in the rotor – about 16 K. Further, the end-windings and slot windings the temperature-rise was abut 13-14 K higher than in the reference test. In the stator iron the additional temperature-rise was much lower, namely about 6-10 K.

Such a significant increase in windings temperature-rise may cause considerable thermal loss of machine life. It should be stressed the temperature is considered the main agent causing aging of the insulation system in low voltage machines. Other agents are vibrations, electrical and mechanical stress. However, for the purpose of estimation of machine life time expectancy these agents are usually omitted [1-3], [5], [7], [10] and the operational life is predicted with Arrhenius' equation. On the basis of Arrhenius' equation [10], the machine lifetime can be assessed with the following expression:

$$L_{\%} = 100\% * 2^{\frac{\vartheta_{rat} - \vartheta_{rwc}}{HIC}} \qquad (1)$$

where $L_{\%}$ - predicted machine life time, in percent of its life time in the nominal work conditions ϑ_{rat} – temperature in the hottest point of the insulation system in the nominal work conditions, ϑ_{rwc} – temperature in the hottest point of the insulation system in the real work conditions, HIC – half interval index [10]

Taking into account that for the applied B class of insulation the half interval index equal to 11 K [10], the 14-K increase in windings temperature-rise due to subharmonic injection reduces the machine life time to merely about 40% of its life time in the nominal work conditions.

To sum up, presence of subharmonics in the supply voltage may result in significant additional temperature-rise inside a machine, and consequently, may cause a considerable reduction of a machine operational life.

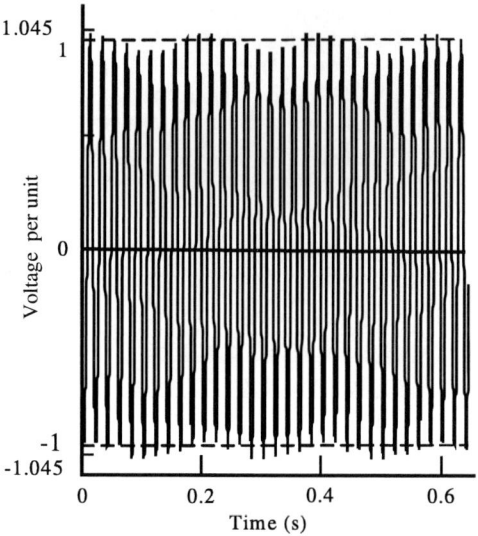

Fig. 3. The recorded voltage waveform (per-unit) for the test with subharmonic injection.

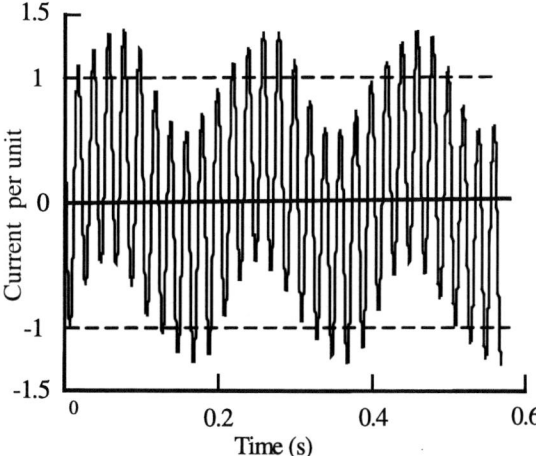

Fig. 4. The recorded current waveform (per-unit) for the test with subharmonic injection.

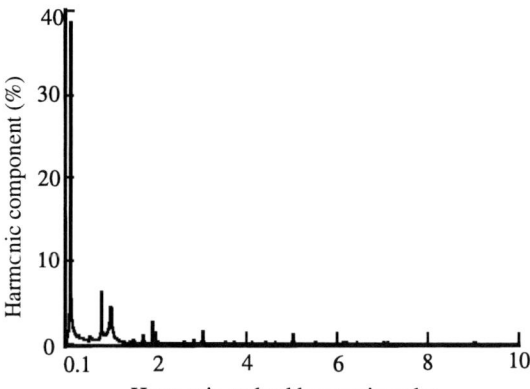

Fig. 5. The subharmonic and harmonic content of the current waveform in Fig. 4.

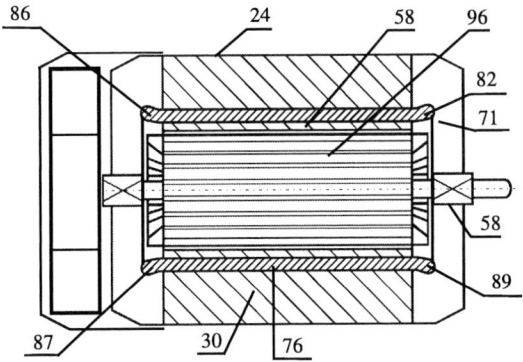

Fig. 6. The measured temperature-rise distribution (in K) in the induction machine mSZJe34a type for the supply voltage without subharmonic injection (a reference test)

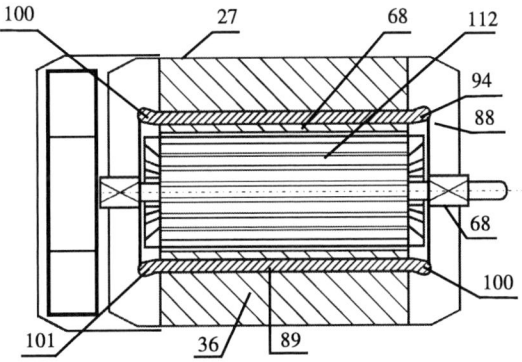

Fig. 7. The measured temperature-rise distribution (in K) in the induction machine mSZJe34a type for the supply voltage containing subharmonic

IV. CONCLUSIONS

The preliminary study shows that voltage subharmonics may lead to a considerable increase in temperature-rise inside an induction cage machine. The increase in windings temperature leads to faster degradation of the insulation system. Consequently, voltage subharmonics may cause a serious reduction in the machine life-time expectancy.

The effect of voltage subharmonics on temperature-rise distribution, life time expectancy, currents and vibration of induction cage machines will be a subject of future detailed investigations.

REFERENCES

[1] J. P. G. de Abreu, A. E. Emanuel, "Induction motor loss of life due to voltage imbalance and harmonics: a preliminary study", in *Proc. of 9th International Conference on Harmonics and Quality of Power*, Orlando, USA, Florida, 1-4 Oct. 2000, vol. 1, pp. 335-40

[2] J. P. G. de Abreu, A. E. Emanuel, "Induction motor thermal aging caused by voltage distortion and imbalance: loss of useful life and its estimated costs", *IEEE Trans. on Industry Applications*, vol. 38, pp. 12-20, Jan./Febr. 2002

[3] J. P. G. de Abreu, A. E. Emanuel, "The need to limit subharmonics injection", in *Proc. of 9th International Conference on Harmonics and Quality of Power*, Orlando, USA, Florida, 1-4 Oct. 2000, vol. 1, pp 251-254

[4] Fuchs E. F., Roesler D. J., Masoum M. A. S. Are harmonics recommendations according to IEEE and IEC too restrictive? *IEEE Trans. on Power Delivery 2004*; 19(4):1775-1786.

[5] P. Gnacinski: "Effect of unbalanced voltage on windings temperature, operational life and load carrying capacity of induction machine" *Energy Conversion and Management*, vol. 49, no 4, pp. 761-770, Apr. 2008, doi. 10.1016/ j.enconman.2007.07.33

[6] P. Gnacinski: "Prediction of windings temperature rise in induction motors supplied with distorted voltage", *Energy Conversion and Management*, , vol. 49, no 4, pp. 707-717, Apr. 2008, doi. 10.1016/ j.enconman.2007.07.23

[7] P. Gnacinski: "Windings temperature and loss of life of an induction machine under voltage unbalance combined with over or undervoltages". *IEEE Trans. on Energy Conversion*, vol. 23, no 2, June 2008, pp. 363-371, DOI 10.1109/TEC.2008.918596

[8] P. Gnacinski, M. Peplinski, M. Szweda: "The effect of subharmonics on the flux in an induction cage machine", proc. of 5th *International Workshop Compatibility in Power Electronics CPE'07* 29.05-1.06.07 Gdansk-Jelitkowo

[9] S.L.Ho, W.N.Fu, "Analysis of indirect temperature-rise tests of induction machines using time stepping finite element method", *IEEE Transaction on Energy Conversion*, vol.16, No. 1, March 2001

[10] P. Pillay, M. Manyage, "Loss of life in induction machines operating with unbalanced supplies", *IEEE Transaction on Energy Conversion* vol. 42, pp. 813-822, Dec. 2006, doi: 10.1109/TEC.2005.853724

[11] M. T. Lazim, W. Shepherd, "Analysis of induction motors subjected to nonsinusoidal voltage containing subharmonics. IEEE Trans. on Industry Applications, vol. IA-21, no 4, Jul/Aug

Influence of Saturation Effects in a Transverse Flux Machine

M. Siatkowski* and B. Orlik

University Bremen, Institute of Electrical Drives, Power Electronics and Devices (IALB), Bremen, Germany,
* e-mail: *siatkowski@ialb.uni-bremen.de*

Abstract—In this paper the torque density of Transverse Flux Machines is investigated against the flux density inside the active parts with analytical calculations and FEM simulations. Especially saturation effects, which reduce the flux linkage with the coil, and core losses, are considered for both, the flux concentrating setup and the surface mounted magnets topology. The calculations are verified by measurements on an existing prototype. Finally new design guidelines are developed for the construction of TFMs with regard to the magnetic circuit.

Keywords—Transversal flux motor, Permanent magnet motor, Flux model.

I. INTRODUCTION

In industrial applications the demand on direct drives is constantly rising, and presently different types of machines are being investigated. Also in the wind energy sector the industry is heading towards gearless permanent magnet generator systems. With its typical characteristics – high torque and low speed – the Transverse Flux Machine (TFM) is particularly suitable for such applications.

Although first concepts for using Transverse Flux technology in machine design have been developed about 80 years ago, only during the last decade first prototypes have been build. The main reasons for this are:

1. The design principle of TFMs is different from the common design methods of conventional machines. There are few references which give type-dependent guidelines in the design of this type of machines. Three-dimensional FEM simulations, which require high computing power, must be used.

2. The construction and building of TFMs is complex and needs to be conducted with the highest possible accuracy in order to achieve the superior advantages.

3. The torque shape of transverse flux machines shows ripples, there are fluctuations in normal force value, and the machine power factor is low.

Finding solutions to overcome these problems is the subject of several research groups worldwide. At the IALB in the University of Bremen there are two main activities: one group deals with the reduction of the torque ripples and normal force fluctuations applying appropriate current waveform [1], and with the control of TFMs in general. Another group works on development of design methods and on optimization of the construction process of permanent magnet excited TFMs [3].

This paper analyzes the flux linkage and the losses in the TFM and compares the relation of torque density to the degree of saturation of the flux conducting elements, also considering the so called flux-concentrating design (see III). In [4] the IALB has already developed some guidelines for designing the pole pitch of TFMs and the ratio between the pole pitch and the iron and magnets in the stator and rotor. In this Paper these guidelines will be extended to some other dimensions in TFMs concerning the saturation of the core elements and if applicable of the flux concentrators. To find these relations, existing analytic models will be used as well as magnetic field simulations with the Finite-Element-Method (FEM) software Flux2D and Flux3D.

A 10-kW-TFM prototype in flux concentrating design was build at the IALB. A second TFM in the power range of 2kW with surface mounted magnets is currently under way. Measurements on the prototype and simulation results from both machines will be taken into account in this paper.

II. STRUCTURE AND PRINCIPLE OF OPERATION OF TFMs

Fig. 1 shows two possible TFM structures. Machine A has C-shaped stator cores which contain the stator winding; the outer rotor is built as a flux concentrating setup (TFM-FCS). The structure is equivalent to the 10kW prototype mentioned above. Machine B has U-shaped stator cores with the windings inside, and the inner rotor is built as a surface mounted magnet topology (TFM-SMM). The structure is equal to the planned 2kW machine. Several other structures, also where the winding is in the rotor part, or the magnets are included in the stator with a passive rotor, can be done.

Unlike conventional machines TFMs do not use a rotating field but an alternating field to generate the torque. With a constant current through the coil the rotor

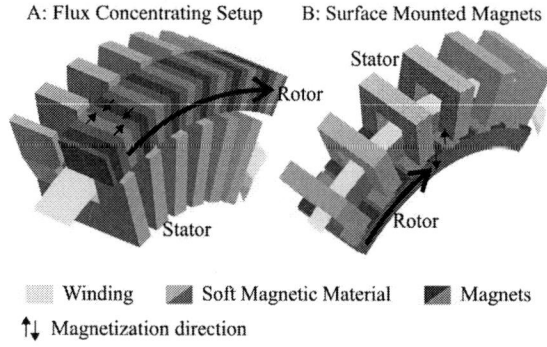

Fig. 1. Two possible TFM structures
Left: Flux Concentrating Setup – Right: Surface Mounted Magnets

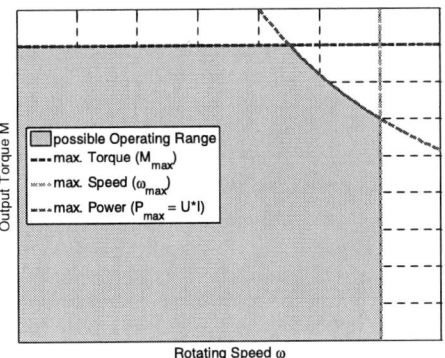

Fig. 2. Operating Range of a TFM

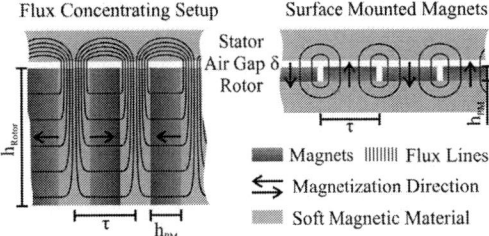

Fig. 3. General structure of the Flux Concentrating Setup in comparison to Surface Mounted Magnets

will find a position, where the stator teeth lie over the rotor poles and the flux produced by the magnets and by the coil will have the same direction. An inversion of the coil current will force the rotor to move to the next pole. An alternating current will lead to a continuous movement. As the direction of this movement is not fixed, and there will be zero-torque-positions in each cycle, TFMs must be built as at least two-phase machines, where the angle between the phases is 90° electrical; they can only run with an inverter.

One advantage of the TFM is that the design of the magnetic and the electric circuit are not coupled. Contrary to conventional topologies the cross section area of the copper winding is not complementary to the cross section of the iron cores. This means that flux density and current density can be adjusted and both, flux and current, can be increased to reach high force densities. The other advantage is the absence of end-windings. It is possible to achieve a very small pole pitch. Along with the flux concentrating setup very high torque-densities can be reached, which can be over four times higher than in conventional high torque machines [2]. TFMs usually operate at low speed with a constant torque, which can sometimes be limited by the maximum power at higher speeds (see Fig. 2), therefore TFMs can also be classified by their torque and not by their power.

III. PRINCIPLE OF THE FLUX CONCENTRATING SETUP

The flux concentrating setup (see Fig. 3) allows a significant increase of the air gap flux density B_δ, which can even rise above the residual flux density B_r of the permanent magnets used in the circuit. The material of the magnetic circuit of the machine can be driven into the saturation region even without electrical current. The flux passing through the permanent magnet's surface $A_{PM} = h_{Rotor} \cdot l$ enters the air gap $A_\delta = \tau \cdot l$, which is smaller than A_{PM}, through the iron surface (l is the axial length of the machine). The air gap flux density is higher, because the flux is *concentrated*. In this setup the iron parts between the magnets become the rotor poles.

In the sketch of the Flux Concentrating setup in Fig. 3 it becomes clear, that it is possible to insert more magnetic material into a machine with the same air gap radius. Especially for machines with a small pole pitch compared to their radius spare space can be better utilized.

The following two equations can be used for a simple calculation of the average air gap flux density B_δ, which does neither consider the reluctance of the magnetic circuit nor the saturation of iron parts. With the conventional surface mounted magnets the air gap flux density will always remain below the residual flux density of the permanent magnets:

$$B_\delta = B_r \frac{h_{PM}}{h_{PM} + \delta} = B_r \frac{1}{1 + \delta/h_{PM}} \qquad (1)$$

where the permanent magnet height is h_{PM}, the air gap length is δ and the relative permeability of the permanent magnet is $\mu_R \approx 1$. As there is significant leakage flux between neighboring magnets, especially in machines with many poles, the air gap flux density will be even lower.

The air gap flux density for the flux concentrating can be calculated as follows:

$$B_\delta = B_r \frac{h_{PM}}{h_{PM}/F_C + \delta} = B_r \frac{F_C}{1 + F_C \cdot \delta/h_{PM}} \qquad (2)$$

with the Concentrating Factor $F_C = A_{PM}/A_\delta = h_{Rotor}/\tau$. It can obviously rise above the residual flux density of the permanent magnets.

Here it must be mentioned, that the definition of the Flux Concentrating Factor in the literature is ambiguous, especially for TFMs, where on the one hand the stator side consists of more than 50% air, and on the other hand the rotor faces stator teeth at two air gaps (see Fig. 4). It may be unclear, which parts of the "air gap" count for the "active air gap". Following the flux path in the TFM with the rotor poles (the flux concentrators) placed over the stator teeth we see that each flux concentrator will face one stator tooth, thus we define the active air gap in the

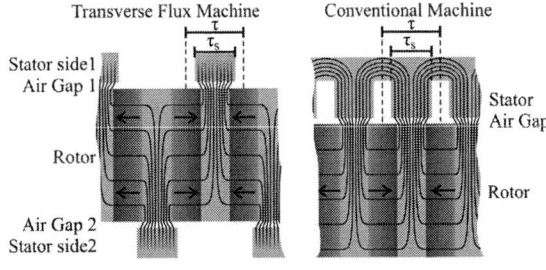

Fig. 4. Flux Concentrating Setup in TFMs and conventional machines

TFM as the sum of the active parts of both air gaps which equals the total area of one air gap.

Anyway, the air gap flux density B_δ is used to calculate machine torque or induced voltage in conventional machines. As there is significant leakage flux in TFMs, B_δ has no pertinence. Instead B_C will be used, which is the flux density inside the stator core. An equivalent average air gap flux density can be given:

$$B_\delta = \tau_s / \tau \cdot B_C \qquad (3)$$

Material saturation can be introduced in a simple way when the flux concentrating setup is used at its limits in a TFM. Generally, when the stator yokes are driven into saturation, additional flux will find paths through the air, and will not be linked with the coil, and there is no possibility, to enforce a higher flux linkage, which will be proved in the next sections. In that way, we can define B_{max} as the maximum flux density inside the stator core to estimate the equivalent average air gap flux density in the TFM:

$$B_\delta = \tau_s / \tau \cdot B_{max} \qquad (4)$$

Also B_{max} can be used with the cross section area of the stator yokes to estimate the flux linkage with the coil in a machine with flux concentration.

IV. CORE LOSSES

Core losses result from the time variation of a magnetic field inside the material. They can be calculated using the loss density model defined in [6], [7] and [8], which is an extension of the Bertotti loss model for non-sinusoidal waveforms. The losses are sub-divided into the hysteresis loss density dP_h, the classical eddy current loss density dP_c and the excess loss density dP_e.

$$dP_{FE} = dP_h + dP_c + dP_e \qquad (5)$$

$$dP_h = k_h B_m^\eta f \qquad (6)$$

$$dP_c = \sigma \frac{d^2}{12} \cdot \frac{1}{T} \int_T \left(\frac{dB(t)}{dt} \right)^2 dt \qquad (7)$$

$$dP_e = k_e \frac{1}{T} \int_T \left(\frac{dB(t)}{dt} \right)^{3/2} dt \qquad (8)$$

with η, k_h and k_e being experimentally determined loss coefficients, B_m, f and $B(t)$ being the amplitude of the flux density with the corresponding frequency and the flux density itself, and σ and d being the conductivity and the lamination thickness.

However, the calculation of core losses is nontrivial. Knowledge of the time variation of the magnetic field in each point of the material is required. The type of magnetic field variation influences the classical eddy current losses. It can be shown, that the losses caused by a rotating field vector can be two times higher, than those caused by an alternating field [9], [10]. Also does (7) not

allow a calculation for a non-laminated material, like a massive aluminum motor housing or permanent magnets which are electrically conducting. In case of hysteresis losses harmonics make the calculation difficult; they can be represented by minor loops. The hysteresis loss coefficient η depends on the degree of material saturation and was determined empirically to $1.6 < \eta < 2.6$ [12].

There is a way of more accurate calculation of core losses using the Loss Surface (LS) model described in the Flux User's Guide [14]. Loss computation is possible as the magnetic behavior of a material is perfectly well defined, having knowledge of a characteristic surface $H(B, dB/dt)$ which is determined experimentally. With the LS model no subdivision of hysteresis, classic and excess losses is possible. Like Bertotti's analytic formula it is only valid for laminated material, but also in saturated material losses can be determined precisely.

In other cases loss computation can be simplified. The conductivity of SMC materials at low frequencies $(f < 1kHz)$ can be assumed to approach zero value. Therefore no eddy current losses will exist, which also eliminates the problems of rotating field vectors in the calculation. The field in laminated steel structure can be considered alternating, so eddy current losses can be calculated with (7).

In [13] PM losses are investigated and it is proved that at low frequencies the hysteresis losses of NdFeB-Magnets are dominant and only for high frequencies eddy currents play an important role.

V. DESCRIPTION OF THE PROTOTYPE

The 10kW Prototype was completed by 2005 and started operation at the beginning of 2006. Along with a special two phase inverter, which also was built at the IALB, the machine has completed many test runs and endurance test. Combined with current- and torque-control algorithms very satisfying results were achieved. Table I gives an overview of the most important machine

TABLE I.
PARAMETERS OF THE PROTOTYPE

Machine Length	270 mm
Diameter	480 mm
Nominal Torque	1000 Nm
Speed	100 rpm
No. of Poles	74
No. of Windings per Phase	66
Armature Current	26 A$_{eff}$
Power Factor (measured)	0.6
Air gap length	0.8 mm
Pole Pitch	13.2 mm
Stator teeth width	25 mm
Magnet surface	907mm²
Weight	165kg

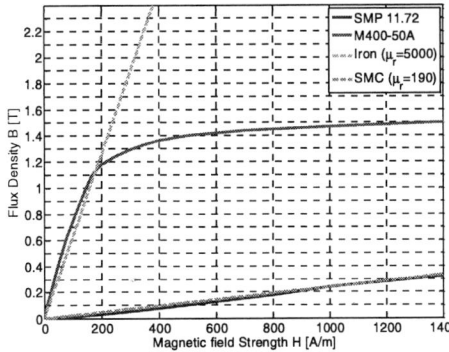

Fig. 5. B-H-Curves for different materials

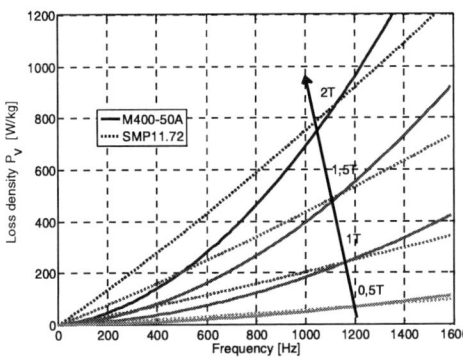

Fig. 6. Core Losses for different flux density peak values

parameters.

The structure of the machine corresponds to Fig. 1 A, which shows only the active parts of one phase of the TFM. The rotor of each phase has two air gaps, one axial and one radial. Two phases, which are shifted 90 degrees electrical against each other, form the main part of the machine.

The prototype was build using M400-50A laminated steel for the stator cores, SMP 11.72 soft magnetic composite (SMC) for the flux concentrators, and NdFeB-magnets with a residual flux density of $B_r = 1.1T$. In Fig. 7 the different material types' magnetic characteristics can be compared to each other. Up to $1.2T$ the M400-50A can be assumed a linear material with $\mu_r = 5000$, for higher values the material is saturated and acts non-linear. Clearly the permeability of the SMC is much lower (approx. $\mu_r = 190$), but it can be assumed as electrically non-conducting. Therefore it was chosen to suppress eddy current losses in the flux concentrating region, where the flux path is not bounded to a spatial plane but is rather three-dimensional. Fig. 6 shows the core losses of both materials for flux density peak values from $0.5T$ to $2T$ assuming a flux in the plane of the lamination if present. The curves were plotted using (5) and values from material data sheets.

The stator and rotor frame and parts of the machine housing are made of aluminum, as it has good mechanical

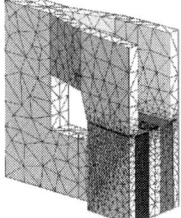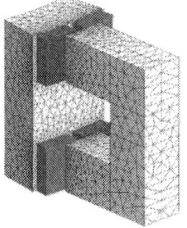

Fig. 7. Two poles of the TFM models in Flux3D with mesh.
Left: 10kW-Prototype - Right: 2kW machine with surface magnets

and thermal properties. The aluminum parts are not illustrated in the figures.

VI. DESCRIPTION OF THE SIMULATION MODELS

The FEM models in Flux3D consist only of the magnetic active parts of the machine. Fig. 7 shows the simulation model geometry for the 10kW prototype as well as for the 2kW machine with surface mounted magnets. To save computation time, only two poles of the geometry are modeled using the periodicity property, as 3D-FEM requires lots of computation time. The mesh density has been adapted to the requirements of the model, and refined especially in the air gap region. It consists of 97799 second order elements for the 10kW simulation model.

The material data for M400-50A has been taken from the Material Database, the characteristic of SMP 11.72 and of the NdFeB-Magnets has been entered according to the manufacturer's data sheets.

Many simulation runs have been performed to adjust the models and to achieve best possible results. Multi-static simulations were used to determine flux distribution and material saturation. For the loss computations transient simulations have been compared to each other, and also values from the multi-static simulations have been taken to estimate the losses using the analytic formulas.

VII. SIMULATION RESULTS

For all simulations the angular position of the rotor α is incrementally changed from $\alpha = 0$ to $\alpha = \tau$. For both, $\alpha = 0$ and $\alpha = \tau$ the flux concentrators or the rotor poles are placed exactly under the stator poles. When $\Theta = 0A$, these are the positions with the maximum Flux through the stator core. Both are stable zero-torque-positions. Due to complex leakage flux paths the torque shape of the cogging torque between these positions is complicated. For $\Theta > 0A$ at $\alpha = 0$ the armature flux and the rotor flux support each other, it is also a stable zero-torque-position. At $\alpha = \tau$ the armature flux opposes the rotor flux, so the core flux is minimal. This zero-torque-position is instable. Fig. 8 shows the torque shape for $\Theta = 0$ and $\Theta = 1700A$. It was extended to $0 < \alpha < 2\tau$.

For the 10kW machine model the effect of the flux concentration is investigated and a focus is put on the saturation of the core material. Fig. 9 shows the total flux through the core compared to the flux produced by the magnets. The magnet size (area perpendicular to the magnetization direction) is the variation parameter which is used to calculate also F_C. Only for a magnet size of

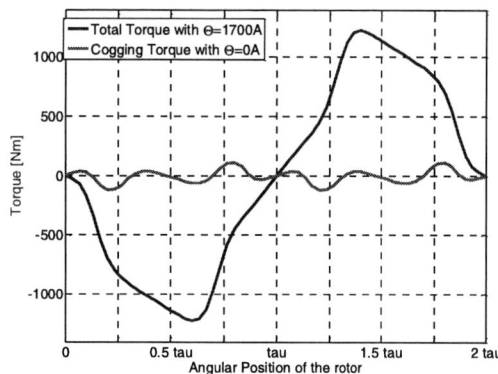

Fig. 8. Torque shape of the 10kW TFM for $\Theta = 0$ and $\Theta = 1700 A$

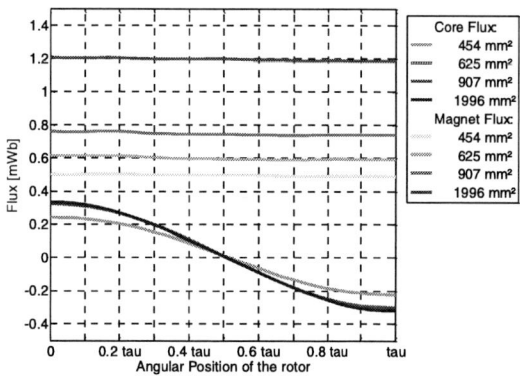

Fig. 9. Flux for $\Theta = 0$

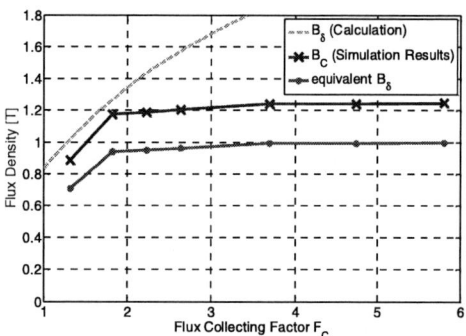

Fig. 10. Flux density over the flux concentrating factor:
Calculation with (2) is compared to simulation results

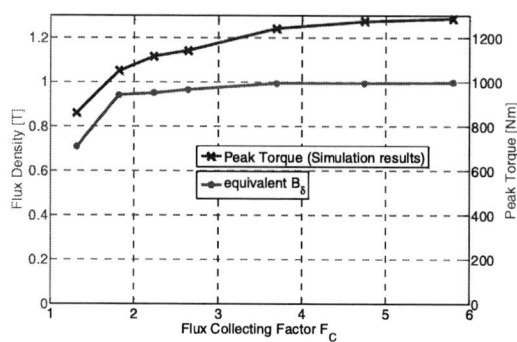

Fig. 11. Comparison of characteristics for different magnet sizes

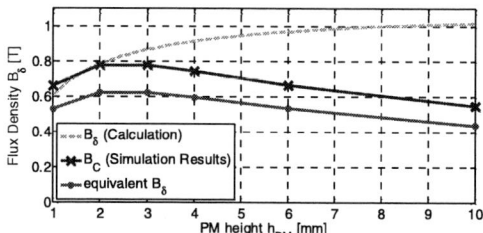

Fig. 12. Flux density for a surface mounted magnet topology:
Calculation with (1) is compared to simulation results

As the torque can be computed from air gap flux density, current and some machine parameters in conventional machines, the impact of F_C on the torque is investigated and compared to the equivalent air gap flux density. We can observe a saturation effect on the peak torque produced by the machine, which is related to the flux density (Fig. 11), although it can still be increased in small portion, while the flux density remains at its maximum level.

For the 2kW model with surface mounted magnets the flux density is investigated against the permanent magnet height and compared to (1) in Fig. 12. We can see that there is a significant deviation, which cannot be explained with material saturation. As illustrated in Fig. 13 and Fig. 14, the leakage flux between two neighboring permanent magnets increases with a growing magnet height. It is

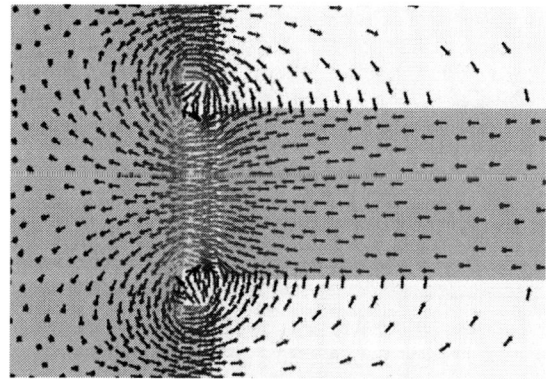

Fig. 13. Flux leakage between
two neighbouring permanent magnets

$454 mm^2$ ($F_C \approx 1.3$) the core flux is reduced significantly compared to the core flux with the prototype's original magnet size of $907 mm^2$ ($F_C \approx 2.65$), while the magnet flux itself has good correlation with the magnets' size. This means that with a rising flux concentrating factor the amount of leakage flux in the rotor area increases dramatically. The linked flux from the FEM simulation has been compared to the calculation with (2) in Fig. 10. Here it becomes clear, that a calculation with (2) is not possible for machines with a big amount of leakage flux, especially when the material reaches the saturation value of $B_{max} = 1.2T$.

834

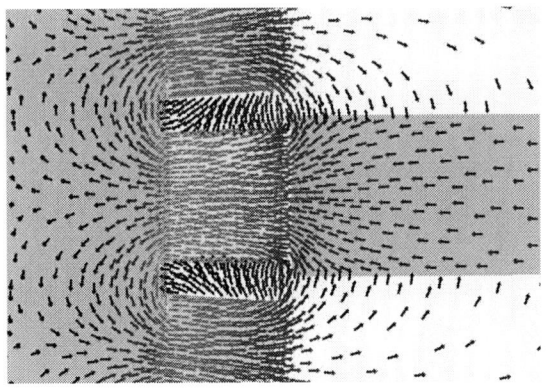

Fig. 14. Flux leakage between
two neighbouring permanent magnets

apparent that the core material in the SMM topology cannot be driven into saturation.

Losses are examined only in the flux concentrating topology, as there is no material saturation in the SMM topology. The core losses in the stator cores are determined with Flux3D using the LS model which allows a precise computation even in saturated material. Losses in the SMC are computed using Bertotti's formula [6] and a conductivity of 0. In the PMs only hysteresis losses can be computed, but in general the PM losses are assumed to be small as there is no significant time variation of the magnetic field as seen in Fig. 9. The computation results are summarized in Table II.

Also in the aluminum frame of the rotor and stator eddy currents can occur. Magnetic fields can only enter the aluminum frame in areas near the active parts, but due to

TABLE II.
LOSSES IN THE FCS-TFM (ONE PHASE) AT 100RPM

Region	Losses [W]			
	P_h	P_c	P_e	P_{tot}
Stator Cores (M400-50A)	-	-	-	17.9
Rotor Cores (SMP 11.72)	66.6	0	0.9	67.5
PMs (NdFeB)	14.2		0.3	14.5
Total				99.9

Fig. 15 Flux density distribution on a cut through
the aluminum stator frame (invisible) close to the air gap

the saturation of the core material there is considerable flux density in the aluminum parts. Fig. 15 shows the cross section through the aluminum stator frame near the air gap. A flux density up to $1T$ can be observed, which can cause significant eddy currents in the non-laminated material. Eddy current losses in the frame will be estimated from measured values.

VIII. MEASUREMENTS ON THE PROTOTYPE

Measurements and simulation results are compared to verify the simulation model for the 10kW-Prototype. As there is no possibility to access the magnetic circuit inside the prototype to perform flux measurements, other quantities must be used for the comparison. The measured induced voltage of both machine phases can be compared to the values computed with the FE-model (Fig. 16). The results match the measurement with good accuracy.

Furthermore, losses have been investigated on the prototype. Fig. 17 shows the measured power loss of the 10kW prototype at different speeds in generator mode at no-load condition. After a subtraction of friction loss the core losses were obtained. At 100rpm the core losses are 430W. The significant difference between measured and calculated losses indicates enormous eddy current losses in the aluminum frames, which could not be calculated. Also in the NdFeB-Magnets eddy currents can occur. As the magnets have a conductivity of $\sigma_{NdFeB} \approx 0.7 \cdot 10^6 \, Sm^{-1}$ compared to the conductivity of aluminum, which is

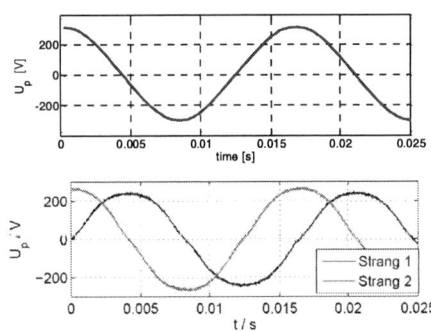

Fig. 16 Comparison of induced voltage from FEM (top) and
measurement on both phases (bottom) at 100rpm

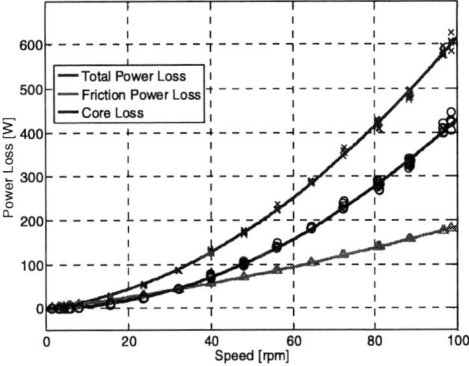

Fig. 17 No-load power loss in the 10kW TFM

$\sigma_{alu} \approx 37 \cdot 10^6 \, Sm^{-1}$, and the time variation of the flux density is much smaller in the PMs, the eddy current losses in the PMs are much smaller. The eddy current losses in the aluminum parts are estimated to be more than 300W.

When the flux concentrating factor is chosen in a way that the stator core material is saturated, more flux is pushed through the aluminum parts which causes these eddy current losses. When designing new machines it is important to limit the flux concentrating factor, so that the core material will not be saturated too much. Saturation results in additional losses, especially in TFMs, where more leakage flux exists in general. The choice of appropriate core materials, particularly for the stator yokes, can increase the performance of TFMs.

IX. Conclusion

In this paper the impact of the flux concentrating setup on the material saturation was investigated and compared to surface mounted magnets in a Transverse Flux Machine. The effects of the saturation value on the flux linkage and leakage flux paths has been described and used to explain the eddy current losses in aluminum parts of the machine.

With the flux concentrating setup the flux density can be increased even above the residual flux density of the permanent magnets. Contrary to a topology with surface mounted magnets with flux concentration it is possible to drive the core material into saturation. In most geometries it is quite easy to adjust the flux concentrating factor without significant changes to other important dimensions, which is one of the great advantages of TFMs. This possibility should be used to choose a factor, which will produce a high flux linkage without pushing the material into saturation too much. This does not only result in a gain of weight and volume of the machine, but also in additional losses.

Improvements can be achieved by the choice of high performance materials introduced in the last few years, which allow higher values for the saturation. Also the construction of the machine housing and the stator and rotor frames can be improved by breaking the eddy current circuits. Special materials or constructions with laminated aluminum can be used.

X. Outlook

Recently a new TFM has been built in the power range of 50kW which makes use of research and analysis results which were achieved in the last years. The stator and rotor frame have been modified to suppress eddy currents. A laminated aluminum rotor frame was developed, which adheres to the electromagnetic requirements as well as to thermal and mechanical conditions. Currently that machine is in the test stand and new results will be available soon.

References

[1] U. Werner, H. Raffel, O.Harling, N. Parspour, B.Orlik, „Strategies to Reduce Torque and Current Ripples of Transverse Flux Permanent Magnet Generators for Wind Turbine Applications," *EPE2003, Touluse*, September 2003

[2] U. Werner, S. Hoffmann, J. Schüttler, M. Vinogradski, B. Orlik „Stromregelung von Permanenterregten Transversalflussmotoren in Servo-Direktantriebsanwendungen," *DFMRS*, February 2006.

[3] A. Babazadeh, N. Parspour,A. Hanifi, „Transverse Flux Machine for Direct Drive Robots: Modelling and Analysis," *IEEE Conference on Robotics, Automation and Mechatronics*, Singapur, 2004

[4] O. Harling, B.Orlik, „Drehmomentsteigerung bei permanentmagneterregten Transversalflussmaschinen durch Optimierung der Rotor-Stator-Längenver-hältnisse", *SPS / IPC / DRIVES2003*, Nürnberg (Germany), November 2003

[5] Y. GUO, „Core Losses in Claw Pole Permanent Magnet Machines With Soft Magnetic Composite Stators", *IEEE Transactions on Magnetics, Vol. 39, No. 5, P. 3199 – 3201*, Sydney, 2003

[6] G. Bertotti, „General properties of power losses in softferromagnetic materials," *IEEE Tranactions. on Magnetics, Vol. 24, No. 1, P. 621–630*, January,1988

[7] F. Magnussen, Y.K. Chin, J. Soulard, A. Broddefalk, S. Eriksson, C. Sadarangani, „Iron losses in salient permanent magnet machines at field-weakening operation," *Proceeding of IAS 2004, Vol. 1, P. 40–47*, October, 2004.

[8] Y. Amara, J. Wang, D. Howe, „Stator iron loss of tubular permanent-magnet machines," *IEEE Trans. on Industry Applications, Vol. 41, No. 4, P. 989–995*, July/August, 2005

[9] Y. Guo, J. Zhu, J. Zhong, W. Wu, "Core losses in claw pole permanent magnet machines with soft magnetic composite stators," *Magnetics Conference, 2003. INTERMAG 2003. IEEE International*, pp. EA-11, 2003

[10] Jian Guo Zhu, Jian Guo Zhu, V. Ramsden, "Improved formulations for rotational core losses in rotating electrical machines," *Magnetics, IEEE Transactions on*, vol. 34, pp. 2234-2242, 1998.

[11] U. Werner, M. Vinogradski, B. Orlik, „Simulation und Entwurf eines analytischen Modells für eine permanentmagneterregte Transversalflussmaschine in Sammler-Bauweise", *VDI/VDE-ETG-Tagung, Elektrisch-mechanische Antriebssysteme*, Fulda 2004

[12] E. Philippow, „Grundlagen der Elektrotechnik", 10.durchgesehene und ergänzte Auflage, Verlag Technik, Berlin, 2000

[13] A. Fukuma, S. Kanazawa, D. Miyagi, N. Takahashi, „Investigation of AC Loss of Permanent Magnet of SPM Motor Considering Hysteresis and Eddy Current Losses", Magnetics, IEEE Transactions on, Vol. 41, No. 5, 1964-1967, May 2005

[14] Cedrat, "Flux 10 User's Guide", Volume 3, Physical applications, p. 147, Aug. 2007

A Model of Semiconductor Converter-Fed Asynchronous Machines Taking into Account Energy Losses and Thermal Processes

M. Pronin[1], O. Shonin[2], Y. Koskin[3], A. Vorontsov[1], P. Kalatchikov[1]

[1] JSC "Power Machines", Branch "Electrosila", Russia, Saint Petersburg,
e-mail: *m_pronin@elsila.spb.ru; a_voroncov@elsila.spb.ru; p_kalachikov@elsila.spb.ru*
[2] Saint-Petersburg Mining Institute, Russia, Saint Petersburg, e-mail: *ninosh_eltech@mail.ru*
[3] Saint-Petersburg Electrotechnical University, Russia, Saint Petersburg, e-mail: *postmaster@raps.etu.spb.ru*

Abstract—A model of a semiconductor converter-fed dual-stator windings induction machine of the adjustable-speed propulsion drive is built up by decomposing the system into sub-circuits with lumped parameters which are coupled together by means of dependent voltage/current sources. Measurements of frequency responses of the driving point impedance of the stator windings have shown a strong impact of the skin effect on machine parameters. In the proposed model, electric and magnetic skin effects are reproduced by means of RL-ladder networks incorporated into the stator and rotor sub-circuits. A thermal model is described in terms of average temperatures and heat flows in the body of an induction machine. The paper concludes by examining the electromechanical and thermal responses of the motor energized by voltage sources of different waveforms.

Keywords— Asynchronous motor, converter machine interactions, thermal design, modeling.

INTRODUCTION

Computer models of electromechanical systems with asynchronous machines (AM) and semiconductor converter (SC) are widely used to facilitate the design procedure of electrical machines, development, research and the debugging of the adjustable-speed drives. The model of the system is expected to properly reproduce not only basic but also specific properties of SC-fed AM, for instance, such as a possible onset of detrimental oscillations of the stator currents in poly-phase AM [1], [2], energy dissipation in SC and AM, dependence of energy loss components on the SC and AM operation modes.

As it is seen from publications [3]-[12], correct identification and evaluation of energy losses components are of great importance because of their affecting the system efficiency and temperature capacity of SC and AM.

Overall energy losses [13] are known to comprise static and dynamic losses in SC, eddy current and hysteresis components of iron losses, joule losses in the stator and the rotor windings, friction and windage components of mechanical losses, and stray load losses due to spatial harmonics of the air gap flux.

In most cases, correct evaluation of energy loss requires taking into account the redistribution of currents due to skin effects. For example, as reported in [14], high-order harmonics of the rotor bar current induced by spatial harmonics of the air gap field give roughly 25% of the bar losses under condition that AM is loaded at $s=0,033$.

A 3D- and t-domain model of AM is the most effective tool at the stage of a machine design. However, less sophisticated models allowing for high speed computing are also in demand, for example, at the stage of the debugging of the microprocessor-based control subsystem of adjustable-speed drive when the model stands for the real SC-fed AM [15].

The redistribution of currents in the rotor, in the first approximation, can be reproduced, as it is shown in [16], by means of *RL*-ladder networks incorporated in the rotor sub-circuits of the AM model.

In increasing the power of SC-fed AM, the electrical skin effect becomes an important factor because of the enlarged cross section of the stator windings conductors. In SC-fed AM, high frequency (HF) losses are defined by high-order harmonics of the voltage spectrum and frequency dependent parameters of AM.

In order to determine AM-parameters versus frequency, measurements of the driving point impedance of the stator windings (see Fig.1) of the induction machine ADR-2000 ($P=2000$ kW, $U=960$ V, $I=700$ A, $f=50$ Hz) have been carried out with a RLC-meter of high accuracy.

Experimental curves $L_{s1}(f)$ and $R(f)$ shown in Fig.2 correlate with data reported in [17] and reveal a strong impact of the skin effect on AM - parameters.

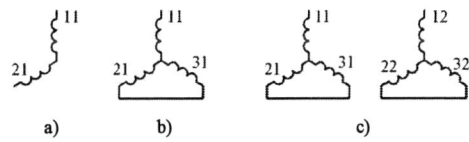

Fig.1. Configurations of the stator windings for measurements of a driving point impedance

978-1-4244-1741-4/08/$25.00 ©2008 IEEE

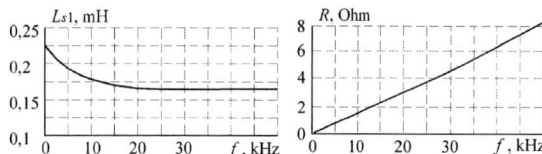

Fig.2. Input inductance Ls1(f) and input resistance R(f) of the stator winding versus frequency

It should be noted that tests of the analyzed motor ADR-2000 powered from frequency converters have shown that the stator windings temperature exceeds the base value by 35°C.

In order to represent interrelated electromechanical and thermal processes, the computer model of SC-fed AM has been extended by building up a thermal model and by taking into account the redistribution of currents under frequency and variation of AM parameters under temperature.

PROPOSED MODEL OF THE SYSTEM WITH SC AND AM

Consider the adjustable-speed propulsion drive based upon the dual-stator windings induction machine ADR-2000 energized by voltage source inverters (VSI) in parallel for each of the stator windings (see Fig.3). The system is decomposed into VSI-subsystem and AM-subsystem which are bound together through phase voltages u_{nm} and phase currents i_{nm}, where $n=1,2,3$ – phase index, $m=1,2,...M$ – stator windings index. The model of VSI-subsystem is reported in [16], [18].

In general, the proposed model allows for describing polyphase machines with M sets of three-phase windings energized from M sources via J_v inverters in parallel for each star. The joint operation of inverters is provided through phase interference–suppressing chokes R_f, L_f with or without magnetic interphase coupling L_{ym}.

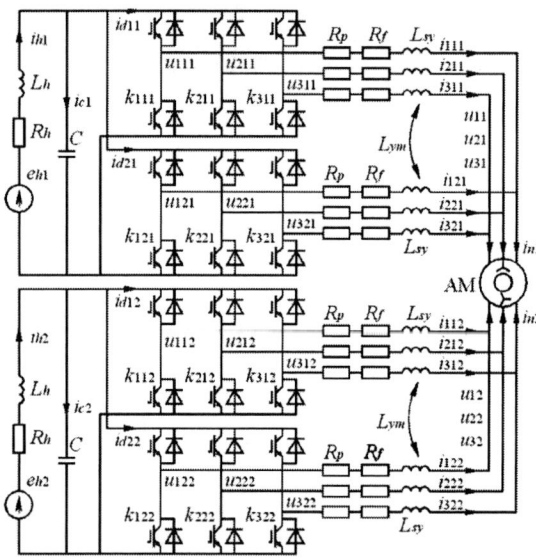

Fig.3 Scheme of adjustable-speed propulsion drive

When building up a model of SC, transistors and anti-parallel diodes are assumed to be ideal valves. Energy losses in SC are taken into account by the resistance R_p. Transistors in a bridge arm are complementary one to the other: if one transistor in an arm is fired, then the other is turned off. Phase voltages and phase currents of an ideal inverter are referred to as U_{njm} and i_{njm}, where $n=1,2,3$ – phase index, $j=1, ...J_v$ –inverter index, $m=1, ..M$ –three phase windings index.

States of semiconductor elements are defined by a discrete function k_{njm}: if an open transistor or a diode connects n-th phase to the positive pole of the capacitor C, then $k_{njm}=1$, and if it connects n-th phase to the negative pole, then $k_{njm}=0$.

The model under consideration is built up by decomposing the system into a SC subsystem and an AM subsystem. The last one in turn is subdivided into the stator and the rotor subcircuits. Subsystems are coupled together through dependent voltage and current sources.

In the subsystem of SC, voltages u_{cm} across the capacitors C and output phase voltages of ideal inverters u_{njm} are determined as follows:

$$u_{cm} = \frac{1}{C} \int i_{cm} dt,$$
$$u_{njm} = \left[k_{njm} - \left(k_{1jm} + k_{2jm} + k_{3jm} \right)/3 \right] u_{cm},$$
$$n = 1,2,3, \quad j = 1,...J_v, \quad m = 1,...M.$$

Current i_{injm} through bridge arms:

$$i_{injm} = k_{njm} i_{njm}.$$

Transistor currents i_{tnjm} and diode currents i_{dnjm}:

$$if \quad i_{injm} > 0, \quad then \quad i_{tnjm} = i_{injm}, \quad i_{dnjm} = 0,$$
$$otherwise \quad i_{tnjm} = 0, \quad i_{dnjm} = -i_{injm}.$$

Inverter input currents:

$$i_{djm} = i_{i1jm} + i_{i2jm} + i_{i3jm}.$$

DC voltage currents i_{hm} are determined from equations:

$$\frac{di_{hm}}{dt} = \frac{e_{hm} - u_{cm} - R_h i_{hm}}{L_h}.$$

Currents i_{cm} through the capacitor:

$$i_{cm} = i_{hm} - \sum_{j=1}^{J_v} i_{djm}, \quad m = 1,...M.$$

Output inverter currents and their derivatives:

$$\frac{di_{njm}}{dt} = \frac{u_{njm} - u_{nm} - \left(R_f + R_p \right) i_{njm} + L_{st} \dfrac{di^*_{njm}}{dt}}{L_f + L_{st}}.$$

A formal inductance L_{st} is incorporated in the model for stabilizing the iterative process of computation, $\dfrac{di^*_{njm}}{dt}(k) = \dfrac{di_{njm}}{dt}(k-1)$, k – iteration index.

SC and AM subsystems are interconnected by means of dependent current sources i_{nm} and voltage sources u_{nm}.

An induction machine is described in different reference frames shown in Fig.4, where ω – angular velocity of the rotor, τ – angle between a fixed axis α and a rotating axis d. In general, the proposed model allows for describing a poly-phase AM with M stator windings

838

shifted by an angle $\pi/3M$ and fed from M DC-sources via several VSI in parallel for each of M stator windings.

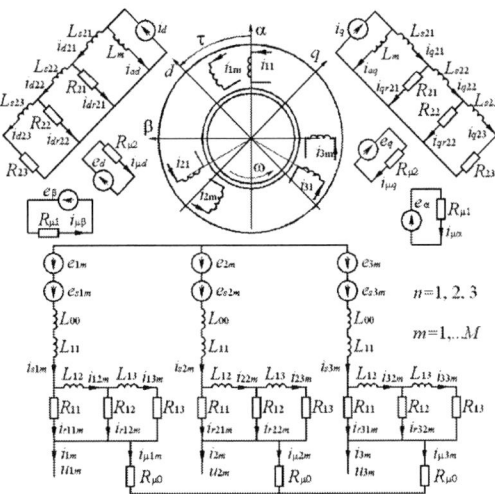

Fig.4. Sub-circuits of a poly-phase AM

AM- phase voltage and phase current references indicated in Fig.4 correspond to a generator mode of operation of AM. In the proposed model, it is adopted as a general case of describing the induction machines.

In the motor mode of operation, AM phase currents i_{nm} and their derivatives are expressed in terms of inverter output currents i_{njm} as follows:

$$\frac{di_{nm}}{dt} = -\sum_{j=1}^{J_v} \frac{di_{njm}}{dt},$$

$$i_{nm} = -\sum_{j=1}^{J_v} i_{njm},$$

$$n = 1,2,3, \quad m = 1,...M.$$

A poly-phase AM is characterized by a magnetizing inductance L_m, a zero-sequence inductance L_0 and leakage inductances L_{s1} and L_{sM} which refer to energizing one set and M sets of the stator windings, correspondingly.

Each of M stator sub-circuits includes a phase emf e_{nm} induced by the air gap flux and emf e_{snm} caused by the inter-phase mutual induction flux along the leakage paths.

The redistribution of currents in the stator windings under frequency is reproduced by a ladder network with inductances L_{00}, L_{1i} and resistances R_{1i}, $i=1,2...J_s$, where J_s – a number of LR sections. An inductance L_{00} is an asymptotic HF value of the function $L_{s1}(f)$ shown in Fig.2. LR parameters of a ladder network are set against a section's index i in the geometrical progression with a common ratio K_s:

$$L_{1i} \approx L_f K_s^{i-1} \Big/ \sum_{i=1}^{J_s} K_s^{i-1}, \qquad R_{1i} \approx R_1 \sum_{i=1}^{J_s} K_s^{i-1} \Big/ K_s^{i-1},$$

where $L_f = (2L_{s1} - L_{sM} - L_{00})$ is an auxiliary parameter referred to a nominal frequency, R_1 – nominal value of the stator winding resistance.

As reported in [19], the stray load losses can be modeled by additional resistance in series with the stator branch of AM equivalent circuit. In the proposed model, the stray load losses are taken into account cumulatively with the iron losses by means of resistances $R_{\mu0}$, $R_{\mu1}$, $R_{\mu2}$ in sub-circuits with dependent voltage sources e_d, e_q, e_α, e_β.

The redistribution of the rotor currents is reproduced by a ladder network in the rotor sub-circuit which also includes a magnetizing inductance L_m and dependent current sources i_d, i_q.

LR parameters are determined as follows:

$$L_{s2i} \approx L_{s2} K_r^{i-1} \Big/ \sum_{i=1}^{J_r} K_r^{i-1}, \qquad R_{2i} \approx R_2 \sum_{i=1}^{J_r} K_r^{i-1} \Big/ K_r^{i-1},$$

where K_r – common ratio of the geometrical progression, i – ladder section index, J_r – a number of the rotor sections, R_2 and L_{s2} – nominal values of the rotor resistance and inductance.

The sub-circuits are bound together by means of dependent voltage source e_{nm}, e_{snm}, e_d, e_q, e_α, e_β and dependent current sources i_{nm}, i_α, i_β, $i_{\mu\alpha}$, $i_{\mu\beta}$, i_d, i_q, $i_{\mu d}$, $i_{\mu q}$.

A numerical solution of a system of differential equations describing electromagnetic processes in AM consists of an iteration cycle to calculate values of the state variable derivatives, values of voltages and emf, and the integration cycle to compute the current values of state variables.

A magnetizing inductance is updated at each step of calculation by referring to the spline-approximated saturation curve. Computational algorithm is realized in C++ Builder.

A system of algebraic equations to be solved at the iteration cycle is formed as follows.

Derivatives of the stator phase current are determined from the expression for phase voltages:

$$u_{nm} = e_{nm} + e_{snm} - R_{11}i_{rnlm} - (L_{00} + L_{11})\frac{di_{snm}}{dt}.$$

Derivatives of currents in $\alpha\beta0$-frame:

$$\frac{di_\alpha}{dt} = \frac{2}{3M}\sum_{m=1}^{M}\sum_{n=1}^{3} c_{nm}\frac{di_{snm}}{dt},$$

$$\frac{di_\beta}{dt} = \frac{2}{3M}\sum_{m=1}^{M}\sum_{n=1}^{3} s_{nm}\frac{di_{snm}}{dt},$$

$$\frac{di_{\alpha0}}{dt} = \sum_{m=1}^{M} c_m \sum_{n=1}^{3}\frac{di_{snm}}{dt},$$

$$\frac{di_{\beta0}}{dt} = \sum_{m=1}^{M} s_m \sum_{n=1}^{3}\frac{di_{snm}}{dt},$$

$$c_{nm} = \cos\left[\frac{2\pi}{3}\left(n-1+\frac{m-1}{2M}\right)\right],$$

$$s_{nm} = \sin\left[\frac{2\pi}{3}\left(n-1+\frac{m-1}{2M}\right)\right],$$

where

$$c_m = \cos\left[\frac{\pi}{M}(m-1)\right],$$

$$s_m = \sin\left[\frac{\pi}{M}(m-1)\right].$$

Emf due to the inter-phase mutual inductance:

$$e_{snm} = -L_a\left(c_{nm}\frac{di_\alpha}{dt} + s_{nm}\frac{di_\beta}{dt}\right) - L_b\left(c_m\frac{di_{\alpha0}}{dt} + s_m\frac{di_{\beta0}}{dt}\right),$$

where $L_a = M(L_{sM} - L_{s1})$, $L_b = (L_0 - L)/3$.

Derivatives of the stator currents in a $dq0$-frame:

$$\frac{di_d}{dt} = \frac{di_\alpha}{dt}\cos\tau + \frac{di_\beta}{dt}\sin\tau - \omega i_q,$$

$$\frac{di_q}{dt} = \frac{di_\alpha}{dt}\sin\tau - \frac{di_\beta}{dt}\cos\tau + \omega i_d.$$

Derivatives of the magnetizing currents i_{ad}, i_{aq}:

$$\frac{di_{ad}}{dt} = \frac{L_{s21}\dfrac{di_d}{dt} + R_{21}i_{dr21}}{L_m + L_{s21}},$$

$$\frac{di_{aq}}{dt} = \frac{L_{s21}\dfrac{di_q}{dt} + R_{21}i_{qr21}}{L_m + L_{s21}}.$$

Components of emf in $dq0$-, $\alpha\beta0$-, phase- frames:

$$e_d = -L_m\left(i_{aq}\omega + \frac{di_{ad}}{dt}\right),$$

$$e_q = L_m\left(i_{ad}\omega - \frac{di_{aq}}{dt}\right),$$

$$e_\alpha = e_d\cos\tau + e_q\sin\tau,$$

$$e_\beta = e_d\sin\tau - e_q\cos\tau,$$

$$e_{nm} = e_\alpha c_{nm} + e_\beta s_{nm}.$$

Calculated values of the derivative of state variables allow for computation of current values of state variables in an integration cycle. Then, other variables are calculated as follows.

Currents in the stator sub-circuits are determined from the next expressions:

$$\frac{di_{n2m}}{dt} = \frac{R_{11}i_{rn1m} - R_{12}i_{rn2m}}{L_{12}},$$

$$\frac{di_{n3m}}{dt} = \frac{R_{12}i_{rn2m} - R_{13}i_{n3m}}{L_{13}}.$$

Currents in the stator sub-circuits:

$$i_{\mu nm} = \frac{u_{nm}}{R_{\mu0}}, \qquad i_{rnm} = i_{snm} - i_{\mu nm},$$

$$i_{rn1m} = i_{snm} - i_{n2m}, \qquad i_{rn2m} = i_{n2m} - i_{n3m},$$

$$i_{\mu\alpha} = \frac{e_\alpha}{R_{\mu1}}, \quad i_{\mu\beta} = \frac{e_\beta}{R_{\mu1}}, \quad i_{\mu d} = \frac{e_d}{R_{\mu2}}, \quad i_{\mu q} = \frac{e_q}{R_{\mu2}}.$$

The stator currents in $dq0$- and $\alpha\beta0$-frames:

$$i_\alpha = \frac{2}{3M}\sum_{m=1}^{M}\sum_{n=1}^{3}c_{nm}i_{snm} + i_{\mu\alpha},$$

$$i_\beta = \frac{2}{3M}\sum_{m=1}^{M}\sum_{n=1}^{3}s_{nm}i_{snm} + i_{\mu\beta},$$

$$i_d = i_\alpha\cos\tau + i_\beta\sin\tau + i_{\mu d},$$

$$i_q = i_\alpha\sin\tau - i_\beta\cos\tau + i_{\mu q}.$$

Derivatives of the rotor currents in a $dq0$-frame:

$$\frac{di_{d22}}{dt} = \frac{R_{21}i_{dr21} - R_{22}i_{dr22}}{L_{s22}},$$

$$\frac{di_{q22}}{dt} = \frac{R_{21}i_{qr21} - R_{22}i_{qr22}}{L_{s22}},$$

$$\frac{di_{d23}}{dt} = \frac{R_{22}i_{dr22} - R_{23}i_{dr23}}{L_{s23}},$$

$$\frac{di_{q23}}{dt} = \frac{R_{22}i_{qr22} - R_{23}i_{qr23}}{L_{s23}}.$$

The rotor currents in a $dq0$-frame:

$$i_{d21} = i_d - i_{ad},$$

$$i_{dr21} = i_{d21} - i_{d22},$$

$$i_{dr22} = i_{d22} - i_{d23},$$

$$i_{q21} = i_q - i_{aq},$$

$$i_{qr21} = i_{q21} - i_{q22},$$

$$i_{qr22} = i_{q22} - i_{q23}.$$

Electromagnetic torque:

$$M_{em} = 1.5ML_m\left(i_{ad}i_q - i_{aq}i_d\right).$$

In the proposed model, mechanical losses are reproduced by additional load torque $M_{mx} = M_{mxn}(\omega/\omega_n)^2$, where, M_{mxn} and ω_n – nominal values. The rotor angular velocity ω and the rotor angular position τ are determined from equations:

$$\frac{d\omega}{dt} = \frac{1}{J}\left(M_{em} - M_c - M_{mx}\right),$$

$$\frac{d\tau}{dt} = \omega,$$

where M_c – load torque, J – inertia coefficient.

Copper losses in the stator windings:

$$p_{m1} = \sum_{m=1}^{M}\sum_{n=1}^{3}\left(R_{11}i_{rn1m}^2 + R_{12}i_{rn2m}^2 + R_{13}i_{rn3m}^2\right).$$

Copper losses in the rotor windings:

$$p_{m2} = \frac{2M}{3}\left[\begin{array}{l}R_{21}\left(i_{dr21}^2 + i_{qr21}^2\right) + \\ + R_{22}\left(i_{dr22}^2 + i_{qr22}^2\right) + R_{23}\left(i_{dr23}^2 + i_{qr23}^2\right)\end{array}\right].$$

Iron and additional losses:

$$p_{\mu0} = 3M\,R_{\mu0}\sum_{m=1}^{M}\sum_{n=1}^{3}i_{\mu0nm}^2,$$

$$p_{\mu1} = 1.5M\,R_{\mu1}\left(i_{\mu\alpha}^2 + i_{\mu\beta}^2\right),$$

$$p_{\mu2} = 1.5M\,R_{\mu2}\left(i_{\mu d}^2 + i_{\mu q}^2\right).$$

Mechanical losses:

$$p_{mx} = M_{mx}\omega^2.$$

Overall losses:

$$p_{am} = p_{m1} + p_{m2} + p_{\mu0} + p_{\mu1} + p_{\mu2} + p_{mx}.$$

AM input power p_1 and shaft power p_2:

$$p_1 = \sum_{m=1}^{M}\sum_{n=1}^{3}u_{nm}i_{nm},$$

$$p_2 = \left(M_{em} - M_{mx}\right)\omega.$$

In a steady-state, active power P is used instead of instantaneous power $p(t)$ introduced above.

Resistances $R_{\mu 0}$, $R_{\mu 1}$, $R_{\mu 2}$ are determined from iron and additional losses related to the stator $P_{\mu 0}$, $P_{\mu 1}$ and to the rotor $P_{\mu 2}$:

$$R_{\mu 0} = \frac{3MU^2}{P_{\mu 0}},$$

$$R_{\mu 1} = \frac{3ME_\delta^2}{P_{\mu 1}}, \qquad R_{\mu 2} = \frac{3ME_\delta^2}{P_{\mu 2}},$$

where U – phase voltage, r.m.s., $E_\delta = I_m L_m \omega_n$, I_m – magnetizing current, r.m.s., L_m – inductance of magnetization, ω_n – nominal angular frequency.

A THERMAL MODEL OF SC

Energy dissipation in SC is represented by means of a resistance R_p, which is determined at each step of calculation on the basis of voltage-current characteristics $u(i)$ of transistors and diodes and characteristics $P_d(i)$, where P_d – dynamic energy losses in a transistor model. These characteristics are available at the temperatures $T_1 = 25°C$ and $T_2 = 125°C$. Calculated currents and $u(i)$ characteristics allow for determining voltage drops across transistor and diodes for indicated temperatures T_1 and T_2, and then for the current temperature. Static energy losses P_s in semiconductor elements are determined as products of current and voltage drops. Resistance R_p is determined from total energy losses $P_p = P_d + P_s$ and inverter phase currents at each moments of time the temperature is updated by thermal calculations.

A THERMAL MODEL OF AN INDUCTION MACHINE

A thermal model of AM is built up on the basis of thermal equations describing heat generation and heat exchange between AM structure elements according to a heat-flow diagram shown in Fig.5.

Heat sources depend on copper and iron energy losses in the rotor P_{m2}, $P_{\mu 2}$ and in the stator P_{m1}, $P_{\mu 0}$, $P_{\mu 1}$. Structure elements of the motor are characterized by average temperatures T_{m1}, T_{m2}, T_{f1}, T_{f2}, T_k, by mass m_{m1}, m_{m2}, m_{f1}, m_{f2}, m_k, and by specific heat capacity c_{m1}, c_{m2}, c_{f1}, c_{f2}, c_k. Indexes m_1 and m_2 refer to the stator and the rotor windings, f_1 and f_2 – to the stator and the rotor iron and index k – to the frame.

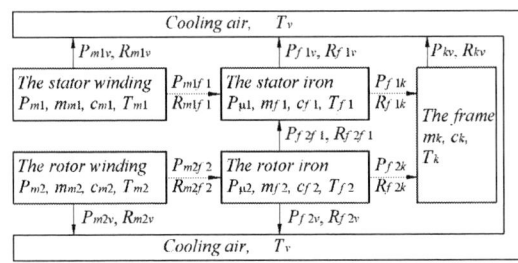

Fig.5. A thermal model scheme

Heat exchange between AM structure elements and cooling air is conditioned by thermal resistances R_{m1v},

R_{m2}, R_{f1v}, R_{f2v}, R_{kv}. Inter-element heat exchange is defined by thermal resistances R_{m1f1}, R_{m2f2}, R_{f1k}, R_{f2k}, R_{f2f1}. Heat flows P_{m1v}, P_{m2v}, P_{f1v}, P_{f2v}, P_{kv}, P_{m1f1}, P_{m2f2}, P_{f1k}, P_{f2k} depend on thermal resistances and temperature differences.

Computation of the temperature is carried out at each instant of time by numerical integration of thermal equations. For example, the temperature of the stator windings is determined as follows:

$$T_{m1} = \frac{1}{m_{m1}c_{m1}} \int_k^{t_{k+1}} \left(P_{m1} - P_{m1v} - P_{m1f1}\right) dt.$$

Heat flows at the same moment of time:

$$P_{m1v} = (T_{m1} - T_v)/R_{m1v}, \dots P_{kv} = (T_k - T_v)/R_{kv}.$$

Under condition that a thermal model is scaled, a computational solution of thermal and electromechanical equations is implemented simultaneously.

Resistances of AM sub-circuits are updated at each step of computation according to the formula:

$$R_T = R_{15}\left[1 + \alpha(T - T_{15})\right],$$

where T and T_{15} – current and reference temperatures, R_{15} – resistance determined at a reference temperature $T_{15} = 15°C$; $\alpha = 0,0038$ and $\alpha = 0,0024$ 1/°C – temperature coefficients of copper and iron.

CHARACTERISTICS OF SC-FED AM UNDER DIFFERENT CONDITIONS OF ENERGIZING

The proposed model has been verified by comparison of simulation results and results of the experimental study of a six phase AM rated as follows: shaft power $P_2 = 2000$ kW, line-to-line voltage $U = 960$ V, a number of three phase windings $M = 2$, power factor $\cos\varphi = 0,867$, frequency $f = 50$ Hz, efficiency $\eta = 95,7\%$, mechanical losses $P_{mf} = 15,4$ kW, the rotor inertia $J = 50$ kgm^2, slip $s = 0,55\%$.

Parameters of the AM model: $L_{sM} = 0,077$ p.u., $L_{s1} = 0,062$ p.u., $L_{00} = 0,049$ p.u., $L_m = 2,55$ p.u., $L_{s2} = 0,084$ p.u., $R_1 = 0,0064$ p.u., $R_2 = 0,0046$ p.u., $R_{\mu 0} = 110$ p.u., $R_{\mu 1} = 200$ p.u., $R_{\mu 2} = 130$ p.u.; ladder network parameters: $J_s = 5$, $K_s = 3,9$ and $J_r = 2$ and $K_r = 1,8$.

Parameters of the thermal model: mass of AM elements $m_{m1} = 597$ kg, $m_{m2} = 342$ kg, $m_{f1} = 2461$ kg, $m_{f2} = 1278$ kg, $m_k = 8332$ kg, specific heat capacity of copper and iron 385 and 481 J/kg°C, thermal resistances $R_{m1v} = 0,0031°C/W$, $R_{m2v} = 0,0049°C/W$, $R_{f1v} = 0,0055°C/W$, $R_{f2v} = 0,004°C/W$, $R_{m1f1} = 0,008°C/W$, $R_{m2f2} = 0,004°C/W$, $R_{f2f1} = 0,1°C/W$, $R_{f1k} = 0,03°C/W$, $R_{f2k} = 0,05°C/W$, $R_{kv} = 0,01°C/W$, input temperature of cooling air $T_v = 25°C$.

Experimental and calculated characteristics of the motor are displayed in Table 1, Fig.6 and Fig.7. Discordance of experiment and computation data, in the worst case, is less than 8%, and may be considered acceptable for adopting the model as adequate.

841

TABLE I. STEADY STATE CHARACTERISTICS OF THE MOTOR
ENERGIZED FROM A SINUSOIDAL VOLTAGE SOURCE

Characteristics	Experiment	Simulation
Starting torque, p.u.	1,25	1,15
Starting current ratio, p.u.	7,88	8,05
Critical torque, p.u.	3,42	3,3
Nominal current, A	724	738
Power factor	0,867	0,851
Magnetizing current, A	274	257
Slip, %	0,55	0,549
The stator copper losses , kW	22,5	23,86
The rotor copper losses, kW	11,1	11,04
Iron losses, kW	39,4	36,5
Mechanical losses, kW	15,4	15,4
Overall losses, kW	88,5	86,8
Efficiency , %	95,76	95,66
Temperature of the stator windings	103°C	104°C

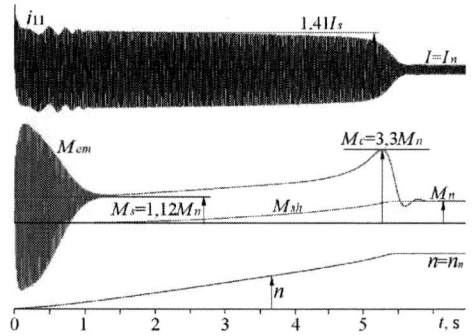

Fig.6. Start of the motor under mechanical load

i11(t) – phase current, Mem(t) – electromagnetic torque, n(t) – the rotor angular velocity, rpm, Is – starting current, r.m.s., Ms and Mc starting and critical torque, In, Mn, nn – nominal values

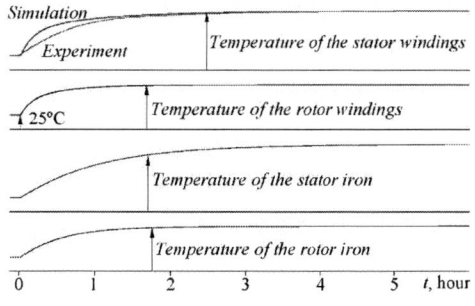

Fig.7. Variation of the motor temperature with time

In order to evaluate the influence of inverter parameters upon AM characteristics several runs of simulation have been carried out. In the first run, PWM triangle signals were set in phase and varied in frequency as follows: f_c=1,18; 2; 3 and 4 kHz. Results of computation are shown in Fig.8, Fig.9 and Table 2.

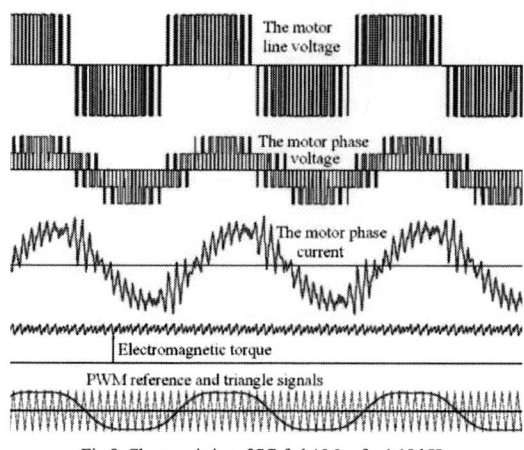

Fig.8. Characteristics of SC-fed AM at fc=1,18 kHz

Fig.9 Graph of the motor phase current at fc=4 kHz

As it is seen from Table 1 and Table 2, going from energizing the motor by a sinusoidal voltage source to energizing the motor by a mono-stage VSI at f_c=1180 Hz results in the distortion of phase currents, an increase in energy losses and a rise in the motor temperature. The increase of PWM carrier signal in frequency from f_c=1,18 kHz up to f_c=4 kHz leads to the decrease of energy losses in the stator windings from 37,1 up to 26,1 kW, reduction of the total harmonic distortion of phase currents from THDI=35 % up to THDI=13 % and lowering the stator windings temperature by 27°C.

In the second run of simulation, inverters in parallel were set in a multistage mode of operation by shifting the PWM triangle signals in time by $T_c/2$, where T_c –period of a triangular carrier wave.

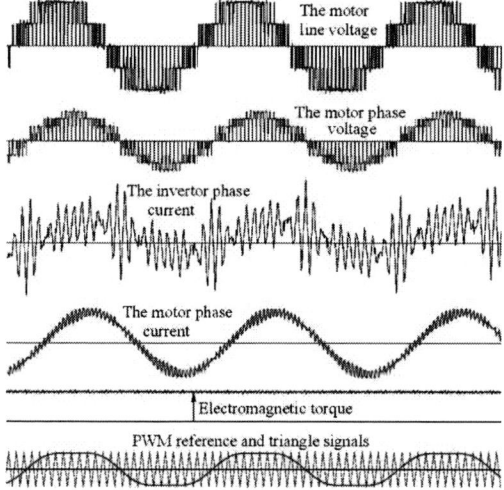

Fig.10 Characteristics of SC-fed AM at a multistage mode of VSI operating at fc=1,18 kHz.

TABLE II. IMPACT OF VSI PARAMETERS ON AM STEADY-STATE CHARACTERISTICS

Parameters	Mono-stage inverter,		Multistage inverter
PWM frequency, kHz	1,18	4	1,18
Fundamental harmonic (50 Hz) of the motor line-to-line voltage, V	959,9	959,8	960,9
The motor phase current, A	781,8	744,3	746.2
Power factor, p.u.	0,85	0,844	0,846
Slip, %	0,544	0,539	0,537
The stator copper losses, kW	37,07	26,09	26,78
The rotor copper losses, kW	11,78	11,11	11,06
Iron losses, kW	36,88	37,28	36,77
Mechanical losses, kW	15,38	15,38	15,38
Overall losses, kW	101,1	89,87	90,01
Temperature of the stator winding, °C	137,5	110,8	112,1
Temperature of the rotor winding, °C	85,3	82,8	82,4
Temperature of the stator iron, °C	131,7	123,1	122,7
Temperature of the rotor iron, °C	90,5	89,5	88,9

Comparison of the data displayed in Table 2 shows that powering the motor by a multistage VSI results in diminishing the motor current distortions and lowering the stator windings temperature by 22°C. The same result can be achieved by the increase in PWM frequency of mono-stage inverters from 1,18 up to 4 kHz.

Energy losses in SC-fed AM P_{li} consist of a high frequency component P_{lh} and a fundamental frequency component P_{n}. Assuming that P_{n} is equal under different conditions of AM energizing, HF losses can be determined as follows: $P_{ef}=P_{li}-P_{ls}$, where P_{ls}-energy losses in the motor energized by a sinusoidal voltage source.

Dependence of HF losses P_{ef} upon a carrier wave's frequency f_c is displayed in Fig.11.

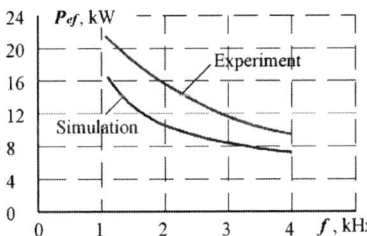

Fig.11. Variation of HF losses Pef against a PWM carrier wave's frequency fc

As reported in [16], HF losses can be evaluated indirectly from the results of direct measurements of a driving point impedance of the stator windings, on one hand, and the results of direct measurements of the inverter voltage spectrum, on the other hand. Fig.11 shows that both simulation and indirect measurement approaches give the same trend of variation of HF losses against a PWM carrier wave's frequency f_c, which correlates with data reported in [20].

CONCLUSION

1. In the proposed model, properties of the low voltage multiphase AM of high power are reproduced by means of the stator and the rotor sub-circuits with lumped parameters. The redistribution of currents under frequency is represented with the ladder sub-circuits. The model takes into account HF losses as well as the influence of temperature-dependent parameters upon electromechanical and thermal processes. The revised model of AM has been incorporated in the model of the frequency-variable propulsion drive with several VSI.

2. Evaluation of the temperature distribution in AM energized under different conditions has shown that powering the motor by a multistage VSI results in a substantial decrease of the stator windings temperature, practically up to the value related to the case of energizing the motor by a sinusoidal voltage source.

REFERENCES

[1] Hadiouche D., Razik H., Rezzoug A. On the Modeling and Design of Dual-Stator Windings to Minimize Circulating Harmonic Currents for VSI Fed AC Machines // IEEE Transactions on Industry Applications, Vol. 40, NO. 2, March/April, 2004.

[2] Pronin M., Vorontsov A. Dependence of current pulsations of multi-phase electrical machine on reduction of winding pitch and scheme of semiconductor converter // EPE-PEMC 2006, Portoroz, Slovenia.

[3] Aoulkadi M., Binder A., Joksimovic G. Additional losses in high-speed induction machine – removed rotor test // EPE 2005, Dresden, Germany.

[4] Cester C., Kedous-Lebouc A., Cornut B. Iron loss under practical working conditions of a PWM powered induction motor // IEEE Transactions on Magnetics. Sep. 1997, Vol. 33, Issue 5, Part 2, pp. 3766-3768.

[5] Ionel D. M., Popescu M., McGilp M. I., Miller T. J. E., Dellinger S. J., Heideman R. J. Computation of Core Losses in Electrical Machines Using Improved Models for Laminated Steel // IEEE Transactions on Industry Applications, Nov.-dec. 2007, Vol. 43, Issue 6, pp. 1554-1564.

[6] Moses A. J., Anayi F. Effect of PWM Voltage Excitation in Iron Loss of Inverter Fed Motors // IEEE Transactions on Magnetics. 2004, Vol. 40 (2), NO. 2, pp. 762-765.

[7] Popescu M., Dorrell D. G., Ionel D. M. A Study of the Engineering Calculations for Iron Losses in 3-phase AC Motor Models // IECON 2007. Nov. 5-8, 2007, Taipei, Taiwan.

[8] Ruderman A., Welch R. Electrical Machine PWM Loss Evaluation Basics // EEMODS 2005, 5–8 September 2005. Heidelberg, Germany.

[9] Roytgarts M., Pronin M. Reducing of induction motors vibrations at drives with semiconductor converters designing // EPE-PEMC 2004, Riga, Lv.

[10] Yamazaki, K., Seto Y. Iron Loss Analysis of Interior Permanent-Magnet Synchronous Motors – Variation of Main Loss Factors Due to Driving Conditions // IEEE Trans. Ind. Applicat., vol. 42, N0 4, 2006, pp. 1045-1052.

[11] Dorell D. G., Staton D.A., Kahout J., Hawking D, McGilp M. I. Linked Electromagnetic and Thermal Modeling of a Permanent Magnet Motor // Proc. of 3rd Int. Conf. on Power Electronics, Machines and Drives, pp 536-540, The Contarf Castle, Dublin, Ireland, Mar. 2006

[12] Munehiro K., Yosihiro K., Takashi K., Nabuyuki, M. Temperature distribution analysis of Permanent Magnet in Interior Permanent Magnet Synchronous Motor Considering PWM Carrier Harmonics // Proc. Of Int. Conf on Electrical Machines and Systems, pp. 2023-2027, Seoul, Oct. 2007

[13] IEC 61972 Standard, "Method for Determining Losses and Efficiency of Three-Phase Cage Induction Motors", 2002

[14] Yamazaki K., Haruishi Y. Stray Load Loss Analysis of Induction Motor-Comparison of Measurement Due to IEEE Standard 112

and Direct Calculation by Finite-Element Method // IEEE Transactions on Industry Applications, Vol. 40, No. 2, March/April, 2004.

[15] Drobkin, B. Z., Vorontsov A. G., Pronin M. V., Krutyakov Y. A., Pavlov P. A. Debugging of microprocessor-based control systems of electric drives using mathematical models // EPE 2003, Toulouse, Fr. –pp. 1-11.

[16] Pronin M., Shonin O., Vorontsov A., Tereschenkov V. Computer model-based evaluation of energy losses components in the systems with asynchronous machines and transistor converters // The 33rd Annual Conference of IEEE Industrial Electronics Society IEEE, IECON 2007, Nov. 5-8, Taiwan.

[17] Holmes D. G., Lipo T. A. Pulse Width Modulation for Power Converter // IEEE Press, USA, 2003.

[18] Pronin M. Modeling and analysis of system with multiphase induction generator and multi-stage active rectifier // Electrotechnika, 2006, №5. –p. 55-61. Russia.

[19] Boglietti A., Cavagnino A., Ferraris L., Lazzari M. Induction Motor Equivalent Circuit Including the Stray Load Losses in the Machine Power Balance //IECON 2007. Nov. 5-8, 2007, Taiwan.

[20] Marcic T., Stumberger G., Hadziselimov M., Zagradisnik I. Analysis of Induction Motor Drive Losses in the Field-Weakening Region // EPE-PEMC 2006, Portoroz, Slovenia.

Use of an AC Self-excited Switched Reluctance Generator as a Battery Charger

Abelardo Martínez*, Estanislao Oyarbide*, Javier Vicuña†, Francisco Perez*, Eduardo Laloya*, Bonifacio Martín-del-Brío*, Tomás Pollán*, Beatriz Sánchez*, and Juan Lladó*

*University of Zaragoza, Dept. of Electronics and Communications Engineering, Zaragoza, Spain. amiturbe@unizar.es
† University of La Rioja. Electrical Engr. Department, Logroño, Spain, e-mail: javier.vicuna@die.unirioja.es

Abstract—We analyzed an AC self-excited switched reluctance generator used as a battery charger, using a second- order nonlinear oscillator model. Generator capacity can be maintained under variable rotor speed by adapting the external capacitance. The AC voltage generated is rectified and adapted with a DC-DC converter to battery voltage. To keep the flux machine under control, the rectified bus voltage should be changed according to oscillating frequency, following a constant voltage/frequency ratio. The simplicity of the system makes it an adequate wind generator for battery charging in isolated locations.

Keywords— Reluctance drive, Generation of electrical energy, Battery charger, Distributed power, Sustainable system/technology.

I. INTRODUCTION

The switched reluctance machine is a highly robust device suitable for power generation in harsh environments. In this paper, the generator is used with a simple drive, where the phase current is alternating current (AC) rather than direct current (DC). Radimov et al [1] has proposed the use of switched reluctance motors as three-phase AC self-excited generators. The phase inductance, AC capacitor, and resistor load are series connected and form a self-excited RLC circuit that exhibits parametric oscillator behaviour [2] where the pumping action affects the variable reluctance [3]. Although many aspects of variable reluctance generator have been studied, not much information is available for switched reluctance machines where there is double saliency in the rotor and stator. Martinez et al [4] have described the working zones in which the machine evolves in the flux-current plane. The steady-state generator behaviour has been analysed in [5] and a simulation model was described in [6]. Although the RLC arrangement in [1] is series connected, in our case the parallel connection shown in Figure 1 is preferred because load modifications can be easily achieved. Subsequently, the resistor load is substituted by a rectifier and a DC bus. The paper is organized as follows: the basic equations are first derived and the resonant frequency is obtained, then, the load resistor is substituted by the rectifier and a DC bus and the behaviour of the new topology is discussed showing the relationships among DC bus voltage, working frequency and maximum flux. Finally, the feasibility of

this arrangement as a variable speed wind generator is outlined.

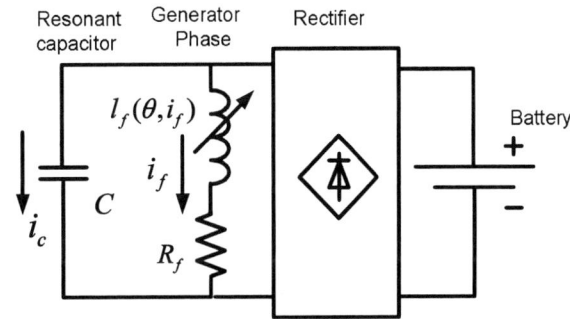

Fig. 1. Single-phase model including the external capacitor, the phase inductance and a rectifier load.

II. CIRCUIT MODEL OF A SINGLE-PHASE GENERATOR

Figure 1 depicts the circuit elements of a single-phase generator. If the rectifier and the battery are initially replaced by a resistor load RL, then the circuit equations are given by the following:

$$U_c - R_f \cdot i_f = \frac{d\Phi(\theta, i_f)}{dt} \tag{1a}$$

$$\Phi(\theta, i_f) = L_f(\theta, i_f) \cdot i_f \tag{1b}$$

$$C \frac{dU_c}{dt} = -\left(i_f + i_{load}\right) \tag{1c}$$

When the Laplace transform is applied to (1c),

$$-\left(\frac{RL}{C \cdot RL \cdot s + 1} + R_f\right) i_f = s \cdot \Phi$$

$$-\left(RL + R_f + C \cdot RL \cdot R_f \cdot s\right) i_f =$$
$$= C \cdot RL \cdot s^2 \cdot \Phi + s \cdot \Phi \tag{2}$$

Flux and current are nonlinearly related as an odd function,

$$i_f = k_1(\theta) \cdot \Phi \cdot \left(1 + k_2(\theta) \cdot \Phi^2\right)$$

$$k_1(\theta) = L_a + L_b \cdot \cos(2 \cdot \theta) \tag{3}$$

Where $k_1(\theta)$ and $k_2(\theta)$ are periodic functions of the rotor angle θ. From (2) and (3), a nonlinear second-order flux equation can be obtained and analysed by describing functions [7]. This equation describes a limit cycle:

$$-\left(RL + R_f\right) \cdot L_a \cdot k_2(\theta) \cdot \Phi^3 -$$
$$\left(RL \cdot R_f \cdot C \cdot L_a \cdot k_2(\theta)\right) \cdot s\left(\Phi^3\right) -$$
$$\left(RL + R_f\right) \cdot L_b \cdot \cos(2 \cdot \omega \cdot t) \cdot \Phi \cdot \left(1 + k_2(\theta) \cdot \Phi^2\right) -$$
$$\left(RL \cdot R_f \cdot C\right) \cdot$$
$$s\left(L_b \cdot \cos(2 \cdot \omega \cdot t) \cdot \Phi \cdot \left(1 + k_2(\theta) \cdot \Phi^2\right)\right) = \tag{4}$$
$$\left(RL \cdot C\right) \cdot s^2 \, \Phi + \left(1 + RL \cdot R_f \cdot C \cdot L_a\right) \cdot s \, \Phi +$$
$$\left(RL + R_f\right) \cdot L_a \cdot \Phi$$

III. GENERATOR BEHAVIOR LOADED WITH A RECTIFIER CHARGING A BATTERY

When the load resistance RL is substituted by a battery-charging rectifier, the resonant frequency changes from:

$$\omega_0 = \sqrt{\frac{\left(RL + R_f\right) \cdot L_a}{RL \cdot C}} \tag{5}$$

to

$$\omega_0 = \sqrt{\frac{L_a}{C}} \tag{6}$$

However, the nonlinearity imposed by the rectifier will allow the generator to work with a wide range of frequencies and DC battery voltages. In the example used, $L_a = 585.88 A/Wb$, C=33uF, $\omega_0 = 720.5 \, rad/s$.

With Nr = 4 poles, the resonant speed is 3440 rpm. The limit cycle is determined when the following condition is met:

$$\frac{1}{2} \cdot C \cdot V_{bus}^2 = \int_0^A i \cdot d\lambda + Joule \, losses_{O-A} \tag{7}$$

where λ is the linked flux and points O and A belong to the solid curve shown in Fig. 2, for a typical $\lambda - i$ trajectory. Point A corresponds to rotor and stator in the aligned position. The generator has the following properties:

A. The size of the limit cycle increases with rotor speed.

The state variables Φ_{max}, i_{max} (core flux and phase current) increase with rotor speed as shown in Figure 3. The capacitor energy and electric work are also shown; the small difference is the Joule losses. Figure 4 shows several limit cycles corresponding to different rotor speeds.

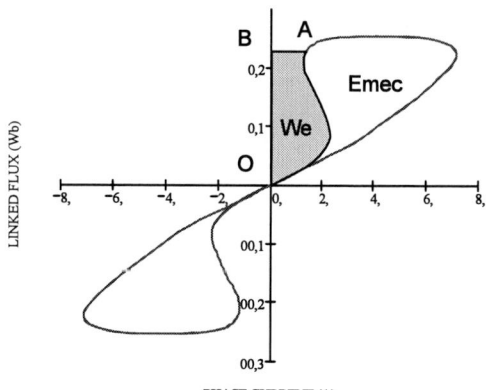

Fig. 2 Area shown as We is the electric work to be supplied by the capacitor. The enclosed area Emec represents half the mechanical energy transformed per cycle (n=3000rpm, Vbus=160V, C=33uF).

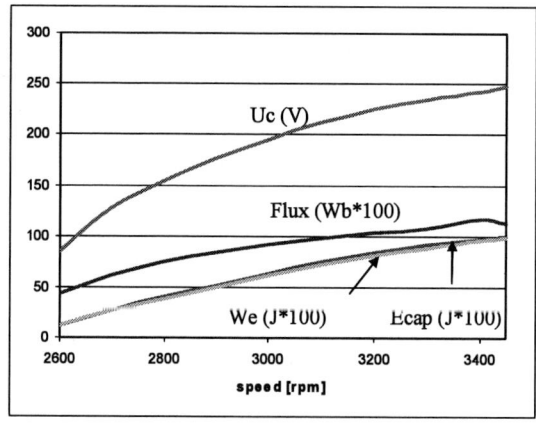

Fig. 3 State variables of the limit cycle versus rotor speed. The electric energy requested and the energy provided by the capacitor are shown below. The small difference is due to Joule losses..

B. The flux amplitude and voltage/speed ratio are maintained for changes in resonant frequency.

This implies that the rotor speed can be changed synchronously with the capacitance to generate at variable speed. The new available voltage has a constant ratio:

$$\left(\frac{C_2}{C_1}\right)^2 = \frac{n_1}{n_2} = \frac{V_{c1\_max}}{V_{c2\_max}} \qquad (7)$$

This result is shown in Figure 5 for a two rotor speeds. The resonant frequency is halved by increasing the capacitance four times. Thus, the speed is also halved. Consequently, the maximum capacitor voltage is halved, with the voltage/speed ratio remaining constant. This property allows the saturation level to be maintained in the machine while it is generating at variable speed.

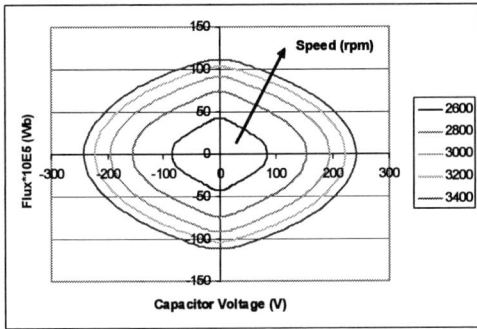

Fig. 4 Increase in the size of the limit cycle as a function of rotor speed.

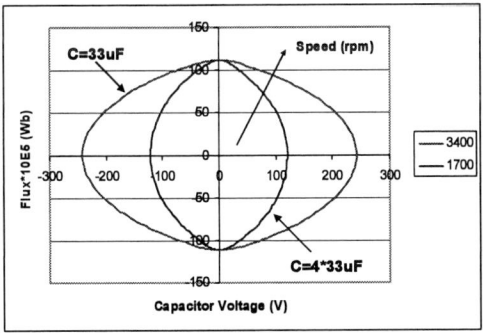

Fig. 5 Limit cycles obtained to show the second property

C. The energy per cycle remains constant with changes in resonant frequency.

Another consequence of the above property is that the energy per cycle remains constant, and thus the power generated at half rotor speed is also halved.

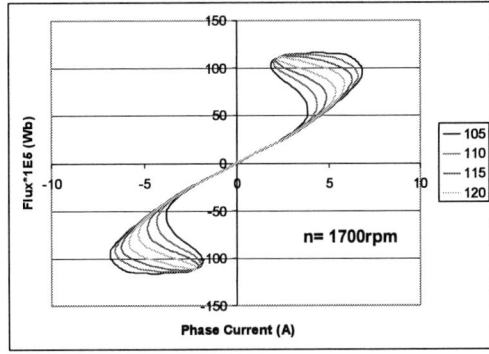

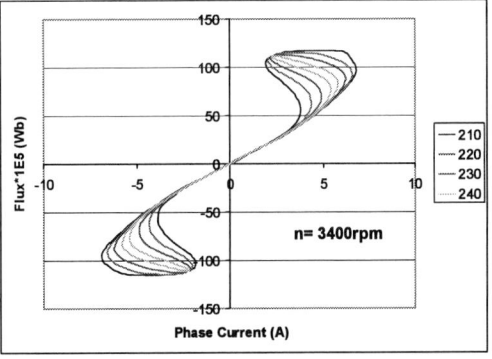

Fig. 6 Limit cycles obtained to show the second property. The speed on the upper picture is half the speed shown on the lower one

This property is shown in Figure 6, which shows the flux/current trajectories for 3400 rpm and 1700 rpm using the rectifier bus voltage as a parameter. In both cases, the trajectories scan the same areas, i.e., the same energy per cycle.

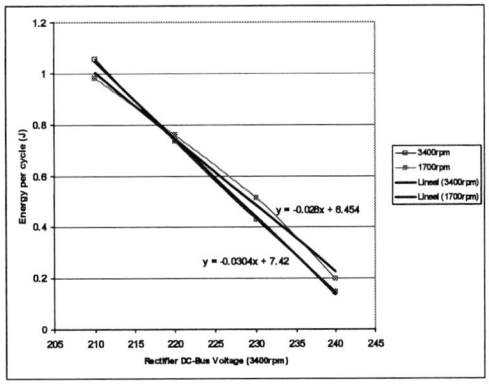

Fig. 7 Energy per cycle conservation. The energy per cycle decreases linearly with the rectifier bus voltage

Figure 7 shows the energy per cycle with results superimposed for 3400 rpm and 1700 rpm. Although the x-axis represents the rectifier DC-bus voltage for 3400 rpm, there is a 1:2 correspondence with the DC-bus voltage for 1700 rpm. Thus, 220V at 3400 rpm corresponds to 110V at 1700 rpm.

The energy per cycle decreases linearly with the rectifier bus voltage in both cases. Further, the linear approximation is quite similar.

D. At constant rotor speed, there is an optional DC/bus voltage with a maximum energy per cycle.

If the rotor speed is held constant, a DC-bus reduction of rectifier output voltage leads first, to some extent, higher flux and energy per cycle, as shown in Figures 8 and 9. When DC-bus is about 70% of the limit cycle, the flux and energy per cycle reach a maximum.

However, further increases will reduce both. As the speed is held constant, the generated power follows the same trend.

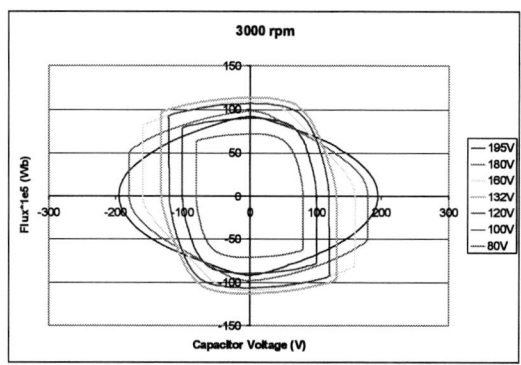

Fig. 8 Energy per cycle conservation. The energy per cycle decreases linearly with the rectifier bus voltage.

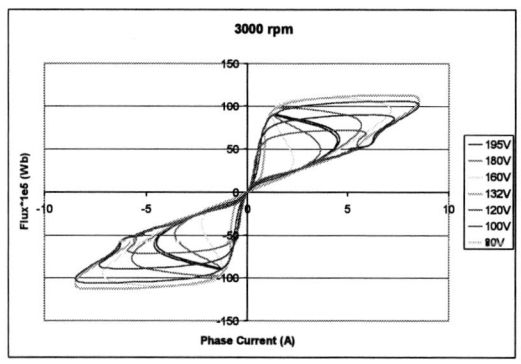

Fig. 9 Flux/current trajectories for different DC-bus voltages. The energy per cycle increases first to a maximum and then decreases.

According to the speed used in the experiment, i.e., the limit cycle selected, the achievable maximum is limited by high phase currents. In this case, the delivered power is controlled by the DC-bus voltage, for example, using a DC-DC buck converter.

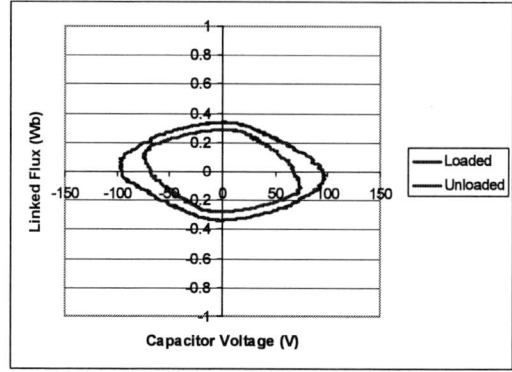

Fig. 10 State variables trajectories at no load and when loaded with a set of five 12V batteries series connected. n=710rpm

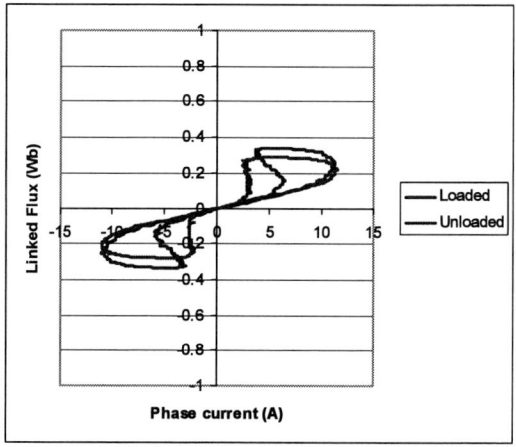

Fig. 11 Flux-phase current trajectories showing the transformed energy at no load and when loaded with a set of five 12V batteries, series connected.

Figure 9 suggests the same conclusion, but using the flux/phase-current information. The energy per cycle first increases to its peak and then decreases.

IV. EXPERIMENTAL RESULTS

A single phase of an 8/6 switched reluctance machine has been used to obtain preliminary results, based on a capacitance of 495 uF. Figure 10 shows the limit cycle at no load and when loaded with a battery bank of five 12V series connected batteries through a bridge rectifier. Figure 11 shows the flux-current enclosed areas in open circuit and when it is loaded with the battery bank. The rotor speed is 710 rpm.

It is important to remark that phase resistance plays an important role in the Joule losses and, consequently, in the efficiency. Its value must be low because resonance implies large phase currents compared to battery current. Battery voltage clamps the natural voltage amplitudes of oscillations to a safe value. To same extend, it runs with variable rotor speed without other control elements than a diode bridge.

V. Conclusions

A switched reluctance machine can be used as an AC generator to charge a battery bank via a bridge rectifier. The nonlinear second-order equation has been used to analyse an oscillator that describes a flux/phase-current limit cycle and its behaviour based on four properties. Simulated and experimental results have been used to illustrate these properties, demonstrating the ability of the machine to generate electricity with variable rotor speed. A DC/DC converter would help control the energy transformed and track the optimal working point. In order to maintain the flux machine under control, the rectified bus voltage should be changed according to oscillating frequency and following a constant voltage/frequency ratio. The simplicity of the system makes it suitable as a wind generator for battery charging in isolated locations.

Acknowledgment

This research was supported by the Spanish Ministry of Science and Technology (Ref. DPI2006-10148).

References

[1] G. Eason, B. Noble, and I. N. Sneddon, "On certain integrals of Lipschitz-Hankel type involving products of Bessel functions," *Phil. Trans. Roy. Soc. London*, vol. A247, pp. 529–551, April 1955.

[2] J. Clerk Maxwell, *A Treatise on Electricity and Magnetism*, 3rd ed., vol. 2. Oxford: Clarendon, 1892, pp.68–73.

[3] I. S. Jacobs and C. P. Bean, "Fine particles, thin films and exchange anisotropy," in *Magnetism*, vol. III, G. T. Rado and H. Suhl, Eds. New York: Academic, 1963, pp. 271–350.

[4] K. Elissa, "Title of paper if known," unpublished.

[5] R. Nicole, "Title of paper with only first word capitalized", *J. Name Stand. Abbrev.*, in press.

[6] Y. Yorozu, M. Hirano, K. Oka, and Y. Tagawa, "Electron spectroscopy studies on magneto-optical media and plastic substrate interface," *IEEE Transl. J. Magn. Japan*, vol. 2, pp. 740–741, August 1987 [Digests 9th Annual Conf. Magnetics Japan, p. 301, 1982].

[7] M. Young, *The Technical Writer's Handbook*. Mill Valley, CA: University Science, 1989.

Direct Thrust Controlled Linear Induction Motor Including End Effect

Berrin Susluoglu*, Vedat M. Karsli[†]

University of Gaziantep, Department of Elect- Electronics Engineering, 27310, Gaziantep, Turkey,
*e-mail: _susluoglu@gantep.edu.tr_, [†] e-mail: _vkarsli@gantep.edu.tr_

Abstract— **As a special feature, the linear induction motor (LIM) has an end effect phenomena causing weakening in airgap flux also in thrust. In this paper, direct thrust control of linear induction motors is improved by considering the end effect in flux and thrust estimator part based on the mathematical model of LIM. To show the effectiveness of the improved system, simulation results are presented using commercially available software package Matlab/Simulink.**

Keywords—**Linear induction motor, end effect, direct thrust control.**

I. INTRODUCTION

The Linear Induction Motor (LIM) has many excellent performance features such as high-starting thrust force, alleviation of gear between motor and the motion devices, reduction of mechanical losses, high speed operation, silence, and so on. Due to these advantages, LIM has been widely used in the field of high-speed transportation systems, elevators, robotic systems and so on [1].

In a LIM, the primary winding corresponds to the stator winding of a rotary induction motor (RIM), while the secondary corresponds to the rotor. The main difference between RIM and LIM is that the primary of the LIM has a finite length. As the primary of LIM moves, a new flux is continuously developed at the entry of the primary side, while existing flux disappears at the exit side. These non-uniform distribution of airgap flux causes the flux and thrust attenuations which become severer as the speed increases. Such a phenomenon is called 'end effect' of LIM [2, 3].

The driving principles of the linear induction motor are similar to the traditional rotary induction motor but its control characteristics are more complicated than RIM. High performance vector control of LIM mostly carried out in secondary flux oriented scheme has been presented in many works [4-7]. However, field oriented control is complicated and requires machine parameters. In contrast to field-oriented control, traditional DTC requires only the accurate knowledge of the stator resistance, the stator flux and torque estimations. The control algorithm of DTC makes it suitable for practical implementation on embedded drive systems [8-10].

Compared with RIM, direct torque control (DTC) for LIM has been less researched. In the papers in literature studying Direct Thrust Control (DTC) for LIM [11, 12], end effect was included in the form of thrust correction coefficient. The thrust correction coefficient was obtained by using the finite element analysis in starting conditions considering only the static end effect. However, in this paper, flux and thrust estimations are improved in a simple manner to reduce the error caused by the end effect by using the equivalent circuit model of LIM rather than using Maxwell equations.

The remainder of paper is organized as follows. At first, the mathematical model of linear induction motor in stationary reference frame is given. Next, the principles of direct thrust control including end effect are introduced. Finally, simulation results and some conclusions are presented.

II. MATHEMATICAL MODEL OF LINEAR INDUCTION MOTOR IN STATIONARY REFERENCE FRAME

The mathematical model of linear induction motor in synchronous reference frame including the end effect was given in literature [2, 3]. To study the dynamic behavior of the linear induction motor both under transient and steady state conditions, the existed model for LIM is well adapted into the stationary reference frame. Fig. 1 shows the d-q equivalent circuit of the LIM including the end effect in stationary frame, in which only the magnetization branch in d-axis equivalent circuit is different from RIM model.

From the dq equivalent circuit of LIM, the primary and secondary voltage and flux linkage equations in stationary reference frame are given by (1), (2) and (3), respectively.

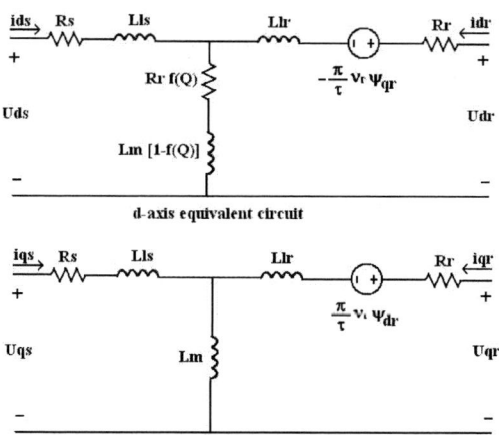

Fig. 1. dq equivalent circuit of the LIM including the end effect in stationary frame

d-q axis primary voltage equations:

$$u_{ds} = R_s i_{ds} + R_r f(Q)(i_{ds} + i_{dr}) + \frac{d}{dt}\psi_{ds}$$

$$u_{qs} = R_s i_{qs} + \frac{d}{dt}\psi_{qs} \qquad (1)$$

d-q axis secondary voltage equations:

$$u_{dr} = R_r i_{dr} + \frac{\pi}{\tau} v_r \psi_{qr} + R_r f(Q)(i_{ds} + i_{dr}) + \frac{d}{dt}\psi_{dr}$$

$$u_{qr} = R_r i_{qr} - \frac{\pi}{\tau} v_r \psi_{dr} + \frac{d}{dt}\psi_{qr} \qquad (2)$$

where the subscript s corresponds to primary, r corresponds to secondary, d corresponds to direct axis, q corresponds to quadrature axis, R_s, R_r the primary and the secondary resistances, τ is pole pitch and v_r is the electrical speed.

Flux linkage equations:

$$\psi_{ds} = L_{ls} i_{ds} + L_m (1 - f(Q)(i_{ds} + i_{dr}))$$

$$\psi_{qs} = L_{ls} i_{qs} + L_m (i_{qs} + i_{qr})$$

$$\psi_{dr} = L_{lr} i_{dr} + L_m (1 - f(Q)(i_{ds} + i_{dr}))$$

$$\psi_{qr} = L_{lr} i_{qr} + L_m (i_{qs} + i_{qr}) \qquad (3)$$

where L_{lr}, L_{ls}, L_m refer to the leakage inductance of secondary, leakage inductance of primary and magnetizing inductance, respectively.

The thrust force is calculated by (4).

$$F_e = \frac{3}{2}\frac{\pi}{\tau}\frac{P}{2}\left(\psi_{ds} i_{qs} - \psi_{qs} i_{ds}\right) \qquad (4)$$

A velocity inverse function given by (5) is used to express the end effect on magnetization factor of the equivalent circuit.

$$f(Q) = \frac{1 - e^{-Q}}{Q} \qquad (5)$$

where $Q = \dfrac{PDR_r}{2v_r L_r}$, D is motor primary length.

The LIM motion dynamics is expressed by (6).

$$F_e - F_L = J\frac{d}{dt}\frac{2}{P}v_r + B\frac{2}{P}v_r \qquad (6)$$

where P is the number of poles, F_e is the electromagnetic thrust, F_L is the external force, J is the mechanical inertia, B is the natural damping.

III. DIRECT THRUST CONTROL FOR LIM WITH END EFFECT

The DTC for LIM can be analyzed in the same way as that for RIM. One problem in the case of LIM is that a new speed dependent resistance and inductance parameters included in the magnetization branch. This makes it necessary to modify the flux and thrust estimation.

The basic idea of DTC is to control stator flux linkage and thrust directly by selecting an optimum inverter voltage vector [8-10].There are eight voltage vectors coming from switching action of a three-phase inverter, six of them are active voltage vectors, U1(100), U2(110), U3(010), U4(011), U5(001), U6(101) and two of them are

zero voltage vectors, U(000), U(111). The voltage vector plane is divided into 6 regions by these active voltage vectors as shown in Fig. 2.

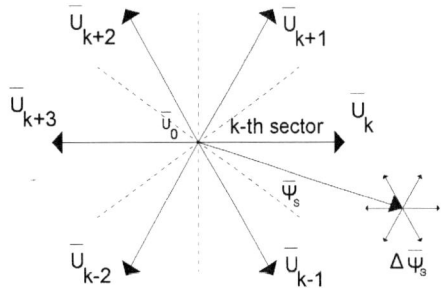

Fig. 2. Voltage vectors and their effects on flux and thrust

TABLE I.
SWITCHING TABLE FOR DTC

Sector		1	2	3	4	5	6
Flux	Thrust						
$e_\psi = 1$	$e_{Fe} = 1$	U2	U3	U4	U5	U6	U1
	$e_{Fe} = 0$	U7	U0	U7	U0	U7	U0
	$e_{Fe} = -1$	U6	U1	U2	U3	U4	U5
$e_\psi = -1$	$e_{Fe} = 1$	U3	U4	U5	U6	U1	U2
	$e_{Fe} = 0$	U0	U7	U0	U7	U0	U7
	$e_{Fe} = -1$	U5	U6	U1	U2	U3	U4

$e_\psi = 1$ to increase flux, $e_\psi = -1$ to decrease flux,

$e_{Fe} = 1$ to increase thrust, $e_{Fe} = 0$ no change of thrust,

$e_{Fe} = -1$ to decrease thrust.

The selection of the voltage vector for the voltage source inverter is made to restrict the flux and torque error within their respective flux and torque hysteresis bands to obtain the fastest torque response at every instant.

The effect of selection of each voltage vector on thrust and flux is given in Table I. The primary flux vector lying on the sector k and rotating anticlockwise can be increased by v_k, v_{k+1}, v_{k-1}; whereas it can be decreased by v_{k+2}, v_{k+3}, v_{k-2}. On the other hand, the electromagnetic thrust can be increased by selecting v_{k+1}, v_{k+2}; whereas it can be decreased by v_{k-1}, v_{k-2}. The states of v_k and v_{k+3} are not considered because of their effects can change depending on the stator flux position at the same sector [10].

For the good performance of DTC, the accurate prediction of the thrust and stator flux is important. However the end effects influences the thrust and flux characteristics of LIM. The application of the traditional stator flux estimation principle to LIM that considers voltage drop only on the stator resistance causes an error on flux and thrust estimations. Due to end effect property of LIM, it is more accurate to use equation (7) for primary flux estimation including the speed dependent coefficient.

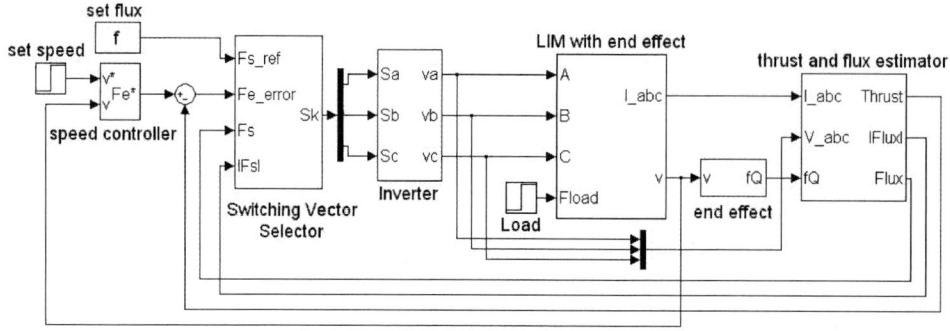

Fig. 3. Block diagram of DTC for LIM including end effect.

$$\psi_{ds} = \int \left(u_{ds} - R_s i_{ds} - R_r f(Q)(i_{ds} + i_{dr})\right) dt \qquad (7)$$

It is reasonable to ignore d-axis rotor current due to its measurement difficulty and less effect on the flux estimation. Then, flux estimate is obtained by (8) considering partial end effect.

$$\psi_{ds} = \int \left(u_{ds} - R_s i_{ds} - R_r f(Q) i_{ds}\right) dt \qquad (8)$$

The thrust value considering end effect is estimated by (9).

$$F_e = \frac{3}{2} \frac{\pi}{\tau} \frac{P}{2} L_m \left(-f(Q) i_{ds} i_{qs} + (1 - f(Q)) i_{dr} i_{qs} - i_{qr} i_{ds}\right) (9)$$

The direct thrust control scheme for LIM drive in Matlab/Simulink is shown in Fig. 3. The overall drive model is implemented by five main blocks: LIM, inverter, thrust and stator flux estimator, speed controller part to obtain thrust reference value and finally switching vector selector to obtain necessary switching state for inverter.

The measured input values for the DTC system are the motor currents, voltages, and speed. The input reference values are the linear speed and the magnitude of the primary flux. The speed control loop is used to produce the thrust reference command.

The end effect factor f(Q), voltage and current signals are inputs through the flux and thrust estimator box which produces the estimated values of primary flux and thrust. In switching vector selector box, three-level thrust and two-level flux hysteresis comparators compare these estimated values with their reference values. Depending on the outputs from these comparators, the switching table logic directly determines the optimum inverter voltage vector and thereby inverter switching directly controls the motor flux and thrust.

IV. SIMULATION RESULTS

The simulation study consists of two parts. In the first part, the validity of the proposed thrust and flux estimators is examined and in the second part, the closed loop performance of DTC employing the improved thrust and flux estimator is shown. Throughout the simulation study, linear induction motor parameters used to test the DTC strategy for LIM are given in Table II. All simulations are done in Matlab/Simulink using SimPowerSystem Toolbox.

In the first part of the simulation study, a step change of thrust command is applied at 0.4 second from zero to 30 Nm and the stator flux reference with amplitude of 0.96 Wb are applied to the LIM drive.

Performance of the estimators without considering end effect, with fully considering end effect and with partially considering end effect is observed in Fig. 4, Fig. 5, Fig. 6, respectively.

TABLE II.
LIM PARAMETERS

Number of Poles-P	2
Primary Resistance-Rs	2.82 Ω
Secondary Resistance-Rr	48.84 Ω
Primary Inductance-Ls	0.0452 H
Secondary Inductance-Lr	0.0301 H
Magnetizing Inductance-Lm	0.0262 H
Pole Pitch-□	0.06 m
Primary Length-D	0.21 m

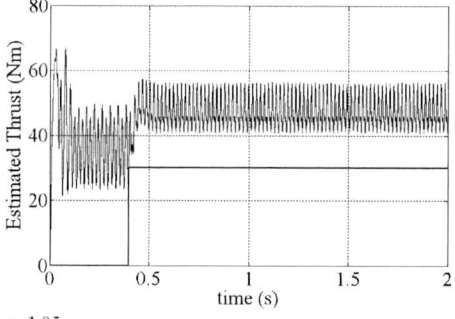

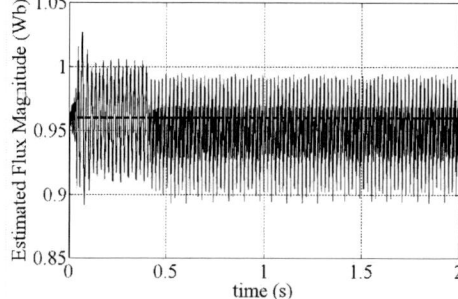

Fig. 4. Estimated thrust and flux waveforms without considering end effect

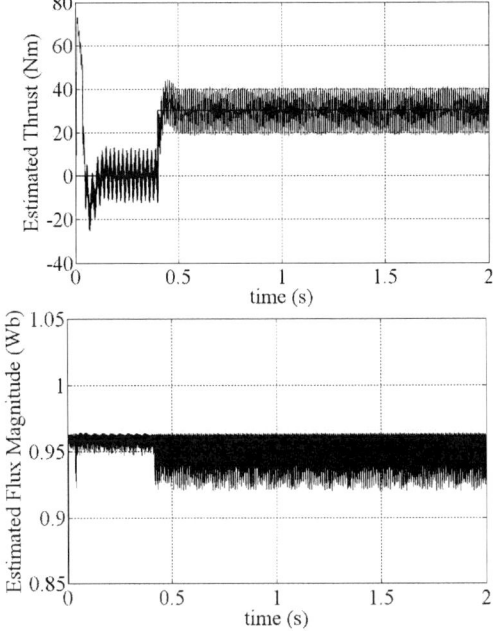

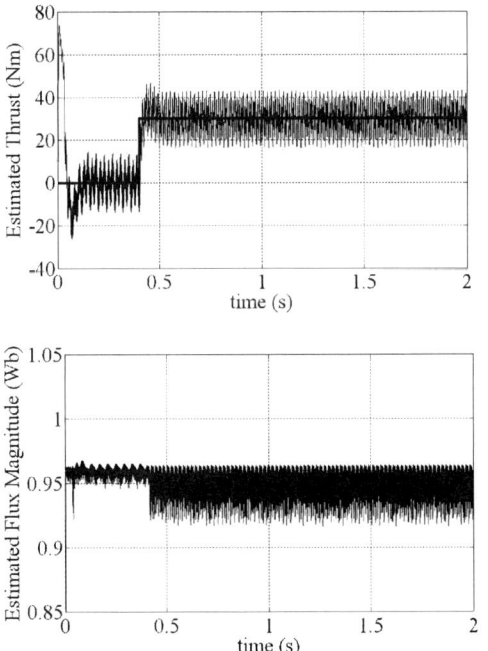

Fig. 5. Improved estimated thrust and flux waveforms
with fully considering end effect

Fig. 6. Estimated thrust and flux waveforms with
partially considering end effect

From Fig. 5 and Fig. 6, thrust and flux estimations including the end effect show satisfactory results. However, without end effect, thrust estimation does not reach the expected value as shown in Fig. 4 which will become worse with the increase of speed.

In the second part of the simulation study, a closed loop performance of DTC strategy with given improvement fully considering end effect is examined by putting a step change of thrust command at 0.4 second from zero to 30 Nm, when the triangular wave speed reference with amplitude of 3.5 m/s and the stator flux reference with amplitude of 0.96 Wb are applied to the LIM drive.

Fig. 7, Fig. 8, Fig. 9 and Fig. 10 show the speed, thrust, flux linkage trajectory and three-phase primary currents, respectively.

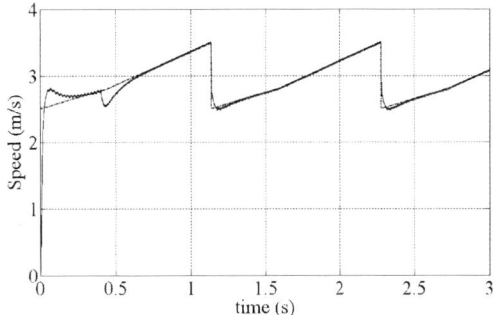

Fig. 7. Speed reference tracking of the DTC in LIM

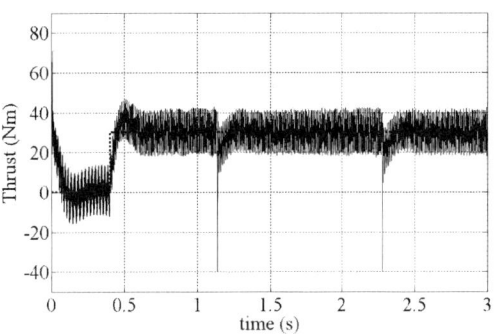

Fig. 8. Thrust response of the DTC in LIM

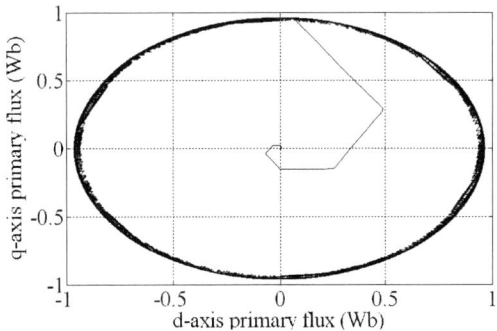

Fig. 9. Primary flux trajectory of the DTC in LIM

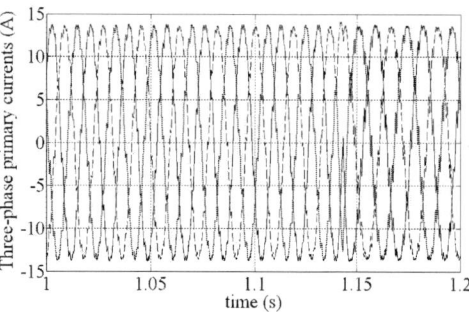

Fig. 10. Three-phase primary currents

It can be seen from simulation results that direct thrust control system follows the imposed references with high performance and fast response. The primary current waveforms are sinusoidal and the primary flux vector locus is a circle. However, the simplicity of the conventional DTC method comes with the disadvantage of thrust ripple. Researchers can choose modified versions of DTC for reducing thrust ripple [13, 14].

According to the simulation results, it is important not to ignore the end effect in flux and thrust estimation part in DTC strategy for LIM.

V. CONCLUSION

Dynamic model and DTC strategy for LIM including end effect are presented. In the case of LIM, the physical dependence between the stator voltage and stator flux is not simple as the case for RIM due to end effect. The thrust and flux estimator for direct thrust control of LIM is improved for considering the end effect. The improved flux and thrust equations require the primary and secondary resistance values. The given improvement is validated by the simulation results. The method used has been satisfactory and can be applied in real situation using a DSP system.

ACKNOWLEDGMENT

The authors wish to express their gratitude to TUBITAK for supporting the project of 104M276.

REFERENCES

[1] J. F. Gieras, *Linear Induction Drives*, Oxford Science Publications, 1994.

[2] G. Kang, K.Nam, "Field-oriented control scheme for linear induction motor with the end effect," *IEE Proceedings on Electric Power Applications*, vol. 152, pp.1565–1572, Nov. 2005.

[3] J.H. Sung; K. Nam, "A new approach to vector control for a linear induction motor considering end effects," *IEEE Industry Applications Conference*, vol. 4, pp.2284–2289, Oct. 1999.

[4] Jianqiang Liu, Fei Lin, Zhongping Yang, Trillion Q. Zheng, "Field Oriented Control of Linear Induction Motor Considering Attraction Force & End-Effects," *IEEE Power Electronics and Motion Control Conference*, vol.1, pp.1–5, 2006.

[5] A. Gastli, "Improved field oriented control of an LIM having joints in its secondary conductors," *IEEE Transactions on Energy Conversion*, vol.17, pp.349–355, Sept. 2002.

[6] E.F. da Silva, C.C. dos Santos, J.W.L. Nerys, "Field oriented control of linear induction motor taking into account end-effects," *IEEE International Workshop on Advanced Motion Control*, pp. 689-694, March 2004.

[7] A.K. Rathore, S.N. Mahendra, "Simulation of secondary flux oriented control of linear induction motor considering attraction force & transverse edge effect," *9th IEEE International Power Electronics Congress*, pp.158-163, 2004.

[8] G. Buja and M. P. Kazmierkowski, "Direct torque control of PWM inverter-fed ac motors-a survey," *IEEE Trans. on Industrial Electronics*, vol. 51-4, pp. 744-757, August 2004.

[9] D. Casadei, G. Serra, A. Tani, and L. Zari, "Assessment of direct torque control for induction motor drives," *Bulletin of the Polish Academy of Technical Sciences*, vol. 54, no. 3, 2006.

[10] Bimal K Bose, *Power Electronics and AC Drives*, 3rd ed., Prentice Hall, 2001.

[11] Byung-Il Kwon, Kyung-Il Woo, Sol Kim, "Finite element analysis of direct thrust-controlled linear induction motor," *IEEE Transactions on Magnetics*, vol. 35, pp.1306–1309, May 1999.

[12] V. Delli Colli, F. Marignetti, M. Scarano, M.M Radulescu, "Implementation of an improved direct thrust and flux control for linear induction motors," *IEEE International Electric Machines and Drives Conference*, vol. 1, pp. 488–493, June 2003.

[13] K. Itoh, H. Kubota, "Thrust ripple reduction of linear induction motor with direct torque control," *Proceedings of the Eighth International Conference on Electrical Machines and Systems*, vol. 1, pp.655–658, Sept. 2005.

[14] Yen-Shin Lai, Jian-Ho Chen, "A new approach to direct torque control of induction motor drives for constant inverter switching frequency and torque ripple reduction," *IEEE Transaction on Energy Conversion*, vol.16, pp. 220-227, Sept. 2001.

Analysis of Short-Circuit Forces at the Top of the Low Voltage U-Type and I-Type Winding in a Power Transformer

Leonardo Štrac[*], Franjo Kelemen[*], Damir Žarko[†]

[*] Končar Power Transformers, Research and Development department, Josipa Mokrovića 6, 10000 Zagreb, Croatia,
e-mail: *leonardo.strac@siemens.com*

[†] University of Zagreb, Faculty of Electrical Engineering and Computing, Department of Electrical Machines, Drives
and Automation, Unska 3, 10000 Zagreb, Croatia, e-mail: *damir.zarko@fer.hr*

Abstract— **The aim of this study was to compare the short-circuit forces in the U-type and I-type low voltage winding of a large power transformer. The comparison of forces calculated using 2D Rabin's method and 3D finite element method is conducted. The influence of the winding helix pitch on the winding forces calculated using 3D finite element method is also investigated.**

Keywords— **Power transformer, short-circuit force, finite element method, Rabin's method.**

I. INTRODUCTION

Every large power transformer can be exposed to short-circuit conditions during its lifecycle. Therefore, large power transformers have to be designed to endure the time limited short circuit. The low-voltage windings are particularly exposed to high forces during short-circuits due to their high current. Two most common types of low voltage windings in a large three phase power transformer are I-type and U-type helix winding. The I-type winding has an entry at the top and an exit at the bottom of the winding. The U-type winding consist actually of two I-type windings connected at the bottom and has an entry and an exit both at the top of the winding. While U-type has some advantages in better cooling and easier connectivity, there is a question which type of winding better handles short-circuit forces. The calculation of radial and axial forces on transformer's winding coils is performed in many papers using different analytical and numerical methods in 2D and 3D geometry [1-2]. The advantages of 2D method are high calculation speed and low consumption of computer resources. While 2D method can describe the transformer windings and related core limbs correctly, it fails to model the transformer yoke properly. This occurs due to the fact that the winding and the core limb on one hand and the yoke on the other hand have mutually perpendicular axes of rotational symmetry. This paper analyzes axial and radial forces on the first wire of the top turn, forces on every turn, force on transformer clamping system as well as the maximal force on the winding. For each case there were three different calculations used: 2D Rabin's method, 3D finite element method on the cylinder-shaped winding and 3D finite element method on the helix-shaped winding. Two different comparisons are actually shown here, one between I-type and U-type of the winding and the other between the three types of calculation. To apply 2D Rabin's method of calculation, Končar Power

Transformer software for force calculation SileKS was used, while 3D calculations were conducted with commercial finite element software Ansoft Maxwell. All calculations are conducted for peak short-circuit current.

II. THEORY

It is well known that axial force is the biggest at the top of the winding, and radial force in the middle. Both forces are calculated on the entire height of the winding with the special attention to the top wire of the top turn. Of course, the relevant value of the radial force is that from the middle of the winding.

Once the magnetic field was calculated using the 3D finite element method, the following equations were used to calculate the force [3]:

The vector force equation is

$$\vec{F} = \vec{J} \times \vec{B}. \qquad (1)$$

The scalar axial force equation is

$$F_a = \vec{F} \cdot \vec{a}_z. \qquad (2)$$

The scalar radial force equation is

$$F_r = \left(\vec{F} \cdot \vec{a}_x\right) \sin\left(\arctan\frac{x}{y}\right) + \left(\vec{F} \cdot \vec{a}_y\right) \cos\left(\arctan\frac{x}{y}\right) \quad (3)$$

The Rabin's method [4] allows analytical formulas to be used for magnetic field calculation on an idealized geometry of a power transformer. The power transformer is presented with a single core limb of an infinite magnetic permeability with a yoke that extends infinitely in the radial direction. Additionally, the sum of winding arrangement ampere-turns must be zero in order for the method to work. Although simplifications impair method's accuracy for the field calculations far away from the winding, it works very well for the field calculations inside the windings or close to the windings. This, in turn makes the method a logic choice for the calculation of forces on the winding.

The Končar Power Transformers software for force calculations SileKS calculates also the short-circuit voltage for a given transformer geometry. According to SileKS, the I-type winding has a bit higher short-circuit voltage than the U-type, 13,57 % versus 13,06 % respectively. Therefore, although the nominal current and flux are the same in both cases, the short-circuit current of

978-1-4244-1741-4/08/$25.00 ©2008 IEEE

I-type winding is going to be smaller because of higher short-circuit impedance. For that reason, our calculations consider that difference in short-circuit current.

III. MODEL

The model is based on the actual 450 MVA transformer with necessary simplifications made for this purpose. From the basic model there were actually six different models derived: the model with the I-type winding in three versions - 2D rotational-symmetry, 3D with the cylinder-shaped winding (fig. 1a) and 3D with the helix-shaped winding (fig. 1b) and the three equivalent models with the U-type winding (figs. 1c and 1d). Each model represents only the middle phase. The core is solid, non-laminated, isotropic and linear in each case. The 2D rotational-symmetry model has a cylinder-shaped winding, proper sized core limb, but the actual yoke is approximated with an infinitely wide yoke. The 3D models have the core of proper shape with the yoke cut on the symmetry of the window. In one 3D model every turn is a ring without a lead, and in the other a turn is a part of a helix and the winding has a lead. The I-type winding has 4 axial parallels versus 8 axial parallels of the U-type winding, but with narrower conductors. The U-type winding is made of 20 turns per layer instead of 20,5 so that the cylinder-shaped winding can be analyzed. The current is recalculated to maintain the correct ampere-turns value.

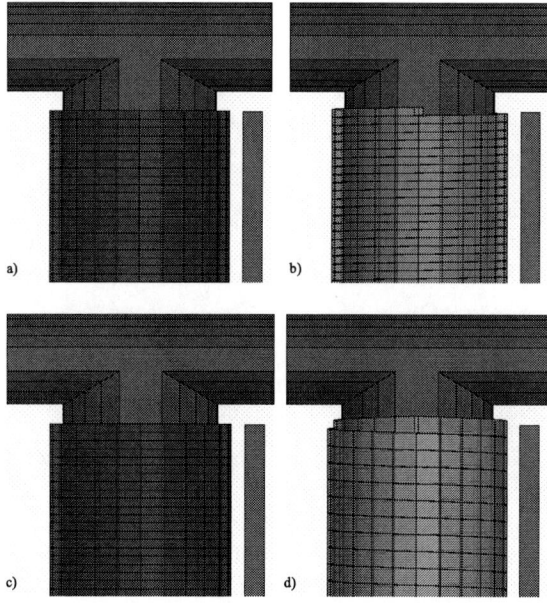

Fig. 1. Cylindrical (a) and helix (b) version of I-type winding in 3D models; cylindrical (c) and helix (b) version of U-type winding in 3D models.

IV. RESULTS

A. Calculation of forces for every turn

Tables 1 and 2 show a comparison of axial and radial force for every turn. The axial force is expectedly the highest at the top and the radial is the highest in the middle. Each of the three applied methods gives very similar results. The largest difference in axial force is in the middle of the winding, but the axial force is very small there. The difference in results of axial force calculation between 2D and 3D methods at the winding top and bottom is 10-20%. The two 3D methods differ insignificantly. However, the difference in results of radial force calculation is much smaller and it is generally about 1%.

TABLE I.
AXIAL AND RADIAL FORCE FOR EVERY TURN OF I-TYPE WINDING

I-type	axial force [kN]			radial force [kN]		
turn	2D	3D cylin.	3D helix	2D	3D cylin.	3D helix
1	1068	1144	1139	920	872	873
2	613	681	677	1136	1095	1096
3	398	433	429	1262	1231	1233
4	261	285	281	1337	1310	1312
5	173	192	189	1376	1356	1358
6	127	131	130	1402	1384	1386
7	84	91	91	1414	1401	1402
8	67	64	65	1426	1412	1413
9	47	46	47	1429	1419	1420
10	36	32	35	1435	1424	1425
11	29	23	26	1436	1427	1428
12	19	16	20	1441	1429	1430
13	19	11	15	1439	1431	1431
14	11	7	11	1442	1432	1432
15	11	5	8	1441	1433	1432
16	7,7	3,3	5,4	1444	1434	1433
17	5,9	2,3	3,4	1441	1434	1433
18	5,8	1,7	1,8	1444	1435	1434
19	2,3	1,4	0,7	1442	1435	1434
20	4,1	1,2	-0,2	1445	1435	1435
21	0,7	0,9	-0,7	1442	1435	1435
22	1,5	0,7	-1,3	1445	1435	1435
23	-0,2	0,3	-2,2	1442	1435	1434
24	-1,4	-0,2	-3,4	1444	1435	1434
25	-1,4	-1,0	-5,0	1442	1434	1433
26	-4,6	-2,2	-7,0	1444	1434	1433
27	-4,4	-3,9	-10	1441	1433	1432
28	-8,1	-6,3	-13	1443	1432	1432
29	-10	-10	-17	1440	1431	1431
30	-14	-14	-22	1441	1429	1430
31	-20	-21	-28	1437	1427	1428
32	-26	-30	-37	1437	1424	1425
33	-36	-43	-49	1432	1419	1420
34	-50	-62	-67	1429	1412	1414
35	-69	-88	-92	1419	1402	1403
36	-101	-127	-130	1408	1385	1387
37	-145	-185	-187	1384	1359	1361
38	-219	-275	-275	1349	1314	1316
39	-343	-419	-418	1281	1237	1240
40	-538	-659	-657	1165	1105	1107
41	-948	-1109	-1105	967	888	892

TABLE II.
AXIAL AND RADIAL FORCE FOR EVERY TURN OF U-TYPE WINDING, FIRST AND SECOND LAYER

U-type	layer 1, axial force [kN]			layer 1, radial force [kN]		
turn	2D	3D cylin.	3D helix	2D	3D cylin.	3D helix
1	946	1025	972	501	468	521
2	369	404	422	665	650	644
3	162	182	184	727	716	714
4	80	86	87	749	742	741
5	43	43	43	757	752	752
6	25	20	22	762	757	757
7	15	6,8	12	763	757	759
8	9,4	8,6	5,4	763	749	760
9	5,2	10,8	2,8	764	760	761
10	2,7	6,9	1,4	765	761	761
11	0,7	4,3	-0,05	765	761	761
12	-1,0	1,2	-1,2	765	761	761
13	-4,1	-2,9	-4,1	764	761	760
14	-8,8	-9,2	-9,3	764	759	759
15	-17	-21	-20	761	757	757
16	-32	-42	-41	759	753	752
17	-63	-84	-83	751	742	742
18	-134	-174	-177	733	714	715
19	-315	-389	-407	675	650	648
20	-836	-993	-935	529	477	527

U-type	layer 2, axial force [kN]			layer 2, radial force [kN]		
turn	2D	3D cylin.	3D helix	2D	3D cylin.	3D helix
1	777	849	798	1790	1725	1635
2	306	331	353	2230	2159	2160
3	142	159	160	2365	2341	2330
4	72	77	77	2414	2397	2390
5	40	39	39	2431	2420	2414
6	24	19	20	2441	2429	2424
7	15	7,8	10	2445	2432	2430
8	8,9	7,7	5,5	2446	2430	2432
9	5,1	9,1	2,5	2446	2436	2433
10	2,5	6,6	1,0	2450	2439	2434
11	0,8	4,4	0,3	2447	2439	2434
12	-0,9	1,4	-1,0	2449	2438	2433
13	-3,7	-2,6	-3,6	2446	2438	2432
14	-8,2	-8,2	-8,3	2446	2435	2429
15	-16	-19	-18	2441	2430	2424
16	-30	-38	-37	2436	2420	2414
17	-57	-75	-74	2418	2398	2391
18	-115	-152	-154	2377	2342	2334
19	-258	-322	-341	2253	2183	2168
20	-684	-824	-768	1851	1744	1661

B. Force on transformer clamping system and maximal force on the winding

Table 3 shows the force on the clamping plate and the maximal force on the winding.

The force on the clamping plate is calculated simply by summing axial forces for every turn from tables 1 and 2. Here we have a significant difference between results obtained using 2D and 3D models and even between the cylindrical and helix versions of the 3D model. The difference is up to 9 times, and appears to be due to the large error in 2D calculation in the middle of the winding. This error probably appears because of the round-off error when calculating small radial field and consequentially small values of axial force using Rabin's method.

The maximal force on the winding is calculated by summing axial forces on one half of the winding. In the middle of the winding the axial force changes direction, so the sum of axial forces for all turns with the same direction of axial force represents the maximal force on the winding. This force is practically identical for both types of winding as well as for all three versions of the model.

TABLE III.
FORCE ON THE CLAMPING SYSTEM AND WINDING

	axial force [kN]		
	2D	3D cylin.	3D helix
I-type: force on the clamping system	451	118	50
I-tip: maximal force in the winding	2990	3174	3170
U-type: force on the clamping system, sum of layer 1 and 2	469	155	134
U-type: maximal force in the winding, sum of layer 1 and 2	3053	3307	3218

In table 4 the maximal radial force in the middle of the winding is given. Since radial force greatly depends on the radius, there is a large difference in the amount of force on the U-type winding, layer 1 and 2. The force on the U-type winding layer 1 is about two times smaller than the force on the I-type, but the force on the layer 2 is almost two times larger.

The difference between the three types of calculation is negligible.

TABLE IV.
MAXIMAL RADIAL FORCE IN THE MIDDLE OF THE WINDING

	maximal radial force [kN]		
	2D	3D cylin.	3D helix
I-type	1444	1435	1435
U-type, layer 1	765	761	761
U-type, layer 2	2450	2439	2434

C. Forces on the first wire of the top turn

Figs. 2 and 3 represent results only for the force on the first wire of the top turn. Since the height of the wire is the same in both I and U-type windings, the wires in both cases collect the similar amount of radial flux, so the axial force is also similar. The radial force depends greatly on the radius, so the force in the second layer of the U-turn winding is much higher than in the first layer.

Fig. 2 represents axial force for the first wire of the top turn per millimeter of circumference for both U and I-type windings. The influence of leads in the helix version of the model can be easily seen as the current is changing direction in that spot. There is also the influence of yoke which is closest to the winding around n = 7 and n = 19. The results in Figs. 2 and 3 show that the 2D model approximates quite well the actual 3D geometry.

Fig. 3 represents radial force for the first wire of the top turn per millimeter of circumference for the I-Type winding and layer 1 and 2 of the U-type winding. Several observations can be made from the figure. The radial force in the layer 2 of the U-type winding is much higher than in the layer 1. The force on the I-type winding is slightly higher than force on the U-type winding, layer 2. All three versions of the model give very similar results with one exception, and that is the helix version of the U-type winding, layer 2. In that case the slope has a large influence on the radial force which is higher in the upper part and lower in the lower part of the turn. In fact, the similar effect is noticeable in the helix version of the U-type winding, layer 1, but it is much less pronounced.

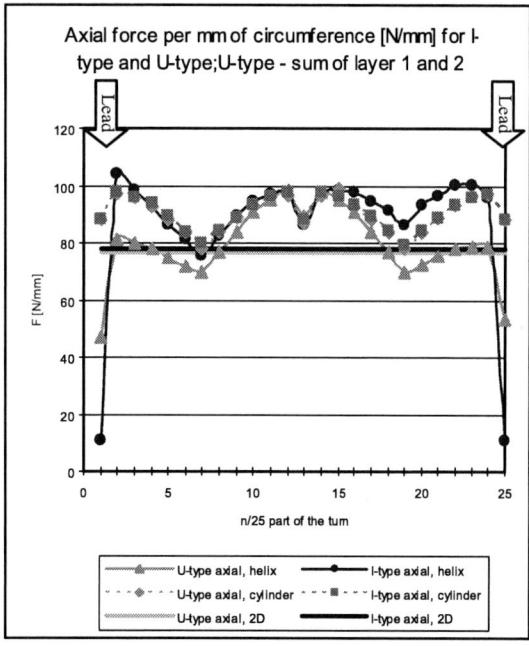

Fig. 2. Axial force on the first wire of the first turn, U and I-type winding

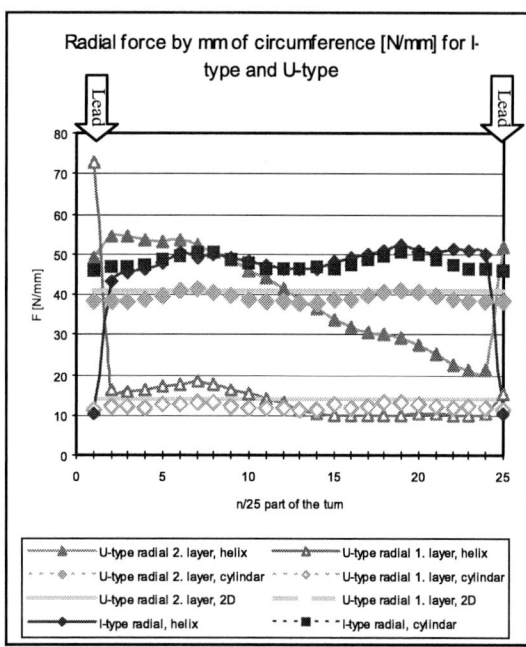

Fig. 3. Radial force on the first wire of the first turn, U and I-type winding

Fig. 3 represents radial force for the first wire of the top turn per millimeter of circumference for the I-Type winding and layer 1 and 2 of the U-type winding. Several

observations can be made from the figure. The radial force in the layer 2 of the U-type winding is much higher than in the layer 1. The force on the I-type winding is slightly higher than force on the U-type winding, layer 2. All three versions of the model give very similar results with one exception, and that is the helix version of the U-type winding, layer 2. In that case the slope has a large influence on the radial force which is higher in the upper part and lower in the lower part of the turn. In fact, the similar effect is noticeable in the helix version of the U-type winding, layer 1, but it is much less pronounced.

V. CONCLUSION

The most important conclusion of this analysis is that 2D analytical calculation of short-circuit forces on low voltage windings of a power transformer using Rabin's method is accurate enough and very close to the calculation performed using 3D finite element method. Since 2D calculation is much quicker and has low demand on computer resources, there is no need to perform 3D calculation except in those cases when very detailed and accurate calculation is required.

The only part where 2D calculation significantly differs from 3D is the calculation of the force on the clamping system. Since 2D calculation gives the largest value among the three cases considered in the analysis, there is no danger for underestimating the forces in the transformer design using this type of calculation. This opens the possibility to lower the number of clamping ties. However, that possibility should be analyzed much more thoroughly than in was done this case.

There is also no significant difference in results between the cylindrical and the helix versions of 3D modeling, so the advice is to use much more complex helix geometry only when it is absolutely necessary.

On the top wire of the top turn where the axial force is the highest, the difference between axial force on the I-type and U-type winding is negligible. These results suggest that there is no significant difference between the two winding types in terms of force on the other vertical segments of the winding as well.

In the middle of the winding where the radial force is the highest, the radial force in the U-type winding, layer 1 is much smaller than the force in the I-type winding. However, the radial force in the U-type winding, layer 2 is much higher than the force in the I-type winding. Hence, it can be concluded that the radial force in the I-type winding can be handled much easier when designing the transformer to mechanically withstand the short-circuit radial forces.

REFERENCES

[1] M. Ardebili, K. Abbaszadeh, S. Jamali, H.A. Toliyat, "Winding Arrangement Effects on Electromagnetic Forces and Short-Circuit Reactance Calculation in Power Transformers via Numerical and Analytical Methods," *12th Biennial IEEE Conference on Electromagnetic Field Computation, 2006.*

[2] A. Benhama, A.C. Williamson, A.B.J. Reece, "Force and torque computation from 2-D and 3-D finite element field solutions," *IEE Proceedings - Electric Power Applications, Jan. 1999.*

[3] Z. Haznadar, Ž. Štih, "Elektromagnetizam," *Zagreb: Školska knjiga, 1997, vol. I & II (in Croatian).*

[4] R.M. Del Vecchio, B. Poulin, P.T. Feghali, D.M. Shah, R.Ahuja, "Transformers Design Principles," *Gordon and Breach Science Publishers, p.149-184.*

Anisotropy Comparison of Reluctance and PM synchronous Machines for Low Speed Position Sensorless Applications

H.W. de Kock and M.J. Kamper
Department of Electrical and Electronic Engineering
University of Stellenbosch
Stellenbosch, South Africa
Email: hugodekock@gmail.com, kamper@sun.ac.za

R.M. Kennel
Electrical Machines and Drives
University of Wuppertal
Wuppertal, Germany
Email: kennel@ieee.org

Abstract—Position sensorless control of reluctance- and PM synchronous machines at zero and low speed is possible using high frequency voltage injection and proper demodulation. The so-called anisotropy position, which is tracked by the HF sensorless scheme, is different to the actual rotor position: the difference contains both offset and time-varying components, which may be explained by carefully considering the high frequency behaviour of the machine and the effect that fundamental excitation and rotor position has upon it. This paper gives insight into the HF behaviour of machines and serves as a practical guide for implementation of stable and robust position estimation at zero and low speed.

I. INTRODUCTION

The focus on high efficiency and cost effective drives, for applications ranging from washing machines to electrical cars, has led to the adoption of certain types of synchronous machines, with control algorithms that avoid the use of expensive sensors, as well as maximize their efficiency. Permanent magnet synchronous machines (PMSM) are widely accepted, due to their high torque to volume ratio. Especially those kind that exhibit rotor saliency characteristics, since they are prime candidates for position sensorless control, even at zero speed [1]–[22]. The reluctance synchronous machine (RSM) represents an alternative to the PMSM for some applications, and the construction cost could be cheaper due to the lack of PMs. The RSM is known for its rotor saliency characteristics and is therefore definitely a candidate for sensorless control [23]–[41]. The introduction of concentrated stator windings, instead of distributed stator windings, represents another cost-saving effort. The implications on the fundamental control, as well as sensorless control, have to be considered [14], [42], [43].

To have field oriented torque control in the entire speed range, rotor position estimation in the entire speed range is necessary. At high speed, back-EMF observers may be used to find the rotor position. At low speed, the lack of back-EMF gives rise to an alternative position estimation scheme, namely one that relies on rotor saliency. Ultimately, a combination of methods may be necessary to have an accurate rotor position estimation in the entire speed range [16], [38], [44].

One method of position estimation at low and zero speed applies a continuous alternating HF voltage with the inverter, and digitally process the resulting HF currents to obtain a saliency dependent position, which may be called the anisotropy position or the saliency position [8], [22], [41]. When there is hardly any rotor saliency, like in surface mount PMSM or induction machines, it is difficult to track the saliency position. In some machines like RSM, PM-assisted RSM and PMSM, the inherent rotor saliency property enables rotor position detection at zero speed [18], [19].

It may be the case that there exists multiple saliencies [2], [22], in which case special care has to be taken to track a useful one. In the case of a single dominant saliency that rotates in the same direction as the rotor and at the electrical speed, the saliency position may be equal to the actual rotor position. However, the fundamental excitation (torque producing current and flux linkage) may have a great impact upon the saliency position: saturation may reduce the effective saliency (reducing signal to noise ratio in the HF current signal) and cross-saturation may cause the saliency position to deviate from the actual rotor position (this is an offset error) [14], [16], [22], [38], [41]. Non-sinusoidal flux distribution, i.e. position dependent inductances in the synchronous reference frame that is caused by e.g. stator slot openings or concentrated windings, also disturbs the position estimation (this is a time-varying error) [41].

In this paper, the effective HF model for a transverse laminated RSM with normal distributed windings is presented based on practical tests on an experimental setup. It is shown how the anisotropy position deviate from the actual rotor position under load (offset error) and the time-varying error due to flux variation is discussed. A simple load dependent compensation function is suggested to compensate the offset error. The experiment is then also repeated for a commercial PMSM. The difference in HF behaviour is pointed out. This paper gives insight into the HF behaviour of synchronous machines and serves as a practical guide to implement a robust and stable position estimation at low and zero speed.

II. MACHINE MODEL

Neglecting core- and endwinding losses, the voltage space vector equation for both RSM and PMSM is given in the stationary $\alpha\beta$ reference frame by (1). The flux linkage vector $\vec{\psi}_s$ is a non-linear function of the magnetizing current vector $\vec{i}_s$ and has in theory a sinusoidal distribution in space, i.e. with respect to the electrical rotor position θ_r. The flux linkage vector $\vec{\psi}_s$ is the total flux linkage, i.e. it includes the leakage flux linkage. The transformation to the synchronously dq reference frame, as in (2), removes the dependency of the flux linkage vector on θ_r, so that $\vec{\psi}_r$ is only a function of $\vec{i}_r$. However, in reality the flux linkage distribution is not perfectly sinusoidal, due to e.g. stator slot openings or concentrated windings, and therefore $\vec{\psi}_r$ nevertheless remain a function of θ_r. As mentioned, $\vec{\psi}_r$ is a non-linear function of $\vec{i}_r$, since in any machine design there is a certain amount of saturation and cross coupling.

The relationship between the flux linkage and the current may be expressed in terms of inductance, and this is best understood in matrix notation as in (4): there are self-inductance terms on the diagonal and mutual-inductance terms on the off-diagonal, where it should be clear that these are differential (also called tangential) inductances, i.e. partial derivatives. Ideally L_d and L_q would be constant, and L_{dq} as well as $\frac{\partial \vec{\psi}_r}{\partial \theta_r}$ would be zero. However, due to the non-linear properties of machines, the inductances are again non-linear functions of $\vec{i}_r$ and θ_r, and there are flux pulsations with movement, i.e. $\frac{\partial \vec{\psi}_r}{\partial \theta_r}$ is not zero.

$$\vec{u}_s = R \cdot \vec{i}_s + \frac{d\vec{\psi}_s}{dt} \tag{1}$$

$$\vec{u}_r = \vec{u}_s \cdot e^{-j\theta_r} \tag{2}$$

$$\vec{u}_r = R \cdot \vec{i}_r + \frac{d\vec{\psi}_r}{dt} + j \cdot \omega_r \cdot \vec{\psi}_r \tag{3}$$

$$\omega_r = \frac{d\theta_r}{dt}$$

$$\begin{bmatrix} \frac{d\psi_d}{dt} \\ \frac{d\psi_q}{dt} \end{bmatrix} = \begin{bmatrix} L_d & L_{dq} \\ L_{dq} & L_q \end{bmatrix} \begin{bmatrix} \frac{di_d}{dt} \\ \frac{di_q}{dt} \end{bmatrix} + \begin{bmatrix} \frac{\partial \psi_d}{\partial \theta_r} \\ \frac{\partial \psi_q}{\partial \theta_r} \end{bmatrix} \frac{d\theta_r}{dt} \tag{4}$$

For a high frequency (HF) model of the machine, consider (3) while applying a HF voltage vector: the resistive loss term and the speed voltage term become insignificant to the derivative of the flux linkage term, as expressed in (5). This equation can further be approximated as (6) if ω_r is sufficiently smaller than ω_{HF}.

$$\vec{u}_r(\omega_{HF}) \approx \frac{d\vec{\psi}_r}{dt} \tag{5}$$

$$\begin{bmatrix} u_d \\ u_q \end{bmatrix} \approx \begin{bmatrix} L_d & L_{dq} \\ L_{dq} & L_q \end{bmatrix} \begin{bmatrix} \frac{di_d}{dt} \\ \frac{di_q}{dt} \end{bmatrix} \tag{6}$$

III. POSITION SENSORLESS CONTROL

Assuming that we have a perfect anisotropy (saliency), i.e. constant L_d and L_q, with $L_d \neq L_d$ and $L_{dq} = 0$, we can superimpose a pulsating HF voltage vector on our fundamental control voltage vector, then filter out the resulting HF current vector from the measured current, and then find the saliency position information within the HF current that allows us to estimate the saliency position. Using the estimated saliency position we can estimate the rotor position, as has been described by numerous authors including [8], [22], [41]. Of course, we would like to use the inherent rotor saliency that is provided by physical rotor design [18], [19], so that the anisotropy position closely resembles the rotor position. So that we don't get confused between the fundamental model and the anisotropy model, we use the subscript A and describe the ideal anisotropy as constant L_{dA} and L_{qA} with $L_{dA} \neq L_{qA}$ and $L_{dqA} = 0$.

The estimation process is described mathematically, in easy to follow logical steps, by (7) through (25) and is illustrated by the block diagrams in Fig. 1 and Fig. 2. The anisotropy position estimation takes place within a reference frame that is aligned with the anisotropy. A pulsating HF voltage is applied in an arbitrary direction δ, as in (8), with respect to the estimated anisotropy position $\hat{\theta}_A$. The common choices are $\delta = 0$ or $\delta = \frac{\pi}{2}$, although it has been suggested that the injection direction should resemble the maximum torque per ampere (MTPA) axis in an effort to minimize the effect of inductance change due to saturation [20].

The HF voltage amplitude U_{HF} is usually chosen as constant and needs to be large enough to give a substantial HF current. If the machine has a very small inductance, then a small voltage amplitude causes a relatively large HF current. It has been suggested that the voltage amplitude be varied in an effort to reduce accoustic noise [21], so that it only has a high value during acceleration or high load.

The chosen frequeny ω_{HF} also largely influences the amplitude of the HF current and the accoustic noise, and needs to be selected so that digital filters can successfully separate the HF current from the fundamental current and so that the fundamental current vector controller does not influence the HF currents much, i.e. a certain amount of frequency separation is needed. In many papers and also in our experience $\omega_{HF} = 2\pi500$ [rad/sec] was a good choise for the machines at hand.

Following (7) and eq:urAHF, the pulsating HF voltage in stationary reference frame is given by (9), and so the pulsating voltage in the reference frame aligned to the actual anisotropy at position θ_A is given by (10). This pulsating voltage causes the HF current in the actual anisotropy reference frame as in (11). Assuming a slowly varying error $\tilde{\theta}_A$, the HF current in the actual anisotropy reference frame may be approximated as in (12). However, since we do not know θ_A and only have access to $\hat{\theta}_A$, we need to find the expression for the HF current in the estimated anisotropy reference frame, as in (13). The useful information, i.e. the modulated amplitude of $\vec{i}_{\hat{A}}$, is expressed by (14).

860

$$\vec{u}_{HF} = U_{HF}\cos(\omega_{HF}t) + j0 \tag{7}$$

$$\vec{u}_{\hat{A}} = \vec{u}_{HF}e^{j\delta} \tag{8}$$

$$\vec{u}_{sHF} = \vec{u}_{\hat{A}}e^{j\hat{\theta}_A} \tag{9}$$

$$\vec{u}_A = \vec{u}_{sHF} \cdot e^{-j\theta_A} = \vec{u}_{\hat{A}}e^{-j\tilde{\theta}_A} \tag{10}$$

$$\tilde{\theta}_A = \theta_A - \hat{\theta}_A$$

$$\begin{bmatrix} i_{dA} \\ i_{qA} \end{bmatrix} = \begin{bmatrix} L_{dA} & 0 \\ 0 & L_{qA} \end{bmatrix}^{-1} \begin{bmatrix} \int u_{dA}dt \\ \int u_{qA}dt \end{bmatrix} \tag{11}$$

$$\vec{i}_A \approx \vec{\psi}_{HF}\left(\frac{1}{L_{dA}}\cos\tilde{\theta}_A - j\frac{1}{L_{qA}}\sin\tilde{\theta}_A\right) \tag{12}$$

$$\vec{\psi}_{HF} = \frac{U_{HF}}{\omega_{HF}}\sin(\omega_{HF}t + \zeta)e^{j\delta}$$

$$\vec{i}_{\hat{A}} = \vec{i}_A e^{j\tilde{\theta}_A} = \vec{\psi}_{HF}\cdot\vec{A} \tag{13}$$

$$\vec{A} = \left(\frac{1}{\Sigma L} + \frac{1}{\Delta L}e^{j2\tilde{\theta}_A}\right) \tag{14}$$

$$\Sigma L = 2\frac{L_{qA}\cdot L_{dA}}{L_{qA} + L_{dA}}$$

$$\Delta L = 2\frac{L_{qA}\cdot L_{dA}}{L_{qA} - L_{dA}}$$

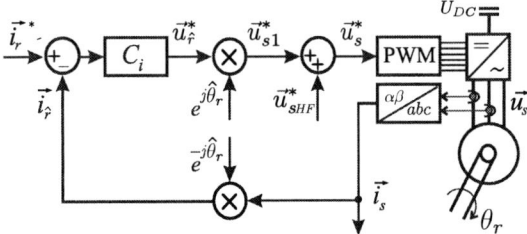

Fig. 1. Fundamental current control.

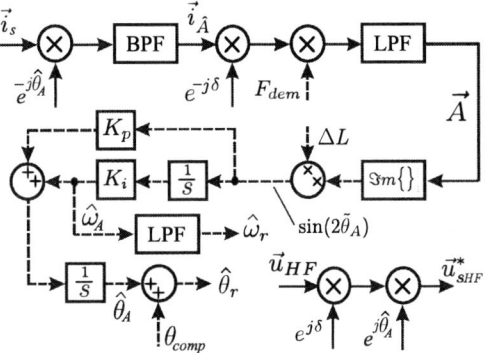

Fig. 2. Rotor position estimation.

The superposition of the HF voltage vector onto the fundamental control voltage vector, as in (15) and shown in Fig. 1, results in a HF current vector that can be separated from the fundamental current vector with a band pass filter (BPF), as in (16). We are only interested in the useful information given by $\vec{A}$ and can get this from $\vec{i}_{\hat{A}}$ using the demodulation given in (17). The demodulation sine wave F_{dem}, as in (18), has the phase shift ζ, which should be zero in theory. In practice the phase shift between HF voltage and HF current could be larger than 90° due to digital system delays. In the demodulation it is therefore important to check that the HF current and the demodulation sine wave F_{dem} is in phase, by setting ζ to an appropriate value.

Now, using the imaginary part of $\vec{A}$ as in (19), as the input signal to a PI controller, i.e. driving this signal to zero, we can assume that $\tilde{\theta}_A$ will be small and (20) becomes valid. Therefore, in this closed loop tracking system (see Eqs. (21), (22) and (23)), or phase locked loop (PLL), we can be sure to track the ansitropy position, and the speed at which it moves, wherever it may be. Important is the normalization with the constant ΔL in (19) and the sign that it has, since negative feedback (stability) has to be ensured.

As the final step, we may estimate the rotor position θ_r from our estimated anisotropy position $\hat{\theta}_A$ as in (24), and in the ideal case θ_{comp} is zero or a constant. We can also estimate the rotor speed using our estimated anisotropy speed, as in (25). Since $\hat{\omega}_A$ is used in part to drive the integrator so that the anisotropy position is tracked, and therefore might have large and fast changes, we might need additional low pass filtering to obtain a reasonable rotor speed estimation.

$$\vec{u}_s^* = \vec{u}_{s1}^* + \vec{u}_{sHF} \tag{15}$$

$$\vec{i}_{\hat{A}} = \text{BPF}\left\{\vec{i}_s e^{-j\hat{\theta}_A}\right\} \tag{16}$$

$$\vec{A} = \text{LPF}\left\{\vec{i}_{\hat{A}}e^{-j\delta}\cdot F_{dem}\right\} \tag{17}$$

$$F_{dem} = 2\frac{\omega_{HF}}{U_{HF}}\sin(\omega_{HF}t + \zeta) \tag{18}$$

$$\sin(2\tilde{\theta}_A) = (\Delta L)\,\Im m\left\{\vec{A}\right\} \tag{19}$$

$$2\tilde{\theta}_A \approx \sin(2\tilde{\theta}_A) \tag{20}$$

$$\tilde{\theta}_A = \theta_A - \hat{\theta}_A \tag{21}$$

$$\hat{\omega}_A = K_i\int\tilde{\theta}_A dt \tag{22}$$

$$\hat{\theta}_A = \int\left(K_p\tilde{\theta}_A + \hat{\omega}_A\right)dt \tag{23}$$

$$\hat{\theta}_r = \hat{\theta}_A + \theta_{comp} \tag{24}$$

$$\hat{\omega}_r = \text{LPF}\{\hat{\omega}_A\} \tag{25}$$

IV. RELUCTANCE SYNCHRONOUS MACHINE

A laboratory 1.5 kW RSM, with a rated current of 3.5 A rms and a rated torque of 10 Nm is used to test the sensorless field oriented control scheme for zero and low speed as described above.

The first test is to find a suitable HF voltage vector. We apply a rotating voltage vector for frequencies ranging from 0 to 1000 Hz, and to look at the current vector response. We expect that at low frequencies the current amplitude will be large, the phase difference between the voltage vector and the current vector will be small and the rotor might turn. Increasing the frequency, we see that the current magnitude reduces and the phase shift between the voltage and the current increases. At high frequencies we start hearing a sound, which could be rather unpleasant. Although the theoretical limit for the phase shift between the voltage and the current vector is 90°, we measure a phase shift greater than 90°, which could be explained by digital system delays. We identify that for this machine, a voltage vector (rotating or pulsating) with a frequency of 508 Hz and an amplitude of 100 V (this represents 30 percent of the total available voltage using space vector PWM and a DC bus voltage of 580 V), results in a HF current vector that has a phase delay of 113° with respect to the voltage vector and an amplitude of 400 mA (about 10 percent of the rated current), and the sound that it produces is bearly noticeable. Therefore, we set $\zeta = 113° - 90° = 23°$ in (18), so that the demodulation sine wave will be in phase with our HF current response. Since we have a sampling frequency of $f_s = 12205$ Hz and the HF waveform of 508 Hz, this phase shift of 23° only means 1 or 2 samples delay.

The second test is to get an idea of the parameters values L_{dA} and L_{qA} at no load. We assume that the anisotropy position θ_A is equal to the rotor position θ_r. We keep the rotor position θ_r at zero and vary the estimated anisotropy position $\hat{\theta}_A$ from 2π to zero, i.e. the error $\theta_r - \hat{\theta}_A$ varies from zero to 2π. We apply the HF voltage like it is shown in Fig. 2 and then we inspect the signal $\vec{A}$. We observe that the frequency component $2\omega_{HF}$ is visible in $\vec{A}$ and that its suppression depends on the cut-off frequency and order of the low pass filter (LPF). We choose a 2nd order LPF that gives us 40 dB suppression at $2\omega_{HF}$. From the real part of $\vec{A}$, i.e. the real part in (14), we have enough information to calculate L_{dA} and L_{qA}. Using the measured values of $\Re\left\{\vec{A}\right\}$ for a complete cycle of $\hat{\theta}_A = 2\pi..0$, the calculation tells us that $L_{dA} = 0.33976$ H and $L_{qA} = 0.26488$ H. Using these values to normalize $\vec{A}$ and to subtract the offset from the real part of $\vec{A}$, we should be able to see the signal $e^{j2\tilde{\theta}}$. Fig 3 shows this measurement, where the top graph is $\tilde{\theta}_A$, the middle graph is $\vec{A}$ and the bottom graph is the normalized $\vec{A}$ with the offset subtracted from the real part.

It has previously been shown in [41] that for this RSM, the anisotropy parameters, or HF inductances, L_{dA} and L_{qA} are load dependent, and that there is a certain amount of mutual inductance. This result is shown here again in Fig. 4. Note that the fundamental current is applied at a current angle of 60°, which for this machine is an approximation to maximum torque per ampere (MTPA).

We expect to see that when the closed loop position estimation is active, there will be a load dependent offset error between the estimated anisotropy position and the actual rotor

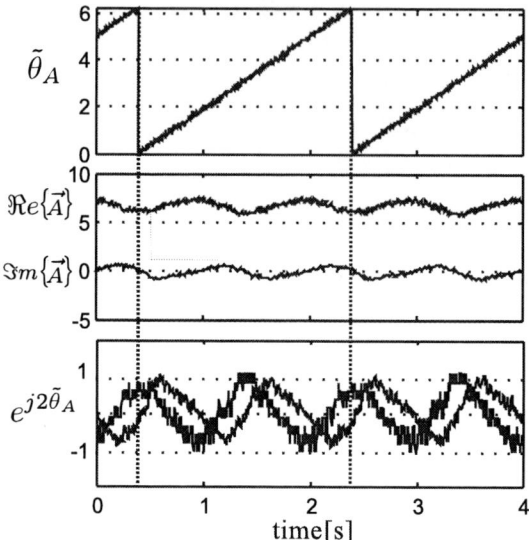

Fig. 3. RSM anisotropy at no load and standstill.

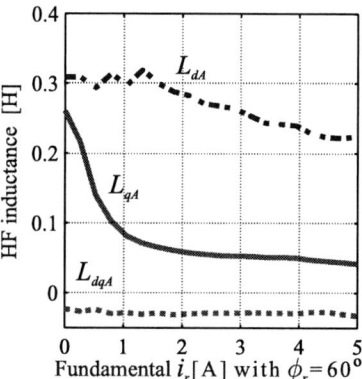

Fig. 4. RSM anisotropy with load at standstill.

position due to the mutual inductance [22], [38], [41]. We also expect that there might be some stability problems in the PLL due to the changing parameters. To maintain stability at all load conditions in this case, it was important to select the L_{dA} and L_{qA} parameters for the worst case scenario: that is when the difference between L_{qA} and L_{dA} is large, since the error signal then also becomes too large and causes instability. Therefore the following constant parameters were chosen $L_{dA} = 0.3$ H and $L_{qA} = 0.05$ H. Of course the parameters can be varied on-line as a function of load, but it has been found that this does not make a big difference to stability or dynamics.

The third test is to look at the dependency of the anisotropy on the actual rotor position. Up to now, all the tests have

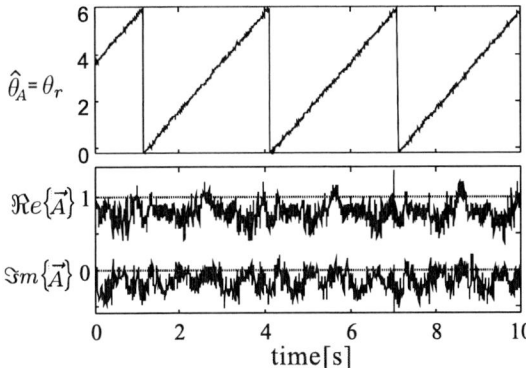

Fig. 5. RSM anisotropy position dependence at no load.

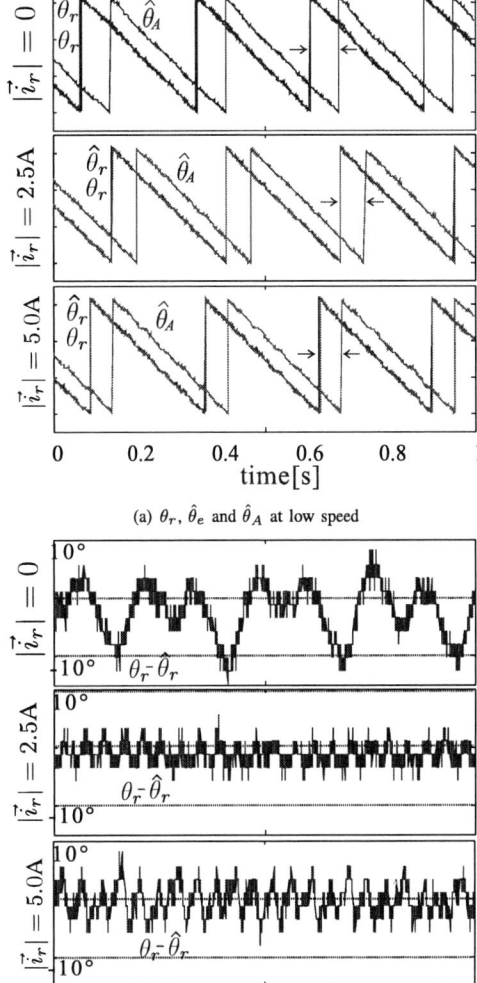

(a) θ_r, $\hat{\theta}_e$ and $\hat{\theta}_A$ at low speed

(b) $\theta_r - \hat{\theta}_r$ at low speed

Fig. 6. RSM sensorless control at zero, half and rated load.

been done with a still standing rotor. We have shown that the parameters L_{dA} and L_{qA} are dependent on the fundamental current vector $\vec{i}_r$ and now we would like to see if these parameters depend on the rotor position θ_r. In this test we set the estimated anisotropy angle equal to the actual rotor position $\hat{\theta}_A = \theta_r$; we inject the HF voltage as indicated in Fig. 2 with $\zeta = 0$; we control the fundamental current to zero and turn the RSM with another machine (mechanically coupled to the RSM) at a low speed; we inspect the HF current amplitudes $\vec{A}$ and normalize it with the parameters L_{dA} and L_{qA} that corresponds with zero fundamental current and subtract the offset from the real part of $\vec{A}$ so that we expect to see $\Re e\left\{\vec{A}\right\} = 1$ and $\Im m\left\{\vec{A}\right\} = 0$. Fig. 5 shows this measurement.

In Fig. 5 we can see that there is a change of $\vec{A}$ with rotor position, but it is not very large. There is a small offset with the respect to the nominal values, and a slight ripple, which might cause an offset in the closed loop estimation of the rotor position and a ripple or time-varying error.

The fourth test is to close the PLL and check the closed loop position estimation at all load conditions, at zero and low speed. For this test, we control the fundamental current using the measured rotor position θ_r and turn the RSM with another machine at a constant speed. We generate the HF voltage using the estimated anisotropy angle $\hat{\theta}_A$. If the normalization constant ΔL is correct, the PLL constants can be chosen as $K_p = 2BW$ and $K_i = BW^2$ to give a closed loop bandwidth of BW and a damped response. We compare $\hat{\theta}_A$ with measured θ_r at all load conditions to see if it is necessary to add compensation θ_{comp}. We find that the compensation function needed is $\theta_{comp} = 1.4 \cdot |\vec{i}_r^*| + 7.2$ [°]. The PLL prefers to track the q-axis of the RSM, but deviates from it when load is applied. This is shown in the final results.

The final step is to use the estimated rotor position $\hat{\theta}_r = \hat{\theta}_A + \theta_{comp}$ for the field orientated control. Fig. 6(a) shows θ_r, $\hat{\theta}_r$ and $\hat{\theta}_A$ for low speed at loads of zero, half load and rated load, while Fig. 6(b) shows the estimation error $\theta_r - \hat{\theta}_r$ that is made. It is clearly seen that the tracking is very good at

any load condition. Attention is drawn to the phase difference between $\hat{\theta}_A$ and θ_r in Fig. 6(a): with zero load the phase difference is about 90°, i.e. the q-axis of the RSM is tracked, but as we increase the load, the phase difference becomes smaller, i.e. the anisotropy position moves towards the d-axis.

The rotor position estimation error that is made due to the position dependency of the anisotropy is not very significant in this RSM, and that is very good, because it is extremely difficult to compensate that kind of error. The rotor position estimation error due to mutual inductance can be successfully compensated, because it is a function of load current and fairly

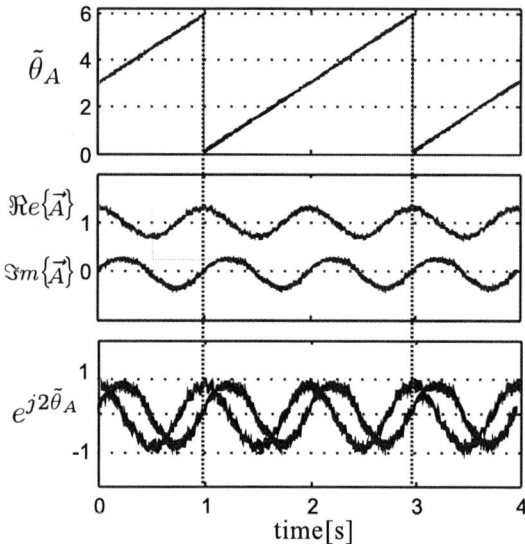

Fig. 7. PMSM anisotropy at no load and standstill.

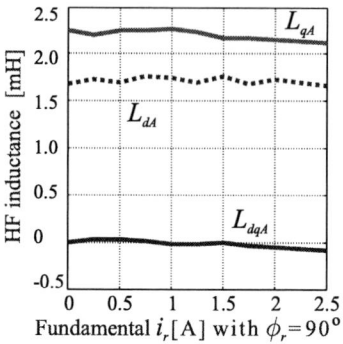

Fig. 8. PMSM anisotropy with load at standstill.

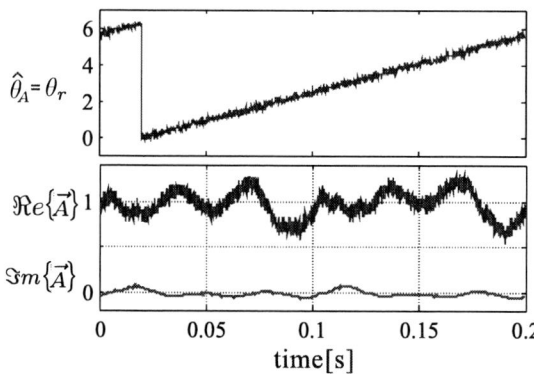

Fig. 9. PMSM anisotropy position dependence at no load.

linear. It is critical to do this compensation, otherwise the wrong position will be used to do the fundamental control, which will lead to loss in torque and instability.

V. COMMERCIAL PMSM

An off-the-shelf 0.5 kW PMSM, with a rated current of 1.6 A rms and a rated torque of 1.5 Nm is used to test the sensorless field orientated scheme for zero and low speed described in this paper and to compare it with the results for the RSM above. The first test is to find a suitable HF voltage: applying $U_{HF} = 20\cos(2\pi 508)$ V results in a HF current $i_{HF} = 0.2\sin(2\pi 508t + 0.4)$, i.e. 6 percent of the available voltage results in a HF current of 9 percent of the rated current. Therefore $\zeta = 0.4\frac{180}{\pi} = 23°$, which represents 2 samples delay with the sampling frequency of $f_s = 12205$ Hz, to align the demodulation sine wave F_{dem} with the HF current. It is noted that much less HF voltage is needed to achieve a HF current of about 10 percent of the rated current, as compared to the RSM. The acoustic noise due to the HF current is minimal.

In the second test we determine the parameters L_{dA} and L_{qA} at no load and find that $L_{dA} = 0.02$ H and $L_{dA} = 0.0255$ H. This measurement is shown in Fig. 7. It is noted that the inductance values are less than 10 times compared with the RSM's inductance values, and also that $L_{dA} < L_{qA}$, where for the RSM it is $L_{dA} > L_{qA}$. We test the dependency of L_{dA} and L_{qA} on the load current $\vec{i}_r$ (current angle of 90°) and find that for this PMSM the parameters stay constant, even up to rated current, shown in Fig. 8. Up to this point, it seems like this machine would be easy to control sensorless and that the compensation function θ_{comp} might not be necessary.

The third test is to look at the dependency of the anisotropy on the actual rotor position. As with the RSM, we set $\hat{\theta}_A = \theta_r$, inject the HF voltage with $\zeta = 0$, and look at the normalized $\vec{A}$, where we expect to see that $\Re\left\{\vec{A}\right\} = 1$ and $\Im\left\{\vec{A}\right\} = 0$. The result is shown in Fig. 9. The anisotropy for this PMSM has a stronger dependency on the rotor position, compared to the result of the RSM in Fig. 5. This test is also performed at half of the rated load and at rated load, and for each load condition the position dependency looks different, although it is always periodic. Simulation was performed using the constant parameters for L_{dA} and L_{qA} to check if the fundamental current vector controller is distorting the HF currents, but it was found that the problem is indeed a rotor position dependent anisotropy and not the current vector controller. So, we expect to find a time-varying and possibly an offset error due to the θ_r dependence of L_{dA} and L_{qA}, where possibly also L_{dqA} is not zero.

In the fourth test we control the fundamental current vector $\vec{i}_r$ using the measured position θ_r and run the anisotropy position estimation in parallel. The result for low speed with zero, half and rated current is shown in Fig. 10(a), with the

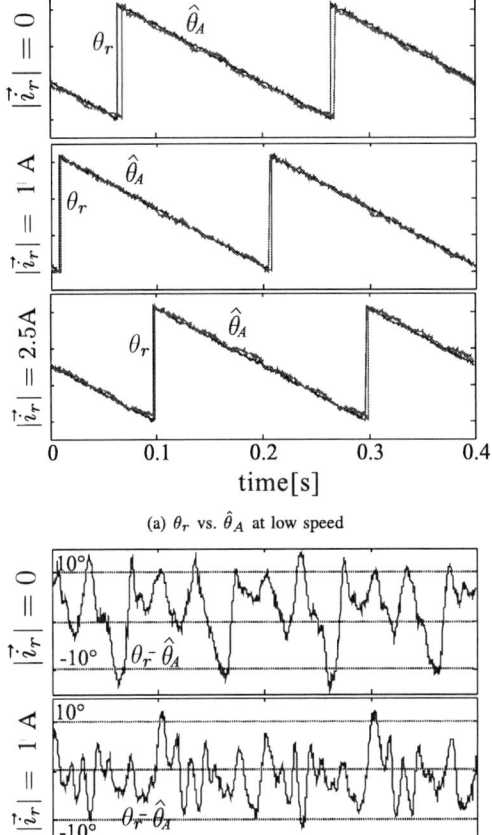

(a) θ_r vs. $\hat{\theta}_A$ at low speed

(b) $\theta_r - \hat{\theta}_A$ at low speed

Fig. 10. θ comparison at zero, half and rated load.

error between actual rotor position and estimated ansitropy position shown in Fig. 10(b). These results can be compared with the results for the RSM in Figs. 6(a) and 6(b).

For this PMSM, the anisotropy angle seems to be aligned with the d-axis of the PMSM (where saturation caused by the PM causes a reduced inductance). At any given rotor position, the parameters L_{dA} and L_{qA} are not influenced so much by the fundamental current, i.e. they are more or less constant. However, the anisotropy is more dependent on the rotor position θ_r, i.e. for some rotor positions, the anisotropy position is equal to the rotor position, and for other rotor

positions not. This kind of error is very difficult to compensate, since it is periodic of nature and also changes its shape for different loads.

VI. CONCLUSION

Position sensorless control of synchronous machines is an important and much researched theme, due to the cost reduction and reliability increase that it brings. A sensorless method has been identified to cope with zero- and low speed requirements, and it involves the application of HF voltages and processing of HF currents that contain the saliency (anisotropy) position information. Under ideal circumstances, the saliency position is equal to the rotor position. However, the estimation algorithm that uses a simplified HF model with constant parameters is plagued by additional saturation caused by the fundamental excitation. The estimation scheme is also disturbed by mutual inductance terms, which cause an offset error in the steady state. The offset error is load dependent, but the problem can easily be solved by using a linearly approximated compensation function. The estimation scheme is further disturbed by the non-sinusoidal flux linkage distribution, which causes a time-varying component in the estimation error signal. This error is difficult to compensate and might lead to large estimation errors, as was shown for the PMSM. This paper has given more insight into the actual HF model of a RSM and PMSM, the effect that the fundamental excitation and rotor position has upon it and how the results can be used to ensure stable sensorless control at zero- and low speed.

REFERENCES

[1] M. Corley and R. Lorenz, "Rotor position and velocity estimation for a salient-pole permanent magnet synchronous machine at standstill and high speeds," *Industry Applications, IEEE Transactions on*, vol. 34, no. 4, pp. 784–789, 1998.

[2] M. Degner and R. Lorenz, "Using multiple saliencies for the estimation of flux, position, and velocity in ac machines," *Industry Applications, IEEE Transactions on*, vol. 34, no. 5, pp. 1097–1104, 1998.

[3] A. Consoli, G. Scarcella, and A. Testa, "Industry application of zero-speed sensorless control techniques for pm synchronous motors," *Industry Applications, IEEE Transactions on*, vol. 37, no. 2, pp. 513–521, 2001.

[4] M. Linke, R. Kennel, and J. Holtz, "Sensorless position control of permanent magnet synchronous machines without limitation at zero speed," in *IECON 02 [Industrial Electronics Society, IEEE 2002 28th Annual Conference of the]*, R. Kennel, Ed., vol. 1, 2002, pp. 674–679 vol.1.

[5] S. Morimoto, K. Kawamoto, M. Sanada, and Y. Takeda, "Sensorless control strategy for salient-pole pmsm based on extended emf in rotating reference frame," *Industry Applications, IEEE Transactions on*, vol. 38, no. 4, pp. 1054–1061, 2002.

[6] J. Faiz and S. Mohseni-Zonoozi, "A novel technique for estimation and control of stator flux of a salient-pole pmsm in dtc method based on mtpf," *Industrial Electronics, IEEE Transactions on*, vol. 50, no. 2, pp. 262–271, 2003.

[7] H. Kim, M. Harke, and R. Lorenz, "Sensorless control of interior permanent-magnet machine drives with zero-phase lag position estimation," *Industry Applications, IEEE Transactions on*, vol. 39, no. 6, pp. 1726–1733, 2003.

[8] M. Linke, R. Kennel, and J. Holtz, "Sensorless speed and position control of synchronous machines using alternating carrier injection," in *Electric Machines and Drives Conference, 2003. IEMDC'03. IEEE International*, vol. 2, 2003, pp. 1211–1217 vol.2.

[9] M. Haque, L. Zhong, and M. Rahman, "A sensorless initial rotor position estimation scheme for a direct torque controlled interior permanent magnet synchronous motor drive," *Power Electronics, IEEE Transactions on*, vol. 18, no. 6, pp. 1376–1383, 2003.

[10] H. Kim, K.-K. Huh, R. Lorenz, and T. Jahns, "A novel method for initial rotor position estimation for ipm synchronous machine drives," *Industry Applications, IEEE Transactions on*, vol. 40, no. 5, pp. 1369–1378, 2004.

[11] Y.-s. Jeong, R. Lorenz, T. Jahns, and S.-K. Sul, "Initial rotor position estimation of an interior permanent-magnet synchronous machine using carrier-frequency injection methods," *Industry Applications, IEEE Transactions on*, vol. 41, no. 1, pp. 38–45, 2005.

[12] O. Wallmark, L. Harnefors, and O. Carlson, "An improved speed and position estimator for salient permanent-magnet synchronous motors," *Industrial Electronics, IEEE Transactions on*, vol. 52, no. 1, pp. 255–262, 2005.

[13] P. Acarnley and J. Watson, "Review of position-sensorless operation of brushless permanent-magnet machines," *Industrial Electronics, IEEE Transactions on*, vol. 53, no. 2, pp. 352–362, 2006.

[14] N. Imai, S. Morimoto, M. Sanada, and Y. Takeda, "Influence of magnetic saturation on sensorless control for interior permanent-magnet synchronous motors with concentrated windings," *Industry Applications, IEEE Transactions on*, vol. 42, no. 5, pp. 1193–1200, 2006.

[15] S. Morimoto, M. Sanada, and Y. Takeda, "Mechanical sensorless drives of ipmsm with online parameter identification," *Industry Applications, IEEE Transactions on*, vol. 42, no. 5, pp. 1241–1248, 2006.

[16] C. Silva, G. Asher, and M. Sumner, "Hybrid rotor position observer for wide speed-range sensorless pm motor drives including zero speed," *Industrial Electronics, IEEE Transactions on*, vol. 53, no. 2, pp. 373–378, 2006.

[17] J. Hu, L. Xu, and J. Liu, "Eddy current effects on rotor position estimation for sensorless control of pm synchronous machine," in *Industry Applications Conference, 2006. 41st IAS Annual Meeting. Conference Record of the 2006 IEEE*, L. Xu, Ed., vol. 4, 2006, pp. 2034–2039.

[18] N. Bianchi and S. Bolognani, "Influence of rotor geometry of an ipm motor on sensorless control feasibility," *Industry Applications, IEEE Transactions on*, vol. 43, no. 1, pp. 87–96, 2007.

[19] N. Bianchi, S. Bolognani, J.-H. Jang, and S.-K. Sul, "Comparison of pm motor structures and sensorless control techniques for zero-speed rotor position detection," *Power Electronics, IEEE Transactions on*, vol. 22, no. 6, pp. 2466–2475, 2007.

[20] H. Hida, Y. Tomigashi, and K. Kishimoto, "Novel sensorless control for pm synchronous motors based on maximum torque control frame," in *Power Electronics and Applications, 2007 European Conference on*, Y. Tomigashi, Ed., 2007, pp. 1–10.

[21] S. Taniguchi, S. Wakao, K. Kondo, and T. Yoneyama, "Position sensorless control of permanent magnet synchronous motor at low speed range using harmonic voltage injection," in *Power Electronics and Applications, 2007 European Conference on*, S. Wakao, Ed., 2007, pp. 1–7.

[22] D. Raca, P. Garcia, D. Reigosa, F. Briz, and R. Lorenz, "A comparative analysis of pulsating vs. rotating vector carrier signal injection-based sensorless control," in *Applied Power Electronics Conference and Exposition, 2008. APEC 2008. Twenty-Third Annual IEEE*, P. Garcia, Ed., 2008, pp. 879–885.

[23] M. Arefeen, M. Ehsani, and A. Lipo, "Sensorless position measurement in synchronous reluctance motor," *Power Electronics, IEEE Transactions on*, vol. 9, no. 6, pp. 624–630, 1994.

[24] R. Lagerquist, I. Boldea, and T. Miller, "Sensorless-control of the synchronous reluctance motor," *Industry Applications, IEEE Transactions on*, vol. 30, no. 3, pp. 673–682, 1994.

[25] M. Schroedl and P. Weinmeier, "Sensorless control of reluctance machines at arbitrary operating conditions including standstill," *Power Electronics, IEEE Transactions on*, vol. 9, no. 2, pp. 225–231, 1994.

[26] P. Jansen and R. Lorenz, "Transducerless position and velocity estimation in induction and salient ac machines," *Industry Applications, IEEE Transactions on*, vol. 31, no. 2, pp. 240–247, 1995.

[27] T. Matsuo and T. Lipo, "Rotor position detection scheme for synchronous reluctance motor based on current measurements," *Industry Applications, IEEE Transactions on*, vol. 31, no. 4, pp. 860–868, 1995.

[28] P. Jansen and R. Lorenz, "Transducerless field orientation concepts employing saturation-induced saliencies in induction machines," *Industry Applications, IEEE Transactions on*, vol. 32, no. 6, pp. 1380–1393, 1996.

[29] M. Jovanovic, R. Betz, and D. Platt, "Sensorless vector controller for a synchronous reluctance motor," *Industry Applications, IEEE Transactions on*, vol. 34, no. 2, pp. 346–354, 1998.

[30] A. Consoli, F. Russo, G. Scarcella, and A. Testa, "Low- and zerospeed sensorless control of synchronous reluctance motors," *Industry Applications, IEEE Transactions on*, vol. 35, no. 5, pp. 1050–1057, 1999.

[31] S.-J. Kang, J.-M. Kim, and S.-K. Sul, "Position sensorless control of synchronous reluctance motor using high frequency current injection," *Energy Conversion, IEEE Transaction on*, vol. 14, no. 4, pp. 1271–1275, 1999.

[32] M.-T. Lin and T.-H. Liu, "Sensorless synchronous reluctance drive with standstill starting," *Aerospace and Electronic Systems, IEEE Transactions on*, vol. 36, no. 4, pp. 1232–1241, 2000.

[33] A. Vagati, M. Pastorelli, F. Scapino, and G. Franceschini, "Impact of cross saturation in synchronous reluctance motors of the transverselaminated type," *Industry Applications, IEEE Transactions on*, vol. 36, no. 4, pp. 1039–1046, 2000.

[34] E. Capecchi, P. Guglielmi, M. Pastorelli, and A. Vagati, "Positionsensorless control of the transverse-laminated synchronous reluctance motor," *Industry Applications, IEEE Transactions on*, vol. 37, no. 6, pp. 1768–1776, 2001.

[35] T. Senjyu, T. Shingaki, and K. Uezato, "Sensorless vector control of synchronous reluctance motors with disturbance torque observer," *Industrial Electronics, IEEE Transactions on*, vol. 48, no. 2, pp. 402–407, 2001.

[36] C.-G. Chen, T.-H. Liu, M.-T. Lin, and C.-A. Tai, "Position control of a sensorless synchronous reluctance motor," *Industrial Electronics, IEEE Transactions on*, vol. 51, no. 1, pp. 15–25, 2004.

[37] P. Guglielmi, M. Pastorelli, G. Pellegrino, and A. Vagati, "Positionsensorless control of permanent-magnet-assisted synchronous reluctance motor," *Industry Applications, IEEE Transactions on*, vol. 40, no. 2, pp. 615–622, 2004.

[38] P. Guglielmi, M. Pastorelli, and A. Vagati, "Impact of cross-saturation in sensorless control of transverse-laminated synchronous reluctance motors," *Industrial Electronics, IEEE Transactions on*, vol. 53, no. 2, pp. 429–439, 2006.

[39] S. Ichikawa, M. Tomita, S. Doki, and S. Okuma, "Sensorless control of synchronous reluctance motors based on extended emf models considering magnetic saturation with online parameter identification," *Industry Applications, IEEE Transactions on*, vol. 42, no. 5, pp. 1264–1274, 2006.

[40] A. Consoli, G. Scarcella, G. Scelba, A. Testa, and D. Triolo, "Sensorless rotor position estimation in synchronous reluctance motors exploiting a flux deviation approach," *Industry Applications, IEEE Transactions on*, vol. 43, no. 5, pp. 1266–1273, 2007.

[41] H. de Kock, M. Kamper, O. Ferreira, and R. Kennel, "Position sensorless control of the reluctance synchronous machine considering high frequency inductances," in *Power Electronics and Drive Systems, 2007. PEDS'07*, 2007.

[42] A. J. Rix, M. J. Kamper, and R.-J. Wang, "Design and performance evaluation of concentrated coil permanent magnet machines for in-wheel drives," in *Electric Machines & Drives Conference, 2007. IEMDC '07. IEEE International*, M. J. Kamper, Ed., vol. 1, 2007, pp. 770–775.

[43] F. Magnussen and H. Lendenmann, "Parasitic effects in pm machines with concentrated windings," *Industry Applications, IEEE Transactions on*, vol. 43, no. 5, pp. 1223–1232, 2007.

[44] O. Wallmark and L. Harnefors, "Sensorless control of salient pmsm drives in the transition region," *Industrial Electronics, IEEE Transactions on*, vol. 53, no. 4, pp. 1179–1187, 2006.

Analysis of VSI-DTC Fed 6-phase Synchronous Machines

Ibrahim Abuishmais*‡, Waqas M. Arshad*, Sami Kanerva†

*ABB Corporate Research, SE-72178 Västerås, Sweden
†ABB Oy, Machines, FI-00380 Helsinki, Finland
‡Norwegian University of Science and Technology O.S. Bragstadsplass 2E, 7491 Trondheim, Norway
Ibrahim.Abuishmais@elkraft.ntnu.no
+47 73594280

Abstract— **High power drives with multiphase machine utilizes paralleled legs converters to realize required power. In addition to the reduced rating of the used semiconductor devices, higher redundancy level could be achieved. Being a relatively new technology for the combination of VSI-fed operation and salient pole synchronous machines, a through analysis is needed. This paper investigates 6-phase synchronous machines with emphasis on redundancy, fault conditions, the machine's behavior under non-sinusoidal voltage profiles and sensitivity of the design parameters. It is shown that new design thinking is required when considering converter-induced machine losses, especially those on the rotor surface. Redundancy study shows the possibility of operating the machine at half load when one supply system is totally or partially lost without exceeding machine's total losses. When studying fault scenarios, the worst case is found to be 4-phase short circuit (equivalent to 2-phase in a 3-phase machine) for the studied cases. Output from sinus supply voltage study was utilized to understand losses distribution inside the machine and establish an effective comparison with different voltage profiles. A study with square wave and DTC supplies shows higher losses in damper bar, when compared to sinus supply. Voltage profiles of two different DTC schemes are studied which show that with particular well-thought control strategies one can improve the machine losses specially those in the damper bars.**

Keywords—Multiphase machines, synchronous machines design, direct torque control.

I. INTRODUCTION

Many industry segments today are increasingly asking for medium-large motor solutions with traditional demands of high efficiency, high power density, reliability and superior control aspects for variable speed operations along with a new demand of redundancy. Dual stator (or 6-phase) motors with 30 degrees displacement between the two 3-phase windings sets (see Fig. 1), provides a solution with this relatively new demand of redundancy. A 6-phase motor satisfies the reliability demands since one of the systems provides a redundancy in case the other fails. Powering the two sets by two different converters; i) adds more redundancy level particularly if one supplying system fails ii) reduces the size of semiconductor devices used to built the converter making the size of large drive system rating feasible and iii) increase the total system efficiency. The choice of synchronous motors fed with voltage source converters and employing direct torque control (DTC) provides a very competitive solution due to

performance advantages of a synchronous motor over an induction motor and the DTC superiority over other converter control schemes. The presented work looks at some of the many aspects associated with this relatively new concept of 6-phase salient-pole synchronous motors fed with DTC-controlled voltage source converters. Issues such as motor design parameters sensitivity, behavior during normal, faulty and partially-faulty (redundancy) operations are addressed by considering the interaction between source's time harmonics and machine's space harmonics and the relevant additional losses.

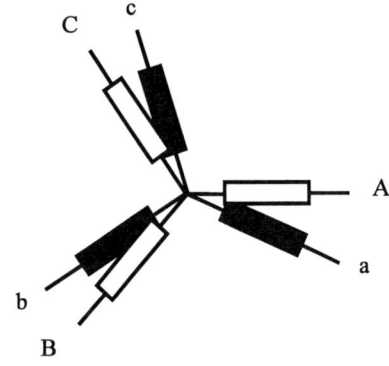

Fig. 1. Schematic drawing of 6-phase machine stator windings, the two winding sets are 30° displaced.

II. LITERATURE SURVEY-A HISTORICAL NOTE

The increase demand on high generation units in late 1920's was restricted by the available circuit breaker capacities at that time. In 1929, design engineers proposed the dual stator windings generator as a solution of the high rated current breaking ability [1] followed by the first addressing of such kind of machines in AIEE in 1930 [2]. The generated voltage in the two windings of separate systems was equal in phase which introduced several discussion issues in generator connections and high voltage switching until P. Robert et al [3] proposal to displace the two windings by 30°. This in addition to solving the connection issues, also resulted in a reduction in stray losses. In 1973, Fuchs and Rosenberg [4] modeled the machine mathematically, using the orthogonal transformation steady state model describing the machine behavior. It is noticed that the used transformation matrix ignores any mutual effect between the two winding sets. In 1974, Nelson and Krause [5] presented a model of multiphase induction machine. Although the proposed

978-1-4244-1741-4/08/$25.00 ©2008 IEEE   867

mathematical model can be useful to gain a better understanding, the authors' assumptions of uniform airgap as well as ignoring the saturation and eddy current effects, makes such model oversimplified to analyze a salient pole machine. However, the authors highlighting of the advantages of the dual stator configuration to achieve higher reliability level and expanded overall drive system availability rate, are valuable. Results of analytical and experimental investigation for 6-phase voltage source inverter driven induction machines were presented in 1983[6],[7]. 6-phase machine model with mutual leakage coupling included was presented by Schiferl and Ong [8], [9]. Back EMF voltage and input current harmonics are also studied. Advantages of multiphase machines, in general, are presented in [10],[4]. A review on multiphase induction machine drives is recently reported in [11].

III. ANALYSIS OF 6-PHASE SYNCHRONOUS MACHINE

A. Harmonic Fields

The nature of synchronous machines and all AC machine designs in general, e.g. rotor saliency and stator's slotting, imposes several deviations from the ideal targeted sinusoidal harmonic free fields and smooth DC torque. The investigation of these phenomena is interesting due to its effect on machine performance and efficiency i.e. total losses and output power.

To analyze harmonic field's causes, analytical approach can be applied. This approach depends on classical analysis of the geometrical parameters of the machine and to determine the resultant flux density. To elaborate more, flux density can be obtained by multiplying permeance waves which are function of stator slotting and rotor saliency by MMF waves see (1) [12]. By applying this approach, the harmonics behavior of the machine can be studied analytically.

$$B(\theta,t) = \frac{\mu_0 A}{\delta(\theta,t)} \frac{1}{A} . MMF(\theta,t) \tag{1}$$
$$= \Lambda(\theta,t). MMF(\theta,t)$$

1) Stator and Rotor Permeance Harmonics

Stator permeance harmonics result mainly due to slotting. For a given number of slots per phase per pole q_S in 6-phase machine stator, total number of stator slots Q_S is given by $Q_S = 12.p.q_S$. Where p is number of pole pairs. Stator spatial harmonics due to slotting can be expressed as

$$h^S = 12.q_S.i \quad i = 1,2,3,... \tag{2}$$

These harmonics are stationary, in other words the wave speed is zero with respect to the synchronous speed. On the other hand, rotor permeance wave depends on pole symmetry. Studied machine has a symmetrical pole pitch geometry which means that permeance wave has following harmonic orders:

$$h^R = 2.j \quad j = 1,2,3,... \tag{3}$$

All of these harmonics pulsate with twice the synchronous speed. Combined effect of rotor and stator permeance results in a waveform with the following harmonic contents:

$$h^{S\&R} = 12.q_S.i \pm 2j \quad i,j = 1,2,3,.. \tag{4}$$

Rotating with speed

$$v^{S\&R} = \frac{\pm 2j}{h^{S\&R}} \quad j = 1,2,3,... \tag{5}$$

2) Stator and Rotor MMF

Stator MMF is a direct result of its winding, in case of 6-phase machines, windings arrangement is interesting as it results in short circuiting some of machine space harmonics for certain time harmonic orders. Windings as a function of space can be expressed as:

$$n_h^x(\theta) = N_h \cos(h\theta - \zeta) \quad h = 1,2,3... \tag{6}$$

Where x index refers to the six phases; A, B, C, a, b, and c, the factor N_h is the windings factor at different harmonic order, while ζ is the phase shift between the two supply systems. The winding functions for different phases can be expressed as [13];

$$n_h^A(\theta) = N_h \cos(h\theta) = \frac{N_h}{2}(e^{jh\theta} + e^{-jh\theta}) \tag{7}$$

$$n_h^B(\theta) = N_h \cos(h\{\theta - \frac{2\pi}{3}\})$$
$$= \frac{N_h}{2}(e^{jh(\theta - \frac{2\pi}{3})} + e^{-jh(\theta - \frac{2\pi}{3})}) = \alpha^{h'} N_h(e^{-jh'})|_{h'=\pm h} \tag{8}$$

$$n_h^c(\theta) = N_h \cos(h\{\theta + \frac{2\pi}{3}\})$$
$$= \frac{N_h}{2}(e^{jh(\theta + \frac{2\pi}{3})} + e^{-jh(\theta + \frac{2\pi}{3})}) = \alpha^{-h'} N_h(e^{-jh'})|_{h'=\pm h} \tag{9}$$

Where $\alpha = e^{j\frac{2\pi}{3}}$. The three remaining functions for the other supply system can be derived in the same manner.

Introducing the operator $\beta = e^{j\frac{\pi}{6}}$, this indicates the physical displacement between the two systems i.e. 30° in the studied case. Winding function of second stator set can be expressed as:

$$n_h^a(\theta) = \beta^{h'} N_h(e^{-jh'})|_{h'=\pm h} \tag{10}$$

$$n_h^b(\theta) = \alpha^{h'}\beta^{h'} N_h(e^{-jh'})|_{h'=\pm h} \tag{11}$$

$$n_h^c(\theta) = \alpha^{-h'}\beta^{h'} N_h(e^{-jh'})|_{h'=\pm h} \tag{12}$$

The resultant magneto motive force is expressed as:

$$MMF^S(\theta,t) = \sum_{x=A,B..} n_h^x(\theta) \times i_k^x(t)$$
$$= \sum_{h'=-\infty}^{\infty}\sum_{k=1}^{S}\{\frac{N_h}{2}((1+\beta^{h'}\beta^{-k})+\alpha^{h'}\alpha^k(1+\beta^{h'}\beta^{-k})$$
$$+ \alpha^{-h'}\alpha^k(1+\beta^{h'}\beta^{-k}))I_k e^{j(k\omega t - h'\theta)}\} \tag{13}$$

Here k is the time harmonic order and the superscript "S" indicates the stator MMF.

Using (13), speeds of airgap space harmonics can be predicted at any time harmonic "k" (k=1 for the fundamental harmonic of supply current) and the space harmonic h'=1,-5, 7,-11, 13 ... etc. It can be noticed that applying different sets of space and time harmonics, results in several harmonics with zero speed, particularly if $h' - k = \pm 6n$ condition is satisfied, here n is an integer. The common term in (13) i.e. $(1 + \beta^{h'} \beta^{-k})$ will be either 0 or 2:

$$(1 + \beta^{h'} \beta^{-k}) = \begin{cases} 0 & \text{if } n \text{ is odd} \\ 2 & \text{if } n \text{ is even} \end{cases} \qquad (14)$$

It is clear that harmonic suppression criterion depends on the phase shift between the two different systems as well as the order of the interacted harmonics.

On the other hand, rotor's excitation creates MMF wave containing harmonics rotate with the synchronous speed and have the order

$$h_m^R = 1, 3, 5 \ldots \qquad (15)$$

The situation for a 6-phase machine is provided in TABLE I showing what harmonics are to be expected due to the interaction between different MMFs and various machine permeances. From this information, a preliminary estimation of the machine behavior for a selected design interacting with a certain converter can be made. This increases machine controllability, knowing the close relation between stator MMF harmonics and the electrical phase shift between the two windings sets. To identify field harmonics; an alternative approach is the FEM analysis employing long time stepping solutions. From the knowledge of fields; torques, losses and damper bars currents developed in the machine can be derived. Fig.2 shows main harmonic contents of the airgap flux density during sinus operation. As can be seen, although the 5th and 7th harmonics are present in the airgap (due to rotor pole saliency) they have no impact on torque pulsations.

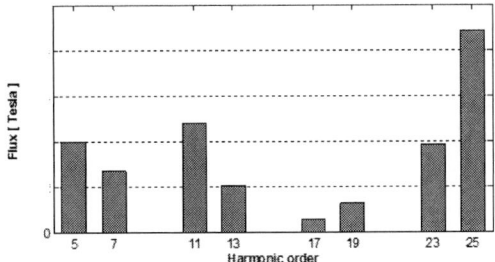

Fig.2. Airgap fields for a sinus-fed 6-phase synchronous motor.

B. Harmonic Torque and Speed

Torque components in synchronous machine are produced mainly by the interaction of stator and rotor fields. Besides the DC torque component that accelerates the machine, pulsating torque components appear at the shaft, these components can be divided into two types, transient and persistent [15].

For persistent non-decaying pulsating components; harmonic orders of developed torque at machine shaft can be predicted utilizing flux linkage components or airgap MMF since it differs only by scale from flux linkage wave. Basically, the produced torque is an interaction between two consecutive flux components resulting in torque with frequency that equals the difference between involved wave's frequencies, taking into account the direction of rotation. For example the 11th MMF component rotates in negative direction with respect to rotor flux rotating direction. It interacts with DC flux component rotating at synchronous speed resulting in {1-(-11)} 12th torque component and same order of the induced damper bars current. See Fig. 3 .

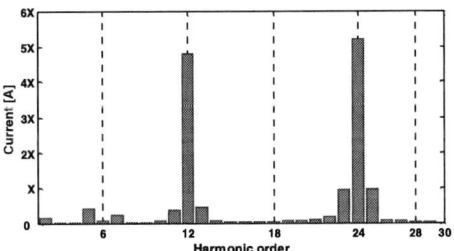

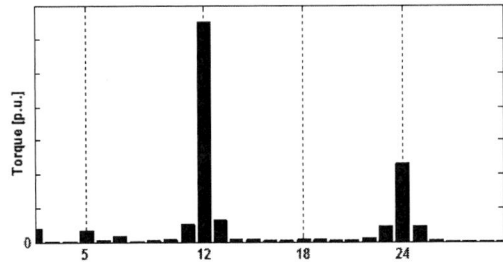

Fig. 3. Correlation between developed torque and damper bars current harmonic spectrums.

In spite of existence of 5th and 7th (17th, 19th ...etc) harmonics in machine's airgap, these harmonics do not contribute in torque production. The same applies to damper bars induced current.

TABLE I.
AIRGAP FLUX DENSITY HARMONICS

Permeance	Stator MMF $h_m^S = 6\,m + 1,$ $m = 0, \pm 2, \pm 4, \pm 6, \ldots$ and $i, j = 1, 2, 3, ..$	Rotor MMF $h_m^R = 1, 3, 5 \ldots$ and $i, j = 1, 2, 3, ..$
Average	h_m^S	h_m^R
Stator slotting	$12.q_s.i \pm h_m^S$	$12.q_s.i \pm h_m^R$
Rotor saliency	$2.j \pm h_m^S$	$2.j \pm h_m^R$
Combined stator and rotor (+)	$12.q_s.i + 2.j \pm h_m'$	$12.q_s.i + 2.j \pm h_m'$
Combined stator and rotor (-)	$12.q_s.i - 2.j \pm h_m'$	$12.q_s.i - 2.j \pm h_m'$

q_S is the number of slots per phase per pole. Superscript "S and R" referred to stator and rotor, respectively.

In 6-phase machine's airgap the fact that the 5th and 7th space harmonics order due to stator MMF are absent results from the combined effect of physical displacement between the two stator windings and electrical phase shift between supplies voltages. It can be concluded that 6th harmonic torque will not appear. Same applies to interpret 18th absence knowing that 17th and 19th MMF components do not exist. On the other hand, airgap fluxes of 6-phase machine contain the same 11th, 13th, 23rd and 25th component's magnitude compared to three-phase machine which means torque pulsations at frequencies 12th and 24th remain the same in the two different machines.

C. 5th, 7th, 17th, 19th Time Harmonics Behavior

The discussed 6-phase machine, differs with a 3-phase machine in terms of its response to 5th, 7th, 17th, 19th,... time harmonics. With 30⁰ degree space shift between the two winding sets and with the same electrical phase shift between the supply voltages, certain harmonics disappear from airgap fluxes. A generalized form of required displacement is π / n for even sets number and $2\pi / n$ for odd number of sets, where n is number of phases [5]. Analytically, analyzing the MMF wave and its interaction with the different permeance functions explains the phenomena. An alternative analytical way is presented by M. Abbas et al. [6] and T. Jahns [14] using the generalized two-phase real component transformation for n-phase machine. The one can conclude that the excitation voltage time harmonics such as 5th, 7th, 17th, 19th ...etc. are prevented from contribution to the airgap flux and torque pulsations while they are contributing in the supply input current. The reactance of these harmonics current paths appears to be relatively small. To verify these analytical conclusions flux paths inside the machine were studied with Finite Element Method (FEM).

Applying an input supply containing only one harmonic order e.g. 5th or 7th at a time, field maps and relative permeability of the rotor region were in focus. Fig. 4 shows field maps in four different cases. 5th and 7th harmonic in supply result in flux which is totally linked in stator region only. This leads to same theoretical conclusions that these components will not contribute in torque pulsations or induce any damper bar currents. It also shows that the stator linkage part of the reactance will disappear making the total path reactance small. The same observation can be noticed in case of 17th and 19th time harmonic orders. On the other hand, 11th, 13th ...etc harmonic results in flux paths linked throughout the rotor region.

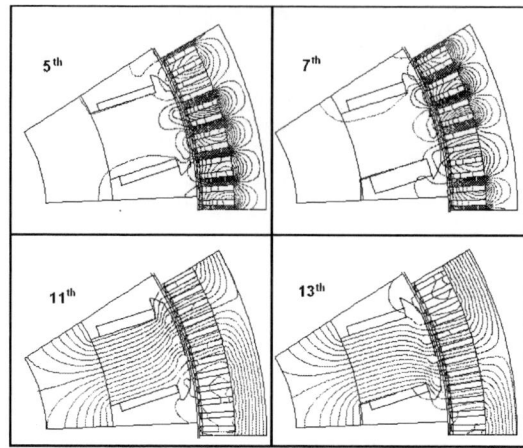

Fig. 4. Field maps for different supply harmonics.

IV. DESIGN PARAMETERS

Sensitivity study for machine's design parameters is also studied, e.g., teeth dimensions, winding layout, wedge material, damper bars and rotor pole shape. The motivation behind this study is to establish a qualitative comparison between the effects of different design parameters on one of the main studied operating aspects, e.g. machine's harmonic fields. It is found that the behavior is not much different to a 3-phase machine. One important difference is to think about pole shape induced 5th, 7th, 17th, 19th etc. harmonics in airgap that induce significant stator harmonic currents and thus affecting machine performance. Such an optimization can be combined with a decrease in the airgap length, which results in a decrease in field windings copper loss at the cost of an increase in damper bar losses.

V. REDUNDANCY ANALYSIS

Using machine with (n) winding systems makes it possible to operate under faulty or partially faulty conditions i.e. open phase or short circuit faults, if each phase's supply or group of phases is being fed from independent supply, like independent inverters. The investigation of the state of being supplied by (n-1) systems is called redundancy analysis. One of the main motivations behind using the 6-phase or double three-phase stator design is to provide a high level of redundancy. Redundancies modes include supply system failure and internal machine faults. In former case losing one supply system means that the machine is driven by three phases out of six in the healthy case. Open phase faults are also considered in this redundancy study.

The simulated cases are:

1. Supplying the 6-phase machine with only one three-phase set, the other three phases are blocked by introducing very high external impedance.
2. Lowering the input current of one system by 25%.
3. Lowering the input current of one system by 65%.

Simulation results for the first redundancy case are provided in Fig. 5, for 50% load torque. Losing three supply phases out of six but developing the same output torque means that the connected phases have to deliver all the required input power. As Fig. 5 shows the increase in the current magnitudes is approximately 100%, it is also noticeable that the waveforms are not identical i.e. they do not contain same harmonic contents. Total harmonic distortion of the input current decreases from 8.5% in case of healthy six phases supply to 6.47% for the three phase supply case. The most significant components are 5th and 7th with big difference in their magnitudes and phase angles. The reason of less distorted input current is the absence of harmonic low impedance paths, in particular for 5th, 7th, 17th and 19th components. As the correspondent counteracting harmonic MMFs which were generated by the adjacent stator set are absent, these fields are able to link through the rotor. Torque pulsations at 12th and 18th are observed in output torque spectrum.

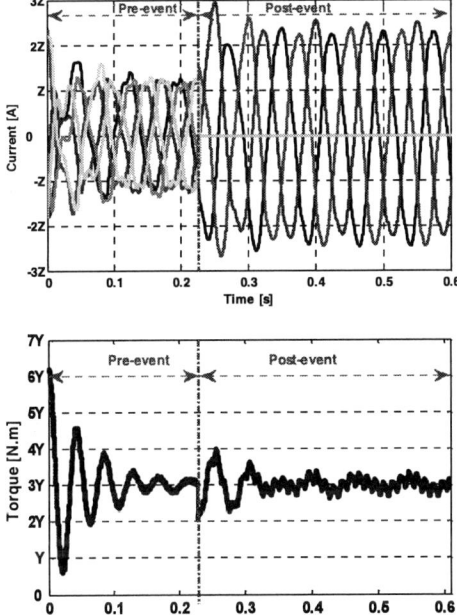

Fig. 5. Input current (above) and developed torque during redundancy event.

During these redundant conditions, input current was limited by introducing a high external resistance. It is found that the best machine behavior is obtained for case 1. Total machine losses are the lowest compared to the other cases.

Input currents for the case 2 and 3 are shown in Fig. 6 and Fig. 7, respectively. for Case 2 where one of the two supplies experiences a partial fault condition causes a reduction of input current approximately by 25%, healthy supply current increases by 68%. Similarly, current reduction in Case 3 by 65% causes an increase in the healthy supply current by 92%. This asymmetry in input currents causes more loss to be dissipated in the two last redundancy cases compared to Case 1. Moreover, torque pulsations like 6th and 18th appear due to asymmetry. Bar losses increase as well.

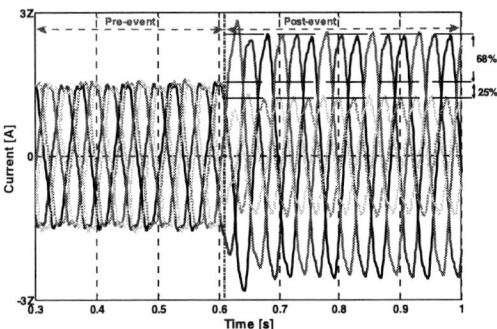

Fig. 6. Input currents during redundancy case 2.

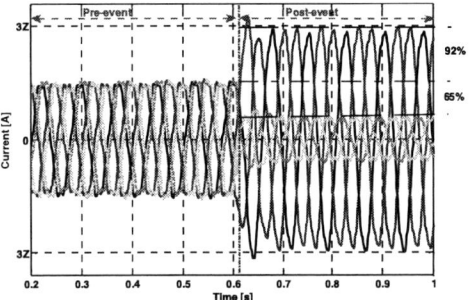

Fig. 7. Input currents during redundancy case 3.

VI. FAULT ANALYSIS

Several different fault cases have been studied, namely; 2×three, 2×two and single phase faults. All short circuits are simulated without introducing any fault impedance. Short circuit currents as well as induced currents in the field windings have been investigated. Results show that the 2×2-phase short circuit results in the highest currents, see Fig. 8. Investigation of flux densities during the different fault conditions show that 2×2-phase case results in the highest stator saturation i.e. lowest machine inductances and hence highest short circuit currents. During single phase faulted phase experiences an increase in the magnitude as expected, but the adjacent phases also experience such an increase.

(a)

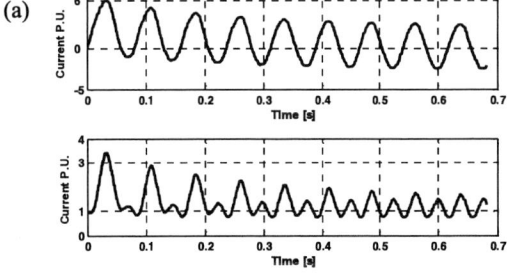

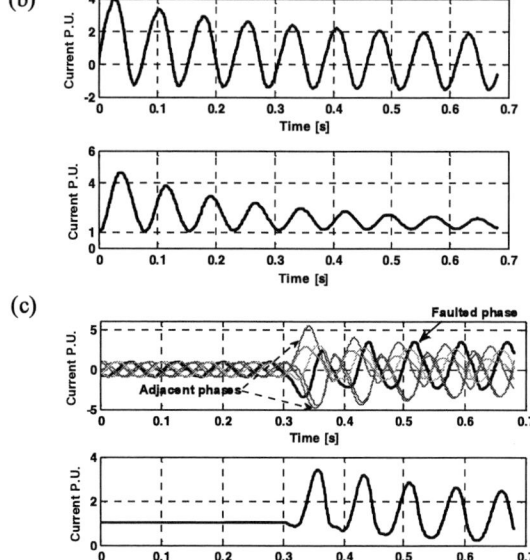

Fig. 8. Per unit phase (above) and field current (down) for: (a) 2×three-phase fault, (b) 2×two phase-fault condition and (c) single phase fault.

VII. Supply Waveforms

Different supply voltage waveforms have been simulated. The aim is to investigate the impact of supply voltage waveform on harmonic fields, machine losses and developed torque. The studied voltage waveforms are:

1. Sinusoidal voltage: the ideal supply waveform.

2. Square wave voltage: With $\pi/4$ P.U. peak and with 1 P.U. peak.

3. Voltage Source Inverter "VSI" waveforms: DTC scheme.

Analyzing the machine under square wave supply facilitate understanding of machine's behavior avoiding the complexity of simulating the stepped voltage waveforms as in DTC case. However, both are presented here. Motivation behind studying two different square waves with different magnitudes is to show the effect of increased voltages i.e. higher flux level, than that the machine is designed for. It is found that the square wave supply results in high magnitudes of THD in stator currents and voltage supply (25-50% higher than sinus cases) and about 5-20% higher losses in total. Instantaneous developed torque in case of sinus and square wave with fundamental component equaling rated voltage (i.e. $\pi/4$ P.U.) peak, are shown in Fig. 9. The torques have same DC value, but higher frequency torque pulsations arise for the later case. An increase of approximately 340% in pulsating magnitude for 12th and 24th components was found.

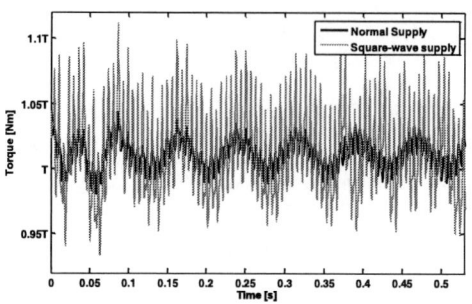

Fig. 9. Developed torque versus time at sinus and square wave supply voltages.

Direct torque control "DTC" scheme embeds the control of the motor and supply inverter together. Switching of inverter devices is dependant upon the electromagnetic state inside the motor. Control signals are produced after comparing reference torque and flux values with real time values and error band limited by hysteresis control method. Such scheme leads to stepped voltage with high harmonic contents and variable switching frequency. The DTC although can have high THD in supply voltage, the magnitude in stator currents is considerably lower and with almost the same total losses as for the sinus case, showing the effectiveness of the DTC solution as far as power density of the motor is concerned. For input voltage and current waveforms see [16]. TABLE II compares the three studied cases.

TABLE II. Comparison between sinus and non-sinus voltage waveform: THD, Losses and main torque pulsation.*

	Sinus	Square wave	DTC
THDi	-	++	+
THDv	-	++	+
Σ Losses	+	++	-
Iron Loss	-	=	=
Field winding loss	+	++	-
6th torque pulsation	=	+	=

* Here "-, +, ++" indicate an increase in magnitude while "=" sign indicates fairly compared figures.

Conclusion

A through analysis of VSI-DTC fed synchronous machines is presented in the paper. Such drive arrangement has been proved to provide a competitive alternative for high power drive systems that demand a high reliability operating conditions. The sensitivity study of design parameters showed the possibility to optimize the machine gaining a higher efficiency level by reducing some of the losses. Redundancy analysis showed the possibility of operating the machine at half load when one supply system is totally or partially lost without exceeding machine's total losses. Saturation level inside the machine varies with fault type that affects fault current as well as field windings induced current during the fault. One natural continuation of this work is to do a multilevel optimization modal where the machine and the converter are optimized together.

ACKNOWLEDGMENT

The authors would like to acknowledge the support from Prof. Tore Undeland (NTNU, Norway), Sonja Lundmark (Chalmers University of Technology, Sweden), Heinz Lendenmann (ABB Corporate Research, Sweden) and Fredrick Kieferndorf (ABB Corporate Research, Switzerland).

REFERENCES

[1] T. F. Baron, "The double winding generator", *General Electric Review*, June 1929. pp. 302-308.

[2] P. L. Alger, E. H. Freburghouse and D. D. Chase, "Double windings for turbine alternators", *AIEE Transaction*, January 1930, pp. 226-244.

[3] R. Robert, J Dispaux and J. Dacier. "Improvement of turbo-alternators efficiency", *Cigre Meeting*, June 8-18, 1966.

[4] E. F. Fuchs, and L. T. Rosenberg, "Analysis of an alternator with two displaced stator windings," *IEEE Transactions on Power Apparatus and Systems*, Vol. 93, No. 6, 1974, pp. 1776-1786.

[5] R. H. Nelson, P. C. Krause, "Induction machine analysis for arbitrary displacement between multiple winding sets", *IEEE Transactions on Power Apparatus and Systems*, Vol. PAS-93, Issue 3. 1974. pp 841-848.

[6] M. Abbas, R. Christen, and T. Jahns, "Six-phase voltage source inverter driven induction motor", *Conference record of 18th annual meeting of IEEE industry application society 1983*, pp 503-511.

[7] M. Abbas and R. Christen, "Characteristics of a high-power density six-phase induction motor", *Conference record of 18th annual meeting of IEEE industry application society 1984*, pp 494-501.

[8] R. F. Schiferl, C. M. Ong., "Six-phase synchronous machine with AC and DC stator connection part I: equivalent circuit representation and steady-state analysis", *IEEE Transactions on Power Apparatus and Systems*, Vol. PAS-102, No. 8, August 1983.

[9] R. F. Schiferl, C. M. Ong., "Six-phase synchronous machine with AC and DC stator connection part II: harmonic study and a proposal uninterruptible power supply scheme", *IEEE Transactions on Power Apparatus and Systems*, Vol. PAS-102, No. 8, August 1983.

[10] L. Paras, "On advantages of multi-phase machines", *31st annual conference of IEEE industrial electronic society record*, IECON 2005. pp 1574-1579.

[11] E. Levi, R. Bojoi, F. Profumo, H.A. Toliyat and S. Williamson, "Multiphase induction motor drives: a technology status review", *IET Electrical power application*, Vol. 1, No. 4, July 2007. pp 489-516.

[12] S. Toader, "Combined analytical and finite element approach to the filed harmonics and magnetic forces in synchronous machines", *ICEM conference record*. Manchester 1994.

[13] D.G. Dorrell, C. Y. Leong, and R. A. McMahon., "Analysis and performance assessment of six-pulse inverter-fed three-phase and six-phase induction machines", *IEEE transaction on industry applications*, Vol. 42, No. 6, November/December 2006.

[14] T. M. Jahns, "Improved reliability in solid state ac drives by means of multiple independent phase-drive units", PhD dissertation at Massachusetts Institute of Technology-MIT. April 1978.

[15] IEEE Std. 1255-2000, "IEEE guide for evaluation of torque pulsations during starting of synchronous motors", *IEEE-SA standards board*, August 2000.

[16] F. Kieferndorf, H. Burzanowska, S. Kanerva and P. Sario, "Modeling of rotor based harmonics in dual-star, wound field, synchronous machines". *Accepted for ICEM 2008*, unpublished.

Optimal Rotor Flux Shape for Multi-phase Permanent Magnet Synchronous Motors

Roberto Zanasi*, Federica Grossi*

*DII, University of Modena and Reggio Emilia, Via Vignolese 905, 41100 Modena, Italy,
e-mail: *roberto.zanasi@unimore.it, federica.grossi@unimore.it*

Abstract—In the paper the Power-Oriented Graphs (POG) technique is used for modeling m-phase permanent magnet synchronous motors and a study on the optimal rotor flux is given. The POG model shows the "power" internal structure of the considered electrical motor: the electric part interacts with the mechanical part by means of a "connection" block which neither stores nor dissipates energy. The dynamic model of the motor is as general as possible and it considers an arbitrary odd number of phases. The rotor flux is analyzed, in particular in order to minimize the currents needed for the torque generation, and its optimal shape is given. The model is finally implemented in Matlab/Simulink and the presented simulation results validate the machine model and the rotor flux choice.

I. INTRODUCTION

The dynamic model of the multi-phase permanent magnet synchronous motors is known in literature obtained using classical mathematical methods. In this paper the dynamic model of this type of motors has been obtained using a Lagrangian approach in the frame of the Power-Oriented Graphs (POG) technique and, for the sake of generality, a generic periodic shape for the rotor flux has been considered. The obtained POG model is very compact, simple and puts in evidence the "power" internal structure of the motor. Using the POG approach and a Concordia-like transformation, the torque vector of the motor assumes a very simple structure which has been analyzed to find the optimal shape of the rotor flux minimizing the electrical power dissipation. The paper is organized as follows. Sec. II gives the basic properties of the POG modeling technique. Sec. III shows the details of POG dynamic model of the m-phase synchronous motors and the optimal shape of the rotor flux. Finally, in Sec. IV some simulation results are reported.

A. Notations

Row matrices will be denoted as follows:

$$\underset{1:n}{\llbracket R_i \rrbracket^i} = \left[\begin{array}{cccc} R_1 & R_2 & \dots & R_n \end{array} \right],$$

column and diagonal matrices as:

$$\underset{1:n}{\llbracket R_i \rrbracket_i} = \left[\begin{array}{c} R_1 \\ R_2 \\ \vdots \\ R_n \end{array} \right], \quad \underset{1:n}{\llbracket R_i \rrbracket^i} = \left[\begin{array}{cccc} R_1 & & & \\ & R_2 & & \\ & & \ddots & \\ & & & R_n \end{array} \right]$$

and full matrices as:

$$\underset{1:n}{\llbracket R_{i,j} \rrbracket_{1:m}^{j}} = \left[\begin{array}{cccc} R_{11} & R_{12} & \cdots & R_{1m} \\ R_{21} & R_{22} & \cdots & R_{2m} \\ \vdots & \vdots & \ddots & \vdots \\ R_{n1} & R_{n2} & \cdots & R_{nm} \end{array} \right]$$

The symbol "$\sum_{n=a:d}^{b} c_n = c_a + c_{a+d} + c_{a+2d} + \dots$" will be used to represent the sum of a succession of numbers c_n where the index n ranges from a to b with increment d that is, using the Matlab symbology, $n = [a : d : b]$.

II. THE BASES OF POWER-ORIENTED GRAPHS

The POG technique [1] is a graphical modeling technique similar to Bond Graph (BG) [2], [3] and Energetic Macroscopic Representation (EMR) [4]. These techniques use the "power interaction" between subsystems as basic element for modeling. The two basic blocks used in the POG technique are shown in Fig. 1: the "elaboration block" (e.b.) and the "connection block" (c.b.). There

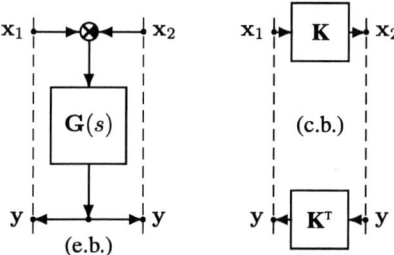

Fig. 1. The POG basic blocks: the elaboration block (e.b.) on the left and the connection block (c.b.) on the right.

is no restriction on the vector variables $\mathbf{x}$ and $\mathbf{y}$ other than the fact that their inner product $\langle \mathbf{x}, \mathbf{y} \rangle = \mathbf{x}^\mathsf{T}\mathbf{y}$ must have the physical meaning of a "power". The e.b. is used for modeling all the physical elements that store and/or dissipate energy (springs, masses, dampers, etc.), i.e. all the 1-port elements (capacitors C, inertias I and resistor R) used in the BG technique. The c.b. is used for modeling all the physical elements that "transform the power" without losses (gear reductors, etc.), i.e. all the Bond Graphs 2-port elements (transformers TR, gyrators GY, modulated transformers MTR and modulated gyrators MGY). The summation element at the top of the e.b. is used for modeling all the 3-port connection elements (0-junction and 1-junction) of the BG technique. More details on Power-Oriented Graphs are reported in [1], [5] and [6].

978-1-4244-1741-4/08/$25.00 ©2008 IEEE

III. ELECTRICAL MOTORS MODELLING

In this paper we will refer only to permanent magnet synchronous electrical motors with an *odd* number m of phases. The electromechanical structure of a seven-phase motor in the case of a single polar expansion ($p = 1$) is shown in Fig. 2. The considered multi-phase electrical motor is characterized by the following parameters:

m : number of motor phases;
p : number of polar expansions;
θ, θ_r : electric and rotor angular positions: $\theta = p\,\theta_r$;
ω, ω_r : electric and rotor angular velocities: $\omega = p\,\omega_r$;
N_c : number of coils for each phase;
R_i : i-th phase resistance ($p = 1$);
L_i : i-th phase self induction coefficient ($p = 1$);
M_{ij} : mutual induction coefficient of i-th phase coupled with j-th phase ($p = 1$);
$\phi(\theta)$: rotor permanent magnet flux;
$\phi_c(\theta)$: total rotor flux chained with stator phase 1;
$\phi_{ci}(\theta)$: total rotor flux chained with stator phase i-th;
φ_r : maximum value of function $\phi(\theta)$;
φ_c : maximum value of function $\phi_c(\theta)$;
J_r : rotor inertia momentum;
b_r : rotor linear friction coefficient;
τ_r : electromotive torque acting on the rotor;
τ_e : external load torque acting on the rotor;
γ : basic angular displacement;

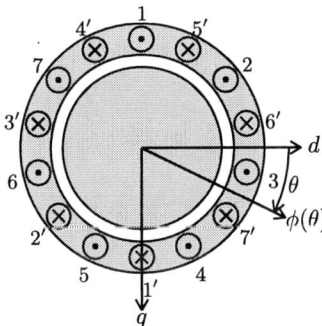

Fig. 2. Structure of a seven-phase motor in the case of single polar expansion ($p = 1$). Vectors d and q denote the "direct" and "quadrature" directions of the flux.

Fluxes $\phi(\theta)$ and $\phi_c(\theta)$ satisfy relations:

$$\phi_c(\theta) = p\,N_c\,\phi(\theta) = p\,N_c\,\varphi_r\,\bar{\phi}(\theta) = \varphi_c\,\bar{\phi}(\theta)$$

where $\varphi_c = p\,N_c\,\varphi_r$ and $\bar{\phi}(\theta)$ is the rotor flux function normalized with respect to its maximum value φ_r.

Let $\gamma = \frac{2\pi}{m}$ denote the basic angular phase displacement for electrical motors with m phases. The following hypotheses are assumed:

H1) Function $\phi(\theta)$ is periodic with period 2π;
H2) Function $\phi(\theta)$ is an even function of θ;
H3) Function $\phi(\theta + \frac{\pi}{2})$ is an odd function of θ;
H4) Flux $\phi_c(\theta)|_{\theta=0}$ chained with phase 1 is maximum;
H5) The motor electrical characteristics are homogeneous.

Let us introduce the state vectors $\dot{\mathbf{q}}$, $\mathbf{q}$ and the gener-

alized flux vector $\mathbf{\Phi}(\mathbf{q})$:

$$\dot{\mathbf{q}} = \begin{bmatrix} \mathbf{I} \\ \omega \end{bmatrix}, \quad \mathbf{q} = \begin{bmatrix} \mathbf{Q} \\ \theta \end{bmatrix}, \quad \mathbf{\Phi}(\mathbf{q}) = \begin{bmatrix} \mathbf{\Phi}_c(\theta) \\ 0 \end{bmatrix}$$

where $\mathbf{I}$ and $\mathbf{\Phi}_c$ are the current and flux vectors:

$$\mathbf{I} = \begin{bmatrix} I_1 \\ I_2 \\ \vdots \\ I_m \end{bmatrix}, \quad \mathbf{\Phi}_c(\theta) = \begin{bmatrix} \phi_{c1}(\theta) \\ \phi_{c2}(\theta) \\ \vdots \\ \phi_{cm}(\theta) \end{bmatrix} = \begin{bmatrix} \phi_c(\theta) \\ \phi_c(\theta - \gamma) \\ \vdots \\ \phi_c(\theta - (m-1)\gamma) \end{bmatrix}.$$

The dynamic equations of the electric motors can be obtained by using the following "Lagrangian" equation:

$$\frac{d}{dt}\left(\frac{\partial K}{\partial \dot{\mathbf{q}}^{\mathrm{T}}}\right) - \frac{\partial K}{\partial \mathbf{q}^{\mathrm{T}}} = \mathbf{V}_e - \mathbf{R}_e\,\dot{\mathbf{q}} \tag{1}$$

where K is the Lagrangian function of the system, $\mathbf{V}_e$ is the extended input vector and $\mathbf{R}_e$ is the extended dissipating matrix. For *multi-phase synchronous motors* the Lagrangian function K has the following structure:

$$K = \frac{1}{2}\dot{\mathbf{q}}^{\mathrm{T}}\mathbf{L}_e\,\dot{\mathbf{q}} - \dot{\mathbf{q}}^{\mathrm{T}}\mathbf{\Phi}(\mathbf{q}) \tag{2}$$

where $\mathbf{L}_e$ is the extended energy matrix of the system. From (1) and (2) one obtains the dynamic equations:

$$\mathbf{L}_e\,\ddot{\mathbf{q}} = \mathbf{V}_e - \mathbf{R}_e\,\dot{\mathbf{q}} - \left[\frac{\partial \mathbf{\Phi}^{\mathrm{T}}}{\partial \mathbf{q}^{\mathrm{T}}} - \frac{\partial \mathbf{\Phi}}{\partial \mathbf{q}}\right]\dot{\mathbf{q}}. \tag{3}$$

The term $\frac{\partial K}{\partial \mathbf{q}^{\mathrm{T}}}$ present in the left part of equation (1) represents the back electromotive voltage generated by the rotor movements. The last term of equation (3) is a skew-symmetric term which represents an internal energy redistribution. From (3) one directly obtains the differential equations of the motor:

$$\underbrace{\left[\begin{array}{c|c} \mathbf{L} & 0 \\ \hline 0 & J_r \end{array}\right]}_{\mathbf{L}_e} \underbrace{\left[\begin{array}{c} \dot{\mathbf{I}} \\ \dot{\omega}_r \end{array}\right]}_{\ddot{\mathbf{q}}} = -\underbrace{\left[\begin{array}{c|c} \mathbf{R} & \mathbf{K}_\tau(\theta) \\ \hline -\mathbf{K}_\tau^{\mathrm{T}}(\theta) & b_r \end{array}\right]}_{\mathbf{R}_e + \mathbf{W}_e} \underbrace{\left[\begin{array}{c} \mathbf{I} \\ \omega_r \end{array}\right]}_{\dot{\mathbf{q}}} + \underbrace{\left[\begin{array}{c} \mathbf{V} \\ -\tau_e \end{array}\right]}_{\mathbf{V}_e} \tag{4}$$

Matrices $\mathbf{L} > 0$, $\mathbf{R}_e$ and $\mathbf{W}_e$ are defined as follows:

$$\mathbf{L} = p \underset{\substack{1:m \ 1:m}}{\left[\! M_{ij} \!\right]^{i \quad j}}, \quad \mathbf{R}_e = \left[\begin{array}{c|c} \mathbf{R} & 0 \\ \hline 0 & b_r \end{array}\right], \quad \mathbf{W}_e = \left[\begin{array}{c|c} 0 & \mathbf{K}_\tau(\theta) \\ \hline -\mathbf{K}_\tau^{\mathrm{T}}(\theta) & 0 \end{array}\right]$$

where $M_{ii} = L_i$, $M_{ij} = M_{ji}$ and terms $\mathbf{K}_\tau(\theta)$ and $\mathbf{R}$ are, respectively, the torque vector and the dissipating matrix:

$$\mathbf{K}_\tau(\theta) = \frac{\partial \mathbf{\Phi}_c^{\mathrm{T}}(\mathbf{q})}{\partial \theta}, \qquad \mathbf{R} = p \underset{1:m}{\left[\! R_i \!\right]^{i}}. \tag{5}$$

Function $\phi_c(\theta)$ is an even and periodic function of period 2π (see H1 and H2) and therefore it can be developed in Fourier series of cosines with only odd harmonics:

$$\phi_c(\theta) = \varphi_c\,\bar{\phi}(\theta) = \varphi_c \sum_{n=1:2}^{\infty} a_n \cos(n\theta). \tag{6}$$

Flux vector $\mathbf{\Phi}_c(\theta)$ can be rewritten in a compact form as:

$$\mathbf{\Phi}_c(\theta) = \varphi_c \underset{0:m-1}{\left[\!\left[\sum_{n=1:2}^{\infty} a_n \cos[n(\theta - h\,\gamma)]\right]\!\right]^{h}}. \tag{7}$$

From (5), the torque vector $\mathbf{K}_\tau(\theta)$ can be expressed as follows:

$$\mathbf{K}_\tau(\theta) = p\,\varphi_c \left[\left[-\sum_{\substack{n=1:2}}^{\infty} n\,a_n \sin[n(\theta - h\,\gamma)] \right]_{0:m-1}\right]. \quad (8)$$

Let us now consider the following orthonormal transformation:

$$
{}^t\mathbf{T}_\omega^{\mathsf{T}} = {}^\omega\mathbf{T}_t = \sqrt{\frac{2}{m}}\left[
\begin{array}{c}
\left[\left[\begin{array}{c} \cos(k\,(\theta - h\,\gamma)) \\ \sin(k\,(\theta - h\,\gamma)) \end{array}\right]_{1:2:m-2}\right]_{0:m-1}^{h} \\[4mm]
\left[\left[\frac{1}{\sqrt{2}}\right]\right]_{0:m-1}^{h}
\end{array}
\right].
$$

Matrix ${}^\omega\mathbf{T}_t$ represents a multi-dimensional rotation in the state space. Matrix ${}^\omega\mathbf{T}_t(\theta)$ is a function of the electrical angle θ and transforms the electric variables $\mathbf{V}$ and $\mathbf{I}$ from the original reference frame Σ_t to a transformed rotating frame Σ_ω. Applying transformation ${}^\omega\mathbf{T}_t$ to matrices $\mathbf{L}$, $\mathbf{R}$ and $\mathbf{K}_\tau(\theta)$, from (4) one obtains the following transformed system:

$$\left[\begin{array}{c|c} {}^\omega\mathbf{L} & 0 \\ \hline 0 & J_r \end{array}\right]\left[\begin{array}{c} {}^\omega\dot{\mathbf{I}} \\ \dot{\omega}_r \end{array}\right] = -\left[\begin{array}{c|c} {}^\omega\mathbf{R}+\mathbf{J}_\omega\,{}^\omega\mathbf{L} & {}^\omega\mathbf{K}_\tau \\ \hline -{}^\omega\mathbf{K}_\tau^{\mathsf{T}} & b_r \end{array}\right]\left[\begin{array}{c} {}^\omega\mathbf{I} \\ \omega_r \end{array}\right] + \left[\begin{array}{c} {}^\omega\mathbf{V} \\ -\tau_e \end{array}\right] \quad (9)$$

where ${}^\omega\mathbf{I} = {}^\omega\mathbf{T}_t\,\mathbf{I}$, ${}^\omega\mathbf{V} = {}^\omega\mathbf{T}_t\,\mathbf{V}$, ${}^\omega\mathbf{R} = {}^\omega\mathbf{T}_t\,\mathbf{R}\,{}^t\mathbf{T}_\omega = \mathbf{R} = p\,R\,\mathbf{I}_m$ and ${}^\omega\mathbf{L} = {}^\omega\mathbf{T}_t\,\mathbf{L}\,{}^t\mathbf{T}_\omega$. Let the self and mutual induction coefficients of matrix $\mathbf{L}$ be defined as:

$$\begin{cases} M_{ij} = M_0 \cos((i-j)\gamma) \\ L_i = \Delta_0 + M_0 \end{cases} \quad \text{for } i,j \in \{1,2,...,m\}.$$

The transformed matrix ${}^\omega\mathbf{L}$ has the following structure:

$${}^\omega\mathbf{L} = p\left[\begin{array}{cccccc} \Delta_0 + \frac{m\,M_0}{2} & 0 & 0 & 0 & \cdots & 0 \\ 0 & \Delta_0 + \frac{m\,M_0}{2} & 0 & 0 & \cdots & 0 \\ 0 & 0 & \Delta_0 & 0 & \cdots & 0 \\ 0 & 0 & 0 & \Delta_0 & \cdots & 0 \\ \vdots & \vdots & \vdots & \vdots & \ddots & \vdots \\ 0 & 0 & 0 & 0 & \cdots & \Delta_0 \end{array}\right],$$

where Δ_0 and M_0 are proper positive parameters. The structure of transformed vector ${}^\omega\mathbf{K}_\tau(\theta)$ is the following:

$${}^\omega\mathbf{K}_\tau(\theta) = {}^\omega\mathbf{T}_t\,\mathbf{K}_\tau(\theta) = -p\,\varphi_c\sqrt{\frac{m}{2}}\,\cdot$$

$$\left[\left[\left[\begin{array}{c} \displaystyle\sum_{n=0:2m}^{\infty}[(n+k)\,a_{n+k} + (n-k)\,a_{n-k}]\sin(n\theta) \\[4mm] \displaystyle\sum_{n=0:2m}^{\infty}[(n+k)\,a_{n+k} - (n-k)\,a_{n-k}]\cos(n\theta) \end{array}\right]_{1:2:m-2}^{k}\right] \\[6mm] -\sqrt{2}\displaystyle\sum_{n=m:2m}^{\infty} n\,a_n \sin(n\theta) \right] \quad (10)$$

Vector ${}^\omega\mathbf{K}_\tau(\theta)$ can be easily computed knowing the coefficients a_n of the Fourier series of the rotor flux,

see eq. (6). Note that vector ${}^\omega\mathbf{K}_\tau(\theta)$ is composed only by the harmonics $\sin(n\theta)$ and $\cos(n\theta)$ where n is an integer number multiple of $2m$. A detailed discussion of the properties of the components of vector ${}^\omega\mathbf{K}_\tau(\theta)$ can be found in [8]. The structure of matrix $\mathbf{J}_\omega$ and vector ${}^\omega\mathbf{I}$ in (9) are the following:

$$\mathbf{J}_\omega = \left[\begin{array}{cc} \left[\left[\begin{array}{cc} 0 & -k\,\omega \\ k\,\omega & 0 \end{array}\right]\right]_{1:2:m-2}^{k} & 0 \\ 0 & 0 \end{array}\right], \quad {}^\omega\mathbf{I} = \left[\begin{array}{c} \left[\left[\begin{array}{c} {}^\omega I_{dk} \\ {}^\omega I_{qk} \end{array}\right]\right]_{1:2:m-2}^{k} \\ {}^\omega I_m \end{array}\right]$$

where $\omega = \dot{\theta}$ is the time-derivative of the electric angle θ and ${}^\omega I_{dk}$, ${}^\omega I_{qk}$ are, respectively, the *direct* and *quadrature* components of the current vector ${}^\omega\mathbf{I}$.

The POG scheme of the multi-phase electrical motor in the transformed space Σ_ω, see eq. (9), is shown in Fig. 3. The elaboration blocks present between the power sections ① and ② represent the *Electrical part* of the system, while the blocks present between sections ③ and ④ represent the *Mechanical part* of the system. The connection blocks present between sections ① and ⓘ and between sections ② and ③ represent, respectively, the state space transformation ${}^t\mathbf{T}_\omega$ between reference frames Σ_t and Σ_ω, and the energy and power conversion (without accumulation nor dissipation) between the electrical and mechanical parts of the motor.

Proposition 1: the torque vector ${}^\omega\mathbf{K}_\tau(\theta)$ in (10) is constant (i.e. it is not function of the electric angle θ) only for the flux functions $\bar{\phi}(\theta)$ which can be expressed in Fourier series as follows:

$$\bar{\phi}(\theta) = \sum_{i=1:2}^{m-2} a_i \cos(i\,\theta) \quad (11)$$

Proof. From (10) it follows that the torque vector ${}^\omega\mathbf{K}_\tau(\theta)$ is constant and different from zero if and only if the following relations hold:

$$\begin{cases} a_{kq}(n,k) \neq 0 & \text{for } n=0 \text{ and } k \in \{1:2:m-2\} \\ \left.\begin{array}{c} a_{kd}(n,k) = 0 \\ a_{kq}(n,k) = 0 \end{array}\right\} \text{ for } \left(\begin{array}{c} n \in \{2m:2m:\infty\} \\ k \in \{1:2:m-2\} \end{array}\right) \\ a_n = 0 & \text{for } n \in \{m:2m:\infty\} \end{cases} \quad (12)$$

where $a_{kq}(n,k) = [(n+k)\,a_{n+k} - (n-k)\,a_{n-k}]$, $a_{kd}(n,k) = [(n+k)\,a_{n+k} + (n-k)\,a_{n-k}]$. Since $n > k$ when $n \neq 0$ and $a_h = 0$ when $h < 0$, relations (12) can be rewritten in the following equivalent form:

$$\begin{cases} a_k \neq 0 & \text{for } k \in \{1:2:m-2\} \\ \left.\begin{array}{c} a_{n+k} = 0 \\ a_{n-k} = 0 \end{array}\right\} \text{ for } \left(\begin{array}{c} n \in \{2m:2m:\infty\} \\ k \in \{1:2:m-2\} \end{array}\right) \\ a_n = 0 & \text{for } n \in \{m:2m:\infty\} \end{cases} \quad (13)$$

that is

$$a_i \neq 0 \qquad \text{for} \qquad i \in \{1:2:m-2\}$$

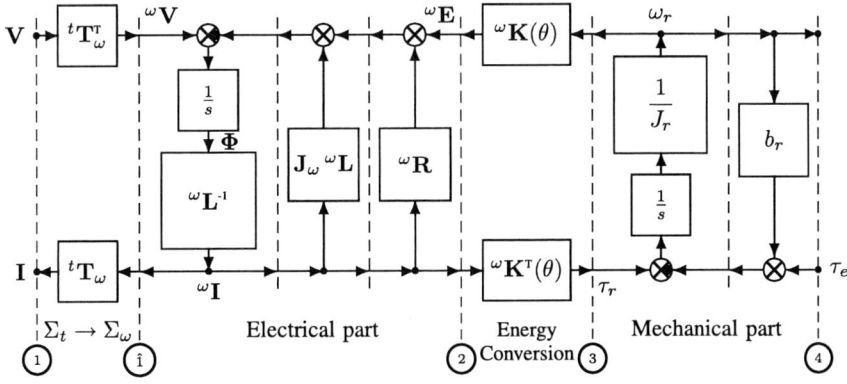

Fig. 3. POG scheme of a multi-phase electrical motor in the transformed space Σ_ω.

as stated in (11), so *Proposition 1* is proved. □

Remark 1: flux function $\bar\phi(\theta)$ in (11) is valid only when the torque vector $^\omega K_\tau(\theta)$ has the form given in (10) which corresponds to the case of m independent motor phases. In the opposite case of star-connected phases, the last term of vector $^\omega K_\tau(\theta)$ is not present and relation (11) must be substituted by $\bar\phi(\theta) = \bar\phi(\theta) + \bar\phi_m(\theta)$ where the new term $\bar\phi_m(\theta) = \sum_{i=m:2m}^\infty a_i \cos(i\,\theta)$ has no influence on the torque generation.

All the possible constant vectors $^\omega K_\tau(\theta)$ are obtained from (10) when $n = 0$:

$$^\omega K_\tau^{\mathsf T} = {^\omega K_\tau^{\mathsf T}(\theta)}|_{n=0} = -\varphi_c\, p\sqrt{\frac{m}{2}}\left[\!\left[\begin{array}{ccc} 0 & k\,a_k & 0 \end{array}\right]\!\right]_{1:2:m-2}^{k}. \quad (14)$$

Main Proposition: among all the fluxes that provide a constant vector $^\omega K_\tau$, see (11), the one that minimizes the current module (and therefore the dissipated power) is given by:

$$\bar\phi(\theta) = a_{m-2}\cos((m-2)\,\theta). \quad (15)$$

Proof. Let $^\omega I_d$ denote the constant desired current. The condition $^\omega I = {^\omega I_d}$ can be achieved by using the following control law:

$$^\omega V = ({^\omega R} + J_\omega\,{^\omega L})\,{^\omega I} + {^\omega K_\tau}\,\omega_r - K_s({^\omega I} - {^\omega I_d}) \quad (16)$$

where $K_s > 0$ is a diagonal matrix used for the tuning of the control. Putting relation (16) in (9) one obtains the dynamics:

$$^\omega L\,{^\omega \dot I} = -K_s({^\omega I} - {^\omega I_d}). \quad (17)$$

Defining $^\omega \tilde I = ({^\omega I} - {^\omega I_d})$ as the current error vector and remembering that $^\omega L\,{^\omega \dot I_d} = 0$, from (17) one obtains

$$^\omega L\,{^\omega \dot{\tilde I}} = -K_s\,{^\omega \tilde I}. $$

With a proper choice of matrix K_s it is possible to give the subsystems the desired dynamics. The time constant τ_i of the i-th subsystem is given by

$$\tau_i = \frac{^\omega L_i}{K_{s_i}}$$

where $^\omega L_i$ and K_{s_i} are the i-th element on the diagonal of matrix $^\omega L$ and matrix K_s, respectively. Chosen the desired time constants, matrix K_s is given by

$$K_s = \left[\!\left[\begin{array}{c} K_{s_i} \end{array}\right]\!\right]_{1:m}^{i}, \quad K_{s_i} = \frac{^\omega L_i}{\tau_i}. \quad (18)$$

Since matrix K_s is positive definite, after a transient the error vector $^\omega \tilde I$ tends asymptotically to zero (i.e. $^\omega I$ tends to $^\omega I_d$) for all the desired constant currents $^\omega I_d$. The desired torque τ_d generated by current vector $^\omega I_d$ is given by relation:

$$\tau_d = {^\omega K_\tau^{\mathsf T}}\,{^\omega I_d}. \quad (19)$$

The set of all the current vectors $^\omega I_d$ satisfying relation (19) is the following:

$$^\omega I_d = {^\omega I_0} + \mathrm{Ker}[{^\omega K_\tau^{\mathsf T}}] \quad (20)$$

where $^\omega I_0$ is a particular solution of system (19) and $\mathrm{Ker}[{^\omega K_\tau^{\mathsf T}}]$ is the kernel of the row matrix $^\omega K_\tau^{\mathsf T}$. Among all the vectors $^\omega I_d$ given by (20) the one which has the minimum modulus is the current vector $^\omega I_d$ which is parallel to vector $^\omega K_\tau$:

$$^\omega I_d = \frac{\tau_d}{|^\omega K_\tau|}\,{^\omega \hat K_\tau} \quad (21)$$

where $^\omega \hat K_\tau$ denotes the versor of vector $^\omega K_\tau$. Note that the modulus of current $^\omega I_d$ is inversely proportional to the modulus of vector $^\omega K_\tau$. So, among all the fluxes given by (11), the one that minimizes the modulus of the current $^\omega I_d$ is the one which maximizes the modulus $|^\omega K_\tau|$ of vector $^\omega K_\tau$. From (14) it follows that the problem of finding the $(m-1)/2$ coefficients a_k that maximize the modulus of vector $^\omega K_\tau$ is equivalent to the problem of maximizing the following functional $F(\mathbf a)$:

$$F(\mathbf a) = \sqrt{\sum_{k=1:2}^{m-2}(k\,a_k)^2}$$

where $\mathbf a = \{a_1, a_3, \ldots, a_{m-2}\}$, under the constraint of unitary maximum amplitude of the flux function $\bar\phi(\theta, \mathbf a)$:

$$\max_\theta \bar\phi(\theta, \mathbf a) = 1. \quad (22)$$

877

This problem admits one single maximum for:

$$a_k = \begin{cases} 0 & \text{for} \quad k = [1 : 2 : m - 4] \\ 1 & \text{for} \quad k = m - 2 \end{cases}$$

which corresponds to the flux shape $\bar{\phi}(\theta) = a_{m-2}\cos((m-2)\,\theta)$ given in (15). So, for constant torque τ_r the rotor flux $\bar{\phi}(\theta)$ which minimizes the modulus of the motor current $^\omega\mathbf{I}_d$ (least power dissipation) is the flux (15). $\square$

In the case of $m = 7$, the functional $F(\mathbf{a})$ is a function of parameters a_1, a_3 and a_5. Since these parameters satisfy constraint (22), then it is always possible to express two of them as a function of the third one. Fig. 4 shows the functional $F(\mathbf{a})$ as a function of the normalized parameters a_1/a_5 and a_3/a_5: the maximum corresponds to solution $a_1 = a_3 = 0$ and $a_5 = 1$.

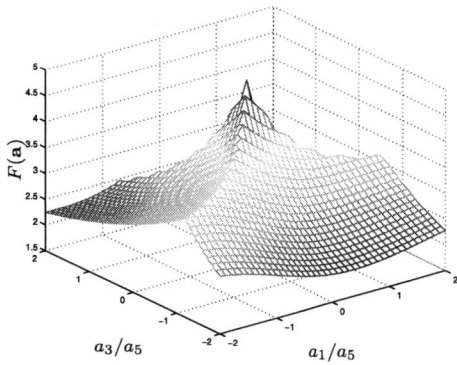

Fig. 4. Functional $F(\mathbf{a})$ as a function of parameters a_k in the case of $m = 7$.

Remark 2: eq. (14) clearly shows that all the odd components of the torque vector $^\omega\mathbf{K}_\tau$ are zero and from (21) it follows that the minimum vector $^\omega\mathbf{I}_d$ that generates the desired torque τ_d is parallel to the torque vector $^\omega\mathbf{K}_\tau$. This leads to conclude that the *direct* components $^\omega I_{dk}$ of vector $^\omega\mathbf{I}_d$ must be zero.

IV. SIMULATIONS

The Simulink scheme of the controlled electric motor is shown in Fig. 5: the main central block corresponds to the POG scheme shown in Fig. 3. The simulation results presented in this Section have been obtained using the following electrical and mechanical parameters: $m = 9$, $p = 1$, $R = 3\,\Omega$, $L_0 = 0.1$ H, $M_0 = 0.08$ H, $N_c = 30$, $\varphi_r = 0.02$ W, $J_r = 0.5$ kg m^2, $b_r = 1.8$ N m s/rad and $\tau_e = 0$ Nm. The odd harmonics $\{1, 3, 5, 7\}$ of the rotor flux $\bar{\phi}(\theta)$ when $m = 9$ and $p = 1$ are shown in Fig. 6. The motor phases are supposed to be star connected. The input vector $^\omega\mathbf{V}$ is given by control law (16) where $^\omega\mathbf{I}_d$ has been calculated using relation (19) considering the desired torque $\tau_d = 10$ Nm for $t \in [0, 1.5]$ s and $\tau_d = 5$ Nm for $t \in [1.5, 3]$ s. The elements K_{s_i} of matrix $\mathbf{K}_s$, see eq. (18), are chosen in order to have the time constants $\tau_i = \{0.33\text{s}, 0.25\text{s}, 0.17\text{s}, 0.09\text{s}\}$ for $i \in \{1, 3, 5, 7\}$. Figures 7÷10 show simulation results

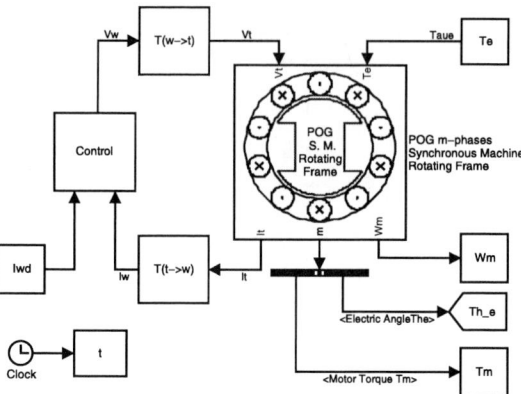

Fig. 5. Simulink scheme of the controlled electric motor.

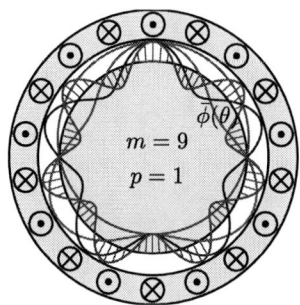

Fig. 6. Shapes of the rotor flux $\bar{\phi}(\theta)$: odd harmonics $\{1, 3, 5, 7\}$ when $m = 9$ and $p = 1$.

obtained for different harmonics of the rotor flux: 1-st harmonic (magenta), 3-rd harmonic (green), 5-th harmonic (blue) and 7-th harmonic (red). Fig. 7 shows the motor velocity ω_r and the rotor torque τ_r: note that the desired value τ_d is reached with the time constants τ_i defined above. Fig. 8 shows the modulus of the phase currents $^\omega\mathbf{I}$: note that the smallest current corresponds to the 7-th harmonic (red line) of the rotor flux, as stated in (15). The quadrature currents $^\omega I_{qk}$, for $k \in \{1, 3, 5, 7\}$, are shown in Fig. 9. Direct currents are not shown because they are equal to zero. Figures 10 and 11 are obtained with the same parameters and same control, but with saturated input voltages $|V_i| \leq 14$ V. Fig. 10 shows the motor velocity ω_r and the rotor torque τ_r: for growing values of the torque vector, the counter electromotive torques increase and the reached velocities decrease. Saturated voltages $\mathbf{V}$ and currents $\mathbf{I}$ in the original reference frame Σ_t for the 7-th harmonic are shown in Fig. 11: note that when the input voltages V_i reach the saturation the desired torque τ_d is no more obtained and a small ripple appears on the final value, see the zoom in Fig. 10.

V. CONCLUSIONS

In this paper a m-phase permanent magnet synchronous motor has been modeled using the Power-Oriented Graphs (POG) technique. The obtained POG model is very compact and can be easily implemented in Simulink.

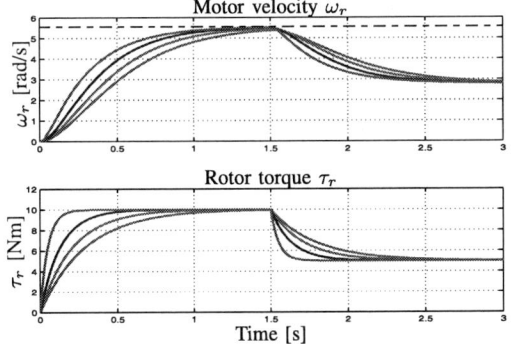

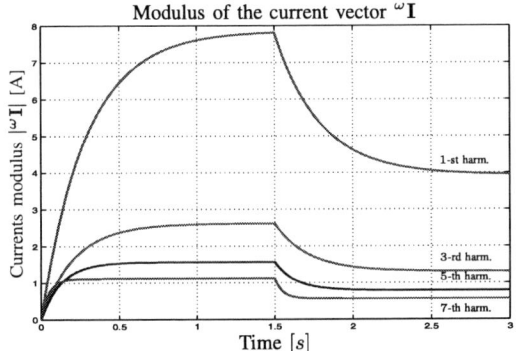

Fig. 7. Motor velocity ω_r and rotor torque τ_r for the odd harmonics $\{1, 3, 5, 7\}$ of the rotor flux $\bar{\phi}(\theta)$.

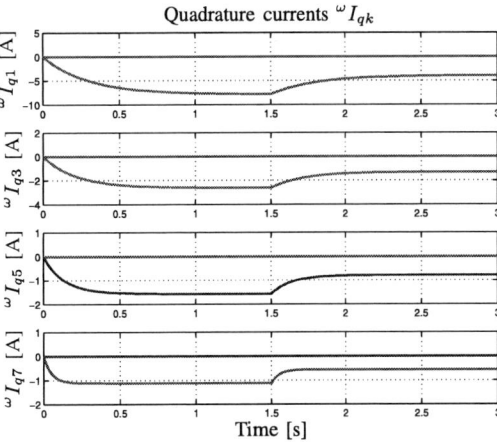

Fig. 9. Quadrature currents $^{\omega}I_{qk}$ in the transformed reference frame Σ_ω for the harmonics $i = \{1, 3, 5, 7\}$ of the rotor flux $\bar{\phi}(\theta)$. Currents $^{\omega}I_{qk}$ are different from zero only when $k = i$.

Fig. 8. Modulus of the phase currents $^{\omega}\mathbf{I}$ in the transformed reference frame Σ_ω.

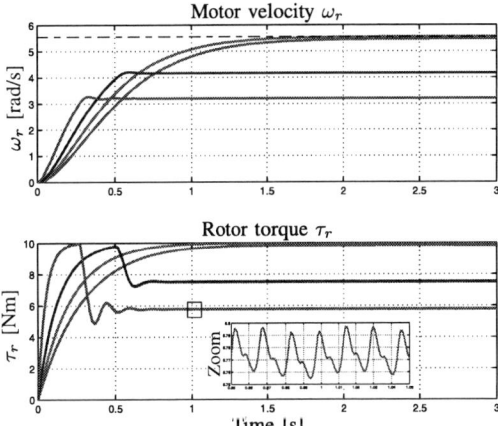

Fig. 10. Motor velocity ω_r and rotor torque τ_r for the odd harmonics $\{1, 3, 5, 7\}$ of the rotor flux $\bar{\phi}(\theta)$ when the input voltage is saturated.

A deep analysis of the torque vector has been carried out in order to find the optimal shape of the rotor flux which minimizes the module of the current vector. Simulation results show the effectiveness of the realized model in the case of a nine-phase star-connected motor.

REFERENCES

[1] R. Zanasi, "Power Oriented Modelling of Dynamical System for Simulation", *IMACS Symp. on Modelling and Control of Technological System*, Lille, France, May 1991.

[2] Paynter, H.M., *Analysis and Design of Engineering Systems*, MIT-press, Camb., MA, 1961.

[3] D. C. Karnopp, D.L. Margolis, R. C. Rosemberg, *System dynamics - Modeling and Simulation of Mechatronic Systems*, Wiley Interscience, ISBN 0-471-33301-8, 3rd ed. 2000.

[4] A. Bouscayrol, B. Davat, B. de Fornel, B. Franois, J. P. Hautier, F. Meibody-Tabar, M. Pietrzak-David, "Multimachine Multiconverter System: application for electromechanical drives", Eur. Physics Journal - Appl. Physics, vol. 10, no. 2, pp. 131-147, 2000.

[5] R. Morselli, R. Zanasi, "Modeling of Automotive Control Systems Using Power Oriented Graphs", 32nd Annual Conference of the IEEE Industrial Electronics Society, IECON 2006, Parigi, 7-10 Novembre, 2006.

[6] R. Zanasi, F. Grossi "The POG technique for modelling multi-phase permanent magnet synchronous motors", 6th EUROSIM Congress on Modelling and Simulation, Ljubljana, 9-13 September 2007.

[7] R. Zanasi, G. H. Geitner, A. Bouscayrol, W. Lhomme, "Different energetic techniques for modelling traction drives", ELECTRIMACS 2008, Québec, Canada, June 2008.

[8] R. Zanasi, F. Grossi, "Multi-phase Synchronous Motors: POG Modeling and Optimal Shaping of the Rotor Flux", ELECTRIMACS 2008, Québec, Canada, June 2008.

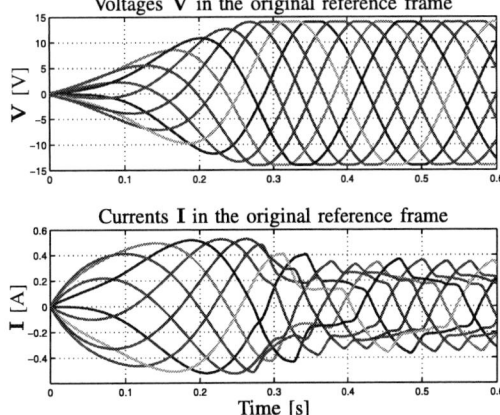

Fig. 11. Voltages $\mathbf{V}$ and currents $\mathbf{I}$ in the original reference frame Σ_t for the 7-th harmonic when the input voltages are saturated to 14 V.

879

Modelling of Electrical Machines Using the Modelica Bond-Graph Library

Mieczysław Ronkowski

Gdansk University of Technology/Department of Power Electronics and Electrical Machines, Gdansk, Poland,
e-mail: *m.ronkowski@ely.pg.gda.pl*

Abstract—The paper presents the bond graphs approach to modelling electrical machines for controlled electromechanical systems applications. The bond graph models of electrical machines are based on the models of ideal couplings: transformer and electromechanical couplings, respectively. Using the couplings models as building blocks the models of electrical machines (brushed and brushless dc machines with permanent magnets, and synchronous machine) have been developed in terms of bond graphs. The models have been elaborated using the Modelica Bond-Graph Library and the DYMOLA package.

Keywords—electrical machine, modelling, simulation.

I. Introduction

Presently, there is an increasing demand for controlled electromechanical (mechatronic) systems with higher performance, more flexibility, and reliability. Such systems are generally multi-physics, i.e., they are made up of a large number of parts, which interact in many ways. Their overall behaviour is derived both from the individual character of components, subsystems and from their interconnection. Moreover, the modern electro-mechanical systems are the result of design activities of not just one discipline engineers.

It is argued that bond graphs based modelling is ideally suited for the description of the energy part of the multi-physics systems [1,4,5,7]. Bond graphs form an modeling tool that enables its user to understand and explain the dynamics of physical processes clearly and concisely. For this reason, bond graphs are suitable both as a didactic and also as a highly practical tool for modelling multi-physics systems. Bond graphs models improves insight and direct feedback on modelling, simulation and design decisions.

In every sort of electromechanical system, which function is power conversion, electrical machines (EMs) play an important role. An EM is an energy-converting link between electrical and mechanical subsystems, and as such it cannot properly be considered in isolation.

The aim of this paper is to present the bond graph approach to development EM models for controlled electro-mechanical systems applications. This paper is one of several resulting from a recent revision of EMs modeling approach at the Faculty of Electrical and Control Engineering, Gdansk University of Technology, to keep pace with the changes in modern technology.

The EM models have been elaborated using the Modelica Bond-Graph Library and the editor of the DYMOLA package. The BondLib is a graphical modelling environment that was developed in the realm of the Modelica/DYMOLA family of modelling tools [2,3,12,13].

II. Basics Of Bond Graphs Notation

An important approach to modeling of engineering systems grew out of the apparent similarities between electrical circuits and other kinds of physical systems. One network-based formalism – the bond graph, was developed by Professor Henry Paynter at MIT [7]. Using this methodology, a physical system is reticulated into a graph by systematic "lumping" of spatially distributed physical quantities, so that the system's dynamic behavior can be simulated and analyzed.

Paynter realized himself that the concept of a 'port', introduced in electrical circuit theory by H.A. Wheeler [11], should be extended to arbitrary power ports that can be applied to any physical domain. Power ports include: mechanical ports, hydraulic ports, thermal ports, electric ports, etc. The port concept allows a domain-independent notation that Paynter called 'bond graphs'.

A bond graph is an example of a more general class of graph-based representation that describe the engineering system functionality and form in qualitative terms and schematically.

Paynter designed the notation using the efficient representation of the relation between two ports by just one line (Fig.1). This line he called a 'bond' (edge). The bond graph notation was completed when Paynter finally introduced the concept of the so-called 'junction structures' which were manifestations of the constraints [7]. Later on, bond graph theory has been further developed by many researchers, like Karnopp, Margolis and Rosenberg, who have worked on extending this modelling technique to mechatronic systems.

A number of researchers have used bond graphs to model and represent the structure of multi-domain systems for the purposes of analysis and design.

An excellent exposure of bond graph modelling can be found in the highly recognised textbook of Karnopp, Margolis and Rosenberg [5] of which the fourth edition was recently published in December 2005. Finally, bond graph modelling is supported by a number of advanced modelling and simulation software packages. A survey compiled by A Samantaray is available at www.bondgraphs.com/software.html. In the bond graph approach, power continuity equations are formulated instead of energy conservation laws. It turns out that, in any physical system, the power balance is a local property, i.e., for modelling the power balance the equations can be expressed for each subsystem separately, and then all the

978-1-4244-1741-4/08/$25.00 ©2008 IEEE

subsystems can be connected as long as the power is also balanced at all the ports between submodels [1,2,3,4,5,7].

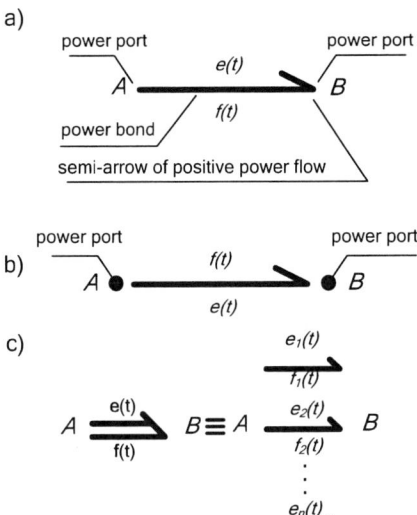

Fig. 1. Variables e(t) and f(t) modelling power flow from element/subsystem A to element/subsystem B: a) model representation in terms of bond graph, b) model representation in terms of bond graph using some software packages (ports are visible) c) model representation in terms of multi bond graphs.

The power in any physical system is written as the product of two conjugate variables (in general as function of time t):

$$p(t) = e(t) \cdot f(t) \qquad (1)$$

called the effort (denoted by e) and the flow (denoted by f) in bond graph terminology. Thus, contrary to block diagrams, in bond graphs the two power conjugated variables e and f are assigned to each bond (edge).

In a bond graph, energy flow (positive) from one port of a element/subsystem A to another B is denoted by a harpoon (a semi-arrow), as shown in Fig.1a.

When similarities in various subsystem components in the model structure can be established, they can be represented in form of a concise notation called vector or multi bond graphs (Fig.1c). The multi bond graphs are useful when initial ideas are being formulated, they may obscure many physical aspects of the system.

In an electrical system, it is customary to select the voltage (or electrical potential) as the effort variable, and the current as the flow variable. In a translational mechanical system, the force will be treated as effort, and the velocity as – flow. In a rotational system, the torque is assumed as the effort, and the angular velocity as flow.

To establish graphically the cause and effect relationships between the factors of power flow in bond graph the causality symbol – causal stroke, has been introduced. The causal stroke indicates the direction in which the effort signal is directed (by implication, the end of the bond that does not have a causal stroke is the end towards which the flow signal is directed). Causality is

simply the specification of which variables are independent and which are dependent. The distinction is quite significant for organizing equations. The basic concept of the causality has been shown in Fig. 2.

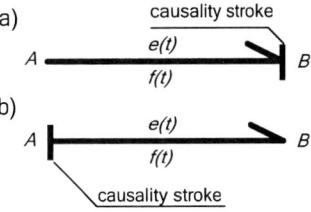

Fig. 2. Causality - the relation of cause and effect between element/subsystem A and element/subsystem B and its representation in terms of bond graph:
a) effort e – cause, flow f – effect; b) flow f – cause, effort e – effect

On the lowest hierarchy level bond graph models are called elements. They represent basic physical processes in which energy is distributed, or transferred from one port to another, or transformed in the same energy domain, or converted into another energy form, in particular into heat, or stored.

Like in physical systems modelling based on networks, bond graph modelling also adopts the abstraction of spatially lumped physical properties. Consequently, the above basic physical processes may be located in space and represented by a node in a bond graph. Since these elementary physical processes are encountered in all energy domains, it is reasonable to represent them by means of a unique mnemonic code that indicates the type of the process and that is the same for all energy domains. Spatial concentration of physical properties means that bond graphs represent so-called lumped parameter models. The above processes are represented by basic bond graph elements.

The fundamental physical processes suggest the introduction of the following classes of basic multi-port elements used for an idealised description of physical processes: i) energy sources and sinks (= negative energy sources), ii) energy stores, iii) dissipators converting energy irreversibly into heat, iv) power couplers and transducers, v) power nodes that instantaneously distribute power.

Energy sources deliver energy into a system, whereas sinks consume energy flowing out of the system. Sources and sinks do not belong to a system. They rather represent boundary conditions of a system embedded into a surroundings.

A bond graph in which bonds connect only nodes that instantaneously transfer or distribute power (without energy conversion into heat), is called generalised junction structure.

A general model structure of hybrid energy systems in terms of bond graphs is shown in Fig. 3a [4].

The model consists of the following elements: i) external elements: *Se* and *Sf* – sources of energy (effort and flow respectively), *M* – sinks of energy; *C* – accumulators of potential energy, *I* – accumulators of kinetic energy, ii) *R* – dissipaters of energy; iii) internal

881

elements: 1, 0 – junctions (nodes), *TF* – energy transformers, *GY* – energy converters.

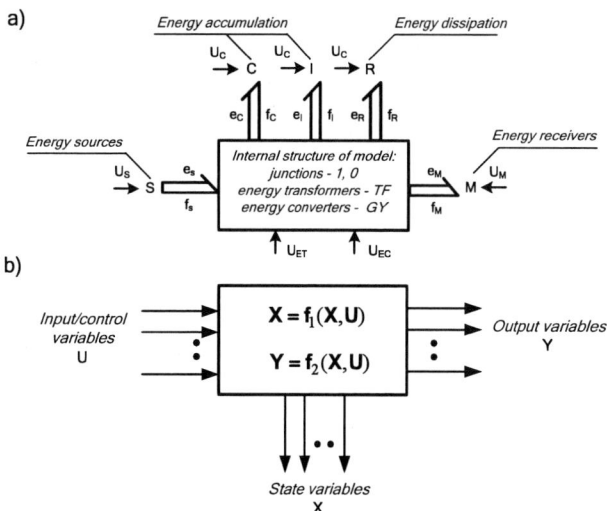

Fig. 3. General structure of hybrid energy system model: a) in terms of bond graphs, b) in terms of state equations.

The ports are classified as 'active ports' and 'passive ports'. The active ports are those, which give reaction to the source. The passive ports are used for the concept of modulated sources of effort and flow, transformers and gyrators. They are denoted respectively as following: *MSe, MSf, MTF* and *MGY, MR, MC* and *MI*. Their energetic parameters either depends upon other model parameters or are independent functions of time *t*. In the later case, the independent functions are defined as control parameters and are denoted by a control vector **U**.

Generally, the control vector consists of the following components: **U**$_S$ – control vector of energy sources, **U**$_M$ – control vector of energy receivers, **U**$_C$, **U**$_I$ – control vectors of energy accumulators, **U**$_R$ – control vector of energy dissipaters, **U**$_{ET}$, **U**$_{EC}$ – control vectors of energy transformers and converters, respectively.

For the purpose of hybrid energy system analysis, identification, and synthesis its mathematical model is setting up.

The type of model that will be found often is described as a "state-determined system". Such a system model often is described by a set of ordinary differential equations in terms of so-called state variables and a set of algebraic equations that relate other system variables of interest to the state variables [4,5]. The system state equations are written usually in the form of vector and matrix notation (Fig.3b).

From the pictorial representation of the bond graph, the derivation of system equations is so systematic that it can be algorithmized [1,4,5].

They can be set up by "hand" on the basis of system bond graph model or set up automatically – using a bond graph oriented simulation packages, like DYMOLA [13] or others (www.bondgraphs.com/software.html).

The model shown in Fig. 3 can be divided into submodels of subsystems and or components that a

considered hybrid energy system has been built up. It should be noticed that the submodels are reusable, and of different levels of accuracy.

A bond graph models of EMs, as an example of an subsystem modelling, are briefly addressed in order to demonstrate the potential of this powerful approach to modelling of controlled electromechanical systems.

III. BASICS OF ELECTRIC MACHINES MODELLING IN TERMS OF BOND GRAPHS

An EM, according to its degree of freedom, has been represented as a multi-port electromechanical transducer (converter) with pair of terminals, which are the winding and shaft terminals (ports). The machine dynamic is described by two power parameters at each pair of terminals. It is assumed that fundamental quantities (variables) of the EM model are: voltages, currents, flux-linkages, rotor angular velocity and rotation torque – electromagnetic and/or load torque.

A general structure of EM model in terms of bond graphs is shown in Fig. 4.

The ports variables are represented by: vectors of voltages ($\mathbf{u}_s$ - stator, $\mathbf{u}_r$ - rotor) and currents ($\mathbf{i}_s$ - stator, $\mathbf{i}_r$ - rotor), rotation torque (T_m) and rotor angular speed (ω_m). For electrical ports the voltages are assumed as efforts, and currents as flows, for mechanical port (denoted by subscript *m*) the torque is assumed to be effort, and the angular velocity as a flow. In turn the elements *C* and *I* are the energy storage elements, and *R* is dissipating energy element. The variables *e* and *f* represent the efforts and flows of these elements, respectively.

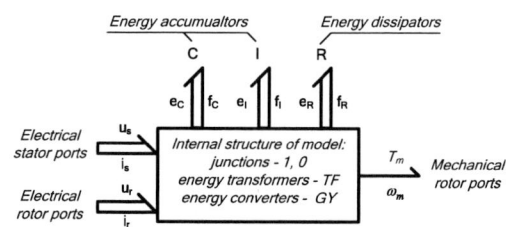

Fig. 4. General structure of electric machine model in terms of bond graphs – power flow directions (arrows directions) for motor operating mode have been assumed.

The internal structure of the model (Fig.4) depends upon the models of the energy transformation and conversion processes in EM. The processes can be modeled in terms of modulated transformers *MTF* and modulated gyrators *MGY* or in terms of *IC* field models [5], i.e., the multiport field and junction structures.

To enhance the analogy between the principles of circuit models and bondgrah models development for all types of EMs, the idea of two ideal couplings has been developed [8].

It has been assumed that generally in all types of EMs two fundamental couplings can be distinguished: referred

882

to as "ideal transformer coupling" and "ideal electromechanical coupling" (Fig. 5).

The couplings represent the processes of energy transformation and energy conversion, respectively. Using the lumped parameter models (bond graph representations) of the couplings as "building blocs" the models of basic EMs can be easy developed. While building EM models in terms of bond graphs these couplings can be modelled in terms of transformers (Fig. 5b) and modulated gyrators (Fig. 5d), respectively.

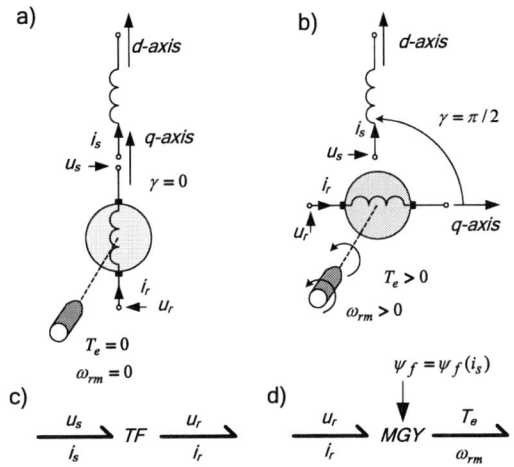

Fig. 5. Idea and models of fundamental couplings in electrical machines – assumed qd reference frame fixed to the stator: a) ideal transformer coupling – energy transformation; b) ideal electromechanical coupling– energy conversion; c) and d) models of fundamental couplings in terms of bond graphs, respectively.

IV. MODELING OF ELECTRICAL MACHINES USING THE MODELICA BOND-GRAPH LIBRARY

A. Brushed DC Motor with Permanent Magnets

As an example a model of a brushed dc motor with permanent magnets has been shown in Fig.6.

The motor model has been built-up using the Modelica Bond-Graph Library [12] and the editor of the DYMOLA package [13]. To build-up the motor model it has been assumed that the eddy currents in the stator and rotor are neglected, i.e., the transformer coupling are not considered. The supply armature voltage Ua is modelled by effort sources Se/Ua, and the load torque Tm is represented by effort source Se/Tm – its value is negative for motor operation. The energy dissipations is represented: in rotor (armature) circuit by resistances Ra, and in mechanical circuit by friction coefficient Bm. The energy accumulation is represented as follows: in rotor circuit by inductance La, and in mechanical circuit by rotor inertia J. The energy conversion from electrical to mechanical (and vice-versa) is represented by gyrator GYq in the q axis (brushes axis). The modulus r of the gyrator is equal to the permanent magnet flux $p\psi_f$ linked with the rotor circuits. Thus, the efforts on the input ports of the gyrator represents the back EMF Ea, and on the output ports the electromagnetic torque Te, respectively.

B. Brushless DC Motor with Permanent Magnets

Using the idea of two ideal electromechanical couplings a brushless DC motor with permanent magnets can be developed in terms of bond graph as shown in Fig.7.

The supply voltages Uqs and Uds in qd axes stator circuits are outputs of the asbscs/qd Park's transformation subsystem. The bond graph model of the Park's transformation subsystem has been developed in [9,10]. The values of the voltages Uqs and Uds are functions of the supply voltages (output inverter voltages) applied across the real terminals as, bs, and cs of the stator winding. In turn, the supply voltage Ufd of the field winding is modelled by a modulated effort source mSE/Ufd. The source mSE/Ufd by output signal of the bloc Ufd_.

The energy dissipations is represented as follows: in stator circuit by resistances Rqs and Rds, in mechanical circuit by friction coefficient Bm. The energy accumulation is represented as follows: in stator circuit by inductances Lqs and Lds, in mechanical circuit by rotor inertia J. The electromechanical energy conversion process from electrical to mechanical (and vice-versa) is represented by modulated gyrators mGYq and mGYd in q and d axes, respectively. The modulus of the gyrators are functions of stator currents Iqs and Ids (measured by sensors Df_Iqs and Df_Ids respectively), the inductances Lqs and Lds, and constant Psi_fr (representing the flux excited by PM and linked with stator winding in the d axis). Thus, the efforts on the input ports of the gyrators represent the back EMFs Eqs and Eds, and on the output ports the

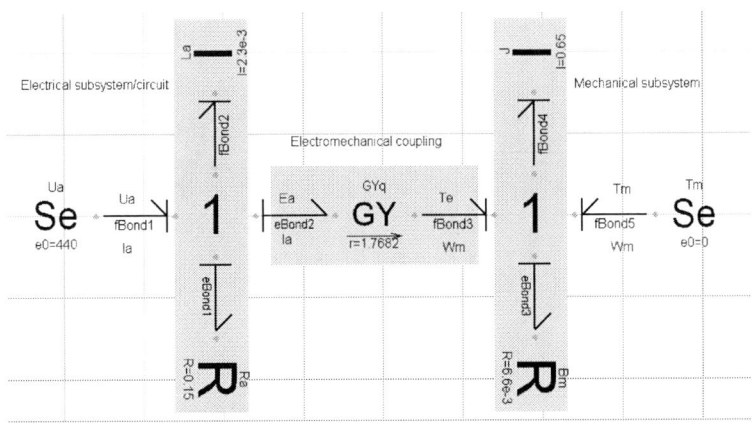

Fig. 6. Model of a brushed dc motor with permanent magnets in terms of bond graphs – Modelica Bond-Graph Library [12] and DYMOLA editor [13] have been applied.

electromagnetic torque components *Teq* and *Ted*, in *qd* axes, respectively.

The load torque *Tm* is represented by effort source Se/*Tm* – its value is negative for motor operation.

The input signal of the *Integrator1* represents the angular speed *Wm*, and the output signal is equal to the electrical rotor position angle *Tetare*. The angle *Tetare* is the input signal to *asbscs/qd* Park's transformation subsystem.

It should be noticed that for a reduced accuracy modelling of a particular electromechanical system it can be assumed that *d* axis component of the supply voltage and current can be assumed as equal to zero, i.e. *Uds*=0 and *Ids*=0. Then the motor model shown in Fig.7 can be reduced to a model shown in Fig.6 – analogous to the classical (brushed) dc motor.

C. Synchronous Machine with Field Winding

The model structure of synchronous machine (SM) depends upon assumed model of the energy conversion process. In [10] the IC multiport field and junction structures have been used to model the transformer coupling between the stator and rotor circuits. In this paper these couplings have been represented by the magnetizing inductances in *q* and *d* axes, respectively [6].

In turn, the electromechanical coupling have been described in terms of the two ideal electromechanical couplings – analogous to the brushless DC motor model. The SM model in terms of bond graphs can be developed as shown in Fig.8.

The supply voltages *Uqs* and *Uds* in *qd* axes stator circuits are outputs of the *asbscs/qd* Park's transformation subsystem – analogous to the brushless DC motor model. The bond graph model of the Park's transformation subsystem has been developed in [7,10].

The energy dissipations is represented as follows: in stator circuit by resistances *Rqs* and *Rds*, in mechanical subsytem by friction coefficient *Bm*. The energy accumulation is represented as follows: in stator circuit by leakage inductances *Llqs* and *Llds*; in rotor circuits by leakage inductances *Llkq* and *Llkd* for the damping windings; leakage inductance *Llfd* for the field winding; magnetizing inductances *Lmq* and *Lmd* for the *q* and *d* axes crcuits; in mechanical circuit by rotor inertia *J*. The electromechanical energy conversion process from electrical to mechanical is represented by modulated gyrators *mGYq* and *mGYd* in *q* and *d* axes, respectively. The modulus of the gyrators are functions of stator currents *Iqs* and *Ids* (measured by sensors *Df_Iqs* and *Df_Ids* respectively), the inductances *Lqs* and *Lds* of stator circuits, and the magnetizing inductances *Lmq* and *Lmd*.

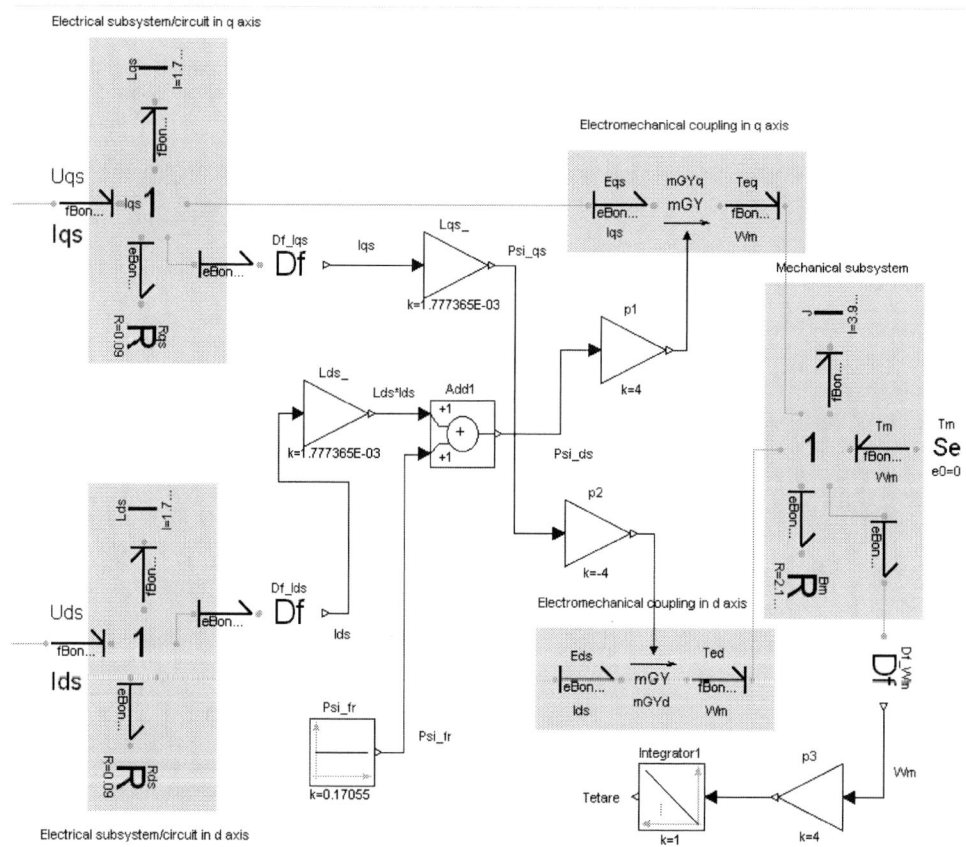

Fig. 7. Two-axes model of brushless DC motor with permanent magnets in terms of bond graphs – assumed *qd* reference frame fixed to the rotor – Modelica Bond-Graph Library [12] and DYMOLA editor [13] have been applied.

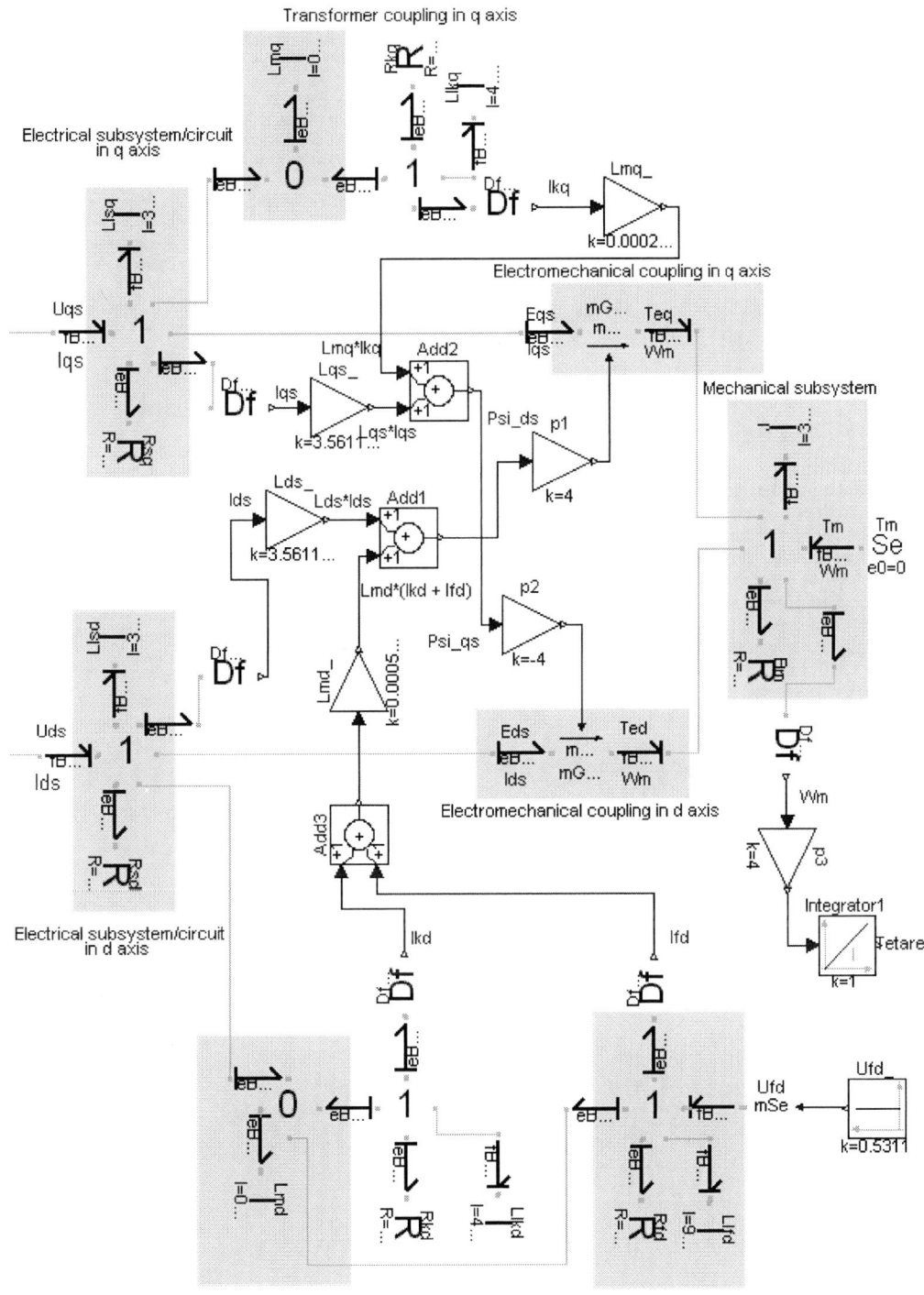

Fig. 8. Two-axes model of synchronous machine with field winding
in terms of bond graphs – assumed *qd* reference frame fixed to the rotor – Modelica Bond-Graph Library [12] and DYMOLA editor [13] have
been applied.

The outputs signals of the adding blocs *Add1* and *Add2* represent the fluxes *Psi_qs* and *Psi_ds* linked with stator winding in the *q* and *d* axes, respectively. Thus, the efforts on the input ports of the gyrators represent the back EMFs *Eqs* and *Ed*s, and on the output ports the electromagnetic torque components *Teq* and *Ted*, in *qd* axes, respectively.

The load torque *Tm* is represented by effort source Se/Tm – its value is negative for motor operation.

The input signal of the *Integrator1* represents the angular speed *Wm*, and the output signal is equal to the electrical rotor position angle *Tetare*. The angle *Tetare* is the input signal to *asbscs/qd* Park's transformation subsystem.

The developed models of induction machine in terms of bond graphs will be presented at the conference.

The elaborated bond graphs EM models can be easy included into the Modelica Bond-Graph Library [12].

V. CONCLUSIONS

A bond graph formalism is an example of a more general class of graph-based representation that describe the engineering system functionality and form in qualitative terms and schematically. This formalism form an modeling tool that enables its user to understand and explain the dynamics of physical processes clearly and concisely. For this reason, bond graphs are suitable both as a didactic and also as a highly practical tool for modelling multi-physics systems. Bond graphs models improves insight and direct feedback on modelling, simulation and design decisions.

The bond graphs approach to modelling electrical machines for controlled electromechanical systems simulations has been presented. The bond graph models of electrical machines are based on the models of ideal couplings: transformer and electromechanical. The models have been elaborated using the Modelica Bond-Graph Library [12] and the editor of the DYMOLA package [13]. The elaborated EM models in terms of bond graphs can be easy included into the Modelica Bond-Graph Library.

REFERENCES

[1] F.E. Cellier, "Hierarchical Non-Linear Bond Graphs: A Unified Methodology for Modeling Complex Physical Systems", *Simulation*, vol. 58, pp. 230-248, No. 4, 1992.

[2] F.E. Cellier, and R.T. McBride, "Object-oriented Modeling of Complex Physical Systems Using the Dymola Bond-graph Library", Proc. ICBGM'03, 6th SCS Intl. Conf. on Bond Graph Modeling and Simulation, Orlando, Florida, pp. 157-162, 2003.

[3] F.E. Cellier,. and A. Nebot, "The Modelica Bond Graph Library", Proc. 4th International Modelica Conference, Hamburg, Germany, vol.1, pp. 57-65, 2005.

[4] M. Cichy, *Modelowanie systemów energetycznych (Modelling of Energy Systems)*, Gdańsk: Wydawnictwo Politechniki Gdańskiej, 2001.

[5] D.C. Karnopp, D.L. Margolis, and R.C. Rosenberg, *System dynamics. Modeling and simulation of mechatronic systems*, 4th ed., New York: J. Wiley & Sons Inc., 2005.

[6] Krause P.C., *Analysis of Electric Machinery*, New York: McGraw-Hill, 1986.

[7] H.M. Paynter, *Analysis and design of engineering systems*, MIT Press, 1961.

[8] M. Ronkowski, and J. Nieznński, "Electrical Machines Modelling – Teaching Aspects", Proc. International Conference on Electrical Machines, ICEM 2000, Espoo, Finland, vol.3, pp. 1256-1260, 2000.

[9] M. Ronkowski, „Modelowanie silnika bezszczotkowego o magnesach trwałych w ujęciu grafów wiązań", Materiały XI Sympozjum Podstawowe Problemy Energoelektroniki I Elektromechaniki, PPEE'2005, Wisła, pp. 45-48, 2005.

[10] Sahm D. A., "Two-Axis Bond Graph Model of the Dynamics of Synchronous Electrical Machines", *Journal Franklin Inst.*, 308, No. 3, 1979, pp. 205–218.

[11] H.A. Wheeler, D. Ettinger: Wheeler Monogr. 9 (1949) 7.

[12] BondLib: Bond Graph Library for Dymola/Modelica, http://www.inf.ethz.ch/personal/fcellier/Soft/BondLib.zip

[13] DYMOLA, Dynamic Modeling Laboratory, http://www.Dynasim.se.

Induction Motor Parameters Identification using Genetic Algorithms for Varying Flux Levels

Konstantinos Kampisios, Pericle Zanchetta, Chris Gerada, Andrew Trentin, Omar Jasim
School of Electrical and Electronic Engineering
University Of Nottingham
Nottingham, NG7 2RD, UK
eexkk4@nottingham.ac.uk; Pericle.zanchetta@nottingham.ac.uk; chris.gerada@nottingham.ac.uk;
Andrew.trentin@nottingham.ac.uk; eexofj@nottingham.ac.uk;

Abstract—This paper describes a novel approach for identifying induction motor electrical parameters in function of flux levels based on experimental transient measurements from a vector controlled Induction Motor (I.M.) drive and using an off line Genetic Algorithm (GA) routine with a linear machine model. The evaluation of the electrical motor parameters is achieved by minimizing the error between experimental and simulation model responses. An accurate and fast estimation of the electrical motor parameters is performed by running a number of optimizations using experimental tests taken under different operating conditions (flux level). Results are verified through a comparison of speed, torque and line current responses between the experimental IM drive and a Matlab – Simulink model.

Keywords-Induction Motor Drives; Genetic Algorithms; System Identification; Vector Control;

I. INTRODUCTION

There is no doubt that the interest for parameter identification in the field orientation control (FOC) drives has been increased in the past few years. Moreover, it is well known that the method of vector control in an induction motor drive allows high – performance control of torque and speed and it can be achieved only if both the electrical and mechanical parameters of the machine are accurately known in all operating conditions [1]. It is clear that the precise knowledge of all IM parameters is very important for indirect rotor field oriented control (IRFO). However, this is hard to achieve due to the variation of parameters at different machine operating points such as the temperature, flux level, torque level and the nonlinearities caused by skin effect and saturation.

There are many different ways to identify I.M. parameters. Traditional ways were used to identify the resistance and inductance of both rotor and stator by performing the locked – rotor and no – load tests [3 – 5]. Nevertheless, one main disadvantage of this method is that the motor has to be locked mechanically and the temperatures of the stator winding and rotor cage have to be measured [6]. Previous works have also discussed the estimation of the machine parameters in standstill [7 – 9]. The advantage of this test is that it can be implemented only adding software routine on the normal control implementation. It also gives effective estimation under any mechanical load [8].

This paper presents an accurate and fast method for evaluation of the electrical induction motor parameters. The motor used for this evaluation test is a 4 kW, 4 – pole I.M. A Matlab model was utilized within the heuristic GA based identification routine.

GAs is a search technique used in many engineering fields to find accurate solutions to large optimization and search problems. The basic concept of GAs is to emulate evolution processes in natural system following the principles first laid down by Charles Darwin of survival of the fittest [10].

The advantage of GAs is that it is a very flexible and intuitive approach to optimization and presents a higher probability of not converging to local optima solutions compared to traditional gradient based methods.

Traditionally GAs optimization has been applied to a variety of power system problems where conventional methods have experienced some difficulties [11 - 12]. One of these main application areas are: (i) unit commitment [13] and generation scheduling [14, 15], (ii) reactive power planning/dispatch [16], (iii) distribution network planning/reconfiguration [17], (iv) load flow [18], (v) alarm processing and fault diagnosis [19], (vi) power system modeling and control [20]. More recently, research work has appeared in the scientific literature about the use of GAs for control design in power Electronics and drives [21 - 22] and general structure identification [23].

The basic idea of this research work is that the evaluation of the electrical motor parameters can be achieved by minimizing, using a GAs approach, the error between the experimental response (speed or current) measured on the experimental motor drive and the respective one obtained by a Matlab-Simulink model implementing the same structure and control of the experimental rig, but with varying electrical parameters [24].

978-1-4244-1741-4/08/$25.00 ©2008 IEEE

II. DESCRIPTION OF THE SYSTEM

Figure 1 represents the optimization experiment set-up. Measures of experimental transient responses of speed and current from the vector controlled electrical drive test rig are stored and used as reference signal for the optimization. A fitness function based on the integral of the absolute error between these reference signals and the ones produced by the simulation model reproducing the experimental rig, will be evaluated by the GAs intelligent search technique which will find the correct system electrical parameters that minimize it.

In the identification experiment the mechanical parameters, moment of inertia and friction (identified using a deceleration test [24]), as well as the resistance of the stator (easy to measure) are supposed to be known - the rest of the electrical parameters have to be identified. Table I presents some of the most important characteristics of the Experimental vector controlled Induction Motor which are also used in the simulation model. The speed and current (Isq) transient responses used as reference signals in the GAs optimization.

The induction machine is modeled using a traditional dynamic model in abc reference frame where the vector control parameters are the same experimental ones shown in table 1 and the simulation is implemented in Matlab – Simulink. The Nameplate data of the examined motor were: P=4 kW, U=415 V, Δ connection, I=8.4 A, n= 1420 rpm. Using the traditional Induction Machine parameters identification based on "no–load" and "locked–rotor" tests [3–5] the following electrical parameters can be obtained: Stator resistance Rs=5.25Ω, Rotor Resistance Rr= 5.19Ω, Magnetizing inductance Lm=0.571H, Stator leakage inductance Lls=0.01816H Rotor leakage inductance Llr=0.01816H. In the following these values will be referred as "standard parameters". A Comparison between modeled and measured time responses is shown in Fig. 2. As it can be noted there is an evident mismatch between the modeled and experimental responses indicating a difference between the experimental and estimated parameters.

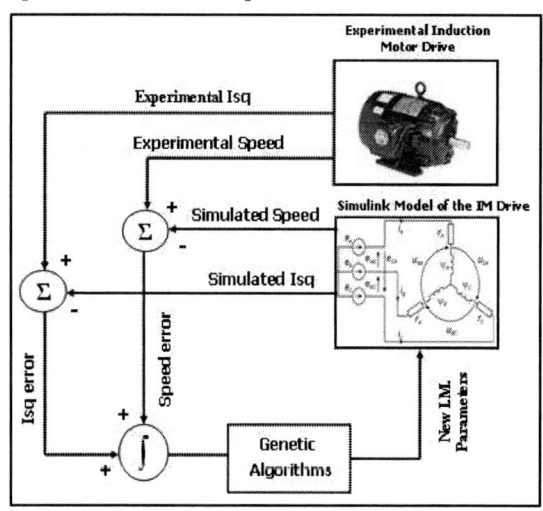

Figure 1. Block diagram representing the experiment.

TABLE I. EXPERIMENTAL VECTOR CONTROLLED IM PARAMETERS

Experimental vector controlled I.M. parameters			
Vector Control Parameters		*Induction Motor Parameters*	
Speed controller	Ksp = 0.6	Inertia J	0.152
	Ksi = 5	Friction	0.0147
Current controller	Kcp = 57.87	Stator Resistance	5.25 Ω
	Kci = 28050	Number of poles	4
Speed reference	250 rad/sec	Power	4 kW
Isd reference	4.9 A	Line voltage	415 V
Rotor time constant	0.1508	Frequency	50 Hz
Current limits	9A		
Voltage limits	1000V		

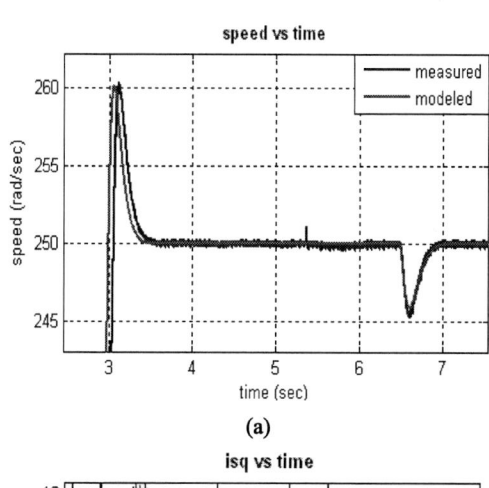

(a)

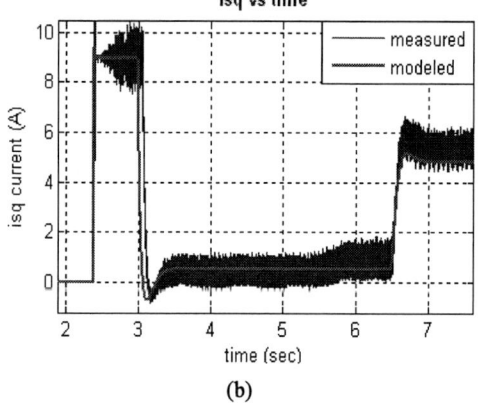

(b)

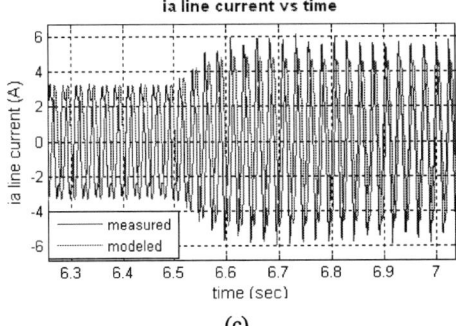

(c)

Figure 2. Measured and modelled speed (a), Isq current (b) and phase line current (c) responses with load and standard parameters.

Figure 3 represents the speed error between the experimental and modeled response by using standard parameters that are calculated using traditional methods. It can be seen that a constant error appears during the speed transient response period.

III. GAs EXPERIMENTAL PARAMETERS IDENTIFICATION

The diagram shown in Fig. 4 explains in details a GAs search procedure. General details on Genetic Algorithms can be found frequently in literature [11 – 24] and therefore will be neglected here. The fitness function plays an important role to the GA optimization procedure as it measures the quality of the represented solution. The fitness function is always problem dependent. An ideal fitness function connects closely with the algorithm's goal and yet may be evaluated quickly because usually a genetic algorithm must be repeated many times in order to achieve the desirable result and the speed of execution might be a very important factor. In some cases, it is even hard to define a fitness function, so an interactive genetic algorithm is used that uses human evaluation. In our problem, the fitness function used is the sum of the integral of the absolute value on both the speed and Isq current error as shown in (1):

$$FF = \int (|speed\ error| + |Isq\ current\ error|) \qquad (1)$$

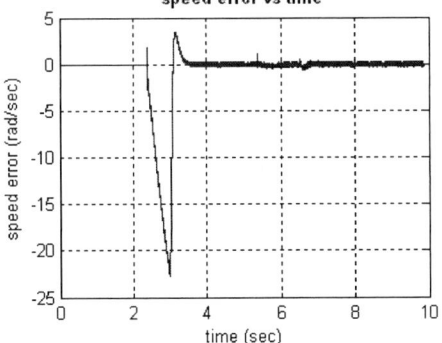

Figure 3. Speed error between measured and modeled speed response

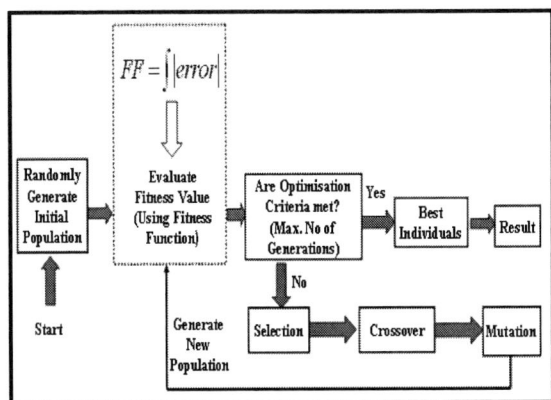

Figure 4. Block diagram representing the GAs routine

It is clear that this is a minimization problem, whereby the accuracy of the identification of the electrical motor parameters improves as the fitness function (FF) approaches zero.

As previously mentioned the unknown electrical parameters of the motor are the rotor resistance (Rr), the magnetizing inductance (Lm), the leakage inductance of the stator (Lls) and of the rotor (Llr). Given a wide range of industrial I.M. (of similar rating of the one used in this study) the bounds for the GA search were considered in a way to create intervals which encompasses most practical parameter values (2):

$$1 \leq R_R \leq 10$$
$$0.1 \leq L_m \leq 1$$
$$0.005 \leq L_{ls} \leq 0.9 \qquad (2)$$
$$0.005 \leq L_{lr} \leq 0.9$$

It has been found by experience that, above all in identification problems, it would be desirable that the GA routine could restrict the parameters search ranges during its optimization. In order to minimize the bounds of the electrical parameters, an identification strategy involving a search space reduction method (SSRM) is used. The aim of this method is to increase the accuracy and reliability of identification by reducing the search space during the algorithm operation. The idea is to let the search space reduce for those parameters that converge quickly and so it is reduced the time wasted looking far outside the area where the optimal solution lies [23].

Five initial runs are performed in order to achieve this using a traditional genetic algorithm with elitist selection. The obtained results are shown in table II.

The optimization routine applies then the search space reduction method (SSRM) in order to estimate the new bounds for the parameters. It can be found that the new limits for the electrical parameters are given as in (3). The GAs optimization settings selected are shown in table III. Finally, applying the last wider genetic algorithms optimization including the new parameters bounds, the final estimation of our I.M. electrical parameters, shown in table IV, can be obtained.

889

TABLE II.
GENETIC ALGORITHM ESTIMATED MOTOR PARAMETERS IN THE INITIAL RUNS

Electrical Parameters	1st run	2nd run	3rd run	4th run	5th run
Rotor Resistance (Rr)	4.1111	4.1111	4.1375	4.1397	4.1488
Magnetizing Inductance (Lm)	0.5476	0.5462	0.5453	0.5453	0.5448
Leakage inductance of stator (Lls)	0.0568	0.0250	0.0429	0.0297	0.0297
Leakage inductance of rotor (Llr)	0.0557	0.0557	0.0558	0.0558	0.0558
Best Objective Value	0.9032	0.8794	0.8524	0.8511	0.8457

$$4.0672 \le R_R \le 4.1920$$
$$0.54188 \le L_m \le 0.54979$$
$$0.01354 \le L_{ls} \le 0.06010 \qquad (3)$$
$$0.05556 \le L_{lr} \le 0.05595$$

TABLE III.
GENETIC ALGORITHM CHARACTERISTICS

GAs parameters		
Generation Gap		0.9
Crossover rate		0.85
Mutation rate		0.15
Initial runs	Maximum Generations	25
	Number of Individuals	20
Final run	Maximum Generations	50
	Number of Individuals	40

TABLE IV.
FINAL GENETIC ALGORITHM ESTIMATED MOTOR PARAMETERS

Electrical Parameters	
Rotor Resistance (Rr)	4.1636
Magnetizing Inductance (Lm)	0.5435
Leakage inductance of stator (Lls)	0.0291
Leakage inductance of rotor (Llr)	0.0556
Best Objective Value	0.8325

A comparison of speed, torque and line current responses between the experimental induction motor drive performance and the Simulink model using the new GA estimated parameters with a load are shown in Fig. 5.

It can be clearly noticed that the system dynamic behavior simulated using the new proposed GA parameters estimation parameters obtained are perfectly matching the experimental measures and show a significant improvement compared with the model obtained using the traditional identification methods. It is to be noticed that in this paper, it is assumed that the machine resistances remain constant since the tests simulation is kept to a maximum value of 10 seconds and so the temperature variation will not affect the value of the resistance. Table V also shows the error introduced by traditional estimation methods.

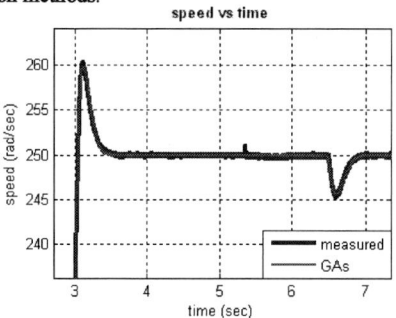

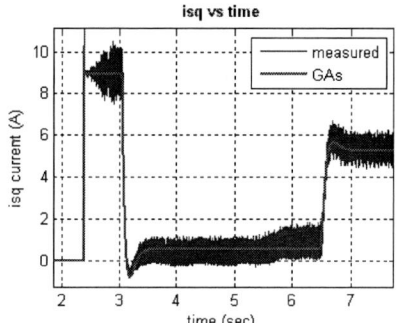

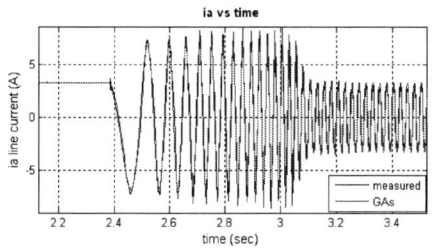

Figure 5. Simulated system behaviours with the new GA evaluated machines electrical parameters against the experimental ones with a load torque of 15.5 Nm.

TABLE V.
COMPARISON BETWEEN THE NEW GENETIC ALGORITHM ESTIMATED MOTOR PARAMETERS AND THE TRADITIONAL ONES

	Traditional Methods	GAs	%error
Rotor Resistance (Rr)	5.19 Ω	4.1636 Ω	19.7
Magnetizing Inductance (Lm)	0.571 H	0.5435 H	4.8
Leakage inductance of stator (Lls)	0.01816 H	0.0291 H	60.24
Leakage inductance of rotor (Llr)	0.01816 H	0.0556 H	206.16

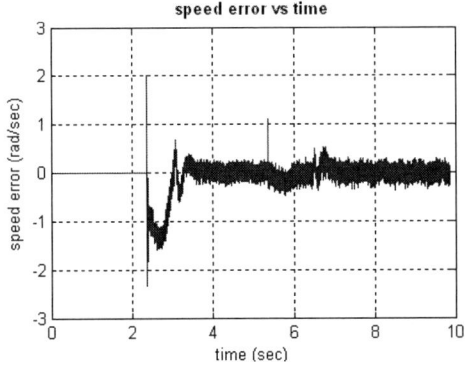

Figure 6.   Speed error between measured and GAs speed response

Figure 6 represents the speed error between the experimental and modeled response behavior with the new GA evaluated electrical parameters.

As it can be seen comparing Fig. 5 and Fig.6, GAs give us less speed error compared with traditional methods, which means more accurate and reliable identified electrical parameters.

Given the nature of the optimization and of the general transient measurement used, the parameters estimation performed has identified a sort of average parameters set. It would be interesting to be able to identify machine parameters variations in function of some relevant quantities. A good exercise to verify this possibility is to apply the same methodology for example to estimate the parameters under different flux levels. The results is a set of electrical parameters for each specific operating conditions from which it is possible to extrapolate their behavior in function of the flux level.

Figure 7 shows the behavior of the magnetizing inductance Lm in function of the Id_ref (flux level); as it can be seen the value of Lm decreases as the flux increases and in low flux level (below 50%) the value of Lm is almost constant as it is expected. This confirms the validity of the proposed identification approach.

Figure 8 shows the behavior of rotor resistance Rr in function of the motor d – axes current Id_ref (flux level); it can be noticed that the value of Rr changes slightly with different flux levels and this is due to either numerical errors or variation of rotor temperature while the pilot experimental measures were taken.

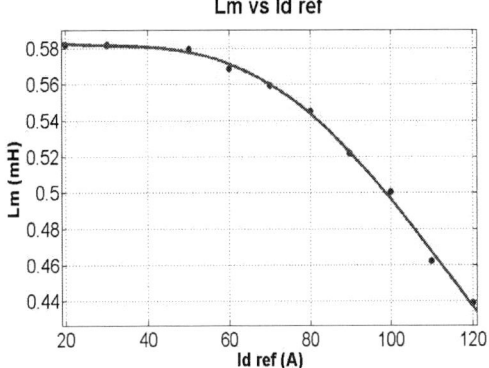

Figure 7.   Values of Lm in function of the Id_ref (different flux levels) found by GA optimization

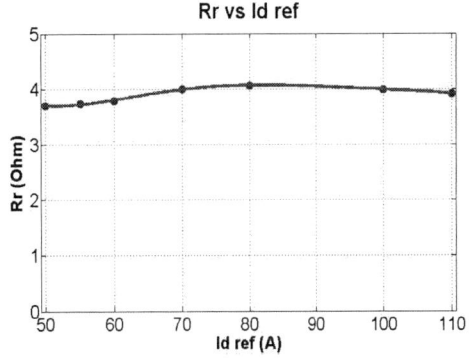

Figure 8.   Values of Rr in function of the Id_ref (different flux levels) found by GA optimization

Conclusion

This paper has presented an intelligent approach to estimate the Induction Machine electrical parameters in a motor drive based on a Genetic Algorithm heuristic optimization approach, a simulation model and experimental transient measurements. Based on both the simulation and experimental measured results it is concluded that the use of the proposed strategy is an effective and reliable method for induction motor parameter identification and for accurately modeling the behavior of the drive system. The reliability of this technique in estimating the machine parameters behavior in function of different operating conditions (flux level) has been presented successfully. The proposed method gives also the basis for an optimized and high performance control design. Having proved this, it is possible in the same way to estimate the motor parameters in function of other relevant variables and therefore to have a better estimation of the electrical parameters for any operating condition.

REFERENCES

[1] W. Leonard: "Control of Electrical Drives", Springer 1997

[2] Marin Despalatovic, martin Jadric, Bozo Terzic, "Identification of Induction Motor Parameters from Free Acceleration and Deceleration Tests", AUTOMATICA 46(2005) 3 – 4, 123 – 128

[3] Charles Kingsley, Stephen D. Umans, "Electric Machinery", Sixth Edition.

[4] Austin Hughes, "Electric Motors and Drives", Third Edition 2006

[5] G.R. Slemon ♦ A. Straughen, "Electric Machines"

[6] Marin Despalatovic, Martin Jadric, Bozo Terzic, "Identification of Induction Motor Parameters from Free Acceleration and Deceleration Tests", ISSN 0005-1144, ATKAAF 46(3-4), 123-128 (2005).

[7] J.R. Willis, G. J. Brock, and J. S. Edmonds, "Derivation of induction motor models from standstill frequency response tests", IEEE Trans. Energy Conv. Vol. 4, pp. 608 – 613, Dec 1989.

[8] J. – K. Seok, S – I. Moon, and S. – K. Sul, "Induction parameter identification using pwm inverter at standstill", IEEE Trans. Energy Conv. Vol. 12, pp. 127 – 132, June 1997.

[9] L. A. de S. Ribeiro, C. B. Jacobina, and A. M. N. Lima, "The influence of the slip and the speed in the parameter estimation of induction machines", in Proc. PESC Conf. Rec., June 1997, pp. 1068 – 1074.

[10] Thomas Back, David B. Fogel, Zbigniew Michalewicz: "Handbook of Evolutionary Computation"

[11] Y.H.Song, C.S.V. Chou: "Advanced engineered – conditioning genetic approach to power economic dispatch", IEE Proc.-Gener. Transm. Distrib., Vol. 144, No 3, May 1997

[12] Srinivasan, D., Wen, F., Chang, C.S., and Liew, A.C.: "A survey of applications of evolutionary computing to power systems", International conference on intelligent system application to power systems, Orlando, 1996

[13] Dasgupta, D., and McGregor, D.R.: "Thermal unit commitment using genetic algorithms", IEE Proc. C., 1994, pp. 459 – 465

[14] Orero, S.O., and Irving, M.R.: "Scheduling of generators with a hybrid genetic algorithm". Proceedings of the first IEE/IEEE international conference on Genetic Algorithms in engineering systems: innovations and applications, 1995, pp. 200 – 206

[15] Song, Y.H., Li, F.R., Morgan, R., and Cheng, D.T.Y.: "Comparison studies of genetic algorithms in power system economic dispatch", J. Power Syst. Technol., 19, (3), pp. 28 – 34

[16] Wu, Q.H., and Ma, J.T.: "Genetic search for optimal reactive power dispatch of power sustems". Proceedings of IEE international conference on Control, 1994, pp. 717 – 722

[17] Nara, K., Shiose, A., and Kitagawa, G.: "Implementation of genetic algorithm for distribution systems loss minimum reconfiguration", IEEE Trans. Power Syst., 1992, pp. 1044 – 1051

[18] Yin, X., and Germay, N.: "Investigation on solving the load flow problem by genetic algorithm", EPRS, 1991, pp. 151 – 163

[19] Wen., F., and Han, Z.: "Optimal fault section estimation in power systems using genetic algorithm and simulated annealing", Proc. CSEE, 1994, pp. 29 – 35

[20] Ju, P., Handschin, E., and Reyer, F.: "Genetic algorithm aided controller design with application to SVC", IEE Proc. C, 1996, pp. 258 – 262

[21] P. Zanchetta, P.W. Wheeler, J.C. Clare, M. Bland, L. Empringham, D. Katsis: "Control Design of a Three-Phase Matrix Converter-Based AC-AC Mobile Utility Power Supply", IEEE Trans. On Industrial Electronics, January 2008.

[22] Cupertino, F.; Mininno, E.; Naso, D.; Turchiano, B.; Salvatore, L.; "On-line genetic design of anti-windup unstructured controllers for electric drives with variable load" IEEE Transactions on Evolutionary Computation, Volume 8, Issue 4, Aug. 2004, Page(s):347 - 364

[23] M.J. Perry, C.G. Koh, Y.S. Choo: "Modified genetic algorithm strategy for structural identification", Elsevier Journal "Computers and Structures" 84 (2006) 529 – 540

[24] Andrew Trentin, Pericle Zanchetta, Patrick Wheeler, Jon Clare : "A New Method for IMs Parameter Estimation Using GAs and Transient Speed Measurements", IAS 2006 conference, Tampa Florida

Study of the sudden symmetrical short-circuit using the mathematical models of the synchronous machine and the numerical methods

Petropol Serb Gabriela*, Petropol Serb Ion[†], Campeanu Aurel*, Sonia Degeratu[†], Anca Petrisor*

* University of Craiova/Faculty of Engineering in Electromechanically, Environment and Industrial Informatics,
Craiova, Romania, gpetropol@em.ucv.ro, [†]/ipetropol@yahoo.com, */acampeanu@em.ucv.ro, [†]/sdegeratu@em.ucv.ro,
*/apetrisor@em.ucv.

Abstract — In the last decade, the field of Fault Detection and Diagnosis in the electric motor manufacturing industry, are becoming more important. The methods used to study the fault detection and diagnoses are based on an explicit mathematical model of the system under test. This article presents, in detail, a method for the study and the simulation of a symmetrical short-circuit of a synchronous machine. It will be calculate and plot the variation of the current of a phase in time, by using the software of Matlab. So, the paper is a theoretical approach on the fault detection and simulation.

Keywords — Fault detection, symmetrical short-circuit, short-circuit current, current plot, Matlab software.

I. INTRODUCTION

In the areas of electrical power systems, a fault is defined as an unusual electrical connection between conductors, this connection being maintained or not in the time. The impedance of the fault is an important parameter in the study of a short circuit.

In the manufacturing environment, unexpected fault of the motor is both undesirable and costly. For that reason, there is a growing demand to improve the reliability of electric motors in general and, especially in industrial applications, to detect forthcoming faults, so the motors can be repaired or replaced during usual maintenance rather than after failure has occurred. It is also desirable to improve stability of electric motors through improved quality control monitoring during manufacture of the electric motors. Newly, fault detection and the methods of diagnosis have been developed by comparing the output signals of complex systems with the output signal obtained from a mathematical model of the fault free system.

Conventional motor quality control equipment utilizes easily measured magnitudes, like steady state current, speed, and input power. These methods utilize indirect information about motor quality and they cannot offer in-depth information about motor faults, which is essential to provide feedback to production process.

The approach based on the modeling of the fault detection use the knowledge of electric motor dynamics to identify, both the symptoms as the causes of faults, irrespective of environmental conditions. The schema of the monitoring system of the operation of a system is given in figure 1. In this context, the aim of this paper is to bring in discussion the models of the synchronous machine to simulate the sudden symmetrical short circuit.

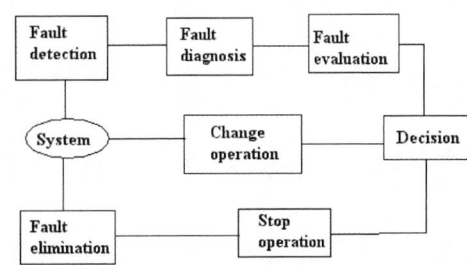

Figure 1: Monitoring schema of the system operation

A. Transient state of the sinchronous machine.

The study of the transient process of synchronous machines, encounter difficulties, because of the electric and magnetic asymmetry in rotor. In the moment when the steady state is interrupted, begin equally, the two kinds of components of windings currents: the periodical one and the no periodical components. These tend to keep unchanged the fluxes in windings. At the moment t=0, the windings behave like over conductors. As a result of currents that have emerged in the coils, the magnetic field configuration is changed, and it corresponds to the parameters of the machine.

To simplify the study of the transient processes in the synchronous machine, it is used the two axes theory with a good precision for the practice [1]. This is a very important process because in the windings there are established transitory currents with high values (multiple of rating value) which are applying electro-dynamical and thermal stress. We considered the following assumptions:

-the armature has 3 identical, symmetrically placed windings: a, b, c;

-the rotor windings E, D, Q are placed in the direction of the two orthogonal axes: d (direct) and q (quadrate);

-the winding E represents the field winding, the windings D and Q are fictitious windings to account for: damper windings and the effects of currents in the iron parts of the rotor;

-the magnetically characteristic is linear;

-the absence of losses iron;

-the torques losses by friction and ventilation are proportional.

978-1-4244-1741-4/08/$25.00 ©2008 IEEE

B. Theoretical problem

We are considering a synchronous generator that is operated initially unloaded, with terminal voltage equal to 1.0 per unit. When in a phase the flux is maxim, a sudden symmetrical short circuit is imposed. During short-circuit the speed is constant and the excitation isn't force. The induced no-load voltage causes a dynamic short-circuit current $i_s(t)$. Due the rotor transients currents the sub transient inductance is active: $L_d^{''} = X_d^{''} / \omega$.

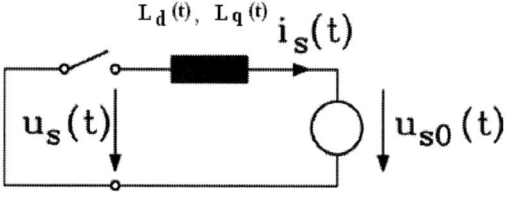

Figure2. Schema of the synchronous machine at the sudden symmetrical short-circuits.

At dynamic operation state the amplitude of stator field is changing rapidly. The linkage of stator field with damper and rotor excitation winding is changing, so in both windings voltage is induced.

The d current i_{sd} excites air gap Φ_{dh}. It links stator winding with rotor excitation winding and damper cage like in a three winding transformer.

The q current i_{sq} excites air gap Φ_{qh}. It links stator winding and rotor damper cage like in a two winding transformer. The q air gap flux Φ_{qh} is not linked with excitation winding (figure 3).

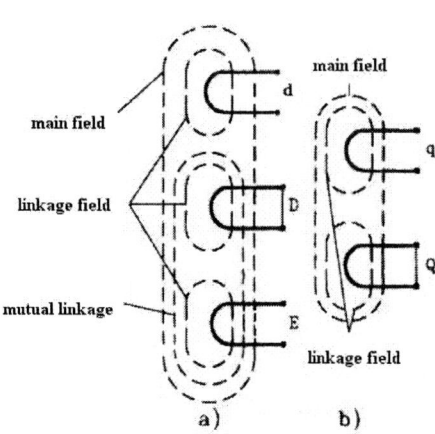

Figure 3. Magnetic fields of the machine in: a). d - axis and b). q - axis.

To simplify calculus, in literature [1], it is supposing at t=0 the superposition principle: in the place of short circuit it is considering that it is applying the stator's voltages - u_{d0},-u_{q0} and the machine isn't excited ($u_E=\Psi_{E0}=\Psi_{D0}=\Psi_{Q0}=0$).

So, the equations for fluxes are:

$$\overline{\Psi}_d = l_d(p)\overline{i}_d. \tag{1}$$

$$\overline{\Psi}_q = l_q(p)\overline{i}_q. \tag{2}$$

Stator's equations are:

$$-u_{d0} = [R_s + pl_d(p)]\overline{i}_d - \omega l_q(p)\overline{i}_q.$$
$$-u_{q0} = \omega l_d(p)\overline{i}_d + [R_s + pl_q(p)]\overline{i}_q \tag{3}$$

Expressions for the images of two axis currents are:

$$\overline{i}_d = \frac{-u_{d0}[R_s + pl_q(p)] - u_{q0}\omega l_q(p)}{[R_s + pl_d(p)][R_s + pl_q(p)] + \omega^2 l_d(p)l_q(p)}.$$
$$\overline{i}_q = \frac{-u_{d0}\omega l_d(p) - u_{q0}[R_s + pl_d(p)]}{[R_s + pl_d(p)][R_s + pl_q(p)] + \omega^2 l_d(p)l_q(p)} \tag{4}$$

If, before short-circuit, the machine operate no load, the phasor of the supply voltage is $\underline{u}_{s0} = j\omega\underline{\Psi}_{s0}$ and $i_{d0} = i_{q0} = 0$. At no load operation, the flux $\underline{\Psi}_{s0}$ is produced only by the field winding, and in the referential rotating with the rotor, we have $\underline{\Psi}_{s0} = \Psi_{s0}$. So, finally we have: $\underline{u}_{s0} = j\omega\Psi_{s0} = jU_{eE}\sqrt{2}$ and $u_{d0} = 0$, $u_{q0} = U_{eE}\sqrt{2}$.

After a substantial amount of algebra [1], arise from the repetitive scheme, we express the stator's current of A phases:

$$i_A(t) = i_d(t)\cos(\omega t + \beta_0) - i_q(t)\sin(\omega t + \beta_0) = -\sqrt{2}U_{eE} \cdot$$

$$\left\{ \left[\frac{1}{X_d} + \left(\frac{1}{X_d^{'}} - \frac{1}{X_d} \right)e^{-\frac{t}{T_d^{'}}} + \left(\frac{1}{X_d^{''}} - \frac{1}{X_d^{'}} \right)e^{-\frac{t}{T_d^{''}}} \right] \cdot \cos(\omega t + \beta_0) - \right.$$

$$\left. \frac{1}{2}\left(\frac{1}{X_d^{''}} + \frac{1}{X_q^{''}} \right)e^{-\frac{t}{T_a}}\cos\beta_0 - \frac{1}{2}\left(\frac{1}{X_d^{''}} - \frac{1}{X_q^{''}} \right)e^{-\frac{t}{T_a}}\cos(2\omega t + \beta_0) \right\}$$

where:

Parameters	Name of parameters
$x_d^{''} = \omega L_d^{''}$	Sub transient reactance on d axes
$x_d^{'} = \omega L_d^{'}$	Transient reactance on d axes
$x_d = \omega L_d$	Synchronous reactance on d axes
$x_q^{''} = \omega L_q^{''}$	Sub transient reactance on q axes
$x_q^{'} = \omega L_q^{'}$	Transient reactance on q axes
$x_q = \omega L_q$	Synchronous reactance on q axes

It is analyzed the equation of the short circuit current in phase A, and the components of the current are:

I. Steady-state short-circuits current;

II. Transient component that is damped by the time constant T'_d;

III. Sub transient component that is damped by the time constant T''_d;

IV. DC component that is damped by the time constant T_a;

V. Double-frequency component that is damped by the time constant T_a.

II. NUMERICAL CALCULATION

We compute and plotted with Matlab the short circuit current's expression and then, we analyzed it.

Using Matlab was simulated the sudden short-circuit at the output of a synchronous generator [3] with: S_N=440MVA, U_N=6300V, n_1=1000rpm, f_1=50Hz, m=3, p=3, $\cos\varphi_N$=0.8 ind., U_{eE}=3928v. The generator is, initial, unload with the terminals voltage equal with 1 r.u.. A symetrical sudden shortcircuit appear when the fluxe is maxim in phase A.

Machine's parameters in phase A are (in r.u.) [3]: synchronous d axis reactance $x_d = 1.4$; synchronous d axis reactance x_q= 0.8; transient d axis reactance x_d'=0.303; subtransient d axis reactance x_d'' =0.16; subtransient d axis reactance x_q''=0.135; transient open circuit time constant T_{do}=8s; subtransient open circuit time constant T_{do}'' =0.00682s; subtransient open circuit time constant T_{qo}'' =0.00682s; inertial constant J=4;armature time constant T_a=0.182.

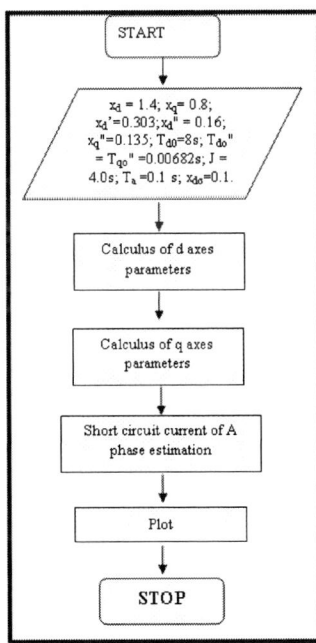

Figure 2 Logical scheme

It was realised four scripts in Matlab:

- a function script to calculate parameters in d axis;

- a function script to calculate parameters in q axis;

- a script to calculate the others parameters of the machine:

```
>> Studying a synchronous generator
Parameters   Relatives units        [Ohm]
xdh               1.3                36.2
xqh               0.7                19.5
xEs               0.241              6.7
xE                1.54               42.9
xDds              0.0852             2.37
xQqs              0.0368             1.03
rE                0.00316            0.0881
rDd               0.134              3.75
rQq               0.344              9.58
xEDs              0.1                2.78
xpEs              0.141              3.91
```

III. SIMULATIONS AND COMMENTS

If the first moment of the short-circuit is for $\beta_0 = \dfrac{\pi}{2}$, will appear in the current of phase A only the periodic components of the clock frequency and the double frequency.

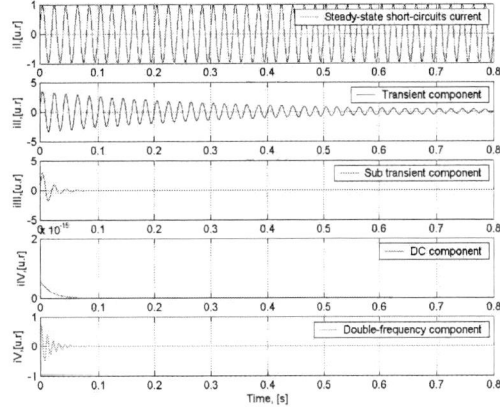

Figure 3. Shortcircuit current's components, $\beta_0 = \dfrac{\pi}{2}$

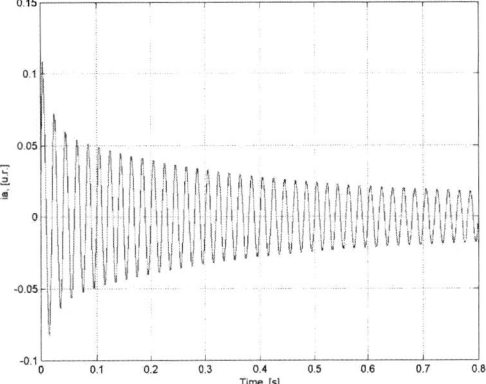

Figure 4. Short-circuit current of phase A, $\beta_0 = \dfrac{\pi}{2}$

Commonly is $X_d'' \ll X_d$. The sudden short-circuit current is outgrowing then the stationary short-circuit current.

If the short circuit is at $\beta_0 \neq \dfrac{\pi}{2}$ the windings currents will have values moreover then rated values. If β_0 is closer to zero, short-circuit is more disadvantageously. If the damper winding is neglected, $X_d'' = X_q''$, the non periodical's amplitude is $I_{max}'' = \dfrac{\sqrt{2}U_e E}{X_d''}$, and theoretical, the amplitude of maxim current could be $2I_{max}''$, at $t = \dfrac{T}{2}$.

In the interval $\left(0, \dfrac{T}{2}\right)$ the stator current's components are attenuated but it is considering that the maxim amplitude could be $(1,8 - 1,9)I''_{max}$.

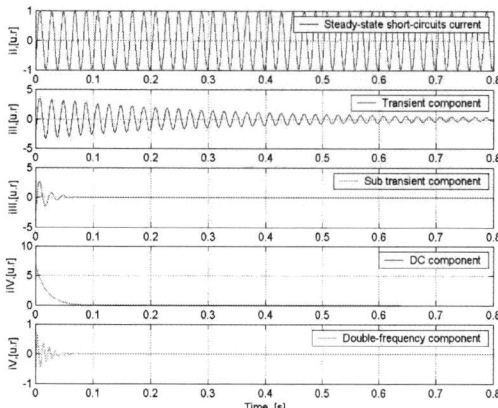

Figure 7. Short-circuit current's components, $\beta_0 = \dfrac{\pi}{4}$

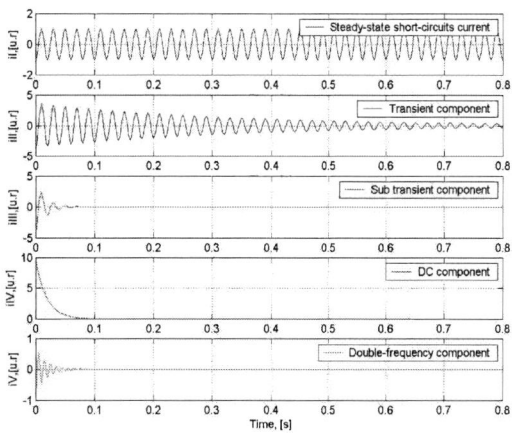

Figure 5. Shortcircuit current's components, $\beta_0 = 0$

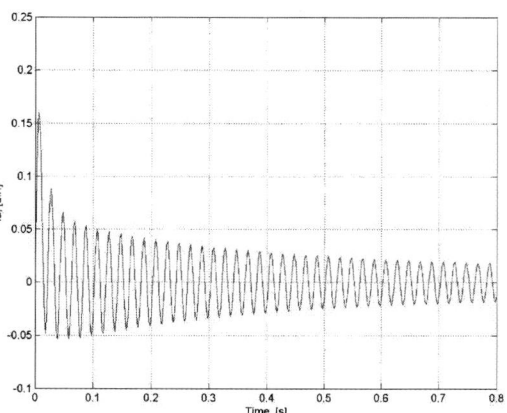

Figure 8. Short-circuit current of phase A, $\beta_0 = \dfrac{\pi}{4}$

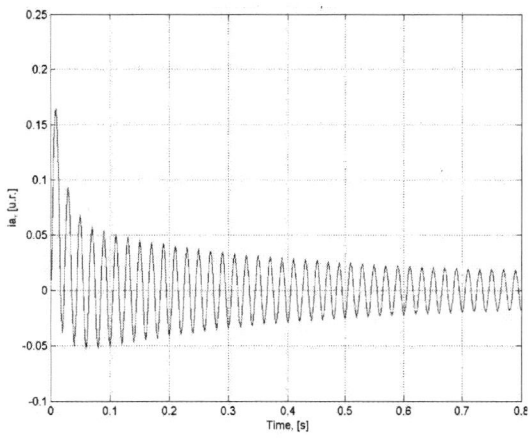

Figure 6. Short-circuit current of phase A, $\beta_0 = 0$

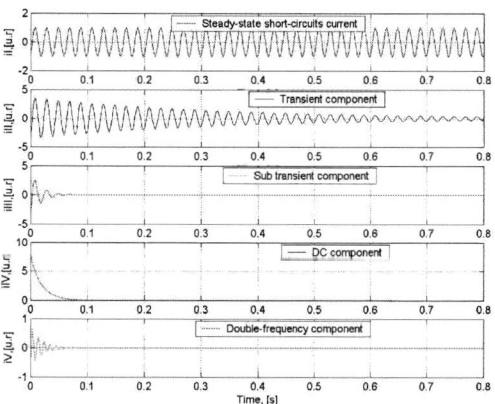

Figure 9. Short-circuit current's components, $\beta_0 = 120^0$

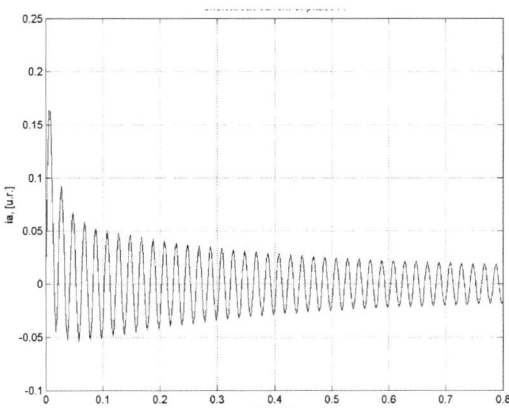

Figure 10. Short-circuit current of phase A, $\beta_0 = 120^0$

Using the equations of voltages and fluxes we obtain the transients currents of the rotors winding.

$$R_E \bar{i}_E + p\left[L_E \bar{i}_E + L_{dh}\left(\bar{i}_d + \bar{i}_D\right) + L_{ED\sigma} i_D\right] = 0$$

$$R_D \bar{i}_D + p\left[L_D \bar{i}_D + L_{dh}\left(\bar{i}_d + \bar{i}_E\right) + L_{ED\sigma} i_E\right] = 0$$

After a substantial among of algebra the equations of current became:

$$\bar{i}_E = \frac{\sqrt{2}U_{eE}T_{Eh}}{\omega L_d} \cdot \frac{p\left(1 + pT_{D\sigma}\right)}{\left(1 + pT_d'\right)\left(1 + pT_d''\right)} \cdot \frac{\omega^2}{p^2 + \omega^2}$$

$$\bar{i}_D = \frac{\sqrt{2}U_{eE}T_{Dh}}{\omega L_d} \cdot \frac{p\left(1 + pT_{E\sigma}\right)}{\left(1 + pT_d'\right)\left(1 + pT_d''\right)} \cdot \frac{\omega^2}{p^2 + \omega^2}$$

$$\bar{i}_Q = -g_Q(p)\bar{i}_q = \frac{\sqrt{2}U_{eE}T_{Qh}}{\omega L_q} \cdot \frac{p}{1 + T_q''} \cdot \frac{p\omega}{p^2 + \omega^2}.$$

By decomposing in simple fractions and neglected some non important terms for the usual synchronous machine $(\frac{T_{d0}''}{T_d'}, \frac{T_d''}{T_d'}$, etc.), we obtain the following expression for the transient current of the field winding:

$$\bar{i}_E(t) = \frac{\sqrt{2}U_{eE}}{\omega L_d}\frac{T_{Eh}}{T_d'}\left[e^{-\frac{t}{T_d'}} - \left(1 - \frac{T_{D\sigma}}{T_d'}\right)e^{-\frac{t}{T_d''}} - \frac{T_{D\sigma}}{T_d''}e^{-\frac{t}{T_a}}\cos\omega t\right]$$

But $\sqrt{2}U_{eE} = \omega L_{dh}i_{E0}$, where i_{E0} is the current of the field winding before the short circuit. The transient current of the field winding to the sudden symmetrical short-circuit is obtained by the superposition of the transient component with i_{E0}. The current became:

$$i_{Esc}(t) = i_{E0} + i_{E0}\frac{X_d - X_d'}{X_d'}\left[e^{-\frac{t}{T_d'}} - \left(1 - \frac{T_{D\sigma}}{T_d''}\right)e^{-\frac{t}{T_d''}} - \frac{T_{D\sigma}}{T_d''}e^{-\frac{1}{T_a}}\cos\omega t\right]$$

Conforming to this equation, in the first moment of the short circuit, the field winding current grow up suddenly with the value $\Delta i_E = \frac{X_d}{X_d'}\frac{T_{D\sigma}}{T_d''}$. Then the current will be damped with the time constant T_d'. Around of this non periodical variation will be established oscillations with the pulsation ω, at the time $t = 0$, with the amplitude Δi_E, damped with the time constant T_a. It is observed that at $t = 0$ and $t = \infty$ it is obtained $i_{Esc}(t) = i_{E0}$ (figure 9).

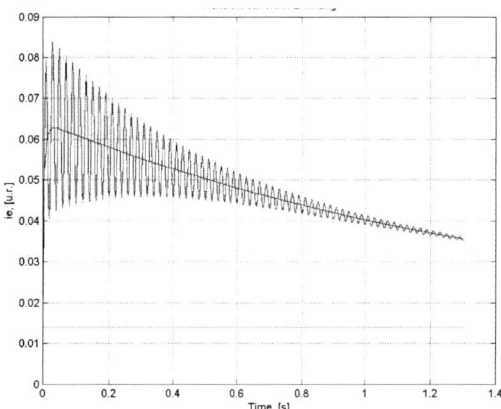

Figure 9. Transient current in E-winding.

Analogous we can obtain the transient current in Q winding (figure 10) and in D winding.

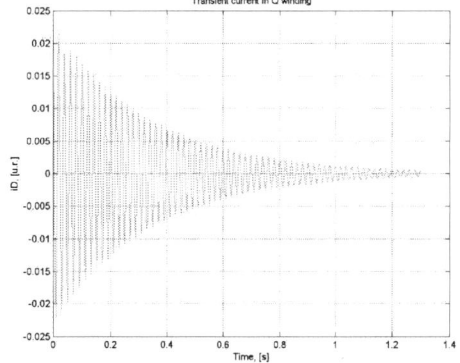

Figure 10. Transient current in Q winding.

IV. SYMBOL AND ABBREVIATIONS

The following symbols are used in this paper: $\omega =$ electrical frequency; u_{d0} =stator's voltages on d axes in the first moment of the short-circuit; u_{q0} =rotor's voltages on d axes in the first moment of the short-circuit; i_A = stator's current of A phases; r_s =stator resistance; r_D =rotor resistance of damper windings; r_Q =rotor resistance of damper windings; r_E =rotor resistance of field windings;

T'_d = transient time constant; T''_d = sub-transient time constant; T_a = time constant, x_s armature leakage reactance, x_d synchronous d- axis reactance, x_d' transient (d-axis) reactance, x_d'' subtransient d- axis reactance, x_q synchronous q- axis reactance.

V. CONCLUSIONS

In concordance with [7], the equivalent schemas in d, q axis shows with bellyful accuracy the reality, ensuring an acceptably concordance between calculated values and those measured during the transient process. Using the software of Matlab the simulations show this concordance.

During a transient state of synchronous machine not the synchronous reactances X_d, X_q are active, but the smaller subtransient reactances: X_d'', X_q''.

Even if $X_d > X_q$ like in the salient poles machines, the subtransient reactances are nearly the same.

The winding resistance: r_s, r_E, r_D, r_Q, cause a decay of transient currents in stator and rotor windings, which flow due induced transient voltages.

Due the small inductance of damper cage the transient current in damper bars decays much faster than in field winding.

Dynamic currents in damper cage have already decayed, whereas in field winding still a big dynamic field current is flowing.

Equivalent reactance for this transient reactance of d axis X_d' is only slightly bigger than subtransient reactance.

After decay all dynamic currents stator reactance becomes again X_d respectively X_q.

REFERENCES

[1] A.,Câmpeanu , "Introducere în dinamica maşinilor electrice de current alternative", Editura Academiei Române, Bucureşti 1998;

[2] A., Câmpeanu, "Maşini electrice. Probleme fundamentale, speciale si de functionare optimală." Editura Scrisul Românesc, Craiova, 1988;

[3] Cioc, I., Nică, C., "Proiectarea maşinilor electrice", Editura Didactică şi Pedagogică R.A., Bucureşti, 1994;

[4] I., Petropol-Şerb, Studiul efectelor parametrilor motorului sincron asupra performanţelor, referat doctorat, Craiova 2005;

[5] G., Petropol-Şerb, I., Petropol-Şerb, Models of the synchronous machine and their numerical calculations, Buletinul Institutului Politehnic Iaşi, 2004,tom L, fasc. 5C, ISSN 1223-8139.

[6] K.,Bonfert, "Betriebsverhalten der Synchronmaschine", Berlin, Springer Verlag,1962.

[7] M.,Canay, "Schemas equivalents de la machine synchrone pour le calcul des constants de la roue polare dans le ces de phenomens non stationaire et de la marche asynchrone", Rev. Brown Bovery, 2, 1969.

[8] Ch., Concordia, "Synchronous machines. Theory and performance. ", New York, John Willley and sons, 1951.

[9] C.V., Jones, "The unified theory of electrical machines", London, Butterworths, 1967.

[10] P., Krause, P., "Analysis of electrical machinery ", New York, McGraw-Hill, 1986.

[11] Th., Laible, "Teoria sinhronnoi masinî pri perehodnîh proţesah ", Moskva-Leningrad, Gosenergoizdat, 1957.

[12] F., Taegen, "Die Glechungen der Synchronmachinen bei dynamischen Vorgangen", Tech. Mitt.AEG – Telefunken, 69,4,1979.

Analytical Method of Calculation of the Current and Torque of a Reluctance Stepper Motor Using Fourier Complex Series

Pavel Záskalický[*], Mária Záskalická[†]

[*]Technical University, KEMPI FEI, Košice, Slovakia, e-mail: *pavel.zaskalicky@tuke.sk*
[†]Technical University, KAMA-SjF, Košice, Slovakia, e-mail: *maria.zaskalicka@tuke.sk*

Abstract—Stepper motors are becoming to be very attractive transducers for conversion of an electric signal to a mechanical position. Due to its simple construction the reluctance machine is considered as a very reliable machine which practically not require any maintenance. The present paper proposes a mathematical analytical method of a calculus of a phase current and electromagnetic torque of the motor via complex Fourier series. Saturation effect of the machine is neglected. Speed of the motor is considered to be constant.

Index Terms—Reluctance stepper motor, analytical method, Fourier series, current and torque calculation.

I. INTRODUCTION

The reluctance stepper motor is an electromagnetic incremental actuator that converts digital pulse sequence on input to analog output shaft position. It is therefore used in digital control systems. A train of pulses turns the shaft of the motor by steps. Neither any position sensor nor any feedback system is normally required for the reluctance stepper motor to make the output response following input command. Typical applications of the reluctance stepper motors requiring incremental motion are printers, desk drives, clocks and robotics.

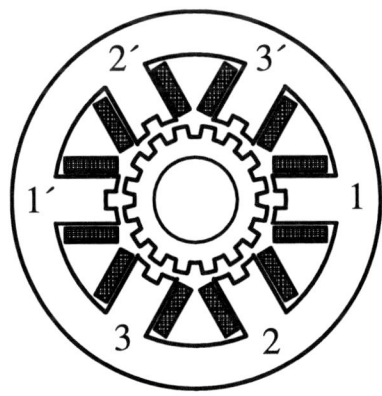

Fig. 1. Design of a reluctance stepper motor.

Due to its simple construction the reluctance machine is considered as a very reliable machine which requiring any maintenance. Typical resolution of commercially available

[0]The financial support of Slovak Research grant No. APVV-0510-06 is acknowledged.

reluctance stepper motor ranges from several steps per revolution to as many as 400 steps per revolution.

A variable reluctance stepper motor can by as a single stack type or a multi stack type constructed. A basic circuit configuration of the three phase single stack reluctance stepper motor is shown on Fig. 1.

When the stator phases are excited with dc current in a proper sequence, the resultant air gap field steps around and the rotor follows the axis of the air gap field by virtue of reluctance torque. The reluctance torque is generated because of the tendency of the ferromagnetic rotor to align itself along with the direction of the resultant magnetic field.

II. MATHEMATICAL MODEL

The analysis of the reluctance stepper motor starts from an electric model which is shown in the Fig. 2. The mutual inductances between the phases are negligible, so it is sufficient to consider the only one phase of the motor. This one phase equivalent circuit comprises ohmic resistance of the coil winding and induced voltage caused by a change of stator inductance.

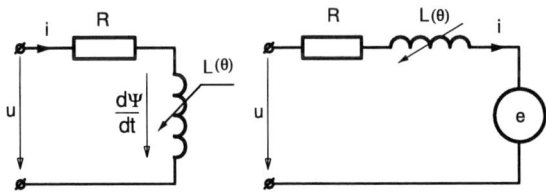

Fig. 2. One phase equivalent circuit of a reluctance stepper motor.

For the one phase equivalent circuit the voltage equation is valid

$$u = Ri + \frac{\mathrm{d}\psi}{\mathrm{d}t} \qquad (1)$$

where ψ is total coil magnetic flux.

Generally, the magnetization curve for nonlineary area is given by

$$\psi = i.L(\theta, i) \qquad (2)$$

In linear area of the magnetization curve the self inductance depends only on the relative position of the stator pole and the rotor tooth

$$\psi = i.L(\theta) \tag{3}$$

The period of the phase inductance has the value $2\pi/N_r$, where N_r is a number of rotor teeth.

Define the electrical angle of rotor position $\theta = N_r\theta_m$, where θ_m is mechanical angle of rotor position.

Similarly the electrical angular velocity is given by $\omega = N_r\omega_m$, where ω_m is the mechanical angular velocity.

Suppose hereafter saturation effect is neglected (the motor is not saturated), so the voltage equation can be writen as

$$u = Ri + L(\theta)\frac{di}{dt} + i\frac{dL(\theta)}{dt} \tag{4}$$

The last term $i\frac{dL(\theta)}{dt} = e$ presents the internal induced electromotive force.

The electrical speed of the motor is given by the change of rotor position in time $\omega = \frac{d\theta}{dt}$. The time in the relation above can be replaced by a rotor position. Finally the voltage equation can be expressed

$$u = Ri + L(\theta)\omega\frac{di}{d\theta} + i\omega\frac{dL(\theta)}{d\theta} \tag{5}$$

Magnetic coenergy of a magnetic circuit is defined

$$W'_{em} = \int_0^{i_0} \psi di \tag{6}$$

The electromagnetic torque of the machine can be calculated from the change of its magnetic coenergy

$$m = \frac{\partial W'_{em}}{\partial \theta} \tag{7}$$

Now, the instantaneous value of the electromagnetic torque may be expressed as

$$m = \frac{1}{2}i^2\frac{dL(\theta)}{d\theta} \tag{8}$$

It is independent from the current polarity.

MODELING OF THE SELF INDUCTANCES

The waveform of the coil self inductance varies with the rotor position. The typical inductance waveform for the non-saturable motor has a form given on the figure 3.

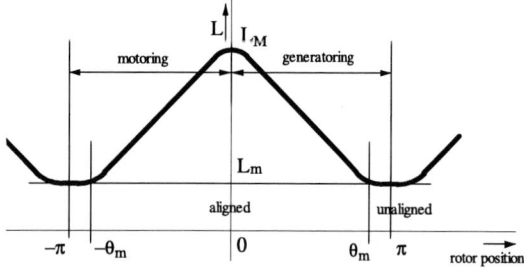

Fig. 3.   Course of the phase inductance versus rotor position.

Mathematically, the waveform of the self inductance can by expressed as a complex Fourier series

$$L(\theta) = \frac{L_0}{2} + \frac{1}{2}\sum_{k=1}^{\infty} L_k\left(e^{jk\theta} + e^{-jk\theta}\right) \tag{9}$$

where L_0, L_k present real Fourier coefficients having the form

$$L_0 = L_m + \frac{L_M - L_m}{2\pi}\theta_m,$$

$$L_k = \frac{2(L_M - L_m)}{k^2\pi\theta_m}\left(1 - \cos k\theta_m\right)$$

The Fourier coefficients are calculated in terms of geometrical and electrical quantities of the motor. L_M is a maximal coil inductance measured in aligned rotor position. L_m is a minimal coil inductance measured in unaligned rotor position. θ_m is a rotor position of approaching of the stator poles and rotor teeth.

III.   SUPPLY CIRCUIT MODELING

The flux in the reluctance stepper motor is not constant, but its must be established from zero in every stroke. In motoring operation the build-up is timed to coincide with the period when rotor poles are approaching the stator poles of the phase. Assuming, each phase is supplied by a circuit of the form shown in Fig. 4.

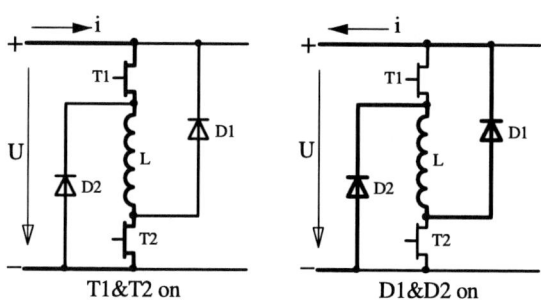

Fig. 4.   Coil supply circuit.

Both transistors T_1, T_2 are simultaneously switched at constant frequency and constant angle of conductance.

The waveform of the supply voltage can be expressed as a Fourier series of the form

$$u = \frac{U}{2}\left\{a_0 + \sum_{k=1}^{\infty}\left[(a_k - jb_k)e^{jk\theta} + (a_k + jb_k)e^{-jk\theta}\right]\right\} \tag{10}$$

a_0, a_k, b_k presents real Fourier coefficients of the form

$$a_0 = \frac{1}{\pi}\left(2\beta - \alpha - \gamma\right)$$

$$a_k = \frac{1}{k\pi}\left(2\sin k\beta - \sin \alpha k - \sin k\gamma\right)$$

$$b_k = \frac{1}{k\pi}\left(\cos \alpha k - 2\cos k\beta + \cos k\gamma\right)$$

The waveform of the suply voltage is given in Fig. 5.

In the Fig. 5 α is turn on angle and β turn off angle of the transistors, γ is turn off angle of the recuperative diodes.

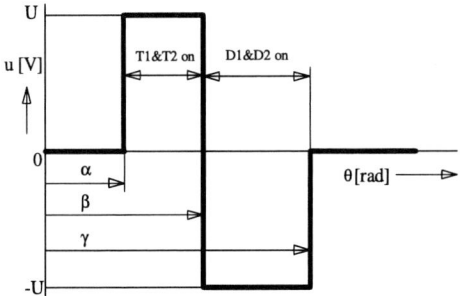

Fig. 5. Waveform of supply voltage.

IV. CURRENT AND TORQUE CALCULATION

The waveform of the supply current is described by the differential voltage equation (5). After introducing (9) and (10) it takes the form:

$$\frac{U}{2}\left\{a_0 + \sum_{k=1}^{\infty}\left[(a_k - jb_k)e^{jk\theta} + (a_k + jb_k)e^{-jk\theta}\right]\right\} =$$
$$Ri + \frac{\omega}{2}\left[L_0 + \sum_{k=1}^{\infty}L_k\left(e^{jk\theta} + e^{-jk\theta}\right)\right]\frac{di}{d\theta} +$$
$$i\omega\sum_{k=1}^{\infty}\frac{jkL_k}{2}\left(e^{jk\theta} - e^{-jk\theta}\right) \tag{11}$$

Equation (11) presents the exact differential equation of the type

$$P\left(\theta,i\right)d\theta + Q\left(\theta,i\right)di = 0 \tag{12}$$

Exact differential equation has an analytical solution provided that

$$\frac{\partial P\left(\theta,i\right)}{\partial i} = \frac{\partial Q\left(\theta,i\right)}{\partial \theta} \tag{13}$$

The condition (13) is fulfilled by neglecting coil resistance ($R = 0$).

The equation (11) takes the form

$$\frac{U}{2}\left\{a_0 + \sum_{k=1}^{\infty}\left[(a_k - jb_k)e^{jk\theta} + (a_k + jb_k)e^{-jk\theta}\right]\right\} =$$
$$\frac{\omega}{2}\left[L_0 + \sum_{k=1}^{\infty}L_k\left(e^{jk\theta} + e^{-jk\theta}\right)\right]\frac{di}{d\theta} +$$
$$\frac{i\omega}{2}\sum_{k=1}^{\infty}jkL_k\left(e^{jk\theta} - e^{-jk\theta}\right) \tag{14}$$

Current analytical solution can by written

$$i = \frac{U}{\omega}\frac{C + a_0\theta + \sum_{k=1}^{\infty}\left(\frac{a_k - jb_k}{jk}e^{jk\theta} - \frac{a_k + jb_k}{jk}e^{-jk\theta}\right)}{L_0 + \sum_{k=1}^{\infty}L_k\left(e^{jk\theta} + e^{-jk\theta}\right)} \tag{15}$$

Integrating constant C can be calculated from the initial condition. For $\theta = \alpha$ is $i = 0$. The final formula for current waveform has the form

$$i = \frac{U}{\omega}\frac{1}{L_0 + \sum_{k=1}^{\infty}L_k(e^{jk\theta} + e^{-jk\theta})}\left\{(\theta - \alpha)a_0 +\right.$$
$$\left. \sum_{k=1}^{\infty}\left[\frac{b_k + ja_k}{k}\left(e^{jk\alpha} - e^{jk\theta}\right) + \frac{b_k - ja_k}{k}\left(e^{-jk\alpha} - e^{-jk\theta}\right)\right]\right\} \tag{16}$$

Instantanous value of the electromagnetic torque in accordance of (8) is given by

$$m = \frac{1}{2}i^2\sum_{k=1}^{\infty}j\omega kL_ke^{jk\theta} \tag{17}$$

V. RESULTS OF THE CALCULUS

To calculate current and torque waveforms, the parameters of a small three phase reluctance motor was used. For the considered machine following values were measured

Inductance in aligned position	$L_M = 100\,mH$
Inductance in unaligned position	$L_m = 10\,mH$
Phase coil resistance	$R = 2,2\,\Omega$
Number of rotor teeth	$N_R = 40$
Number of stator pole	$N_S = 6$
Number of the teeth per pole	$N_{St} = 2$
Stator supply voltage	$U = 5\,V$

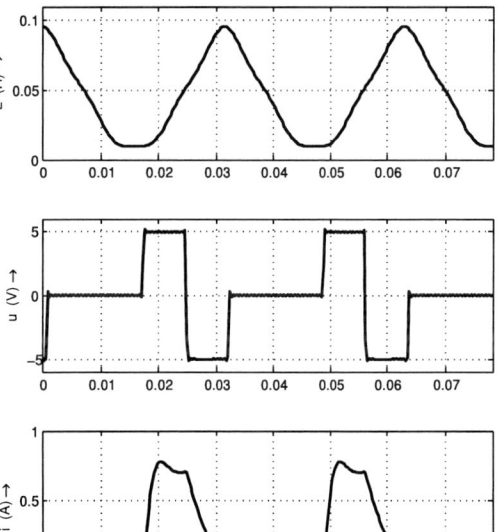

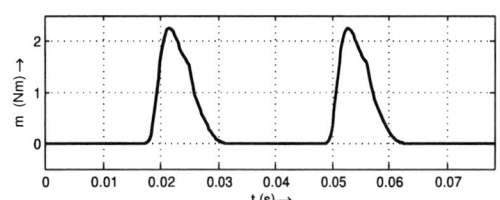

Fig. 6. Calculated quantities for motoring.

In the Fig. (6) there are show a waveforms of the phase current, voltage and torque for machine which works as a motor. Phase coil is excited in interval of positive derivate of the inductance. Turn off angle of the recuperate diodes γ must be calculated by the iteration.

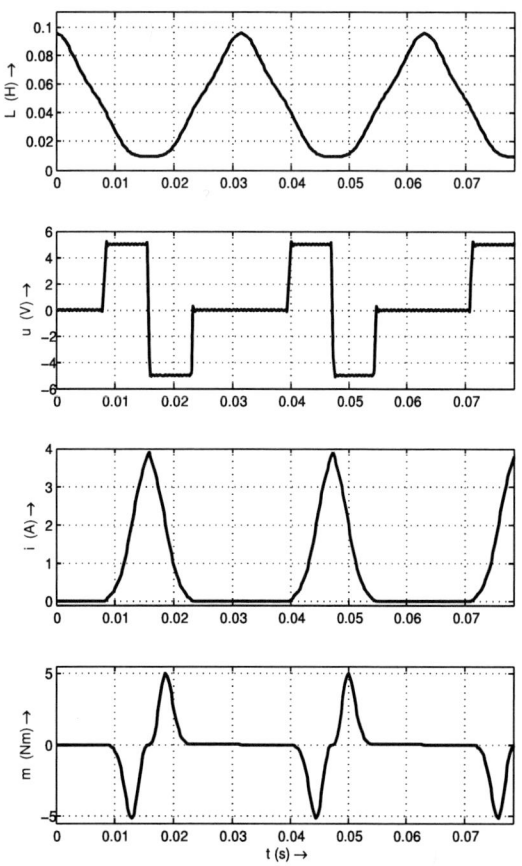

Fig. 7. Calculated quantities for no loaded machine.

Fig. (7) shows the waveforms for no loaded machine. The average induced torque is equal zero.

In the Fig. (8) there are shown the waveforms of the electrical quantities for the generating area of the stepper motor. Motor is excited in interval of negative derivate of the inductances. The induced torque becomes negative.

CONCLUSION

In present paper there was shown a mathematical method for analytical calculus of the phase current and electromagnetic torque of the reluctance stepper motor. Presented method is based on the analytical formularization of the waveform of motor inductance and supply voltage using complex Fourier series. Calculated motor quantities confirm accuracy of the method.

REFERENCES

[1] T. J. E. Miler, "Switched Reluctance Motors and their Control" *Magne Physics Publishing*, Oxford: Clarendon, 1993

[2] F. M. Sargos, E.J. Gudefin, P. Záskalický, "Etudes analytique du fonctionnement des moteurs à reluctance alimenté à frequence variable," *Journal Physique III*, vol.5, 339–354, March 1995, France

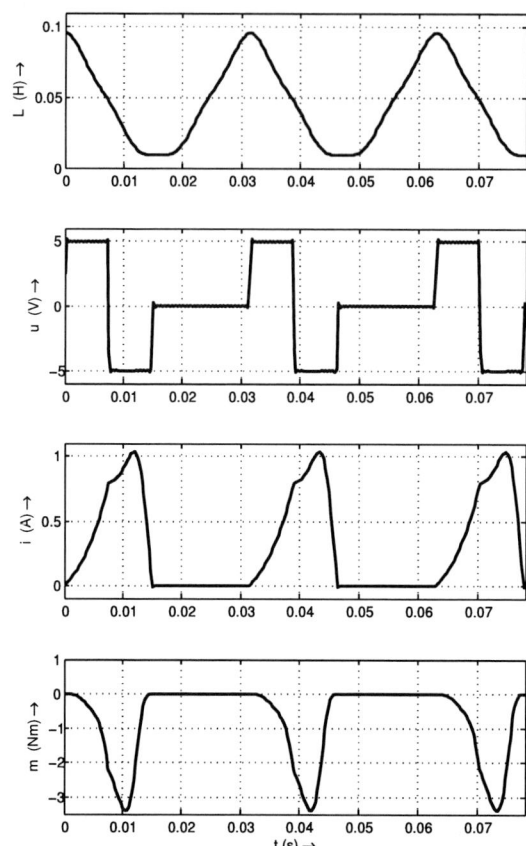

Fig. 8. Calculated quantities for generating area.

[3] J. Kudla, R. Miksiewicz , "Field-circuit model of switched reluctance motor, at single pulse supply," in *Low voltage electric machines, Joint Czech-Polish Conference*, Nov. 14-15,2005, pp.58–64, Brna, Czech Republic.

[4] L. Schreier, M. Chomát, I. Doležel, "Influence of magnetic circuit of synchronous reluctance machine on harmonic content in electrical quantities," in *Symposium ČSAV*, September, 20, 2001, pp.51–59, Prague, Czech Republic.

Bearing Damage Analysis by Calculation of Capacitive Coupling between Inner and Outer Races of a Ball Bearing

Jafar Adabi[*], Firuz Zare[*], Gerard Ledwich[*], Arindam Ghosh[*], Robert D.Lorenz[†]

[*] Queensland University of Technology, School of Electrical Engineering, GPO Box 2434, Brisbane, QLD, 4001, Australia, adabi.jafar@student.qut.edu.au

[†] University of Wisconsin-Madison, Depts. of ME and ECE,1513 University Avenue, Madison, WI 53706, lorenz@engr.wisc.edu

Abstract— bearing damage in modern inverter-fed AC drive systems is more common than in motors working with 50 or 60 Hz power supply. Fast switching transients and common mode voltage generated by a PWM inverter cause unwanted shaft voltage and resultant bearing currents. Parasitic capacitive coupling creates a path to discharge current in rotors and bearings. In order to analyze bearing current discharges and their effect on bearing damage under different conditions, calculation of the capacitive coupling between the outer and inner races is needed. During motor operation, the distances between the balls and races may change the capacitance values. Due to changing of the thickness and spatial distribution of the lubricating grease, this capacitance does not have a constant value and is known to change with speed and load. Thus, the resultant electric field between the races and balls varies with motor speed. The lubricating grease in the ball bearing cannot withstand high voltages and a short circuit through the lubricated grease can occur. At low speeds, because of gravity, balls and shaft voltage may shift down and the system (ball positions and shaft) will be asymmetric. In this study, two different asymmetric cases (asymmetric ball position, asymmetric shaft position) are analyzed and the results are compared with the symmetric case. The objective of this paper is to calculate the capacitive coupling and electric fields between the outer and inner races and the balls at different motor speeds in symmetrical and asymmetrical shaft and balls positions. The analysis is carried out using finite element simulations to determine the conditions which will increase the probability of high rates of bearing failure due to current discharges through the balls and races.

Keywords— bearing failure, capacitive coupling, discharge current, shaft voltage, symmetrical and asymmetrical shaft position

I. INTRODUCTION

Nowadays, modern AC motor drive systems are widely used in industrial and commercial applications. Due to rapid developments of IGBT technology, switching times have decreased to a fraction of a micro second and as a result, the switching frequency has dramatically increased. Fig.1.a shows the structure of a modern power electronic drive consisting a filter, a rectifier, a dc link capacitor, an inverter and an AC motor. It also shows that many parasitic capacitive couplings exist which may be neglected in low frequency analysis but the conditions are completely different in high frequencies. In high

switching frequencies, a low impedance path is created for the current to flow through these capacitors [1-2].

Fig.1.b shows the different forms of capacitive coupling in an induction motor, where C_{WR} is the capacitive coupling between the stator winding and rotor, C_{WS} is the capacitive coupling between the stator winding and stator, C_{SR} is the capacitive coupling between the rotor and stator frame.

In principle, all inverters generate common mode voltages relative to the earth ground due to coupling through the parasitic capacitances [1]. Fig.2 shows a simple equivalent circuit model of an AC motor which depicts the main high frequency coupling capacitances [2-3].

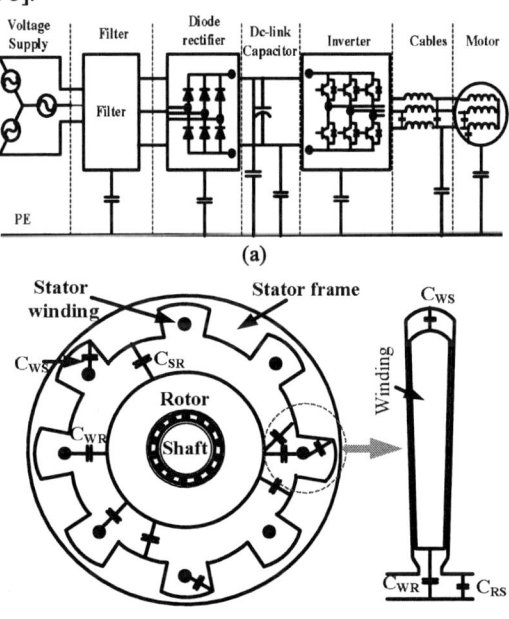

(a)

(b)

Fig.1. Capacitance coupling in an induction motor and a view of stator slot

High dv/dt (fast switching transients) and common mode voltage generated by a PWM inverter can cause unwanted problems such as shaft voltage and resultant bearing currents [4-7]. Fig.3 shows the general structure of ball bearings and shaft in an AC machine. As shown in this figure, there are balls between outer and inner

978-1-4244-1741-4/08/$25.00 ©2008 IEEE

races with lubricating grease between balls and the races. There is a capacitive coupling between the outer and inner races.

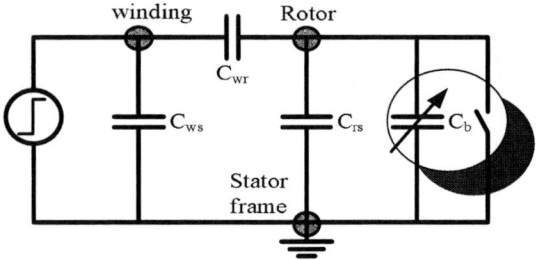

Fig.2. High frequency model of an induction motor

During operation, the distances between the balls and races may change and vary the capacitance and resultant electric field between the races and balls. This capacitance has a nonlinear relationship with load and speed. Lubricating grease in the ball bearing cannot withstand high voltages and a short circuit through the lubricated grease can occur. This breakdown phenomenon can be modeled as a switch.

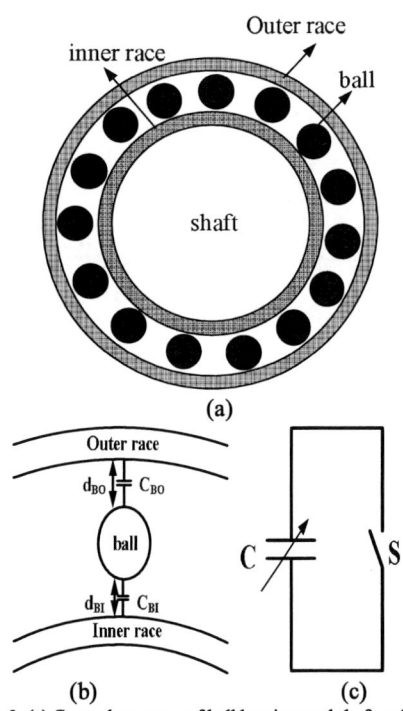

Figure.3. (a) General structure of ball bearings and shaft and outer and inner race of an AC machine (b) a view of ball, outer and inner races and capacitive couplings (c) simple model of ball bearing

This paper focuses on calculation of capacitive coupling between ball bearing and inner and outer races using finite element simulations to analyze the probability of increased bearing failure rates under different conditions.

II. DISCHARGE CURRENT PATHS BY CALCULATION OF CAPACITIVE COUPLINGS FIGURES AND TABLES

2-D Finite element simulations are carried out based on the proposed structure in order to calculate capacitive

coupling terms between the inner and outer races and balls in low and high speeds in symmetrical and asymmetrical positions. For the test case bearing as shown in Fig.3, there are 15 balls with the diameter of 20 mm, shaft diameter is 80 mm and three ranges of 1mm, 0.1mm, 0.01mm oil thickness were simulated. The objective of the simulation is to calculate the electric fields between the outer race and balls (d_{BO}) and between the inner race and balls (d_{BI}) at different motor speeds which cause symmetrical and asymmetrical shaft and ball positions. Analyses are carried out in order to determine the conditions under which the probability of bearing failure rate due to discharging current through the balls and races are very high. Several conditions are simulated based on balls and shaft positions in low and high speeds.

A. SymmetricCase

At high speed, balls and shaft positions are considered symmetric and the distances between the inner race and balls (d_{BI}) and between outer races and balls (d_{BO}) are assumed to be equal. Also the shaft position is not changed and the shaft and outer race are concentric. Table I shows the capacitive coupling between the inner and outer races and the ball, the electric field in the area between the inner race and ball (E_{BI}) and the outer race (E_{BO}) assuming a typical 100 volts voltage across the races.

TABLE I. CAPACITIVE COUPLING TERMS, VOLTAGE AND ELECTRIC FIELDS IN THE SYMMETRIC CASE

d_{BO} (mm)	d_{BI} (mm)	C_{BO} (nF)	C_{BI} (nF)	E_{BO} (V/mm)	E_{BI} (V/mm)
0.5	0.5	1.010	0.807	88.87	111.13
0.05	0.05	3.540	2.890	899.47	1100.53
0.005	0.005	11.300	9.020	8881.72	11118.28

As depicted in Fig.4, if a short circuit (breakdown) occurs, then a discharge current will be divided into several paths and the probability of bearing damage is decreased.

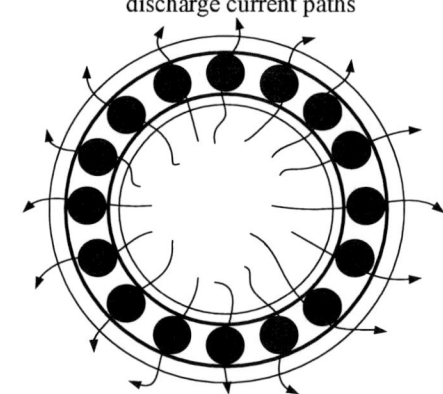

Fig.4. Possible discharge current paths in the symmetric case

B. Asymmetric case

At low speeds, because of gravity, balls and shaft may shift down and the system (balls position and shaft) will be asymmetrical. In this study, two different cases (asymmetric ball positions, asymmetric shaft position) are

904

analyzed. Fig.5 shows these two types of asymmetries. As shown in Fig.5.a, in this asymmetric case, the upper and lower side balls are shifted down because of gravity but the separations between the inner and outer races with other balls can approximately be considered as symmetric. As shown as in Fig.5.b, at lower speeds, an asymmetric shaft position may occur, which is more common than other cases.

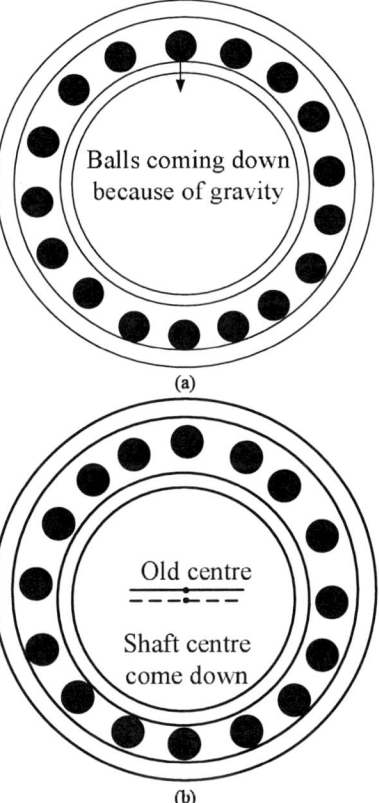

(a)

(b)

Fig.5. Asymmetric (a) ball positions (b) shaft position

•Asymmetric ball positions

As shown in Table. II, several distances are simulated to compare the capacitive couplings (C_{BO}, C_{BI}) and electric fields (E_{BO}, E_{BI}) for each of them. Simulations are carried out for oil thicknesses of 1mm, 0.1mm, and 0.01mm. As shown in Fig.6.a, in the asymmetrical balls case, balls come down and the region between the upper ball and shaft (see Fig.6.b) and the lower ball and shaft (see Fig.6.c) are more important than other areas.

From the results in Table II, the electric field is increased when d_{BI} or d_{Bo} are decreased but the electric field between the inner race and upper ball (E) is more than the electric field between the outer race and lower ball (E') for the same rate of change in distances. The capacitive coupling terms and resultant electric fields for $d_{BI1}=d_{BO2}=0.001$ mm & $d_{BI2}=d_{BO1}=0.009$ mm as shown in Table III. However d_{BO2} & d_{BI1} are equal, because of different positions of balls and races (which is shown in Fig.6.b&c), capacitive coupling terms and electric fields are different (E_{BI1} is 50% more than E_{BO2}).

TABLE II. CAPACITIVE COUPLING TERMS AND ELECTRIC FIELDS IN AN ASYMMETRICAL BALL POSITION

Oil Thickness (mm)	d_{BO} (mm)	d_{BI} (mm)	C_{BO} (nF)	C_{BI} (nF)	E_{BO} (V/mm)	E_{BI} (V/mm)
1	0.1	0.9	2.490	0.616	198.71	89.03
1	0.3	0.7	1.400	0.710	111.97	94.87
1	0.5	0.5	1.010	0.807	88.86	111.13
1	0.7	0.3	0.893	1.130	79.82	147.08
1	0.9	0.1	0.778	2.020	80.21	278.05
0.1	0.01	0.09	7.760	2.130	2155.97	871.56
0.1	0.03	0.07	4.570	2.430	1157.93	932.31
0.1	0.05	0.05	3.540	2.890	899.46	1100.53
0.1	0.07	0.03	2.980	3.750	796.03	1475.91
0.1	0.09	0.01	2.620	6.530	792.73	2865.36
0.01	0.001	0.009	26.200	6.890	20821.87	8797.57
0.01	0.003	0.007	13.100	7.800	12443.87	8952.63
0.01	0.005	0.005	11.300	9.020	8881.72	11118.28
0.01	0.007	0.003	9.140	11.800	8048.07	14554.51
0.01	0.009	0.001	8.150	18.700	7736.50	30371.47

TABLE III. CAPACITIVE COUPLING TERMS AND ELECTRIC FIELDS IN OIL THICKNESS OF 0.001 MM

ball	Oil Thickness (mm)	d_{BO} (mm)	d_{BI} (mm)	C_{BO} (nF)	C_{BI} (nF)	E_{BO} (V/mm)	E_{BI} (V/mm)
1	0.01	0.009	0.001	8.150	18.700	7736.50	30371.47
2	0.01	0.001	0.009	26.20	6.890	20821.87	8797.57

Thus, increasing the electric field between inner race and balls at upper side will create a path to discharge current. In other words, if a short circuit (breakdown) occurs at these balls, the probability of dividing the discharge current into other paths will decrease and the upper ball near the inner race (ball 1 in Fig.6.a) is the highest probability candidate to create a path for discharging current. If the voltage breakdown occurs, a bearing damage problem could occur at this area (position A in Fig.7). If the damage occurs at this position, the same problem will happen at the distance between ball and outer race (position A' in Fig.7).

•Asymmetric shaft position

An asymmetry in the shaft position is analyzed via simulations. The simulations are carried out to find the capacitive coupling terms and electric field in three separation ranges: 1mm, 0.1mm, and 0.001mm. In this case, shaft position is shifted down corresponding to 20%, 40% and 60% grease thickness. Table IV shows the capacitive coupling terms, voltage and electric fields with respect to different variables associated with the balls position assuming the inner and outer distances in each side are equal.

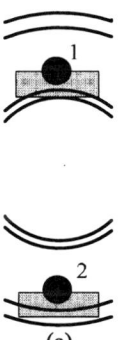

(a)

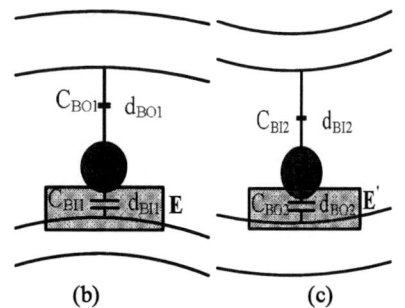

(b)   (c)

Fig 6. (a) Asymmetric ball positions (b) upper side ball (c) lower side ball

TABLE IV. CAPACITIVE COUPLING TERMS AND ELECTRIC FIELDS IN AN ASYMMETRIC SHAFT POSITION

Shift in Shaft center (mm)	d_{BO} (mm)	d_{BI} (mm)	C_{BO} (nF)	C_{BI} (nF)	E_{BO} (V/mm)	E_{BI} (V/mm)
0.2	0.4	0.4	1.21	0.967	111.21	138.79
0.4	0.3	0.3	1.41	1.130	148.54	184.79
0.6	0.2	0.2	1.74	1.400	223.22	276.78
0.02	0.04	0.04	4.01	3.240	1117.02	1382.99
0.04	0.03	0.03	4.64	3.750	1489.93	1843.40
0.06	0.02	0.02	5.71	4.610	2233.12	2766.88
0.002	0.004	0.004	13.20	9.960	10767.64	14232.36
0.004	0.003	0.003	17.90	10.200	12121.21	21212.12
0.006	0.002	0.002	24.10	11.600	16308.64	33691.36

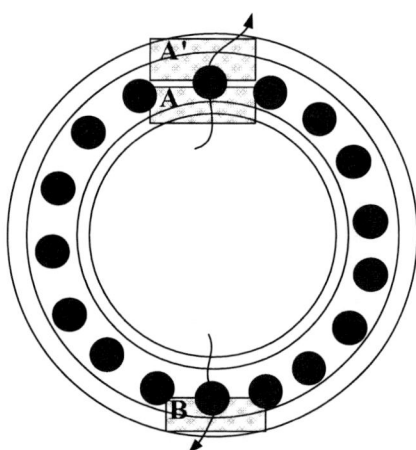

Fig.7. Discharge current paths for asymmetric ball positions

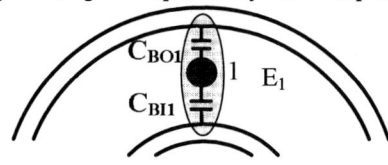

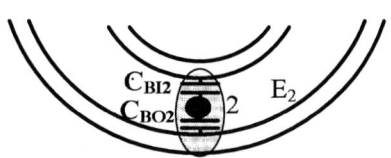

Fig.8. Capacitive coupling terms between upper and lower balls and races for an asymmetric shaft position

According to simulation results, electric field between the lower ball (ball 2 in Fig.8) and the inner race is more than other separations. In other words, if a breakdown occurs in this area, the probability of division of the discharge current into other paths will decrease and ball 2 is the highest probability candidate to create a path for the discharge current. In this case, the distance between ball 1 and the races is more than the distance between ball 2 and races. Thus, capacitance and the resultant electric field in the upper side is less than in the lower side ($E_1<E_2$ as shown in Fig.8). In the lower side, because of different positions of ball 2 and the races, the electric field is different while the distance between ball and races are the same (for instance, at $d_{BI2}=d_{BO2}=.002$ mm, E_{BI2} is 40% more than E_{BO2}). As shown in Fig.9, if the breakdown voltage is exceeded, a bearing damage problem may occur at this area (position C in Fig.9). If the damage happens at this position, the same problem will happen at the distance between ball and outer race (position C' in Fig.9). This may cause multiple bearing damage sites.

.

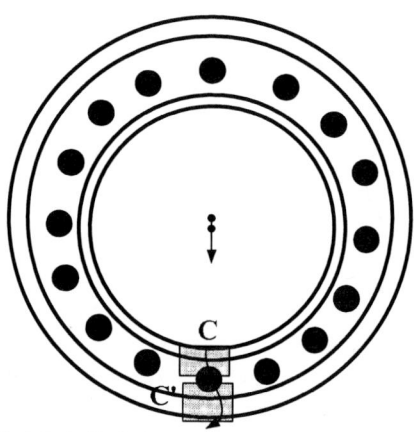

Fig.9. Probable discharge current paths for an asymmetric shaft position

III. CONCLUSIONS

Based on the simulation and analysis which are presented in this paper, during motor operation, the distances between the balls and races may change the capacitance values. At a high speed, balls and shaft positions are considered symmetrical and the distances between the inner race and balls (d_{BI}) and between outer races and balls (d_{BO}) are assumed to be equal. Also the shaft position is not changed and the centers of the shaft and the outer race are the same (symmetrical position). In a low speed case, because of gravity, balls and shaft voltage may shift down and the system (balls position and shaft) will be in an asymmetric shape. In this study, two different asymmetric cases (asymmetric ball positions, asymmetric shaft position) are analyzed and the results are compared with the symmetrical case to determine the probability of bearing damage. Several distances are simulated to compare the capacitive couplings between ball bearing and inner and outer races (C_{BO}, C_{BI}) and electric fields (E_{BO}, E_{BI}) for each of them. Simulations are carried out for oil thicknesses of 1mm, 0.1mm, and 0.01mm for both symmetrical and asymmetrical cases to determine the conditions which will increase the probability of high rates of bearing failure due to current discharges through the balls and races.

ACKNOWLEDGEMENT

The authors thank the Australian Research Council (ARC) for the financial support for this project through the ARC Discovery Grant DP0774497

REFERENCES

[1] S. Chen, T. A. Lipo, and D. Fitzgerald, "Modeling of motor bearing currents in PWM inverter drives," *Proc. of the 30th Annual IEEE Industry Applications Conference*, vol.32, issue 6, pp. 1365-1370, 1995.

[2] S. Chen, T. A. Lipo, and D. Fitzgerald, "Source of induction motor bearing currents caused by PWM inverters" *IEEE Transactions on Energy Conversion*, vol. 11, pp. 25-32, 1996.

[3] A. Muetze and A. Binder, "Calculation of Circulating Bearing Currents in Machines of Inverter-Based Drive Systems" *IEEE Transactions on Industrial Electronics*, vol. 54, pp. 932-938, 2007.

[4] ABB Technical guide No.5 'bearing currents in modern AC Drive systems", Helsinki, 1999

[5] A. Muetze and A. Binder, "Practical Rules for Assessment of Inverter-Induced Bearing Currents in Inverter-Fed AC Motors up to 500 kW," *IEEE Transactions on Industrial Electronics*, vol. 54, pp. 1614-1622, 2007.

[6] J. M. Erdman, R. J. Kerkman, D. W. Schlegel, and G. L. Skibinski, "Effect of PWM inverters on AC motor bearing currents and shaft voltages," *IEEE Transactions on Industry Applications*, vol. 32, pp. 250-259, 1996.

[7] Michael J. Devaney and Levent Eren, "Detecting motor bearing faults", *IEEE Instrumentation & Measurement Magazine*, Volume 7, Issue 4, pp. 30-50, Dec 2004.

The Model of the Squirrel Cage AC Motor including Rotor Slot Harmonics

Eleonora Darie[*], Costin Cepişcă[†] and Emanuel Darie[**]

[*] Technical University of Civil Engineering/ Electrotechnical Department, Bucharest, Romania, e-mail:
eleonora_darie@yahoo.com
[†] University Politechnica of Bucharest/ Electrotechnical Department, Bucharest, Romania, e-mail: *costin@wing.ro*
[**] Police Academy Bucharest/ Engineering Department, Bucharest, Romania, e-mail: *e_darie@yahoo.com*

Abstract—The model of the squirrel cage AC Motor is based on multiple coupled circuits and takes into account the geometry and winding layout of the machine. All inductances are derived by means of the winding function approach and are integrated with the decomposition into their Fourier series. An important issue in such effort is the modeling of the induction motor including rotor slot harmonics (RSH) under symmetrical and asymmetrical conditions, with minimum computational complexity.

Keywords— Diagnostics, Harmonics, Induction Motor,, Modeling, Simulation, Signal Processing.

I. INTRODUCTION

The majority models of the squirrel cage AC Motor are based on the simple idealist machine without tacking into account the physical layout of the stator and rotor windings. One example is the conventional Park model and the development of its current, torque and power relationships are based on the assumptions that the rotating magneto motive force (MMF) produced by stator winding excitation is sinusoidal distributed in space and that the rotor MMF due to the slip frequency induced currents is similarly distributed. It is apparent that these models are not suitable for diagnosis and/or sensorless speed estimation by investigating the rotor slot harmonics (RSH). Therefore, there is a real need to derive an accurate model which can take into account the effect of the field harmonics both in time and in space.

In this work, an alternative way for formulating the suitable model is analyzed, using the decomposition into Fourier series of the mutual inductance matrix and the presentation of the induction motor in Park frame.

The model can be extended to the solution of a wide variety of fault and predict the squirrel induction motor response in transient as well as in steady-state modes of operation [1], [2], [3].

II. THEORETICAL MODEL OF THE SQUIRREL CAGE AC MOTOR

A. Rotor Slot Harmonics

The air gap field of an induction motor fed by a sinusoidal voltage supply waveform comprises a wide range of different space harmonics. The following analysis assumes that these air-gap flux harmonics are a result of the interaction of air-gap permeance and harmonic MMF waves. Only harmonics due to slotting are considered here (rotor slot harmonics).

The rotor slot harmonics (RSH) are generated in the stator line current for healthy machine at frequencies given by the following equation [3], [5]:

$$f_{sh} = \left[\frac{\lambda N_r}{p} \left(1 - s\right) \pm 1 \right] f \cdot \tag{1}$$

The rotor slot harmonic frequencies on calculated by the equation:

$$f_{hk} = \left[\frac{\lambda N_r}{p} \left(1 - s\right) \pm 1 \pm 2 k s \right] f_s, \tag{2}$$

where: p – is the number of pole pairs; s – is the slip; f_s – is the fundamental supply frequency; N_r – is the number of rotor slot; λ – is a positive integer.

B. Electrical Equations

On considered a squirrel cage induction motor (Figure 1) what have three identical and symmetrical phases in the stator.

The rotor cage has ($N_r = n$) bars is viewed as n identical spaced loops and the currents distribution can be specified in term of $n+1$ independent rotor currents.

These currents are formed of n rotor loop currents (i_{nr}) plus a circulating current in one of the end rings i_e.

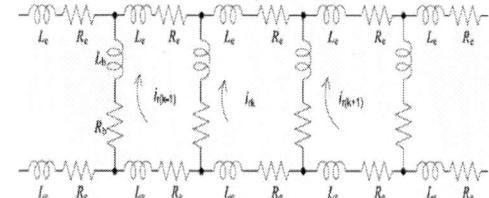

Fig. 1. The Rotor loops in AC motor.

The mesh model is based on a coupled magnetic circuits approach and by making the following assumptions: 1) the state of operation remains far from magnetic saturation; 2) the magnetic permeability of iron is considered to be infinite and the air gap is very small and smooth.

The stator voltage equation can be written as:

$$[V_{3s}] = [R_s][i_{3s}] + \frac{d}{dt}[\psi_{3s}],\qquad (3)$$

where: $[V]$ is the voltage, $[i]$ the current matrix, $[R_s]$ and $[R_r]$ are the stator and rotor resistance matrixes respectively, $[\psi_s]$ and $[\psi_r]$ are the stator and rotor flux linkage matrixes respectively.

C. Analysis of Inductances

The inductances of the above system of equations on calculated using the winding method, from which inductance between any two windings i and j in any electric machine can be computed by the equation:

$$L_{ij}(\varphi) = \mu_0 L r \int_0^{2\pi} \frac{n_i(\varphi,\theta) N_j(\varphi,\theta)}{e(\varphi,\theta)} d\theta,\qquad (4)$$

where: φ is the angular position of the rotor with respect to some stator reference, θ is the particular angular position along the stator inner surface, e is the air gap function, L is the length of stack and r is the average radius of air gap. $n_i(\varphi,\theta)$ is the winding distribution of coil i (it was introduced to describe the considered coil), $N_j(\varphi,\theta)$ is the winding distribution of coil j (it represents the magneto motive force of the air gap produced by unit current flowing in the considered coil).

The AC motor used for simulations used in this work is a three phases, P_n= 2.2 kW, 50 Hz, 2 poles, 36 stator slots and 28 rotor bars. The mutual inductances between stator and rotor are considered to be time-varying, the others inductances are pre-calculated and treated constant because of the round stator and rotor structure.

The Figure 2 shows the turn function or the MMF distribution of the stator phase and the winding function of the rotor sloop 1.

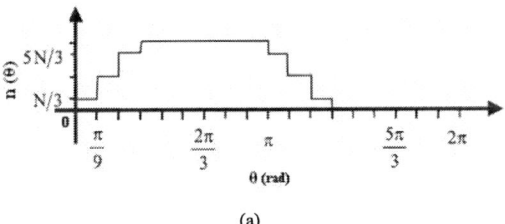

(a)

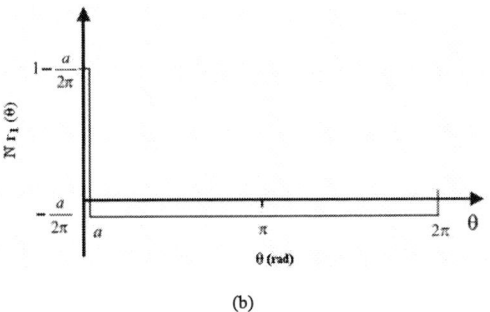

(b)

Fig. 2. Winding distribution of stator phase (a), Winding function of rotor loop 1 (b).

The electrical angle of a rotor loop is notated with a,

$$a = p\frac{2\pi}{N_r}$$

The $[L_r]$ is n by n symmetrical matrix, where L_b is the rotor leakage inductance, the rotor end ring leakage inductance and L_e is the mutual inductance. All inductances on derived using (4).

$$L_r = \begin{bmatrix} L_{rp} + 2L_b + 2\dfrac{L_e}{N_r} & \cdots & M_{rr} & \cdots \\ \vdots & & \vdots & \\ M_{rr} & & M_{rr} - L_b & \\ \vdots & & \vdots & \\ M_{rr} - L_b & \cdots & M_{rr} & \\ M_{rr} & & M_{rr} & \\ \vdots & & \vdots & \\ L_{rp} + 2L_b + 2\dfrac{L_e}{N_r} & & M_{rr} - L_b & \cdots \\ \vdots & & \vdots & \\ M_{rr} & & M_{rr} & \end{bmatrix}.(5)$$

The space distribution of the mutual inductance is not sinusoidal. This implies that the mutual inductances matrix presents harmonics with respect to the electrical angle θ. This matrix can be resolved into its Fourier series [12]:

$$[M_{sr}] = \sum_{h=1}^{\infty} M_{srh} \begin{bmatrix} \cdots & \cos[h(\theta + \varphi_h + ka)] & \cdots \\ \cdots & \cosh\left[(\theta + \varphi_h + ka) - \dfrac{2\xi_h\pi}{3}\right] & \cdots \\ \cdots & \cosh\left[(\theta + \varphi_h + ka) + \dfrac{2\xi_h\pi}{3}\right] & \cdots \end{bmatrix},(6)$$

where ξ_h is the initial phase angle, $k = 0, 1, 2, \dots, N_r$,

$$\xi_h = \begin{cases} +1 & if\ h \in F \\ -1 & if\ h \in B \end{cases},\qquad (7)$$

and $F = \{1, 7, 13, 19, \dots\}$ is the set of forward harmonic component, $B = \{5, 11, 17, \dots\}$ is the set of backward harmonic component.

III. THE PARK TRANSFORMATION

The Park Transformation equation (the three phases to two phase transformation equation in machine analysis) is of the form:

$$[X_{sodq}] = [X_{so}\quad X_{sd}\quad X_{sq}]^T = [P_3(\theta_s)]^T X_{3s},\ (8)$$

where the transformation matrix is defined in (9) and θ_s is the angular displacement between the Park reference and the first phase of the stator [7], [8].

$$[P_3(\theta_s)] = \sqrt{\frac{2}{3}} \begin{bmatrix} \frac{1}{\sqrt{2}} & \cos(\theta_s) & -\sin(\theta_s) \\ \frac{1}{\sqrt{2}} & \cos\left(\theta_s - \frac{2\pi}{3}h\right) & -\sin\left(\theta_s - \frac{2\pi}{3}\right) \\ \frac{1}{\sqrt{2}} & \cos\left(\theta_s - \frac{2\pi}{3}h\right) & -\sin\left(\theta_s + \frac{2\pi}{3}\right) \end{bmatrix}.(9)$$

The relation between θ_s and θ is:

$$\theta_s = \theta_r + \theta, \qquad (10)$$

where θ_r is the angular displacement between the Park reference frame and the first rotor loop.

IV. THE MECHANICAL EQUATIONS

The mechanical equations are:

$$\begin{cases} \dfrac{d\omega}{dt} = \dfrac{p}{J}(\Gamma_e - \Gamma_r) \\ \dfrac{d\theta}{dt} = \omega \end{cases}, \qquad (11)$$

where: Γ_e is the electromagnetic torque produced by the motor, Γ_r is the load torque and ω is the rotor speed.

V. NUMERICAL SIMULATIONS

A. Analysis of Healthy AC motor

The AC motor used for simulations used in this work is a three phases, 2.2 kw, 50 Hz, 2 poles, 36 stator slots and 28 rotor bars.

For purposes of comparison the healthy machine on first modeled using the conventional Park d-q model (Figure 3). The same machine on simulated next using the model analyzed in this work. The simulations results are carrier out a slip around 0.0035 (Figure 4).

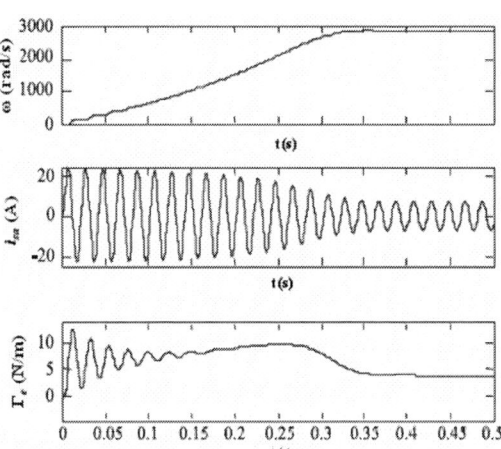

Fig. 3. The acceleration transient using conventional d-q mesh model under sinusoidal voltage excitation.

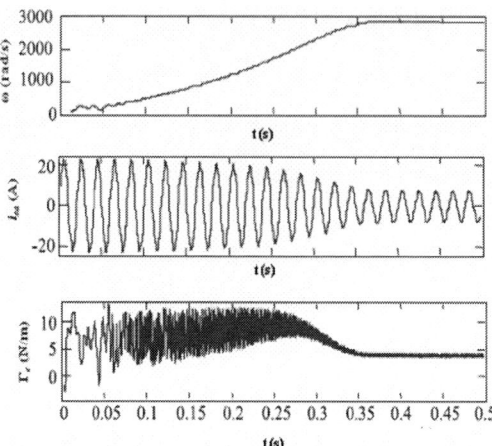

Fig. 4. The acceleration transient using the analyzed model under sinusoidal voltage excitation.

Comparison of the two simulations traces shows very good correlation. The effects of RSH can be observed to have a more significant effect on the electromagnetic torque [6].

The Figure 5 shows the Fast Fourier Transform (FFT) normalized to the fundamental of the line current for the conventional and the analyzed model.

The Fast Fourier Transform analysis is usually applied when the motor is operating in the steady-state [8], [9].

Then, a set of measurements is taken over a period of time and is analyzed to obtain the signal frequency components with any desired frequency resolution.

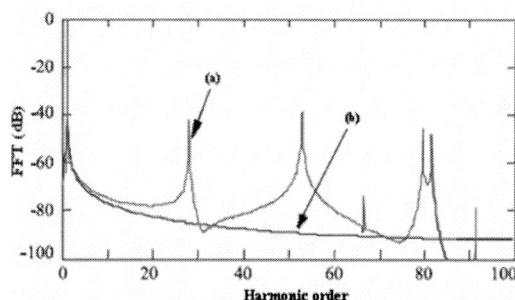

Fig. 5. The Normalized FFT of the stator current in quasi steady state. Conventional model d-q meshes model (a), analyzed model (b).

On can observer the presence of the RSH in the case of analyzed model.

B. Analysis of AC motor with broken rotor bars

On analyzed AC motor have incipient broken bars under similar load and inertia conditions. For simulate the broken rotor bar, on increase its resistance by a coefficient such as the current bar is closest to zero. The simulations are shown in the Figure 6. The broken rotor bar related harmonic components are clearly located around the fundamental (Figure 7) [10], [11].

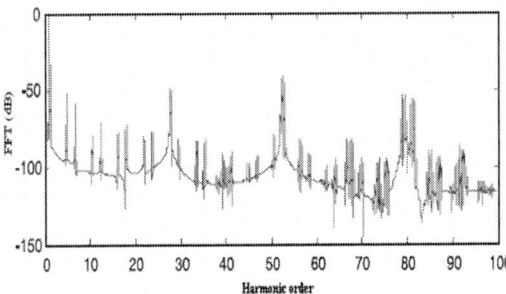

Fig. 6. The line current spectra of machine with 4 broken rotor bars.

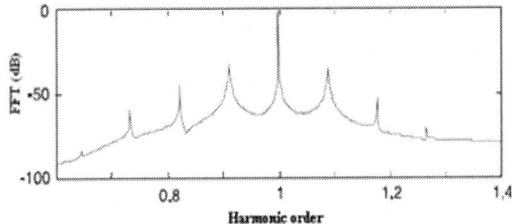

Fig. 7. The zoomed spectrum of the stator current around the fundamental.

These classical twice slip frequency sidebands are not the only effect due to rotor broken bars. There are other frequencies induced around all rotor slot harmonics (Figure 8).

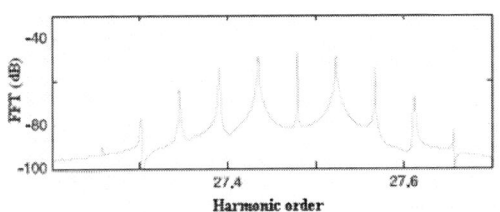

Fig. 8. The zoomed spectrum of the stator current around the first RSH.

These harmonic components come from the mathematical relations given in (3). They can give additional information about the rotor asymmetry and its gravity.

Broken rotor bars [3], [4] are one of the easiest AC motor faults to detect using steady-state stator current condition monitoring. This is based on monitoring the amplitudes of the double slip frequency sidebands of the fundamental supply frequency in the current spectrum [1]. It has shown that the greater the rotor bar fault severity, the higher is the amplitude of these sidebands. However, the sideband amplitudes are also sensitive to motor loading [3]. For example at no load or at light loads, these broken rotor bar sidebands are undetectable due to the small rotor currents under this condition.

The broken rotor bar fault at an early stage or partially broken rotor bars, which can lead to a larger failure or even be catastrophic, may not be detectable even under full load conditions. Therefore, there is a strong need to develop condition monitoring techniques to address these issues to allow earlier detection of rotor faults.

VI. CONCLUSION

The analyzed model of squirrel cage AC motor is based on multiple coupled circuits and takes into account the geometry and winding layout of the machine, without complexity in equations formulations or long computation.

The model equations are directly extracted by the decomposition into Fourier series of the mutual inductance matrix and the presentation of the induction motor in d-q frame. This model is helpful in quantifying the rotor slot harmonics under healthy as well as faulty condition.

This model is used in quantifying the RSH under healthy as well as faulty condition.

This model indicates that: 1) The reverse rotating field caused by the supply unbalance induces some of space harmonics in the stator current and 2) For machines with broken rotor bars; the stator current spectrum contains other significant harmonics than $(1 \pm 2\ ks)\ f_s$. These harmonics can be located around all rotor slot harmonics.

REFERENCES

[1] M. Boucerma, A. Khezzar, "Park Model of Squirrel Cage Induction Machine including Space Harmonics Effects", *Journal of Electrical Engineering*, vol. 57, No. 4, pp. 193-199, 2006.

[2] A. Bellini, G. Franceschini, "Quantitative evaluation of induction motor broken bars by means of electrical signature analysis", *Proc. IEEE. Trans. Indust. Electronics*, vol. 37, No. 2, pp. 1248-1245, 2003.

[3] V. Devanneux, V. Dagues "An Accurate Model of Squirrel Cage Induction Motor under Stator Faults", *Proceedings of Mathematics and Computers in Simulations*, vol. 63, pp. 377-391, 2003.

[4] T.J.E. Miller, "A Review of Induction Motors Signature Analysis as a Medium for Faults Detection", *Proc. IEEE. Trans. Indust. Electronics*, vol. 47, No. 4, pp. 463-464, 2002.

[5] S. Nandi, S. Ahmed "Detection of Rotor slot and Other Eccentricity Related Harmonics in a Three Phase Induction Motor with Different Rotor Cages", *IEEE Transactions on Energy*, vol. 16, No.3, pp. 253-260, 2001.

[6] A. A. Da Silva, "Rotating machinery monitoring and diagnosis using short-time Fourier transform and wavelet techniques", *Proc. Int. Conf. Maintenance and Reliability*, Knoxville, vol. 1, pp. 1401-1415, 2001.

[7] I. Daubechies, "The wavelet transform, time-frequency localization and signal analysis" *IEEE Trans. Indust. Electron.* Vol. 36, No. 3 pp. 961-1005, 2001.

[8] G. Jocsimovic "The detection of Inter-turn Short-circuits in the Stator Windings of Operating Motors", *IEEE Transactions on Industriel Electronics*, vol. 47, No. 5, pp. 1078-1084, 2000.

[9] W. Deleroi, "Broken bars in squirrel cage rotor of an induction motor-Part 1: Description by superimposed fault currents" *Proc. Int. Conf. (in German) Arch. Elektrotech.*, vol. 67, pp. 91-99, 2000.

[10] M. E. H. Benbouzid., *Induction motor faults detection using advanced spectral analysis technique*, Proc. Int. Conf. Electrical Machines, vol. 3, Istanbul, Turkey, pp. 1849-1854, 1998.

[11] Eleonora Darie, "A Wavelet Analysis model for detection of broken rotor bars in Induction Machines", Proc. of 4 th International Conference: Metrology & Measurement Systems - METSIM 2007, University of Politechnica Bucharest, 2007.

Identification of mathematical model induction motor's parameters with using evolutionary algorithm and multiple criteria of quality

dr inż. Hudy Wiktor*, dr hab. inż. Jaracz Kazimierz, prof. AP†

* Institute of Technology, Pedagogical University of Cracow, ul. Podchorążych 2, 30 – 084 Kraków, Poland, e-mail:
whudy@ap.krakow.pl
† Institute of Technology, Pedagogical University of Cracow, ul. Podchorążych 2, 30 – 084 Kraków, Poland, e-mail:
jaracz@ap.krakow.pl

Abstract— **In this paper identification's method of mathematical model induction motor's parameters was shown. This identification is based on evolutionary algorithm. The criterion, which is the functional of four quality indexes, was applied. This criterion was calculated by finding a common one from four quality indexes, which is the functional of absolute value of four quality indexes product. The identification was calculated by finding the minimum of last functional. Four factors of product were the difference between values of dynamic characteristics and characteristics at present calculated on the basis of individuals. Mathematical model parameters of induction motor were values of individuals' parameters. On the basis of calculated parameters' values characteristics generated were compared with experimental characteristics. The identification was made with using Tamel Sg90L-6 induction motor.**

Keywords— **induction motor, evolutionary algorithm**

I. Prefatory

At present stage of development induction motors find application in electric driving systems.

In research of dynamic of these systems the basic task is describing mathematical model of induction motor. The next step is identification of parameters of this model. The work of mathematical model of induction motor reflects transitory process, which occurs in motor with some approximation. This model shouldn't be too complicated, but should reflect these processes precisely enough. In this paper the commonly used simplifications are assumed [1, 2, 3, 4, 5, 6, 7, 8, 10, 12, 13].

To identify mathematical model' parameters of induction motor were used the evolutionary algorithm (AE), which has the basic attitudes [11]:

- manipulates many potential solutions at the same time – individuals,

- imitates natural evolution,

- is easy in implementation,

- it can not be assumed the received result is optimal, but it can use techniques, which allow a decrease in final mistake ex. by using progressive mutation and two kinds of selection.

TABLE I.
EVOLUTIONARY ALGORITHM'S PARAMETERS

Generation amount	100 000
Individual amount in population	200
Cross amount	80
Mutation amount	80
Progressive mutation amount	from 30 to 100. Every 1000 generations it was increased amount mutations about 1 to border 100.
Amount of points, in which there are calculated criterion (P)	150

For selection there was used a method, which takes advantage of two ways of natural selection: tournament method and deterministic method [5].

Measurements of Sg90L-6 induction motor's characteristics were made on Department of Electrical Machines with using DAMOT program. Measurements of starting characteristics were made for five values of power voltages (among it for the power U_n, which was used in identification process) and steady state's characteristics (the mechanical characteristic and the function of phase current in the slip function).

In this paper a method of identification of mathematical model induction motor's parameters with use of multi-criteria evolutionary algorithm. Four quality indexes were used (K_1, K_2, K_3, K_4) and they were calculated as following:

- K_1 – as the sum of squares of errors in discrete moments of time between measured values of registered during starting rotational speed and values, which were calculated on the basis of individual;

- K_2 – as the sum of squares of errors in discrete moments of time between measured root mean square values of phase current registered during starting and values, which were calculated on the basis of individual;

- K_3 – as the sum of squares of errors between measured values of electromagnetically moment and values which was calculated on the basis of individual in measurement points;

- K_4 – as the sum of squares of errors between measured values of phase current and values, which were calculated on the basis of the individual in selected measurement points.

In [3, 4, 5, 6, 9, 12, 13, 14, 15] paper the same assumptions were assumed, with regarding to mathematic model, evolutionary algorithm's parameters and partial index of quality similar as in this paper. Distinctively to this paper final index of quality was assumed, which was the sum of partial indexes of quality.

The final index is the product fitness function (1) with assuming real-valued representation of task. After taken simplified assumption the individual is seven-dimensional vector at coordinates $\{R_1, R_2, L_1, L_2, L_{12}, J, D\}$, where R_1 – stator resistance, R_2 – rotor resistance, L_{12} – magnetizing inductance, L_1 – stator leakage inductance, L_2 – rotor leakage inductance, J – moment of inertia, D – coefficient of friction. In this situation final index has a form:

$$F = \prod_{i=1}^{k} \frac{1}{N_r^2} \sum_{j=1}^{P} \left(w_{zi,j} - w_{oi,j} \right)^2 \tag{1}$$

where:
k – number of criteria ($k=4$),
N_r – maximum value received for i-th measure characteristics,
$w_{zi,j}$ – measure value of i-th criterion in j-th time's moment,
$w_{oi,j}$ – the value of i-th criterion calculated in j-th time's moment.

II. Identification of mathematical model induction motor's parameters

There were many processes of simulation made based on evolutionary algorithm. Evolution results chosen for made assumptions are in table 2. Values of fitness function (F) are unitless quantities.

TABLE II.
RESULTS OF EVOLUTION FOR SG90L-6 INDUCTION MOTOR

Lp.	R_1 [Ω]	R_2 [Ω]	L_1 [H]	L_2 [H]	L_{12} [H]	D [Nms/rad]	J [kg m²]	F
1.	5,03858	3,42817	0,20297	0,10028	0,08056	0,00609	0,01360	0,01200
2.	5,02335	3,43469	0,19985	0,10191	0,08080	0,00603	0,01384	0,00075
3.	5,02197	3,43951	0,20150	0,10083	0,08085	0,00597	0,01384	0,00150
4.	5,02710	3,43345	0,19812	0,10311	0,08083	0,00602	0,01404	0,00056
5.	5,02107	3,43550	0,20057	0,10149	0,08076	0,00599	0,01378	0,00101
6.	5,01527	3,43361	0,20122	0,10128	0,08078	0,00603	0,01375	0,00152
7.	5,01479	3,42843	0,20446	0,10045	0,08092	0,00608	0,01342	0,00022
8.	5,02177	3,43749	0,19826	0,10291	0,08077	0,00597	0,01370	0,00065
9.	5,01471	3,42506	0,20085	0,10124	0,08077	0,00596	0,01372	0,00099
10.	5,01988	3,43788	0,20117	0,10120	0,08083	0,00601	0,01351	0,00143

The best individual (it is the solution of identification problems) is '7' individual from table 2. This individual represents identified mathematical model's parameters of induction motor. On drawings 1 a, b, c, d there is measurable data and characteristics calculated base on optimal individual number '7' from table 2.

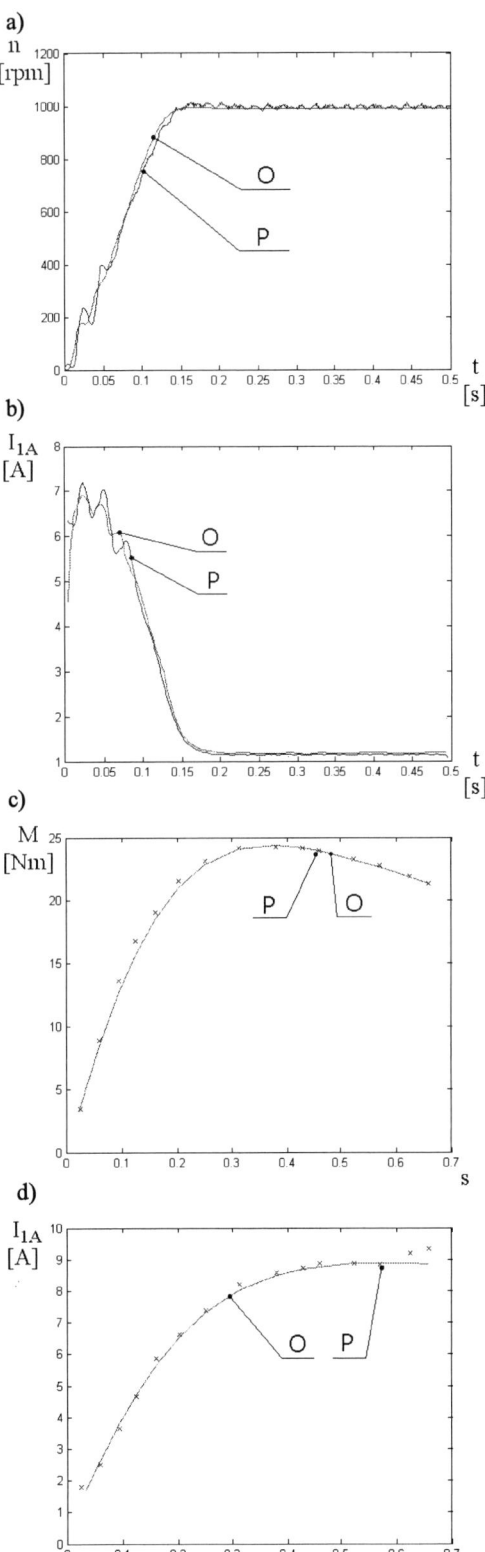

Figure 1. Rotational speed characteristic (a) and current first phase stator (b), mechanical characteristic (c) and first phase current in the slip function (d) for measuring data and individual '7' from table 2

III. VERIFICATION OF CALCULATED PARAMETERS

Measurable characteristics for various values of power voltage were using to verification of counted parameters of mathematical model induction motor. Figures. 2a,b, 3a,b, 4a,b show measurable characteristics (P) and counted characteristics (O) based on individual '7'.

a)

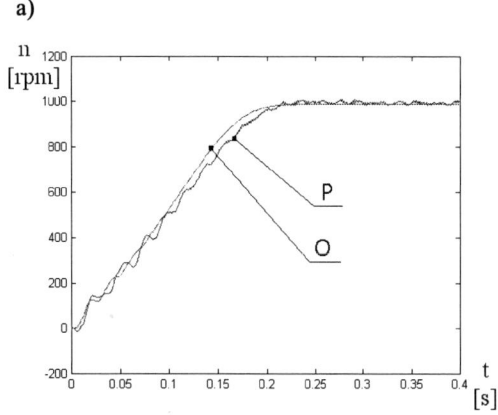

b)

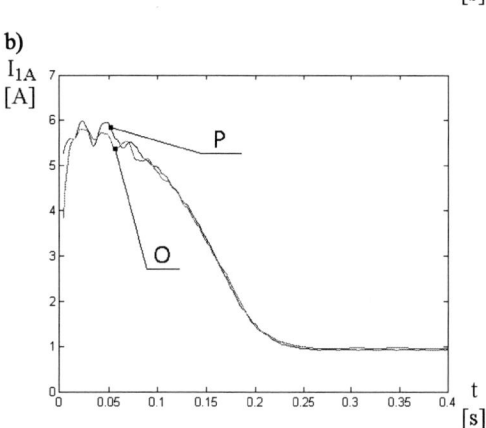

Figure 2. Rotational speed characteristic – (a) and first phase current – (b) start motor Sg90L-6 for $U_{z,sk}$ = 252 V

a)

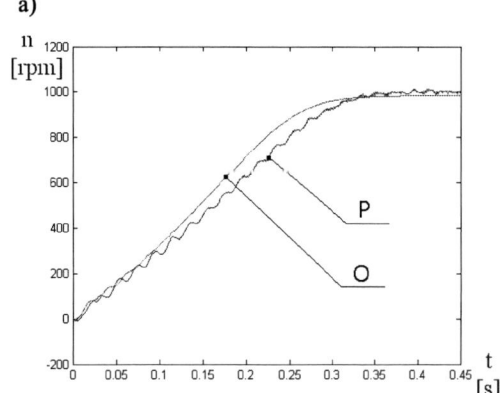

b)

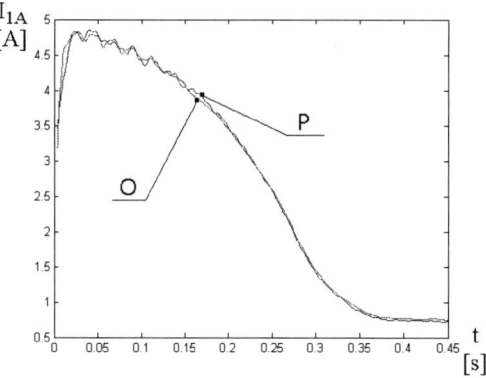

Figure. 3. Rotational speed characteristic – (a) and first phase current – (b) start motor Sg90L-6 for $U_{z,sk}$ = 203.7 V

a)

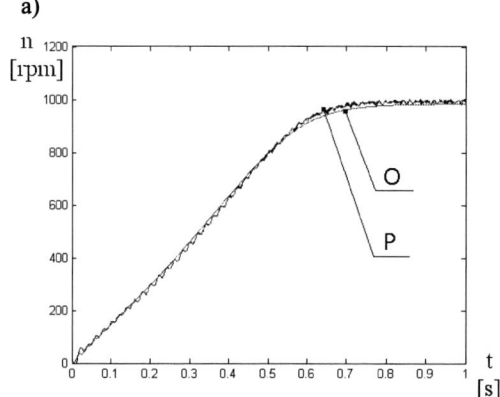

b)

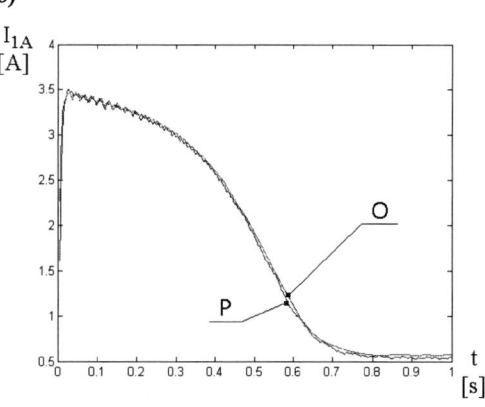

Figure. 4. Rotational speed characteristic – (a) and first phase current – (b) start motor Sg90L-6 for $U_{z,sk}$ = 150 V

IV. RESUME

As follows from table 2 received results are recurring. On the basis of drawings 2, 3, 4 it can be affirmed that assumed model and received results of mathematical model's parameters imitate processes occurred into induction motor in a good way. The result from '7' position from table 2 is the best one. On the basis of received results it is possible to affirm that using the evolutionary algorithm, which uses 4-th quality indexes and a product quality index, gave good results. These results differ insignificantly from results received in paper [3]. Using the final criterion, which is product of partly criterions, insignificantly shorten time of calculations. Received results are usable as a input degree to other optimization methods.

REFERENCES

[1] Chrzan P.J.: *Identyfikacja parametrów silnika indukcyjnego w układzie polowozorientowanym podczas postoju.* I Krajowa Konferencja Użytkowników MATLAB'a, AGH – Kraków, 14-15 listopada 1995

[2] Henri Arnold: *Ein Beitrag zur Identifikation der Parameter der Asynchronmaschine im geregelten Betrieb.* Dissertation, Technischen Universität Bergakademie Freiberg, Deutschland 2005

[3] Hudy W.: *Projektowanie układu sterowania oraz identyfikacja parametrów silnika indukcyjnego z zastosowaniem algorytmów ewolucyjnych i różnych kryteriów jakości,* Rozprawa Doktorska, WEAIiE AGH, Kraków 2007

[4] Hudy W., Jaracz K.: *Identyfikacja parametrów silnika indukcyjnego przy wykorzystaniu zasady optymalności w sensie Pareto i zastosowaniu algorytmów ewolucyjnych.* Materiały Konferencyjne SENE 2005, Łódź 23-25 listopada 2005

[5] Hudy W., Jaracz K.: *Wielokryterialna identyfikacja parametrów silnika indukcyjnego przy zastosowaniu algorytmu ewolucyjnego.* 33 Konferencja Automatyka, Telekomunikacja, Informatyka – ATI 2005, Szczyrk, 1-3 czerwca 2005

[6] Jaracz K., Hudy W.: *Criteria Identification of Inductive Motor's Parameters with using Evolutionary Algorithm, Pareto's Optimal Rule and Steady State's Characteristics.* International Carpathian Control Conference ICCC' 2006 Roznov pod Radhostem, Czech Republic, 29-31 May 2006

[7] Jadżyński W.: *Identification of a model of induction motor with function parameters.* ICEMS2003: proceedings of the sixth International Conference on Electrical Machines and Systems: November 9–11, Beijing, China 2003

[8] Jelonkiewicz J., Przybył A., Zawirski K.: *Projektowanie obserwatorów stanu w oparciu o algorytmy ewolucyjne na przykładzie silnika indukcyjnego,* VIII Konferencja Naukowo-Techniczna "Zastosowania Komputerów w Elektrotechnice - ZKwE 2003", Poznań 2003

[9] Kovacs K.P., Racz I.: Transiente Vorgange in Wechselstrommaschinen, Vol 1 & 2, Verlag Der Ungarischen Akademie Der Wissenschaften, Budapest 1959

[10] Macek-Kamińska K.: *Estymacja parametrów modeli matematycznych silników indukcyjnych dwuklatkowych i głębokożłobkowych.* Wyd. Wyższej Szkoły Inżynieryjnej w Opolu, Opole 1992

[11] Michalewicz Z.: *Genetic Algorithms + Data Structures = Evolution Programs.* 3rd, rev. and extended ed. 1996. Corr. 2nd printing, 1998

[12] Orłowska-Kowalska T., Lis J.: *Identyfikacja parametrów silnika indukcyjnego w stanie zatrzymanym za pomocą algorytmu ewolucyjnego.* Prace Naukowe Instytutu Maszyn, Napędów i Pomiarów Elektrycznych Politechniki Wrocławskiej Nr 56, Wrocław 2004

[13] Orłowska-Kowalska T., Szabat K., Ritter W.: *Zastosowanie algorytmów genetycznych do identyfikacji parametrów silnika indukcyjnego.* X Sympozjum PPEE 2003 – Wisła 7-10 grudnia 2003

[14] Orłowska-Kowalska T., Szabat K., Ritter W.: *Identyfikacja parametrów silnika indukcyjnego za pomocą algorytmów genetycznych.* Prace Naukowe Instytutu Maszyn, Napędów i

Pomiarów Elektrycznych Politechniki Wrocławskiej, Nr 54, Wrocław 2003

[15] Rutczyńska-Wdowiak K.: *Algorytmy Genetyczne w zastosowaniu do identyfikacji parametrycznej modelu matematycznego silnika indukcyjnego – aspekty obliczeniowe.* VI Krajowa Konferencja Naukowa, SENE 2003, Łódź 19-21 listopada 2003

Simulation Study on Control of Ultrahigh Speed Drives in Waste Energy Recovery Systems

Péter Stumpf (student), Miklós G. Simon (student), Rafael K. Járdán, István Nagy

* Budapest University of Technology and Economics, Faculty of Electrical Engineering and Informatics
Budafoki út 8., H-1111 Budapest, Hungary, e-mail: *nagy@elektro.get.bme.hu*

Abstract— **Special problem arising from the unique characteristics of the ultrahigh-speed induction generator applied in a system developed to generate electric power by utilizing waste and/or renewable energy sources. The studied system is built up from a high-speed turbine coupled to a three-phase induction generator, an AC/AC converter and a supervisory control unit. The purpose of the system is to utilize the energy content of a working medium by the help of a special high-speed turbine converting the energy content of the medium into electric energy with a generator and an AC/AC converter. Many different energy sources can be used with the application of the system, e.g. the energy resulting from the process of pressure reduction in steam, gas or fluid networks. The generated electric power can be fed to the utility mains or it can be used in stand-alone mode. The special properties of the system are studied only in respect of the unusual behavior of the ultrahigh-speed induction generator.**

Keywords— **Induction generators, Renewable Energy, Distributed generation.**

I. INTRODUCTION

The system described in this paper has been developed for the utilization of renewable or waste energy sources [2,3,4]. The possible fields of application of the system include most of the alternative, renewable energy sources that are capable of producing saturated or superheated steam with low or medium pressure. In most cases the gas can be produced in co-generation or less frequently in direct generation process. The alternative renewable energy sources are considered to be solar energy (direct solar steam system and binary cycle system), geothermal energy or energy obtained from biomass. The system can be applied with the use of various waste energy sources also. The example of energy gain during the process of pressure reduction in gas or fluid systems can also be utilized with this system. The generation of power is used instead of the original pressure reduction method which uses a special throttling valve, thereby missing the opportunity for an environmental friendly mode of power generation. The basic idea of the subject is elaborated and the research work is described in more detail in the literature [1.].

The system that is still under development applies ultrahigh-speed turbine-generator machine set. The solution offers the advantages of reduced weight and size as well as higher efficiency.

The technology used for the construction and development of these machines is most up-to-date, leading to great challenges in the construction and also in control and power electronics. The description and analysis of the turbine is out of the scope of the paper. We are concentrating to the study of the speed control of the high

sped machine set. Instead of using a permanent magnet synchronous generator in the turbine-generator set for electromechanical energy conversion as in previous solutions, this task is done by an induction machine. The characteristics of the induction machine need to be taken into account during the development of the control and power electronic system. The problems caused by the special properties of the induction machine will mainly be discussed in the paper in connection with the Turbine-Generator- Converter system, studying the dynamic behavior of the machine set.

II. DESCRIPTION OF THE SYSTEM

Considering that the system is to be used in a steam or gas network for pressure reduction to produce electric energy, the basic construction and operation of the system is described briefly using a simplified block diagram shown in Fig. 1. The system is connected in parallel with the utility network.

The torque of the turbine *T* is produced by the working medium. The turbine drives an induction generator *IG*. The electric power produced by the generator has a varying voltage level and frequency. The power is fed to the utility grid through a DC link AC/AC converter. It consists of two high frequency **PWM** converters *CONV1* and *CONV2*.

The functions of control, protection, fault diagnostics, monitoring, data acquisition and storage are handled by a microprocessor controller. The system uses two basic control loops.

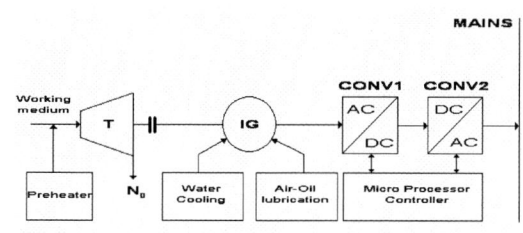

Fig. 1. Simplified block diagram of the system

A Pressure controller loop is used to keep the outlet pressure (N0) of the working medium at a constant level which is set by a reference signal. The applied control cannot only manipulate the angle of the blades in the turbine to adjust the pressure drop, but it can also change the input temperature of the medium by the use of a preheating system.

The generated mechanical power is proportional to the pressure drop in the turbine. Due to the fluctuation in the

978-1-4244-1741-4/08/$25.00 ©2008 IEEE

inlet pressure the power that can be fed to the utility mains is changed as well.

The power balance can be ensured by controlling the speed of the turbine-generator set by keeping the speed constant [5.,6.,7.,8.,9.,16.]. Both **CONV1** and **CONV2** are three phase PWM converters. The speed controller defines the fundamental frequency of **IG** via **CONV1** that furthermore ensures constant main flux for the machine The speed control loop varies the torque and power of **IG** and ensures the torque and power balance between **T** and **IG**. **CONV2** is synchronized to the utility mains and makes it possible to feed power to the utility mains.

Auxiliary equipment are used for the supporting of operation of the machine (Water cooling, Minimal air-oil lubrication as the machine has ceramic bearings). The additional equipment increases the complexity of the system.

III. SIMULATION OF ULTRAHIGH SPEED TURBINE-GENERATOR SYSTEM

The start-up, steady-state and transient operation of the ultrahigh speed turbine – generator set furnished with speed control loop have been studied by simulation using MatLab/Simulink program. Ultrahigh speed induction machine (I.M.) is applied as a generator. The plate data of the machine can be found in Appendix A. The I.M. is connected to a three-phase PWM converter generates symmetrical three-phase voltage with variable angular frequency Ω_1. The harmonics are neglected since we are interested in the performance of the speed control loop. Study of the additional losses in the I.M. is not our aim in the current simulation. An other assumption is that the converter is changing the amplitude of the stator voltages with the variable frequency Ω_1 in a manner to keep the main flux of I.M. constant, i.e. in steady-state the speed-torque $\Omega - \tau$ characteristics of the machine can be used. By changing the stator frequency Ω_1, the characteristics Ω versus τ is shifted along the axis in parallel to itself. Due to the high electromechanical time constant of the turbine-generator set the electromagnetic transients in I.M. are neglected. Under the assumptions described, the behaviour of I.M. is approximated in the simulation by the Kloss formula (see later).

Two kinds of control loop were studied. In the first one the rotor frequency Ω_2 was not limited, while in the second control loop a saturation block is applied to limit Ω_2.

IV. SPEED CONTROL CONFIGURATION WITHOUT LIMITATION

Fig.2. shows the block diagram of the Matlab/Simulink model without applying limitation on the maximum value of the rotor frequency. The input of the system is the reference speed Ω_r. To control the output mechanical speed Ω a PI controller is applied. The input of the controller is the error signal $\Omega_\varepsilon = \Omega_r - \Omega$. The output signal of the controller is the synchronous speed Ω_1. The "Converter + I.M." block calculates first the slip. The torque τ is determined from the slip by the Kloss formula is written in the form:

$$\tau = \frac{2\tau_b \cdot s_b \cdot s \cdot (1+\varepsilon)}{s^2 + s_b^2 + 2\varepsilon \cdot s \cdot s_b} \qquad (1)$$

where τ_b is the maximum or breaking torque in pu, s_b is the slip at the maximum torque and $\varepsilon = R_1 / \sqrt{R_1^2 + (L_r'\Omega_1)^2}$, where R_1 the stator resistance and L_r' the rotor transient inductance (Appendix A).

The difference of the electrical torque τ and the turbine torque τ_{turb} accelerates or decelerates the machine. The equation of motion:

$$T_m \frac{d\Omega}{dt} = \tau - \tau_{turb} \qquad (2)$$

where T_m is the electromechanical time constant.

Due to the special construction required by the ultrahigh speed some basic properties of the machine are different from those of standard induction machines. The slope $d\tau / ds$ is very steep around the synchronous speed Ω_1 and the rated and maximum slip are very small (Appendix A).

The system has two operation regions.

1., The I.M. operates in its motoring quadrat ($\Omega_1 > \Omega$). The supplying converter is increasing the angular stator frequency Ω_1 from zero up to its rated value with constant slope $d\Omega_r / dt = const$. The power is drawn from the main. After reaching the steady-state, the inlet valve of the turbine is open and the turbine starts driving I.M. that enters its generating quadrant ($\Omega > \Omega_1$).

2., The system works in the second region when ($\Omega > \Omega_1$) to supply electric energy through the converter. The power is delivered by the turbine.

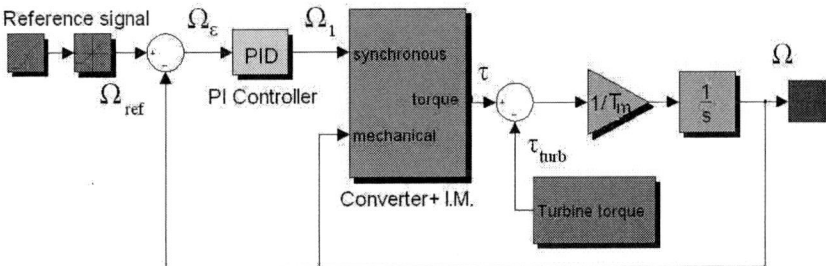

Fig.2. Block diagram of the speed control loop without limitation

A large sudden turbine torque increase might result in a serious operation failure. During acceleration the slip of the machine can exceed s_b in a short time, thus the drive gets into the unstable part of the speed-torque characteristics, where τ decreases with speed. When $|\tau_{turb}| - |\tau| > 0$, the speed Ω of the turbine-generator set can reach dangerous levels. Protection against overspeed has to turn-off the system.

There are three main targets for the simulation study.

    a) To find the configuration and the parameters of the speed control loop.

    b) To study the start-up and shut-down transient of the system.

    c) To investigate the dynamic process caused by sudden turbine torque change to avoid overspeeding.

A. Stability consideration

The nonlinear system shown in Fig.2. is approximated by a linear model [14,15,17,18]. It is assumed that the working point of I.M. remains always on the "linear part" of the Ω versus τ characteristic near the synchronous speed Ω_1.

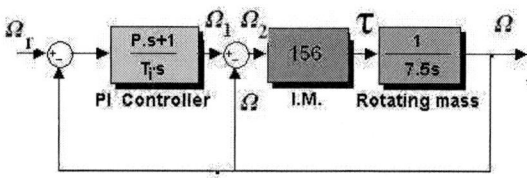

Fig. 3.   Approximate block diagram of the speed control loop

In other words the nonlinear equation is replaced by a straight line. Its slope is $\tau_{rate} / \Omega_{2rate} = 1 / 0.00642 \cong 156$ pu. The approximate block diagram is presented in Fig.3a. Here the electromechanical time constant $T_m \cong 7.5s$. The resultant transfer function of the inner loop is

$$\frac{\Omega}{\Omega_1} = \frac{1}{1 + s / 20.8}$$

The amplitude Bode diagram is depicted in Fig.3b. for two parameter sets of PI controller:

    a) $P = 1, T_i = 1s$, without limiter

    b) $P = 10, T_i = 0.1s$, with limiter

The cut-off frequency for the first set is $\omega_{ca} = 4.57$ rad/s and for the second set $\omega_{cb} = 2000$ rad/s. As we neglected the electromagnetic transients in the I.M. and all transients in the inverter ω_{cb} must be considerably reduced, in the real system it could hardly be realized. It can be concluded that the level of the cut-off frequency has to be selected between these two values likely near to ω_{ca} by choosing the appropriate P and T_i values.

We will present the simulation results first by using the PI controller with $P = 1, T_i = 1s$ for block diagram shown in Fig.2. and after the second set of parameters

$P = 10, T_i = 0.1s$ is applied for the block diagram having a limiter shown in Fig.15.

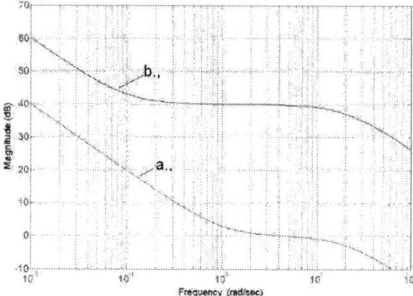

Fig.3.b Amplitude Bode diagrams, P=1,Ti=1s (curve a.,),
P=10,Ti=0.1s (curve b.,)

Fig.3c presents the amplitude Bode diagram for three other parameter sets of the PI controller. Only the start up transient processes are shown later with these sets using the speed control loop of Fig.2. and the conclusion will be drawn there.

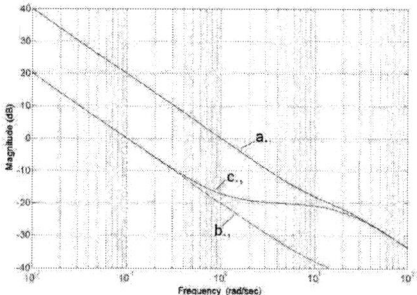

Fig.3.c Amplitude Bode diagrams, P=0.1,Ti=1s (curve a.,),
P=0.1,Ti=10s (curve b.,), P=1,Ti=10s (curve c.,)

V. SIMULATION RESULTS WITHOUT LIMITATION

A. Start-up and Shut-down

During the study of the dynamic behaviour of the system in order to satisfy the assumption of neglecting electromechanical transients, the acceleration of the I.M. is limited. For this reason the speed reference signal Ω_r is a ramp signal with different slopes instead of a stepwise change. The three different slopes of the reference signal applied are (0.05, 0.1 and 0.2) pu/s. During the start-up the turbine torque is supposed to be zero. The I.M. torque accelerates the machine set drawing power from mains via the DC link of the converters.

In the next figures samples of the simulation results are presented. The PI controller parameters are $P = 1$, $T_i = 1s$.

Fig.4. shows the time functions of $\Omega_r(t)$, $\Omega_1(t)$ and $\Omega(t)$. The last two ones, due to the low slip levels, are almost the same, difference is not visible in the diagram.

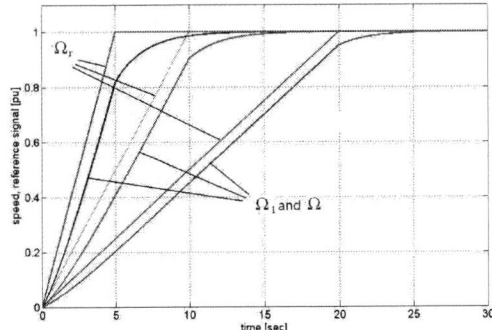

Fig. 4. Ω_r, Ω_1 and Ω versus time during start up

Both the synchronous and the mechanical speed follow the reference signal in a linear fashion. After the reference signals settle at their rated value, the actual mechanical speed and stator frequency (synchronous speed) follow themcorrectly.

The time function of the slip and the torque (Fig.5.) are almost the same, they differ only in magnitude. It means that during start-up the motor is working in the linear part of the $\tau - s$ characteristics.

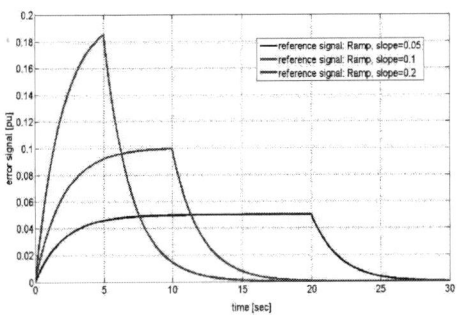

Fig. 5. Torque versus time during start up

Knowing the slope $d\Omega / dt \cong 0.05\,pu / s$ and $T_m \cong 7.5s$, the average value of the torque $\tau_{ave} = 0.375\,pu$. The same τ_{ave} can be read from Fig.5.

Similar results were obtained for shut-down as for start-up (Fig.6.) The absolute values of the slopes of Ω_r were the same as in Fig.4.

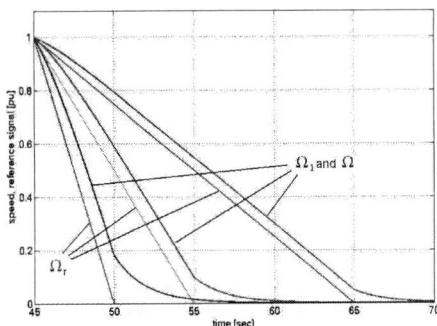

Fig. 6. Ω_r, Ω_1 and Ω versus time during shutdown

The simulation was performed with three other parameter values of the PI controller listed in the caption

in Fig.3c. (Fig.7.). The result obtained with $P = 0.1, T_i = 1s$ is similar what was found in Fig.4. The transient response is very slow with the other two parameter sets, since their cut-off frequency is one decade lower.

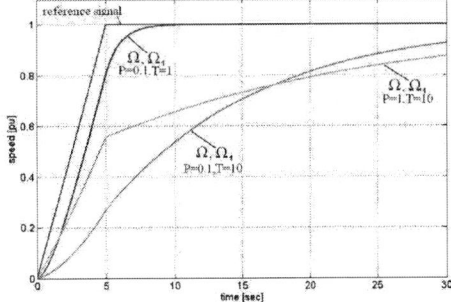

Fig. 7. Response using different PI parameters, $d\Omega/dt=0.2$pu/s

B. Transients in generating Region

Next the transient processes of the system are shown for sudden change in the turbine torque, lasting from $t = 35s$ to $t = 40s$. The PI controller parameters were again $P = 1, T_I = 1$ s.

Stepwise change of the turbine torque -1 pu was applied in *Fig.8*. The torque of the I.M. almost perfectly copies the time function of the turbine torque, but of course its sign is opposite. The small difference between the torques starts accelerating the machine. The higher speed Ω results in negative error signal, causing a reduction in the stator frequency (synchronous speed Ω_1) (Fig.9.). The reverse of this transient takes place when the turbine torque falls back to zero at $t = 40s$.

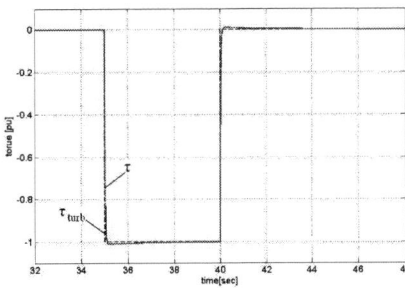

Fig. 8. Torques during transient process

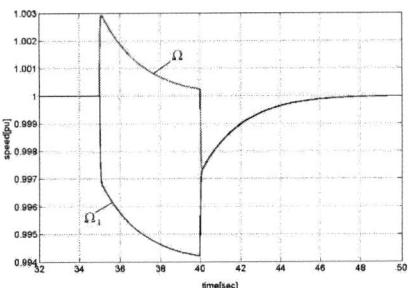

Fig. 9. Speed response during transient process

Fig.10. plots the enlarged torque-time function in the vicinity of $t = 35s$.

The time function of the machine torque is approximated by exponential function with time constant $T = 0.025\ s$. On the basis of (2)

$$\int_{t_1}^{t_2}(1-(1-e^{\frac{-t}{T}}))dt = T_m\int_{\Omega_1}^{\Omega_2}d\Omega = T_m\Delta\Omega .$$

Substituting $T_m = 7.5s$ and the time $t_1 = 35s, t_2 = 35.1s$, the calculated result is $\Delta\Omega = 0.0032\ pu$.

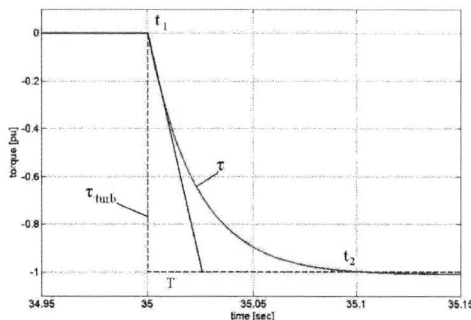

Fig. 10. Enlarged torque transient

The same result can be read from Fig.9. Because $P = 1$, the synchronous speed change is the same but with negative sign near at $t = 35s$.

Applying a high turbine torque $\tau_{turb} = -3.5\ pu$ but lower than τ_b, the controller keeps the overspeed within quite acceptable value (Fig.11,12.)

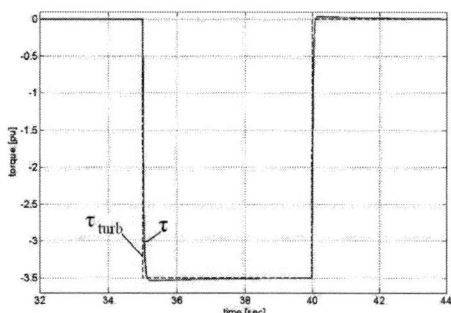

Fig. 11. Torque at $\tau_{turb} = -3.5\,\mathrm{pu}$

Increasing the turbine torque over the breaking torque of I.M. in generator mode (Fig.13.) results in dangerous overspeed. The machine "flips over" and starts accelerating to very high speeds.

Fig.14. shows the time functions of the synchronous Ω_1 and mechanical speed Ω together with Ω_2 and Ω_ε, at applied turbine torque $-5\ pu$. The operation point moves to the unstable region of the characteristic Ω versus τ where τ is decreasing with increasing Ω. The system loses its stability.

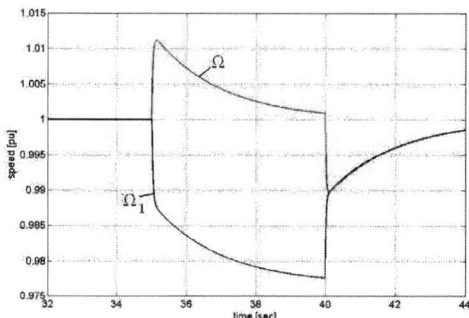

Fig. 12. Speed response at $\tau_{turb} = -3.5\,\mathrm{pu}$

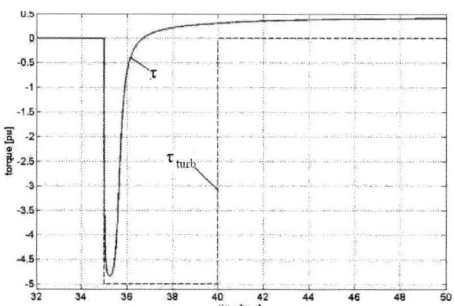

Fig. 13. $|\tau_{turb}|$ is higher than the breaking torque

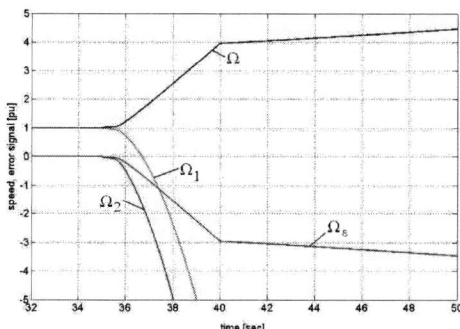

Fig. 14. Speeds at high acceleration turbine torque

VI. SPEED CONTROL CONFIGURATION WITH LIMITATION OF ROTOR FREQUENCY

Now the output signal of the PI controller is considered to be the rotor frequency. One way for trying to maintain system stability is to keep the operation point in the "linear part" of $\Omega - \tau$ characteristics. It can be achieved by limiting the value $|\Omega_2|$. In the next control configuration the rotor frequency Ω_2^* is kept between two limit values (Fig.15.). The upper and lower limit of Ω_2 are set as three times of the rated rotor frequency. $\Omega_{2upper}^* = -\Omega_{2lower}^* = 0.01914[pu] = 1722.6[rpm]$

Substituting the lower and upper limit in to (1) results in the limit values of the torque. The maximum torque of I.M. in motoring $\tau_{mot} = 2.58pu$ and in generating mode $\tau_{gen} = -2.956pu$, respectively.

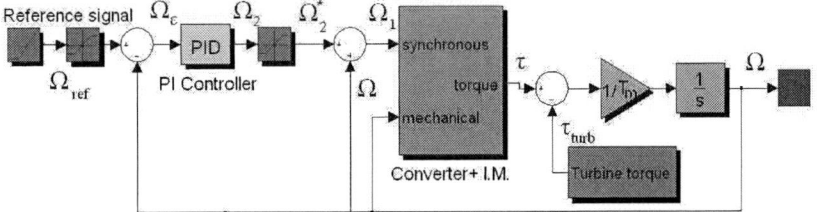

Fig.15. Block diagram of the speed control loop without limitation

VII. SIMULATION RESULTS WITH ROTOR FREQUENCY LIMITATION

A. Start-up and Shut-down

All simulation results presented from now on are obtained by using $P = 10$, $T_I = 0.1\ s$. Fig.16. shows the time functions of the speed Ω, Ω_1 and their reference signals. The traces of the mechanical speed Ω and synchronous speed Ω_1 almost perfectly coincide with the reference signal. The difference remains within the thickness of the line. In the real system this performance cannot be achieved, due to the small time constants neglected in our model, the loop gain and the cut-off frequency have to be reduced.

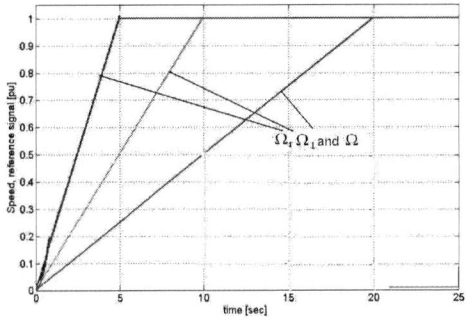

Fig. 16. Ω_r, Ω_1 and Ω versus time during start up

The shape of $s(t)$ and $\tau(t)$ (Fig.14.) are almost the same, they differ only in magnitude. It means again that the motor during the start-up is working in the linear part of the $s - \tau$ characteristic curve.

Comparing the torque time functions with the result of the previous model (Fig.4.) shows that the speed of response is much better.

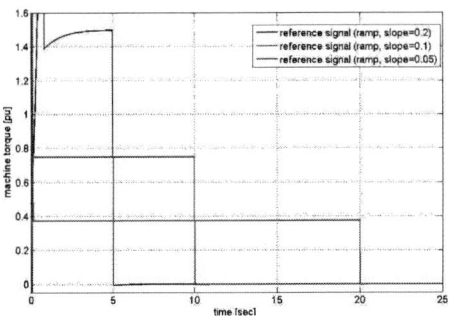

Fig. 17. Torque versus time during start up

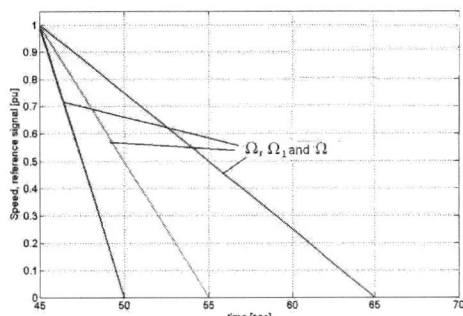

Fig. 18. Ω_r, Ω_1 and Ω versus time during shutdown

During shut-down the reverse process of the start up happens (Fig.18.).

B. Transients in Generating Region

Next the transient processes of the system is shown for sudden stepwise turbine torque change in generating mode of the induction machine starting from $t = 35s$ to $t = 40s$. Fig.19 presents the machine and turbine torque time function. The applied turbine torque value is $-1pu$. The torque of the induction machine almost perfectly copies the time function of the turbine torque.

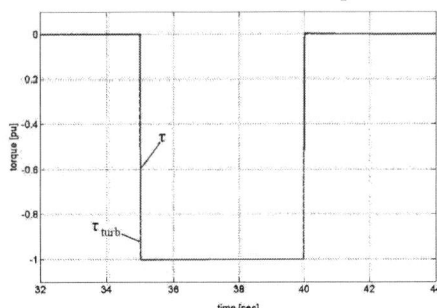

Fig. 19. Torques during transient process

Thanks to the high performance of the speed control loop, the overshoot in the speed response is much lower than that of in Fig.9.

The same calculation carried out earlier in connection with Fig.9. and 10. verified the results shown in Fig.20.

As mentioned previously, the rotor frequency Ω_2 is controlled in such a way that its highest possible value limited below the set level, in our case three times of the rated Ω_2.

921

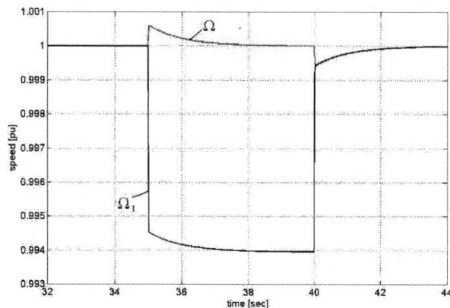

Fig. 20. Speed during transient process

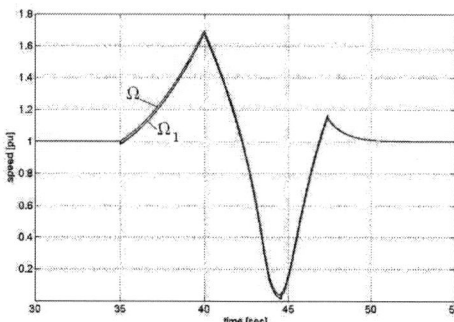

Fig. 21. τ_{turb}=-3.5 pu in 35s$\leq$t$\leq$40s

Applying $\tau_{turb} = -3.5\ pu$ from $t = 35s$ to $t = 40s$ the $\Omega(t)$ and $\Omega_1(t)$ are shown in Fig.21. τ_{turb} is considerable higher than $|\tau_{turb}| = 2.956$. It produces a very high overspeed $\Omega_{max} = 1.7\,pu$ although now the system does not loose its stability, it settles down after oscillation to its original state. Of course 70% overspeed is unacceptable, the overspeed protection would intervene.

VIII. CONCLUSION

It can be concluded that the configuration without limiter (Fig.2.) provides better performance compared to the one with limiter as far as the dynamic response caused by large sudden turbine torque pulses τ_{turb} is concerned. The controller can keep speed Ω within acceptable border at higher τ_{turb} pulse than the rated one (Fig.11.,12. and 21.). The performance of the two configurations would be closer to each other if the rotor frequency limiter would be set near to the maximum toque.

The dynamic performance for small signal disturbances is the same for the two configurations since the optimal parameter set for the PI controller can be the same.

IX. ACKNOWLEDGEMENT

The authors wish to thank the Hungarian Research Fund (OTKA K72338, OTKA T049640), the Control Research Group of the Hungarian Academy of Sciences (HAS) for their financial support.

REFERENCES

[1] Jardan, R.K.,Nagy,I.:"High Speed Turbine Induction Generator System for Utilization of Renewable and Waste Energies," 38th Annual IEEE Power Electronics Specialists Conference, PESC07,

June 17-21, 2007, Orlando, Florida, US. pp. 2639-2645. CD Rom ISBN 1-4244-0655-2.

[2] Martin Olejar, Vladimir Ruŝĉin, Milan Leko, Jaroslav Dudrik,"Bidirectional DC/DC Converter for hybrid battery", EDPE'07, High Tatras, Slovakia, 24-26 september, 2007, CDROm ISBN: 978-80-8073-868-6

[3] V. Oleschuk, F Blaabjerg, "Novel Simplifying Approach for Analysis and Synthesis of Space Vector PWM Algorithms", EPE 2003 Toulouse, France 1-4 September, CD Rom

[4] Wlodzimierz Koczara, Nazar –AlKhayat, "Variable Speed Integrated Generator VSIG as a Modern Controlled and Decoupled Generation System of Electrical Power", EPE 2005, Dresden, Germany, 11-14 September 2005, CD Rom ISBN: 90-7581-08-05

[5] M. Orabi, T. Ninomiya, "Analysis of PEFC Converter Stability Using Energy Balance Theory", IECON'03, 2-6 November, 2003 Roanoke, Virginia, USE, Pp.544-549

[6] Dudrík, J: soft Switching PWM DC-DC Converters for High Power Applications, Proc. Of the Int. Conf. IC-SPETO 2003, Gliwice-Niedzica, Poland, 2003, Vol.1, pp 11-11a-11f-12

[7] M. Cirrincione, M. Pucci, G. Vitele, „Direct Control of Three-Phase VSIs for fte Minimalization of Common-Mode Emissions in Distributed Generation Systems" ISIE'07, Vigo, Spain, 4-7, June, 2007, pp. 2532 – 2539, CD Rom ISBN: 1-4244-0755-9

[8] V. Ruscin, M. Lacko, M. Olejar, J.Dudrik, "Soft Switching PS-PWM DC-DC ConverterControlled by Microprocessors", EDPE'07, High Tatras, Slovakia, 24 -26 Sept, 2007, CD rom ISBN: 978-80-8073-868-6

[9] J. Leuchter, P. Bauer, P. Bojda, V. Rerucha, „Bi-Directional DC-DC Converters for Supercapacitor Based Energy Buffer for Electrical Gen-Sets" EPE'07, Aalborg, Denmark, 2-5 szept, 2007, CD Rom ISBN: 9789075815108

[10] S. Ryvkin, D. Izosimov, S. Belkin, "Commutation Laws Transfer Strategy for the Feedforward Switching Losses Optimal PWM for Three-phases voltage Inverter" IEEE Industrial Electronics Society, IECON'98, 31 Aug-4 Spet 1998, Vol.2, pp. 768-773 *IEEE Xplore

[11] V. Oleschuk, F. Profumo, A. Tenconi. „Simplifying Approach for Analysis of Space-Vector PWM for Three-Phase and Multiphase Converters" EPE'07, Aalborg, Denmark, 2-5 szept, 2007, CD Rom ISBN: 9789075815108

[12] H.Abdulrahman, P.Bauer, " Flicker Mitigation with the Smarttrafo",Dresden, Germany, 11-14 September, 2005, CD Rom ISBN: 9-7581-08-05

[13] B. Grzesik, M. Stepien, "Coaxial HF Power Transformer with Turbular Linear Windings-FEM Results vs. Laboratory Test" EPE-PEMC'06, Portoroz, Slovenia, August 30- September 01, 2006, pp. 1313-1317, CD Rom ISBN: 1-4244-0121-6

[14] Abdelali El Aroudi, Bruno Robert, "Stability Analysis of a Voltage Mode Controlled Two-Cells DC-DC Buck Converter", IEEE 36th Annual Power Electronis Specialists Conference, Recife, Brasil, 12-16 June, 2005, CD Rom ISBN: 0-7803-9034-2, pp.1057-1061, IEEE Xplore

[15] A. Aroudi, B. Robert, L. Martinez-Salamero, „Bifurcation Behavior of a Three Cell DC-DC Buck Converter", EPE-PEMC'06, Portoroz, Slovenia, August 30- September 01, 2006, pp. 1994-2001, CD Rom ISBN: 1-4244-0121-6

[16] M.R. Baiju, K.K. Mohapatra, R.S. Kanchan, P.N. Tekwani, K. Gopakumar "A Space Phasor based Current Hysteresis Controller Using Adjacent Inverter Voltage Vectors with Smooth Transition to Six Step Operation for a Three Phase Voltage Source Inverter" EPE Journal, Brussels, Belgium, February 2005, Vol.15.,No.1,pp.36-47

[17] V. Moreno, L. Benadero," Investigating Stabilty of a Single Inductor Current Mode Controlled Dual Switching DC-DC Converter" EPE 2005, Dresden, Germany, 11-14, September, 2005, CD Rom, ISBN: 90-7581-08-05

[18] M.Z. Youssef, H. Pinherio, P.K. Jain, „Analysis & modeling of a self-sustained oscillation series-parallel resonant converter with capacitive output filter using sampled-data analytical technique", Telecommunications Energy Conference, INETELEC'03, 19-23 Oct, 2003, pp.282 - 289, IEEE Xplore

APPENDIX A

Induction machine Rated Data

Type: PSSG102P, Squirrel cage

$V_n = 380[V]; I_n = 9[A]; f_{1n} = 1500[Hz]; P_n = 4500[W]; p = 1;$

$\cos \varphi = 0.8$

Symbol	Name	Absolute value	pu value
τ_n	Rated torque	0.4805 [Nm]	0.7644
τ_b	Maximum torque	2.4545 [Nm]	3.9052
τ_i	Starting torque	0.284 [Nm]	0.452
s_n	Rated slip	0.642 [%]	0.00642
s_b	Slip at max. torque	5.31 [%]	0.0531

Adaptive Back EMF Parameter Adjustment of Simplified Vector Control for Position Sensorless Permanent Magnet Synchronous Motors

Kiyoshi Sakamoto[*], Yoshitaka Iwaji[*], Daigo Kaneko[*], Toshihiro Takeuchi[**],

Tsunehiro Endo[***], and Atsuo Kawamura[†]

[*] Hitachi Research Laboratory / Hitachi, Ltd., Hitachi, Japan
[**] Hitachi Information and Control Solutions, Ltd., Hitachi, Japan
[***] Power and Industrial Systems Division / Hitachi, Ltd., Hitachi, Japan
[†] Yokohama National University, Yokohama, Japan

Abstract— A previously reported simplified vector control for position-sensorless permanent magnet synchronous motors can be affected by errors in setting the motor parameters, because no feedback compensator is used. We examine the effects of errors on the steady-state performance of these motors in this paper. It was found that the current amplitude increases when a parameter error is given and the effect of a back EMF parameter error is especially large. We propose a new back EMF parameter adjustment method based on these results. We determined from our simulation and experimental results that the proposed method can compensate for the current increase caused by the parameter error. A theoretical analysis showed that the proposed method is affected by only the *d*-axis inductance parameter L_d. Therefore, even if there are errors in the other motor parameters, the motor current of the constant load condition is minimized by using the proposed method.

Keywords— AC machine, Permanent magnet motor, Variable speed drive, Vector control, Sensorless control.

I. INTRODUCTION

A position-sensorless trapezoidal current drive, i.e. a six-step commutation brushless-dc-motor drive, used as a drive method for Permanent Magnet Synchronous Motors (PMSMs) for home appliance motor drive systems, has been widely used because of its simplicity and low-cost implementation. However, a distorted current waveform generates pulsating torque in the motor and motor core loss. Therefore, sinusoidal current drives are increasingly used.

A major example of a sinusoidal current drive is the position-sensorless vector control. Many research papers have focused on this control [1]-[4]. However, the position estimation methods that were proposed, such as the Kalman filter, the state observer, and the disturbance observer, are relatively complicated and their calculation requirements are large. Furthermore, the vector control, which includes a speed-control loop and a current-control loop, requires a short interval of control. Therefore, a high-performance micro-controller unit (MCU) or a

digital signal processor (DSP) is needed for the implementation of the control. The use of such an expensive controller/processor is unrealistic for the motor drives of electrical household appliances.

In a previous paper [11]-[13], the authors proposed a simplified vector-control method that was suitable for implementation with a low-cost typical MCU. However, its performance can be affected by errors in setting the motor parameters because no feedback compensator was used.

As the steady-state performance, we focus on the motor current amplitude under a constant speed and constant load torque condition. The reason we chose the criteria is that the capabilities of the minimized current drive is the most important feature of the motor drives of electrical appliances, such as air conditioners.

For this paper, the effects of motor parameter errors on the steady-state performance were examined under various error conditions using a computer simulation. We are now proposing a new adaptive back EMF parameter adjustment method based on these simulation results. A theoretical analysis showed that the proposed method has the capability to compensate for a current amplitude increase that was caused by a motor parameter error. Finally, the proposed control method was verified by using experimentation.

II. SIMPLIFIED VECTOR CONTROL OF POSITION SENSORLESS PMSM

A. Rotational reference frame and motor model

The proposed position sensorless control method was developed on an assumed rotor reference frame, named the "*dc-qc* axis" in this paper. The *dc-qc* axis is controlled so that it synchronizes with the real rotor reference frame. Fig. 1 shows the two rotating axes of the PMSM, where the assumed *dc-qc* axis is shifted from the *d-q* axis. The difference between the two axes is defined as the position error $\Delta\theta$ given in (1):

$$\Delta\theta = \theta_{dc} - \theta_d,\qquad(1)$$

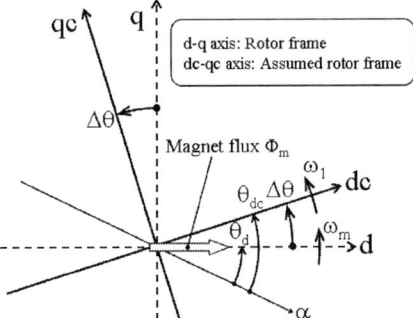

Fig. 1 Definition of Two Rotational Reference Frames

where θ_d is the real angular position of the d-axis and θ_{dc} is the assumed angular position of the dc-axis relative to the stationary α-axis. The time derivatives θ_d and θ_{dc} are expressed as follows:

$$\frac{d}{dt}\theta_d = \omega_m = \frac{P}{2}\omega_r \text{, and} \quad (2)$$

$$\frac{d}{dt}\theta_1 = \omega_1 , \quad (3)$$

where ω_m is the rotational angular velocity of the d-q axis, ω_1 is the rotational angular velocity of the dc-qc axis, ω_r is the rotational angular velocity of the mechanical shaft of the motor, and P is the pole number. The relation of dc-qc axis value and the d-q axis value is given by

$$\begin{bmatrix} i_{dc} \\ i_{qc} \end{bmatrix} = \begin{bmatrix} \cos\Delta\theta & \sin\Delta\theta \\ -\sin\Delta\theta & \cos\Delta\theta \end{bmatrix}\begin{bmatrix} i_d \\ i_q \end{bmatrix} . \quad (4)$$

The voltage equation of a salient pole PMSM in a synchronous reference frame is known as

$$\begin{bmatrix} v_d \\ v_q \end{bmatrix} = R\begin{bmatrix} i_d \\ i_q \end{bmatrix} + p\begin{bmatrix} L_d i_d \\ L_q i_q \end{bmatrix} + \omega_m\begin{bmatrix} -L_q i_q \\ L_d i_d \end{bmatrix} + \begin{bmatrix} 0 \\ K_E \omega_m \end{bmatrix} , \quad (5)$$

where R is the stator winding resistance, L_d, L_q are the d and q axes inductances, v_d, v_q are the d and q axes voltages; i_d, i_q are the d and q axes currents; p is the differential operator with respect to time; and K_E is the parameter of the back electromotive force (EMF).

In the assumed rotor frame (dc-qc axis), voltage equation (5) is easily transformed using the extended EMF idea [5] [6], which results in

$$\begin{aligned} \begin{bmatrix} v_{dc} \\ v_{qc} \end{bmatrix} &= R\begin{bmatrix} i_{dc} \\ i_{qc} \end{bmatrix} + pL_d\begin{bmatrix} i_{dc} \\ i_{qc} \end{bmatrix} + \omega_m L_q\begin{bmatrix} -i_{qc} \\ i_{dc} \end{bmatrix} \\ &+ \frac{d\Delta\theta}{dt}L_d\cdot\begin{bmatrix} -i_{qc} \\ i_{dc} \end{bmatrix} + E_{0x}\begin{bmatrix} \sin\Delta\theta \\ \cos\Delta\theta \end{bmatrix} . \end{aligned} \quad (6)$$

The E_{0x} in (6) is the amplitude of the extended EMF given by

$$E_{0x} = K_E\omega_m + p\left(L_q - L_d\right)\cdot i_q + \omega_m\left(L_d - L_q\right)\cdot i_d .$$

B. Proposed Control method

Fig. 2 shows a block diagram of the simplified vector control for position sensorless PMSMs [11]. The proposed method is characterized by eliminating the automatic

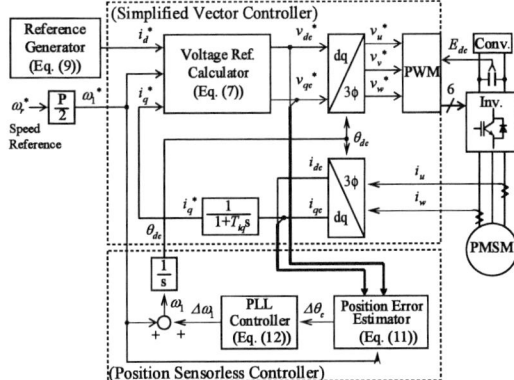

Fig. 2 Control diagram of simplified vector controller for position sensorless PMSM

speed regulator and automatic current regulator that are used in the conventional vector control.

The proposed controller can be divided to two parts. In the following paragraphs, we will briefly explain each part.

In the vector control part, the voltage command v_{dc}^* and v_{qc}^* are calculated by using (7). This equation is obtained by neglecting the time derivative term of (5):

$$\begin{bmatrix} v_{dc}^* \\ v_{qc}^* \end{bmatrix} = R^*\begin{bmatrix} i_d^* \\ i_q^* \end{bmatrix} + \omega_1^*\cdot\begin{bmatrix} -L_q^*\cdot i_q^* \\ L_d^*\cdot i_d^* \end{bmatrix} + \begin{bmatrix} 0 \\ K_E^*\cdot\omega_1^* \end{bmatrix} , \quad (7)$$

where ω_1^* is the angular frequency command, i_d^*, i_q^* are the current command values, and R^*, L_d^*, L_q^*, and K_E^* are the motor parameters of controller. In this paper, a motor parameter with an asterisk represents a motor parameter of the controller.

The qc-axis current command i_q^* in (7) is obtained from the detected qc-axis motor current through the LPF as follows:

$$i_q^* = \frac{1}{1+T_{iq}s}i_{qc} . \quad (8)$$

However, the dc-axis current command i_d^* is obtained by

$$i_d^* = \frac{K_E^*}{2(L_q^* - L_d^*)} - \sqrt{\left[\frac{K_E^*}{2(L_q^* - L_d^*)}\right]^2 + \left(i_q^{**}\right)^2} . \quad (9)$$

Equation (9) is known as the relationship between i_d and i_q under the maximum torque per current amplitude condition [7] [8]. However, in this paper, the motor parameters of the controller are used for the calculation.

We set the speed of the qc-axis current value i_q^{**} slower than that for i_q^* for stabilization reasons.

$$i_q^{**} = \frac{1}{1+T_{iqL}s}i_{qc} . \quad (10)$$

In the position sensorless control part, an estimated position error $\Delta\theta_c$ is estimated by using the following equation [10].

$$\Delta\theta_c = \tan^{-1}\left[\frac{v_{dc}^* - R^*\cdot i_{dc} + \omega_1 L_q^*\cdot i_{qc}}{v_{qc}^* - R^*\cdot i_{qc} - \omega_1 L_q^*\cdot i_{dc}}\right] . \quad (11)$$

925

Note that (11) uses only L_q^* parameter because (11) is derived from (6).

Using $\Delta\theta_e$, the angular velocity ω_1 and phase θ_{dc} are obtained by calculating the phase locked loop (PLL) in the following way:

$$\Delta\omega_1 = -K_{ps}\Delta\theta_e , \tag{12}$$

$$\omega_1 = \omega_1^* + \Delta\omega_1 , \text{ and} \tag{13}$$

$$\theta_{dc} = \int \omega_1 dt , \tag{14}$$

where K_{ps} is the proportional gain of the PLL loop and ω_1^* is the angular velocity command. Note that the proposed method has to use a ramp variation function for ω_1^*. The acceleration of ω_1^* can be designed with an inertia moment of the rotor shaft.

The proposed position sensorless algorithm is based on the back EMF voltage information. Thus, low-speed or standstill operations of the motor are difficult because the EMF amplitude is very small. To drive the motor from a standstill to approximately 10% of the rated speed, we used an open-loop start-up method, i.e. the synchronous drive method.

III. PARAMETER ERROR EFFECT AND A NEW COMPENSATION METHOD

The proposed method described in the previous section uses motor parameters for calculation in (7), (9), and (11), but does not use any feedback compensator. Therefore, the drive performance of the proposed method can be affected by errors in setting the motor parameters. In this section, we examine the effects of errors on the steady-state performance. In particular, we investigate the motor current amplitude under a constant speed and constant load torque condition. The reason we chose the criteria is that the capabilities of the minimized current drive is the most important feature of motor drives of electrical appliances, such as air conditioners.

A. Analysis of parameter error effect

Under a steady-state condition (a constant speed and constant load torque condition), the average values for the motor voltage equation (6) and voltage command equation (7) are almost the same. Assuming that $\omega_1^* = \omega_1 = \omega_m$, we obtain

$$\begin{bmatrix} v_{dc} \\ v_{qc} \end{bmatrix} = R \begin{bmatrix} i_{dc} \\ i_{qc} \end{bmatrix} + \omega_m L_q \begin{bmatrix} -i_{qc} \\ i_{dc} \end{bmatrix} + E_{0x} \cdot \begin{bmatrix} \sin\Delta\theta \\ \cos\Delta\theta \end{bmatrix}, \text{ and} \tag{15}$$

$$where \quad E_{0x} = K_E \omega_m + \omega_m \left(L_d - L_q \right) i_d ,$$

$$\begin{bmatrix} v_{dc}^* \\ v_{qc}^* \end{bmatrix} = R^* \begin{bmatrix} i_d^* \\ i_{qc} \end{bmatrix} + \omega_m \begin{bmatrix} -L_q^* \cdot i_{qc} \\ L_d^* \cdot i_d^* \end{bmatrix} + \begin{bmatrix} 0 \\ K_E^* \cdot \omega_m \end{bmatrix}. \tag{16}$$

Note that we substitute i_{qc} for i_q^* in (15) and (16). (The proposed method calculated i_q^* by using (8), thus $i_q^* = i_{qc}$ in the steady-state.)

From the v_{qc} in (15) and the v_{qc}^* in (16), we obtain

$$i_{dc} = \frac{L_d^*}{L_d} \cdot i_d^* + \frac{1}{L_d}\left(K_E^* - K_E \cos\Delta\theta\right)$$
$$+ \frac{\left(L_d - L_q\right)}{L_d} \cdot i_q \sin\Delta\theta + \frac{\left(R^* - R\right)i_{qc}}{\omega_m L_d} . \tag{17}$$

From the v_{dc} in (15) and the v_{dc}^* in (16), we obtain

$$\sin\Delta\theta = \frac{-\left(L_q^* - L_q\right)i_{qc} + \dfrac{R^* i_d^* - R i_{dc}}{\omega_m}}{K_E + \left(L_d - L_q\right) \cdot i_d} . \tag{18}$$

Equations (17) and (18) help to prove that the motor current i_{dc} varies from i_d^* depending on the motor parameter error.

In the proposed method shown in Fig. 2, the i_{dc}^* is calculated by using (9) in order to minimize the motor current. However, if a parameter error exists, a minimized current drive is not achieved.

By neglecting the stator resistance term, we can obtain the following approximated equations:

$$i_{dc} \cong \frac{L_d^*}{L_d} \cdot i_d^* + \frac{1}{L_d}\left(K_E^* - K_E \cos\Delta\theta\right) + \frac{\left(L_d - L_q\right)}{L_d} \cdot i_q \sin\Delta\theta ; \tag{19}$$

$$\sin\Delta\theta \cong \frac{-\left(L_q^* - L_q\right)i_{qc}}{K_E + \left(L_d - L_q\right) \cdot i_d} . \tag{20}$$

We were able to find from using (20) that a position error $\Delta\theta$ is generated by an error in the L_q^*. As a result, the motor current i_{dc} varies from the i_d^* depending on the third term of right side of (19).

Furthermore, assuming $L_q^* = L_q$, we obtained the simple approximate formula

$$i_{dc} \cong \frac{L_d^*}{L_d} \cdot i_d^* + \frac{1}{L_d}\left(K_E^* - K_E\right). \tag{21}$$

We were able to find from using (21) that the motor current i_{dc} varies from the i_d^* depending on errors in the L_d^* and K_E^*.

B. Numerical examples of parameter error effect

We have used a computer simulation to examine the effect under various error conditions. The specifications of the test motor are listed in Table 1. In this simulation, the test motor was driven with 100% rated-load torque. In each simulation, the steady-state motor current amplitude was recorded. The motor speeds that were selected were either 100 % speed (rated speed) or 10% speed.

The motor parameters were individually varied in this simulation. (In other words, when parameter K_E^* was set

Table 1. Test motor specifications.

Output power	3.7 kW
Maximum speed	6900 r/min
Rated torque	5.1 Nm
L_d	1.4 mH
L_q	3.8 mH
R	0.1 Ω
K_E	0.071 V/(rad/s)

to various values, other parameters, R^*, L_d^*, L_q^*, were set to real motor values.)

Fig. 3 shows the relationship between the motor parameter error and the motor current amplitude. In the case of a no error condition, i.e. $R^*/R = 100\%$, $L_d^*/L_d = 100\%$, $L_q^*/L_q = 100\%$, and $K_E^*/K_E = 100\%$, the motor current amplitude was 94%. The test motor rotor was salient, and a 94% motor current was the minimum value under the 100% rated-load torque condition.

It was found that the current amplitude increases when a parameter error was given. However, the error effect differed depending on the parameter type. The error effect of L_d^* was very small in this simulation. The error effect

of R^* differed depending on the speed of rotation. The motor current amplitude increased because of an R^* error in the case of 10% speed. The error effects of K_E^* and L_q^* were large compared with those of R^* and L_d^*.

C. A new compensation method

As a result of the simulation results mentioned above, we found that the error effect of the back EMF parameter setting error is especially large. In this case, a K_E^* setting error of about 10% makes the motor current 10% bigger. Generally, the magnetic flux density of PMSM decreases when the temperature of the permanent magnet is increased. As a result, the back EMF parameter changes while driving. Therefore, the proposed simplified vector control needs to dynamically compensate for the K_E variation.

In this paper, we propose a new back EMF parameter adjustment method. Fig. 4 shows a block diagram of the entire system of the proposed control. The proposed compensation method is shown within the heavy dashed line in the diagram. Compensator G is used to dynamically adjust K_E^* according to the dc-axis current error, i.e. $i_d^* - i_{dc}$. Note that K_{E0}^* is a constant value of the initial setting. The proposed adjustment method is activated while driving the motor.

Assuming $L_d^* = L_d$ in (21), an approximate relation between K_E^* and i_{dc} can be obtained as follows:

$$i_{dc} \simeq i_d^* + \frac{1}{L_d}\left(K_E^* - K_E\right). \tag{22}$$

Based on approximation (22), the dc-axis current model is expressed as shown in Fig. 5. An integral compensator is

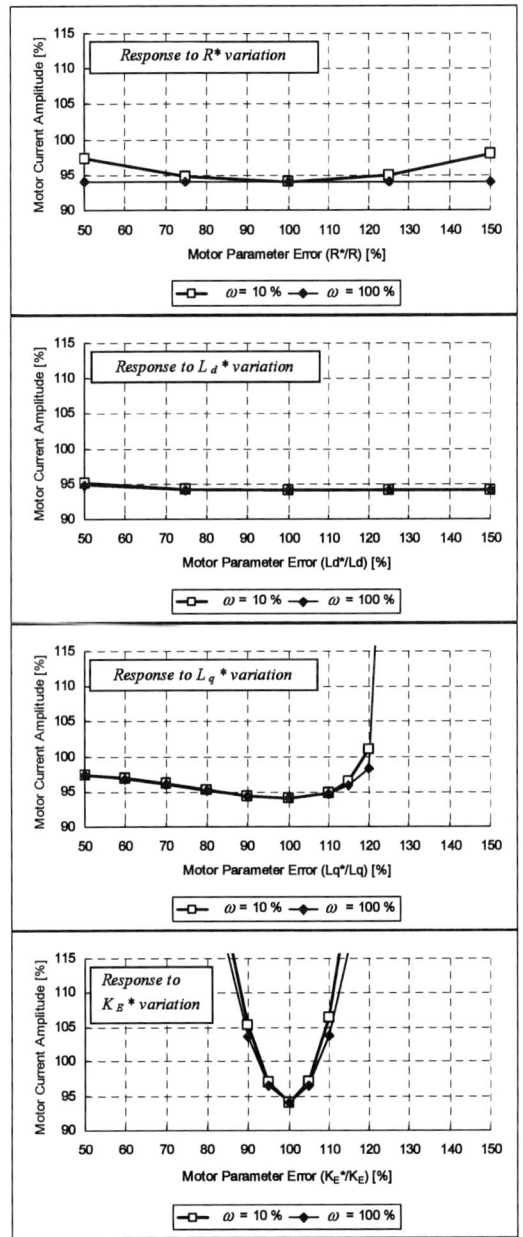

Fig. 3 Relation between motor parameter error and motor current.

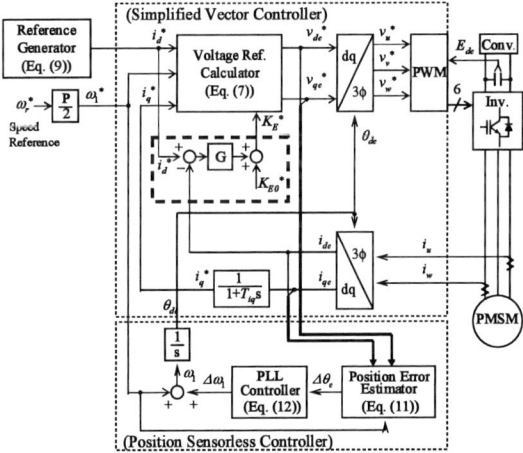

Fig. 4 Simplified vector controller using back EMF parameter adjustment.

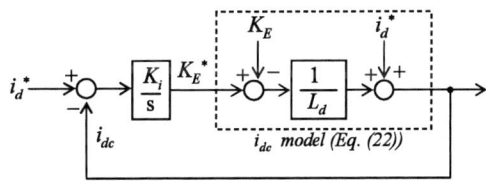

Fig. 5 dc-axis current i_{dc} model.

927

selected as compensator G in order to make i_{dc} to first-order delay response.

The integral gain K_i can be determined by

$$K_i = \omega_c L_d^{\ *}, \tag{23}$$

where ω_c is a response angular velocity from $i_d^{\ *}$ to i_{dc}. If the approximate relation in (22) can be formed, the control system becomes a 1st-order system and it is stable.

In principle, the proposed method is similar to the EMF constant identification proposed in [9]. However, the control model which uses adjusted EMF parameter in the proposed method differs from that in [9].

IV. VALIDATION OF PROPOSED COMPENSATION METHOD

A. Simulations

The validity of the simplified vector control for position sensorless PMSM with back EMF parameter adjustment was confirmed by using a computer simulation. In this simulation, the compensator gain K_i was designed by using $\omega_c = 20$ (rad/s).

First of all, we examined the transient response of the proposed control. Fig. 6 shows the adjustment response of the back EMF parameter. In this simulation, the K_E of the PMSM dropped to 10% lower value in a single step at 0.1 (s). It was found that $K_E^{\ *}$ is adjusted after the K_E drop.

Fig. 7 shows the load torque response at a 100% rated speed. Fig. 7 (a) shows the response of the control with a back EMF parameter adjustment; Fig. 7 (b) shows the response of the control without the adjustment. In this simulation, the $K_E^{\ *}$ transiently varies after the load torque is applied, although $K_E^{\ *}$ is set to K_E. Since the two responses in Fig. 7 are almost the same, it was found that activating the proposed method while the motor is in operation does not affect the response.

Then, we used the same simulation as in Fig. 3 to investigate the steady-state performance. Fig. 8 shows the relation between the motor parameter error and the motor current. Under each simulation condition, the motor current amplitude was 94%. It was found that the motor current amplitude was suppressed even when there were errors in the motor parameters of controller.

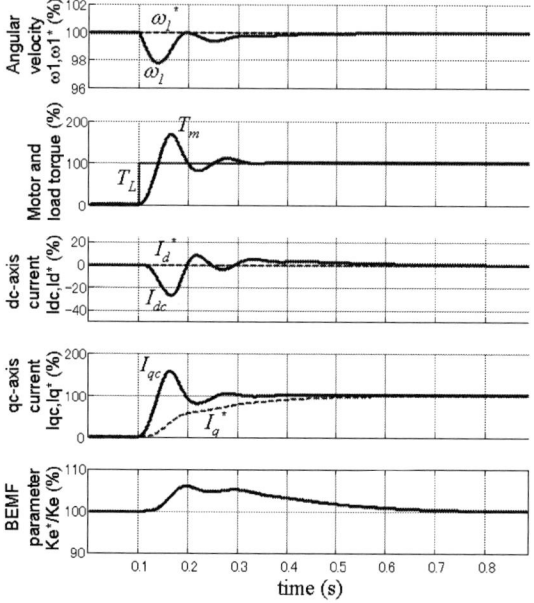

(a) Proposed method

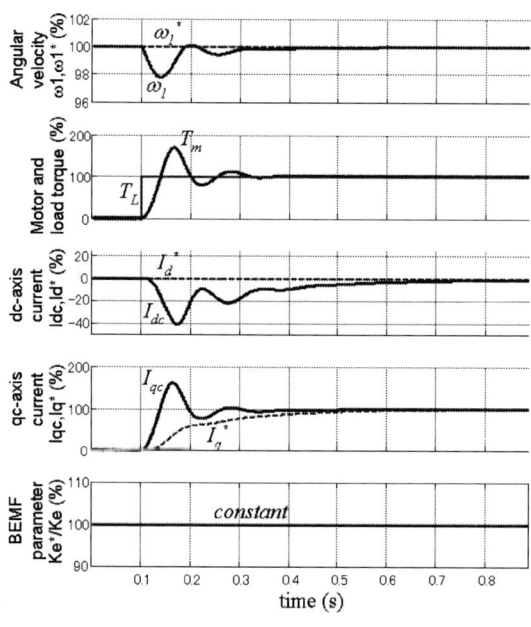

(b) Without $K_E^{\ *}$ adjustment

Fig. 7 Load torque response (simulation results)

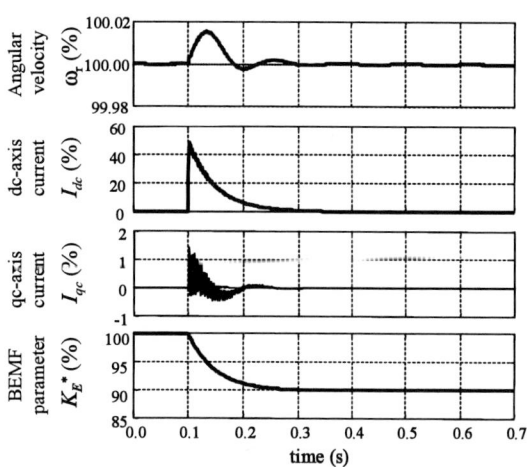

Fig. 6 Adjustment response of BEMF constant $K_E^{\ *}$ (simulation result)

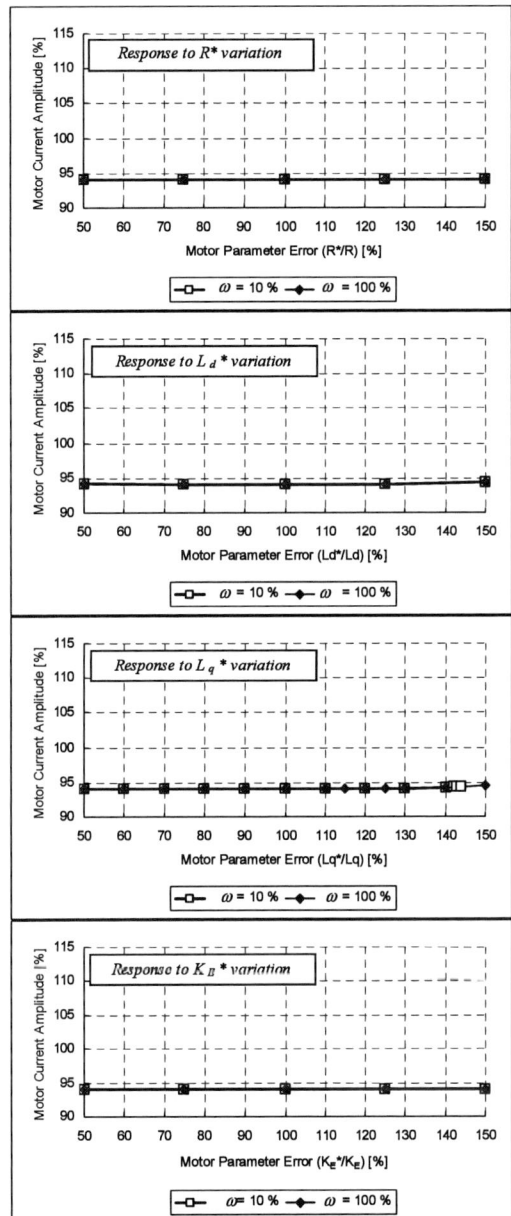

Fig. 8 Relation between motor parameter error and motor current. (Proposed method)

B. Theoretical analysis

The proposed back EMF parameter adjustment method is designed to compensate for the back EMF parameter error. However, the simulation results show that the proposed method can compensate for a current increase caused by the other motor parameter error. In this subsection, we try to clear up the principle of the unexpected effect.

The relationship between i_d and i_q under the maximum torque per current amplitude condition is known as that given by (9). However, (9) was derived from the voltage equation of PMSM on the rotor reference frame [7] [8]. In

this paper, the relationship is derived from the voltage equation on the assumed rotor reference frame.

Assuming $\omega_1{}^* = \omega_1 = \omega_m$, $i_q{}^* = i_{qc}$, and $i_d{}^* = i_{dc}$ from (7), the voltage command of the proposed method can be expressed as follows:

$$\begin{bmatrix} v_{dc}{}^* \\ v_{qc}{}^* \end{bmatrix} = R^* \begin{bmatrix} i_{dc} \\ i_{qc} \end{bmatrix} + \omega_m \cdot \begin{bmatrix} -L_q{}^* \cdot i_{qc} \\ L_d{}^* \cdot i_{dc} \end{bmatrix} + \begin{bmatrix} 0 \\ K_E{}^* \cdot \omega_m \end{bmatrix}. \qquad (24)$$

The instantaneous active power of (24) can be calculated by using

$$\begin{aligned} P_{ins} &= v_{dc}{}^* i_{dc} + v_{qc}{}^* i_{qc} \\ &= R^* i_{dc}{}^2 + R^* i_{qc}{}^2 + \omega_m \left(L_d{}^* - L_q{}^* \right) i_{dc} i_{qc} + K_E{}^* \omega_m i_{qc}. \end{aligned} \qquad (25)$$

By subtracting the copper loss from P_{ins}, the motor output torque T_m is obtained by using the following equations.

$$T_m = \frac{1}{\omega_m} \left[P_{ins} - R \cdot i_{dc}{}^2 - R \cdot i_{qc}{}^2 \right]. \qquad (26)$$

$$T_m = \left(L_d{}^* - L_q{}^* \right) i_{dc} i_{qc} + K_E{}^* i_{qc} + \frac{\left(R^* - R \right) \left(i_{dc}{}^2 + i_{qc}{}^2 \right)}{\omega_m}. \qquad (27)$$

We use the Lagrange multiplier method to solve the maximum torque per current amplitude problem [8]. By using a Lagrange multiplier λ, function L is defined by

$$L \left(i_{dc}, i_{qc}, \lambda \right) = R^* \left(i_{dc}{}^2 + i_{qc}{}^2 \right) + \lambda \left(T_m - T_{load} \right), \qquad (28)$$

where T_{load} is the motor load torque. On the right side of (28), the first term represents the copper loss and the second term represents the torque condition of the constraint.

The condition for minimizing the copper loss is obtained by

$$\frac{\partial L}{\partial i_{dc}} = \frac{\partial L}{\partial i_{qc}} = 0. \qquad (29)$$

After solving this equation, the condition for minimizing the copper loss can be expressed by using (30):

$$i_{dc}{}^2 + \frac{K_E{}^*}{\left(L_d{}^* - L_q{}^* \right)} i_{dc} - \left[1 + \frac{1}{\left(L_d{}^* - L_q{}^* \right)} \frac{\partial K_E{}^*}{\partial i_{dc}} \right] i_{qc}{}^2 \\ + \frac{1}{\left(L_d{}^* - L_q{}^* \right)} \frac{\partial K_E{}^*}{\partial i_{qc}} i_{dc} i_{qc} = 0 \qquad (30)$$

The proposed method uses the back EMF parameter adjustment method. Therefore, the partial derivative of $K_E{}^*$ is not zero. From (21), parameter $K_E{}^*$ can be approximated by using

$$K_E{}^* \simeq \left(L_d - L_d{}^* \right) \cdot i_{dc} + K_E. \qquad (31)$$

Thus, partial derivatives can be approximated by using

$$\frac{\partial K_E{}^*}{\partial i_{dc}} \simeq L_d - L_d{}^*, \quad \frac{\partial K_E{}^*}{\partial i_{qc}} \simeq 0. \qquad (32)$$

Substituting (32) with (30) results in

$$i_{dc}^2 + \frac{K_E^*}{L_d^* - L_q^*} i_{dc} - \left[1 + \frac{L_d - L_d^*}{L_d^* - L_q^*}\right] \cdot i_{qc}^2 = 0. \quad (33)$$

Finally, the relationship between i_{dc} and i_{qc} under the maximum torque per current amplitude condition is obtained in the following way.

$$i_{dc} = \frac{K_E^*}{2(L_q^* - L_d^*)} - \sqrt{\left[\frac{K_E^*}{2(L_q^* - L_d^*)}\right]^2 + \left[1 + \frac{L_d - L_d^*}{L_d^* - L_q^*}\right] \cdot i_{qc}^2}$$

(34)

Equation (34) is similar to the conventional equation reported in [7] and [8]. However, as a real motor parameter, only L_d is used in (34). Note that the other motor parameters, R, L_q, and K_E are not included in this equation. In other words, (34) proves that the proposed method is only affected by an error in the d-axis inductance parameter L_d. Therefore, even if there are errors in the other motor parameters, the motor current under a constant load condition is minimized by using the proposed method.

V. Experiments

The experiments we carried out were conducted to confirm the simulation results and our theoretical analysis.

For the experiments, we used a test motor with the parameters shown in Table 2. The test motor was driven by four different parameter conditions written in Fig. 9 and motor phase current amplitude was measured in the steady-state. We set a motor load torque to 30 % value because of mechanical constraint of the experimental equipment.

Fig. 9 shows the results of measurement. In the case of no compensation, motor current amplitude was about 2.3A. In the case of proposed method, we confirmed that motor current amplitude decreased to 1.9A, even if there was an error in parameter settings.

Fig. 10 shows the measured characteristics of dc-axis current command i_d^* and motor current amplitude. We chose the same motor torque condition as in Fig. 9. In this experiments, i_d^* was set to a constant value, but was not calculated by (9). It was found that motor current was minimized by setting $i_d^* = 0.5A$ and the minimum current amplitude was 1.9A.

Therefore, we are able to consider that the current value 1.9A in Fig. 9 is the minimum motor current of the motor torque condition of these experiments. We prove that the motor current of the constant load condition is minimized by using the proposed method.

VI. Conclusion

We proposed a new adaptive back EMF parameter adjustment method for the simplified vector control. We proved through the use of a simulation and experimentation that the proposed method can compensate for the motor parameter error, and has sufficient robustness in the steady-state.

A theoretical analysis showed that the proposed method is affected by only the d-axis inductance parameter L_d. Therefore, even if there are errors in the other motor

Table 2. Specifications of the test motor for the experiments.

L_d	10 mH
L_q	12 ~15 mH
R	0.5 Ω
K_E	0.155 V/(rad/s)

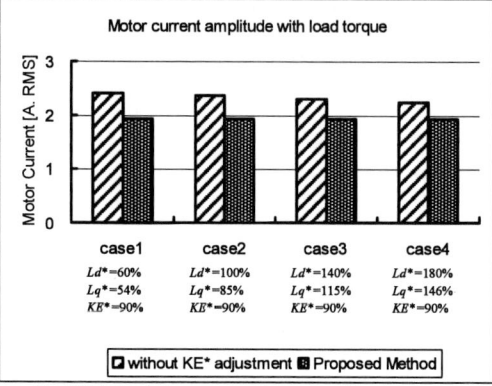

Fig. 9 Relation between parameter error and motor current. (experiment)

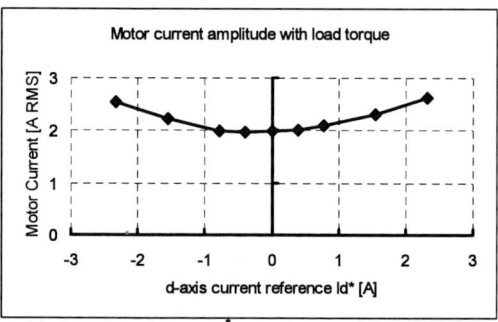

Fig. 10 Relation between i_d^* and motor current. (experiment).

parameters, the motor current of the constant load condition is minimized by using the proposed method.

References

[1] P. P. Acarnley; J. F. Watson, "Review of position-sensorless operation of brushless permanent-magnet machines," *IEEE Trans. Ind. Electron.*, vol. 53, no. 2, pp. 352- 362, Apr. 2006.

[2] T. Takeshita, M. Ichikawa, J-S Lee, and N. Matsui, "Back EMF estimation-based sensorless salient-pole brushless DC Motor Drives," *Trans. IEE Japan*, vol. 117-D, no. 1 pp. 98-104, Jan. 1997 (in Japanese).

[3] L. A. Jones and J. H. Lang, "A state observer for the permanent magnet synchronous motor," *IEEE Trans. Ind. Electron.*, vol. 36, no. 3, pp. 374-382, Aug. 1989.

[4] S. Bolognani, R. Oboe, and M. Zigliotto, "Sensorless full-digital PMSM drive with EKF estimation of speed and rotor position," *IEEE Trans. Ind.l Electron.*, vol. 46, no. 1, pp. 184-191, Feb. 1999.

[5] S. Ichikawa, Z. Chen, M. Tomita, S. Doki, and S. Okuma, "Sensorless controls of salient-pole permanent magnet synchronous motors using extended electromotive force models," *Electr. Eng. Japan*, vol. 146, no. 3, pp.55-64, 2004 [*Trans. IEE Japan*, vol. 122-D, no. 12 pp. 1088-1096, Dec. 2002].

[6] Z. Chen, M. Tomita, S. Doki, and S. Okuma, "An extended electromotive force model for sensorless control of interior permanent-magnet synchronous motors," *IEEE Trans. Ind. Electron.*, vol. 50, no. 2, pp. 288-295, Apr. 2003.

[7] S. Morimoto, M. Sanada, Y. Takeda, "Wide-speed operation of interior permanent magnet synchronous motors with high-performance current regulator," *IEEE Trans. Ind. Appl.*, vol. 30, no. 4, pp. 920-926, July 1994.

[8] S. Shinnaka: "A vector-signal-based analysis for salient synchronous motors oriented to energy-efficient current controls," *Trans. IEE Japan*, vol.119-D, no.5, pp. 648-658, 1999 (in Japanese).

[9] T. Takeshita and N. Matsui, "Sensorless brushless dc motor drive with EMF constant identifier," *Proc. of 1994 IECON (Industrial Electronics Conference)*, vol. 1, pp. 14-19, 1994

[10] K. Sakamoto, Y. Iwaji, T. Endo, and Y. Takakura, "Position and Speed Sensorless Control for PMSM Drive Using Direct Position Error Estimation," *Proc. of 2001 IECON (Industrial Electronics Conference)*, vol.3, pp.1680-1685, 2001.

[11] K. Sakamoto, Y. Iwaji, and T. Endo, "A simplified vector control of position sensorless permanent magnet synchronous motor for electrical household appliances," *Trans. IEE Japan*, vol. 124-D, no. 11, pp. 1133-1140 2004 (in Japanese).

[12] D. Li, T. Suzuki, K. Sakamoto, Y. Notohara, T. Endo, C. Tanaka, T. Ando, "Sensorless Control and PMSM Drive System for Compressor Applications," *Proc. of Power Electronics and Motion Control Conference*, vol. 2, pp. 1-5, Aug. 2006

[13] K. Sakamoto, Y. Iwaji, T. Endo, T. Taniguchi, T. Niki, M. Kawamata, A. Kawamura, "Position Sensorless Vector Control of Permanent Magnet Synchronous Motors for Electrical Household Appliances," *Proc. of Power Conversion Conference - Nagoya, 2007*, pp. 1119-1125, April 2007

Identification and Control of Precision XY Stages with Active Vibration Suppression System

Mayumi Nitta[*] and Seiji Hashimoto[†]

[*] Gunma University/ Electronic Engineering, Kiryu, Gunma, e-mail: *m07e642@gs.eng.gunma-u.ac.jp*
[†] Gunma University/ Electronic Engineering, Kiryu, Gunma, e-mail: *seijiha@el.gunma-u.ac.jp*

Abstract— In this paper, a design method of the reaction force control system for ultra-precision stage which equips vibration control system, is proposed. Taking the application to semiconductor manufacturing into consideration, the designed system deals with both the reaction force due to the stage movement and the vibration from the external environment. The experimental setup for a 6-degree of freedom (DOF) motion control system is firstly designed. The system is controlled by two counter weights that are placed eccentric from the center of gravity, i.e. two weights are not co-axial. Next, the system identification experiments based on a subspace method are performed to build the multi-input multi-output state-space model. The reaction force and vibration control systems are designed using the identified model based on bilateral control strategies. The main advantages of the proposed identification and control approach are that the experimental time for accurate identification is quite short and the control system can be systematically designed. Finally, the effectiveness of the proposed control system is verified through the vibration control experiments.

Keywords— reaction force control, vibration suppression, bilateral system, ultra-precision stage, precision positioning.

I. INTRODUCTION

For the future application of an electron beam apparatus for the semiconductor industry, a nonresonant ultrasonic motor is the most attractive device for a stage system instead of an electromagnetic motor, because the power source of the stage system is required for non-magnetic and vacuum applications. The authors have proposed the control methods of the nonresonant ultrasonic motor especially focusing on the continuous movement and the precision positioning. In order to meet the future demand for a high throughput in the semiconductor manufacturing, the precision stage should overcome difficulties related to the high torque and high response for feeding a heavy stage.

Present control system is classified into the following two methods. One is the force control system that counterbalances driving reaction force of the stage by setting up the counter-balance weights so that the devices do not generate the force. The other is the vibration control system which suppress the vibration using the vibration isolation equipment.

In this paper, the reaction force control system integrated with the vibration control based on bilateral control method is newly proposed. The effectiveness of the proposed reaction force control system is verified through the constructed 6DOF experimental setup.

II. CONTROLLED OBJECT

The precision stage with reaction force control equipment is typically constructed of the X-Y stage and the linear actuator-driven counter weight as shown in Fig. 1. Two counter weights symmetrically placed for one axis can be moved to balance the driven force of the stage.

In the proposed reaction force control system, the counter weights are eccentrically placed on each axis, i.e. one counter weight for one direction, for simplicity of the construction and the breadth of industrial applications. Developed 6DOF control equipment is shown in Fig. 2. Configurations of the system are represented in Table I. The linear actuator-driven counter weights are placed along X-Y axis. The isolator in order to suppress the vertical vibration is driven by the voice coil motors.

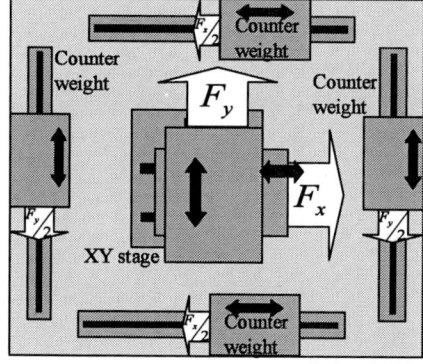

Fig.1 Example of reaction force control system.

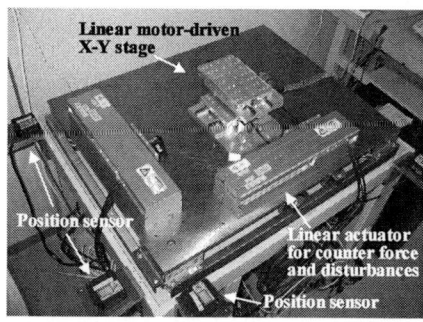

Fig. 2. Constructed 6DOF experimental system.

TABLE I. SPECIFICATION OF 6DOF EXPERIMENTAL SYSTEM

Actuator for counter weight	Techno Hands / SGL80-S3-200-0.1S
Stroke	200 mm
Moving mass	0.92 kg
Rated force	33 N
Sensor resolution	100 nm
Peak force	132 N
Actuator for XY stage	Tech concierge Kumamoto / SPIDER
Stroke	100 mm
Moving mass	1.0 kg
Rated force	13 N
Sensor resolution	100 nm
Position sensor	SUNX / ANR1250
Measure range	±10 mm
Sensor resolution	10 μm

III. REACTION FORCE CONTROL SYSTEM

A. Identification of the controlled object

Before constructing the position and reaction force control systems, the modeling of the controlled object is investigated. In the modeling, since the object is the multi-input and multi-output (MIMO) system, a subspace identification method is introduced [1], [2]. The subspace identification method involves matrices obtained from output and input data just by once identification experiment. Thus, the MIMO system model is obtained in state-space form using these data matrices. The pseudo-random binary signal, PRBS which is a pseudo-white signal, is used as an excitation signal in the experiments. Sampling period of the PRBS is decided considering the frequency range to be identified.

B. Design of the position and reaction force control system

Based on the identified MIMO model, a position control and reaction force control systems are designed. The block diagram of the proposed control system with the transfer function matrix is shown in Fig. 3. In Fig. 3, the plant P represents the plant transfer function matrix, which is composed of the primary mode to be controlled and the coupling term between the stage and the vibration suppression system. C describes the controller transfer function matrix whose factors are constructed of the reaction force controllers, vibration suppression controllers and the position controller. The position control system is constructed not only for the stage but also the counter weights.

In the design of the position control system for the XY stage, the disturbance observer-based internal model control method (DIMC) is utilized [3], [4]. The block diagram of the DIMC system is shown in Fig. 4. In Fig. 4, $P(s)$ and $P_n(s)$ are the controlled plant and its nominal model. $F(s)$ is a low pass filter with a steady state gain of one. The transfer function $F(s) \cdot P_n^{-1}(s)$ becomes proper. $F_d(s)$ is assumed to be the same as the IMC filter $F(s)$ ($F_d(s) = F(s)$). The controller's parameters for DIMC are only the parameters of the plant model and control bandwidth. Therefore, the position controller can be systematically designed by using the identified model.

For the design of the reaction force control system by the counter weights, the decoupling control method and the direct velocity feedback control (DVFC) method are introduced [5]. Both methods do not require the additional control parameters other than that of the identified plant and control band. The advantage to utilize the DVFC are that the system is stable even if the modeling error exists in the design, and all observable vibration modes are improved since the effectiveness of DVFC is similar to a mechanical damper. Therefore, the DVFC system is constructed for both axes mainly to compensate the modeling error and to suppress the vibration due to external disturbance.

To compensate the coupling term due to the eccentric placement of the counter weights as well as the coupling term among the stage, the counter weight and the active vibration isolator, the decoupling control is implemented. By the proposed control method, both the vibration due to the driving force and the vibration due to the external disturbance can be suppressed simultaneously. In addition to the control methods, the position control system with narrow control bandwidth is constructed for the linear-motor driven counter weights in order to utilize its stroke effectively.

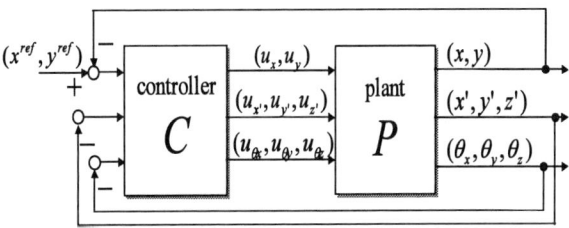

Fig. 3. Block diagram of reaction force control system.

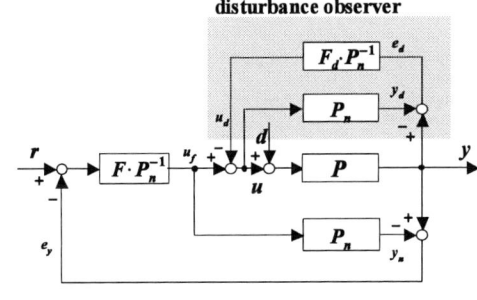

Fig. 4. Block diagram of DIMC-based position control for XY stage.

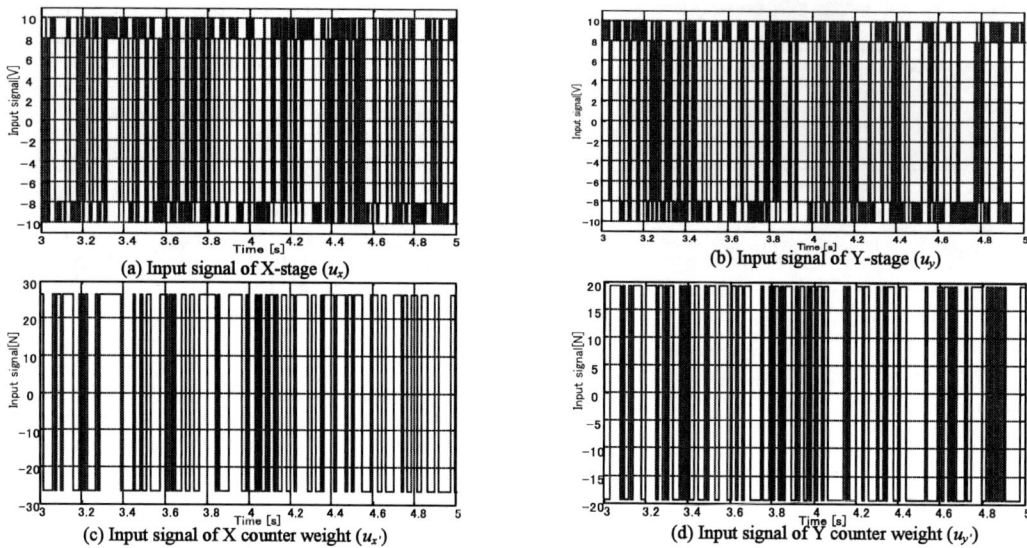

(a) Input signal of X-stage (u_x)

(b) Input signal of Y-stage (u_y)

(c) Input signal of X counter weight ($u_{x'}$)

(d) Input signal of Y counter weight ($u_{y'}$)

Fig. 5. Input signals for system identification experiment.

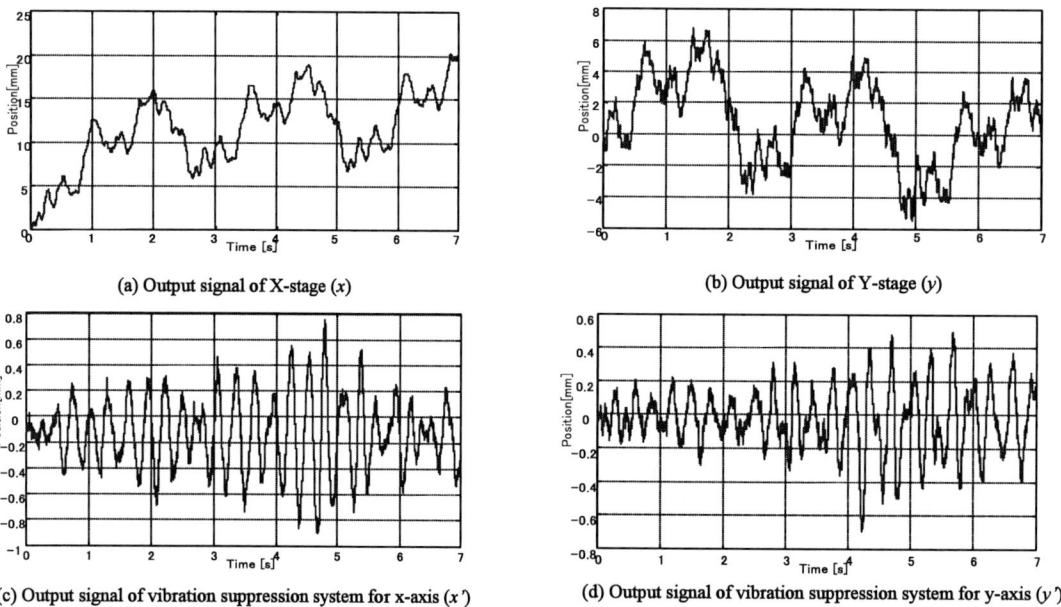

(a) Output signal of X-stage (x)

(b) Output signal of Y-stage (y)

(c) Output signal of vibration suppression system for x-axis (x')

(d) Output signal of vibration suppression system for y-axis (y')

Fig. 6. Output signals for system identification experiment.

IV. EXPERIMENTAL RESULTS

A. System identification experiments

In order to the MIMO model P, the system identification experiment previously described was carried out. In this paper, since the main object is the MIMO identification base on the subspace method, the input and output signals are simply focused on the stage's position and position of the vibration suppression system. Four input signals u_x, u_y, $u_{x'}$, and $u_{y'}$ mainly drive the stage of x-direction, the stage of y-direction, the counter weight for x-direction and the counter weight for y-direction, respectively. The input signals are PRBS, and are applied at the same time. The sampling time for the experiment was 1 ms. Taking the frequency bandwidth of the identified plant into consideration, the clock period of PRBS was selected. The clock period for $u_{x'}$ and $u_{y'}$ was 10 times of the sampling time. The period of the PRBS was 1023. For u_x and u_y, the input signals are the convolution of two signals. One is the signal with the clock period of 10 times, which is for identifying the dynamics from u_x (u_y)to $x'(y')$. The other is the signal with the clock period of 4 times, which is for identifying the dynamics from u_x (u_y)to x (y). In order not to make correlation between these inputs, the initial value for PRBS is differently selected. The outputs x, y, x' and y' are the position of x-stage, position of y-stage, position of the basement for x-direction, and that for y-direction, respectively. The measured input and output signals are shown in Figs. 5 and 6. As can be seen in Fig. 4, the output signals are well excited by the input signals. The coherence between these inputs is shown in Fig. 7. It is verified that the correlation between the input signals are enough small at the frequency range to be identified. Moreover, the coherence between the inputs and outputs are represented in Fig. 8 and 9. In order to improve the coherence in low frequency range which relates to the control bands, the decimation filter consisting of the low pass filtering and the downsampling was applied to the data. The thick solid line shows the coherence for diagonal factor of the identified plant model. Since the coherence is comparatively high, the accurate identification can be expected. Moreover, the mid thick solid line represents the cross-coupling term from the input u_x, u_y to x' and y'. It can be noticed that these coupling terms cannot be ignored since the coherence is high.

With this, the system identification based on the subspace method is investigated. A 45th order model was assumed. The order of the model is decided using trial and error techniques. The system matrices, can be identified simultaneously by the subspace method. The System Identification Toolbox of MATLAB is used for the calculation. The identification results are shown in Figs. 10 and 11. It can be noticed that the gain of the diagonal factor, which is represented by the thick solid line, is high comparing to that of the off-diagonal factors. According to this model, the frequency of the first vibration mode is 21 rad/s, and that of the second mode is about 32 rad/s. In order to verify identification accuracy, the typical least-square method based on the ARX model to obtain the SISO model was also applied. The order of the model was assumed to 30th. The gains of the diagonal factor are quite similar to those of the ARX model. Therefore, the accuracy of the MIMO model can be confirmed.

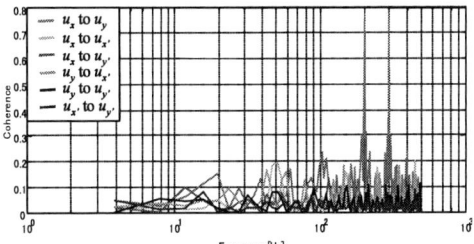

Fig. 7. Coherence between input signals.

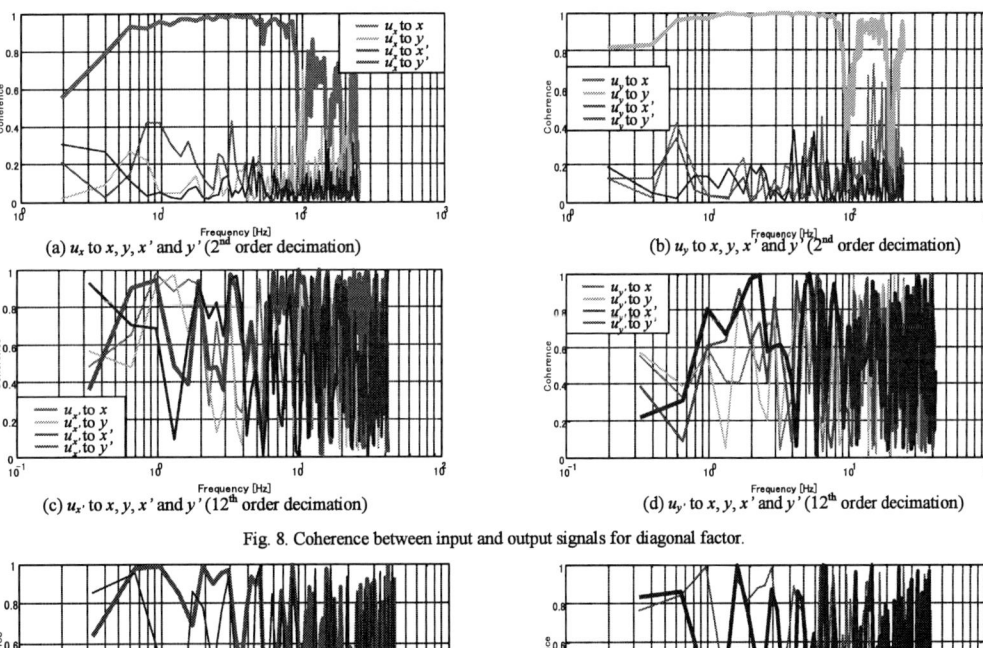

(a) u_x to x, y, x' and y' (2nd order decimation)

(b) u_y to x, y, x' and y' (2nd order decimation)

(c) $u_{x'}$ to x, y, x' and y' (12th order decimation)

(d) $u_{y'}$ to x, y, x' and y' (12th order decimation)

Fig. 8. Coherence between input and output signals for diagonal factor.

(a) u_x to x' and y' (12th order decimation)

(b) u_y to x' and y' (12th order decimation)

Fig. 9. Coherence between input and output signals for off-diagonal factor.

935

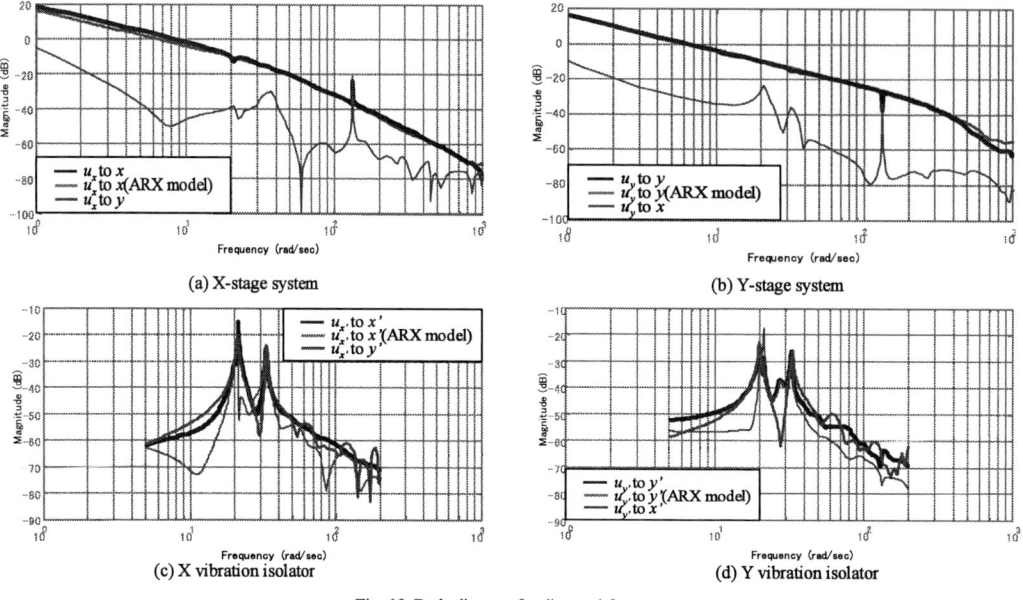

(a) X-stage system   (b) Y-stage system

(c) X vibration isolator   (d) Y vibration isolator

Fig. 10. Bode diagram for diagonal factor.

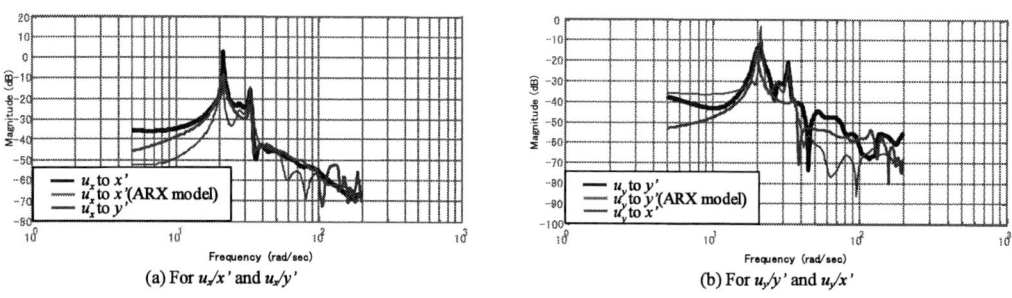

(a) For u_x/x' and u_x/y' (b) For u_y/y' and u_y/x'

Fig. 11. Bode diagram for off-diagonal factor.

B. Control experiments

Based on the identified models, the proposed reaction force control system described in previous section is designed. Here, the control experiments to the step movement of the precision stage with the amplitude of 10 mm were investigated. As an example, the result for X-direction is represented in Fig.12. Fig. 12 (a) and (b) show the transient state and steady state responses of the isolator's position x', respectively.

For the comparison, 1) without control, 2) decoupling control, 3) DVFC and 4) proposed reaction force control were examined.

From Fig.12, the proposed control system well suppresses the vibration for both transient state which is mainly due to the reaction force and steady state which is mainly due to the external vibration. The quantitative results for the X-direction experiments are compared in Table II. Maximum displacement when the stage moves was reduced to 36.7% for X-direction. Moreover, the summation of the squared error for 6 seconds was decreased to 2.7% for X-direction.

V. CONCLUSION

In the present paper, a reaction force control problem for the precision X-Y stage is considered. In addition, a modeling procedure and a control method for the system have been proposed. The subspace method is verified to be helpful for modeling the system. The reaction force and vibration control method employing DVFC and decoupling control is verified to be suitable for the control problem. The performance of the design control system is experimentally demonstrated by using the constructed 6DOF active vibration isolator.

ACKNOWLEDGMENT

This work was partially supported by the Ministry of Education, Science, Sports and Culture, Grant-in-Aid for Young Scientists.

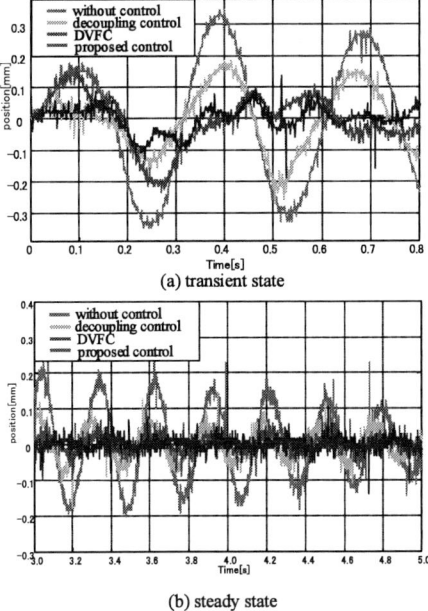

(a) transient state

(b) steady state

Fig. 12. Comparison of time response of position for X vibration
suppression system.

TABLE II. VIBRATION OF POSITION FOR X VIBRATION SUPPRESSION
SYSTEM

	max. displacement [mm]	squared error
non control	0.349(100%)	100%
decoupling control	0.274(78.5%)	21.6%
DVFC	0.235(67.3%)	7.3%
bilateral control	0.128(36.7%)	2.7%

REFERENCES

[1] S. Adachi, "Advanced system identification for control by MATLAB," Tokyo Electrical Engineering College Publications Service, 2004.

[2] S. Adachi, "System identification for control by MATLAB," Tokyo Electrical Engineering College Publications Service, 1996.

[3] M. Kigure, S. Hashimoto, T. Ishikawa, "A Fast-Response and High-Precision Control Method Based on Internal Model Control," Meeting on Industrial Instrumentation and Control, IEE Japan, pp.31-36, 2006.

[4] M. Kigure, S, Hashimoto, T. Ishikawa, "Disturbance Observer-Based Internal Model Control With An Adaptive Mechanism for Linear Actuators," SICE Annual Conference 2007.

[5] S. Hashimoto, H. Funato, K. Hara, K. Kamiyama, "DVFC-Based Identification and Its Application to Robust Vibration Suppression Control of Flexible Structures," Trans. IEE of Japan, Vol. 119-D, No. 6, June, 1999.

[6] Graham C. Goodwin, Stefan F. Graebe, Mario E. Salgado, "Control System Design," Prentice Hall, 2001.

[7] S. Hashimoto, T. Ishikawa, S. Adachi, H. Funao, K. Kamiyama, A. Isojima, "Multi-Decimation Identification-Based Modeling Method of Flexible Structures for Robust Vibration Control," IEEJ Trans. IA, Vol.124, No.5, pp.471-478, 2004.

[8] K. Wei, G. Meng, S. Zhou, J. Liu, "Vibration control of variable speed/acceleration rotating beams using smart materials," Journal of Sound and Vibration 298, pp.1150-1158, 2006.

[9] Chih-Chergn Ho, Chih-Kao Ma, "Active vibration control of structural systems by a combination of the linear quadratic Gaussian and input estimation approaches," Journal of Sound and Vibration 301, pp.429-449 2007.

[10] T. Shimono, R. Kubo, K. Ohnishi, S. Katsura, K. Ohishi : "Design of Haptic Transmission Ratio for Multilateral Control", IEEE Japan Ind. Appl. Soc. Annual Conf.,No.126, 2006.

[11] Seok-Jun Moon, Chae-Wook Lim, Byung-Hyun Kim, Youngjin Park, "Structural vibration control using linear magnetostrictive actuators," Journal of Sound and Vibration 302, pp.875-891, 2006.

[12] Yuichi Chida, Yoshiyuki Ishihara, Takuya Okina, Tomoaki Noshimura, Koichi Ohtomi, Ryo Fukukawa, "Identification and Frequency Shaping Control of a Vibration Isolation System", Control Engineering Practice16, pp.711-723, 2008.

[13] Yongjun Chen, Changsheng Zhu, "Active Vibration Control Based on Linear matrix Inequality for Rotor System under Seismic Excitation," Journal of Sound and Vibration 314, pp.53-69, 2008.

[14] Shinji Wakui, "Incline compensation control using an air-spring type active isolated apparatus," Precision Engineering 27, pp.170-174, 2003.

[15] Shinji Wakui, Azusa Noda, Takeshi Akiyama, Masato Takahashi, "Development of velocity sensor with high frequency band and its application to a vibration isolate table," Precision Engineering 31, pp.146-155, 2007.

[16] S. Wakui, "Construction of 6-DOF Attitude Control System Using Linear Motor," the 17th Symposium on Electromagnetics and Dynamics, pp.213-216, 2005.

Sensitivity of the Currents Input-Output Decoupling Vector Control of the DFIM versus Current Sensors Fault

Meriem ABDELLATIF[*†], Maria PIETRZAK-DAVID[†], Ilhem SLAMA-BELKHODJA[*]

*Laboratoire des Systèmes Electriques (LSE), Ecole Nationale d'Ingénieurs de Tunis (ENIT),
e-mail: *meriem.abdellatif@laplace.univ-tlse.fr* , *ilhem.slama@enit.rnu.tn*
[†] Université de Toulouse; Laplace; INPT, UPS; Toulouse, France,
e-mail: *maria.david@laplace.univ-tlse.fr*

Abstract— The goal of this paper is a Doubly Fed Induction Machine (DFIM) working in motor mode and supplied by two voltages PWM inverters, in stator and rotor sides. Effects of current sensors faults on the currents input-output decoupling in the Stator Flux Oriented Vector Control (SFOVC) are analyzed. A sensitivity study is carried out considering both stator and rotor current sensors failure, under different working conditions and a large speed range. Stator flux coefficient sensitivities emphasize the required accuracy of the currents sensors to guaranty the correct angle orientation and the required level of control performances. Matlab-Simulink simulation results are given and illustrate this sensitivity analysis.

Keywords— Doubly Fed Induction Motor, Vector Control, Sensor Fault, Reliability, Robustness, Simulation.

I. INTRODUCTION

The SFOVC is nowadays widespread for high dynamic performances of AC drives [1]. However, these control static and dynamic performances and robustness depend strongly on the accuracy of the current sensors measured signals. They are present in all system controls and also in the stator flux orientation. So, as reliability is a criterion of outstanding importance for satisfactory operation drive, faulty sensors investigations finds an increasing interest and different works in this domain are presented in the literature. Some authors investigate the effect of sensor faults on system performances [2], [3]; others analyze the fault detection and isolation of a faulty sensor [4], [5]. Some solutions are proposed to improve the system operation results when failure occurs in current sensor. The first is the integration in the control of sensors errors compensation [6], the second is direct sensors drifts compensation [7]. The reconfiguration strategy is also proposed in order to guarantee a continuous working when fault occurs in current sensors [8], [9].

In this paper, a sensitivity study of currents input-output decoupling is introduced in the SFOVC of the DFIM drive system working in motor mode and fed by two voltages PWM inverters, in stator and rotor sides. This control method insures an effective currents decoupling control. After this introduction the authors propose, the studied system presentation. So, a brief description of its control is given to emphasize its current sensor dependence. Then they analyze this dependency using the sensitivity

coefficients of stator flux relatively versus the stator and rotor line current sensors. Simulation results are carried out to highlight the current sensor failures consequences and to validate the sensitivity analysis.

II. SYSTEM PRESENTATION

The general scheme of the power part of the DFIM drive installation is shown in the Fig.1. Two power electronic devices, each composed by a Rectifier, a DC bus and a PWM Voltage Inverter, are used to feed this 850 kW motor and its load.

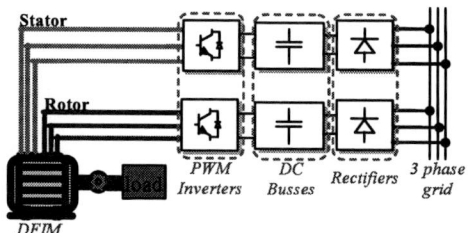

Fig. 1. General scheme of DFIM drive installation.

The control strategy is determined from the DFIM model expressed in the «d,q» rotating reference frame. As it is illustrated in Fig.2, the stator flux vector is oriented with the direct axis ($\Phi s = \Phi sd$ and $\Phi sq = 0$) and rotates at $\omega dq = d\theta dq/dt$ speed. The general model is expressed by (1). In steady state, $\omega dq = \omega s$ and $\theta dq = \theta s$, and by imposing $isdref = 0$ to ensure a unitary power factor working [11], this model can be simplified and written in (2). So, we obtain the null direct stator voltage component too ($Vsd = 0$).

978-1-4244-1741-4/08/$25.00 ©2008 IEEE

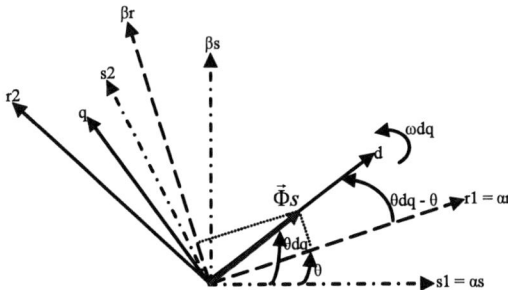

Fig. 2. Stator flux vector linked to the direct axis of the "d,q" frame.

General DFIM model

$$V_{sd} = R_s.i_{sd} + \frac{d\Phi_{sd}}{dt} - \left(\frac{d\theta_{dq}}{dt}\right)\Phi_{sq}$$

$$V_{sq} = R_s.i_{sq} + \frac{d\Phi_{sq}}{dt} + \left(\frac{d\theta_{dq}}{dt}\right)\Phi_{sd}$$

$$V_{rd} = R_r.i_{rd} + \frac{d\Phi_{rd}}{dt} - \left(\frac{d(\theta_{dq}-\theta)}{dt}\right)\Phi_{rq} \quad (1)$$

$$V_{rq} = R_r.i_{rq} + \frac{d\Phi_{rq}}{dt} + \left(\frac{d(\theta_{dq}-\theta)}{dt}\right)\Phi_{rd}$$

Steady state DFIM model

$$V_{sd} = 0$$
$$V_{sq} = R_s.i_{sq} + \omega_s.\Phi_{sd}$$
$$V_{rd} = R_r.i_{rd} + \omega_r.\Phi_{rq} \quad (2)$$
$$V_{rq} = R_r.i_{rq} - \omega_r.\Phi_{rd}$$

The electromagnetic torque is expressed by (3).

$$T_{em} = p.M_{sr}.i_{sq}.i_{rd} = p\Phi_{sd}.i_{sq} \quad (3)$$

The DFIM vector control is designed with an input-output current decoupling strategy which allows an independent control of the four current components, *Isd, Isq, Ird* and *Irq*. All details of this method are presented in [10]. This decoupling strategy is based on state space DFIM modeling as fellows in (4):

$$\dot{x} = Ax + Bu, y = Cx \quad (4)$$

$$x = \begin{bmatrix} i_{sd} & i_{sq} & i_{rd} & i_{rq} \end{bmatrix}^T \in IR^n$$

the state space vector,

$$u = \begin{bmatrix} V_{sd} & V_{sq} & V_{rd} & V_{rq} \end{bmatrix}^T \in IR^m$$

the input vector

$$y = \begin{bmatrix} i_{sd} & i_{sq} & i_{rd} & i_{rq} \end{bmatrix}^T \in IR^p$$

With n, state variables number, m, inputs number, p, outputs number. So, the different matrices of the state space equation are as below:

$$A = \begin{bmatrix} -a_1.I_2 + (a.\omega + \omega_s)J_2 & a_3.I_2 + a_5.\omega.J_2 \\ a_4.I_2 - a_6.\omega.J_2 & -a_2.I_2 + \left(-\frac{\omega}{\sigma} + \omega_s\right).J_2 \end{bmatrix}$$

the dynamic matrix,

$$B = \begin{bmatrix} b_1.I_2 & -b_3.I_2 \\ -b_3.I_2 & b_2.I_2 \end{bmatrix}$$

the control matrix,

$$C = I_4$$

the output matrix,

With $I_2 = \begin{bmatrix} 1 & 0 \\ 0 & 1 \end{bmatrix}$, $J_2 = \begin{bmatrix} 0 & 1 \\ -1 & 0 \end{bmatrix}$, $I_4 = \begin{bmatrix} 1 & 0 & 0 & 0 \\ 0 & 1 & 0 & 0 \\ 0 & 0 & 1 & 0 \\ 0 & 0 & 0 & 1 \end{bmatrix}$

Where:

$$a = \frac{1-\sigma}{\sigma}, \ a_1 = \frac{R_s}{\sigma.L_s}, \ , \ a_2 = \frac{R_r}{\sigma.L_r}, \ a_3 = \frac{R_r.M_{sr}}{\sigma.L_s.L_r}, \ a_4 = \frac{R_s.M_{sr}}{\sigma.L_s.L_r},$$

$$a_5 = \frac{M_{sr}}{\sigma.L_s}, \ a_6 = \frac{M_{sr}}{\sigma.L_r}, \ b_1 = \frac{1}{\sigma.L_s}, \ b_2 = \frac{1}{\sigma.L_r}, \ b_3 = \frac{M_{sr}}{\sigma.L_s.L_r},$$

and $\sigma = 1 - \frac{M_{sr}^2}{L_r L_s}$

Consequently, the general scheme of applied decoupling current method is presented in Fig.3.

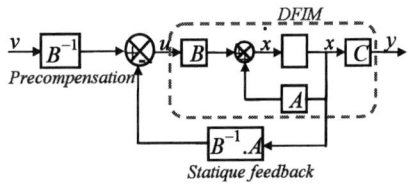

Fig. 3. Decoupling current feedback for DFIM.

The new input vector *v* is imposed to obtain the decoupling current control [10] associated with SFOVC strategy. The general block diagram of this control is shown in Fig.4. We have chosen the same controllers acting on all system variables, i.e. rotation speed, stator flux and stator and rotor currents. Therefore we applied the Integral-Proportional without a zero structure, called IP controller.

In order to verify the static and dynamic performances of the developed system control, we introduce the following operation. Firstly, the DFIM is magnetized at zero speed. Then, at *t=3s*, the reference speed is set to its nominal value *Wn*. The reversal speed rotation direction is applied at *t=9s* and then the reference speed is set to 25% *Wn* at *t=16s*. We impose the load torque null during DFIM starting phase (*Tl=0*). Next, we set it equal to the DFIM nominal torque *Tl=Tnom* at *t=7s*. Then we modify its sign at *t=13,5s* and finally we impose *Tl=Tnom* at *t=20s*. The MATLAB-SIMULIK simulation results permit to validate the global system operation. This specific cycle is shown in Fig.5.a.

The four currents (Fig.5.b) and the stator flux responses (5.c) confirm also the good static and dynamic behavior of this drive when its control strategy is realized with healthy current sensors. A gain error is then introduced on respectively rotor and stator line current sensors. Simulations carried out with 1% gain error value show weak variations on stator flux components (Fig.6) but severe and unacceptable performances deterioration current controls (Fig.7).

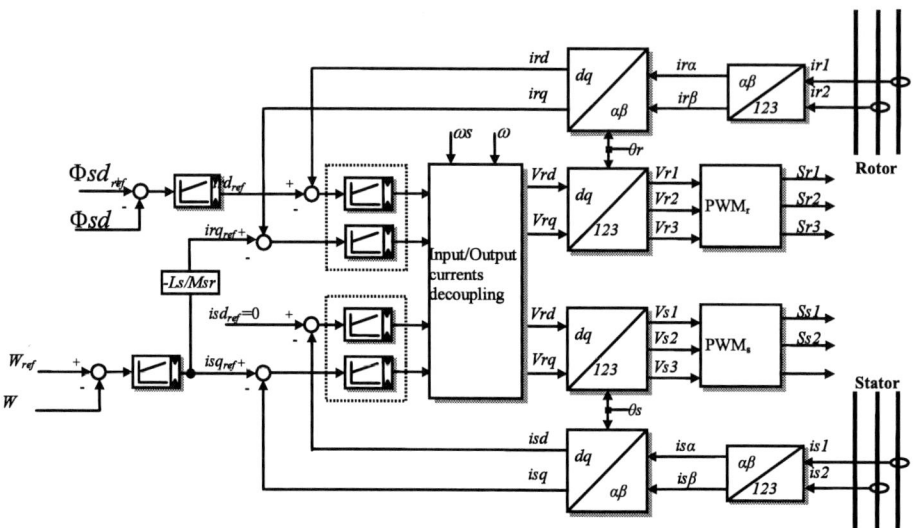

Fig. 4. DFIM developed control.

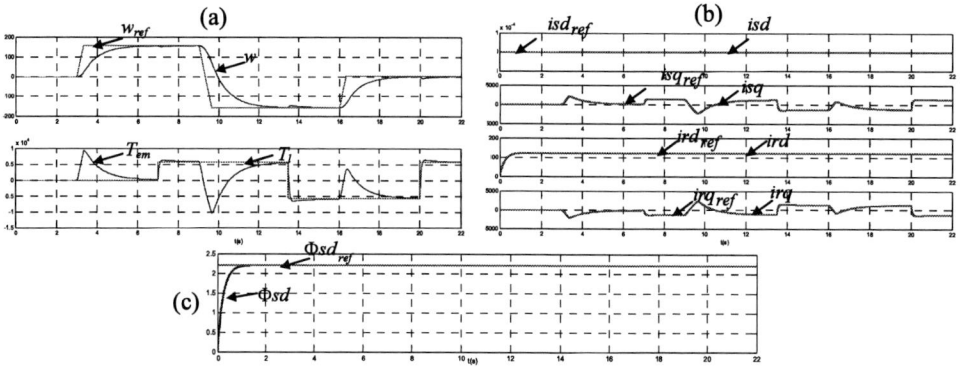

Fig. 5. DFIM control current responses with safety current sensors.

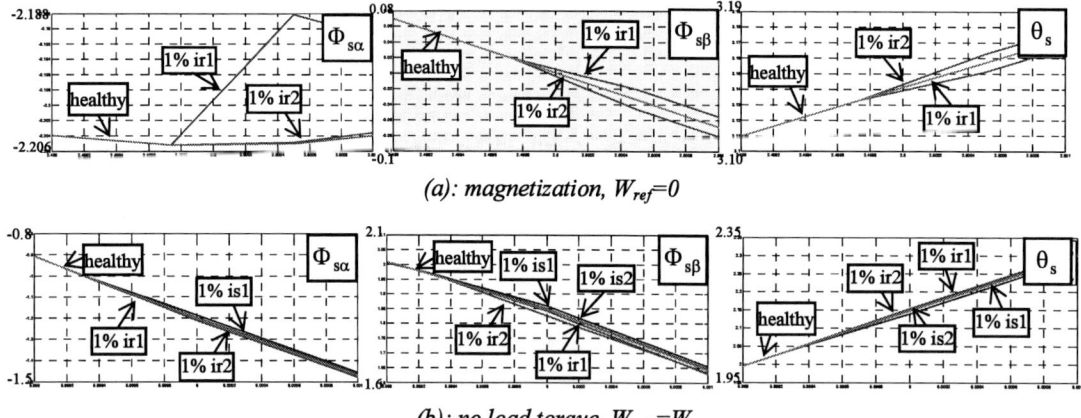

(a): magnetization, $W_{ref}=0$

(b): no load torque, $W_{ref}=W_n$

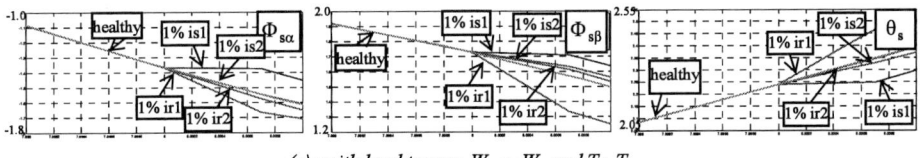

(c): with load torque, $W_{ref}=W_n$ and $T_l=T_{nom}$

Fig. 6. DFIM control with faulty current sensors at different operation zones (gain faults of 1% applied after safety operation).

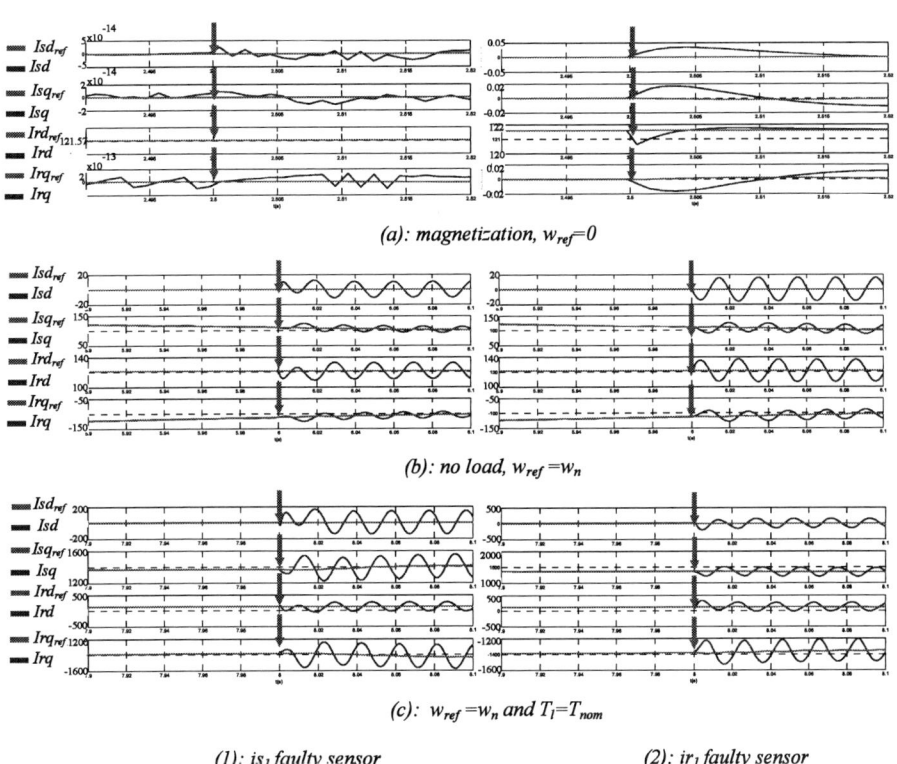

(a): magnetization, $w_{ref}=0$

(b): no load, $w_{ref}=w_n$

(c): $w_{ref}=w_n$ and $T_l=T_{nom}$

(1): is_1 faulty sensor　　　　　　　　*(2): ir_1 faulty sensor*

Fig. 7. DFIM control current responses with faulty current sensor at different DFIM operation zones (gain faults of 1% applied after safety operation).

III. SENSITIVITY STUDY

A. Analysis of control algorithm current sensor dependence

The first use of current measurement is Concordia transformation which provides stator and rotor currents in « α_x,β_x » frame transformation as expressed in (5). Then, the Park transformation is applied as written in (6). This transformation requires exact angle calculation. The synchronous angle θ_s is determined by (7). The rotor current Park transformations are performed using θ_r obtained with the measured electrical angle θ ($\theta_r = \theta_s - \theta$).

$$ix\alpha = \sqrt{\frac{3}{2}}.ix1$$

$$ix\beta = \frac{\sqrt{2}}{2}.ix1 + \sqrt{2}ix2 \tag{5}$$

$$\begin{cases} ixd = ix\alpha.\cos\theta x + ix\beta.\sin\theta x \\ ixq = -ix\alpha.\sin\theta x + ix\beta.\cos\theta x \end{cases} \tag{6}$$

With $x=s$ for stator variables and $x=r$ for rotor ones. The stator flux components needed for synchronous angle are determined by (8).

941

$$\theta_s = \tan^{-1}\frac{\Phi_{s\beta}}{\Phi_{s\alpha}} \qquad (7)$$

$$\begin{cases} \Phi_{s\alpha} = Ls.is\alpha + Msr.[ir\alpha.\cos\theta - ir\beta.\sin\theta] \\ \Phi_{s\beta} = Ls.is\beta + Msr.[ir\alpha.\sin\theta + ir\beta.\cos\theta] \end{cases} \qquad (8)$$

So, high accuracy is required for stator flux components determination since stator angle depends directly on it and rotor angle indirectly. In order to analyze the effect of faulty current sensor on stator flux components, these components are expressed directly with the line currents. So, in their expressions (7), $is\alpha$ and $is\beta$ are replaced by stator line current (8) and $ir\alpha$ and $ir\beta$ are replaced by rotor line current (9) for the study of faulty rotor current sensor effect as below.

The stator flux expressions with stator line currents are as in (9)

$$\begin{cases} \Phi_{s\alpha} = Ls.\sqrt{\frac{3}{2}}is1 + Msr.[ir\alpha.\cos\theta - ir\beta.\sin\theta] \\ \Phi_{s\beta} = Ls.\sqrt{2}\left(\frac{1}{2}.is1 + is2\right) + Msr.[ir\alpha.\sin\theta + ir\beta.\cos\theta] \end{cases} \qquad (9)$$

and d and q stator currents components are calculated:

$$\begin{cases} isd = is1.\left[\sqrt{\frac{3}{2}}.\cos\theta_s + \frac{\sqrt{2}}{2}.\sin\theta_s\right] + is2.\sqrt{2}.\sin\theta_s \\ isq = is1.\left[-\sqrt{\frac{3}{2}}.\sin\theta_s + \frac{\sqrt{2}}{2}.\cos\theta_s\right] + is2.\sqrt{2}.\cos\theta_s \end{cases} \qquad (10)$$

The rotor flux expressions with rotor line currents are as in (11)

$$\begin{cases} \Phi_{s\alpha} = Ls.is\alpha + Msr.\left[ir1.\left(\sqrt{\frac{3}{2}}.\cos\theta - \frac{\sqrt{2}}{2}.\sin\theta\right) - \sqrt{2}.ir2.\sin\theta\right] \\ \Phi_{s\beta} = Ls.is\beta + Msr.\left[ir1.\left(\sqrt{\frac{3}{2}}.\sin\theta + \frac{\sqrt{2}}{2}.\cos\theta\right) + \sqrt{2}.ir2.\cos\theta\right] \end{cases} \qquad (11)$$

and d and q rotor currents components are calculated:

$$\begin{cases} ird = ir1.\left[\sqrt{\frac{3}{2}}.\cos\theta_r + \frac{\sqrt{2}}{2}.\sin\theta_r\right] + ir2.\sqrt{2}.\sin\theta_r \\ irq = ir1.\left[-\sqrt{\frac{3}{2}}.\sin\theta_r + \frac{\sqrt{2}}{2}.\cos\theta_r\right] + ir2.\sqrt{2}.\cos\theta_r \end{cases} \qquad (12)$$

The study of the fault sensor effect is carried out using sensitivity coefficient, the sensitivity of X related to Y is defined as fellows:

$$S_X^Y = \frac{\partial X}{\partial Y}.\frac{Y}{X} \qquad (13)$$

B. Sensitivity to stator currents faults

Calculation of stator flux components sensitivity versus the stator currents leads to the expression resumed in Table I. These expressed are approximated considering each stator flux component equal to the nominal flux value. They are noted $S_{X^a}^Y$. The approximated values are simulated when the control is working with faulty sensor, in the whole range of operating cycle of the DFIM drive. The different coefficients have the same evolution but their magnitude varies, as shown in Fig.8. The higher sensitivity values are obtained when a load torque is applied and at reversal speed direction. The lower values are obtained during magnetization and at nominal rotation speed but without load torque ($Tl=0$).

Beside these last operating conditions, any stator current sensor fault would deteriorate system performances. Gain fault occurring in the $is1$ current sensor leads to bad control results of both the DFIM stator flux and its angular rotation speed inducing important torque oscillations and the drive should be stopped. However, an $is2$ current sensor gain fault induces lower oscillations. This is due to the fact that $\Phi_{s\alpha}$ presents no sensitivity versus this fault. Consequently, the flux orientation is not disturbed by this failure.

According to the sensitivity coefficients expression (Table I), the flux component $\Phi_{s\alpha}$ is not sensitive to any $is2$ sensor fault. As the current isq is the rotation speed controller output and ird is the stator flux controller one (Fig.4), during the magnetization phase, the machine rotation speed is null and the stator current too. So, the control is not sensitive to stator current sensor fault during this working step. In the other working point any stator sensor fault leads to the deterioration of the control system efficiency. These results are validated by simulations shown in Fig.5.

TABLE I. STATOR FLUX SENSITIVITIES EXPRESSIONS RELATED TO STATOR CURRENTS AND THEIR OVERESTIMATIONS

	$S_{\Phi_{s\alpha}}^{is1}$	$S_{\Phi_{s\alpha}}^{is2}$	$S_{\Phi_{s\beta}}^{is1}$	$S_{\Phi_{s\beta}}^{is2}$
S_X^Y	$Ls.\sqrt{\frac{3}{2}}.\frac{is1}{\Phi_{s\alpha}}$	0	$Ls.\frac{\sqrt{2}}{2}.\frac{is1}{\Phi_{s\beta}}$	$Ls.\sqrt{2}.\frac{is2}{\Phi_{s\beta}}$
$S_{X^a}^Y$	$0.0225.\frac{is_{eff}}{\Phi_{Sn}}$	0	$0.013.\frac{is_{eff}}{\Phi_{Sn}}$	$0.026.\frac{is_{eff}}{\Phi_{Sn}}$

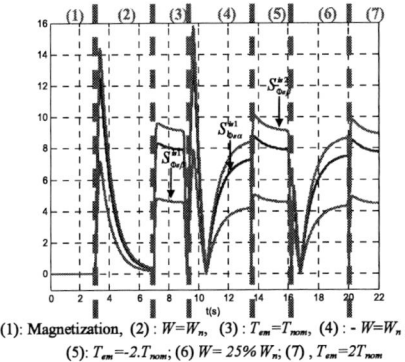

(1): Magnetization, (2) : $W=W_n$, (3) : $T_{em}=T_{nom}$, (4) : - $W=W_n$;
(5): T_{em}=-2.T_{nom}; (6) W= 25% W_n; (7) , T_{em}=2T_{nom}

Fig. 8. Stator flux sensitivity behaviors relatively to stator currents.

C. Sensitivity to rotor currents faults

At the present, the stator flux sensitivity expressions are calculated considering rotor currents faults. The obtained expressions, resumed in Table II, depend on the angle position θ. In magnetization zone, θ is equal to zero, while it varies in the others zones. In order to simplify this study, the θ values leading to a maximum of the sensitivities expressions are found out. The resulting approximations are reported in Table II. The sensitivity coefficients of stator flux components $\Phi_{s\alpha}$ and the flux $\Phi_{s\beta}$ relatively to ir_1 and ir_2 have similar magnitude.

TABLE II. STATOR FLUX SENSITIVITIES EXPRESSIONS RELATED TO ROTOR CURRENTS AND THEIR OVERESTIMATIONS

S_X^Y	$S_{\Phi s\alpha}^{ir1}$	$S_{\Phi s\alpha}^{ir2}$	$S_{\Phi s\beta}^{ir1}$	$S_{\Phi s\beta}^{ir2}$
	$Msr.\left(\sqrt{\dfrac{3}{2}}.\cos\theta - \dfrac{\sqrt{2}}{2}.\sin\theta\right).\dfrac{ir_1}{\Phi s\alpha}$	$-\sqrt{2}.Msr.\sin\theta.\dfrac{ir_2}{\Phi s\alpha}$	$Msr.\left(\sqrt{\dfrac{3}{2}}.\sin\theta + \dfrac{\sqrt{2}}{2}.\cos\theta\right).\dfrac{ir_1}{\Phi s\beta}$	$Msr.\sqrt{2}.\cos\theta.\dfrac{ir_2}{\Phi s\beta}$
S_{Xa}^{is1}	$0.025.\dfrac{ir_{eff}}{\Phi_{Sn}}$	$0.0257.\dfrac{ir_{eff}}{\Phi_{Sn}}$	$0.025.\dfrac{ir_{eff}}{\Phi_{Sn}}$	$0.0257.\dfrac{ir_{eff}}{\Phi_{Sn}}$

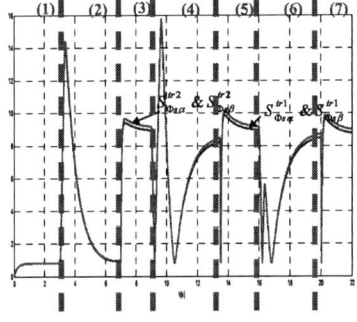

Fig. 9. Stator flux sensitivity behavior versus rotor currents in different zones.

As illustrated in Fig.9, even in magnetization zone, $\Phi_{s\alpha}$ and $\Phi_{s\beta}$ point out sensitivities versus rotor currents. Consequently, any rotor current sensor fault appeared in any operating zone will deteriorate the good system working.

IV. CONCLUSION

In this paper, a sensitivity study of currents input-output decoupling introduced in SFOVC is presented for a DFIM drive system working in motor mode and supplied by two voltages PWM inverters in stator and rotor sides. This study shows that the current sensor sensitivity varies according to the operating zone of the system and also even though there is a stator current measure fault or a rotor current one. Theoretical development of the sensitivity analysis is made and it is validate by the Matlab-Simulink simulation results. The different scenarios of the reconfiguration control algorithms are investigated by the authors in order to ensure the continuous system operating with high performances even with faulty current sensors.

V. ANNEXE

DFIM PARAMETERS

Coefficient	Value	Unit
Stator resistance Rs	0.016	Ω
Inductance Ls	18.39	mH
Mutual inductance Msr	18.14	mH
Friction f	1	Kg.m².s⁻¹
Nominal power P_n	850	KW
Rotor nominal voltage Vr_n	400	V
Rotor nominal current Ir_n	850	A
Stator nominal frequency f_n	50	Hz
rotor resistance Rr	0.012	Ω
Inductance Lr	18.53	mH
Inertia J	60	Kg.m2
Dispersion coefficient σ	3.43%	-
Stator nominal voltage Vsn	400	V
Stator nominal current Isn	850	A
Nominal rotation speed Nn	1420	rpm
Pair poles p	2	

References

[1] S. Bolognani, M.Zigliotto, "Essentials of IM Parameters Measurement for FOC Drives Tuning", *ICEM 2002, Bruges, Belgium*, august 2002, 26-28

[2] S. Skander Mustapha, I. Slama-Belkhodja "Current sensor failure in a DFIG Wind-Turbine: Effect analysis, detection and control reconfiguration", *International Review of Electrical Engineering (I.R.E.E.)*, ISSN1827-660, Vol.1n°3, July-august 2006-pp.426-433

[3] L. Laurila, P. Kurronen, M. Niemela and J. Pyrhonen, "Effect of unideal current sensors in direct torque controlled PMSM drives", NORpi2002, *Nordic Workshop on Power and Industrial Electronics, Stockholm, Sweden*, August 2002.

[4] K. Rothenhagen and F.W. Fuches "Current sensor Fault Detection by bilinear Observer for a Doubly Fed Induction generator" *IECON'2006, the 32nd Conference on the IEEE Industrial Electronics Society, Paris, France*, November 6-10, 2006

[5] I. Bahri, M-W. Naouar, I. Slama-Belkhodja, E. Monmasson "FPGA-based FDI of faulty current sensor in current controlled PWM converters", *IEEE-EUROCON 2007, Warsaw , Poland*, September 9-12, 200,

[6] L Galotto, B.K. Bose, C.Leite, J.O.Pereira Pinto, L.E.Borges da Silva, G. Lambert-Torres, "Auto-Associative Neural Network Based Sensor Drift Compensation in Indirect Vector Controlled Drive System" , *The 33rd Annual Conference of the IEEE Industrial Electronics Society (IECON)*, , *Taipei, Taiwan* Nov. 5-8, 2007

[7] Han-Su Jung, Jang-Mok Kim, Cheul-U Kim, Cheol Choi, "Diminution of Current Measurement Error For Vector Controlled AC Motor Drives", *Industry Applications, Electric Machines and Drives, 2005 IEEE International Conference*, 15-18 May 2005 Page(s): 551 - 557

[8] C. Klumpner, F.Blaabjerg, "Measuring with only one Current Sensor all the Load Currents in a Multiple Drive System based on a Two-Stage Direct Power Electronic Conversion Topology" , *EPE 2003 – Toulouse*

[9] M.Abedllatif, I.Slama-Belkhodja, M.Pietrzak-David " Reconfiguration de la commande en présence d'un défaut capteur de courant " *EF'07, Conférence Electrotechnique du Futur, EF 2007 - Toulouse, France,*6-7 Septembre 2007

[10] G. Salloum, R. Ghosn, M. Pietrzak-David, B. de Fornel, " Robustness of currents input-output decoupling in vector control of a doubly fed induction machine ", *EPE-PEMC, Riga Latvia* -2,4 Septembre 2004-

[11] S. Khojet El Khil, I.Slama-Belkhodja, M. Pietrzak-David, B. de Fornel, "Power Distribution Law in a Doubly Fed Induction Machine", *Special Issue of the Transactions of IMACS on Mathematics and Computers in Simulation*. Vol.71 -issue 4-6 ISSN 0378-4754-pp.360-368, Edition Elsevier, 2006

Extended Back EMF model for PM synchronous machines with different inductances in d- and q-axis

Andreas Eilenberger*, Manfred Schroedl†

*Institute of Electrical Drives and Machines, Vienna, Austria, e-mail: *andreas.eilenberger@tuwien.ac.at*
†Institute of Electrical Drives and Machines, Vienna, Austria, e-mail: *manfred.schroedl@tuwien.ac.at*

Abstract—This paper discusses a reluctance-dependent back electromotive force (EMF) model for encoderless vector driven permanent magnet synchronous machines (PMSM) compared to well known, non reluctance-dependent back EMF models. Established PMSM could have considerable varieties in direct and quadrature inductances. So it makes sense to consider this behaviour in an extended back EMF model. On closer examination it will turn out that a reluctance dependent EMF model has a better behaviour at low speed as the standard model. Also the derivation of the extended model will be illustrated. Furthermore some simulation results and practical measurements will be discussed.

Keywords—Electrical Drive, Permanent magnet motor, Synchronous motor, Reluctance drive, Sensorless control.

I. INTRODUCTION

There are a lot of publications on the field of sensorless driven PMSM, which means replacing the mechanical position encoder by mathematical models. Low speed models combined with a back EMF model for higher speed enables an operation in the whole speed range. The lower the speed the more sensitive the parameters of the EMF model are. Such parameters are stator inductance and stator resistance. In most cases there is a significant difference between direct and quadrature inductance. Hence, this paper describes an extended EMF model with regard to the saliency to achieve highly dynamical operation in a wide speed range, especially at low speed.

II. THE CLASSICAL BACK EMF MODEL

In the subsequent analyses for determining the operational behaviour of the given machine, steady-state operation in conjunction with constant rotor angular velocity ω_K is supposed. All data are given in normalized values. The following calculations deal with the classical back EMF model [1],[2] and [6]. The stator flux linkage ψ_s is derived from the stator voltage equation

$$\underline{u}_s = r_s \cdot \underline{i}_s + \frac{d}{d\tau} \underline{\psi}_s + \jmath \cdot \omega_K \cdot \underline{\psi}_s \quad (1)$$

by integration

$$\underline{\psi}_s(\tau) = \int \left[\underline{u}_s(\tau) - r_s \cdot \underline{i}_s(\tau) \right] d\tau.$$

Furthermore flux linkage $\underline{\psi}_m$ due to permanent magnets is calculated from stator flux linkage $\underline{\psi}_s$

$$\underline{\psi}_s = l_s \cdot \underline{i}_s + \underline{\psi}_m \quad (2)$$

and yields in the $\alpha\beta$ stator-oriented reference frame to equations

$$\psi_{m\alpha}(\tau) = \int \left[u_{s\alpha}(\tau) - r_s \cdot i_{s\alpha}(\tau) \right] d\tau - l_s \cdot i_{s\alpha}(\tau)$$

$$\psi_{m\beta}(\tau) = \int \left[u_{s\beta}(\tau) - r_s \cdot i_{s\beta}(\tau) \right] d\tau - l_s \cdot i_{s\beta}(\tau).$$

Normally, the integration is stabilized by a certain feedback. The argument of flux linkage vector ψ_m is the searched rotor position γ

$$\gamma = \arg[\underline{\psi}_m] = \arctan \left[\frac{\psi_{m\beta}}{\psi_{m\alpha}} \right] = \quad (3)$$

$$= \arctan \left\{ \frac{\int \left[u_{s\alpha}(\tau) - r_s i_{s\alpha}(\tau) \right] d\tau - l_s i_{s\alpha}(\tau)}{\int \left[u_{s\beta}(\tau) - r_s i_{s\beta}(\tau) \right] d\tau - l_s i_{s\beta}(\tau)} \right\}.$$

III. EXTENDED BACK EMF MODEL

Using well known two-axis theory [4], [5] the flux linkage $\underline{\psi}_s$ in the dq rotor-oriented reference frame is given by

$$\underline{\psi}_{s,dq} = \underline{\psi}_{m,dq} + l_d i_d + \jmath l_q i_q \quad | \cdot e^{\jmath\gamma} \quad (4)$$

$$\underline{\psi}_{s,\alpha\beta} = \underline{\psi}_{m,\alpha\beta} + (l_d i_d + \jmath l_q i_q) \cdot (\cos\gamma + \jmath\sin\gamma)$$

$$\underline{\psi}_{s,\alpha\beta} = \underline{\psi}_{m,\alpha\beta} + l_d i_d \cos\gamma - l_q i_q \sin\gamma$$
$$+ \jmath \left(l_q i_q \cos\gamma + l_d i_d \sin\gamma \right)$$

which yields to

$$\psi_{m\alpha} = \psi_{s\alpha} - l_d i_d \cos\gamma + l_q i_q \sin\gamma \quad (5)$$

$$\psi_{m\beta} = \psi_{s\beta} - l_q i_q \cos\gamma - l_d i_d \sin\gamma. \quad (6)$$

The rotor position γ follows with simplified notation to

$$\arctan \left\{ \frac{\int (u_\alpha - r_s i_\alpha) d\tau - l_d i_d \cos\gamma + l_q i_q \sin\gamma}{\int (u_\beta - r_s i_\beta) d\tau - l_q i_q \cos\gamma - l_d i_d \sin\gamma} \right\}. \quad (7)$$

The block diagram of the realized extended EMF model is shown in fig. 3. Furthermore an extended observer structure as discussed in [3] is implemented.

978-1-4244-1741-4/08/$25.00 ©2008 IEEE

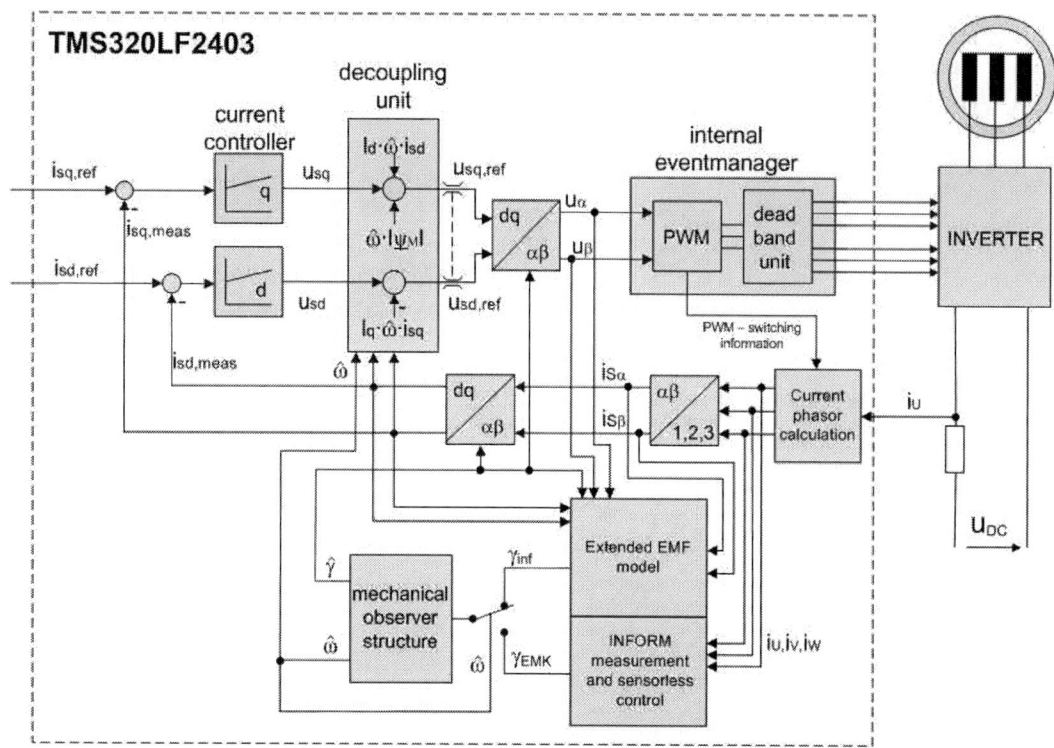

Fig. 1. Structure of the realized encoderless drive with an outer-rotor PMSM

TABLE I
CHARACTERISTICS OF PMSM WITH OUTER ROTOR

continuous torque:	24 Nm
steady state current:	19 A rms
reference voltage level:	18.5 V
battery voltage:	46 V
rated output power:	750 W
rated velocity:	300 rpm
number of pole pairs:	p=12
stator resistance:	$r_s = 0.2$
direct inductance:	$l_d \approx 0.11$
quadrature inductance:	$l_q \approx 0.16$

IV. EXPERIMENTAL SETUP

For verification of the novel back EMF model, an outer-rotor PMSM was used, shown in fig. 2. Furthermore a voltage source inverter with only a DC-link measurement is used. A short overview of the given hardware is shown in figure 1. The subsequent table I specifies the characteristics of the used PMSM.

Following fig. 2 shows the used outer-rotor PMSM with assembled encoder and belt to load-machine.

V. SIMULATION RESULTS

For simulation of both back EMF models the simulation tool MATLAB-Simulink is used. In the following, two scenarios were simulated. First scenario shows results of the calculated rotor position with the well known EMF model with constant inductances and extended

Fig. 2. Picture of the used outer-rotor PMSM with encoder and load-machine

EMF model at constant quadrature-axis current $i_{q,ref}$. To verify results, the reference rotor position is mapped. This simulation starts at standstill and accelerates rotor speed slowly. As be can seen in fig. 4 the extended back EMF model provides a useful angular rotor position at first while increasing the rotor speed. At higher speed, the standard back EMF model also yields an angular rotor position. The significant offset between reference and calculated EMF angular rotor position is a result of the integrator feedback. An offset on voltage space phasor $u_{s\alpha}$ is used in the simulation representing voltage measurement errors of the experimental setup.

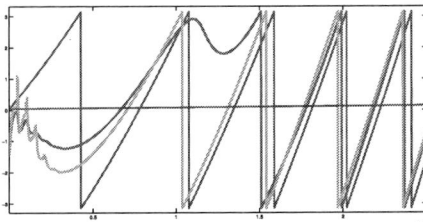

Fig. 4. Characteristics of calculated rotor position with standard back EMF model (blue) and extended back EMF model (yellow), furthermore encoder position (red) and rising rotor velocity (green) with $i_{q,ref} = 0.8$.

The second scenario shows the behaviour of the EMF models (fig. 5) at a constant rotor speed $\omega = 0.2$ with a rising quadrature current $i_{q,ref}$.

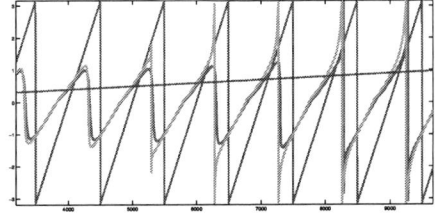

Fig. 5. Characteristics of calculated rotor position with standard back EMF model (blue) and extended back EMF model (yellow), encoder position (red) and the rising quadrature current $i_{q,ref}$(green) with constant $\omega = 0.2$.

VI. MEASUREMENTS RESULTS

The measured results verify the results of the two simulated scenarios. First diagram (fig. 6) depicts the measurement according to fig. 4 with increasing rotor speed. Due to an offset error of the measured voltage space phasor $u_{s,\alpha\beta}$ the calculated rotor position with standard back EMF model has a visible positive offset.

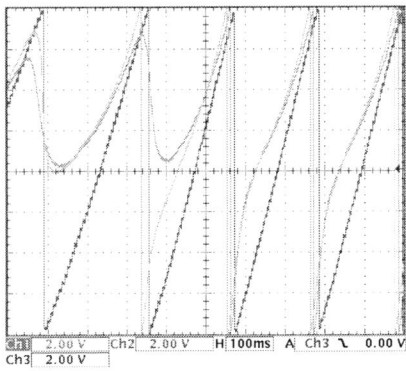

Fig. 6. Measured characteristics with a rising rotor speed and constant $i_{q,ref} = 0.5$, Ch1: extended model (22.5/Div.), Ch2: novel model (22.5/Div.), Ch3: encoder position (22.5/Div.)

Subsequent fig. 7 depicts the calculated angular rotor position with the mentioned two back EMF models at constant rotor speed and quadrature current. The extended model shows a better behaviour as the novel back EMF model.

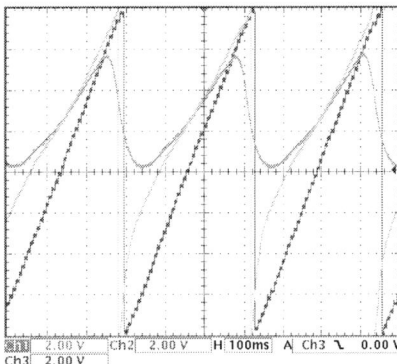

Fig. 7. Measured characteristics with $i_{q,ref} = 1$ and $\omega \approx 0.15$, Ch1: extended model (22.5/Div.), Ch2: novel model (22.5/Div.), Ch3: encoder position (22.5/Div.)

VII. CONCLUSION AND OUTLOOK

This paper demonstrates an extended back EMF model with improved low-speed properties compared to the well known back EMF model with constant inductances, both with an observer structure. It is also shown that measured results with the experimental setup shows expected simulation results. Furthermore, the extended EMF model was successfully implemented in a drive for light vehicles.

ACKNOWLEDGMENT

The authors are very much indebted to the Austrian Science Foundation (FWF) which generously supports the work at the Institute of Electrical Drives and Machines at the Vienna University of Technology.

REFERENCES

[1] Schroedl M., Hofer M., Staffler W. "Combining INFORM method, Voltage model and mechanical observer for sensorless control of PM Synchronous Motors in the whole speed range including standstill", *PCIM, Nuernberg (Germany)*, (2006).

[2] Rieder U.-H.: "Optimierung der sensorlosen Regelung von permanentmagneterregten Aussenlaeufer-Synchronmaschinen", *PhD thesis*, Vienna University of Technology, 2005.

[3] Schroedl M., Hofer M., Staffler W. "Sensorless control of PM Synchronous Motors in the whole speed range including standstill using a combined INFORM/EMF model", *EPE-PEMC, Portoroz (Slovenia)*, (2006).

[4] Park R.H., "Two-Reaction Theory of synchronous machines, Generalized Method of Analysis", *AIEE Trans.*, pt. I, Vol.48, pp. 716, July 1929.

[5] Jones C. V., "The unified theory of electrical machines", *Butterworth & Co. (Publishers) Ltd.*, pt. II, London, 1967.

[6] T.M. Wolbank, M.A. Vogelsberger, R. Stumberger, "Adaptive Flux model for commissioning of signal injection based zero speed sensorless flux control of induction machines" *Proceedings of IEEE International Conference on Power Electronics and Drive Systems*, Bankok (2007)

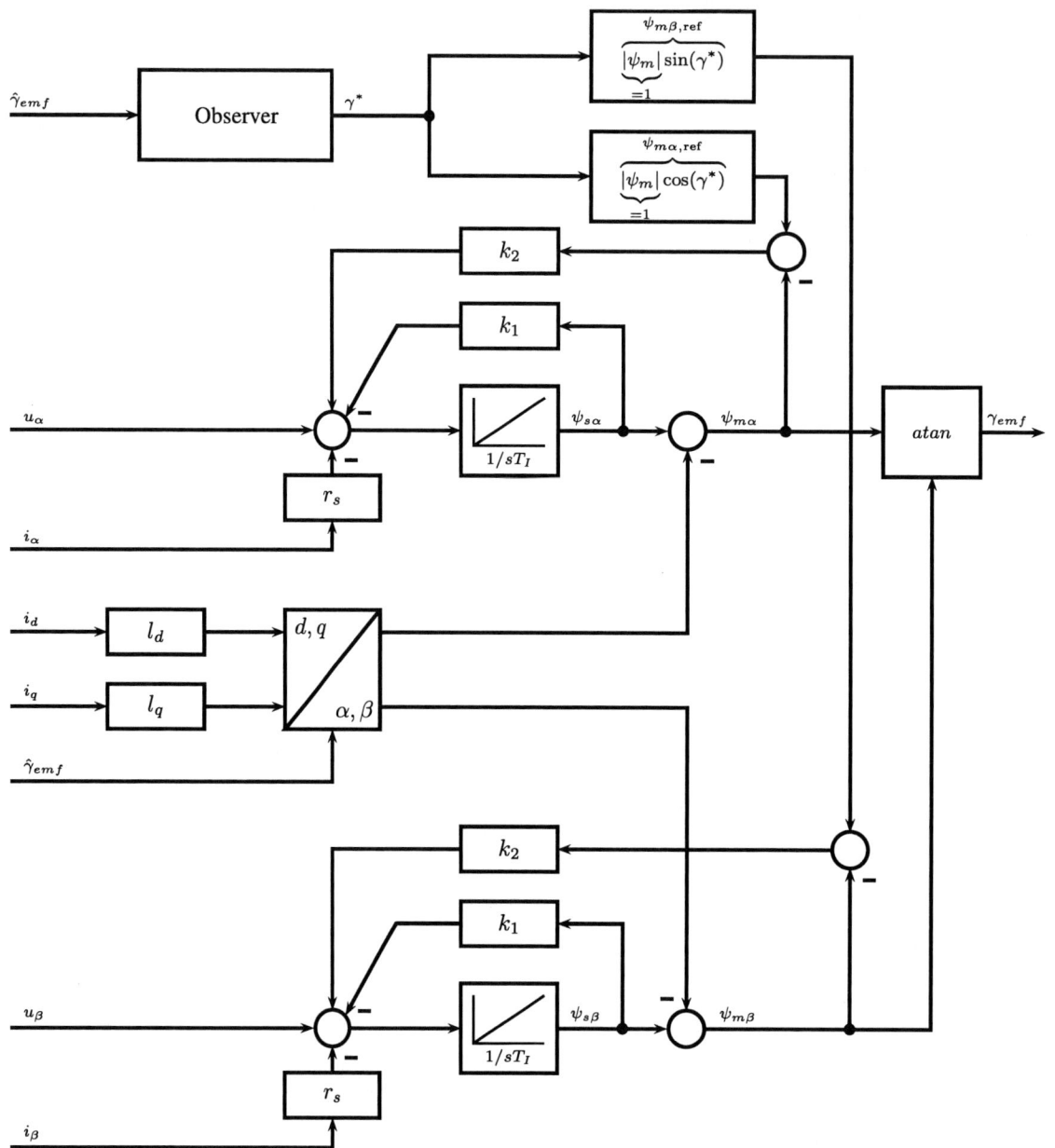

Fig. 3.   Schematic of the realized extended back EMF model with extended observer structure [3]

Gait generation of a two-legged robot by using adaptive network based fuzzy logic control

Umit ONEN[1], Mete KALYONCU[2], Mustafa TINKIR[3], Fatih M. BOTSALI[4]
Department of Mechanical Engineering, Faculty of Engineering and Architecture, University of Selçuk
Alaeddin Keykubat Campus, 42079 Konya, TURKEY
[1]uonen@selcuk.edu.tr, [2]mkalyoncu@selcuk.edu.tr, [3]mtinkir@selcuk.edu.tr, [4]fbotsali@selcuk.edu.tr

Abstract — In this paper, a control strategy is proposed for gait generation for a two-legged humanoid robot. The two-legged robot is assumed as a 3-dimensional robot with 5-links. Gait generation is performed by assuming motions in the saggital and lateral planes. Dynamic model of the robot is obtained by using MATLAB®/SimMechanics Toolbox. A fuzzy logic controller (FLC) is established for gait generation. The rule-base of the controller is optimized offline by using artificial neural network (NN). The neural networks are trained by using the reference joint trajectories obtained from clinical gait analysis (CGA) [13].

Keywords — Robotics, modelling, fuzzy logic based control, motion control, neural network, humanoid robot, gait generation

I. INTRODUCTION

The design and control of bipedal robots are among the more challenging topics in the field of robotics and were treated by a great number of researchers since many years. The bipedal humanoid robots are intended to replace human beings for interventions in hostile environments or to help human beings in the daily tasks. In addition to problems related to autonomy, determination of the essential locomotion task is also a big challenge. Although some advanced prototypes like ASIMO [1], and HPR-2 [2] are produced, performance of these prototypes are still far from the human locomotion process. Among the different kinds of human movements, bipedal walking has absorbed a great attention due to its different nature. More than 200 degrees of freedom, unstable nature, and periodicity are the main features of walking. Generally, keeping the balance for upward standing and walking control are two different aspects of human locomotion, which have to be considered in any control strategy which is proposed to model this control system. In fact, in order to get an efficient walking gait for biped robots, it is necessary to solve some problems. First of all, dynamic stability of the robot has to be maintained. Computing time must be decreased in order to obtain an efficient control algorithm. Adaptive control strategies should make it possible to modify the gait generation process with regard to the environment. Lastly, energy cost must be minimized for improving autonomy.

II. DYNAMIC MODEL OF TWO-LEGGED ROBOT

The model of the two-legged robot is obtained by using MATLAB®/SimMechanics Toolbox. The robot model

consists of 5 limbs (2 thigh, 2 shank and a trunk) attached to each other with revolute joints.

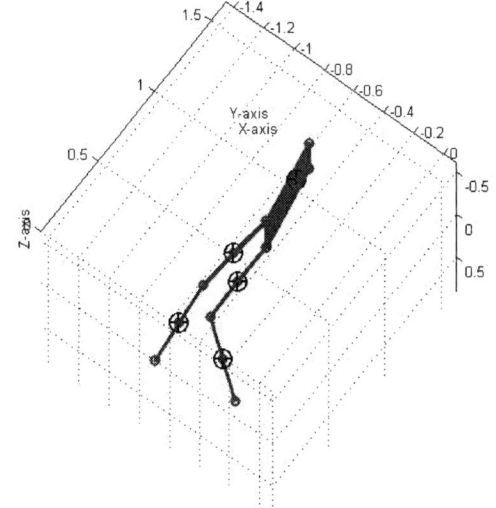

Fig. 1. Two-legged robot model

Joints are located at the hip, at the knee and at the spine. It is assumed that thighs and shanks are rotating in the saggital plane and trunk is rotating in the lateral plane. The model of the two-legged robot is shown in Fig. 1. Physical characteristics of the members (masses and lengths of the limbs) are summarized in Table 1.

TABLE I.
MASSES AND LENGTHS OF THE LIMBS

Limb	Weigh (kg)	Length (m)
Trunk	30	0.5
Thigh	8	0.5
Shank	4	0.5

III. ADAPTIVE NETWORK BASED FUZZY LOGIC CONTROLLER DESIGN (ANFLC)

In this study, adaptive network based fuzzy logic controllers are applied for motion control of two-legged humanoid robots. Established adaptive network based fuzzy inference system (ANFIS) uses a hybrid learning algorithm to identify parameters of Sugeno-type fuzzy inference system. It applies a combination of the least-squares method and the back propagation gradient descent method for training fuzzy

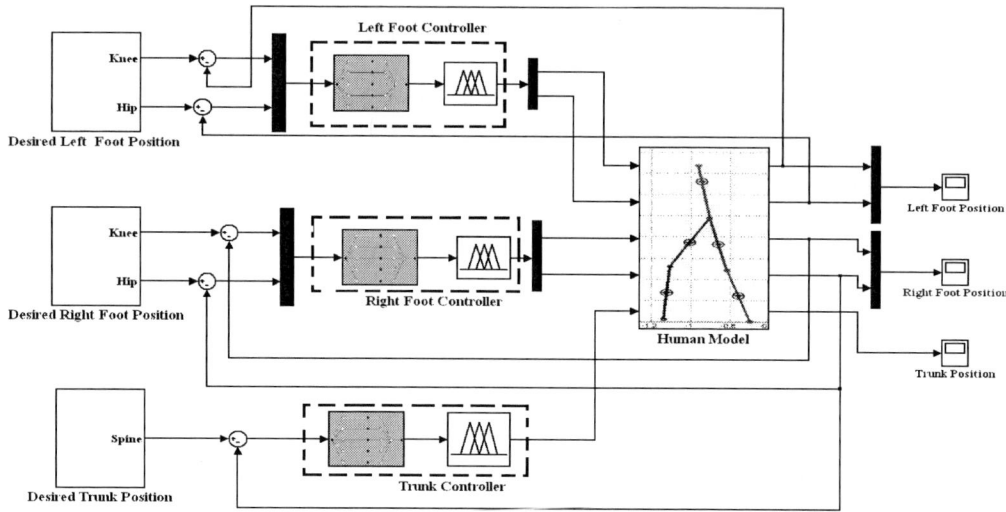

Fig. 2. Adaptive Network Based Fuzzy Logic Controller (ANFLC)

inference system (FIS) membership function parameters to emulate a given training data set. The control algorithm proposed to be used in control of the two legged-robot is shown in Fig. 2. Three fuzzy logic controllers (FLC) are designed for motion control of the left foot, right foot and trunk. Inputs of the controllers are the angles of each joint (hips, knees and spine). Left and right foot fuzzy logic controllers have two inputs and two outputs so they are called as multi inputs multi outputs (MIMO) controllers. Trunk fuzzy logic controller is called as single input single output (SISO) controller.

Before adaptation to the system, the fuzzy logic controller rule-base is optimized offline by using artificial neural network (ANN). These controllers are designed by training and checking data sets that are obtained from PID control of the system.

First of all hierarchical PID controllers applied to the two legged-robot and then their inputs and outputs data are obtained to set up the adaptive network based fuzzy inference system (ANFIS). After training the network, fuzzy inference system structure is established. The network base is trained offline. Number of membership functions and type of membership functions are selected. A few tests are done to obtain appropriate fuzzy inference system for motion control of the two legged-robot system. Control torqueses are evaluated for left foot, right foot and trunk controllers separately. In this way, designing of the controller becomes easier.

A. Adaptive Neural Network

Neural networks are composed of simple elements operating in parallel. These elements are inspired from biological nervous systems. As in nature, the network function is determined largely by the connections between elements. One can train a neural network to perform a particular function by adjusting the values of the connections (weights) between elements. Commonly,

neural networks are adjusted, or trained, so that a particular input leads to a specific target output. Such a situation is shown in Fig. 3. The network is adjusted by comparing the output and the target, until the network output matches the target.

Typically, many such input/target pairs are needed to train a network. Neural networks have been trained to perform complex functions in various fields, including pattern recognition, identification, classification, speech, vision, and control systems.

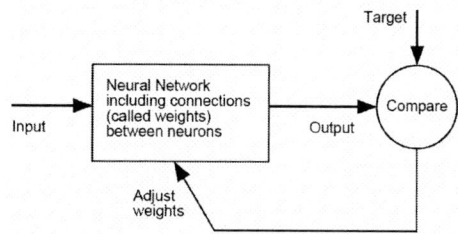

Fig. 3. Neural network structure

ANFIS architecture consists of five layers with the output of the nodes in each respective layer is represented by $O_{i,l}$ where i is the ith node of layer l.

Layer 1: Generate the membership grades

$$O_{i,l} = \mu_{A_i}(x), \ i = 1, 2 \tag{1}$$

or

$$O_{i,l} = \mu_{B_{i,2}}(y), \ i = 3, 4 \tag{2}$$

Where x (or y) is the input to the node and A_i (or B_{i-2}) is the fuzzy set associated with this node.

950

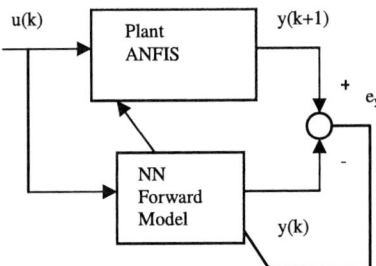

Fig. 4. Training of NN forward model

Layer 2: Generate the firing strengths by multiplying the incoming signals and outputs the t-norm operator result, e.g.

$$O_{2,i} = w_i = \mu_{Ai}(x) \times \mu_{Bi}(y) , \ i = 1,2 \tag{3}$$

Layer 3: Normalize the firing strengths

$$O_{3,i} = \overline{w}_i = \frac{w_i}{w_1 + w_2} , \ i = 1,2 \tag{4}$$

Layer 4: Calculate rule outputs based on the consequent parameters $\{p_i, q_i, r_i\}$

$$O_{4,i} = \overline{w}_i f_i = \overline{w}_i (p_i x + q_i y + r_i) \tag{5}$$

Layer 5: Computes the overall outputs as the summation of incoming signals

$$O_{5,l} = \sum_i \overline{w}_i f_i = \frac{\sum_i w_i f}{\sum_i w_i} \tag{6}$$

We follow these steps for creating of ANFIS controller shown in below:

- 600 training and checking data (from hierarchical PID control) are obtained for neural network of ANFIS control.
- The number and type of membership functions are determined.
- Hybrid learning algorithm and 40 epochs is chosen to train network.

B. Hybrid Learning Algorithm

In this study, forward hybrid learning algorithm is used for neural network part of ANFIS controllers shown in Fig. 4. Nearly 40 epochs later error rate close to 10^{-1}. In the forward pass of the hybrid learning algorithm, node outputs go forward until layer 4 and the consequent are identified by the least-squares method. When the values of the premise parameters are fixed, the overall output can be expressed as a linear combination of the consequent parameters

$$
\begin{aligned}
f &= \frac{w_1}{w_1 + w_2} f_1 + \frac{w_2}{w_1 + w_2} f_2 \\
&= \overline{w}_1 f_1 + \overline{w}_2 f_2 \\
&= (\overline{w}_1 x) p_1 + (\overline{w}_1 y) q_1 + (\overline{w}_1) r_1 + (\overline{w}_2 x) p_2 + (\overline{w}_1 y) q_2 + (\overline{w}_1) r_2
\end{aligned} \tag{7}
$$

which is linear in the consequent parameters p_1, q_1, r_1, p_2, q_2 and r_2,

$$f = XW \tag{8}$$

If X matrix is invertible then

$$W = X^{-1} f \tag{9}$$

otherwise a pseudo-inverse is used to solve for W.

$$W = (X^T X)^{-1} X^T f \tag{10}$$

Due to the adaptive capability of ANFIS, their applications to adaptive and learning control are immediate. The most common design techniques for ANFIS controllers are derived directly from neural networks counterpart methodologies. However certain design techniques apply exclusively to ANFIS.

C. Left Foot Adaptive Network Based Fuzzy Logic Controller

Left foot adaptive network based fuzzy logic controller is designed using the rotation (θ) of knee and hip to implement the control methodology. Hence left foot control torque is suitable output in the left foot fuzzy logic control. Rotation of the knee and hip is about applied control torque. In left foot adaptive network based fuzzy logic control multi inputs and multi outputs (MIMO) occur.

D. Right Foot and Trunk Adaptive Network Based Fuzzy Logic Controller

The right foot and the trunk adaptive network based fuzzy logic controllers are designed by using the rotation (θ) of right knee, hip and spine as left foot controller design. These controllers' outputs are control torques applied to knee, hip and spine. Only one difference between these controllers is that trunk controller has single input and single output (SISO).

E. Membership Functions and Fuzzy Rules

Two of the difficulties with the design or any fuzzy control systems are the shape of the membership functions and the choice of the fuzzy rules. In fact, the decision-making logic is the way in which the controller output is generated. It uses the input fuzzy sets, and the decision is taking according to the values of the inputs. Moreover, the knowledge base comprises knowledge of application domain and the attendant control goals. It consists of a database and a fuzzy control rule base.

A control system is said to be an adaptive fuzzy control system if either a set of fuzzy rules are used to modify or change an existing fuzzy controller's architecture, i.e., membership functions and/or rules. The fuzzification uses membership functions to determine the degree of inputs. The aim of control action is to minimize the rotation error. The higher the error, the higher the control input. However, the rate of change of error also affects the value of the control input. In left foot adaptive network based fuzzy logic controller, error is used in control rules as linguistic variables.

This is defined as:

$$U_{\theta e}=\text{Angular Displ. Error} = (\text{Desired} - \text{Actual}) \qquad (11)$$

In adaptive network based fuzzy inference system (ANFIS), number and type of membership functions are created by user. Then fuzzy inference system rule base is obtained automatically by ANFIS.

Left foot adaptive network based fuzzy logic controller has five and four membership functions for each input as hip and knee errors. Triangular type membership functions are used for fuzzy inputs in fuzzification process of the left foot controller. Sugeno type fuzzy inference system is adopted so these controllers' outputs are not fuzzy. ANFIS determine these outputs by given training and checking data sets. Figs. 5 shows that membership functions and control surfaces of left foot adaptive network based fuzzy logic controller.

Right foot adaptive network based fuzzy logic controller has five and three membership functions for each inputs as hip and knee errors like left foot controller. Triangular type membership functions are used for fuzzy inputs in fuzzification process of right foot controller. Figs. 6 shows that membership functions and control surfaces of right foot adaptive network based fuzzy logic controller.

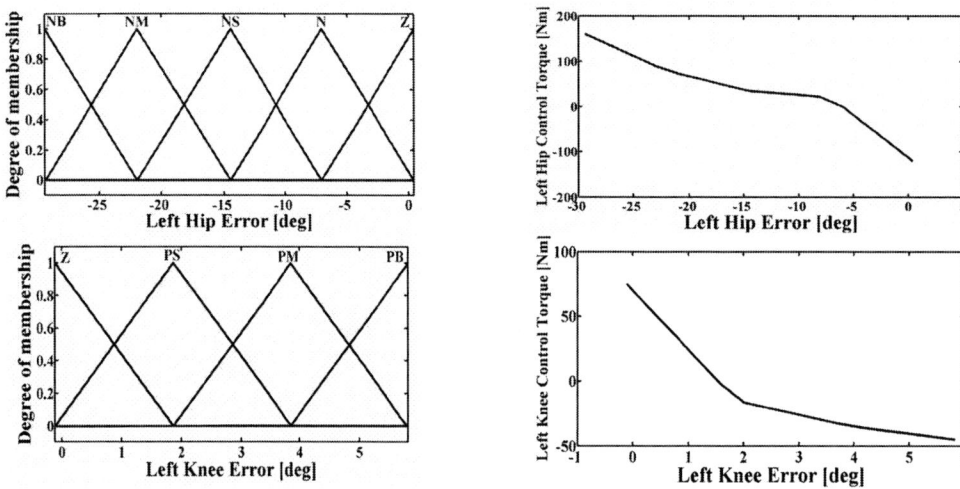

Fig. 5. Membership functions and control torques of left foot adaptive network based fuzzy logic controller

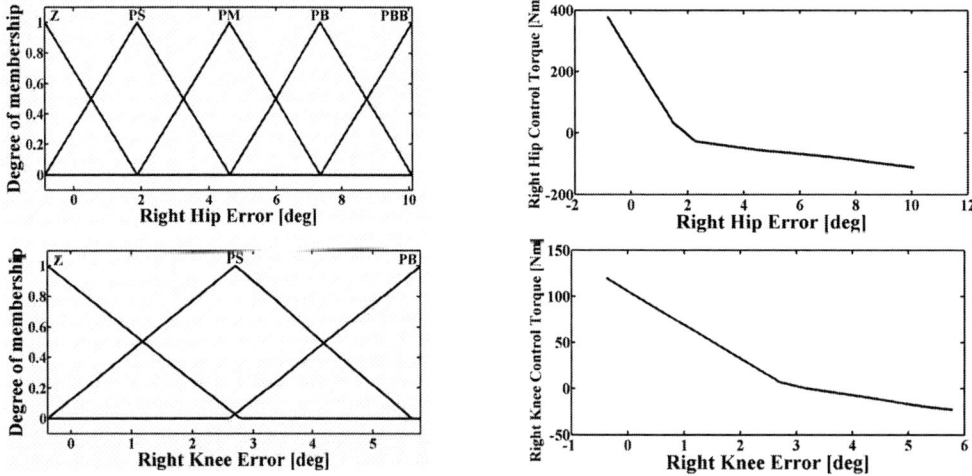

Fig. 6. Membership functions and control torques of right foot adaptive network based fuzzy logic controller.

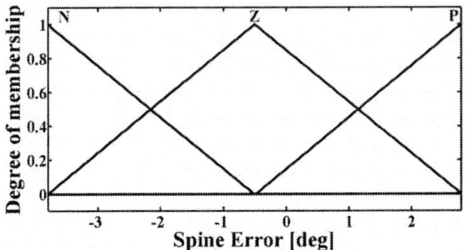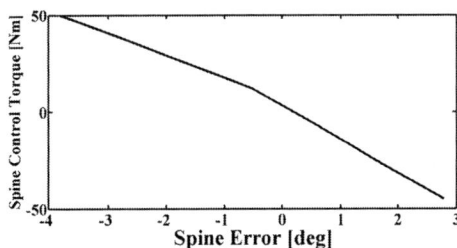

Fig. 7. Membership functions and control torques of trunk adaptive network based fuzzy logic controller.

Trunk adaptive network based fuzzy logic controller has three membership functions for spine error. Triangular type membership functions are used for fuzzy inputs in fuzzification process of trunk controller. Figs. 7 shows that membership functions and control surfaces of trunk adaptive network based fuzzy logic controller.

F. Adaptive Network Based Fuzzy Logic Controller Rule Bases

The fuzzy rule base can be illustrated as a look-up table. Left and right foot adaptive network based fuzzy logic controllers' look-up tables are divided in two parts as vectors of five, four and three rows and one column. Left foot and right foot controllers rule bases are constituted nine and eight rules respectively shown in Table 2-5. Trunk controller has three rules as shown in Table 6. There are nine fuzzy membership functions, that correspond to, *negative(N), negative small(NS), negative medium (NM), negative big(NB), zero (Z), positive small (PS), positive medium (PM) positive big (PB)* and *positive big big (PBB)* values of hip and knee errors. The control laws of the fuzzy controller consist of a complete set of these control rules defined as:

If $\theta_{Lhiperror}$ = Z then T_{Lhip} = Constant Torque 1,

If $\theta_{Rhiperror}$ =PS then T_{Rhip} =Constant Torque 2,

...

TABLE II.
LEFT FOOT ADAPTIVE NETWORK BASED FUZZY LOGIC
CONTROLLER RULE BASE FOR HIP ERROR

	Left Foot Hip Error($\theta_{Lhiperror}$)				
	Z	N	NS	NM	NB
Control Torque (T_{Lhip})	-120	20	35	78	160

TABLE III.
LEFT FOOT ADAPTIVE NETWORK BASED FUZZY LOGIC
CONTROLLER RULE BASE FOR KNEE ERROR

	Left Foot Knee Error($\theta_{Lkneeerror}$)			
	Z	PS	PM	PB
Control Torque (T_{Lknee})	75	-15	-34	-45

TABLE IV.
RIGHT FOOT ADAPTIVE NETWORK BASED FUZZY LOGIC
CONTROLLER RULE BASE FOR HIP ERROR

	Right Foot Hip Error($\theta_{Rhiperror}$)				
	Z	PS	PM	PB	PBB
Control Torque (T_{Rhip})	380	-25	-56	-80	-112

TABLE V.
RIGHT FOOT ADAPTIVE NETWORK BASED FUZZY LOGIC
CONTROLLER RULE BASE FOR KNEE ERROR

	Right Foot Knee Error($\theta_{Rkneeerror}$)		
	Z	PS	PB
Control Torque (T_{Rknee})	120	5	-23

TABLE VI.
TRUNK ADAPTIVE NETWORK BASED FUZZY LOGIC
CONTROLLER RULE BASE FOR KNEE ERROR

	Spine Error($\theta_{Spineerror}$)		
	N	Z	P
Control Torque (T_{Spine})	50	12	-45

G. Defuzzification Process

Once the fuzzy controller is activated, rule evaluation is performed and all the true rules are fired. Utilizing the true output membership functions, defuzzification is then applied to determine a crisp control action. The defuzzification is to transform the control signal into an exact control output. For Sugeno-style inference, we have to choose between wtaver (weighted average) or wtsum (weighted sum) defuzzification method. In defuzzification process of the adaptive network based fuzzy logic control, the method of weighted average (wtaver) is used:

$$u = \frac{\sum_{i=1}^{N} w_i z_i}{\sum_{i=1}^{N} w_i} \tag{12}$$

953

IV. RESULTS AND DISCUSSION

In this study, a control strategy is proposed for gait generation for a two-legged humanoid robot. The robot is assumed as a 3-dimensional robot with 5-links moving in the saggital and lateral planes. Dynamic model of the robot is obtained by using MATLAB®/SimMechanics Toolbox. A fuzzy logic controller (FLC) is established for gait generation. The rule-base of the controller is optimized offline by using artificial neural network (NN).

Effectiveness of the proposed algorithm is tested through computer simulations. The successive movements of the robot for a walking velocity of 1 m/s are given in Fig. 8. The reference and measured joint trajectories when the robot is walking with a speed of 1 m/s are shown in Figs. 9-13. The control torques for the left leg, right leg and the spine are given in Figs. 14-18.

Obtained robot model can provide successive walking movements very similar to human as can be seen from Fig. 8. It can also be seen from Figs. 9-13 that the proposed control algorithm can control the joint trajectories effectively for walking.

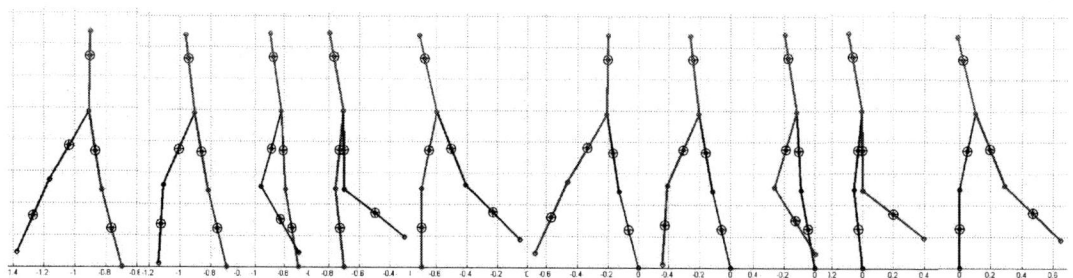

Fig. 8. The successive movements of the robot

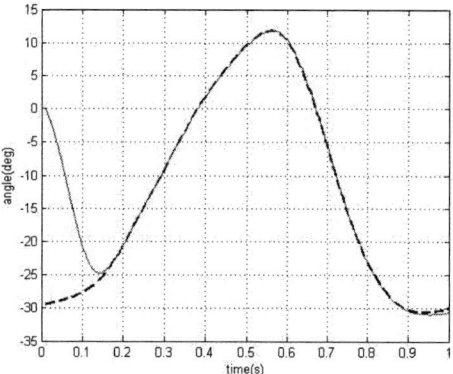

Fig. 9. Left hip angle Reference trajectory (dashed),
ANFLC trajectory (solid)

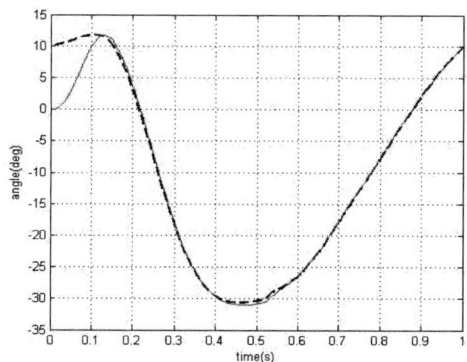

Fig. 11. Right hip angle Reference trajectory (dashed),
ANFLC trajectory (solid)

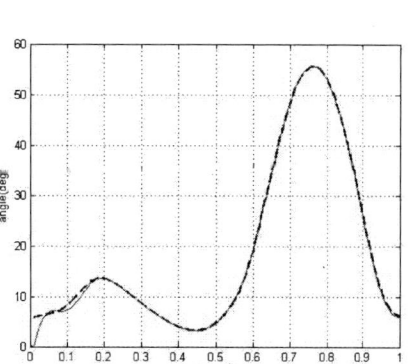

Fig. 10. Left knee angle Reference trajectory (dashed),
ANFLC trajectory (solid)

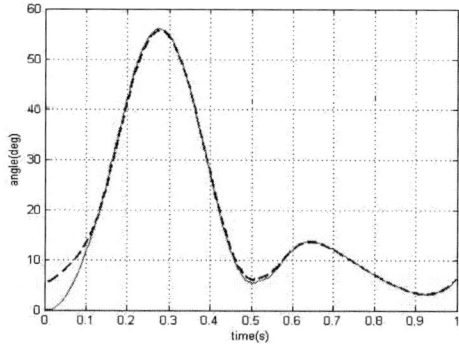

Fig. 12. Right knee angle Reference trajectory (dashed),
ANFLC trajectory (solid)

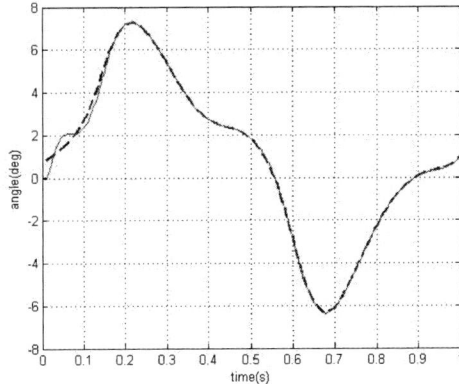

Fig. 13. Trunk angle Reference trajectory (dashed) ANFLC trajectory (solid)

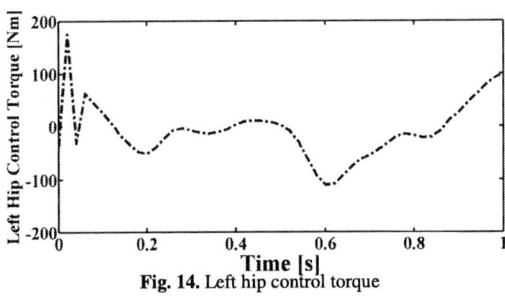

Fig. 14. Left hip control torque

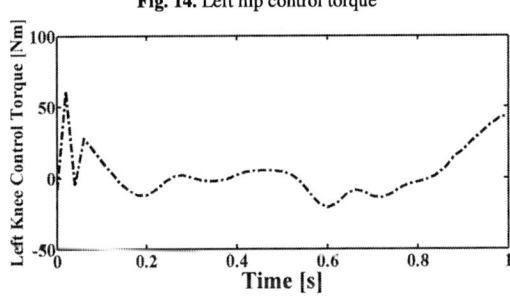

Fig. 15. Left knee control torque

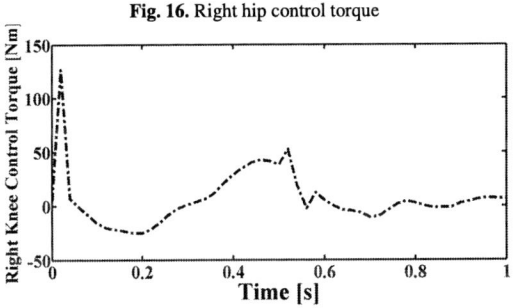

Fig. 16. Right hip control torque

Fig. 17. Right knee control torque

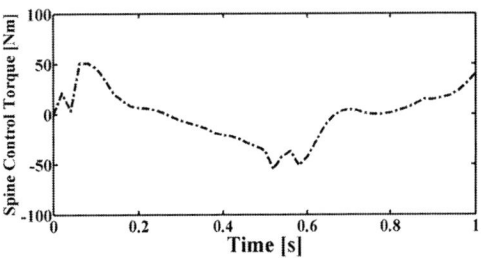

Fig. 18. Spine control torque

REFERENCES

[1] Y. Sakagami, et al., The intelligent ASIMO: system overview and integration, Proc. IEEE Conf. Robot. Autom. (2002) 2478-2483.

[2] F Kaneko et al., Humanoid robot HRP-2, Proc. IEEE Conf. Robot. Autom. (2004) 1083-1090.

[3] K. Hirai, M. Hirose, Y. Haikawa, T. Takensa, The development of Honda humanoid robot, Proc. IEEE Conf. Robot. Autom. (1998) 1321-1326.

[4] J. Pratt, G. Pratt, Virtuel model control: an intuitive approach for bipedal locomotion, Int. J. Robot. Res. 20 (2) (2001) 129-143.

[5] J. Forret, O. Bruneau, J. G. Fontaine, Unified approach for m-stability analysis and control of legged robots, Proc. IEEE Conf. Intell. Robots Syst. (2003) 106-111.

[6] M. Vukabratovic, B. Bocovac, D. Surla, D. Stokic, Biped locomotion Scientific Fundamentals of Robotics, vol.7, Springer Verlah, 1990.

[7] M.A. Lewis, L. S. Simo, Certain principles of biomorphic robots, Auton. Robots 11 (3) (2001) 211-216.

[8] J. H. Park, Fuzzy-logic zero-moment-point trajectory generation for reduced trunk motions of biped robots, Fuzzy Sets Syst, 134 (2003) 169-187

[9] W. T. Miller, Real-time neural network control of a biped walking robot, IEEE Contr. Syst. (1994) 41-48.

[10] Y. J. Seo, Y. S. Yoon, Design of a robust dynamic gait of the biped using the concept of dynamic stability margin, Robotica 13 (1994) 461-468.

[11] C. Sabourin, O. Bruneau, Robustness of the dynamic walk of a biped robot subjected to disturbing external forces by using CMAC neural networks, Robot. and Auton. Syst. 51 (2005) 81-99.

[12] R. K. Jha, B. Singh, D. K. Pratihar, On-line stabil gait generation of a two-legged robot using a genetic-fuzzy system, Robot. and Auton. Syst. 53 (2005) 15-35.

[13] C. Kirtley. CGA Normative Gait Database, Hong Kong Polytechnic University, 10 Young Adults. Available: http://guardian.curtin.edu.au/cga/data/

Walking robot HEXOR® II – a versatile platform for engineering education

Sajkowski M. *†, Stenzel T. *†, Grzesik B. *

* Silesian University of Technology, Department of Power Elekctronics, Electrical Drives and Robotics KENER, Gliwice, Poland, e-mail: *tomasz.stenzel@polsl.pl*

* Silesian University of Technology, Department of Power Elekctronics, Electrical Drives and Robotics KENER, Gliwice, Poland, e-mail: *maciej.sajkowski@polsl.pl*

* Silesian University of Technology, Department of Power Elekctronics, Electrical Drives and Robotics KENER, Gliwice, Poland, e-mail: *boguslaw.grzesik@polsl.pl*

† STENZEL Ltd., Gliwice, Poland, e-mail: *stenzel@stenzel.com.pl*

Abstract— The paper describes walking robot Hexor®II that has been designed as versatile platform for engineering education. It was designed, manufactured, tested and introduced for teaching process by Silesian University of Technology and Stenzel Ltd. The project of robot was undertaken basing on the previous long experience of above partners in the field of advanced power electronics. The robot allows for teaching the knowledge that embraces many fields such as mechanical engineering, electrical engineering, electronics, power electronics, control engineering, microprocessor technique, elements of telecommunications, mechatronics and even also a certain knowledge of biology (use of biological prototypes). Four years of experience proved that described robot is extremely useful teaching aid, giving student great amount of interdisciplinary knowledge that reaches far beyond accepted standards.

Keywords—mobile robot, teaching aid, mechatronics, education.

I. INTRODUCTION

"Using robots for educational purposes is not an entirely new idea. Designing, building, programming and testing an autonomous mobile device – one that walks and talks in addition to engaging in acrobatics and doing dance steps – helps young people to become familiar with the development of technical systems in their pre-adolescent years, while allowing them to have fun in the process. A one-day programme enables the children to learn basic notions of science, technology, electrical engineering, mechanics, robotics and information technology." [3]. It is a comment on European project *Roberta Goes EU*. Similar experiences are obtained by many university and secondary schools teachers [2], [4], [6].

Authors of this work describe walking robot HEXOR of their own design and construction that is dedicated for teaching purposes. Our adventure begun from the lecture of Prof. Marek Perkowski from Portland State University when he delivered important lecture "Intelligent Toys". We found that all activities and experience we had at that time could allow us to attack the area of mobile robots. We had experience in power electronics that was accumulated during long time of cooperation with

industry that resulted in number of complex power electronic systems based on microprocessor control. After short analysis and discussion, during which Prof Perkowski encouraged us for undertaking research of walking robots, the decision had been made. Than the small group of researches undertook the research work on walking robots. The next stage of the development of Hexor®II was designing of its mechanical part, its controls and telecommunications. The prototype of Hexor was completed in the beginning of 2003 [7], [8].

II. OVERVIEW OF MOBILE ROBOTS DEDICATED FOR EDUCATION

There are many mobile robots that are dedicated for education. Five of them are representative for this class. They are: i) Paralax and Lynxmotion [5], ii) Rugwarior [1], iii) Lego Mindstorms [13], iv) Activemedia [14], v) Kephera and Koala [15]. Due to many experts, there is no universal mobile robot that could be used for teaching and research. Students experience lack of useful hardware/software platform for learning and experimenting. For example, Lego Mindstorms have some drawbacks of programming language; it is impossible to use sensors and additional actuators. Another example of the mobile robot that is dedicated for teaching as well as research is Kephera and Koala. They have also the same shortcomings, they are not equipped with high-level language, they do not allow for higher number of sensors, and they are design for relatively simple environment. The above information was taken into account before decision concerning building and development of Hexor®II was made.

III. HEXOR®II – DESIGN AND CHARACTERISTICS [9]

The photograph of the robot is given in Fig. 1. It is hexapod type. The mobile robot Hexor®II is of a modular construction imitating the a scorpion. in respect of its anatomy and behaviors. It has been optimized for maximum possible reliability, assuming the smallest number of servomotors. Its energy consumption is minimal and it is easy to use and handling.

Fig. 1. Photograph of Hexor®II

Fig. 3. Pan/tilt camera and sonar head

There are five essential modules described below. They are:

A. The walking mechanism

The walking mechanism is mounted on a duralumin carrying flat construction. It has the form of openwork, having six legs driven by three DC pulse servomotors. Each leg has one degree of freedom. Front and rear legs move only in horizontal plane and the middle ones in vertical plane. The robot stability and balance during movement is assured by three points of support.

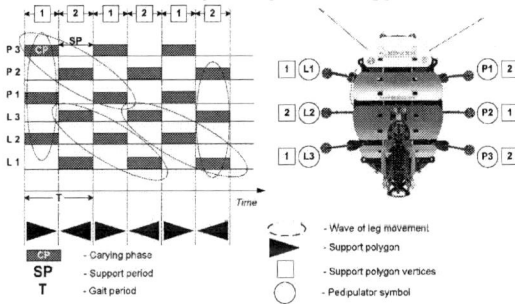

Fig. 2. Schematic diagram of robot Hexor®II fast tripod gait

B. The camera head

The camera head (robot's eye) is mounted on to of the metal tail. High position of the eye guarantees wide perspective of environment observation. The azimuth angle of the head can change ±90° and elevation angle can change also ±90°. Two DC pulse servomotors, placed on the top of tail drive the head.

C. The sonar

The sonar used in the robot has two ultrasonic converters, receiver and transmitter. It is also equipped with an integrated microchip, which allows measurement of distance from vertical obstacles in range 3cm ÷ 3m with accuracy of 1 cm. The sonar is placed above the camera so it can measure the distance to visible object to the camera.

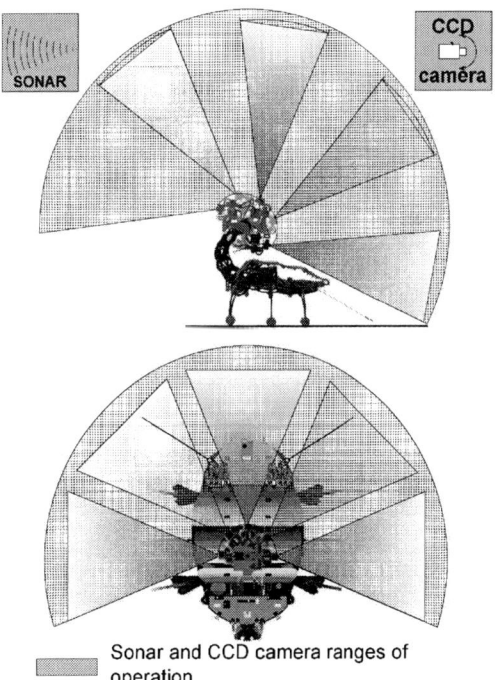

Fig. 4. Ultrasonic distance sensor range of operation in horizontal and vertical plane

957

D. Tactile sensor

Tactile sensors are mounted on both sides of the board, imitating feelers. When touching an obstacle, they close switches, which are connected to microcontroller inputs.

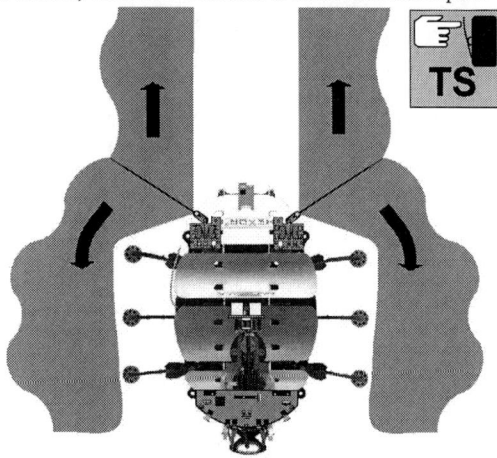

Tactile sensors operetional range
Sensors dead zone

Fig. 5. Range of operation of tactile sensors in case of forward movement and right or left rotation

E. Infra Read Proximity Detector IRPD

There are five pairs of IR sensors (transmitter and receiver). Four of them are mounted in the rear part of the robot board. They detect obstacles behind the robot. One another pair of IR sensors is mounted in the front of the board, between the tactile sensors.

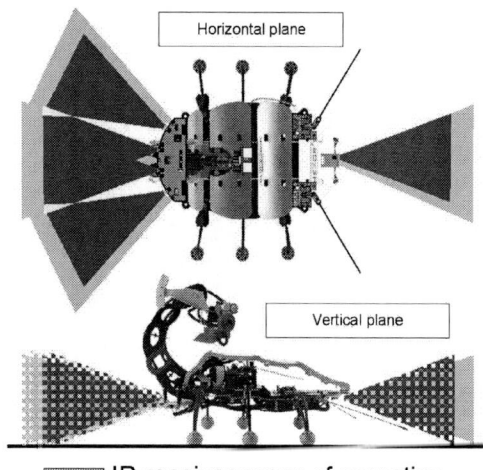

IR receiver range of operation
IR transmitter range of operation

Fig. 6. Front and rear infra red proximity detectors range of operation in horizontal and vertical plane

The IR sensors allow detecting obstacles by means of reflected IR rays. The detailed data characterizing the robot is given in table 1.

TABLE I.
TECHNICAL DATA OF THE ROBOT

1	Dimensions	Length 38 cm Width 47 cm Height 35 cm Weight 4.9 kg
2	Power supply	Gel battery 3.2 Ah Voltage supply 6V Working time of the battery 1,5h–3h Charging time ca. 1h
3	Connection of the robot to a PC computer	USB connection – radio base module Video Input of a TV Card – Audio/Video receiver Wireless connection Digital transmission – two-way, at 433 MHz - (commands and data) Analog transmission - Audio/Video signal at 2.4 GHz (image from the camera)
4	CCD camera	B&W or color, wide angle lens, IR light in option
5	Microcontroller	Main CPU AVR Mega 128 User program memory 128 kB Main board programming language - Basic (Bascom package), assembler or C i.e. GCC compiler
6	Sensors set:	Ultrasonic distance sensor (sonar) – distance measurement from 3 cm up to 3m, Close-up sensors based on IR light Touch sensors – „feelers"

F. Overall structure

Overall structure of the Hexor system is depicted in figure 2. Main µP is the central part of the system. It controls servomotors (servos). Three essential sensors, sonar, IRPD and TS are connected to microprocessor input and output lines. There is a possibility for connection of additional sensors via I²C bus. The mentioned bus is utilized for operating the temperature sensor (°C) and configuration EEPROM memory. It is possible to apply voice recognition module (VC), that can be connected to µP by SPI bus. Exchange digital data between robot's µP and host computer is realized by wireless 433 MHz base radio module (BRM). It is also possible to use analog subsystem that is connected to µP by ADC port. Besides, the system includes watchdog (WD).

One of the tasks of µP is power management and control. Power unit is based on single 6 V battery. The unit delivers 3.3 V, 5 V and 12 V voltage.

The last part of the system is designed for audio and video signals wireless transmission, operating at 2.4 GHz. It consists of color or b/w video camera (camera), transmitter (A/V transmit) and receiver (A/V receive).

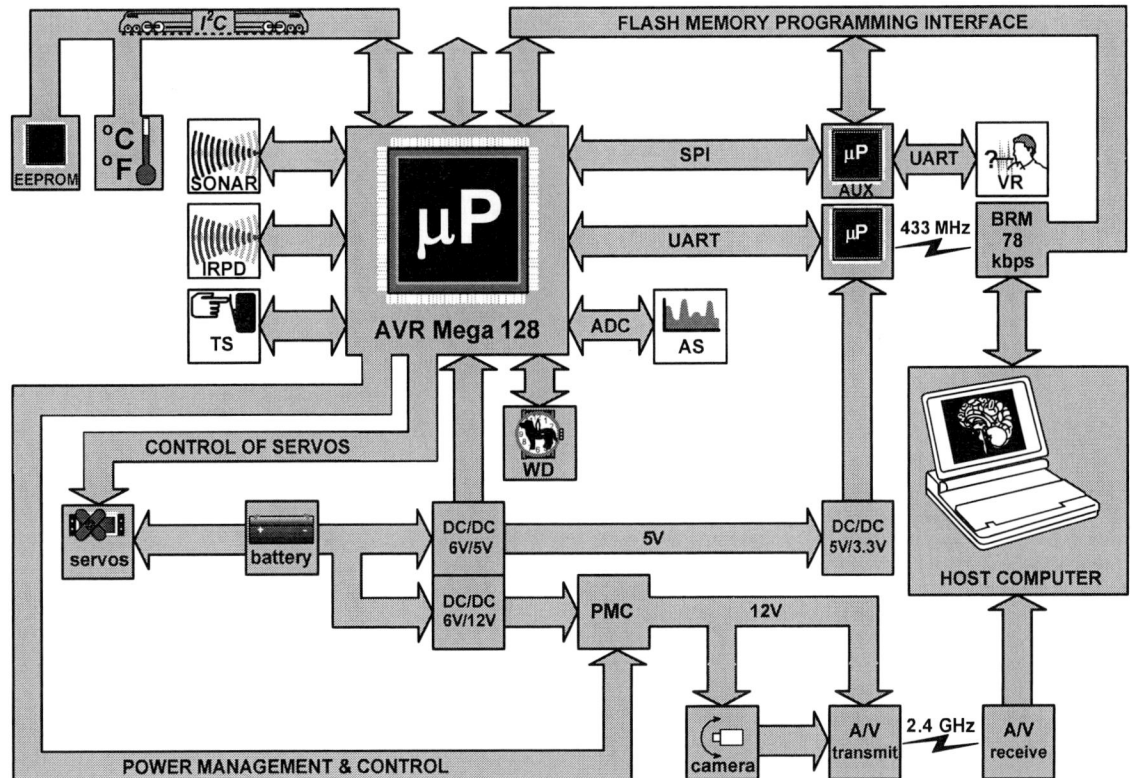

Fig. 7. Block diagram of HEXOR®II–system

IV. HOW WE USE ROBOT HEXOR IN TEACHING

Robot is used for the following teaching activities: i) Lectures (demonstrations), ii) Laboratory exercises – for university and secondary schools (mainly developing : control software, mechanical parts, gait control algorithms and improving efficiency of electrical energy usage), iii) Diploma thesis for MSc and BSc – for university (e.g. solar system for robot), iv) Projects aimed at development of Hexor® II system - for university (e.g. energy consumption by robot, image analysis, voice control and recognition) [11], [12].

The HEXOR robot control system was designed according to the previously assumed following key issues:

- The control board is based on popular commercial microcontroller including AVR RISC microcontrollers and based on 51 architecture CC1010 wireless transceiver;

- The machine behavior can be programmed using the commercially and free available software like: assembler language implementation suitable for applied microcontrollers, C language commercial and free compilers including KEIL or IAR commercial compilers and GCC free programming tool. It is also possible to use PASCAL and BASIC programming languages compilers;

- The elements of sensory system of the robot are models of the sensors used in industrial conditions. It is possible to learn the rule of operation of the sensors

and test the reactions after the sensor parameters modification.

Thanks to fulfillment of the assumptions listed above, students can apply the experiences and knowledge learnt with HEXOR in their future professional work. They develop programming skills, learn how to handle real microcontroller architecture features like timers or interrupts and also discover the industrial sensors from the inside. The technology of the machine including communication and power system is also based on the industrially applied solutions.

Walking robot is also very good example of the demonstrational equipment for teaching theoretical mechanics introducing kinematic and dynamic fundamental notions.

During five years of teaching with the robot as an aid, authors find it extremely useful and observe that students much easier become familiar with systems like Hexor. It means that they easily acquire the advanced knowledge and skills, embracing mechanical systems, electronics, telecommunications, radioelectronics, TV technology and problems of energy usage.

The robot was used for teaching the students associated in BOARD OF EUROPEAN STUDENST OF TECHNOLOGY (BEST) in their Mobile Robots Workshop [10]. Application of the HEXOR as main teaching aid for the mentioned course was real proof, that, it is possible to develop students team working skills by mobile robot projects. Similar experience comes from the student projects, exercises and seminars based on the

HEXOR®II. Walking robot HEXOR®II as an aid in teaching is a very complex device. To reach the satisfying results it is reasonable to divide the process of teaching into parts in accordance with the interest of particular students. Individual student's work is strongly recommended. This way of teaching is very efficient and satisfies not only students but also management of many engineering companies, what resulted in sponsorship of the robot project for secondary schools. Educational advantages of walking robots were appreciated by VATTENFALL POLAND corporation and Polish company Southern Energy Concern (PKE). The two companies decided to fund HEXOR walking robots as a teaching aid for ten technical colleges, which are very important elements in educational chain under the university level.

Introducing of HEXOR II to secondary schools influenced the quality of teaching, noticed also by engineering companies employing graduates from these schools. The same effect was observed in case of graduates from Silesian University of Technology. When discussing the role of mobile robots in education, it can be pointed out, that some scientific and educational solutions should be presented to potential technical university candidates in order to direct their interests to area of modern technology. Basing on the authors experience, HEXOR is a very good tool to attract candidates to become a student of electrical engineering faculty. The machine is successfully promoting the university and faculty during the educational fairs, scientific festivals and other university events.

Fig. 8. Group of students programming robot HEXOR®II and verifying the results in laboratory

What is more, students are very satisfied as they are aware of the enormous amount of knowledge they acquire in relatively short period of time.

CONCLUSIONS

Four years of using HEXOR®II for teaching confirmed the expectations of the authors – the knowledge obtained by students reaches far beyond standards. Using robots improves quality of teaching. It is so as student deals with real problems observing immediately the results of their software operation. Very frequently student is faced with mechanical engineering that is not common for electrical and electronics engineers. The teaching using robots is the teaching through work and play. About eighty robots has been delivered to Polish secondary schools and one to abroad secondary school where they are tested in real

environment. They are also used by several university centers and one abroad. These universities use our robots for experiments with their own software.

The robot has entirely been designed at our University and is manufactured by a small company Stenzel Ltd., which cooperates with the Silesian University of Technology. It is relatively cheap even in Poland.

The following problems are foreseen as a continuation of this work: i) how the reliability can be increased, ii) how to increase efficiency of electrical energy usage, iii) design analysis of various gaits, iv) optimization of walking mechanism and relevant control algorithm, v) possibility of increasing of leg degree-of-freedom and vi) the problem of how open mechanical system should be designed. It is also planned to use this machine for robots navigation and motion planning algorithms development.

REFERENCES

[1] Perkowski M.: Intelligent Toys, Lecture at Department of Electrical Engineering, Silesian University of Technology, Gliwice, Poland May 18, 2001.

[2] Claessens M.: Missing links (Editorial), The Magazine of the European Research Area, research*eu, Special issue, June 2007. http://ec.europa.eu/research/research-eu/special_education/02/article_edu16_en.html#article

[3] Roberta. Robots and girls. The Magazine of the European Research Area, research*eu, Special issue, June 2007. http://ec.europa.eu/research/research-eu/special_education/02/article_edu16_en.html#article

[4] Lima P.U.: Robotics Educational Activities in Portugal: A Motivating Experience [Education] Robotics & Automation Magazine, IEEE, June 2007, Volume: 14, Issue: 2, pp. 16-17.

[5] Jones J.L., Flynn A. M., Sieger B.A.: Mobile robots. Inspiration to Implementation. 2nd ed., A.K. Peters Natick, Massachusetts 1999.

[6] Zielinska T.: Walking machines. Fundamentals, Designing, Control and Biological Prototypes, Polish Scientific Publishers PWN, Warszawa 2003 (in Polish: Maszyny kroczące).

[7] Sajkowski M. et al.: Mobile Robots as an Educational Tool in Engineering Education, Scientific Periodical of Silesian University of Technology (Gliwice, Poland) ELEKTRYKA, Z. 187, 2003, pp. 87-94.

[8] Grzesik B., Sajkowski M., Stenzel T.: Which software architecture is applicable for mobile robot? Analysis on the Hexor robot example, Przegląd Elektrotechniczny 9'2004, pp. 813-816 (in Polish).

[9] Tze-wen Wang et al.: An Inexpensive LISP-Based Educational Platform to Teach Humanoid Robotics and Artificial Intelligence, International Conference on Engineering Education, ICEE 2005, Gliwice, Poland, pp.310-318.

[10] Zielińska T., Sajkowski M., Stenzel T., Grzesik B.: Lectures and laboratory exercises on mobile robots, BEST'06-SPRING Mobile Robots Workshop, organized by Board of European Students of Technology, Gliwice, Poland, May 8-12, 2006.

[11] Brown Ch. et al.: Hexor, a Walking and Talking Robot with Quantum and Fuzzy Inference, 15th International Workshop on Post-Binary ULSI Systems, May 17, 2006, Nanyang Technological University Singapore.

[12] Hwa Kim D. et al.: Artificial Immune-Neuro-Fuzzy-System to Control a Walking Robot Hexor. 15th International Workshop on Post-Binary ULSI Systems, May 17, 2006, Nanyang Technological University Singapore.

[13] Horwill I. A.: Laboratory Course in Behavior Based Robotics, IEEE Intelligent System Magazine, November/December 2000, pp 16 – 21.

[14] Ashraf Saad: Mobile Robotics As The Platform For Undergraduate Capstone Electrical And Computer Engineering Design Projects, http://fie.engrng.pitt.edu/fie2004/papers

[15] Dertien E.: Design of a miniature autonomous mobile robot (Individual design assignment), University of Twente Department of Electrical Engineering, April 2002G.

Motion Control of Steel Sheet Shears with Rocking Knife Mechanism

Jan Fetyko, Frantisek Durovsky and Viliam Fedak

Technical University of Kosice, Dept. of Electrical, Mechatronic and Industrial Engn., Kosice, Slovak Republic,
e-mail: *jan.fetyko@tuke.sk, frantisek.durovsky@tuke.sk, viliam.fedak@tuke.sk*

Abstract—The paper deals with development and realisation of the motion control for a very precise mechanism of a rocking knife driven by an electrical drive. The rocking knife provides transversal cutting of a moving steel strip into flat sheets of various, adjustable lengths. After development of mathematical model of the controlled mechanism there is designed the trajectory of the mechanism and based on this, a control law for the electrical drive is derived. The correctness of the design has been verified on a real mechanism in a hot rolling mill plant.

Keywords—motion control, machine tool drive.

1. INTRODUCTION

Cutting of a moving steel strip on the shearing line in a metallurgical plant is based on control of the drive moving the mechanism of the shears. In the real shearing line the strip speed reaches 1 m/s. To get flat sheets of the pre-set length, a precise control of the cutting mechanism with precise trajectory of the cutting knife during shearing and thus correct control of the electrical drive is required.

The body of the cutting knife performs a pendulous movement and is driven by an electrical drive through a crank mechanism. From detailed analysis it is known that the cutting blade trajectory is in form of an arch. To get a precise cutting, the drive of the cutting mechanism must perform the required trajectory. A proper time dependency of the trajectory requires a precise speed control regardless the mass of the mechanism and additive centrifugal forces that in the real mechanism cause disturbances in the calculated trajectory. To avoid their influence, they must be compensated by proper torque correction.

2. KINEMATIC SCHEME AND PRINCIPLE OF OPERATION

The rocking knife is located in the output part of the shearing line. The kinematic scheme of the mechanism is shown in Fig. 1.

The body of the cutting mechanism performs a pendulous movement and it resembles to an inverse pendulum. The bottom blade is fixed and the upper one presents a moving knife. The mechanism is driven by a DC drive that acts to the body of the knife through a gear and the crank mechanism. In that arrangement, the crank rotational movement is transformed to a pendulous movement of the rocking knife. One rotation of the crank corresponds to one stroke of the knife body.

The starting shears position is perpendicular to the cut steel strip. The upper knife is driven by an asynchronous motor with a flywheel that is connected with the crank

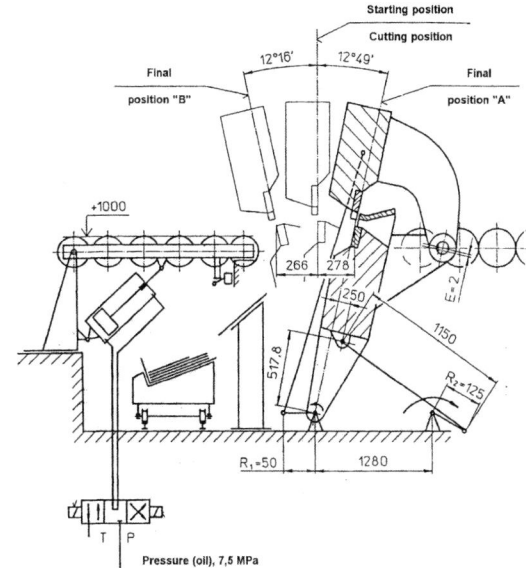

Fig. 1 Principal kinematic diagram of the rocking knife mechanism (cutting position = shears position).

mechanism through a pneumatic clutch and it causes movement of the upper knife.

The process of cutting is as follows:

1) In the defined time instant before the cut spot on the strip arrives under the knife, the drive of the rocking knife starts its activity. The body (so called „tower") moves from the starting position firstly backwards until it reaches the position of the rear dead-centre and afterwards it starts to move forward. In the cutting position the speed of the knife is synchronised with the speed of the strip and the spot of the cutting (i.e. position of the cutting) lies on the blade edge of the knives.

2) In the lead time the upper knife starts its movement by coupling its crank mechanism to the drive through a pneumatic clutch, so that the edges of the knives would cut the material in the required cut position. After cutting off the strip, the upper knife is disconnected from the drive and is braked by an electro-mechanic brake so it will remain to stay in the upper position.

The body of the knives continues with movement forwards up to the front dead-lock of position and afterwards it moves to the starting position.

3. SHEARS MECHANISM MATHEMATICAL MODEL AND CONTROL LAW PRINCIPLE

The mechanism of the rocking knife consists of multi-joint kinematic chain with four rotating joints. The task solution consists in a kinematic analysis of the knife mechanism body result of which is calculation of the angle position of the knives blade edge as a function of the crank angle position. Let's denote joint variables in the sequence starting from the crank joint up to the joint of knives body as follows: q_1, q_2, q_3, q_4.

At control of the knives movement the following conditions should be met:

1) In the cutting position, the circumferential speed of the knives must be equal to the strip speed.

2) The position of the cut spot has to lay down precisely to the bottom knife edge.

The position of the crank determines uniquely the position of the knives according to the nonlinear function shown below. This is reason why we have utilised the sensors of the crank position and speed for controlling the knives movement. Resolving the inverse kinematic task, the pendulous movement of the knives is transformed into a rotational movement of the crank. Now it is easy to control the knives movement on a principle of control of rotating shears. As a result of the analysis we have derived the diameter of a fictive cylinder with the axis identical to the crank joint where the circumferential speed in the cutting position is accurately the same like the circumferential speed of the edge of knives and is equal to the strip speed.

The principle of the fictive drum control (Fig. 3):

1) The zero position of the crank $q_1=0°$ corresponds to the shears position and 180° to the starting position of the knives.

2) After half revolution of the crank, the knives move by half stroke and they are in the cutting position. During this half revolution the mechanism of knives is synchronised to the strip speed.

3) In surroundings of the shears position $q_1=0°$ there is marked out a narrow band AX and AY where the circumferential speed of a fictive cylinder v_b has to be synchronised with the strip speed.

4) In the shears position the own shearing is preformed and simultaneously the position of the reference cylinder point B is reset. To this position of the reference points there corresponds the shears position of blade edges of the knives.

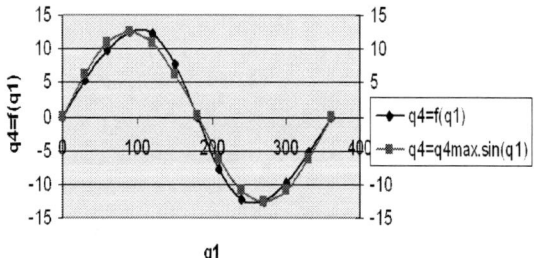

Fig. 2 Curve of angle of the cutting knife blade.

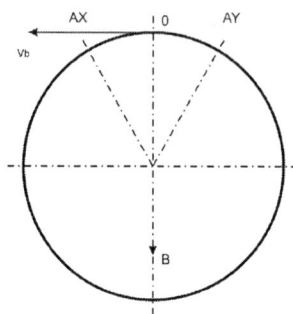

Fig. 3 Principle of cylinder shears operation.

5) According to the cutting length the shears stop (long sheets), or their speed decreases (shorter sheets).

For precise control of the rocking knife it is necessary to calculate the motor load torque. Then, the torque correction component is fed into the control circuit. We can easy derive the following set of equations:

The correction torque T_{corr} consists of 4 components:

$$T_{corr} = T_{dm} + T_{dshear} + T_{centr} + T_g \qquad (1)$$

where T_{dm} is dynamic torque of the motor with gear:

$$T_{dm} = K_{dm}(J_m + J_{gear})\frac{d\dot{q}_{ref}}{dt} \qquad (2)$$

The dynamic torque for the shears T_{dshear} follows up from the kinematic analysis:

$$T_{shear} = K_{dshear} \cos^2 \varphi_{ref} \cdot \frac{d\dot{q}_{ref}}{dt}, \qquad (3)$$

the torque T_{centr} involves a correction of centrifugal forces on the shear:

$$T_{centr} = -K_{centr} \omega_{ref}^2 \sin \varphi_{ref} \cos \varphi_{ref}, \qquad (4)$$

And the torque T_g involves the gravitational forces correction:

$$T_g = K_g m_{shear} \cos \varphi_{ref} \qquad (5)$$

where K_{dm}, K_{dshear}, K_{centr}, and K_g are constants that can be derived from a detailed analysis.

The proposed control structure is shown in Fig. 4.

4. REALIZATION

For realization the drive system in industry we used components from SIEMENS company. A DC motor fed by Simoreg 6RA70 DC Master drove the mechanism of the shear. The motor speed controller gets a feedback from the speed sensor mounted on the motor shaft. To sense position of the shears we used the second sensor mounted behind the gear - on the axis of the crankshaft driven the tower.

This arrangement enabled to increase dynamics and stability in the speed loop, also with regards to a possible backlash in the gear. The material speed was sensed by an incremental sensor mounted on a levelling machine. The technological module T400 with the software package SPS450 was used for controlling the shears movement.

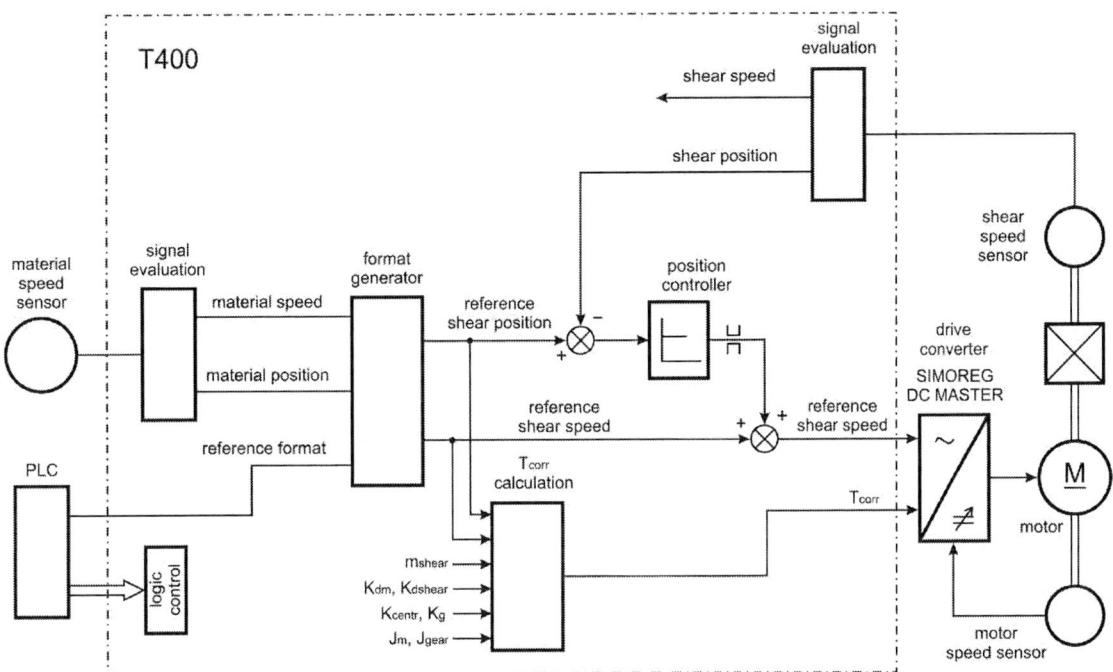

Fig. 4 Principle of the control structure.

The module contains a powerful processor enabling to perform fast calculations in the floating point arithmetic. The module ensures the logical control of the shears, handling the signals from sensors of the material and shears positions, calculation of reference shears speed towards the material movement (called format generator), a communication with superordinated control system and with subordinated converter. Except of this, it contains free blocks enabling to adapt pre-set control structure according to user needs. The communication structure is based on the BICO technology that is known from SIEMENS high performance converters. It is also possible to perform parametrisation of the module without substantial changes in the control structure or to programme it in the CFC environment using the software package Step 7. The module has pre-set functions for the drum shears and flying shears that is fixed on a travel moving in parallel with the strip.

In our case, for the pendulum shears we had to modify the control structure and complete it by a block that calculates corrections of the torque. For these purposes, we utilised advantageously the free blocks that are available in the module. Superodrinated logical control, control of the shears mechanism and visualisation was ensured by the PLC SIEMENS S7-300.

The control structure is shown in Fig. 4. The signals from incremental sensors of tower and material positions are fed into the module and processed in the blocks "**signal evaluation**". The output signal contains information about actual positions of the tower and material. Reference values of the speed and tower position are generated in the block "**format generator**" from the actual speed and actual position of the material and from the required length of the cutting (format). The reference speed of the tower is corrected by the signal from position controller and it is fed into the converter.

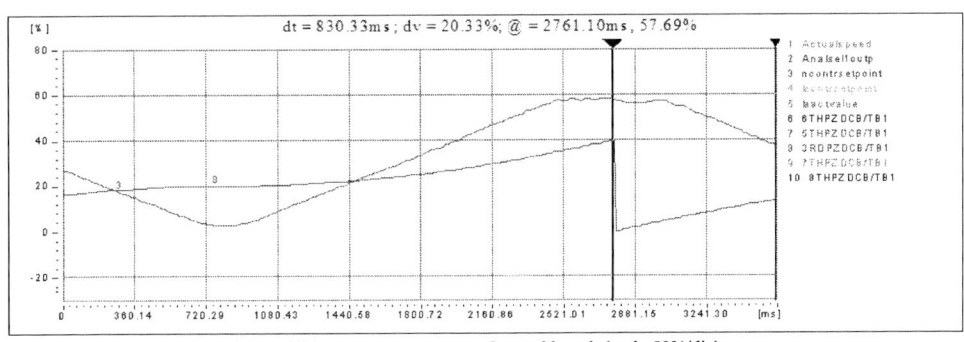

3 reference speed (material) (scale 20%/div)   8 actual length (scale 50%/div)

Fig. 5 Trace of speed and position of a fictive drum taken from real equipment.

5. EXPERIMENTAL RESULTS

Basic philosophy of the control system was elaborated in detail and realised using the SIEMENS equipment in a steel industry company. Fig. 5 shows courses of speed and position of the fictive cylinder of shears. The shears at the prescribed sheet length start running from non-zero speed, they are synchronised to the strip speed and in the time instant of cursor, the own shearing starts and the required position of the cylinder reference point is reset.

Fig. 6 shows courses of reference and real shears speeds without torque compensation. Simultaneously here is the shears position error and I-component of shears speed controller. Fig. 7 shows courses with correction of the centrifugal force compensation component. Fig. 8 shows the gravitational force compensation component. In both cases there was not connected any compensation to the control .Fig. 9 shows courses o f reference and real shears speed with full compensation. Form the courses it follows that the shears position error in time of cutting and the I-component of the shears speed controller are lower that in case without compensation what serves as an example of properly introduced compensation to the drive of the mechanism. The cursor shows time instant of the cutting.

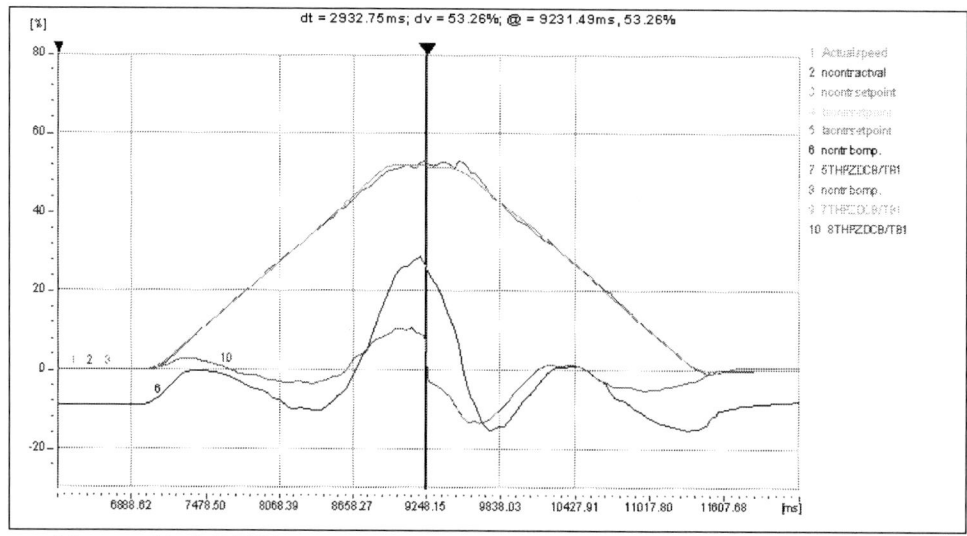

1 - actual shears speed (scale 20%/div)
2 - actual shears speed after filtering (scale 20%/div)
3 - reference speed (material) (scale 20%/div)
6 - output of speed controller (I-component) - (scale 20%/div)
10 - shears position error (scale 0,1%/div)

Fig. 6 The shears speed without compensation.

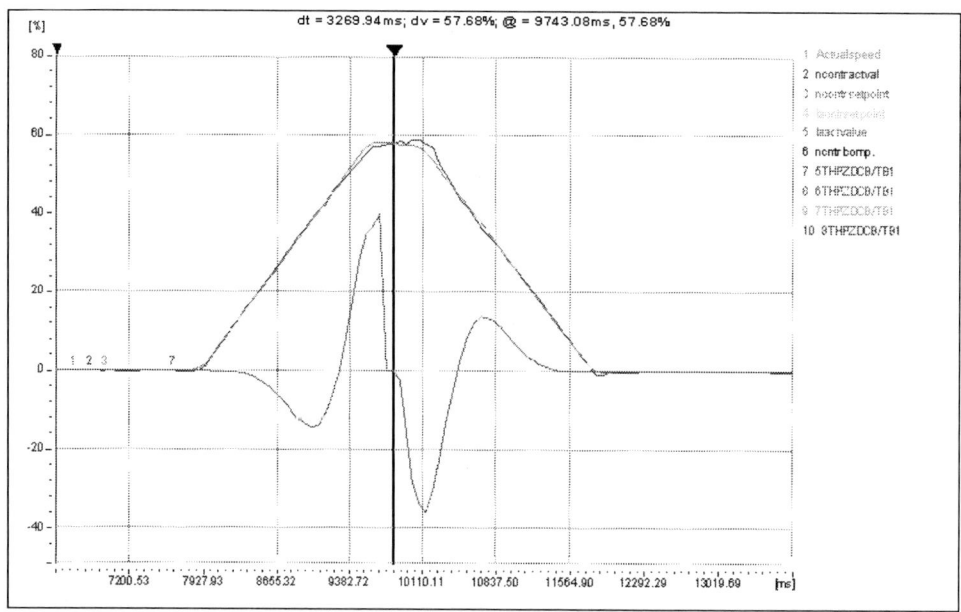

1 - actual shears speed (scale 20%/div)
2 - actual shears speed after filtering (scale 20%/div)
3 - reference speed (material) (scale 20%/div)
7 - torque compensation (scale 50%/div) - centrifugal force compensation component only

Fig. 7 The time courses in case of the torque compensation by centrifugal force compensation component (the shears without compensation).

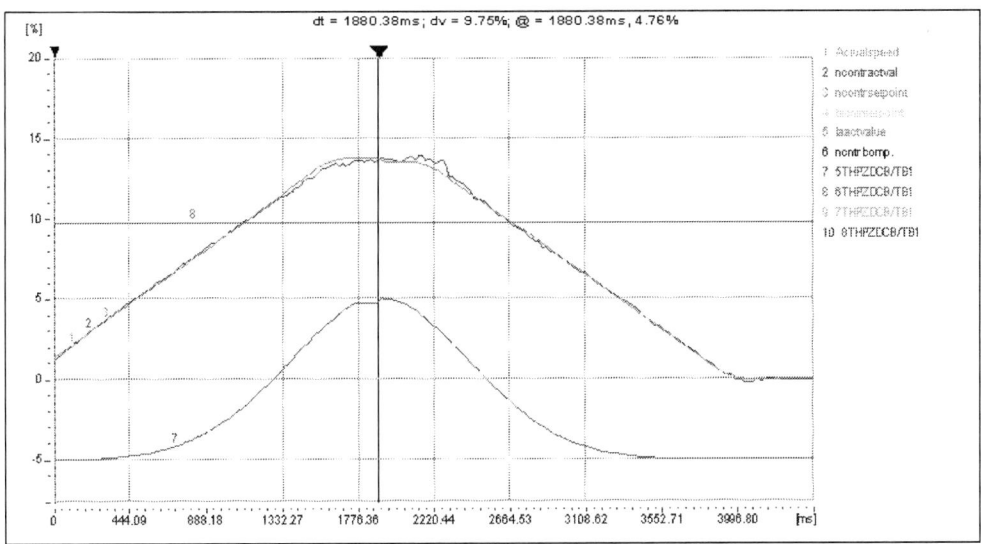

Legend:
1 - actual shears speed (scale 20%/div) 3 - reference speed (material) (scale 20%/div)
2 - actual shears speed after filtering (scale 20%/div) 7 - torque compensation (scale 5 %/div) - gravitational force compensation component only

Fig. 8 The course of the gravitational force compensation component (the shears without compensation).

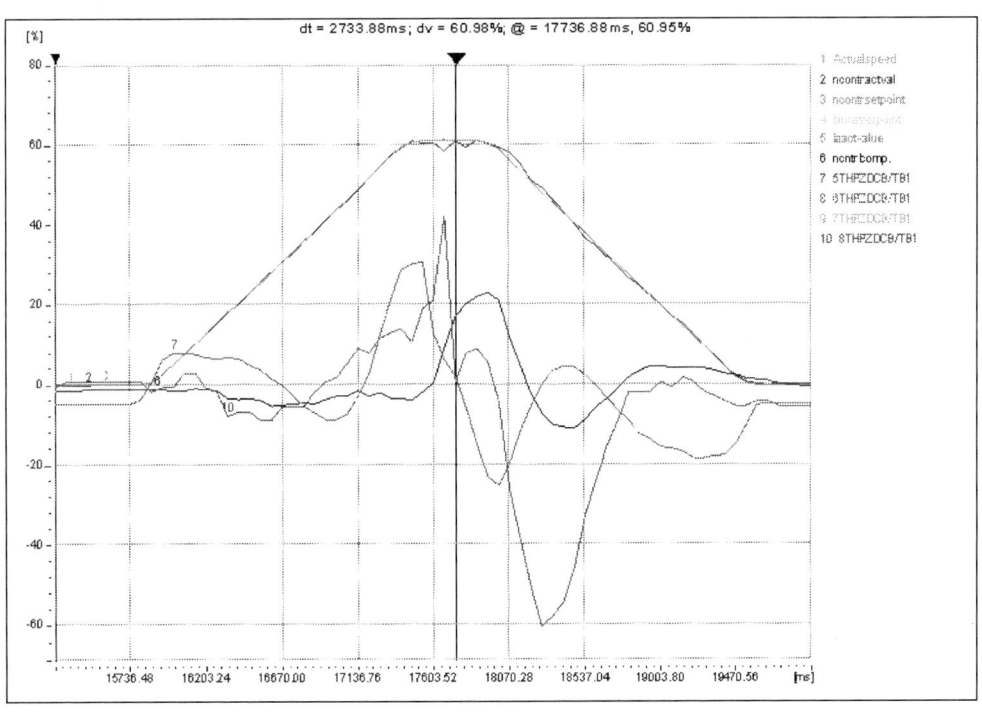

Legend:
1 - actual shears speed (scale 20%/div) 6 - output of speed controller (I-component) - (scale 20%/div)
2 - actual shears speed after filtering (scale 20%/div) 7 - torque compensation (scale 20%/div) – full torque compensation
3 - reference speed (material) (scale 20%/div) 10 - shears position error (scale 0,1%/div)

Fig. 9 The time courses of the controlled variables in case of the shears with full compensation.

6. Conclusions

The paper describes derivation of the control algorithm for the electrical drive driving the rocking knife mechanism in the shearing line for cutting the moving steel strip into sheets of a required length. The derivation starts from the kinematic diagram of the multi-joint mechanism. Based on the strip speed and required length of the sheet, the shearing control system generates the reference speed and position of the fictive cylinder shears. The control of the fictive cylinder speed is ensured by control system of the crank.

The control algorithm was realised in a steel industry company using the components from SIEMENS. In the real control system there is necessary to eliminate disturbances caused by moving mass of the equipment at acceleration of the mechanism and also centrifugal forces.

The time courses of the controlled variables prove evidence of improvement tower drive behaviour and higher precision in following the reference position at cutting.

Acknowledgment

The authors wish to thank to Slovak Grant Agency KEGA for its support of the project N° č. 3/5240/05 "Virtual Laboratory of Mechatronic Systems Control".

References

[1] Yaodong Cui, Ling Huang, „Dynamic programming algorithms for generating optimal strip layouts. *Computational Optimization and Applications.* v33, 2006: pp. 287-301.
www.springerlink.com/content/ y8975k4562642116/

[2] Shearing Line for Edge-Trimming and Cutting Metal Sheets and Plates. United States Patent 3555951.
www.freepatentsonline.com/3555951.html

[3] ISO Standard 77.140.50: Flat steel products and semi-products.
www.iso.org/iso/iso_catalogue/

[4] Sheet-Cutter/Cut to Length for T400 Technology Module. Manual SIEMENS, Ord. No. 6DD1903-0DB0

[5] SIMOREG DC-Master 6RA70 Series. Documentation. SIEMENS. Ord. No. 6RX1700-0DA64.

Intelligent Adaptive Control and Monitoring Of Band Sawing

İlhan ASİLTÜRK*, Ali ÜNÜVAR[†]

* Department of Mechanical Education, Faculty of Technical Education, Konya, Turkey,
e-mail: *iasilturk@selcuk.edu.tr*
[†] Department of Mechanical Engineering, Faculty of Engineering and Architecture, Konya, Turkey,
e-mail: *aunuvar@selcuk.edu.tr*

Abstract— In this paper, we propose that a neuro-fuzzy based adaptive controller for control of band sawing process. The system composed of two different kinds of back propagation networks and a fuzzy logic controller. The first network accomplishes the reference forces that are compared with the measured cutting force values in the real time. The required feed rate and cutting speed are adjusted by the proposed controller. The cutting parameters are continuously updated by a secondary neural network model to compensate the disturbances. The system provides that possibility of identification of material to be cut. Material identification is determined by the measured cutting forces while cutting operation. Experimental results show that the performance of the proposed controller is very well as adaptation of the cutting speed and feed rate during band sawing in the real time.

Keywords— Fuzzy control; neural network; adaptive bandsaw control; neuro fuzzy.

I. INTRODUCTION

Bandsawing is a parting-off method that has become more popular as material and manufacturing cost increase. Bandsawing is a multi-tooth cutting process and factors influencing the cutting performance of bandsawing operations should be controlled. In bandsaw machines, it is desired to feed the bandsaw blade into the workpiece with an appropriate feeding force in order to perform an efficient cutting operation. This can be accomplished by controlling the feed rate and thrust force by accurately detecting the cutting resistance against the bandsaw blade during cutting operation. In the process metal removal is accomplished by forcing a multi-toothed tool against the workpiece. The amount of metal removed by each tooth is dependent primarily on how well the blade transmits the applied pressure to the workpiece and also on the penetration ability of the cutting teeth. Cutting forces generating during a sawing operation are therefore found to have greater significance than in other chip forming machining processes. It has been found that thrust and cutting loads per tooth per unit thickness reduced with an increase in cutting speed. A reduction in the thrust force will cause a reduction in the depth of cut taken by the engaged teeth. An increase in the feed rate causes a substantial increase in both and thrust loads per tooth. In bandsawing, the thrust load is normally constant along the workpiece breadth. When sawing round sections the width of the workpiece changes within the cut, the cut width increases as the blade moves towards the centre and decreases as the cut is being finished.

Artificial Intelligent methods are used in every stage of manufacture. Machining is the basic manufacturing system in the industry. The selected machining parameters are usually conservative to avoid machining failure. To ensure the quality of machining products, to reduce the machining costs and increase the machining efficiency, it is necessary to adjust the machining parameters in real-time and to optimize machining process at that time. Adaptive control provides on-line adjustment of the operating conditions. Therefore, parameter adaptive control techniques for machining processes have been developed to adjust the feed rate automatically to maintain a constant cutting force. Applications of these techniques successfully increased both the metal removal rate and tool life.

The conventional PID feedback control system has been used in controlling machining processes by numerous researchers [1- 4]. The main problem with the fixed gain Adaptive Control Constraint (ACC) system is the one that produce poor performance and may become unstable during the time-varying machining process. Model reference adaptive control based ACC systems (MRAC) have been developed by some researchers [1-3,4].These studies found that MRAC perform control duties better than fixed gain controllers. Recently, many studies have been devoted to the theory of fuzzy control and its application to machining processes. Tarng et al [5,6] developed a Fuzzy Logic based Controller (FLC) for adaptive control of turning operations. The developed FLC can adjust feed rate on-line so as to reduce machining time and maintain constant force. In the experimental studies of Zhang et al [7] it is shown that process optimization is possible by online monitoring and controlling of the machining process. This eliminates the effect of disturbances caused by operator. An online monitoring system was designed by Ordonez et al [8] by using artificial intelligence based on sensors. Signals which were taken from sensors are used in AI decision making during the cutting process. The real time signals obtained through force transducers and estimated cutting forces obtained by using NN were compared. An adaptive control approach was suggested by Rodolfo et al [9] for maintaining the cutting force constant, in the milling process. Tsai et al [10] observed that, surface roughness can experimentally be determined by one or more quantitative measurements. An adaptive controller with optimization was designed based on two kinds of NN by Liu and Wang [11] for milling process. A modified back propagation NN was proposed adjusting its learning rate and adding dynamic factor in the learning process, and was used for the online modeling of the milling system. Adaptive Control Constraint (ACC) is one of the methods

978-1-4244-1741-4/08/$25.00 ©2008 IEEE

used in controlling machining processes. Force control algorithms have been developed and evaluated by numerous researchers. Among the most common is the fixed gain proportional integral (PI) controller, originally proposed for milling by Kim et al [12].

The essential aim of the neural network based controller is to construct a reverse function for the machining system using the NN so that the output of the machining system approaches to the desired output. Machining process can usually be controlled by adjusting the feed rate or spindle speed. The neural network based ACC system has been applied to machining process control by [13-14].

In a study by Liu et al [15], the major adaptive control constraint systems were discussed based on the feedback control, parameter adaptive control/self-tuning control, model reference adaptive control, variable structure control/sliding mode control, neural network control, and fuzzy control. Their typical applications to constant cutting force control system are also described, and some recent experiments results were presented.

Online method of achieving optimal settings of a fuzzy-neural network has been developed by Sandak and Tanaka. Results of the cutting experiments using several wood species show that the fuzzy-neural system developed performs well in online feed rate optimization during band sawing, while maintaining saw deviation within specified limits [16].

Zuperl et al [17] discussed the application of fuzzy adaptive control strategy to the problem of cutting force control in high end milling operations. In their AC system, the feed rate is adjusted on-line in order to maintain a constant cutting force in spite of variations in cutting conditions. They developed a simple fuzzy control strategy in the intelligent system and carried out some experimental simulations with the fuzzy control strategy.

The effect of cutting speed, feed rate and workpiece geometry in band sawing were investigated by Ahmad et al [18]. In the experimental studies, reduction in the thrust force and cutting force per teeth for unit thickness were observed, as the cutting speed increased.

In this paper, we propose an intelligent adaptive control for band sawing by using a fuzzy-neural system. The proposed adaptive control system can be applied effectively in various cutting situations. Band sawing system is shown in Fig. 1.

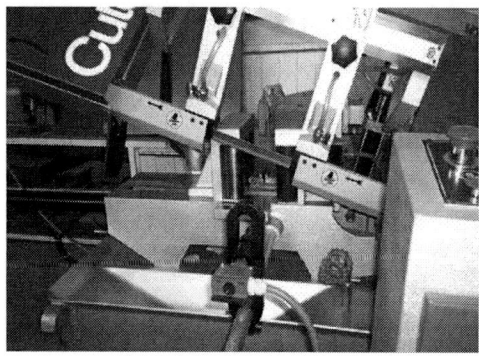

Fig. 1. Band Sawing System.

II. ADAPTIVE CONTROL OF MACHINING PROCESSES

Intelligent machining system improves the machining operations. In this system, process related data is acquired, and then process is controlled. The sensing of the machining process is much more comprehensive and complex. Many papers have been prepared about monitoring and control of the machine tool.

Researchers and industrialists concerned with tool monitoring and adaptive control. One of the most important functions of the intelligent control is the provision of required action in the unknown or indefinite ambient processes. Machine tool and cutting tools are protected by the monitoring system. Tool changing cost, scrap rate and production cost is reduced by real time tool wear measurement. Thus, full capacities of the machine tools are maintained.

In the machining processes, feed rate is continuously adjusted for keeping on the process with constant reference force in the adaptive control systems. Consequently, tool life increases with the restricted load application. In the case of depth or width of cut, feed rate are usually adjusted to compensate for the variability. This type of variability is often encountered in profile milling or contouring operations. When hard spots or the areas of difficulty to machine are encountered in the workpiece, either speed or feed is reduced to avoid premature failure of the tool. If the workpiece deflects as a result of insufficient rigidity in the setup, the feed rate must be reduced in order to maintain accuracy in the process. As the tool begins to dull, the cutting forces increase. The adaptive controller will typically respond to tool dulling by reducing the feed rate. The workpiece geometry may contain shaped sections where no machining needs to be performed. If the tool were to continue feeding through these so-called air gaps at the same rate, time would be lost. Accordingly, the typical procedure is to increase the feed rate, by a factor of two or three, when air gaps are encountered. These sources of variability present themselves as time-varying and, for the most part, unpredictable changes in the machining process. It should be examined how adaptive control can be used to compensate for these changes.

III. ADAPTIVE NEURO-FUZZY CONTROL OF THE BAND SAWING SYSTEM

In metal cutting, it is necessary to control the machining process online to provide stable cutting and to select the appropriate machining parameters depending on material characteristics. Also the machining parameters must be adjusted in real-time. Due to the machining process being a nonlinear and time-varying process with random disturbance, traditional adaptive controller depends greatly on the accurate model of the machining process. Consequently it is difficult for the traditional adaptive controller to realize effective control in the machining process. Adaptive control constraint (ACC) is one of the effective methods of solving the above problems. ACC controls the machining parameters to maintain the maximum working conditions during the time-varying machining process. In this work, to avoid aforementioned difficulties, a neural-fuzzy based adaptive controller with constraints for band sawing was established and model outputs were compared with measured cutting force in real time, then according to difference between them, processes were maintained with fuzzy logic controller in appropriate conditions. Cutting parameters were continuously updated by a secondary neural network, to compensate for the effect of environmental disturbances. Required feed rate and cutting speed were continuously adjusted by fuzzy logic controller. A new control scheme which is called adaptive neural-fuzzy control is developed by using NNs with back propagation algorithm and fuzzy logic controller. The objective of this control is to keep the metal removal rate as high as possible and maintain cutting and trust forces as close as possible to a given reference value. Furthermore, the amount of computational task and time can be reduced as compared to classical or modern control theory.

The training and testing of NNs and also the operation of machining system require a number of experimental data. For this reason, an experimental design was created for band sawing. In order to establish 162 different cutting states in band sawing, the specimens which are 18 different type samples in three kinds of material groups were prepared. And then normalization was made at 880°C for the homogeneity of samples. Data acquisition was made by the software called as "ilhan_daq_v01". A data acquisition with this software was made with 16 channels real time data acquisition hardware. The constants and cutting parameters were entered to the interface. Outputs were measured as 80 samples / sec. and their average values were recorded as one datum. Circular and square cross sectional samples which have three different hardness and three different diameters, were machined at three different cutting speeds, and three different feed rates. Consequently, tests were performed 162 experimental runs [19]. The block diagram of the experiment set is shown in Fig. 2.

Filtered data have been combined and data files have been composed for NN. Training and test processes were A supervised multilayer back propagation network

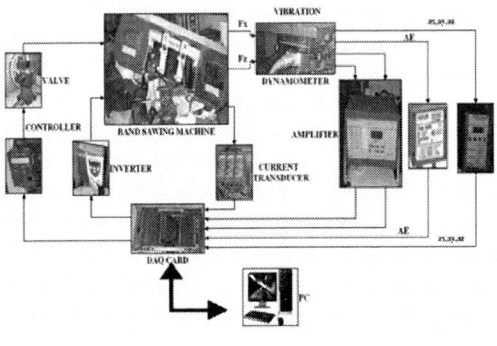

Fig. 2. Block diagram of the experiment set[19].

algorithm was used as NN. made with this data and suitable NN model were determined.

In this adaptive control system, recorded data are evaluated by NN and FL and then these data are transformed into adaptive control decision. Consequently, the operation of the band sawing machine is improved to operate in required cutting conditions.

The model is composed of three intelligent models: 1- Neural Network (NN_1), 2- Fuzzy logic control (FLC) 3- Neural Network (NN_2). Where; NN_1 is used for estimated cutting force in process variations. FLC produces control outputs by the difference between estimated cutting force and measured cutting force. NN_2 is used for upgrading of the cutting force model. Control block diagram of the band sawing system is shown in Fig. 3.

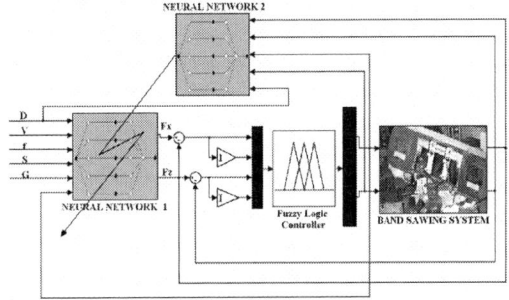

Fig. 3. Control block diagram of the band sawing IACS.

NN_1 is used to learn the appropriate mappings between the input and output variables of the machining process. Cutting forces (Fx and Fz) is accepted as measurable output parameters with respect to input parameters of our system consisting of cutting speed (V), feed rate (f), diameter (D), hardness (H), material group no (G), instantaneous feed rate (fi). Data consisting of 4482 lines and 8 columns extracted from the experiments is source of input data to NN_1. Data is normalized between the ranges 0-1 and network is chosen as multilayered feed forward network and transfer function is either tangent sigmoid or logarithmic sigmoid. Moreover gradient descent optimization method is chosen as training method function and weights and bias values are updated accordingly. NN_1 system is shown in Fig. 4.

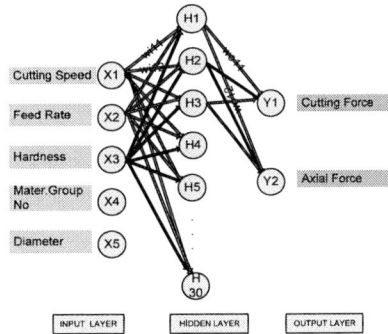

Fig. 4. Basic Structure of Cutting Forces to Predict Neural Network Model.(NN_1)

For NN_1, a matrix with a dimension of (3000x6) is used as input and a matrix with a dimension of (3000x2) is used as output. To test the system, a data matrix with a dimension of (1482x6) is used as input and a data matrix with a dimension of (1482x2) is used as output matrix. Both two matrices are the remaining data which is not used for training. Modeling, training and test of the NN_1 are realized by using MATLAB™ Neural Network Toolbox.

To determine the appropriate network structure, performance of the network consisting of constant number of input and output layer neuron is measured with respect to changing number of hidden layer neurons which are 2, 3, 4, 5, 6, 10, 15, 20, 25, 30, 35, 40, 50. The network with 6x30x2 neurons per layer which has MSE (Mean Squared Error) of 0.0020407 at the 6000th epoch is chosen. The graphic of the data which is part of the test process by using aforementioned network structure are given below.

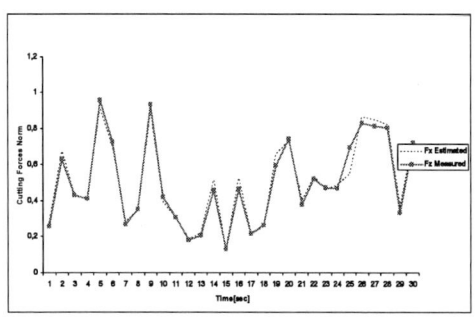

Fig. 5. The comparison of measured and estimated by NN cutting forces.

FLC is a multi input/multi output fuzzy logic algorithm which functions as a controller which produces output parameters for the process network. FLC modeling is realized by using a MATLAB Fuzzy Logic Toolbox. FLC is a controller structure by using fuzzy logic. For this reason, it tries to minimize the difference between the real time measured values and estimated values produced by NN1. Expert opinions and experiences are used to build FLC and to determine FLC parameters. Error values and their integrations are used as inputs, and feed rate and cutting speed driving the system are used as output. Since the model used here is nonlinear, Mamdani fuzzy processing type is preferred. The number of input variables is 4, the number of output variables is 2, the degree of fuzziness is 3 and the number of the rules is 36 for the fuzzy controller. The center of gravity is also used as max-min inference method and defuzzification methods, respectively. The scaling of Fuzzy regions is made by named 1=low, 2=medium, 3=high for both input variables and output variables. Triangular shape was preferred as membership function in the study. Table of rules are prepared to realize all probabilities for the input variables and the output variables that are assigned the required values by the expert in it. The structure of rules is created in the term of "IF.... THEN" in the designed fuzzy controller. Thirty six rules are constructed based on the knowledge extracted the errors and the changes of errors.

NN_2 drives the system to the optimum state against the instantaneous changes in the process or outside distorting variables. NN_2 changes reference model according to the effect of disturbances by real time monitoring of the system and thus NN_1 can estimate the true outcome. The measured inputs of the system are cutting speed, feed rate, instantaneous feed rate, speed, diameter, and coordinate/diameter values. Material hardness is the only output. The experiments to gather data for training and test purposes result with 5487 lines by 7 columns data matrix which is normalized to the values ranging from 0 to 1. For the training phase, a data matrix having a dimension of (3800 x 6) is used as input and a data matrix having a dimension of (3800 x 1) is used as output.

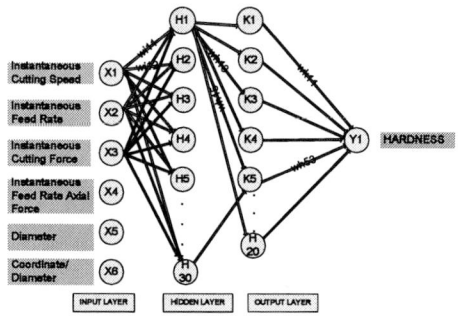

Fig. 6. Basic Structure of Hardness to Predict Neural Network Model.(NN_2)

To test the system, a data matrix with a dimension of (1686 x 6) is used as input and a data matrix with a dimension of (1686 x 2) is used as output matrix. Both of the matrices are the remaining data which is not used for training. Modeling, training and test of the NN_2 are realized by using MATLAB™ Neural Network Toolbox. To determine the appropriate number of layers and neurons per layers, the following network structures are experimented. Performance of the network consists of constant number of input and output layer neuron is measured with respect to changing number of two hidden layer neurons which are 3, 6, 10, 15, 20, 25, 30, and 40. The network structure with 6x30x20x1 neurons per layer which has MSE of 0.0133115 at the 5000th epoch is chosen.

IV. EXPERIMENTAL WORK

A. Experimental Conditions

Minimum and maximum cutting speed and feed rate values used at experimental works were determined by using the mean values obtained from the band manufacturers' handbooks, and Median Values were determined by using Full Factorial Experiment calculations. A semi automatic type bandsaw machine which has pulling cut (IMAS 280) was used. To control the machine by a computer, hydraulic, electrical, electronics and mechanic revisions are made on the machine. An experiment set which is able to measure the cutting forces, current, vibration, AE, speed and feed rate

Fuzzy İnterface System inputs	Name	Number of Membership functions	Membership geometry	MF1 Membership interval	MF2 Membership interval	MF3 Membership interval
Differences of cutting forces	efx	3	trimf	[-18 -10 -2]	[-8 0 8]	[2 10 18]
Integral of cutting forces differences	iefx	3	trimf	[-18 -10 -2]	[-8 0 8]	[2 10 18]
Differences of Feed forces	efz	3	trimf	[-18 -10 -2]	[-8 0 8]	[2 10 18]
Integral of differences of the feed forces	iefz	3	trimf	[-18 -10 -2]	[-8 0 8]	[2 10 18]

Table 1. Input data of the selected fuzzy logic architecture

from the process was set up. The cutting tests were carried out on a IMAS 280 Band Saw Machine. Experimental conditions are given below;

Cutting Speed levels: 40, 63, 100 [m/min],

Feed Rate: 35; 66; 125 [mm/min],

Workpiece diameter: 30, 46, 70 [mm]

Workpiece material: C1015, C1040 and C4140

B. Experiment Set Elements

Analog data was collected for the different cutting parameters and materials by using sensors. Force measurements were realized by using Kistler 9257B dynamometer and Kistler 5019B Charge Amplifier. Acceleration measurements were made by using Kistler 8792A500. Acoustic emission measurements were made by using Kistler 8152B111, the valve which adjust the feed rate is proportional valve Rexroth 2FRE 6 B-2X/K4RV. Cutting speed is adjusted by using Telemecanique inventor, and National Instrument 6221 is used as DAQ card. By using the data collection card, analog data is converted into digital values for different cutting parameters and materials and stored in the computer. When the data collection experiment results are investigated; it can be seen that cutting forces decreases as cutting speed decreases, and increases as the feed rate increases. These results are considered carefully when the rule table for control structure is generated. This database is used both for training and test process.

V. EXPERIMENTAL RESULTS AND DISCUSSION

In the beginning of experimental work, the material is clamped and entered the diameter value to the system.

Process starts cutting with minimum federate, after 100 data sampling print, chooses the appropriate federate according to NN_2 results. Material hardness, material group info and proper reference model is determined as soon as the material is started to be cut. Process adjusts the cutting speed and feed rate based on the principle of constant cutting force. When the situations like chipping and workpiece local hardness arise, system updates itself, the reference forces changes instantaneously, and process is regulated according to the parameters accommodating the new conditions.

The determined network parameters and the results of the constructed network structure were compared. ANN model of 6 x 30 x 2 which has the least MSE was chosen as the most appropriate model. NN estimated the reference forces with 97.13% and 97.05%.ratios for training and test results, respectively.

First ANN system continuously produces new reference force by using the feedback from the second ANN regarding to hardness change and then sends it to the fuzzy controller. New cutting parameters produced by fuzzy controller according to the new situation depending on the reference cutting forces are produced.

To test the system, the workpieces having round cross section were used. After the band saw contacts with the workpiece, the cutting forces increase as the cutting length increases. As the cutting forces increases, feed rate is decreased by ANN Fuzzy System. During the process, with feed rate and cutting speed changing, second ANN (NN_2) produces output and the hardness changes and this parameter is feed back to the ANN (NN_1), as a result of this process it is produced a new reference force value by the first ANN (NN_1). Consequently, the system adjusts itself to the new situation and cuts the material by using new parameters. Graphics of changing of cutting forces vs. cutting time is given in the Figs. 5-8, both for ANN fuzzy adaptive control system and uncontrolled system. At the start of the cutting, cutting parameters are entered according to the known material type and workpiece diameter. Cutting operation starts by using entered parameters, as the cutting operation progresses, the controller determines the values of cutting parameters. The controller tries to maintain the cutting forces steady at determined reference values by changing the feed rate and cutting speed. Hardness information is extracted from cutting forces feed back and then reference force model is updated.

Graphics of results of experiment on SAE 1040 diameter of 46 mm. materials for both normal and adaptive IACS cutting conditions are as follows:

As seen in Fig.7 and Fig.8, at the beginning of cutting, to maintain the chosen cutting force reference level, feed rate is decreased and cutting speed is increased. The system acquires real time material hardness value based on cutting forces and cutting force model is updated. Any local force increase during the cutting session updates the reference force model. The system reacts by increasing cutting speed and decreasing the feed rate in the case of increasing cutting forces.

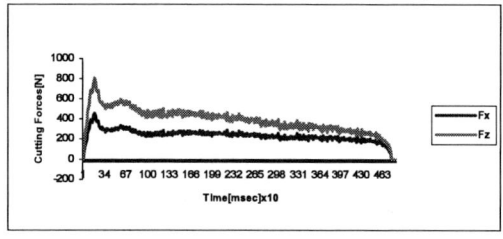

Fig. 7. Cutting forces during the cutting of the material (SAE 1040 Φ46 mm) by IACS.

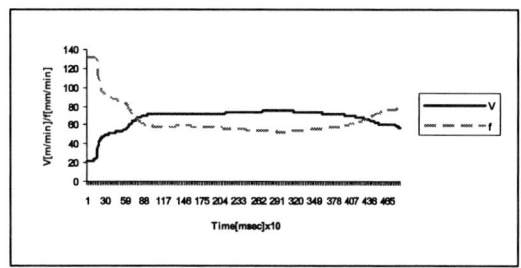

Fig. 8. Adaptive change of feed rate and cutting speed while the material (SAE 1040_ Φ 46mm) cutting by IACS.

A material hardness recognition system is developed based on cutting forces by means of NN2. Hardness of

material SAE 1040 Φ46 mm workpiece is obtained in the real time as shown in the Fig 9.

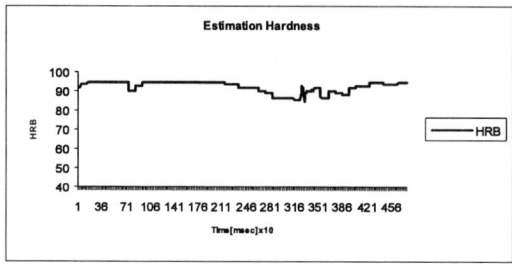

Fig. 9. Recognition of the material hardness of the material (SAE 1040_ Φ 46 mm) by IACS during the cutting process

In Fig.10, Force change in the process of the cutting of SAE 1040_ Φ46 mm material by using a conventional band saw machine is shown. As seen in the graphic, at the beginning of cutting the cutting forces are to be low according to the number of the teeth which are in touch with the material as the band saw progresses, the cutting forces increase as the cutting length increases. To decrease the thrust forces, the depth of cut is decreased by the system toward the workpiece center. In cutting circular workpieces by using a band saw with constant feed rate, the forces increase as the cross-sectional length increases. In Fig 8, it can be seen that feed rate is constant in band saw cutting of SAE 1040 Φ46 mm. workpiece by using a conventional band saw machine.

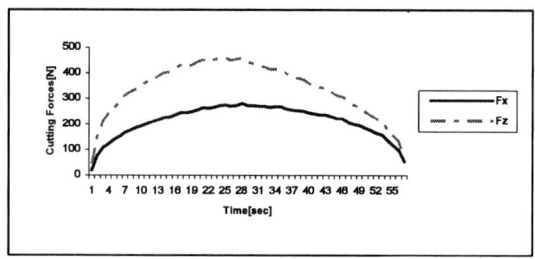

Fig. 10. Cutting forces during the cutting of the material (SAE 1040 Φ46mm) by conventional band saw machine.

Proposed system estimates the material hardness consistently with real cutting forces and produces cutting parameters accordingly. Proposed system estimates the real cutting forces consistently with the material hardness and produces cutting parameters accordingly. The system provides quick response to drive the system to appropriate state and results with regular updates in cutting parameters and reference model with respect to conventional systems.

A neuro-fuzzy based force model of band sawing system is obtained and outputs of the model are compared with measured cutting force values in real time. Fuzzy logic controller is designed to control process while it works. Cutting parameters are continuously updated by a secondary neural network to compensate the effect of environmental disturbances. Required feed rate and cutting speed are adjusted by using fuzzy logic controller. Material identification is determined by measured cutting forces while cutting operation. Experimental results show that performance of the improved neuro-fuzzy control system is very well in real time cutting speed and feed rate during band sawing in Fig 7-9.

VI. 6. CONCLUSION

The experimental results shown that the hybrid ANN-Fuzzy adaptive controller decreases the tool costs by working the machine tool in appropriate feed rate and cutting speed and provides less duration than conventional controllers. This study shows that hybrid artificial intelligence method can be used successfully to control of machine tools adaptively. ANFIS was not used in the system, because the system has two real time input-output variables. Metal removal problem which is very hard to model mathematically is solved easily by using these methods. There are many parameters affecting the band sawing process. Therefore control of this cutting process can not be achieved effectively by using conventional methods. In this study, the system has been modeled by using an artificial intelligence model. Effect of operator intervention has been set to minimum by the help of developed monitoring and control system. Designed and realized system increases effectiveness in production speed and energy saving. The positive effect of the proposed system on the machine tools is demonstrated theoretically by working them in appropriate and safe cutting speeds and feed rates against the changes in cutting forces resulting from local variations in the hardness of workpiece. Additionally, the scientific point of view and expertise in hardware and software development in this work can be used in different applications.

ACKNOWLEDGMENT

The authors would like to thank to Selcuk University Scientific Research Project Coordinatorship (Project No: 06101008) and executives of IMAS (Integrated Cutting Systems) because of their sponsorship.

REFERENCES

[1] O. Masory, Y. Koren, "Adaptive control system for turning", Annals *CIRP, 29(1)*, 1980, pp. 281-284.

[2] O. Masory, Y. Koren, "Stability analysis of a constant force adaptive control system for turning", *Transactions ASME Journal of Engineering for Industry*, 107, 1985, pp. 295-300

[3] L. K. Lauderbaugh, A.G. Ulsoy, "Model reference adaptive force control in milling", *Transactions ASME Journal of Engineering for Industry*, 111, 1989, pp. 14-21

[4] Y. Koren, "Adaptive control systems for machining", *Proceedings of the American Control Conference*, 1988, pp. 1161–1167,

[5] Y. S. Tarng, S. T. Cheng , "Fuzzy control of feed rate in end milling operation", *Int. J. Mach. Tools & Manufact.*, Vol 33(4), 1993, pp 643-650.

[6] Y. S. Tarng, Y. S. Wang , "An adaptive fuzzy control system for turning operations", *Int. J. Mach. Tools & Manufact.*, Vol 33(6), 1993, pp 761-771.

[7] G. Zhang, R, G. Khanchustambham, "Neural Network Applications in On-line Monitoring of a Turning Process",

Neural networks in design and manufacturing, 1993, 247 - 268, ISBN:981-02-1281-X .

[8] R. Ordonez, J. Zumberge, J.T. Spooner, K.M.O Passino, "Adaptive Fuzzy Control: Experiments and Analyses", *IEEE Transactions on Fuzzy Systems*, Vol. 5, No. 2, 1997, pp.167-188.

[9] E.H. Rodolfo, R. Clodeinir, A. Angel, R. Salvador, G.Carlos, "Toward Intelligent Machining: Hierarchical Fuzzy Control for the End Milling Process", *IEEE Transactions On Control Systems Technology*, Vol. 6, No. 2, 1998, .pp. 188-200.

[10] Y. H. Tsai, J. C. Chen, S. J. Lou, "An In-Process Surface Recognition, System Based on Neural Networks in End Milling Cutting Operation", *Int. J. Mach. Tools Manuf.*, 39, 1999, pp. 583–605.

[11] Y. Liu, C. Wang, "Neural Network Based Adaptive Control and Optimization in the Milling Process", *Journal of Advanced Manufacturing Technology*, 15, 1999, pp. 791-795.

[12] M.K. Kim, M.W. Cho, K Kim, "Application of the fuzzy control strategy to adaptive control of non-minimum phase end milling operations", *Int. Journal of Advanced Manufacturing Technology*, 17, 1999, 791 – 795.

[13] Y. M Liu, L. Zuo, C. J. Wang, "Intelligent Adaptive Control in the Milling Process", *Int. Journal of Computer Integrated Manufacturing*, 12(5), 1999, pp. 453-460

[14] S.J. Hang, K. C. Chiou , "Application of neural networks in self tuning constant force control", *Int. J. Mach. Tools & Manufact.*, Vol 36, 1996, pp 17-31.

[15] Y. Liu, T. Cheng, L. Zuo, "Adaptive Control Constraint of Machining Processes", *Journal of Advanced Manufacturing Technology*, 17, 2001,720 – 726.

[16] J. Sandak, C. Tanaka, "Online adaptive control of bandsaw feed speed using a fuzzy-neural system", *Forest products journal*, ISSN 0015-7473, vol. 53, 2003, pp. 36-43.

[17] U. Zuperl, F. Cus, M. Milfelner, "Fuzzy control strategy for adaptive force control in end-milling", *Journal of Materials processing Technology*, 164-165, 2005, 1472-1478.

[18] M.M. Ahmad, B. Hogan, E. Goode, "Identification of variables to improve cutting performance of a band sawing process", *4th. I.M.C. Conf.* 1987.

[19] İ. Asiltürk, "Adaptive Control Application Based On Artificial Intelligence In The Bandsawing Process", *Ph.D. These, Selcuk University, F.B.E*, 2007, Konya, Turkey.

Nomenclature:

Symbol	Name
F_{out}	Process output;
F_{meas}	Measured force;
F_{ref}	Reference force;
f	Feed rate;
L	Instantaneous height
vi	Instantaneous cutting speed
fi	Instantaneous feed rate
e	Output error
IACS	Intelligent Adaptive Control System of the Band Sawing.
D	Diameter
H	Hardness
G	Material group no
I	Integral of error

973

Hierarchical adaptive network based fuzzy logic controller design for a single flexible link robot manipulator

Mete KALYONCU[*], Mustafa TINKIR[†]

Department of Mechanical Engineering, Faculty of Engineering and Architecture, University of Selçuk,
Alaeddin Keykubat Campus, 42079 Konya, TURKEY,
[*] e-mail: mkalyoncu@selcuk.edu.tr   [†] email: mtinkir@selcuk.edu.tr

Abstract — In this study, hierarchical adaptive network based fuzzy logic controller for a single flexible link robot manipulator is designed. This study is aimed to get the end of the flexible robot manipulator to desired position and terminate vibrations on arm while it moves under the action of an external driving torque. The proposed controller has two subsystems. The fast-subsystem reduces the vibration of the flexible robot manipulator end point, and the other slow-subsystem controller controls the desired position. The performance of the proposed control system is evaluated on the basis of the experimental results.

Keywords — Fuzzy control, neural network, adaptive control, motion control, robotics, modelling

I. INTRODUCTION

Robots have been shown to be useful in performing some repetitious, labour-intensive, dangerous, monotonous, and tedious jobs in many industrial and aerospace applications. Robot are also useful in hazardous or hostile environments such as arc welding shops, contaminated areas, nuclear reactors, outer space, and under water. For many tasks, the presently available robots are not fast enough, economical, or sufficiently accurate. Flexible-link robotic manipulators have many advantages with respect to conventional rigid robots. Future robots will be designed with lighter weight, more accurate position control, and larger payloads, so more flexible arms must be developed. With additional requirements for higher speed, better system performance, lower power consumption, and cheaper operating cost on practical robots, the structural flexibility must be considered in the design of robotic control systems. The dynamic analysis and control of flexible-link manipulators is much more complex than the analysis and control of the equivalent rigid manipulators. Therefore, research on the dynamic modelling [1-8] and control of flexible robots [9-19] has received increased attention in the last decades due to these advantages. A first step towards designing an efficient control strategy for these manipulators must be aimed at developing accurate dynamic models that can characterize the flexibility of the links. Controller design solutions to minimize the effects of the flexible displacements in light robots are highly demanded in the industrial and space applications which require accurate trajectory control. Controlled robot manipulators are usually designed to reach a target or to follow a trajectory. In the first case, a short settling time is expected while in the tracking condition a high speed robot arm displacement is planned. Thus, strong control actions are applied and, as a result, undesired controlled system features could appear if hidden vibrating modes are excited enough.

Existing studies on flexible robot manipulator can be divided into two groups: those on dynamic, and those on control. Some of these investigations on the control of flexible robot manipulators consider hierarchical fuzzy logic control [20-35] due to its several advantages over other control techniques. Adaptive network based fuzzy logic control was not taken into consideration for single link flexible robot manipulator in most of the cited investigations despite its some advantage to be indicated in this study.

Akbarzadeh-Totonchi et al developed a two-level hierarchical fuzzy controller for a single flexible link. The second level of hierarchy monitored the behaviour of the robot arm and extracts features such as Straight, Oscillatory, Gently Curved [20]. In references [21, 23, 26], a two-time scale fuzzy logic controller was applied to the flexible-link robot arm. A singular perturbation approach was introduced to derive the slow and fast subsystems. The fast subsystem controller damped out the vibration of the flexible structure by two hierarchical fuzzy logic controllers. The slow subsystem fuzzy controller dominated the tracking of the trajectory. Caswara and Unbehauen used a neurofuzzy controller as a nonlinear compensator for a flexible four-link manipulator. Two classes of neurofuzzy models, the Takagi-Sugeno fuzzy model and the rectangular local linear model network were applied as a feedforward controller to compensate the nonlinearities [22]. Lin applied a multi-time-scale fuzzy logic controller to a robotic manipulator with link structural flexibility. The large-scale system was decomposed into a finite number of reduced-order subsystems using the singular perturbation approach. A hierarchical ordering of fuzzy rules was used to reduce the size of the inference engine [24]. Chiang and Lin applied the hybrid position/force control for flexible link robot arm. Therefore, a multi-time scale fuzzy logic controller was applied for a flexible link robot arm. Using this methodology, the control of the force and the position of the end point were possible while the end effector moves on the constraint surface [25]. Emara and Elshafei considered robust control of robots including motor dynamics based on a dynamic game approach [27]. Lin developed a novel approach for

978-1-4244-1741-4/08/$25.00 ©2008 IEEE

achieving a high-performance active piezoelectric absorber of a smart panel using adaptive networks in hierarchical fuzzy control [28]. Jnifene and Andrews concerned with the design and implementation of active vibration control based on fuzzy logic and neural networks (NNs). The controller was used to damp the end-point vibration in a single-link flexible manipulator mounted on a two degrees-of-freedom platform. The inputs to the fuzzy logic controller (FLC) were the angular position of the hub and the end point deflection of the flexible beam [29]. Kamalasadan and Ghandakly proposed a neural network (NN)-based intelligent adaptive controller that introduces a new concept of intelligent supervisory loop. The scheme consists of an online radial basis-function NN (RBFNN) in parallel with a model reference adaptive controller (MRAC) and uses a growing dynamic RBFNN to augment MRAC [30]. Karimi and Yazdanpanah proposed a new methodology for modelling a single-link flexible manipulator with an arbitrarily large (infinite) number of deflection modes based on the singular perturbation method [31, 32]. Lin et al dealt with active damping control problems of robot manipulators with oscillatory bases. A first investigation of two-time scale fuzzy logic controller with vibration stabilizer for such structures was proposed, where the dynamics of a robotic system was strongly affected by disturbances due to the base oscillation [33, 34].

The aim of this study is investigation of hierarchical adaptive network based fuzzy logic control of a flexible arm. The tip end of a single flexible link robot manipulator traces a desired position signal under the action of an external driving torque. Considered robot manipulator consists of a single flexible link without tip mass and a revolute joint. The flexible link is assumed to be an Euler-Bernoulli beam. Equations of motion of the flexible manipulator are obtained by using Lagrange's equation of motion. Using different type of applications of fuzzy logic controller in the studies is aimed to get the end of the flexible robot manipulator to desired position and terminate vibrations on arm while it moves. A two-time-scale adaptive network based fuzzy logic controller is applied to control the system in this study. Adaptive network based fuzzy inference system (ANFIS) uses a hybrid learning algorithm to identify parameters of Sugeno-type fuzzy inference systems. It applies a combination of the least-squares method and the back propagation gradient descent method for training fuzzy inference system (FIS) membership function parameters to emulate a given training data set. The controller has two subsystems. The fast-subsystem reduces the vibration of the flexible robot manipulator end point, and the other slow-subsystem controller controls desired position. Hierarchical fuzzy rules are used to reduce size of the inference engine. The main contribution of this work is the derivation of the hierarchical adaptive fuzzy algorithm to avoid the rule explosion phenomenon that characterizes traditional fuzzy systems. We also prove that the proposed control algorithm is locally stable. The control of the vibration and position of a single flexible link robot manipulator end point are realized successfully by such this approach while the end-effector traces desired position signal. Experimental results obtained by using experiment system of a single flexible link robot arm are presented and physical trend of the results are discussed. The performance of the hierarchical adaptive network based fuzzy control system is evaluated on the basis of the experimental results.

II. DYNAMIC MODELLING OF THE FLEXIBLE ROBOT MANIPULATOR

The schematic diagram of the flexible robot manipulator is given in Fig. 1. This considered robot arm consists of a flexible beam with a distributed mass. The mass and flexible properties are assumed to be distributed uniformly along the flexible arm. The flexible arm is assumed as an Euler-Bernoulli beam. θ represents the angular position of the equivalent rigid link with respect to the fixed frame XY. D represents the position of the end point on the flexible arm with respect to the equivalent rigid link. Assuming small deflection of the link, an approximate linear time invariant dynamic model is derived using Lagrangian formulation and the dynamic equation is represented in matrix form as:

$$M\ddot{q} + C\dot{q} + Kq = F \qquad (1)$$

where the vector q is generalized coordinates. $q=[\theta \;\; \alpha]^T$, M is the mass matrix, C is the damping matrix, K is the stiffness matrix, and $F=[T\;\;0]^T$. T is the input torque applied at the joint. The variable α ($\alpha=D/L$) represents the slope at the free end of the flexible link.

The dynamic model can be represented in state-space form as:

$$\begin{aligned} x &= A\,\dot{x} + Bu \\ y &= Cx \end{aligned} \qquad (2)$$

the vector $x = \begin{bmatrix} \theta & \alpha & \dot{\theta} & \dot{\alpha} \end{bmatrix}^T$, the matrices A, B, and C are given by

$$A = \begin{bmatrix} 0_{2x2} & I_{2x2} \\ -M^{-1}K & -M^{-1}C \end{bmatrix}, \; B = \begin{bmatrix} 0_{2x1} \\ -M^{-1} \end{bmatrix}, \; C = I_{2x4} \qquad (3)$$

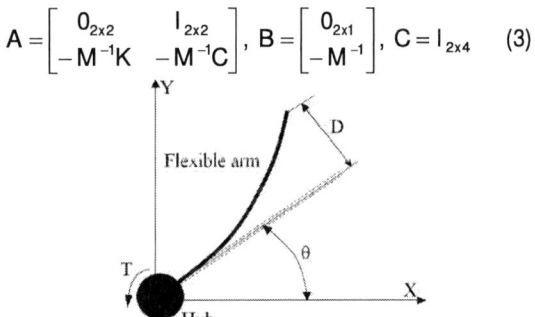

Fig. 1. Schematic diagram of the flexible arm

III. HIERARCHICAL ADAPTIVE NETWORK BASED FUZZY LOGIC CONTROLLER DESIGN

A two-time-scale adaptive network based fuzzy logic controller is applied to control the system in this study. Adaptive network based fuzzy inference system (ANFIS) uses a hybrid learning algorithm to identify parameters of Sugeno-type fuzzy inference systems. It applies a combination of the least-squares method and the back propagation gradient descent method for training fuzzy inference system (FIS) membership function parameters to emulate a given training data set. Two adaptive networks based fuzzy logic controllers are used in hierarchical mean control strategy. These controllers are designed by training and checking data sets that are obtained from hierarchical PID control of the system.

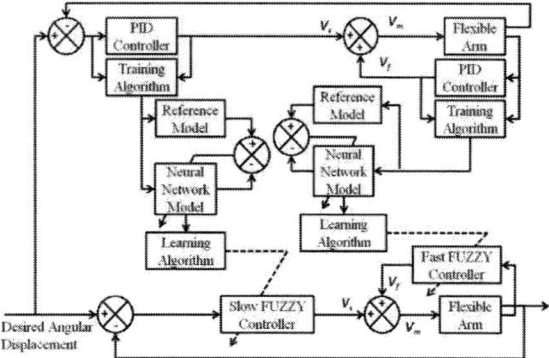

Fig. 2. Detailed block diagram of the proposed control system

First of all hierarchical PID controllers applied to a single flexible link robot arm and then their input and output data are obtained to set up network base of the adaptive network based fuzzy inference system (ANFIS). After trained network, fuzzy inference system structure is built. The network base was trained offline. Number of membership functions and type of membership functions are decided. A few experiments are done to obtain appropriate fuzzy inference system for angular position and deflection control of a single flexible link robot manipulator. Fig. 2 shows that the proposed control system closed loop block diagram.

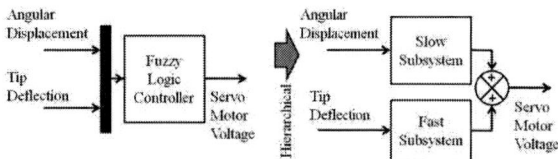

Fig. 3. Hierarchical control structure

Control voltages are evaluated by slow and fast controllers separately. In this way, designing of the controllers becomes easy. If every two variables could be combined, then for an even number of variables, the reduction is even more pronounced. For a fuzzy system with two variables (θ, α) and five fuzzy sets (labels), then the rules would reduce from $5^2 = 25$ to $5^1 = 5$, a reduction 80%. Clearly, depending on how many outputs can be fused and in what order, when they are put into a hierarchical structure, the size of the rule base would be reduced differently. It is illustrated the fuzzy logic controller's rule-base reduction for such case in the Fig. 3.

A. Adaptive Neural Network

Neural networks are composed of simple elements operating in parallel. These elements are inspired by biological nervous systems. As in nature, the network function is determined largely by the connections between elements. We can train a neural network to perform a particular function by adjusting the values of the connections (weights) between elements. Commonly neural networks are adjusted, or trained, so that a particular input leads to a specific target output. Such a situation is shown in Fig. 4. There, the network is adjusted, based on a comparison of the output and the target, until the network output matches the target.

Typically many such input/target pairs are needed to train a network. Neural networks have been trained to perform complex functions in various fields, including pattern recognition, identification, classification, speech, vision, and control systems.

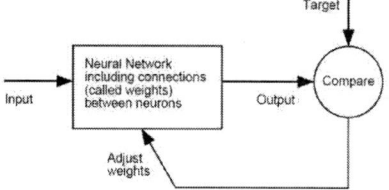

Fig. 4. Neural network structure

ANFIS architecture consists of five layers with the output of the nodes in each respective layer is represented by $O_{i,l}$ where i is the ith node of layer l.

Layer 1: Generate the membership grades

$$O_{i,l} = \mu_A(x) , \ i = 1,2 \tag{4}$$

or

$$O_{i,l} = \mu_{B_{i-2}}(y) , \ i = 3,4 \tag{5}$$

Where x (or y) is the input to the node and A_i (or B_{i-2}) is the fuzzy set associated with this node.

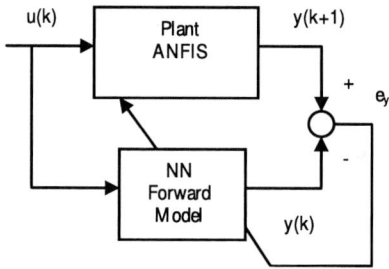

Fig. 5. Training of NN forward model

Layer 2: Generate the firing strengths by multiplying the incoming signals and outputs the t-norm operator result, e.g.

$$O_{2,i} = w_i = \mu_{Ai}(x) \times \mu_{Bi}(y) , \ i = 1,2 \tag{6}$$

Layer 3: Normalize the firing strengths

$$O_{3,i} = \overline{w}_i = \frac{w_i}{w_1 + w_2} , \ i = 1,2 \tag{7}$$

Layer 4: Calculate rule outputs based on the consequent parameters $\{p_i, q_i, r_i\}$

$$O_{4,i} = \overline{w}_i f_i = \overline{w}_i (p_i x + q_i y + r_i) \tag{8}$$

Layer 5: Computes the overall outputs as the summation of incoming signals

$$O_{5,i} = \sum_i \overline{w}_i f_i = \frac{\sum_i w_i f}{\sum_i w_i} \tag{9}$$

We follow these steps for creating of ANFIS controller shown in below:

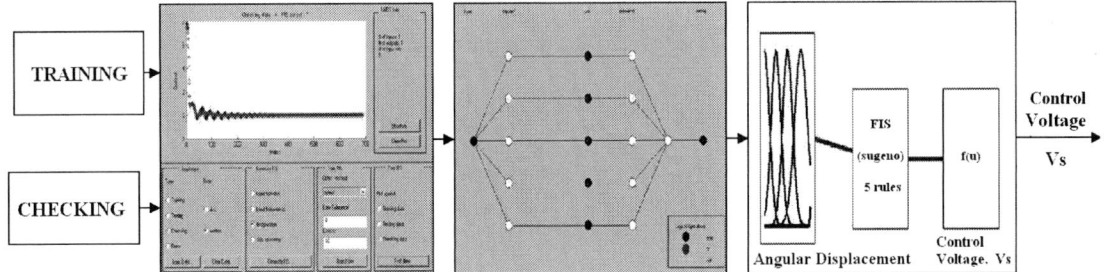

Fig. 6. Slow adaptive network based fuzzy logic controller scheme

- 3000 training and checking data (from hierarchical PID control) are obtained for neural network of ANFIS control.
- The number and type of membership functions are determined.
- Hybrid learning algorithm and 60 epochs is chosen to train network.

B. Hybrid Learning Algorithm

In this study, forward hybrid learning algorithm is used for neural network part of ANFIS controllers shown in Fig. 5. Nearly 60 epochs later error rate close to 10^{-6}. In the forward pass of the hybrid learning algorithm, node outputs go forward until layer 4 and the consequent are identified by the least-squares method. When the values of the premise parameters are fixed, the overall output can be expressed as a linear combination of the consequent parameters

$$f = \frac{w_1}{w_1 + w_2} f_1 + \frac{w_2}{w_1 + w_2} f_2$$
$$= \overline{w_1} f_1 + \overline{w_2} f_2$$
$$= \left(\overline{w_1}x\right)p_1 + \left(\overline{w_1}y\right)q_1 + \left(\overline{w_1}\right)r_1 + \left(\overline{w_2}x\right)p_2 + \left(\overline{w_1}y\right)q_2 + \left(\overline{w_1}\right)r_2$$
(10)

which is linear in the consequent parameters p_1, q_1, r_1, p_2, q_2 and r_2,

$$f = XW \qquad (11)$$

If X matrix is invertible then

$$W = X^{-1}f \qquad (12)$$

otherwise a pseudo- inverse is used to solve for W.

$$W = \left(X^T X\right)^{-1} X^T f \qquad (13)$$

Due to the adaptive capability of ANFIS, their applications to adaptive and learning control are immediate. The most common design techniques for ANFIS controllers are derived directly from neural networks counterpart methodologies. However certain design techniques apply exclusively to ANFIS.

C. The Slow Subsystem Adaptive Network Based Fuzzy Logic Controller

The slow subsystem fuzzy logic controller is designed using the rotation (θ) to implement the control methodology. Hence V_s is suitable output in the slow subsystem. Rotation of the arm is about applied control voltage. In slow subsystem adaptive network based fuzzy logic control single input and single output (SISO) occur. The slow adaptive network based fuzzy logic controller is shown in Fig. 6.

D. The Fast Subsystem Adaptive Network Based Fuzzy Logic Controller

The control scheme of the system must be able to control the tip deflection of the arm. In this study, the fast subsystem adaptive network based fuzzy logic controller are designed to control deflection (α) of the arm motion. There is no desired deflection signal to compare with actual deflection of the system. The fast subsystem adaptive network based fuzzy logic controller's input and output are tip deflection and fast control voltage (V_f) respectively. Aim of this control strategy is that using controller on feedback line. The fast adaptive network based fuzzy logic controller is shown in Fig. 7.

E. Membership Functions and Fuzzy Rules

Two of the difficulties with the design or any fuzzy control systems are the shape of the membership functions and choice of the fuzzy rules. In fact, the decision-making logic is the way in which the controller output is generated. It uses the input fuzzy sets, and the decision is taking according to the values of the inputs. Moreover, the knowledge base comprises knowledge of application domain and the attendant control goals. It consists of a database and a fuzzy control rule base.

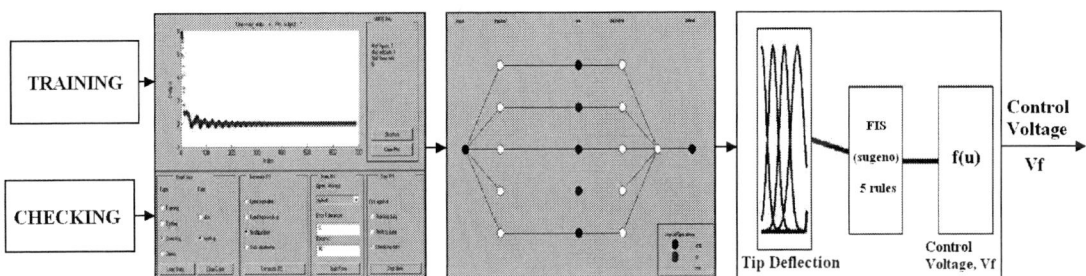

Fig. 7. Fast adaptive network based fuzzy logic controller scheme

977

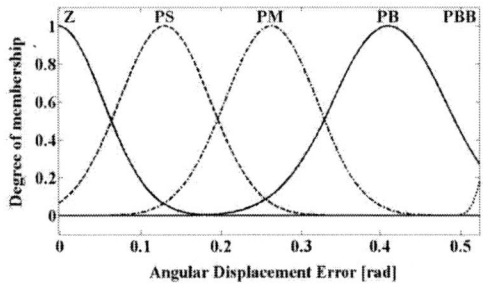

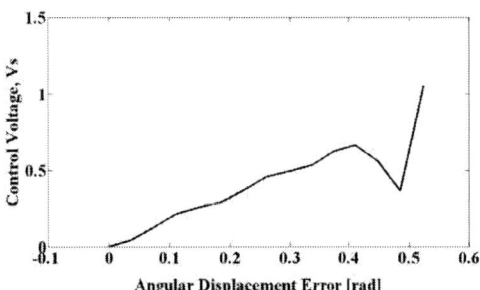

Fig. 8. Membership functions and control voltage of slow subsystem adaptive network based fuzzy logic controller

A control system is said to be an adaptive fuzzy control system if either a set of fuzzy rules are used to modify or change an existing fuzzy controller's architecture, i.e., membership functions and/or rules. The fuzzification uses membership functions to determine the degree of inputs. The aim of control action is to minimize the rotation error. The higher the error, the higher the control input. However, the rate of change of error also affects the value of the control input. In slow subsystem fuzzy logic controller, error is used in control rules as linguistic variables.

This is defined as:

$U_{\theta e}$=Angular Displacement Error=(Desired − Actual) (14)

In adaptive network based fuzzy inference system (ANFIS), number and type of membership functions are created by user. Then fuzzy inference system rule base is obtained automatically by ANFIS. Slow and fast adaptive network based fuzzy logic controllers have five membership functions for each inputs and outputs. Gauss type membership functions are used for fuzzy inputs in fuzzification process of slow and fast subsystem control. Sugeno type fuzzy inference system is adopted so these controllers' outputs are not fuzzy. ANFIS determine these outputs by given training and checking data sets. Figs. 8 and 9 show that membership functions and control surfaces of slow and fast subsystem adaptive network based fuzzy logic controllers.

F. Adaptive Network Based Fuzzy Logic
Controllers' Rule Bases

The fuzzy rule base can be illustrated as a look-up table. Slow subsystem adaptive network based fuzzy logic controller's look-up table is a vector of five rows and one column. In slow subsystem rule base is constituted five rules. Slow subsystem controller rule base is shown in Table 1. The following example demonstrates the use of the natural language modelling

approach for slow subsystem fuzzy control decision-making. There are five fuzzy membership functions, that correspond to, zero (Z), positive small (PS), positive medium (PM) positive big (PB) and positive big big (PBB) values of rotation. The control laws of the fuzzy controller consist of a complete set of these control rules defined as:

If $U_{\theta e}$=Z then V_s=Constant Voltage 1,

If $U_{\theta e}$=PS then V_s=Constant Voltage 3,

...

Fast subsystem adaptive network based fuzzy logic controllers' has one look-up table which is made up of five rules as slow subsystem. Fast subsystem adaptive network based fuzzy logic controllers' rule base is shown in Table 2. There are five fuzzy membership functions, that correspond to, negative big (NB), negative medium (NM), negative small (NS), zero (Z) and positive small (PS) values of tip deflection. The control laws of the fast subsystem fuzzy controller are defined as:

If α=NB then V_f= Constant Voltage 2,

If α=NM then V_f= Constant Voltage 4,

...

G. Defuzzification Process

Once the fuzzy controller is activated, rule evaluation is performed and all the rules which are true are fired. Utilizing the true output membership functions, defuzzification is then applied to determine a crisp control action.The defuzzification is to transform the control signal into an exact control output. For Sugeno-style inference, we have to choose between wtaver (weighted average) or wtsum (weighted sum) defuzzification method. In defuzzification process both slow and fast subsystem adaptive network based fuzzy logic control, the method of weighted average (wtaver) is used:

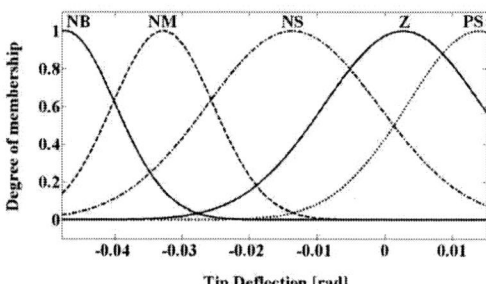

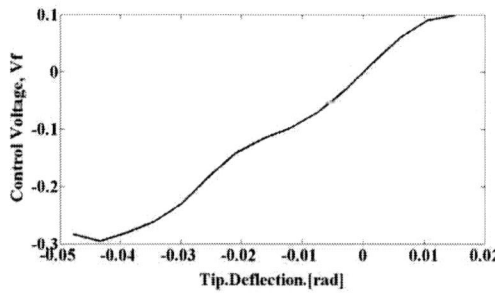

Fig. 9. Membership functions and control voltage of fast subsystem adaptive network based fuzzy logic controller

$$U = \frac{\sum\limits_{i=1}^{N} W_i Z_i}{\sum\limits_{i=1}^{N} W_i} \qquad (15)$$

Slow and fast subsystem adaptive network based fuzzy logic controllers' control surfaces are shown in Figs. 13 and 14. Because of these controllers are SISO system, their control output graphics are not surface.

TABLE I.
SLOW SUBSYSTEM ADAPTIVE NETWORK BASED FUZZY LOGIC CONTROLLER RULE BASE

	Angular Displacement Error ($U_{\theta e}$)				
	Z	PS	PM	PB	PBB
Control Voltage (V_s)	-0.0143	0.2165	0.3752	0.6487	4.7680

TABLE II.
FAST SUBSYSTEM ADAPTIVE NETWORK BASED FUZZY LOGIC CONTROLLER RULE BASE

	Tip Deflection (α)				
	NB	NM	NS	Z	PS
Control Voltage (V_f)	-0.2552	-0.1771	-0.1097	0.0198	0.0969

IV. EXPERIMENTAL RESULTS

The effectiveness of the proposed control schemes has been tested by means of real time experiments on a laboratory single flexible link robot manipulator. The photograph of the flexible robot manipulator considered in this study is given in Fig. 10. The flexible link robot is a Quanser FLEXGAGE thin aluminium beam with strain gage mounted at its base. The gage is calibrated to output 1 volt per 1 inch of tip deflection. The robot consists of a DC servo and a flexible link attached to the motor's shaft. The servo is voltage driven Quanser SRV02-ET model equipped with angular speed and position sensors. Interaction with the controlled plant is provided by Quanser UPM 2405 power module and by Quanser MultiQ PCI card, as shown in Fig. 10. WinCon software developed by Quanser allows communication with Simulink. This makes control by Matlab possible. The schematic diagram of the system considered in this study is given in Fig. 11.

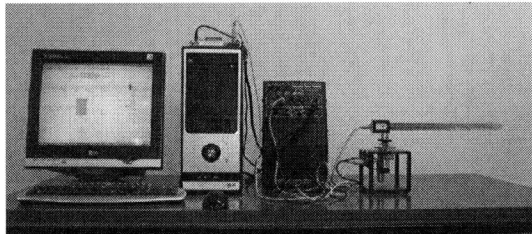

Fig. 10. Photograph of the single flexible link robot manipulator experimental set up

The objective in this experiment is to tune a controller for the rotary flexible link robot manipulator. The proposed controller allows us to command a desired tip angle position, and eliminates the link's vibrations while maintaining a fast response.

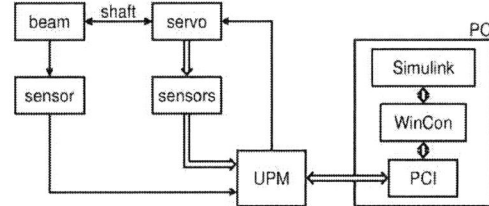

Fig. 11. Schematic diagram of the system

TABLE III.
PHYSICAL AND GEOMETRIC CHARACTERISTICS OF FLEXIBLE ROBOT MANIPULATOR

Symbol	Parameter	Value	Unit
J_h	Motor inertia	0.0020	kg.m^2
L	Flexible Arm length	0.4500	m
h	Flexible Arm height	0.0200	m
d	Flexible Arm thickness	0.0008	m
ρ	Linear density	0.1333	kg/m
EI	Flexural rigidity	0.1621	N.m^2

Substituting the physical parameters in Table 3 of the flexible arm and the other components of the system into the dynamic model equations (3), the state-space form of the system can be obtained as:

$$\begin{bmatrix} \dot{\theta} \\ \dot{\alpha} \\ \ddot{\theta} \\ \ddot{\alpha} \end{bmatrix} = \begin{bmatrix} 0 & 0 & 1 & 0 \\ 0 & 0 & 0 & 1 \\ 0 & 566.46 & -37.02 & 0 \\ 0 & -921.77 & 37.02 & 0 \end{bmatrix} \begin{bmatrix} \theta \\ \alpha \\ \dot{\theta} \\ \dot{\alpha} \end{bmatrix} + \begin{bmatrix} 0 \\ 0 \\ 65.11 \\ -65.11 \end{bmatrix} V_m$$
$$(16)$$

where V_m is the servomotor voltage. This basic equation was used to simulate the control system for a trial before experiments.

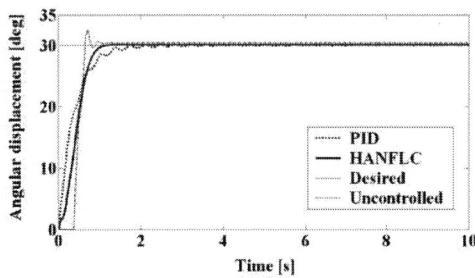

Fig. 12. Performance of the angular displacement control

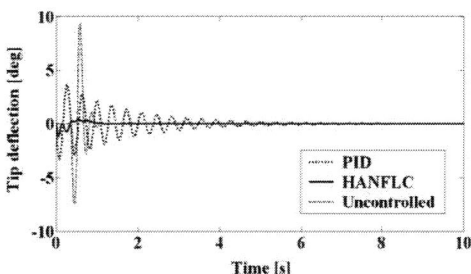

Fig. 13. Performance of the tip deflection control

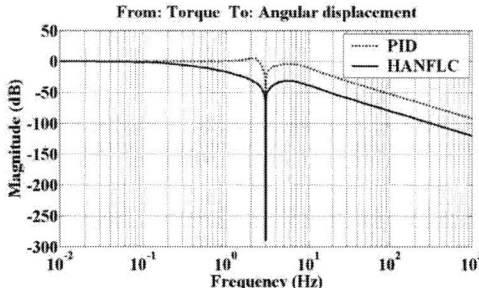

Fig. 14. Frequency responses of the controllers for the angular displacement

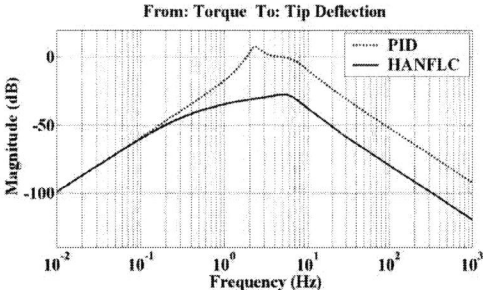

Fig. 15. Frequency responses of the controllers for the tip deflection

First of all, hierarchical PID controllers are applied to a single flexible link robot manipulator, and then their input and output datas are obtained to set up the network base of the adaptive network based fuzzy inference system (ANFIS). The gains of the PID controller are tuned by using the constants determined through a trial and error method, (Kp=15; Ki=0.3; Kd=5 for slow subsystem), (Kp=1.5; Ki=0.3; Kd=1 for fast subsystem). 3000 training and checking datas are used for training of artificial neural network (Optimization method is hybrid method, Error tolerance is zero, Epoch number is 60). After training the network, fuzzy inference system structure is built up. Number of membership functions and type of membership functions are decided. A few experiments are carried out obtaining appropriate fuzzy inference system for position and deflection control of a single flexible link robot manipulator. The rule base of both fast subsystem and slow subsystem consists of 5 (IF ... THEN) rules.

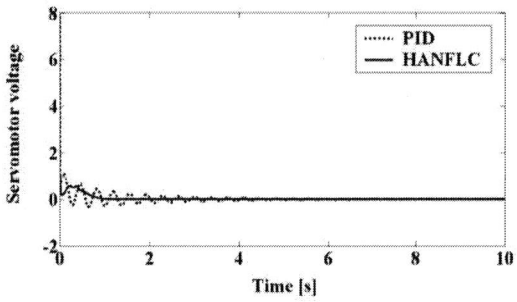

Fig. 16. Servomotor control voltages of the controllers

In experimental studies, performances of PID control and hierarchical adaptive network based fuzzy logic control (HANFLC) are obtained and compared with each other. For desired rotation motion of thirty degree,

controllers' performances are applied to a single flexible link robot arm. When robot arm moves, tip deflection of the arm must be close to zero degree. For these desire, angular displacement and tip deflection graphics are obtained. In Figs. 12 and 13, it is shown the comparison of the PID, the proposed HANFLC controllers with an uncontrolled response about angular displacement and tip deflection control of flexible link. From these results, it can be said the HANLC control is more effective than the PID control. Because of the HANFLC controller's base is depend on the PID controller training data sets. The HANFLC control increased the PID control performance on a single flexible link robot arm. Tip deflection comparisons of controllers show that the fast adaptive network based fuzzy logic controller influences deflection of the arm very good mean. The PID controller effect to tip deflection of the arm is worse than its angular displacement control. Maximum deflection 4 degrees greater than the desired value. The experimental results show the effectiveness of the proposed controller in reducing the end-point vibration of the flexible link and tracking the desired angular displacement.

Frequency responses of the controllers for angular displacement and the tip deflection are given in Figs. 14 and 15. It can be seen from the figures; the HANFLC controllers increase the system damping; thus, increasing the isolation in the region of the system resonance, and producing higher isolation at higher frequencies, lower isolation at lower frequencies for both angular displacement and tip deflection compared to the PID controller.

The comparison of the control voltages for both controllers is given in Fig. 16. The proposed controller requires a smaller voltage compared to the control voltage required in PID controller. The proposed controller shows smaller fluctuation in the control voltage compared to the PID controller.

V. CONCLUSION

Hierarchical adaptive network based fuzzy logic control (HANFLC) of a single flexible link robot manipulator with revolute joint is investigated in this study. The tip end of a single flexible link robot manipulator traces a desired position signal under the action of an external driving torque. Considered robot manipulator consists of a single flexible link without tip mass. Using different type of applications of fuzzy logic controller in the studies is aimed to get the end of the flexible robot manipulator to desired position and terminate vibrations on arm while it moves. A two-time-scale adaptive network based fuzzy logic controller is applied to control the system in this study. Adaptive network based fuzzy inference system (ANFIS) uses a hybrid learning algorithm to identify parameters of Sugeno-type fuzzy inference systems. It applies a combination of the least-squares method and the back propagation gradient descent method for training fuzzy inference system (FIS) membership function parameters to emulate a given training data set. The controller has two subsystems. The fast-subsystem reduces the vibration of the flexible robot manipulator end point, and the other slow-subsystem controller controls desired position. Hierarchical fuzzy rules are used to reduce size of the inference engine. The control of the vibration and position of a single flexible link robot manipulator end point are realized successfully by such

980

this approach while the end-effector traces desired position signal. Experimental results obtained by using experiment system of a single flexible link robot arm are presented and physical trend of the results are discussed. Experimental results of the flexible manipulator illustrate a successful performance of the proposed controller. The HANFLC has been optimized to reduce the rotation and tip deflection error. Results obtained for the HANFLC are in good agreement with the calculated reference input for the rotation and tip deflection. The proposed controller has good performance on a single flexible link robot manipulator.

ACKNOWLEDGMENT

This work is supported by the Coordinatorship of Selçuk University's Scientific Research Projects (Project no.: 06401020).

REFERENCES

[1] M. Kalyoncu and F. M. Botsalı, "Effect of axial shortening on the vibration of elastic robot arm", International Symposium, Second Turkish-German Joint Computer Application Days, Konya, Turkey, 1998, pp.313-324.

[2] Z. X. Shi, Eric H. K. Fung and Y. C. Li, "Dynamic modelling of a rigid-flexible manipulator for constrained motion task control", Applied Math. Modelling, vol.23, 1999, 509-525.

[3] J. Knani, "Dynamic modelling of flexible robotic mechanisms and adaptive robust control of trajectory computer simulation–Part I", Applied Mathematical Modelling, vol.26, 2002, pp.1113-1124.

[4] A. Ankaralı, M. Kalyoncu, F. M. Botsalı and T. Şişman, "Mathematical modelling and simulation of a flexible shaft-flexible link system with end mass", Mathematical and Computer Modelling of Dynamical System, vol.10, no.3-4, 2004, pp.187-200.

[5] J. Lin, F. L. Lewis, "A symbolic formulation of dynamic equations for a manipulator with rigid and flexible links", Internat. J. Robotics Res., vol.13, no.5, 1994, pp.454-466.

[6] M. Kalyoncu and F. M. Botsalı, "Dynamic modelling of a robot manipulator with a translating and rotating elastic arm (in Turkish)", 3rd International Mechatronics Design and Modelling Workshop, Ankara, Turkey, 1997, pp.237-247.

[7] M. Kalyoncu, F. M. Botsalı, "Vibration analysis of an elastic robot manipulator with prismatic joint and a time varying end mass", The Arabian Journal For Science And Engineering, vol.29, no.1C, 2004, pp.27-38.

[8] S. K. Dwivedy, P. Eberhard, "Dynamic analysis of flexible manipulators, a literature review", Mechanism and Machine Theory, vol.41, 2006, pp.749–777.

[9] M. Kalyoncu, "Mathematical modelling and dynamic response of a multi-straight-line path tracing flexible robot manipulator with rotating-prismatic joint", Applied Mathematical Modelling, vol.32, no.6, 2007, pp. 1087-1098.

[10] C.A. Monje, F. Ramos, V. Feliu, B.M. Vinagre, "Tip position control of a lightweight flexible manipulator using a fractional order controller", IET Control Theory and Applications, vol.1, no.5, 2007, pp.1451-1460.

[11] A. Sanz and V. Etxebarria, "Experimental control of a single-link flexible robot arm using energy shaping", International Journal of Systems Science, Vol. 38, No. 1, 2007, pp.61–71

[12] B. Subudhi, A.S.Morris, "Fuzzy and neuro-fuzzy approaches to control a flexible single-link manipulator", Proceedings of the Institution of Mechanical Engineers. Part I: Journal of Systems and Control Engineering, vol.217, no.5, 2003, pp.387-400

[13] M. A. Arteaga, B. Siciliano, "On tracking control of flexible robot arms", IEEE Transactions on Automatic Control, vol.45, no.3, 2000, pp.520-527.

[14] I. Dul and I. Karcz–Dul, "Algorithms of trajectory planning with constrained deviation from a given end-effector path", Robotica, vol.22, 2004, pp.633-642.

[15] A. A. Ata and T. R. Myo, "Optimal point-to-point trajectory tracking of redundant manipulators using Generalized Pattern Search", International Journal of Advanced Robotic Systems, vol.2, no.3, 2005, pp.239-244.

[16] R. Burkan and İ. Uzmay, "A model of parameter adaptive law with time varying function for robot control", Applied Mathematical Modelling, vol.29, 2005, pp.361–371.

[17] P. Huang, Y. Xu and B. Liang, "Tracking trajectory planning of space manipulator for capturing operation", Int. Journal of Advanced Robotic Systems, vol.3, no.3, 2006, pp.211-218.

[18] W. S. Owen, E. A. Croft and B. Benhabib, "Real-time trajectory resolution for a two-manipulator machining system", J. of Robotic Systems, vol.22 (Supplement), 2006, pp.S51–S63.

[19] T.H. Lee, S.S. Ge, Z.P. Wang, "Adaptive robust controller design for multi-link flexible robots", Mechatronics, vol.11, no.8, 2001, pp.951-967.

[20] M. R Akbarzadeh-Totonchi, M. Jamshidi, N. Vadiee, "Hierarchical fuzzy controller using line-curvature feature extraction for a single link flexible arm", IEEE International Conference on Fuzzy Systems, vol.1, 1994, pp.524-529.

[21] J. Lin, "Flexible link robot arm control by a hierarchical fuzzy logic approach", IEEE International Conference on Plasma Science, vol.2, 2002, pp.1138-1143.

[22] F.M. Caswara, H. Unbehauen, "A neurofuzzy approach to the control of a flexible-link manipulator", IEEE Transactions on Robotics and Automation, vol.18, no.6, 2002, pp.932-944.

[23] K. Y. Kuo, J. Lin, "Fuzzy logic control for flexible link robot arm by singular perturbation approach", Applied Soft Computing, vol.2, 2002, pp.24–38.

[24] J. Lin, "Hierarchical fuzzy logic controller for a flexible link robot arm performing constrained motion tasks", IEE Proc. - Control Theory Appl., vol.150, no.4, 2003, pp.355-364.

[25] T. S. Chiang, J. Lin, "A new design of hierarchical fuzzy hybrid position/force control for flexible link robot arm", Proceedings of the American Control Conference, vol.6, 2003, pp.5239-5244.

[26] J. Lin, F. L. Lewis, "Two-time scale fuzzy logic controller of flexible link robot arm", Fuzzy Sets and Systems, vol.139, no.1, 2003, pp.125-149.

[27] H. Emara, A. L. Elshafei, "Robust robot control enhanced by a hierarchical adaptive fuzzy algorithm", Engineering Appl. of Artificial Intelligence, vol.17, 2004, pp.187–198.

[28] J. Lin, "A vibration absorber of smart structures using adaptive networks in hierarchical fuzzy control", Journal of Sound and Vibration, vol.287, no.4-5, 2005, pp.683-705.

[29] A. Jnifene, W. Andrews, "Experimental study on active vibration control of a single-link flexible manipulator using tools of fuzzy logic and neural networks", IEEE Transactions on Inst. and Measurement, vol.54, no.3, 2005, pp.1200-1208.

[30] S. Kamalasadan, A.A. Ghandakly, "A neural network parallel adaptive controller for dynamic system control", IEEE Transactions on Instrumentation and Measurement, vol.56, no.5, October, 2007, pp.1786-1796.

[31] H.R. Karimi, M.J. Yazdanpanah, "A new modeling approach to single-link flexible manipulator using singular perturbation method", Electrical Eng., vol.88, no.5, 2006, pp.375-382.

[32] H.R. Karimi, M.J. Yazdanpanah, R.V. Patel, K. Khorasani, "Modeling and control of linear two-time scale systems: Applied to single-link flexible manipulator", Journal of Intelligent and Robotic Systems: Theory and Applications, vol.45, no.3, 2006, pp.235-265.

[33] J. Lin, Z. Z. Huang, P. H. Huang, "An active damping control of robot manipulators with oscillatory bases by singular perturbation approach", Journal of Sound and Vibration, vol.304, 2007, pp.345–360.

[34] J. Lin, Z. Z. Huang, "A hierarchical fuzzy approach to supervisory control of robot manipulators with oscillatory bases", Mechatronics, vol.17, no.10, 2007, pp.589-600.

Digital Controlled High Speed Synchronous Motor

Zdeněk Čeřovský[1], Jaroslav Novák[2], Martin Novák [2], Marek Čambál[2]

Czech Technical University in Prague, Prague, Czech Republic

[1]Faculty of Electrical Engineering, [1]Department of Electrical Drives and Traction

[1]e-mail: *Cerovsky@fel.cvut.cz*

[2]Faculty of Mechanical Engineering, [2]Department of Instrumentation and Control Engineering

[2]e-mail: *Jaroslav.Novak@fs.cvut.cz* , *Martin.Novak2@fs.cvut.cz* , *Marek.Cambal@fs.cvut.cz*

Abstract— Contribution deals with predictive torque control of the high speed permanent magnet synchronous motor. Such motors are used for electrically driven compressors (E-booster) on turbocharged gasoline or diesel engines. Theory and experimental results from the laboratory are revealed.

Keywords— Adjustable speed drive, Automotive electronics, Adaptive control, Synchronous motor, Control of Drive

I. Introduction

Permanent magnet synchronous machines are broadly used in the propulsion technique. Synchronous motors are very robust and can be used in almost all hard conditions. They are reliable especially in high speed drives. Compressors for turbocharged gasoline or diesel engines are such cases.

Turbocharged combustion engines are produced for applications where the mass of the engine is one of many very important parameters. It is so in railway, ship and car transportation. The green house gases production especially the production of carbon oxide and nitrogen oxides is very important parameter recently. Compressors of turbocharged combustion motors are usually driven by gas turbine taking the power from exhausted gas. Compressor blows the air into inlet tube with compression of 4,5 up to 5,5 .

Turbocharger benefits:

- Possibility to increase fuel quantity into cylinder
- Increase engine power with 50% or more
- Fuel consumption diminishing
- Fuel burning time in cylinder prolongation
- Green house gases emission diminishing. Diminishing especially of carbon oxide and nitrogen oxides emissions.
- Better intake valves cooling

Disadvantage of compressors driven by a turbine is the fact that they are unable to deliver needed air amount for the combustion engine in the full range of engine revolutions and power. Following the construction of the turbine and compressor it happens that in low revolutions of the engine the turbine does not give enough power and compressor does not give enough air pressure. In high revolutions the turbine power may be higher then needed on the contrary. In this case it would be possible to drive with the overflowing power an electric generator and to supply the gained power into the internal vehicle network.

The turbine driven compressors have relatively low dynamic. The reason lies in the fact that after fuel addition it takes time to raise the engine revolutions to make higher the exhaust of gas, to accelerate the turbocharger, to raise pressure on the compressor and to apply it into inlet tube.

The ideal compressor drive should be therefore controlled with respect to the fuel quantity and engine revolutions. Its dynamics should quickly follow the engine fuel control.

Turbo compressors are driven mostly by turbine today. It can be added either an electric motor mounted on the shaft between turbine and compressor or the separate additional compressor driven by special electric motor.

In the first case the velocity of the motor is the same as the velocity of the turbo compressor. In the second case it is possible to compromise the optimal velocity for the compressor with the optimal velocity for the electric motor.

New control manner of the torque of an electric permanent magnet synchronous motor for combustion engine is described in this paper.

II. Discussion of Posiible Aproaches to High Speed Permanent Magnet Synchronous Motor Torque Control

From the hardware point of view, the standard solution for powering a permanent magnet synchronous motor is a three phase bridge inverter. In the case of automotive industry, the most perspective element is a power FET, considering that the motor is working with a small voltage, maximally 48V, and a switching frequency over 20 kHz with minimal losses in the semiconductor element. At our test stand, the motor has a nominal voltage of 400V and we are therefore using an IGBT inverter.

The choice of controller structures for high speed permanent magnet synchronous motors can be based on several criteria. In the case of a high speed motor for automotive industry, the goals are maximum efficiency, robustness and minimal price while maintaining reasonable but not extreme control qualities. Also the question of EMC is important.

Three control strategies come in question for the torque control of high speed permanent magnet synchronous motor: direct non-linear torque control, simple electronic commutation methods and linear vector control.

Direct non-linear control promises very good control loop dynamics, but requires high frequency of changes in the controlled system and high computational power of the controller. For the high speed motor, this would mean an increase in the switching frequency of the inverter, meaning higher switching looses and lower efficiency. The demands for high computational power of the controller would also mean its increase in price. It would also mean less favorable conditions from the EMC point of view. For this reasons, we are not considering direct non-linear control in this paper. Only some experiments with on-off control of instantaneous phase current values were made while synchronizing with the rotors position. However, to achieve sufficient control quality, it would be necessary to increase the inverter's switching frequency so this way was abandoned.

Simple methods that use electronically commutated motors are easily implemented even in the cheapest control units; there are even application oriented programmable circuits for this area. The methods are based on the production of square waves of the motor's phase currents whose value is changing 6 times per period. These methods allow to use simple and cheap position sensors with only 6 positions per revolution or to implement sensor less approaches based on measuring of the induced voltage. These methods are mainly used for low power high speed motors and can also be used in the replacement of some non electric drives in automotive industry with higher power motors. The disadvantage of these methods is generally the lower quality of control and in some cases also the impossibility to use generator braking. In the future, we would like to focus also on these methods.

Linear control methods can be divided into two groups: control of currents in transformed rectangular coordinate system in the d,q axes and control of instantaneous phase current values with linear controllers. In this case, the requested value of phase currents is synchronized with the rotors instantaneous position. The control in the d,q transformed coordinate system can give in principle higher quality of control, especially for high speed motors, but the algorithm is more complex and demands higher computational power from the control system. On the contrary, the control of instantaneous current values is simpler, but it has much higher demands on the controller operation as it has to work with changing values of the controlled value.

In our work we want to focus on a comparison between the properties of both methods of linear control of high speed motors. The comparison will be done from the point of view of control quality, algorithm complexity demands, necessary PWM switching frequency and required number of measured rotor positions.

The work was initiated by implementing methods based on instantaneous current values control.

III. TORQUE CONTROL DESCRIPTION

Torque control uses the basic formula for torque calculation:

$$T = 1.5 \cdot p_p \cdot (F_d \cdot i_q - F_q \cdot i_d) \tag{1}$$

F_d is the d-axis magnetic linkage component, F_q is the q-axis magnetic linkage component, i_d is the d-axis stator current component, i_q is the q-axis stator current component, p_p is pole pair number. F_d, F_q, i_d, i_q are instantaneous values. The coordinate system is firmly coupled on the magnetic rotor flux in no load and that means that it is firmly coupled on the space rotor position.

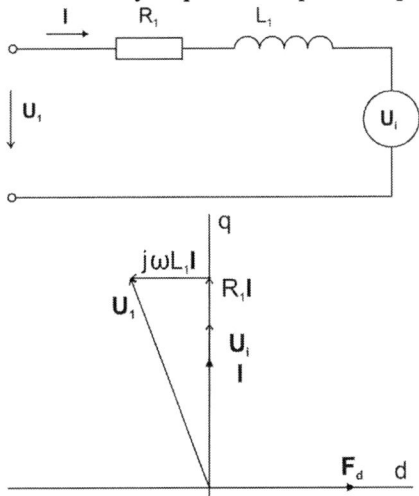

Fig.1 Vector diagram for torque controlled synchronous motor with permanent magnets excitation and equivalent scheme.

At the torque control of the motor with permanent magnets mounted on the rotor the motor does not need magnetic excitation. We do not want to change the magnetic flux with rotor current and therefore the i_d current component is not required. The remaining i_q current component is controlled to lead 90^0 the space rotor position. Therefore it is in phase with the induced voltage U_i. Under this presumption it holds for the torque:

$$T = 1.5 \cdot p_p \cdot F_d \cdot I \tag{2}$$

On Fig.1 the U_1 is terminal stator voltage, and U_i is the induced stator voltage component. Magnetic flux weakening is not assumed so far. Therefore demagnetizing stator current component i_d is zero.

IV. FEEDBACK TORQUE CONTROL STRUCTURE

Designed new torque control structure rises from previously published papers but was enhanced with respect to recent experiments performed in Research Center of Combustion Motors and Automobiles at Czech Technical University in Prague, [4], [5] , [6], [7].

Aim of the paper is to improve features of phase current control of high speed synchronous motors with permanent magnets on the rotor. To use on-off controllers [4] is disadvantageous because of the high frequency of needed calculations in the controller and high switching frequency of power electronic devices in the converter which can cause electromagnetic compatibility difficulties and lower converter efficiency. Linear control has simple algorithms [4], [5] and [8]. Therefore it is often used on simple, cheap universal microcontrollers. Its computing demanding is low. This feature is advantageous in high revolution drives.

Linear control with controllers in all three phases, with the outgoing signals u_{Ar} u_{Br} u_{Cr} proportional to the phase voltage from controllers to control the PWM modulator has following disadvantages:

First disadvantage is, that when the integral component of the current controller is used, then the zero mean values of the per unit voltages u_{AR}, u_{BR}, u_{CR} set as input on the PWM modulator, are not guaranteed. The time curves of u_{AR}, u_{BR}, u_{CR}, can be shifted to the control zone margin. This disadvantage can be eliminated by using only two per unit voltages of u_{AR}, u_{BR}, u_{CR} and by calculating the third from the condition $u_{AR} + u_{BR} + u_{CR} = 0$.

Important improvement can be reached by modification of the control structure and calculation of each per unit voltages u_{AR}, u_{BR}, u_{CR} as sum of the controller output $\overline{u}_{PI}$ and induced p.u. phase voltage $\overline{u}_i$ as gives Eq (3):

$$\overline{u_R} = \overline{u}_{PI} + \overline{u_i} \qquad (3)$$

$\overline{u_i}$ represents the compensation of induced stator voltage U_i shown on Fig. 1. The is denoted in the scheme on Fig.2 as u_{comp}. The use of this method improves markedly the dynamics of the response by which the controller reaches the requested current value.

The controller controls then the motor terminal voltage for which it holds

$$u_A = R_1 \cdot i + L_1 \cdot di / dt + u_i \qquad (4)$$

R_1 and L_1 are resistance and inductance of the stator winding ($L_d = L_q = L_1$), u_i is the induced voltage and i is the motor current. Performed experiments [5] proved the improvement. Nevertheless the control system had at stable proportional and integral controller constants at different revolutions different features. Therefore the adaptation of constants for different instantaneous current derivation and for instantaneous induced voltage was implemented. This adaptation depends on motor revolutions and was implemented by using following equations.

$$K_P = K_{P0} + C_{P1} \cdot \Delta i_{set} / \Delta t + C_{Pu} \cdot u_i$$
$$K_I = K_{I0} + C_{I1} \cdot \Delta i_{set} / \Delta t + C_{Iu} \cdot u_i \qquad (5)$$

K_P is actual controller proportional constant,

K_I is actual controller integral constant,

K_{P0} is actual controller proportional constant corresponding to optimal proportional constant at low revolutions,

K_{I0} is actual controller integral constant corresponding to optimal integral constant at low revolutions,

C_{P1} and C_{I1} are constants corresponding to the weight of current derivation in the proportional respectively integral constant calculation,

Δi_{set} is the changed requested current value during period of controller calculation,

Δt is period of controller calculation

C_{Pu} and C_{Iu} are constants corresponding to the weight of instantaneous induced phase voltage u_i calculated in controller units.

Experiments were performed with this arrangement on synchronous motor 4kW, 1500min-1, 50Hz. Sufficient control quality was reached on high speed motor with revolutions 5000min-1.

To develop the control system for higher motor revolutions a new testing working place was established.

V. NEW TESTING PLACE

New testing place for testing static and dynamic features of high speed synchronous motors with permanent magnets on the rotor was built in Research Center of Combustion Motors and Automobile Technique Josef Božek at the Czech Technical University in Prague.
Working place consists of:
Frequency controlled asynchronous dynamometer 2,3kW, 350V, 70 000min-1 , 0,3Nm. (Fig.3)

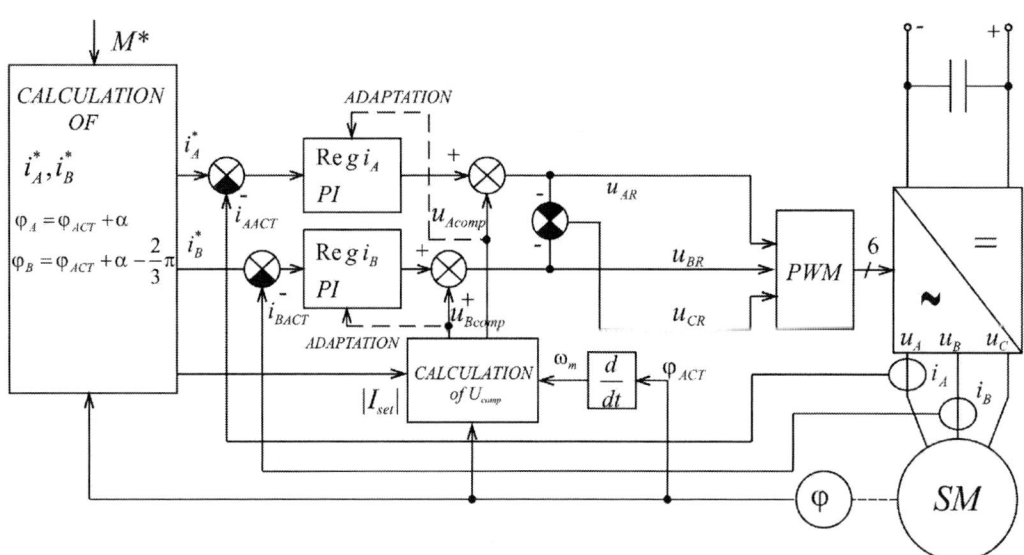

Fig. 2 Block scheme of the control system

984

Fig.3 High speed dynamometer with tested Synchronous motor

Two pole high speed synchronous motor 2,9kW, 400V, 40 000min⁻¹, 0,7Nm with permanent magnet rotor (Fig.3). IGBT converter with microelectronic controller.

Motor was supplied from its separate IGBT converter. Chokes in series with stator winding and inductance of 2,4mH were implemented to diminish stator current pulsation. Synchronous motor has two pole resolver integrated to determine the space rotor position.

Special electronic unit was developed to supply the resover with exciting signal 10kHz and evaluates the signals from the resolver. The unit has 12 bites resolution and discriminates 4096 positions in one motor revolution. The information on rotor absolute space position is transferred via parallel bus after reset. Information on relative rotor position is transferred in sensor signal form IRC. More information is in [3].

System DSP TMS320F240 for torque control was used in experiments. Switching frequency was 5 kHz and discrimination of 4096 positions in one motor revolution was used. The calculation power of the used DSP has limited the motor revolutions on the testing place. It was possible to control the motor up to 11 000 revolution min⁻¹ with good results. But later on experiments on new working place showed that at higher revolutions a new problem occurs.

The actual current shapes were delayed with respect to the reference values. A high control error between reference and actual current amplitudes was measured.

It was not possible to improve the control quality by no controller constants modification nor by adaptation of C_{P1}, C_{I1}, C_{Pu}, C_{Iu} constants.

This problem could be solved by quicker controller. Such a controller is prepared for the future. Higher control signal frequency for PWM will be also possible when the new controller will be used.

In the mean time a new solution was found.

The frequency of stator current is changing in broad limits in case of high speed motor. The same holds on stator reactance and additional phase reactance. The voltage drop on these reactance's is changing in broad limits in case of high speed motor too. The voltage drop $\Delta U = j\omega(L_1+L_{pr})I$ on the Fig.1 is very high on high speed motor. Therefore the control structure was completed with the compensation of total inductance voltage drop. The

claims on the controller work were notably reduced using this strategy. This idea was implemented by changing the calculation of u_{comp} in the modified structure. Its amplitude is calculated as sooner from the curve of induced voltage but the induced voltage is introduced into the control structure as compensating voltage in such a manner that it leads the induced voltage with the angel φ_{comp}. Practically it is in such case in phase with the motor terminal voltage. The conditions for the work of the controller are enhanced a lot by this procedure because the controller does not need to control the whole motor terminal voltage U_1 but only the difference between it and the compensating voltage u_{comp}. The calculation of the angel φ_{comp} uses following idea:

At high revolutions there is possible to omit the influence of the winding resistance in the vector diagram. The induced voltage equals $\omega\Psi_d$, where Ψ_d is magnetic stator linkage. Induced voltage is turned with the angle φ_{comp} and used as compensation voltage.

We calculate the angle for turning the induced voltage from the Eq. (6):

$$tg\,\varphi_{comp} = \frac{\omega L\;I}{\omega F_d} = \frac{LI}{F_d} \qquad (6)$$

When we assume constant magnetic stator linkage it holds

$$\varphi_{comp} = arctg\left(kL\;I\right) \qquad (7)$$

Block scheme of control system is on Fig. 2. The symmetric asynchronous PWM with pulse modulation frequency of 5kHz was used for converter control. Calculation tact in the control structure was synchronized with the pulse modulation frequency. Sampling period of the controller calculation matches with the period at which the actual values are transferred in the PWM modulator. These periods are 100µs. In the same period is the actual rotor space position evaluated.

VI. RESULTS OF TESTING EXPERIMENTS

Many testing experiments were performed on the testing place in steady state and transient conditions. Only tests in motor mode were performed. The dynamometer during testing was in speed control mode and that means that it worked in constant given speed. Tested motor worked in torque mode and that means that its torque remained constant, respectively was changed in a step. Tests were performed up to 10800min⁻¹ revolutions.

Detail of measured values is shown in Fig.5. Current reference phase value is (Ibzad) and actual value is (Ibskut). Compensating phase voltage value is (Ubcomp) at revolutions 10800 min-1 and torque 0,16Nm. Reference current value is violet, actual current value is blue and compensating voltage is red. In curves is leading of compensating voltage good seen. On following figures Fig. 6 and Fig. 7 are dynamic curves of reference and actual current values after step by step torque command seen. Reference current curve is marked (Ibzad), actual current curve is marked (Ibskut). Reference current amplitude is marked |I|zad, actual current amplitude is marked |I|skut.

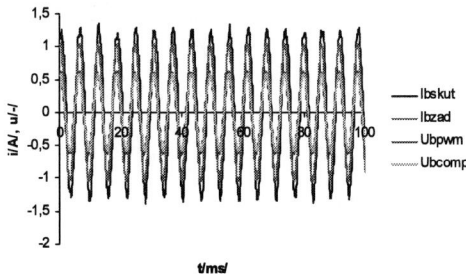

Fig. 4. Curves of actual (blue) and reference (violet) current phase values, voltage u_R given in PWM modulator (red) and compensation voltage (light blue) without compensation of inductance influence, n=10000min⁻¹, T=0,08Nm

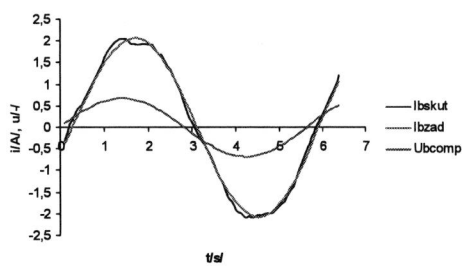

Fig.5. Detail curves of synchronous motor reference and measured actual currents and of compensation voltage at n=10 800 min⁻¹

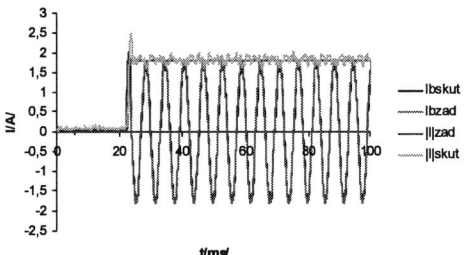

Fig. 6a. Reference current step from zero to 1,8 A (T=0,14Nm) at revolutions n=10 000 min⁻¹ and actual current response.

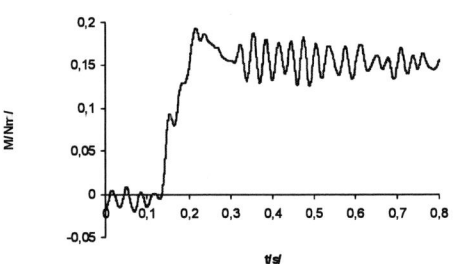

Fig. 6b. Step response to the change of requested current amplitude from 0 to 1,8A (0,14Nm) by n=10 000 min⁻¹ – torque waveform measured with a tenzometric sensor on dynamometer

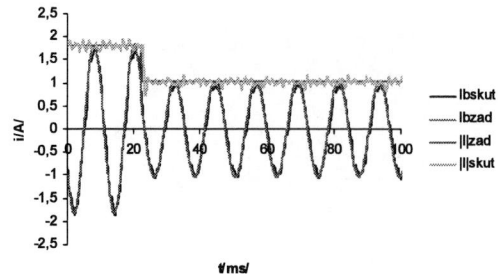

Fig. 7a. Reference current step from 1,8A to 1A, (T=0,14->0,08Nm) at revolutions n=5000 min⁻¹ and actual current response.

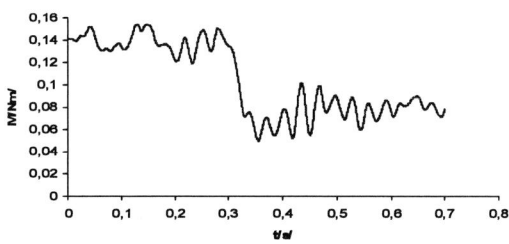

Fig. 7b. Step response to the change of requested current amplitude from 1,8A to 1,0A (M=0,14->0,08Nm) by n=5000min⁻¹ - torque waveform measured with a tenzometric sensor on dynamometer

In following Fig. 8 curves of actual current (blue), reference current amplitude (red), actual current amplitude (light blue) and revolutions violet are shown at no load accelerating with torque 0,14Nm. The diagram ends at 7680 min⁻¹ revolutions.

Actual curves of phase current (blue), reference curves of phase current (violet), voltage u_r that is transferred on the PWM modulator (red) and compensation voltage (light blue) on steady state operation 5000 min⁻¹and torque 0,14Nm are shown in Fig.9. With respect to low revolutions and low load the leading of the compensation voltage before the current is not visible.

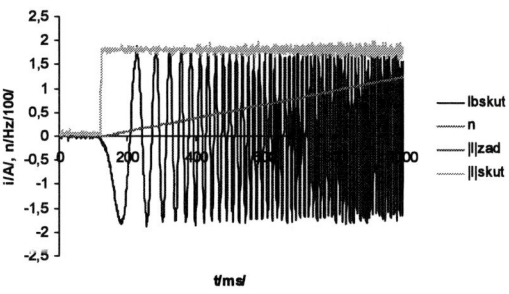

Fig. 8. No load speeding up with torque 0,14Nm up to 7680 min⁻¹ revolutions.

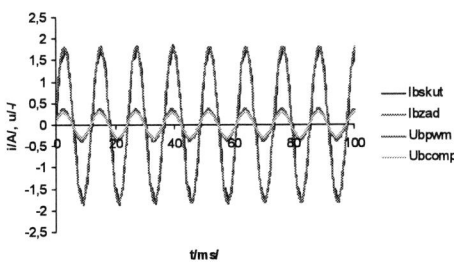

Fig. 9 Reference and actual current curves, voltage ur on the PWM modulator input and compensating voltage on steady state operation at 5000 min⁻¹ and 0,14Nm.

VII. CONCLUSIONS

As it is clear from the measured waveforms, the torque control quality is good in both steady states and transition states while maintaining a simple algorithm. In transition states a new stable value is reached in some milliseconds. An increased ripple of the current (current is proportional to internal motor torque) for increased speed is evident from figure 8. It is a principal property given by a decreasing ratio between the period of fundamental harmonic of the inverter output voltage and PWM period. The controller structure is in principle based on the prediction of instantaneous phase voltage values calculated from rotor position, speed and requested torque. The current controllers in two phases of the circuit are then only working with a small control deviation corrections. The described method has a simple algorithm and good control quality, but can be used only in areas where generator braking is not necessary. For our area of high speed non-electric motor replacement this is however not an issue.

At the present time, we are implementing a new, more powerful controller based on TMS320F2812 (Fig. 10,11) which should allow us to increase the PWM switching frequency to 10-20 kHz and to test the motor for speeds up to 40 000 min⁻¹. We will also focus on experimentally comparing the properties of the described simple controller structure with a d,q rectangular co-ordinate system control for high speed synchronous motor torque control. Next area will be to optimize the rotor position measurement system with regard to reduce the number of measured positions per one revolution while maintaining good torque control quality or to use a sensor less approach.

Fig. 10. Experimental inverter for high speed motor

Fig. 11. Drive controller with TMS320F2812 and position measurement add-on

ACKNOWLEDGMENT

Authors would like to thank the Josef Božek Research Center for the financial support of the work – project 1M0568.

REFERENCES

[1] Novotny D.W., Lipo T.A.:Vector Control and Dynamics of AC Drives, Oxford Science Publications Nr 41, 1996

[2] ⬜emus J., Hamata V.: Transient Stability Analysis of Synchronous Motors. Czechoslovak Academy of Sciences, Academia/Prague 1990

[3] ⬜ambál, M. - Novák, M. - Novák, J.: Study of Synchronous Motor Rotor Position Measuring Methods. In 13th International Conference on Electrical Drivers and Power Electronics. Zagreb, Croatia: KoREMA, 2005, p. 62-66. ISBN 953-6037-42-4.

[4] Novák, M. - ⬜ambál, M. - Novák, J.: Application of Sinusoidal Phase Current Control for Synchronous Drives. In ISIE 2006 International Symposium on Industrial Electronic [CD-ROM]. Montreal, Canada: IEEE Industrial Electronic Society, 2006, ISBN 1-4244-0497-5.

[5] ⬜ambál M. - Novák M. - Novák J.: Possibilities to Increase the Quality of Phase Current Control for Synchronous Motors, The 15th Mediterranean conference on Control and Automation - MED 07 - 27. - 29.6. 2007, Athina, Grece

[6] Uhlí⬜ I. - ⬜ambál, M. - Novák, M. - Novák, J.: Synchronous Motor Current Controler Quality Augmentation with Adaptive Control, The 33rd Annual Conference of the IEEE Industrial Electronics Society. Taipei, Taiwan: IEEE Industrial Electronics Society, 2007

[7] ⬜ambál, M. - Novák, M. - Novák, J.: Synchronous Motors Phase Current Adaptive Control, Proceedings of the 8th International Carpathian Control Conference. Košice: Technical University , BERG Faculty, 2007

[8] Šimánek, J. – Doleček, R., - ⬜erný, O.: PMSM Drive Control based on Sinusoidal Commutation Control of Brushless Motor, Proceedings of the International Carpathian Control Conference ICCC 08, Sinaia, Romania, May 2008

Analysis of combustion engine - electric Linear generator set operation

Jiří Pavelka

CTU Prague, Faculty of Electrical Engineering /Department of Electric Drives and Traction, Prague, Czech Republic,
e-mail: *pavelka@fel.cvut.cz*

Abstract— This paper analyses a power system formed by a combustion engine and an electric linear generator. Such system allows to transmit the power from a combustion engine directly to an electric generator without a crank mechanism and a fly wheel. The classical solution is compared with new system. Results of calculations are presented. They show that some optimum speed exists for the concrete aggregate and that increasing of speed requires uneconomic increasing of the linear generator size. The prototype of such aggregate was designed, constructed and is now tested.

Keywords— Adjustable speed generation system, Linear drive, Design, Automotive application, Efficiency

I. INTRODUCTION

Transport of persons, materials and goods represents significant part of human working and energy consummation. A combustion engine is used in a great deal of applications. A hybrid drive that allows an operation with a higher energetic efficiency is now intensively studied in many research institutions.

The hybrid drive combines a combustion engine with an electric generator and variable speed electric motor for wheel driving. The combustion engine operates with optimal constant load and speed. A demanded variable load and speed of wheels is controlled in the electric power part and the control part of variable speed drive.

A classical solution is realized by a DC or AC rotating electric generator. The theory of electric machines knows also linear machines. Linear motors are used in many different drive units. Application of a linear generator is not so frequent. This paper analyzes one of possible applications of a linear generator in a combination with a combustion engine.

II. DESCRIPTION OF A CLASSICAL POWER SYSTEM

Pistons of a combustion engine run in a linear direction between two dead points. The volume V, pressure p, entropy s and temperature T change in the piston workspace during this motion. Two diagrams – the working diagram p = f(V) and the heat diagram T = f(s) – are used for a quantitative engine description. A typical working diagram of one combustion two stroke engine in p.u. system $p_{pu} = f(V_{pu})$ is shown in the Fig. 1.

A crank mechanism is used in a classical combustion engine – electric generator machine unit for the transfer of the piston linear motion to the generator shaft rotating motion. A typical diagram $p_{pu} = f(\varphi_{Cpu})$ is in the Fig. 2 where φ_{Cpu} is the crank angular displacement.

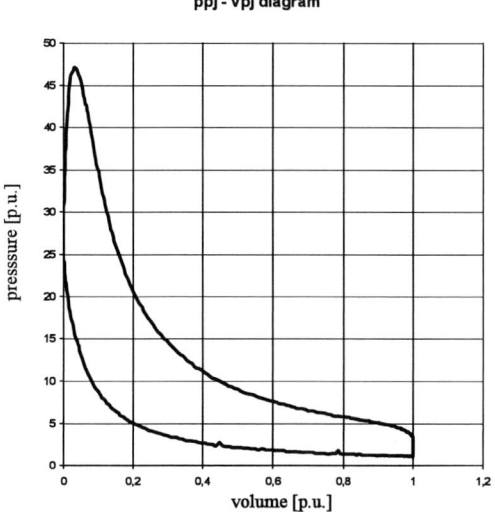

Fig. 1 Typical working diagram ppj = f(Vpj) for one piston in p.u. system

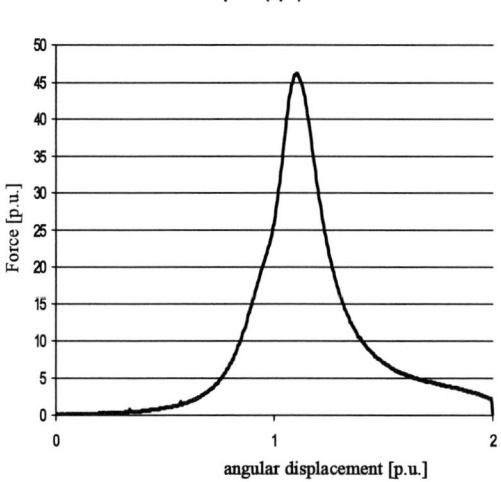

Fig. 2 Typical diagram $p_{pu} = f(\varphi_{pu})$

978-1-4244-1741-4/08/$25.00 ©2008 IEEE

The piston speed changes from zero value in one dead point to the maximum speed value v_{Pmaxpu} and back to the zero value in the opposite dead point during one half of the shaft revolution. An acceleration and a deceleration of the piston, the piston rod and the part of crank mass m_{compj} requires the force F_{accpj} given by the next equation

$$F_{accpj} = m_{compj} * \frac{dv_{Ppu}}{dt_{pu}} = m_{compj} * \frac{d^2 h_{Ppu}}{dt_{pu}^2} \qquad (1)$$

The motion of the generator shaft is given by the motion equation in p.u. system as follows:

$$M_{tan\,pj}\left(\varphi_{Cpu}, \Omega_{pu}, t_{pu}\right) - M_{Lpu}\left(\varphi_{Cpu}, \Omega_{pu}, t_{pu}\right) =$$
$$= J_{celpu} * \frac{d^2 \varphi_{pu}\left(\Omega_{pu}, t_{pu}\right)}{dt_{pu}^2} \qquad (2)$$

where M_{tanpj} is a tangential torque caused by the sum of the piston force F_{Ppj} and the acceleration force F_{accpj} in p.u. system

M_{Lpj} is the load torque of the electric generator in p.u. system

Ω_{pu} is the angular speed of shaft in p.u. system

J_{compu} is the moment of inertia of a rotor in p.u. system

The relation between the tangential torque M_{tanpj} and the total force ($F_{Ppu} + F_{accpu}$) in the piston motion direction represents a complicated function of the shaft angle φ_{Cpu} as follow

$$M_{tan\,pj} = -F_{Ppj} * \frac{h_{P\max pj}}{2} \sin\left(\pi * \varphi_{pj}\right) * \left(1 - \frac{\cos\left(\pi * \varphi_{pj}\right)}{\sqrt{4 * \chi^2 - \sin^2\left(\pi * \varphi_{pj}\right)}}\right)$$
$$(3)$$

where h_{Pmaxpj} is the stroke otherwise distance between both piston dead points in p.u. system

χ is the ratio of the piston rod length to the stroke

The integral of M_{tanpj} within one shaft revolution represents the active energy of the piston and it must equal to the integral of M_{Lpu} during the same time in the steady operation. A dynamic torque tangential component ΔM_{tanpj} (the difference between the instantaneous torque M_{tanpj} and the average torque $M_{tan(AV)}$) causes a change of the angular speed Ω_{pj} . The integral of ΔM_{tanpj} during one shaft revolution is zero in the steady state operation. The dynamic component causes only a change of a shaft angular speed Ω_{pj} in some range around the average shaft speed $\Omega_{(AV)pj}$. If the average shaft angular speed increases then the value of the acceleration force F_{accpj} increases with square of the angular speed. The acceleration force can therefore represent a main part of the torque in the high speed operation.

The differential of an energy ($\Delta M_{tanpj} * d\varphi_{Cpj}$) is accumulated in the rotor moment of inertia J_{compu} and causes the instantaneous change of the angular speed ($J_{compu} * (d\Omega_{pu}/dt_{pu})$). The electric generator loads the shaft only with the average torque $M_{tan(AV)pj}$.

In practice it is used more than one piston and the crankshaft with shifted positions. This allows to decrease the dynamic tangential torque ΔM_{tanpj} and to minimize the angular speed deviation.

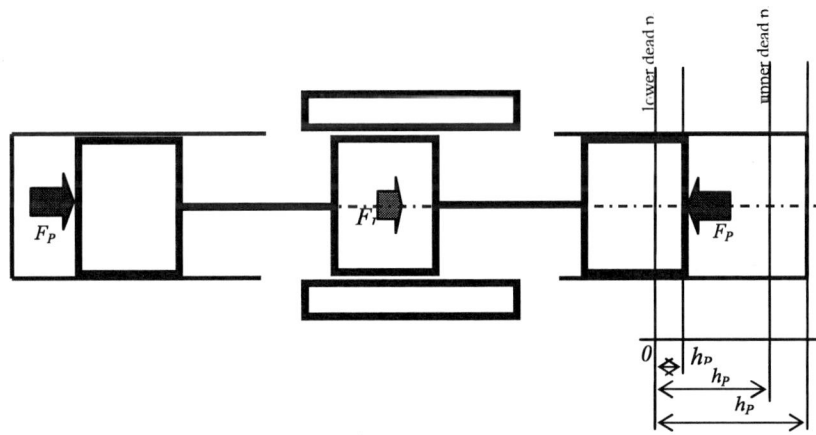

Fig. 3 Mechanical diagram of an analyzed power system

III. DESCRIPTION OF THE ANALYZED NEW POWER SYSTEM

The mechanical diagram of an analyzed new power system with a linear generator is in the Fig. 3. Two pistons move between their lower dead points and upper dead points. The result force $F_{Ppu} = (F_{P1pu} - F_{P2pu})$ (see Fig. 4) acts directly against the linear generator force F_{Lpu}. The motion of a moving part is given by the motion equation in p.u. system as follow:

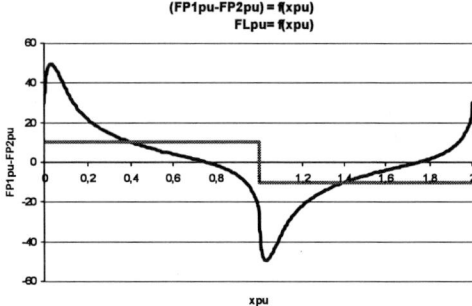

Fig. 4 Resulting force of pistons (F_{P1pu}-F_{P2pu}) (black) and average generator force F_{Lpu} (red) as function of the position x_{pu}

$$F_{Lpu}(h_{Ppu}) - F_{Ppu}(h_{Ppu}) = F_{upu}(h_{Ppu}) = \tag{4}$$
$$= m_{Ppu} * h_{PZpu} * \Omega_{Ppu}^2 * K_2 * \frac{d^2 h_{Ppu}}{dt_{pu}^2}$$

where m_{Ppu} is the total moving mass in p.u. system

h_{PZpu} is the length of the stroke

Ω_{Ppu} is "the angular speed" of the piston in p.u. system ($\Omega_{Ppu} = f_{Ppu} = n_{Ppu}$)

K_2 is the constant defined as

$$K_2 = \frac{m_0 * h_{PZ0}}{F_0 * T_0^2}$$

m_0 is the base mass in [kg]

h_{PZ0} is the base length of the stroke in [m]

F_0 is the base force in [N]

T_0 is the base time of one half-cycle in [s]

Dependence of p.u. linear generator force on p.u. piston position

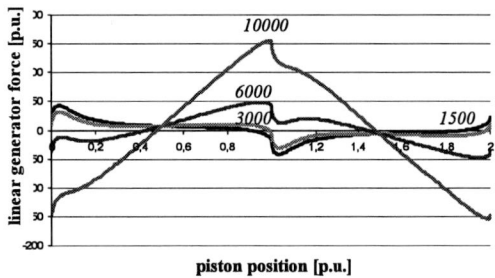

Fig. 5 Demanded force F_{Lpu} as function of the position x_{pu} for different ppm. (Black-1500, green-3000, blue-6000, red-10000)

The forces F_{Ppu} and F_{Lpu} are functions of the piston position h_{Ppu} and therefore the equation (4) must be rearranged to the form

$$F_{Lpu}(h_{Ppu}) - F_{Ppu}(h_{Ppu}) = F_{upu}(h_{Ppu}) = \tag{5}$$
$$= m_{Ppu} * h_{PZpj} * \Omega_{Ppj}^2 * K_2 * \frac{d(v_{pu}^2)}{2 * dh_{Ppu}}$$

We assume that the average piston force $F_{P(AV)pu}$ is equal to the linear generator load F_{Lpu}. The solution of the equation (5) gives only one "natural" value of the vibration period $T_{natur} = 1/f_{natur}$ and therefore also one "natural" value of periods per minute n_{natur} for concrete parameters. If we can decrease or increase the number of periods per minute n we must use the linear generator for creation of the demanded force. The results of the calculation are shown in the Fig. 5. It can be seen that the maximum generator force F_{Lmaxpj} changes its direction and increases rapidly. Some results of the calculation for concrete parameters are given also in following table. The "natural" value of periods per minute ("natural speed") is 1870 ppm for this concrete case.

TABLE I.

periods per minute	[min⁻¹]	1500	3000	6000	10000
vibration period	[ms]	40	20	10	6
maximal force. F_{Lmaxpu}	[p.u.]	42,27	31,27	48,17	154,53
effective force F_{Lefpu}	[p.u.]	15,82	12,19	25,82	88,37
F_{Lmaxpj} / $F_{L(AV)pu}$	[-]	4,14	3,06	4,72	15,14
F_{Lmaxpj} / F_{Lefpu}	[-]	2,67	2,56	1,87	1,75

The effective value of the linear generator force F_{Lmaxpu} increases rapidly with the "speed". This effective value bears a proportion to the linear generator size. But greater generator size requires larger moving mass and that requires a greater generator force. It indicates an existence of some optimal speed for concrete parameters of the system and the additional speed increase leads to uneconomic or unreal solutions.

IV. PROTOTYPE OF LINEAR GENERATOR

Fig. 6 View on a prototype of the combustion engine with the electric linear generator

A prototype of this power system was designed and produced in the Faculty of Electrical Engineering in Prague [1]. A view on this prototype is in Fig. 6 and principal block diagram of the control loops is in Fig. 7. This prototype is now tested. Test results confirmed existence of "natural speed" and existence of "speed" limit for realization of the high speed power system.

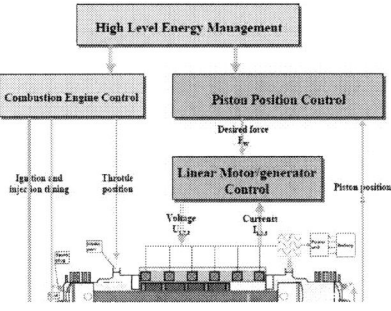

Fig. 7 Principal block diagram of the control loops

The prototype of the linear generator was realized on the base of a standard synchronous linear motor L3SK075P-2415 manufactured by company VUES Brno Czech Republic [2]. This linear motor has a three phase winding on the stationary part and permanent magnets on the moving part. The schematic cross section of this generator can be seen in Fig. 8. The stationary part has 24 slots with 12 coils. The slot pitch is 15 mm and therefore the length of the stationary active part is 360 mm. The moving part has the permanent magnet pitch equal to 24/22*15=16,36 mm. It must be longer for about the stroke 40 mm i.e. 400 mm. The nearest suitable standard moving part is manufactured with 32 permanent magnets in the length 512 mm and therefore a special moving part was designed and manufactured with 25 permanent magnets.

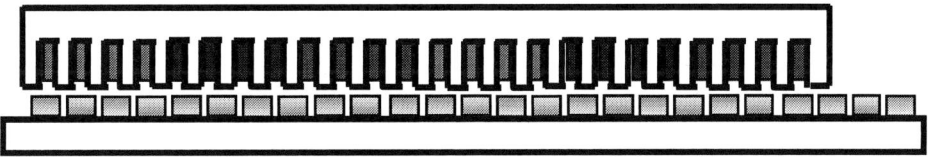

Fig. 8 Schematic cross section of the linear generator

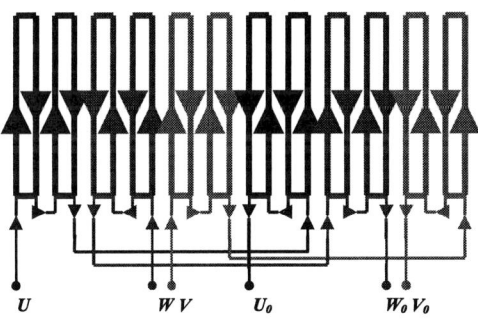

Fig. 9 Interconnection of coils and connection to the three phase power supply

The interconnection of coils and a connection to the three phase supply source is shown in Fig. 9. Each coil is put in two adjoining slots and therefore adjoining teeth are magnetized with opposite magnetic polarity. The result of different pitches of stationary and moving parts is the cause of an electromagnetic force. The position of moving part can be controlled by the amplitude and the phase shift of supplied currents. One period of the supply current represents the displacement of two permanent magnet pitches i.e. 32,72 mm. Move

of about 40 mm requires 40/32,72=1,22 period of the supplied current.

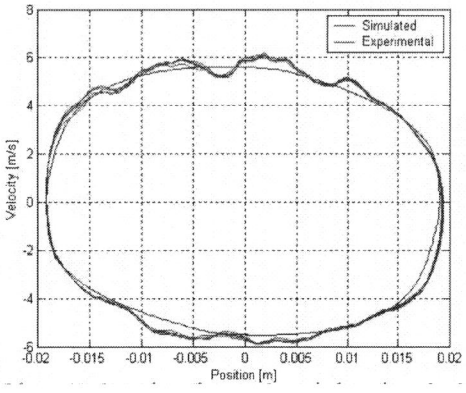

Fig. 10 Comparison of desired and real values of the piston velocity as a function of its position

The control system operates with the main position loop. The output of the position controller is a desired value of the linear generator force for the slave controller of the linear motor. It transforms this desired generator force to desired values of the phase currents. An oscillogram of a desired and a real value of the piston velocity as the function of the position can be seen in Fig. 10.

V. WAYS FOR REDUCTION OF THE DESIRED LINEAR GENERATOR FORCE

It can be seen from the equation (5) that a relative needed force for an acceleration and a deceleration F_{upu} of the studied system moving part depends linearly on its moving mass m_{Ppj} and on square of its "angular speed" Ω_{Ppj}. A reduction of moving mass m_{Ppj} is one way for increasing the system "angular speed". Above described standard solution of a linear motor needs a ferromagnetic circuit on its moving part. The use of two stationary parts located on both sides of the moving part allows removing this ferromagnetic circuit from the

moving part and to reduce its mass approximately to one half. It allows increasing of the system "angular speed" in the ratio of $\sqrt{2}$. The schematic cross section of a linear generator with the moving part without ferromagnetic circuit is shown in Fig. 11.

This solution is used in the second prototype that is now being manufactured.

The consumed work of the needed force for the acceleration and the deceleration F_{upu} during one period of the piston move is equal zero. This work is accumulated in the moving mass during acceleration and returned from it to the linear generator during the deceleration. The linear generator and its power supply operate then as accumulator of energy. The physics knows also others energy accumulators as steel springs, pneumatic springs or electric condensers.

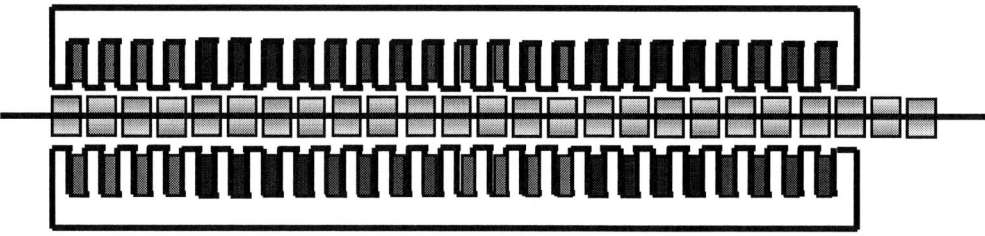

Fig. 11 Schematic cross section of a linear generator with the moving part without ferromagnetic circuit

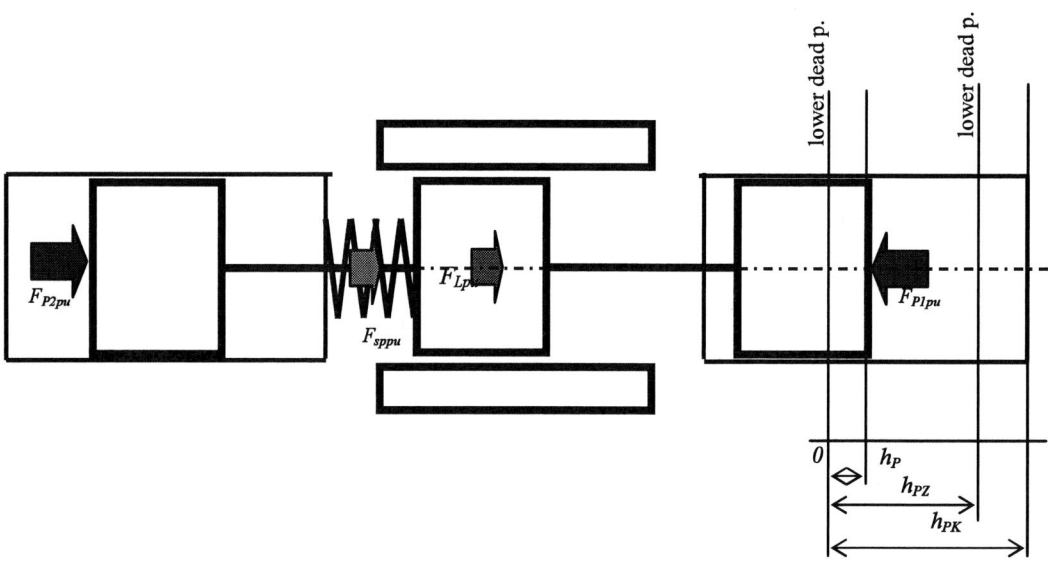

Fig. 12 A mechanical diagram of combusted engine-linear generator with a spiral spring

A. Steel spring accumulator

A mechanical diagram of a combusted engine-linear generator with a spiral spring is shown in Fig. 12.

The dependence of the steel spring force F_{sppu} is a linear function of its deflection and can be written in the form

$$F_{sppu} = K_{sppu} * \left(\frac{1}{2} - h_{pj} \right) \qquad (6)$$

Elasticity factor K_{sppu} is a function of the spring sizes. Its change influences a time of the piston oscillation and maximum piston speed. The linear generator operates only with the average power of combustion engine. A possibility to use a flat spring or a spiral spring is studied in the present time.

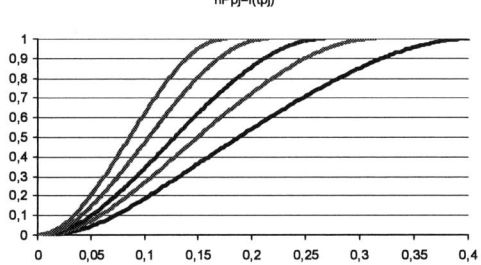

hPpj=f(tpj)

Fig. 13 Dependence of the relative piston position h_{Ppj} on the relative time t_{pj}

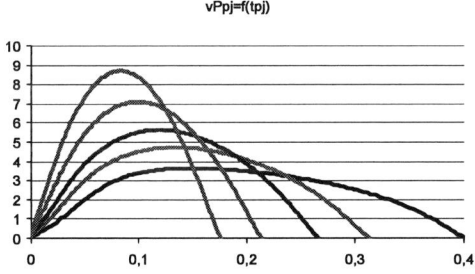

vPpj=f(tpj)

Fig. 14 Dependence of the relative piston speed v_{Ppj} on the relative time t_{pj}

Gas spring

A gas spring makes use a relation between changes of pressure p_{sp} in a closed room and its volume V_{sp}. This relation is called polytrophic compression and is described by the equation

$$V_{sp}^{n} * p_{sp} = const \qquad (7)$$

Compared with a steel spring the function $F_{sppu} = F(h_{Ppu})$ is nonlinear in this case. A possibility to use this type of spring is also studied in the present time.

B. Force of an electromagnet

Magnetic field in an air gap produces attractive force between its both faces. If two electromagnets are located in the opposite sides of the moving part then resulting force is their difference and both direction of the resulting force can be obtained. But this solution requires separate magnetic circuits for both electromagnets and magnetic material on the moving part. It requires sizable increasing of the moving mass. Preliminary calculations indicate that this type of an accumulator is practically unapplicable.

VI. CONCLUSION

The results of calculations and analyses showed that a direct transmission of the power from a combustion engine to a linear generator is possible and that some "natural" frequency exists for concrete parameters. These analyses also demonstrated that a high speed aggregate cannot be realized with linear generator only but it requires the use of some mechanical accumulator.

ACKNOWLEDGMENT

This paper summarizes results of theoretical studies, mathematical calculations, simulations and practical realization that were down in the "Josef Bozek Research Center of Engine and Automotive Engineering", workplace at CTU Prague, FEE.

REFERENCES

[1] P.Němeček, M.Šindelka, and O.Vysoký, "Control of Two-Stroke Free-Piston Generator" in *Proceeding of the 6th Asian Control Conference (IFAC) Bandung*, Bali 2006.

[2] "Linear motors series L3S and L3SK", in *Catalogues of VUES Company*, http://www.vues.cz/en/linear-motors.html

Closed Loop Control of AC Drive with LC Filter[1]

Jaroslaw Guzinski

Gdansk University of Technology/Faculty of Electrical and Control Eng., Gdansk, Poland, *jarguz@pg.gda.pl*,

Abstract— In electric drives with frequency converter a LC filter is used to smooth the voltages and currents supplying the AC motors. Unfortunately the LC filter introduces the phase shift and voltage drop on input and output of the filter. It makes difficult to implement precise control of the electric motors especially in the case of speed sensorless control methods. It is necessary to change the used control system and the state variables calculate block in case of the drive with the filter. In this paper a method for an asynchronous motor speed sensorless control in the system with inverter and LC output filter is presented. In the proposed system a nonlinear control method of the asynchronous motor and the speed and flux observer structure were modified. The system does not need any additional sensors except these which were previously used in the system without LC filter. The proposed electric drive was realized in simulation and experiment for 1.5kW asynchronous motor.

Keywords— asynchronous motor, passive filter, sensorless control, control of drive.

I. INTRODUCTION

High switching frequency in the frequency converters is a source of the numerous problems which appear in the modern electric drives [3]. In such drives the most meaningful problems are: currents which flow through motor parasitic capacitances into ground, faster motor bearings degradation, motor terminal overvoltages and disturbing of other electric equipment. That problems are specially serious in case when long cable to the motor is used.

To prevent these problems the LC filters are used [1], [2]. The LC filter is an additional circuit consisted of: inductors, capacitors and resistors placed between inverter output terminals and motor input terminals. The main purpose of that filter is to smooth currents and voltages waveform supplied motor and to eliminate of the unwanted high frequency harmonics which are results of high switching frequency of the transistors. On the output of the filter the voltage has sinusoidal shape. With the LC filter is possible to use the long cable connection to the motor - Fig. 1.

When inverter output filter is included into the electric drive the voltage drops and phase shift between filter input and output voltages and currents appear. Because of that in some modern sophisticated electric drives can appear problems with control of the motor. That problem is for example important when speed sensorless control methods are used. The principles of that control is that inverter output voltages and currents are equal to the motor input voltages and currents. In case of voltages and current discrepancy the motor region of the properly work could

be decreased or deleted. Because of that in a control method of electric drives with inverter output filter the important modifications are indispensable in measurement circuits or in control algorithms [4]-[8].

The simplest way to improve electric drive with inverter output filter is to introduce the additional sensors for motor voltages and current measurement [4], [5]. Unfortunately that solution is practically not accepted because LC filter is an external element in relation to the converter. More practically accepted solution is to leave inverter structure without changes and to provide some changes in the control and estimation algorithms [6]-[8].

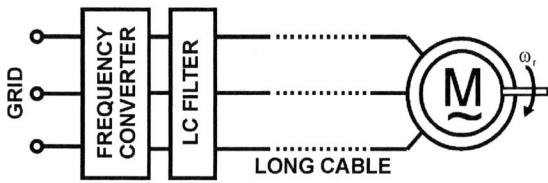

Fig. 1. Electric drive with frequency converter, LC filter and long cable connection to the AC motor

In the paper the solution is proposed for the asynchronous motor electric drive with the inverter output LC filter.

II. MODEL OF THE SYSTEM

A. LC Filter

In the investigated system the inverter output LC filter designed for use in system with 1.5kW motor and inverter was applied. Structure of the filter is presented in Fig. 2.

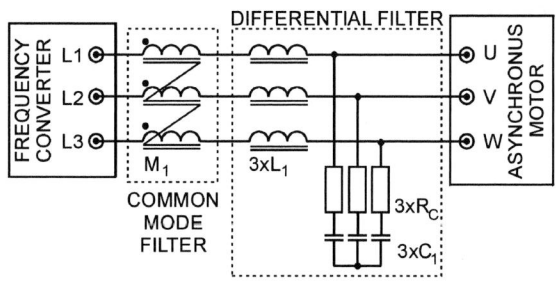

Fig. 2. Structure of the inverter output LC filter

The presented filter structure is connection of differential and common mode filter. Differential filter which smoothing motor voltages and current consist of L1, C1 and RC elements. Common mode filter which

---

[1] The scientific work funded as a development project by finance for scientific research in the years 2007 – 2009.

includes M1 inductor prevents leakage current flow. More detailed description of that filter was presented in [2].

In the investigated system the filter with parameter presented in Table I was used.

Table I.
LC FILTER PARAMETERS

Denotation	Value
L_1	4 mH
R_1	0,05 Ω
C_1	3 µF
R_C	1 Ω
M_1	14 mH
$L_{\sigma M1}$	0,7 mH

where L_1, C_1, R_C, and M_1 are denoted in Fig. 1, R_1 is resistance of the L_1 and M_1 chokes and $L_{\sigma M1}$ is leakage inductance of the M_1 choke.

For practical reasons, in the system, model the filter equivalent circuit (Fig. 3) was obtained after variable transformations form three phase system into orthogonal two phase system αβ. The zero component in that transformation was omitted because it has no influence on investigated control algorithm.

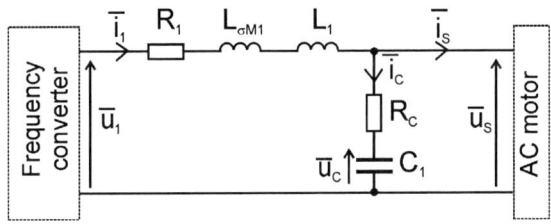

Fig. 3. Equivalent circuit of the differential inverter output LC filter

The model of the LC filer could be described using equations as follows:

$$\frac{du_{c\alpha}}{d\tau} = \frac{i_{c\alpha}}{C_1}, \qquad (1)$$

$$\frac{di_{1\alpha}}{d\tau} = \frac{u_{1\alpha} - R_1 i_{1\alpha} - R_c i_{c\alpha} - u_{c\alpha}}{(L_1 + L_{\sigma M1})}, \qquad (2)$$

$$i_{c\alpha} = i_{1\alpha} - i_{s\alpha}, \qquad (3)$$

$$u_{s\alpha} = R_c(i_{1\alpha} - i_{s\alpha}) + u_{c\alpha}, \qquad (4)$$

$$\frac{du_{c\beta}}{d\tau} = \frac{i_{c\beta}}{C_1}, \qquad (5)$$

$$\frac{di_{1\beta}}{d\tau} = \frac{u_{1\beta} - R_1 i_{1\beta} - R_c i_{c\beta} - u_{c\beta}}{(L_1 + L_{\sigma M1})}, \qquad (6)$$

$$i_{c\beta} = i_{1\beta} - i_{s\beta}, \qquad (7)$$

$$u_{s\beta} = R_c(i_{1\beta} - i_{s\beta}) + u_{c\beta}, \qquad (8)$$

where $\tau = 2\pi f_o t$ and f_o - electrical grid frequency.

An examples of commanded voltage of the inverter and real voltages and currents waveforms on input and output of the filter are presented in Fig. 4 - 6.

In case of the motor nominal load the phase shift of the voltages is less than 5 deg and filter voltage drop is about 6 V for 50 Hz of the motor fundamental voltage.

All the frequencies close to the filter resonant

frequency are dangerous for the inverter. During voltage modulation except switching frequency and fundamental frequency also other frequencies appear. This other frequencies are results of inaccuracy of PWM generation, dead time effects and nonlinearities of inverter transistors and diodes. Therefore to assure safe system operation, is necessary to use passive or active system for a resonance suppressing [15], [16]. In the system presented in this paper, more frequently used solution, with passive suppressing of the resonance was used.

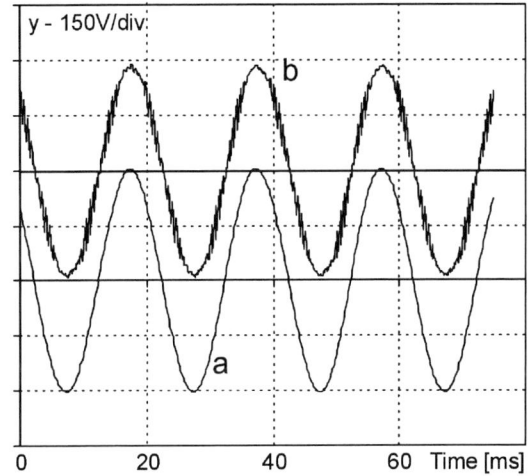

Fig. 4. Waveforms of the (a) inverter output commanded voltage and (b) output LC filter voltages

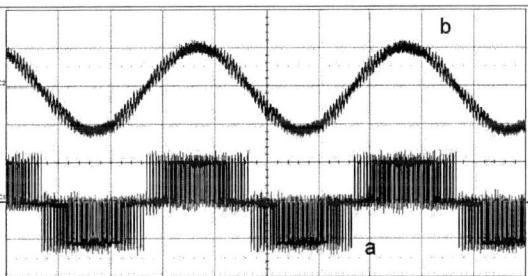

Fig. 5. Waveforms of the (a) input and (b) output LC filter voltages (a-500V/div, b-360V/div, 5ms/div)

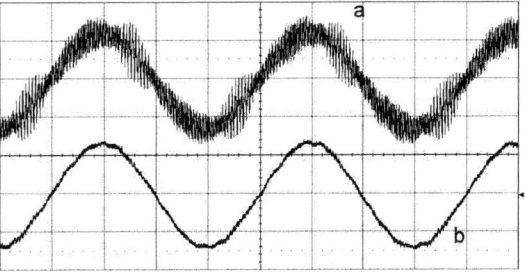

Fig. 6. Waveforms of the (a) input and (b) output LC filter currents (5A/div, 5ms/div)

B. Asynchronous motor

The nominal power, voltage and speed and other parameters of the tested motor are presented in Table II.

TABLE II
MOTOR NOMINAL PARAMETERS

Parameter	Value	Description
P_n	1.5 kW	power
U_n	300 V	phase to phase voltage
I_n	4.7 A	current
n_n	1420 rpm	speed
p	2	number of poles
f_n	50 Hz	supply voltage frequency
J	0.0038 kg·m^2	inertia
R_s	2.1 Ω	stator resistance
R_r	2.1 Ω	rotor resistance
L_s	0.263 H	stator inductance (leakage + mutual)
L_r	0.263 H	rotor inductance (leakage + mutual)
L_m	0.254 H	mutual inductance

For the asynchronous motor a mathematical model in the stationary rectangular coordinates $\alpha\beta$ was used [9]. As the state variables were chosen: stator current and rotor flux components and motor mechanical speed noted as $i_{s\alpha}$, $i_{s\beta}$, $\psi_{r\alpha}$, $\psi_{r\beta}$ and ω_r respectively. The equations of the motor model in per unit system (Table III) are as follows [9]-[12]

$$\frac{di_{s\alpha}}{d\tau} = -\frac{R_s L_r^2 + R_r L_m^2}{L_r w_\sigma} i_{s\alpha} + \frac{R_r L_m}{L_r w_\sigma} \psi_{r\alpha} + \\ + \omega_r \frac{L_m}{w_\sigma} \psi_{r\beta} + \frac{L_r}{w_\sigma} u_{s\alpha} \tag{9}$$

$$\frac{di_{s\beta}}{d\tau} = -\frac{R_s L_r^2 + R_r L_m^2}{L_r w_\sigma} i_{s\beta} + \frac{R_r L_m}{L_r w_\sigma} \psi_{r\beta} + \\ - \omega_r \frac{L_m}{w_\sigma} \psi_{r\alpha} + \frac{L_r}{w_\sigma} u_{s\beta} \tag{10}$$

$$\frac{d\psi_{r\alpha}}{d\tau} = -\frac{R_r}{L_r} \psi_{r\alpha} - \omega_r \psi_{r\beta} + R_r \frac{L_m}{L_r} i_{s\alpha} , \tag{11}$$

$$\frac{d\psi_{r\beta}}{d\tau} = -\frac{R_r}{L_r} \psi_{r\beta} + \omega_r \psi_{r\alpha} + R_r \frac{L_m}{L_r} i_{s\beta} , \tag{12}$$

$$\frac{d\omega_r}{d\tau} = \frac{L_m}{L_r J} \left(\psi_{r\alpha} i_{s\beta} - \psi_{r\beta} i_{s\alpha} \right) - \frac{1}{J_M} T_L , \tag{13}$$

where: R_r, R_s, L_r, L_s, L_m – motor equivalent circuit parameters, T_L – load torque, J_M – motor inertia and $w_\sigma = L_r L_s - L_m^2$.

III. CONTROL SYSTEM

In the investigated drive with the LC filter one of the modern control methods for asynchronous motor was used. That control method presented first time in [9] is the method of nonlinear control of the asynchronous motor and is noted as multiscalar model based (MMB) method.

In the presented control system only three sensors were used: two current sensors for inverter output currents an one voltage sensor for inverter input voltage measurement. Speed sensor was not used in control system so presented system is speed sensorless.

TABLE III
DEFINITION OF PER UNIT VALUES

Definition	Description
$U_b = \sqrt{3} U_n$	base voltage
$I_b = \sqrt{3} I_n$	base current
$Z_b = U_b / I_b$	base impedance
$m_b = (U_b I_b p)/\omega_0$	base torque
$\Psi_b = U_b / \omega_0$	base flux
$\omega_b = \omega_0 / p$	base mechanical speed
$L_b = \Psi_b / I_b$	base inductance
$J_b = m_b /(\omega_b \omega_0)$	base inertia
$\tau = \omega_0 t$	relative time

A. Nonlinear control method

The principle of the MMB method is to use only four new state variables x_{11}, x_{12}, x_{21} and x_{22} in order to describe asynchronous motor model. That MMB variables are defined as follows:

$$x_{11} = \omega_r , \tag{14}$$

$$x_{12} = \psi_{r\alpha} i_{s\beta} - \psi_{r\beta} i_{s\alpha} , \tag{15}$$

$$x_{21} = \psi_{r\alpha}^2 + \psi_{r\beta}^2 , \tag{16}$$

$$x_{22} = \psi_{r\alpha} i_{s\alpha} + \psi_{r\beta} i_{s\beta} . \tag{17}$$

In the MMB control the compensation of motor nonlinearities and internal motor state variables coupling is realized in decoupling block:

$$u_1 = \frac{w_\delta}{L_r} [x_{11}(x_{22} + \frac{L_m}{w_\delta} x_{21}) + m_1] , \tag{18}$$

$$u_1 = \frac{w_\delta}{L_r} (-x_{11}x_{12} - \frac{R_r L_m}{L_r w_\delta} x_{21} - \frac{R_r L_m}{L_r} \frac{x_{12}^2 + x_{22}^2}{x_{21}} + m_2) , \tag{19}$$

$$u_{s\alpha}^{com} = \frac{\psi_{r\alpha} u_2 - \psi_{r\beta} u_1}{\psi_r^2} , \tag{20}$$

$$u_{s\beta}^{com} = \frac{\psi_{r\alpha} u_1 - \psi_{r\beta} u_2}{\psi_r^2} . \tag{21}$$

With multiscalar variables the nonlinear asynchronous motor control structure is divided to two partially independent subsystems: mechanical and electromagnetic.

The multiscalar model of induction motor has been determined from the derivations of the new MM variables (14)-(17) taking into account the differential equations of stator current and rotor flux vectors (9)-(13) as the next [9]:

Mechanical subsystem:

$$\frac{dx_{11}}{d\tau} = \frac{L_m}{J_M L_r} x_{12} - \frac{1}{J_M} T_L , \tag{22}$$

$$\frac{dx_{12}}{d\tau} = -\frac{1}{T_v} x_{12} + u_1 , \tag{23}$$

Electromagnetic subsystem:

$$\frac{dx_{21}}{d\tau} = -2 \frac{R_r}{L_r} x_{21} + 2R_r \frac{L_m}{L_r} x_{22} , \tag{24}$$

$$\frac{dx_{22}}{dt} = -\frac{1}{T_v}x_{22} + u_2. \qquad (25)$$

where: $\dfrac{1}{T_v} = \dfrac{R_r w_\sigma + R_s L_r^2 + R_r L_m^2}{w_\sigma L_r}$

More detailed information about MMB control are presented in [9], [13], [14].

The structure of the base MMB control of the motor state variables is presented in Fig. 6.

In Fig. 7 variables denoted with ^ are variables estimated in the calculation and observer block.

B. Modification of the MMB control method in the system with LC filter

MMB method assures precise control of asynchronous motor in steady state and in transients. However including of inverter LC filter into drive system is a reason of necessarily of control algorithm change. In the MMB control system presented in the chapter III.A the control system sets the motor supply voltages $u_{s\alpha}^{com}$ and $u_{s\beta}^{com}$.

Unfortunately, in case when LC filter is inserted, the inverter is unable to generate that voltages on input of the filter. On output of the filter the different voltage appears.

For full control of the system with LC filter it is necessary to control the LC filter state variables also [4]. The solution is to use additional controllers subordinated to the superior controllers. That idea was presented in literature for other control principle: for field oriented control [3], [6], [7]. The solution proposed there is applicable to the MMB control method also.

In Fig. 8 a block scheme of the MMB control system which includes LC filter state variables control is presented.

In the electric drive with inverter output filter an additional state variables which have to be controlled are motor stator voltage u_s and inverter output current i_1 [4], [6]. That variables are controlled by proportional-integrating (PI) controllers. In order to omit phase shift of the PI controllers new state variables are controlled in the dq rotating frame of references where position of d axis is linked to the motor stator voltage vector u_s. Angle position of d axis is noted as θ_{US}.

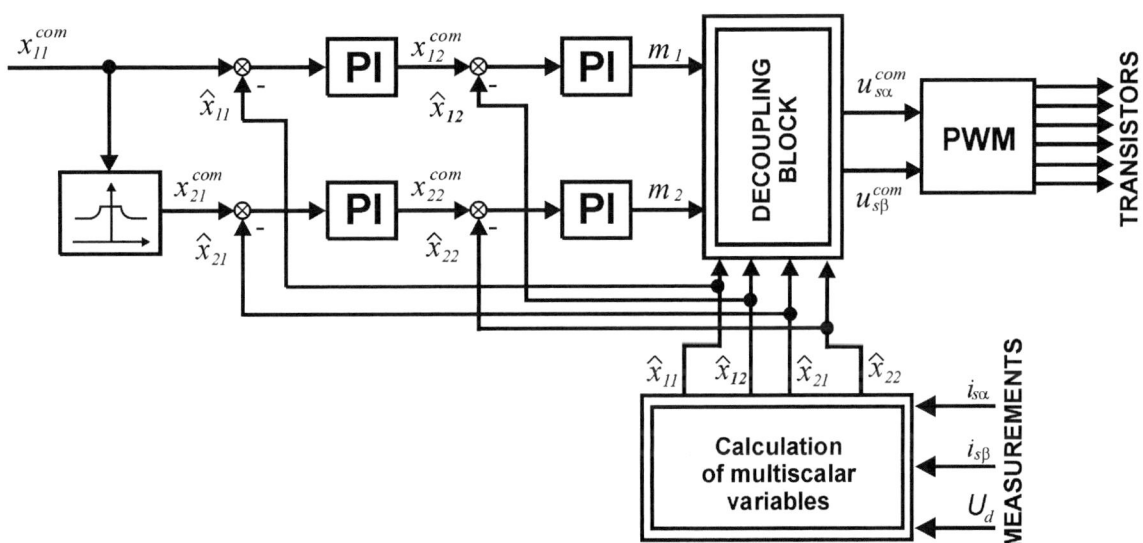

Fig. 7. Control system with multiscalar variables controllers will be presented in the final paper

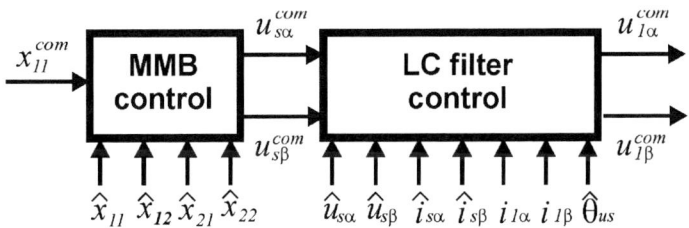

Fig. 8. Block diagram of the MMB control structure in the system with LC filter

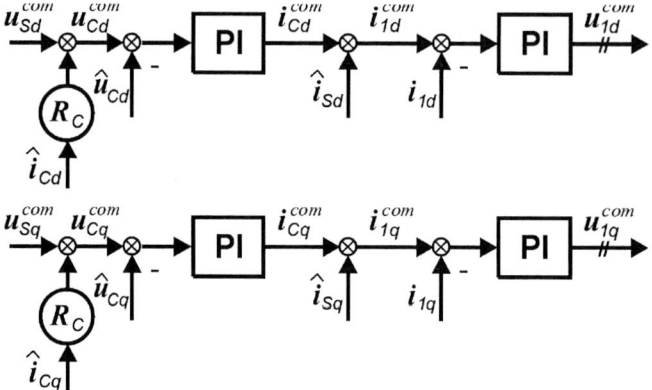

Fig. 9. Control system for LC filter state variables

Transformation of arbitrary variables x from stationary $\alpha\beta$ frame of references into rotating dq frame of references is specified as follows:

$$\begin{bmatrix} x_d \\ x_q \end{bmatrix} = \begin{bmatrix} \cos\theta_{US} & \sin\theta_{US} \\ -\sin\theta_{US} & \cos\theta_{US} \end{bmatrix} \begin{bmatrix} x_\alpha \\ x_\beta \end{bmatrix}, \tag{26}$$

and reverse transformation is as follows:

$$\begin{bmatrix} x_\alpha \\ x_\beta \end{bmatrix} = \begin{bmatrix} \cos\theta_{US} & -\sin\theta_{US} \\ \sin\theta_{US} & \cos\theta_{US} \end{bmatrix} \begin{bmatrix} x_d \\ x_q \end{bmatrix}, \tag{27}$$

where: x – adequately vector of stator voltage u_S or current i_1.

The structure of the LC filter control is presented in Fig. 9.

In the LC filter control structure control the influence of the damping resistance R_C was considered according to the filter model:

$$u_{Cd}^{com} = u_{Sd}^{com} + i_{Cd}R_c \tag{28}$$

$$u_{Cq}^{com} = u_{Sq}^{com} + i_{Cq}R_c \tag{29}$$

For the proper work of the control system the value of the damping resistance should be known.

IV. STATE VARIABLES OBSERVER

A. Base observer structure

Operating of modern complex asynchronous motor electric drives requires application of the calculating system which compute on-line the all state variables used in control system. Nowadays frequent solution in state variables calculation is to use a state observer. In the proposed, in the paper, control system the state observer presented earlier in [10], [11] was used. For stationary frame of references $\alpha\beta$ equations for that state observer are as follows:

$$\frac{d\hat{i}_{s\alpha}}{d\tau} = -\frac{R_s L_r^2 + R_r L_m^2}{L_r w_\delta}\hat{i}_{s\alpha} + \frac{R_r L_m}{L_r w_\delta}\psi_{r\alpha} + \frac{L_m}{w_\delta}\xi_\beta + \\ + \frac{L_r}{w_\delta}u_{s\alpha}^{com} + k_3\left(i_{s\alpha} - \hat{i}_{s\alpha}\right) \tag{30}$$

$$\frac{d\hat{i}_{s\beta}}{d\tau} = -\frac{R_s L_r^2 + R_r L_m^2}{L_r w_\delta}\hat{i}_{s\beta} + \frac{R_r L_m}{L_r w_\delta}\psi_{r\beta} - \frac{L_m}{w_\delta}\xi_\alpha + \\ + \frac{L_r}{w_\delta}u_{s\beta}^{com} + k_3\left(i_{s\beta} - \hat{i}_{s\beta}\right) \tag{31}$$

$$\frac{d\psi_{r\alpha}}{d\tau} = -\frac{R_r}{L_r}\psi_{r\alpha} - \xi_\beta + R_r\frac{L_m}{L_r}\hat{i}_{s\alpha} - k_2 S_b\psi_{r\alpha} + \\ + Sk_2 k_3\psi_{r\beta}(S_b - S_{bF}) + \\ + Sk_5\left((S_x - S_{xF})\psi_{r\alpha} - (S_b - S_{bF})\psi_{r\beta}\right) \tag{32}$$

$$\frac{d\psi_{r\beta}}{d\tau} = -\frac{R_r}{L_r}\psi_{r\beta} + \xi_\alpha + R_r\frac{L_m}{L_r}\hat{i}_{s\beta} - k_2 S_b\psi_{r\beta} + \\ - Sk_2 k_3\psi_{r\alpha}(S_b - S_{bF}) + \\ + Sk_5\left(-(S_x - S_{xF})\psi_{r\beta} - (S_b - S_{bF})\psi_{r\alpha}\right) \tag{33}$$

$$\frac{d\xi_\alpha}{d\tau} = -\omega_{\psi r}\xi_\beta - k_1\left(i_{s\beta} - \hat{i}_{s\beta}\right), \tag{34}$$

$$\frac{d\xi_\beta}{d\tau} = \omega_{\psi r}\xi_\alpha + k_1\left(i_{s\alpha} - \hat{i}_{s\alpha}\right), \tag{35}$$

$$\frac{dS_{bF}}{d\tau} = \frac{1}{T_{Sb}}\left(S_b - S_{bF}\right), \tag{36}$$

$$\frac{d\omega_{rF}}{d\tau} = \frac{1}{T_{KT}}\left(\omega_r - \omega_{rF}\right), \tag{37}$$

$$\frac{dS_{xF}}{d\tau} = \frac{1}{T_{Sx}}\left(S_x - S_{xF}\right), \tag{38}$$

$$S = \begin{cases} 1 & if \quad \hat{\omega}_{\psi r} > 0 \\ -1 & if \quad \hat{\omega}_{\psi r} \leq 0 \end{cases}, \tag{39}$$

$$S_x = \xi_\alpha\psi_{r\alpha} + \xi_\beta\psi_{r\beta}, \tag{40}$$

$$S_b = \xi_\alpha\psi_{r\beta} - \xi_\beta\psi_{r\alpha}, \tag{41}$$

$$\omega_{\psi r} = \omega_{rF} + R_r\frac{L_m}{L_r}\left(\frac{\psi_{r\alpha}\hat{i}_{s\beta} + \psi_{r\beta}\hat{i}_{s\alpha}}{\psi_{r\alpha}^2 + \psi_{r\beta}^2}\right), \tag{42}$$

$$\omega_r = \frac{\zeta_\alpha\psi_{r\alpha} + \zeta_\alpha\psi_{r\alpha}}{\psi_{r\alpha}^2 + \psi_{r\beta}^2}, \tag{43}$$

where: ξ_α, ξ_β - components of the motor electromotive forces, $\omega_{\psi r}$ – angular speed of the motor flux vector, k_1, k_2, k_3, k_4, k_5 – observer gains, S_x, S_{xF}, S_b, S_{bF} – additional variables used to stabilize the observer work, T_{Sb}, T_{KT}, T_{Sx} – time constants of the filters, S – sign of the rotor flux speed.

B. Observer modification

State observer described in previous chapter IV.A was modified in order to assure electric drive proper work according to conception presented in [8]. But in [8] the other observer were used. Presented in [8] observer is Kubota adaptation observer [19]. Used in this paper observer has different conception and has better than Kubota observer properties [10, 11].

In the proposed observer the equations (30)-(43) were extended with LC filter model equations (1)-(8). The correction parts in (30) and (31) were changed to the difference between measured and calculated inverter output current. Instead of commanded motor voltages in equations (30) and (31) the commanded inverter output voltages were used in equations (51) and (52). In the equations (2) and (6) of the LC filter model the new correction terms were added.

Equations of the modified state observer for electric drives with inverter output filter are as follows:

$$\frac{d\hat{i}_{s\alpha}}{dt} = -\frac{R_s L_r^2 + R_r L_m^2}{L_r w_\delta}\hat{i}_{s\alpha} + \frac{R_r L_m}{L_r w_\delta}\psi_{r\alpha} + \frac{L_m}{w_\delta}\xi_\beta + \\ + \frac{L_r}{w_\delta}\hat{a}_{s\alpha} + k_3(i_{1\alpha} - \hat{i}_{1\alpha}) \tag{44}$$

$$\frac{d\hat{i}_{s\beta}}{dt} = -\frac{R_s L_r^2 + R_r L_m^2}{L_r w_\delta}\hat{i}_{s\beta} + \frac{R_r L_m}{L_r w_\delta}\psi_{r\beta} - \frac{L_m}{w_\delta}\xi_\alpha + \\ + \frac{L_r}{w_\delta}\hat{a}_{s\beta} + k_3(i_{1\beta} - \hat{i}_{1\beta}) \tag{45}$$

$$\frac{d\psi_{r\alpha}}{dt} = \frac{R_r}{L_r}\psi_{r\alpha} + \xi_\beta + R_r\frac{L_m}{L_r}\hat{i}_{s\alpha} + k_2 S_b \psi_{r\alpha} + \\ + Sk_2 k_3 \psi_{r\beta}(S_b - S_{bF}) + \\ + Sk_5((S_x - S_{xF})\psi_{r\alpha} - (S_b - S_{bF})\psi_{r\beta}) \tag{46}$$

$$\frac{d\psi_{r\beta}}{dt} = -\frac{R_r}{L_r}\psi_{r\beta} + \xi_\alpha + R_r\frac{L_m}{L_r}\hat{i}_{s\beta} - k_2 S_b \psi_{r\beta} + \\ - Sk_2 k_3 \psi_{r\alpha}(S_b - S_{bF}) + \\ + Sk_5(-(S_x - S_{xF})\psi_{r\beta} - (S_b - S_{bF})\psi_{r\alpha}) \tag{47}$$

$$\frac{d\xi_\alpha}{d\tau} = -\omega_{\psi r}\xi_\beta - k_1(i_{1\beta} - \hat{i}_{1\beta}) \tag{47}$$

$$\frac{d\xi_\beta}{d\tau} = \omega_{\psi r}\xi_\alpha + k_1(i_{1\alpha} - \hat{i}_{1\alpha}) \tag{48}$$

$$\frac{d\hat{a}_{c\alpha}}{d\tau} = \frac{i_{1\alpha} - \hat{i}_{s\alpha}}{C_1} \tag{49}$$

$$\frac{d\hat{a}_{c\beta}}{d\tau} = \frac{i_{1\beta} - \hat{i}_{s\beta}}{C_1} \tag{50}$$

$$\frac{d\hat{i}_{1\alpha}}{d\tau} = \frac{u_{1\alpha}^{com} - \hat{a}_{s\alpha}}{L_1} + k_A(i_{1\alpha} - \hat{i}_{1\alpha}) - k_B(i_{1\beta} - \hat{i}_{1\beta}) \tag{51}$$

$$\frac{d\hat{i}_{1\beta}}{d\tau} = \frac{u_{1\beta}^{com} - \hat{a}_{s\beta}}{L_1} + k_A(i_{1\beta} - \hat{i}_{1\beta}) + k_B(i_{1\alpha} - \hat{i}_{1\alpha}) \tag{52}$$

$$\hat{a}_{s\alpha} = \hat{a}_{c\alpha} + (i_{1\alpha} - \hat{i}_{s\alpha})R_c \tag{53}$$

$$\hat{a}_{s\beta} = \hat{a}_{c\beta} + (i_{1\beta} - \hat{i}_{s\beta})R_c \tag{54}$$

$$\frac{dS_{bF}}{d\tau} = \frac{1}{T_{Sb}}(S_b - S_{bF}) \tag{55}$$

$$\frac{d\hat{\omega}_{rF}}{d\tau} = \frac{1}{T_{KT}}(\hat{\omega}_r - \hat{\omega}_{rF}) \tag{56}$$

$$\frac{dS_{xF}}{d\tau} = \frac{1}{T_{Sx}}(S_x - S_{xF}) \tag{57}$$

$$S = \begin{cases} 1 & \text{if} \quad \hat{\omega}_{\psi r} > 0 \\ -1 & \text{if} \quad \hat{\omega}_{\psi r} \leq 0 \end{cases} \tag{58}$$

$$S_x = \xi_\alpha \psi_{r\alpha} + \xi_\beta \psi_{r\beta} \tag{59}$$

$$S_b = \xi_\alpha \psi_{r\beta} - \xi_\beta \psi_{r\alpha} \tag{60}$$

$$\hat{\omega}_{\psi r} = \hat{\omega}_{rF} + R_r \frac{L_m}{L_r}\left(\frac{\psi_{r\alpha}\hat{i}_{s\beta} + \psi_{r\beta}\hat{i}_{s\alpha}}{\psi_{r\alpha}^2 + \psi_{r\beta}^2}\right) \tag{61}$$

$$\hat{\omega}_r = \frac{\zeta_\alpha \psi_{r\alpha} + \zeta_\alpha \psi_{r\alpha}}{\psi_{r\alpha}^2 + \psi_{r\beta}^2} \tag{62}$$

where k_A, k_B – additional state observer gains.

The small resistance R_1 and leakage inductance $L_{\sigma M1}$ were omitted in the modified observer procedure.

V. INVESTIGATIONS

In the experiment inverter output filter with parameters presented in Table I was used. Investigations of the proposed control method were made for a drive with the asynchronous motor with parameters presented in Table II.

In Fig. 10 are presented results of simulation investigations for the full sensorless control system.

Without proposed modification the asynchronous motor sensorless control with MMB principle system was unable to work with LC filter. When the proposed modification in the control and observer procedure were add the closed loop control of the asynchronous motor the system started to work.

In Fig. 10 the step change of the motor and speed reverse for the motor without the load presented. After the speed reverse the load torque is changed. MMB variable x_{11} which is equal to motor mechanical speed was measured but was not used in control system. The estimated speed was calculated properly except the change of the speed direction when system crosses the regenerative, unstable region of the work. In that region the high frequency changes of LC filter control variables appear. In spite of that the full system works stable.

The observer gains and the internal filter time constants should be properly tuned to assure a fast estimation of the variables and simultaneously to assure a stable work of the observer. Also the PI controllers coefficient should be tuned carefully. In the presented paper the observer's gains and time constants and all PI controllers gains and time constants were tuned by set of simulation tests. The author next step of the work with proposed system is to tune all the gains etc. with use of the random weight change algorithm (RWC) [17], [18].

In the test bench microprocessor control board type SH65L with floating point digital signal processor ADSP21065L and programmable circuit EPF6016QC were used – Fig. 11.

In Fig. 12-15 are presented results of experimental results.

In Fig. 12 system was tested during increasing and decreasing of the commanded speed and during changes of the motor flux. In Fig 13, for constant motor speed, the motor flux was decreased and increased twice. In Fig. 14 the load torque was changed. And in Fig. 15 the motor reverse results are presented.

The experimental system works stable. Both the commanded speed $x_{11}{}^{com}$ and commanded flux $x_{21}{}^{com}$ were set to the commanded level. In case of motor reverse the calculated motor speed has increasing error near zero speed. In spite of that for dynamic change of

the motor speed direction the system work properly – similar as in the simulation. That range of the very low speed for the proposed system is going to be improved in the future works.

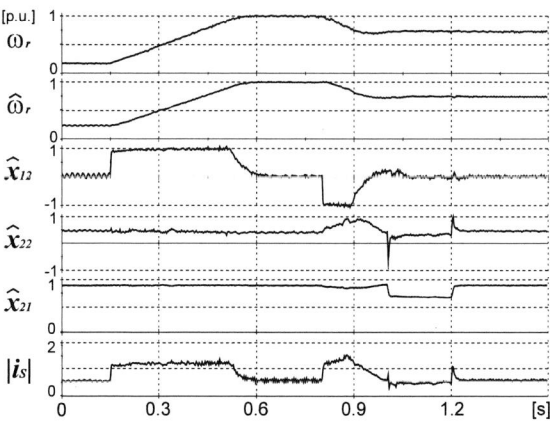

Fig. 12. Experimental results of the proposed system during changes of motor speed and motor flux

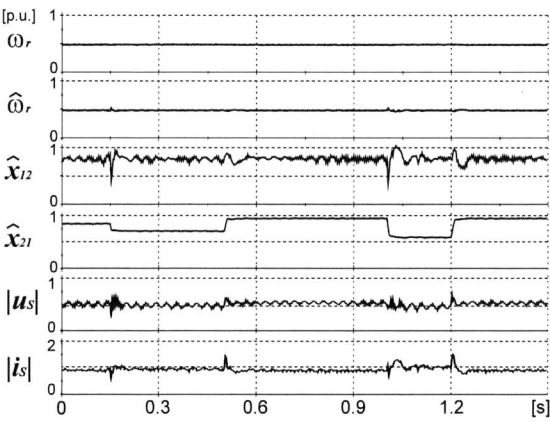

Fig. 13. Experimental results of the proposed system during changes of motor flux for the loaded motor

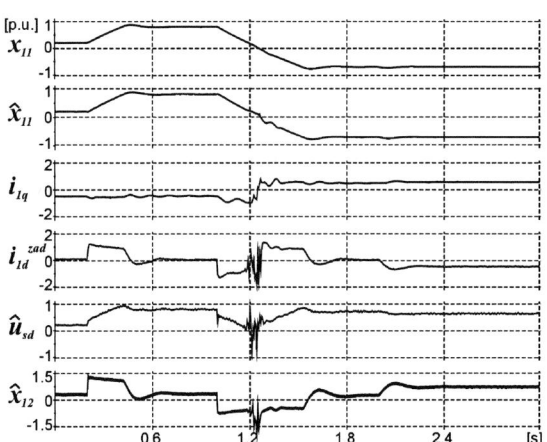

Fig. 10. Simulation results of the proposed system during changes of speed and load torque

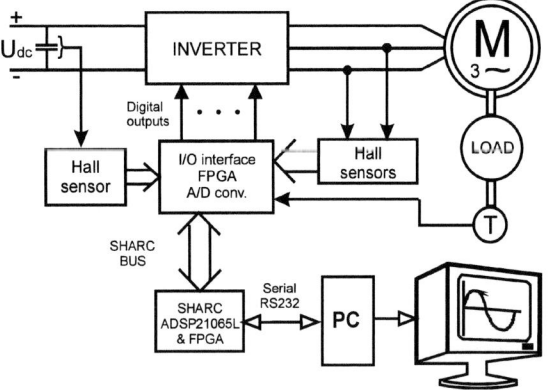

Fig. 11. Experimental setup with SH65L board

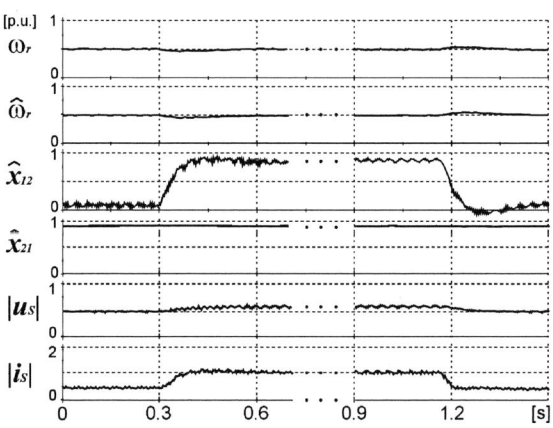

Fig. 14. Experimental results of the proposed system during changes of load torque

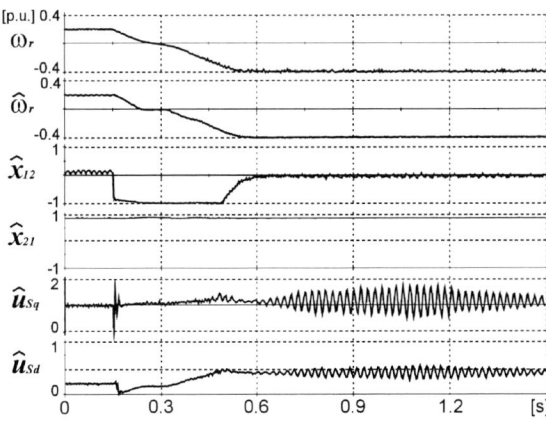

Fig. 15. Experimental results of the proposed system during motor speed reverse

VI. CONCLUSIONS

The use of inverter output filter in an electric drive make difficult precise control of the asynchronous motor. In the sophisticated control algorithm the motor parameters should be precisely known. If inverter output filter is used the parameters of the system changes. Additional inductances, capacitances and resistances should be considered for precise control of the motor in such system.

System proposed in the paper assures stable work for the drive with inverter output filter. That filter was included into control algorithm and into state observer. In the control system additional PI controllers were add to control inverter output current and motor supply voltage. State observer which was used for both flux and speed estimation was extended to motor supply voltage estimation.

The author is going to continue the work to obtain higher dynamics of the system using tuned observer and controllers gains.

The obtained hardware structure of the frequency converter stay without changes, no additional sensor are needed.

ACKNOWLEDGEMENTS

The author thank Prof. Zbigniew Krzeminski and colleagues from the research group from the Department of Automatic Control of Electric Drives, Faculty of Electrical and Control Engineering, Gdansk University of Technology, Poland for their remarks and help with preparing the simulation and experimental tests.

REFERENCES

[1] Akagi H., Hasegawa H., Doumoto T.: Passive EMI Filter for Use With a Voltage-Source PWM Inverter Having Sinusoidal Output Voltage and Zero Common-Mode. IEEE Trans. on Power Electronics, 19 (2004), n.4, 1069-1076.

[2] Krzemiński Z., Guziński J.: Output filter for voltage source inverter supplying induction motor. International Conference on Power Electronics, Intelligent Motions and Power Quality PCIM, (2005), Nuremberg, Germany.

[3] Salomaki J., Kerkman R., Schlegel D., Skibinski G.: Effect of PWM Inverters on AC Motor Bearing Currents and Shaft Voltages. IEEE APEC Conference, (1995), Dallas, USA

[4] Seliga R., Koczara W.: Instantaneous Current and Voltage Control Strategy in Low-Pass Filter Based Sine-Wave Voltage DC/AC Converter Topology for Adjustable Speed PWM Drive System. IEEE International Symposium on Industrial Electronics ISIE, (2002), L'Aquila, Italy.

[5] Kojima M., Hirabayashi K., Kawabata Y., Eijogu E. C., Kawabata T.: Novel Vector Control System Using Deadbeat-Controlled PWM Inverter With Output LC Filter. IEEE Trans. on Industry Applications, 40 (2004), n.1, 162-169.

[6] Salomaki J., Luomi J.: Vector Control of an Induction Motor Fed by a PWM Inverter with Output LC Filter. Proceedings of the 4th Nordic Workshop on Power and Industrial Electronics, NORPIE (2004), Trondheim, Norway.

[7] Salomaki J., Hikkanen M., Luomi J.: Sensorless Control of Induction Motor Drives Equipped With Inverter Output Filter. IEEE International Conference on Electric Machines and Drives, (2005), San Antonio, USA.

[8] Guziński J., Włas M.: "Closed Loop Control of Speed Sensorless Drive with Motor Filter". International Conference on Power Electronics, Intelligent Motions and Power Quality PCIM (2006), Nuremberg, Germany.

[9] Krzeminski Z.: Nonlinear Control of Induction Motor. 10th World Congress on Automatic Control, IFAC (1987), Munich, German

[10] Krzemiński Z: Sensorless Control of the Induction Motor Based on New Observer. International Conference on Power Conversion and Intelligent Motion PCIM, (2000), Nuremberg, Germany.

[11] Krzeminski Z.: Structure of Induction Motor Speed Observer with Disturbance Model. 4th Polish Conference on Modeling and Simulation, MiS-4 (2006), Koscielisko, Poland – in Polish.

[12] Adamowicz M., Guziński J.: "Control of Sensorless Electric Drive with Inverter Output Filter". 4th International Symposium on Automatic Control AUTSYM 2005. 22-23 September 2005. Wismar, Germany.

[13] Krzemiński Z.: Asynchronous machines digital control. Gdansk University of Technology Publication. Poland, Gdansk 2001. (in Polish).

[14] Abu-Rub H., Krzemiński Z., Guziński J.: Nonlinear control of induction motor. Idea and application. 9th International Conference and Exhibition on Power Electronics and Motion Control EPE-PEMC 2000. 5-7 September 2000, Koszyce, Slovakia.

[15] J. Pontt, J. Rodriguez, M. Rotella, "Output sinus filter for medium voltage drive with direct torque control", in Proc. *40th Annual Meeting, Industry Applications Society, IAS 2005*, Hong Kong, Chine, 2005.

[16] B.-H. Bae, B.-H. Cho, S.-K. Sul, "Damping Control Strategy for the Vector Controlled Traction Drive", in Proc. *9th European Conference on Power Electronics and Applications EPE'2001*, Graz, Austria, 2001.

[17] B. Burton, F. Kamran, R. G. Harley, T. G. Habetler, M. A. Brooke, R. Poddar, "Identification and Control of Induction Motor Stator Currents Using Fast On-Line Random Training of a Neural Network" Trans. on Industry Applications, vol. 33, no. 3, pp. 697-704, May/June 1997.

[18] T. Pajchrowski, K. Urbański "DSP application to robust speed control of PMSM by means of Artificial Neural Network," In Proc. Scientific Conference SENE'2001, Lodz-Arturowek, Poland (in Polish). pp 647-652.

[19] Kubota H., Matsuse K., Nakano. T.: DSP-Based Speed Adaptive Flux Observer of Induction Motor. IEEE Transactions on Idustry Applications, March/April 1993.

Sensorless IPMSM based drive for reciprocating compressor

Anton Dianov, *Member IEEE*, Kim Young-Kwan, Lee Sang-Joon, Lee Sang-Taek, Yoon Tae-Ho

Samsung Electronics / Digital Appliance Division, Suwon, Korea, e-mail: anton.dianov@samsung.com

Abstract— This paper presents a new algorithm for the position sensorless vector control of the Interior Permanent Magnets Synchronous Motors (IPMSM), which is based on the well-known approach for the Permanent Magnets Synchronous Motors (PMSM), when rotor position and speed information is obtained by using current error between actual and estimated currents. Estimated current is calculated using motor model, which is written in the synchronous reference frame *dq*. The current difference is decomposed into two components. One of them is used for the motor back-emf and speed estimation and another one is used as a correction term. Rotor position is calculated as an integral of the estimated speed. Utilization of two current error components allows to build reliable system with low estimation error, where one current error component is used for estimation in static modes and another one is used for estimation in dynamic mode. This paper shows that sensorless algorithm for the PMSM can be spread also on the IPMSM and it works perfectly even under the difficult load conditions such as reciprocating compressor. Robustness of the proposed algorithm and its sensitivity to the motor parameters variations are also described. This paper also pays attention to the drive starting procedure in the sensorless mode.

Keywords— Electrical drives, synchronous motors, sensorless control.

I. INTRODUCTION

Permanent magnet synchronous motors (PMSM) are widely used for small and low power applications due to high efficiency, high torque to current ration, simplicity and reliability. However, the most important problems for such applications are cost and reliability of the drive system, especially in the field of household appliances. During last years there was made progress in the field of materials, power electronics and microcontrollers, which became cheaper and better. With the increase in performance of the low-cost microcontrollers it became possible to simplify hardware structure of the electric drive at the price of software sophistication. The main modification of the traditional vector control scheme is elimination of the position sensor. It simplifies drive structure, increases its reliability and decreases total cost. However it needs developing of the speed and position estimation algorithm for the vector sensorless control system.

This paper presents algorithm, which uses motor back-emf estimation to estimate speed and position of the rotor. Such method is more preferable to injecting-based methods, since it produces less noises and vibration, which is important for home appliances. The estimation algorithm was originally developed for the surface-

mounted PMSM [1, 2], which have symmetrical structure along direct and quadrature axes, however it is shown in the paper that such approach can also be used for the IPMSM.

It is known that back-emf based estimation methods have pure performance at low speed due to the considerable estimation error. So they require developing of the starting algorithm, which accelerates motor to the speed, where position estimation error is small enough. Such algorithm was also developed and described in the paper. This paper is continuation of the work published in [4], which focused on the surface-mounted PMSM.

The proposed control scheme was verified by experiments using 180 W IPMSM built in the reciprocating compressor for refrigerator.

II. ESTIMATION ALGORITHM

Proposed algorithm is based on the idea published in [1, 2, 4], which describe estimation algorithm for the PMSM. This paper shows that such approach can be also spread on the IPMSM. The idea of the speed and position estimation is to estimate any measuring value for the next step of calculation using motor model in synchronous reference frame *dq*. Then, on the next calculation step, the difference between estimated and measured values is used for calculation and correction of the estimated parameters.

Fig. 1 shows scheme of the PM machine. d_sq_s is stationary reference frame coupled with stator, dq is rotational reference frame, coupled with rotor, where d axis coincides with the flux of the rotor, θ is the position of the rotor, $\gamma\delta$ is estimated rotational reference frame with estimated position $\hat{\theta}$, and $\Delta\theta$ is position estimation error.

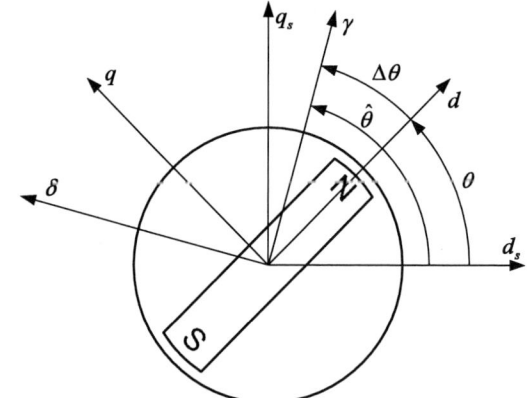

Fig. 1. IPMSM reference frames

For ideal asymmetrical synchronous machine its electrical equations in synchronous reference frame can be written as:

$$\begin{bmatrix} u_d \\ u_q \end{bmatrix} = \begin{bmatrix} R + L_d p & -\omega L_q \\ \omega L_d & R + L_q p \end{bmatrix} \begin{bmatrix} i_d \\ i_q \end{bmatrix} + \begin{bmatrix} 0 \\ E \end{bmatrix}. \quad (1)$$

where:

R phase resistance,

L_d, L_q inductances along direct and quadrature axes respectively,

E motor back-EMF, which is equal to $k_e \omega$, where k_e is motor back-EMF coefficient.

Rearranging (1) regarding to the current derivatives gives current model of the motor:

$$p \begin{bmatrix} i_d \\ i_q \end{bmatrix} = \begin{bmatrix} L_d^{-1} & 0 \\ 0 & L_q^{-1} \end{bmatrix} \left(\begin{bmatrix} u_d \\ u_q \end{bmatrix} - \begin{bmatrix} R & -\omega L_q \\ \omega L_d & R \end{bmatrix} \begin{bmatrix} i_d \\ i_q \end{bmatrix} - \begin{bmatrix} 0 \\ E \end{bmatrix} \right). (2)$$

Assuming that sampling period is short enough, current derivatives can be substituted with first order differences:

$$p \begin{bmatrix} i_d \\ i_q \end{bmatrix} = \frac{1}{T_s} \left(\begin{bmatrix} i_d^{n+1} \\ i_q^{n+1} \end{bmatrix} - \begin{bmatrix} i_d^n \\ i_q^n \end{bmatrix} \right). \quad (3)$$

where superscript is used to denote calculation step.

Since currents and voltages can be measured only in stationary reference frame $d_s q_s$ and exact rotor position is unknown, all calculations are performed in the estimated reference frame $\gamma\delta$. Then the Park transformation is used to change the stationary reference frame to the rotational one:

$$\begin{bmatrix} i_\gamma^n \\ i_\delta^n \end{bmatrix} = \begin{bmatrix} \cos(\hat\theta) & \sin(\hat\theta) \\ -\sin(\hat\theta) & \cos(\hat\theta) \end{bmatrix} \begin{bmatrix} i_{ds}^n \\ i_{qs}^n \end{bmatrix}. \quad (4)$$

Supposing that $\hat\theta = \theta + \Delta\theta$, (4) can be rewritten:

$$\begin{bmatrix} i_\gamma^n \\ i_\delta^n \end{bmatrix} = \begin{bmatrix} \cos(\Delta\theta) & \sin(\Delta\theta) \\ -\sin(\Delta\theta) & \cos(\Delta\theta) \end{bmatrix} \begin{bmatrix} i_d^n \\ i_q^n \end{bmatrix}. \quad (5)$$

As it was mentioned earlier motor current model can be used to estimate currents for the next calculation step, using measured currents and reconstructed voltages for the current calculation step:

$$\begin{bmatrix} \hat i_\gamma^{n+1} \\ \hat i_\delta^{n+1} \end{bmatrix} = \begin{bmatrix} i_\gamma^n \\ i_\delta^n \end{bmatrix} + T_s \begin{bmatrix} L_d^{-1} & 0 \\ 0 & L_q^{-1} \end{bmatrix} \times$$
$$\times \left(\begin{bmatrix} u_\gamma^n \\ u_\delta^n \end{bmatrix} - \begin{bmatrix} R & -\hat\omega L_q \\ \hat\omega L_d & R \end{bmatrix} \begin{bmatrix} i_\gamma^n \\ i_\delta^n \end{bmatrix} - \begin{bmatrix} 0 \\ \hat E \end{bmatrix} \right) . \quad (6)$$

Assuming that sampling period is short enough and motor speed does not change significantly, currents measured at the next calculation step can be expressed as:

$$\begin{bmatrix} i_\gamma^{n+1} \\ i_\delta^{n+1} \end{bmatrix} = \begin{bmatrix} \cos(\hat\theta + \hat\omega T_s) & \sin(\hat\theta + \hat\omega T_s) \\ -\sin(\hat\theta + \hat\omega T_s) & \cos(\hat\theta + \hat\omega T_s) \end{bmatrix} \begin{bmatrix} i_{ds}^{n+1} \\ i_{qs}^{n+1} \end{bmatrix} =$$
$$= \begin{bmatrix} \cos(\Delta\theta) & \sin(\Delta\theta) \\ -\sin(\Delta\theta) & \cos(\Delta\theta) \end{bmatrix} \begin{bmatrix} i_d^{n+1} \\ i_q^{n+1} \end{bmatrix} \quad (7)$$

Subtraction (6) from (7) results current estimation errors. Expression for the γ-axis current error is:

$$\Delta i_\gamma^{n+1} = i_\gamma^{n+1} - \hat i_\gamma^{n+1} = -\frac{T_s k_e \omega}{L_d} \sin(\Delta\theta) - \Delta\omega T_s i_\delta^n +$$
$$+ \sin(\Delta\theta) \frac{(L_q - L_d)}{L_d} \left(\hat\omega T_s i_d^n - (i_q^{n+1} - i_q^n) \right) \approx \quad . \quad (8)$$
$$- \frac{T_s k_e \omega}{L_d} \Delta\theta - \Delta\omega T_s i_\delta^n \frac{L_q}{L_d} +$$
$$+ \frac{(L_q - L_d)}{L_d} \left(\omega T_s i_d^n - (i_q^{n+1} - i_q^n) \right) \Delta\theta$$

And expression for the δ-axis current error is:

$$\Delta i_\delta^{n+1} = i_\delta^{n+1} - \hat i_\delta^{n+1} = \frac{T_s}{L_q} k_e \hat\omega - \frac{T_s}{L_q} k_e \omega \cos(\Delta\theta) +$$
$$+ \Delta\omega T_s i_\gamma^n - \sin(\Delta\theta) \frac{(L_q - L_d)}{L_q} \left(\hat\omega T_s i_q^n + (i_d^{n+1} - i_d^n) \right) \approx \quad . \quad (9)$$
$$\approx \frac{T_s}{L_q} k_e \Delta\omega + \Delta\omega T_s i_\gamma^n \frac{L_d}{L_q} -$$
$$- \frac{(L_q - L_d)}{L_q} \left(\omega T_s i_q^n + (i_d^{n+1} - i_d^n) \right) \Delta\theta$$

It can be seen from (8) that first two terms of the γ-current estimation error are proportional to the position and speed errors and they are equal to the expression for the PMSM [4]. But unlike expression for PMSM (8) contains additional term caused by the motor asymmetry, which is also proportional to the position error, but has another sign. However third term does not significantly affect the sign of the current estimation error since d-current of the motor is controlled to be zero or negative.

It is also evident that first two terms in (9) are proportional to the back-EMF estimation error and they are equal to the δ-current estimation error of PMSM [4]. However (9) contains third term, caused by the motor asymmetry, which is proportional to the position estimation error and which sign differs from the first two terms. It causes inevitable error in the back-emf estimation scheme, which utilizes (9), so speed estimation algorithm must contain correction branch. This branch can use γ-current estimation error (8), which is not considerably impact by the motor asymmetry.

Considered above expressions for the current estimation errors can be used for designing of the estimation scheme. In this paper we propose scheme, which is shown in the Fig. 2. It uses integral type controller for the back-EMF estimation and PI type controller for the correction branch. It also contains low pass filters for estimated parameters, which decrease ripples. The feature of the algorithm is fact that i_γ based branch works faster, than i_δ based branch. Since I-controller has large time constant, it decreases influence of the current error ripples on the estimated back-EMF, but in transients it also has considerable phase delay. Slow response of the I-controller is compensated by the fast PI-controller, which guarantees stability in static and good performance in dynamic for the estimation

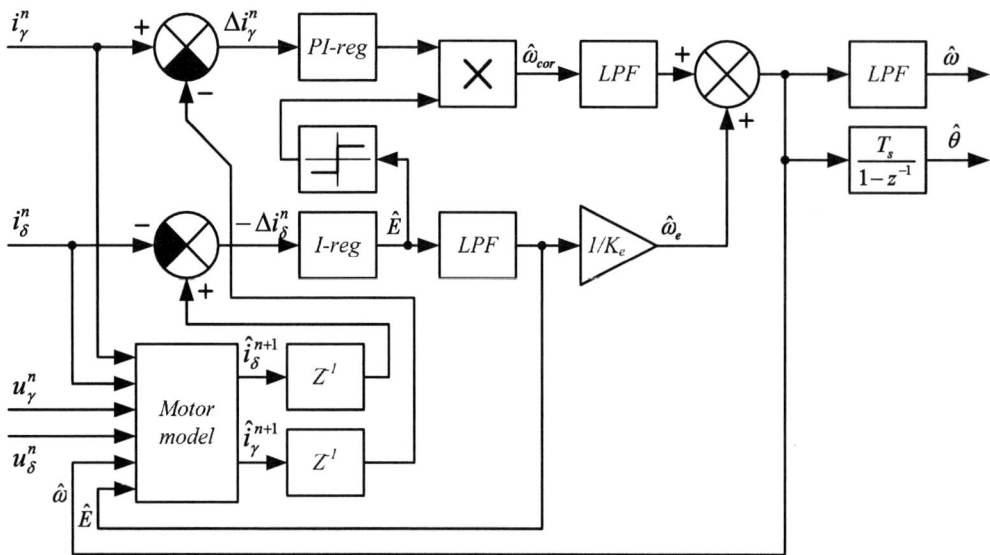

Fig. 2. Scheme of the proposed estimation algorithm

algorithm. Furthermore i_γ based branch by means of correction term compensates inaccuracy of the motor's parameters and increases robustness of the proposed algorithm.

Fig. 3 illustrates work of the proposed algorithm. It shows that estimated speed, based on the i_δ estimation error, is changing slowly, but it is compensated by the correction term, which improves performance in dynamic. Fig. 3 also shows that there is a small gap between estimated speed and speed estimated in δ-current branch even after end of transient, which is caused by the third term in (9). However this estimation error is compensated by the correction term.

III. ROBUSTNESS OF THE ESTIMATION ALGORITHM

Suppose that exact values of the motor parameters are unknown and estimation algorithm operates with rough values. It should be also noted that in properly designed estimator, where position estimation error is less 90 degrees, average value of the speed error is zero (if it's not equal to zero the position estimation error increases and

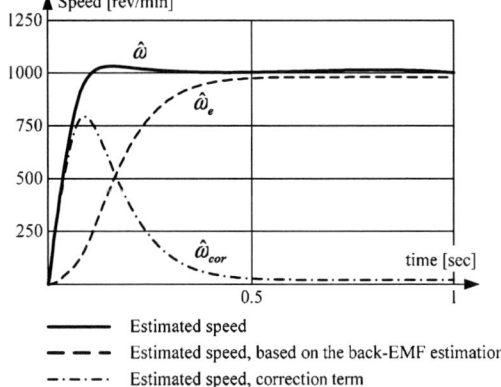

Fig. 3. Estimated speed

— Estimated speed
- - - Estimated speed, based on the back-EMF estimation
-·-·- Estimated speed, correction term

algorithm fails).

Assume that motor resistance in the estimation algorithm is $r + \Delta r$, where Δr is resistance error. In that case current estimation errors (8) and (9) are transformed into:

$$\Delta i_\gamma^{n+1} \approx -\frac{T_s k_e \omega}{L_d} \Delta\theta - \Delta\omega T_s i_\delta^n \frac{L_q}{L_d} + \tag{10}$$
$$+ \frac{\left(L_q - L_d\right)}{L_d}\left(\omega T_s i_d^n - \left(i_q^{n+1} - i_q^n\right)\right)\Delta\theta + \frac{T_s}{L_d} i_\gamma^n \Delta r$$

$$\Delta i_\delta^{n+1} \approx \frac{T_s}{L_q}\left(\hat{E} - E\right) + \Delta\omega T_s i_\gamma^n \frac{L_d}{L_q} - \tag{11}$$
$$- \frac{\left(L_q - L_d\right)}{L_q}\left(\omega T_s i_q^n + \left(i_d^{n+1} - i_d^n\right)\right)\Delta\theta + \frac{T_s}{L_q} i_\delta^n \Delta r$$

Additional term in (11) caused by the resistance error is compensated by the back-EMF estimation error, so motor speed, estimated by i_δ based branch of the algorithm contain error. This error is compensated by the i_γ based branch of the algorithm and position estimation error is:

$$\Delta\theta \approx \frac{i_\gamma^n \Delta r}{\left(k_e \omega - \left(L_q - L_d\right)\left(\omega i_d^n - \frac{\left(i_q^{n+1} - i_q^n\right)}{T_s}\right)\right)} \tag{12}$$

It is proportional to the γ-axis current and reverse proportional to the speed of the motor. However i_γ is usually controlled to be zero, so position estimation error is small and proposed algorithm is robust to the stator resistance variations.

Suppose that motor inductances L_d and L_q are not known exactly and these parameters in the estimation algorithm contain error: $L_d + \Delta L_d$ and $L_q + \Delta L_q$ respectively. In that case position estimation errors are:

$$\Delta\theta \approx \frac{\Delta L_d\left(i_d^{n+1}-i_d^n\right)}{k_e\omega T_s-\left(L_q-L_d\right)\omega T_s i_d^n+\left(L_d-L_q+\Delta L_d\right)\left(i_q^{n+1}-i_q^n\right)} \quad (13)$$

for L_d, and:

$$\Delta\theta \approx \frac{-\omega i_\delta^n \Delta L_q}{k_e\omega-\left(L_q-L_d\right)\left(\omega i_d^n+\dfrac{i_q^{n+1}-i_q^n}{T_s}\right)} \quad (14)$$

for L_q.

These expressions show that estimation algorithm is robust to the L_d variation and is more sensitive to the L_q variation. Furthermore, position error caused by the L_q error does not almost depend on the motor speed and increases with the load.

If the back-EMF coefficient in the estimation algorithm contains error $k_e + \Delta k_e$, it affects only i_δ based branch of the algorithm and (9) transforms into:

$$\Delta i_\delta^{n+1} \approx \frac{T_s}{L_q}k_e\Delta\omega+\Delta\omega T_s i_\gamma^n\frac{L_d}{L_q}- $$
$$-\frac{\left(L_q-L_d\right)}{L_q}\left(\omega T_s i_q^n+\left(i_d^{n+1}-i_d^n\right)\right)\Delta\theta+\frac{T_s}{L_q}\Delta k_e\widehat{\omega} \quad (15)$$

However error of the motor back-EMF coefficient does not impact on i_γ based branch of the algorithm, which compensates speed estimation error in the i_δ based branch.

IV. STARTING PROCEDURE

The most important problem for sensorless vector control system is to design algorithm, which works at low speed. Since most of the sensorless algorithms use back-EMF for position and speed estimation, low speed operations are difficult. But many electric drives are not supposed to operate at low speed, so this problem is important only for starting.

There are some ways to start drive in sensorless control system. There can be used algorithms based on motor speed-independent parameters (inductances, steel saturation, etc.). But the simplest way is to use open-loop control, accelerate motor to the speed, where sensorless control algorithm works well and then use closed-loop control.

Since proposed sensorless control algorithm uses back-

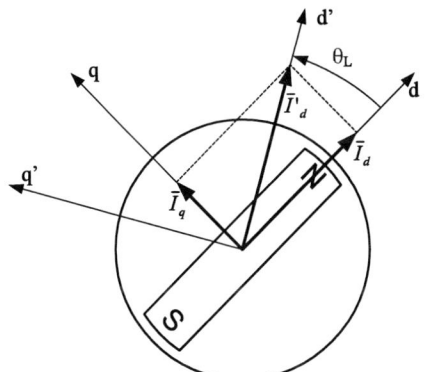

Fig. 4. Reference frames at starting

EMF for position and speed estimation, starting algorithm has to be developed. This paper proposes starting algorithm, which contains the following steps: initial rotor positioning; starting in open-loop system and closing of the control system. Since the position of the rotor reference frame dq is not exactly known, starting algorithm uses starting reference frame $d'q'$ which is shifted from dq at load angle θ_L.

Initial rotor positioning is used to set rotor in the predefined position to start from. Usually it corresponds to 0 °, which is aligned with the phase "A". Initial positioning is performed in two steps: At first, position of the $d'q'$ reference frame is set to -90 ° and then d-current is linearly increased form zero to its maximum value. At this step rotor is aligning with the current vector and dq reference frame coincides with $d'q'$. After that $d'q'$ reference frame is rotated through 90 ° to its initial position of 0 °.

Such two-stepped positioning is used to prevent a fault, which can occur if the difference between rotor initial position (before starting) and desired one is close to 180 degrees. In that case applied current produces torque, which is close to zero, so the rotor can be immovable. Using slow increasing magnitude of the starting current vector and its slow rotation removes undesirable noises during starting.

Starting in open-loop system is used to accelerate rotor to the speed, where position estimator can operate. At this step system is accelerated with a constant acceleration to the predefined speed and then it is closed.

After initial positioning current vector $\bar{I}'_d$ is applied along d'-axis, which coincides with d-axis. At this step starting reference frame $d'q'$ is rotated with constant acceleration and d'-current is controlled to have maximum value. If the sum of the load torque and dynamic torque is less than maximum torque of the motor, the rotor rotates with the same average acceleration, but the real dq reference frame is shifted from $d'q'$ reference frame at load angle θ_L (Fig. 4). At the same time position estimation algorithm begins to work, but estimated parameters are ignored, because the estimation errors are unacceptable huge. Estimated parameters are used only as inputs for the estimation algorithm for its synchronization. With the increase of the motor speed the back-EMF increases and estimation errors decrease to the level of acceptance. And at the predefined speed, which was determined at the experiments, the control system is closed.

Since estimation algorithm is working during acceleration it calculates the speed and position of the rotor and at the moment of closing, where estimation error is supposed to be convenient, the real rotor speed and position are known. So from the moment of closing estimated parameters are used in the control system.

However it should be noted that speed PI controller is not initialized, since it was not used during acceleration and current regulators was set to regulate currents in the starting reference frame $d'q'$, where d'-axis current was set to maximum current and q'-axis current was set to zero. So, outputs of the regulators and their integral part have to be set to the proper values.

Suppose that conventional control system with speed and currents PI controllers is used. So at the moment of closing commanded and actual speed have to be set to the closing speed, proportional component have to be set to zero and its integral component and output have to be set to the estimated value of the q-current $i_q = I_q$. Inputs of the current controllers and their proportional components have to be set to zero; their integral components and outputs have to be set to the estimated values of the voltages u_d and u_q.

Proposed closing algorithm allows to change system structure without any problems and avoid current peaks at the moment of closing.

V. EXPERIMENTAL SYSTEM

The experimental system is shown in Fig. 5. It contains reciprocating compressor with built in IPMSM, which parameters are given in the Table I. The load for the used system is pressure accumulator with needle valve equipped with a pressure metering system. Motor control hardware includes two boards: power board and control board. The control board is placed over the power board and they are connected with a cable. Power board is designed to be supplied from 220 V (50 – 60) Hz source.

It contains "Fairchild" inverter FSBS5CH60AS (5 A/600 V), which is used for controlling of the experimental motor. Power board also comprises two hall-effect current sensors in the motor phases and dc-link voltage sensor. Motor phase voltages are reconstructed by the control system using dc-link voltage and modulation coefficients.

Control board is based on the 60 MIPS "Matsushita" microcontroller MN103SFA7K. Three analog signal from the sensors are converted by the built-in 10-bit ADC of microcontroller with sampling time of 250 µs. The sampling time was selected after energy consumption tests, which showed that the best carrier frequency for the PWM is 4 kHz. Control board also contains four DAC, which are used for monitoring of the internal variables such as estimated values.

However position sensor cannot be mounted inside the reciprocating compressor, so for the verification of the estimated position there was used dynamo set, which contained IPMSM, similar to the built in the compressor, quadrature encoder (4096 pulses per revolution) and brushed DC motor as a load.

The structure of the control system is shown in Fig. 6.

Fig. 5. Experimental system

TABLE I.
MOTOR PARAMETERS

Parameter	Value
Number of poles	2p = 6
Rated current, A	1
Rated speed, rpm	3000
Phase resistance, Ω	2.85
L_d inductance, mH	80
L_q inductance, mH	120
Back-EMF constant, V·s/rad	0.12

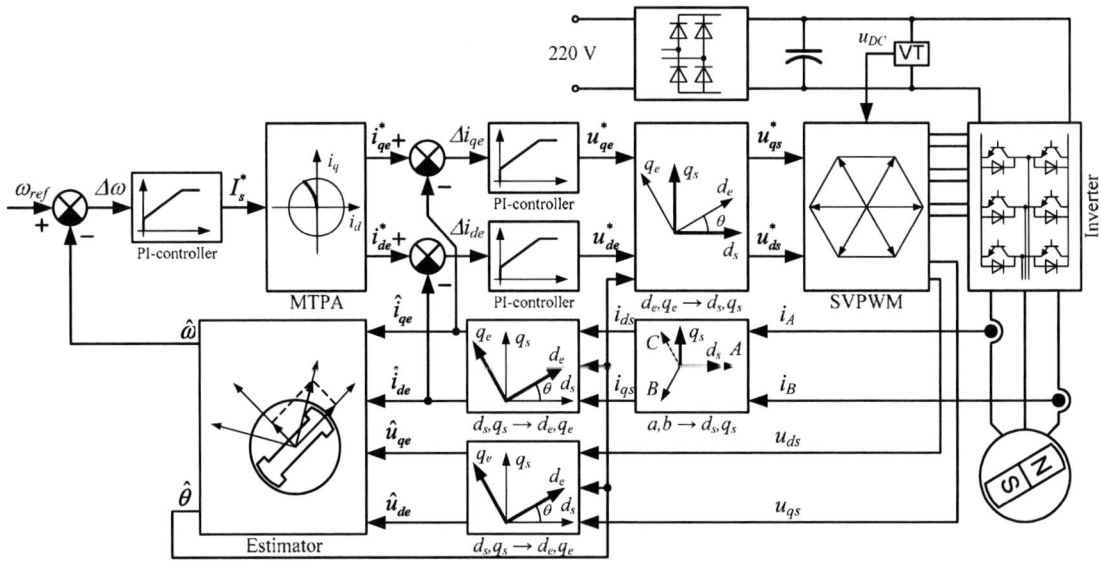

Fig. 6. Control system

VI. EXPERIMENTAL RESULTS

Fig. 7 shows starting of the motor under sensorless control at 0.25 Nm load torque. As it was stated above, encoder could not be mounted into reciprocating compressor, so motor set with dc load and quadrature encoder has been used for the for the experiment. Fig 8 illustrates closing of the control system. It shows that there are no current peaks at the moment of closing and control system accurately regulates motor currents after closing.

Minimum speed, where the stable rotation can be obtained, was also determined. For the used dynamo set it is 30 rpm. However for the motor mounted into compressor the minimum speed is 400 rpm. It is caused by the irregular load of the motor and resonances in the system motor–compressor. So the motor has to be accelerated with a maximum possible acceleration to that speed and then control system can be closed.

Fig. 9 shows the steady state operation of the motor at 2800 rpm. The load of the motor is 0.5 Nm. Fig 10 illustrates performance of the estimation algorithm. Position estimation error is one calculation steps, which

approximately corresponds to 11° el. This error can be significantly decreased by increasing the PWM frequency, but as it was said earlier, that PWM frequency was selected from the point of view of the drive efficiency.

Currents of the motor built in the compressor are shown in Fig. 11. The difference between suction pressure and discharge pressure is 4 kgf/cm^2. Oscillations of the motor currents with a period three times more than the electrical one are caused by the irregularity of the load torque, which is the nature of the reciprocating compressor.

The sensitivity of the developed algorithm to the motor parameters variation has also been tested. It is almost insensitive to the stator resistance variations ±50% and it can compensate ±50% error of the d-axis inductance L_d. However sensorless algorithm is more sensitive to the q-axis inductance L_q deviation and it can function only with ±25% error. As it was stated above, if the back-EMF constant k_e of the motor model differs from the real one, estimation algorithm compensates the difference by the speed correction term. Fig. 12 illustrates response of the estimation algorithm to the 50% increase of the motor model parameter k_e.

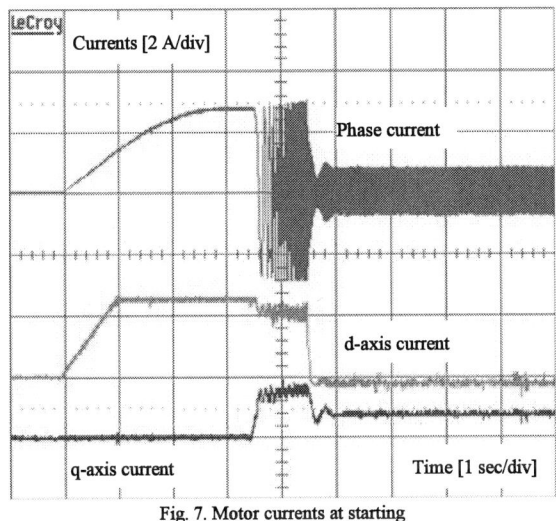

Fig. 7. Motor currents at starting

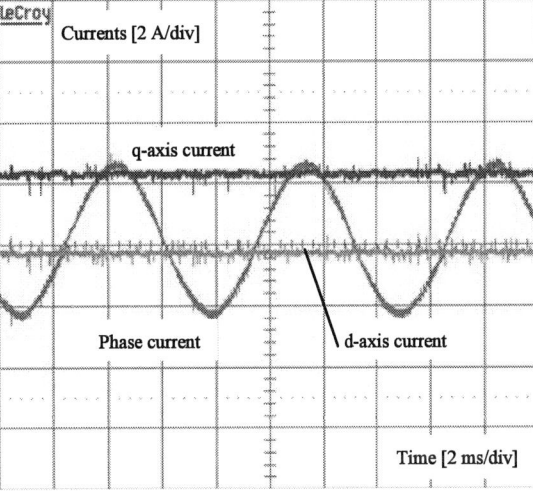

Fig. 9. Motor currents at 2800 rpm

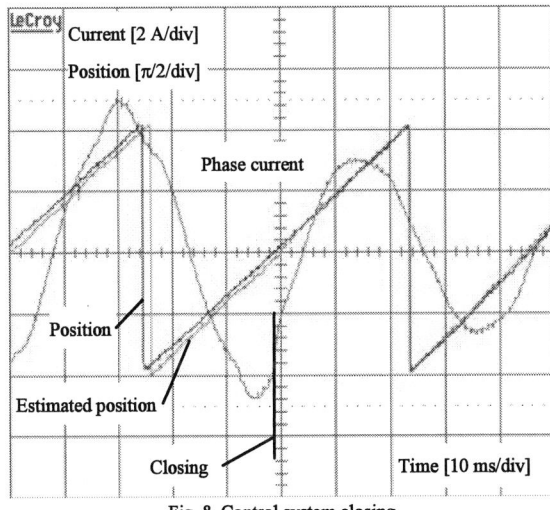

Fig. 8. Control system closing

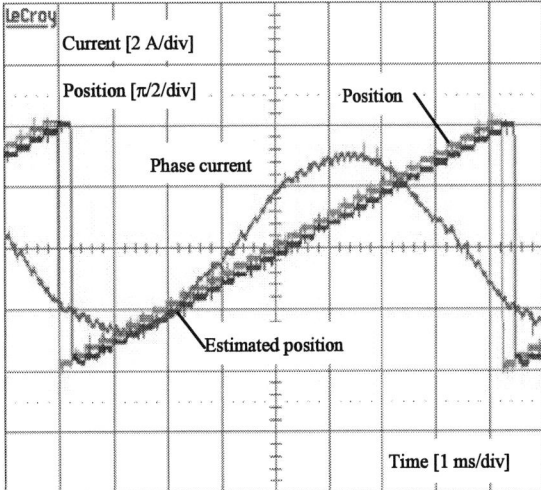

Fig. 10. Position estimation error

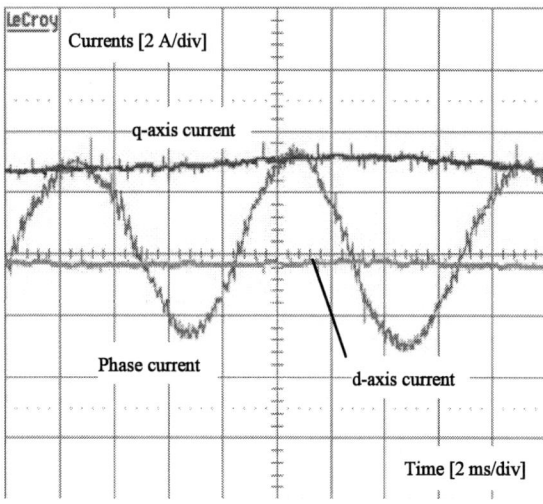

Fig. 11. Motor currents

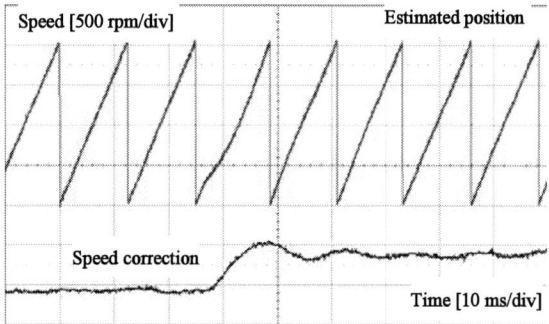

Fig. 12. Correction of the k_e increase

VII. CONCLUSIONS

The paper proposed new speed and position estimation algorithm for position sensorless vector control system for the IPMSM-based drive. The basic idea of the estimation is to use difference between actual motor currents and calculated using motor model in synchronous reference frame. Current error is decomposed into two components; one of them is used for back-EMF and speed estimation and another one is used for correction of the estimated value. Then position is calculated as an integral of the estimated speed. This paper also pays attention to the robustness of the algorithm and its sensitivity to the motor parameters variation.

This paper also presents starting algorithm, which accelerates motor in open-loop till the predefined speed and then closes control system. New closing algorithm is suggested in this paper. Simulation and experiments verified effectiveness of the proposed algorithm and proved its feasibility.

REFERENCES

[1] N. Matsui. "Sensorless PM Brushless DC Motor Drives", IEEE Trans. on Industrial Electronics, Vol. 43, No. 2, Apr, 1996, pp. 300-308.

[2] N. Matsui, T. Takeshita, K. Yasuda "A new sensorless drive of brushless DC motor", in Proc. IECON'92, 1992, pp. 430-435.

[3] U.-H. Rieder, M. Schroedl, A. Ebner, "Sensorless control of an external rotor PMSM in the whole speed range including standstill using DC-link measurements only", Power Electronics Specialists Conference, 2004. PESC 04. 2004 IEEE 35th Annual, Vol. 2, June 2004, pp. 1280 – 1285.

[4] Dianov Anton, Choi Jae-Young, Lee Kwang-Woon, Lee Joon-Hwan, "Sensorless Vector Controlled Drive For Reciprocating Compressor", Power Electronics Specialists Conference, 2007. PESC 2007. In Proc. PESC 2007, pp. 580 – 586.

Controlling system of electrodynamic drive

Josef Černohorský*

* Technical university of Liberec, Faculty of Mechatronics and Interdisciplinary Engineering Studies, Hálkova 6,
Liberec, Czech Republic, e-mail: *josef.cernohorsky@tul.cz*

Abstract— The winding and unwinding process is very common in textile industry. Nowadays solutions of winding drive with centralized drive and mechanical cams do not fulfil requirement on production of modern textile machine. Therefore we develop low-power model of linear electrodynamic drive. This article is aimed at controlling system based on ADSP 21992 and at mathematical modelling. For power electronics design we make mathematical model of electrodynamic motor. We also made a mathematical model of traverse bar for emulating of load during mathematical simulation of electrodynamic drive. The results of mathematical simulation and real drive parameters are discussed in conclusion.

Keywords— Permanent magnet motor, Electrical Drive, Highly dynamic drive, Simulation, Industrial application.

I. INTRODUCTION

The spinning machine Fig.1 makes a yarn of prefabricated cotton or synthetic materials. Our topic is modernization of delivery drive. Basically it is a positioning system which deploys the yarn on spool. The delivery drive is situated at the end of the machine in a cabinet and the motion is distributed by a traverse bar.

Fig. 1. Block diagram of DSP control program

On delivery drive there are requirements on dynamic and precision, especially during reverzation. Due to textile technology requirements the motion is complex with modification of end positions and velocity during positioning. This modification is particularly made by external mechanism out of main CAM and particularly by controlling system of the drive. The aim of the modernization is full electronic control [1] of positioning parameters and removing mechanical cam and additional mechanism.

II. MATHEMATIC MODEL OF DRIVE CONTROL

Suitable mathematic model is basic precondition to investigate the properties and behaviour of drive by simulation. In developing mathematical model of electrodynamic motor we can use the analogy with electric DC drive with separate excitation [2]. The model is based on mechanical equitation (1):

$$m\ddot{x} + D\dot{x} + F_z(t) = F_m(t) \qquad (1)$$

and electrical equitation (2):

$$u(t) = L_s \frac{di}{dt} - U_e(t) + R_s i(t). \qquad (2)$$

Where U_e is back EMF voltage, in case of electrodynamic drive described by (3):

$$U_e = B \cdot l \cdot \dot{x}. \qquad (3)$$

Force of electrodynamic drive could be described by (4):

$$F_m(t) = B \cdot i(t) \cdot l. \qquad (4)$$

Where l is an active part of whole winding, part which is in magnetic field B, between pitch poles. The substitute circuit of electrodynamic drive is shown on fig.

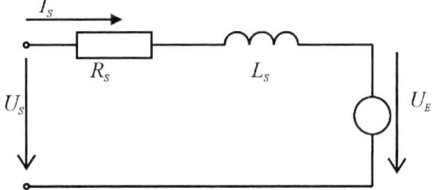

Fig. 2. Substitute circuit of electrodynamic drive

The model of the drive is shown on Fig. 3.

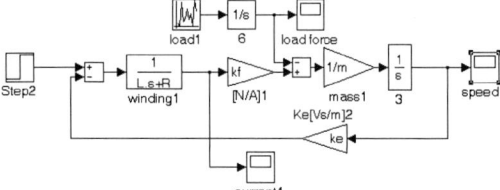

Fig. 3. Model of electrodynamic drive

Suitable mathematic model is basic precondition to investigate the properties and behaviour of drive by simulation. Therefore is made continuous model with current controller, speed and position controller. The model is nonlinear according to H-bridge transfer function (5).

$$F(s) = \frac{\dfrac{T_0^2}{12}s^2 - \dfrac{T_0}{2}s + 1}{\dfrac{T_0^2}{12}s^2 + \dfrac{T_0}{2}s + 1} \qquad (5)$$

The T_0 is a period of PWM modulation. The model only with current controller is shown on Fig.

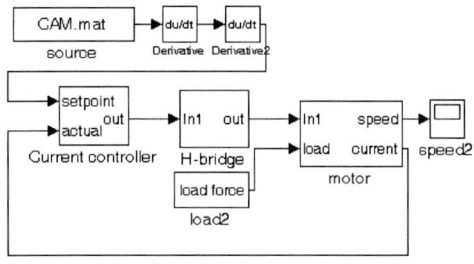

Fig. 4. Drive with current feedback

As an excitation signal of the model is used CAM definition signal, the description of real mechanical CAM. Like a disturbance signal, force, we used once integrated noise generator. This model gives us enough information for power electronics board design.

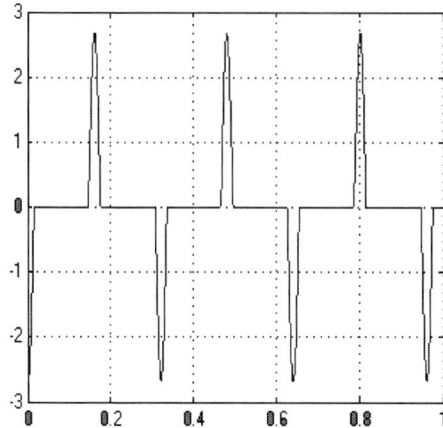

Fig. 5. Current [A] vs. time (no-load)

During controller tuning we have to respect current and voltage limitation of our power electronic design. Therefore was to the controller structure added saturation block.

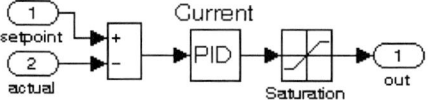

Fig. 6. Current controller block internal structure

The parameters of each controller are recalculated to discrete time due to implementation to digital signal processor.

III. CALCULATION OF CONTROLLING PARAMETERS

During controlling parameter calculation we used time mechanical (6) and electrical time constant (7) for motor description. The nonlinear blocks in the model, H-bridge of transfer function is removed according to high chopping frequency 20kHz.

$$T_m = \frac{m \cdot R}{K_f \cdot K_e}. \qquad (6)$$

$$T_e = \frac{L}{R}. \qquad (7)$$

$$\Omega_m = \frac{1}{\sqrt{T_m \cdot T_e}}. \qquad (8)$$

$$\varsigma_m = \frac{m \cdot R}{K_f \cdot K_e}. \qquad (9)$$

The transfer function of motor is described by (10)

$$F_{M0} = \frac{\dfrac{m}{K_e \cdot K_f} \cdot s}{\dfrac{s^2}{\Omega_m^2} + \dfrac{2\varsigma}{\Omega_m} + 1} \qquad (10)$$

and transfer function of the PI controller is (11)

$$R = Kp_i \frac{Tn_i \cdot s + 1}{Tn_i \cdot s} \qquad (11)$$

The open loop of current controller (12)

$$F_{i0} = R \cdot F_{M0} = Kp_i \frac{Tn_i \cdot s + 1}{Tn_i \cdot s} \cdot \frac{\dfrac{m}{K_e \cdot K_f} \cdot s}{\dfrac{s^2}{\Omega_m^2} + \dfrac{2\varsigma}{\Omega_m} + 1} \qquad (12)$$

The roots of equation are equal to (13)

$$s_{1,2} = -\varsigma_m \Omega_m \pm \sqrt{-\Omega_m^2 + \varsigma_m^2 \Omega_m^2} \;. \qquad (13)$$

The root trajectory is shown on Fig. 7.

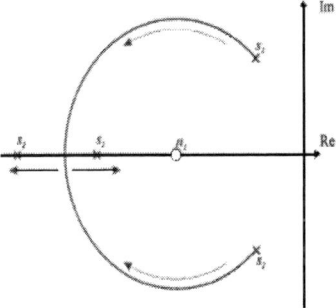

Fig. 7. Root trajectory of idealized current loop with increasing gain

With increasing gain the roots follow the trajectory; one of the roots is by the zero of the numerator. The influence of this root is rejected by the zero. The second root is far enough from real axis. Therefore we can use simplification for speed controller calculation; the transfer function of well tuned current controller is one.

The speed controller open loop is described by (14)

$$F_{\omega 0} = R \cdot 1 \cdot \frac{K_f}{m \cdot s} = Kp_\omega \frac{Tn_\omega \cdot s + 1}{Tn_\omega \cdot s} \cdot \frac{K_f}{m \cdot s} \;. \qquad (14)$$

The closed loop transfer function is (15)

$$F_\omega = \frac{K_f \cdot Kp_\omega (Tn_\omega \cdot s + 1)}{Tn_\omega \cdot m \cdot s^2 + K_f \cdot Kp_\omega \cdot s + K_f \cdot Kp_\omega} \;. \qquad (15)$$

Roots of this equatiton are equal (16)

$$s_{1,2} = \frac{-K_f Kp_\omega - \sqrt{(K_f Kp_\omega)^2 - 4 \cdot Tn_\omega \cdot m \cdot K_f Kp_\omega}}{2 \cdot Tn_\omega \cdot m} \;. \qquad (16)$$

By the condition (17) the complex root become real and with increasing gain the trajectory is shown on Fig. 8.

$$\frac{K_f \cdot Kp_\omega \cdot Tn_\omega}{m} \geq 4 \;. \qquad (17)$$

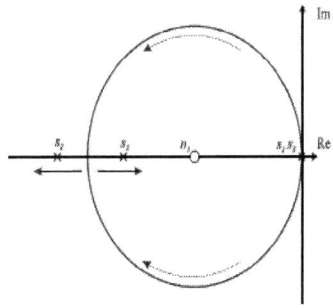

Fig. 8. Root trajectory of idealized speed controller with increasing gain

With increasing gain, one of the roots is close to zero of numerator.

The position controller is represented by simple proportional gain. The open loop of position controller is described by (18)

$$F_{p0} = Kp_p \cdot \frac{K_f \cdot Kp_\omega (Tn_\omega \cdot s + 1)}{Tn_\omega \cdot m \cdot s^2 + K_f \cdot Kp_\omega \cdot s + K_f \cdot Kp_\omega} \cdot \frac{1}{s} \;. \qquad (18)$$

The solution of characteristic equation is three roots, two complex numbers and one negative real. The root trajectory is shown on picture

1011

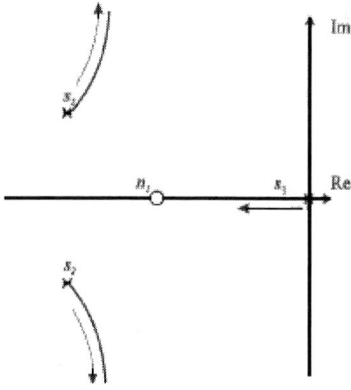

Fig. 9. Root trajectory of position control loop with increasing gain

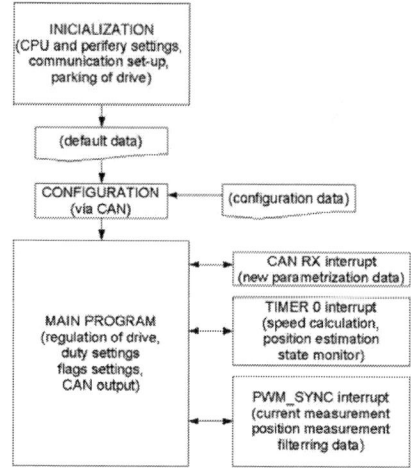

Fig. 10. Block diagram of DSP control program

The calculation of root locus diagrams gives us information about system setting and proper gain of each loop.

IV. IMPLEMENTATION OF CONTROLLING SYSTEM

The controlling system of the drive consist of two layers, one of them is operating level in digital signal processor necessary to control the drive, the higher level is implemented in PC and it is useful for debugging, drive parameterization and configuration.

A. Low level control

According to implementation control algorithm to DPS we have to split the control algorithm to separate time slots and operations. The basic program structure is shown in Fig. 1.

During the initialization there are made DSP clock setting, initialization of CAN controller, ADC and PWM controller. The next operation is loading the default data for positioning, end points, speed of positioning and default controller settings. After this software and hardware initialization we can modify the positioning data and controller setting by high level control system via CAN as well.

In the main program loop there are made calculations of controlling parameters and there are settings of drive status and flags. The process values are transmitted via CAN to higher level of control. The most important result of the controlling is a duty register setting which is a new setpoint value of the drive current.

The interrupts have a different priority. CAN messages have the highest priority, because we want to control the drive via CAN. But only a flag of a received message sets this interrupt, because transmitting process values for visualization are not as important as a quick stop from the higher control system.

The Timer0 interrupt has a lower priority. During Timer interrupt the speed controller calculation is made. The measured value of the drive is an actual position, because of the used IRC sensor. Therefore we have to calculate actual speed value like a numerical derivation of position.

The PWM sync interrupt has the lowest priority. Basically there are measurements of actual current and position of the drive, data preprocessing and filtering of the measured values. This interrupt is the most frequent with 20 kHz.

B. Current, speed and position control loops

For current loop sensing was used in first version sensing resistor and in-built current amplifier in H-bridge driver with analogue low pass filter. This method has one big disadvantage; the information about the current value is given, but not about the direction. The current direction must be calculated by internal model, form information about actual speed and several samples of current. Later was used current transformer to get information about direction as well as amount. There is also simplification in analogue low pass filter and amplifier.

The measured current is filtered by sliding average and final value is over scaled to Amperes. The synchronization of measurement to PWM sync interrupt is very important, because the current is measured in same time after PWM duty register setting. Due to transients on inductance, we get valid current information.

With current loop measurement is also measured the position by magnetostriction sensor, because the channels on ADCs are multiplexed.

The velocity control is based on Timer 0 interrupt. To make shorter duration of the interrupt we split internal structure of the interrupt into twice. Each even interrupt we make calculation of velocity and calculate the velocity controller. In odd interrupt is updated data for visualization. In each interrupt is calculated new setpoint values of sepoint generator, for generating smooth setpoint function.

The position control is calculated in main loop. But the most of the time the controller calculates the interrupts. Therefore the position control has lowest priority, despite of calculation in main control loop. To prevent unexpected functionality the nested interrupts are not allowed.

C. Configuration and visualization

Each modern servo drive has options for visualization and parameterization of the drive. Thanks to CAN interface on ADPS 21192 we used CAN 2.0b to obtain data for visualization and writing the parameters.

Creating the simple graphical user interface, shown on Fig. 11 was next step. There is several tab sheets to make CAN settings, Drive Control and Monitoring. The very important is Expert monitor to detect internal states of DSP program, important internal values and communication errors.

The events on buttons for PWM enable Stop and release, Stop and park/hold, Stop and hold is transmitted via CAN to ADPS, therefore the CAN Rx interrupt must have the highest priority to enable handling the drive any time.

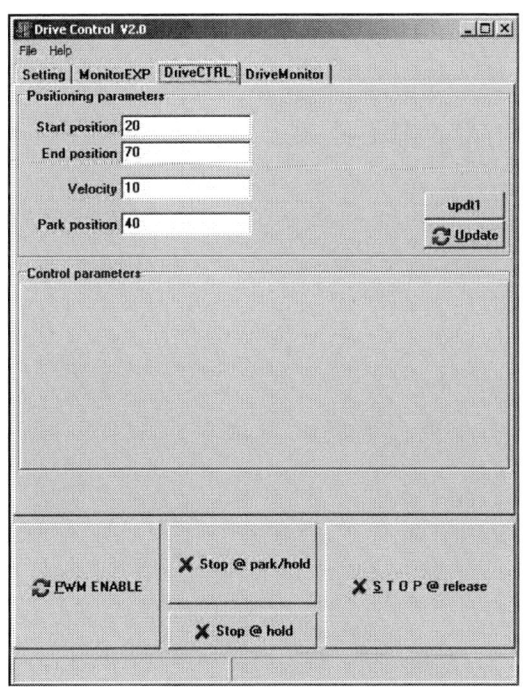

Fig. 11. GUI for motor commissioning and configuration

The important feature of each high level system is visualization. The visualization gives us information about actual position, speed and current via built in oscilloscope.

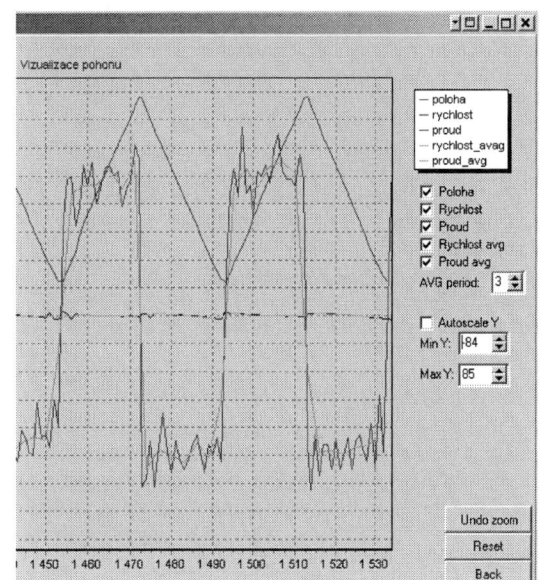

Fig. 12. Visualization of actual position, speed and current

There is also option to disable some lines in chart or add averaged lines speed and current.

V. POWER BOAD OF INVERTER

The power board of inverter is based on H-bridge driver IR2132 and power HexFet transistors IRFZ 44. The schematic of power board is shown on Fig. 14. The commissioning of power board is made without DPS control unit, but with generators and with intensive current measurement. During functionality validation of power board is necessary to test the board with inductive load. Therefore was used single phase transformer with a light bulb, Fig 13.

Fig. 13. Layout of power board and commissioning

A. Possibilities of over-current protection

The current measurement is very important with regard to power electronic overload in one hand and motor dynamic in the other. Modern power semiconductors allow multiple short time overloads that improves dynamic of the motor. In practical way we can solve the over current protection in two ways.

1013

One way is in hardware layer, because in bridge driver is a comparator with internal reference and current sense input from sensing resistor. The output of the comparator signal Fault can be connected with digital input of processor, but only for information of over current. The PWM is already disabled by bridge driver. Benefit of this hardware solution is in reliability. The disadvantage of this solution is problematic short time overload solution, by external low pass filter and setting the proper filter parameters.

The other way is based on software solution of this problem. The current is measured by the transformer and output voltage is evaluated with processor. This solution is reliable only with proper code in processor.

B. Short and long time over-current protection

The basic idea of our circuit solution is that we have to allow short time over current protection in one hand, but not allow overheat the transistor, by long time overload.

Therefore we used hardware solution set up near to maximal current of the transistors. This improves the dynamics of the motor thru the peak force.

The current measurement for regulation is realized with current transformer. This transformer was used for software solution of continuous over current protection. In processor was simplification of device duty cycle calculation. The continuous current value vas set by the external amplifier to saturation of the ADC. If the ADC is in saturation, the saturation time is measured by the timer. If timer overflow the PWM unit of the controller is disabled. The new enable of the PWM unit is possible after several seconds, when the heat from semiconductor is lead off.

The power supplies and analog converter amplifiers is not shown in this schematic. The driver IR2132 has in built current monitor. But due to noise in chopped current signal we used the LEM transformer to measure actual current.

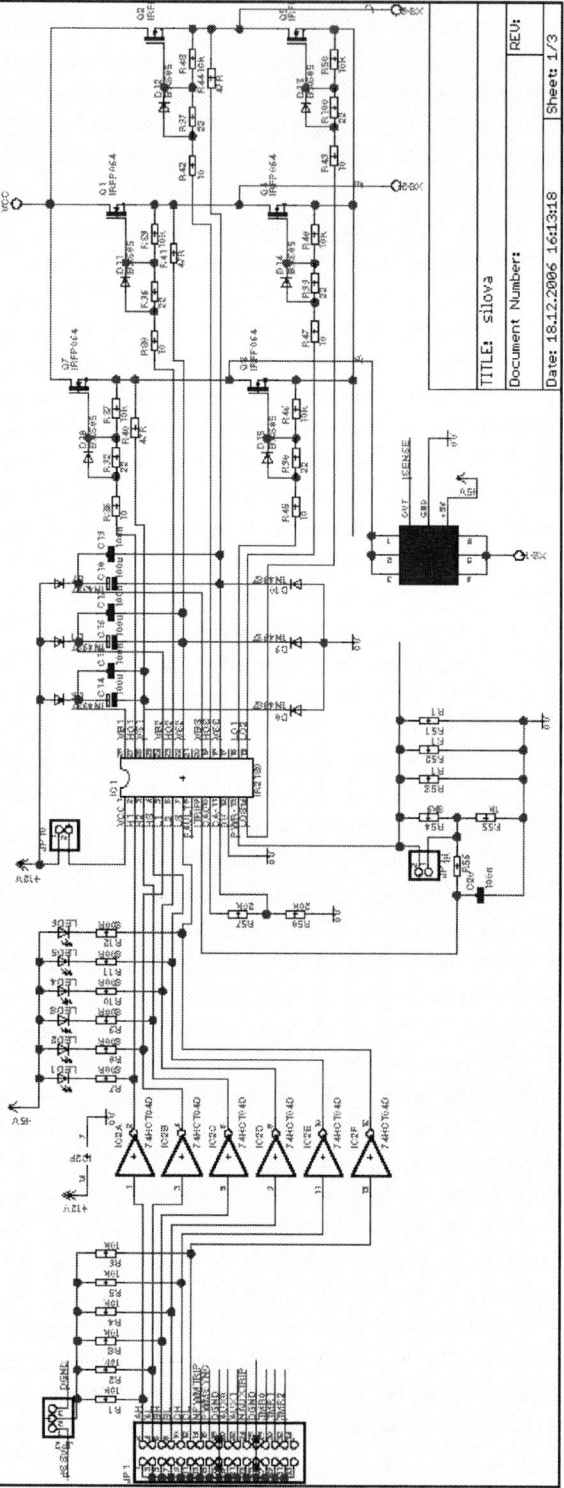

Fig. 14. Schematic of power stage

VI. MECHANICAL CONSTRUCTION

The mechanical arrangement of the final low power model of electrodynamic drive is shown in Fig 15.

Fig. 15. Mechanical construction of electrodynamic drive

The frame of electrodynamic drive is made of two coplanar desks. On the top desk, there is gantry arrangement of linear bearing. The stator with two rings of permanent magnets is mounted to lower desk. The coil is coaxial to orthogonal axis of permanent magnet polarization, fig. 16.

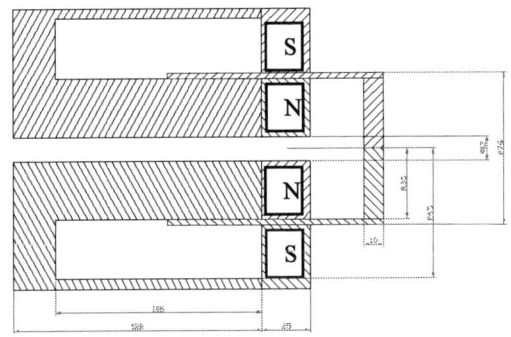

Fig. 16. Stator with permanent magnets and coil

To the gantry of linear bearing is connected to the flange of coil. These parts create moving part of the motor.

For position measurement there are two sensors. The linear IRC sensor is used for positioning and velocity measurement. The absolute magnetostriction sensor is used for motor parking and referencing to machine zero.

VII. MOTOR DATA AND PERFORMANCE

Peak Force @ 0.02s	111.25	[N]
Continuous Force	53.4	[N]
Peak Current @ 0.02s	25	[A]
Continuous Current	12	[A]
Max. Supply Voltage	36	[V]
Typ. Sypply Voltage	24	[V]
Force constant	4.45	[N/A]
Electric constant	4.46	[V/m/s]

Winding resistance @ 21°C	0.4	[Ohm]
Winding resistance @ 100°C	0.54	[Ohm]
Winding Inductance @ 1kHz	27	[mH]
Max. Winding Temperature	115	[°C]
Max. Theoretical Speed	5.38	[m/s]
Max. Stoke	120	[mm]
Moving Parts Mass	0.300	[kg]

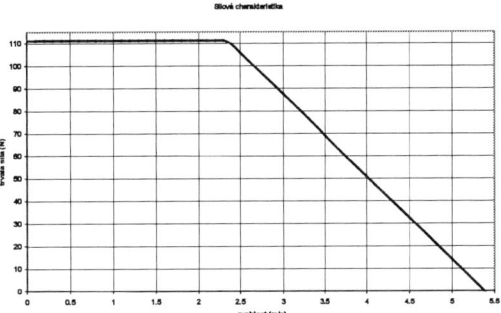

Fig. 17. Force vs. Speed chart

VIII. CONCLUSION

The mathematical model of the electrodynamic motor gives us good information for power electronics board design and power electronic module sizing. The mathematical model of electrodynamic drive including control structure of PI controllers provides us particular information for controller initial setting. With voltage and current limitation we can achieve better precision and find out the non-limiting setting of controllers.

The real drive does not achieve simulated performance. The main problems are especially in speed controller setting. The absolute measurement with magnetostriction principle is distorted by electromagnetic field of a main coil. With optical measurement by IRC sensor, we can get better speed controller setting. The other problems are caused by EMC with current measurement. Using the shield cables and better grounding of the power electronics and the analogue circuits reduce the problems with EMC.

The electrodynamic drive is easy to control due to one winding, one current controller and nearly invariant force during a whole stoke. But for the higher power ranges, for example real unwinding drive, this principle can not reach performance parameters of nowadays linear tubular drives.

ACKNOWLEDGMENT

Thanks to Research Centre Textile 1M0553 and Ministry of Industry and Trade of the Czech Republic project FT-TA3/017.

REFERENCES

[1] J. □ernohorský, P. Střeštík, Possibilities of winding parameter improvement. Liberec: Research Institute of Textile Machines, 2003 (in Czech).

[2] P. Souček, Servomechanismy ve výrobních strojích: vydavatelství □VUT : Praha, 2004, ISBN 80-01-02902-6 (in Czech).

[3] J. □ernohorský, Possibilities of electrodynamic drive usage on unwinding machines: Technical university of Liberec: Liberec, 2006, (in Czech).

Expert System for Electric Drive Design

Juhan Laugis* and Valery Vodovozov†

* Tallinn University of Technology / Institute if Electric Drive and Power Electronics, Tallinn, Estonia,
e-mail: *laugis@cc.ttu.ee*
† State Electrotechnical University / Dept. of Electric Drive and Robotics, St. Petersburg, Russia,
e-mail: *edrive@narod.ru*

Abstract— The problem of electric drive design concerns the multi-criteria non-linear problems having a number of solutions. In daily practice the search of optimum solution bases a great deal on the designer intuition and experience. Therefore, multiple obtained results have a low degree of equipment use, unwarranted complexity, or excessive cost. Proposed approach is devoted to implementation of the novel expert system into the design stages concerned the work with databases of the drive components. Examples of effective algorithms are given to choose an optimum ratio of electrical and mechanical components. The soft tool is developed to generate and edit structured query language sentences, to extract information from databases, as well as to suit multitude sets for particular applications.

Keywords — Design, electrical drive, expert system, modelling, simulation, software.

I. INTRODUCTION

An electric drive belongs to inhomogeneous, nonlinear, and non-stationary class of systems [1]. Object heterogeneity reflects in various natures of structural components and assemble of informational and energetic processes. Heterogeneity deals with analog and discrete nature of different inputs and variables. Power pulse generation with continuous electromechanical transformation and discontinuous informational conversion is referred to the same processes as well. When the reaction on different inputs is discussed, an electromechanical system is devoted to non-linear group. Non-linearity is produced by the limited controllers and sensors code lengths, as well as by technical and program non-linear chains and blocks. They course amplitude modularity of digit-to-analog transformation, non-regular signal shapes, and semi-controllability of power converters, saturation in magnetic chains of electric machines, plugs, gaps, backslashes, Coulomb and fluid friction of mechanical gears and actuators. A non-stationary property of electric drive is a result of changing parameters of the system and varying load torques and moments of inertia of mechanical system. This feature is taking into account in description of objects behavior as time-dependent statements.

Therefore, an electric drive design procedure involves a number of complex problems [2]. Some of them are:

- timing calculation and the mechanism travel diagram construction;
- computation of mechanism forces and torque/power patterns synthesis;
- gear dimensioning and selection;
- motor dimensioning and selection;
- optimum motor-gear composition and checking;
- power electronic converter dimensioning and selection or design a novel one;
- design of controllers for multi-loop control and adjustment;
- process simulation with building of steady-state diagrams and transients;
- development of the drive documentation;
- economical basing and efficiency evaluation.

Effective assembling of the drive systems meets with the problems of such kind owing to rather complicated algorithms [3]. Today, the three ways of the drive design are used in the designers' community.

The first of them starts from the selection of a manufacturer; further the manufacturer design technology and recommendations is followed. To compute and select equipment, the leading companies have developed their specific technologies. Examples are the guides and software of "Siemens" [4], "Omron", "Sew Eurodrive" [5], "Maxon Motors", "Mitsubishi Electric", etc. Such approach is usual for the majority of firms that carry out project designs and have rich experience in acceptance of the decisions based on extensive computer databases, coming up to numerous catalogue archives and "absorbing" their contents and structures. The main its drawback is the technological restriction and data limiting that deprive a designer of optimum way in the project. It is especially important in the first stage, when the most responsible decisions are decided.

The novel way has been developed and described in [6], [7]; it starts from the traditional load computation using the designers' own experience and methods. Further a designer selects a group of gear families and types of different companies using the load calculation results. Through these gears, the forces and mechanism speed found before will be converted to the equivalent values on motor shaft. Then, for each gear type the own motor type is to be picked out, counted on the converted forces and speed. Thus, hundreds and thousands sets of possible drive variants are generated. Once the equipment framework will be found, the new problem appears: which of the suitable motor-gear combinations are optimum? To find a solution, a designer can form appropriate criteria and sort them. It may be a criterion of maximum accuracy or speed, minimum weight, power, or inertia, highest rigidity, etc. This way, the full scale of the electromechanical and electronic properties is gathered, from which the choice is done based on judgments about preferences of that or other criterion.

An approach proposed in this paper bases on the automatic multitude simultaneous gear-motor-converter assembles calculation and selection with analyzing and sorting them on the judgments about the criterion preferences.

II. QUERY LANGUAGE TO PROCESS DATABASES OF THE DRIVE COMPONENTS

Different manufacturers worked out hundreds and thousands of databases that help in providing most of stages described above, short of first and last one. A specialist in electric drive design receives next input specifications from these data sheets:

- Parameters of motors with various action and construction principles, powers and speeds. Each motor is specified by its rated torque, moment of inertia, mass, efficiency and a list of electrical values.

- Specifications of transmissions distinguished by design and construction, power and ratio. Each gear-box has its rated and maximum torque, primary and secondary speeds, moment of inertia, mass and efficiency.

- The nameplate data of industrial electrical power converters with ac and dc principles and different powers. Each record includes rated and maximum voltages, currents, frequencies, and efficiency.

All these data sheets are prepared by various manufactures using different languages and operation systems. They have multiple structures, data types, data fields, keys, names, and inter-sheet links. Not all required data are published into the sheets; some of them present the know-how of manufacturers hided from users. Often, information carriers advertising nature, which is not needed to derive a design. Some data are published in Internet; other is obtainable through printed materials and journals.

To overcome these problems and obtain an access to the required information, a novel expert system based on the universal database has been proposed and described data sources were connected to this joined database. Unlike the known simulation tools, for example [8], the proposed package is deeply integrated with database equipment. For the selected connection, specific software has been developed [9]. The uniform structure, data types, keys, and tables were prepared also. Then using structured query language (SQL), a number of templates were developed, in which calculated data should be substituted to find the sets of the drive components.

A query language gives a set of rules and tools that helps user to search required information into the databases. As a query answer, a database management system replies a result in the form of screen (virtual) table displayed a recordset of a certain structure, which may be used to build a real summary table, report, or a new query. Today, SQL is an international standard of query languages used in all databases used all-round the world. The data search is not the only operation of SQL. The language provides tables' creation, control, connection, and edition as well.

SQL grammar bases on SQL statements, each included variables, which are the names and values of particular tables, fields, and keys. For example:

SELECT * FROM csCONTROLTECHNIQUES
WHERE Pc_W > 900 AND Ic_A > 21
ORDER BY Pc_W

This statement searches all records in the table csCONTROLTECHNIQUES, which satisfy the requirement that the candidate converter power must exceed 900 and current must exceed 21 whilst the obtained records have to be ordered alphabetically on the screen.

The next statement searches information in three tables simultaneously with online calculation of dependant parameters. Here, gears, motors, and converters are selected together matching powers (P_W), torques (M_Nm), rotational and linear speeds (n_rpn, vg_cms), moments of inertia (J_kgcm2), and currents (Ic_A) together:

SELECT * FROM gs, msSEW, csCONTROLTECH-NIQUES
WHERE vg_cms / 100 >= 0.1
AND Pg_W * 100 / vg_cms >= 671 / 0.1
AND M_Nm > 671 / ng_rpm * 10 / [%g] * 100
AND J_kgcm2 / 10000 > 0.004 / 10
AND n_rpm / 10 > 59.5
AND Pc_W > 15 * 60
AND Ic_A > 14.9 * I_A / M_Nm
ORDER BY Pg_W

III. DATABASE MANAGEMENT SYSTEM

Further, the soft tool has been developed to build the necessary SQL statement templates and to insert the arguments into this templates. The proposed software includes the comfortable Windows interface with dialog boxes and SQL edition environment. The required databases link to the package automatically or manually. The first tab of the database window shown in Fig. 1 involves the list of tables used to find data as well as the detailed data sheets with the drive components' records. Figure 2 depicts the SQL editor page having all the needed instruments to write, copy, cut, paste, delete, insert, and check the language components and SQL statements. Further the combined result of the query running is generated, where the matched sets of motors, gears, and converters are ranged on the preferred order.

All information selected using SQL may be inserted into the drive model to suit their correctness and efficiency in the common gear-motor-converter dataset. To execute simulation, the three tabs display simulation result: Dynamics, Statics, and Analysis. Any change of model provides the result updating when it activates. Here, the tab Dynamics represents the speed transients of the mechanism and the torque or the current transients of the motor. The values of variables one may measure in the areas where the curves are crossed by the vertical line of cursor. The program indicates them in the status bar. The tab Statics displays the idealized speed-current or speed-torque relations in the open-loop system. Simulation analyses screen shown in Fig.4 includes:

- table of torque, current, and speed instantaneous values per time;

- summary of maximum, minimum, and steady-state values of variables;

- the data, calculated in the process of simulation;

- summary of service factors of the drive equipment that is calculated as the steady-to-rated (in the case of the step signal) or rms-to-rated (in other cases) values ratio:

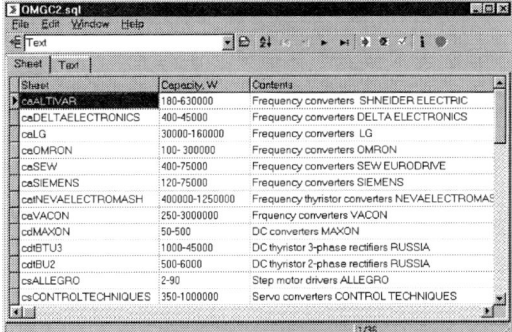

Fig. 1. The tab with table of contents on the database screen

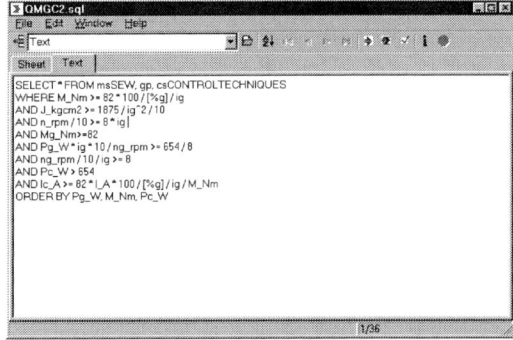

Fig. 2. SQL editor tab of the database screen

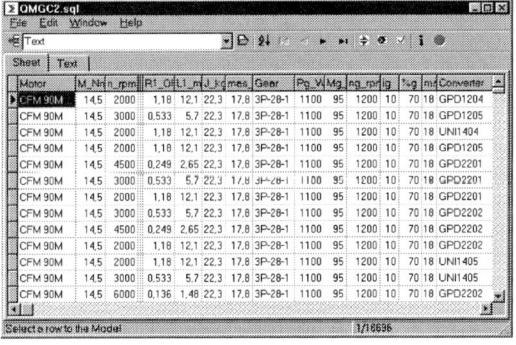

Fig. 3. Example of an SQL query running where almost 20000 possible drive variants are sorted.

- if the motor rated power has been specified in the database, the motor power service factor is obtained as the product of motor torque and motor speed divided by motor power, and the motor torque service factor is obtained as the ratio of steady (rms) torque value and the quotient obtained when rated power is divided by motor speed;

- if the motor rated torque has been specified in the database, the motor torque service factor is obtained as the ratio of the steady (rms) torque and the rated torque values;

- if the motor rated current has been specified in the database, the motor current service factor is obtained as the ratio of the steady (rms) current and the rated current values;

- if the converter rated power has been specified in the database, the converter power service factor is obtained the ratio of the motor torque-speed product and the rated converter power;

- if the gear rated power has been specified in the database, the gear power service factor is obtained as the ratio of the load torque-speed product and the gear rated power;

- if the gear transmission ratio and rated torque have been specified in the database, the gear torque service factor is obtained as the product of the steady (rms) torque and transmission ratio divided by the gear rated torque.

The selected components may be used in the project since the simulation results satisfied the designed requirements. If the results are unsatisfactory, new SQL queries have to be created to continue the search of optimum drive components.

IV. CONCLUSION

A novel expert system to search the optimum components of the designed motor drive has been proposed. This approach is based on the implementation of the SQL into the design stages concerned the work with databases. Examples of effective algorithms are given to choose an optimum ratio of electrical and mechanical components. The soft tool is described to generate and edit SQL sentences, to extract information from databases, as well as to test different component sets for particular applications. Obtained results help to increase the equipment service factors and to avoid unwarranted complexity or excessive motor drive cost.

REFERENCES

[1] V. Vodovozov, "Theory and systems of electric drives", ETU Publishing, St.Petersburg, 2004.

[2] S. Cetinkunt, "Optimal design issues in high-speed high-precision motion servo systems", Mechatronics, 1991, 1(2), 187–201.

[3] F. Roos, H. Johansson, and J. Wikander, "Optimum selection of motor and gearhead in mechatronic applications", Mechatronics, 2006, 16(1), 63-72.

[4] "Siemens standard drives. Application handbook", Congleton, 1997.

[5] "Drive engineering — Practical implementation", Volume 1. Drive arrangements with SEW geared motors. Calculation methods and examples. SEW Eurodrive, 1998.

[6] V. Vodovozov and A. Loparev, "Simulation tools for design and testing of electric drives", 10th International Power Electronics and Motion Control Conference EPE-PEMC-2004, Riga, Latvia, 2004, Paper DS 7.16.

[7] V. Vodovozov and J. Laugis, "Object-oriented electric drive development technology", IEEE International Electric Machines and Drives Conference IEMDC'07, Antalya, Turkey, 2007, Paper AF000434.

[8] N. Ergodian, H. Hemao, and R. Grisel, "A proposed technique for simulating the complete electric drive systems with a complex kinematics chain", IEEE International Electric Machines and Drives Conference IEMDC'07, Antalya, Turkey, 2007, Paper AF014095.

[9] V. Vodovozov, E. Vodovozova, and E. Tsvetikov, "Object-oriented models of electromechanical systems", 8th European Simulation Symposium "Simulation in Industry", Italy, 1996.

Improvement of Moving Characteristics of Cableless Micro-actuator and Consideration of Reversible Motion

Hiroyuki Yaguchi [1], Kazumi Ishikawa [2], Toshihiro Zamma, Koichi Funayama

[1] Tohoku Gakuin University/Faculty of Engineering, Tagajo, Miyagi, Japan, yaguchi@tjcc.tohoku-gakuin.ac.jp
[2] Tohoku Gakuin University/Faculty of Engineering, Tagajo, Miyagi, Japan, ishikawa@tjcc.tohoku-gakuin.ac.jp

Abstract—We previously proposed a novel cableless micro-actuator that provides propulsion using the mechanical resonance energy of a system excited by an electromagnetic force. However, it was possible for that the actuator to move only one direction and comparative low speed locomotion. This paper proposes a cableless micro-actuator that exhibits a high speed locomotion and considers reversible motion of the cableless actuator due to control of a magnetic field by using coils outside the pipe. This actuator contains a mechanical inverter that directly transforms DC from button batteries into AC. The mechanical DC-AC inverter incorporates a one-degree-of-freedom-model and a curved beam with a concentrated mass that switches under the electromagnetic force. The actuator is moved by inertia force of the one-degree-of-freedom-model due to mechanical resonance energy. Experimental results show that the actuator can move upwards at a speed of 16.5 mm/s by using 10 button batteries when pulling 10 g load mass. The results hold great promise for the creation of highly mobile actuators capable of moving in thin pipes with diameters of under 8 mm.

Keywords— cableless micro-actuator, mechanical DC-AC inverter, inertia force, reversible motion.

I . INTRODUCTION

For inspection of pipes used in nuclear power and chemical plants, it is very important to find damage from inside pipes. Safety issues have encouraged the growing use of such robots in recent years for a variety of duties, not only inspections, maintenance and cleaning in the pipe. Some papers have addressed possible mechanisms for moving within piping [1-9]. We have proposed [10] a new cableless micro-actuator to provide locomotion for a maintenance robot in pipes. This paper proposes a new type of actuator which exhibits a very high thrusting force and is capable of extension to locomotion range. The propulsion module of the actuator is remodel in order to reinforce a magnetic force due to reconsideration of arrangement of permanent magnet and electromagnet. For that reason, a new type of mechanical DC-AC inverter is developed also. The actuator is based on a small, mechanical inverter which directly converts the DC from button batteries to AC. Switching is performed by the vibration of a one-degree-of-freedom-model and a curved beam with a concentrated copper mass at tip. Propulsion is provided by the mechanical resonance energy of the

one-degree-of-freedom-model suspended by using double exciting due to arrange of two permanent magnets. A prototype actuator has been fabricated containing a mechanical DC-AC inverter and is able to move through a pipe of 8 mm inner diameter. This paper demonstrates the good performance over the range of the cableless micro-actuator. We consider a reversible motion of the actuator due to control of a magnetic field by using coils outside the pipe.

II . CABLELESS MICRO-ACTUATOR FOR A MOVEABLE IN-PIPING

Fig.1 is a diagram of the cableless micro-actuator capable of moving within 8 mm inner diameter pipe. The actuator consists of two permanent magnets to reinforce the magnetic force due to double exciting, a translational spring, and an excitation electromagnet and the mechanical DC- AC inverter due to the curved beam switching system. The permanent magnets are cylindrical NdFeB and are magnetized in the axial direction. These are 6 mm in diameter and height is 3 mm and 5 mm. The surface magnetic flux density is 350 mT and 400 mT, respectively. The translational spring is the stainless steel compression coil type with an outer diameter of 6 mm,

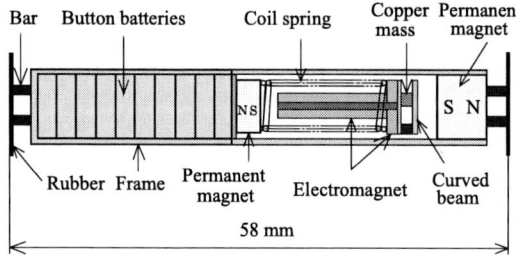

Fig. 1 Structure of the cableless micro-actuator.

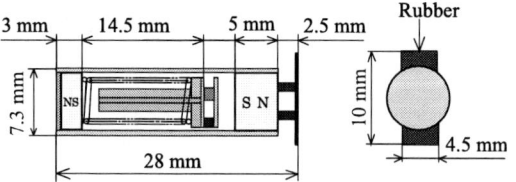

Fig. 2. Propulsion module of the actuator.

free length of 14 mm, and a spring constant $k = 632$ N/m. The excitation electromagnet with a stepped diameter consists of an iron core 1.7 mm in diameter with 2550 turns of 0.07 mm diameter copper wire. The stepped diameter is 6 mm and 4 mm, respectively. The electrical resistance of the electromagnet is 160 Ω. This electromagnet also achieves duty of mass of the one-degree-of-freedom-model by combining together with a coil spring. There is the curved beam system with the copper mass upon the electromagnet. As shown in Fig.2, the leg supporting the actuator is a natural rubber of two sheets of 10 mm length, 4.5 mm width and 0.5 mm thickness. The actuator is 58 mm length and it has the same length and total mass compared with previous work [10] when the actuator has 10 button batteries. This actuator converts the mechanical resonance energy stimulated by the AC electrical power source into locomotive power.

III. PRINCIPLE OF LOCOMOTION

The two rubbers transform as shown in Fig.3 (a) when the actuator is inserted in the pipe. We assume that the mass-spring system of the actuator vibrates with pure translational vibration inside coordinate x. The principle of locomotion is as follows: 1) As shown in Fig. 3 (a), the supporting force of two rubbers are not changed due to inertia force of a vibrating mass when the mass of the mass-spring system vibrates at direction of coordinate x. Therefore, the actuator can slide when the inertia force of the vibrating mass is bigger than a frictional force between two rubbers and wall of the pipe. 2) The other hand, as shown in Fig. 3 (b), tip of the rubber is locked wall of the pipe when the mass vibrates at opposite direction of coordinate x. In this case, the frictional force between two rubbers and wall of the pipe becomes quite big. 3) Thus, the frictional force between the rubber and wall of the pipe inside alternately changes during one period of the vibration. As the results, the actuator can move only at direction of coordinate x utilizing inertia force of the vibrating mass of the mass-spring system.

IV. STRUCTURE AND OPERATING PRINCIPLES OF MECHANICAL DC-AC INVERTER

Fig.4 (a) and (b) show outline of the new mechanical DC-AC inverter proposed in this paper to realize the cableless actuator and dimension of the curved beam system with the concentrated copper mass at tip. The inverter is composed of a switching system due to the permanent magnets, the copper circular plate and the curved beam upon the electromagnet, and a mass-spring system. The curved beam is two layers structure due to permanent magnets, the copper circular plate and the curved beam upon the electromagnet, and a mass-spring system. The curved beam is two layers structure due to plastic and steel material, curved free length of 10.6 mm, and the tip copper mass m is 0.043 g. The copper plate and copper mass are connected to a DC source such as a battery in Fig.4. The mass-spring system vibrates and the conductors make and break contact, it is cycled on and off. As a result, the DC voltage is converted into a square alternating waveform and then the magnetic force operates on the mass-spring system.

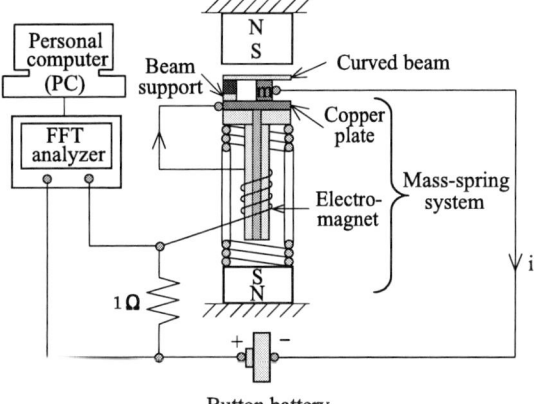

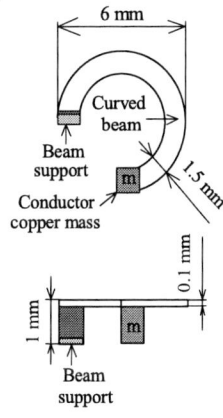

(a) Connection of the two vibration systems.

(b) The curved beam with the copper mass at tip.

Fig.4. Principle of the mechanical DC-AC inverter and dimension of the curved beam system.

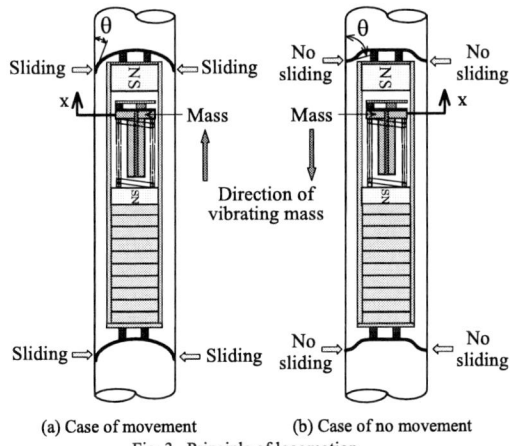

(a) Case of movement   (b) Case of no movement

Fig. 3. Principle of locomotion.

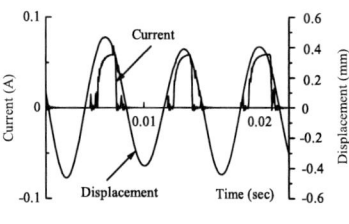

(a) Initial contact force F = 0.36 mN

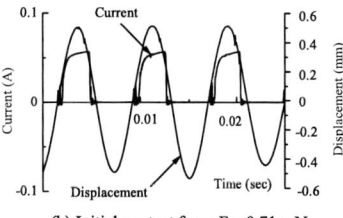

(b) Initial contact force F = 0.71 mN

Fig. 5 Waveform of the current and displacement due to initial contact force.

It was anticipated that the two vibration systems would restrict each other's displacement of vibration when there was an initial contact force between them, increasing their contact time and in consequence, the duty factor. The initial contact force can be adjusted by bending the beam support. An experiment was conducted to measure the duty factor of the generated square wave due to change of the initial contact force. The DC source for the experiment was a pack of six alkaline batteries of LR6 type. The current waveform generated during switching of the two vibration systems was stored in a PC via a fast Fourier transform (FFT) analyzer. Fig. 5 shows the displacement of the mass-spring system and the variations in produced current with initial contact force *F*.

V. LOCOMOTION CHARACTERISTICS OF CABLELESS MICRO-ACTUATOR

As the power source for the actuator, Maxell SR621W silver-oxide button batteries were used. Each battery is 6.8 mm in diameter and 2.15 mm thick, with a mass of

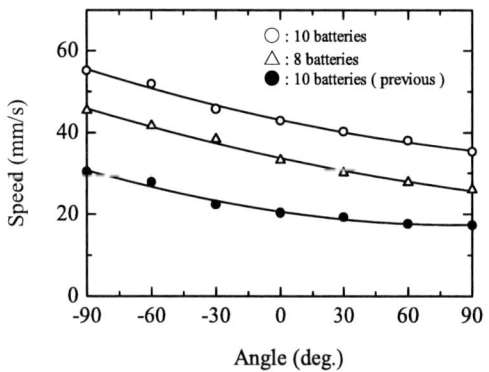

Fig. 6. Relationship between angle and speed.

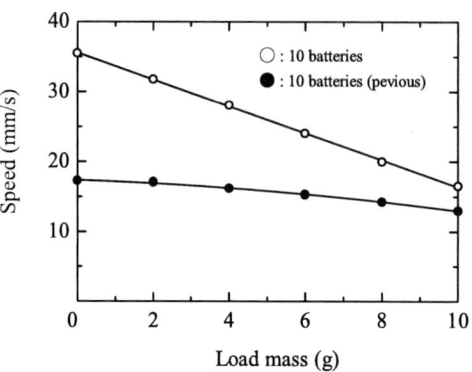

Fig. 7 Relationship between load mass and vertical straight up speed.

TABLE 1.
Motion time and range.

	Present actuator	Present actuaor	Previous actuator
Batteries	10	8	10
Motion time (min)	42	42	62
Horizontal speed (mm/s)	42.8	33.3	20.4
Vertical speed (mm/s)	35.3	26	17.3
Horizontal motion (m)	108	84	76
Vertical motion (m)	89	66	64

0.3 g and a capacity of 18 mAh at a nominal output of 1.55 V. The battery pack was mounted on a frame. The completely fabricated actuator was then inserted into an acrylic pipe with an inner diameter of 8 mm. Harmonic excitation tuned at eigenfrequency of the mass-spring system was then induced using a coil outside the pipe in order to excite vibration of the two vibration systems. This initiated action in the inverter, in turn initiating motion of the actuator. Fig.6 shows the relationship between speed of the actuator and the angle α by using 10 and 8 button batteries. The angle was varied from α = -90° (straight down) to α = 90° (straight up). The speed decreased in a roughly linear manner with increasing angle α. The maximum vertical straight up speed was 35.3 mm/s with 10 button batteries and 26 mm/s with 8 button batteries. This is quite high performance for the cableless micro-actuator. As shown in Fig.6, this actuator demonstrates speed of about twice compared with previous actuator [10] having the same dimension and total mass. Fig.7 shows the relationship between load mass and speed when moving straight up using 10 button batteries. The figure indicates that the actuator could climb at 16.5 mm/s when pulling a load mass of 10 g, a moderately high performance. Table 1 shows the continuous motion time and range. This actuator demonstrates extension of locomotive range about 140 percent compared with previous work [10] due to reinforcement of the magnetic force and improvement of the new mechanical DC-AC inverter. Fig. 8 shows the

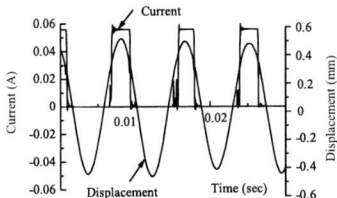

(a) 10 button batteries

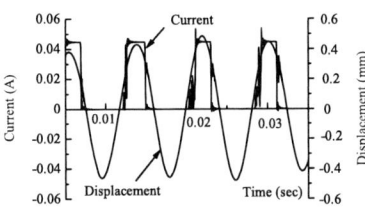

(b) 8 button batteries

Fig. 8. Waveform of the current and displacement.

124 Hz current square wave produced by the inverter module when the initial contact force in the two vibration systems was set at 0.5 mN. This figure indicates that the duty factor of this waveform was 30 %.

VI. CONSIDERATION OF REVERSIBLE MOTION

Fig.9 (a) shows outline of the reversible motion of the micro-actuator. As shown in Fig.9 (b), moving direction of the actuator can reverse due to control of the magnetic field by using an arranged coil A, coil B and coil C outside the pipe. The operation procedure of the reversible motion is as follows: 1) The actuator starts to move by using an excitation coil. 2) The coil C generates the

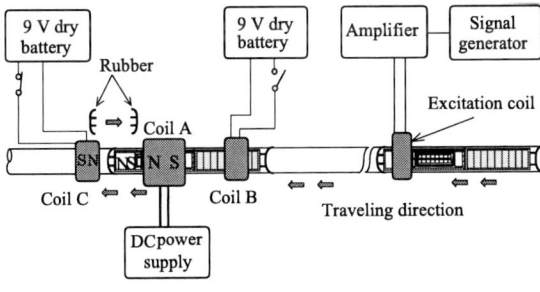

(a) Outline of reversible motion.

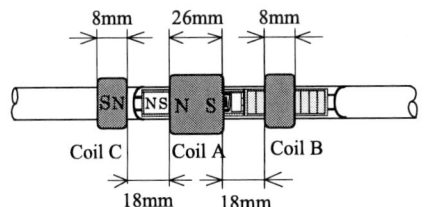

(b) Arrangement and dimension of the coils.
Fig. 9 Principle of the reversible motion.

TABLE 2.
Specification of the coils and magnetic field.

	Diameter of wire (mm)	Number of turns	Width (mm)	Input current (A)	Magnetic flux density (mT)
Reversible Coil A	0.5	200	10	5	46
Stopping Coil B, Coil C	0.2	96	6	0.8	4.8

repulsion magnetic field against the permanent magnet of the actuator due to input current of 0.8 A, and the magnetic field of the coil A and coil B is zero due to no input current. As the results, the actuator can be through the coil A and coil B but it stops around in the coil C. 3) When the coil A generates the attractive magnetic field for the permanent magnet of the actuator due to input of electrical current of 5 A, as shown in Fig.9 (a), The two rubbers of the actuator are reversed, and the actuator can move to another direction. Table 2 shows the properties of the coils and the relationship between the input current and magnitude of the magnetic field generated at center of the coils.

VII. CONCLUSION

A cableless micro-actuator powered by a mechanical DC-AC inverter and capable of moving within a pipe has been proposed and tested. When double exciting by using two permanent magnets, the actuator displayed speeds of 55.0 mm/s moving straight down, 42.8 mm/s moving horizontally, and 35.3 mm/s moving straight up, Since it moves downwards much more quickly than it climbs, the actuator must be lightened. The total operating range of this actuator is 108 m in the horizontal direction or 89 m in the vertical direction over an operating time of 42 min, when powered by 10 button batteries. Result demonstrates reversible motion of the actuator due to control of a magnetic field by using coils outside the pipe. If the power requirement of the propulsion module can be lowered and the generation of high-frequency harmonics of square current wave suppressed, it certainly seems possible to extend the range to several hundred meters.

REFERENCES

[1] H. Yaguchi, S. Sasaki : *Trans. Magn. Soc. Jpn*, 5, 1, pp. 43-48 (2005).

[2] H. Yaguchi, K. Tsurumoto : *The 2005 IEEE International Magnetics Conference*, FT-12(2005)

[3] H. Yaguchi : *J. Magn. Soc. Jpn.*, 27, 4, pp. 509-512 (2003).

[4] K. Tsuruta, T. Shibata, N. Mitsumoto, T. Sasaya and M. Kawahara : *Trans. IEE Jpn.*, 122, 2 (2002)

[5] Y. Kondo, S. Yokota : *Trans. J. Soc. Mec . Eng.*, 64, 617 (1998).

[6] S. Yamamoto, K. Sato, K. Kikuchi and T. Matsu : *Trans. Japan Soc. Mec. Eng.*, 57, 2675 (1991).

[7] H. Saito, K. Sato, K. Kudo and K. Sato : *Trans. Japan Soc. Mec. Eng.*, 66, 346 (2000).

[8] S. Aoshima, T. Morimitsu and K. Tujimura ： *Trans. J. Soc. Mec. Eng.*, 55, 516 (1989).

[9] T. Fukuda : *Trans. J. Soc. Mec. Eng.*, 52, 477 (1986).

[10] H. Yaguchi, Y. Nanjo and K. Ishikawa : *The 12th European Conference on Power Electronics and Applications*,0646(2007).

Sensorless Control of AC Machines using High-Frequency Excitation

Heiko Zatocil

Chair of Electrical Drives, University of Erlangen-Nuremberg, Erlangen, Germany
e-mail: Zatocil@eas.eei.uni-erlangen.de

Abstract—In the present paper a new method for sensorless control is presented, which can handle the presence of multiple non-separatable saliencies. Because of the inimitability of the measured impedance curves, the method is based on the minimum error between the measured curves during operation and precommissioned reference curves. The method can handle alternating and rotating test signals. The capacity of the proposed method is tested on two different standard induction machines, which are not especially designed for sensorless control. Operation at standstill, and even at zero-frequency, is possible.

Keywords—Sensorless control, induction motor, signal processing, vector control

I. INTRODUCTION

For the last few years many researchers have been working in the field of sensorless control in order to reduce costs, increase reliability and decrease the need for maintenance. There are two very promising methods for realising a drive control without a mechanical transducer. The first approach evaluates the fundamental model of the induction machine as defined in the classical space vector theory. For example, this can be realised by a Model Reference Adaptive System (MRAS) [1], an observer (e.g. Kalman filter, Luenberger observer) [2]–[4] and so on. In the case of zero stator frequency, no voltage will be induced in the stator and the observability of the machine is lost. A sensorless control that deals with such an approach will therefore only perform satisfactorily until a minimum frequency - close to zero - is reached.

The second opportunity for implementing a transducerless drive control is based on the exploitation of anisotropies that can depend on saturation or slotting [5] but also on the magnetic anisotropy of the lamination [6]. In most cases, these inherent saliencies are detected by additional high-frequency signal injection [7]. The high-frequency excitation can be realised by an alternating, rotating or transient signal, as presented in [8].

The test signal response is influenced by several saliencies. The separation of each anisotropy is therefore hard to implement but fundamental. For the estimation of the mechanical speed only the slotting effects must be exploited, whereas for the detection of the flux angle only a special saturation-induced saliency carries the needed information. All non-exploited signal components act therefore as disturbances and thus must be eliminated.

At present, there are many approaches for separating the signals and using them in a sensorless field-oriented control [9]–[14]. Most of them are based on the idea of rotating, clearly-defined saliencies. But in practice a saliency is influenced by different factors like saturation, load or mechanical angle. For an exact physical understanding a detailed study of the high-frequency behaviour of the machine is necessary, which can be found in [15]–[22].

In the present paper, a new method of measuring the mechanical speed of the rotor with the help of a test signal is presented; this can handle the presence of multiple saliencies not by separating them, but by exploring them simultaneously. The method as published in [15] is extended to rotating test signals in order to reduce the precomissioning effort. The capacity of the proposed estimation technique is tested on two different standard induction machines, which are not especially designed for sensorless control.

II. TEST SIGNAL

To enable examination of every single spatial angle, the machine is excited by an alternating test signal. This excitation signal can be described in the stator fixed coordinate system by the equation

$$\vec{U}_{hf} = \hat{U}_{hf} \cdot \cos(\omega_{hf} t) \cdot e^{j\alpha} \qquad (1)$$

wherein $\hat{U}_{hf}$ is the amplitude, ω_{hf} is the frequency and α is the spatial orientation of the test signal. The equation can also be expressed in real (α-part) and imaginary (β-part) components:

$$\vec{U}_{hf} = (\hat{u}_{hf\alpha} + j \cdot \hat{u}_{hf\beta}) \cdot \cos(\omega_{hf} t) \qquad (2)$$

The response to such a high-frequency voltage space vector will be a high-frequency current space vector $\vec{I}_{hf}$, which describes an ellipse, owing to the machine's imbalances. The angle of its principal axis can be different from the angle α of the stimulating voltage. A complex impedance can be defined as follows:

$$\underline{Z} = \frac{\hat{u}_{hf\alpha} + j \cdot \hat{u}_{hf\beta}}{\hat{i}_{hf\alpha} + j \cdot \hat{i}_{hf\beta}} \qquad (3)$$

wherein $\hat{u}_{hf\alpha}$ and $\hat{u}_{hf\beta}$ describe the amplitudes of real and imaginary parts of the test signal, and $\hat{i}_{hf\alpha}$ and $\hat{i}_{hf\beta}$ the amplitudes of the associated complex test signal current.

III. CURRENT SEPARATION

In practice, the test signal (1) is added to the reference voltage space vector demanded for normal operation, e.g. from a field-oriented control. The resulting stator current

978-1-4244-1741-4/08/$25.00 ©2008 IEEE

space vector $\vec{I}_{res}$ consists of the fundamental wave current $\vec{I}_1$ and the high-frequency current $\vec{I}_{hf}$ as the consequence of the test signal.

$$\vec{I}_{res} = \vec{I}_1 + \vec{I}_{hf} \tag{4}$$
$$= (I_{1\alpha} + j \cdot I_{1\beta}) + (I_{hf\alpha} + j \cdot I_{hf\beta}) \tag{5}$$

For calculation of the complex impedance and decoupling of the test signal from the field-oriented control, these two currents have to be separated.

The extraction of the high-frequency current can be made by a Fast Fourier Transformation (FFT) of $\vec{I}_{res}$. The FFT demands a high computational effort and delivers the amplitude and the phase angle of every single frequency in the current spectrum, although only the values for the test signal frequency are needed. The Goertzel algorithm is therfore implemented, which only computes phase and amplitude of a requested harmonic. If the number of computed harmonics M is less than $log_2 N$, (N is the number of samples), the Goertzel algorithm has less computational effort than a Fast Fourier Transformation [23], but delivers the same results for the requested harmonic. If the amplitude and phase angle of more than one signal are needed, as in the present case, this advantage of the Goertzel algorithm increases.

The Goertzel algorithm is applied to the real and imaginary part of $\vec{I}_{res}$ (Fig. 1). It delivers the amplitudes $\hat{i}_{hf\alpha}$ and $\hat{i}_{hf\beta}$ of the corresponding signals, as well as the associated phase angles (not depicted). In order to calculate the fundamental waveform current $\vec{I}_1$, the high-frequency current waveforms $I_{hf\alpha}$ and $I_{hf\beta}$ are reconstructed. According to (5) the subtraction of the measured current space vector $\vec{I}_{res}$ and the test signal current $\vec{I}_{hf}$ provides the desired current $\vec{I}_1$.

$$I_{1\alpha} = I_{res\alpha} - I_{hf\alpha} \tag{6}$$
$$I_{1\beta} = I_{res\beta} - I_{hf\beta} \tag{7}$$

Consequently, the test signal injection and analysis are decoupled from the heterodyned field-oriented control.

IV. DEPENDENCIES OF COMPLEX HF-IMPEDANCE

Detailed investigations have shown that the measured complex impedance depends on numerous impacts [15]. The research was made on a $15\,\mathrm{kW}$ standard induction motor with two pole pairs, forty-eight slots in the stator and thirty-six rotor slots, whereof twelve are opened and

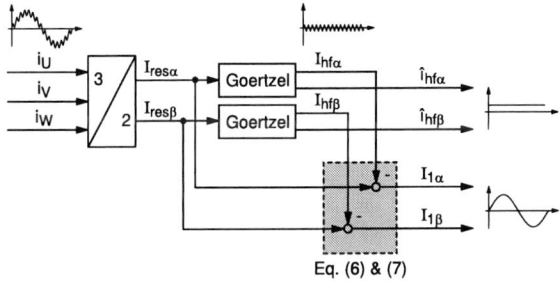

Fig. 1: Topology of the current separation scheme

twenty-four are closed. The machine design was changed in neither the stator nor in the rotor. The investigations concluded FE calculations which were made with the FE program FLUX2D [24]. A magnetic equivalent circuit for the high-frequency flux, including the skewing of the rotor, was also derived in order to investigate the machine's high-frequency behaviour in detail.

The investigations have shown that the measured complex high-frequency impedance, as defined in (3), depends on the mechanical angle of the rotor φ_{mech}, the slip frequency f_2, the angle of the stator current space vector in the stator fixed coordinate system φ_1 and the test signal angle α. In the case of nominal flux every impedance curve is unique, and can therefore be measured and described as

$$\underline{Z} = f\left(\varphi_1, \alpha, f_2, \varphi_{mech}\right) \tag{8}$$

These reference impedance curves can be measured while the induction motor is magnetised by a DC stator current and the mechanical velocity is determined by a speed-controlled machine, which is coupled on the motor shaft. A field-oriented control provides through the DC current nominal flux in the induction motor. The slip frequency f_2 can be calculated via:

$$f_2 = f_1 - p \cdot n \tag{9}$$

wherein f_1 is the stator frequency, p is the number of pole pairs and n is the mechanical velocity. Thus, in the presence of DC stator current ($f_1 = 0\,\mathrm{Hz}$), the slip frequency f_2 and the mechanical angle φ_{mech} are determined by the load machine and can be measured directly, whereas the parameters φ_1 and α can be chosen.

Fig. 2 shows a measured impedance curve of the investigated machine at standstill and nominal flux ($\varphi_1 = 0°$). The dependency on the mechanical angle and the test signal angle, as well as the $30°$-periodicity with reference to φ_{mech} owing to the opened and closed rotor slots, becomes visible. Fig. 3 shows clearly, that even a slight change in the mechanical angle of the rotor φ_{mech} leads to a different impedance curve. The magnitude value as well as the angle at which this maximum appears changes.

Because of the large number of impacts on the complex impedance, a new estimation method for the mechanical speed of the induction motor was developed.

V. ESTIMATION METHOD

As shown above, the complex impedance is also addicted to the slip frequency f_2. Consequently, the slip frequency of the induction machine can be calculated via a comparison between the measured reference curves in the presence of DC magnetisation as described in section IV and the actual measured impedance values during the machine's operation. By the way of explanation the induction motor should first be magnetised by a DC current to its nominal rotor flux level and operate at a constant mechanical speed. Later, it will be shown that the consideration of a rotating field is only a small step.

During the machine's operation the test signal as described by (1) is added to the demanded voltages of the

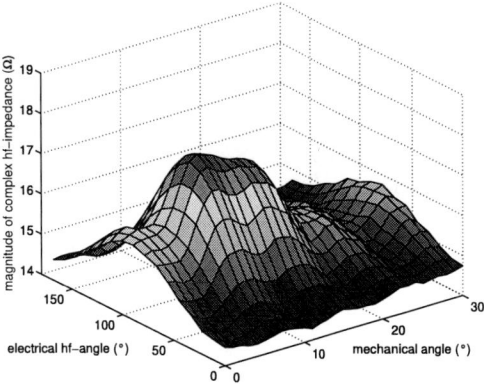

Fig. 2: Measured hf-impedance at standstill and magnetisation to nominal rotor flux in α-direction

Fig. 3: Curves of measured hf-impedance of Fig. 2 for different mechanical angles

field-oriented control as illustrated in Fig. 4. The resulting test signal current is extracted for the calculation of the complex impedance $\underline{Z}$ as explained in section III. The measured value of the complex impedance is stored in the array $\underline{Z}_{meas}$ with k elements, wherein the currently measured value is on the first position, and the latest on the last. Although the test signal has an alternating nature, it rotates in the stator fixed coordinate system with a fixed frequency, in order to measure the complex impedance for different test signal angles.

A measured impedance curve while the machine is in operation is shown in Fig. 5. The machine was, as described above, magnetised by a DC current. While the impedance was measured, the machine rotated with $n = 6\,\text{min}^{-1}$ and the angle of the stator current φ_1 was zero (α-direction). Between the measurement of a single impedance value a small time delay Δt was inserted.

The angle of the magnetising DC current space vector in the stator can be measured easily. The angle of the test signal of every measured value and the time space between the single measurements Δt are also known. For

constant speed, only the mechanical angle φ_{mech} of the last measured impedance value and the slip frequency f_2 are therefore the unknown parameters.

$$\underline{Z} = f\Big(\underbrace{\varphi_1}_{\text{measured}} , \overbrace{\alpha}^{\text{given}} , \underbrace{f_2, \varphi_{mech}}_{\text{unknown}} \Big) \tag{10}$$

For a fixed pair of a proposed mechanical angle φ'_{mech} and a proposed slip frequency f'_2, the mechanical angles of all measured impedance values can be computed. From (9) follows:

$$\varphi_{mech} = 2\pi \int \left(\frac{f_1 - f_2}{p} \right) dt \tag{11}$$

The changing of the mechanical angle as a consequence of the slip frequency can be described by

$$\Delta\varphi'_2 (\nu) = \nu \cdot \frac{2\pi \cdot f'_2}{p} \cdot \Delta t \tag{12}$$

with ν as the counting value starting with 0 (current measured value). According to (11), the mechanical angles of the measured values in the absence of a rotating field ($f_1 = 0\,\text{Hz}$) can be calculated via

$$\varphi'_{mech} (\nu) = \varphi'_{mech} + \Delta\varphi'_2 (\nu) \tag{13}$$

Afterwards, the differences between the magnitude of the measured values and their opposites in the belonging reference curve are calculated. An error value is defined as follows:

$$\epsilon \left(\varphi'_{mech}, f'_2 \right) = \sum_{\nu} \big| |\underline{Z}_{meas} (\nu)| - |\underline{Z}_{ref}| \big| \tag{14}$$

with

$$\underline{Z}_{ref} = \underline{Z}_{ref} \left(\varphi_1 (\nu), \alpha (\nu), f'_2, \varphi'_{mech} (\nu) \right) \tag{15}$$

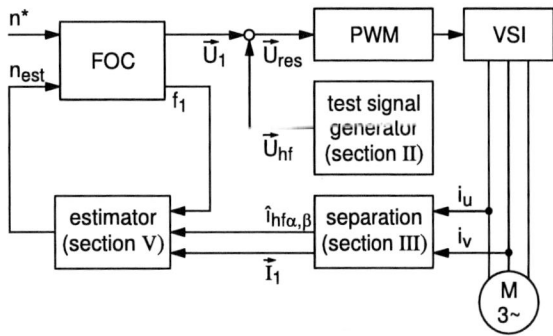

Fig. 4: Topology of the sensorless field-oriented control (FOC) with alternating test signal

This error value is calculated for several combinations of φ'_{mech} and f'_2. The combination of φ'_{mech} and f'_2 for which

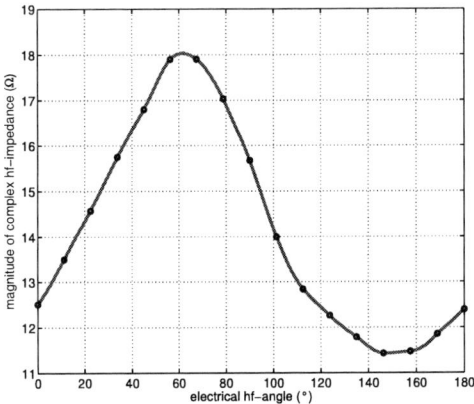

Fig. 5: Measured hf-impedance values (marker) at $n = 6\,\text{min}^{-1}$, DC nominal flux and $\varphi_1 = 0°$ while machine is in operation

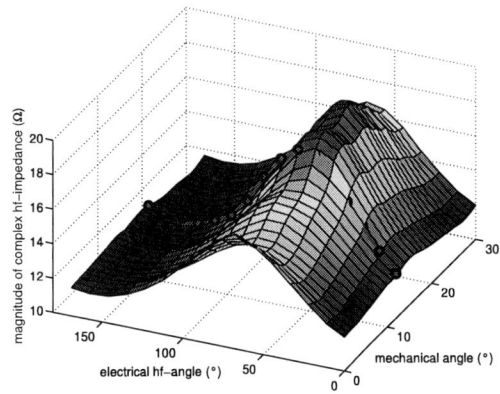

Fig. 6: Reference impedance face for $f_2 = -0,2\,\text{min}^{-1}$, DC nominal flux and $\varphi_1 = 0°$ and measured values of Fig. 5 (red)

the error ϵ has a minimum, corresponds to the current mechanical values of the machine, and represents therefore the estimated values.

$$\min\left(\epsilon\left(\varphi'_{mech}, f'_2\right)\right) \to \varphi_{mech}, f_2 \qquad (16)$$

In the case of the measured values depicted in Fig. 5, the associated combination is $\varphi_{mech} = 17°$ and $f_2 = -0,2\,\text{Hz}$ as shown in Fig. 6.

In the next step the rotating magnetic field has to be taken into account. It is clear that it has to be considered in the selection of $\underline{Z}_{ref}$ (15), but also in (13). Note that the equation only takes the slip frequency f_2 into account, and not the rotating field. The changing of the angle of the rotating field relative to the actual value can be described by:

$$\Delta\varphi_1\left(\nu\right) = \frac{\varphi_1\left(0\right) - \varphi_1\left(\nu\right)}{p} \qquad (17)$$

Thus, the proposed mechanical angle of the measured impedance values can be calculated via

$$\varphi'_{mech}\left(\nu\right) = \varphi'_{mech} + \Delta\varphi'_2\left(\nu\right) - \Delta\varphi_1\left(\nu\right) \qquad (18)$$

The described method delivers the mechanical angle φ_{mech} and the slip frequency f_2, which are two independent values of the algorithm. In order to get the demanded mechanical velocity, (9) can be evaluated, wherein f_1 is the frequency of the rotor flux, which is a known value in a field-oriented control.

$$n = \frac{1}{p} \cdot \left(\underbrace{f_1}_{\text{known}} - \underbrace{f_2}_{\text{estimated}}\right) \qquad (19)$$

Another method is to differentiate the estimated mechanical angle. In practice the differentiation doesn't lead to satisfactory results because of the non-smooth nature of the estimated angle. The first option (evaluation of (19)) is therefore used.

VI. ROTATING TEST SIGNAL

When the machine is excited by a test signal with alternating nature (1), every change in the test signal angle α induces a transient response of the system. In order to get the steady-state value of the complex impedance (3), a delay time Δt must be inserted between the angle's changing and the measurement of $\underline{Z}$. This delay time can be reduced by use of a rotating test signal. The measurement effort for the reference impedance curves can also be reduced.

The alternating test signal (1) can be described by two rotating signals, a positive and a negative phase-sequence voltage, $\vec{U}_{hf}^+$ and $\vec{U}_{hf}^-$.

$$\vec{U}_{hf} = \hat{U}_{hf} \cdot \cos\left(\omega_{hf}t\right) \cdot e^{j\alpha} \qquad (20)$$
$$= \vec{U}_{hf}^+ + \vec{U}_{hf}^- \qquad (21)$$

The voltages follow the equations:

$$\vec{U}_{hf}^+ = \frac{\hat{U}_{hf}}{2} \cdot e^{j(+\omega_{hf}t+\alpha)} \qquad (22)$$
$$\vec{U}_{hf}^- = \frac{\hat{U}_{hf}}{2} \cdot e^{j(-\omega_{hf}t+\alpha)} \qquad (23)$$

All values in the positive phase-sequence system are marked by a '+', those of the negative phase-sequence system with a '−' as exponent.

The positive and negative phase-sequence space vector voltages are generated, according to the space vector theory, by the superposition of the phase-to-neutral voltages u_U, u_V and u_W:

$$u_U^\pm = \frac{\hat{U}_{hf}}{2} \cdot \cos\left(\pm\omega_{hf}t + \alpha\right) \qquad (24)$$
$$u_V^\pm = \frac{\hat{U}_{hf}}{4}\left(\begin{array}{l}+\sqrt{3}\sin\left(\pm\omega_{hf}t+\alpha\right) + \\ -\cos\left(\pm\omega_{hf}t+\alpha\right)\end{array}\right) \qquad (25)$$
$$u_W^\pm = \frac{\hat{U}_{hf}}{4}\left(\begin{array}{l}-\sqrt{3}\sin\left(\pm\omega_{hf}t+\alpha\right) + \\ -\cos\left(\pm\omega_{hf}t+\alpha\right)\end{array}\right) \qquad (26)$$

wherein $u_U^\pm$, $u_V^\pm$ and $u_W^\pm$ describe the positive and negative phase-sequence phase-to-neutral voltages. Note that the test signal angle α acts in all equations like a time-shift.

$$\alpha = \omega_{hf}t_0 \qquad (27)$$

At high-frequency excitation the corresponding hf-flux does not penetrate the rotor cage, but distributes on the rotor's surface. The vector of the phase-to-neutral voltages at the machine terminals $\boldsymbol{u_m}$ can therefore be described as follows:

$$\boldsymbol{u_m} = \boldsymbol{R} \cdot \boldsymbol{i} + \boldsymbol{M} \cdot \boldsymbol{i'} \qquad (28)$$

with $\boldsymbol{i'}$ as the derivation of $\boldsymbol{i}$ and

$$\boldsymbol{u_m} = \begin{bmatrix} u_{Um} \\ u_{Vm} \\ u_{Wm} \end{bmatrix} \qquad (29)$$

$$\boldsymbol{i} = \begin{bmatrix} i_U \\ i_V \\ i_W \end{bmatrix} \qquad (30)$$

$$\boldsymbol{R} = \begin{bmatrix} R_U & 0 & 0 \\ 0 & R_V & 0 \\ 0 & 0 & R_W \end{bmatrix} \qquad (31)$$

$$\boldsymbol{M} = \begin{bmatrix} L_U & M_{UV} & M_{UW} \\ M_{VU} & L_V & M_{VW} \\ M_{WU} & M_{WV} & L_W \end{bmatrix} \qquad (32)$$

wherein R_U, R_V and R_W are the resistances and L_U, L_V and L_W are the self-inductances of the associated phase. The couplings between the phases are considered by M.

Note that the phase-to-neutral voltages of the machine may differ from the inverter voltages as described in (24) - (26), because of the unsymmetrical hf-behaviour of the machine. Thus, in the case of a star-connected machine, a voltage u_0 between the two neutral points can be measured. The machine voltages (marked by the index m) can be expressed by the inverter voltages and the voltage between the neutral points of the machine and the inverter u_0 (Fig. 7).

$$u_{Um}^\pm = u_U^\pm - u_0 \qquad (33)$$
$$u_{Vm}^\pm = u_V^\pm - u_0 \qquad (34)$$
$$u_{Wm}^\pm = u_W^\pm - u_0 \qquad (35)$$

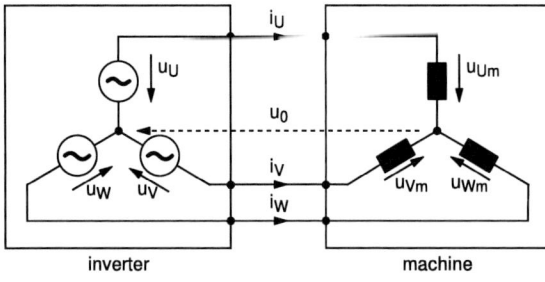

Fig. 7: Voltages and currents of the inverter and the machine

Thus, equation (28) can be expressed by:

$$\begin{bmatrix} u_U^\pm - u_0 \\ u_V^\pm - u_0 \\ u_W^\pm - u_0 \end{bmatrix} = \boldsymbol{R} \cdot \boldsymbol{i^*} + \boldsymbol{M} \cdot (\boldsymbol{i^*})' \qquad (36)$$

with

$$\boldsymbol{i^*} = \begin{bmatrix} i_U^\pm \\ i_V^\pm \\ -i_U^\pm - i_V^\pm \end{bmatrix} \qquad (37)$$

Finally, the voltages of the two meshes in Fig. 7 can be calculated by:

$$u_U^\pm - u_V^\pm = R_U \cdot i_U^\pm - R_V \cdot i_V^\pm + \\ + \frac{di_U^\pm}{dt} \cdot (L_U - M_{VU} - M_{UW} + M_{VW}) \\ + \frac{di_V^\pm}{dt} \cdot (M_{UV} - L_V - M_{UW} + M_{VW}) \quad (38)$$

and

$$u_V^\pm - u_W^\pm = R_W \cdot i_U^\pm + (R_V + R_W) \cdot i_V^\pm + \\ + \frac{di_U^\pm}{dt} \cdot (M_{VU} - M_{WU} - M_{VW} + L_W) \\ + \frac{di_V^\pm}{dt} \cdot (L_V - M_{WV} - M_{VW} + L_W) \quad (39)$$

wherein the voltages are described by (24) - (26) and the resistances and reactances are determined by the machine.

The phase currents $i_U^\pm$, $i_V^\pm$ and $i_W^\pm$ can be calculated by the application of the Laplace transformation on (38) and (39). In the following, only the Laplace transformed $I_U^+(s)$ of the positive phase-sequence line current i_U^+ is examined, but the results can also be assigned to the other currents. The transient part of the current i_U^+ is also neglected, since the test signal is applied during the whole machine's operation and thus the general solution is zero.

In the case of steady-state behaviour, the test signal angle α can be neglected for the transformation of the phase voltages $u_U^\pm$, $u_V^\pm$ and $u_W^\pm$, since it acts like a time-shift (27). This time-shift, and thus the test signal angle, can be considered in the current's solution. The steady-state behaviour of i_U^+ in the Laplace transformation follows the equation:

$$I_U^+(s) = c_1 \cdot \frac{s + c_2}{s^2 + \omega_{hf}^2} \qquad (40)$$

wherein c_1 and c_2 are constants, which depend on the test signal parameters as well as on the machine's resistances and reactances. The transformation of (40) to the time domain and the consideration of the test signal angle (27) leads to:

$$i_U^+(t) = c_0 \cdot \sin(\omega_{hf}t + \phi + \alpha) \qquad (41)$$

wherein

$$c_0 = c_1 \cdot \sqrt{1 + \frac{c_2^2}{\omega_{hf}^2}} \qquad (42)$$

$$\tan\phi = \frac{\omega_{hf}}{c_2} \qquad (43)$$

$$c_1 = f(\omega_{hf}, R_U, ...) \qquad (44)$$

$$c_2 = f(\omega_{hf}, R_U, ...) \qquad (45)$$

1028

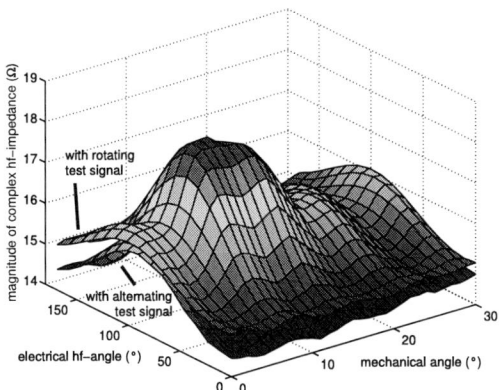

Fig. 8: Measured impedance curves for standstill, nominal DC flux with alternating (Fig. 2) and rotating test signal

Note that c_0 and ϕ are measured values (e.g. FFT or Goertzel of i_U^+), and that the test signal angle α is a choosable input value. Thus, with a single measurement of c_0 and ϕ, i_U^+ can be calculated for every test signal angle.

From (41) - (45), it becomes obvious that the amplitude c_0 as well as the phase angle ϕ of the phase current also depends on the test signal frequency and amplitude. Because of the frequency dependency of c_1 and c_2 the negative phase-sequence current cannot be exactly calculated if only the positive phase-sequence current is applied to the machine. But in practice these differences are small and thus the amplitude c_0 and the phase angle ϕ are similar for i_U^+ and i_U^-.

Fig. 8 depicts the complex impedance curve while the measurement was made with a rotating test signal. Only the positive phase-sequence test signal voltage was applied to the machine. Only the corresponding positive phase-sequence current was therefore measured. The calculation of the negative phase-sequence current was made on the assumption that c_1 and c_2, and consequently c_0 and ϕ, are equal for both currents.

$$i_U^-(t) = c_0 \cdot \sin\left(-\omega_{hf}t + \phi + \alpha\right) \quad (46)$$

The addition of the measured positive phase-sequence current space vector and the calculated negative phase-sequence current leads to a fictitious current which corresponds to a test signal as described by (21).

$$\vec{I}_{hf}^{\pm} = \frac{2}{3} \cdot \left(i_U^{\pm} + \underline{a} \cdot i_V^{\pm} + \underline{a}^2 \cdot i_W^{\pm}\right) \quad (47)$$

$$\vec{I}_{hf} = \vec{I}_{hf}^+ + \vec{I}_{hf}^- \quad (48)$$

As already indicated, the calculated negative phase-sequence current is different from that actually measured with an alternating test signal, because of the different values of c_1 and c_2 for i_U^+ and i_U^-. The resulting high-frequency current of (48) is also therefore not equal to that actually measured.

The comparison of the shape of the complex impedance, measured with an alternating test signal, with

the shape for the measurement with the rotating test signal shows that the error is very small (Fig. 8). Indeed, the absolute values are slightly smaller in the case of the alternating test signal but, and this is remarkable, the shape of the impedance curve is preserved. Thus, if the reference impedance curves are measured in the described manner, the proposed method (section V) also works with the rotating test signal.

The advantage of the rotating test signal is the strong reduced measurement effort for the reference curves. With the alternating test signal, the complex impedance for every combination of test signal angle α and mechanical angle φ_{mech} for a fixed combination of φ_1 and f_2 have to be measured. With the rotating test signal, only c_0 and ϕ has to be measured for every mechanical angle. The curve over the test signal angle α can be calculated as described above. When the complex impedance is measured for every $10°$ of α, the measurement amount is reduced to the seventeenth part when the rotating test signal is used.

VII. EXPERIMENTAL RESULTS

The proposed method using the rotating test signal was implemented on a dSPACE-System and tested on two different induction machines (technical data in Tab. I). The detailed investigations were made on machine 1. Machine 2 is an induction machine, of which no further technical data (shape of lamination, number of slots, etc.) are known. Both machines are standard induction motors, which are not specially designed for sensorless control, and which are not manipulated in order to show special behaviour.

The topology of the sensorless control is depicted in Fig. 9. Instead of the alternating test signal, only the positive phase-sequence voltage

$$\vec{U}_{hf}^+ = \frac{\hat{U}_{hf}}{2} \cdot e^{j(\omega_{hf}t)} \quad (49)$$

is added to the voltage $\vec{U}_1$, which is demanded by the field-oriented control. Note that α does not appear in (49), because it only acts as a time-shift. The test signal angle is important for the calculation of the alternating high-frequency current $\vec{I}_{hf}$ as described in section VI. The

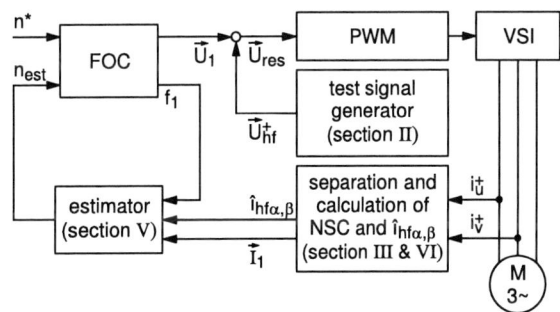

Fig. 9: Topology of the sensorless field-oriented control (FOC) with rotating test signal and calculation of negative phase-sequence current (NSC)

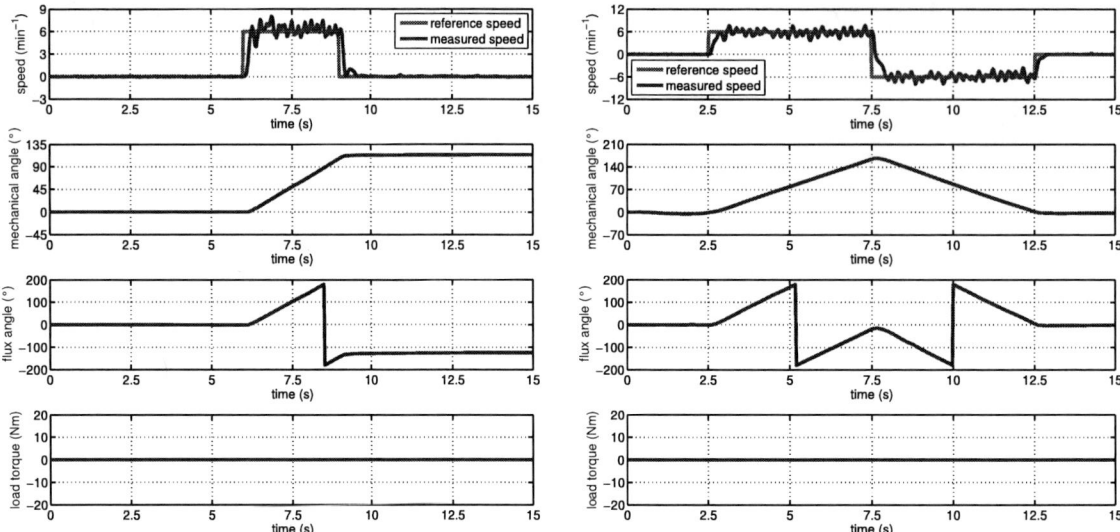

Fig. 10: **Machine 1**, sensorless control: response to steps of the reference speed at no-load operation

Fig. 11: **Machine 1**, sensorless control: response to steps of the reference speed while operating at no-load

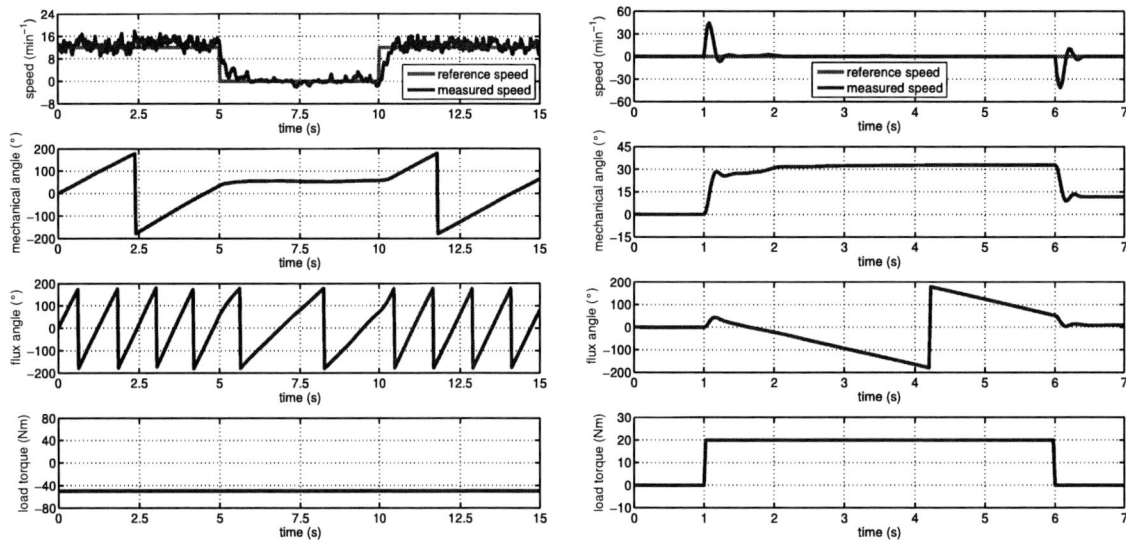

Fig. 12: **Machine 1**, sensorless control: response to steps of the reference speed while operating at 50 Nm (50% of nominal torque)

Fig. 13: **Machine 1**, sensorless control: response to steps of the load torque (20% of nominal torque) while operating at zero speed

results using the alternating test signal were published in [15].

A. Machine 1

Figs. 10 - 13 show the behaviour of the sensorless control of machine 1. The responses to steps in the reference speed at no-load operation are depicted in Figs. 10 and 11. It becomes obvious that the proposed method allows save operation even at standstill and zero frequency.

In addition, the sensorless control can handle loads up to 50% of the nominal torque (Fig. 12), and also load steps at standstill can be controlled precisely (Fig. 13).

B. Machine 2

The step responses of the sensorless control of machine 2 are shown in Figs. 14 - 17. In Figs. 14 and 15 the machine operated at no-load. The behaviour under constant loaded conditions and under load steps is shown in

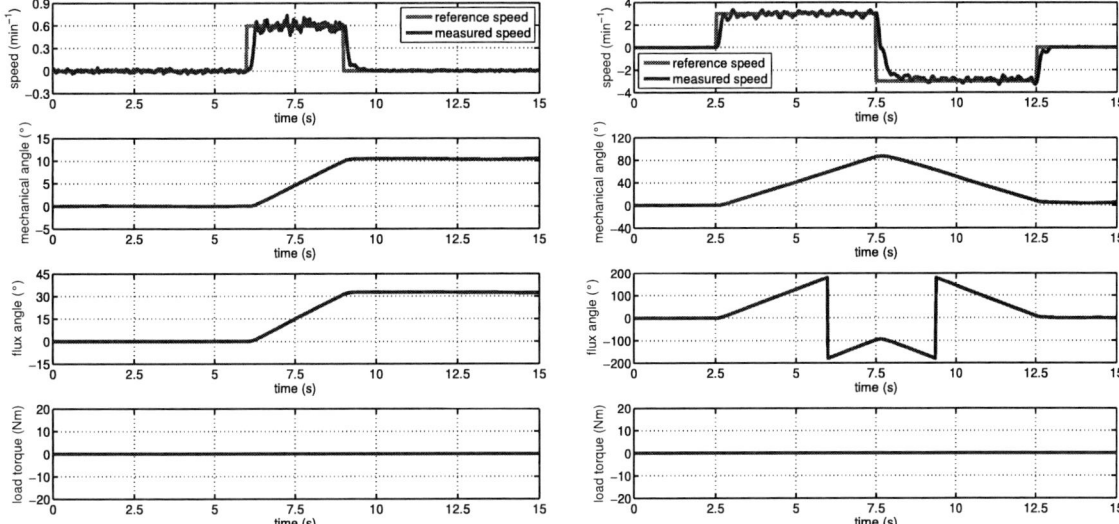

Fig. 14: **Machine 2**, sensorless control: response to steps of the reference speed at no-load operation

Fig. 15: **Machine 2**, sensorless control: response to steps of the reference speed while operating at no-load

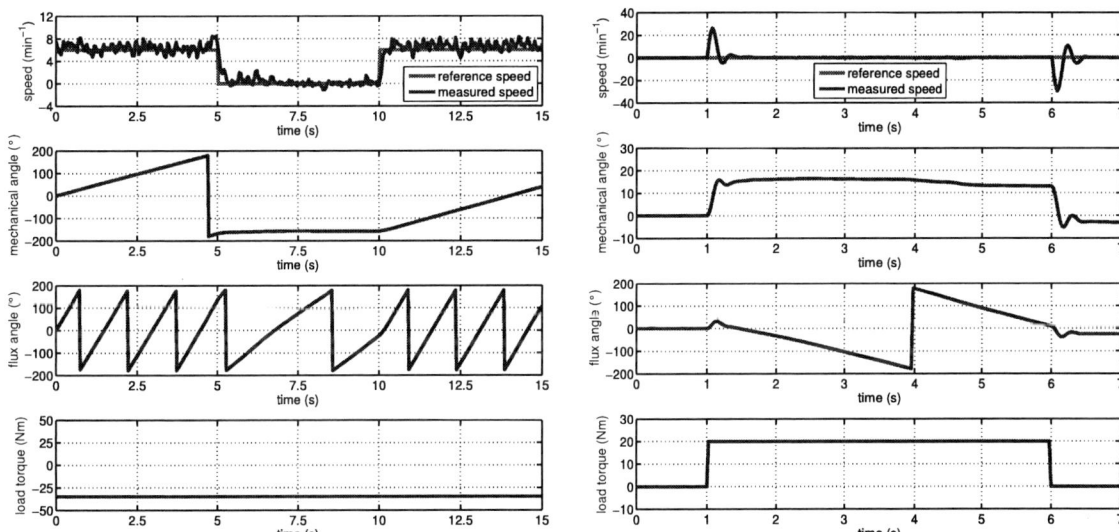

Fig. 16: **Machine 2**, sensorless control: response to steps of the reference speed while operating at 35 Nm (40% of nominal torque)

Fig. 17: **Machine 2**, sensorless control: response to steps of the load torque (20% of nominal torque) while operating at zero speed

Figs 16 and 17. Thus, in the case of the second machine also, the proposed method allows save and precise machine operation even at standstill and zero frequency.

VIII. CONCLUSION

With the help of a test signal a complex impedance can be measured which depends on numerous impacts. Hitherto, the separation of these phenomena was fundamental in order to extract e.g. the mechanical speed or the flux angle. Since separation is not always possible, a new method was developed that allows sensorless field-oriented control of machines with multiple non-separable or single saliencies without the introduction of an additional sensor.

Since the measured impedance curves are unique, the method is based on the comparison of the impedance curve measured during the machine's operation with precommissioned reference impedance curves. The combination of the slip frequency f_2 and the mechanical angle φ_{mech} for which the error between these two curves is the lowest represents the actual values. The proposed method allows save machine operation at standstill (loaded and unloaded) and even at zero frequency. It was tested on two different induction machines.

In order to reduce the precommissioning effort, the alternating test signal was exchanged for a rotating test signal. Only the positive phase-sequence voltage of the test signal with alternating nature is applied to the machine, and the negative phase-sequence current is calculated. The calculated negative phase-sequence current differs from that actually measured with the alternating test signal, but only slightly in the absolute value of the complex impedance. The shape of the complex impedance curve therefore does not change. It was shown that the method works also with the rotating test signal.

The author is now testing the performance of the presented sensorless control on different induction machines from a few hundred W to a few kW. The proposed method will also be tested on synchronous motors with permanent magnets (burried and surface mounted), as well as on synchronous motors with tooth coils, because the dependence of the complex impedance on the load and the mechanical angle also exists in these machines. The first measurements are very promising.

APPENDIX

TABLE I: Technical data of the used induction motors

	machine 1	machine 2
nominal power, P_N	15 kW	11 kW
nominal voltage, U_N	400 V, Y	400 V, Δ
nominal current, I_N	28, 6 A	22 A
nominal frequency, f_N	50 Hz	60 Hz
number of pole pairs, p	2	3

REFERENCES

[1] C. Schauder, "Adaptive speed identification for vector control of induction motors without rational transducers," *IEEE Transactions on Industry Applications*, vol. 28, no. 5, pp. 1054–1061, 1992.

[2] K. Ohyama, G. M. Asher, and M. Sumner, "Comparative analysis of experimental perfomance and stability of sensorless induction motor drives," *IEEE Transactions on Industrial Electronics*, vol. 53, no. 1, pp. 178–186, 2006.

[3] C. Lascu, I. Boldea, and F. Blaabjerg, "Comparative study of adaptive and inherently sensorless observers for variable-speed induction-motor drives," *IEEE Transactions on Industrial Electronics*, vol. 53, no. 1, pp. 57–65, 2006.

[4] T. Orlowska-Kowalska, "Rotor flux observer and speed estimators for sensorless induction motor drives - comparative study," *Proceedings of International Power Electronics and Motion Control Conference (EPE-PEMC)*, Dubrovnik and Cavtat, Croatia, 2002.

[5] P. Vas, *Sensorless vector and direct torque control*. Oxford: Oxford University Press, 1998.

[6] T. M. Wolbank, J. L. Machl, H. Hauser, and P. Macheiner, "Anisotropy in induction machine lamination and its influence on mechanical sensorless control and condition monitoring," *Proceedings of European Conference on Power Electronics and Applications (EPE)*, Toulouse, France, 2003.

[7] J. Holtz, "Sensorless control of induction machines - with or without signal injection?," *IEEE Transactions on Industrial Electronics*, vol. 53, no. 1, pp. 7–30, 2006.

[8] M. Schrödl, "Sensorless control of ac machines at low speed and standstill based on the "inform" method," *Conference Record of the 1996 IEEE Industry Applications Conference*, pp. 270–277, 1996.

[9] C. S. Staines, G. M. Asher, and M. Sumner, "Rotor-position estimation for induction machines at zero and low frequency utilizing zero-sequence currents," *IEEE Transactions on Industry Applications*, vol. 42, no. 1, pp. 105–112, 2006.

[10] C. Caruana, G. M. Asher, and M. Sumner, "Performance of hf signal injection techniques for zero-low-frequency vector control of induction machines under sensorless condition," *IEEE Transactions on Industrial Electronics*, vol. 53, no. 1, pp. 225–238, 2006.

[11] A. Consoli, G. Scarcella, G. Bottiglieri, G. Scelba, A. Testa, and D. A. Triolo, "Low-frequency signal-demodulation-based sensorless technique for induction motor drives at low speed," *IEEE Transactions on Energy Conversion*, vol. 53, no. 1, pp. 20–215, 2006.

[12] P. L. Jansen and R. D. Lorenz, "Transducerless field orientation concepts employing saturation-induced saliencies in induction machines," *IEEE Transactions on Industry Applications*, vol. 32, no. 6, pp. 1380–1393, 1996.

[13] M. L. Aime, M. W. Degner, and R. D. Lorenz, "Saturation measurements in ac machines using carrier signal injection," *IEEE Industry Application Society Annual Meeting*, 1998.

[14] R. D. Lorenz, "Practical issues and research opportunities when implementing zero speed sensorless control," *Proceedings of International Conference on Electric Machines and Systems (ICEMS)*, Shenyang, China, 2001.

[15] H. Zatocil, "Physical understanding of multiple saliencies in induction motors and their impact on sensorless control," *SPEEDAM*, Ischia, Italy, 2008.

[16] A. Consoli, G. Scarcella, and A. Testa, "An alternative to high frequency current detection techniques for zero speed sensorless control of ac motor drives," *Proceedings of International Power Electronics and Motion Control Conference (EPE-PEMC)*, Dubrovnik and Cavtat, Croatia, 2002.

[17] A. Consoli, G. Scarcella, G. Bottiglieri, and A. Testa, "Harmonic analysis of voltage zero-sequence-based encoderless techniques," *IEEE Transactions on Industry Applications*, vol. 42, no. 6, pp. 1548–1557, 2006.

[18] J.-I. Ha, S.-K. Sul, K. Ide, I. Murokita, and K. Sawamura, "Physical understanding of high frequency injection method to sensorless drives of an induction machine," *IEEE Industry Application Society Annual Meeting*, 2000.

[19] A. Consoli, G. Scarcella, G. Tutino, and A. Testa, "Finite element analysis of flux angle estimation techniques based on high frequency signal injection," *Proceedings of International Symposium on Power Electronics, Electrical Drives, Automation & Motion (SPEEDAM)*, pp. 1–6, Ischia, Italy, 2002.

[20] D. Drevensek, D. Zarko, and T. A. Lipo, "A study of sensorless control of induction motor at zero speed utilizing high frequency voltage injection," *Proceedings of International Power Electronics and Motion Control Conference (EPE-PEMC)*, Dubrovnik and Cavtat, Croatia, 2002.

[21] T. M. Wolbank, R. Wöhrnschimmel, and J. L. Machl, "Slot geometry - an important design parameter for zero speed sensorless control of standard induction machines," *Proceedings of International Power Electronics and Motion Control Conference (EPE-PEMC)*, Dubrovnik and Cavtat, Croatia, 2002.

[22] T. M. Wolbank and R. Wöhrnschimmel, "Influence of rotor design on sensorless control for induction motors," *EPE*, Graz, Austria, 2001.

[23] A. V. Oppenheim and R. W. Schafer, *Discrete-time signal processing*. New Jersey: Prentice Hall, 2 ed., 1999.

[24] FLUX2D, *User's guide*, vol. 1-5. Grenoble: CEDRAT, 2006.

Author Index

A

Abbatelli, L. 61
Abbey, Chad 2178
Abdelhamid, Tamer H. 606
Abdellatif, Meriem 938
Abe, Seiya..393
Abourida, Simon 1077
Abroshan, Mohammad 1117
Abroushan, Mohammad................................365
Abuishmais, Ibrahim....................................867
Abu-Rub, Haithem 1084, 1382
Adabi, Jafar..718, 903
Adamidis, Georgios 1840
Adamowicz, Marek......................................1729
Adzic, Evgenije...1957
Ahmadi, Muhammad 1847
Ahmed, M.M.R. 1866, 2472
Ahn, Jonng-Bo..2524
Ahn, ong-Bo...2492
Ait-Ahmed, Mourad 1740
Akhondi, Hamidreza.....................................2071
Alarcón, E. 2108
Albert, Laurent..2037
Al-Diab, Ahmad 1710
Alexandrov, Alexandar..................................787
Al-Khayat, Nazar 2150
Allard, Bruno 2457
Al-Othman, A. K. 606
Amelon, Nicolas 1740
Anaya-Lara, O. 1784, 1941
Andersen, Michael A.E..................................127
Ando, Kenji 614
Andrzejewski, Andrzej 1090
Areerak, K-N..2049
Arellano-Padilla, J. 1173
Arellano-Padilla, Jesus..................................769
Armstrong, S. 1688
Aroudi, A. El 2108
Aroudi, Abdelali El...........................2115, 2120
Arshad, Waqas M. 867
Asher, G. M. 2261
Asher, G.M. 2049
Asher, Greg..2300
Asiltürk, Ilhan...967
Aurel, Campeanu 893
Averberg, Andreas.......................................213

B

Baalbergen, Freek J.F. 2170
Baghaee, H.R.......................................313, 629, 750
Bahri, I. 1365
Bailey, Chris 76
Bakas, Panagiotis.......................................1840
Balazovic, Peter 1402

Balouktsis, Anastasios 1840
Baluta, Gh. 2043
Ban, Drago...818
Barai, Mukti 674
Barakat, Georges 1834, 810
Baranowski, Jerzy.............................1432, 1446
Barbosa, Fabián H. 637
Barlik, Roman..84
Barrero, R. 1512
Bartelt, R..521
Baskys, Algirdas 1140
Bastiani, A. 1293
Baszynski, Marcin 1779
Bauer, Pavel...422
Bauer, Pavol 2170, 2354, 2368, 2371
Beck, Hans-Peter 1243
Bekbudov, Radiy 337
Bekishev, Anatoly 663
Bélanger, Jean.................................1077, 1475
Belfkira, Rachid..1834
Belkhodja, I. Slama 1149
Bellini, Armando 490
Bellmunt, Oriol 731
Belter, D. 1044
Benadero, Luis 2115
Bendkowski, Lukas 250
Benecke, Marcel 1280
Benkhoris, Mohamed-Fouad 1740
Bennani, A.Ben Abdelghani 1149
Beran, Leos 782
Bergas-Jane, Joan 731
Bergogne, Dominique...................................2457
Berthon, A. 1542
Bertoluzzo, Manuele....................................1491
Bertram, Torsten 1215
Betz, R.E...1293
Bevilacqua, Pascal 2457
Bifaretti, Stefano 1771, 490, 561
Binder, Andreas.................................1625, 2385
Binkowski, T. 714
Birolleau, Damien 2037
Biswas, Jayanta...674
Bizon, Nicu 621
Blahník, Vojtech 1535
Blanco, M. 2481
Blazic, B. 2510
Böcker, J. 1598
Böcker, Joachim 159
Bodora, A. 326
Bogalecka, Elzbieta 1975, 804
Bojda, Petr 422
Bolgov, Viktor 154
Bolognani, Silverio 1097
Boora, Arash A 468, 723
Bossche, A. Van den....................................1326
Botan, Corneliu...1111

Author Index

Botsali, FatihM. 949
Bouafia, Abdelouahab 703
Boucherit, M.S. 1987
Bouhalli, Nadia 281
Bozhko, S.V. 2049
Brand1tetter, Pavel 1375
Braslavsky, I.Ya. 1050
Breban, Stefan 1896
Brown, Neil L. 2150
Bruno, Francois 2205
Bucher, Alexander 244, 250
Buja, Giuseppe 1491
Bukatov, Alexander 1872
Bulic, Neven 556
Buonomo, S. 61
Buss, Martin 2312

C

C., Ilioudis Vasilios 1105
Caballero, M. 1555
Cabrita, Carlos M. P. 1646
Calado, Maria R. A. 1646
Camara, M.B. 1542
Cambál, Marek 982
Candusso, Denis 734
Cartes, D.A. 793
Case, Michael James 1798
Castaing, Ambroise 2464
Catalão, J. P. S. 1682
Cédl, Marek 1593, 372
Ceglia, Gerardo 268
Cepisca, Costin 1963, 908
Cernat, Mihai 1748
Cernohorský, Josef 1009
Cerovský, Zdenk 982
Cha, Gil-Ro 383
Champenois, Gérard 2015
Chan, Paul K.W. 1688
Chang, Hao-Chi 1652
Chang, Lon-Kou 320
Chang, Yuan-Chih 456
Chante, Jean-Pierre 2457
Charaabi, L. 1365
Chekhet, Eduard 307
Chen, Anyuan 799
Chen, Junling 2000
Chen, Yonggang 1981, 2000, 405, 515
Chen, Zhe 2325
Chen, Zong-Jie 1704
Cheng, K.W.E. 576
Cherif, M. Ghodbane 1149
Cheung, N. C. 1221
Chien, Sywe-Bin 1652
Chillet, Christian 2037
Chimento, F. 61

Chlodnicki, Zdzislaw 2150
Choi, Heung-Kwan 2524
Choi, Jaeho 2498
Choi, Uk-Don 1421
Chou, Ming-Chang 1652
Chrenko, Daniela 2156
Chrzan, Piotr J. 144
Chudzik, Piotr 1568
Chun, Tae-Won 1421
Clare, J. 1326
Clare, Jon C 207
Clare, Jon C. 229, 561
Clare, Jon 1771, 307
Comnac, Vasile 1748
Cook, B.J. 1293
Cook, D. 1326
Coquery, Gérard 2192
Correa, Pablo 451, 699
Courtecuisse, Vincent 1896, 2184
Cousineau, Marc 281
Cuk, Vladimir 1426
Cychowski, Marcin 2241
Czapp, Stanislaw 2059

D

Dabroom, A.M. 1337
Dakyo, B. 1911
Dannehl, J. 444
Darie, Eleonora 1963, 908
Darie, Emanuel 1963, 908
Davey, J. 1918
De Bernardinis, Alexandre 2192
De Castro, M.R. 2126
De Gersem, Herbert 2385
de Kock, H.W. 859
De Souza, Kleber C.A. 1951
Deaconu, Sorin 1409
Debowski, Andrzej 1568, 2289
Degeratu, Sonia 893
Delaney, Kieran 2241
Demenko, Andrzej 2412
Denny, Ernest Edward 1798
Depernet, Daniel 734
Derbel, Nabil 2120
Deskur, Jan 1204, 2227
Deuse, Jacques 2184
Dheilly, Nicolas 2457
Di, Lu 2205
Dianov, Anton 1002
Díaz, Nelson L. 637
Diblík, Martin 1676
Diguet, Marc 1382
Dilevs, Guntis 1811
Dimitrakakis, Georgios S. 1301
Dinkhauser, Vincenz 1819

Author Index

Dobrucky, Branislav 1402
Dockhorn, Matthias 1734
Dodds, Stephen J. 2551, 2559
Dodds, Stephen James 2543
Doebbelin, Reinhard 1280
Doi, Nobuaki 744
Dong, P. 576
Dong, Wei.. 1716
Dontchev, Dimitar 787
Drábek, Pavel..................................... 1593, 372
Draganov, Denis 1610
Dubowski, Marian Roch 1090
Dudak, Juraj.. 2368
Dudrik, Jaroslav 295
Duerbaum, Thomas.............................. 244, 250
Dufour, Christian 1077, 1475
Duke, Richard 528
Dumur, Guillaume 1475
Durovsky, Frantisek....................................... 961
Dybkowski, Mateusz 2211, 2306
Dzieniakowski, Maciej A. 2082

E

Eberhard, Andreas 1371
Eckel, Hans-Guenter 48
Edrington, C.S. 793
Egan, Michael G. 1249
Egorov, Mikhail 1257
Ehsan, Mehdi 1847
Eilenberger, Andreas 945
Elmoctar, Mohamed Y. Ould......................... 810
Empringham, Lee 207, 229, 388
Endo, Tsunehiro... 924
Eno, Otu A. 114
Erceg, Gorislav 556
Etxeberria-Otadui, I. 1555

F

Fabianowski, Jan... 2082
Fabijanski, Pawel...................... 1040, 2055, 2087
Fahrni, C. 256
Fakham, Hicham 2142
Fan, Yue.. 1771
Farhangi, Sh. 173
Farshad, Siamak.. 1575
Fedák, Viliam 2354
Fedak, Viliam 961
Fedyczak, Zbigniew............................. 165, 236
Feki, Moez 2120
Fernández, Herman 1947
Fernandez-Mola, Josep-Maria....................... 731
Ferreira, Jan Abraham 187
Ferreira, Luís António Fialho Marcelino 2076
Fetyko, Jan... 961
Filchev, T.. 1326

Filho, Braz Jesus Cardoso........................... 1345
Filka, Roman 1402
Fisher, R. 1293
Fleisch, Karl 48
Fodor, D.. 2096
Foft, Jiří 1593
Foo, Gilbert... 2269
Forster, Stefan... 2420
Fotouhi, Reza.. 1575
Francois, Bruno 2142, 2184
Franke, W. Toke... 69
Franko, Marek... 2538
Friedli, T. 27
Fröhleke, Norbert.. 159
Fuchs, F.W. 444
Fuchs, Friedrich W. 1390, 1819, 69
Fujita, Y. 275
Fukushima, Kentaro.................................... 148
Funabashi, Toshihisa 2478, 2487
Funato, Hirohito 479
Funayama, Koichi....................................... 1020
Futami, Motoo 2337

G

Gabriela, Petropol Serb................................. 893
Gan, W. C. 1221
Gao, Fanqiang... 515
Gao, Q. 2261
Gao, Qiang.. 1058
García-Tabarés, L. 2481
Gardecki, Arkadiusz 1193
Gasiewski, Marcin 1562
Gaubert, Jean-Paul 703
Gavranic, Ivica 818
Gaztañaga, H. 1555
Gelezevicius, Vilius Antanas........................ 1144
Gennadevich, Kiselev Michail......................... 428
Gennadevich, Lepanov Michail........................ 428
Gerada, C. 1173
Gerada, Chris 1058, 388, 769, 887
Ghaedi, Azam 1054
Gharehpetian, G.B. 313, 629, 750
Ghosh, Arindam 468, 723, 903
Gímenez, María Isabel................................. 1947
Giménez, María 268
Giral, Roberto 2115
Gizinski, Zygmunt 1562
Glasberger, Tomál 1268
Glavin, M.E. 1688
Glushkin, Evgeny 1872
Gnacinski, Piotr 826
Gobis, Vitoldas.. 1140
Goeldel, C... 2126
Gomis-Bellmunt, Oriol 1670
González-Hernández, S. 1784

Author Index

Gorbounov, Yassen...787
Goto, Hiroki..1163, 1168
Grabic, Stevan...1957
Grad, M..714
Grecki, Filip..1440
Grigaitis, Arunas...1144
Grigans, Linards 2066
Grossi, Federica...874
Grzesiak, Lech M...1071
Grzesik, B...956
Gualous, H..1542
Guo, Hai-Jiao..1163, 1168
Gustin, F...1542
Gustin, Frederic 734
Guy, Owen J..2464
Guzinski, Jaroslaw......................................1382, 994
Guzmán, Víctor...1947, 268
Gwózdz, Michal..728

H

Haan, Sjoerd de..187
Habetler, Thomas G..21
Hadas, Zdenek...1665
Hadjov, Kliment..787
Hájek, Vítezslav..2371
Halasz, S...682
Halgos, Jan..2368
Hamada, Tomoyuki 1884
Hamar, J...1755
Hameyer, K..2393
Hameyer, Kay..2412
Harada, Yosuke..148
Hartansky, Rene...2368
Hartnett, Kevin J..1249
Hasegawa, Masaru...614
Hashimoto, Seiji...932
Hayashi, Kenta...589
Hayashi, Yusuke...2445
Hayes, John G...1249
Heising, C...521
Helmut, Weiss...1722, 1934
Henrotte, F..2393
Henze, Olaf..2385
Hercog, Darko...2349
Hicham, Fakham...2205
Himmelstoss, Felix. A..331
Hiraki, Eiji...119, 1877
Hirokawa, Masahiko..393
Hissel, Daniel..2156
HISSEL, Daniel...734
Hmasic, N...2134
Ho, S.L..576
Hoffmann, Frank...1215
Hõimoja, H..2005
Hõimoja, Hardi...1581

Hojo, Masahide..2487
Holtz, Joachim...1084
Holub, Marcin..195
Horen, Yoram..776
Horga, V..2043
Horga, Vasile..1111
Hrasko, Martin...2538
Hu, Weihao..2325
Hubik, Vladimir..1620
Huiqing, Wen...1518, 417
Hurley, W.G...1688
Huttin, N..1523

I

I., Margaris Nikolaos...1105
Iannuzzi, Diego..1469
Ibach, Robert 2082
Ibáñez, Fernando 268
Ichinokura, Osamu.............................1163, 1168, 758
Ichinose, Masaya 2337
Ide, Kazumasa 2337
Igic, P. M...2464
Iida, Takahiko..595
Ikeda, Yoshiko...498
Ikhouane, Faycal...1670
Iman-Eini, H...173
Inoue, Yukinori..1859
Ion, Petropol Serb...893
Iov, Florin..1771, 561
Ishikawa, Kazumi..1020
Iskhakov, Albert..663
Ito, Fumio 1309
Itoh, Jun-ichi..581
Itoi, M..275
Ivanovic, Zoran..1957
Iwaji, Yoshitaka..924
Iwanski, Grzegorz...1440, 2164
Iwase, Yuta..2487
Izadbakhsh, Alireza..2102

J

Jalakas, T...1263
Jalakas, Tanel 1257
Jan, Mucko 1316
Ján, Vittek 2219
Jang, Gil-Soo 2498
Jang, Su-Jin...1924
Jansen, Uwe..88
Janson, Kuno 154
Járdán, Rafael K..916
Jardan, Rafael Kalman...2360
Jasim, O..1173
Jasim, Omar...887
Jasinski, Marek..1904
Javurek, Jiri...1465

Author Index

Jedryczka, Cezary 2406
Jennings, Michael R. 2464
Jeon, Jin-Hong 2492, 2524
Jezernik, Karel 2283, 2349, 432
Ji, Young-Hyok 1929
Jian, Xiao 1722
Jin, Zhao 1128
Johnson, C Mark 76
Joós, Géza 2178
Joost, M. 1064
Judek, Slawomir 1497
Jufer, Marcel 1
Jun, Liu 1518, 417
Jung, Doo-Yong 1929
Jung, Yong-Chae 181, 1924, 1929, 383

K

Kalatchikov, P. 837
Kalisiak, Stanislaw 195
Kallaste, Ants 154
Kallenbach, E. 1598
Kalyoncu, Mete 1132, 949, 974
Kamata, Yuki 498
Kamiski, Bartlomiej 2378
Kamper, M.J. 859
Kampisios, Konstantinos 887
Kaneko, Daigo 924
Kanerva, Sami 867
Kaplon, Andrzej 377
Karaffy, Z. 2096
Karsli, Vedat M. 850
Karwowski, Krzysztof 1497
Kasa, Nobuyuki 595
Kasinski, A. 1044
Kasprowicz, Andrzej 1332
Katic, Vladimir 1957
Kato, Koji 581
Katsura, Seiichiro 1187, 1604, 1614
Kawamura, Atsuo 7, 924
Kayhan, Ince 1934
Kazimierz, Jaracz 912
Kazmierkowski, Marian P. 1548, 1904
Kelemen, Franjo 855
Kennel, R.M. 859
Kennel, Ralph 1239
Khaldi, B.S 1987
Kim, Eel-Hwan 2498
Kim, Heung-Gun 1421
Kim, Jae-Hong 2498
Kim, Jae-Hyung 1924, 1929
Kim, Jong-Yul 2492
Kim, Se-Ho 2498
Kim, Seul-Ki 2492, 2524
Kimura, Kensuke 1168
Kimura, Noriyuki 1884

Kinoshita, Hirotaka 2337
Kireev, V. 1598
Klimczak, Pawel 108
Klug, O. 2096
Klyachko, Leonid 663
Klytta, Marius 165
Knop, André 69
Kobayashi, Yukinori 479
Kobougias, Ioannis C. 1274
Koczara, Wlodzimierz 1440, 2150, 2164, 2254
Koda, Noriaki 1877
Kolar, J. W. 27
Kolesnikov, Artem 1872
Kolomeitsev, L. 1598
Kompa, K. 695
Komura, Akiyoshi 2337
Kondo, Masaki 1614
Koneke, Thies 1458
Kong, S.T. 43
Konstantinovich, Rozanov Yurie 428
Korondi, Peter 2360
Korotyeyev, Igor 236
Koskin, Y. 837
Kosmecki, Michal 1975
Kostylev, A.V. 1050
Kotodziejek, Piotr 804
Kouzou, A. 1987
Kowalski, Czeslaw T. 1359
Kraeftner, Wilhelm 331
Kraynov, D. 1598
Krettek, Johannes 1215
Krim, Fateh 703
Krismer, F. 27
Krykowski, K. 326
Krystkowiak, Michal 728
Krzeminski, Zbigniew 1382, 2294
Kubiak, Andrzej 2452
Kubin, Jiri 1815
Kubota, Sachio 1309
Kuchta, Jozef 2538
Kudarauskas, Sigitas 2200
Kuebrich, Daniel 244
Kuhn, Harald 1458
Kuisma, M. 1233
Kulka, Arkadiusz 657
Kumar, Dinesh 207
Kuperman, Alon 776
Kurokawa, Fujio 2434, 2504
Kürschner, Daniel 1696, 1734
Kuß, H. 695
Kusserow, Wolf 1239
Kütt, Lauri 154
kuwata, M. 275
Kyritsis, A.C. 1287

Author Index

L

Laczynski, Tomasz	569, 649
Lafoz, M.	2481
Lagoda, Ryszard	1040, 2055, 2087
Laloya, Eduardo	845
Lange, E.	2393
Lapointe, Vincent	1077
Lastowiecki, Jozef	1440
Latka, M.	714
Latkovskis, Leonards	2066
Laugis, J.	1263
Laugis, Juhan	1017
Laur, R.	1064
Lazar, C.	2043
Lazar, Mihai	2457
Ledwich, Gerard	468, 723, 903
Lee, Joo-Hyuk	1924
Lee, Tzung-Lin	1704
Lehtla, Madis	1581
Lehtla, T.	2011
Leidhold, Roberto	1353
Leszek, Szychta	2091
Leuchter, Jan	422
Levins, Nikolajs	1811
Lewandowski, Daniel	2289, 669
Lewicki, Arka diusz	1382
Lewis, A.W.	1790
Leyva, R.	2108
Li, Kaihang	97
Li, Rongyuan	159
Li, Yaohua	1981, 2000, 405, 515
Li, Zixin	1981, 2000, 405, 515
Liaw, Chang-Ming	1652, 456
Lie, Xu	229
Liffran, Florent	409
Lillo, Liliana de	388
Lindemann, Andreas	1280, 2420
Lingemann, M.	2134
Lis, Jacek D.	1359
Lisik, Zbigniew	2452
Lisowski, Grzegorz	669
Liu, Congwei	405
Liu, Li	793
Lladó, Juan	845
Lodzinski, Michal	2464
Lopez-de-Heredia, A.	1555
Lorenz, Robert D.	903
LU, Di	2142
Lu, Hua	76
Lu, Y.	1221
Luft, Miroslaw	463
Luiz, Alex-Sander Amavel	1345
Luniewski, Piotr	88
Lyons, Brendan J.	1249
Lyskawinski, Wieslaw	2406

M

Macek-Kaminska, Krystyna	1193
Madawala, U. K.	139
Madawala, Udaya K.	1918
Maga, Dusan	2368
Mahmoudi, M.O.	1987
Mahyob, Amin	810
Mailat, Adrian	1748
Majidi, Behrooz	763
Maksimovic, Dragan	498
MAKYS, Pavol	2538
Malekian, Kaveh	1117, 1123, 2071, 365, 763
Malska, W.	714
Man, T.K.	400, 475
Mandache, Lucian	1585
Mandra, Slawomir	1071
Mandrek, Slawomir	144
Marek, Stulrajter	2219
Margaliot, M.	260
Mariano, Sílvio José Pinto Simões	2076
Marouchos, Christos	1967
Martín-del-Brío, Bonifacio	845
Martínez, Abelardo	1947, 845
Martinez, Itziar	437
Martins, Denizar C.	1951
Masada, E.	1755
Mascibrodzki, Ireneusz	1562
Mathis, W.	132
Matsui, Keiju	614
Matsui, Nobumasa	2504
Mawby, P.A.	2472
Mawby, Philip A.	2464
McEachern, Alex	1371
Mecke, Rudolf	1734
Melício, R.	1682
Mendes, V. M. F.	1682
Mertens, A.	132
Mertens, Axel	1458, 213, 569, 649
Meuret, R.	1523
Meynard, Thierry	281
Michalík, Jan	1535, 550
Michalke, N.	695
Mierlo, J. Van	1512
Milanovic, Miro	301
Milimonfared, Jafar	1117, 2071, 365, 763
Mimura, Yasuhiro	2434
Mirsalim, M.	313, 629, 750
Mirsalim, Mojtaba	1123
Mirzaeva, G.	1155
Mishima, Tomokazu	119
Mitani, Tetsuya	2428
Mladenovic, I.	2022
Mohd, A.	2134
Mokrovica, Josipa	855
Mõlder, Heigo	154

Author Index

Molinas, Marta...2318
Möller, T..2005
Mollov, Stefan V..350
Molnár, Jan..1535, 550
Mondzik, Andrzej...345
Monmasson, E...1365
Montesinos-Miracle, Daniel................1670, 731
Morel, Herve..2457
Moreno-Font, Vanessa....................................2115
Moreno-Goytia, E............................1784, 1941
Morimoto, Shigeo...1859
Morino, Kimio...2478
Morizane, Toshimitsu.....................................1884
Morton, D...2134
Mouni, Emile...2015
Mukhopadhyay, Siddhartha.............................485
Munk-Nielsen, Stig..108
Murata, Toshiaki...2337
Musallam, Mahera..76
Mustonen, P...1233
Musumeci, S..61
Muszynski, Roman..2227
Mutschler, Peter...1353
Müür, M...2005
Mysinski, Wojciech...1321

N

Nagy, I...1755
Nagy, Istvan...2360
Nagy, István...916
Naka, Toshiyuki...498
Nakagawa, Akio...498
Nakamura, Kazutoshi.......................................498
Nakamura, Kenji..758
Nakaoka, M...275
Nakaoka, Mutsuo...119
Nakayama, Hiroaki...1877
Nanakos, Anastasios Ch.................................1827
Naouar, M-W..1365
Narayanan, E.M. Sankara.................................43
Narjiss, Abdellah..734
Nasser, Mehdi...1896
Navarro, Daniel..437
Nawaz, Muhammed...2472
Nekoui, Mohammad Ali...................................1054
Ngwendson, L..43
Ni, Bingchang..2331
Nichita, C...1911
Nichita, Cristian..1834
Nicolae, Ileana-Diana......................................1585
Nicolae, Petre Marian......................................1585
Nicolae, Petre-Marian.....................................1181
Niechaj, Marek..1890
Niemelä, Markku...1763
Nikolic, Aleksandar...1426

Nilssen, Robert..799
Ninomiya, Tamotsu..................................148, 393
Nishida, Yasuyuki...2530
Nishikata, Shoji..2343
Nishimiya, Ayumu...1163
Nishioka, Kunihiro..1309
Nitta, Mayumi...932
Noda, Shuji...1877
Norigoe, Isami..148
Novák, Jaroslav..982
Novák, Martin...982
Nowak, Lech..2400
Nowak, Mietek...84
Numata, Shigeo...2478
Nuutinen, Pasi..1763
Nyczkowski, Lukasz..740
Nymand, Morten..127
Nysveen, Arne..799

O

O'Sullivan, D.L...1790
Ogiwara, H..275
Ohashi, Hiromichi.....................2428, 2445, 54
Ohishi, Kiyoshi....................1187, 1604, 1614
Ohsaki, H...1755
Ohyama, Kazuhiro...2300
Okamatsu, Masashi..2434
Oleschuk, Valentin...1548
Omari, O...2134
OMORI, Hideki..2530
Ondrusek, Cestmir..1665
ONEN, Umit...949
OPROESCU, Mihai..621
Orlik, B..1064, 830
Orlowska-Kowalska, Teresa...............2211, 2306
Ortjohann, E..2134
Oyarbide, Estanislao.......................................845

P

Pacas, Mario...2248
Pajchrowski, Tomasz.....................1198, 1204
Pakhomin, S...1598
Palis, Frank..1610
Palis, Stefan...1660
Panoiu, Caius...1409
Panoiu, Manuela...1409
Papanikolaou, N.P..1287
Papic, I..2510
Paquin, Jean-Nicolas......................................1475
Park, JuneHo..2492
Park, Sang-Hoon......................................181, 383
Park, So-Ri...181
Parkatti, P..201
Parker-Allotey, N-A..2472
Patel, N. D...139

Author Index

Patra, Pradipta 485
Patra", Amit 485
Pavelka, Jiri 221
Pavelka, Jirí 988
Pavlitov, Constantin 787
Pavlovsky, Martin 7
Pavol, Makys 2219
Pavoni, Alessandro 1491
Peftitsis, Dimosthenis 1840
Peltoniemi, Pasi 1763
Peplinski, Marcin 826
Pera, Marie-Cecile 2156
Perez, Francisco 845
Perez-Tomas, Amador 2464
Peric, Nedjeljko 2235
Peroutka, Zdenek 1268, 1529, 1535, 550
Peter, Bris 2219
Peter, Zaucher 1722
Petit, Marc 2184
Petrella, Roberto 1097
Petrisor, Anca 893
Piatek, Pawel 1446
Pietrzak-David, Maria 938
Piróg, Stanislaw 1779
Pittermann, Martin 1593, 372
Planson, Dominique 2457
Poljugan, Alen 1058
Pollán, Tomás 845
Popa, Anca Sorana 1225
Popa, Mircea 1225
Porada, Ryszard 740
Pospelov, Vladimir 663
Pronin, M. 837
Pugachevs, Vladislavs 1811
Pyrhönen, Juha 1763

Q

Quiroga, J. 793

R

Rabkowski, Jacek 84
Raciti, A. 61
Radomski, Grzegorz 504
Raducu, Marian 621
Radulescu, Mircea M. 1896
Rafecas-Sabate, Josep 731
Rafiei, S.M.R. 2102
Rahman, M.F. 2269
Rahnamaee, Arash 1117, 365
Rao, Sachit 2312
Rathge, Christian 1696
Ratoi, Marcel 1111
Rawicki, Stanislaw 1481
Raynaud, Christophe 2457
Rednov, F. 1598

Reghem, Pascal 1834, 810
Rerucha, Vladimir 422
Rezaei, Mohammad Mehdi 1123
Reznikov, B. 260
Ribickis, Leonids 1811
Richter, F. 1398
Riipinen, T. 1233
Risteiu, Mircea 1243
Riz, A. 2096
Roasto, I. 2011
Robert, B.G.M. 2126
Robert, Bruno Gerard Michel 2120
Robinson, Jonathan 2178
Robyns, B. 1523
Robyns, Benoît 1896
Robyns, Benoit 2184
Rodic, Miran 2283
Rodriguez, E. 2108
Rodriguez, Jose 451, 699
Rojas, A. 1155
Rojko, Andreja 2349
Rolek, Jaroslaw 377
Rompelman, Otto 2354
Ronkowski, Mieczyslaw 880
Rosin, A. 2005
Rothenhagen, Kai 1390, 1904
Round, S. D. 27
Ru1scin, Vladimír 295
Ruderman, A. 260
Ruderman, Michael 1215
Rufer, A. 256
Ruger, N. E. 132
Rusinov, Radoslav 787
Rylko, Marek S. 1249
Ryvkin, Sergey 1505
Rzasa, Janina 357

S

Saadi, S. 1987
Sabirin, Chip Rinaldi 1625
Saito, Makoto 2439
Saito, Tsuyoshi 744
Sajkowski, M. 956
Sakamoto, Kiyoshi 924
Sakamoto, Yosei 288
Salo, M. 201
Salonen, Pasi 1763
Samanta, Susovon 485
Samuelsen, Dag 1416
Sanada, Masayuki 1859
Sánchez, Beatriz 845
Sánchez, Carlos 268
Sang-Joon, Lee 1002
Sang-Taek, Lee 1002
Sanjari, M. J. 313, 629, 750

Author Index

San-Sebastian, J. 1555
Santo, António Espírito............................ 1646
Sarraute, Emmanuel 281
Sasaki, Masahiro 2434
Sato, Muneo 1309
Saudemont, C. 1523
Sayed, Mahmoud A. 542
Sayeef, S. 2269
Schallschmidt, Thomas 1610, 1660
Schanen, JL. 173
Schmelter, A. 2134
Schmid, Markus 244, 250
Schmidt, Istvan 1803
Schmidt-Obermoeller, Richard.................. 1505
Schmitt, Günter 1239
Schneider, T. 1598
Schnick, O. 132
Schrödl, Manfred 2275
Schroedl, Manfred 945
Schuffenhauer, U. 695
Scollo, R. 61
Sengupta, Sabyasachi 674
Seppä, L. 1233
Shao, S. 1293
Shapoval, Ivan 307
Sharma, R. 1918
She, X. 710
She, Yun 710
Shieh, Fa-Hwa 1652
Shimaoka, Yoshihiro 1309
Shimizu, Takaaki 498
Shimizu, Toshihisa 2428, 2445, 288, 600
Shimoda, Eisuke 2478
Shiraishi, Keiichi 2504
Shonin, O. 837
Shoyama, Masahito 148, 393
Shyu, Juei Lung 643
Siatkowski, M. 830
Silea, Ioan 1225
Silventoinen, P. 1233
Simetzberger, Christian 2275
Simon, Miklós G. 916
Singule, Vladislav 1620, 1665
Sinsukthavorn, W. 2134
Siostrzonek, Tomasz 1779
Sîrbu, Ioana-Gabriela 1181, 1585
Siroky, Peter 2368
Sitar, Jan 2368
Sivkov, Oleg 221
Skovpen, Sergey 663
Skuta, Ondřej 1375
Slama-Belkhodja, I. 1365
Slama-Belkhodja, Ilhem 938
Smet, Bart 102
Sobczuk, Dariusz 2378
Sobczynski, D. 714

Sochacki, Mariusz 2452
Soltani, Hamid 718
Song, Sang-Hoon 383
Song, Seung-Ho 2498
Soroudi, Alireza 1847
Sosa-Ruiz, J. 1941
Souad, Rafa 1209
Sourkounis, C. 1398
Sourkounis, Constantinos 1633, 1710, 2331
Sozanski, Krzysztof Piotr 1995
Stadler, Paul Andreas 2543
Stala, Robert 1852, 345
Stamann, Mario 1660
Stanescu, Dan-Gabriel 1181
Staudt, V. 521
Staudt, Volker 2371
Stefanutti, Fabio 1097
Steimel, A. 521
Steimel, Andreas 1505, 2371
Stenzel, T. 956
Stepanyuk, D.P. 1050
Stepien, P. 1293
Stocco, Piero 1097
Strac, Leonardo 855
Strzelecki, Ryszard Michal 1332
Strzelecki, Ryszard 1729
Stumpf, P. 1755
Stumpf, Péter 916
Sugai, T. 275
Sugimasa, Junji Tamura Masatoshi 2337
Suissa, Uri 776
Sulkowski, Waldemar 1416
Sumida, Yuichi 2434
Sumina, Damir 556
Sumiyoshi, Shinichiro 2530
Summers, T.J. 1293
Sumner, M. 1173, 2261
Sumner, Mark 1058, 2300, 769
Sun, Z. G. 1221
Susluoglu, Berrin 850
Suul, Jon Are 2318
Sveda, Martin 1620
Sweet, M. 43
Sykulski, Jan K. 2383
Szabat, Krzysztof 2211, 2241
Szamel, Laszlo 1033
Szczeniak, Pawel 165
Szczepankowski, Pawel 1332
Szczesniak, Pawel 236
Szelag, Wojciech 2406
Sziebig, Gabor 2360
Szmidt, Jan 2452
Szubert, Krzysztof 536
Szweda, Mariusz 826
Szychta, Elzbieta 463
Szychta, Leszek 463

A-9

Author Index

Szymanski, B. J. 695

T

Tackoen, X. 1512
Tae-Ho, Yoon 1002
Taguchi, Toyoki 498
Takahashi, Nobuo 1877
Takahashi, Rion 2337
Takao, Kazuto 2445, 54
Takeshita, Takaharu 542
Takeuchi, Nobuhito 614
Takeuchi, Toshihiro 924
Tan, Longcheng 1981, 2000, 405, 515
Tanabe, Takayuki 2478
Tanaka, Toshihiko 1877
Taniguchi, Katsunori 1884
Taniguchi, Satoshi 600
Tankari, A.M. 1911
Tao, Zhou 2205
Tapuchi, Saad 776
Tarczewski, Tomasz 1071
Tatakis, E.C. 1287
Tatakis, Emmanuel C. 1274, 1301, 1827
Tatsuta, Fujio 2343
Theodoridis, Michael P. 350
Thomas, D.W.P. 2049
Thomas, David W.P. 1716
Thompson, David S. 114
Tinkir, Mustafa 1132, 949, 974
Tnani, Slim 2015
Tournier, Dominique 2457
Tran, Quang-Vinh 1421
Trentin, Andrew. 887
Trujillo, Cesar L. 637
Tsai, Jih-Run 1652
Tseng, K.J. 2516
Tsukakoshi, Kenta 148
Tsuruta, Yukinori. 7
Tulbure, Adrian. 1243
Turner, Robert W. 528
Tutaj, Andrzej 1432
Tuusa, H. 201

U

Ueda, Yoshinobu 2478, 2487
Ummaneni, Ravindra. B. 799
Undeland, Tore 2318, 657
Ünüvar, Ali 967
Urabe, R. 275
Urbanski, Konrad 1454
Utkin, Vadim 2312, 512

V

Väisänen, V. 1233

Valchev, V. 1326
van Duivenbode, Jeroen 102
Vasak, Mario 2235
Vedrana, Jerkovic 690
Vekic, Marko 1957
Vergnol, Arnaud 1896
Veszpremi, Karoly 1803
Vicuña, Javier 845
Villanueva, Elena. 451
Villwock, Sebastian 2248
Vinnikov, D. 1263, 2011
Vinnikov, Dmitri 1257
Viscarret, U. 1555
Vittek, Jan 2551
Vladimír, Vavrus 2219
Vodovozo, Valery 1017
Vorontsov, A. 837
Vrana, Petr 1465

W

Wada, Keiji 2428, 288, 600
Walas, K. 1044
Walter, Julio 268
Walton, Simon 528
Wang, Ping 1981, 2000, 405, 515
Wang, Yi. 187
Wang, Yue 2325
Wang, Zhaoan 2325
Weidinger, Thomas 2028
Weiland, Thomas 2385
Weindl, Ch. 2022
Werner, Timur 649
Wheeler, P. 1326
Wheeler, Patrick W. 207
Wheeler, Patrick W. 229
Wheeler, Patrick 388
Wiktor, Hudy 912
Willis, K. 1293
Winternheimer, Stefan 1872
Wisniewski, Janusz 2254
Wlas, Miroslaw 1084
Won, Chung-Yuen 181, 1924, 1929, 383
Wong, L.K. 400, 475
Wu, Dongming 97

X

XiaoyanHuang, ...388
Xu, Wei 2000, 405
Xuhui, Wen 1518, 417
Xuhui, Zhang 1518, 417

Y

Yaguchi, Hiroyuki 1020
Yamanouchi, Wataru 1187

Author Index

Yang, Lingling 97
Yang, Liu 1128
Yang, Ru-Shiuan 320
Yin, Chunyan 76
Yokokura, Yuki 1187, 1604
Yokoyama, Tomoki 589, 744
Young-Kwan, Kim 1002
Yousefi, Ashkan 1847

Z

Zakrzewski, Zbigniew 1332
Zamma, Toshihiro 1020
Zanasi, Roberto 874
Zanchetta, Pericle 1716, 1771, 561, 887
Zare, Firuz 468, 718, 723, 903
Zaring, Carina 2472
Zarko, Damir 855
Zarko, Damirarko 818
Zaskalicka, Maria 899
Zaskalicky, Pavel 899
Zatocil, Heiko 1024
Zawirski, Krzysztof 1198, 1204, 1454
Zdenek, Jiri 1638
Zdravko, Valter 690
Zeljko, Spoljaric 690
Zeman, Karel 1529
Zeroug, Houcine 1209
Zhang, H. 1523
Zhang, S. 2516
Zhao, S. W. 1221
Zhou, Tao 2142
Zhu, Haibin 515
Zielinski, K. 1064
Zigic, Aleksandar 1426
Zinoviev, Genady Stepanovic 1332
Zlosnikas, Valerijus 1140
Zouhar, Jan 1665
Zulawnik, Marcin 1562
Zych, Michal 1562
Zymmer, Krzysztof 1332, 1562

CURRAN ASSOCIATES INC.
proceedings
.com

9781424417414